For Extra Practice

The following sections feature extra practice problems available when you register for MyMathLab.com. For your convenience, these exercises are also printed in the *Instructor's Resource Guide*. Please contact your instructor for assistance.

Intermediate Algebra

CONCEPTS AND APPLICATIONS

Intermediate Algebra

SEVENTH EDITION

CONCEPTS AND APPLICATIONS

MARVIN L. BITTINGER

Indiana University Purdue University Indianapolis

DAVID J. ELLENBOGEN

Community College of Vermont

PEARSON

Addison
Wesley

Boston   San Francisco   New York
London   Toronto   Sydney   Tokyo   Singapore   Madrid
Mexico City   Munich   Paris   Cape Town   Hong Kong   Montreal

Publisher	Greg Tobin
Editor in Chief	Maureen O'Connor
Acquisitions Editor	Jennifer Crum
Associate Editor	Katie Nopper
Editorial Assistant	Elizabeth Bernardi
Managing Editor	Ron Hampton
Production Supervisor	Kathleen A. Manley
Editorial and Production Services	Martha K. Morong/Quadrata, Inc.
Art Editor and Photo Researcher	Geri Davis/The Davis Group, Inc.
Compositor	Beacon Publishing Services
Software Development	David Malone and Rebecca Williams
Marketing Manager	Dona Kenly
Marketing Coordinator	Tracy Rabinowitz
Prepress Supervisor	Caroline Fell
Manufacturing Buyer	Evelyn Beaton
Design Supervisor/Cover Designer	Dennis Schaefer
Text Designer	Geri Davis/The Davis Group, Inc.
Cover Photograph	Water Lily; © H. Kehrer/Masterfile

For permission to use copyrighted material, grateful acknowledgment is made to the copyright holders on page 760, which is hereby made part of this copyright page.

Library of Congress Cataloging-in-Publication Data
Bittinger, Marvin L.
   Intermediate algebra: concepts and applications.– 7th ed. / Marvin L.
     Bittinger, David J. Ellenbogen. p. cm.
   Includes indexes.
   ISBN 0-321-23386-7 (SE)
   1. Algebra—Textbooks. I. Ellenbogen, David. II. Title.
QA154.3.B58 2004
512.9—dc22
                                    2004062480

3 4 5 6 7 8 9 10—VH—09 08 07 06

For Kit, Tony, Kristin, and Max

Contents

Systems of Linear Equations and Problem Solving

Inequalities and Problem Solving

Polynomials and Polynomial Functions

**Rational Expressions,
Equations, and
Functions**

**Exponents and
Radicals**

Quadratic Functions and Equations

Exponential and Logarithmic Functions

Conic Sections

11

Sequences, Series, and the Binomial Theorem

Preface

I t is with great pleasure that we introduce you to the seventh edition of *Intermediate Algebra: Concepts and Applications*. Every time we work on a new edition, it's a balancing act. On the one hand, we want to preserve the features, applications, and explanations that faculty have come to rely on and expect. On the other hand, we want to blend our own ideas for improvement with the many insights that we receive from faculty and students throughout North America. The result is a living document in which new features and applications are developed at the same time that successful features and popular applications from previous editions are updated and refined. Our goal, as always, is to present content that is easy to understand and has the depth required for success in this and future courses.

Appropriate for a one-term course in intermediate algebra, this text is intended for those students who have a firm background in elementary algebra. It is one of three texts in an algebra series that also includes *Elementary Algebra: Concepts and Applications*, Seventh Edition, by Bittinger/Ellenbogen, and *Elementary and Intermediate Algebra: Concepts and Applications*, Fourth Edition, by Bittinger/Ellenbogen/Johnson.

Approach

Our goal, quite simply, is to help today's students both learn and retain mathematical concepts. To achieve this goal, we feel that we must prepare developmental-mathematics students for the transition from "skills-oriented" elementary and intermediate algebra courses to more "concept-oriented" college-level mathematics courses. This requires that we teach these same students critical thinking skills: to reason mathematically, to communicate mathematically, and to identify and solve mathematical problems. Following are three aspects of our approach that we use to help meet the challenges we all face when teaching developmental mathematics.

Problem Solving

One distinguishing feature of our approach is our treatment of and emphasis on problem solving. We use problem solving and applications to motivate the material wherever possible, and we make use of real-life applications and problem-solving techniques throughout the text. Problem solving not only encourages students to think about how mathematics can be used, it also helps to prepare them for more advanced material in future courses.

In Chapter 1, we introduce our five-step process for solving problems: (1) Familiarize, (2) Translate, (3) Carry out, (4) Check, and (5) State the answer. These steps are then used consistently throughout the text whenever we encounter a problem-solving situation. Repeated use of this problem-solving strategy provides students with a starting point for any type of problem they encounter, and frees them to focus on the unique aspects of each situation. We often use estimation and carefully checked guesses to help with the *Familiarize* and *Check* steps (see pp. 172 and 391–392).

Applications

Interesting applications of mathematics help motivate both students and instructors. Solving applied problems gives students the opportunity to see their conceptual understanding put to use in a real way. In the Seventh Edition of *Intermediate Algebra: Concepts and Applications*, we have increased the number of applications, the number of real-data problems, and the number of reference lines that specify the sources of the real-world data. As in the past, art is integrated into the applications and exercises to aid the student in visualizing the mathematics. (See pp. 126, 140, 178, and 566.)

Pedagogy

New!   **CONCEPT REINFORCEMENT EXERCISES.**   This feature is designed to help students build their confidence and comprehension through true/false, matching, and fill-in-the-blank exercises at the beginning of most exercise sets. Whenever possible, special attention is devoted to increasing student understanding of the new vocabulary and notation developed in that section. (See pp. 93, 129, 241, and 285.)

STUDY SKILLS.   These remarks, located in the margin near the beginning of each section, provide suggestions for successful study habits that can be applied to both this and other college courses. Ranging from ideas for better time management to suggestions for test preparation, these comments can be useful to even experienced college students. (See pp. 10, 99, and 151.)

New!   **STUDENT NOTES.**   These comments, strategically located in the margin within each section, are specific to the mathematics appearing on that page. Remarks are more casual in format than the typical exposition and range from suggestions on how to avoid common mistakes to how to best read new mathematical notation. (See pp. 90, 238, and 314.)

New!   **STUDY SUMMARY.**   Each chapter closes with a Study Summary in which the authors have written notes that highlight the most important concepts and terminology from that chapter. Page numbers are provided so students can reference where the terminology is introduced and where important properties and formulas are listed. Examples of specific types of problems are often included. In short, the Study Summary provides a terrific point from which to begin reviewing for a chapter test. (See pp. 214, 499, and 578.)

New! **THE ALGEBRAIC–GRAPHICAL CONNECTION.** Because many of us master concepts more easily when provided with visual information, we now include a feature that offers students the opportunity to visualize algebraic concepts that might otherwise prove elusive. (See pp. 506 and 574.)

CONNECTING THE CONCEPTS. To help students understand the "big picture," Connecting the Concepts subsections within each chapter (and highlighted in the table of contents) relate the concept at hand to previously learned and upcoming concepts. Because students may occasionally "lose sight of the forest because of the trees," this feature helps them to better keep their bearings as they encounter new material. (See pp. 28, 164, and 475.)

AHA! EXERCISES. Designated by Aha!, these exercises can be solved quickly if the student has the proper insight. Without the insight, the problem can be solved like the others around it. The Aha! designation is used the first time a new insight can be used on a particular type of exercise and tells students that there is a simpler way to complete the exercise than the more routine approach. It's then up to the student to find the simpler approach and, in subsequent exercises, to determine if and when that particular insight can be used again. The Aha! exercises are not more difficult than the neighboring exercises, but are designed to reward students who "look before they leap" into a problem. (See pp. 49, 130, and 194.)

TECHNOLOGY CONNECTIONS. Throughout each chapter, optional Technology Connection boxes help students use graphing calculator technology to better visualize a concept that they have just learned. To connect this feature to the exercise sets, certain exercises, marked with a graphing calculator icon, reinforce the use of this optional technology. (See pp. 79, 100, 155, and 338.)

SKILL MAINTENANCE EXERCISES. Retention of skills is critical to a student's success in this and future courses. To this end, beginning in Section 1.2, every exercise set includes Skill Maintenance exercises that review skills and concepts from preceding sections of the text. Whenever possible, these six to eight exercises provide extra practice with specific skills that may be "rusty" but are needed for the next section of the text. (See pp. 242, 424, and 531.)

SYNTHESIS EXERCISES. Following the Skill Maintenance section, every exercise set ends with a group of Synthesis exercises that offer opportunities for students to synthesize skills and concepts from earlier sections with the present material, and often provide students with deeper insights into the current topic. Synthesis exercises are generally more challenging than those in the main body of the exercise set and occasionally include Aha! exercises. (See pp. 83, 180, and 252.)

WRITING EXERCISES. Every set of exercises includes at least four writing exercises. Two of these are more basic and appear just before the Skill Maintenance exercises. The other writing exercises are more challenging and appear as Synthesis exercises. All writing exercises are marked with 📄 and require answers that are one or more complete sentences. Because some instructors may collect answers to writing exercises, and because more than one answer may be correct, answers to writing exercises are listed at the back of the text only when they are within Review Exercises. Writing exercises have been found to aid in student comprehension, critical thinking, and conceptualization. (See pp. 50, 158, 345, and 399.)

COLLABORATIVE CORNERS.   Studies have shown that students who study together generally outperform those who do not. Throughout the text, we provide optional Collaborative Corner activities that require students to work in groups to explore and solve mathematical problems. There are typically two to three Collaborative Corners per chapter, each one appearing after the appropriate exercise set. (See pp. 42, 167, and 426.)

CUMULATIVE REVIEW.   After Chapters 3, 6, 9, and 11, we have included a Cumulative Review, which reviews skills and concepts from all preceding chapters of the text. (See pp. 219, 433, 654, and 743.)

What's New in the Seventh Edition?

We have rewritten many key topics in response to user and reviewer feedback and have made significant improvements in design, art, pedagogy, and an expanded supplements package. Detailed information about the content changes is available in the form of a conversion guide. Please ask your local Addison-Wesley sales consultant for more information. Following is a list of the major changes in this edition.

New Design

While incorporating a new layout, a fresh palette of colors, and new features, we have maintained the larger page dimension for an open look, and a typeface that is easy to read. As always, it is our goal to make the text look mature without being intimidating. In addition, we continue to pay close attention to the pedagogical use of color to make sure that it is used to present concepts in the clearest possible manner.

Content Changes

A variety of content changes have been made throughout the text. Some of the more significant changes are listed below.

- Chapter 1 now includes improved discussion on the use of scientific notation.
- Chapter 6 now places greater emphasis on identifying the domain of a rational function.
- Chapter 8 now makes greater use of graphs as a way to visualize quadratic equations and functions.
- Throughout the text, there is an increased emphasis on students learning how to distinguish between equivalent expressions and equivalent equations.

Ancillaries

The following ancillaries are available to help both instructors and students use this text more effectively.

STUDENT SUPPLEMENTS	INSTRUCTOR SUPPLEMENTS
Student's Solutions Manual • By Judith A. Penna, *Indiana University Purdue University Indianapolis* • Contains completely worked-out solutions with step-by-step annotations for all the odd-numbered exercises in the text, with the exception of the writing exercises. ISBN: 0-321-27822-4 **Videotapes** • Present a series of lectures correlated directly to the content of each section of the text. • Feature an engaging team of instructors, including one of the authors, who present material using examples and exercises often from the text in a format that stresses student interaction. • New! A videotape guide correlated directly to the text provides practice exercises for each video segment so students can assess their understanding. ISBN: 0-321-27816-X **Digital Video Tutor** • Complete set of digitized videos on CD-ROMs for student use at home or on campus. • Ideal for distance learning or supplemental instruction. • New! Practice Problems and their solutions now follow each lesson. • Videotape guide available. ISBN: 0-321-27823-2	**Instructor's Solutions Manual** • By Judith A. Penna, *Indiana University Purdue University Indianapolis* • Contains fully worked-out solutions to the odd-numbered exercises and brief solutions to the even-numbered exercises in the exercise sets. ISBN: 0-321-27821-6 **Answer Book** • By Judith A. Penna, *Indiana University Purdue University Indianapolis* • Contains answers to all the section exercises in the text. ISBN: 0-321-27815-1 **Printed Test Bank/Instructor's Resource Guide** • By Patricia A. Slipher • Contains two multiple-choice tests per chapter; six free-response tests per chapter; eight final exams; extra practice exercises on selected topics from the text; correlation guide; video index; and transparency masters. ISBN: 0-321-27819-4 **New! *Adjunct Support Manual*** • Includes resources designed to help both new and adjunct faculty with course preparation and classroom management. • Offers helpful teaching tips. ISBN: 0-321-27818-6

(continued)

STUDENT SUPPLEMENTS	INSTRUCTOR SUPPLEMENTS
Addison-Wesley Math Tutor Center • The Addison-Wesley Math Tutor Center is staffed by qualified mathematics instructors who provide students with tutoring on examples and odd-numbered exercises from the textbook. Tutoring is available via toll-free telephone, toll-free fax, e-mail, or the Internet. White Board technology allows tutors and students to actually see problems worked while they "talk" in real time over the Internet during tutoring sessions. For more information, go to www.aw-bc.com/tutorcenter **_MathXL® Tutorials on CD_** • Provides algorithmically generated practice exercises that correlate to exercises at the end of sections. • Every exercise is accompanied by an example and a guided solution, and selected exercises may also include a video clip. • The software recognizes student errors and provides feedback. It can also generate printed summaries of students' progress. ISBN: 0-321-27824-0	**_TestGen with Quizmaster_** • Enables instructors to build, edit, print, and administer tests. • Features a computerized bank of questions developed to cover all text objectives. • Available on a dual-platform Windows/Macintosh CD-ROM. ISBN: 0-321-27820-8

MathXL® www.mathxl.com

MathXL is a powerful online homework, tutorial, and assessment system that accompanies your Addison-Wesley textbook. With MathXL, instructors can create, edit, and assign online homework and tests using algorithmically generated exercises correlated to your textbook. All student work is tracked in MathXL's online gradebook. Students can take chapter tests in MathXL and receive personalized study plans based on their test results. The study plan diagnoses weaknesses and links students directly to tutorial exercises for the objectives they need to study and retest. Students can also access supplemental video clips and animations directly from selected exercises. MathXL is available to qualified adopters. For more information, visit our website at www.mathxl.com, or contact your Addison-Wesley sales representative.

MyMathLab

MyMathLab is a series of text-specific, easily customizable online courses for Addison-Wesley textbooks in mathematics and statistics. MyMathLab is powered by CourseCompass™—Pearson Education's online teaching and learning environment—and by MathXL®—our online homework, tutorial, and assessment system. MyMathLab gives instructors the tools they need to deliver all or a portion of their course online, whether students are in a lab setting or working from home. MyMathLab provides a rich and flexible set of course materials, featuring free-response exercises that are algorithmically generated for unlimited practice and mastery. Students can also use online tools, such as video lectures, animations, and a multimedia textbook, to independently improve their understanding and performance. Instructors can use MyMathLab's homework and test managers to select and assign online exercises correlated directly to the textbook, and they can import TestGen tests into MyMathLab for added flexibility. MyMathLab's online gradebook—designed specifically for mathematics and statistics—automatically tracks students' homework and test results and gives the instructor control over how to calculate final grades. Instructors can also add offline (paper-and-pencil) grades to the gradebook. MyMathLab is available to qualified adopters. For more information, visit our Web site at www.mymathlab.com or contact your Addison-Wesley sales representative.

InterAct Math® Tutorial Web site www.interactmath.com

Get practice and tutorial help online! This interactive tutorial Web site provides algorithmically generated practice exercises that correlate directly to the exercises in your textbook. You can retry an exercise as many times as you like with new values each time for unlimited practice and mastery. Every exercise is accompanied by an interactive guided solution that gives you helpful feedback if you enter an incorrect answer, and you can also view a worked-out sample problem that steps you through an exercise similar to the one you're working on.

Acknowledgments

No book can be produced without a team of professionals who take pride in their work and are willing to put in long hours. Barbara Johnson, in particular, deserves extra thanks for her work as developmental editor. Without Barbara's tireless devotion and her many fine suggestions, this book simply would not have been possible. Laurie Hurley, Bob Maynard, Elina Niemelä, Dawn Mulheron, Vince Koehler, Jeremy Pletcher, and Louise Jarvis also deserve special thanks for their careful accuracy checks, well-thought-out suggestions, and uncanny eye for detail. Judy Penna's outstanding work in organizing and preparing the printed supplements amounts to an inspection of the text that goes far beyond the call of duty and for which we are always extremely grateful. Thanks to Patty Slipher for authoring the *Printed Test Bank.*

We are also indebted to Chris Burditt and Jann MacInnes for their many fine ideas that appear in our Collaborative Corners and Vince McGarry and Janet Wyatt for their recommendations for Teaching Tips featured in the *Annotated Instructor's Edition.*

Martha Morong, of Quadrata, Inc., provided editorial and production services of the highest quality imaginable—she is amazing and a joy to work with. Geri Davis, of the Davis Group, Inc., performed superb work as designer, art editor, and photo researcher, and is always a pleasure to work with. Network Graphics generated the graphs, charts, and many of the illustrations. Not only are the people at Network reliable, but they clearly take pride in their work. The many illustrations appear thanks to Jim Bryant and Bill Melvin, both of whom are artists with insights and creativity.

Our team at Addison-Wesley deserves special thanks. Associate Editor Katie Nopper and Editorial Assistant Elizabeth Bernardi helped with a variety of jobs and always in a timely way. Senior Acquisitions Editor Jenny Crum provided many fine suggestions and is superb at remaining involved and helpful throughout the project. Senior Production Supervisor Kathy Manley exhibited patience when others would have shown frustration. Designer Dennis Schaefer's willingness to listen and then creatively respond resulted in a book that is beautiful to look at. Senior Marketing Manager Dona Kenly and Marketing Coordinator Tracy Rabinowitz skillfully kept us in touch with the needs of faculty. Our Editor in Chief, Maureen O'Connor, and our Publisher, Greg Tobin, deserve credit for assembling this fine team.

We also thank the students at the Community College of Vermont and the following professors for their thoughtful reviews and insightful comments.

Prerevision Diary Reviewers (Sixth Edition)

Barbara Armenta, *Pima Community College—East Campus*
Debi McCandrew, *Florence–Darlington Technical College*
Timothy McKenna, *University of Michigan—Dearborn*

Manuscript Reviewers

Marie Aratari, *Oakland Community College—Orchard Ridge Campus*
Douglas Brozovic, *University of North Texas*
Barbara Burke, *Hawaii Pacific University*
Sharon Edgmon, *Bakersfield College*
Karen Ernst, *Hawkeye College*
Kathy Garrison, *Clayton College and State University*
Shirley Goos, *Cameron University*
Tracey L. Johnson, *University of Georgia*
Joanne Kawczenski, *Luzerne County Community College*
Rachel Lamp, *North Iowa Community College*
Doug Mace, *Kirtland Community College*
Bob McCarthy, *Community College of Allegheny County—South Campus*
Rhea Meyerholtz, *Indiana State University*
Kausha Miller, *Lexington Community College*
Rebecca Parrish, *Ohio University*
Debra Pharo, *Northwestern Michigan College*
Terry Reeves, *Red Rock Community College*
Kathy Rod, *Wharton County Junior College*
Nicole Saporito, *Luzerne Community College*
Donald Soloman, *University of Wisconsin—Milwaukee*
Fran Smith, *Oakland Community College*

Finally, a special thank you to all those who so generously agreed to discuss their professional use of mathematics in our chapter openers. These dedicated people all share a desire to make math more meaningful to students. We cannot imagine a finer set of role models.

M.L.B.
D.J.E.

Feature Walkthrough

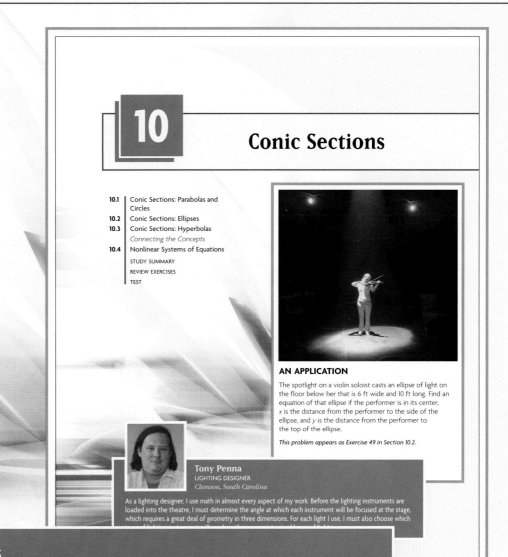

AN APPLICATION

The spotlight on a violin soloist casts an ellipse of light on the floor below her that is 6 ft wide and 10 ft long. Find an equation of that ellipse if the performer is in its center, x is the distance from the performer to the side of the ellipse, and y is the distance from the performer to the top of the ellipse.

This problem appears as Exercise 49 in Section 10.2.

Tony Penna
LIGHTING DESIGNER
Clemson, South Carolina

As a lighting designer, I use math in almost every aspect of my work. Before the lighting instruments are loaded into the theatre, I must determine the angle at which each instrument will be focused at the stage, which requires a great deal of geometry in three dimensions. For each light I use, I must also choose which

Chapter Openers

Each chapter opens with a list of sections covered and a real-life application that includes a testimonial from a person in that field to show how important mathematics is in everyday problem solving. Real data often appear in these applications, in many other exercises, and in "on the job" examples similar to what students might find in the workplace.

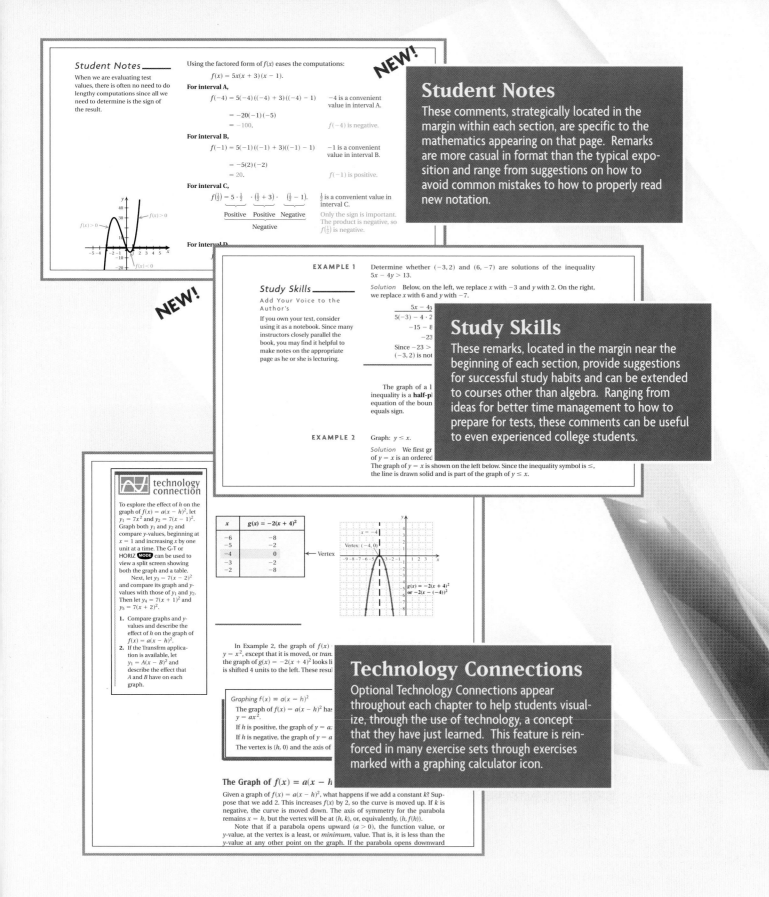

and a tie is worth 1 point. Graph a system of inequalities that describes the situation. (*Hint:* Let $w =$ the number of wins and $t =$ the number of ties.)

73. *Elevators.* Many elevators have a capacity of 1 metric ton (1000 kg). Suppose that c children, each weighing 35 kg, and a adults, each 75 kg, are on an elevator. Graph a system of inequalities that indicates when the elevator is overloaded.

74. *Widths of a basketball floor.* Sizes of basketball floors vary due to building sizes and other constraints such as cost. The length L is to be at most 94 ft and the width W is to be at most 50 ft. Graph a system of inequalities that describes the possible dimensions of a basketball floor.

c) $13x - 25y + 10 \le 0$
d) $2x + 5y > 0$

 **76.** Use a graphing calculator to check your answers to Exercises 35–48. Then use INTERSECT to determine any point(s) of intersection.

CORNER

COLLABORATIVE

How Old Is Old Enough?

Focus: Linear inequalities
Time: 15–25 minutes
Group size: 2

It is not unusual for the ages of a bride and groom to differ significantly. Yet is it possible for the difference in age to be too great? In answer to this question, the following rule of thumb has emerged: *The younger spouse's age should be at least seven more than half the age of the older spouse.* (*Source:* http://home.earthlink.net/ ~mybrainhurts/2002_06_01_archive.html)

ACTIVITY

1. Let $b =$ the age of the bride, in years, and $g =$ the age of the groom, in years. One group member should write an equation for

calculating the bride's minimum age if the groom's age is known. The other group member should write an equation for finding the groom's minimum age if the bride's age is known. The equations should look similar.

2. Convert each equation into an inequality by selecting the appropriate symbol from $<$, $>$, $\le$, and $\ge$. Be sure to reflect the rule of thumb stated above.

3. Graph both inequalities from step 2 as a system of linear inequalities. What does the solution set represent?

4. If your group feels that a minimum or maximum age for marriage should exist, adjust your graph accordingly.

5. Compare your finished graph with those of other groups.

, AND LINEAR EQUATIONS

can be used to predict the height, in centimeters, of a woman whose humerus (the bone from the elbow to the shoulder) is x cm long. Predict the height of a woman whose humerus is the length given.

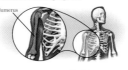

Humerus

57. 32 cm   **58.** 35 cm

Chemistry. The function F described by
$$F(C) = \tfrac{9}{5}C + 32$$
gives the Fahrenheit temperature corresponding to the Celsius temperature C.

59. Find the Fahrenheit temperature equivalent to $-10°C$.

60. Find the Fahrenheit temperature equivalent to $5°C$.

Heart attacks and cholesterol. For Exercises 61 and 62, use the following graph, which shows the annual heart attack rate per 10,000 men as a function of blood cholesterol level.*

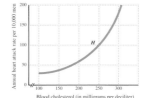

61. Approximate the annual heart attack rate for those men whose blood cholesterol level is 225 mg/dl. That is, find $H(225)$.

*Copyright 1989, CSPI. Adapted from *Nutrition Action Healthletter* (1875 Connecticut Avenue, N.W., Suite 300, Washington, DC 20009-5728. $24 for 10 issues).

62. Approximate the annual heart attack rate for those men whose blood cholesterol level is 275 mg/dl. That is, find $H(275)$.

Voting attitudes. For Exercises 63 and 64, use this graph, which shows the percentage of people responding yes to the question, "If your (political) party nominated a generally well-qualified person for president who happened to be a woman, would you vote for that person?" **Source:** www.gallup.com

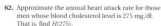

63. Approximate the percentage of Americans willing to vote for a woman for president in 1960. That is, find $P(1960)$.

64. Approximate the percentage of Americans willing to vote for a woman for president in 2000. That is, find $P(2000)$.

Energy-saving lightbulbs. Compact fluorescent (CFL) lightbulbs can be used in most places that conventional incandescent bulbs are, but they use a fraction of the electricity. The table below lists the CFL wattage and the incandescent wattage required to create the same amount of light. **Source:** Westinghouse Lighting Corporation

Input, CFL Wattage	Output, Wattage of Incandescent Equivalent
7	25
20	75
30	120

65. Use the data in the figure above to draw a graph and to estimate the wattage of an incandescent bulb that creates light equivalent to a 15-watt CFL bulb. Then predict the wattage of an incandescent bulb that creates light equivalent to a 35-watt CFL bulb.

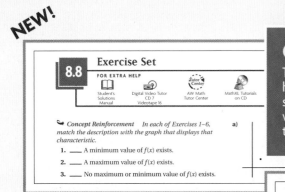

8.8 Exercise Set

FOR EXTRA HELP

Student's Solutions Manual | Digital Video Tutor CD 7 Videotape 16 | AW Math Tutor Center | MathXL Tutorials on CD

Concept Reinforcement In each of Exercises 1–6, match the description with the graph that displays that characteristic.

1. ____ A minimum value of $f(x)$ exists.

2. ____ A maximum value of $f(x)$ exists.

3. ____ No maximum or minimum value of $f(x)$ exists.

a)

Concept Reinforcement Exercises

This feature is designed to help students build their confidence and comprehension through true/false, matching, and fill-in-the-blank exercises at the start of nearly all exercise sets. Where appropriate, special attention is devoted to increasing student understanding of the new vocabulary and notation developed in that section.

Connecting the Concepts

This feature highlights the importance of connecting concepts and invites students to pause and check that they understand the "big picture." This helps assure that students understand how concepts work together in several sections at once.

CONNECTING THE CONCEPTS

We now have three different methods for solving systems of equations. Each method has certain strengths and weaknesses, as outlined below.

Method	Strengths	Weaknesses
Graphical	Solutions are displayed graphically. Can be used with any system that can be graphed.	Inexact when solutions involve numbers that are not integers. Solution may not appear on the part of the graph drawn.
Substitution	Yields exact solutions. Easy to use when a variable is alone on one side.	Introduces extensive computations with fractions when solving more complicated systems. Solutions are not displayed graphically.
Elimination	Yields exact solutions. Easy to use when fractions or decimals appear in the system. The preferred method for systems of 3 or more equations in 3 or more variables (see Section 3.4)	Solutions are not displayed graphically.

Exercises

75. $|4 - 3y| > 8$

76. $|7 - 2y| < -6$

Aha! 77. $|5 - 4x| < -6$

78. $7 + |4a - 5| \leq 26$

79. $\left|\dfrac{2 - 5x}{4}\right| \geq \dfrac{2}{3}$

80. $\left|\dfrac{1 + 3x}{5}\right| > \dfrac{7}{8}$

81. $|m + 3| + 8 \leq 14$

82. $|t - 7| + 3 \geq 4$

83. $25 - 2|a + 3| > 19$

84. $30 - 4|a + 2| > 12$

85. Let $f(x) = |2x - 3|$. Find all x for which $f(x) \leq 4$.

86. Let $f(x) = |5x + 2|$. Find all x for which $f(x) \leq 3$.

87. Let $f(x) = 5 + |3x - 4|$. Find all x for which $f(x) \geq 16$.

88. Let $f(x) = |2 - 9x|$. Find all x for which $f(x) \geq 25$.

89. Let $f(x) = 7 + |2x - 1|$. Find all x for which $f(x) < 16$.

90. Let $f(x) = 5 + |3x + 2|$. Find all x for which $f(x) < 19$.

91. Explain in your own words why -7 is not a solution of $|x| < 5$.

92. Explain in your own words why $[6, \infty)$ is only part of the solution of $|x| \geq 6$.

SKILL MAINTENANCE

Solve using substitution or elimination. [3.2]

93. $2x - 3y = 7,$
    $3x + 2y = -10$

94. $3x - 5y = 9,$
    $4x - 3y = 1$

95. $x = -2 + 3y,$
    $x - 2y = 2$

96. $y = 3 - 4x,$
    $2x - y = -9$

Solve graphically. [3.1]

97. $x + 2y = 9,$
    $3x - y = -1$

98. $2x + y = 7,$
    $-3x - 2y = 10$

SYNTHESIS

99. Is it possible for an equation in x of the form $|ax + b| = c$ to have exactly one solution? Why or why not?

100. Explain why the inequality $|x + 5| \geq 2$ can be interpreted as "the number x is at least 2 units from -5."

101. From the definition of absolute value, $|x| = x$ only when $x \geq 0$. Solve $|3t - 5| = 3t - 5$ using this same reasoning.

102. $|3x - 5| = x$

103. $|x + 2| > x$

104. $2 \leq |x - 1| \leq 5$

105. $|5t - 3| = 2t + 4$

106. $t - 2 \leq |t - 3|$

Find an equivalent inequality with absolute value.

107. $-3 < x < 3$

108. $-5 \leq y \leq 5$

109. $x \leq -6$ or $6 \leq x$

110. $x < -4$ or $4 < x$

111. $x < -8$ or $2 < x$

112. $-5 < x < 1$

113. x is less than 2 units from 7.

114. x is less than 1 unit from 5.

Write an absolute-value inequality for which the interval shown is the solution.

115. [number line from -7 to 7]

116. [number line from -5 to 9]

117. [number line from -7 to 7]

118. [number line from 0 to 14]

119. *Bungee jumping.* A bungee jumper is bouncing up and down so that her distance d above a river satisfies the inequality $|d - 60 \text{ ft}| \leq 10 \text{ ft}$ (see the figure below). If the bridge from which she jumped is 150 ft above the river, how far is the bungee jumper from the bridge at any given time?

150 ft
d
60 ft
d

SKILL MAINTENANCE EXERCISES Skill Maintenance exercises appear in all exercise sets as a means of keeping past concepts fresh and as a way of reviewing previously covered skills to prepare students for upcoming topics.

SYNTHESIS EXERCISES Synthesis exercises guarantee an extensive and wide-ranging variety of problems in every exercise set. The Synthesis exercises allow students to combine concepts from more than one section and provide challenge for even the strongest students.

WRITING EXERCISES Writing exercises, indicated by ▨, provide opportunities for students to answer problems with one or more sentences. Often, these questions have more than one correct response and ask students to explain *why* a certain concept works as it does.

AHA! EXERCISES In many exercise sets, students will see the **Aha!** icon. This icon indicates that there is a simpler way to complete the exercise without going through a lengthy computation. It's then up to the student to discover that simpler approach. The **Aha!** icon appears the first time a new insight can be used on a particular type of exercise. After that, the student must determine if and when that particular insight can be reused.

3 Study Summary

Because so many real-world problems translate into two or more equations in two or more variables, **systems of equations** are studied in great detail (p. 151). Most of the systems studied in this chapter are **consistent**, meaning that they have at least one solution, although we also studied some **inconsistent** systems, for which there is no solution (p. 155). The equations in the systems we solved were **independent**, except for those cases in which one equation could be written as a multiple and/or sum of the other equation(s) (such equations are called **dependent**) (p. 155).

Three methods—graphing, substitution, and elimination—can be used to solve systems of equations (pp. 153, 159, 161). Of the three methods, graphing is the easiest to visualize.

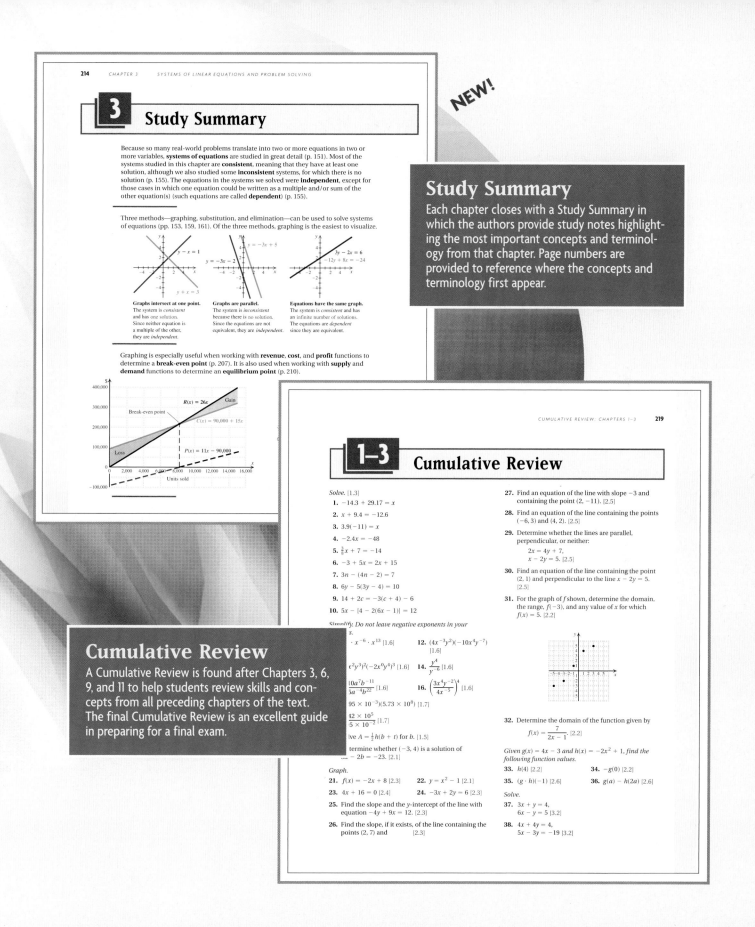

Graphs intersect at one point.
The system is *consistent* and has one solution. Since neither equation is a multiple of the other, they are *independent*.

Graphs are parallel.
The system is *inconsistent* because there is no solution. Since the equations are not equivalent, they are *independent*.

Equations have the same graph.
The system is *consistent* and has an infinite number of solutions. The equations are *dependent* since they are equivalent.

Graphing is especially useful when working with **revenue**, **cost**, and **profit** functions to determine a **break-even point** (p. 207). It is also used when working with **supply** and **demand** functions to determine an **equilibrium point** (p. 210).

Study Summary

Each chapter closes with a Study Summary in which the authors provide study notes highlighting the most important concepts and terminology from that chapter. Page numbers are provided to reference where the concepts and terminology first appear.

1–3 Cumulative Review

Solve. [1.3]

1. $-14.3 + 29.17 = x$

2. $x + 9.4 = -12.6$

3. $3.9(-11) = x$

4. $-2.4x = -48$

5. $\frac{3}{8}x + 7 = -14$

6. $-3 + 5x = 2x + 15$

7. $3n - (4n - 2) = 7$

8. $6y - 5(3y - 4) = 10$

9. $14 + 2c = -3(c + 4) - 6$

10. $5x - [4 - 2(6x - 1)] = 12$

Simplify. Do not leave negative exponents in your [answers].

11. $\cdot x^{-6} \cdot x^{13}$ [1.6]

12. $(4x^{-3}y^2)(-10x^4y^{-7})$ [1.6]

13. $(x^2y^3)^2(-2x^0y^4)^3$ [1.6]

14. $\dfrac{y^4}{y^{-6}}$ [1.6]

15. $\dfrac{10a^7b^{-11}}{5a^{-4}b^{22}}$ [1.6]

16. $\left(\dfrac{3x^4y^{-2}}{4x^{-5}}\right)^4$ [1.6]

17. $95 \times 10^{-3})(5.73 \times 10^8)$ [1.7]

18. $\dfrac{42 \times 10^5}{5 \times 10^{-2}}$ [1.7]

19. [Sol]ve $A = \frac{1}{2}h(b + t)$ for b. [1.5]

20. [De]termine whether $(-3, 4)$ is a solution of [$\ldots$] $- 2b = -23$. [2.1]

Graph.

21. $f(x) = -2x + 8$ [2.3]

22. $y = x^2 - 1$ [2.1]

23. $4x + 16 = 0$ [2.4]

24. $-3x + 2y = 6$ [2.3]

25. Find the slope and the *y*-intercept of the line with equation $-4y + 9x = 12$. [2.3]

26. Find the slope, if it exists, of the line containing the points $(2, 7)$ and [$\ldots$] [2.3]

27. Find an equation of the line with slope -3 and containing the point $(2, -11)$. [2.5]

28. Find an equation of the line containing the points $(-6, 3)$ and $(4, 2)$. [2.5]

29. Determine whether the lines are parallel, perpendicular, or neither:
$$2x = 4y + 7,$$
$$x - 2y = 5. \; [2.5]$$

30. Find an equation of the line containing the point $(2, 1)$ and perpendicular to the line $x - 2y = 5$. [2.5]

31. For the graph of *f* shown, determine the domain, the range, $f(-3)$, and any value of *x* for which $f(x) = 5$. [2.2]

32. Determine the domain of the function given by
$$f(x) = \frac{7}{2x - 1}. \; [2.2]$$

Given $g(x) = 4x - 3$ and $h(x) = -2x^2 + 1$, find the following function values.

33. $h(4)$ [2.2]

34. $-g(0)$ [2.2]

35. $(g \cdot h)(-1)$ [2.6]

36. $g(a) - h(2a)$ [2.6]

Solve.

37. $3x + y = 4,$
$6x - y = 5$ [3.2]

38. $4x + 4y = 4,$
$5x - 3y = -19$ [3.2]

Cumulative Review

A Cumulative Review is found after Chapters 3, 6, 9, and 11 to help students review skills and concepts from all preceding chapters of the text. The final Cumulative Review is an excellent guide in preparing for a final exam.

Intermediate Algebra

CONCEPTS AND APPLICATIONS

1

Algebra and Problem Solving

AN APPLICATION

In 2003, painting the interior of a house increased the average sale price of the home by about $2500. This was approximately $\frac{5}{3}$ of what the average paint job cost. (*Source*: Based on information from HomeGain.com) What was the cost of the average paint job?

This problem appears as Exercise 24 in Section 1.4.

Marilyn Lewis
CONTRACTOR
Kansas City, Missouri

Anyone entering a building trade definitely needs math to figure such things as rafter length and square footage. As a contractor, I use math to watch for waste, to check invoices, to compare budgeted costs with actual costs, and to figure costs in advance.

*T*he principal theme of this text is problem solving in algebra. An overall strategy for solving problems is presented in Section 1.4. Additional and increasing emphasis on problem solving appears throughout the book. This chapter begins with a short review of algebraic symbolism and properties of numbers. As you will see, the manipulations of algebra, such as simplifying expressions and solving equations, are based on the properties of numbers.

1.1   Some Basics of Algebra

Algebraic Expressions and Their Use • Translating to Algebraic Expressions • Evaluating Algebraic Expressions • Sets of Numbers

The primary difference between algebra and arithmetic is the use of *variables*. In this section, we will see how variables can be used to represent various situations. We will also examine the different types of numbers that will be represented by variables throughout this text.

Algebraic Expressions and Their Use

We are all familiar with expressions like

$$95 + 21, \qquad 57 \times 34, \qquad 9 - 4, \quad \text{and} \quad \frac{35}{71}.$$

In algebra, we use these as well as expressions like

$$x + 21, \qquad l \cdot w, \qquad 9 - s, \quad \text{and} \quad \frac{d}{t}.$$

A letter that can be any one of various numbers is called a **variable**. If a letter always represents a particular number that never changes, it is called a **constant**. Let d = the number of hours it takes the moon to orbit the earth. Then d is a constant. If a = the age of a baby chick, in minutes, then a is a variable since a changes as time passes.

An **algebraic expression** consists of variables, numbers, and operation signs. All of the expressions above are examples of algebraic expressions. When an equals sign is placed between two expressions, an **equation** is formed.

Algebraic expressions and equations arise frequently in problem-solving situations. Suppose, for example, that we want to determine by how much the number of DVD–video copies has increased in North America from 2001 to 2003.

North American DVD–Video Replication

Source: The International Recording Media Association

By using *x* to represent the increase in copying, in millions, we can form an equation:

Millions of copies made in 2001	plus	increase in copying	is	millions of copies made in 2003.
636	+	x	=	1225.

To find a **solution**, we can subtract 636 from both sides of the equation:

$$x = 1225 - 636$$
$$x = 589.$$

We see that the number of DVDs copied in North America increased by 589 million from 2001 to 2003.

Translating to Algebraic Expressions

To translate problems to equations, we need to know which words correspond to which symbols:

Key Words

Addition	Subtraction	Multiplication	Division
add	subtract	multiply	divide
sum of	difference of	product of	divided by
plus	minus	times	quotient of
increased by	decreased by	twice	ratio
more than	less than	of	per

When the value of a number is not given, we represent that number with a variable.

Phrase	Algebraic Expression
Five *more than* some number	$n + 5$
Half *of* a number	$\frac{1}{2}t,$ or $\frac{t}{2}$
Five *more than* three *times* some number	$3p + 5$
The *difference of* two numbers	$x - y$
Six *less than* the *product of* two numbers	$rs - 6$
Seventy-six percent *of* some number	$0.76z,$ or $\frac{76}{100}z$

Note that an expression like *rs* represents a product and can also be written as $r \cdot s$, $r \times s$, or $(r)(s)$. The multipliers *r* and *s* are called **factors**.

EXAMPLE 1   Translate to an algebraic expression:

Five less than forty-three percent of the quotient of two numbers.

Solution   We let *r* and *s* represent the two numbers.

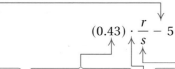

$$(0.43) \cdot \frac{r}{s} - 5$$

Five less than forty-three percent of the quotient of two numbers

Evaluating Algebraic Expressions

When we replace a variable with a number, we say that we are **substituting** for the variable. The calculation that follows the substitution is called **evaluating the expression**.

EXAMPLE 2   Evaluate the expression $3xy + z$ for $x = 2$, $y = 5$, and $z = 7$.

Solution   We substitute and carry out the multiplication and addition:

$$3xy + z = 3 \cdot 2 \cdot 5 + 7 \qquad \text{We use color to highlight the substitution.}$$
$$= 30 + 7$$
$$= 37.$$

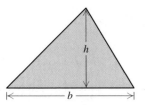

Geometric formulas are often evaluated. In the next example, we use the formula for the area *A* of a triangle with a base of length *b* and a height of length *h*. This is an important formula that is worth remembering:

$$A = \tfrac{1}{2} \cdot b \cdot h.$$

EXAMPLE 3

4 m

3.1 m

The base of a triangular sail is 3.1 m and the height is 4 m. Find the area of the sail.

Solution   We substitute 3.1 for b and 4 for h and multiply:

$$\frac{1}{2} \cdot b \cdot h = \frac{1}{2} \cdot 3.1 \cdot 4$$
$$= 6.2 \text{ square meters (sq m)}.$$

Before evaluating other algebraic expressions, we need to develop *exponential notation*. Many different kinds of numbers can be used as *exponents*. Here we establish the meaning of a^n when n is a counting number, 1, 2, 3, ...

Exponential Notation

The expression a^n, in which n is a counting number, means

$$\underbrace{a \cdot a \cdot a \cdot \cdots \cdot a \cdot a}_{n \text{ factors.}}$$

In a^n, a is called the *base* and n is the *exponent*, or *power*. When no exponent appears, the power is assumed to be 1. Thus, $a^1 = a$.

The expression a^n is read "a raised to the nth power" or simply "a to the nth." We read s^2 as "s-squared" and x^3 as "x-cubed." This terminology comes from the fact that the area of a square of side s is $s \cdot s = s^2$ and the volume of a cube of side x is $x \cdot x \cdot x = x^3$.

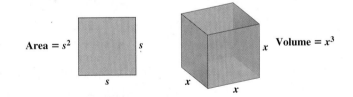

Area $= s^2$ s x **Volume** $= x^3$

s x x

Exponential notation tells us that 5^2 means $5 \cdot 5$, or 25, but what does $1 + 2 \cdot 5^2$ mean? If we add 1 and 2 and multiply by 25, we get 75. If we multiply 2 times 5^2, or 25, and add 1, we get 51. A third possibility is to square $2 \cdot 5$ to get 100 and then add 1. The following convention indicates that only the second of these approaches is correct: We square 5, then multiply, and then add.

Student Notes

Note that rule 3 states that when division precedes multiplication, the division is performed first. Thus, $20 \div 5 \cdot 2$ represents $4 \cdot 2$, or 8. Similarly, $9 - 3 + 1$ represents $6 + 1$, or 7.

Rules for Order of Operations

1. Simplify within any grouping symbols.
2. Simplify all exponential expressions.
3. Perform all multiplication and division, as either occurs, working from left to right.
4. Perform all addition and subtraction, as either occurs, working from left to right.

EXAMPLE 4   Evaluate $5 + 2(a - 1)^2$ for $a = 4$.

Solution

$$
\begin{aligned}
5 + 2(a - 1)^2 &= 5 + 2(4 - 1)^2 && \text{Substituting} \\
&= 5 + 2(3)^2 && \text{Working within parentheses first} \\
&= 5 + 2(9) && \text{Simplifying } 3^2 \\
&= 5 + 18 && \text{Multiplying} \\
&= 23 && \text{Adding}
\end{aligned}
$$

Step (3) in the rules for order of operations tells us to divide before we multiply when division appears first, reading left to right. This means that an expression like $6 \div 2x$ should be thought of as $(6 \div 2)x$.

EXAMPLE 5   Evaluate $9 - x^3 + 6 \div 2y^2$ for $x = 2$ and $y = 5$.

Solution

$$
\begin{aligned}
9 - x^3 + 6 \div 2y^2 &= 9 - 2^3 + 6 \div 2(5)^2 && \text{Substituting} \\
&= 9 - 8 + 6 \div 2 \cdot 25 && \text{Simplifying } 2^3 \text{ and } 5^2 \\
&= 9 - 8 + 3 \cdot 25 && \text{Dividing} \\
&= 9 - 8 + 75 && \text{Multiplying} \\
&= 1 + 75 && \text{Subtracting} \\
&= 76 && \text{Adding}
\end{aligned}
$$

Sets of Numbers

When evaluating algebraic expressions, and in problem solving in general, we often must examine the *type* of numbers used. For example, if a formula is used to determine an optimal class size, any fraction results must be rounded up or down, since it is impossible to have a fraction part of a student. Three frequently used sets of numbers are listed below.

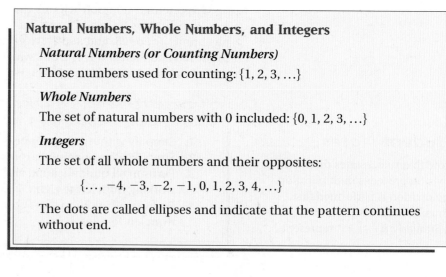

Natural Numbers, Whole Numbers, and Integers

Natural Numbers (or Counting Numbers)
Those numbers used for counting: $\{1, 2, 3, \ldots\}$

Whole Numbers
The set of natural numbers with 0 included: $\{0, 1, 2, 3, \ldots\}$

Integers
The set of all whole numbers and their opposites:

$$\{\ldots, -4, -3, -2, -1, 0, 1, 2, 3, 4, \ldots\}$$

The dots are called ellipses and indicate that the pattern continues without end.

The integers correspond to the points on a number line as follows:

To fill in the numbers between these points, we must describe two more sets of numbers. This requires us to first discuss set notation.

The set containing the numbers -2, 1, and 3 can be written $\{-2, 1, 3\}$. This way of writing a set is known as **roster notation**. Roster notation was used for the three sets listed above. A second type of set notation, **set-builder notation**, specifies conditions under which a number is in the set. The following example of set-builder notation is read as shown:

Set-builder notation is generally used when it is difficult to list a set using roster notation.

EXAMPLE 6   Using both roster notation and set-builder notation, represent the set consisting of the first 15 even natural numbers.

Solution

Using roster notation: $\{2, 4, 6, 8, 10, 12, 14, 16, 18, 20, 22, 24, 26, 28, 30\}$

Using set-builder notation: $\{n \mid n \text{ is an even number between 1 and 31}\}$

The symbol $\in$ is used to indicate that an element or member belongs to a set. Thus if $A = \{2, 4, 6, 8\}$, we can write $4 \in A$ to indicate that 4 *is an element of A*. We can also write $5 \notin A$ to indicate that 5 *is not an element of A*.

EXAMPLE 7   Classify the statement $8 \in \{x \mid x \text{ is an integer}\}$ as true or false.

Solution   Since 8 *is* an integer, the statement is true. In other words, since 8 is an integer, it belongs to the set of all integers.

With set-builder notation, we can describe the set of all *rational numbers*.

Rational Numbers

Numbers that can be expressed as an integer divided by a nonzero integer are called *rational numbers*:

$$\left\{ \frac{p}{q} \,\middle|\, p \text{ is an integer, } q \text{ is an integer, and } q \neq 0 \right\}.$$

Rational numbers can be written using fraction or decimal notation. *Fraction notation* uses symbolism like the following:

$$\frac{5}{8}, \quad \frac{12}{-7}, \quad \frac{-17}{15}, \quad -\frac{9}{7}, \quad \frac{39}{1}, \quad \frac{0}{6}.$$

In *decimal notation*, rational numbers either *terminate* (end) or *repeat*.

EXAMPLE 8   When written in decimal form, does each of the following numbers terminate or repeat? **(a)** $\frac{5}{8}$; **(b)** $\frac{6}{11}$.

Solution

a) Since $\frac{5}{8}$ means $5 \div 8$, we perform long division to find that $\frac{5}{8} = 0.625$, a decimal that ends. Thus, $\frac{5}{8}$ can be written as a terminating decimal.

b) Using long division, we find that $6 \div 11 = 0.5454\ldots$, so we can write $\frac{6}{11}$ as a repeating decimal. Repeating decimal notation can be abbreviated by writing a bar over the repeating part—in this case, $0.\overline{54}$.

Many numbers, like π, $\sqrt{2}$, and $-\sqrt{15}$, are not rational numbers. For example, $\sqrt{2}$ is the number for which $\sqrt{2} \cdot \sqrt{2} = 2$. A calculator's representation of $\sqrt{2}$ as 1.414213562 is only an approximation since $(1.414213562)^2$ is not exactly 2. Note that $\sqrt{2}$ does not repeat or terminate when written in decimal form and cannot be written as a fraction.

To see that $\sqrt{2}$ is a "real" point on the number line, it can be shown that when a right triangle has two legs of length 1, the remaining side has length $\sqrt{2}$. Thus we can "measure" $\sqrt{2}$ units and locate $\sqrt{2}$ on a number line.

Numbers like π, $\sqrt{2}$, and $-\sqrt{15}$ are said to be **irrational**. Decimal notation for irrational numbers neither terminates nor repeats.

The set of all rational numbers, combined with the set of all irrational numbers, gives us the set of all **real numbers**.

Real Numbers

Numbers that are either rational or irrational are called *real numbers*:

$$\{x \,|\, x \text{ is rational or } x \text{ is irrational}\}.$$

Every point on the number line represents some real number and every real number is represented by some point on the number line.

The following figure shows the relationships among various kinds of numbers, along with examples of how real numbers can be sorted.

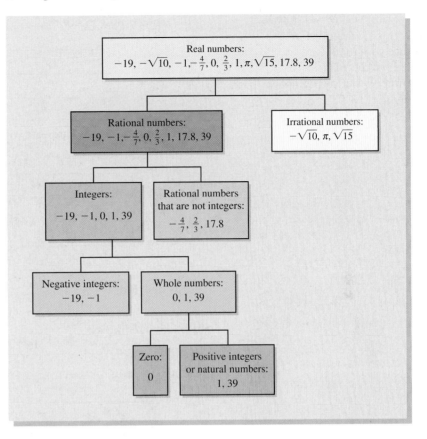

EXAMPLE 9   Which numbers in the following list are **(a)** whole numbers? **(b)** integers? **(c)** rational numbers? **(d)** irrational numbers? **(e)** real numbers?

$$-29, \quad -\frac{7}{4}, \quad 0, \quad 2, \quad 3.9, \quad \sqrt{42}, \quad 78$$

Solution

a) 0, 2, and 78 are whole numbers.

b) -29, 0, 2, and 78 are integers.

c) -29, $-\dfrac{7}{4}$, 0, 2, 3.9, and 78 are rational numbers.

d) $\sqrt{42}$ is an irrational number.

e) -29, $-\dfrac{7}{4}$, 0, 2, 3.9, $\sqrt{42}$, and 78 are all real numbers.

When all members of one set are members of a second set, the first set is a **subset** of the second set. Thus if $A = \{2, 4, 6\}$ and $B = \{1, 2, 4, 5, 6\}$, we write $A \subseteq B$ to indicate that *A is a subset of B*. Similarly, if $\mathbb{N}$ represents the set of all natural numbers and $\mathbb{Z}$ the set of all integers, we can write $\mathbb{N} \subseteq \mathbb{Z}$. Additional statements can be made using other sets in the diagram above.

Study Skills _____

Throughout this textbook, you will find a feature called *Study Skills*. These tips are intended to help improve your math study skills. On the first day of class, we recommend that you complete this chart.

Instructor:

Name _____

Office hours and location _____

Phone number _____

Fax number _____

E-mail address _____

Find the names of two students whom you could contact for information or study questions:

1. Name _____

   Phone number _____

   E-mail address _____

2. Name _____

   Phone number _____

   E-mail address _____

Math lab on campus:

Location _____

Hours _____

Phone _____

Tutoring:

Campus location _____

Hours _____

Addison-Wesley Tutor Center _____

To order, call _____

(See the preface for important information concerning this tutoring.)

Important supplements:

(See the preface for a complete list of available supplements.)

Supplements recommended by the instructor.

_____

_____

Exercise Set

1.1

FOR EXTRA HELP

Student's Solutions Manual · Digital Video Tutor CD 1 Videotape 1 · AW Math Tutor Center · MathXL Tutorials on CD · MathXL · MyMathLab

⮕ *Concept Reinforcement   In each of Exercises 1–10, fill in the blank with the appropriate word or words.*

1. A letter representing a specific number that never changes is called a(n) _____ .

2. A letter that can be any one of a set of numbers is called a(n) _____ .

3. In the expression $7y$, the multipliers 7 and y are called _____ .

4. When all variables in a variable expression are replaced by numbers and a result is calculated, we say that we are _____ the expression.

5. When no grouping symbols, exponents, division, or multiplication appear, we subtract before we add, provided the subtraction appears to the _____ of any addition.

6. In a^b, the a is called the _____ and the b is called the _____ .

7. A number that can be written in the form a/b, where a and b are integers (with $b \neq 0$), is said to be a(n) _____ number.

8. A real number that cannot be written as a quotient of two integers is an example of a(n) _____ number.

9. Division can be used to show that $\frac{7}{40}$ can be written as a(n) _____ decimal.

10. Division can be used to show that $\frac{13}{7}$ can be written as a(n) _____ decimal.

Use mathematical symbols to translate each phrase.

11. Six less than some number

12. Four more than some number

13. Twelve times a number

14. Twice a number

15. Sixty-five percent of some number

16. Thirty-nine percent of some number

17. Nine more than twice a number

18. Six less than half of a number

19. Eight more than ten percent of some number

20. Five less than six percent of some number

21. One less than the difference of two numbers

22. One more than the product of two numbers

23. Ninety miles per every four gallons of gas

24. One hundred words per every sixty seconds

To the student and the instructor: Throughout this text, selected exercises are marked with the icon **Aha!** *. These "Aha!" exercises can be answered quite easily if the student pauses to inspect the exercise rather than proceed mechanically. This is done to discourage rote memorization. Some "Aha!" exercises are left unmarked to encourage students to always pause before working a problem.*

Evaluate each expression using the values provided.

25. $7x + y$, for $x = 3$ and $y = 4$

26. $6a - b$, for $a = 5$ and $b = 3$

27. $2c \div 3b$, for $b = 2$ and $c = 6$

28. $3z \div 2y$, for $y = 1$ and $z = 6$

29. $25 + r^2 - s$, for $r = 3$ and $s = 7$

30. $n^3 + 2 - p$, for $n = 2$ and $p = 5$

Aha! 31. $3n^2p - 3pn^2$, for $n = 5$ and $p = 9$

32. $2a^3b - 2b^2$, for $a = 3$ and $b = 7$

33. $5x \div (2 + x - y)$, for $x = 6$ and $y = 2$

34. $3(m + 2n) \div m$, for $m = 7$ and $n = 0$

35. $[10 - (a - b)]^2$, for $a = 7$ and $b = 2$

36. $[17 - (x + y)]^2$, for $x = 4$ and $y = 1$

37. $[5(r + s)]^2$, for $r = 1$ and $s = 2$

38. $[3(a - b)]^2$, for $a = 7$ and $b = 5$

39. $m^2 - [2(m - n)]^2$, for $m = 7$ and $n = 5$

40. $x^2 - [3(x - y)]^2$, for $x = 6$ and $y = 4$

41. $(r - s)^2 - 3(2r - s)$, for $r = 11$ and $s = 3$

42. $(m - 2n)^2 - 2(m + n)$, for $m = 8$ and $n = 1$

In Exercises 43–46, find the area of a triangular window with the given base and height.

43. Base = 5 ft, height = 7 ft

44. Base = 2.9 m, height = 2.1 m

45. Base = 7 m, height = 3.2 m

46. Base = 3.6 ft, height = 4 ft

Use roster notation to write each set.

47. The set of all vowels in the alphabet

48. The set of all days of the week

49. The set of all odd natural numbers

50. The set of all even natural numbers

51. The set of all natural numbers that are multiples of 5

52. The set of all natural numbers that are multiples of 10

Use set-builder notation to write each set.

53. The set of all odd numbers between 10 and 20

54. The set of all multiples of 4 between 22 and 35

55. $\{0, 1, 2, 3, 4\}$

56. $\{-3, -2, -1, 0, 1, 2\}$

57. The set of all multiples of 5 between 7 and 79

58. The set of all even numbers between 9 and 99

*In Exercises 59–62, determine which numbers in the list provided are **(a)** whole numbers? **(b)** integers? **(c)** rational numbers? **(d)** irrational numbers? **(e)** real numbers?*

59. $-8.7, -3, 0, \dfrac{2}{3}, \sqrt{7}, 6$

60. $-\dfrac{9}{2}, -4, -1.2, 0, \sqrt{5}, 3$

61. $-17, -4.13, 0, \dfrac{5}{4}, 3, \sqrt{77}$

62. $-9.1, -2, 0, 4, \sqrt{17}, \dfrac{99}{2}$

Classify each statement as true or false. The following sets are used:

$\mathbb{N}$ = the set of natural numbers;

$\mathbb{W}$ = the set of whole numbers;

$\mathbb{Z}$ = the set of integers;

$\mathbb{Q}$ = the set of rational numbers;

$\mathbb{H}$ = the set of irrational numbers;

$\mathbb{R}$ = the set of real numbers.

63. $5.1 \in \mathbb{N}$

64. $\mathbb{N} \subseteq \mathbb{W}$

65. $\mathbb{W} \subseteq \mathbb{Z}$

66. $\sqrt{8} \in \mathbb{Q}$

67. $\frac{2}{3} \in \mathbb{H}$

68. $\mathbb{H} \subseteq \mathbb{R}$

69. $\sqrt{10} \in \mathbb{R}$

70. $4.3 \notin \mathbb{Z}$

71. $\mathbb{Z} \nsubseteq \mathbb{N}$

72. $\mathbb{Q} \subseteq \mathbb{R}$

73. $\mathbb{Q} \subseteq \mathbb{Z}$

74. $9 \in \mathbb{N}$

To the student and the instructor: The icon ▨ is used to denote writing exercises. These exercises are meant to be answered with one or more English sentences. Because many writing exercises have a variety of correct answers, these solutions are not listed in the answers at the back of the book.

75. What is the difference between rational numbers and integers?

76. Werner insists that $15 - 4 + 1 \div 2 \cdot 3$ is 2. What error is he making?

SYNTHESIS

To the student and the instructor: *Synthesis exercises are designed to challenge students to extend the concepts or skills studied in each section. Many synthesis exercises require the assimilation of skills and concepts from several sections.*

77. Is the following true or false, and why?
$$\{2, 4, 6\} \subseteq \{2, 4, 6\}$$

78. On a quiz, Francesca answers $6 \in \mathbb{Z}$ while Jacob writes $\{6\} \in \mathbb{Z}$. Jacob's answer does not receive full credit while Francesca's does. Why?

Translate to an algebraic expression.

79. The quotient of the sum of two numbers and their difference

80. Three times the sum of the cubes of two numbers

81. Half of the difference of the squares of two numbers

82. The product of the difference of two numbers and their sum

Use roster notation to write each set.

83. The set of all whole numbers that are not natural numbers

84. The set of all integers that are not whole numbers

85. $\{x \mid x = 5n, n \text{ is a natural number}\}$

86. $\{x \mid x = 3n, n \text{ is a natural number}\}$

87. $\{x \mid x = 2n + 1, n \text{ is a whole number}\}$

88. $\{x \mid x = 2n, n \text{ is an integer}\}$

89. Draw a right triangle that could be used to measure $\sqrt{13}$ units.

1.2 Operations and Properties of Real Numbers

Absolute Value • Inequalities • Addition, Subtraction, and Opposites • Multiplication, Division, and Reciprocals • The Commutative, Associative, and Distributive Laws

In this section, we review addition, subtraction, multiplication, and division of real numbers. We also study important rules for manipulating algebraic expressions. First, however, we must discuss absolute value and inequalities.

Absolute Value

It is convenient to have a notation that represents a number's distance from zero on the number line. Note that distance is never negative.

Absolute Value

The notation $|a|$, read "the absolute value of a," represents the number of units that a is from zero.

EXAMPLE 1

Find the absolute value: **(a)** $|-3|$; **(b)** $|2.5|$; **(c)** $|0|$.

Solution

a) $|-3| = 3$ -3 is 3 units from 0.

b) $|2.5| = 2.5$ 2.5 is 2.5 units from 0.

c) $|0| = 0$ 0 is 0 units from itself.

Note that whereas the absolute value of a nonnegative number is the number itself, to find the absolute value of a negative number, we must make it positive.

Inequalities

For any two numbers on the number line, the one to the left is said to be less than, or smaller than, the one to the right. The symbol $<$ means "is less than" and the symbol $>$ means "is greater than." The symbol $\leq$ means "is less than or equal to" and the symbol $\geq$ means "is greater than or equal to." These symbols are used to form **inequalities**.

As shown in the figure below, $-6 < -1$ (since -6 is to the left of -1) and $|-6| > |-1|$ (since 6 is to the right of 1).

EXAMPLE 2

Write out the meaning of each inequality and determine whether it is a true statement.

a) $-7 < -2$ **b)** $4 > -1$ **c)** $-3 \geq -2$

d) $5 \leq 6$ **e)** $6 \leq 6$

Solution

Inequality	*Meaning*
a) $-7 < -2$	"-7 is less than -2" is true because -7 is to the left of -2.
b) $4 > -1$	"4 is greater than -1" is true because 4 is to the right of -1.
c) $-3 \geq -2$	"-3 is greater than or equal to -2" is false because -3 is to the left of -2.
d) $5 \leq 6$	"5 is less than or equal to 6" is true because $5 < 6$ is true.
e) $6 \leq 6$	"6 is less than or equal to 6" is true because $6 = 6$ is true.

Addition, Subtraction, and Opposites

We are now ready to review the addition of real numbers.

Addition of Two Real Numbers

1. *Positive numbers*: Add the numbers. The result is positive.
2. *Negative numbers*: Add absolute values. Make the answer negative.
3. *A negative and a positive number*: If the numbers have the same absolute value, the answer is 0. Otherwise, subtract the smaller absolute value from the larger one:
   a) If the positive number has the greater absolute value, make the answer positive.
   b) If the negative number has the greater absolute value, make the answer negative.
4. *One number is zero*: The sum is the other number.

EXAMPLE 3   Add: **(a)** $-9 + (-5)$; **(b)** $-3.24 + 8.7$; **(c)** $-\frac{3}{4} + \frac{1}{3}$.

Solution

a) $-9 + (-5)$ — We add the absolute values, getting 14. The answer is *negative*, -14.

b) $-3.24 + 8.7$ — The absolute values are 3.24 and 8.7. Subtract 3.24 from 8.7 to get 5.46. The positive number is further from 0, so the answer is *positive*, 5.46.

c) $-\frac{3}{4} + \frac{1}{3} = -\frac{9}{12} + \frac{4}{12}$ — The absolute values are $\frac{9}{12}$ and $\frac{4}{12}$. Subtract to get $\frac{5}{12}$. The negative number is further from 0, so the answer is *negative*, $-\frac{5}{12}$.

When numbers like 7 and -7 are added, the result is 0. Such numbers are called **opposites**, or **additive inverses**, of one another.

The Law of Opposites

For any two numbers a and $-a$,

$$a + (-a) = 0.$$

(When opposites are added, their sum is 0.)

EXAMPLE 4   Find the opposite: **(a)** -17.5; **(b)** $\frac{4}{5}$; **(c)** 0.

Solution

a) The opposite of -17.5 is 17.5 because $-17.5 + 17.5 = 0$.

b) The opposite of $\frac{4}{5}$ is $-\frac{4}{5}$ because $\frac{4}{5} + \left(-\frac{4}{5}\right) = 0$.

c) The opposite of 0 is 0 because $0 + 0 = 0$.

To name the opposite, we use the symbol "−" and read the symbolism −a as "the opposite of a."

Caution!   −a does not necessarily denote a negative number. In particular, when a is *negative*, −a is *positive*.

EXAMPLE 5

Find −x for the following: **(a)** $x = -2$; **(b)** $x = \frac{3}{4}$.

Solution

a) If $x = -2$, then $-x = -(-2) = 2$.   The opposite of −2 is 2.

b) If $x = \frac{3}{4}$, then $-x = -\frac{3}{4}$.   The opposite of $\frac{3}{4}$ is $-\frac{3}{4}$.

---

technology connection

Technology Connections highlight situations in which calculators (primarily graphing calculators) or computers can be used to enrich the learning experience. Most Technology Connections present information in a generic form—consult an outside reference for specific keystrokes.

Graphing calculators have different keys for subtracting and writing negatives. The key labeled (-) is used to create a negative sign, whereas − is used for subtraction.

1. Use a graphing calculator to check Example 6.

2. Calculate: $-3.9 - (-4.87)$.

Using the notation of opposites, we can formally define absolute value.

Absolute Value

$$|x| = \begin{cases} x & \text{if } x \geq 0, \\ -x & \text{if } x < 0 \end{cases}$$

(When x is nonnegative, the absolute value of x is x. When x is negative, the absolute value of x is the opposite of x. Thus, $|x|$ is never negative.)

A negative number is said to have a negative "sign" and a positive number a positive "sign." To subtract, we can add an opposite. This can be stated as: "Change the sign of the number being subtracted and then add."

EXAMPLE 6

Subtract: **(a)** $5 - 9$; **(b)** $-1.2 - (-3.7)$; **(c)** $-\frac{4}{5} - \frac{2}{3}$.

Solution

a) $5 - 9 = 5 + (-9)$ Change the sign and add.

  $= -4$

b) $-1.2 - (-3.7) = -1.2 + 3.7$ Instead of *subtracting* −3.7, we *add* 3.7.

  $= 2.5$

c) $-\frac{4}{5} - \frac{2}{3} = -\frac{4}{5} + \left(-\frac{2}{3}\right)$ Finding a common denominator

  $= -\frac{12}{15} + \left(-\frac{10}{15}\right)$

  $= -\frac{22}{15}$

Multiplication, Division, and Reciprocals

Multiplication of real numbers can be regarded as repeated addition or as repeated subtraction that begins at 0. For example,

$$3 \cdot (-4) = 0 + (-4) + (-4) + (-4) = -12$$

and

$$(-2)(-5) = 0 - (-5) - (-5) = 0 + 5 + 5 = 10.$$

Multiplication of Two Real Numbers

1. To multiply two numbers with *unlike signs*, multiply their absolute values. The answer is *negative*.
2. To multiply two numbers having the *same sign*, multiply their absolute values. The answer is *positive*.

Thus we have

$$(-4)9 = -36 \quad \text{and} \quad \left(-\frac{2}{3}\right)\left(-\frac{3}{7}\right) = \frac{2}{7}.$$

To divide, recall that the quotient $a \div b$ (also denoted a/b) is that number c for which $c \cdot b = a$. For example, $10 \div (-2) = -5$ since $(-5)(-2) = 10$; $(-12) \div 3 = -4$ since $(-4)3 = -12$; and $-18 \div (-6) = 3$ since $3(-6) = -18$. Thus the rules for division are just like those for multiplication.

Division of Two Real Numbers

1. To divide two numbers with *unlike signs*, divide their absolute values. The answer is *negative*.
2. To divide two numbers having the *same sign*, divide their absolute values. The answer is *positive*.

Thus we have

$$\frac{-45}{-15} = 3 \quad \text{and} \quad \frac{20}{-4} = -5.$$

Note that since

$$\frac{-8}{2} = \frac{8}{-2} = -\frac{8}{2} = -4,$$

we have the following generalization.

The Sign of a Fraction

For any number a and any nonzero number b,

$$\frac{-a}{b} = \frac{a}{-b} = -\frac{a}{b}.$$

Recall that

$$\frac{a}{b} = \frac{a}{1} \cdot \frac{1}{b} = a \cdot \frac{1}{b}.$$

That is, rather than divide by b, we can multiply by $1/b$. Provided that b is not 0, the numbers b and $1/b$ are called **reciprocals**, or **multiplicative inverses**, of each other.

The Law of Reciprocals

For any two numbers a and $1/a$ $(a \neq 0)$,

$$a \cdot \frac{1}{a} = 1.$$

The product of two reciprocals is 1.

Every number except 0 has exactly one reciprocal. The number 0 has no reciprocal.

EXAMPLE 7   Find the reciprocal: **(a)** $\frac{7}{8}$; **(b)** $-\frac{3}{4}$; **(c)** -8.

Solution

a) The reciprocal of $\frac{7}{8}$ is $\frac{8}{7}$ because $\frac{7}{8} \cdot \frac{8}{7} = 1$.

b) The reciprocal of $-\frac{3}{4}$ is $-\frac{4}{3}$.

c) The reciprocal of -8 is $\frac{1}{-8}$, or $-\frac{1}{8}$.

To divide, we can multiply by a reciprocal. We sometimes say that we "invert and multiply."

EXAMPLE 8   Divide: **(a)** $-\frac{1}{4} \div \frac{3}{5}$; **(b)** $-\frac{6}{7} \div (-10)$.

Solution

a) $-\frac{1}{4} \div \frac{3}{5} = -\frac{1}{4} \cdot \frac{5}{3}$ "Inverting" $\frac{3}{5}$ and changing division to multiplication

$$= -\frac{5}{12}$$

b) $-\frac{6}{7} \div (-10) = -\frac{6}{7} \cdot \left(-\frac{1}{10}\right) = \frac{6}{70}$, or $\frac{3}{35}$

Thus far, we have never divided by 0 or, equivalently, had a denominator of 0. There is a reason for this. Suppose 5 were divided by 0. The answer would have to be a number that, when multiplied by 0, gave 5. But any number times 0 is 0. Thus we cannot divide 5 or any other nonzero number by 0.

What if we divide 0 by 0? In this case, our solution would need to be some number that, when multiplied by 0, gave 0. But then *any* number would work as a solution to $0 \div 0$. This could lead to contradictions so we agree to exclude division of 0 by 0 also.

Division by Zero

We never divide by 0. If asked to divide a nonzero number by 0, we say that the answer is *undefined*. If asked to divide 0 by 0, we say that the answer is *indeterminate*.

The rules for order of operations discussed in Section 1.1 apply to *all* real numbers, regardless of their signs.

EXAMPLE 9   Simplify: $7 - 5^2 + 6 \div 2(-5)^2$.

Solution

$$7 - 5^2 + 6 \div 2(-5)^2 = 7 - 25 + 6 \div 2 \cdot 25 \qquad \text{Simplifying } 5^2 \text{ and } (-5)^2$$

$$= 7 - 25 + 3 \cdot 25 \qquad \text{Dividing}$$

$$= 7 - 25 + 75 \qquad \text{Multiplying}$$

$$= -18 + 75 \qquad \text{Subtracting}$$

$$= 57 \qquad \text{Adding}$$

Besides parentheses, brackets, and braces, groupings may be indicated by a fraction bar, absolute-value symbol, or radical sign ($\sqrt{}$).

EXAMPLE 10   Calculate: $\dfrac{12\,|7 - 9| + 4 \cdot 5}{(-3)^4 + 2^3}$.

Solution   We simplify the numerator and the denominator and divide the results:

$$\frac{12\,|7 - 9| + 4 \cdot 5}{(-3)^4 + 2^3} = \frac{12\,|-2| + 20}{81 + 8}$$

$$= \frac{12(2) + 20}{89}$$

$$= \frac{44}{89}. \qquad \text{Multiplying and adding}$$

The Commutative, Associative, and Distributive Laws

When a pair of real numbers are added or multiplied, the order in which the numbers are written does not affect the result.

The Commutative Laws

For any real numbers a and b,

$$a + b = b + a; \qquad a \cdot b = b \cdot a.$$
(for Addition) $\qquad$ (for Multiplication)

The commutative laws provide one way of writing *equivalent expressions*.

Equivalent Expressions

Two expressions that have the same value for all possible replacements are called *equivalent expressions*.

Much of this text is devoted to finding equivalent expressions.

EXAMPLE 11 Use a commutative law to write an expression equivalent to $7x + 9$.

Solution   Using the commutative law of addition, we have

$$7x + 9 = 9 + 7x.$$

We can also use the commutative law of multiplication to write

$$7 \cdot x + 9 = x \cdot 7 + 9.$$

The expressions $7x + 9$, $9 + 7x$, and $x \cdot 7 + 9$ are all equivalent. They name the same number for any replacement of x.

The *associative laws* also enable us to form equivalent expressions.

The Associative Laws

For any real numbers a, b, and c,

$$a + (b + c) = (a + b) + c; \qquad a \cdot (b \cdot c) = (a \cdot b) \cdot c.$$
(for Addition) $\qquad$ (for Multiplication)

EXAMPLE 12 Write an expression equivalent to $(3x + 7y) + 9z$, using the associative law of addition.

Solution   We have

$$(3x + 7y) + 9z = 3x + (7y + 9z).$$

The expressions $(3x + 7y) + 9z$ and $3x + (7y + 9z)$ are equivalent. They name the same number for any replacements of x, y, and z.

EXAMPLE 13   Use the commutative and associative laws to write an expression equivalent to

$$\frac{5}{x} \cdot (yz).$$

Solution   Answers may vary. We use the associative law first, then the commutative law, although the reverse sequence could also be used.

$$\frac{5}{x} \cdot (yz) = \left(\frac{5}{x} \cdot y\right) \cdot z \qquad \text{Using the associative law of multiplication}$$

$$= \left(y \cdot \frac{5}{x}\right) \cdot z \qquad \text{Using the commutative law inside the parentheses}$$

The *distributive law* that follows provides still another way of forming equivalent expressions. In essence, the distributive law allows us to rewrite the *product* of a and $b + c$ as the *sum* of ab and ac.

The Distributive Law

For any real numbers a, b, and c,

$$a(b + c) = ab + ac.$$

EXAMPLE 14   Obtain an expression equivalent to $5x(y + 4)$ by multiplying.

Solution   We use the distributive law to get

$$5x(y + 4) = 5x \cdot y + 5x \cdot 4 \qquad \text{Using the distributive law}$$

$$= 5xy + 5 \cdot 4 \cdot x \qquad \text{Using the commutative law of multiplication}$$

$$= 5xy + 20x. \qquad \text{Simplifying}$$

The expressions $5x(y + 4)$ and $5xy + 20x$ are equivalent. They name the same number for any replacements of x and y.

Student Notes ⎯⎯⎯⎯

The commutative, associative, and distributive laws are used so often in this course that it is worth the effort to memorize them.

When we reverse what we did in Example 14, we say that we are **factoring** an expression. This allows us to rewrite a sum or a difference as a product.

EXAMPLE 15   Obtain an expression equivalent to $3x - 6$ by factoring.

Solution   We use the distributive law to get

$$3x - 6 = 3(x - 2).$$

In Example 15, since the product of 3 and $x - 2$ is $3x - 6$, we say that 3 and $x - 2$ are **factors** of $3x - 6$. Thus the word "factor" can act as a noun or as a verb.

Exercise Set

1.2

↪ *Concept Reinforcement* *Classify each of the following as either true or false.*

1. The sum of two negative numbers is always negative.

2. The product of two negative numbers is always negative.

3. The product of a negative number and a positive number is always negative.

4. The sum of a negative number and a positive number is always negative.

5. The sum of a negative number and a positive number is always positive.

6. If a and b are negative, with $a < b$, then $|a| > |b|$.

7. If a and b are positive, with $a < b$, then $|a| > |b|$.

8. The commutative law of addition states that for all real numbers a and b, $a + b$ and $b + a$ are equivalent.

9. The associative law of multiplication states that for all real numbers a, b, and c, $(ab)c$ is equivalent to $a(bc)$.

10. The distributive law states that the order in which two numbers are multiplied does not change the result.

Find each absolute value.

11. $|-9|$ 12. $|-7|$ 13. $|6|$

14. $|47|$ 15. $|-6.2|$ 16. $|-7.9|$

17. $|0|$ 18. $\left|3\frac{3}{4}\right|$ 19. $\left|1\frac{7}{8}\right|$

20. $|7.24|$ 21. $|-4.21|$ 22. $|-5.309|$

Write the meaning of each inequality, and determine whether it is a true statement.

23. $-6 \leq -2$ 24. $-1 \leq -5$

25. $-9 > 1$ 26. $7 \geq -2$

27. $3 \geq -5$ 28. $9 \leq 9$

29. $-8 < -3$ 30. $7 \geq -8$

31. $-4 \geq -4$ 32. $2 < 2$

33. $-5 < -5$ 34. $-2 > -12$

Add.

35. $3 + 9$ 36. $8 + 3$

37. $-4 + (-7)$ 38. $-8 + (-3)$

39. $-3.9 + 2.7$ 40. $-1.9 + 7.3$

41. $\frac{2}{7} + \left(-\frac{3}{5}\right)$ 42. $\frac{3}{8} + \left(-\frac{2}{5}\right)$

43. $-3.26 + (-5.8)$ 44. $-2.1 + (-7.5)$

45. $-\frac{1}{9} + \frac{2}{3}$ 46. $-\frac{1}{2} + \frac{4}{5}$

47. $0 + (-4.5)$ 48. $-3.19 + 0$

49. $-7.24 + 7.24$ 50. $-9.46 + 9.46$

51. $15.9 + (-22.3)$ 52. $21.7 + (-28.3)$

Find the opposite, or additive inverse.

53. 3.14 54. 5.43 55. $-4\frac{1}{3}$

56. $2\frac{3}{5}$ 57. 0 58. $-2\frac{3}{4}$

Find $-x$ for each of the following.

59. $x = 9$ 60. $x = 3$

61. $x = -2.7$ 62. $x = -1.9$

63. $x = 1.79$ 64. $x = 3.14$

65. $x = 0$ 66. $x = -7$

Subtract.

67. $8 - 5$ 68. $10 - 3$

69. $5 - 8$ 70. $3 - 10$

71. $-5 - (-12)$ 72. $-3 - (-9)$

73. $-5 - 14$ 74. $-9 - 8$

75. $2.7 - 5.8$ 76. $3.7 - 4.2$

77. $-\frac{3}{5} - \frac{1}{2}$

78. $-\frac{2}{3} - \frac{1}{5}$

79. $-3.9 - (-6.8)$

80. $-5.4 - (-4.3)$

81. $0 - (-7.9)$

82. $0 - 5.3$

Multiply.

83. $(-5)6$

84. $(-4)7$

85. $(-4)(-9)$

86. $(-7)(-8)$

87. $(4.2)(-5)$

88. $(3.5)(-8)$

89. $\frac{3}{7}(-1)$

90. $-1 \cdot \frac{2}{5}$

91. $(-17.45) \cdot 0$

92. 15.2×0

93. $(-3.2) \times (-1.7)$

94. $(1.9) \cdot (4.3)$

Divide.

95. $\frac{-10}{-2}$

96. $\frac{-15}{-3}$

97. $\frac{-100}{20}$

98. $\frac{-50}{5}$

99. $\frac{73}{-1}$

100. $\frac{-62}{1}$

101. $\frac{0}{-7}$

102. $\frac{0}{-11}$

Find the reciprocal, or multiplicative inverse, if it exists.

103. 4

104. 3

105. -9

106. -5

107. $\frac{2}{3}$

108. $\frac{4}{7}$

109. $-\frac{3}{11}$

110. 0

Divide.

111. $\frac{2}{3} \div \frac{4}{5}$

112. $\frac{2}{7} \div \frac{6}{5}$

113. $-\frac{3}{5} \div \frac{1}{2}$

114. $\left(-\frac{4}{7}\right) \div \frac{1}{3}$

115. $\left(-\frac{2}{9}\right) \div (-8)$

116. $\left(-\frac{2}{11}\right) \div (-6)$

Aha! **117.** $-\frac{12}{7} \div \left(-\frac{12}{7}\right)$

118. $\left(-\frac{2}{7}\right) \div (-1)$

Calculate using the rules for order of operations. If an expression is undefined, state this.

119. $9 - (8 - 3 \cdot 2^3)$

120. $19 - (4 + 2 \cdot 3^2)$

121. $\frac{5 \cdot 2 - 4^2}{27 - 2^4}$

122. $\frac{7 \cdot 3 - 5^2}{9 + 4 \cdot 2}$

123. $\frac{3^4 - (5-3)^4}{8 - 2^3}$

124. $\frac{4^3 - (7-4)^2}{3^2 - 7}$

125. $\frac{(2-3)^3 - 5|2-4|}{7 - 2 \cdot 5^2}$

126. $\frac{8 \div 4 \cdot 6|4^2 - 5^2|}{9 - 4 + 11 - 4^2}$

127. $|2^2 - 7|^3 + 4$

128. $|-2 - 3| \cdot 4^2 - 3$

129. $32 - (-5)^2 + 15 \div (-3) \cdot 2$

130. $43 - (-9 + 2)^2 + 18 \div 6 \cdot (-2)$

131. $17 - \sqrt{11 - (3+4)} \div [-5 - (-6)]^2$

132. $15 - 1 + \sqrt{5^2 - (3+1)^2}(-1)$

Write an equivalent expression using a commutative law. Answers may vary.

133. $4a + 7b$

134. $6 + xy$

135. $(7x)y$

136. $-9(ab)$

Write an equivalent expression using an associative law.

137. $(3x)y$

138. $-7(ab)$

139. $x + (2y + 5)$

140. $(3y + 4) + 10$

Write an equivalent expression using the distributive law.

141. $2(a + 5)$

142. $8(x + 1)$

143. $4(x - y)$

144. $9(a - b)$

145. $-5(2a + 3b)$

146. $-2(3c + 5d)$

147. $9a(b - c + d)$

148. $5x(y - z + w)$

Find an equivalent expression by factoring.

149. $5x + 15$

150. $7a + 7b$

151. $3p - 9$

152. $15x - 3$

153. $7x - 21y + 14z$

154. $6y - 9x - 3w$

Aha! **155.** $255 - 34b$

156. $132a + 33$

157. Describe in your own words a method for determining the sign of the sum of a positive number and a negative number.

158. What is the difference between the associative law of multiplication and the distributive law?

SKILL MAINTENANCE

To the student and the instructor: Exercises included for Skill Maintenance review skills previously studied in the text. Usually these exercises provide preparation for the next section of the text. The section(s) in which these types of exercises first appeared is shown in brackets. Answers to all Skill Maintenance exercises appear at the back of the book.

Evaluate. [1.1]

159. $2(x + 5)$ and $2x + 10$, for $x = 3$

160. $2a - 3$ and $a - 3 + a$, for $a = 7$

SYNTHESIS

161. Explain in your own words why 7/0 is undefined.

162. Write a sentence in which the word "factor" appears once as a verb and once as a noun.

Insert one pair of parentheses to convert each false statement into a true statement.

163. $8 - 5^3 + 9 = 36$

164. $2 \cdot 7 + 3^2 \cdot 5 = 104$

165. $5 \cdot 2^3 \div 3 - 4^4 = 40$

166. $2 - 7 \cdot 2^2 + 9 = -11$

167. Find the greatest value of a for which $|a| \geq 6.2$ and $a < 0$.

168. Use the commutative, associative, and distributive laws to show that $5(a + bc)$ is equivalent to $c(b \cdot 5) + a \cdot 5$. Use only one law in each step of your work.

169. Are subtraction and division commutative? Why or why not?

170. Are subtraction and division associative? Why or why not?

1.3 Solving Equations

Equivalent Equations • The Addition and Multiplication Principles • Combining Like Terms • Types of Equations

Solving equations is an essential part of problem solving in algebra. In this section, we review and practice solving basic equations.

Equivalent Equations

In Section 1.1, we saw that the solution of $636 + x = 1225$ is 589. That is, when x is replaced with 589, the equation $636 + x = 1225$ is a true statement. Although this solution may seem obvious, it is important to know how to find such a solution using the principles of algebra. These principles are used to produce *equivalent equations* from which solutions are easily found.

> **Equivalent Equations**
>
> Two equations are *equivalent* if they have the same solution(s).

EXAMPLE 1   Determine whether $4x = 12$ and $10x = 30$ are equivalent equations.

Solution   The equation $4x = 12$ is true only when x is 3. Similarly, $10x = 30$ is true only when x is 3. Since both equations have the same solution, they are equivalent.

EXAMPLE 7   Simplify: $3x + 2[4 + 5(x + 2y)]$.

Solution

$$3x + 2[4 + 5(x + 2y)] = 3x + 2[4 + 5x + 10y] \qquad \text{Using the distributive law}$$

$$= 3x + 8 + 10x + 20y \qquad \text{Using the distributive law again}$$

$$= 13x + 8 + 20y \qquad \text{Combining like terms}$$

The product of a number and -1 is its opposite, or additive inverse. For example,

$$-1 \cdot 8 = -8 \qquad \text{(the opposite of 8).}$$

Thus we have $-8 = -1 \cdot 8$, and in general, $-x = -1 \cdot x$. We can use this fact along with the distributive law when parentheses are preceded by a negative sign or subtraction.

EXAMPLE 8   Simplify $-(a - b)$, using multiplication by -1.

Solution   We have

$$-(a - b) = -1 \cdot (a - b) \qquad \text{Replacing} - \text{with multiplication by} -1$$

$$= -1 \cdot a - (-1) \cdot b \qquad \text{Using the distributive law}$$

$$= -a - (-b) \qquad \text{Replacing } -1 \cdot a \text{ with } -a \text{ and } (-1) \cdot b \text{ with } -b$$

$$= -a + b, \text{ or } b - a. \qquad \text{Try to go directly to this step.}$$

The expressions $-(a - b)$ and $b - a$ are equivalent. They represent the same number for all replacements of a and b.

Example 8 illustrates a useful shortcut worth remembering:

The opposite of $a - b$ is $-a + b$,   or $b - a$.

EXAMPLE 9   Simplify: $9x - 5y - (5x + y - 7)$.

Solution

$$9x - 5y - (5x + y - 7) = 9x - 5y - 5x - y + 7 \qquad \text{Using the distributive law}$$

$$= 4x - 6y + 7 \qquad \text{Combining like terms}$$

CONNECTING THE CONCEPTS

It is important to distinguish between *equivalent expressions* and *equivalent equations*.

In Examples 4 and 5, we used the addition and multiplication principles to write a sequence of equivalent equations that led to an equation for which the solution was clear. Because the equations were equivalent, the solution of the last equation was also a solution of the original equation.

In Examples 6–9, we used the commutative, associative, and distributive laws to write a sequence of equivalent, and increasingly simple, expressions that take on the same value when the variables are replaced with numbers. This is how we "simplify" an expression.

Equivalent Equations

$$
\begin{aligned}
y - 4.7 &= 13.9 \\
y - 4.7 + 4.7 &= 13.9 + 4.7 \\
y + 0 &= 13.9 + 4.7 \\
y &= 18.6
\end{aligned}
$$

Each line here is a complete equation. Because they are equivalent, all four equations share the same solution.

Equivalent Expressions

$$
\begin{aligned}
&3x + 2[4 + 5(x + 2y)] \\
&= 3x + 2[4 + 5x + 10y] \\
&= 3x + 8 + 10x + 20y \\
&= 13x + 8 + 20y
\end{aligned}
$$

Each line here is an expression that is equivalent to those written above or below. There is no equation to "solve."

Often, as in the next example, we merge these ideas by forming an equivalent equation by replacing part of an equation with an equivalent expression.

EXAMPLE 10   Solve: $5x - 2(x - 5) = 7x - 2$.

Solution

$5x - 2(x - 5) = 7x - 2$	
$5x - 2x + 10 = 7x - 2$	Using the distributive law
$3x + 10 = 7x - 2$	Combining like terms
$3x + 10 - 3x = 7x - 2 - 3x$	Using the addition principle; adding $-3x$, the opposite of $3x$, to both sides
$10 = 4x - 2$	Combining like terms
$10 + 2 = 4x - 2 + 2$	Using the addition principle
$12 = 4x$	Simplifying
$\frac{1}{4} \cdot 12 = \frac{1}{4} \cdot 4x$	Using the multiplication principle; multiplying both sides by $\frac{1}{4}$, the reciprocal of 4
$3 = x$	Using the law of reciprocals; simplifying

Check:

$$
\begin{array}{c|c}
\multicolumn{2}{c}{5x - 2(x - 5) = 7x - 2} \\
\hline
5 \cdot 3 - 2(3 - 5) & 7 \cdot 3 - 2 \\
15 - 2(-2) & 21 - 2 \\
15 + 4 & 19 \\
19 \stackrel{?}{=} 19 & \text{TRUE}
\end{array}
$$

The solution is 3.

Types of Equations

In Examples 4, 5, and 10, we solved *linear equations*. A **linear equation** in one variable—say, x—is an equation equivalent to one of the form $ax = b$ with a and b constants and $a \neq 0$. The variable in a linear equation is always raised to the first power.

Don't rush to solve equations in your head. Work neatly, keeping in mind that the number of steps in a solution is less important than producing a simpler, yet equivalent, equation in each step.

Every equation falls into one of three categories. An **identity** is an equation, like $x + 5 = 3 + x + 2$, that is true for all replacements. A **contradiction** is an equation, like $n + 5 = n + 7$, that is *never* true. A **conditional equation**, like $2x + 5 = 17$, is sometimes true and sometimes false, depending on what the replacement of x is. Most of the equations examined in this text are conditional.

EXAMPLE 11   Solve each of the following equations and classify the equation as an identity, a contradiction, or a conditional equation.

a) $2x + 7 = 7(x + 1) - 5x$

b) $3x - 5 = 3(x - 2) + 4$

c) $3 - 8x = 5 - 7x$

Solution

a) $2x + 7 = 7(x + 1) - 5x$

$\quad 2x + 7 = 7x + 7 - 5x$ Using the distributive law

$\quad 2x + 7 = 2x + 7$ Combining like terms

The equation $2x + 7 = 2x + 7$ is true regardless of what x is replaced with, so all real numbers are solutions. Note that $2x + 7 = 2x + 7$ is equivalent to $2x = 2x$, $7 = 7$, or $0 = 0$. All real numbers are solutions and the equation is an identity.

Student Notes _____

The addition and multiplication principles can be used even when 0 appears on one side of an equation. Thus to solve $3x - 5 = 0$, we would begin by adding 5 to both sides.

b) $\qquad\quad 3x - 5 = 3(x - 2) + 4$

$\qquad\quad 3x - 5 = 3x - 6 + 4$ Using the distributive law

$\qquad\quad 3x - 5 = 3x - 2$ Combining like terms

$\quad -3x + 3x - 5 = -3x + 3x - 2$ Using the addition principle

$\qquad\qquad\quad -5 = -2$

Since the original equation is equivalent to $-5 = -2$, which is false for any choice of x, the original equation has no solution. There is no choice of x that will solve $3x - 5 = 3(x - 2) + 4$. The equation is a contradiction.

c) $\qquad\quad 3 - 8x = 5 - 7x$

$\quad 3 - 8x + 7x = 5 - 7x + 7x$ Using the addition principle

$\qquad\quad 3 - x = 5$ Simplifying

$\quad -3 + 3 - x = -3 + 5$ Using the addition principle

$\qquad\quad -x = 2$ Simplifying

$\qquad\quad x = \dfrac{2}{-1}, \text{ or } -2$ Dividing both sides by -1 or multiplying both sides by $\dfrac{1}{-1}$, or -1

There is one solution, -2. For other choices of x, the equation is false. This equation is conditional since it can be true or false, depending on the replacement for x.

We will sometimes refer to the set of solutions, or **solution set**, of a particular equation. Thus the solution set for Example 11(c) is $\{-2\}$. The solution set for Example 11(a) is simply $\mathbb{R}$, the set of all real numbers, and the solution set for Example 11(b) is the **empty set**, denoted $\varnothing$ or $\{\ \}$. As its name suggests, the empty set is the set containing no elements.

Exercise Set

1.3

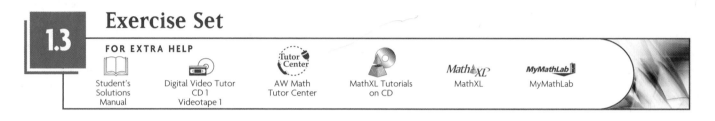

FOR EXTRA HELP

Student's Solutions Manual   Digital Video Tutor CD 1 Videotape 1   AW Math Tutor Center   MathXL Tutorials on CD   MathXL   MyMathLab

🖙 *Concept Reinforcement   Classify each of the following as either a pair of equivalent equations or a pair of equivalent expressions.*

1. $2x - 10,\ 2(x - 5)$

2. $2(x - 5) = 19,\ 2x - 10 = 19$

3. $3(t + 4) = 25,\ 3t + 12 = 25$

4. $3(t + 4),\ 3t + 12$

5. $4x - 9 = 7,\ 4x = 16$

6. $4x - 9,\ 5x - 9 - x$

7. $8t + 5 - 2t + 1,\ 6t + 6$

8. $5t - 2 + t = 8,\ 6t = 10$

9. $6x - 3 = 10x + 5,\ -8 = 4x$

10. $9x - 2,\ 13x - 6 - 4x + 4$

Determine whether the two equations in each pair are equivalent.

11. $t + 5 = 11$ and $3t = 18$

12. $t - 3 = 7$ and $3t = 24$

13. $12 - x = 3$ and $2x = 20$

14. $3x - 4 = 8$ and $3x = 12$

15. $5x = 2x$ and $\dfrac{4}{x} = 3$

16. $6 = 2x$ and $5 = \dfrac{2}{3 - x}$

Solve. Be sure to check.

17. $x - 2.9 = 13.4$

18. $y + 4.3 = 11.2$

19. $8t = 72$

20. $9t = 63$

21. $4x - 12 = 60$

22. $4x - 6 = 70$

23. $\frac{3}{5}n + 2 = 17$

24. $\frac{2}{7}n + 1 = 9$

25. $2y - 11 = 37$

26. $3x - 13 = 29$

Simplify to form an equivalent expression by combining like terms. Use the distributive law as needed.

27. $3x + 7x$

28. $9x + 3x$

29. $7rt - 9rt$

30. $3ab + 7ab$

31. $9t^2 + t^2$

32. $7a^2 + a^2$

33. $12a - a$

34. $15x - x$

35. $n - 8n$

36. $x - 6x$

37. $5x - 3x + 8x$

38. $3x - 11x + 2x$

39. $4x - 2x^2 + 3x$

40. $9a - 5a^2 + 4a$

41. $4x - 7 + 18x + 25$

42. $13p + 5 - 4p + 7$

43. $-7t^2 + 3t + 5t^3 - t^3 + 2t^2 - t$

44. $-9n + 8n^2 + n^3 - 2n^2 - 3n + 4n^3$

45. $7a - (2a + 5)$

46. $x - (5x + 9)$

47. $m - (6m - 2)$

48. $5a - (4a - 3)$

49. $3d - 7 - (5 - 2d)$

50. $8x - 9 - (7 - 5x)$

51. $-2(x + 3) - 5(x - 4)$

52. $9y - 4(5y - 6)$

53. $9a - [7 - 5(7a - 3)]$

54. $12b - [9 - 7(5b - 6)]$

55. $5\{-2a + 3[4 - 2(3a + 5)]\}$

56. $7\{-7x + 8[5 - 3(4x + 6)]\}$

57. $2y + \{7[3(2y - 5) - (8y + 7)] + 9\}$

58. $7b - \{6[4(3b - 7) - (9b + 10)] + 11\}$

Solve. Be sure to check.

59. $6x + 3x = 54$

60. $3x + 7x = 150$

61. $\frac{2}{3}y - \frac{1}{4}y = 5$

62. $\frac{3}{5}t - \frac{1}{2}t = 3$

63. $5t - 13t = -32$

64. $-9y - 5y = 28$

65. $3(x + 4) = 7x$

66. $3(y + 5) = 8y$

67. $70 = 10(3t - 2)$

68. $27 = 9(5y - 2)$

69. $1.8(n - 2) = 9$

70. $2.1(x - 3) = 8.4$

71. $5y - (2y - 10) = 25$

72. $8x - (3x - 5) = 40$

73. $7y - 1 = 23 - 5y$

74. $14t + 20 = 8t - 22$

75. $\frac{1}{5} + \frac{3}{10}x = \frac{4}{5}$

76. $-\frac{5}{2}x + \frac{1}{2} = -18$

77. $\frac{9}{10}y - \frac{7}{10} = \frac{21}{5}$

78. $\frac{4}{5}t - \frac{3}{10} = \frac{2}{5}$

79. $7r - 2 + 5r = 6r + 6 - 4r$

80. $9m - 15 - 2m = 6m - 1 - m$

81. $\frac{2}{3}(x - 2) - 1 = \frac{1}{4}(x - 3)$

82. $\frac{1}{4}(6t + 48) - 20 = -\frac{1}{3}(4t - 72)$

83. $5 + 2(x - 3) = 2[5 - 4(x + 2)]$

84. $3[2 - 4(x - 1)] = 3 - 4(x + 2)$

Find each solution set. Then classify each equation as a conditional equation, an identity, or a contradiction.

85. $7x - 2 - 3x = 4x$

86. $3t + 5 + t = 5 + 4t$

87. $2 + 9x = 3(4x + 1) - 1$

88. $4 + 7x = 7(x + 1)$

Aha! **89.** $-9t + 2 = -9t - 7(6 \div 2(49) + 8)$

90. $-9t + 2 = 2 - 9t - 5(8 \div 4(1 + 3^4))$

91. $2\{9 - 3[-2x - 4]\} = 12x + 42$

92. $3\{7 - 2[7x - 4]\} = -40x + 45$

93. As the first step in solving
$$2x + 5 = -3,$$
Pat multiplies both sides by $\frac{1}{2}$. Is this incorrect? Why or why not?

94. Explain how an identity can be easily altered so that it becomes a contradiction.

SKILL MAINTENANCE

Translate to an algebraic expression. [1.1]

95. Nine more than twice a number

96. Forty-two percent of half of a number

SYNTHESIS

97. Explain how the distributive and commutative laws can be used to rewrite $3x + 6y + 4x + 2y$ as $7x + 8y$.

98. Explain the difference between equivalent expressions and equivalent equations.

Solve and check. The symbol ▤ *indicates an exercise designed to be solved with a calculator.*

99. $4.23x - 17.898 = -1.65x - 42.454$

100. $-0.00458y + 1.7787 = 13.002y - 1.005$

101. $8x - \{3x - [2x - (5x - (7x - 1))]\} = 8x + 7$

102. $6x - \{5x - [7x - (4x - (3x + 1))]\} = 6x + 5$

103. $17 - 3\{5 + 2[x - 2]\} + 4\{x - 3(x + 7)\}$
$= 9\{x + 3[2 + 3(4 - x)]\}$

104. $23 - 2\{4 + 3[x - 1]\} + 5\{x - 2(x + 3)\}$
$= 7\{x - 2[5 - (2x + 3)]\}$

105. Create an equation for which it is preferable to use the multiplication principle *before* using the addition principle. Explain why it is best to solve the equation in this manner.

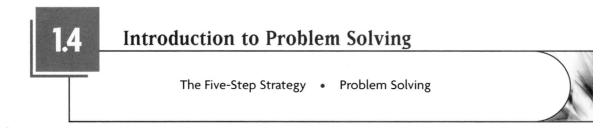

1.4   Introduction to Problem Solving

The Five-Step Strategy   •   Problem Solving

We now begin to study and practice the "art" of problem solving. Although we are interested mainly in using algebra to solve problems, much of what we say here applies to solving all kinds of problems.

What do we mean by a *problem*? Perhaps you've already used algebra to solve some "real-world" problems. What procedure did you use? Is there an approach that can be used to solve problems of a more general nature? These are questions that we will answer in this section.

In this text, we do not restrict the use of the word "problem" to computations involving arithmetic or algebra, such as $589 + 437 = a$ or $3x + 5x = 9$. In this text, a problem is simply a question to which we wish to find an answer. Perhaps this can best be illustrated with some sample problems:

1. Can I afford to rent a bigger apartment?
2. If I exercise twice a week and eat 3000 calories a day, will I lose weight?
3. Do I have enough time to take 4 courses while working 20 hours a week?
4. My fishing boat travels 12 km/h in still water. How long will it take me to cruise 25 km upstream if the river's current is 3 km/h?

Although these problems differ, there is a strategy that can be applied to all of them.

The Five-Step Strategy

Since you have already studied some algebra, you have some experience with problem solving. The following steps constitute a strategy that you may already have used and comprise a sound strategy for problem solving in general.

Five Steps for Problem Solving with Algebra

1. *Familiarize* yourself with the problem.
2. *Translate* to mathematical language.
3. *Carry out* some mathematical manipulation.
4. *Check* your possible answer in the original problem.
5. *State* the answer clearly.

Of the five steps, probably the most important is the first: becoming familiar with the problem situation. Here are some ways in which this can be done.

The First Step in Problem Solving with Algebra

To familiarize yourself with the problem:

1. If the problem is written, read it carefully. Then read it again, perhaps aloud. Verbalize the problem to yourself.
2. List the information given and restate the question being asked. Select a variable or variables to represent any unknown(s) and clearly state what each variable represents. Be descriptive! For example, let t = the flight time, in hours; let p = Paul's weight, in pounds; and so on.
3. Find additional information. Look up formulas or definitions with which you are not familiar. Geometric formulas appear on the inside back cover of this text; important words appear in the index. Consult an expert in the field or a reference librarian.
4. Create a table, using variables, in which both known and unknown information is listed. Look for possible patterns.
5. Make and label a drawing.
6. Estimate or guess an answer and check to see whether it is correct.

EXAMPLE 1   How might you familiarize yourself with the situation of Problem 1: "Can I afford to rent a bigger apartment?"

Solution   Clearly more information is needed to solve this problem. You might:

a) Estimate the rent of some apartments in which you are interested.

b) Examine what your savings are and how your income is budgeted.

c) Determine how much rent you can afford and whether you would consider having a roommate.

When enough information is known, it might be wise to make a chart or table to help you reach an answer.

EXAMPLE 2   How might you familiarize yourself with Problem 4: "How long will it take the boat to cruise 25 km upstream?"

Solution   First read the question *very* carefully. This may even involve speaking aloud. You may need to reread the problem several times to fully understand what information is given and what information is required. A sketch or table is often helpful.

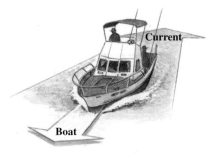

Distance to be Traveled	25 km
Speed of Boat in Still Water	12 km/h
Speed of Current	3 km/h
Speed of Boat Upstream	?
Time Required	?

Student Notes

It is extremely helpful to write down exactly what each variable represents before attempting to form an equation.

To gain more familiarity with the problem, we should determine, possibly with the aid of outside references, what relationships exist among the various quantities in the problem. With some effort it can be learned that the current's speed should be subtracted from the boat's speed in still water to determine the boat's speed going upstream. We also need to find or recall an extremely important formula:

Distance = Speed × Time.   It is important to remember this equation.

We rewrite part of the table, letting t = the number of hours required for the boat to cruise 25 km upstream.

Distance to be Traveled	25 km
Speed of Boat Upstream	12 − 3 = 9 km/h
Time Required	t

At this point we might try a guess. Suppose the boat traveled upstream for 2 hr. The boat would have then traveled

$$9\,\frac{km}{hr} \times 2\,hr = 18\,km. \qquad \text{Note that } \frac{km}{hr} \cdot hr = km.$$

$$\uparrow \qquad \uparrow \qquad \uparrow$$

Speed × Time = Distance

Since 18 ≠ 25, our guess is wrong. Still, examining how we checked our guess sheds light on how to translate the problem to an equation. A better guess, when multiplied by 9, would yield a number closer to 25.

The second step in problem solving is to translate the situation to mathematical language. In algebra, this often means forming an equation.

> *The Second Step in Problem Solving with Algebra*
>
> Translate the problem to mathematical language. This is sometimes done by writing an algebraic expression, but most often in this text is done by translating to an equation.

In the third step of our process, we work with the results of the first two steps. Often this requires us to use the algebra that we have studied.

> *The Third Step in Problem Solving with Algebra*
>
> Carry out some mathematical manipulation. If you have translated to an equation, this means to solve the equation.

To complete the problem-solving process, we should always **check** our solution and then **state** the solution in a clear and precise manner. To check, we make sure that our answer is reasonable and that all the conditions of the original problem are satisfied. If our answer checks, we write a complete English sentence stating the solution. The five steps are listed again below. Try to apply them regularly in your work.

> **Five Steps for Problem Solving with Algebra**
>
> 1. *Familiarize* yourself with the problem.
> 2. *Translate* to mathematical language.
> 3. *Carry out* some mathematical manipulation.
> 4. *Check* your possible answer in the original problem.
> 5. *State* the answer clearly.

Problem Solving

At this point, our study of algebra has just begun. Thus we have few algebraic tools with which to work problems. As the number of tools in our algebraic "toolbox" increases, so will the difficulty of the problems we can solve. For now our problems may seem simple; however, to gain practice with the problem-solving process, you should try to use all five steps. Later some steps may be shortened or combined.

EXAMPLE 3   Purchasing.   Maya pays $581.94 for a home theater system. If the price paid includes a 6% sales tax, what is the price of the system itself?

Solution

1. Familiarize.   First, we familiarize ourselves with the problem. Note that tax is calculated from, and then added to, the item's price. Let's guess that the

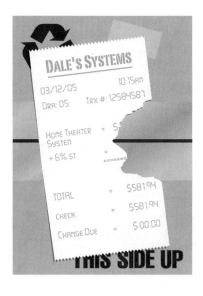

home theater system's price is $500. To check the guess, we calculate the amount of tax, $(0.06)(\$500) = \$30$, and add it to $500:

$$\$500 + (0.06)(\$500) = \$500 + \$30$$
$$= \$530. \qquad \$530 \neq \$581.94$$

Our guess was too low, but the manner in which we checked the guess will guide us in the next step. We let

$h =$ the home theater system's price, in dollars.

2. Translate. Our guess leads us to the following translation:

Rewording:

The system's price	plus	6% sales tax	is	the price with sales tax.

Translating: $h \quad + \quad (0.06)h \quad = \quad \$581.94$

3. Carry out. Next, we carry out some mathematical manipulation:

$$h + (0.06)h = 581.94$$
$$1.06h = 581.94 \qquad \text{Combining like terms}$$
$$\frac{1}{1.06} \cdot 1.06h = \frac{1}{1.06} \cdot 581.94 \qquad \text{Using the multiplication principle}$$
$$h = 549.$$

4. Check. To check the answer in the original problem, note that the tax on a home theater system costing $549 would be $(0.06)(\$549) = \$32.94$. When this is added to $549, we have

$$\$549 + \$32.94, \quad \text{or} \quad \$581.94.$$

Thus, $549 checks in the original problem.

5. State. We clearly state the answer: The home theater system itself costs $549.

EXAMPLE 4

Home maintenance. In an effort to make their home more energy-efficient, Alma and Drew purchased 200 in. of 3M Press-In-Place™ window glazing. This will be just enough to outline their two square skylights. If the length of the sides of the larger skylight is $1\frac{1}{2}$ times the length of the sides of the smaller one, how should the glazing be cut?

Solution

1. Familiarize. Note that the *perimeter* of (distance around) each square is four times the length of a side. Furthermore, if s represents the length of a side of the smaller square, then $\left(1\frac{1}{2}\right)s$ represents the length of a side of the larger square. We make a drawing and note that the two perimeters must add up to 200 in.

Perimeter of a square $= 4 \cdot$ *length of a side*

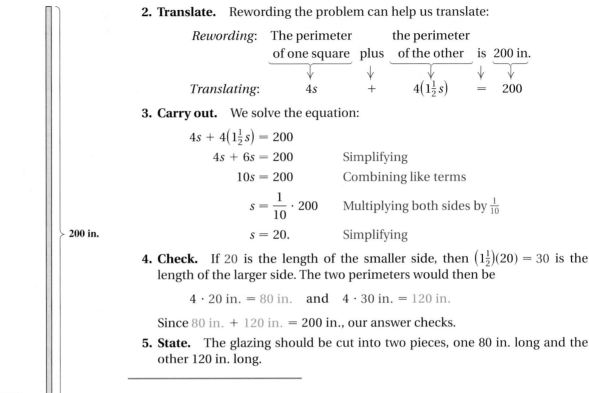

2. Translate.   Rewording the problem can help us translate:

Rewording: The perimeter   the perimeter
of one square   plus   of the other   is   200 in.

Translating: $4s$ $+$ $4\left(1\frac{1}{2}s\right)$ $=$ 200

3. Carry out.   We solve the equation:

$$4s + 4\left(1\frac{1}{2}s\right) = 200$$

$$4s + 6s = 200 \qquad \text{Simplifying}$$

$$10s = 200 \qquad \text{Combining like terms}$$

$$s = \frac{1}{10} \cdot 200 \qquad \text{Multiplying both sides by } \tfrac{1}{10}$$

$$s = 20. \qquad \text{Simplifying}$$

4. Check.   If 20 is the length of the smaller side, then $\left(1\frac{1}{2}\right)(20) = 30$ is the length of the larger side. The two perimeters would then be

$$4 \cdot 20 \text{ in.} = 80 \text{ in.} \quad \text{and} \quad 4 \cdot 30 \text{ in.} = 120 \text{ in.}$$

Since 80 in. + 120 in. = 200 in., our answer checks.

5. State.   The glazing should be cut into two pieces, one 80 in. long and the other 120 in. long.

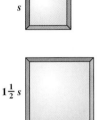

We cannot stress too much the importance of labeling the variables in your problem. In Example 4, solving for s is not enough: We need to find $4s$ and $4\left(1\frac{1}{2}s\right)$ to determine the numbers we are after.

EXAMPLE 5   Three numbers are such that the second is 6 less than three times the first and the third is 2 more than two-thirds the first. The sum of the three numbers is 150. Find the largest of the three numbers.

Solution   We proceed according to the five-step process.

1. Familiarize.   We need to find the largest of three numbers. We list the information given in a table in which x represents the first number.

First Number	x
Second Number	6 less than 3 times the first
Third Number	2 more than $\frac{2}{3}$ the first

First + Second + Third = 150

Try to check a guess at this point. We will proceed to the next step.

2. **Translate.**   Because we wish to write an equation in just one variable, we need to express the second and third numbers using x ("in terms of x"). To do so, we expand the table:

First Number	x	x
Second Number	6 less than 3 times the first	$3x - 6$
Third Number	2 more than $\frac{2}{3}$ the first	$\frac{2}{3}x + 2$

We know that the sum is 150. Substituting, we obtain an equation:

$$\underbrace{\text{First}} + \underbrace{\text{second}} + \underbrace{\text{third}} = 150.$$
$$x + (3x - 6) + \left(\tfrac{2}{3}x + 2\right) = 150$$

3. **Carry out.**   We solve the equation:

$$x + 3x - 6 + \tfrac{2}{3}x + 2 = 150 \qquad \text{Leaving off unnecessary parentheses}$$

$$\left(4 + \tfrac{2}{3}\right)x - 4 = 150 \qquad \text{Combining like terms}$$

$$\tfrac{14}{3}x - 4 = 150$$

$$\tfrac{14}{3}x = 154 \qquad \text{Adding 4 to both sides}$$

$$x = \tfrac{3}{14} \cdot 154 \qquad \text{Multiplying both sides by } \tfrac{3}{14}$$

$$x = 33. \qquad \text{Remember, } x \text{ represents the first number.}$$

Going back to the table, we can find the other two numbers:

Second: $3x - 6 = 3 \cdot 33 - 6 = 93$;
Third: $\frac{2}{3}x + 2 = \frac{2}{3} \cdot 33 + 2 = 24$.

4. **Check.**   We return to the original problem. There are three numbers: 33, 93, and 24. Is the second number 6 less than three times the first?

$$3 \times 33 - 6 = 99 - 6 = 93$$

The answer is *yes*.
        Is the third number 2 more than two-thirds the first?

$$\tfrac{2}{3} \times 33 + 2 = 22 + 2 = 24$$

The answer is *yes*.
        Is the sum of the three numbers 150?

$$33 + 93 + 24 = 150$$

The answer is *yes*. The numbers do check.

5. **State.**   The problem asks us to find the largest number, so the answer is: "The largest of the three numbers is 93."

> *Caution!*   In Example 5, although the equation $x = 33$ enables us to find the largest number, 93, the number 33 is *not* the solution of the problem. By clearly labeling our variable in the first step, we can avoid thinking that the variable always represents the solution of the problem.

Exercise Set

1.4

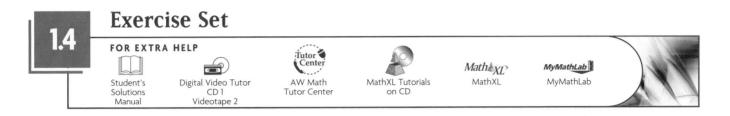

FOR EXTRA HELP

Student's Solutions Manual | Digital Video Tutor CD 1 Videotape 2 | AW Math Tutor Center | MathXL Tutorials on CD | MathXL | MyMathLab

For each problem, familiarize yourself with the situation. Then translate to mathematical language. You need not actually solve the problem; just carry out the first two steps of the five-step strategy. You will be asked to complete some of the solutions as Exercises 31–38.

1. The sum of two numbers is 65. One of the numbers is 7 more than the other. What are the numbers?

2. The sum of two numbers is 83. One of the numbers is 11 more than the other. What are the numbers?

3. *Swimming.* An olympic swimmer can swim at a sustained rate of 5 km/h in still water. The Lazy River flows at a rate of 2.3 km/h. How long would it take an olympic swimmer to swim 1.8 km upstream?

4. *Boating.* The Delta Queen is a paddleboat that tours the Mississippi River near New Orleans, Louisiana. It is not uncommon for the Delta Queen to run 7 mph in still water and for the Mississippi to flow at a rate of 3 mph. At these rates, how long will it take the boat to cruise 2 mi upstream?
Source: Delta Queen information

5. *Moving sidewalks.* The moving sidewalk in O'Hare Airport is 300 ft long and moves at a rate of 5 ft/sec. If Alida walks at a rate of 4 ft/sec, how long will it take her to walk the length of the moving sidewalk?

4 ft/sec

5 ft/sec

6. *Aviation.* A Cessna airplane traveling 390 km/h in still air encounters a 65-km/h headwind. How long will it take the plane to travel 725 km into the wind?

7. *Angles in a triangle.* The degree measures of the angles in a triangle are three consecutive integers. Find the measures of the angles.

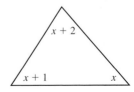

$x + 2$

$x + 1$ x

8. *Pricing.* Becker Lumber gives contractors a 10% discount on all orders. After the discount, a contractor's order cost $279. What was the original cost of the order?

9. *Pricing.* The Sound Connection prices packages of CDRs by raising the wholesale price 50% and adding $1.50. What must a package's wholesale price be if it is being sold for $22.50?

10. *Pricing.* Miller Oil offers a 5% discount to customers who pay promptly for an oil delivery. The Blancos promptly paid $142.50 for their December oil bill. What would the cost have been had they not promptly paid?

11. *Cruising altitude.* A Boeing 747 has been instructed to climb from its present altitude of 8000 ft to a cruising altitude of 29,000 ft. If the plane ascends at a rate of 3500 ft/min, how long will it take to reach the cruising altitude?

12. A piece of wire 10 m long is to be cut into two pieces, one of them $\frac{2}{3}$ as long as the other. How should the wire be cut?

13. *Angles in a triangle.* One angle of a triangle is three times as great as a second angle. The third angle measures 12° less than twice the second angle. Find the measures of the angles.

14. *Angles in a triangle.* One angle of a triangle is four times as great as a second angle. The third angle measures 5° more than twice the second angle. Find the measures of the angles.

15. Find two consecutive even integers such that two times the first plus three times the second is 76.

16. Find three consecutive odd integers such that the sum of the first, twice the second, and three times the third is 70.

17. A steel rod 90 cm long is to be cut into two pieces, each to be bent to make an equilateral triangle. The length of a side of one triangle is to be twice the length of a side of the other. How should the rod be cut?

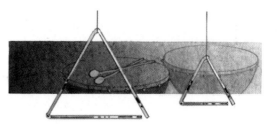

18. A piece of wire 100 cm long is to be cut into two pieces, and those pieces are each to be bent to make a square. The area of one square is to be 144 cm^2 greater than that of the other. How should the wire be cut? (*Remember*: Do not solve.)

19. *Rescue calls.* Rescue crews working for Stockton Rescue average 3 calls per shift. After his first four shifts, Brian had received 5, 2, 1, and 3 calls. How many calls will Brian need on his next shift if he is to average 3 calls per shift?

20. *Test scores.* Deirdre's scores on five tests are 93, 89, 72, 80, and 96. What must the score be on her next test so that the average will be 88?

Solve each problem. Use all five problem-solving steps.

21. *Pricing.* The price that Ruth paid for her graphing calculator, $84, is less than what Tony paid by $13. How much did Tony pay for his graphing calculator?

22. *Class size.* The number of students in James' class, 35, is greater than the number in Rose's class by 12. How many students are in Rose's class?

23. *Public health.* In 2050, the number of diagnosed cases of diabetes in the United States is predicted to reach 29 million. This is approximately $\frac{13}{5}$ of the number of cases in 2000. How many cases had been diagnosed in 2000?
Source: U.S. Centers for Disease Control

24. *Home improvement.* In 2003, painting the interior of a house increased the average sale price of the

home by about $2500. This was about $\frac{5}{3}$ of what the average paint job cost. What was the cost of the average paint job?

Source: Based on information from HomeGain.com

25. Officer Reid wrote up 9 more tickets than Officer Schultz did. Together they wrote up 35 tickets. How many tickets did Officer Reid write?

26. On an evening shift, Lily waited on 7 fewer tables than Keith did. Together they waited on 25 tables. How many tables did Lily wait on?

27. The length of a rectangular mirror is three times its width and its perimeter is 120 cm. Find the length and the width of the mirror.

28. The length of a rectangular tile is twice its width and its perimeter is 21 cm. Find the length and the width of the tile.

29. The width of a rectangular greenhouse is one-fourth its length and its perimeter is 130 m. Find the length and the width of the greenhouse.

30. The width of a rectangular garden is one-third its length and its perimeter is 32 m. Find the dimensions of the garden.

31. Solve the problem of Exercise 3.

32. Solve the problem of Exercise 4.

33. Solve the problem of Exercise 14.

34. Solve the problem of Exercise 13.

35. Solve the problem of Exercise 10.

36. Solve the problem of Exercise 8.

37. Solve the problem of Exercise 9.

38. Solve the problem of Exercise 15.

39. Write a problem for a classmate to solve. Devise the problem so that the solution is "The material should be cut into two pieces, one 30 cm long and the other 45 cm long."

40. Write a problem for a classmate to solve. Devise the problem so that the solution is "The first angle is 40°, the second angle is 50°, and the third angle is 90°."

SKILL MAINTENANCE

Solve. [1.3]

41. $7 = \dfrac{2}{3}(x + 6)$

42. $9 = \dfrac{x}{4}$

43. $8 = \dfrac{5 + t}{3}$

44. $6t - 8 = 0$

SYNTHESIS

45. How can a guess or estimate help prepare you for the *Translate* step when solving problems?

46. Why is it important to check the solution from step 3 (*Carry out*) in the original wording of the problem being solved?

47. *Test scores.* Tico's scores on four tests are 83, 91, 78, and 81. How many points above his current average must Tico score on the next test in order to raise his average 2 points?

48. *Geometry.* The height and sides of a triangle are four consecutive integers. The height is the first integer, and the base is the third integer. The perimeter of the triangle is 42 in. Find the area of the triangle.

49. *Home prices.* Panduski's real estate prices increased 6% from 2001 to 2002 and 2% from 2002 to 2003. From 2003 to 2004, prices dropped 1%. If a house sold for $117,743 in 2004, what was its worth in 2001? (Round to the nearest dollar.)

50. *Adjusted wages.* Blanche's salary is reduced $n\%$ during a period of financial difficulty. By what number should her salary be multiplied in order to bring it back to where it was before the reduction?

CORNER

COLLABORATIVE

Who Pays What?

Focus: Problem solving

Time: 15 minutes

Group size: 5

Suppose that two of the five members in each group are celebrating birthdays and the entire group goes out to lunch. Suppose further that each member whose birthday it is gets treated to his or her lunch by the other *four* members. Finally, suppose that all meals cost the same amount and that the total bill is $40.00.*

*This activity was inspired by "The Birthday-Lunch Problem," *Mathematics Teaching in the Middle School*, vol. 2, no. 1, September–October 1996, pp. 40–42.

ACTIVITY

1. Determine, as a group, how much each group member should pay for the lunch described above. Then explain how this determination was made.
2. Compare the results and methods used for part (1) with those of the other groups in the class.
3. If the total bill is $65, how much should each group member pay? Again compare results with those of other groups.
4. If time permits, generalize the results of parts 1–3 for a total bill of x dollars.

1.5 Formulas, Models, and Geometry

Solving Formulas • Mathematical Models

Study Skills

Avoiding Temptation

Be kind to yourself when studying—in a supportive way. For example, sit in a comfortable chair in a well-lit location, but stay away from a coffee shop where friends may stop by. Once you begin studying, let the answering machine answer the phone, shut off any cell phones, and do not check e-mail during your study session.

A **formula** is an equation that uses letters to represent a relationship between two or more quantities. For example, in Section 1.4, we made use of the formula $P = 4s$, where P represents the perimeter of a square and s the length of a side. Other important geometric formulas are $A = \pi r^2$ (for the area A of a circle of radius r), $C = \pi d$ (for the circumference C of a circle of diameter d), and $A = b \cdot h$ (for the area A of a parallelogram of height h and base length b).* A more complete list of geometric formulas appears at the very end of this text.

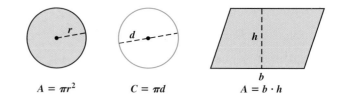

$$A = \pi r^2 \qquad C = \pi d \qquad A = b \cdot h$$

---

*The Greek letter π, read "pi," is *approximately* 3.14159265358979323846264. Often 3.14 or 22/7 is used to approximate π when a calculator with a π key is unavailable.

Solving Formulas

Suppose we know the floor area and the width of a rectangular room and want to find the length. To do so, we could "solve" the formula $A = l \cdot w$ (Area = Length · Width) for l, using the same principles that we use for solving equations.

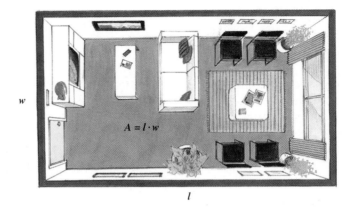

$$A = l \cdot w$$

EXAMPLE 1   Area of a rectangle.   Solve the formula $A = l \cdot w$ for l.

Solution

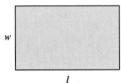

$$A = l \cdot w \qquad \text{We want this letter alone.}$$

$$\frac{A}{w} = \frac{l \cdot w}{w} \qquad \begin{array}{l}\text{Dividing both sides by } w, \text{ or} \\ \text{multiplying both sides by } 1/w\end{array}$$

$$\left.\begin{array}{l}\dfrac{A}{w} = l \cdot \dfrac{w}{w} \\[2ex] \dfrac{A}{w} = l\end{array}\right\} \quad \begin{array}{l}\text{Simplifying by removing a factor} \\[1ex] \text{equal to 1: } \dfrac{w}{w} = 1\end{array}$$

Thus to find the length of a rectangular room, we can divide the area of the floor by its width. Were we to do this calculation for a variety of rectangular rooms, the formula $l = A/w$ would be more convenient than repeatedly substituting into $A = l \cdot w$ and then dividing.

EXAMPLE 2   Simple interest.   The formula $I = Prt$ is used to determine the simple interest I earned when a principal of P dollars is invested for t years at an interest rate r. Solve this formula for t.

Solution

$$I = Prt \qquad \text{We want this letter alone.}$$

$$\frac{I}{Pr} = \frac{Prt}{Pr} \qquad \text{Dividing both sides by } Pr, \text{ or multiplying both sides by } \frac{1}{Pr}$$

$$\frac{I}{Pr} = t \qquad \text{Simplifying by removing a factor equal to 1: } \frac{Pr}{Pr} = 1$$

EXAMPLE 3

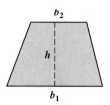

Area of a trapezoid.   A trapezoid is a geometric shape with four sides, exactly two of which, the bases, are parallel to each other. The formula for calculating the area A of a trapezoid with bases b_1 and b_2 (read "b sub one" and "b sub two") and height h is given by

$$A = \frac{h}{2}(b_1 + b_2),$$ A derivation of this formula is outlined in Exercise 86 of this section.

where the *subscripts* 1 and 2 distinguish one base from the other. Solve for b_1.

Solution   There are several ways to "remove" the parentheses. We could distribute $h/2$, but an easier approach is to multiply both sides by the reciprocal of $h/2$.

$$A = \frac{h}{2}(b_1 + b_2)$$

$$\frac{2}{h} \cdot A = \frac{2}{h} \cdot \frac{h}{2}(b_1 + b_2)$$ Multiplying both sides by $\frac{2}{h}\left(\text{or dividing by } \frac{h}{2}\right)$

$$\frac{2A}{h} = b_1 + b_2$$ Simplifying. The right side is "cleared" of fractions.

$$\frac{2A}{h} - b_2 = b_1$$ Adding $-b_2$ on both sides

The similarities between solving formulas and solving equations can be seen below. In (a), we solve as we did before; in (b), we do not carry out all calculations; and in (c), we cannot carry out all calculations because the numbers are unknown. The same steps are used each time.

a) $9 = \frac{3}{2}(x + 5)$

$\frac{2}{3} \cdot 9 = \frac{2}{3} \cdot \frac{3}{2}(x + 5)$

$6 = x + 5$

$1 = x$

b) $9 = \frac{3}{2}(x + 5)$

$\frac{2}{3} \cdot 9 = \frac{2}{3} \cdot \frac{3}{2}(x + 5)$

$\frac{2 \cdot 9}{3} = x + 5$

$\frac{2 \cdot 9}{3} - 5 = x$

c) $A = \frac{h}{2}(b_1 + b_2)$

$\frac{2}{h} \cdot A = \frac{2}{h} \cdot \frac{h}{2}(b_1 + b_2)$

$\frac{2A}{h} = b_1 + b_2$

$\frac{2A}{h} - b_2 = b_1$

EXAMPLE 4

Accumulated simple interest.   The formula $A = P + Prt$ gives the amount A that a principal of P dollars will be worth in t years when invested at simple interest rate r. Solve the formula for P.

Student Notes ———

As is often the case, the material in this section builds on the material in the preceding sections. If you experience difficulty in this section, before seeking help make certain that you have a thorough understanding of the material in Sections 1.4 and (especially) 1.3.

Solution

$$A = P + Prt \qquad \text{We want this letter alone.}$$

$$A = P(1 + rt) \qquad \text{Factoring (using the distributive law) to combine like terms}$$

$$\frac{A}{1 + rt} = \frac{P(1 + rt)}{1 + rt} \qquad \text{Dividing both sides by } 1 + rt, \text{ or multiplying both sides by } \frac{1}{1 + rt}$$

$$\frac{A}{1 + rt} = P \qquad \text{Simplifying}$$

This last equation can be used to determine how much should be invested at interest rate r in order to have A dollars t years later.

Note in Example 4 that the factoring enabled us to write P once rather than twice. This is comparable to combining like terms when solving an equation like $16 = x + 7x$.

You may find the following summary useful.

To Solve a Formula for a Specified Letter

1. Get all terms with the letter being solved for on one side of the equation and all other terms on the other side, using the addition principle. To do this may require removing parentheses.

   - To remove parentheses, either divide both sides by the multiplier in front of the parentheses or use the distributive law.

2. When all terms with the specified letter are on the same side, factor (if necessary) so that the variable is written only once.
3. Solve for the letter in question by dividing both sides by the multiplier of that letter.

Mathematical Models

The above formulas from geometry and economics are examples of *mathematical models*. A **mathematical model** can be a formula, or set of formulas, developed to represent a real-world situation. In problem solving, a mathematical model is formed in the *Translate* step.

EXAMPLE 5   Body mass index.   Shaquille O'Neal of the Miami Heat is 7 ft 1 in. tall and has a body mass index of approximately 31.2. What is his weight?

Solution

1. **Familiarize.**   From an outside source, we find that body mass index I depends on a person's height and weight and is found using the formula

$$I = \frac{704.5W}{H^2},$$

where W is the weight, in pounds, and H is the height, in inches (*Source*: National Center for Health Statistics).

2. **Translate.**   Because we are interested in finding O'Neal's weight, we solve for W:

$$I = \frac{704.5W}{H^2} \qquad \text{We want this letter alone.}$$

$$I \cdot H^2 = \frac{704.5W}{H^2} \cdot H^2 \qquad \text{Multiplying both sides by } H^2 \text{ to clear the fraction}$$

$$IH^2 = 704.5W \qquad \text{Simplifying}$$

$$\frac{IH^2}{704.5} = \frac{704.5W}{704.5} \qquad \text{Dividing by 704.5}$$

$$\frac{IH^2}{704.5} = W.$$

3. **Carry out.**   The model

$$W = \frac{IH^2}{704.5}$$

can be used to calculate the weight of someone whose body mass index and height are known. Using the information given in the problem, we have

$$W = \frac{31.2 \cdot 85^2}{704.5} \qquad \text{7 ft 1 in. is 85 in.}$$

$$\approx 320. \qquad \text{Using a calculator}$$

4. **Check.**   We could repeat the calculations or substitute in the original formula and then solve for W. The check is left to the student.

5. **State.**   Shaquille O'Neal weighs about 320 lb.

EXAMPLE 6

Density.   A collector suspects that a silver coin is not solid silver. The density of silver is 10.5 grams per cubic centimeter (g/cm^3) and the coin is 0.2 cm thick with a radius of 2 cm. If the coin is really silver, how much should it weigh?

Solution

1. **Familiarize.**   From an outside reference, we find that density depends on mass and volume and that, in this setting, mass means weight. A formula for the volume of a right circular cylinder appears at the very end of this text.

2. **Translate.**   We need to use two formulas:

$$D = \frac{m}{V} \quad \text{and} \quad V = \pi r^2 h,$$

where D is the density, m the mass, V the volume, r the length of the radius, and h the height of a right circular cylinder. Since we need a model relating mass to the measurements of the coin, we solve for m and then substitute for V:

$$D = \frac{m}{V}$$

$$V \cdot D = V \cdot \frac{m}{V} \qquad \text{Multiplying by } V$$

$$V \cdot D = m \qquad \text{Simplifying}$$

$$\pi r^2 h \cdot D = m. \qquad \text{Substituting}$$

3. **Carry out.**   The model $m = \pi r^2 hD$ can be used to find the mass of any right circular cylinder for which the dimensions and the density are known:

$$m = \pi r^2 hD$$
$$= \pi(2)^2(0.2)(10.5) \qquad \text{Substituting}$$
$$\approx 26.3894. \qquad \text{Using a calculator with a } \pi \text{ key}$$

4. **Check.**   To check, we could repeat the calculations. We might also check the model by examining the units:

$$\pi r^2 h \cdot D = \text{cm}^2 \cdot \text{cm} \cdot \frac{\text{g}}{\text{cm}^3} = \text{cm}^3 \cdot \frac{\text{g}}{\text{cm}^3} = \text{g}. \qquad \pi \text{ has no unit.}$$

Since g (grams) is the unit in which the mass, m, is given, we have at least a partial check.

5. **State.**   The coin, if it is indeed silver, should weigh about 26 grams.

Exercise Set

1.5

FOR EXTRA HELP

Student's Solutions Manual | Digital Video Tutor CD 1 Videotape 2 | AW Math Tutor Center | MathXL Tutorials on CD | MathXL | MyMathLab

---

↪ *Concept Reinforcement   Complete each of the following statements.*

1. A formula is a(n) _____ that uses letters to represent a relationship between two or more quantities.

2. The formula $A = \pi r^2$ is used to calculate the _____ of a circle.

3. The formula $C = \pi d$ is used to calculate the _____ of a circle.

4. The formula _____ is used to calculate the perimeter of a rectangle of length l and width w.

5. The formula _____ is used to calculate the area of a parallelogram of height h and base length b.

6. The formula $l = A/w$ can be used to determine the _____ of a rectangle, given its area and width.

7. In the formula for the area of a trapezoid, $A = \dfrac{h}{2}(b_1 + b_2)$, the numbers 1 and 2 are referred to as _____ .

8. When two or more terms on the same side of a formula contain the letter for which we are solving, we can _____ so that the letter is only written once.

Solve.

9. $d = rt$, for r (a distance formula)

10. $d = rt$, for t

11. $F = ma$, for a (a physics formula)

12. $A = lw$, for w (an area formula)

13. $W = EI$, for I (an electricity formula)

14. $W = EI$, for E

15. $V = lwh$, for h (a volume formula)

16. $I = Prt$, for r (a formula for interest)

17. $L = \dfrac{k}{d^2}$, for k
    (a formula for intensity of sound or light)

18. $F = \dfrac{mv^2}{r}$, for m (a physics formula)

19. $G = w + 150n$, for n
(a formula for the gross weight of a bus)

20. $P = b + 0.5t$, for t (a formula for parking prices)

21. $2w + 2h + l = p$, for l
(a formula used when shipping boxes)

22. $2w + 2h + l = p$, for w

23. $nl + nm = k$, for n

24. $ba + bc = d$, for b

25. $yx + zx = w$, for x

26. $sr + tr = u$, for r

27. $Ax + By = C$, for y (a formula for graphing lines)

28. $P = 2l + 2w$, for l (a perimeter formula)

29. $C = \frac{5}{9}(F - 32)$, for F (a temperature formula)

30. $T = \frac{3}{10}(I - 12{,}000)$, for I (a tax formula)

31. $V = \frac{4}{3}\pi r^3$, for r^3
(a formula for the volume of a sphere)

32. $V = \frac{4}{3}\pi r^3$, for π

33. $ab = d - ac$, for a

34. $mn = p - mr$, for m

35. $xy = w + zy$, for y

36. $rs = t + ps$, for s

37. $A = \dfrac{q_1 + q_2 + q_3}{n}$, for n (a formula for averaging)
(*Hint*: Multiply by n to "clear" fractions.)

38. $g = \dfrac{km_1m_2}{d^2}$, for d^2 (Newton's law of gravitation)

39. $v = \dfrac{d_2 - d_1}{t}$, for t (a physics formula)

40. $v = \dfrac{s_2 - s_1}{m}$, for m

41. $v = \dfrac{d_2 - d_1}{t}$, for d_1

42. $v = \dfrac{s_2 - s_1}{m}$, for s_1

43. $r - m = mnp$, for m

44. $p - x = xyz$, for x

45. $y - ab = ac^2$, for a

46. $d - mn = mp^3$, for m

47. *Investing.* Janos has $2600 to invest for 6 months. If he needs the money to earn $156 in that time, at what rate of simple interest must Janos invest?

48. *Banking.* Yvonne plans to buy a one-year certificate of deposit (CD) that earns 7% simple interest. If she needs the CD to earn $110, how much should Yvonne invest?

49. *Geometry.* The area of a parallelogram is 78 cm^2. The base of the figure is 13 cm. What is the height?

50. *Geometry.* The area of a parallelogram is 72 cm^2. The height of the figure is 6 cm. How long is the base?

For Exercises 51–58, make use of the formulas given in Examples 1–6.

51. *Body mass index.* Arnold Schwarzenegger, the current governor of California and a former bodybuilder, is 6 ft 2 in. tall and has a body mass index of 30.2. How much does he weigh?

52. *Body mass index.* Kate Winslet, star of the movie *Titanic*, has a body mass index of 21 and a height of 5 ft 6 in. What is her weight?

53. *Weight of salt.* The density of salt is 2.16 g/cm^3 (grams per cubic centimeter). An empty cardboard salt canister weighs 28 g, is 13.6 cm tall, and has a 4-cm radius. How much will a filled canister weigh?

54. *Weight of a coin.* The density of gold is 19.3 g/cm^3. If the coin in Example 6 were made of gold instead of silver, how much more would it weigh?

55. *Gardening.* A garden is being constructed in the shape of a trapezoid. The dimensions are as shown in the figure. The unknown dimension is to be such that the area of the garden is 90 ft^2. Find that unknown dimension.

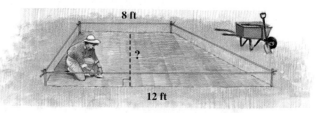

56. *Pet care.*   A rectangular kennel is being constructed, and 76 ft of fencing is available. The width of the kennel is to be 13 ft. What should the length be, in order to use just 76 ft of fence?

Aha! **57.** *Investing.*   Bok Lum Chan is going to invest $1000 at simple interest at 9%. How long will it take for the investment to be worth $1090?

58. Rik is going to invest $950 at simple interest at 7%. How long will it take for his investment to be worth $1349?

Chess ratings.   The formula

$$R = r + \frac{400(W - L)}{N}$$

is used to establish a chess player's rating R, after he or she has played N games, where W is the number of wins, L is the number of losses, and r is the average rating of the opponents.
Source: U.S. Chess Federation

59. Ulana's rating is 1305 after winning 5 games and losing 3 in tournament play. What was the average rating of her opponents? (Assume there were no draws.)

60. Vladimir's rating fell to 1050 after winning twice and losing 5 times in tournament play. What was the average rating of his opponents? (Assume there were no draws.)

Female caloric needs.   The number of calories K needed each day by a moderately active woman who weighs w pounds, is h inches tall, and is a years old can be estimated by the formula

$$K = 917 + 6(w + h - a).$$

Source: M. Parker, *She Does Math.* Mathematical Association of America, p. 96

61. Julie is moderately active, weighs 120 lb, and is 23 years old. If Julie needs 1901 calories per day to maintain her weight, how tall is she?

62. Tawana is moderately active, 31 years old, and 5 ft 4 in. tall. If Tawana needs 1901 calories per day to maintain her weight, how much does she weigh?

Male caloric needs.   The number of calories K needed each day by a moderately active man who weighs w kilograms, is h centimeters tall, and is a years old can be estimated by the formula

$$K = 19.18w + 7h - 9.52a + 92.4.$$

Source: M. Parker, *She Does Math.* Mathematical Association of America, p. 96

63. One of the authors of this text, Marv Bittinger, is moderately active, weighs 87 kg, and is 185 cm tall. If Marv needs 2476 calories per day to maintain his weight, how old is he?

64. One of the authors of this text, David Ellenbogen, is moderately active, weighs 82 kg, and is 46 years old. If David needs 2523 calories per day to maintain his weight, how tall is he?

Projected birth weight.   *Ultrasonic images of 29-week-old fetuses can be used to predict weight. One model, developed by Thurnau,[*] is P = 9.337da − 299; a second model, developed by Weiner,[†] is P = 94.593c + 34.227a − 2134.616. For both formulas, P represents the estimated fetal weight in grams, d the diameter of the fetal head in centimeters, c the circumference of the fetal head in centimeters, and a the circumference of the fetal abdomen in centimeters.*

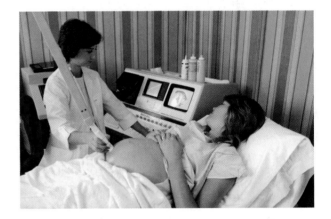

65. Solve Thurnau's model for d and use that equation to estimate the diameter of a fetus' head at 29 weeks when the estimated weight is 1614 g and the circumference of the fetal abdomen is 24.1 cm.

66. Solve Weiner's model for c and use that equation to estimate the circumference of a fetus' head at 29 weeks when the estimated weight is 1277 g and the circumference of the fetal abdomen is 23.4 cm.

---

[*]Thurnau, G. R., R. K. Tamura, R. E. Sabbagha, et al. *Am. J. Obstet Gynecol* 1983; **145**: 557.

[†]Weiner, C. P., R. E. Sabbagha, N. Vaisrub, et al. *Obstet Gynecol* 1985; **65**: 812.

Waiting time. *In an effort to minimize waiting time for patients at a doctor's office without increasing a physician's idle time, Michael Goiten of Massachusetts General Hospital has developed a model. Goiten suggests that the interval time I, in minutes, between scheduled appointments be related to the total number of minutes T that a physician spends with patients in a day and the number of scheduled appointments N according to the formula $I = 1.08(T/N)$.***

67. Dr. Cruz determines that she has a total of 8 hr a day to see patients. If she insists on an interval time of 15 min, according to Goiten's model, how many appointments should she make in one day?

68. A doctor insists on an interval time of 20 min and must be able to schedule 25 appointments a day. According to Goiten's model, how many hours a day should the doctor be prepared to spend with patients?

69. Is every rectangle a trapezoid? Why or why not?

70. Predictions made using the models of Exercises 65 and 66 are often off by as much as 10%. Does this mean the models should be discarded? Why or why not?

SKILL MAINTENANCE

Use the associative and commutative laws to write two equivalent expressions for each of the following. Answers may vary. [1.2]

71. $(7a)(3a)$

72. $(4y)(xy)$

SYNTHESIS

73. Which would you expect to have the greater density, and why: cork or steel?

74. Both of the models used in Exercises 65 and 66 have *P* alone on one side of the equation. Why?

75. The density of platinum is 21.5 g/cm^3. If the ring shown in the figure below is crafted out of platinum, how much will it weigh?

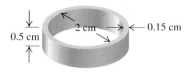

76. The density of a penny is 8.93 g/cm^3. The mass of a roll of pennies is 177.6 g. If the diameter of a penny is 1.85 cm, how tall is a roll of pennies?

77. See Exercises 59 and 60. Suppose Heidi plays in a tournament in which all of her opponents have the same rating. Under what circumstances will playing to a draw help or hurt her rating?

Solve.

78. $s = v_i t + \frac{1}{2}at^2$, for *a*

79. $A = 4lw + w^2$, for *l*

80. $\dfrac{P_1 V_1}{T_1} = \dfrac{P_2 V_2}{T_2}$, for T_2

81. $\dfrac{P_1 V_1}{T_1} = \dfrac{P_2 V_2}{T_2}$, for T_1

82. $\dfrac{b}{a - b} = c$, for *b*

83. $m = \dfrac{(d/e)}{(e/f)}$, for *d*

84. $\dfrac{a}{a + b} = c$, for *a*

Aha! **85.** $s + \dfrac{s + t}{s - t} = \dfrac{1}{t} + \dfrac{s + t}{s - t}$, for *t*

86. To derive the formula for the area of a trapezoid, consider the area of two trapezoids, one of which is upside down.

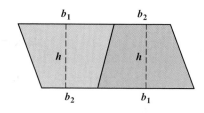

Explain why the total area of the two trapezoids is given by $h(b_1 + b_2)$. Then explain why the area of a trapezoid is given by $\dfrac{h}{2}(b_1 + b_2)$.

**New England Journal of Medicine*, 30 August 1990, pp. 604–608.

Properties of Exponents

The Product and Quotient Rules • The Zero Exponent •
Negative Integers as Exponents • Simplifying $(a^m)^n$ •
Raising a Product or a Quotient to a Power

Study Skills

Seeking Help Off Campus

Are you aware of all the supplements that exist for this textbook? See the preface for a description of each supplement: the *Student Solutions Manual*, a complete set of lessons in video or CD form, tutorial software, and the new Web site at www.MyMathLab.com. Through this Web site, you can find additional practice resources, helpful ideas, and links to other learning resources.

In Section 1.1, we discussed how whole-number exponents are used. We now develop rules for manipulating exponents and determine what zero and negative integers will mean as exponents.

The Product and Quotient Rules

Note that the expression $x^3 \cdot x^4$ can be rewritten as follows:

$$x^3 \cdot x^4 = \underbrace{x \cdot x \cdot x}_{3 \text{ factors}} \cdot \underbrace{x \cdot x \cdot x \cdot x}_{4 \text{ factors}}$$

$$= \underbrace{x \cdot x \cdot x \cdot x \cdot x \cdot x \cdot x}_{7 \text{ factors}}$$

$$= x^7.$$

This result is generalized in the *product rule*.

Multiplying with Like Bases: The Product Rule

For any number a and any positive integers m and n,

$$a^m \cdot a^n = a^{m+n}.$$

(When multiplying, if the bases are the same, keep the base and add the exponents.)

EXAMPLE 1

Multiply and simplify: **(a)** $m^5 \cdot m^7$; **(b)** $(5a^2b^3)(3a^4b^5)$.

Solution

a) $m^5 \cdot m^7 = m^{5+7} = m^{12}$ Multiplying powers by adding exponents

b) $(5a^2b^3)(3a^4b^5) = 5 \cdot 3 \cdot a^2 \cdot a^4 \cdot b^3 \cdot b^5$ Using the associative and commutative laws

$$= 15a^{2+4}b^{3+5}$$ Multiplying; using the product rule

$$= 15a^6b^8$$

Student Notes

Be careful to distinguish between how coefficients and exponents are handled. For example, in Example 1(b), the product of the coefficients 5 and 3 is 15, whereas the product of b^3 and b^5 is b^8.

Caution! $5^8 \cdot 5^6 = 5^{14}$; $5^8 \cdot 5^6 \neq 25^{14}$.

Next, we simplify a quotient:

$$\frac{x^8}{x^3} = \frac{x \cdot x \cdot x \cdot x \cdot x \cdot x \cdot x \cdot x}{x \cdot x \cdot x}$$ ⟵ 8 factors
   ⟵ 3 factors

$$= \frac{x \cdot x \cdot x}{x \cdot x \cdot x} \cdot x \cdot x \cdot x \cdot x \cdot x$$ Note that x^3/x^3 is 1.

$$= x \cdot x \cdot x \cdot x \cdot x$$ ⟵ 5 factors

$$= x^5.$$

The generalization of this result is the *quotient rule*.

Dividing with Like Bases: The Quotient Rule

For any nonzero number a and any positive integers m and n, $m > n$,

$$\frac{a^m}{a^n} = a^{m-n}.$$

(When dividing, if the bases are the same, keep the base and subtract the exponent of the denominator from the exponent of the numerator.)

EXAMPLE 2   Divide and simplify: **(a)** $\dfrac{r^9}{r^3}$; **(b)** $\dfrac{10x^{11}y^5}{2x^4y^3}$.

Solution

a) $\dfrac{r^9}{r^3} = r^{9-3} = r^6$ Using the quotient rule

b) $\dfrac{10x^{11}y^5}{2x^4y^3} = 5 \cdot x^{11-4} \cdot y^{5-3}$ Dividing; using the quotient rule

$$= 5x^7y^2$$

Caution! $\dfrac{7^8}{7^2} = 7^6$; $\dfrac{7^8}{7^2} \neq 7^4$.

The Zero Exponent

Suppose now that the bases in the numerator and the denominator are identical and are both raised to the same power. On the one hand, any (nonzero) expression divided by itself is equal to 1. For example,

$$\frac{t^5}{t^5} = 1 \quad \text{and} \quad \frac{6^4}{6^4} = 1.$$

On the other hand, if we continue subtracting exponents when dividing powers with the same base, we have

$$\frac{t^5}{t^5} = t^{5-5} = t^0 \quad \text{and} \quad \frac{6^4}{6^4} = 6^{4-4} = 6^0.$$

This suggests that t^5/t^5 equals both 1 *and* t^0. It also suggests that $6^4/6^4$ equals both 1 *and* 6^0. This leads to the following definition.

> **The Zero Exponent**
>
> For any nonzero real number a,
>
> $$a^0 = 1.$$
>
> (Any nonzero number raised to the zero power is 1. 0^0 is undefined.)

EXAMPLE 3   Evaluate each of the following for $x = 2.9$: **(a)** x^0; **(b)** $-x^0$; **(c)** $(-x)^0$.

Solution

a) $x^0 = 2.9^0 = 1$ Using the definition of 0 as an exponent

b) $-x^0 = -2.9^0 = -1$ The exponent 0 pertains only to the 2.9.

c) $(-x)^0 = (-2.9)^0 = 1$ Because of the parentheses, the base here is -2.9.

Parts (b) and (c) of Example 3 illustrate an important result:

Since $-a^n$ means $-1 \cdot a^n$, in general, $-a^n \neq (-a)^n$.*

Negative Integers as Exponents

Later in this text we will explain what numbers like $\frac{2}{9}$ or $\sqrt{2}$ mean as exponents. Until then, integer exponents will suffice.

To develop a definition for negative integer exponents, we simplify $5^3/5^7$ two ways. First we proceed as in arithmetic:

$$\frac{5^3}{5^7} = \frac{5 \cdot 5 \cdot 5}{5 \cdot 5 \cdot 5 \cdot 5 \cdot 5 \cdot 5 \cdot 5} = \frac{5 \cdot 5 \cdot 5 \cdot 1}{5 \cdot 5 \cdot 5 \cdot 5 \cdot 5 \cdot 5 \cdot 5}$$

$$= \frac{5 \cdot 5 \cdot 5}{5 \cdot 5 \cdot 5} \cdot \frac{1}{5 \cdot 5 \cdot 5 \cdot 5}$$

$$= \frac{1}{5^4}.$$

---

* When n is odd, it *does* follow that $-a^n = (-a)^n$. However, when n is even, we always have $-a^n \neq (-a)^n$, since $-a^n$ is always negative and $(-a)^n$ is always positive. We assume $a \neq 0$.

Were we to apply the quotient rule, we would have

$$\frac{5^3}{5^7} = 5^{3-7} = 5^{-4}.$$

These two expressions for $5^3/5^7$ suggest that

$$5^{-4} = \frac{1}{5^4}.$$

This leads to the definition of integer exponents, which includes negative exponents.

Integer Exponents

For any real number a that is nonzero and any integer n,

$$a^{-n} = \frac{1}{a^n}.$$

(The numbers a^{-n} and a^n are reciprocals of each other.)

The definitions above preserve the following pattern:

$$4^3 = 4 \cdot 4 \cdot 4,$$
$$4^2 = 4 \cdot 4, \qquad\qquad \text{Dividing both sides by 4}$$
$$4^1 = 4, \qquad\qquad \text{Dividing both sides by 4}$$
$$4^0 = 1, \qquad\qquad \text{Dividing both sides by 4}$$
$$4^{-1} = \frac{1}{4}, \qquad\qquad \text{Dividing both sides by 4}$$
$$4^{-2} = \frac{1}{4 \cdot 4} = \frac{1}{4^2}. \qquad \text{Dividing both sides by 4}$$

Caution! A negative exponent does not, in itself, indicate that an expression is negative. As shown above,

$$4^{-2} \neq 4(-2).$$

EXAMPLE 4

Express using positive exponents and, if possible, simplify.

a) 3^{-2} **b)** $5x^{-4}y^3$ **c)** $\dfrac{1}{7^{-2}}$

Solution

a) $3^{-2} = \dfrac{1}{3^2} = \dfrac{1}{9}$

b) $5x^{-4}y^3 = 5\left(\dfrac{1}{x^4}\right)y^3 = \dfrac{5y^3}{x^4}$

c) Since $\dfrac{1}{a^n} = a^{-n}$, we have

$$\frac{1}{7^{-2}} = 7^{-(-2)} = 7^2, \text{ or } 49. \qquad \text{Remember: } n \text{ can be a negative integer.}$$

The result from part (c) above can be generalized.

Factors and Negative Exponents

For any nonzero real numbers a and b and any integers m and n,

$$\frac{a^{-n}}{b^{-m}} = \frac{b^m}{a^n}.$$

(A factor can be moved to the other side of the fraction bar if the sign of the exponent is changed.)

EXAMPLE 5   Write an equivalent expression without negative exponents:

$$\frac{vx^{-2}y^{-5}}{z^{-4}w^{-3}}.$$

Solution   We can move each factor to the other side of the fraction bar if we change the sign of each exponent:

$$\frac{vx^{-2}y^{-5}}{z^{-4}w^{-3}} = \frac{vz^4w^3}{x^2y^5}.$$

The product and quotient rules apply for all integer exponents.

EXAMPLE 6   Simplify: **(a)** $7^{-3} \cdot 7^8$;   **(b)** $\dfrac{b^{-5}}{b^{-4}}$.

Solution

a) $7^{-3} \cdot 7^8 = 7^{-3+8}$ $\qquad$ Using the product rule

$\qquad\qquad\quad = 7^5$

b) $\dfrac{b^{-5}}{b^{-4}} = b^{-5-(-4)} = b^{-1}$ $\qquad$ Using the quotient rule

$\qquad\quad = \dfrac{1}{b}$ $\qquad$ Writing the answer without a negative exponent

Example 6(b) can also be simplified as follows:

$$\frac{b^{-5}}{b^{-4}} = \frac{b^4}{b^5} = b^{4-5} = b^{-1} = \frac{1}{b}.$$

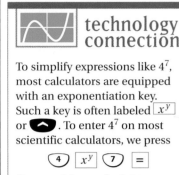

Simplifying $(a^m)^n$

Next, consider an expression like $(3^4)^2$:

$$(3^4)^2 = (3^4)(3^4)$$ We are raising 3^4 to the second power.

$$= (3 \cdot 3 \cdot 3 \cdot 3)(3 \cdot 3 \cdot 3 \cdot 3)$$
$$= 3 \cdot 3 \cdot 3 \cdot 3 \cdot 3 \cdot 3 \cdot 3 \cdot 3$$ Using the associative law
$$= 3^8.$$

Note that in this case, we could have multiplied the exponents:

$$(3^4)^2 = 3^{4 \cdot 2} = 3^8.$$

Likewise, $(y^8)^3 = (y^8)(y^8)(y^8) = y^{24}$. Once again, we get the same result if we multiply the exponents:

$$(y^8)^3 = y^{8 \cdot 3} = y^{24}.$$

The Power Rule

For any real number a and any integers m and n,

$$(a^m)^n = a^{mn}.$$

(To raise a power to a power, multiply the exponents.)

EXAMPLE 7

Simplify: **(a)** $(3^5)^4$; **(b)** $(y^{-5})^7$; **(c)** $(a^{-3})^{-7}$.

Solution

a) $(3^5)^4 = 3^{5 \cdot 4} = 3^{20}$ **b)** $(y^{-5})^7 = y^{-5 \cdot 7} = y^{-35} = \dfrac{1}{y^{35}}$

c) $(a^{-3})^{-7} = a^{(-3)(-7)} = a^{21}$

Raising a Product or a Quotient to a Power

When an expression inside parentheses is raised to a power, the inside expression is the base. Let's compare $2a^3$ and $(2a)^3$.

$$2a^3 = 2 \cdot a \cdot a \cdot a; \qquad (2a)^3 = (2a)(2a)(2a)$$
$$= 2 \cdot 2 \cdot 2 \cdot a \cdot a \cdot a$$
$$= 2^3 a^3 = 8a^3$$

We see that $2a^3$ and $(2a)^3$ are *not* equivalent. Note also that to simplify $(2a)^3$ we can raise each factor to the power 3. This leads to the following rule.

Raising a Product to a Power

For any integer n, and any real numbers a and b for which $(ab)^n$ exists,

$$(ab)^n = a^n b^n.$$

(To raise a product to a power, raise each factor to that power.)

EXAMPLE 8   Simplify: **(a)** $(-2x)^3$; **(b)** $(-3x^5y^{-1})^{-4}$.

Solution

a) $(-2x)^3 = (-2)^3 \cdot x^3$ Raising each factor to the third power

$\qquad\quad = -8x^3$

b) $(-3x^5y^{-1})^{-4} = (-3)^{-4}(x^5)^{-4}(y^{-1})^{-4}$ Raising each factor to the negative fourth power

$\qquad\qquad\qquad = \dfrac{1}{(-3)^4} \cdot x^{-20}y^4$ Multiplying powers; writing $(-3)^{-4}$ as $\dfrac{1}{(-3)^4}$

$\qquad\qquad\qquad = \dfrac{1}{81} \cdot \dfrac{1}{x^{20}} \cdot y^4$

$\qquad\qquad\qquad = \dfrac{y^4}{81x^{20}}$

There is a similar rule for raising a quotient to a power.

Raising a Quotient to a Power

For any integer n, and any real numbers a and b for which a/b, a^n, and b^n exist,

$$\left(\frac{a}{b}\right)^n = \frac{a^n}{b^n}.$$

(To raise a quotient to a power, raise both the numerator and the denominator to that power.)

EXAMPLE 9   Simplify: **(a)** $\left(\dfrac{x^2}{2}\right)^4$; **(b)** $\left(\dfrac{y^2z^3}{5}\right)^{-3}$.

Solution

a) $\left(\dfrac{x^2}{2}\right)^4 = \dfrac{(x^2)^4}{2^4} = \dfrac{x^8}{16}$ $\begin{array}{l}\longleftarrow 2 \cdot 4 = 8 \\ \longleftarrow 2^4 = 16\end{array}$

b) $\left(\dfrac{y^2z^3}{5}\right)^{-3} = \dfrac{(y^2z^3)^{-3}}{5^{-3}}$

$\qquad\qquad\qquad = \dfrac{5^3}{(y^2z^3)^3}$ Moving factors to the other side of the fraction bar and reversing the sign of those powers

$\qquad\qquad\qquad = \dfrac{125}{y^6z^9}$

The rule for raising a quotient to a power allows us to derive a useful result for manipulating negative exponents:

$$\left(\frac{a}{b}\right)^{-n} = \frac{a^{-n}}{b^{-n}} = \frac{b^{n}}{a^{n}} = \left(\frac{b}{a}\right)^{n}.$$

Using this result, we can simplify Example 9(b) as follows:

$$\left(\frac{y^2z^3}{5}\right)^{-3} = \left(\frac{5}{y^2z^3}\right)^3 \qquad \text{Taking the reciprocal of the base and changing the exponent's sign}$$

$$= \frac{5^3}{(y^2z^3)^3} = \frac{125}{y^6z^9}.$$

Definitions and Properties of Exponents

The following summary assumes that no denominators are 0 and that 0^0 is not considered. For any integers m and n,

1 as an exponent:	$a^1 = a$
0 as an exponent:	$a^0 = 1$
Negative exponents:	$a^{-n} = \dfrac{1}{a^n}$
	$\dfrac{a^{-n}}{b^{-m}} = \dfrac{b^m}{a^n}$
	$\left(\dfrac{a}{b}\right)^{-n} = \left(\dfrac{b}{a}\right)^n$
The Product Rule:	$a^m \cdot a^n = a^{m+n}$
The Quotient Rule:	$\dfrac{a^m}{a^n} = a^{m-n}$
The Power Rule:	$(a^m)^n = a^{mn}$
Raising a product to a power:	$(ab)^n = a^n b^n$
Raising a quotient to a power:	$\left(\dfrac{a}{b}\right)^n = \dfrac{a^n}{b^n}$

1.6 ## Exercise Set

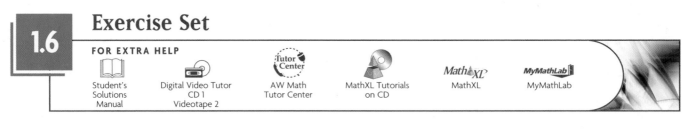
↩ *Concept Reinforcement   In each of Exercises 1–10, state whether the equation is an example of the product rule, the quotient rule, the power rule, raising a product to a power, or raising a quotient to a power.*

1. $(a^6)^4 = a^{24}$

2. $\left(\dfrac{5}{7}\right)^4 = \dfrac{5^4}{7^4}$

3. $(5x)^7 = 5^7 x^7$

4. $\dfrac{m^9}{m^3} = m^6$

5. $m^6 \cdot m^4 = m^{10}$

6. $(5^2)^7 = 5^{14}$

7. $\left(\dfrac{a}{4}\right)^7 = \dfrac{a^7}{4^7}$

8. $(ab)^{10} = a^{10} b^{10}$

9. $\dfrac{x^{10}}{x^2} = x^8$ **10.** $r^5 \cdot r^7 = r^{12}$

Multiply and simplify. Leave the answer in exponential notation.

11. $2^8 \cdot 2^4$ **12.** $7^3 \cdot 7^4$

13. $5^6 \cdot 5^4$ **14.** $6^3 \cdot 6^5$

15. $m^9 \cdot m^0$ **16.** $x^0 \cdot x^5$

17. $6x^5 \cdot 3x^2$ **18.** $4a^3 \cdot 2a^7$

19. $(-2m^4)(-8m^9)$ **20.** $(-2a^5)(7a^4)$

21. $(x^3y^4)(x^7y^6z^0)$ **22.** $(m^6n^5)(m^4n^7p^0)$

Divide and simplify.

23. $\dfrac{a^9}{a^3}$ **24.** $\dfrac{x^{12}}{x^3}$

25. $\dfrac{12t^7}{4t^2}$ **26.** $\dfrac{20a^{20}}{5a^4}$

27. $\dfrac{m^7n^9}{m^2n^5}$ **28.** $\dfrac{m^{12}n^9}{m^4n^6}$

29. $\dfrac{32x^8y^5}{8x^2y}$ **30.** $\dfrac{35x^7y^8}{7xy^2}$

31. $\dfrac{28x^{10}y^9z^8}{-7x^2y^3z^2}$ **32.** $\dfrac{18x^8y^6z^7}{-3x^2y^3z}$

Evaluate each of the following for $x = -2$.

33. $-x^0$ **34.** $(-x)^0$ **35.** $(4x)^0$ **36.** $4x^0$

Simplify.

37. $(-2)^4$ **38.** $(-3)^4$ **39.** -2^4

40. -3^4 **41.** $(-4)^{-2}$ **42.** $(-5)^{-2}$

43. -4^{-2} **44.** -5^{-2} **45.** -2^{-4}

46. -5^{-3} **47.** -2^{-6} **48.** -1^{-8}

Write an equivalent expression without negative exponents and, if possible, simplify.

49. a^{-3} **50.** n^{-6} **51.** $\dfrac{1}{5^{-3}}$

52. $\dfrac{1}{2^{-6}}$ **53.** $8x^{-3}$ **54.** $7x^{-3}$

55. $3a^8b^{-6}$ **56.** $5a^{-7}b^4$ **57.** $\dfrac{z^{-4}}{3x^5}$

58. $\dfrac{y^{-5}}{x^{-3}}$ **59.** $\dfrac{x^{-2}y^7}{z^{-4}}$ **60.** $\dfrac{y^4z^{-3}}{x^{-2}}$

Write an equivalent expression with negative exponents.

61. $\dfrac{1}{8^4}$ **62.** $\dfrac{1}{9^2}$ **63.** $\dfrac{1}{(-12)^2}$

64. $\dfrac{1}{(-5)^6}$ **65.** x^5 **66.** n^3

67. $4x^2$ **68.** $-4y^5$ **69.** $\dfrac{1}{(5y)^3}$

70. $\dfrac{1}{(5x)^5}$ **71.** $\dfrac{1}{3y^4}$ **72.** $\dfrac{1}{4b^3}$

Simplify. Should negative exponents appear in the answer, write a second answer using only positive exponents.

73. $8^{-2} \cdot 8^{-4}$ **74.** $9^{-1} \cdot 9^{-6}$

75. $b^2 \cdot b^{-5}$ **76.** $a^4 \cdot a^{-3}$

77. $a^{-3} \cdot a^4 \cdot a^2$ **78.** $x^{-8} \cdot x^5 \cdot x^3$

79. $(4mn^3)(-2m^3n^2)$ **80.** $(6x^6y^{-2})(-3x^2y^3)$

81. $(-2x^{-3})(7x^{-8})$ **82.** $(6x^{-4}y^3)(-4x^{-8}y^{-2})$

83. $(5a^{-2}b^{-3})(2a^{-4}b)$ **84.** $(3a^{-5}b^{-7})(2ab^{-2})$

85. $\dfrac{10^{-3}}{10^6}$ **86.** $\dfrac{12^{-4}}{12^8}$

87. $\dfrac{2^{-7}}{2^{-5}}$ **88.** $\dfrac{9^{-4}}{9^{-6}}$

89. $\dfrac{y^4}{y^{-5}}$ **90.** $\dfrac{a^3}{a^{-2}}$

91. $\dfrac{24a^5b^3}{-8a^4b}$ **92.** $\dfrac{9a^2}{3ab^3}$

93. $\dfrac{14a^4b^{-3}}{-8a^8b^{-5}}$ **94.** $\dfrac{-24x^6y^7}{18x^{-3}y^9}$

95. $\dfrac{-6x^{-2}y^4z^8}{24x^{-5}y^6z^{-3}}$ **96.** $\dfrac{8a^6b^{-4}c^8}{32a^{-4}b^5c^9}$

97. $(x^4)^3$ **98.** $(a^3)^2$

99. $(9^3)^{-4}$ **100.** $(8^4)^{-3}$

101. $(t^{-8})^{-5}$ **102.** $(x^{-4})^{-3}$

103. $(5xy)^2$ **104.** $(5ab)^3$

105. $(a^3b)^4$ **106.** $(x^3y)^5$

107. $\dfrac{(x^5)^2(x^{-3})^4}{(x^2)^3}$ **108.** $\dfrac{(a^{-2})^3(a^4)^2}{(a^3)^{-3}}$

109. $\dfrac{(2a^3)^4 4a^{-3}}{(a^2)^5}$

110. $\dfrac{(3x^2)^3 2x^{-4}}{(x^4)^2}$

Aha! **111.** $(8x^{-3}y^2)^{-4}(8x^{-3}y^2)^4$

112. $(2a^{-1}b^3)^{-2}(2a^{-1}b^3)^{-2}$

113. $\dfrac{(3x^3y^4)^3}{6xy^3}$

114. $\dfrac{(5a^3b)^2}{10a^2b}$

115. $\left(\dfrac{-4x^4y^{-2}}{5x^{-1}y^4}\right)^{-4}$

116. $\left(\dfrac{2x^3y^{-2}}{3y^{-3}}\right)^3$

117. $\left(\dfrac{6a^{-2}b^6}{8a^{-4}b^0}\right)^{-2}$

118. $\left(\dfrac{21x^5y^{-7}}{14x^{-2}y^{-6}}\right)^0$

Aha! **119.** $\left(\dfrac{4a^3b^{-9}}{6a^{-2}b^5}\right)^0$

120. $\left(\dfrac{5x^0y^{-7}}{2x^{-2}y^4}\right)^{-2}$

 **121.** Explain why $(-1)^n = 1$ for any even number n.

 **122.** Explain why $(-17)^{-8}$ is positive.

SKILL MAINTENANCE

Evaluate. [1.1], [1.2]

123. $4.9t^2 + 3t$, for $t = -3$

124. $16t^2 + 10t$, for $t = -2$

SYNTHESIS

 **125.** Explain in your own words why, for $a \neq 0$, a^0 is defined to be 1.

 **126.** Is the following true or false, and why?

$$5^{-6} > 4^{-9}$$

Simplify. Assume that all variables represent nonzero integers.

127. $\dfrac{8a^{x-2}}{2a^{2x+2}}$

128. $[7y(7-8)^{-4} - 8y(8-7)^{-2}]^{(-2)^2}$

129. $\dfrac{(2^{-2})^a \cdot (2^b)^{-a}}{(2^{-2})^{-b}(2^b)^{-2a}}$

130. $\{[(8^{-a})^{-2}]^b\}^{-c} \cdot [(8^0)^a]^c$

131. $(3^{a+2})^a$

132. $\dfrac{-28x^{b+5}y^{4+c}}{7x^{b-5}y^{c-4}}$

133. $(7^{3-a})^{2b}$

134. $\dfrac{4x^{2a+3}y^{2b-1}}{2x^{a+1}y^{b+1}}$

135. $\dfrac{3^{q+3} - 3^2(3^q)}{3(3^{q+4})}$

136. $\dfrac{25x^{a+b}y^{b-a}}{-5x^{a-b}y^{b+a}}$

137. $\left[\left(\dfrac{a^{-2c}}{b^{7c}}\right)^{-3}\left(\dfrac{a^{4c}}{b^{-3c}}\right)^2\right]^{-a}$

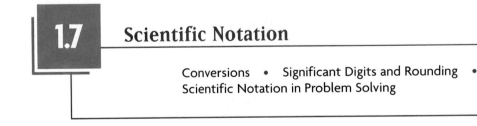

Scientific Notation

Conversions   •   Significant Digits and Rounding   •   Scientific Notation in Problem Solving

There is a variety of symbolism, or *notation*, for numbers. You are already familiar with fraction notation, decimal notation, and percent notation. We now study **scientific notation**, so named because of its usefulness in work with the very large and very small numbers that occur in science.

The following are examples of scientific notation:

$$7.2 \times 10^5 \quad \text{means} \quad 720{,}000;$$
$$3.4 \times 10^{-6} \quad \text{means} \quad 0.0000034;$$
$$4.89 \times 10^{-3} \quad \text{means} \quad 0.00489.$$

Scientific Notation

Scientific notation for a number is an expression of the form $N \times 10^m$, where N is in decimal notation, $1 \le N < 10$, and m is an integer.

Conversions

Note that $10^b/10^b = 10^b \cdot 10^{-b} = 1$. To convert to scientific notation, we can multiply by 1, writing 1 in the form $10^b/10^b$ or $10^b \cdot 10^{-b}$.

EXAMPLE 1   Population projections.   It has been estimated that in the year 2020, the world population will be approximately 7,516,000,000 (*Source*: U.S. Bureau of the Census, International Data Base). Write scientific notation for this number.

Solution   To write 7,516,000,000 as 7.516×10^m for some integer m, we must move the decimal point in 7,516,000,000 to the left 9 places. This can be accomplished by dividing—and then multiplying—by 10^9:

$$7{,}516{,}000{,}000 = \frac{7{,}516{,}000{,}000}{10^9} \cdot 10^9 \qquad \text{Multiplying by 1: } \frac{10^9}{10^9} = 1$$

$$= 7.516 \times 10^9. \qquad \text{This is scientific notation.}$$

EXAMPLE 2   Write scientific notation for the mass of a grain of sand:

0.0648 gram (g).

Solution   To write 0.0648 as 6.48×10^m for some integer m, we must move the decimal point 2 places to the right. To do this, we multiply—and then divide—by 10^2:

$$0.0648 = \frac{0.0648 \times 10^2}{10^2} \qquad \text{Multiplying by 1: } \frac{10^2}{10^2} = 1$$

$$= \frac{6.48}{10^2}$$

$$= 6.48 \times 10^{-2}\,\text{g}. \qquad \text{Writing scientific notation}$$

Try to make conversions to and from scientific notation mentally if possible. In doing so, remember that negative powers of 10 are used when representing small numbers and positive powers of 10 are used to represent large numbers.

EXAMPLE 3   Convert mentally to decimal notation: **(a)** 4.371×10^7; **(b)** 1.73×10^{-5}.

Solution

a) $4.371 \times 10^7 = 43{,}710{,}000$ Moving the decimal point 7 places to the right

b) $1.73 \times 10^{-5} = 0.0000173$ Moving the decimal point 5 places to the left

EXAMPLE 4   Convert mentally to scientific notation: **(a)** 82,500,000; **(b)** 0.0000091.

Solution

a) $82{,}500{,}000 = 8.25 \times 10^7$ *Check*: Multiplying 8.25 by 10^7 moves the decimal point 7 places to the right.

b) $0.0000091 = 9.1 \times 10^{-6}$ *Check*: Multiplying 9.1 by 10^{-6} moves the decimal point 6 places to the left.

Significant Digits and Rounding

In the world of science, it is important to know just how accurate a measurement is. For example, the measurement 5.12×10^3 km is more precise than the measurement 5.1×10^3 km. We say that 5.12×10^3 has three **significant digits** whereas 5.1×10^3 has only two significant digits. If 5.1×10^3, or 5100, includes no rounding in the tens column, we would indicate that by writing 5.10×10^3.

> When two or more measurements written in scientific notation are multiplied or divided, the result should be rounded so that it has the same number of significant digits as the measurement with the fewest significant digits. Rounding should be performed at the *end* of the calculation.

Thus,

$$\underbrace{(3.1 \times 10^{-3}\,\text{mm})}_{2\ \text{digits}}\underbrace{(2.45 \times 10^{-4}\,\text{mm})}_{3\ \text{digits}} = 7.595 \times 10^{-7}\,\text{mm}^2$$

should be rounded to

$$\underbrace{7.6 \times 10^{-7}}_{2\ \text{digits}}\,\text{mm}^2.$$

> When two or more measurements written in scientific notation are added or subtracted, the result should be rounded so that it has as many decimal places as the measurement with the fewest decimal places.

For example,

$$\underset{\text{5 digits}}{\underline{1.6354}} \times 10^4 \text{ km} + \underset{\text{4 digits}}{\underline{2.078}} \times 10^4 \text{ km} = 3.7134 \times 10^4 \text{ km}$$

should be rounded to

$$\underset{\text{4 digits}}{\overbrace{3.713}} \times 10^4 \text{ km}.$$

EXAMPLE 5

Multiply and write scientific notation for the answer:

$$(7.2 \times 10^5)(4.3 \times 10^9).$$

Solution   We have

$$(7.2 \times 10^5)(4.3 \times 10^9) = (7.2 \times 4.3)(10^5 \times 10^9) \qquad \text{Using the commutative and associative laws}$$

$$= 30.96 \times 10^{14} \qquad \text{Adding exponents}$$

To find scientific notation for this result, we convert 30.96 to scientific notation and simplify:

$$30.96 \times 10^{14} = (3.096 \times 10^1) \times 10^{14}$$

$$= 3.096 \times 10^{15}$$

$$\approx 3.1 \times 10^{15}. \qquad \text{Rounding to 2 significant digits}$$

technology connection

Both graphing and scientific calculators allow expressions to be entered using scientific notation. To do so, a key normally labeled EE or EXP is used. Often this is a secondary function and a key labeled SHIFT or 2ND must be pressed first. To check Example 5, we press 7.2 EE 5 × 4.3 EE 9. When we then press ENTER or =, the result 3.096E15 or 3.096 15 appears. We must interpret this result as 3.096×10^{15}.

EXAMPLE 6

Divide and write scientific notation for the answer:

$$\frac{3.48 \times 10^{-7}}{4.64 \times 10^6}.$$

Solution

$$\frac{3.48 \times 10^{-7}}{4.64 \times 10^6} = \frac{3.48}{4.64} \times \frac{10^{-7}}{10^6} \qquad \text{Separating factors. Our answer must have 3 significant digits.}$$

$$= 0.75 \times 10^{-13} \qquad \text{Subtracting exponents; simplifying}$$

$$= (7.5 \times 10^{-1}) \times 10^{-13} \qquad \text{Converting 0.75 to scientific notation}$$

$$= 7.50 \times 10^{-14} \qquad \text{Adding exponents. We write 7.50 to indicate 3 significant digits.}$$

Scientific Notation in Problem Solving

Scientific notation can be useful in problem solving.

EXAMPLE 7   *Information technology.*   In 2002, approximately 5.0×10^{18} megabytes of information were generated by the worldwide population of 6.3 billion people (*Source: The Indianapolis Star*, 11/2/03). Find the average amount of information generated by each person in 2002.

Solution

1. **Familiarize.**   If necessary, we can consult an outside reference to confirm that one billion is 1,000,000,000, or 10^9. Thus, 6.3 billion is 6.3×10^9. Note also that to find an average we need to divide. We let $a =$ the average number of megabytes of information generated, per person, in 2002.

2. **Translate.**   To find the average amount of information generated per person, we divide the total amount of information generated by the number representing the world population:
$$a = \frac{5.0 \times 10^{18} \text{ megabytes}}{6.3 \times 10^9 \text{ people}}.$$

3. **Carry out.**   We calculate and write scientific notation for the result:
$$a = \frac{5.0 \times 10^{18} \text{ megabytes}}{6.3 \times 10^9 \text{ people}}$$
$$= \frac{5.0}{6.3} \times \frac{10^{18} \text{ megabytes}}{10^9 \text{ people}}$$
$$\approx 0.79 \times 10^9 \text{ megabytes/person} \qquad \text{Rounding to 2 significant digits}$$
$$\approx 7.9 \times 10^{-1} \times 10^9 \text{ megabytes/person} \left.\vphantom{\begin{matrix}a\\a\end{matrix}}\right\} \quad \text{Writing scientific}$$
$$\approx 7.9 \times 10^8 \text{ megabytes/person.} \qquad\qquad\quad \text{notation}$$

4. **Check.**   To check, we multiply our answer, the average number of megabytes per person, by the worldwide population:
$$\underbrace{(7.9 \times 10^8 \text{ megabytes per person})}_{\substack{\text{Average amount of} \\ \text{information generated}}}\underbrace{(6.3 \times 10^9 \text{ people})}_{\text{Worldwide population}}$$
$$= (7.9 \times 6.3)(10^8 \times 10^9)\frac{\text{megabytes}}{\text{person}} \cdot \text{people}$$
$$= 49.77 \times 10^{17} \text{ megabytes}$$
$$\approx 5.0 \times 10^1 \times 10^{17} \text{ megabytes} \qquad \text{Rounding to 2 significant digits}$$
$$\approx 5.0 \times 10^{18} \text{ megabytes.}$$

Our answer checks.

5. **State.**   An average of 7.9×10^8 megabytes of information was generated by each person in the world in 2002.

EXAMPLE 8   Telecommunications.   A fiber-optic wire will be used for 375 km of transmission line. The wire has a diameter of 1.2 cm. What is the volume of wire needed for the line?

Solution

1. **Familiarize.**   Making a drawing, we see that we have a cylinder (a very *long* one). Its length is 375 km and the base has a diameter of 1.2 cm.

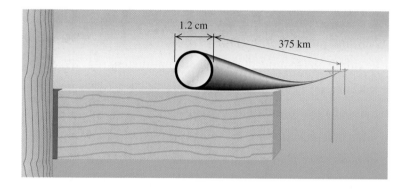

Recall that the formula for the volume of a cylinder is

$$V = \pi r^2 h,$$

where r is the radius and h is the height (in this case, the length of the wire).

2. **Translate.**   We will use the volume formula, but it is important to make the units consistent. Let's express everything in meters:

Length: 375 km = 375,000 m, or 3.75×10^5 m;
Diameter: 1.2 cm = 0.012 m, or 1.2×10^{-2} m.

The radius, which we will need in the formula, is half the diameter:

Radius: 0.6×10^{-2} m, or 6×10^{-3} m.

We now substitute into the above formula:

$$V = \pi (6 \times 10^{-3} \text{ m})^2 (3.75 \times 10^5 \text{ m}).$$

3. **Carry out.**   We do the calculations:

$$\begin{aligned}
V &= \pi \times (6 \times 10^{-3} \text{ m})^2 (3.75 \times 10^5 \text{ m}) \\
&= \pi \times 6^2 \times 10^{-6} \text{ m}^2 \times 3.75 \times 10^5 \text{ m} \quad &&\text{Using the properties of exponents} \\
&= (\pi \times 6^2 \times 3.75) \times (10^{-6} \times 10^5) \text{ m}^3 \\
&\approx 423.9 \times 10^{-1} \text{ m}^3 \quad &&\text{Using 3.14 for } \pi \\
&\approx 42 \text{ m}^3. \quad &&\text{Rounding 42.39 to 2 significant digits because of the 1.2}
\end{aligned}$$

4. **Check.**   About all we can do here is recheck the translation and calculations.

5. **State.**   The volume of the wire is about 42 m^3 (cubic meters).

Exercise Set

1.7

FOR EXTRA HELP

Student's Solutions Manual

Digital Video Tutor CD 1 Videotape 2

Tutor Center AW Math Tutor Center

MathXL Tutorials on CD

MathXL MathXL

MyMathLab MyMathLab

◄ *Concept Reinforcement* State whether scientific notation for each of the following numbers would include a positive or a negative power of 10.

1. The length of an Olympic marathon, in centimeters

2. The thickness of a cat's whisker, in meters

3. The mass of a hydrogen atom, in grams

4. The mass of a pickup truck, in grams

5. The time between leap years, in seconds

6. The time between a bird's heartbeats, in hours

Convert to decimal notation.

7. 5×10^{-4}

8. 5×10^{-5}

9. 9.73×10^{8}

10. 9.24×10^{7}

11. 4.923×10^{-10}

12. 7.034×10^{-2}

13. 9.03×10^{10}

14. 1.01×10^{12}

15. 4.037×10^{-8}

16. 3.007×10^{-9}

17. 7.01×10^{12}

18. 9.001×10^{10}

Convert to scientific notation.

19. 83,000,000,000

20. 2,600,000,000,000

21. 863,000,000,000,000,000

22. 572,000,000,000,000,000

23. 0.000000016

24. 0.000000263

25. 0.00000000007

26. 0.00000000009

27. 803,000,000,000

28. 3,090,000,000,000

29. 0.000000904

30. 0.00000000802

31. 431,700,000,000

32. 953,400,000,000

Simplify and write scientific notation for the answer. Use the correct number of significant digits.

33. $(2.3 \times 10^{6})(4.2 \times 10^{-11})$

34. $(6.5 \times 10^{3})(5.2 \times 10^{-8})$

35. $(2.34 \times 10^{-8})(5.7 \times 10^{-4})$

36. $(4.26 \times 10^{-6})(8.2 \times 10^{-6})$

37. $(5.2 \times 10^{6})(2.6 \times 10^{4})$

38. $(6.11 \times 10^{3})(1.01 \times 10^{13})$

39. $(7.01 \times 10^{-5})(6.5 \times 10^{7})$

40. $(4.08 \times 10^{-10})(7.7 \times 10^{5})$

Aha! 41. $(2.0 \times 10^{6})(3.02 \times 10^{-6})$

42. $(7.04 \times 10^{-9})(9.01 \times 10^{-7})$

43. $\dfrac{5.1 \times 10^{6}}{3.4 \times 10^{3}}$

44. $\dfrac{8.5 \times 10^{8}}{3.4 \times 10^{5}}$

45. $\dfrac{7.5 \times 10^{-9}}{2.5 \times 10^{-4}}$

46. $\dfrac{4.0 \times 10^{-6}}{8.0 \times 10^{-3}}$

47. $\dfrac{3.2 \times 10^{-7}}{8.0 \times 10^{8}}$

48. $\dfrac{12.6 \times 10^{8}}{4.2 \times 10^{-3}}$

49. $\dfrac{9.36 \times 10^{-11}}{3.12 \times 10^{11}}$

50. $\dfrac{2.42 \times 10^{5}}{1.21 \times 10^{-5}}$

51. $\dfrac{6.12 \times 10^{19}}{3.06 \times 10^{-7}}$

52. $\dfrac{4.7 \times 10^{-9}}{2.0 \times 10^{-9}}$

53. $4.6 \times 10^{-9} + 3.2 \times 10^{-9}$

54. $2.9 \times 10^{15} + 4.6 \times 10^{15}$

55. $5.9 \times 10^{23} + 6.3 \times 10^{23}$

56. $7.8 \times 10^{-34} + 5.4 \times 10^{-34}$

Solve.

57. *Radioactivity.* The lightest known particle in the universe, a neutrino has a maximum mass of 1.8×10^{-36} kg. What is the smallest number of neutrinos that could have the same mass as an alpha particle of mass 3.62×10^{-27} kg that results from the decay of radon?

Source: *Guinness Book of World Records* 2004

58. *Printing and engraving.* A ton of five-dollar bills is worth $4,540,000. How many pounds does a five-dollar bill weigh?

59. *High-tech fibers.* A carbon nanotube is a thin cylinder of carbon atoms that, pound for pound, is stronger than steel and may one day be used in clothing. With a diameter of about 4.0×10^{-10} in., a fiber can be made 100 yd long. Find the volume of such a fiber.
Source: *The Indianapolis Star*, 6/15/03

60. *Finance.* A *mil* is one thousandth of a dollar. The taxation rate in a certain school district is 5.0 mils for every dollar of assessed valuation. The assessed valuation for the district is 13.4 million dollars. How much tax revenue will be raised?

61. *Home maintenance.* The thickness of a sheet of plastic is measured in *mils*, where 1 mil $= \frac{1}{1000}$ in. To help conserve heat, the foundation of a 24-ft by 32-ft rectangular home is covered with a 4-ft high sheet of 8-mil plastic. Find the volume of plastic used.

62. *Office supplies.* A ream of copier paper weighs 2.25 kg. How much does a sheet of copier paper weigh?

For Exercises 63 and 64, use the fact that 1 *light year* $= 5.88 \times 10^{12}$ *mi.*

63. *Astronomy.* The brightest star in the night sky, Sirius, is about 4.704×10^{13} mi from the earth. How many light years is it from the earth to Sirius?

64. *Astronomy.* The diameter of the Milky Way galaxy is approximately 5.88×10^{17} mi. How many light years is it from one end of the galaxy to the other?

Named in tribute to Anders Ångström, a Swedish physicist who measured light waves, 1 Å *(read "one Angstrom") equals* 10^{-10} *meters. One parsec is about* 3.26 *light years, and one light year equals* 9.46×10^{15} *meters.*

65. How many Angstroms are in one parsec?

66. How many kilometers are in one parsec?

For Exercises 67 and 68, use the approximate average distance from the earth to the sun of 1.50×10^{11} *meters.*

67. Determine the volume of a cylindrical sunbeam that is 3 Å in diameter.

68. Determine the volume of a cylindrical sunbeam that is 5 Å in diameter.

69. *Biology.* An average of 4.55×10^{11} bacteria live in each pound of U.S. mud. There are 60.0 drops in one teaspoon and 6.0 teaspoons in an ounce. How many bacteria live in a drop of U.S. mud?
Source: *Harper's Magazine*, April 1996, p. 13

70. *Astronomy.* If a star 5.9×10^{14} mi from the earth were to explode today, its light would not reach us for 100 years. How far does light travel in 13 weeks?

71. *Astronomy.* The diameter of Jupiter is about 1.43×10^{5} km. A day on Jupiter lasts about 10 hr. At what speed is Jupiter's equator spinning?

72. *Astronomy.* The average distance of the earth from the sun is about 9.3×10^{7} mi. About how far does the earth travel in a yearly orbit about the sun? (Assume a circular orbit.)

73. Write a problem for a classmate to solve. Design the problem so the solution is "The area of the pinhead is 3.14×10^{-4} cm^2."

74. List two advantages of using scientific notation. Answers may vary.

SKILL MAINTENANCE

Evaluate.

75. $3x - 7y$, for $x = 5$ and $y = 1$ [1.1]

76. $2a - 5b$, for $a = 1$ and $b = -6$ [1.1], [1.2]

SYNTHESIS

77. A criminal claims to be carrying $5 million in twenty-dollar bills in a briefcase. Is this possible? Why or why not? (*Hint*: See Exercise 58.)

78. When a calculator indicates that $5^{17} = 7.629394531 \times 10^{11}$, an approximation is being made. How can you tell? (*Hint*: Examine the ones digit.)

79. The Sartorius Microbalance Model 4108 can weigh objects to an accuracy of 3.5×10^{-10} oz. A chemical compound weighing 1.2×10^{-9} oz is split in half and weighed on the microbalance. Give a weight range for the actual weight of each half.
Source: *Guinness Book of World Records* 2000

80. Write the reciprocal of 2.57×10^{-17} in scientific notation.

81. Compare $8 \cdot 10^{-90}$ and $9 \cdot 10^{-91}$. Which is the larger value? How much larger is it? Write scientific notation for the difference.

82. Write the reciprocal of 8.00×10^{-23} in scientific notation.

83. Evaluate: $(4096)^{0.05}(4096)^{0.2}$.

84. What is the ones digit in 513^{128}?

85. A grain of sand is placed on the first square of a chessboard, two grains on the second square, four grains on the third, eight on the fourth, and so on. Without a calculator, use scientific notation to approximate the number of grains of sand required for the 64th square. (*Hint*: Use the fact that $2^{10} \approx 10^3$.)

CORNER

Paired Problem Solving

COLLABORATIVE

Focus: Problem solving, scientific notation, and unit conversion

Time: 15–25 minutes

Group size: 3

ACTIVITY

Given that the earth's average distance from the sun is 1.5×10^{11} meters, determine the earth's orbital speed around the sun in miles per hour. Assume a circular orbit and use the following guidelines.

1. Each group should spend about 10 minutes attempting to solve this problem. Do not worry if the solution is not found.

2. Group members should each describe the interactions that took place, answering these three questions:

  a) What successful strategies were used?
  b) What unsuccessful strategies were used?
  c) What recommendations can you make for students working together to solve a problem?

3. Each group should then share their observations with each other and report to the class as a whole what they feel are their most significant observations.

1 Study Summary

The study of algebra consists largely of learning how to write **equivalent expressions** and **equivalent equations** (pp. 20, 24).

Equivalent Expressions

An **algebraic expression** consists of **variables**, numbers or **constants**, and operation signs. There are many ways in which we can form equivalent expressions:

- Any **addition**, **subtraction**, **multiplication**, or **division** of **real numbers** that appears within the expression can be calculated (pp. 15–18).
- The **rules for order of operations** may be useful for calculating certain combinations of operations involving real numbers (p. 5).
- The **commutative**, **associative**, and **distributive laws** can be used (pp. 20–21).
- **Like**, or **similar**, **terms** can be **combined** (p. 26).
- The **definitions** and **rules for exponents** can be used (p. 58).

A quick partial check to see whether two expressions are equivalent is to **evaluate** both expressions by replacing variables with numbers (p. 58).

Equivalent Equations

When an equals sign is placed between two expressions, an **equation** is formed (p. 2). Although we will sometimes discuss **identities** (like $2x + 3 = x + 3 + x$) or **contradictions** (like $t + 5 = t + 3$), most of the equations in this text are **conditional** (like $2a + 9 = 4$) (p. 29). Most equations are solved by writing a sequence of equivalent equations, the last of which isolates the variable on one side and a constant on the other. There are several techniques for creating equivalent equations:

- The **addition principle for equations** can be used, followed by the **law of opposites**: If $x - 9 = -4$, then $x - 9 + 9 = -4 + 9$, and $x + 0 = 5$ (pp. 15, 25).
- The **multiplication principle for equations** can be used, followed by the **law of reciprocals**: If $\frac{2}{3}x = -5$, then $\frac{3}{2} \cdot \frac{2}{3}x = \frac{3}{2}(-5)$, and $1x = -\frac{15}{2}$ (pp. 18, 25)
- All of the methods for writing equivalent expressions can be used on either side of the equation (see above).

The same techniques are used to solve a **formula** for a specified letter (p. 42).

Algebraic concepts are often described using different notations. We use both **roster** and **set-builder notation** (p. 7) to write sets. **Fraction, decimal,** percent, and **scientific notation** (pp. 8, 60) are all used when writing numbers.

We can approach problems of any kind with the following **five-step problem-solving strategy** (p. 32):

1. *Familiarize* yourself with the problem.
2. *Translate* to mathematical language.
3. *Carry out* some mathematical manipulation.
4. *Check* your possible answer in the original problem.
5. *State* the answer clearly.

We use the following rules to manipulate exponents (p. 58):

1 as an exponent:	$a^1 = a$
0 as an exponent:	$a^0 = 1$
Negative exponents:	$a^{-n} = \dfrac{1}{a^n}$
	$\dfrac{a^{-n}}{b^{-m}} = \dfrac{b^m}{a^n}$
	$\left(\dfrac{a}{b}\right)^{-n} = \left(\dfrac{b}{a}\right)^n$
The Product Rule:	$a^m \cdot a^n = a^{m+n}$
The Quotient Rule:	$\dfrac{a^m}{a^n} = a^{m-n}$
The Power Rule:	$(a^m)^n = a^{mn}$
Raising a product to a power:	$(ab)^n = a^n b^n$
Raising a quotient to a power:	$\left(\dfrac{a}{b}\right)^n = \dfrac{a^n}{b^n}$

The following important formulas will be used throughout the study of algebra:

Area of a rectangle:	$A = lw$
Area of a square:	$A = s^2$
Area of a parallelogram:	$A = bh$
Area of a trapezoid:	$A = \dfrac{h}{2}(b_1 + b_2)$
Area of a triangle:	$A = \frac{1}{2}bh$
Area of a circle:	$A = \pi r^2$
Circumference of a circle:	$C = \pi d$
Volume of a cube:	$V = s^3$
Volume of a right circular cylinder:	$V = \pi r^2 h$
Perimeter of a square:	$P = 4s$
Distance traveled:	$d = rt$
Simple interest:	$I = Prt$

1 Review Exercises

The following review exercises are for practice. Answers are at the back of the book. If you need to, restudy the section indicated alongside the answer.

☛ *Concept Reinforcement*   In each of Exercises 1–10, match the expression or equation with an equivalent expression or equation from the column on the right.

1. ____ $2x - 1 = 9$ [1.3]

2. ____ $2x - 1$ [1.2]

3. ____ $\frac{3}{4}x = 5$ [1.3]

4. ____ $\frac{3}{4}x - 5$ [1.2]

5. ____ $2(x + 7)$ [1.2]

6. ____ $2(x + 7) = 6$ [1.3]

7. ____ $4x - 3 + 2x = 5$ [1.3]

8. ____ $4x - 3 + 2x$ [1.2]

9. ____ $6 + 2x$ [1.2]

10. ____ $6 = 2x$ [1.3]

a) $2 + \frac{3}{4}x - 7$

b) $2x + 14 = 6$

c) $6x - 3$

d) $2(3 + x)$

e) $2x = 10$

f) $6x - 3 = 5$

g) $5x - 1 - 3x$

h) $3 = x$

i) $2x + 14$

j) $\frac{4}{3} \cdot \frac{3}{4}x = \frac{4}{3} \cdot 5$

11. Translate to an algebraic expression: Five less than the quotient of two numbers. [1.1]

12. Evaluate
$$7x^2 - 5y \div zx$$
for $x = -2$, $y = 3$, and $z = -5$. [1.1]

13. Name the set consisting of the first five odd natural numbers using both roster notation and set-builder notation. [1.1]

14. Find the area of a triangular flag that has a base of 50 cm and a height of 70 cm. [1.1]

Find the absolute value. [1.2]

15. $|-9.3|$ **16.** $|4.09|$ **17.** $|0|$

Perform the indicated operation. [1.2]

18. $-6.5 + (-3.7)$ **19.** $\left(-\frac{4}{5}\right) + \left(\frac{1}{7}\right)$

20. $\left(-\frac{1}{3}\right) + \frac{4}{5}$ **21.** $-7.9 - 3.6$

22. $-\frac{2}{3} - \left(-\frac{1}{2}\right)$ **23.** $12.5 - 17.9$

24. $(-4.2)(-3)$ **25.** $\left(-\frac{2}{3}\right)\left(\frac{5}{8}\right)$

26. $(1.2)(-4)$ **27.** $\frac{-18}{-3}$

28. $\frac{72.8}{-8}$ **29.** $-7 \div \frac{4}{3}$

30. Find $-a$ if $a = -4.01$.

Use a commutative law to write an equivalent expression. [1.2]

31. $9 + a$ **32.** $7y$ **33.** $5x + y$

Use an associative law to write an equivalent expression. [1.2]

34. $(4 + a) + b$ **35.** $(xy)3$

36. Obtain an expression that is equivalent to $7mn + 14m$ by factoring. [1.2]

37. Combine like terms: $3x^3 - 6x^2 + x^3 + 5$. [1.3]

38. Simplify: $7x - 4[2x + 3(5 - 4x)]$. [1.3]

Solve. If the solution set is $\varnothing$ or $\mathbb{R}$, classify the equation as a contradiction or as an identity. [1.3]

39. $x - 8.9 = 2.7$

40. $\frac{2}{3}a = 9$

41. $-9x + 4(2x - 3) = 5(2x - 3) + 7$

42. $3(x - 4) + 2 = x + 2(x - 5)$

43. $5t - (7 - t) = 4t + 2(9 + t)$

44. Translate to an equation but do not solve: 15 more than twice a number is 21. [1.4]

45. A number is 19 less than another number. The sum of the numbers is 115. Find the smaller number. [1.4]

46. One angle of a triangle measures three times the second angle. The third angle measures twice the second angle. Find the measures of the angles. [1.4]

47. Solve for m: $P = m/S$. [1.5]

48. Solve for x: $c = mx - rx$. [1.5]

49. The volume of a film canister is 28.26 cm^3. If the radius of the canister is 1.5 cm, determine the height. [1.5]

50. Multiply and simplify: $(6a^2b^7)(-2a^4b)$. [1.6]

51. Divide and simplify: $\dfrac{12x^3y^8}{3x^2y^2}$. [1.6]

52. Evaluate a^0, a^2, and $-a^2$ for $a = -5.3$. [1.6]

Simplify. Do not use negative exponents in the answer. [1.6]

53. $3^{-4} \cdot 3^9$

54. $(5a^2)^3$

55. $(-2a^{-3}b^2)^{-3}$

56. $\left(\dfrac{x^2y^3}{z^4}\right)^{-2}$

57. $\left(\dfrac{2a^{-2}b}{4a^3b^{-3}}\right)^4$

Simplify. [1.2]

58. $\dfrac{4(9 - 2 \cdot 3) - 3^2}{4^2 - 3^2}$

59. $1 - (2 - 5)^2 + 5 \div 10 \cdot 4^2$

60. Convert 0.000000103 to scientific notation. [1.7]

61. One *parsec* (a unit that is used in astronomy) is 30,860,000,000,000 km. Write scientific notation for this number. [1.7]

Simplify and write scientific notation for each answer. Use the correct number of significant digits. [1.7]

62. $(8.7 \times 10^{-9}) \times (4.3 \times 10^{15})$

63. $\dfrac{1.2 \times 10^{-12}}{6.1 \times 10^{-7}}$

64. A sheet of plastic shrink wrap has a thickness of 0.00015 mm. The sheet is 1.2 m by 79 m. Use scientific notation to find the volume of the sheet. [1.7]

SYNTHESIS

65. Describe a method that could be used to write equations that have no solution. [1.3]

66. Explain how the distributive law can be used when combining like terms. [1.3]

67. If the smell of gasoline is detectable at 3 parts per billion, what percent of the air is occupied by the gasoline? [1.7]

68. Evaluate $a + b(c - a^2)^0 + (abc)^{-1}$ for $a = 3$, $b = -2$, and $c = -4$. [1.1], [1.6]

69. What's a better deal: a 13-in. diameter pizza for \$8 or a 17-in. diameter pizza for \$11? Explain. [1.4], [1.5]

70. The surface area of a cube is 486 cm^2. Find the volume of the cube. [1.5]

71. Solve for z: $m = \dfrac{x}{y - z}$. [1.5]

72. Simplify: $\dfrac{(3^{-2})^a \cdot (3^b)^{-2a}}{(3^{-2})^b \cdot (9^{-b})^{-3a}}$. [1.6]

73. Each of Ray's test scores counts three times as much as a quiz score. If after 4 quizzes Ray's average is 82.5, what score does he need on the first test in order to raise his average to 85? [1.4]

74. Fill in the following blank so as to ensure that the equation is an identity. [1.3]
$$5x - 7(x + 3) - 4 = 2(7 - x) + \underline{\hspace{1.5cm}}$$

75. Replace the blank with one term to ensure that the equation is a contradiction. [1.3]
$$20 - 7[3(2x + 4) - 10] = 9 - 2(x - 5) + \underline{\hspace{1.5cm}}$$

76. Use the commutative law for addition once and the distributive law twice to show that [1.2]
$$a2 + cb + cd + ad = a(d + 2) + c(b + d).$$

77. Find an irrational number between $\frac{1}{2}$ and $\frac{3}{4}$. [1.1]

Chapter Test

1. Translate to an algebraic expression: Four more than the product of two numbers.

2. Evaluate $a^3 - 5b + b \div ac$ for $a = -2$, $b = 6$, and $c = 3$.

3. A triangular stamp's base measures 3 cm and its height 2.5 cm. Find its area.

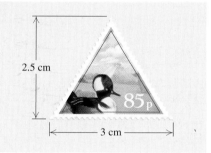

Perform the indicated operation.

4. $-15 + (-16)$

5. $-7.5 + 3.8$

6. $3.21 + (-8.32)$

7. $29.5 - 43.7$

8. $-16.8 - 26.4$

9. $-6.4(5.3)$

10. $-\frac{7}{3} - \left(-\frac{3}{4}\right)$

11. $-\frac{2}{7}\left(-\frac{5}{14}\right)$

12. $\frac{-42.6}{-7.1}$

13. $\frac{2}{5} \div \left(-\frac{3}{10}\right)$

14. Simplify: $7 + (1 - 3)^2 - 9 \div 2^2 \cdot 6$.

15. Use a commutative law to write an expression equivalent to $7x + y$.

Combine like terms.

16. $3y - 9y + 5y$

17. $6a^2b - 5ab^2 + ab^2 - 5a^2b + 2$

18. Simplify: $2x - 3(2x - 5) - 7$.

Solve. If the solution set is $\mathbb{R}$ or $\varnothing$, classify the equation as an identity or a contradiction.

19. $10x - 7 = 38x + 49$

20. $13t - (5 - 2t) = 5(3t - 1)$

21. Solve for P_2: $\dfrac{P_1 V_1}{T_1} = \dfrac{P_2 V_2}{T_2}$.

22. Linda's scores on five tests are 84, 80, 76, 96, and 80. What must Linda score on the sixth test so that her average will be 85?

23. Find three consecutive odd integers such that the sum of four times the first, three times the second, and two times the third is 167.

Simplify. Do not use negative exponents in the answer.

24. $-5(x - 4) - 3(x + 7)$

25. $6b - [7 - 2(9b - 1)]$

26. $(7x^{-4}y^{-7})(-6x^{-6}y)$

27. -3^{-2}

28. $(-6x^2y^{-4})^{-2}$

29. $\left(\dfrac{2x^3 y^{-6}}{-4y^{-2}}\right)^2$

30. $(7x^3y)^0$

Simplify and write scientific notation for the answer. Use the correct number of significant digits.

31. $(9.05 \times 10^{-3})(2.22 \times 10^{-5})$

32. $\dfrac{5.6 \times 10^7}{2.8 \times 10^{-3}}$

33. $\dfrac{1.8 \times 10^{-4}}{4.8 \times 10^{-7}}$

Solve.

34. Binary stars orbit around each other and can be as close to each other as the earth is to the sun. If a pair of binary stars are 1.5×10^8 km apart, what is the length of their orbit? (Assume the orbit is circular.)
Source: Abell, George O., *Exploration of the Universe*, 3rd ed. New York: Holt, Rinehart and Winston, 1975

SYNTHESIS

Simplify. Do not use negative exponents in the answer.

35. $(2x^{3a}y^{b+1})^{3c}$

36. $\dfrac{-27a^{x+1}}{3a^{x-2}}$

37. $\dfrac{(-16x^{x-1}y^{y-2})(2x^{x+1}y^{y+1})}{(-7x^{x+2}y^{y+2})(8x^{x-2}y^{y-1})}$

2

Graphs, Functions, and Linear Equations

AN APPLICATION

Under one particular CIGNA health-insurance plan, an individual pays the first $100 of each hospital stay plus $\frac{1}{5}$ of all charges in excess of $100. By approximately how much did Gerry's hospital bill exceed $100, if the hospital stay cost him a total of $2200? (*Source*: VT State Chamber of Commerce membership information from VACE, 2004)

This problem appears as Exercise 67 in Section 2.4.

Akin B. Aboyadé-Cole
SMALL-BUSINESS OWNER
Elmsford, New York

As a materials and environmental engineering technology firm, we cannot deliver services to our clients without using mathematics. Besides engineering calculations, we use math to calculate the healthcare costs of participating employees, which helps us determine the company's overhead and fringe-benefit costs.

Graphs are important because they allow us to see relationships. For example, the graph of an equation in two variables helps us see how those two variables are related. Graphs are useful in a wide variety of settings. In this chapter, we emphasize graphs of equations.

A certain kind of relationship between two variables is known as a function. Functions are very important in mathematics in general, and in problem solving in particular. In this chapter, we will explain what a function is as well as how it can be used in problem solving.

2.1 Graphs

Points and Ordered Pairs • Quadrants • Solutions of Equations • Nonlinear Equations

It has often been said that a picture is worth a thousand words. As we turn our attention to the study of graphs, we discover that in mathematics this is quite literally the case. Graphs are a compact means of displaying information and provide a visual approach to problem solving.

Points and Ordered Pairs

Whereas on a number line, each point corresponds to a number, on a plane, each point corresponds to a pair of numbers. The idea of using two perpendicular number lines, called **axes** (pronounced ak-sēz; singular, **axis**) to identify points in a plane is commonly attributed to the great French mathematician and philosopher René Descartes (1596–1650). In honor of Descartes, this representation is also called the **Cartesian coordinate system**. The variable x is most often used for the horizontal axis and the variable y for the vertical axis, so we will also refer to the **x, y-coordinate system**.

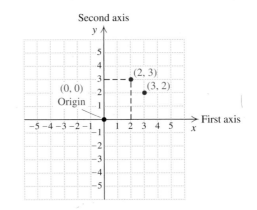

To label a point that appears on the *x, y*-coordinate system, we write a pair of numbers in the form (*x, y*), where *x* is the number located on the *x*-axis directly above or below the point and *y* is the number located on the *y*-axis to the left or right of the point (see the dashed lines in the figure on p. 76). Thus, as shown on p. 76, (2, 3) and (3, 2) are different points. Because the order in which the numbers are listed is important, these are called **ordered pairs**.

The numbers in an ordered pair are called **coordinates**. In (−4, 3), the *first coordinate* is −4 and the *second coordinate** is 3. The point with coordinates (0, 0) is called the **origin**.

EXAMPLE 1   Plot the points (−4, 3), (−5, −3), (0, 4), and (2.5, 0).

Solution   To plot (−4, 3), note that the first coordinate, −4, tells us the distance in the first, or horizontal, direction. We go 4 units *left* of the origin. From that location, we go 3 units *up*. The point (−4, 3) is then marked, or "plotted." The points (−5, −3), (0, 4), and (2.5, 0) are also plotted below.

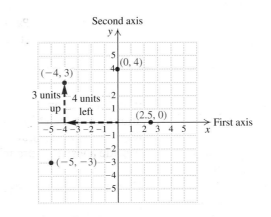

Quadrants

The axes divide the plane into four regions called **quadrants**, as shown here.

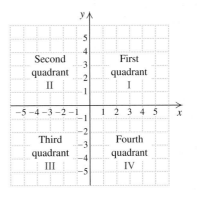

*The first coordinate is sometimes called the **abscissa** and the second coordinate the **ordinate**.

In region I (the *first* quadrant), both coordinates of a point are positive. In region II (the *second* quadrant), the first coordinate is negative and the second coordinate is positive. In the third quadrant, both coordinates are negative, and in the fourth quadrant, the first coordinate is positive while the second coordinate is negative.

Points with one or more 0's as coordinates, such as $(0, -6)$, $(4, 0)$, and $(0, 0)$, are on axes and *not* in quadrants.

Solutions of Equations

If an equation has two variables, its solutions are pairs of numbers. When such a solution is written as an ordered pair, the first number listed in the pair generally replaces the variable that occurs first alphabetically.

EXAMPLE 2   Determine whether the pairs $(4, 2)$, $(-1, -4)$, and $(2, 5)$ are solutions of the equation $y = 3x - 1$.

Solution   To determine whether each pair is a solution, we replace x with the first coordinate and y with the second coordinate. When the replacements make the equation true, we say that the ordered pair is a solution.

$$
\begin{array}{c|c}
\multicolumn{2}{l}{y = 3x - 1} \\
\hline
2 & 3(4) - 1 \\
 & 12 - 1 \\
\end{array}
\qquad
2 \overset{?}{=} 11
$$

$$
\begin{array}{c|c}
\multicolumn{2}{l}{y = 3x - 1} \\
\hline
-4 & 3(-1) - 1 \\
 & -3 - 1 \\
\end{array}
\qquad
-4 \overset{?}{=} -4
$$

$$
\begin{array}{c|c}
\multicolumn{2}{l}{y = 3x - 1} \\
\hline
5 & 3(2) - 1 \\
 & 6 - 1 \\
\end{array}
\qquad
5 \overset{?}{=} 5
$$

Since $2 = 11$ is *false*, the pair $(4, 2)$ is *not* a solution.

Since $-4 = -4$ is *true*, the pair $(-1, -4)$ *is* a solution.

Since $5 = 5$ is *true*, the pair $(2, 5)$ *is* a solution.

In fact, there is an infinite number of solutions of $y = 3x - 1$. We can use a graph as a convenient way of representing these solutions. Thus to **graph** an equation means to make a drawing that represents all of its solutions.

EXAMPLE 3   Graph the equation $y = x$.

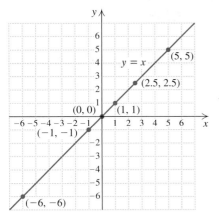

Solution   We label the horizontal axis as the x-axis and the vertical axis as the y-axis.

Next, we find some ordered pairs that are solutions of the equation. In this case, no calculations are necessary. Here are a few pairs that satisfy the equation $y = x$:

$$(0, 0), \qquad (1, 1), \qquad (5, 5), \qquad (-1, -1), \qquad (-6, -6).$$

Plotting these points, we can see that if we were to plot a hundred solutions, the dots would appear to form a line. Observing the pattern, we can draw the line with a ruler. The line is the graph of the equation $y = x$. We label the line $y = x$.

Note that the coordinates of *any* point on the line—for example, $(2.5, 2.5)$—satisfy the equation $y = x$. The line continues indefinitely in both directions, so we draw it to the edge of the grid.

EXAMPLE 4 Graph the equation $y = 2x$.

Solution We find some ordered pairs that are solutions. This time we list the pairs in a table. To find an ordered pair, we can choose *any* number for x and then determine y. For example, if we choose 3 for x, then $y = 2 \cdot 3 = 6$ (substituting into $y = 2x$). We choose some negative values for x, as well as some positive ones (generally, we avoid selecting values that are beyond the edge of the graph paper). Next, we plot these points. If we plotted *many* such points, they would appear to make a solid line. We draw the line with a ruler and label it $y = 2x$.

x	y	(x, y)
0	0	$(0, 0)$
1	2	$(1, 2)$
3	6	$(3, 6)$
-2	-4	$(-2, -4)$
-3	-6	$(-3, -6)$

Choose any x.
Compute y.
Form the pair.
Plot the points and draw the line.

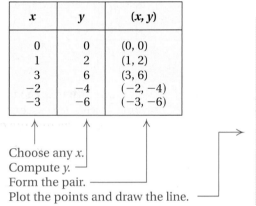

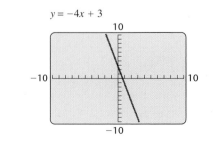technology connection

The window of a graphing calculator is the rectangular portion of the screen in which a graph appears. Windows are described by four numbers of the form [L, R, B, T], representing the left and right endpoints of the x-axis and the bottom and top endpoints of the y-axis. Below, we have graphed $y = -4x + 3$, using the "standard" $[-10, 10, -10, 10]$ window.

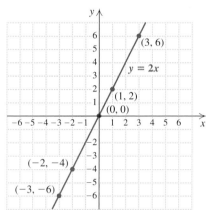

$y = -4x + 3$

When (TRACE) is pressed, a cursor can be moved along the graph while its coordinates appear. To find the y-value that is paired with a particular x-value, we simply key in that x-value and press **ENTER**.

1. Graph $y = -4x + 3$ using a $[-10, 10, -10, 10]$ window. Then TRACE to find coordinates of several points, including the points with the x-values -1.5 and 1.

EXAMPLE 5   Graph the equation $y = -\frac{1}{2}x$.

Solution   By choosing even integers for x, we can avoid fractions when calculating y. For example, if we choose 4 for x, we get $y = \left(-\frac{1}{2}\right)(4)$, or -2. When x is -6, we get $y = \left(-\frac{1}{2}\right)(-6)$, or 3. We find several ordered pairs, plot them, and draw the line.

Student Notes

Your choice of values for x can vary from what other students select. Certain choices will help you avoid fractions or points that lie off the edge of your grid.

x	y	(x, y)
4	-2	$(4, -2)$
-6	3	$(-6, 3)$
0	0	$(0, 0)$
2	-1	$(2, -1)$

Choose any x.
Compute y.
Form the pair.
Plot the points and draw the line.

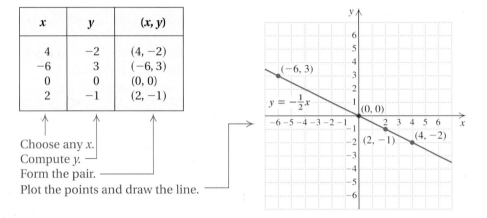

As you can see, the graphs in Examples 3–5 are straight lines. We will refer to any equation whose graph is a straight line as a **linear equation**. Linear equations will be discussed in more detail in Sections 2.3–2.5 at which time we will develop other methods for graphing them. To graph a line, be sure to plot at least two points, using a third point as a check.

Nonlinear Equations

There are many equations for which the graph is not a straight line. Graphing these **nonlinear equations** often requires plotting many points in order to see the general shape of the graph.

EXAMPLE 6   Graph: $y = |x|$.

Solution   We select numbers for x and find the corresponding values for y. For example, if we choose -1 for x, we get $y = |-1| = 1$. Several ordered pairs are listed in the table below.

x	y
-3	3
-2	2
-1	1
0	0
1	1
2	2
3	3

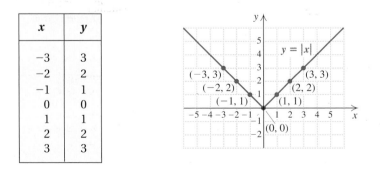

We plot these points, noting that the absolute value of a positive number is the same as the absolute value of its opposite. Thus the x-values 3 and -3 both are paired with the y-value 3. The graph is V-shaped, as shown above.

Curves similar to the one in Example 7 below are studied in Chapter 8.

EXAMPLE 7 Graph: $y = x^2 - 5$.

Solution We select numbers for x and find the corresponding values for y. For example, if we choose -2 for x, we get $y = (-2)^2 - 5 = 4 - 5 = -1$. The table lists several ordered pairs.

x	y
0	-5
-1	-4
1	-4
-2	-1
2	-1
-3	4
3	4

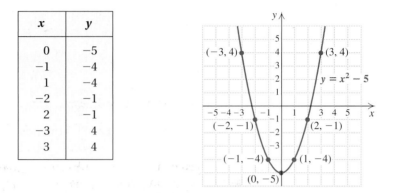

Next, we plot the points. The more points plotted, the clearer the shape of the graph becomes. Since the value of $x^2 - 5$ grows rapidly as x moves away from the origin, the graph rises steeply on either side of the y-axis.

---

technology connection

Often, graphs are drawn with the aid of a graphing calculator. As shown above, determining just a few ordered pairs can be quite time-consuming. With a graphing calculator, thousands of ordered pairs can be graphed at the touch of a button. Most graphing calculators can also set up a table of pairs for any equation that is entered. By pressing **2ND** (TBLSET), we can control the smallest x-value listed (using TblStart) and the difference between successive x-values (using ΔTbl). Setting Indpnt and Depend both to Auto allows the calculator to complete a table. For the table shown, we used $y_1 = -4x + 3$, with TblStart $= 1.4$ and ΔTbl $= .1$. To view the table, we press **2ND** (TABLE).

Graph each of the following equations using a $[-10, 10, -10, 10]$ window. Then create a table of ordered pairs in which the x-values start at -1 and are 0.1 unit apart.

TblStart $= 1.4$ ΔTbl $= .1$

X	Y1
1.4	-2.6
1.5	-3
1.6	-3.4
1.7	-3.8
1.8	-4.2
1.9	-4.6
2	-5

X = 1.4

1. $y = 5x - 3$ **2.** $y = x^2 - 4x + 3$
3. $y = (x + 4)^2$ **4.** $y = \sqrt{x + 2}$
5. $y = |x + 2|$

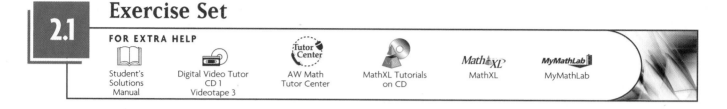

Exercise Set

2.1

FOR EXTRA HELP

Student's Solutions Manual | Digital Video Tutor CD 1 Videotape 3 | AW Math Tutor Center | MathXL Tutorials on CD | MathXL | MyMathLab

To the student and the instructor: **Throughout this text, selected exercises are marked with the icon Aha! . These "Aha!" exercises can be answered quite easily if the student pauses to inspect the exercise rather than proceed mechanically. This is done to discourage rote memorization. Some "Aha!" exercises are left unmarked to encourage students to always pause before working a problem.**

↩ *Concept Reinforcement   Complete each of the following.*

1. In the fourth quadrant, a point's first coordinate is always positive and its second coordinate is always _____.

2. In the third quadrant, a point's first coordinate is always _____ and its second coordinate is always negative.

3. The two perpendicular number lines that are used for graphing are called _____.

4. Because the order in which the numbers are listed is important, numbers listed in the form (x, y) are called _____ pairs.

5. To graph an equation means to make a drawing that represents all _____ of the equation.

6. An equation whose graph is a straight line is said to be a(n) _____ equation.

Give the coordinates of each point.

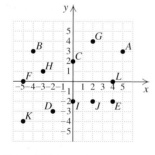

7. A, B, C, D, E, and F

8. G, H, I, J, K, and L

Plot the points. Label each point with the indicated letter.

9. $A(3, 0)$, $B(4, 2)$, $C(5, 4)$, $D(6, 6)$, $E(3, -4)$, $F(3, -3)$, $G(3, -2)$, $H(3, -1)$

10. $A(1, 1)$, $B(2, 3)$, $C(3, 5)$, $D(4, 7)$, $E(-2, 1)$, $F(-2, 2)$, $G(-2, 3)$, $H(-2, 4)$, $J(-2, 5)$, $K(-2, 6)$

11. Plot the points $M(2, 3)$, $N(5, -3)$, and $P(-2, -3)$. Draw $\overline{MN}$, $\overline{NP}$, and $\overline{MP}$. ($\overline{MN}$ means the line segment from M to N.) What kind of geometric figure is formed? What is its area?

12. Plot the points $Q(-4, 3)$, $R(5, 3)$, $S(2, -1)$, and $T(-7, -1)$. Draw $\overline{QR}$, $\overline{RS}$, $\overline{ST}$, and $\overline{TQ}$. What kind of figure is formed? What is its area?

Name the quadrant in which each point is located.

13. $(-4, 1)$

14. $(2, 17)$

15. $(-6, -7)$

16. $(4, -8)$

17. $\left(3, \frac{1}{2}\right)$

18. $(-1, -7)$

19. $(6.9, -2)$

20. $(-4, 31)$

Determine whether each ordered pair is a solution of the given equation. Remember to use alphabetical order for substitution.

21. $(1, -1)$; $y = 3x - 4$

22. $(2, 5)$; $y = 4x - 3$

23. $(2, 4)$; $5s - t = 8$

24. $(1, 3)$; $4p - q = 1$

25. $(3, 5)$; $4x - y = 7$

26. $(2, 7)$; $5x - y = 3$

27. $\left(0, \frac{3}{5}\right)$; $6a + 5b = 3$

28. $\left(0, \frac{3}{2}\right)$; $3f + 4g = 6$

29. $(2, -1)$; $4r - 2s = 10$

30. $(2, -4)$; $5w + 2z = 2$

31. $(5, 3)$; $x - 3y = -4$

32. $(1, 2); 2x - 5y = -6$

33. $(3, -1); y = 3x^2$

34. $(2, 4); 2r^2 - s = 5$

35. $(2, 3); 5s^2 - t = 7$

36. $(2, 3); y = x^3 - 5$

Graph.

37. $y = x + 4$ **38.** $y = x + 3$

39. $y = -x$ **40.** $y = 3x$

41. $y = 3x - 1$ **42.** $y = -4x + 1$

43. $y = -2x + 3$ **44.** $y = -3x + 1$

Aha! **45.** $y + 2x = 3$ **46.** $y + 3x = 1$

47. $y = -\frac{3}{2}x + 5$ **48.** $y = -\frac{2}{3}x - 2$

49. $y = \frac{3}{4}x + 1$ **50.** $y = \frac{3}{4}x + 2$

51. $y = |x| + 2$ **52.** $y = |x| + 1$

53. $y = |x| - 2$ **54.** $y = |x| - 3$

55. $y = x^2 + 2$ **56.** $y = x^2 + 1$

57. $y = x^2 - 2$ **58.** $y = x^2 - 3$

59. What can be said about the location of two points that have the same first coordinates and second coordinates that are opposites of each other?

60. Examine Example 6 and explain why it is unwise to draw a graph after plotting just two points.

SKILL MAINTENANCE

Evaluate.

61. $5s - 3t$, for $s = 2$ and $t = 4$ [1.1]

62. $2a + 7b$, for $a = 3$ and $b = 1$ [1.1]

63. $(3x - y)^2$, for $x = 4$ and $y = 2$ [1.1]

64. $(2m + n)^2$, for $m = 3$ and $n = 1$ [1.1]

Aha! **65.** $(5 - x)^4(x + 2)^3$, for $x = -2$ [1.2]

66. $2x^2 + 4x - 9$, for $x = -3$ [1.2]

SYNTHESIS

67. Without making a drawing, how can you tell that the graph of $y = x - 30$ passes through three quadrants?

68. At what point will the line passing through $(a, -1)$ and $(a, 5)$ intersect the line that passes through $(-3, b)$ and $(2, b)$? Why?

69. Using the same set of axes, graph $y = 2x$, $y = 2x - 3$, and $y = 2x + 3$. Describe the pattern relating each line to the number that is added to $2x$.

70. Graph $y = 6x, y = 3x, y = \frac{1}{2}x, y = -6x, y = -3x$, and $y = -\frac{1}{2}x$ using the same set of axes and compare the slants of the lines. Describe the pattern that relates the slant of the line to the multiplier of x.

71. Match each sentence with the most appropriate of the four graphs shown below.

  a) Carpooling to work, Terry spent 10 min on local streets, then 20 min cruising on the freeway, and then 5 min on local streets to his office.

  b) For her commute to work, Sharon drove 10 min to the train station, rode the express for 20 min, and then walked for 5 min to her office.

  c) For his commute to school, Roger walked 10 min to the bus stop, rode the express for 20 min, and then walked for 5 min to his class.

  d) Coming home from school, Kristy waited 10 min for the school bus, rode the bus for 20 min, and then walked 5 min to her house.

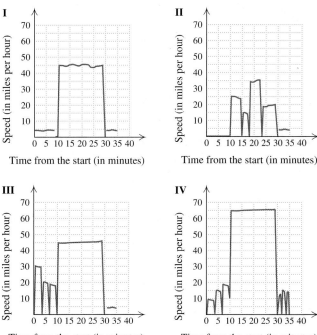

72. Match each sentence with the most appropriate of the four graphs shown below.

a) Roberta worked part time until September, full time until December, and overtime until Christmas.

b) Clyde worked full time until September, half time until December, and full time until Christmas.

c) Clarissa worked overtime until September, full time until December, and overtime until Christmas.

d) Doug worked part time until September, half time until December, and full time until Christmas.

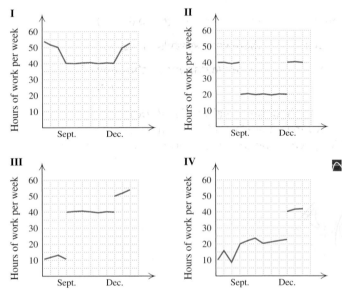

73. Which of the following equations have $\left(-\frac{1}{3}, \frac{1}{4}\right)$ as a solution?

a) $-\frac{3}{2}x - 3y = -\frac{1}{4}$

b) $8y - 15x = \frac{7}{2}$

c) $0.16y = -0.09x + 0.1$

d) $2(-y + 2) - \frac{1}{4}(3x - 1) = 4$

74. If $(-10, -2)$, $(-3, 4)$, and $(6, 4)$ are the coordinates of three vertices of a parallelogram, determine the coordinates of three different points that could serve as the fourth vertex.

75. If $(2, -3)$ and $(-5, 4)$ are the endpoints of a diagonal of a square, what are the coordinates of the other two vertices? What is the area of the square?

📟 *Graph each equation after plotting at least 10 points.*

76. $y = 1/x$; use x-values from -4 to 4

77. $y = 1/(x - 2)$; use values of x from -2 to 6

78. $y = \sqrt{x}$; use values of x from 0 to 10

79. $y = 1/x^2$; use values of x from -4 to 4

80. $y = \frac{1}{2}x^2$; use x-values from -4 to 4

81. $y = \sqrt{x} + 1$; use x-values from 0 to 10

82. $y = \frac{1}{x} + 3$; use x-values from -4 to 4

83. $y = x^3$; use x-values from -2 to 2

84. $y = x^3 - 5$; use x-values from -2 to 2

Note: Throughout this text, the icon 〰 *indicates exercises designed for graphing calculators.*

〰 *In Exercises 85 and 86, use a graphing calculator to draw the graph of each equation. For each equation, select a window that shows the curvature of the graph and create a table of ordered pairs in which x-values extend, by tenths, from 0 to 0.6.*

85. a) $y = 0.375x^3$

b) $y = -3.5x^2 + 6x - 8$

c) $y = (x - 3.4)^3 + 5.6$

86. a) $y = 2.3x^4 + 3.4x^2 + 1.2x - 4$

b) $y = -0.25x^2 + 3.7$

c) $y = 3(x + 2.3)^2 + 2.3$

2.2   Functions

Functions and Graphs   •   Function Notation and Equations   •
Applications: Interpolation and Extrapolation

We now develop the idea of a *function*—one of the most important concepts in mathematics. In much the same way that the ordered pairs of Section 2.1 formed correspondences between first coordinates and second coordinates, a function is a special kind of correspondence between two sets. For example,

To each person in a class	there corresponds	a date of birth.
To each bar code in a store	there corresponds	a price.
To each real number	there corresponds	the cube of that number.

In each example, the first set is called the **domain**. The second set is called the **range**. For any member of the domain, there is *exactly one* member of the range to which it corresponds. This kind of correspondence is called a **function**.

EXAMPLE 1   Determine whether each correspondence is a function.

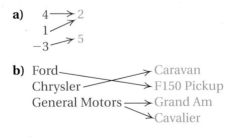

Solution

a) The correspondence *is* a function because each member of the domain corresponds to *exactly one* member of the range.

b) The correspondence *is not* a function because a member of the domain (General Motors) corresponds to more than one member of the range.

> **Function**
>
> A *function* is a correspondence between a first set, called the *domain*, and a second set, called the *range*, such that each member of the domain corresponds to *exactly one* member of the range.

EXAMPLE 2   Determine whether each correspondence is a function.

Domain	*Correspondence*	*Range*
a) An elevator full of people	Each person's weight	A set of positive numbers
b) $\{-2, 0, 1, 2\}$	Each number's square	$\{0, 1, 4\}$
c) Authors of best-selling books	The titles of books written by each author	A set of book titles

Solution

a) The correspondence *is* a function, because each person has *only one* weight.

b) The correspondence *is* a function, because every number has *only one* square.

c) The correspondence *is not* a function, because some authors have written *more than one* book.

Study Skills

The *Student's Solutions Manual* is an excellent resource if you need additional help with an exercise in the exercise sets. It contains step-by-step solutions for the odd-numbered exercises in each exercise set.

Functions and Graphs

The functions in Examples 1(a) and 2(b) can be expressed as sets of ordered pairs. Example 1(a) can be written $\{(-3, 5), (1, 2), (4, 2)\}$ and Example 2(b) can be written $\{(-2, 4), (0, 0), (1, 1), (2, 4)\}$. We can graph these functions as follows.

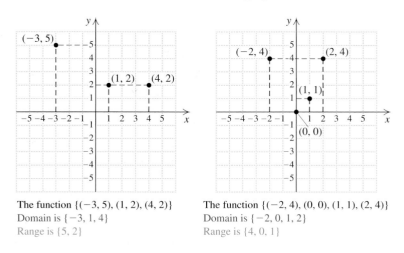

The function $\{(-3, 5), (1, 2), (4, 2)\}$
Domain is $\{-3, 1, 4\}$
Range is $\{5, 2\}$

The function $\{(-2, 4), (0, 0), (1, 1), (2, 4)\}$
Domain is $\{-2, 0, 1, 2\}$
Range is $\{4, 0, 1\}$

When a function is given as a set of ordered pairs, the domain is the set of all first coordinates and the range is the set of all second coordinates. Function names are generally represented by lower- or upper-case letters.

EXAMPLE 3   Find the domain and the range of the function *f* below.

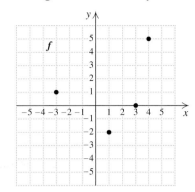

Solution   Here *f* can be written $\{(-3, 1), (1, -2), (3, 0), (4, 5)\}$. The domain is the set of all first coordinates, $\{-3, 1, 3, 4\}$, and the range is the set of all second coordinates, $\{1, -2, 0, 5\}$. We can also find the domain and the range directly, without first writing *f*.

EXAMPLE 4   For the function *f* represented below, determine each of the following.

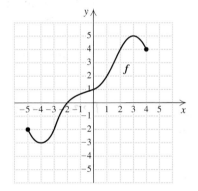

a) What member of the range is paired with 2

b) The domain of *f*

c) What member of the domain is paired with −3

d) The range of *f*

Solution

a) To determine what member of the range is paired with 2, we locate 2 on the horizontal axis (this is where the domain is located). Next, we find the point directly above 2 on the graph of *f*. From that point, we can look to the vertical axis to find the corresponding *y*-coordinate, 4. The "input" 2 has the "output" 4.

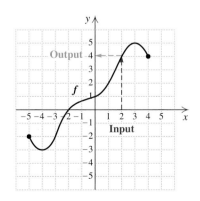

b) The domain of f is the set of all x-values that are used in the points on the curve. Because there are no breaks in the graph of f, these extend continuously from -5 to 4 and can be viewed as the curve's shadow, or *projection*, on the x-axis. Thus the domain is $\{x \mid -5 \leq x \leq 4\}$.

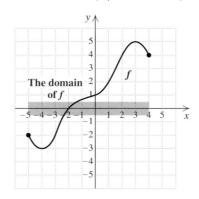

c) To determine what member of the domain is paired with -3, we locate -3 on the vertical axis. (This is where the range is located; see below.) From there we look left and right to the graph of f to find any points for which -3 is the second coordinate. One such point exists, $(-4, -3)$. We observe that -4 is the only element of the domain paired with -3.

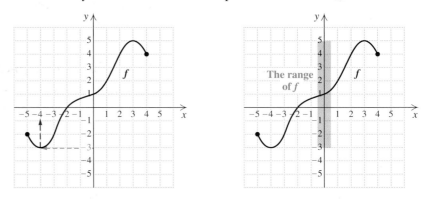

d) The range of f is the set of all y-values that are used in the points on the curve. (See the graph on the right above.) These extend continuously from -3 to 5, and can be viewed as the curve's projection on the y-axis. Thus the range is $\{y \mid -3 \leq y \leq 5\}$.

Note that if a graph contains two or more points with the same first coordinate, that graph cannot represent a function (otherwise one member of the domain would correspond to more than one member of the range). This observation is the basis of the *vertical-line test*.

The Vertical-Line Test

If it is possible for a vertical line to cross a graph more than once, then the graph is not the graph of a function.

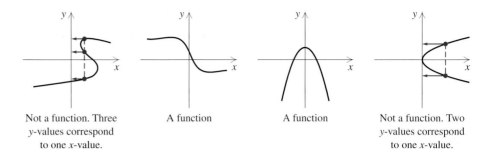

Not a function. Three
y-values correspond
to one x-value.

A function

A function

Not a function. Two
y-values correspond
to one x-value.

Graphs that do not represent functions still do represent *relations*.

Relation

A *relation* is a correspondence between a first set, called the *domain*, and a second set, called the *range*, such that each member of the domain corresponds to *at least one* member of the range.

Thus, although the correspondences and graphs above are not all functions, they *are* all relations.

Function Notation and Equations

To understand function notation, it helps to imagine a "function machine." Think of putting a member of the domain (an *input*) into the machine. The machine is programmed to produce the appropriate member of the range (the *output*).

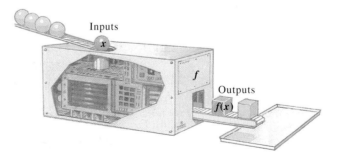

The function pictured has been named f. Here x represents an arbitrary input, and $f(x)$—read "f of x," "f at x," or "the value of f at x"—represents the corresponding output. You should check that in Example 3, $f(-3)$ is 1, $f(4)$ is 5, and so on. Similarly, in Example 4, we showed that $f(2) = 4$.

Caution! $f(x)$ *does not mean f times x.*

Most functions are described by equations. For example, $f(x) = 2x + 3$ describes the function that takes an input x, multiplies it by 2, and then adds 3.

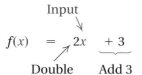

To calculate the output $f(4)$, we take the input 4, double it, and add 3 to get 11. That is, we substitute 4 into the formula for $f(x)$:

$$f(4) = 2 \cdot 4 + 3$$
$$= 11.$$

Sometimes, in place of $f(x) = 2x + 3$, we write $y = 2x + 3$, where it is understood that the value of y, the *dependent variable*, depends on our choice of x, the *independent variable*. To understand why $f(x)$ notation is so useful, consider two equivalent statements:

a) If $f(x) = 2x + 3$, then $f(4) = 11$.
b) If $y = 2x + 3$, then the value of y is 11 when x is 4.

The notation used in part (a) is far more concise and emphasizes that x is the independent variable.

EXAMPLE 5   Find each indicated function value.

a) $f(5)$, for $f(x) = 3x + 2$
b) $g(-2)$, for $g(r) = 5r^2 + 3r$
c) $h(4)$, for $h(x) = 7$
d) $F(a + 1)$, for $F(x) = 3x + 2$

Solution   Finding function values is much like evaluating an algebraic expression.

a) $f(5) = 3 \cdot 5 + 2 = 17$
b) $g(-2) = 5(-2)^2 + 3(-2)$
$$= 5 \cdot 4 - 6 = 14$$

c) For the function given by $h(x) = 7$, all inputs share the same output, 7. Therefore, $h(4) = 7$. The function h is an example of a *constant function*.

d) $F(a + 1) = 3(a + 1) + 2$
$$= 3a + 3 + 2 = 3a + 5$$

Student Notes

In Example 5(d), it is important to note that the parentheses on the left are for function notation, whereas those on the right indicate multiplication.

Note that whether we write $f(x) = 3x + 2$, or $f(t) = 3t + 2$, or $f(\square) = 3\square + 2$, we still have $f(5) = 17$. Thus the independent variable can be thought of as a *dummy variable*. The letter chosen for the dummy variable is not as important as the algebraic manipulations to which it is subjected.

When a function is described by an equation, the domain is often unspecified. In such cases, the domain is the set of all numbers for which function values can be calculated. If an x-value is not in the domain of a function, the graph of that function will not include any point above or below that x-value.

EXAMPLE 6

For each equation, determine the domain of f.

a) $f(x) = |x|$

b) $f(x) = \dfrac{7}{2x - 6}$

Solution

a) We ask ourselves, "Is there any number x for which we cannot compute $|x|$?" Since we can find the absolute value of *any* number, the answer is no. Thus the domain of f is $\mathbb{R}$, the set of all real numbers.

b) Is there any number x for which $\dfrac{7}{2x - 6}$ cannot be computed? Since $\dfrac{7}{2x - 6}$ cannot be computed when $2x - 6$ is 0, the answer is yes. To determine what x-value would cause $2x - 6$ to be 0, we set up and solve an equation:

$$2x - 6 = 0 \qquad \text{Setting the denominator equal to 0}$$
$$2x = 6 \qquad \text{Adding 6 to both sides}$$
$$x = 3. \qquad \text{Dividing both sides by 2}$$

Thus, 3 is *not* in the domain of f, whereas all other real numbers are. The domain of f is $\{x \,|\, x$ is a real number *and* $x \neq 3\}$.

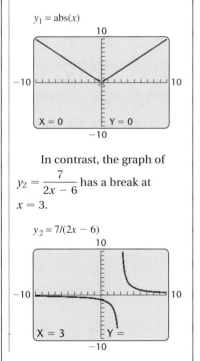

technology connection

To visualize Example 6, note that the graph of $y_1 = |x|$ (which is entered $y_1 = \text{abs}(x)$, using the NUM option of the MATH menu) appears without interruption for any piece of the x-axis that we examine.

$y_1 = \text{abs}(x)$

In contrast, the graph of $y_2 = \dfrac{7}{2x - 6}$ has a break at $x = 3$.

$y_2 = 7/(2x - 6)$

Applications: Interpolation and Extrapolation

Function notation is often used in formulas. For example, to emphasize that the area A of a circle is a function of its radius r, instead of

$$A = \pi r^2,$$

we can write

$$A(r) = \pi r^2.$$

When a function is given as a graph in a problem-solving situation, we are often asked to determine certain quantities on the basis of the graph. Later in this text, we will develop models that can be used for calculations. For now we simply use the graph to estimate the coordinates of an unknown point by using other points with known coordinates. When the unknown point is *between* the known points, this process is called **interpolation**. If the unknown point extends *beyond* the known points, the process is called **extrapolation**.

EXAMPLE 7

Elementary school math proficiency.   According to the National Assessment of Educational Progress, the percentage of fourth-graders who are proficient in math has grown from 18% in 1992 to 24% in 2000 and 32% in 2003 (*Source: National Council of Teachers of Mathematics News Bulletin* Jan./Feb. 2004). Estimate the percentage of fourth-graders who showed proficiency in 1996 and predict the percentage who will demonstrate proficiency in 2007.

Solution

1. and **2. Familiarize.** and **Translate.**   The given information enables us to plot and connect three points. We let the horizontal axis represent the year and the vertical axis the percentage of fourth-graders demonstrating mathematical proficiency. We label the function itself *P*.

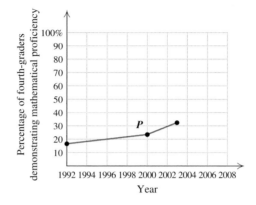

3. Carry out.   To estimate the percentage of fourth-graders showing mathematical proficiency in 1996, we locate the point directly above the year 1996. We then estimate its second coordinate by moving horizontally from that point to the *y*-axis. Although our result is not exact, we see that $P(1996) \approx 20$.

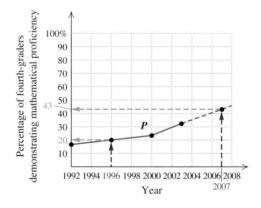

To predict the percentage of fourth-graders showing proficiency in 2007, we extend the graph and extrapolate. It appears that $P(2007) \approx 43$.

4. Check.   A precise check requires consulting an outside information source. Since 20% is between 18% and 24% and 43% is greater than 32%, our estimates seem plausible.

5. State.   In 1996, about 20% of all fourth-graders showed proficiency in math. By 2007, that figure is predicted to grow to 43%.

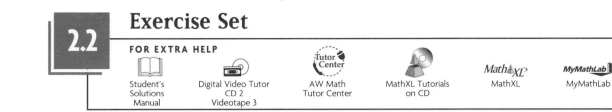

Exercise Set

2.2

✏ *Concept Reinforcement*   *Complete each of the following sentences.*

1. For any function, the set of all inputs, or first values, is called the _____.

2. For any function, the set of all outputs, or second values, is called the _____.

3. In any function, each member of the domain is paired with _____ one member of the range.

4. In a relation, a member of the domain may be paired with _____ than one member of the range.

5. When a function is graphed, members of the domain are located on the _____ axis.

6. When a function is graphed, members of the range are located on the _____ axis.

7. The notation $f(3)$ is read _____.

8. The vertical-line test can be used to determine whether or not a graph represents a(n) _____.

Determine whether each correspondence is a function.

9. $2 \longrightarrow a$
   $4 \longrightarrow b$
   $6 \longrightarrow c$
   $8 \longrightarrow d$
   10

10. $3 \longrightarrow 9$
   $6 \longrightarrow 8$
   $9 \longrightarrow 7$
   $12 \longrightarrow 6$

11.

Girl's age (in months)	Average daily weight gain (in grams)
2	21.8
9	11.7
16	8.5
23	7.0

Source: *American Family Physician*, December 1993, p. 1435

12.

Boy's age (in months)	Average daily weight gain (in grams)
2	24.3
9	11.7
16	8.2
23	7.0

Source: *American Family Physician*, December 1993, p. 1435

13. *Celebrity*   *Birthday*

Sigourney Weaver
Jesse Jackson ⟶ October 8
Chevy Chase
Muhammad Ali
Jim Carrey ⟶ January 17

Source: www.kidsparties.com

14. *Predator*   *Prey*

cat   dog
fish   worm
dog   cat
tiger   fish
bat ⟶ mosquito

15. *Birthday*   *Celebrity*

          Jennifer Lopez
July 24   Barry Bonds
          Pam Tillis
          Vin Diesel
July 18   John Glenn
          Nelson Mandela

Source: www.leannesbirthdays.com and www.kidsparties.com

16. *State*   *Neighboring state*

Texas   Oklahoma
          New Mexico
Colorado   Arkansas
          Louisiana

Determine whether each of the following is a function.
Identify any relations that are not functions.

Domain	Correspondence	Range
17. A yard full of pumpkins	The price of each pumpkin	A set of prices
18. The members of a rock band	An instrument the person can play	A set of instruments
19. The players on a team	The uniform number of each player	A set of numbers
20. A set of triangles	The area of each triangle	A set of numbers

For each graph of a function, determine **(a)** $f(1)$; **(b)** *the domain;* **(c)** *any x-values for which* $f(x) = 2$; *and* **(d)** *the range.*

21.
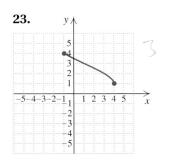

22.
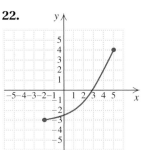

23.
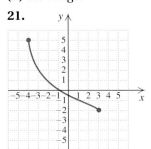

24.
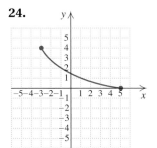

25.
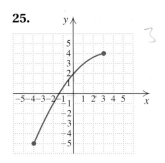

26.

27.
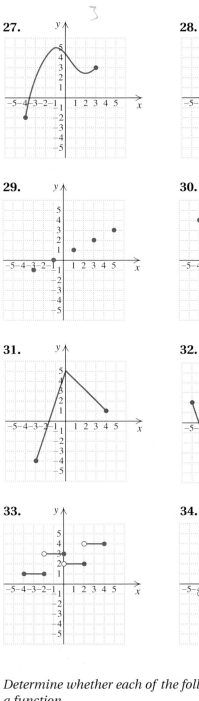

28.
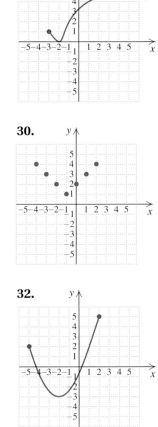

29.

30.

31.

32.

33.

34.
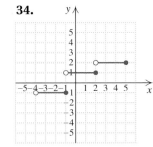

Determine whether each of the following is the graph of a function.

35.

36.
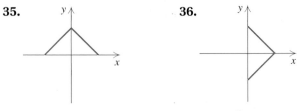

37.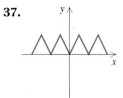

38.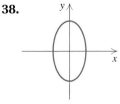

39.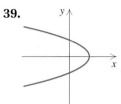

40.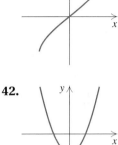

41.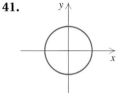

42.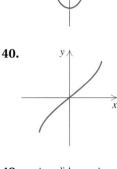

Find the function values.

43. $g(x) = x + 5$

    a) $g(0)$ **b)** $g(-4)$ **c)** $g(-7)$
    d) $g(8)$ **e)** $g(a + 2)$ **f)** $g(a) + 2$

44. $h(x) = x - 2$

    a) $h(4)$ **b)** $h(8)$ **c)** $h(-3)$
    d) $h(-4)$ **e)** $h(a - 1)$ **f)** $h(a) + 3$

45. $f(n) = 5n^2 + 4n$

    a) $f(0)$ **b)** $f(-1)$ **c)** $f(3)$
    d) $f(t)$ **e)** $f(2a)$ **f)** $f(3) - 9$

46. $g(n) = 3n^2 - 2n$

    a) $g(0)$ **b)** $g(-1)$ **c)** $g(3)$
    d) $g(t)$ **e)** $g(2a)$ **f)** $g(3) - 4$

47. $f(x) = \dfrac{x - 3}{2x - 5}$

    a) $f(0)$ **b)** $f(4)$ **c)** $f(-1)$
    d) $f(3)$ **e)** $f(x + 2)$

48. $s(x) = \dfrac{3x - 4}{2x + 5}$

    a) $s(10)$ **b)** $s(2)$ **c)** $s\left(\frac{1}{2}\right)$
    d) $s(-1)$ **e)** $s(x + 3)$

49. Find the domain of f.

    a) $f(x) = \dfrac{5}{x - 3}$ **b)** $f(x) = \dfrac{7}{6 - x}$
    c) $f(x) = 2x + 1$ **d)** $f(x) = x^2 + 3$
    e) $f(x) = \dfrac{3}{2x - 5}$ **f)** $f(x) = |3x - 4|$

50. Find the domain of g.

    a) $g(x) = \dfrac{3}{x - 1}$ **b)** $g(x) = |5 - x|$
    c) $g(x) = \dfrac{9}{x + 3}$ **d)** $g(x) = \dfrac{4}{3x + 4}$
    e) $g(x) = x^3 - 1$ **f)** $g(x) = 7x - 8$

The function A described by $A(s) = s^2 \dfrac{\sqrt{3}}{4}$ gives the area of an equilateral triangle with side s.

51. Find the area when a side measures 4 cm.

52. Find the area when a side measures 6 in.

The function V described by $V(r) = 4\pi r^2$ gives the surface area of a sphere with radius r.

53. Find the surface area when the radius is 3 in.

54. Find the surface area when the radius is 5 cm.

55. *Pressure at sea depth.* The function $P(d) = 1 + (d/33)$ gives the pressure, in *atmospheres* (atm), at a depth of d feet in the sea. Note that $P(0) = 1$ atm, $P(33) = 2$ atm, and so on. Find the pressure at 20 ft, 30 ft, and 100 ft.

56. *Melting snow.* The function $W(d) = 0.112d$ approximates the amount, in centimeters, of water that results from d centimeters of snow melting. Find the amount of water that results from snow melting from depths of 16 cm, 25 cm, and 100 cm.

Archaeology.   *The function H described by*

$$H(x) = 2.75x + 71.48$$

can be used to predict the height, in centimeters, of a woman whose humerus (the bone from the elbow to the shoulder) is x cm long. Predict the height of a woman whose humerus is the length given.

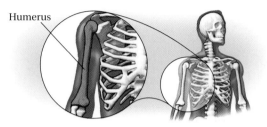

Humerus

57. 32 cm   **58.** 35 cm

Chemistry.   *The function F described by*

$$F(C) = \frac{9}{5}C + 32$$

gives the Fahrenheit temperature corresponding to the Celsius temperature C.

59. Find the Fahrenheit temperature equivalent to $-10°C$.

60. Find the Fahrenheit temperature equivalent to 5°C.

Heart attacks and cholesterol.   *For Exercises 61 and 62, use the following graph, which shows the annual heart attack rate per 10,000 men as a function of blood cholesterol level.**

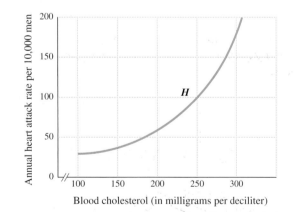

61. Approximate the annual heart attack rate for those men whose blood cholesterol level is 225 mg/dl. That is, find $H(225)$.

*Copyright 1989, CSPI. Adapted from *Nutrition Action Healthletter* (1875 Connecticut Avenue, N.W., Suite 300, Washington, DC 20009-5728. $24 for 10 issues).

62. Approximate the annual heart attack rate for those men whose blood cholesterol level is 275 mg/dl. That is, find $H(275)$.

Voting attitudes.   *For Exercises 63 and 64, use this graph, which shows the percentage of people responding yes to the question, "If your (political) party nominated a generally well-qualified person for president who happened to be a woman, would you vote for that person?"*
Source: www.gallup.com

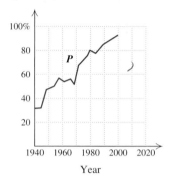

63. Approximate the percentage of Americans willing to vote for a woman for president in 1960. That is, find $P(1960)$.

64. Approximate the percentage of Americans willing to vote for a woman for president in 2000. That is, find $P(2000)$.

Energy-saving lightbulbs.   *Compact fluorescent (CFL) lightbulbs can be used in most places that conventional incandescent bulbs are, but they use a fraction of the electricity. The table below lists the CFL wattage and the incandescent wattage required to create the same amount of light.*
Source: Westinghouse Lighting Corporation

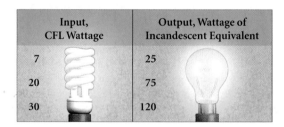

Input, CFL Wattage	Output, Wattage of Incandescent Equivalent
7	25
20	75
30	120

65. Use the data in the figure above to draw a graph and to estimate the wattage of an incandescent bulb that creates light equivalent to a 15-watt CFL bulb. Then predict the wattage of an incandescent bulb that creates light equivalent to a 35-watt CFL bulb.

66. Use the graph from Exercise 65 to estimate the wattage of an incandescent bulb that creates light equivalent to a 26-watt CFL bulb. Then predict the wattage of an incandescent bulb that creates light equivalent to a 40-watt CFL bulb.

Blood alcohol level.   *The following table can be used to predict the number of drinks required for a person of a specified weight to be considered legally intoxicated (blood alcohol level of 0.08 or above). One 12-oz glass of beer, a 5-oz glass of wine, or a cocktail containing 1 oz of a distilled liquor all count as one drink. Assume that all drinks are consumed within one hour.*

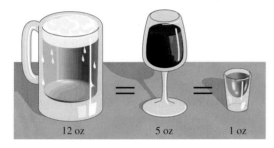

Input, Body Weight (in pounds)	Output, Number of Drinks
100	2.5
160	4
180	4.5
200	5

67. Use the data in the table above to draw a graph and to estimate the number of drinks that a 140-lb person would have to drink to be considered intoxicated. Then predict the number of drinks it would take for a 230-lb person to be considered intoxicated.

68. Use the graph from Exercise 67 to estimate the number of drinks that a 120-lb person would have to drink to be considered intoxicated. Then predict the number of drinks it would take for a 250-lb person to be considered intoxicated.

Incidence of AIDS.   *The following table indicates the number of cases of AIDS reported in each of several years.*

Input, Year	Output, Number of Cases Reported
1994	77,103
1996	66,497
1998	48,269
2000	40,282
2002	42,745

Source: U.S. Centers for Disease Control and Prevention

69. Use the data in the table above to draw a graph and to estimate the number of cases of AIDS reported in 1997. Then predict the number of cases that will be reported in 2007.

70. Use the graph from Exercise 69 to estimate the number of cases of AIDS reported in 1999. Then predict the number of cases that will be reported in 2008.

71. *Retailing.*   Shoreside Gifts is experiencing constant growth. They recorded a total of $250,000 in sales in 1999 and $285,000 in 2004. Use a graph that displays the store's total sales as a function of time to estimate sales for 2000 and for 2007.

72. Use the graph in Exercise 71 to estimate sales for 2002 and for 2008.

Researchers at Yale University have suggested that the following graphs* may represent three different aspects of love.

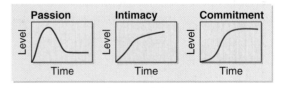

73. In what unit would you measure time if the horizontal length of each graph were ten units? Why?

74. Do you agree with the researchers that these graphs should be shaped as they are? Why or why not?

*From "A Triangular Theory of Love," by R. J. Sternberg, 1986, *Psychological Review*, **93**(2), 119–135. Copyright 1986 by the American Psychological Association, Inc. Reprinted by permission.

SKILL MAINTENANCE

Simplify. [1.2]

75. $\dfrac{10 - 3^2}{9 - 2 \cdot 3}$

76. $\dfrac{2^4 - 10}{6 - 4 \cdot 3}$

77. The surface area of a rectangular solid of length l, width w, and height h is given by $S = 2lh + 2lw + 2wh$. Solve for l. [1.5]

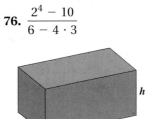

78. Solve the formula in Exercise 77 for w. [1.5]

Solve for y. [1.5]

79. $2x + 3y = 6$

80. $5x - 4y = 8$

SYNTHESIS

81. What would be wrong with using the data for only the years 1994, 1998, and 2002 to answer Exercises 69 and 70?

82. Which would you trust more and why: estimates made using interpolation or those made using extrapolation?

For Exercises 83 and 84, let $f(x) = 3x^2 - 1$ and $g(x) = 2x + 5$.

83. Find $f(g(-4))$ and $g(f(-4))$.

84. Find $f(g(-1))$ and $g(f(-1))$.

85. If f represents the function in Exercise 14, find $f(f(f(f(\text{tiger}))))$.

Pregnancy. *For Exercises 86–89, use the following graph of a woman's "stress test." This graph shows the size of a pregnant woman's contractions as a function of time.*

86. How large is the largest contraction that occurred during the test?

87. At what time during the test did the largest contraction occur?

88. On the basis of the information provided, how large a contraction would you expect 60 seconds after the end of the test? Why?

89. What is the frequency of the largest contraction?

90. The *greatest integer function* $f(x) = [\![x]\!]$ is defined as follows: $[\![x]\!]$ is the greatest integer that is less than or equal to x. For example, if $x = 3.74$, then $[\![x]\!] = 3$; and if $x = -0.98$, then $[\![x]\!] = -1$. Graph the greatest integer function for $-5 \le x \le 5$. (The notation $f(x) = \text{INT}(x)$ is used in many graphing calculators and computer programs.)

91. Suppose that a function g is such that $g(-1) = -7$ and $g(3) = 8$. Find a formula for g if $g(x)$ is of the form $g(x) = mx + b$, where m and b are constants.

92. *Energy expenditure.* On the basis of the information given below, what burns more energy: walking $4\frac{1}{2}$ mph for two hours or bicycling 14 mph for one hour?

Approximate Energy Expenditure by a 150-Pound Person in Various Activities

Activity	Calories per Hour
Walking, $2\frac{1}{2}$ mph	210
Bicycling, $5\frac{1}{2}$ mph	210
Walking, $3\frac{3}{4}$ mph	300
Bicycling, 13 mph	660

Source: Based on material prepared by Robert E. Johnson, M.D., Ph.D., and colleagues, University of Illinois.

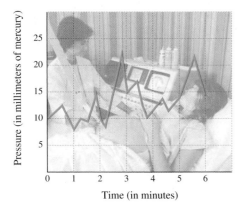

2.3   Linear Functions: Slope, Graphs, and Models

Slope–Intercept Form of an Equation   •   Applications

Different functions have different graphs. In this section, we examine functions and real-life models with graphs that are straight lines. Such functions and their graphs are called *linear* and can be written in the form $f(x) = mx + b$, where m and b are constants.

Slope–Intercept Form of an Equation

Examples 3–5 in Section 2.1 suggest that for any number m, the graph of $y = mx$ is a straight line passing through the origin. What will happen if we add a number b on the right side and graph the equation $y = mx + b$?

EXAMPLE 1   Graph $y = 2x$ and $y = 2x + 3$, using the same set of axes.

Solution   We first make a table of solutions of both equations.

x	y $y = 2x$	y $y = 2x + 3$
0	0	3
1	2	5
−1	−2	1
2	4	7
−2	−4	−1
3	6	9

We then plot these points. Drawing a blue line for $y = 2x + 3$ and a red line for $y = 2x$, we observe that the graph of $y = 2x + 3$ is simply the graph of $y = 2x$ shifted, or *translated*, 3 units up. The lines are parallel.

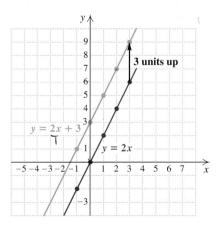

EXAMPLE 2   Graph $f(x) = \frac{1}{3}x$ and $g(x) = \frac{1}{3}x - 2$, using the same set of axes.

Solution   We first make a table of solutions of both equations. By choosing multiples of 3, we can avoid fractions.

	$f(x)$	$g(x)$
x	$f(x) = \frac{1}{3}x$	$g(x) = \frac{1}{3}x - 2$
0	0	-2
3	1	-1
-3	-1	-3
6	2	0

We then plot these points. Drawing a blue line for $g(x) = \frac{1}{3}x - 2$ and a red line for $f(x) = \frac{1}{3}x$, we see that the graph of $g(x) = \frac{1}{3}x - 2$ is simply the graph of $f(x) = \frac{1}{3}x$ shifted, or translated, 2 units down.

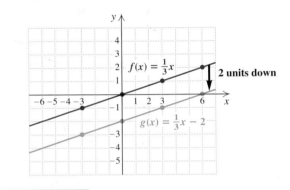

To explore the effect of b when graphing $y = mx + b$, begin with the graph of $y_1 = x$. On the same set of axes, graph the lines $y_2 = x + 3$ and $y_3 = x - 4$. How do these lines differ from $y_1 = x$? What do you think the line $y = x - 5$ will look like? Try drawing lines like $y = x + \frac{1}{4}$ and $y = x + (-3.2)$ and describe what happens to the graph of $y = x$ when a number b is added to x. By creating a table of values, explain how the values of y_2 and y_3 differ from y_1.

technology connection

Note that the graph of $y = 2x + 3$ passed through the point $(0, 3)$ and the graph of $g(x) = \frac{1}{3}x - 2$ passed through the point $(0, -2)$. In general, the graph of $y = mx + b$ is a line parallel to $y = mx$, passing through the point $(0, b)$. **The point $(0, b)$ is called the y-intercept.** Often, for convenience, we refer to the number b as the y-intercept.

EXAMPLE 3   For each equation, find the y-intercept.

a) $y = -5x + 4$ **b)** $f(x) = 5.3x - 12$

Solution

a) The y-intercept is $(0, 4)$, or simply 4.

b) The y-intercept is $(0, -12)$, or simply -12.

In examining the graphs in Examples 1 and 2, note that the slant of the red lines seems to match the slant of the blue lines. Note too that the slant of the lines in Example 1 differs from the slant of the lines in Example 2. This leads us to suspect that it is the number m, in the equation $y = mx + b$, that is responsible for the slant of the line. The following definition enables us to visualize this slant, or *slope*, as a ratio of two lengths.

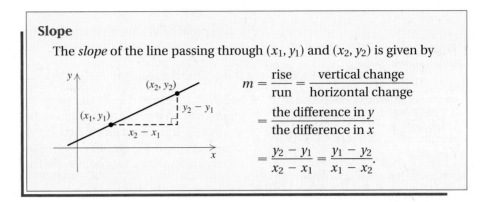

Slope

The *slope* of the line passing through (x_1, y_1) and (x_2, y_2) is given by

$$m = \frac{\text{rise}}{\text{run}} = \frac{\text{vertical change}}{\text{horizontal change}}$$

$$= \frac{\text{the difference in } y}{\text{the difference in } x}$$

$$= \frac{y_2 - y_1}{x_2 - x_1} = \frac{y_1 - y_2}{x_1 - x_2}.$$

In the definition above, (x_1, y_1) and (x_2, y_2)—read "x sub-one, y sub-one and x sub-two, y sub-two"—represent two different points on a line. It does not matter which point is considered (x_1, y_1) and which is considered (x_2, y_2) so long as coordinates are subtracted in the same order in both the numerator and the denominator.

The letter m is traditionally used for slope. This usage has its roots in the French verb *monter*, to climb.

EXAMPLE 4

Find the slope of the lines drawn in Examples 1 and 2.

Solution   To find the slope of a line, we can use the coordinates of any two points on that line. We use $(1, 5)$ and $(2, 7)$ to find the slope of the blue line in Example 1:

$$\text{Slope} = \frac{\text{rise}}{\text{run}} = \frac{\text{difference in } y}{\text{difference in } x}$$

$$= \frac{y_2 - y_1}{x_2 - x_1} = \frac{7 - 5}{2 - 1} = 2. \qquad \begin{array}{l}\textit{Any} \text{ pair of points on the line will}\\ \text{give the same slope.}\end{array}$$

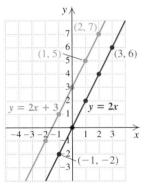

To find the slope of the red line in Example 1, we use $(-1, -2)$ and $(3, 6)$:

$$\text{Slope} = \frac{\text{rise}}{\text{run}} = \frac{\text{difference in } y}{\text{difference in } x} = \frac{6 - (-2)}{3 - (-1)} = \frac{8}{4} = 2. \qquad \begin{array}{l}\textit{Any} \text{ pair of}\\ \text{points on}\\ \text{the line will}\\ \text{give the}\\ \text{same slope.}\end{array}$$

We can use $(3, -1)$ and $(6, 0)$ to find the slope of the blue line in Example 2:

$$\text{Slope} = \frac{\text{rise}}{\text{run}} = \frac{\text{difference in } y}{\text{difference in } x} = \frac{0 - (-1)}{6 - 3} = \frac{1}{3}. \qquad \begin{array}{l}\textit{Any} \text{ pair of}\\ \text{points on the}\\ \text{line will give the}\\ \text{same slope.}\end{array}$$

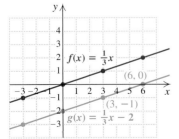

You should confirm that the red line in Example 2 also has a slope of $\frac{1}{3}$.

In Example 4, we found that the lines given by $y = 2x + 3$, $y = 2x$, $g(x) = \frac{1}{3}x - 2$, and $f(x) = \frac{1}{3}x$ have slopes 2, 2, $\frac{1}{3}$, and $\frac{1}{3}$, respectively. This supports (but does not prove) the following:

The slope of any line written in the form $y = mx + b$ is m.

A proof of this result is outlined in Exercise 101 on p. 111.

EXAMPLE 5

Determine the slope of the line given by $y = \frac{2}{3}x + 4$, and graph the line.

Solution   Here $m = \frac{2}{3}$, so the slope is $\frac{2}{3}$. This means that from *any* point on the graph, we can locate a second point by simply going *up* 2 units (the *rise*) and *to the right* 3 units (the *run*). Where do we start? Because the y-intercept, $(0, 4)$, is known to be on the graph, we calculate that $(0 + 3, 4 + 2)$, or $(3, 6)$, is also on the graph. Knowing two points, we can draw the line.

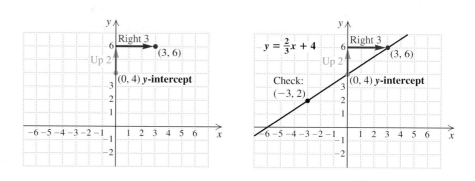

Important! To check the graph, we use some other value for x, say -3, and determine y (in this case, 2). We plot that point and see that it *is* on the line. Were it not, we would know that some error had been made.

> **Slope–Intercept Form**
>
> Any equation $y = mx + b$ has a graph that is a straight line. It goes through the y-intercept $(0, b)$ and has slope m.
>   Any equation of the form
>
> $$y = mx + b$$
>
> is said to be written in *slope–intercept form*.

EXAMPLE 6

Determine the slope and the y-intercept of the line given by $y = -\frac{1}{3}x + 2$.

Solution   The equation $y = -\frac{1}{3}x + 2$ is written in the form $y = mx + b$, with $m = -\frac{1}{3}$ and $b = 2$. Thus the slope is $-\frac{1}{3}$ and the y-intercept is $(0, 2)$.

EXAMPLE 7   Find a linear function whose graph has slope 3 and y-intercept $(0, -1)$.

Solution   We use the slope–intercept form, $f(x) = mx + b$:

$$f(x) = 3x - 1. \qquad \text{Substituting 3 for } m \text{ and } -1 \text{ for } b$$

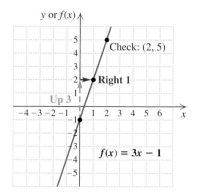

To graph $f(x) = 3x - 1$, we regard the slope of 3 as $\frac{3}{1}$. Then, beginning at the y-intercept $(0, -1)$, we count *up* 3 units and *to the right* 1 unit, as shown at left.

In Examples 1, 2, and 5, the lines have positive slopes and slant upward from left to right. When the slope is negative, lines slant downward from left to right.

EXAMPLE 8   Graph: $f(x) = -\frac{1}{2}x + 5$.

Student Notes

Slope–intercept form is not only important—it is convenient to use. The sooner you feel comfortable writing, reading, and graphing slope–intercept form, the easier your work will become.

Solution   The y-intercept is $(0, 5)$. The slope is $-\frac{1}{2}$, or $\frac{-1}{2}$. From the y-intercept, we go *down* 1 unit and *to the right* 2 units. That gives us the point $(2, 4)$. We can now draw the graph.

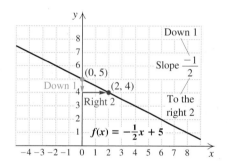

As a new type of check, we rewrite the slope in another form and find another point:

$$-\frac{1}{2} = \frac{1}{-2}.$$

Thus we can go *up* 1 unit and then *to the left* 2 units. This gives the point $(-2, 6)$. Since $(-2, 6)$ is on the line, we have a check.

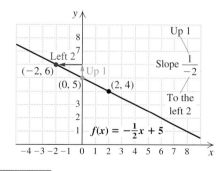

Often the easiest way to graph an equation is to rewrite it in slope–intercept form and then proceed as in Examples 5 and 8 above.

EXAMPLE 9   Determine the slope and the y-intercept for the equation $5x - 4y = 8$. Then graph.

Solution   We convert to slope–intercept form:

$$5x - 4y = 8$$
$$-4y = -5x + 8 \qquad \text{Adding } -5x \text{ to both sides}$$
$$y = -\tfrac{1}{4}(-5x + 8) \qquad \text{Multiplying both sides by } -\tfrac{1}{4}$$
$$y = \tfrac{5}{4}x - 2. \qquad \text{Using the distributive law}$$

Because we have an equation of the form $y = mx + b$, we know that the slope is $\frac{5}{4}$ and the y-intercept is $(0, -2)$. We plot $(0, -2)$, and from there we go *up* 5 units and *to the right* 4 units, and plot a second point at $(4, 3)$. We then draw the graph.

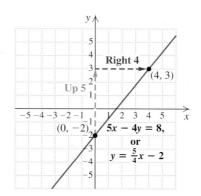

To check that the line is drawn correctly, we calculate the coordinates of another point on the line. For $x = 2$, we have

$$5 \cdot 2 - 4y = 8$$
$$10 - 4y = 8$$
$$-4y = -2$$
$$y = \frac{1}{2}.$$

Thus, $\left(2, \frac{1}{2}\right)$ should appear on the graph. Since it *does* appear to be on the line, we have a check.

Applications

Because slope is a ratio that indicates how a change in the vertical direction corresponds to a change in the horizontal direction, it has many real-world applications. Foremost is the use of slope to represent a *rate of change*.

EXAMPLE 10   First-class mail.   The number of pieces of priority mail has been decreasing steadily since 2000. (This is probably due to the increasing use of e-mail.) Use the graph on the following page to find the rate at which this number is changing. Note that the jagged "break" on the vertical axis is used to avoid including a large portion of unused grid.

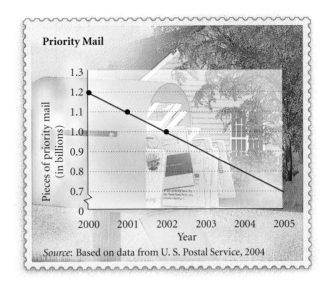

Priority Mail

Source: Based on data from U. S. Postal Service, 2004

Solution   Since the graph is linear, we can use any pair of points to determine the rate of change. We choose the points (2000, 1.2 billion) and (2001, 1.1 billion), which gives us

$$\text{Rate of change} = \frac{1.1 \text{ billion} - 1.2 \text{ billion}}{2001 - 2000}$$

$$= \frac{-0.1 \text{ billion}}{1 \text{ year}} = -0.1 \text{ billion per year.}$$

The number of pieces of priority mail sent each year is changing at a rate of -0.1 billion pieces per year.

EXAMPLE 11   Cell-phone use.   In 2000, there were approximately 109.5 million cell-phone customers in the United States. By 2002, the figure had grown to 140.8 million (*Source*: Based on data from the Cellular Telecommunications Industry Association). At what rate did the number of cell-phone customers change?

Solution   The rate at which the number of customers changed is given by

$$\text{Rate of change} = \frac{\text{change in number of customers}}{\text{change in time}}$$

$$= \frac{140.8 \text{ million customers} - 109.5 \text{ million customers}}{2002 - 2000}$$

$$= \frac{31.3 \text{ million customers}}{2 \text{ years}}$$

$$= 15.65 \text{ million customers per year.}$$

Between 2000 and 2002, the number of cell-phone customers grew at a rate of approximately 15.65 million, or 15,650,000 customers per year.

EXAMPLE 12   *Running speed.*   Stephanie runs 10 km during each workout. For the first 7 km, her pace is twice as fast as it is for the last 3 km. Which of the following graphs best describes Stephanie's workout?

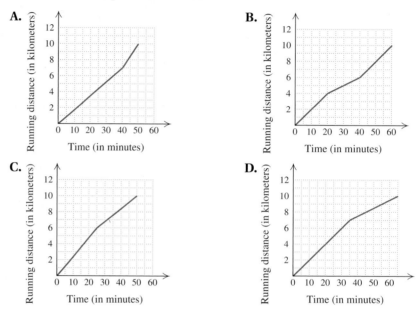

Solution   The slopes in graph A increase as we move to the right. This would indicate that Stephanie ran faster for the *last* part of her workout. Thus graph A is not the correct one.

The slopes in graph B indicate that Stephanie slowed down in the middle of her run and then resumed her original speed. Thus graph B does not correctly model the situation either.

According to graph C, Stephanie slowed down not at the 7-km mark, but at the 6-km mark. Thus graph C is also incorrect.

Graph D indicates that Stephanie ran the first 7 km in 35 min, a rate of 0.2 km/min. It also indicates that she ran the final 3 km in 30 min, a rate of 0.1 km/min. This means that Stephanie's rate was twice as fast for the first 7 km, so graph D provides a correct description of her workout.

Linear functions arise continually in today's world. As with the rate problems appearing in Examples 10–12, it is critical to use proper units in all answers.

EXAMPLE 13   *Salvage value.*   Tyline Electric uses the function $S(t) = -700t + 3500$ to determine the *salvage value $S(t)$*, in dollars, of a color photocopier t years after its purchase.

a) What do the numbers -700 and 3500 signify?

b) How long will it take the copier to *depreciate* completely?

c) What is the domain of S?

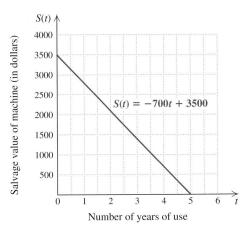

S(t) = −700t + 3500

Salvage value of machine (in dollars)

Number of years of use

Solution   Drawing, or at least visualizing, a graph can be useful here.

a) At time $t = 0$, we have $S(0) = -700 \cdot 0 + 3500 = 3500$. Thus the number 3500 signifies the original cost of the copier, in dollars.

This function is written in slope–intercept form. Since the output is measured in dollars and the input in years, the number -700 signifies that the value of the copier is declining at a rate of $700 per year.

b) The copier will have depreciated completely when its value drops to 0. To learn when this occurs, we determine when $S(t) = 0$:

$$S(t) = 0 \qquad \text{A graph is not always available.}$$
$$-700t + 3500 = 0 \qquad \text{Substituting } -700t + 3500 \text{ for } S(t)$$
$$-700t = -3500 \qquad \text{Subtracting 3500 from both sides}$$
$$t = 5. \qquad \text{Dividing both sides by } -700$$

The copier will have depreciated completely in 5 yr.

c) Neither the number of years of service nor the salvage value can be negative. In part (b), we found that after 5 yr the salvage value will have dropped to 0. Thus the domain of S is $\{t \mid 0 \le t \le 5\}$. The graph at left serves as a visual check of this result.

Exercise Set

2.3

FOR EXTRA HELP

Student's Solutions Manual

Digital Video Tutor CD 2 Videotape 3

Tutor Center

AW Math Tutor Center

MathXL Tutorials on CD

Math XL MathXL

MyMathLab MyMathLab

↪ *Concept Reinforcement*   *In each of Exercises 1–6, match the word with the most appropriate choice from the column on the right.*

1. _____ Rise

2. _____ Run

3. _____ Slope

4. _____ y-intercept

5. _____ Slope–intercept form

6. _____ Translated

a) $y = mx + b$

b) Shifted

c) $\dfrac{\text{Difference in } y}{\text{Difference in } x}$

d) Difference in x

e) Difference in y

f) $(0, b)$

Graph.

7. $f(x) = 2x - 7$

8. $g(x) = 3x - 7$

9. $g(x) = -\frac{1}{3}x + 2$

10. $f(x) = -\frac{1}{2}x - 5$

11. $h(x) = \frac{2}{5}x - 4$

12. $h(x) = \frac{4}{5}x + 2$

Determine the y-intercept.

13. $y = 3x + 4$

14. $y = 4x - 9$

15. $g(x) = -4x - 3$

16. $g(x) = -5x + 7$

17. $y = -\frac{3}{8}x - 4.5$

18. $y = \frac{15}{7}x + 2.2$

19. $f(x) = 2.9x - 9$

20. $f(x) = -3.1x + 5$

21. $y = 37x + 204$

22. $y = -52x + 700$

For each pair of points, find the slope of the line containing them.

23. $(6, 9)$ and $(4, 5)$

24. $(8, 7)$ and $(2, -1)$

25. $(3, 8)$ and $(9, -4)$

26. $(17, -12)$ and $(-9, -15)$

27. $\left(-\frac{2}{3}, \frac{1}{2}\right)$ and $\left(-\frac{4}{5}, \frac{4}{5}\right)$

28. $\left(\frac{3}{4}, -\frac{2}{5}\right)$ and $\left(\frac{1}{3}, -\frac{1}{4}\right)$

29. $(-9.7, 43.6)$ and $(4.5, 43.6)$

30. $(-2.8, -3.1)$ and $(-1.8, -2.6)$

Determine the slope and the y-intercept. Then draw a graph. Be sure to check as in Example 5 or Example 8.

31. $y = \frac{5}{2}x + 3$ **32.** $y = \frac{2}{5}x + 4$

33. $f(x) = -\frac{5}{2}x + 2$ **34.** $f(x) = -\frac{2}{5}x + 3$

35. $2x - y = 5$ **36.** $2x + y = 4$

37. $F(x) = \frac{1}{3}x + 2$ **38.** $g(x) = -3x + 6$

39. $6y + 4x = 6$ **40.** $4y + 20 = x$

Aha! **41.** $g(x) = -0.25x$ **42.** $F(x) = 1.5x - 3$

43. $4x - 5y = 10$ **44.** $5x + 4y = 4$

45. $f(x) = \frac{5}{4}x - 2$ **46.** $f(x) = \frac{4}{3}x + 2$

47. $12 - 4f(x) = 3x$ **48.** $15 + 5f(x) = -2x$

Aha! **49.** $g(x) = 4.5$ **50.** $g(x) = \frac{3}{4}x$

Find a linear function whose graph has the given slope and y-intercept.

51. Slope $\frac{2}{3}$, y-intercept $(0, -9)$

52. Slope $-\frac{3}{4}$, y-intercept $(0, 12)$

53. Slope -6, y-intercept $(0, 2)$

54. Slope 2, y-intercept $(0, -3)$

55. Slope $-\frac{7}{9}$, y-intercept $(0, 5)$

56. Slope $-\frac{4}{11}$, y-intercept $(0, 9)$

57. Slope 5, y-intercept $\left(0, \frac{1}{2}\right)$

58. Slope 6, y-intercept $\left(0, \frac{2}{3}\right)$

For each graph, find the rate of change. Remember to use appropriate units. See Example 10.

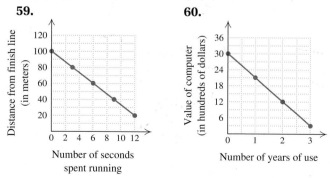

59.

60.

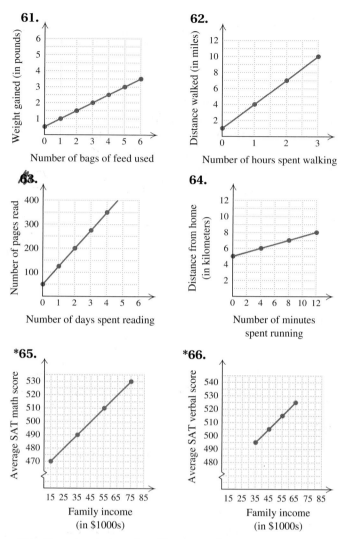

61.

62.

63.

64.

***65.**

***66.**

67. *Running rate.* An ultra-marathoner passes the 15-mi point of a race after 2 hr and reaches the 22-mi point 56 min later. Assuming a constant rate, find the speed of the marathoner.

68. *Skiing rate.* A cross-country skier reaches the 3-km mark of a race in 15 min and the 12-km mark 45 min later. Assuming a constant rate, find the speed of the skier.

*Based on data from the College Board Online.

69. *Work rate.* As a painter begins work, one-fourth of a house has already been painted. Eight hours later, the house is two-thirds done. Calculate the painter's work rate.

70. *Rate of computer hits.* At the beginning of 2002, Starfarm.com had already received 80,000 hits at their website. At the beginning of 2004, that number had climbed to 430,000. Calculate the rate at which the number of hits is increasing.

71. *Rate of descent.* A plane descends to sea level from 12,000 ft after being airborne for $1\frac{1}{2}$ hr. The entire flight time is 2 hr and 10 min. Determine the average rate of descent of the plane.

72. *Growth in overseas travel.* In 1997, the number of U.S. visitors overseas was about 21.6 million. In 2000, the number grew to 26.9 million. Determine the rate at which the number of U.S. visitors overseas was growing.
Source: *Statistical Abstract of the United States*

73. *Nursing.* Match each sentence with the most appropriate of the four graphs shown.
a) The rate at which fluids were given intravenously was doubled after 3 hr.
b) The rate at which fluids were given intravenously was gradually reduced to 0.
c) The rate at which fluids were given intravenously remained constant for 5 hr.
d) The rate at which fluids were given intravenously was gradually increased.

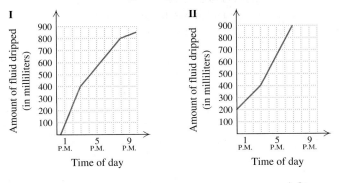

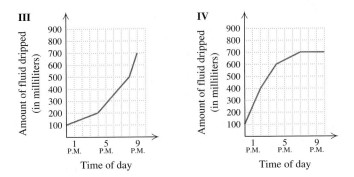

74. *Market research.* Match each sentence with the most appropriate of the four graphs shown.
a) After January 1, daily sales continued to rise, but at a slower rate.
b) After January 1, sales decreased faster than they ever grew.
c) The rate of growth in daily sales doubled after January 1.
d) After January 1, daily sales decreased at half the rate that they grew in December.

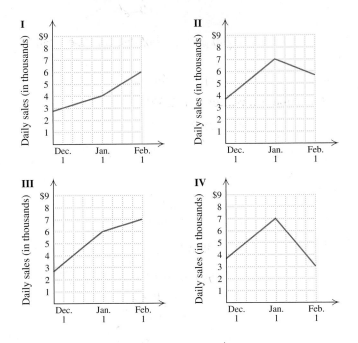

In Exercises 75–84, each model is of the form
$f(x) = mx + b$. *In each case, determine what m and b signify.*

75. *Catering.* When catering a party for x people, Jennette's Catering uses the formula $C(x) = 25x + 75$, where $C(x)$ is the cost of the party, in dollars.

76. *Weekly pay.*   Each salesperson at Knobby's Furniture is paid $P(x)$ dollars, where $P(x) = 0.05x + 200$ and x is the value of the salesperson's sales for the week.

77. *Hair growth.*   After Oscar gets a "buzz cut," the length $L(t)$ of his hair, in inches, is given by $L(t) = \frac{1}{2}t + 1$, where t is the number of months after he gets the haircut.

78. *Natural gas demand.*   The demand, in quadrillions of Btu's, for natural gas is approximated by $D(t) = \frac{9}{25}t + 22$, where t is the number of years after 2000.
Source: Based on data from the U.S. Energy Information Administration

79. *Life expectancy of American women.*   The life expectancy of American women t years after 1970 is given by $A(t) = \frac{1}{7}t + 75.5$.

80. *Landscaping.*   After being cut, the length $G(t)$ of the lawn, in inches, at Great Harrington Community College is given by $G(t) = \frac{1}{8}t + 2$, where t is the number of days since the lawn was cut.

81. *Cost of a movie ticket.*   The average price $P(t)$, in dollars, of a movie ticket is given by $P(t) = 0.227t + 4.29$, where t is the number of years since 1995.

82. *Sales of cotton goods.*   The function given by $f(t) = -0.5t + 7$ can be used to estimate the cash receipts, in billions of dollars, from farm marketing of cotton t years after 1995.

83. *Cost of a taxi ride.*   The cost, in dollars, of a taxi ride in New York City is given by $C(d) = 2d + 2.5$,* where d is the number of miles traveled.

84. *Cost of renting a truck.*   The cost, in dollars, of a one-day truck rental is given by $C(d) = 0.3d + 20$, where d is the number of miles driven.

85. *Salvage value.*   Green Glass Recycling uses the function given by $F(t) = -5000t + 90,000$ to determine the salvage value $F(t)$, in dollars, of a waste removal truck t years after it has been put into use.
  a) What do the numbers -5000 and $90,000$ signify?
  b) How long will it take the truck to depreciate completely?
  c) What is the domain of F?

86. *Salvage value.*   Consolidated Shirt Works uses the function given by $V(t) = -2000t + 15,000$ to determine the salvage value $V(t)$, in dollars, of a color separator t years after it has been put into use.
  a) What do the numbers -2000 and $15,000$ signify?
  b) How long will it take the machine to depreciate completely?
  c) What is the domain of V?

87. *Trade-in value.*   The trade-in value of a Homelite snowblower can be determined using the function given by $v(n) = -150n + 900$. Here $v(n)$ is the trade-in value, in dollars, after n winters of use.
  a) What do the numbers -150 and 900 signify?
  b) When will the trade-in value of the snowblower be $300?
  c) What is the domain of v?

88. *Trade-in value.*   The trade-in value of a John Deere riding lawnmower can be determined using the function given by $T(x) = -300x + 2400$. Here $T(x)$ is the trade-in value, in dollars, after x summers of use.
  a) What do the numbers -300 and 2400 signify?
  b) When will the value of the mower be $1200?
  c) What is the domain of T?

*Rates are higher between 4 and 8 P.M. (*Source*: Based on data in *The Burlington Free Press* 3/31/04)

89. *Economics.*   In March of 2004, the federal debt could be modeled using $D(t) = mt + 7.1$, where $D(t)$ is in trillions of dollars t months after March and m is some constant. If you were president of the United States, would you want m to be positive or negative? Why?
Source: www.publicdebt.treas.gov

90. *Economics.*   Examine the function given in Exercise 89. What units of measure must be used for m? Why?

SKILL MAINTENANCE

Solve. [1.3]

91. $2x - 5 = 7x + 3$

92. $4t + 9 = t - 6$

93. $\frac{1}{5}x + 7 = 2$

94. $-\frac{2}{3}t + 4 = t - 1$

95. $3 \cdot 0 - 2y = 9$

96. $4x - 7 \cdot 0 = 3$

SYNTHESIS

97. Hope Diswerkz claims that her firm's profits continue to go up, but the rate of increase is going down.
   a) Sketch a graph that might represent her firm's profits as a function of time.
   b) Explain why the graph can go up while the rate of increase goes down.

98. Belly Up, Inc., is losing $1.5 million per year while Spinning Wheels, Inc., is losing $170 an hour. Which firm would you rather own and why?

In Exercises 99 and 100, assume that r, p, and s are constants and that x and y are variables. Determine the slope and the y-intercept.

99. $rx + py = s - ry$

100. $rx + py = s$

101. Let (x_1, y_1) and (x_2, y_2) be two distinct points on the graph of $y = mx + b$. Use the fact that both pairs are solutions of the equation to prove that m is the slope of the line given by $y = mx + b$. (*Hint*: Use the slope formula.)

Given that $f(x) = mx + b$, classify each of the following as true or false.

102. $f(c + d) = f(c) + f(d)$

103. $f(cd) = f(c)f(d)$

104. $f(kx) = kf(x)$

105. $f(c - d) = f(c) - f(d)$

106. Find k such that the line containing $(-3, k)$ and $(4, 8)$ is parallel to the line containing $(5, 3)$ and $(1, -6)$.

107. Match each sentence with the most appropriate of the four graphs shown.
   a) Annie drove 2 mi to a lake, swam 1 mi, and then drove 3 mi to a store.
   b) During a preseason workout, Rico biked 2 mi, ran for 1 mi, and then walked 3 mi.
   c) James bicycled 2 mi to a park, hiked 1 mi over the notch, and then took a 3-mi bus ride back to the park.
   d) After hiking 2 mi, Marcy ran for 1 mi before catching a bus for the 3-mi ride into town.

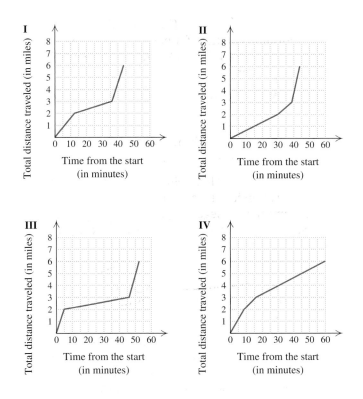

108. *Cost of a speeding ticket.* The penalty schedule shown below is used to determine the cost of a speeding ticket in certain states. Use this schedule to graph the cost of a speeding ticket as a function of the number of miles per hour over the limit that a driver is going.

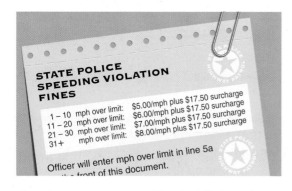

109. Find the slope of the line that contains the given pair of points.

a) $(5b, -6c), (b, -c)$
b) $(b, d), (b, d + e)$
c) $(c + f, a + d), (c - f, -a - d)$

110. Graph the equations

$$y_1 = 1.4x + 2, \qquad y_2 = 0.6x + 2,$$
$$y_3 = 1.4x + 5, \quad \text{and} \quad y_4 = 0.6x + 5$$

using a graphing calculator. If possible, use the SIMULTANEOUS mode so that you cannot tell which equation is being graphed first. Then decide which line corresponds to each equation.

111. A student makes a mistake when using a graphing calculator to draw $4x + 5y = 12$ and the following screen appears. Use algebra to show that a mistake has been made. What do you think the mistake was?

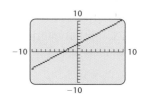

112. A student makes a mistake when using a graphing calculator to draw $5x - 2y = 3$ and the following screen appears. Use algebra to show that a mistake has been made. What do you think the mistake was?

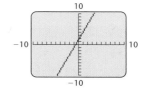

2.4 Another Look at Linear Graphs

Zero Slope and Lines with Undefined Slope • Graphing Using Intercepts • Solving Equations Graphically • Recognizing Linear Equations

In Section 2.3, we graphed linear equations using slopes and y-intercepts. We now graph lines that have slope 0 or that have an undefined slope. We also graph lines using both x- and y-intercepts and learn how to use graphs to solve certain equations.

Zero Slope and Lines with Undefined Slope

If two different points have the same second coordinate, what is the slope of
the line joining them? In this case, we have $y_2 = y_1$, so

$$m = \frac{y_2 - y_1}{x_2 - x_1} = \frac{0}{x_2 - x_1} = 0.$$

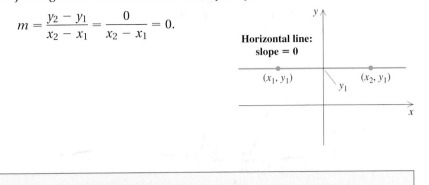

Slope of a Horizontal Line

Every horizontal line has a slope of 0.

EXAMPLE 1

Use slope–intercept form to graph $f(x) = 3$.

Solution   Recall from Example 5(c) in Section 2.2 that a function of this type
is called a *constant function*. Writing $f(x)$ in slope–intercept form,

$$f(x) = 0 \cdot x + 3,$$

we see that the y-intercept is $(0, 3)$ and the slope is 0. Thus we can graph f by
plotting $(0, 3)$ and, from there, determining a slope of 0. Because $0 = 0/2$ (any
nonzero number could be used in place of 2), we can draw the graph by going
up 0 units and to the right 2 units. As a check, we also find some ordered pairs.
Note that for any choice of x-value, $f(x)$ must be 3.

x	$f(x)$
-1	3
0	3
2	3

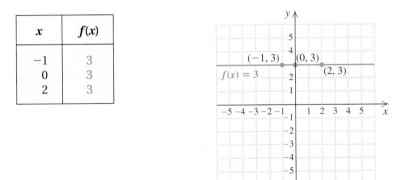

We see from Example 1 the following:

**The graph of any function of the form $f(x) = b$ or
$y = b$ is a horizontal line that crosses the y-axis at $(0, b)$.**

Suppose that two different points are on a vertical line. They then have the same first coordinate. In this case, we have $x_2 = x_1$, so

$$m = \frac{y_2 - y_1}{x_2 - x_1} = \frac{y_2 - y_1}{0}.$$

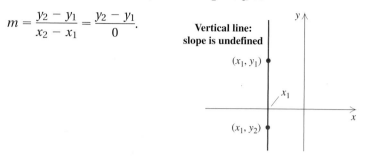

Since we cannot divide by 0, this is undefined. Note that when we say that $(y_2 - y_1)/0$ is undefined, it means that we have agreed to not attach any meaning to that expression.

Slope of a Vertical Line

The slope of a vertical line is undefined.

EXAMPLE 2   Graph: $x = -2$.

Solution   With y missing, no matter which value of y is chosen, x must be -2. Thus the pairs $(-2, 3)$, $(-2, 0)$, and $(-2, -4)$ all satisfy the equation. The graph is a line parallel to the y-axis. Note that since y is missing, this equation cannot be written in slope–intercept form.

x	y
-2	3
-2	0
-2	-4

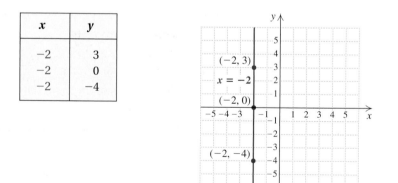

Example 2 shows the following:

**The graph of any equation of the form $x = a$
is a vertical line that crosses the x-axis at $(a, 0)$.**

EXAMPLE 3   Find the slope of each given line. If the slope is undefined, state this.

a) $3y + 2 = 14$ **b)** $2x = 10$

Solution

a) We solve for y:

$$3y + 2 = 14$$
$$3y = 12 \qquad \text{Subtracting 2 from both sides}$$
$$y = 4. \qquad \text{Dividing both sides by 3}$$

The graph of $y = 4$ is a horizontal line. Since $3y + 2 = 14$ is equivalent to $y = 4$, the slope of the line $3y + 2 = 14$ is 0.

b) When y does not appear, we solve for x:

$$2x = 10$$
$$x = 5. \qquad \text{Dividing both sides by 2}$$

The graph of $x = 5$ is a vertical line. Since $2x = 10$ is equivalent to $x = 5$, the slope of the line $2x = 10$ is undefined.

Graphing Using Intercepts

Any line that is not horizontal or vertical will cross both the x- and y-axes. We have already seen that the point at which a line crosses the y-axis is called the *y-intercept*. Similarly, the point at which a line crosses the x-axis is called the *x-intercept*. When the x- and y-intercepts are not both $(0, 0)$, they are two distinct points and can thus be used to draw the graph of the line. Recall that to find the y-intercept, we replace x with 0 and solve for y. To find the x-intercept, we replace y with 0 and solve for x.

> **To Determine Intercepts**
>
> The x-intercept is of the form $(a, 0)$. To find a, let $y = 0$ and solve the original equation for x.
>
> The y-intercept is of the form $(0, b)$. To find b, let $x = 0$ and solve the original equation for y.

EXAMPLE 4   Graph the equation $3x + 2y = 12$ by using intercepts.

Solution   To find the y-intercept, we let $x = 0$ and solve for y:

$$3 \cdot 0 + 2y = 12$$
$$2y = 12$$
$$y = 6.$$

The y-intercept is $(0, 6)$.

To find the x-intercept, we let y = 0 and solve for x:

$$3x + 2 \cdot 0 = 12$$
$$3x = 12$$
$$x = 4.$$

The *x*-intercept is (4, 0).

We plot the two intercepts and draw the line. A third point could be calculated and used as a check.

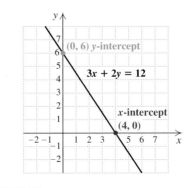

EXAMPLE 5

Graph $f(x) = 2x + 5$ by using intercepts.

Solution   Because the function is in slope–intercept form, we know that the *y*-intercept is (0, 5). To find the *x*-intercept, we replace $f(x)$ with 0 and solve for *x*:

$$0 = 2x + 5$$
$$-5 = 2x$$
$$-\tfrac{5}{2} = x.$$

The *x*-intercept is $\left(-\tfrac{5}{2}, 0\right)$.

We plot the intercepts (0, 5) and $\left(-\tfrac{5}{2}, 0\right)$ and draw the line. As a check, we can calculate the slope:

$$m = \frac{5 - 0}{0 - \left(-\tfrac{5}{2}\right)}$$
$$= \frac{5}{\tfrac{5}{2}}$$
$$= 5 \cdot \frac{2}{5}$$
$$= 2.$$

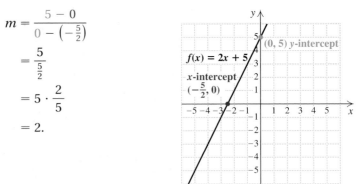

The slope is 2, as expected.

technology connection

One way to find intercepts on a graphing calculator is to graph the function and then make use of **2ND** (CALC). This menu allows you to select VALUE and then enter 0 for *x*. The same menu permits us to select ZERO. The *zero* or *root* of a function is the value for which $f(x) = 0$. To find this, we enter a value less than the *x*-intercept and then a value greater than the *x*-intercept as left and right bounds, respectively. We next enter a guess and the calculator then finds the value of the intercept.

Solving Equations Graphically

Note in Example 5 that the x-intercept, $-\frac{5}{2}$, is the solution of $2x + 5 = 0$. Visually, $-\frac{5}{2}$ is the x-coordinate of the point at which the graphs of $f(x) = 2x + 5$ and $h(x) = 0$ intersect. Similarly, we can solve $2x + 5 = -3$ by finding the x-coordinate of the point at which the graphs of $f(x) = 2x + 5$ and $g(x) = -3$ intersect. Careful inspection suggests that -4 is that x-value. To check, note that $f(-4) = 2(-4) + 5 = -3$.

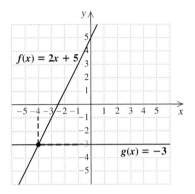

EXAMPLE 6   Solve graphically: $\frac{1}{2}x + 3 = 2$.

Solution   To find the x-value for which $\frac{1}{2}x + 3$ will equal 2, we graph $f(x) = \frac{1}{2}x + 3$ and $g(x) = 2$ on the same set of axes. Since the intersection appears to be $(-2, 2)$, the solution is apparently -2.

Student Notes

Remember that it is only the first coordinate of the point of intersection that is the solution of the equation.

Check:

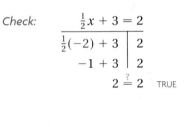

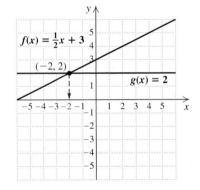

The solution is -2.

EXAMPLE 7   Cost projections.   Cleartone Communications charges $50 for a cell phone and $40 per month for calls made under its Call Anywhere plan. Formulate and graph a mathematical model for the cost. Then use the model to estimate the time required for the total cost to reach $250.

Solution

1. **Familiarize.**   The problem describes a situation in which a monthly fee is charged after an initial purchase has been made. After 1 month of service,

the total cost will be $50 + $40 = $90. After 2 months, the total cost will be $50 + $40 · 2 = $130. This can be generalized in a model if we let $C(t)$ represent the total cost, in dollars, for t months of service.

2. **Translate.**   We rephrase and translate as follows:

$$
\begin{array}{cccccc}
& & & \text{the cost of} & & \\
\textit{Rephrasing:} & \underbrace{\text{The total cost}} & \text{is} & \underbrace{\text{the phone}} & \text{plus} & \underbrace{\$40 \text{ per month.}} \\
& \downarrow & \downarrow & \downarrow & \downarrow & \downarrow \\
\textit{Translating:} & C(t) & = & 50 & + & 40 \cdot t
\end{array}
$$

where $t \geq 0$ (since there cannot be a negative number of months).

3. **Carry out.**   Before graphing, we rewrite the model in slope–intercept form: $C(t) = 40t + 50$. We see that the vertical intercept is (0, 50) and the slope— or rate—is $40 per month.

We plot (0, 50) and, from there, count up $40 and to the right 1 month. This takes us to (1, 90). We then draw a line passing through both points. Counting by 20's on the vertical axis allows us to graph both this line *and* the line $y = 250$.

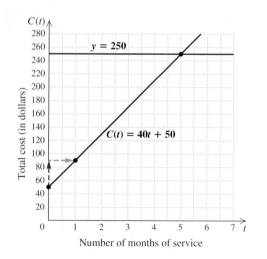

To estimate the time required for the total cost to reach $250, we are estimating the solution of

$$40t + 50 = 250. \qquad \text{Replacing } C(t) \text{ with } 250$$

We do this by graphing $y = 250$ and looking for the point of intersection. This point appears to be $(5, 250)$. Thus we estimate that it takes 5 months for the total cost to reach $250.

4. **Check.**   We evaluate:

$$C(5) = 40 \cdot 5 + 50 = 200 + 50 = 250.$$

Our estimate turns out to be precise.

5. **State.**   It takes 5 months for the total cost to reach $250.

There are limitations to solving equations graphically, as the next example illustrates.

EXAMPLE 8

Solve graphically: $-\frac{3}{4}x + 6 = 2x - 1$.

Solution   We graph $f(x) = -\frac{3}{4}x + 6$ and $g(x) = 2x - 1$ on the same set of axes. It *appears* that the lines intersect at $(2.5, 4)$. This would mean that $f(2.5)$ and $g(2.5)$ are identical. Let's check.

Check:

$$
\begin{array}{c|c}
-\frac{3}{4}x + 6 = 2x - 1 & \\
\hline
-\frac{3}{4}(2.5) + 6 & 2(2.5) - 1 \\
-1.875 + 6 & 5 - 1 \\
4.125 \overset{?}{=} 4 & \qquad \text{FALSE}
\end{array}
$$

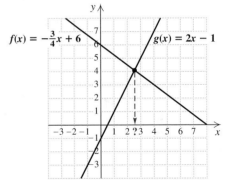

Our check shows that 2.5 is *not* the solution, although it may not be off by much. To find the exact solution, we need either a more precise way of determining the point of intersection (see the Technology Connection) or an algebraic approach:

$$
\begin{aligned}
-\tfrac{3}{4}x + 6 &= 2x - 1 & \\
-\tfrac{3}{4}x + 7 &= 2x & \text{Adding 1 to both sides} \\
7 &= \tfrac{11}{4}x & \text{Adding } \tfrac{3}{4}x \text{ to both sides} \\
\tfrac{28}{11} &= x. & \text{Multiplying both sides by } \tfrac{4}{11}
\end{aligned}
$$

The solution is $\frac{28}{11}$, or about 2.55. A check of this answer is left to the student.

Caution!   When using a graph to solve an equation, it is important to use graph paper and to work as neatly as possible.

technology connection

To solve Example 8 with a graphing calculator, we can graph each side of the equation and then select the INTERSECT option of the CALC menu. Once this is done, we locate the cursor on each line and press **ENTER**. Finally, we enter a guess, and the calculator determines the coordinates of the intersection.

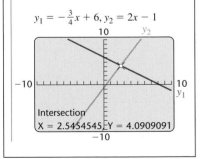

Recognizing Linear Equations

Is every equation of the form $Ax + By = C$ linear? To find out, suppose that A and B are nonzero and solve for y:

$$Ax + By = C \qquad \text{A, B, and C are constants.}$$
$$By = -Ax + C \qquad \text{Adding } -Ax \text{ to both sides}$$
$$y = -\frac{A}{B}x + \frac{C}{B}. \qquad \text{Dividing both sides by B}$$

Since the last equation is a slope–intercept equation, we see that $Ax + By = C$ is a linear equation when $A \neq 0$ and $B \neq 0$.

But what if A or B (but not both) is 0? If A is 0, then $By = C$ and $y = C/B$. If B is 0, then $Ax = C$ and $x = C/A$. In the first case, the graph is a horizontal line; in the second case, the line is vertical. In either case, $Ax + By = C$ is a linear equation when A or B (but not both) is 0. We have now justified the following result.

Standard Form of a Linear Equation

Any equation $Ax + By = C$, where A, B, and C are real numbers and A and B are not both 0, has a graph that is a straight line.

Any equation of the form

$$Ax + By = C$$

is said to be written in *standard form*.

EXAMPLE 9   Determine whether the equation $y = x^2 - 5$ is linear.

Solution   We attempt to put the equation in standard form:

$$y = x^2 - 5$$
$$-x^2 + y = -5. \qquad \text{Adding } -x^2 \text{ to both sides}$$

This last equation is not linear because it has an x^2-term. We graphed this equation as Example 7 in Section 2.1.

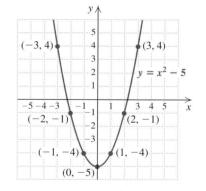

Only linear equations have graphs that are straight lines. Also, only linear graphs have a constant slope. Were you to try to calculate the slope between several pairs of points in Example 9, you would find that the slopes vary.

Exercise Set

2.4

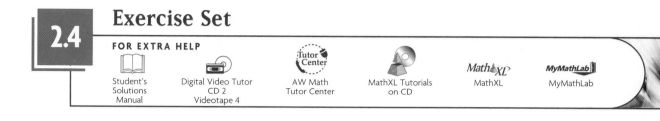

↪ *Concept Reinforcement*   *Complete each of the following statements.*

1. Every _____ line has a slope of 0.

2. The slope of a vertical line is _____.

3. The graph of any equation of the form $x = a$ is a _____ line that crosses the x-axis at $(a, 0)$.

4. The graph of any function of the form $f(x) = b$ is a horizontal line that crosses the _____ at $(0, b)$.

5. To find the x-intercept, we let $y =$ _____ and solve the original equation for _____.

6. To find the y-intercept, we let $x =$ _____ and solve the original equation for _____.

7. To solve $3x - 5 = 7$, we can graph $f(x) = 3x - 5$ and $g(x) = 7$ and find the x-value at the point of _____.

8. An equation like $4x + 3y = 8$ is said to be written in _____ form.

9. Only _____ equations have graphs that are straight lines.

10. Using graphs to solve an equation does not always yield an _____ solution.

For each equation, find the slope. If the slope is undefined, state this.

11. $y - 9 = 3$

12. $x + 1 = 7$

13. $8x = 6$

14. $y - 3 = 5$

15. $3y = 28$

16. $19 = -6y$

17. $9 + x = 12$

18. $2x = 18$

19. $2x - 4 = 3$

20. $5y - 1 = 16$

21. $5y - 4 = 35$

22. $2x - 17 = 3$

23. $3y + x = 3y + 2$

24. $x - 4y = 12 - 4y$

25. $5x - 2 = 2x - 7$

26. $5y + 3 = y + 9$

Aha! 27. $y = -\frac{2}{3}x + 5$

28. $y = -\frac{3}{2}x + 4$

Graph.

29. $y = 5$

30. $x = -1$

31. $x = 3$

32. $y = 2$

33. $4 \cdot f(x) = 20$

34. $6 \cdot g(x) = 12$

35. $3x = -15$

36. $2x = 10$

37. $4 \cdot g(x) + 3x = 12 + 3x$

38. $3 - f(x) = 2$

Find the intercepts. Then graph by using the intercepts, if possible, and a third point as a check.

39. $x + y = 4$

40. $x + y = 5$

41. $y = 2x + 6$

42. $y = 3x + 9$

43. $3x + 5y = -15$

44. $5x - 4y = 20$

45. $2x - 3y = 18$

46. $3x + 2y = -18$

47. $3y = 6x$

48. $5y = 15x$

49. $f(x) = 3x - 7$

50. $g(x) = 2x - 9$

51. $1.4y - 3.5x = -9.8$

52. $3.6x - 2.1y = 22.68$

53. $5x + 2g(x) = 7$

54. $3x - 4f(x) = 11$

Solve each equation graphically. Then check your answer by solving the same equation algebraically.

55. $x - 2 = 6$

56. $x + 4 = 6$

57. $3x - 4 = -1$

58. $2x + 1 = 7$

59. $\frac{1}{2}x + 3 = 5$

60. $\frac{1}{3}x - 2 = 1$

61. $x - 8 = 3x - 5$

62. $x + 3 = 5 - x$

63. $3 - x = \frac{1}{2}x - 3$

64. $5 - \frac{1}{2}x = x - 4$

65. $2x + 1 = -x + 7$

66. $-3x + 4 = 3x - 4$

Use a graph to estimate the solution in each of the following. Be sure to use graph paper and a straightedge.

67. *Healthcare costs.*   Under one particular CIGNA health-insurance plan, an individual pays the first $100 of each hospital stay plus $\frac{1}{5}$ of all charges in excess of $100. By approximately how much did Gerry's hospital bill exceed $100, if the hospital stay cost him a total of $2200?
Source: VT State Chamber of Commerce membership information from VACE, 2004

68. *Indiana Toll Road.*   To drive an automobile on the Indiana Toll Road costs approximately 21¢ plus 3¢ per mile. Estimate how far Gina has driven on the Toll Road if her toll is $2.50.
Source: www.in.gov/dot/motoristinfo

69. *Telephone charges.*   Skytone Calling charges $20 for a telephone and $25 per month under its economy plan. Estimate the time required for the total cost to reach $145.

70. *Cell-phone charges.*   The Cellular Connection charges $30 for a cell phone and $40 per month under its economy plan. Estimate the time required for the total cost to reach $190.

71. *Parking fees.*   Karla's Parking charges $3.00 to park plus 50¢ for each 15-min unit of time. Estimate how long someone can park for $7.50.*

72. *Cost of a road call.*   Dave's Foreign Auto Village charges $50 for a road call plus $15 for each 15-min unit of time. Estimate the time required for a road call that cost $140.*

73. *Cost of a FedEx delivery.*   In 2004, for Standard delivery to the closest zone of packages weighing from 100 to 499 lb, FedEx charged $109 plus $1.09 per pound. Estimate the weight of a package that cost $305.20 to ship.
Source: www.fedex.com

74. *Copying costs.*   For each copy of a town report, a FedEx Kinko's Office and Print Center charged $3.45 for a coil binding and 15¢ for each page that was a double-sided copy. Estimate the number of pages in a coil-bound town report that cost $6.90 per copy. Assume that every page in the report was a double-sided copy.
Source: FedEx Kinko's price list brochure

Determine whether each equation is linear. Find the slope of any nonvertical lines.

75. $5x - 3y = 15$

76. $3x + 5y + 15 = 0$

77. $16 + 4y = 10$

78. $3x - 12 = 0$

79. $4g(x) = 6x^2$

80. $2x + 4f(x) = 8$

81. $3y = 7(2x - 4)$

82. $2(5 - 3x) = 5y$

---

*More precise, nonlinear models of Exercises 71 and 72 appear in Exercises 107 and 106, respectively.

83. $g(x) - \dfrac{1}{x} = 0$

84. $f(x) + \dfrac{1}{x} = 0$

85. $\dfrac{f(x)}{5} = x$

86. $\dfrac{g(x)}{2} = 3 + x^2$

87. *Engineering.* Wind friction, or *air resistance*, increases with speed. Following are some measurements made in a wind tunnel. Plot the data and explain why a linear function does or does not give an approximate fit.

Velocity (in kilometers per hour)	Force of Resistance (in newtons)
10	3
21	4.2
34	6.2
40	7.1
45	15.1
52	29.0

88. *Meteorology.* Wind chill is a measure of how cold the wind makes you feel. Below are some measurements of wind chill for a 15-mph breeze. How can you tell from the data that a linear function will give an approximate fit?

Temperature	15-mph Wind Chill
30°F	19°F
25°F	13°F
20°F	6°F
15°F	0°F
10°F	−7°F
5°F	−13°F
0°F	−19°F

Source: National Oceanic & Atmospheric Administration, as reported in USA TODAY.com, 2004

SKILL MAINTENANCE

Simplify.

89. $-\dfrac{3}{7} \cdot \dfrac{7}{3}$ [1.2]

90. $\dfrac{5}{4}\left(-\dfrac{4}{5}\right)$ [1.2]

91. $-5[x - (-3)]$ [1.3]

92. $-2[x - (-4)]$ [1.3]

93. $\dfrac{2}{3}\left[x - \left(-\dfrac{1}{2}\right)\right] - 1$ [1.3]

94. $-\dfrac{3}{2}\left(x - \dfrac{2}{5}\right) - 3$ [1.3]

SYNTHESIS

95. Jim tries to avoid fractions as often as possible. Under what conditions will graphing using intercepts allow him to avoid fractions? Why?

96. Under what condition(s) will the *x*- and *y*-intercepts of a line coincide? What would the equation for such a line look like?

97. Give an equation, in standard form, for the line whose *x*-intercept is 5 and whose *y*-intercept is −4.

98. Find the *x*-intercept of $y = mx + b$, assuming that $m \neq 0$.

In Exercises 99–102, assume that r, p, and s are nonzero constants and that x and y are variables. Determine whether each equation is linear.

99. $rx + 3y = p^2 - s$

100. $py = sx - r^2y - 9$

101. $r^2x = py + 5$

102. $\dfrac{x}{r} - py = 17$

103. Suppose that two linear equations have the same *y*-intercept but that equation A has an *x*-intercept that is half the *x*-intercept of equation B. How do the slopes compare?

Consider the linear equation

$$ax + 3y = 5x - by + 8.$$

104. Find *a* and *b* if the graph is a horizontal line passing through (0, 4).

105. Find *a* and *b* if the graph is a vertical line passing through (4, 0).

106. (Refer to Exercise 72.) A 32-min road call with Dave's costs the same as a 44-min road call. Thus the linear graph drawn in the solution of Exercise 72 is not a precise representation of the situation. Draw a graph with a series of "steps" that more accurately reflects the situation.

107. (Refer to Exercise 71.) It costs as much to park at Karla's for 16 min as it does for 29 min. Thus the linear graph drawn in the solution of Exercise 71 is not a precise representation of the situation. Draw a graph with a series of "steps" that more accurately reflects the situation.

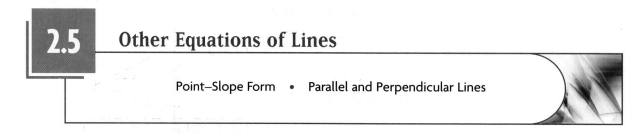 *Solve graphically and then check by solving algebraically.*

108. $5x + 3 = 7 - 2x$ **109.** $4x - 1 = 3 - 2x$

110. $3x - 2 = 5x - 9$ **111.** $8 - 7x = -2x - 5$

Solve using a graphing calculator.

112. Weekly pay at Bikes for Hikes is \$219 plus a 3.5% sales commission. If a salesperson's pay was \$370.03, what did that salesperson's sales total?

113. It costs Bert's Shirts \$38 plus \$2.35 a shirt to print tee shirts for a day camp. Camp Weehawken paid Bert's \$623.15 for shirts. How many shirts were printed?

2.5   Other Equations of Lines

Point–Slope Form   •   Parallel and Perpendicular Lines

Specifying the slope of a line and a point through which the line passes enables us to draw the line. We now study how this information can be used to write an *equation* of the line. This skill is important in more advanced courses.

Point–Slope Form

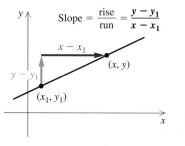

Suppose that a line of slope m passes through the point (x_1, y_1). For any other point (x, y) to lie on this line, we must have

$$\frac{y - y_1}{x - x_1} = m.$$

It is tempting to use this equation as an equation for the line of slope m that passes through (x_1, y_1). The problem with doing so is that when x and y are replaced with x_1 and y_1, we have $\frac{0}{0} = m$, a false equation. To avoid this difficulty, we multiply both sides by $x - x_1$ and simplify:

$$(x - x_1)\frac{y - y_1}{x - x_1} = m(x - x_1) \qquad \text{Multiplying both sides by } x - x_1$$

$$y - y_1 = m(x - x_1). \qquad \text{Removing a factor equal to 1:}$$
$$\frac{x - x_1}{x - x_1} = 1$$

This is the *point–slope* form of a linear equation.

> **Point–Slope Form**
>
> Any equation $y - y_1 = m(x - x_1)$ has a graph that is a straight line.
> It passes through (x_1, y_1) and has slope m.
>   Any equation of the form
> $$y - y_1 = m(x - x_1)$$
> is said to be written in *point–slope form*.

EXAMPLE 1   Find and graph an equation of the line passing through (3, 4) with slope $-\frac{1}{2}$.

Solution   We substitute in the point–slope equation:

$$y - y_1 = m(x - x_1)$$
$$y - 4 = -\tfrac{1}{2}(x - 3). \qquad \text{Substituting}$$

To graph this point–slope equation, we count off a slope of $-\frac{1}{2}$, starting at (3, 4). Then we draw the line.

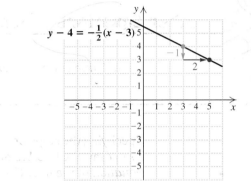

Student Notes _____

To help remember point–slope form, many students begin by writing the equation for slope and then clearing fractions. The equation in Example 1 would then be found as

$$\frac{y - 4}{x - 3} = -\frac{1}{2},$$

and

$$y - 4 = -\tfrac{1}{2}(x - 3).$$

EXAMPLE 2   Find a linear function that has a graph passing through the points $(-1, -5)$ and $(3, -2)$.

Solution   We first determine the slope of the line and then write an equation in point–slope form. Note that

$$m = \frac{-5 - (-2)}{-1 - 3} = \frac{-3}{-4} = \frac{3}{4}.$$

Since the line passes through $(3, -2)$, we have

$$y - (-2) = \tfrac{3}{4}(x - 3) \qquad \text{Substituting into } y - y_1 = m(x - x_1)$$
$$y + 2 = \tfrac{3}{4}x - \tfrac{9}{4}. \qquad \text{Using the distributive law}$$

Before using function notation, we isolate y:

$$y = \tfrac{3}{4}x - \tfrac{9}{4} - 2 \qquad \text{Subtracting 2 from both sides}$$
$$y = \tfrac{3}{4}x - \tfrac{17}{4} \qquad -\tfrac{9}{4} - \tfrac{8}{4} = -\tfrac{17}{4}$$
$$f(x) = \tfrac{3}{4}x - \tfrac{17}{4}. \qquad \text{Using function notation}$$

You can check that using $(-1, -5)$ as (x_1, y_1) in $y - y_1 = \frac{3}{4}(x - x_1)$ will yield the same expression for $f(x)$.

EXAMPLE 3

Tattoo removal.   In 1996, an estimated 275,000 Americans visited a doctor for tattoo removal. That figure was expected to grow to 410,000 in 2000 (*Source*: Mike Meyers, staff writer, *Star-Tribune Newspaper of the Twin Cities Minneapolis–St. Paul*, copyright 2000). Assuming constant growth since 1995, how many people will visit a doctor for tattoo removal in 2008?

Solution

1. **Familiarize.**   Constant growth indicates a constant rate of change, so a linear relationship can be assumed. If we let n represent the number of people, in thousands, who visit a doctor for tattoo removal and t the number of years since 1995, we can form the pairs (1, 275) and (5, 410). After choosing suitable scales on the two axes, we draw the graph.

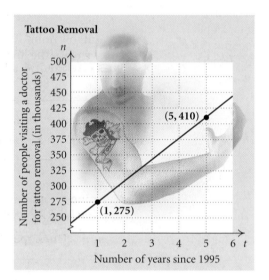

2. **Translate.**   To find an equation relating n and t, we first find the slope of the line. This corresponds to the *growth rate*:

$$m = \frac{410 \text{ thousand people} - 275 \text{ thousand people}}{5 \text{ years} - 1 \text{ year}}$$

$$= \frac{135 \text{ thousand people}}{4 \text{ years}}$$

$$= 33.75 \text{ thousand people per year.}$$

Next, we write point–slope form and solve for n:

$n - 275 = 33.75(t - 1)$	Writing point–slope form
$n - 275 = 33.75t - 33.75$	Using the distributive law
$n = 33.75t + 241.25.$	Adding 275 to both sides

3. **Carry out.**   Using function notation, we have

$$n(t) = 33.75t + 241.25.$$

To predict the number of people who will visit a doctor for tattoo removal in 2008, we find

$$n(13) = 33.75 \cdot 13 + 241.25 \qquad \text{2008 is 13 years from 1995.}$$
$$= 680. \qquad\qquad\qquad \text{This represents 680,000 people.}$$

4. **Check.**   To check, we can repeat our calculations. We could also extend the graph to see if (13, 680) appears to be on the line.

5. **State.**   Assuming constant growth, there will be about 680,000 people visiting a doctor for tattoo removal in 2008.

Parallel and Perpendicular Lines

Two lines are parallel if they lie in the same plane and do not intersect no matter how far they are extended. If two lines are vertical, they are parallel. How can we tell if nonvertical lines are parallel? The answer is simple: We look at their slopes (see Examples 1 and 2 in Section 2.3).

Slope and Parallel Lines

Two lines are parallel if they have the same slope.

EXAMPLE 4   Determine whether the line passing through (1, 7) and (4, −2) is parallel to the line given by $f(x) = -3x + 4.2$.

Solution   The slope of the line passing through (1, 7) and (4, −2) is given by

$$m = \frac{7 - (-2)}{1 - 4} = \frac{9}{-3} = -3.$$

Since the graph of $f(x) = -3x + 4.2$ also has a slope of −3, the lines are parallel.

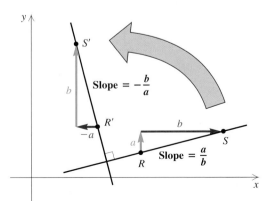

Two lines are perpendicular if they intersect at a right angle. If one line is vertical and another is horizontal, they are perpendicular. There are other instances in which two lines are perpendicular.

Consider a line $\overleftrightarrow{RS}$, as shown at left, with slope a/b. Then think of rotating the figure 90° to get a line $\overleftrightarrow{R'S'}$ perpendicular to $\overleftrightarrow{RS}$. For the new line, the rise and the run are interchanged, but the run is now negative. Thus the slope of the new line is $-b/a$. Let's multiply the slopes:

$$\frac{a}{b}\left(-\frac{b}{a}\right) = -1.$$

This can help us determine which lines are perpendicular.

Slope and Perpendicular Lines

Two lines are perpendicular if the product of their slopes is -1 or if one line is vertical and the other is horizontal.

Thus, if one line has slope m ($m \neq 0$), the slope of a line perpendicular to it is $-1/m$. That is, we take the reciprocal of m ($m \neq 0$) and change the sign.

EXAMPLE 5

Consider the line given by the equation $8y = 7x - 24$.

a) Find an equation for a parallel line passing through $(-1, 2)$.

b) Find an equation for a perpendicular line passing through $(-1, 2)$.

Solution   Both parts (a) and (b) require us to find the slope of the line given by $8y = 7x - 24$. To do so, we solve for y to find slope–intercept form:

$$8y = 7x - 24$$
$$y = \tfrac{7}{8}x - 3. \qquad \text{Multiplying both sides by } \tfrac{1}{8}$$
$$\text{The slope is } \tfrac{7}{8}.$$

a) The slope of any parallel line will be $\tfrac{7}{8}$. The point–slope equation yields

$$y - 2 = \tfrac{7}{8}[x - (-1)] \qquad \text{Substituting } \tfrac{7}{8} \text{ for the slope and } (-1, 2) \text{ for the point}$$

$$y - 2 = \tfrac{7}{8}[x + 1]$$
$$y = \tfrac{7}{8}x + \tfrac{7}{8} + 2 \qquad \text{Using the distributive law and adding 2 to both sides}$$

$$y = \tfrac{7}{8}x + \tfrac{23}{8}.$$

b) The slope of a perpendicular line is given by the opposite of the reciprocal of $\tfrac{7}{8}$, or $-\tfrac{8}{7}$. The point–slope equation yields

$$y - 2 = -\tfrac{8}{7}[x - (-1)] \qquad \text{Substituting } -\tfrac{8}{7} \text{ for the slope and } (-1, 2) \text{ for the point}$$

$$y - 2 = -\tfrac{8}{7}[x + 1]$$
$$y = -\tfrac{8}{7}x - \tfrac{8}{7} + 2 \qquad \text{Using the distributive law and adding 2 to both sides}$$

$$y = -\tfrac{8}{7}x + \tfrac{6}{7}.$$

---

technology connection

To check that the graphs of $y = \tfrac{7}{8}x - 3$ and $y = -\tfrac{8}{7}x + \tfrac{6}{7}$ are perpendicular, we use the ZSQUARE option of the ZOOM menu to create a "squared" window. This corrects for distortion that would result from the window being wider than it is tall.

1. Show that the graphs of
   $y = \tfrac{3}{4}x + 2$ and $y = -\tfrac{4}{3}x - 1$
   appear to be perpendicular.
2. Show that the graphs of
   $y = -\tfrac{2}{5}x - 4$ and $y = \tfrac{5}{2}x + 3$
   appear to be perpendicular.
3. To see that this type of check is not foolproof, graph
   $y = \tfrac{31}{40}x + 2$ and $y = -\tfrac{40}{30}x - 1$.
   Are the lines perpendicular? Why or why not?

CONNECTING THE CONCEPTS

We have now studied the slope–intercept, point–slope, and standard forms of a linear equation. These are the most common ways in which linear equations are written. Depending on what information we are given and what information we are seeking, one form may be more useful than the others. A referenced summary is given below.

Slope–intercept form, $y = mx + b$ or $f(x) = mx + b$	• Useful when an equation is needed and the slope and y-intercept are given. See Example 7 on p. 103. • Useful when a line's slope and y-intercept are needed. See Example 6 on p. 102. • Useful when solving equations graphically. See Example 6 on p. 117. • Commonly used for linear functions.
Standard form, $Ax + By = C$	• Allows for easy calculation of intercepts. See Example 4 on pp. 115–116. • Will prove useful in future work. See Sections 3.1–3.3.
Point–slope form, $y - y_1 = m(x - x_1)$	• Useful when an equation is needed and the slope and a point on the line are given. See Example 1 on p. 125. • Useful when a linear function is needed and two points on its graph are given. See Example 2 on p. 125. • Will prove useful in future work with curves and tangents in calculus.

Exercise Set

2.5

FOR EXTRA HELP

Student's Solutions Manual | Digital Video Tutor CD 2 Videotape 4 | Tutor Center AW Math Tutor Center | MathXL Tutorials on CD | Math XL MathXL | MyMathLab MyMathLab

🔖 *Concept Reinforcement*   *Classify each statement as either true or false.*

1. The equation $y - 5 = -3(x - 7)$ is written in point–slope form.

2. The equation $y = -2x + 6$ is written in point–slope form.

3. Knowing the coordinates of just one point on a line is enough to write an equation of the line.

4. Knowing the coordinates of just two points on a line is enough to write an equation of the line.

5. If two lines are perpendicular, then they have the same slope.

6. If the product of the slopes of two lines is 1, then the lines are perpendicular.

7. If two nonvertical lines are parallel, then they have the same slope.

8. If two nonvertical lines are perpendicular, then the product of their slopes is -1.

9. Point–slope form can be used with either point that is used to calculate the slope of that line.

10. There are situations for which point–slope form is more convenient to use than slope–intercept form.

Find an equation in point–slope form of the line having the specified slope and containing the point indicated. Then graph the line.

11. $m = -2, (1, 4)$ **12.** $m = 5, (3, 1)$

13. $m = 3, (5, 2)$ **14.** $m = 2, (7, 3)$

15. $m = \frac{1}{2}, (-2, -4)$ **16.** $m = 1, (-5, -7)$

17. $m = -1, (8, 0)$ **18.** $m = -3, (-2, 0)$

For each point–slope equation listed, state the slope and a point on the graph.

19. $y - 9 = \frac{2}{7}(x - 8)$ **20.** $y - 3 = 9(x - 2)$

21. $y + 2 = -5(x - 7)$ **22.** $y - 4 = -\frac{2}{9}(x + 5)$

23. $y - 4 = -\frac{5}{3}(x + 2)$ **24.** $y + 7 = -4(x - 9)$

Aha!   **25.** $y = \frac{4}{7}x$ **26.** $y = 3x$

Find an equation of the line having the specified slope and containing the indicated point. Write your final answer as a linear function in slope–intercept form. Then graph the line.

27. $m = 4, \ (2, -3)$ **28.** $m = -4, \ (-1, 5)$

29. $m = -\frac{3}{5}, \ (-4, 8)$ **30.** $m = -\frac{1}{5}, \ (-2, 1)$

31. $m = -0.6, \ (-3, -4)$ **32.** $m = 2.3, \ (4, -5)$

Aha!   **33.** $m = \frac{2}{7}, \ (0, -6)$ **34.** $m = \frac{1}{4}, \ (0, 3)$

35. $m = \frac{3}{5}, \ (-4, 6)$ **36.** $m = -\frac{2}{7}, \ (6, -5)$

Find an equation of the line containing each pair of points. Write your final answer as a linear function in slope–intercept form.

37. (1, 4) and (5, 6)   **38.** (2, 6) and (4, 1)

39. (2.5, −3) and (6.5, 3)   **40.** (2, −1.3) and (7, 1.7)

Aha!   **41.** (1, 3) and (0, −2)   **42.** (−3, 0) and (0, −4)

43. (−2, −3) and (−4, −6)

44. (−4, −7) and (−2, −1)

In Exercises 45–56, assume that a constant rate of change exists for each model formed.

45. *Dietary trends.* In 1971, the average American woman consumed 1542 calories per day. By 2000, the figure had risen to 1877 calories per day. Let $C(t)$ represent the average number of calories consumed per day by an American woman t years after 1971.

Source: Centers for Disease Control and Prevention

a) Find a linear function that fits the data.

b) Use the function from part (a) to predict the average number of calories consumed per day by an American woman in 2009.

c) When will the average number of calories consumed per day reach 2000?

46. *Dietary trends.* In 1971, the average American man consumed 2450 calories per day. By 2000, the figure had risen to 2618 calories per day. Let $C(t)$ represent the average number of calories consumed per day by an American man t years after 1971.

Source: Centers for Disease Control and Prevention

a) Find a linear function that fits the data.

b) Use the function from part (a) to predict the average number of calories consumed per day by an American man in 2008.

c) When will the average number of calories consumed per day reach 2750?

47. *Life expectancy of males in the United States.* In 1990, the life expectancy of males was 71.8 yr. In 2000, it was 74.1 yr. Let $E(t)$ represent life expectancy and t the number of years since 1990.

Source: *Statistical Abstract of the United States,* 2002

a) Find a linear function that fits the data.

b) Use the function of part (a) to predict the life expectancy of males in 2009.

48. *Life expectancy of females in the United States.* In 1990, the life expectancy of females was 78.8 yr. In 2000, it was 79.5 yr. Let $E(t)$ represent life expectancy and t the number of years since 1990.

Source: *Statistical Abstract of the United States,* 2002

a) Find a linear function that fits the data.

Aha!   **b)** Use the function of part (a) to predict the life expectancy of females in 2010.

49. *PAC contributions.* In 1992, Political Action Committees (PACs) contributed $178.6 million to congressional candidates. In 2002, the figure rose to $282 million. Let $A(t)$ represent the amount of PAC contributions, in millions, and t the number of years since 1992.

Source: Congressional Research Service and Federal Election Commission

PAC Contributions to Congressional candidates

$178.6 million in 1992
$282 million in 2002

FOR SALE ?

Congressional Research Service & Federal Election Commission

a) Find a linear function that fits the data.
b) Use the function of part (a) to predict the amount of PAC contributions in 2008.

50. *Consumer demand.*   Suppose that 6.5 million lb of coffee are sold when the price is $8 per pound, and 4.0 million lb are sold when it is $9 per pound.

a) Find a linear function that expresses the amount of coffee sold as a function of the price per pound.
b) Use the function of part (a) to predict how much consumers would be willing to buy at a price of $6 per pound.

51. *Recycling.*   In 1996, Americans recycled 57.3 million tons of solid waste. In 2000, the figure grew to 69.9 million tons. Let $N(t)$ represent the number of tons recycled, in millions, and t the number of years since 1996.
Source: *Statistical Abstract of the United States*, 2002

a) Find a linear function that fits the data.
b) Use the function of part (a) to predict the amount recycled in 2008.

52. *Seller's supply.*   Suppose that suppliers are willing to sell 5.0 million lb of coffee at a price of $8 per pound and 7.0 million lb at $9 per pound.

a) Find a linear function that expresses the amount suppliers are willing to sell as a function of the price per pound.
b) Use the function of part (a) to predict how much suppliers would be willing to sell at a price of $6 per pound.

53. *Online travel plans.*   In 1999, about 48 million Americans used the Internet to find travel information. By 2003, that number had grown to about 64 million. Let $N(t)$ represent the number of

Americans using the Internet for travel information, in millions, t years after 1999.
a) Find a linear function that fits the data.
b) Use the function of part (a) to predict the number of Americans who will find travel information on the Internet in 2009.
c) In what year will 104 million Americans find travel information on the Internet?

54. *Records in the 100-meter run.*   In 1991, the record for the 100-m run was 9.86 sec. In 2002, it was 9.78 sec. Let $R(t)$ represent the record in the 100-m run and t the number of years since 1991.
Source: runnersworld.com

a) Find a linear function that fits the data.
b) Use the function of part (a) to predict the record in 2008 and in 2015.
c) When will the record be 9.6 sec?

55. *National Park land.*   In 1994, the National Park system consisted of about 74.9 million acres. By 2000, the figure had grown to 78.2 million acres. Let $A(t)$ represent the amount of land in the National Park system, in millions of acres, t years after 1994.
Source: *Statistical Abstract of the United States*, 2002

a) Find a linear function that fits the data.
b) Use the function of part (a) to predict the amount of land in the National Park system in 2009.

56. *Pressure at sea depth.*   The pressure 100 ft beneath the ocean's surface is approximately 4 atm (atmospheres), whereas at a depth of 200 ft, the pressure is about 7 atm.
   a) Find a linear function that expresses pressure as a function of depth.
   b) Use the function of part (a) to determine the pressure at a depth of 690 ft.

Without graphing, tell whether the graphs of each pair of equations are parallel.

57. $x + 8 = y,$
   $y - x = -5$

58. $2x - 3 = y,$
   $y - 2x = 9$

59. $y + 9 = 3x,$
   $3x - y = -2$

60. $y + 8 = -6x,$
   $-2x + y = 5$

61. $f(x) = 3x + 9,$
   $2y = 8x - 2$

62. $f(x) = -7x - 9,$
   $-3y = 21x + 7$

Write an equation of the line containing the specified point and parallel to the indicated line.

63. $(4, 7),\ x + 2y = 6$

64. $(1, 3),\ 3x - y = 7$

65. $(2, -6),\ 5x - 3y = 8$

66. $(-4, -5),\ 2x + y = -3$

Aha!   **67.** $(0, -7),\ y = 2x + 1$

68. $(0, 4),\ y = -x - 5$

69. $(-2, -3),\ 2x + 3y = -7$

70. $(3, -4),\ 5x - 6y = 4$

71. $(-6, 2),\ 3x - 9y = 2$

72. $(-7, 0),\ 5x + 2y = 6$

73. $(5, -4),\ x = 2$

74. $(-3, 6),\ y = 7$

Without graphing, tell whether the graphs of each pair of equations are perpendicular.

75. $f(x) = 4x - 3,$
   $4y = 7 - x$

76. $2x - 5y = -3,$
   $2x + 5y = 4$

77. $x + 2y = 7,$
   $2x + 4y = 4$

78. $y = -x + 7,$
   $f(x) = x + 3$

Write an equation of the line containing the specified point and perpendicular to the indicated line.

79. $(2, 5),\ x + 2y = -7$

80. $(4, 0),\ x - 3y = 0$

81. $(3, -2),\ 3x + 6y = 5$

82. $(-3, -5),\ 4x - 2y = 4$

83. $(0, 9),\ 2x + 5y = 7$

84. $(-3, -4),\ -7x + 6y = 2$

85. $(-4, -7),\ 3x - 5y = 6$

86. $(-4, 5),\ 7x - 2y = 1$

Aha!   **87.** $(0, 6),\ 2x - 5 = y$

88. $(0, -7),\ 4x + 3 = y$

89. $(-3, 7),\ y = 5$

90. $(4, -2),\ x = 1$

91. Rosewood Graphics recently promised its employees 6% raises each year for the next five years. Amy currently earns $30,000 a year. Can she use a linear function to predict her salary for the next five years? Why or why not?

92. If two lines are perpendicular, does it follow that the lines have slopes that are negative reciprocals of each other? Why or why not?

SKILL MAINTENANCE

Simplify. [1.3]

93. $(3x^2 + 5x) + (2x - 4)$

94. $(5t^2 - 2t) - (4t + 3)$

Evaluate.

95. $\dfrac{2t - 6}{4t + 1}$, for $t = 3$ [1.1]

96. $(3x - 1)(4t + 20)$, for $t = -5$ [1.2]

97. $2x - 5y$, for $x = 3$ and $y = -1$ [1.2]

98. $3x - 7y$, for $x = -2$ and $y = 4$ [1.2]

SYNTHESIS

99. In 1998, Political Action Committees contributed $206.8 million to congressional candidates. Does this information make your answer to Exercise 49(b) seem too low or too high? Why?

100. On the basis of your answers to Exercises 47 and 48, would you predict that at some point in the future the life expectancy of males will exceed that of females? Why or why not?

For Exercises 101–105, assume that a linear equation models each situation.

101. *Temperature conversion.* Water freezes at 32° Fahrenheit and at 0° Celsius. Water boils at 212°F and at 100°C. What Celsius temperature corresponds to a room temperature of 70°F?

102. *Depreciation of a computer.* After 6 mos of use, the value of Pearl's computer had dropped to $900. After 8 mos, the value had gone down to $750. How much did the computer cost originally?

103. *Cell-phone charges.* The total cost of Mel's cell-phone was $230 after 5 mos of service and $390 after 9 mos. What costs had Mel already incurred when his service just began? Assume that Mel's monthly charge is constant.

104. *Operating expenses.* The total cost for operating Ming's Wings was $7500 after 4 mos and $9250 after 7 mos. Predict the total cost after 10 mos.

105. Based on the information given in Exercises 50 and 52, at what price will the supply equal the demand?

106. Specify the domain of your answer to Exercise 50(a).

107. Specify the domain of your answer to Exercise 52(a).

108. For a linear function g, $g(3) = -5$ and $g(7) = -1$.
  a) Find an equation for g.
  b) Find $g(-2)$.
  c) Find a such that $g(a) = 75$.

109. Find the value of k such that the graph of $5y - kx = 7$ and the line containing the points $(7, -3)$ and $(-2, 5)$ are parallel.

110. Find the value of k such that the graph of $7y - kx = 9$ and the line containing the points $(2, -1)$ and $(-4, 5)$ are perpendicular.

111. When several data points are available and they appear to be nearly collinear, a procedure known as *linear regression* can be used to find an equation for the line that most closely fits the data.
  a) Use a graphing calculator with a LINEAR REGRESSION option and the table that follows to find a linear function that predicts the wattage of an incandescent (conventional)

lightbulb as a function of the wattage of a Compact Fluorescent (CFL) energy-saving bulb of equivalent brightness.

Energy Conservation

CFL Wattage	Incandescent Equivalent
7 W	25 W
20 W	75 W
23 W	90 W
28 W	110 W
30 W	120 W
40 W	135 W

Source: Westinghouse Lighting Corporation

  b) Use the function from part (a) to estimate the incandescent wattage that is equivalent to a 35-watt CFL bulb. Then compare your answer with the corresponding answer to Exercise 65 in Section 2.2. Which answer seems more reliable? Why?

112. Use linear regression (see Exercise 111) to find a linear function that predicts a woman's life expectancy as a function of the year in which she was born. Then use the function to predict the life expectancy in 2010 and compare this with the corresponding answer to Exercise 48 of this exercise set. Which answer seems more reliable? Why?

Life Expectancy of Women

Year, x	Life Expectancy, y (in years)
1920	54.6
1930	61.6
1940	65.2
1950	71.1
1960	73.1
1970	74.7
1980	77.5
1990	78.8
2000	79.5

Sources: Statistical Abstract of the United States 2002 and The World Almanac 1999

113. Use a graphing calculator with a *squared* window to check your answers to Exercises 75–88.

2.6   The Algebra of Functions

The Sum, Difference, Product, or Quotient of Two Functions   •
Domains and Graphs

We now return to the idea of a function as a machine and examine four ways in which functions can be combined.

The Sum, Difference, Product, or Quotient of Two Functions

Suppose that a is in the domain of two functions, f and g. The input a is paired with $f(a)$ by f and with $g(a)$ by g. The outputs can then be added to get $f(a) + g(a)$.

EXAMPLE 1   Let $f(x) = x + 4$ and $g(x) = x^2 + 1$. Find $f(2) + g(2)$.

Solution   We visualize two function machines. Because 2 is in the domain of each function, we can compute $f(2)$ and $g(2)$.

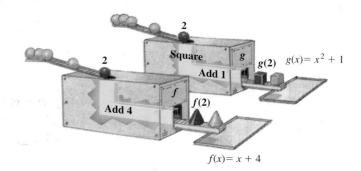

Since

$$f(2) = 2 + 4 = 6 \quad \text{and} \quad g(2) = 2^2 + 1 = 5,$$

we have

$$f(2) + g(2) = 6 + 5 = 11.$$

In Example 1, suppose that we were to write $f(x) + g(x)$ as $(x + 4) + (x^2 + 1)$, or $f(x) + g(x) = x^2 + x + 5$. This could then be regarded as a "new" function. The notation $(f + g)(x)$ is generally used to denote a function formed in this manner. Similar notations exist for subtraction, multiplication, and division of functions.

> **The Algebra of Functions**
>
> If f and g are functions and x is in the domain of both functions, then:
>
> **1.** $(f + g)(x) = f(x) + g(x)$;
> **2.** $(f - g)(x) = f(x) - g(x)$;
> **3.** $(f \cdot g)(x) = f(x) \cdot g(x)$;
> **4.** $(f/g)(x) = f(x)/g(x)$, provided $g(x) \neq 0$.

EXAMPLE 2   For $f(x) = x^2 - x$ and $g(x) = x + 2$, find the following.

a) $(f + g)(3)$

b) $(f - g)(x)$ and $(f - g)(-1)$

c) $(f/g)(x)$ and $(f/g)(-4)$

d) $(f \cdot g)(3)$

Solution

a) Since $f(3) = 3^2 - 3 = 6$ and $g(3) = 3 + 2 = 5$, we have

$$(f + g)(3) = f(3) + g(3)$$
$$= 6 + 5 \qquad \text{Substituting}$$
$$= 11.$$

Alternatively, we could first find $(f + g)(x)$:

$$(f + g)(x) = f(x) + g(x)$$
$$= x^2 - x + x + 2$$
$$= x^2 + 2. \qquad \text{Combining like terms}$$

Thus,

$$(f + g)(3) = 3^2 + 2 = 11. \qquad \text{Our results match.}$$

b) We have

$$(f - g)(x) = f(x) - g(x)$$
$$= x^2 - x - (x + 2) \qquad \text{Substituting}$$
$$= x^2 - 2x - 2. \qquad \begin{array}{l}\text{Removing parentheses and}\\\text{combining like terms}\end{array}$$

Thus,

$$(f - g)(-1) = (-1)^2 - 2(-1) - 2 \qquad \begin{array}{l}\text{Using } (f - g)(x) \text{ is faster than}\\\text{using } f(x) - g(x).\end{array}$$
$$= 1. \qquad \text{Simplifying}$$

c) We have

$$(f/g)(x) = f(x)/g(x)$$
$$= \frac{x^2 - x}{x + 2}. \qquad \text{We assume that } x \neq -2.$$

Thus,

$$(f/g)(-4) = \frac{(-4)^2 - (-4)}{-4 + 2} \qquad \text{Substituting}$$
$$= \frac{20}{-2} = -10.$$

d) Using our work in part (a), we have

$$(f \cdot g)(3) = f(3) \cdot g(3)$$
$$= 6 \cdot 5$$
$$= 30.$$

It is also possible to compute $(f \cdot g)(3)$ by first multiplying $x^2 - x$ and $x + 2$ using methods we will discuss in Chapter 5.

Domains and Graphs

Although applications involving products and quotients of functions rarely appear in newspapers, situations involving sums or differences of functions often do appear in print. For example, the following graphs are similar to those published by the California Department of Education to promote breakfast programs in which students eat a balanced meal of fruit or juice, toast or cereal, and 2% or whole milk. The combination of carbohydrate, protein, and fat gives a sustained release of energy, delaying the onset of hunger for several hours.

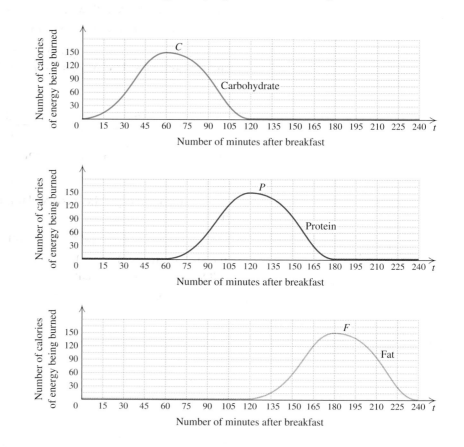

When the three graphs are superimposed, and the calorie expenditures added, it becomes clear that a balanced meal results in a steady, sustained supply of energy.

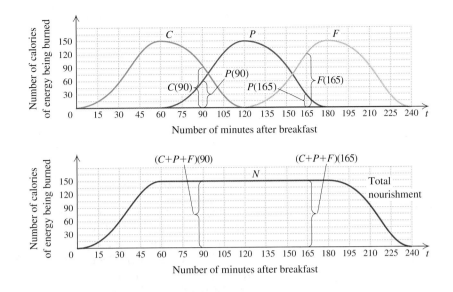

Note that for $t > 120$, we have $C(t) = 0$; for $t < 60$ or $t > 180$, we have $P(t) = 0$; and for $t < 120$, we have $F(t) = 0$. For any point $(t, N(t))$, we have

$$N(t) = (C + P + F)(t) = C(t) + P(t) + F(t).$$

To find $(f + g)(a)$, $(f - g)(a)$, $(f \cdot g)(a)$, or $(f/g)(a)$, we must first be able to find $f(a)$ and $g(a)$. This means a must be in the domain of both f and g.

EXAMPLE 3   Let

$$f(x) = \frac{5}{x} \quad \text{and} \quad g(x) = \frac{2x - 6}{x + 1}.$$

Find the domain of $f + g$, the domain of $f - g$, and the domain of $f \cdot g$.

Solution   Note that because division by 0 is undefined, we have

Domain of $f = \{x \,|\, x \text{ is a real number } and \, x \neq 0\}$

and

Domain of $g = \{x \,|\, x \text{ is a real number } and \, x \neq -1\}$.

In order to find $f(a) + g(a)$, $f(a) - g(a)$, or $f(a) \cdot g(a)$, we must know that a is in *both* of the above domains. Thus,

Domain of $f + g =$ Domain of $f - g =$ Domain of $f \cdot g$

$$= \{x \,|\, x \text{ is a real number } and \, x \neq 0 \, and \, x \neq -1\}.$$

Suppose in Example 3 that we want to find $(f/g)(3)$. Finding $f(3)$ and $g(3)$ poses no problem:

$$f(3) = \frac{5}{3} \quad \text{and} \quad g(3) = \frac{2 \cdot 3 - 6}{3 + 1} = 0;$$

but then

$$(f/g)(3) = f(3)/g(3)$$
$$= \tfrac{5}{3} \, / \, 0 \, . \qquad \text{Division by 0 is undefined.}$$

Thus, although 3 is in the domain of both f and g, it is not in the domain of f/g.

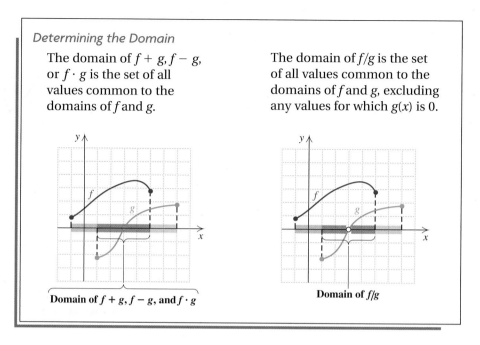

Determining the Domain

The domain of $f + g, f - g,$ or $f \cdot g$ is the set of all values common to the domains of f and g.

The domain of f/g is the set of all values common to the domains of f and g, excluding any values for which $g(x)$ is 0.

Domain of $f + g, f - g,$ and $f \cdot g$

Domain of f/g

EXAMPLE 4

Given $f(x) = 1/x$ and $g(x) = 2x - 7$, find the domains of $f + g, f - g, f \cdot g,$ and f/g.

Solution   The domain of f is $\{x \mid x \neq 0\}$ or $\{x \mid x$ is a real number *and* $x \neq 0\}$. The domain of g is $\mathbb{R}$. Thus the domains of $f + g, f - g,$ and $f \cdot g$ are the set of all elements common to both the domain of f and the domain of g. We have

the domain of $f + g =$ the domain of $f - g =$ the domain of $f \cdot g$

$$= \{x \mid x \text{ is a real number } and \ x \neq 0\}.$$

Student Notes _____

The concern over a denominator being 0 arises throughout this course. Try to develop the habit of checking for any possible input-values that would create a denominator of 0 whenever you work with functions.

To find the domain of f/g, note that

$$(f/g)(x) = \frac{f(x)}{g(x)} = \frac{1/x}{2x - 7} \quad \text{cannot be evaluated if } x = 0 \text{ or if } 2x - 7 = 0.$$

This means the domain of f/g is $\{x \mid x$ is a real number *and* $x \neq 0\}$, *with the additional restriction* that $g(x) \neq 0$. To determine what x-values would make $g(x) = 0$, we solve:

$$2x - 7 = 0 \qquad \text{Replacing } g(x) \text{ with } 2x - 7$$
$$2x = 7$$
$$x = \tfrac{7}{2}.$$

Since $g(x) = 0$ for $x = \tfrac{7}{2}$,

the domain of $f/g = \left\{x \mid x \text{ is a real number } and \ x \neq 0 \ and \ x \neq \tfrac{7}{2}\right\}.$

---

technology connection

A partial check of Example 4 can be performed by setting up a table so the TABLE MINIMUM is 0 and the increment of change (ΔTbl) is 0.7. (Other choices, like 0.1, will also work.) Next, we let $y_1 = 1/x$ and $y_2 = 2x - 7$. Using the Y-VARS key to write $y_3 = y_1 + y_2$ and $y_4 = y_1/y_2$, we can create the table of values shown here. Note that when x is 3.5, a value for y_3 can be found, but y_4 is undefined. When setting up these functions, you may wish to enter y_1 and y_2 without "selecting" either. Otherwise, the table's columns must be scrolled to display y_3 and y_4.

X	Y₃	Y₄
0	ERROR	ERROR
.7	−4.171	−.2551
1.4	−3.486	−.1701
2.1	−2.324	−.1701
2.8	−1.043	−.2551
3.5	.28571	ERROR
4.2	1.6381	.17007
X = 0		

Use a similar approach to partially check Example 3.

Division by 0 is not the only condition that can force restrictions on the domain of a function. In Chapter 7, we will examine functions similar to that given by $f(x) = \sqrt{x}$, for which the concern is taking the square root of a negative number.

Exercise Set

2.6

FOR EXTRA HELP

Student's Solutions Manual · Digital Video Tutor CD 2 Videotape 4 · Tutor Center AW Math Tutor Center · MathXL Tutorials on CD · MathXL MathXL · MyMathLab MyMathLab

---

✎ *Concept Reinforcement*   *Make each of the following sentences true by selecting the correct word for each blank.*

1. If f and g are functions and x is in the _____ of both functions, then
range/domain
$(f + g)(x) = f(x) + g(x)$.

2. One way to compute $(f - g)(2)$ is to _____ $g(2)$ from $f(2)$.
erase/subtract

3. One way to compute $(f - g)(2)$ is to simplify $f(x) - g(x)$ and then _____ the
evaluate/substitute
result for $x = 2$.

4. The domain of $f + g$, $f - g$, and $f \cdot g$ is the set of all values common to the _____
domains/ranges
of f and g.

5. The domain of f/g is the set of all values common to the domains of f and g, _____
including/excluding
any values for which $g(x)$ is 0.

6. The height of $(f + g)(a)$ on a graph is the _____ of the heights of $f(a)$ and $g(a)$.
product/sum

Let $f(x) = -3x + 1$ and $g(x) = x^2 + 2$. Find the following.

7. $f(2) + g(2)$

8. $f(-1) + g(-1)$

9. $f(5) - g(5)$

10. $f(4) - g(4)$

11. $f(-1) \cdot g(-1)$

12. $f(-2) \cdot g(-2)$

13. $f(-4)/g(-4)$

14. $f(3)/g(3)$

15. $g(1) - f(1)$

16. $g(2)/f(2)$

17. $(f + g)(x)$

18. $(g - f)(x)$

Let $F(x) = x^2 - 2$ and $G(x) = 5 - x$. Find the following.

19. $(F + G)(x)$

20. $(F + G)(a)$

21. $(F + G)(-4)$

22. $(F + G)(-5)$

23. $(F - G)(3)$

24. $(F - G)(2)$

25. $(F \cdot G)(-3)$

26. $(F \cdot G)(-4)$

27. $(F/G)(x)$

28. $(G - F)(x)$

29. $(F/G)(-2)$

30. $(F/G)(-1)$

In 2004, a study comparing high doses of the cholesterol-lowering drugs Lipitor and Pravachol indicated that patients taking Lipitor were significantly less likely to have heart attacks or require angioplasty or surgery.

In the graph below, L(t) is the percentage of patients on Lipitor (80 mg) and P(t) is the percentage of patients on Pravachol (40 mg) who suffered heart problems or death t years after beginning to take the medication.
Source: *New York Times*, March 9, 2004

31. Use estimates of $P(2)$ and $L(2)$ to estimate $(P - L)(2)$.

32. Use estimates of $P(1)$ and $L(1)$ to estimate $(P - L)(1)$.

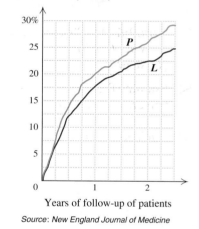

Years of follow-up of patients

Source: New England Journal of Medicine

The following graph shows the number of women, in millions, who had a child the previous year. Here W(t) represents the number of women under 30 who gave birth in year t, R(t) the number of women 30 and older who gave birth in year t, and N(t) the total number of women who gave birth in year t.

33. Use estimates of $R(2000)$ and $W(2000)$ to estimate $N(2000)$.

34. Use estimates of $R(1990)$ and $W(1990)$ to estimate $N(1990)$.

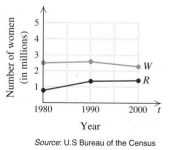

Source: U.S Bureau of the Census

Often function addition is represented by stacking the individual functions directly on top of each other. The graph below indicates how the three major airports servicing New York City have been utilized. The braces indicate the values of the individual functions.

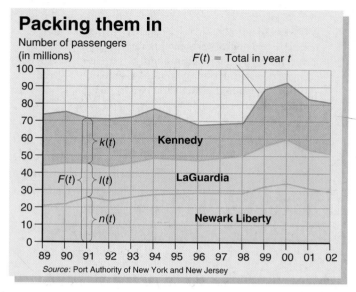

Packing them in

Number of passengers (in millions)

$F(t) = $ Total in year t

Source: Port Authority of New York and New Jersey

35. Estimate $(n + l)('98)$. What does it represent?

36. Estimate $(k + l)('98)$. What does it represent?

37. Estimate $F('02)$. What does it represent?

38. Estimate $F('01)$. What does it represent?

39. Estimate $(F - k)('02)$. What does it represent?

40. Estimate $(F - k)('01)$. What does it represent?

For each pair of functions f and g, determine the domain of the sum, difference, and product of the two functions.

41. $f(x) = x^2,$
$g(x) = 7x - 4$

42. $f(x) = 5x - 1,$
$g(x) = 2x^2$

43. $f(x) = \dfrac{1}{x - 3},$
$g(x) = 4x^3$

44. $f(x) = 3x^2,$
$g(x) = \dfrac{1}{x - 9}$

45. $f(x) = \dfrac{2}{x},$
$g(x) = x^2 - 4$

46. $f(x) = x^3 + 1,$
$g(x) = \dfrac{5}{x}$

47. $f(x) = x + \dfrac{2}{x - 1}$,

$g(x) = 3x^3$

48. $f(x) = 9 - x^2$,

$g(x) = \dfrac{3}{x - 6} + 2x$

49. $f(x) = \dfrac{3}{x - 2}$,

$g(x) = \dfrac{5}{4 - x}$

50. $f(x) = \dfrac{5}{x - 3}$,

$g(x) = \dfrac{1}{x - 2}$

For each pair of functions f and g, determine the domain of f/g.

51. $f(x) = x^4$,

$g(x) = x - 3$

52. $f(x) = 2x^3$,

$g(x) = 5 - x$

53. $f(x) = 3x - 2$,

$g(x) = 2x - 8$

54. $f(x) = 5 + x$,

$g(x) = 6 - 2x$

55. $f(x) = \dfrac{3}{x - 4}$,

$g(x) = 5 - x$

56. $f(x) = \dfrac{1}{2 - x}$,

$g(x) = 7 - x$

57. $f(x) = \dfrac{2x}{x + 1}$,

$g(x) = 2x + 5$

58. $f(x) = \dfrac{7x}{x - 2}$,

$g(x) = 3x + 7$

For Exercises 59–66, consider the functions F and G as shown.

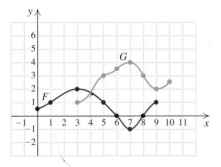

59. Determine $(F + G)(5)$ and $(F + G)(7)$.

60. Determine $(F \cdot G)(6)$ and $(F \cdot G)(9)$.

61. Determine $(G - F)(7)$ and $(G - F)(3)$.

62. Determine $(F/G)(3)$ and $(F/G)(7)$.

63. Find the domains of F, G, $F + G$, and F/G.

64. Find the domains of $F - G$, $F \cdot G$, and G/F.

65. Graph $F + G$.

66. Graph $G - F$.

In the following graph, W(t) represents the number of gallons of whole milk, L(t) the number of gallons of lowfat milk, and S(t) the number of gallons of skim milk consumed by the average American in year t.*

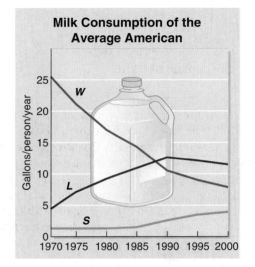

67. From 1970 to 2000, did American milk consumption increase or decrease? Explain how you determined this.

68. Examine the graphs in Exercises 35–40. To what do you attribute the decline in $F(t)$ from 2000 to 2002?

SKILL MAINTENANCE

Solve. [1.5]

69. $4x - 7y = 8$, for x

70. $3x - 8y = 5$, for y

71. $5x + 2y = -3$, for y

72. $6x + 5y = -2$, for x

Translate each of the following. Do not solve. [1.4]

73. Five more than twice a number is 49.

74. Three less than half of some number is 57.

75. The sum of two consecutive integers is 145.

76. The difference between a number and its opposite is 20.

---

*Sources: Copyright 1990, CSPI. Adapted from *Nutrition Action Healthletter* (1875 Connecticut Avenue, N.W., Suite 300, Washington, DC 20009-5728. $24.00 for 10 issues); USDA Agricultural Fact Book 2000, USDA Economic Research Service.

SYNTHESIS

77. If $f(x) = c$, where c is some positive constant, describe how the graphs of $y = g(x)$ and $y = (f + g)(x)$ will differ.

78. Examine the graphs following Example 2 and explain how they might be modified to represent the absorption of 200 mg of Advil® taken four times a day.

79. Find the domain of f/g, if
$$f(x) = \frac{3x}{2x + 5} \quad \text{and} \quad g(x) = \frac{x^4 - 1}{3x + 9}.$$

80. Find the domain of F/G, if
$$F(x) = \frac{1}{x - 4} \quad \text{and} \quad G(x) = \frac{x^2 - 4}{x - 3}.$$

81. Sketch the graph of two functions f and g such that the domain of f/g is
$$\{x \mid -2 \le x \le 3 \text{ and } x \ne 1\}.$$

82. Find the domains of $f + g$, $f - g$, $f \cdot g$, and f/g, if
$$f = \{(-2, 1), (-1, 2), (0, 3), (1, 4), (2, 5)\}$$
and
$$g = \{(-4, 4), (-3, 3), (-2, 4), (-1, 0), (0, 5), (1, 6)\}.$$

83. Find the domain of m/n, if
$$m(x) = 3x \text{ for } -1 < x < 5$$
and
$$n(x) = 2x - 3.$$

84. For f and g as defined in Exercise 82, find $(f + g)(-2)$, $(f \cdot g)(0)$, and $(f/g)(1)$.

85. Write equations for two functions f and g such that the domain of $f + g$ is
$$\{x \mid x \text{ is a real number } and \ x \ne -2 \ and \ x \ne 5\}.$$

86. Let $y_1 = 2.5x + 1.5$, $y_2 = x - 3$, and $y_3 = y_1/y_2$. Depending on whether the CONNECTED or DOT mode is used, the graph of y_3 appears as follows. Use algebra to determine which graph more accurately represents y_3.

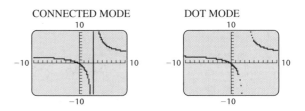

87. Using the window $[-5, 5, -1, 9]$, graph $y_1 = 5$, $y_2 = x + 2$, and $y_3 = \sqrt{x}$. Then predict what shape the graphs of $y_1 + y_2$, $y_1 + y_3$, and $y_2 + y_3$ will take. Use a graphing calculator to check each prediction.

88. Use the TABLE feature on a graphing calculator to check your answers to Exercises 45, 47, 55, and 57. (See the Technology Connection on p. 139.)

CORNER

Time On Your Hands

COLLABORATIVE

Focus: The algebra of functions

Time: 10–15 minutes

Group size: 2–3

The graph and the data at right chart the average retirement age $R(x)$ and life expectancy $E(x)$ of U.S. citizens in year x.

ACTIVITY

1. Working as a team, perform the appropriate calculations and then graph $E - R$.

2. What does $(E - R)(x)$ represent? In what fields of study or business might the function $E - R$ prove useful?

3. Should E and R really be calculated separately for men and women? Why or why not?

4. What advice would you give to someone considering early retirement?

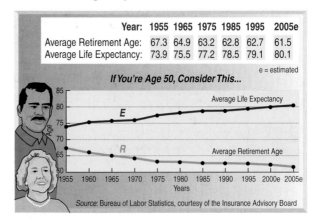

Year:	1955	1965	1975	1985	1995	2005e
Average Retirement Age:	67.3	64.9	63.2	62.8	62.7	61.5
Average Life Expectancy:	73.9	75.5	77.2	78.5	79.1	80.1

e = estimated

If You're Age 50, Consider This...

Source: Bureau of Labor Statistics, courtesy of the Insurance Advisory Board

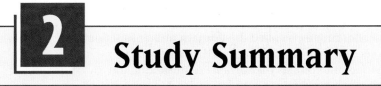

Study Summary

Our study of graphing began by developing the **Cartesian coordinate system,** which is formed by drawing two **axes** that intersect at a right angle (p. 76). **Ordered pairs** can then be graphed by reading their **coordinates** and locating the appropriate points on the plane (p. 77). The axes form four **quadrants,** as shown below.

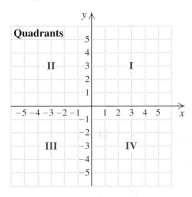

Equations containing two variables, like $2x + 3y = 9$ or $y = 2x^2$, have too many solutions to list. By graphing the solutions of such equations, we can represent the solutions visually. Equations like $2x + 3y = 9$, $y = -4x + 1$, or $y - 5 = \frac{1}{2}(x + 6)$ are classified as **linear** because their graphs are lines (p. 80). Equations like $y = x^2$ or $y = 1/x$ are classified as **nonlinear** because their graphs are not lines (p. 80). Linear equations are usually written one of three ways:

Slope–intercept form: $y = mx + b$ (p. 102);

Point–slope form: $y - y_1 = m(x - x_1)$ (p. 125);

Standard form: $Ax + By = C$ (p. 120).

An important characteristic of any nonvertical line is its **slope** (p. 101). A line's slope measures the rate at which the vertical coordinates change with respect to their respective horizontal coordinates:

$$\text{Slope} = m = \frac{\text{rise}}{\text{run}} = \frac{\text{difference in } y}{\text{difference in } x} = \frac{y_2 - y_1}{x_2 - x_1}.$$

Graphs provide a useful way to represent **functions** (p. 85). A function is a set of ordered pairs for which no two different ordered pairs have the same first coordinate. The **vertical-line test** provides a quick way to determine whether a graph represents a function: If no vertical line crosses the graph more than once, then the graph is a function. A **linear function** is a function that has a straight line as its graph (p. 99). A **constant function** is given by an equation of the form $f(x) = c$ and has a horizontal line as its graph (p. 113). All horizontal lines have a **slope of 0**, whereas the slope of a vertical line is **undefined** (pp. 113–114).

The set of all of a function's first coordinates is called the **domain** of that function (p. 85). The set of all of a function's second coordinates is called the **range** of the function (p. 85). When an equation of a function has the variable representing the second coordinate alone on one side of an equation, it appears as the **dependent variable** (p. 90). This means that its value is calculated after a choice has been made for the replacement of the other, **independent**, **variable** (p. 90). Numbers replacing the independent variable are often called **inputs** and the resulting values of the dependent variable are often called **outputs** (p. 89).

Linear functions are often used to estimate the coordinates of points that appear between known data points. This process is known as **interpolation** (p. 91). When a linear function is used to predict future coordinates that appear beyond known data points, we have what is known as **extrapolation** (p. 91).

When the graphs of two linear functions are **parallel** to each other, their slopes are equal (p. 127). When the graphs of two linear functions are **perpendicular** to each other, the product of their slopes is -1 (p. 128).

Functions can be added, subtracted, multiplied, and divided as follows:

1. $(f + g)(x) = f(x) + g(x)$
2. $(f - g)(x) = f(x) - g(x)$
3. $(f \cdot g)(x) = f(x) \cdot g(x)$
4. $(f/g)(x) = f(x)/g(x)$, provided $g(x) \neq 0$ (p. 135).

The domain of $f + g, f - g$, and $f \cdot g$ is the set of all values common to the domains of f and g (p. 138).

The domain of f/g is the set of all values common to the domains of f and g, excluding any values for which $g(x)$ is 0 (p. 138).

2 Review Exercises

↬ *Concept Reinforcement*   *Classify each statement as either true or false.*

1. A line's slope is a measure of how the line is slanted or tilted. [2.3]

2. Every line has a y-intercept. [2.4]

3. Every line has an x-intercept. [2.4]

4. No member of a function's range can be used in two different ordered pairs. [2.2]

5. The horizontal-line test is a quick way to determine whether a graph represents a function. [2.2]

6. The slope of a vertical line is undefined. [2.4]

7. The slope of the graph of a constant function is 0. [2.4]

8. Extrapolation is done to predict future values. [2.2]

9. $(f + g)(x) = f(x) + g(x)$ is not an example of the distributive law when f and g are functions. [2.6]

10. In order for $(f/g)(a)$ to exist, we must have $g(a) \neq 0$. [2.6]

Determine whether the ordered pair is a solution of the given equation. [2.1]

11. $(3, 7),\ 4p - q = 5$

12. $(-2, 8),\ x = 2y + 12$

13. $\left(0, -\frac{1}{2}\right),\ 3a - 4b = 2$

14. $(8, -2),\ 3c + 2d = 28$

Graph.

15. $y = -3x + 2$ [2.1], [2.3]

16. $y = 6$ [2.4]

17. $y + 1 = \frac{3}{4}(x - 5)$ [2.5]

18. $8x + 32 = 0$ [2.4]

19. $y = -x^2 + 1$ [2.1]

20. For the following graph of f, determine **(a)** $f(2)$; **(b)** the domain of f; **(c)** any x-values for which $f(x) = 2$; and **(d)** the range of f. [2.2]

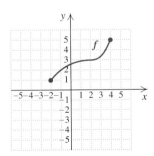

21. The function $A(t) = 0.233t + 5.87$ can be used to estimate the median age of cars in the United States t years after 1990. (In this context, a median age of 3 yr means that half the cars are more than 3 yr old and half are less.) Predict the median age of cars in 2010. [2.2]
Source: The Polk Co.

Find the slope and the y-intercept. Then draw the graph. [2.3]

22. $g(x) = -4x - 9$

23. $-6y + 2x = 14$

24. Find the rate of change for the graph shown. Be sure to use appropriate units. [2.3]

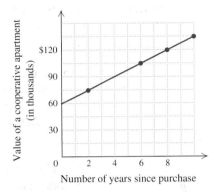

Number of years since purchase

The following table shows the U.S. minimum hourly wage. [2.2]

Input, Year	Output, U.S. Minimum Hourly Wage
1955	$0.75
1980	3.10
2004	5.15

25. Use the data in the table to draw a graph and to estimate the U.S. minimum hourly wage in 1985. [2.2]

26. Use the graph from Exercise 25 to estimate the U.S. hourly minimum wage in 2009. [2.2]

Find the slope of each line. If the slope is undefined, state this.

27. Containing the points $(4, 5)$ and $(-3, 1)$ [2.3]

28. Containing the points $(-16.4, 2.8)$ and $(-16.4, 3.5)$ [2.4]

29. By March 1, 2003, U.S. builders had begun construction of 227,500 homes. By July 1, 2003, the year-to-date number of homes started had grown to 865,400. Calculate the rate at which new homes were being started. [2.3]
Source: www.census.gov

30. The average cost of tuition at a state university t years after 1997 can be estimated by $C(t) = 645t + 9800$. What do the numbers 645 and 9800 signify? [2.3]

31. Find a linear function whose graph has slope $\frac{2}{7}$ and y-intercept $(0, -6)$. [2.3]

32. Graph using intercepts: $-2x + 4y = 8$. [2.4]

33. Solve $2 - x = 4 + x$ graphically. Then check your answer by solving the equation algebraically. [2.4]

34. To join the Family Fitness Center, it costs $75 plus $15 a month. Use a graph to estimate the time required for the total cost to reach $180. [2.4]

Determine whether each of these is a linear equation. [2.4]

35. $2x - 7 = 0$ **36.** $3x - 8f(x) = 7$

37. $2a + 7b^2 = 3$ **38.** $2p - \dfrac{7}{q} = 1$

39. Find an equation in point–slope form of the line with slope -2 and containing $(-3, 4)$. [2.5]

40. Using function notation, write a slope–intercept equation for the line containing $(2, 5)$ and $(-4, -3)$. [2.5]

Determine whether each pair of lines is parallel, perpendicular, or neither. [2.5]

41. $y + 5 = -x$,
    $x - y = 2$

42. $3x - 5 = 7y$,
    $7y - 3x = 7$

43. *Records in the 200-meter run.* In 1983, the record for the 200-m run was 19.75 sec.* In 2003, it was 19.32 sec. Let $R(t)$ represent the record in the 200-m run and t the number of years since 1983. [2.5]
Source: runnersworld.com

   a) Find a linear function that fits the data.
   b) Use the function of part (a) to predict the record in 2008 and in 2013.

Find an equation of the line. [2.5]

44. Containing the point $(2, -5)$ and parallel to the line $3x - 5y = 9$

45. Containing the point $(2, -5)$ and perpendicular to the line $3x - 5y = 9$

Let $g(x) = 3x - 6$ and $h(x) = x^2 + 1$. Find the following.

46. $g(0)$ [2.2]   **47.** $h(-5)$ [2.2]

48. $(g \cdot h)(4)$ [2.6]   **49.** $(g - h)(-2)$ [2.6]

50. $(g/h)(-1)$ [2.6]   **51.** $g(a + b)$ [2.2]

52. The domains of $g + h$ and $g \cdot h$ [2.6]

53. The domain of h/g [2.6]

SYNTHESIS

54. Explain why every function is a relation, but not every relation is a function. [2.2]

55. Explain why the slope of a vertical line is undefined whereas the slope of a horizontal line is 0. [2.4]

56. Find the y-intercept of the function given by
$$f(x) + 3 = 0.17x^2 + (5 - 2x)^x - 7. \; [1.6], [2.4]$$

57. Determine the value of a such that the lines
$$3x - 4y = 12 \quad \text{and} \quad ax + 6y = -9$$
are parallel. [2.5]

58. Homespun Jellies charges $2.49 for each jar of preserves. Shipping charges are $3.75 for handling, plus $0.60 per jar. Find a linear function for determining the cost of x jars of preserves, including shipping and handling. [2.5]

_____

*Records are for elevations less than 1000 m.

59. Match each sentence with the most appropriate of the four graphs at right. [2.3]

a) Joni walks for 10 min to the train station, rides the train for 15 min, and then walks 5 min to the office.

b) During a workout, Phil bikes for 10 min, runs for 15 min, and then walks for 5 min.

c) Sam pilots his motorboat for 10 min to the middle of the lake, fishes for 15 min, and then motors for another 5 min to another spot.

d) Patti waits 10 min for her train, rides the train for 15 min, and then runs for 5 min to her job.

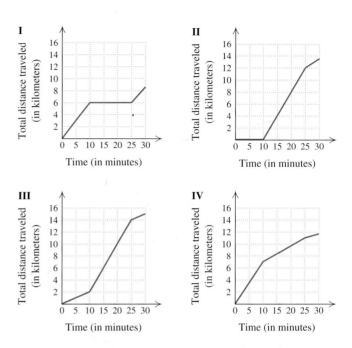

2 Chapter Test

Determine whether the ordered pair is a solution of the given equation.

1. $(12, -3)$, $x + 4y = -20$

2. $(1, 4)$, $-2p + 5q = 18$

Graph.

3. $y = -3x + 4$

4. $f(x) = x^2 + 3$

5. $y - 1 = -\frac{1}{2}(x + 4)$

6. $3 - x = 9$

7. For the following graph of f, determine **(a)** $f(-2)$; **(b)** the domain of f; **(c)** any x-value for which $f(x) = \frac{1}{2}$; and **(d)** the range of f.

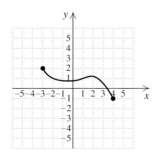

8. The function $S(t) = 1.7t + 26.2$ can be used to estimate the total U.S. sales of books, in billions of dollars, t years after 1995.
Source: Based on data from *Statistical Abstract of the United States*, 2002

a) Predict the total U.S. sales of books in 2009.

b) What do the numbers 1.7 and 26.2 signify?

9. There were 48.5 million international visitors to the United States in 1999 and 41.9 million in 2002. Draw a graph and estimate the number of international visitors in 2001.
Source: Tourism Industries, International Trade Administration, U.S. Department of Commerce

Find the slope and the y-intercept.

10. $f(x) = -\frac{3}{5}x + 12$

11. $-5y - 2x = 7$

Find the slope of the line containing the following points. If the slope is undefined, state this.

12. $(-2, -2)$ and $(6, 3)$

13. $(-3.1, 5.2)$ and $(-4.4, 5.2)$

14. Find the rate of change for the graph below. Use appropriate units.

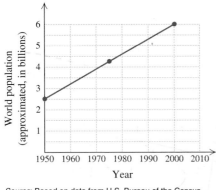

Source: Based on data from U.S. Bureau of the Census

15. Find a linear function whose graph has slope -5 and y-intercept $(0, -1)$.

16. Graph using intercepts: $-2x + 5y = 12$.

17. Solve $x + 3 = 2x$ graphically. Then check your answer by solving the equation algebraically.

18. Which of these are linear equations?
   a) $8x - 7 = 0$
   b) $4b - 9a^2 = 2$
   c) $2x - 5y = 3$

19. Find an equation in point–slope form of the line with slope 4 and containing $(-2, -4)$.

20. Using function notation, write a slope–intercept equation for the line containing $(3, -1)$ and $(4, -2)$.

Determine without graphing whether each pair of lines is parallel, perpendicular, or neither.

21. $4y + 2 = 3x$,
$-3x + 4y = -12$

22. $y = -2x + 5$,
$2y - x = 6$

Find an equation of the line.

23. Containing $(-3, 2)$ and parallel to the line $2x - 5y = 8$

24. Containing $(-3, 2)$ and perpendicular to the line $2x - 5y = 8$

25. Find the following, given that $g(x) = -3x - 4$ and $h(x) = x^2 + 1$.
   a) $h(-2)$
   b) $(g \cdot h)(3)$
   c) The domain of h/g

26. If you rent a truck for one day and drive it 250 mi, the cost is \$100. If you drive it 300 mi, the cost is \$115. Let $C(m)$ represent the cost, in dollars, of driving m miles.
   a) Find a linear function that fits the data.
   b) Use the function to determine how much it will cost to rent the truck for one day and drive it 500 mi.

SYNTHESIS

27. The function $f(t) = 5 + 15t$ can be used to determine a bicycle racer's location, in miles from the starting line, measured t hours after passing the 5-mi mark.
   a) How far from the start will the racer be 1 hr and 40 min after passing the 5-mi mark?
   b) Assuming a constant rate, how fast is the racer traveling?

28. The graph of the function $f(x) = mx + b$ contains the points $(r, 3)$ and $(7, s)$. Express s in terms of r if the graph is parallel to the line $3x - 2y = 7$.

29. Given that $f(x) = 5x^2 + 1$ and $g(x) = 4x - 3$, find an expression for $h(x)$ so that the domain of $f/g/h$ is $\left\{x \mid x \text{ is a real number } and \ x \neq \frac{3}{4} and \ x \neq \frac{2}{7}\right\}$.

Answers may vary.

3

Systems of Linear Equations and Problem Solving

AN APPLICATION

King Street Printing recently charged 1.9¢ per sheet of paper, but 2.4¢ per sheet for paper made of recycled fibers. Darren's bill for 150 sheets of paper was $3.41. How many sheets of each type were used?

This problem appears as Exercise 15 in Section 3.3.

Nancy Plunkett
WASTE REDUCTION MANAGER
Williston, Vermont

I regularly use algebra whenever I create spreadsheets to track the costs and performance of recycling and composting programs and to compile survey data. I also use it to convert volumes of different materials to weight and to calculate rates of change in program participation.

*T*he most difficult part of problem solving is almost always translating the problem situation to mathematical language. Once a problem has been translated, the rest is usually straightforward. In this chapter, we study systems of equations *and how to solve them using graphing, substitution, elimination, and matrices. Systems of equations often provide the easiest way to model real-world situations in fields such as psychology, sociology, business, education, engineering, and science.*

3.1 Systems of Equations in Two Variables

Translating • Identifying Solutions • Solving Systems Graphically

Translating

Problems involving two unknown quantities are often solved most easily if we can first translate the situation to two equations in two unknowns.

EXAMPLE 1

Bottled-water consumption. Americans are buying more bottled water than ever before. At the time of this writing, the average American buys more bottled water than milk, coffee, or beer. In 2003, the average American purchased 76.4 gal of soft drinks and water. The amount of water was only 4.3 gal less than half of the amount of soft drinks. (*Source*: Based on data from www.beveragemarketing.com) How many gallons of water and how many of soft drinks did the average American buy in 2003?

Solution

1. Familiarize. We have already seen problems in which we need to look up certain formulas or the meaning of certain words. Here we simply observe that the words *per person* mean the same thing as the amount consumed by the *average American.*

Often problems contain information that has no bearing on the situation being discussed. In this case, the fact that more water is bought than milk, coffee, or beer is irrelevant to the question being asked. Instead we focus on the amounts of water and soft drinks consumed. Rather than guess and check, let's proceed to the next step, using w to represent the average number of gallons of bottled water purchased and s for the average number of gallons of soft drinks purchased in 2003.

2. **Translate.** There are two statements to translate. First we look at the total number of gallons purchased by the average American:

Rewording: The amount of $\underbrace{\text{water purchased}}$ plus the amount of $\underbrace{\text{soft drinks purchased}}$ was 76.4.

Translating: w $+$ s $=$ 76.4

The second statement compares the two amounts, w and s:

Rewording: The amount of $\underbrace{\text{water purchased}}$ was $\underbrace{\text{4.3 gal less than half the amount of soft drinks purchased.}}$

Translating: w $=$ $\dfrac{1}{2}s - 4.3$

We have now translated the problem to a pair, or **system, of equations**:

$$w + s = 76.4,$$
$$w = \frac{1}{2}s - 4.3.$$

> **System of Equations**
>
> A *system of equations* is a set of two or more equations, in two or more variables, for which a common solution is sought.

Problems like Example 1 *can* be solved using one variable; however, as problems become complicated, you will find that using more than one variable (and more than one equation) is often the preferable approach.

EXAMPLE 2   Purchasing.   Recently the Woods County Art Center purchased 120 stamps for $33.90. If the stamps were a combination of 23¢ postcard stamps and 37¢ first-class stamps, how many of each type were bought?

Solution

1. **Familiarize.**   To familiarize ourselves with this problem, let's guess that the art center bought 60 stamps at 23¢ each and 60 stamps at 37¢ each. The total cost would then be

$$60 \cdot \$0.23 + 60 \cdot \$0.37 = \$13.80 + \$22.20, \text{ or } \$36.00.$$

Since $\$36.00 \neq \$33.90$, our guess is incorrect. Rather than guess again, let's see how algebra can be used to translate the problem.

2. Translate. We let $p =$ the number of postcard stamps and $f =$ the number of first-class stamps. The information can be organized in a table, which will help with the translating.

Type of Stamp	Postcard	First-class	Total	
Number Sold	p	f	120	→ $p + f = 120$
Price	$0.23	$0.37		
Amount	$0.23p	$0.37f	$33.90	→ $0.23p + 0.37f = 33.90$

The first row of the table and the first sentence of the problem indicate that a total of 120 stamps were bought:

$$p + f = 120.$$

Since each postcard stamp cost $0.23 and p stamps were bought, $0.23p$ represents the amount paid, in dollars, for the postcard stamps. Similarly, $0.37f$ represents the amount paid, in dollars, for the first-class stamps. This leads to a second equation:

$$0.23p + 0.37f = 33.90.$$

Multiplying both sides by 100, we can clear the decimals. This gives the following system of equations as the translation:

$$p + f = 120,$$
$$23p + 37f = 3390.$$

We will complete the solutions of Examples 1 and 2 in Section 3.3.

Identifying Solutions

A *solution* of a system of two equations in two variables is an ordered pair of numbers that makes *both* equations true.

EXAMPLE 3   Determine whether $(-4, 7)$ is a solution of the system

$$x + y = 3,$$
$$5x - y = -27.$$

Solution   As discussed in Chapter 2, unless stated otherwise, we use alphabetical order of the variables. Thus we replace x with -4 and y with 7:

$$\frac{x + y = 3}{-4 + 7 \;\vert\; 3}$$
$$3 \stackrel{?}{=} 3 \quad \text{TRUE}$$

$$\frac{5x - y = -27}{5(-4) - 7 \;\vert\; -27}$$
$$-20 - 7 \;\vert\;$$
$$-27 \stackrel{?}{=} -27 \quad \text{TRUE}$$

The pair $(-4, 7)$ makes both equations true, so it is a solution of the system. We can also describe the solution by writing $x = -4$ and $y = 7$. Set notation can also be used to list the solution set $\{(-4, 7)\}$.

Solving Systems Graphically

Recall that the graph of an equation is a drawing that represents its solution set. If we graph the equations in Example 3, we find that $(-4, 7)$ is the only point common to both lines. Thus one way to solve a system of two equations is to graph both equations and identify any points of intersection. The coordinates of each point of intersection represent a solution of that system.

$$x + y = 3,$$
$$5x - y = -27$$

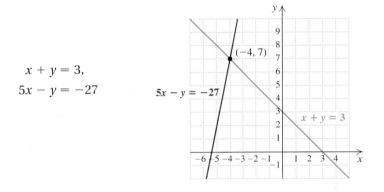

Most pairs of lines have exactly one point in common. We will soon see, however, that this is not always the case.

EXAMPLE 4   Solve each system graphically.

a) $y - x = 1,$
     $y + x = 3$

b) $y = -3x + 5,$
     $y = -3x - 2$

c) $3y - 2x = 6,$
      $-12y + 8x = -24$

Solution

a) We graph each equation using any method studied in Chapter 2. All ordered pairs from line L_1 are solutions of the first equation. All ordered pairs from line L_2 are solutions of the second equation. The point of intersection has coordinates that make *both* equations true. Apparently, $(1, 2)$ is the solution. Graphs are not always accurate, so solving by graphing may yield approximate answers. Our check below shows that $(1, 2)$ is indeed the solution.

$$y - x = 1,$$
$$y + x = 3$$

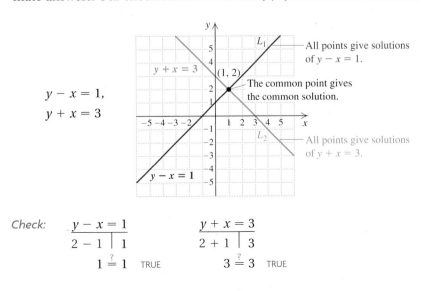

Check:

$$\begin{array}{c|c} y - x = 1 \\ \hline 2 - 1 & 1 \\ 1 \overset{?}{=} 1 & \text{TRUE} \end{array}$$

$$\begin{array}{c|c} y + x = 3 \\ \hline 2 + 1 & 3 \\ 3 \overset{?}{=} 3 & \text{TRUE} \end{array}$$

b) We graph the equations. The lines have the same slope, -3, and different y-intercepts, so they are parallel. There is no point at which they cross, so the system has no solution.

$$y = -3x + 5,$$
$$y = -3x - 2$$

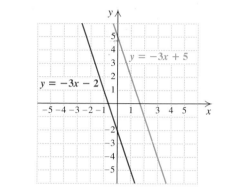

c) We graph the equations and find that the same line is drawn twice. Thus any solution of one equation is a solution of the other. Each equation has an infinite number of solutions, so the system itself has an infinite number of solutions. We check one solution, $(0, 2)$, which is the y-intercept of each equation.

Student Notes _____

Although the system in Example 4(c) is true for an infinite number of ordered pairs, those pairs must be of a certain form. Only pairs that are solutions of $3y - 2x = 6$ or $-12y + 8x = -24$ are solutions of the system. It is incorrect to think that *all* ordered pairs are solutions.

$$3y - 2x = 6,$$
$$-12y + 8x = -24$$

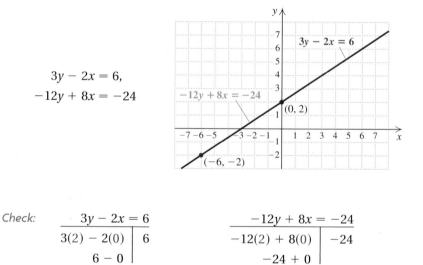

Check:

$3y - 2x = 6$		$-12y + 8x = -24$	
$3(2) - 2(0)$	6	$-12(2) + 8(0)$	-24
$6 - 0$		$-24 + 0$	
$6 \overset{?}{=} 6$	TRUE	$-24 \overset{?}{=} -24$	TRUE

You can check that $(-6, -2)$ is another solution of both equations. In fact, any pair that is a solution of one equation is a solution of the other equation as well. Thus the solution set is

$$\{(x, y) \mid 3y - 2x = 6\}$$

or, in words, "the set of all pairs (x, y) for which $3y - 2x = 6$." Since the two equations are equivalent, we could have written instead $\{(x, y) \mid -12y + 8x = -24\}$.

technology connection

On most graphing calculators, an INTERSECT option allows you to find the coordinates of the intersection directly. This is especially useful when equations contain fractions or decimals or when the coordinates of the intersection are not integers. To illustrate, consider the following system:

$$3.45x + 4.21y = 8.39,$$
$$7.12x - 5.43y = 6.18.$$

After solving for y in each equation, we obtain the graph below. Using INTERSECT, we see that, to the nearest hundredth, the coordinates of the intersection are $(1.47, 0.79)$.

$$y_1 = (8.39 - 3.45x)/4.21,$$
$$y_2 = (6.18 - 7.12x)/(-5.43)$$

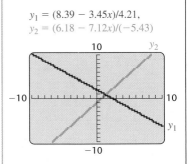

Use a graphing calculator to solve each of the following systems. Make sure that all x- and y-coordinates are correct to the nearest hundredth.

1. $y = -5.43x + 10.89,$
   $y = 6.29x - 7.04$
2. $y = 123.52x + 89.32,$
   $y = -89.22x + 33.76$
3. $2.18x + 7.81y = 13.78,$
   $5.79x - 3.45y = 8.94$
4. $-9.25x - 12.94y = -3.88,$
   $21.83x + 16.33y = 13.69$

When we graph a system of two linear equations in two variables, one of the following three outcomes will occur.

1. The lines have one point in common, and that point is the only solution of the system (see Example 4a). Any system that has *at least* one solution is said to be **consistent**.

2. The lines are parallel, with no point in common, and the system has no solution (see Example 4b). This type of system is called **inconsistent**.

3. The lines coincide, sharing the same graph. Because every solution of one equation is a solution of the other, the system has an infinite number of solutions (see Example 4c). Since it has at least one solution, this type of system is also consistent.

When one equation in a system can be obtained by multiplying both sides of another equation by a constant, the two equations are said to be **dependent**. Thus the equations in Example 4(c) are dependent, but those in Examples 4(a) and 4(b) are **independent**. For systems of three or more equations, the definitions of dependent and independent will be slightly modified.

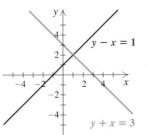
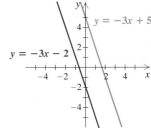
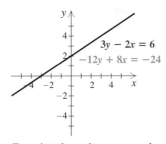

Graphs intersect at one point.
The system is *consistent* and has one solution. Since neither equation is a multiple of the other, they are *independent*.

Graphs are parallel.
The system is *inconsistent* because there is no solution. Since the equations are not equivalent, they are *independent*.

Equations have the same graph.
The system is *consistent* and has an infinite number of solutions. The equations are *dependent* since they are equivalent.

Graphing is helpful when solving systems because it allows us to "see" the solution. It can also be used on systems of nonlinear equations, and in many applications, it provides a satisfactory answer. However, graphing often lacks precision, especially when fraction or decimal solutions are involved. In Section 3.2, we will develop two algebraic methods of solving systems. Both methods produce exact answers.

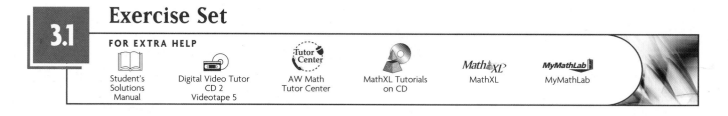

Exercise Set
3.1

FOR EXTRA HELP

Student's Solutions Manual | Digital Video Tutor CD 2 Videotape 5 | Tutor Center AW Math Tutor Center | MathXL Tutorials on CD | MathXL | MyMathLab

↩ *Concept Reinforcement   Classify each statement as either true or false.*

1. Solutions of systems of equations in two variables are ordered pairs.

2. Every system of equations has at least one solution.

3. It is possible for a system of equations to have an infinite number of solutions.

4. The graphs of the equations in a system of two equations may coincide.

5. The graphs of the equations in a system of two equations could be parallel lines.

6. Any system of equations that has at most one solution is said to be consistent.

7. Any system of equations that has more than one solution is said to be inconsistent.

8. If one equation in a system can be obtained by multiplying both sides of another equation in that system by a constant, the two equations are said to be dependent.

Determine whether the ordered pair is a solution of the given system of equations. Remember to use alphabetical order of variables.

9. $(1, 2)$; $4x - y = 2$,
$\qquad 10x - 3y = 4$

10. $(-1, -2)$; $2x + y = -4$,
$\qquad x - y = 1$

11. $(2, 5)$; $y = 3x - 1$,
$\qquad 2x + y = 4$

12. $(-1, -2)$; $x + 3y = -7$,
$\qquad 3x - 2y = 12$

13. $(1, 5)$; $x + y = 6$,
$\qquad y = 2x + 3$

14. $(5, 2)$; $a + b = 7$,
$\qquad 2a - 8 = b$

Aha! 15. $(3, 1)$; $3x + 4y = 13$,
$\qquad 6x + 8y = 26$

16. $(4, -2)$; $-3x - 2y = -8$,
$\qquad 8 = 3x + 2y$

Solve each system graphically. Be sure to check your solution. If a system has an infinite number of solutions, use set-builder notation to write the solution set. If a system has no solution, state this.

17. $x - y = 3$,
$\qquad x + y = 5$

18. $x + y = 4$,
$\qquad x - y = 2$

19. $3x + y = 5$,
$\qquad x - 2y = 4$

20. $2x - y = 4$,
$\qquad 5x - y = 13$

21. $4y = x + 8$,
$\qquad 3x - 2y = 6$

22. $4x - y = 9$,
$\qquad x - 3y = 16$

23. $x = y - 1$,
$\qquad 2x = 3y$

24. $a = 1 + b$,
$\qquad b = 5 - 2a$

25. $x = -3$,
$\qquad y = 2$

26. $x = 4$,
$\qquad y = -5$

27. $t + 2s = -1$,
$\qquad s = t + 10$

28. $b + 2a = 2$,
$\qquad a = -3 - b$

29. $2b + a = 11$,
$\qquad a - b = 5$

30. $y = -\frac{1}{3}x - 1$,
$\qquad 4x - 3y = 18$

31. $y = -\frac{1}{4}x + 1$,
$\qquad 2y = x - 4$

32. $6x - 2y = 2$,
$\qquad 9x - 3y = 1$

33. $y - x = 5$,
$\qquad 2x - 2y = 10$

34. $y = -x - 1$,
$\qquad 4x - 3y = 24$

35. $y = 3 - x$,
$\qquad 2x + 2y = 6$

36. $2x - 3y = 6$,
$\qquad 3y - 2x = -6$

37. For the systems in the odd-numbered exercises 17–35, which are consistent?

38. For the systems in the even-numbered exercises 18–36, which are consistent?

39. For the systems in the odd-numbered exercises 17–35, which contain dependent equations?

40. For the systems in the even-numbered exercises 18–36, which contain dependent equations?

Translate each problem situation to a system of equations. Do not attempt to solve, but save for later use.

41. The sum of two numbers is 50. The first number is 25% of the second number. What are the numbers?

42. The sum of two numbers is 40. The first number is 60% of the second number. What are the numbers?

43. *Nontoxic furniture polish.*   A nontoxic wood furniture polish can be made by mixing mineral (or olive) oil with vinegar. To make a 16-oz batch for a squirt bottle, Mabel uses an amount of mineral oil that is 4 oz more than twice the amount of vinegar. How much of each ingredient is required?
Sources: Based on information from Chittenden Solid Waste District and *Clean House, Clean Planet* by Karen Logan

44. *Scholastic Aptitude Test.*   Many high-school students take the Scholastic Aptitude Test. Each student receives two scores, a *verbal* score and a *math* score. In 2002–2003, the average total score of students was 1026, with the average math score exceeding the verbal score by 12 points. What was the average verbal score and what was the average math score?
Source: College Entrance Examination Board

45. *Geometry.*   Two angles are supplementary.* One angle is 3° less than twice the other. Find the measures of the angles.

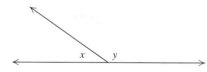
Supplementary angles

46. *Geometry.*   Two angles are complementary.† The sum of the measures of the first angle and half the second angle is 64°. Find the measures of the angles.

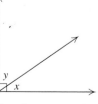
Complementary angles

47. *Basketball scoring.*   Wilt Chamberlain once scored 100 points, setting a record for points scored in an NBA game. Chamberlain took only two-point shots and (one-point) foul shots and made a total of 64 shots. How many shots of each type did he make?

48. *Basketball scoring.*   The Fenton College Cougars made 40 field goals in a recent basketball game, some 2-pointers and the rest 3-pointers. Altogether the 40 baskets counted for 89 points. How many of each type of field goal was made?

49. *Retail sales.*   Paint Town sold 45 paintbrushes, one kind at $8.50 each and another at $9.75 each. In all, $398.75 was taken in for the brushes. How many of each kind were sold?

50. *Retail sales.*   Mountainside Fleece sold 40 neckwarmers. Polarfleece neckwarmers sold for $9.90 each and wool ones sold for $12.75 each. In all, $421.65 was taken in for the neckwarmers. How many of each type were sold?

51. *Sales of pharmaceuticals.*   In 2004, the Diabetic Express charged $27.06 for a vial of Humulin insulin and $34.39 for a vial of Novolin Velosulin insulin. If a total of $1565.57 was collected for 50 vials of insulin, how many vials of each type were sold?

52. *Fundraising.*   The St. Mark's Community Barbecue served 250 dinners. A child's plate cost $3.50 and an adult's plate cost $7.00. A total of $1347.50 was collected. How many of each type of plate was served?

53. *Court dimensions.*   The perimeter of a standard basketball court is 288 ft. The length is 44 ft longer than the width. Find the dimensions.

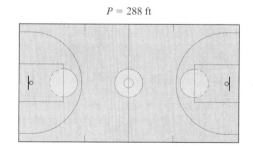
$P = 288$ ft

54. *Court dimensions.*   The perimeter of a standard tennis court used for doubles is 228 ft. The width is 42 ft less than the length. Find the dimensions.

55. Write a problem for a classmate to solve that requires writing a system of two equations. Devise the problem so that the solution is "The Lakers made 6 three-point baskets and 31 two-point baskets."

---

*The sum of the measures of two supplementary angles is 180°.
†The sum of the measures of two complementary angles is 90°.

56. Write a problem for a classmate to solve that can be translated into a system of two equations. Devise the problem so that the solution is "Arnie took five 3-credit classes and two 4-credit classes."

SKILL MAINTENANCE

Solve. [1.3]

57. $2(4x - 3) - 7x = 9$ **58.** $6y - 3(5 - 2y) = 4$

59. $4x - 5x = 8x - 9 + 11x$

60. $8x - 2(5 - x) = 7x + 3$

Solve. [1.5]

61. $3x + 4y = 7$, for y

62. $2x - 5y = 9$, for y

SYNTHESIS

Presidential primaries.   For Exercises 63 and 64, consider the following graph showing the results of a poll in which Iowans were asked which Democrat candidate for president they favored.

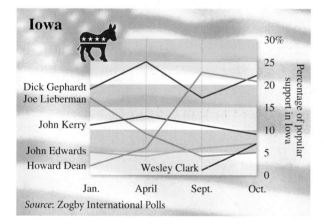

Source: Zogby International Polls

63. At what point in time could it have been said that no one was in fourth place? Explain.

64. At what point in time was there no clear leader? Explain how you reach this conclusion.

65. For each of the following conditions, write a system of equations.
   a) $(5, 1)$ is a solution.
   b) There is no solution.
   c) There is an infinite number of solutions.

66. A system of linear equations has $(1, -1)$ and $(-2, 3)$ as solutions. Determine:
   a) a third point that is a solution, and
   b) how many solutions there are.

67. The solution of the following system is $(4, -5)$. Find A and B.
$$Ax - 6y = 13,$$
$$x - By = -8.$$

Translate to a system of equations. Do not solve.

68. *Ages.*   Burl is twice as old as his son. Ten years ago, Burl was three times as old as his son. How old are they now?

69. *Work experience.*   Lou and Juanita are mathematics professors at a state university. Together, they have 46 years of service. Two years ago, Lou had taught 2.5 times as many years as Juanita. How long has each taught at the university?

70. *Design.*   A piece of posterboard has a perimeter of 156 in. If you cut 6 in. off the width, the length becomes four times the width. What are the dimensions of the original piece of posterboard?

$P = 156$ in.

71. *Nontoxic scouring powder.*   A nontoxic scouring powder is made up of 4 parts baking soda and 1 part vinegar. How much of each ingredient is needed for a 16-oz mixture?

Solve graphically.

72. $y = |x|,$ **73.** $x - y = 0,$
   $x + 4y = 15$ $y = x^2$

In Exercises 74–77, use a graphing calculator to solve each system of linear equations for x and y. Round all coordinates to the nearest hundredth.

74. $y =$ $8.23x + 2.11,$ **75.** $y = -3.44x - 7.72,$
   $y = -9.11x - 4.66$ $y =$ $4.19x - 8.22$

76. $14.12x + 7.32y = 2.98,$
   $21.88x - 6.45y = -7.22$

77. $5.22x -$ $8.21y = -10.21,$
   $-12.67x + 10.34y = 12.84$

3.2 Solving by Substitution or Elimination

The Substitution Method • The Elimination Method •
Comparing Methods

The Substitution Method

Algebraic (nongraphical) methods for solving systems are often superior to graphing, especially when fractions are involved. One algebraic method, the *substitution method*, relies on having a variable isolated.

EXAMPLE 1

Solve the system

$$x + y = 4, \quad (1)$$
$$x = y + 1. \quad (2)$$

For easy reference, we have numbered the equations.

Solution   Equation (2) says that x and $y + 1$ name the same number. Thus we can substitute $y + 1$ for x in equation (1):

$$x + y = 4 \qquad \text{Equation (1)}$$
$$(y + 1) + y = 4. \qquad \text{Substituting } y + 1 \text{ for } x$$

We solve this last equation, using methods learned earlier:

$$(y + 1) + y = 4$$
$$2y + 1 = 4 \qquad \text{Removing parentheses and combining like terms}$$
$$2y = 3 \qquad \text{Subtracting 1 from both sides}$$
$$y = \tfrac{3}{2}. \qquad \text{Dividing by 2}$$

We now return to the original pair of equations and substitute $\tfrac{3}{2}$ for y in either equation so that we can solve for x. For this problem, calculations are slightly easier if we use equation (2):

$$x = y + 1 \qquad \text{Equation (2)}$$
$$= \tfrac{3}{2} + 1 \qquad \text{Substituting } \tfrac{3}{2} \text{ for } y$$
$$= \tfrac{3}{2} + \tfrac{2}{2} = \tfrac{5}{2}.$$

We obtain the ordered pair $\left(\tfrac{5}{2}, \tfrac{3}{2}\right)$. A check ensures that it is a solution:

Check:

$$\begin{array}{c|c} x + y = 4 \\ \hline \tfrac{5}{2} + \tfrac{3}{2} & 4 \\ \tfrac{8}{2} & \\ 4 \overset{?}{=} 4 & \text{TRUE} \end{array} \qquad \begin{array}{c|c} x = y + 1 \\ \hline \tfrac{5}{2} & \tfrac{3}{2} + 1 \\ & \tfrac{3}{2} + \tfrac{2}{2} \\ \tfrac{5}{2} \overset{?}{=} \tfrac{5}{2} & \text{TRUE} \end{array}$$

Since $\left(\tfrac{5}{2}, \tfrac{3}{2}\right)$ checks, it is the solution.

The exact solution to Example 1 is difficult to find graphically because it involves fractions. Despite this, the graph shown does serve as a check and provides a visualization of the problem.

Study Skills————

Learn from Your Mistakes

Immediately after each quiz or test, write out a step-by-step solution to any questions you missed. Visit your professor during office hours or consult with a tutor for help with problems that are still giving you trouble. Misconceptions tend to resurface if they are not corrected as soon as possible.

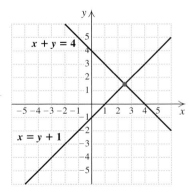

A visualization of Example 1.
Note that the coordinates of the intersection are not obvious.

If neither equation in a system has a variable alone on one side, we first isolate a variable in one equation and then substitute.

EXAMPLE 2   Solve the system

$$2x + y = 6, \qquad (1)$$
$$3x + 4y = 4. \qquad (2)$$

Solution   First, we select an equation and solve for one variable. To isolate y, we can subtract $2x$ from both sides of equation (1):

$$2x + y = 6 \qquad \qquad (1)$$
$$y = 6 - 2x. \qquad (3) \qquad \text{Subtracting } 2x \text{ from both sides}$$

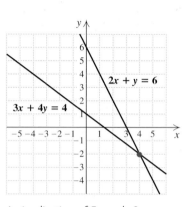

A visualization of Example 2

Next, we proceed as in Example 1, by substituting:

$$3x + 4(6 - 2x) = 4 \qquad \qquad \text{Substituting } 6 - 2x \text{ for } y \text{ in equation (2).}$$
$$\text{Use parentheses!}$$
$$3x + 24 - 8x = 4 \qquad \qquad \text{Distributing to remove parentheses}$$
$$3x - 8x = 4 - 24 \qquad \qquad \text{Subtracting 24 from both sides}$$
$$-5x = -20$$
$$x = 4. \qquad \qquad \text{Dividing both sides by } -5$$

Next, we substitute 4 for x in either equation (1), (2), or (3). It is easiest to use equation (3) because it has already been solved for y:

$$y = 6 - 2x$$
$$= 6 - 2(4)$$
$$= 6 - 8 = -2.$$

The pair $(4, -2)$ appears to be the solution. We check in equations (1) and (2).

Check:

$$\begin{array}{c|c} 2x + y = 6 \\ \hline 2(4) + (-2) & 6 \\ 8 - 2 & \\ & 6 \overset{?}{=} 6 \quad \text{TRUE} \end{array} \qquad \begin{array}{c|c} 3x + 4y = 4 \\ \hline 3(4) + 4(-2) & 4 \\ 12 - 8 & \\ & 4 \overset{?}{=} 4 \quad \text{TRUE} \end{array}$$

Since $(4, -2)$ checks, it is the solution.

Some systems have no solution, as we saw graphically in Section 3.1. How do we recognize such systems if we are solving by an algebraic method?

EXAMPLE 3   Solve the system

$$y = -3x + 5, \qquad (1)$$
$$y = -3x - 2. \qquad (2)$$

Solution   We solved this system graphically in Example 4(b) of Section 3.1, and found that the lines are parallel and the system has no solution. Let's now

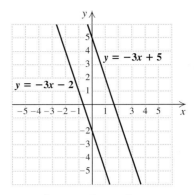

A visualization of Example 3

try to solve the system by substitution. Proceeding as in Example 1, we substitute $-3x - 2$ for y in the first equation:

$$-3x - 2 = -3x + 5 \qquad \text{Substituting } -3x - 2 \text{ for } y \text{ in equation (1)}$$
$$-2 = 5. \qquad \text{Adding } 3x \text{ to both sides; } -2 = 5 \text{ is a contradiction.}$$

When we add $3x$ to get the x-terms on one side, the x-terms drop out and the result is a contradiction—an equation that is always false. When solving algebraically yields a contradiction, we state that the system has no solution.

The Elimination Method

The *elimination method* for solving systems of equations makes use of the *addition principle*: If $a = b$, then $a + c = b + c$. Consider the following system:

$$2x - 3y = 0, \qquad (1)$$
$$-4x + 3y = -1. \qquad (2)$$

To see why the elimination method works well with this system, notice the $-3y$ in one equation and the $3y$ in the other. These terms are opposites. If we add all terms on the left side of the equations, the sum of $-3y$ and $3y$ is 0, so in effect, the variable y is "eliminated."

To use the addition principle with a system, note that according to equation (2), $-4x + 3y$ and -1 are the same number. Thus we can work vertically and add $-4x + 3y$ to the left side of equation (1) and -1 to the right side:

$$2x - 3y = 0 \qquad (1)$$
$$\underline{-4x + 3y = -1} \qquad (2)$$
$$-2x + 0y = -1. \qquad \text{Adding}$$

This eliminates the variable y, and leaves an equation with just one variable, x, for which we solve:

$$-2x = -1$$
$$x = \tfrac{1}{2}.$$

Next, we substitute $\tfrac{1}{2}$ for x in equation (1) and solve for y:

$$2 \cdot \tfrac{1}{2} - 3y = 0 \qquad \text{Substituting. We also could have used equation (2).}$$
$$1 - 3y = 0$$
$$-3y = -1, \text{ so } y = \tfrac{1}{3}.$$

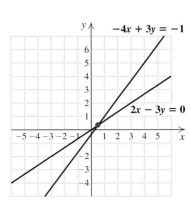

A visualization of the solution of
$$2x - 3y = 0,$$
$$-4x + 3y = -1$$

Check:

Since $\left(\tfrac{1}{2}, \tfrac{1}{3}\right)$ checks, it is the solution. See also the graph at left.

To eliminate a variable, we must sometimes multiply before adding.

EXAMPLE 4

Solve the system

$$5x + 4y = 22, \quad (1)$$
$$-3x + 8y = 18. \quad (2)$$

Student Notes

It is wise to double-check each step of your work as you go along, rather than checking all steps after reaching the end of a problem. Finding and correcting an error as it occurs will save you time in the long run. One common error is to forget to multiply *both* sides of the equation when you use the multiplication principle.

Solution If we add the left sides of the two equations, we will not eliminate a variable. However, if the $4y$ in equation (1) were changed to $-8y$, we would. To accomplish this change, we multiply both sides of equation (1) by -2:

$$
\begin{array}{ll}
-10x - 8y = -44 & \text{Multiplying both sides of equation (1) by } -2 \\
\underline{-3x + 8y = 18} & \\
-13x + 0 = -26 & \text{Adding} \\
x = 2. & \text{Solving for } x
\end{array}
$$

Then

$$
\begin{array}{ll}
-3 \cdot 2 + 8y = 18 & \text{Substituting 2 for } x \text{ in equation (2)} \\
-6 + 8y = 18 & \\
\left. \begin{array}{l} 8y = 24 \\ y = 3. \end{array} \right\} & \text{Solving for } y
\end{array}
$$

We obtain $(2, 3)$, or $x = 2$, $y = 3$. We leave it to the student to confirm that this checks and is the solution.

Sometimes we must multiply twice in order to make two terms become opposites.

EXAMPLE 5

Solve the system

$$2x + 3y = 17, \quad (1)$$
$$5x + 7y = 29. \quad (2)$$

Solution We multiply so that the x-terms are eliminated.

$$
\begin{array}{lll}
2x + 3y = 17, & \xrightarrow{\text{Multiplying both sides by 5}} & 10x + 15y = 85 \\
5x + 7y = 29 & \xrightarrow{\text{Multiplying both sides by } -2} & \underline{-10x - 14y = -58} \\
& & 0 + y = 27 \quad \text{Adding} \\
& & y = 27
\end{array}
$$

Next, we substitute to find x:

$$
\begin{array}{ll}
2x + 3 \cdot 27 = 17 & \text{Substituting 27 for } y \text{ in equation (1)} \\
2x + 81 = 17 & \\
\left. \begin{array}{l} 2x = -64 \\ x = -32. \end{array} \right\} & \text{Solving for } x
\end{array}
$$

Check:

$$
\begin{array}{c|c}
\multicolumn{2}{l}{2x + 3y = 17} \\
\hline
2(-32) + 3(27) & 17 \\
-64 + 81 & \\
17 \overset{?}{=} 17 & \text{TRUE}
\end{array}
\qquad
\begin{array}{c|c}
\multicolumn{2}{l}{5x + 7y = 29} \\
\hline
5(-32) + 7(27) & 29 \\
-160 + 189 & \\
29 \overset{?}{=} 29 & \text{TRUE}
\end{array}
$$

We obtain $(-32, 27)$, or $x = -32$, $y = 27$, as the solution.

EXAMPLE 6

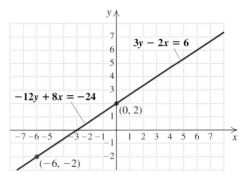

A visualization of Example 6

Solve the system

$$3y - 2x = 6, \qquad (1)$$
$$-12y + 8x = -24. \qquad (2)$$

Solution   We graphed this system in Example 4(c) of Section 3.1, and found that the lines coincide and the system has an infinite number of solutions. Suppose we were to solve this system using the elimination method:

$$\begin{aligned} 12y - 8x &= 24 \qquad \text{Multiplying both sides of equation (1) by 4} \\ -12y + 8x &= -24 \\ \hline 0 &= 0. \qquad \text{We obtain an identity; } 0 = 0 \text{ is always true.} \end{aligned}$$

Note that both variables have been eliminated and what remains is an identity—that is, an equation that is always true. Any pair that is a solution of equation (1) is also a solution of equation (2). The equations are dependent and the solution set is infinite:

$$\{(x, y) \,|\, 3y - 2x = 6\}, \quad \text{or equivalently,} \quad \{(x, y) \,|\, -12y + 8x = -24\}.$$

Rules for Special Cases

When solving a system of two linear equations in two variables:

1. If an identity is obtained, such as $0 = 0$, then the system has an infinite number of solutions. The equations are dependent and, since a solution exists, the system is consistent.*
2. If a contradiction is obtained, such as $0 = 7$, then the system has no solution. The system is inconsistent.

Should decimals or fractions appear, it often helps to *clear* before solving.

EXAMPLE 7

Solve the system

$$0.2x + 0.3y = 1.7,$$
$$\tfrac{1}{7}x + \tfrac{1}{5}y = \tfrac{29}{35}.$$

Solution   We have

$$0.2x + 0.3y = 1.7, \quad \rightarrow \text{Multiplying both sides by 10} \rightarrow \quad 2x + 3y = 17$$
$$\tfrac{1}{7}x + \tfrac{1}{5}y = \tfrac{29}{35} \quad \rightarrow \text{Multiplying both sides by 35} \rightarrow \quad 5x + 7y = 29.$$

We multiplied both sides of the first equation by 10 to clear the decimals. Multiplication by 35, the least common denominator, clears the fractions in the second equation. The problem now happens to be identical to Example 5. The solution is $(-32, 27)$, or $x = -32, y = 27$.

---

*Consistent systems and dependent equations are discussed in greater detail in Section 3.4.

Comparing Methods

The following table is a summary that compares the graphical, substitution, and elimination methods for solving systems of equations.

CONNECTING THE CONCEPTS

We now have three different methods for solving systems of equations. Each method has certain strengths and weaknesses, as outlined below.

Method	Strengths	Weaknesses
Graphical	Solutions are displayed graphically. Can be used with any system that can be graphed.	Inexact when solutions involve numbers that are not integers. Solution may not appear on the part of the graph drawn.
Substitution	Yields exact solutions. Easy to use when a variable is alone on one side.	Introduces extensive computations with fractions when solving more complicated systems. Solutions are not displayed graphically.
Elimination	Yields exact solutions. Easy to use when fractions or decimals appear in the system. The preferred method for systems of 3 or more equations in 3 or more variables (see Section 3.4).	Solutions are not displayed graphically.

Before selecting a method to use, try to remember the strengths and weaknesses of each method. If possible, begin solving the system mentally to help discover the method that seems best suited for that particular system. Selecting the "best" method for a problem is a bit like selecting one of three different saws with which to cut a piece of wood. The "best" choice depends on what kind of wood is being cut and what type of cut is being made, as well as your skill level with each saw.

Note that each of the three methods was introduced using a rather simple example. As the examples became more complicated, additional steps were required in order to "turn" the new problem into a more familiar format. This is a common approach in mathematics: We perform one or more steps to make a "new" problem resemble a problem we already know how to solve.

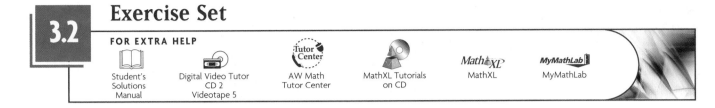

Exercise Set

3.2

✏ *Concept Reinforcement   In each of Exercises 1–6, match the system listed with the choice from the column on the right that would be a subsequent step in solving the system.*

1. ____ $-2x + 3y = 7,$
$\qquad 2x + 5y = -8$

2. ____ $x = 4y - 5,$
$\qquad 5x + 7y = 2$

3. ____ $3x + 4y = 13,$
$\qquad 4x - 2y = 5$

4. ____ $8x + 6y = -15,$
$\qquad 5x - 3y = 8$

5. ____ $y = 4x - 7,$
$\qquad 6x + 3y = 19$

6. ____ $y = \frac{2}{3}x + 4,$
$\qquad y = -\frac{1}{5}x + 4$

a) $3x + 4y = 13,$
$\quad 8x - 4y = 10$

b) The lines intersect at $(0, 4)$.

c) $6x + 3(4x - 7) = 19$

d) $8y = -1$

e) $5(4y - 5) + 7y = 2$

f) $8x + 6y = -15,$
$\quad 10x - 6y = 16$

For Exercises 7–54, if a system has an infinite number of solutions, use set-builder notation to write the solution set. If a system has no solution, state this.

Solve using the substitution method.

7. $y = 5 - 4x,$
$\quad 2x - 3y = 13$

8. $2y + x = 9,$
$\quad x = 3y - 3$

9. $3x + 5y = 3,$
$\quad x = 8 - 4y$

10. $9x - 2y = 3,$
$\quad 3x - 6 = y$

11. $3s - 4t = 14,$
$\quad 5s + t = 8$

12. $m - 2n = 16,$
$\quad 4m + n = 1$

13. $4x - 2y = 6,$
$\quad 2x - 3 = y$

14. $t = 4 - 2s,$
$\quad t + 2s = 6$

15. $-5s + t = 11,$
$\quad 4s + 12t = 4$

16. $5x + 6y = 14,$
$\quad -3y + x = 7$

17. $2x + 2y = 2,$
$\quad 3x - y = 1$

18. $4p - 2q = 16,$
$\quad 5p + 7q = 1$

19. $x - 4y = 3,$
$\quad 5x + 3y = 4$

20. $2a + 2b = 5,$
$\quad 3a - b = 7$

21. $2x - 3 = y,$
$\quad y - 2x = 1$

22. $a - 2b = 3,$
$\quad 3a = 6b + 9$

Solve using the elimination method.

23. $x + 3y = 7,$
$\quad -x + 4y = 7$

24. $2x + y = 6,$
$\quad x - y = 3$

25. $2x - y = -3,$
$\quad x + y = 9$

26. $x - 2y = 6,$
$\quad -x + 3y = -4$

27. $9x + 3y = -3,$
$\quad 2x - 3y = -8$

28. $6x - 3y = 18,$
$\quad 6x + 3y = -12$

29. $5x + 3y = 19,$
$\quad 2x - 5y = 11$

30. $3x + 2y = 3,$
$\quad 9x - 8y = -2$

31. $5r - 3s = 24,$
$\quad 3r + 5s = 28$

32. $5x - 7y = -16,$
$\quad 2x + 8y = 26$

33. $6s + 9t = 12,$
$\quad 4s + 6t = 5$

34. $10a + 6b = 8,$
$\quad 5a + 3b = 2$

35. $\frac{1}{2}x - \frac{1}{6}y = 3,$
$\quad \frac{2}{5}x + \frac{1}{2}y = 2$

36. $\frac{1}{3}x + \frac{1}{5}y = 7,$
$\quad \frac{1}{6}x - \frac{2}{5}y = -4$

37. $\dfrac{x}{2} + \dfrac{y}{3} = \dfrac{7}{6},$
$\quad \dfrac{2x}{3} + \dfrac{3y}{4} = \dfrac{5}{4}$

38. $\dfrac{2x}{3} + \dfrac{3y}{4} = \dfrac{11}{12},$
$\quad \dfrac{x}{3} + \dfrac{7y}{18} = \dfrac{1}{2}$

Aha! **39.** $12x - 6y = -15,$
$\quad -4x + 2y = 5$

40. $8s + 12t = 16,$
$\quad 6s + 9t = 12$

41. $0.2a + 0.3b = 1,$
$\quad 0.3a - 0.2b = 4$

42. $-0.4x + 0.7y = 1.3,$
$\quad 0.7x - 0.3y = 0.5$

Solve using any appropriate method.

43. $a - 2b = 16,$
$\quad b + 3 = 3a$

44. $5x - 9y = 7,$
$\quad 7y - 3x = -5$

45. $10x + y = 306,$
$\quad 10y + x = 90$

46. $3(a - b) = 15,$
$\quad 4a = b + 1$

47. $3y = x - 2,$
$\quad x = 2 + 3y$

48. $x + 2y = 8,$
$\quad x = 4 - 2y$

49. $3s - 7t = 5,$
$\quad 7t - 3s = 8$

50. $2s - 13t = 120,$
$\quad -14s + 91t = -840$

51. $0.05x + 0.25y = 22,$
$0.15x + 0.05y = 24$

52. $2.1x - 0.9y = 15,$
$-1.4x + 0.6y = 10$

53. $13a - 7b = 9,$
$2a - 8b = 6$

54. $3a - 12b = 9,$
$14a - 11b = 5$

55. Describe a procedure that can be used to write an inconsistent system of equations.

56. Describe a procedure that can be used to write a system that has an infinite number of solutions.

SKILL MAINTENANCE

Solve. [1.4]

57. The fare for a taxi ride from Johnson Street to Elm Street is $5.20. If the rate of the taxi is $1.00 for the first $\frac{1}{2}$ mi and 30¢ for each additional $\frac{1}{4}$ mi, how far is it from Johnson Street to Elm Street?

58. A student's average after 4 tests is 78.5. What score is needed on the fifth test in order to raise the average to 80?

59. *Home remodeling.* In a recent year, Americans spent $35 billion to remodel bathrooms and kitchens. Twice as much was spent on kitchens as on bathrooms. How much was spent on each?
Source: *Indianapolis Star*

60. A 480-m wire is cut into three pieces. The second piece is three times as long as the first. The third is four times as long as the second. How long is each piece?

61. *Car rentals.* Badger Rent-A-Car rents a compact car at a daily rate of $34.95 plus 10¢ per mile. A sales representative is allotted $80 for car rental for one day. How many miles can she travel on the $80 budget?

62. *Car rentals.* Badger rents midsized cars at a rate of $43.95 plus 10¢ per mile. A tourist has a car-rental budget of $90 for one day. How many miles can he travel on the $90?

SYNTHESIS

63. Some systems are more easily solved by substitution and some are more easily solved by elimination. What guidelines could be used to help someone determine which method to use?

64. Explain how it is possible to solve Exercise 39 mentally.

65. If $(1, 2)$ and $(-3, 4)$ are two solutions of $f(x) = mx + b$, find m and b.

66. If $(0, -3)$ and $\left(-\frac{3}{2}, 6\right)$ are two solutions of $px - qy = -1$, find p and q.

67. Determine a and b for which $(-4, -3)$ is a solution of the system
$$ax + by = -26,$$
$$bx - ay = 7.$$

68. Solve for x and y in terms of a and b:
$$5x + 2y = a,$$
$$x - y = b.$$

Solve.

69. $\dfrac{x + y}{2} - \dfrac{x - y}{5} = 1,$
$\dfrac{x - y}{2} + \dfrac{x + y}{6} = -2$

70. $3.5x - 2.1y = 106.2,$
$4.1x + 16.7y = -106.28$

Each of the following is a system of nonlinear equations. However, each is reducible to linear, since an appropriate substitution (say, u for $1/x$ and v for $1/y$) yields a linear system. Make such a substitution, solve for the new variables, and then solve for the original variables.

71. $\dfrac{2}{x} + \dfrac{1}{y} = 0,$
$\dfrac{5}{x} + \dfrac{2}{y} = -5$

72. $\dfrac{1}{x} - \dfrac{3}{y} = 2,$
$\dfrac{6}{x} + \dfrac{5}{y} = -34$

73. A student solving the system
$$17x + 19y = 102,$$
$$136x + 152y = 826$$
graphs both equations on a graphing calculator and gets the following screen. The student then (incorrectly) concludes that the equations are dependent and the solution set is infinite. How can algebra be used to convince the student that a mistake has been made?

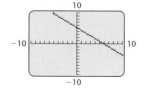

CORNER

How Many Two's? How Many Three's?

Focus: Systems of linear equations

Time: 20 minutes

Group size: 3

The box score at right, from the 2004 NBA All-Star game, contains information on how many field goals and free throws each player attempted and made. For example, the line "Kidd 4-6 3-4 14" means that the East's Jason Kidd made 4 field goals out of 6 attempts and 3 free throws out of 4 attempts, for a total of 14 points. (Each free throw is worth 1 point and each field goal is worth either 2 or 3 points, depending on how far from the basket it was shot.)

ACTIVITY

1. Work as a group to develop a system of two equations in two unknowns that can be used to determine how many 2-pointers and how many 3-pointers were made by the West.
2. Each group member should solve the system from part (1) in a different way: one person

algebraically, one person by making a table and methodically checking all combinations of 2- and 3-pointers, and one person by guesswork. Compare answers when this has been completed.
3. Determine, as a group, how many 2- and 3-pointers the East made.

East (132)
Iverson 1-6 1-4 3, McGrady 5-11 2-4 13, J.O'Neal 7-13 2-4 16, Carter 5-7 0-0 11, Wallace 2-5 0-0 4, Martin 8-10 1-2 17, Kidd 4-6 3-4 14, Magloire 9-16 1-2 19, Artest 3-5 1-2 7, Davis 3-9 0-0 7, Redd 5-12 0-0 13, Pierce 4-8 0-0 8
Totals 56-108 11-22 132

West (136)
Bryant 9-12 0-1 20, Francis 6-9 0-0 13, Garnett 6-14 0-0 12, Duncan 6-11 2-4 14, Yao Ming 8-14 0-0 16, S.O'Neal 12-19 0-1 24, Allen 6-13 3-4 16, Cassell 2-3 0-0 4, Nowitzki 1-3 0-0 4, Stojakovic 2-5 0-0 5, Kirilenko 1-3 0-0 2, Brad Miller 4-5 0-0 8
Totals 63-111 5-10 136

East	33	31	37	31	—	132
West	31	27	45	33	—	136

3.3 Solving Applications: Systems of Two Equations

Total-Value and Mixture Problems • Motion Problems

You are in a much better position to solve problems now that you know how systems of equations can be used. Using systems often makes the translating step easier.

EXAMPLE 1 Bottled-water consumption. Americans are buying more bottled water than ever before. At the time of this writing, the average American buys more bottled water than milk, coffee, or beer. In 2003, the average American purchased 76.4 gal of soft drinks and water. The amount of water was only 4.3 gal less than

Study Skills _____

Expect to be Challenged

Do not be surprised if your success rate drops some as you work on real-world problems. *This is normal.* Your success rate will increase as you gain experience with these types of problems and use some of the study skills already listed.

half of the amount of soft drinks. (*Source*: Based on data from www.beveragemarketing.com) How many gallons of water and how many of soft drinks did the average American buy in 2003?

Solution The *Familiarize* and *Translate* steps have been done in Example 1 of Section 3.1. The resulting system of equations is

$$w + s = 76.4,$$
$$w = \frac{1}{2}s - 4.3,$$

where w is the number of gallons of water and s is the number of gallons of soft drinks purchased by the average American in 2003.

3. Carry out. We solve the system of equations. Since one equation already has a variable isolated, let's use the substitution method:

$$w + s = 76.4$$

$$\frac{1}{2}s - 4.3 + s = 76.4 \qquad \text{Substituting } \frac{1}{2}s - 4.3 \text{ for } w$$

$$\frac{3}{2}s - 4.3 = 76.4 \qquad \text{Combining like terms}$$

$$\frac{3}{2}s = 80.7 \qquad \text{Adding 4.3 to both sides}$$

$$s = \frac{2}{3} \cdot 80.7 \qquad \text{Multiplying both sides by } \frac{2}{3}: \frac{2}{3} \cdot \frac{3}{2} = 1$$

$$s = 53.8. \qquad \text{Simplifying}$$

Next, using either of the original equations, we substitute and solve for w:

$$w = \frac{1}{2} \cdot 53.8 - 4.3 = 26.9 - 4.3 = 22.6.$$

4. Check. The sum of 53.8 and 22.6 is 76.4, so the total consumed is correct. Since 4.3 less than half of 53.8 is $26.9 - 4.3$, or 22.6, the numbers check.

5. State. In 2003, the average American purchased 22.6 gal of bottled water and 53.8 gal of soft drinks.

Student Notes _____

It is very important that you clearly label precisely what each variable represents. Not only will this assist you in writing equations, but it will help you to identify and state solutions.

Total-Value and Mixture Problems

EXAMPLE 2

Purchasing. Recently the Woods County Art Center purchased 120 stamps for $33.90. If the stamps were a combination of 23¢ postcard stamps and 37¢ first-class stamps, how many of each type were bought?

Solution The *Familiarize* and *Translate* steps were completed in Example 2 of Section 3.1.

3. Carry out. We are to solve the system of equations

$$p + f = 120, \qquad (1)$$
$$23p + 37f = 3390, \qquad (2) \qquad \text{Working in cents rather than dollars}$$

where p is the number of postcard stamps bought and f is the number of first-class stamps bought. Because both equations are in the form

Ax + *By* = *C*, let's use the elimination method to solve the system. We can eliminate *p* by multiplying both sides of equation (1) by −23 and adding them to the corresponding sides of equation (2):

$$-23p - 23f = -2760 \qquad \text{Multiplying both sides of equation (1) by } -23$$
$$\underline{23p + 37f = 3390}$$
$$14f = 630 \qquad \text{Adding}$$
$$f = 45. \qquad \text{Solving for } f$$

To find *p*, we substitute 45 for *f* in equation (1) and then solve for *p*:

$$p + f = 120 \qquad \text{Equation (1)}$$
$$p + 45 = 120 \qquad \text{Substituting 45 for } f$$
$$p = 75. \qquad \text{Solving for } p$$

We obtain (45, 75), or *f* = 45 and *p* = 75.

4. **Check.**   We check in the original problem. Recall that *f* is the number of first-class stamps and *p* the number of postcard stamps.

Number of stamps: *f* + *p* = 45 + 75 = 120

Cost of first-class stamps: $0.37*f* = 0.37 × 45 = $16.65

Cost of postcard stamps: $0.23*p* = 0.23 × 75 = $17.25

Total = $33.90

The numbers check.

5. **State.**   The art center bought 45 first-class stamps and 75 postcard stamps.

Example 2 involved two types of items (first-class stamps and postcard stamps), the quantity of each type bought, and the total value of the items. We refer to this type of problem as a *total-value problem*.

EXAMPLE 3   Blending teas.   Sonya's House of Tea sells loose Lapsang Souchong tea for 95¢ an ounce and Assam Gingia for $1.43 an ounce. Sonya wants to make a 20-oz mixture of the two types, called Dragon Blend, that sells for $1.10 an ounce. How much tea of each type should Sonya use?

Solution

1. **Familiarize.**   This problem is similar to Example 2. Rather than postcard stamps and first-class stamps, we have ounces of Assam Gingia and ounces of Lapsang Souchong. Instead of a different price for each type of stamp, we

have a different price per ounce for each type of tea. Finally, rather than knowing the total cost of the stamps, we know the weight and the price per ounce of the mixture. Thus we can find the total value of the blend by multiplying 20 ounces times $1.10, or 110¢ per ounce. Although we could make and check a guess, we proceed to let l = the number of ounces of Lapsang Souchong and a = the number of ounces of Assam Gingia.

2. Translate.   Since a 20-oz batch is being made, we must have

$$l + a = 20.$$

To find a second equation, note that the total value of the 20-oz blend must match the combined value of the separate ingredients:

Rewording: The value   the value   the value
of the   of the   of the
Lapsang Souchong   plus   Assam Gingia   is   Dragon Blend.

Translating: $l \cdot 95$ $+$ $a \cdot 143$ $=$ $20 \cdot 110$

These equations can also be obtained from a table.

	Lapsang Souchong	Assam Gingia	Dragon Blend
Number of Ounces	l	a	20
Price per Ounce	95¢	143¢	110¢
Value of Tea	$95l$	$143a$	$20 \cdot 110$, or 2200¢

$\longrightarrow l + a = 20$

$\longrightarrow 95l + 143a = 2200$

We have translated to a system of equations:

$$l + a = 20, \qquad (1)$$
$$95l + 143a = 2200. \qquad (2)$$

3. Carry out.   We can solve using substitution. When equation (1) is solved for l, we have $l = 20 - a$. Substituting $20 - a$ for l in equation (2), we find a:

$95(20 - a) + 143a = 2200$ Substituting

$1900 - 95a + 143a = 2200$ Using the distributive law

$48a = 300$ Combining like terms; subtracting 1900 from both sides

$a = 6.25.$ Dividing both sides by 48

We have $a = 6.25$ and, from equation (1) above, $l + a = 20$. Thus, $l = 13.75$.

4. Check.   If 13.75 oz of Lapsang Souchong and 6.25 oz of Assam Gingia are combined, a 20-oz blend will result. The value of 13.75 oz of Lapsang Souchong is 13.75(95¢) or 1306.25¢. The value of 6.25 oz of Assam Gingia is 6.25(143¢), or 893.75¢. Thus the combined value of the blend is 1306.25¢ + 893.75¢, or 2200¢, which is $22. A 20-oz blend priced at $1.10 an ounce would also be worth $22, so our answer checks.

5. State.   The Dragon Blend should be made by combining 13.75 oz of Lapsang Souchong with 6.25 oz of Assam Gingia.

EXAMPLE 4   Student loans.   Ranjay's student loans totaled $9600. Part was a Perkins loan made at 5% interest and the rest was a Stafford Loan made at 8% interest. After one year, Ranjay's loans accumulated $633 in interest. What was the original amount of each loan?

Solution

1. **Familiarize.**   We begin with a guess. If $7000 was borrowed at 5% and $2600 was borrowed at 8%, the two loans would total $9600. The interest would then be 0.05($7000), or $350, and 0.08($2600), or $208, for a total of only $558 in interest. Our guess was wrong, but checking the guess familiarized us with the problem. More than $2600 was borrowed at the higher rate.

2. **Translate.**   We let $p =$ the amount of the Perkins loan and $f =$ the amount of the Stafford Loan. Next, we organize a table in which the entries in each column come from the formula for simple interest:

$$Principal \cdot Rate \cdot Time = Interest.$$

	Perkins Loan	Stafford Loan	Total	
Principal	p	f	$9600	$\rightarrow p + f = 9600$
Rate of Interest	5%	8%		
Time	1 yr	1 yr		
Interest	$0.05p$	$0.08f$	$633	$\rightarrow 0.05p + 0.08f = 633$

The total amount borrowed is found in the first row of the table:

$$p + f = 9600.$$

A second equation, representing the accumulated interest, can be found in the last row:

$$0.05p + 0.08f = 633, \quad or \quad 5p + 8f = 63{,}300. \qquad \text{Clearing decimals}$$

3. **Carry out.**   The system can be solved by elimination:

$$\begin{array}{l} p + f = 9600, \\ 5p + 8f = 63{,}300. \end{array} \quad \begin{array}{c} \longrightarrow \text{Multiplying both} \\ \text{sides by } -5 \end{array} \quad \begin{array}{c} \longrightarrow \\ \end{array} \quad \begin{array}{l} -5p - 5f = -48{,}000 \\ \underline{5p + 8f = 63{,}300} \\ 3f = 15{,}300 \end{array}$$

$$p + f = 9600 \longleftarrow \qquad f = 5100$$
$$p + 5100 = 9600$$
$$p = 4500.$$

We find that $p = 4500$ and $f = 5100$.

4. **Check.**   The total amount borrowed is $4500 + $5100, or $9600. The interest on $4500 at 5% for 1 yr is 0.05($4500), or $225. The interest on $5100 at 8% for 1 yr is 0.08($5100), or $408. The total amount of interest is $225 + $408, or $633, so the numbers check.

5. **State.**   The Perkins loan was for $4500 and the Stafford loan was for $5100.

Before proceeding to Example 5, briefly scan Examples 2–4 for similarities. Note that in each case, one of the equations in the system is a simple sum while the other equation represents a sum of products. Example 5 continues this pattern with what is commonly called a *mixture problem*.

Problem-Solving Tip

When solving a problem, see if it is patterned or modeled after a problem that you have already solved.

EXAMPLE 5

Mixing fertilizers.   Sky Meadow Gardening, Inc., carries two brands of fertilizer containing nitrogen and water. "Gently Green" is 5% nitrogen and "Sun Saver" is 15% nitrogen. Sky Meadow Gardening needs to combine the two types of solutions in order to make 90 L of a solution that is 12% nitrogen. How much of each brand should be used?

Solution

1. **Familiarize.** We make a drawing and then make a guess to gain familiarity with the problem.

Suppose that 40 L of Gently Green and 50 L of Sun Saver are mixed. The resulting mixture will be the right size, 90 L, but will it be the right strength? To find out, note that 40 L of Gently Green would contribute $0.05(40) = 2$ L of nitrogen to the mixture while 50 L of Sun Saver would contribute $0.15(50) = 7.5$ L of nitrogen to the mixture. The total amount of nitrogen in the mixture would then be $2 + 7.5$, or 9.5 L. But we want 12% of 90, or 10.8 L, to be nitrogen. Thus our guess of 40 L and 50 L is incorrect. Still, checking our guess has familiarized us with the problem.

2. **Translate.** Let $g =$ the number of liters of Gently Green and $s =$ the number of liters of Sun Saver. The information can be organized in a table.

	Gently Green	Sun Saver	Mixture	
Number of Liters	g	s	90	→ $g + s = 90$
Percent of Nitrogen	5%	15%	12%	
Amount of Nitrogen	$0.05g$	$0.15s$	0.12×90, or 10.8 liters	→ $0.05g + 0.15s = 10.8$

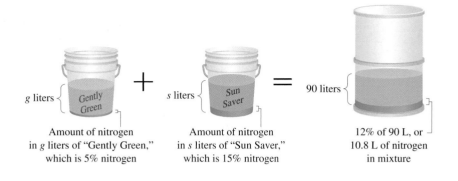

| Amount of nitrogen in *g* liters of "Gently Green," which is 5% nitrogen | Amount of nitrogen in *s* liters of "Sun Saver," which is 15% nitrogen | 12% of 90 L, or 10.8 L of nitrogen in mixture |

If we add *g* and *s* in the first row, we get one equation. It represents the total amount of mixture: $g + s = 90$.

If we add the amounts of nitrogen listed in the third row, we get a second equation. This equation represents the amount of nitrogen in the mixture: $0.05g + 0.15s = 10.8$.

After clearing decimals, we have translated the problem to the system

$$g + s = 90 \qquad (1)$$
$$5g + 15s = 1080. \qquad (2)$$

3. Carry out.   We use the elimination method to solve the system:

$$
\begin{array}{ll}
-5g - 5s = -450 & \text{Multiplying both sides of} \\
\underline{5g + 15s = 1080} & \text{equation (1) by } -5 \\
10s = 630 & \text{Adding} \\
s = 63; & \text{Solving for } s
\end{array}
$$

$$
\begin{array}{ll}
g + 63 = 90 & \text{Substituting into equation (1)} \\
g = 27. & \text{Solving for } g
\end{array}
$$

4. Check.   Remember, *g* is the number of liters of Gently Green and *s* is the number of liters of Sun Saver.

Total amount of mixture:	$g + s = 27 + 63 = 90$
Total amount of nitrogen:	5% of 27 + 15% of 63 = 1.35 + 9.45 = 10.8
Percentage of nitrogen in mixture:	$\dfrac{\text{Total amount of nitrogen}}{\text{Total amount of mixture}} = \dfrac{10.8}{90} = 12\%$

The numbers check in the original problem.

5. State.   Sky Meadow Gardening should mix 27 L of Gently Green with 63 L of Sun Saver.

Motion Problems

When a problem deals with distance, speed (rate), and time, recall the following.

Distance, Rate, and Time Equations

If r represents rate, t represents time, and d represents distance, then:

$$d = rt, \qquad r = \frac{d}{t}, \quad \text{and} \quad t = \frac{d}{r}.$$

Be sure to remember at least one of these equations. The others can be obtained by multiplying or dividing on both sides as needed.

EXAMPLE 6   *Train travel.*   A Vermont Railways freight train, loaded with logs, leaves Boston, heading to Washington D.C. at a speed of 60 km/h. Two hours later, an Amtrak® Metroliner leaves Boston, bound for Washington D.C., on a parallel track at 90 km/h. At what point will the Metroliner catch up to the freight train?

Solution

1. **Familiarize.**   Let's make a guess—say, 180 km—and check to see if it is correct. The freight train, traveling 60 km/h, would travel 180 km in $\frac{180}{60} = 3$ hr. The Metroliner, traveling 90 km/h, would cover 180 km in $\frac{180}{90} = 2$ hr. Since 3 hr is *not* two hours more than 2 hr, our guess of 180 km is incorrect. Although our guess is wrong, we see that the time that the trains are running and the point at which they meet are both unknown. We let $t =$ the number of hours that the freight train is running before they meet and $d =$ the distance at which the trains meet. Since the freight train has a 2-hr head start, the Metroliner runs for $t - 2$ hours before catching up to the freight train, at which point both trains have traveled the same distance.

60 km/h
d kilometers
t hours

90 km/h
d kilometers
$t - 2$ hours

Trains meet here

2. **Translate.**   We can organize the information in a chart. Each row is determined by the formula *Distance = Rate · Time*.

	Distance	Rate	Time	
Freight Train	d	60	t	$\rightarrow d = 60t$
Metroliner	d	90	$t - 2$	$\rightarrow d = 90(t - 2)$

Using *Distance = Rate · Time* twice, we get two equations:

$$d = 60t, \qquad (1)$$
$$d = 90(t - 2). \qquad (2)$$

3. **Carry out.**   We solve the system using substitution:

$$60t = 90(t - 2) \qquad \text{Substituting } 60t \text{ for } d \text{ in equation (2)}$$
$$60t = 90t - 180$$
$$-30t = -180$$
$$t = 6.$$

The time for the freight train is 6 hr, which means that the time for the Metroliner is 6 − 2, or 4 hr. Remember that it is distance, not time, that the problem asked for. Thus for $t = 6$, we have $d = 60 \cdot 6 = 360$ km.

4. **Check.**   At 60 km/h, the freight train will travel 60 · 6, or 360 km, in 6 hr. At 90 km/h, the Metroliner will travel 90 · (6 − 2) = 360 km in 4 hr. The numbers check.

5. **State.**   The freight train will catch up to the Metroliner at a point 360 km from Boston.

EXAMPLE 7   Jet travel.   A Boeing 747-400 jet flies 4 hr west with a 60-mph tailwind. Returning *against* the wind takes 5 hr. Find the speed of the plane with no wind.

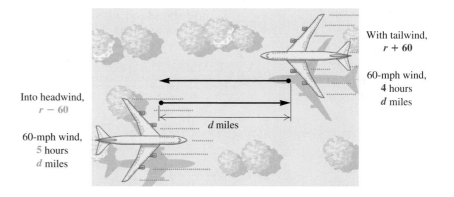

Solution

1. **Familiarize.**   We imagine the situation and make a drawing. Note that the wind *speeds up* the jet on the outbound flight, but *slows down* the jet on the return flight. Since the distances traveled each way must be the same, we can check a guess of the jet's speed with no wind. Suppose the speed of the jet with no wind is 400 mph. The jet would then fly 400 + 60 = 460 mph with the wind and 400 − 60 = 340 mph into the wind. In 4 hr, the jet would travel 460 · 4 = 1840 mi with the wind and 340 · 5 = 1700 mi against the wind. Since 1840 ≠ 1700, our guess of 400 mph is incorrect. Rather than guess again, let's have $r =$ the speed, in miles per hour, of the jet in still air. Then $r + 60 =$ the jet's speed with the wind and $r - 60 =$ the jet's speed against the wind. We also let $d =$ the distance traveled, in miles.

2. Translate. The information can be organized in a chart. The distances traveled are the same, so we use *Distance = Rate* (or *Speed*) · *Time*. Each row of the chart gives an equation.

	Distance	Rate	Time	
With Wind	d	$r + 60$	4	$\rightarrow d = (r + 60)4$
Against Wind	d	$r - 60$	5	$\rightarrow d = (r - 60)5$

The two equations constitute a system:

$$d = (r + 60)4, \quad (1)$$
$$d = (r - 60)5. \quad (2)$$

3. Carry out. We solve the system using substitution:

$(r - 60)5 = (r + 60)4$ Substituting $(r - 60)5$ for d in equation (1)

$5r - 300 = 4r + 240$ Using the distributive law

$r = 540.$ Solving for r

4. Check. When $r = 540$, the speed with the wind is $540 + 60 = 600$ mph, and the speed against the wind is $540 - 60 = 480$ mph. The distance with the wind, $600 \cdot 4 = 2400$ mi, matches the distance into the wind, $480 \cdot 5 = 2400$ mi, so we have a check.

5. State. The speed of the plane with no wind is 540 mph.

Tips for Solving Motion Problems

1. Draw a diagram using an arrow or arrows to represent distance and the direction of each object in motion.
2. Organize the information in a chart.
3. Look for times, distances, or rates that are the same. These often can lead to an equation.
4. Translating to a system of equations allows for the use of two variables.
5. Always make sure that you have answered the question asked.

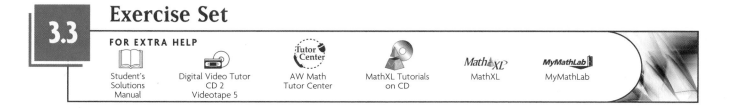

Exercise Set

3.3

FOR EXTRA HELP

Student's Solutions Manual | Digital Video Tutor CD 2 Videotape 5 | AW Math Tutor Center | MathXL Tutorials on CD | MathXL | MyMathLab

1.–14. For Exercises 1–14, solve Exercises 41–54 from p. 157.

15. *Printing.* King Street Printing recently charged 1.9¢ per sheet of paper, but 2.4¢ per sheet for paper made of recycled fibers. Darren's bill for 150 sheets of paper was $3.41. How many sheets of each type were used?

16. *Photocopying.* Quick Copy recently charged 6¢ a page for copying pages that can be machine-fed and 18¢ a page for copying pages that must be hand-placed on the copier. If Lea's bill for 90 copies was $9.24, how many copies of each type were made?

17. *Lighting.* Booth Bros. Hardware charges $7.50 for a General Electric Biax Energy Saver light bulb and $5 for an SLi Lighting Cool White Energy Saver bulb. If Paul County Hospital purchased 200 such bulbs for $1150, how many of each type did they purchase?

18. *Office supplies.* Barlow's Office Supply charges $16.75 for a box of Erase-A-Gel™ pens and $14.25 for a box of Icy™ automatic pencils. If Letsonville Community College purchased 120 such boxes for $1790, how many boxes of each type did they purchase?

19. *Sales.* Staples® recently sold a black Apple Stylewriter II ink cartridge for $30.86 and a black HP Designjet 10PS cartridge for $43.58. At the start of a recent fall semester, a total of 50 of these cartridges was sold for a total of $1733.80. How many of each type were purchased?

20. *Sales.* Staples® recently sold a wirebound graph-paper notebook for $2.50 and a college-ruled note-book made of recycled paper for $2.30. At the start of a recent spring semester, a combination of 50 of these notebooks was sold for a total of $118.60. How many of each type were sold?

21. *Blending coffees.* The Bean Counter charges $9.00 per pound for Kenyan French Roast coffee and $8.00 per pound for Sumatran coffee. How much of each type should be used to make a 20-lb blend that sells for $8.40 per pound?

22. *Mixed nuts.* Oh Nuts! sells cashews for $6.75 per pound and Brazil nuts for $5.00 per pound. How much of each type should be used to make a 50-lb mixture that sells for $5.70 per pound?

23. *Catering.* Casella's Catering is planning a wedding reception. The bride and groom would like to serve a nut mixture containing 25% peanuts. Casella has available mixtures that are either 40% or 10% peanuts. How much of each type should be mixed to get a 20-lb mixture that is 25% peanuts?

24. *Ink remover.* Etch Clean Graphics uses one cleanser that is 25% acid and a second that is 50% acid. How many liters of each should be mixed to get 30 L of a solution that is 40% acid?

25. *Blending granola.* Deep Thought Granola is 25% nuts and dried fruit. Oat Dream Granola is 10% nuts and dried fruit. How much of Deep

Thought and how much of Oat Dream should be mixed to form a 20-lb batch of granola that is 19% nuts and dried fruit?

26. *Livestock feed.*   Soybean meal is 16% protein and corn meal is 9% protein. How many pounds of each should be mixed to get a 350-lb mixture that is 12% protein?

27. *Student loans.*   Lomasi's two student loans totaled $12,000. One of her loans was at 6% simple interest and the other at 9%. After one year, Lomasi owed $855 in interest. What was the amount of each loan?

28. *Investments.*   An executive nearing retirement made two investments totaling $15,000. In one year, these investments yielded $1432 in simple interest. Part of the money was invested at 9% and the rest at 10%. How much was invested at each rate?

29. *Automotive maintenance.*   "Arctic Antifreeze" is 18% alcohol and "Frost No-More" is 10% alcohol. How many liters of each should be mixed to get 20 L of a mixture that is 15% alcohol?

30. *Chemistry.*   E-Chem Testing has a solution that is 80% base and another that is 30% base. A technician needs 150 L of a solution that is 62% base. The 150 L will be prepared by mixing the two solutions on hand. How much of each should be used?

31. *Octane ratings.*   The octane rating of a gasoline is a measure of the amount of isooctane in the gas. The 2002 Dodge Neon RT requires 91-octane gasoline. How much 87-octane gas and 93-octane gas should Kasey mix in order to make 12 gal of 91-octane gas for her Neon RT?
Sources: Champlain Electric and Petroleum Equipment; Goss Dodge

32. *Octane ratings.*   The octane rating of a gasoline is a measure of the amount of isooctane in the gas. The 2005 Chrysler Crossfire requires 93-octane gasoline. How much 87-octane gas and 95-octane gas should Ken mix in order to make 10 gal of 93-octane gas for his Crossfire?
Sources: Champlain Electric and Petroleum Equipment; Freedom Chrysler Plymouth

33. *Food science.*   The following bar graph shows the milk fat percentages in three dairy products. How many pounds each of whole milk and cream should be mixed to form 200 lb of milk for cream cheese?

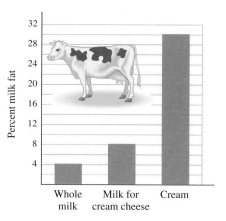

34. *Food science.*   How much lowfat (1% fat) milk and how much whole milk (4% fat) should be mixed to make 5 gal of reduced fat (2% fat) milk?

35. *Train travel.*   A train leaves Danville Junction and travels north at a speed of 75 km/h. Two hours later, an express train leaves on a parallel track and travels north at 125 km/h. How far from the station will they meet?

36. *Car travel.*   Two cars leave Salt Lake City, traveling in opposite directions. One car travels at a speed of 80 km/h and the other at 96 km/h. In how many hours will they be 528 km apart?

37. *Boating.*   Mia's motorboat took 3 hr to make a trip downstream with a 6-mph current. The return trip against the same current took 5 hr. Find the speed of the boat in still water.

38. *Canoeing.*   Alvin paddled for 4 hr with a 6-km/h current to reach a campsite. The return trip against the same current took 10 hr. Find the speed of Alvin's canoe in still water.

39. *Point of no return.*   A plane flying the 3458-mi trip from New York City to London has a 50-mph tailwind. The flight's *point of no return* is the point at which the flight time required to return to New York is the same as the time required to continue to London. If the speed of the plane in still air is 360 mph, how far is New York from the point of no return?

40. *Point of no return.*   A plane is flying the 2553-mi trip from Los Angeles to Honolulu into a 60-mph headwind. If the speed of the plane in still air is 310 mph, how far from Los Angeles is the plane's point of no return? (See Exercise 39.)

41. *Architecture.*   The rectangular ground floor of the John Hancock building has a perimeter of 860 ft. The length is 100 ft more than the width. Find the length and the width.

$x + 100$

x

42. *Real estate.*   The perimeter of a rectangular oceanfront lot is 190 m. The width is one fourth of the length. Find the dimensions.

43. *Real estate.*   In 1996, the Simon Property Group and the DeBartolo Realty Corporation merged to form the largest real estate company in the United States, owning 183 shopping centers in 32 states. Prior to merging, Simon owned twice as many properties as DeBartolo. How many properties did each company own before the merger?

The Indianapolis Star
REAL ESTATE GIANTS MERGE

DeBartolo shareholders would receive 0.68 share of Simon common stock for each share of DeBartolo common stock. Simon also would agree to repay $1.5 billion in DeBartolo debt. At Tuesday's closing price of $23.625 a share for common stock, the transaction is valued at roughly $3 billion.

Executives say the proposed company, Simon DeBartolo Group, would be the largest real estate company in the United States, worth $7.5 billion. **Not included in the deal:** DeBartolo's ownership stake in the San Francisco 49ers, or the Indiana Pacers, owned separately by the Simon

44. *Hockey rankings.*   Hockey teams receive 2 points for a win and 1 point for a tie. The Wildcats once won a championship with 60 points. They won 9 more games than they tied. How many wins and how many ties did the Wildcats have?

45. *Radio airplay.*   Roscoe must play 12 commercials during his 1-hr radio show. Each commercial is either 30 sec or 60 sec long. If the total commercial time during that hour is 10 min, how many commercials of each type does Roscoe play?

46. *Video rentals.*   J. P.'s Video rents general-interest films for $3.00 each and children's films for $1.50 each. In one day, a total of $213 was taken in from the rental of 77 videos. How many of each type of video was rented?

47. *Making change.*   Cecilia makes a $9.25 purchase at the bookstore with a $20 bill. The store has no bills and gives her the change in quarters and fifty-cent pieces. There are 30 coins in all. How many of each kind are there?

48. *Teller work.*   Ashford goes to a bank and gets change for a $50 bill consisting of all $5 bills and $1 bills. There are 22 bills in all. How many of each kind are there?

49. In what ways are Examples 3 and 4 similar? In what sense are their systems of equations similar?

50. Write at least three study tips of your own for someone beginning this exercise set.

SKILL MAINTENANCE

Evaluate.

51. $2x - 3y + 12$, for $x = 5$ and $y = 2$ [1.1]

52. $7x - 4y + 9$, for $x = 2$ and $y = 3$ [1.1]

53. $5a - 7b + 3c$, for $a = -2$, $b = 3$, and $c = 1$ [1.1], [1.2]

54. $3a - 8b - 2c$, for $a = -4$, $b = -1$, and $c = 3$ [1.1], [1.2]

55. $4 - 2y + 3z$, for $y = \frac{1}{3}$ and $z = \frac{1}{4}$ [1.1]

56. $3 - 5y + 4z$, for $y = \frac{1}{2}$ and $z = \frac{1}{5}$ [1.1]

SYNTHESIS

57. Suppose that in Example 3 you are asked only for the amount of Assam Gingia needed for the Dragon Blend. Would the method of solving the problem change? Why or why not?

58. Write a problem similar to Example 2 for a classmate to solve. Design the problem so that the solution is "The florist sold 14 hanging plants and 9 flats of petunias."

59. *Recycled paper.*   Unable to purchase 60 reams of paper that contains 20% post-consumer fiber, the Naylor School bought paper that was either 0% post-consumer fiber or 30% post-consumer fiber. How many reams of each should be purchased in order to use the same amount of post-consumer fiber as if the 20% post-consumer fiber paper were available?

60. *Retail.*   Some of the world's best and most expensive coffee is Hawaii's Kona coffee. In order for coffee to be labeled "Kona Blend," it must contain at least 30% Kona beans. Bean Town Roasters has 40 lb of Mexican coffee. How much Kona coffee must they add if they wish to market it as Kona Blend?

61. *Automotive maintenance.*   The radiator in Michelle's car contains 6.3 L of antifreeze and water. This mixture is 30% antifreeze. How much of this mixture should she drain and replace with pure antifreeze so that there will be a mixture of 50% antifreeze?

62. *Exercise.*   Natalie jogs and walks to school each day. She averages 4 km/h walking and 8 km/h jogging. From home to school is 6 km and Natalie makes the trip in 1 hr. How far does she jog in a trip?

63. *Book sales.*   A limited edition of a book published by a historical society was offered for sale to members. The cost was one book for $12 or two books for $20 (maximum of two per member). The society sold 880 books, for a total of $9840. How many members ordered two books?

64. The tens digit of a two-digit positive integer is 2 more than three times the units digit. If the digits are interchanged, the new number is 13 less than half the given number. Find the given integer. (*Hint*: Let x = the tens-place digit and y = the units-place digit; then $10x + y$ is the number.)

65. *Wood stains.*   Williams' Custom Flooring has 0.5 gal of stain that is 20% brown and 80% neutral. A customer orders 1.5 gal of a stain that is 60% brown and 40% neutral. How much pure brown stain and how much neutral stain should be added to the original 0.5 gal in order to make up the order?*

66. *Train travel.*   A train leaves Union Station for Central Station, 216 km away, at 9 A.M. One hour later, a train leaves Central Station for Union Station. They meet at noon. If the second train had started at 9 A.M. and the first train at 10:30 A.M., they would still have met at noon. Find the speed of each train.

67. *Fuel economy.*   Grady's station wagon gets 18 miles per gallon (mpg) in city driving and 24 mpg in highway driving. The car is driven 465 mi on 23 gal of gasoline. How many miles were driven in the city and how many were driven on the highway?

68. *Biochemistry.*   Industrial biochemists routinely use a machine to mix a buffer of 10% acetone by adding 100% acetone to water. One day, instead of adding 5 L of acetone to create a vat of buffer, a machine added 10 L. How much additional water was needed to bring the concentration down to 10%?

---

*This problem was suggested by Professor Chris Burditt of Yountville, California.

69. See Exercise 65 above. Let $x =$ the amount of pure brown stain added to the original 0.5 gal. Find a function $P(x)$ that can be used to determine the percentage of brown stain in the 1.5-gal mixture. On a graphing calculator, draw the graph of P and use INTERSECT to confirm the answer to Exercise 65.

70. *Gender.* Phil and Phyllis are siblings. Phyllis has twice as many brothers as she has sisters. Phil has the same number of brothers as sisters. How many girls and how many boys are in the family?

3.4 Systems of Equations in Three Variables

Identifying Solutions • Solving Systems in Three Variables • Dependency, Inconsistency, and Geometric Considerations

CONNECTING THE CONCEPTS

As often happens in mathematics, once an idea is thoroughly understood, it can be extended to increasingly more complicated problems. This is precisely the situation for the material in Sections 3.4–3.7: We will extend the elimination method of Section 3.2 to systems of three equations in three unknowns. Although we will not do so in this text, the approach that we use can be further extended to systems with four equations in four unknowns, five equations in five unknowns, and so on.

Another common occurrence in mathematics is the streamlining of a sequence of steps that are used repeatedly. In Sections 3.6 and 3.7, we develop different notations that streamline the calculations of Sections 3.2 and 3.4.

Some problems translate directly to two equations. Others more naturally call for a translation to three or more equations. In this section, we learn how to solve systems of three linear equations. Later, we will use such systems in problem-solving situations.

Identifying Solutions

A **linear equation in three variables** is an equation equivalent to one in the form $Ax + By + Cz = D$, where A, B, C, and D are real numbers. We refer to the form $Ax + By + Cz = D$ as *standard form* for a linear equation in three variables.

A solution of a system of three equations in three variables is an ordered triple (x, y, z) that makes *all three* equations true.

EXAMPLE 1   Determine whether $\left(\frac{3}{2}, -4, 3\right)$ is a solution of the system

$$4x - 2y - 3z = 5,$$
$$-8x - y + z = -5,$$
$$2x + y + 2z = 5.$$

Solution   We substitute $\left(\frac{3}{2}, -4, 3\right)$ into the three equations, using alphabetical order:

$$
\begin{array}{r|l}
\multicolumn{2}{l}{4x - 2y - 3z = 5} \\
\hline
4 \cdot \frac{3}{2} - 2(-4) - 3 \cdot 3 & 5 \\
6 + 8 - 9 & \\
5 \stackrel{?}{=} 5 & \text{TRUE}
\end{array}
\qquad
\begin{array}{r|l}
\multicolumn{2}{l}{-8x - y + z = -5} \\
\hline
-8 \cdot \frac{3}{2} - (-4) + 3 & -5 \\
-12 + 4 + 3 & \\
-5 \stackrel{?}{=} -5 & \text{TRUE}
\end{array}
$$

$$
\begin{array}{r|l}
\multicolumn{2}{l}{2x + y + 2z = 5} \\
\hline
2 \cdot \frac{3}{2} + (-4) + 2 \cdot 3 & 5 \\
3 - 4 + 6 & \\
5 \stackrel{?}{=} 5 & \text{TRUE}
\end{array}
$$

The triple makes all three equations true, so it is a solution.

Solving Systems in Three Variables

Graphical methods for solving linear equations in three variables are problematic, because a three-dimensional coordinate system is required and the graph of a linear equation in three variables is a plane. The substitution method *can* be used but becomes very cumbersome unless one or more of the equations has only two variables. Fortunately, the elimination method allows us to manipulate a system of three equations in three variables so that a simpler system of two equations in two variables is formed. Once that simpler system has been solved, we can substitute into one of the three original equations and solve for the third variable.

EXAMPLE 2   Solve the following system of equations:

$$
\begin{aligned}
x + y + z &= 4, & (1) \\
x - 2y - z &= 1, & (2) \\
2x - y - 2z &= -1. & (3)
\end{aligned}
$$

Solution   We select *any* two of the three equations and work to get one equation in two variables. Let's add equations (1) and (2):

$$
\begin{array}{ll}
x + y + z = 4 & (1) \\
\underline{x - 2y - z = 1} & (2) \\
2x - y \phantom{{}+ z} = 5. & (4) \qquad \text{Adding to eliminate } z
\end{array}
$$

Next, we select a different pair of equations and eliminate the *same variable* that we did above. Let's use equations (1) and (3) to again eliminate z. Be careful here! A common error is to eliminate a different variable in this step.

$$
\begin{array}{ll}
x + y + z = 4, \\
2x - y - 2z = -1
\end{array}
\quad
\xrightarrow[\text{of equation (1) by 2}]{\text{Multiplying both sides}}
\quad
\begin{array}{ll}
2x + 2y + 2z = 8 \\
\underline{2x - y - 2z = -1} \\
4x + y \phantom{{}+ 2z} = 7 \qquad (5)
\end{array}
$$

Now we solve the resulting system of equations (4) and (5). That solution will give us two of the numbers in the solution of the original system.

$$2x - y = 5 \qquad (4)$$
$$\underline{4x + y = 7} \qquad (5)$$
$$6x \quad\;\; = 12 \qquad \text{Adding}$$
$$x = 2$$

Note that we now have two equations in two variables. Had we not eliminated the same variable in both of the above steps, this would not be the case.

We can use either equation (4) or (5) to find y. We choose equation (5):

$$4x + y = 7 \qquad (5)$$
$$4 \cdot 2 + y = 7 \qquad \text{Substituting 2 for } x \text{ in equation (5)}$$
$$8 + y = 7$$
$$y = -1.$$

We now have $x = 2$ and $y = -1$. To find the value for z, we use any of the original three equations and substitute to find the third number, z. Let's use equation (1) and substitute our two numbers in it:

$$x + y + z = 4 \qquad (1)$$
$$2 + (-1) + z = 4 \qquad \text{Substituting 2 for } x \text{ and } -1 \text{ for } y$$
$$1 + z = 4$$
$$z = 3.$$

We have obtained the triple $(2, -1, 3)$. It should check in *all three* equations:

$$\frac{x + y + z = 4}{2 + (-1) + 3 \;\big|\; 4}$$
$$4 \overset{?}{=} 4 \quad \text{TRUE}$$

$$\frac{x - 2y - z = 1}{2 - 2(-1) - 3 \;\big|\; 1}$$
$$1 \overset{?}{=} 1 \quad \text{TRUE}$$

$$\frac{2x - y - 2z = -1}{2 \cdot 2 - (-1) - 2 \cdot 3 \;\big|\; -1}$$
$$-1 \overset{?}{=} -1 \quad \text{TRUE}$$

The solution is $(2, -1, 3)$.

Solving Systems of Three Linear Equations

To use the elimination method to solve systems of three linear equations:

1. Write all equations in the standard form $Ax + By + Cz = D$.
2. Clear any decimals or fractions.
3. Choose a variable to eliminate. Then select two of the three equations and work to get one equation in which the selected variable is eliminated.
4. Next, use a different pair of equations and eliminate the same variable that you did in step (3).
5. Solve the system of equations that resulted from steps (3) and (4).
6. Substitute the solution from step (5) into one of the original three equations and solve for the third variable. Then check.

EXAMPLE 3   Solve the system

$$4x - 2y - 3z = 5, \qquad (1)$$
$$-8x - y + z = -5, \qquad (2)$$
$$2x + y + 2z = 5. \qquad (3)$$

Student Notes _____

Because solving systems of three equations can be lengthy, it is important that you use plenty of paper, work in pencil, and double-check each step as you proceed.

Solution

1., 2. The equations are already in standard form with no fractions or decimals.

3. Next, select a variable to eliminate. We decide on y because the y-terms are opposites of each other in equations (2) and (3). We add:

$$-8x - y + z = -5 \qquad (2)$$
$$\underline{2x + y + 2z = 5} \qquad (3)$$
$$-6x \qquad + 3z = 0. \qquad (4) \qquad \text{Adding}$$

4. We use another pair of equations to create a second equation in x and z. That is, we eliminate the same variable, y, as in step (3). We use equations (1) and (3):

$$
\begin{array}{l}
4x - 2y - 3z = 5, \\
2x + y + 2z = 5
\end{array}
\quad
\xrightarrow[\text{of equation (3) by 2}]{\text{Multiplying both sides}}
\quad
\begin{array}{l}
4x - 2y - 3z = 5 \\
\underline{4x + 2y + 4z = 10} \\
8x \qquad + z = 15. \qquad (5)
\end{array}
$$

5. Now we solve the resulting system of equations (4) and (5). That allows us to find two parts of the ordered triple.

$$
\begin{array}{l}
-6x + 3z = 0, \\
8x + z = 15
\end{array}
\quad
\xrightarrow[\text{of equation (5) by } -3]{\text{Multiplying both sides}}
\quad
\begin{array}{l}
-6x + 3z = 0 \\
\underline{-24x - 3z = -45} \\
-30x \qquad = -45 \\
x = \frac{-45}{-30} = \frac{3}{2}
\end{array}
$$

We use equation (5) to find z:

$$8x + z = 15$$
$$8 \cdot \tfrac{3}{2} + z = 15 \qquad \text{Substituting } \tfrac{3}{2} \text{ for } x$$
$$12 + z = 15$$
$$z = 3.$$

6. Finally, we use any of the original equations and substitute to find the third number, y. We choose equation (3):

$$2x + y + 2z = 5 \qquad (3)$$
$$2 \cdot \tfrac{3}{2} + y + 2 \cdot 3 = 5 \qquad \text{Substituting } \tfrac{3}{2} \text{ for } x \text{ and 3 for } z$$
$$3 + y + 6 = 5$$
$$y + 9 = 5$$
$$y = -4.$$

The solution is $\left(\tfrac{3}{2}, -4, 3\right)$. The check was performed as Example 1.

Sometimes, certain variables are missing at the outset.

EXAMPLE 4   Solve the system

$$x + y + z = 180, \quad (1)$$
$$x \quad\;\; - z = -70, \quad (2)$$
$$2y - z = \quad 0. \quad (3)$$

Solution

1., 2. The equations appear in standard form with no fractions or decimals.

3., 4. Note that there is no y in equation (2). Thus, at the outset, we already have y eliminated from one equation. We need another equation with y eliminated, so we use equations (1) and (3):

$$
\begin{array}{l}
x + y + z = 180, \\
2y - z = \quad 0
\end{array}
\xrightarrow[\text{of equation (1) by } -2]{\text{Multiplying both sides}}
\begin{array}{r}
-2x - 2y - 2z = -360 \\
2y - \;z = \quad 0 \\
\hline
-2x \quad\;\; - 3z = -360. \quad (4)
\end{array}
$$

5., 6. Now we solve the resulting system of equations (2) and (4):

$$
\begin{array}{l}
x - \;z = \;-70, \\
-2x - 3z = -360
\end{array}
\xrightarrow[\text{of equation (2) by } 2]{\text{Multiplying both sides}}
\begin{array}{r}
2x - 2z = -140 \\
-2x - 3z = -360 \\
\hline
-5z = -500 \\
z = \quad 100.
\end{array}
$$

Continuing as in Examples 2 and 3, we get the solution (30, 50, 100). The check is left to the student.

Dependency, Inconsistency, and Geometric Considerations

Each equation in Examples 2, 3, and 4 has a graph that is a plane in three dimensions. The solutions are points common to the planes of each system. Since three planes can have an infinite number of points in common or no points at all in common, we need to generalize the concept of *consistency*.

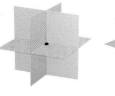  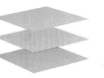

Planes intersect at one point. System is *consistent* and has one solution.

Planes intersect along a common line. System is *consistent* and has an infinite number of solutions.

Three parallel planes. System is *inconsistent;* it has no solution.

Planes intersect two at a time, with no point common to all three. System is *inconsistent;* it has no solution.

Consistency

A system of equations that has at least one solution is said to be **consistent**.

A system of equations that has no solution is said to be **inconsistent**.

EXAMPLE 5   Solve:

$$y + 3z = 4, \qquad (1)$$
$$-x - y + 2z = 0, \qquad (2)$$
$$x + 2y + z = 1. \qquad (3)$$

Solution   The variable x is missing in equation (1). By adding equations (2) and (3), we can find a second equation in which x is missing:

$$-x - y + 2z = 0 \qquad (2)$$
$$\underline{x + 2y + z = 1} \qquad (3)$$
$$y + 3z = 1. \qquad (4) \qquad \text{Adding}$$

Equations (1) and (4) form a system in y and z. We solve as before:

$$
y + 3z = 4, \quad \xrightarrow[\text{of equation (1) by } -1]{\text{Multiplying both sides}} \quad -y - 3z = -4
$$
$$
y + 3z = 1 \qquad\qquad\qquad\qquad\qquad \underline{y + 3z = \ \ 1}
$$
$$
\text{This is a contradiction.} \ \longrightarrow \ 0 = -3. \qquad \text{Adding}
$$

Since we end up with a *false* equation, or contradiction, we know that the system has no solution. It is *inconsistent*.

The notion of *dependency* from Section 3.1 can also be extended.

EXAMPLE 6   Solve:

$$2x + y + z = 3, \qquad (1)$$
$$x - 2y - z = 1, \qquad (2)$$
$$3x + 4y + 3z = 5. \qquad (3)$$

Solution   Our plan is to first use equations (1) and (2) to eliminate z. Then we will select another pair of equations and again eliminate z:

$$2x + y + z = 3$$
$$\underline{x - 2y - z = 1}$$
$$3x - y \qquad = 4. \qquad (4)$$

Next, we use equations (2) and (3) to eliminate z again:

$$
x - 2y - z = 1, \quad \xrightarrow[\text{of equation (2) by 3}]{\text{Multiplying both sides}} \quad 3x - 6y - 3z = 3
$$
$$
3x + 4y + 3z = 5 \qquad\qquad\qquad\qquad \underline{3x + 4y + 3z = 5}
$$
$$
6x - 2y \qquad = 8. \qquad (5)
$$

We now try to solve the resulting system of equations (4) and (5):

$$3x - y = 4, \quad \xrightarrow{\text{Multiplying both sides of equation (4) by } -2} \quad -6x + 2y = -8$$
$$6x - 2y = 8 \qquad\qquad\qquad\qquad\qquad \underline{6x - 2y = 8}$$
$$0 = 0. \quad (6)$$

Equation (6), which is an identity, indicates that equations (1), (2), and (3) are *dependent*. This means that the original system of three equations is equivalent to a system of two equations. One way to see this is to observe that two times equation (1), minus equation (2), is equation (3). Thus removing equation (3) from the system does not affect the solution of the system.* In writing an answer to this problem, we simply state that "the equations are dependent."

Recall that when dependent equations appeared in Section 3.1, the solution sets were always infinite in size and were written in set-builder notation. There, all systems of dependent equations were *consistent*. This is not always the case for systems of three or more equations. The following figures illustrate some possibilities geometrically.

The planes intersect along a common line. The equations are *dependent* and the system is *consistent*. There is an infinite number of solutions.

The planes coincide. The equations are *dependent* and the system is *consistent*. There is an infinite number of solutions.

Two planes coincide. The third plane is parallel. The equations are *dependent* and the system is *inconsistent*. There is no solution.

Exercise Set

3.4

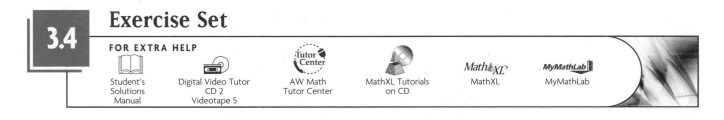

FOR EXTRA HELP

Student's Solutions Manual

Digital Video Tutor CD 2 Videotape 5

AW Math Tutor Center

MathXL Tutorials on CD

MathXL

MyMathLab

🔖 *Concept Reinforcement   Classify each statement as either true or false.*

1. $3x + 5y + 4z = 7$ is a linear equation in three variables.

2. It is not difficult to solve a system of three equations in three unknowns by graphing.

3. Every system of three equations in three unknowns has at least one solution.

4. If, when we are solving a system of three equations, a false equation results from adding a multiple of one equation to another, the system is inconsistent.

---

*A set of equations is dependent if at least one equation can be expressed as a sum of multiples of other equations in that set.

5. If, when we are solving a system of three equations, an identity results from adding a multiple of one equation to another, the equations are dependent.

6. Whenever a system of three equations contains dependent equations, there is an infinite number of solutions.

7. Determine whether $(2, -1, -2)$ is a solution of the system

$$x + y - 2z = 5,$$
$$2x - y - z = 7,$$
$$-x - 2y + 3z = 6.$$

8. Determine whether $(1, -2, 3)$ is a solution of the system

$$x + y + z = 2,$$
$$x - 2y - z = 2,$$
$$3x + 2y + z = 2.$$

Solve each system. If a system's equations are dependent or if there is no solution, state this.

9. $2x - y + z = 10,$
$4x + 2y - 3z = 10,$
$x - 3y + 2z = 8$

10. $x + y + z = 6,$
$2x - y + 3z = 9,$
$-x + 2y + 2z = 9$

11. $x - y + z = 6,$
$2x + 3y + 2z = 2,$
$3x + 5y + 4z = 4$

12. $2x - y - 3z = -1,$
$2x - y + z = -9,$
$x + 2y - 4z = 17$

13. $6x - 4y + 5z = 31,$
$5x + 2y + 2z = 13,$
$x + y + z = 2$

14. $2x - 3y + z = 5,$
$x + 3y + 8z = 22,$
$3x - y + 2z = 12$

15. $x + y + z = 0,$
$2x + 3y + 2z = -3,$
$-x - 2y - z = 1$

16. $3a - 2b + 7c = 13,$
$a + 8b - 6c = -47,$
$7a - 9b - 9c = -3$

17. $2x + y - 3z = -4,$
$4x - 2y + z = 9,$
$3x + 5y - 2z = 5$

18. $4x + y + z = 17,$
$x - 3y + 2z = -8,$
$5x - 2y + 3z = 5$

19. $2x + y + 2z = 11,$
$3x + 2y + 2z = 8,$
$x + 4y + 3z = 0$

20. $2x + y + z = -2,$
$2x - y + 3z = 6,$
$3x - 5y + 4z = 7$

21. $-2x + 8y + 2z = 4,$
$x + 6y + 3z = 4,$
$3x - 2y + z = 0$

22. $x - y + z = 4,$
$5x + 2y - 3z = 2,$
$4x + 3y - 4z = -2$

23. $4x - y - z = 4,$
$2x + y + z = -1,$
$6x - 3y - 2z = 3$

24. $a + 2b + c = 1,$
$7a + 3b - c = -2,$
$a + 5b + 3c = 2$

25. $r + \frac{3}{2}s + 6t = 2,$
$2r - 3s + 3t = 0.5,$
$r + s + t = 1$

26. $5x + 3y + \frac{1}{2}z = \frac{7}{2},$
$0.5x - 0.9y - 0.2z = 0.3,$
$3x - 2.4y + 0.4z = -1$

27. $4a + 9b = 8,$
$8a + 6c = -1,$
$6b + 6c = -1$

28. $3p + 2r = 11,$
$q - 7r = 4,$
$p - 6q = 1$

29. $x + y + z = 57,$
$-2x + y = 3,$
$x - z = 6$

30. $x + y + z = 105,$
$10y - z = 11,$
$2x - 3y = 7$

31. $a - 3c = 6,$
$b + 2c = 2,$
$7a - 3b - 5c = 14$

32. $2a - 3b = 2,$
$7a + 4c = \frac{3}{4},$
$2c - 3b = 1$

Aha! **33.** $x + y + z = 83,$
$y = 2x + 3,$
$z = 40 + x$

34. $l + m = 7,$
$3m + 2n = 9,$
$4l + n = 5$

35. $x + z = 0,$
$x + y + 2z = 3,$
$y + z = 2$

36. $x + y = 0,$
$x + z = 1,$
$2x + y + z = 2$

37. $x + y + z = 1,$
$-x + 2y + z = 2,$
$2x - y = -1$

38. $y + z = 1,$
$x + y + z = 1,$
$x + 2y + 2z = 2$

39. Describe a method for writing an inconsistent system of three equations in three variables.

40. Abbie recommends that a frustrated classmate double- and triple-check each step of work when attempting to solve a system of three equations. Is this good advice? Why or why not?

SKILL MAINTENANCE

Translate each sentence to mathematics. [1.1]

41. One number is twice another.

42. The sum of two numbers is three times the first number.

43. The sum of three consecutive numbers is 45.

44. One number plus twice another number is 17.

45. The sum of two numbers is five times a third number.

46. The product of two numbers is twice their sum.

SYNTHESIS

🗒 **47.** Is it possible for a system of three linear equations to have exactly two ordered triples in its solution set? Why or why not?

🗒 **48.** Describe a procedure that could be used to solve a system of four equations in four variables.

Solve.

49. $\dfrac{x+2}{3} - \dfrac{y+4}{2} + \dfrac{z+1}{6} = 0,$

$\dfrac{x-4}{3} + \dfrac{y+1}{4} - \dfrac{z-2}{2} = -1,$

$\dfrac{x+1}{2} + \dfrac{y}{2} + \dfrac{z-1}{4} = \dfrac{3}{4}$

50. $w + x + y + z = 2,$
$w + 2x + 2y + 4z = 1,$
$w - x + y + z = 6,$
$w - 3x - y + z = 2$

51. $w + x - y + z = 0,$
$w - 2x - 2y - z = -5,$
$w - 3x - y + z = 4,$
$2w - x - y + 3z = 7$

For Exercises 52 and 53, let u represent 1/x, v represent 1/y, and w represent 1/z. Solve for u, v, and w, and then solve for x, y, and z.

52. $\dfrac{2}{x} - \dfrac{1}{y} - \dfrac{3}{z} = -1,$

$\dfrac{2}{x} - \dfrac{1}{y} + \dfrac{1}{z} = -9,$

$\dfrac{1}{x} + \dfrac{2}{y} - \dfrac{4}{z} = 17$

53. $\dfrac{2}{x} + \dfrac{2}{y} - \dfrac{3}{z} = 3,$

$\dfrac{1}{x} - \dfrac{2}{y} - \dfrac{3}{z} = 9,$

$\dfrac{7}{x} - \dfrac{2}{y} + \dfrac{9}{z} = -39$

Determine k so that each system is dependent.

54. $x - 3y + 2z = 1,$
$2x + y - z = 3,$
$9x - 6y + 3z = k$

55. $5x - 6y + kz = -5,$
$x + 3y - 2z = 2,$
$2x - y + 4z = -1$

In each case, three solutions of an equation in x, y, and z are given. Find the equation.

56. $Ax + By + Cz = 12;$
$\left(1, \frac{3}{4}, 3\right), \left(\frac{4}{3}, 1, 2\right),$ and $(2, 1, 1)$

57. $z = b - mx - ny;$
$(1, 1, 2), (3, 2, -6),$ and $\left(\frac{3}{2}, 1, 1\right)$

58. Write an inconsistent system of equations that contains dependent equations.

CORNER

Finding the Preferred Approach

COLLABORATIVE

Focus: Systems of three linear equations
Time: 10–15 minutes
Group size: 3

Consider the six steps outlined on p. 183 along with the following system:

$$2x + 4y = 3 - 5z,$$
$$0.3x = 0.2y + 0.7z + 1.4,$$
$$0.04x + 0.03y = 0.07 + 0.04z.$$

ACTIVITY

1. Working independently, each group member should solve the system above. One person should begin by eliminating *x*, one should first eliminate *y*, and one should first eliminate *z*. Write neatly so that others can follow your steps.

2. Once all group members have solved the system, compare your answers. If the answers do not check, exchange notebooks and check each other's work. If a mistake is detected, allow the person who made the mistake to make the repair.

3. Decide as a group which of the three approaches above (if any) ranks as easiest and which (if any) ranks as most difficult. Then compare your rankings with the other groups in the class.

3.5 Solving Applications: Systems of Three Equations

Applications of Three Equations in Three Unknowns

Solving systems of three or more equations is important in many applications. Such systems arise in the natural and social sciences, business, and engineering. In mathematics, purely numerical applications also arise.

EXAMPLE 1 The sum of three numbers is 4. The first number minus twice the second, minus the third is 1. Twice the first number minus the second, minus twice the third is −1. Find the numbers.

Solution

1. **Familiarize.** There are three statements involving the same three numbers. Let's label these numbers x, y, and z.

2. **Translate.** We can translate directly as follows.

$$\underbrace{\text{The sum of the three numbers}}_{x + y + z} \;\; \underbrace{\text{is}}_{=} \;\; \underbrace{4.}_{4}$$

$$\underbrace{\text{The first number}}_{x} \;\; \underbrace{\text{minus}}_{-} \;\; \underbrace{\text{twice the second}}_{2y} \;\; \underbrace{\text{minus}}_{-} \;\; \underbrace{\text{the third}}_{z} \;\; \underbrace{\text{is}}_{=} \;\; \underbrace{1.}_{1}$$

$$\underbrace{\text{Twice the first number}}_{2x} \;\; \underbrace{\text{minus}}_{-} \;\; \underbrace{\text{the second}}_{y} \;\; \underbrace{\text{minus}}_{-} \;\; \underbrace{\text{twice the third}}_{2z} \;\; \underbrace{\text{is}}_{=} \;\; \underbrace{-1.}_{-1}$$

We now have a system of three equations:

$$x + y + z = 4,$$
$$x - 2y - z = 1,$$
$$2x - y - 2z = -1.$$

3. **Carry out.** We need to solve the system of equations. Note that we found the solution, $(2, -1, 3)$, in Example 2 of Section 3.4.

4. **Check.** The first statement of the problem says that the sum of the three numbers is 4. That checks, because $2 + (-1) + 3 = 4$. The second statement says that the first number minus twice the second, minus the third is 1: $2 - 2(-1) - 3 = 1$. That checks. The check of the third statement is left to the student.

5. **State.** The three numbers are 2, −1, and 3.

EXAMPLE 2   Architecture.   In a triangular cross section of a roof, the largest angle is 70° greater than the smallest angle. The largest angle is twice as large as the remaining angle. Find the measure of each angle.

Solution

1. Familiarize.   The first thing we do is make a drawing, or a sketch.

Student Notes ____

It is quite likely that you are expected to remember that the sum of the measures of the angles in any triangle is 180°. You may want to ask your instructor which other formulas from geometry and elsewhere you are expected to know.

Since we don't know the size of any angle, we use x, y, and z to represent the three measures, from smallest to largest. Recall that the measures of the angles in any triangle add up to 180°.

2. Translate.   This geometric fact about triangles gives us one equation:

$$x + y + z = 180.$$

Two of the statements can be translated almost directly.

The largest angle   is   70° greater than the smallest angle.
$$z \quad = \quad x + 70$$

The largest angle   is   twice as large as the remaining angle.
$$z \quad = \quad 2y$$

We now have a system of three equations:

$$
\begin{aligned}
x + y + z &= 180, & \quad x + y + z &= 180, \\
x + 70 &= z, & \text{or} \quad x \quad\; - z &= -70, \\
2y &= z; & 2y - z &= 0.
\end{aligned}
$$
Rewriting in standard form

3. Carry out.   The system was solved in Example 4 of Section 3.4. The solution is (30, 50, 100).

4. Check.   The sum of the numbers is 180, so that checks. The measure of the largest angle, 100°, is 70° greater than the measure of the smallest angle, 30°, so that checks. The measure of the largest angle is also twice the measure of the remaining angle, 50°. Thus we have a check.

5. State.   The angles in the triangle measure 30°, 50°, and 100°.

EXAMPLE 3

Cholesterol levels. Recent studies indicate that a child's intake of cholesterol should be no more than 300 mg per day. By eating 1 egg, 1 cupcake, and 1 slice of pizza, a child consumes 302 mg of cholesterol. A child who eats 2 cupcakes and 3 slices of pizza takes in 65 mg of cholesterol. By eating 2 eggs and 1 cupcake, a child consumes 567 mg of cholesterol. How much cholesterol is in each item?

Solution

1. **Familiarize.** After reading the problem, it becomes clear that an egg contains considerably more cholesterol than the other foods. Let's guess that one egg contains 200 mg of cholesterol and one cupcake contains 50 mg. Because of the second sentence in the problem, it would follow that a slice of pizza contains 52 mg of cholesterol since $200 + 50 + 52 = 302$.

   To see if our guess satisfies the other statements in the problem, we find the amount of cholesterol that 2 cupcakes and 3 slices of pizza would contain: $2 \cdot 50 + 3 \cdot 52 = 256$. Since this does not match the 65 mg listed in the fourth sentence of the problem, our guess was incorrect. Rather than guess again, we examine how we checked our guess and let g, c, and $s =$ the number of milligrams of cholesterol in an egg, a cupcake, and a slice of pizza, respectively.

2. **Translate.** Rewording some of the sentences, we can translate as follows:

   The amount of cholesterol in 1 egg | plus | the amount of cholesterol in 1 cupcake | plus | the amount of cholesterol in 1 slice of pizza | is | 302 mg.

   $$g + c + s = 302$$

   The amount of cholesterol in 2 cupcakes | plus | the amount of cholesterol in 3 slices of pizza | is | 65 mg.

   $$2c + 3s = 65$$

   The amount of cholesterol in 2 eggs | plus | the amount of cholesterol in 1 cupcake | is | 567 mg.

   $$2g + c = 567$$

   We now have a system of three equations:

   $$g + c + s = 302,$$
   $$2c + 3s = 65,$$
   $$2g + c = 567.$$

3. **Carry out.** We solve and get $g = 274$, $c = 19$, and $s = 9$.

4. **Check.** The sum of 274, 19, and 9 is 302 so the total cholesterol in 1 egg, 1 cupcake, and 1 slice of pizza checks. Two cupcakes and three slices of pizza would contain $2 \cdot 19 + 3 \cdot 9 = 65$ mg, while two eggs and one cupcake would contain $2 \cdot 274 + 19 = 567$ mg of cholesterol. The answer checks.

5. **State.** An egg contains 274 mg of cholesterol, a cupcake contains 19 mg of cholesterol, and a slice of pizza contains 9 mg of cholesterol.

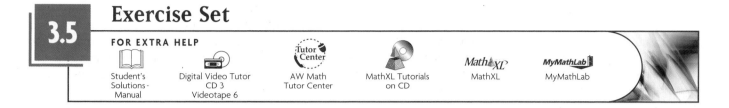

3.5 Exercise Set

Solve.

1. The sum of three numbers is 57. The second is 3 more than the first. The third is 6 more than the first. Find the numbers.

2. The sum of three numbers is 5. The first number minus the second plus the third is 1. The first minus the third is 3 more than the second. Find the numbers.

3. The sum of three numbers is 26. Twice the first minus the second is 2 less than the third. The third is the second minus three times the first. Find the numbers.

4. The sum of three numbers is 105. The third is 11 less than ten times the second. Twice the first is 7 more than three times the second. Find the numbers.

5. *Geometry.*   In triangle *ABC*, the measure of angle *B* is three times that of angle *A*. The measure of angle *C* is 20° more than that of angle *A*. Find the angle measures.

6. *Geometry.*   In triangle *ABC*, the measure of angle *B* is twice the measure of angle *A*. The measure of angle *C* is 80° more than that of angle *A*. Find the angle measures.

7. *Health insurance.*   In 2004, UNICARE® health insurance for adults under the age of 30 cost $121/month for a couple, $107/month for an adult with one child, and $164/month for a couple with one child. On the basis of the information given, find the monthly rate for an individual adult, a spouse, and a child.
   Source: UNICARE Life and Health Insurance Company® advertisement

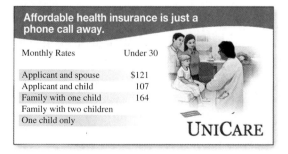

8. *Health insurance.*   In 2004, UNICARE® health insurance cost $160/month for a 35–39-year-old adult and spouse, $145/month for a 35–39-year-old adult with one child, and $245/month for a 39-year-old adult with a spouse and child. On the basis of the information given, find the monthly rates for a 39-year-old adult, a 39-year-old's spouse, and a 39-year-old's child.
   Source: UNICARE Life and Insurance Company® advertisement

9. *Nutrition.*   Most nutritionists now agree that a healthy adult diet should include 25–35 g of fiber each day. A breakfast of 2 bran muffins, 1 banana, and a 1-cup serving of Wheaties® contains 9 g of fiber; a breakfast of 1 bran muffin, 2 bananas, and a 1-cup serving of Wheaties® contains 10.5 g of fiber; and a breakfast of 2 bran muffins and a 1-cup serving of Wheaties® contains 6 g of fiber. How much fiber is in each of these foods?
   Sources: usda.gov and InteliHealth.com

10. *Nutrition.*   Refer to Exercise 9. A breakfast consisting of 2 pancakes and a 1-cup serving of strawberries contains 4.5 g of fiber, whereas a breakfast of 2 pancakes and a 1-cup serving of Cheerios® contains 4 g of fiber. When a meal consists of 1 pancake, a 1-cup serving of Cheerios®, and a 1-cup serving of strawberries, it contains 7 g of fiber. How much fiber is in each of these foods?
    Source: InteliHealth.com

Aha! **11.** *Automobile pricing.*   The basic model of a 2004 Jeep Grand Cherokee Laredo (2WD) with a power sunroof cost $25,495. When equipped with 4WD and a sunroof, the vehicle's price rose to $27,465. The cost of the basic model with 4WD was $26,665. Find the basic price, the cost of 4WD, and the cost of a sunroof.

12. *Lens production.*   When Sight-Rite's three polishing machines, A, B, and C, are all working, 5700 lenses can be polished in one week. When only A and B are working, 3400 lenses can be polished in one week. When only B and C are working, 4200 lenses can be polished in one week. How many lenses can be polished in a week by each machine?

13. *Welding rates.*   Elrod, Dot, and Wendy can weld 74 linear feet per hour when working together. Elrod and Dot together can weld 44 linear feet per hour, while Elrod and Wendy can weld 50 linear feet per hour. How many linear feet per hour can each weld alone?

14. *Telemarketing.*   Sven, Tillie, and Isaiah can process 740 telephone orders per day. Sven and Tillie together can process 470 orders, while Tillie and Isaiah together can process 520 orders per day. How many orders can each person process alone?

15. *Coffee prices.*   Roz works at a Starbucks® coffee shop where a 12-oz cup of coffee costs $1.40, a 16-oz cup costs $1.60, and a 20-oz cup costs $1.70. During one busy period, Roz served 55 cups of coffee, emptying six 144-oz "brewers" while collecting a total of $85.90. How many cups of each size did Roz fill?

| 12 oz | 16 oz | 20 oz |
| $1.40 | $1.60 | $1.70 |

16. *Advertising.*   In a recent year, U.S. companies spent a total of $106.5 billion on newspaper, television, and radio ads. The total amount spent on television and radio ads was $18.7 billion more than the amount spent on newspaper ads alone. The amount spent on newspaper ads was $28.8 billion more than what was spent on radio ads. How much was spent on each form of advertising? **Sources**: NAA (newspapers); McCann–Erickson Inc. (television and radio)

17. *Restaurant management.*   McDonald's® recently sold small soft drinks for $1, medium soft drinks for $1.15, and large soft drinks for $1.30. During a lunch-time rush, Chris sold 40 soft drinks for a total of $45.25. The number of small and large drinks, combined, was 10 fewer than the number of medium drinks. How many drinks of each size were sold?

18. *Investments.*   A business class divided an imaginary investment of $80,000 among three mutual funds. The first fund grew by 10%, the second by 6%, and the third by 15%. Total earnings were $8850. The earnings from the first fund were $750 more than the earnings from the third. How much was invested in each fund?

19. *Nutrition.*   A dietician in a hospital prepares meals under the guidance of a physician. Suppose that for a particular patient a physician prescribes a meal to have 800 calories, 55 g of protein, and 220 mg of vitamin C. The dietician prepares a meal of roast beef, baked potatoes, and broccoli according to the data in the following table.

Serving Size	Calories	Protein (in grams)	Vitamin C (in milligrams)
Roast Beef, 3 oz	300	20	0
Baked Potato, 1	100	5	20
Broccoli, 156 g	50	5	100

How many servings of each food are needed in order to satisfy the doctor's orders?

20. *Nutrition.*   Repeat Exercise 19 but replace the broccoli with asparagus, for which a 180-g serving contains 50 calories, 5 g of protein, and 44 mg of vitamin C. Which meal would you prefer eating?

21. *World population growth.*   The world population is projected to be 9.1 billion in 2050. At that time, there are expected to be approximately 3 billion more people in Asia than in Africa. The population for the rest of the world will be approximately 0.1 billion more than half the population of Asia. Find the projected populations of Asia, Africa, and the rest of the world in 2050.
Sources: U.S. Bureau of the Census; *Burlington Free Press* 3/23/04

22. *Crying rate.*   The sum of the average number of times a man, a woman, and a one-year-old child cry each month is 56.7. A woman cries 3.9 more times than a man. The average number of times a one-year-old cries per month is 43.3 more than the average number of times combined that a man and a woman cry. What is the average number of times per month that each cries?

23. *Basketball scoring.*   The New York Knicks recently scored a total of 92 points on a combination of 2-point field goals, 3-point field goals, and 1-point foul shots. Altogether, the Knicks made 50 baskets and 19 more 2-pointers than foul shots. How many ·shots of each kind were made?

24. *History.*   Find the year in which the first U.S. transcontinental railroad was completed. The following are some facts about the number. The sum of the digits in the year is 24. The ones digit is 1 more than the hundreds digit. Both the tens and the ones digits are multiples of 3.

25. Problems like Exercises 15 and 17 could be classified as total-value problems. How do these problems differ from the total-value problems of Section 3.3?

26. Write a problem for a classmate to solve. Design the problem so that it translates to a system of three equations in three variables.

SKILL MAINTENANCE

Simplify. [1.2]

27. $5(-3) + 7$

28. $-4(-6) + 9$

29. $-6(8) + (-7)$

30. $7(-9) + (-8)$

31. $-7(2x - 3y + 5z)$

32. $-6(4a + 7b - 9c)$

33. $-4(2a + 5b) + 3a + 20b$

34. $3(2x - 7y) + 5x + 21y$

SYNTHESIS

35. Consider Exercise 23. Suppose there were no foul shots made. Would there still be a solution? Why or why not?

36. Consider Exercise 15. Suppose Roz collected $46. Could the problem still be solved? Why or why not?

37. *Health insurance.*   In 2004, UNICARE® health insurance for a 35–39-year-old and his or her spouse cost $160/month. That rate increased to $203/month if a child were included and $243/month if two children were included. The rate dropped to $145/month for just the applicant and one child. Find the separate costs for insuring the applicant, the spouse, the first child, and the second child.
Source: UNICARE Life and Health Insurance Company® advertisement

38. Find a three-digit positive integer such that the sum of all three digits is 14, the tens digit is 2 more than the ones digit, and if the digits are reversed, the number is unchanged.

39. *Ages.*   Tammy's age is the sum of the ages of Carmen and Dennis. Carmen's age is 2 more than the sum of the ages of Dennis and Mark. Dennis's age is four times Mark's age. The sum of all four ages is 42. How old is Tammy?

40. *Ticket revenue.* A magic show's audience of 100 people consists of adults, students, and children. The ticket prices are $10 for adults, $3 for students, and 50¢ for children. The total amount of money taken in is $100. How many adults, students, and children are in attendance? Does there seem to be some information missing? Do some more careful reasoning.

41. *Sharing raffle tickets.* Hal gives Tom as many raffle tickets as Tom first had and Gary as many as Gary first had. In like manner, Tom then gives Hal and Gary as many tickets as each then has. Similarly, Gary gives Hal and Tom as many tickets as each then has. If each finally has 40 tickets, with how many tickets does Tom begin?

42. Find the sum of the angle measures at the tips of the star in this figure.

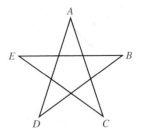

3.6 Elimination Using Matrices

Matrices and Systems • Row-Equivalent Operations

In solving systems of equations, we perform computations with the constants. The variables play no important role until the end. Thus we can simplify writing a system by omitting the variables. For example, the system

$$3x + 4y = 5,$$
$$x - 2y = 1$$

simplifies to

$$\begin{matrix} 3 & 4 & 5 \\ 1 & -2 & 1 \end{matrix}$$

if we do not write the variables, the operation of addition, and the equals signs.

Matrices and Systems

In the example above, we have written a rectangular array of numbers. Such an array is called a **matrix** (plural, **matrices**). We ordinarily write brackets around matrices. The following are matrices:

$$\begin{bmatrix} -3 & 1 \\ 0 & 5 \end{bmatrix}, \begin{bmatrix} 2 & 0 & -1 & 3 \\ -5 & 2 & 7 & -1 \\ 4 & 5 & 3 & 0 \end{bmatrix}, \begin{bmatrix} 2 & 3 \\ 7 & 15 \\ -2 & 23 \\ 4 & 1 \end{bmatrix}$$

The individual numbers are called *elements* or *entries*.

The **rows** of a matrix are horizontal, and the **columns** are vertical.

Let's see how matrices can be used to solve a system.

EXAMPLE 1   Solve the system

$$5x - 4y = -1,$$
$$-2x + 3y = 2.$$

As an aid for understanding, we list the corresponding system in the margin.

$$5x - 4y = -1,$$
$$-2x + 3y = 2$$

Solution   We write a matrix using only coefficients and constants, listing x-coefficients in the first column and y-coefficients in the second. Note that in each matrix a dashed line separates the coefficients from the constants:

$$\begin{bmatrix} 5 & -4 & | & -1 \\ -2 & 3 & | & 2 \end{bmatrix}.$$

Consult the notes in the margin for further information.

Our goal is to transform

$$\begin{bmatrix} 5 & -4 & | & -1 \\ -2 & 3 & | & 2 \end{bmatrix} \quad \text{into the form} \quad \begin{bmatrix} a & b & | & c \\ 0 & d & | & e \end{bmatrix}.$$

The variables x and y can then be reinserted to form equations from which we can complete the solution.

We do calculations that are similar to those that we would do if we wrote the entire equations. The first step is to multiply and/or interchange the rows so that each number in the first column below the first number is a multiple of that number. Here that means multiplying Row 2 by 5. This corresponds to multiplying both sides of the second equation by 5.

$$5x - 4y = -1,$$
$$-10x + 15y = 10$$

$$\begin{bmatrix} 5 & -4 & | & -1 \\ -10 & 15 & | & 10 \end{bmatrix}$$

New Row 2 = 5(Row 2 from above)

Next, we multiply the first row by 2, add this to Row 2, and write that result as the "new" Row 2. This corresponds to multiplying the first equation by 2 and adding the result to the second equation in order to eliminate a variable. Write out these computations as necessary—we perform them mentally.

$$5x - 4y = -1,$$
$$7y = 8$$

$$\begin{bmatrix} 5 & -4 & | & -1 \\ 0 & 7 & | & 8 \end{bmatrix}$$

$2(5 \quad -4 \; | \; -1) = (10 \quad -8 \; | \; -2)$ and
$(10 \quad -8 \; | \; -2) + (-10 \quad 15 \; | \; 10) = (0 \quad 7 \; | \; 8)$
New Row 2 = 2(Row 1) + (Row 2)

If we now reinsert the variables, we have

$$5x - 4y = -1, \qquad (1)$$
$$7y = 8. \qquad (2)$$

We can now proceed as before, solving equation (2) for y:

$$7y = 8 \qquad (2)$$
$$y = \tfrac{8}{7}.$$

Next, we substitute $\tfrac{8}{7}$ for y in equation (1):

$$5x - 4y = -1 \qquad (1)$$
$$5x - 4 \cdot \tfrac{8}{7} = -1 \qquad \text{Substituting } \tfrac{8}{7} \text{ for } y \text{ in equation (1)}$$
$$x = \tfrac{5}{7}. \qquad \text{Solving for } x$$

The solution is $\left(\tfrac{5}{7}, \tfrac{8}{7}\right)$. The check is left to the student.

EXAMPLE 2 Solve the system

$$2x - y + 4z = -3,$$
$$x \quad\;\; - 4z = 5,$$
$$6x - y + 2z = 10.$$

Solution We first write a matrix, using only the constants. Where there are missing terms, we must write 0's:

$$2x - y + 4z = -3,$$
$$x \quad\;\; - 4z = 5,$$
$$6x - y + 2z = 10$$

$$\begin{bmatrix} 2 & -1 & 4 & | & -3 \\ 1 & 0 & -4 & | & 5 \\ 6 & -1 & 2 & | & 10 \end{bmatrix}$$

Our goal is to transform the matrix to one of the form

$$ax + by + cz = d,$$
$$ey + fz = g,$$
$$hz = i$$

$$\begin{bmatrix} a & b & c & | & d \\ 0 & e & f & | & g \\ 0 & 0 & h & | & i \end{bmatrix}.$$

A matrix of this form can be rewritten as a system of equations that is equivalent to the original system, and from which a solution can be easily found.

The first step is to multiply and/or interchange the rows so that each number in the first column is a multiple of the first number in the first row. In this case, we do so by interchanging Rows 1 and 2:

$$x \quad\;\; - 4z = 5,$$
$$2x - y + 4z = -3,$$
$$6x - y + 2z = 10$$

$$\begin{bmatrix} 1 & 0 & -4 & | & 5 \\ 2 & -1 & 4 & | & -3 \\ 6 & -1 & 2 & | & 10 \end{bmatrix}$$

This corresponds to interchanging the first two equations.

Next, we multiply the first row by -2, add it to the second row, and replace Row 2 with the result:

$$x \quad\;\; - 4z = 5,$$
$$-y + 12z = -13,$$
$$6x - y + 2z = 10$$

$$\begin{bmatrix} 1 & 0 & -4 & | & 5 \\ 0 & -1 & 12 & | & -13 \\ 6 & -1 & 2 & | & 10 \end{bmatrix}.$$

$-2(1 \;\; 0 \;\; -4 \;| \; 5) = (-2 \;\; 0 \;\; 8 \; | \; -10)$ and
$(-2 \;\; 0 \;\; 8 \; | \; -10) + (2 \;\; -1 \;\; 4 \; | \; -3) = $
$(0 \;\; -1 \;\; 12 \; | \; -13)$

Now we multiply the first row by -6, add it to the third row, and replace Row 3 with the result:

$$x \quad\;\; - 4z = 5,$$
$$-y + 12z = -13,$$
$$-y + 26z = -20$$

$$\begin{bmatrix} 1 & 0 & -4 & | & 5 \\ 0 & -1 & 12 & | & -13 \\ 0 & -1 & 26 & | & -20 \end{bmatrix}.$$

$-6(1 \;\; 0 \;\; -4 \; | \; 5) = (-6 \;\; 0 \;\; 24 \; | \; -30)$ and
$(-6 \;\; 0 \;\; 24 \; | \; -30) + (6 \;\; -1 \;\; 2 \; | \; 10) = $
$(0 \;\; -1 \;\; 26 \; | \; -20)$

Next, we multiply Row 2 by -1, add it to the third row, and replace Row 3 with the result:

$$x \quad\;\; - 4z = 5,$$
$$-y + 12z = -13,$$
$$14z = -7$$

$$\begin{bmatrix} 1 & 0 & -4 & | & 5 \\ 0 & -1 & 12 & | & -13 \\ 0 & 0 & 14 & | & -7 \end{bmatrix}.$$

$-1(0 \;\; -1 \;\; 12 \; | \; -13) = (0 \;\; 1 \;\; -12 \; | \; 13)$
and $(0 \;\; 1 \;\; -12 \; | \; 13) + (0 \;\; -1 \;\; 26 \; | \; -20) = $
$(0 \;\; 0 \;\; 14 \; | \; -7)$

Reinserting the variables gives us

$$x \quad\;\; - 4z = 5,$$
$$-y + 12z = -13,$$
$$14z = -7.$$

We now solve this last equation for z and get $z = -\frac{1}{2}$. Next, we substitute $-\frac{1}{2}$ for z in the preceding equation and solve for y: $-y + 12\left(-\frac{1}{2}\right) = -13$, so $y = 7$. Since there is no y-term in the first equation of this last system, we need only substitute $-\frac{1}{2}$ for z to solve for x: $x - 4\left(-\frac{1}{2}\right) = 5$, so $x = 3$. The solution is $\left(3, 7, -\frac{1}{2}\right)$. The check is left to the student.

The operations used in the preceding example correspond to those used to produce equivalent systems of equations. We call the matrices **row-equivalent** and the operations that produce them **row-equivalent operations**.

Row-Equivalent Operations

Row-Equivalent Operations

Each of the following row-equivalent operations produces a row-equivalent matrix:

a) Interchanging any two rows.
b) Multiplying all elements of a row by a nonzero constant.
c) Replacing a row with the sum of that row and a multiple of another row.

Student Notes ———

Try to remember that row-equivalent matrices are not *equal*. It is the solutions of the corresponding systems that are the same.

The best overall method for solving systems of equations is by row-equivalent matrices; even computers are programmed to use them. Matrices are part of a branch of mathematics known as linear algebra. They are also studied in many courses in finite mathematics.

---

technology connection

Row-equivalent operations can be performed on a graphing calculator. For example, to interchange the first and second rows of the matrix, as in step (1) of Example 2 above, we enter the matrix as matrix **A** and select "rowSwap" from the MATRIX MATH menu. Some graphing calculators will not automatically store the matrix produced using a row-equivalent operation, so when several operations are to be performed in succession, it is helpful to store the result of each operation as it is produced. In the window at right, we see both the matrix produced by the rowSwap operation and the indication that this matrix is stored, using **STO**, as matrix **B**.

```
rowSwap([A],1,2)→[B]
[[1  0 -4  5]
 [2 -1  4 -3]
 [6 -1  2 10]]
```

1. Use a graphing calculator to proceed through all the steps in Example 2.

Exercise Set

3.6

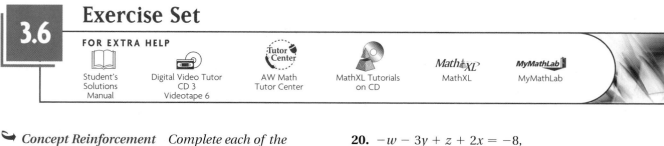
⤺ *Concept Reinforcement*   *Complete each of the following statements.*

1. The rows of a matrix are _____ and the _____ are vertical.

2. Multiplying the numbers in a row of a matrix by a constant corresponds to multiplying both sides of a(n) _____ by a constant.

3. Each number in a matrix is called a(n) _____ or element.

4. The plural of the word matrix is _____ .

5. To solve a system using matrices, we can replace any row by the sum of that row and a(n) _____ of another row.

6. In the final step of solving a system of equations, the leftmost column has zeros in all rows except the _____ one.

Solve using matrices.

7. $9x - 2y = 5,$
$3x - 3y = 11$

8. $4x + y = 7,$
$5x - 3y = 13$

9. $x + 4y = 8,$
$3x + 5y = 3$

10. $x + 4y = 5,$
$-3x + 2y = 13$

11. $6x - 2y = 4,$
$7x + y = 13$

12. $3x + 4y = 7,$
$-5x + 2y = 10$

13. $3x + 2y + 2z = 3,$
$x + 2y - z = 5,$
$2x - 4y + z = 0$

14. $4x - y - 3z = 19,$
$8x + y - z = 11,$
$2x + y + 2z = -7$

15. $p - 2q - 3r = 3,$
$2p - q - 2r = 4,$
$4p + 5q + 6r = 4$

16. $x + 2y - 3z = 9,$
$2x - y + 2z = -8,$
$3x - y - 4z = 3$

17. $3p + 2r = 11,$
$q - 7r = 4,$
$p - 6q = 1$

18. $4a + 9b = 8,$
$8a + 6c = -1,$
$6b + 6c = -1$

19. $2x + 2y - 2z - 2w = -10,$
$w + y + z + x = -5,$
$x - y + 4z + 3w = -2,$
$w - 2y + 2z + 3x = -6$

20. $-w - 3y + z + 2x = -8,$
$x + y - z - w = -4,$
$w + y + z + x = 22,$
$x - y - z - w = -14$

Solve using matrices.

21. *Coin value.*   A collection of 42 coins consists of dimes and nickels. The total value is $3.00. How many dimes and how many nickels are there?

22. *Coin value.*   A collection of 43 coins consists of dimes and quarters. The total value is $7.60. How many dimes and how many quarters are there?

23. *Mixed granola.*   Grace sells two kinds of granola. One is worth $4.05 per pound and the other is worth $2.70 per pound. She wants to blend the two granolas to get a 15-lb mixture worth $3.15 per pound. How much of each kind of granola should be used?

24. *Trail mix.*   Phil mixes nuts worth $1.60 per pound with oats worth $1.40 per pound to get 20 lb of trail mix worth $1.54 per pound. How many pounds of nuts and how many pounds of oats should be used?

25. *Investments.*   Elena receives $212 per year in simple interest from three investments totaling $2500. Part is invested at 7%, part at 8%, and part at 9%. There is $1100 more invested at 9% than at 8%. Find the amount invested at each rate.

26. *Investments.*   Miguel receives $306 per year in simple interest from three investments totaling $3200. Part is invested at 8%, part at 9%, and part at 10%. There is $1900 more invested at 10% than at 9%. Find the amount invested at each rate.

27. Explain how you can recognize dependent equations when solving with matrices.

28. Explain how you can recognize an inconsistent system when solving with matrices.

SKILL MAINTENANCE

Simplify. [1.2]

29. $5(-3) - (-7)4$

30. $8(-5) - (-2)9$

31. $-2(5 \cdot 3 - 4 \cdot 6) - 3(2 \cdot 7 - 15) + 4(3 \cdot 8 - 5 \cdot 4)$

32. $6(2 \cdot 7 - 3(-4)) - 4(3(-8) - 10) + 5(4 \cdot 3 - (-2)7)$

SYNTHESIS

33. If the matrices

$$\begin{bmatrix} a_1 & b_1 & | & c_1 \\ d_1 & e_1 & | & f_1 \end{bmatrix} \quad \text{and} \quad \begin{bmatrix} a_2 & b_2 & | & c_2 \\ d_2 & e_2 & | & f_2 \end{bmatrix}$$

share the same solution, does it follow that the corresponding entries are all equal to each other $(a_1 = a_2, b_1 = b_2, \text{etc.})$? Why or why not?

34. Explain how the row-equivalent operations make use of the addition, multiplication, and distributive properties.

35. The sum of the digits in a four-digit number is 10. Twice the sum of the thousands digit and the tens digit is 1 less than the sum of the other two digits. The tens digit is twice the thousands digit. The ones digit equals the sum of the thousands digit and the hundreds digit. Find the four-digit number.

36. Solve for x and y:
$$ax + by = c,$$
$$dx + ey = f.$$

3.7   Determinants and Cramer's Rule

Determinants of 2 × 2 Matrices   •   Cramer's Rule: 2 × 2 Systems   •   Cramer's Rule: 3 × 3 Systems

Determinants of 2 × 2 Matrices

When a matrix has m rows and n columns, it is called an "m by n" matrix. Thus its *dimensions* are denoted by $m \times n$. If a matrix has the same number of rows and columns, it is called a **square matrix**. Associated with every square matrix is a number called its **determinant**, defined as follows for 2 × 2 matrices.

> **2 × 2 Determinants**
>
> The determinant of a two-by-two matrix $\begin{bmatrix} a & c \\ b & d \end{bmatrix}$ is denoted $\begin{vmatrix} a & c \\ b & d \end{vmatrix}$ and is defined as follows:
>
> $$\begin{vmatrix} a & c \\ b & d \end{vmatrix} = ad - bc.$$

EXAMPLE 1   Evaluate: $\begin{vmatrix} 2 & -5 \\ 6 & 7 \end{vmatrix}$.

Solution   We multiply and subtract as follows:

$$\begin{vmatrix} 2 & -5 \\ 6 & 7 \end{vmatrix} = 2 \cdot 7 - 6 \cdot (-5) = 14 + 30 = 44.$$

Cramer's Rule: 2 × 2 Systems

One of the many uses for determinants is in solving systems of linear equations in which the number of variables is the same as the number of equations and the constants are not all 0. Let's consider a system of two equations:

$$a_1 x + b_1 y = c_1,$$
$$a_2 x + b_2 y = c_2.$$

If we use the elimination method, a series of steps can show that

$$x = \frac{c_1 b_2 - c_2 b_1}{a_1 b_2 - a_2 b_1} \quad \text{and} \quad y = \frac{a_1 c_2 - a_2 c_1}{a_1 b_2 - a_2 b_1}.$$

These fractions can be rewritten using determinants.

Cramer's Rule: 2 × 2 Systems

The solution of the system

$$a_1 x + b_1 y = c_1,$$
$$a_2 x + b_2 y = c_2,$$

if it is unique, is given by

$$x = \frac{\begin{vmatrix} c_1 & b_1 \\ c_2 & b_2 \end{vmatrix}}{\begin{vmatrix} a_1 & b_1 \\ a_2 & b_2 \end{vmatrix}}, \quad y = \frac{\begin{vmatrix} a_1 & c_1 \\ a_2 & c_2 \end{vmatrix}}{\begin{vmatrix} a_1 & b_1 \\ a_2 & b_2 \end{vmatrix}}.$$

These formulas apply only if the denominator is not 0. If the denominator *is* 0, then one of two things happens:

1. If the denominator is 0 and the numerators are also 0, then the equations in the system are dependent.
2. If the denominator is 0 and at least one numerator is not 0, then the system is inconsistent.

To use Cramer's rule, we find the determinants and compute x and y as shown above. Note that the denominators are identical and the coefficients of x and y appear in the same position as in the original equations. In the numerator of x, the constants c_1 and c_2 replace a_1 and a_2. In the numerator of y, the constants c_1 and c_2 replace b_1 and b_2.

EXAMPLE 2 Solve using Cramer's rule:

$$2x + 5y = 7,$$
$$5x - 2y = -3.$$

Solution We have

$$x = \frac{\begin{vmatrix} 7 & 5 \\ -3 & -2 \end{vmatrix}}{\begin{vmatrix} 2 & 5 \\ 5 & -2 \end{vmatrix}} \qquad \text{Using Cramer's rule}$$

$$= \frac{7(-2) - (-3)5}{2(-2) - 5 \cdot 5} = -\frac{1}{29}$$

and

$$y = \frac{\begin{vmatrix} 2 & 7 \\ 5 & -3 \end{vmatrix}}{\begin{vmatrix} 2 & 5 \\ 5 & -2 \end{vmatrix}} \qquad \text{Using Cramer's rule}$$

$$= \frac{2(-3) - 5 \cdot 7}{-29} = \frac{41}{29}. \qquad \text{The denominator is the same as in the expression for } x.$$

The solution is $\left(-\frac{1}{29}, \frac{41}{29}\right)$. The check is left to the student.

Cramer's Rule: 3 × 3 Systems

Cramer's rule can be extended for systems of three linear equations. However, before doing so, we must define what a 3 × 3 determinant is.

> **3 × 3 Determinants**
>
> The determinant of a three-by-three matrix is defined as follows:
>
> $$\begin{vmatrix} a_1 & b_1 & c_1 \\ a_2 & b_2 & c_2 \\ a_3 & b_3 & c_3 \end{vmatrix} = a_1 \begin{vmatrix} b_2 & c_2 \\ b_3 & c_3 \end{vmatrix} \overset{\text{Subtract.}}{\underset{}{-}} a_2 \begin{vmatrix} b_1 & c_1 \\ b_3 & c_3 \end{vmatrix} \overset{\text{Add.}}{\underset{}{+}} a_3 \begin{vmatrix} b_1 & c_1 \\ b_2 & c_2 \end{vmatrix}$$

Note that the a's come from the first column. Note too that the 2 × 2 determinants above can be obtained by crossing out the row and the column in which the a occurs.

Student Notes

Cramer's rule and the evaluation of determinants rely on patterns. The specific formulas are less important than the patterns that they represent.

For a_1:   For a_2:   For a_3:

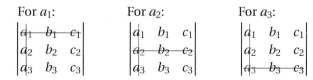

EXAMPLE 3   Evaluate:

$$\begin{vmatrix} -1 & 0 & 1 \\ -5 & 1 & -1 \\ 4 & 8 & 1 \end{vmatrix}.$$

Solution   We have

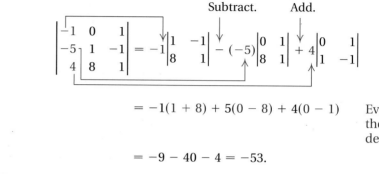

$$= -1(1 + 8) + 5(0 - 8) + 4(0 - 1) \qquad \text{Evaluating the three determinants}$$

$$= -9 - 40 - 4 = -53.$$

Cramer's Rule: 3 × 3 Systems

The solution of the system

$$a_1x + b_1y + c_1z = d_1,$$
$$a_2x + b_2y + c_2z = d_2,$$
$$a_3x + b_3y + c_3z = d_3$$

can be found using the following determinants:

$$D = \begin{vmatrix} a_1 & b_1 & c_1 \\ a_2 & b_2 & c_2 \\ a_3 & b_3 & c_3 \end{vmatrix}, \qquad D_x = \begin{vmatrix} d_1 & b_1 & c_1 \\ d_2 & b_2 & c_2 \\ d_3 & b_3 & c_3 \end{vmatrix},$$

D contains only coefficients. In D_x, the d's replace the a's.

$$D_y = \begin{vmatrix} a_1 & d_1 & c_1 \\ a_2 & d_2 & c_2 \\ a_3 & d_3 & c_3 \end{vmatrix}, \qquad D_z = \begin{vmatrix} a_1 & b_1 & d_1 \\ a_2 & b_2 & d_2 \\ a_3 & b_3 & d_3 \end{vmatrix}.$$

In D_y, the d's replace the b's. In D_z, the d's replace the c's.

If a unique solution exists, it is given by

$$x = \frac{D_x}{D}, \qquad y = \frac{D_y}{D}, \qquad z = \frac{D_z}{D}.$$

EXAMPLE 4   Solve using Cramer's rule:

$$x - 3y + 7z = 13,$$
$$x + y + z = 1,$$
$$x - 2y + 3z = 4.$$

Determinants can be evaluated on most graphing calculators using **2ND** **MATRIX**. After entering a matrix, we select the determinant operation from the MATRIX MATH menu and enter the name of the matrix. The graphing calculator will return the value of the determinant of the matrix. For example, if

$$\mathbf{A} = \begin{bmatrix} 1 & 6 & -1 \\ -3 & -5 & 3 \\ 0 & 4 & 2 \end{bmatrix},$$

we have

det([A])
           26

1. Confirm the calculations in Example 4.

Solution   We compute D, D_x, D_y, and D_z:

$$D = \begin{vmatrix} 1 & -3 & 7 \\ 1 & 1 & 1 \\ 1 & -2 & 3 \end{vmatrix} = -10; \qquad D_x = \begin{vmatrix} 13 & -3 & 7 \\ 1 & 1 & 1 \\ 4 & -2 & 3 \end{vmatrix} = 20;$$

$$D_y = \begin{vmatrix} 1 & 13 & 7 \\ 1 & 1 & 1 \\ 1 & 4 & 3 \end{vmatrix} = -6; \qquad D_z = \begin{vmatrix} 1 & -3 & 13 \\ 1 & 1 & 1 \\ 1 & -2 & 4 \end{vmatrix} = -24.$$

Then

$$x = \frac{D_x}{D} = \frac{20}{-10} = -2;$$

$$y = \frac{D_y}{D} = \frac{-6}{-10} = \frac{3}{5};$$

$$z = \frac{D_z}{D} = \frac{-24}{-10} = \frac{12}{5}.$$

The solution is $\left(-2, \frac{3}{5}, \frac{12}{5}\right)$. The check is left to the student.

In Example 4, we need not have evaluated D_z. Once x and y were found, we could have substituted them into one of the equations to find z.

To use Cramer's rule, we divide by D, provided $D \neq 0$. If $D = 0$ and at least one of the other determinants is not 0, then the system is inconsistent. If *all* the determinants are 0, then the equations in the system are dependent.

Exercise Set

3.7

FOR EXTRA HELP

Student's Solutions Manual | Digital Video Tutor CD 3 Videotape 6 | AW Math Tutor Center | MathXL Tutorials on CD | MathXL | MyMathLab

↩ *Concept Reinforcement*   *Classify each of the following as either true or false.*

1. A square matrix has the same number of rows and columns.

2. A 3×4 matrix has 3 rows and 4 columns.

3. Cramer's rule exists only for 2×2 systems.

4. Whenever Cramer's rule yields a denominator that is 0, the system has no solution.

5. Whenever Cramer's rule yields a numerator that is 0, the equations are dependent.

6. Cramer's rule allows us to solve some systems that could not be solved any other way.

Evaluate.

7. $\begin{vmatrix} 5 & 1 \\ 2 & 4 \end{vmatrix}$ **8.** $\begin{vmatrix} 3 & 2 \\ 2 & -3 \end{vmatrix}$

9. $\begin{vmatrix} 6 & -9 \\ 2 & 3 \end{vmatrix}$ **10.** $\begin{vmatrix} 3 & 2 \\ -7 & 5 \end{vmatrix}$

11. $\begin{vmatrix} 1 & 4 & 0 \\ 0 & -1 & 2 \\ 3 & -2 & 1 \end{vmatrix}$ **12.** $\begin{vmatrix} 3 & 0 & -2 \\ 5 & 1 & 2 \\ 2 & 0 & -1 \end{vmatrix}$

13. $\begin{vmatrix} -1 & -2 & -3 \\ 3 & 4 & 2 \\ 0 & 1 & 2 \end{vmatrix}$

14. $\begin{vmatrix} 1 & 2 & 2 \\ 2 & 1 & 0 \\ 3 & 3 & 1 \end{vmatrix}$

15. $\begin{vmatrix} -4 & -2 & 3 \\ -3 & 1 & 2 \\ 3 & 4 & -2 \end{vmatrix}$

16. $\begin{vmatrix} 2 & -1 & 1 \\ 1 & 2 & -1 \\ 3 & 4 & -3 \end{vmatrix}$

Solve using Cramer's rule.

17. $5x + 8y = 1,$
$3x + 7y = 5$

18. $3x - 4y = 6,$
$5x + 9y = 10$

19. $5x - 4y = -3,$
$7x + 2y = 6$

20. $-2x + 4y = 3,$
$3x - 7y = 1$

21. $3x - y + 2z = 1,$
$x - y + 2z = 3,$
$-2x + 3y + z = 1$

22. $3x + 2y - z = 4,$
$3x - 2y + z = 5,$
$4x - 5y - z = -1$

23. $2x - 3y + 5z = 27,$
$x + 2y - z = -4,$
$5x - y + 4z = 27$

24. $x - y + 2z = -3,$
$x + 2y + 3z = 4,$
$2x + y + z = -3$

25. $r - 2s + 3t = 6,$
$2r - s - t = -3,$
$r + s + t = 6$

26. $a \quad\quad - 3c = 6,$
$\quad b + 2c = 2,$
$7a - 3b - 5c = 14$

27. What is it about Cramer's rule that makes it useful?

28. Which version of Cramer's rule do you find more useful: the version for 2×2 systems or the version for 3×3 systems? Why?

SKILL MAINTENANCE

Solve. [1.3]

29. $0.5x - 2.34 + 2.4x = 7.8x - 9$

30. $5x + 7x = -144$

31. A piece of wire 32.8 ft long is to be cut into two pieces, and those pieces are each to be bent to make a square. The length of a side of one square is to be 2.2 ft greater than the length of a side of the other. How should the wire be cut? [1.4]

32. *Inventory.* The Freeport College store paid $1728 for an order of 45 calculators. The store paid $9 for each scientific calculator. The others, all graphing calculators, cost the store $58 each. How many of each type of calculator was ordered? [3.3]

33. *Insulation.* The Mazzas' attic required three and a half times as much insulation as did the Kranepools'. Together, the two attics required 36 rolls of insulation. How much insulation did each attic require? [3.3]

34. *Sales of food.* High Flyin' Wings charges $12 for a bucket of chicken wings and $7 for a chicken dinner. After filling 28 orders for buckets and dinners, High Flyin' Wings had collected $281. How many buckets and how many dinners did they sell? [3.3]

SYNTHESIS

35. Cramer's rule states that if $a_1x + b_1y = c_1$ and $a_2x + b_2y = c_2$ are dependent, then
$$\begin{vmatrix} a_1 & b_1 \\ a_2 & b_2 \end{vmatrix} = 0.$$
Explain why this will always happen.

36. Under what conditions can a 3×3 system of linear equations be consistent but unable to be solved using Cramer's rule?

Solve.

37. $\begin{vmatrix} y & -2 \\ 4 & 3 \end{vmatrix} = 44$

38. $\begin{vmatrix} 2 & x & -1 \\ -1 & 3 & 2 \\ -2 & 1 & 1 \end{vmatrix} = -12$

39. $\begin{vmatrix} m+1 & -2 \\ m-2 & 1 \end{vmatrix} = 27$

40. Show that an equation of the line through (x_1, y_1) and (x_2, y_2) can be written
$$\begin{vmatrix} x & y & 1 \\ x_1 & y_1 & 1 \\ x_2 & y_2 & 1 \end{vmatrix} = 0.$$

3.8 Business and Economic Applications

Break-Even Analysis • Supply and Demand

Break-Even Analysis

When a company manufactures x units of a product, it spends money. This is **total cost** and can be thought of as a function C, where $C(x)$ is the total cost of producing x units. When the company sells x units of the product, it takes in money. This is **total revenue** and can be thought of as a function R, where $R(x)$ is the total revenue from the sale of x units. **Total profit** is the money taken in less the money spent, or total revenue minus total cost. Total profit from the production and sale of x units is a function P given by

Profit = Revenue − Cost, or $P(x) = R(x) - C(x)$.

If $R(x)$ is greater than $C(x)$, there is a gain and $P(x)$ is positive. If $C(x)$ is greater than $R(x)$, there is a loss and $P(x)$ is negative. When $R(x) = C(x)$, the company breaks even.

There are two kinds of costs. First, there are costs like rent, insurance, machinery, and so on. These costs, which must be paid whether a product is produced or not, are called *fixed costs*. When a product is being produced, there are costs for labor, materials, marketing, and so on. These are called *variable costs*, because they vary according to the amount being produced. The sum of the fixed cost and the variable cost gives the *total cost* of producing a product.

> *Caution!* Do not confuse "cost" with "price." When we discuss the *cost* of an item, we are referring to what it costs to produce the item. The *price* of an item is what a consumer pays to purchase the item and is used when calculating revenue.

EXAMPLE 1   Manufacturing lamps.   Ergs, Inc., is planning to make a new lamp. Fixed costs will be $90,000, and it will cost $15 to produce each lamp (variable costs). Each lamp sells for $26.

a) Find the total cost $C(x)$ of producing x lamps.

b) Find the total revenue $R(x)$ from the sale of x lamps.

c) Find the total profit $P(x)$ from the production and sale of x lamps.

d) What profit will the company realize from the production and sale of 3000 lamps? of 14,000 lamps?

e) Graph the total-cost, total-revenue, and total-profit functions using the same set of axes. Determine the break-even point.

Solution

a) Total cost is given by

$$C(x) = (\text{Fixed costs}) \text{ plus } (\text{Variable costs}),$$

or $C(x) = \quad 90{,}000 \quad + \quad 15x,$

where x is the number of lamps produced.

b) Total revenue is given by

$R(x) = 26x.$ $26 times the number of lamps sold. We assume that every lamp produced is sold.

c) Total profit is given by

$$P(x) = R(x) - C(x) \qquad \text{Profit is revenue minus cost.}$$
$$= 26x - (90{,}000 + 15x)$$
$$= 11x - 90{,}000.$$

d) Profits will be

$$P(3000) = 11 \cdot 3000 - 90{,}000 = -\$57{,}000$$

when 3000 lamps are produced and sold, and

$$P(14{,}000) = 11 \cdot 14{,}000 - 90{,}000 = \$64{,}000$$

when 14,000 lamps are produced and sold. Thus the company loses money if only 3000 lamps are sold, but makes money if 14,000 are sold.

e) The graphs of each of the three functions are shown below:

$R(x) = 26x,$ This represents the revenue function.

$C(x) = 90{,}000 + 15x,$ This represents the cost function.

$P(x) = 11x - 90{,}000.$ This represents the profit function.

$R(x)$, $C(x)$, and $P(x)$ are all in dollars.

The revenue function has a graph that goes through the origin and has a slope of 26. The cost function has an intercept on the $-axis of 90,000 and has a slope of 15. The profit function has an intercept on the $-axis of −90,000 and has a slope of 11. It is shown by the dashed line. The red dashed line shows a "negative" profit, which is a loss. (That is what is known as "being in the red.") The black dashed line shows a "positive" profit, or gain. (That is what is known as "being in the black.")

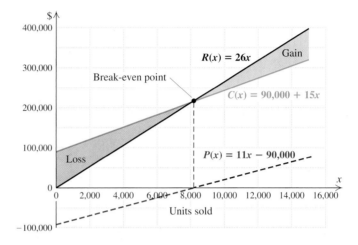

If you plan to study business or economics, you may want to consult the material in this section when these topics arise in your other courses.

Gains occur where the revenue is greater than the cost. Losses occur where the revenue is less than the cost. The **break-even point** occurs where the graphs of R and C cross. Thus to find the break-even point, we solve a system:

$$R(x) = 26x,$$
$$C(x) = 90,000 + 15x.$$

Since both revenue and cost are in *dollars* and they are equal at the break-even point, the system can be rewritten as

$$d = 26x, \qquad (1)$$
$$d = 90,000 + 15x \qquad (2)$$

and solved using substitution:

$$26x = 90,000 + 15x \qquad \text{Substituting } 26x \text{ for } d \text{ in equation (2)}$$
$$11x = 90,000$$
$$x \approx 8181.8.$$

The firm will break even if it produces and sells about 8182 lamps (8181 will yield a tiny loss and 8182 a tiny gain), and takes in a total of $R(8182) = 26 \cdot 8182 = \$212,732$ in revenue. Note that the x-coordinate of the break-even point can also be found by solving $P(x) = 0$. The break-even point is (8182 lamps, $212,732).

Supply and Demand

As the price of coffee varies, the amount sold varies. The table and graph below show that *consumers will demand less as the price goes up.*

Demand Function, *D*

Price, *p*, per Kilogram	Quantity, *D(p)* (in millions of kilograms)
$ 8.00	25
9.00	20
10.00	15
11.00	10
12.00	5

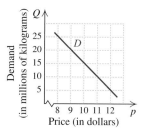

As the price of coffee varies, the amount available varies. The table and graph below show that *sellers will supply more as the price goes up.*

Supply Function, *S*

Price, *p*, per Kilogram	Quantity, *S(p)* (in millions of kilograms)
$ 9.00	5
9.50	10
10.00	15
10.50	20
11.00	25

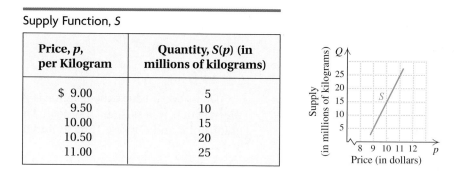

Let's look at the above graphs together. We see that as price increases, demand decreases. As price increases, supply increases. The point of intersection is called the **equilibrium point**. At that price, the amount that the seller will supply is the same amount that the consumer will buy. The situation is analogous to a buyer and a seller negotiating the price of an item. The equilibrium point is the price and quantity that they finally agree on.

Any ordered pair of coordinates from the graph is (price, quantity), because the horizontal axis is the price axis and the vertical axis is the quantity axis. If *D* is a demand function and *S* is a supply function, then the equilibrium point is where demand equals supply:

$$D(p) = S(p).$$

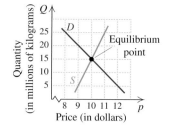

EXAMPLE 2   Find the equilibrium point for the demand and supply functions given:

$$D(p) = 1000 - 60p, \qquad (1)$$
$$S(p) = \ \ 200 + \ \ 4p. \qquad (2)$$

Solution   Since both demand and supply are *quantities* and they are equal at the equilibrium point, we rewrite the system as

$$q = 1000 - 60p, \qquad (1)$$
$$q = \ \ 200 + \ \ 4p. \qquad (2)$$

We substitute $200 + 4p$ for q in equation (1) and solve:

$200 + 4p = 1000 - 60p$	Substituting $200 + 4p$ for q in equation (1)
$200 + 64p = 1000$	Adding $60p$ to both sides
$64p = 800$	Adding -200 to both sides
$p = \frac{800}{64} = 12.5.$	

Thus the equilibrium price is $12.50 per unit.

To find the equilibrium quantity, we substitute $12.50 into either $D(p)$ or $S(p)$. We use $S(p)$:

$$S(12.5) = 200 + 4(12.5) = 200 + 50 = 250.$$

Thus the equilibrium quantity is 250 units, and the equilibrium point is ($12.50, 250).

Exercise Set

3.8

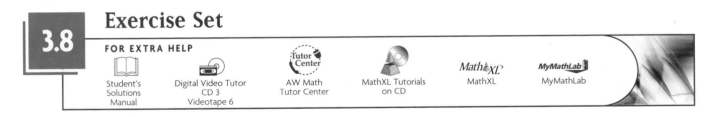

FOR EXTRA HELP

Student's Solutions Manual   Digital Video Tutor CD 3 Videotape 6   Tutor Center AW Math Tutor Center   MathXL Tutorials on CD   Math XL MathXL   MyMathLab MyMathLab

↩ *Concept Reinforcement*   *In each of Exercises 1–8, match the word or phrase with the most appropriate choice from the column on the right.*

1. ___ Total cost

2. ___ Total revenue

3. ___ Total profit

4. ___ Fixed costs

5. ___ Variable costs

6. ___ Break-even point

7. ___ Equilibrium point

8. ___ Price

a) The amount of money that a company takes in

b) The sum of fixed costs and variable costs

c) The point at which total revenue equals total cost

d) What consumers pay per item

e) The difference between total revenue and total cost

f) What companies spend whether or not a product is produced

g) The point at which supply equals demand

h) The costs that vary according to the number of items produced

For each of the following pairs of total-cost and total-revenue functions, find **(a)** *the total-profit function and* **(b)** *the break-even point.*

9. $C(x) = 45x + 300,000;$
$R(x) = 65x$

10. $C(x) = 25x + 270,000;$
$R(x) = 70x$

11. $C(x) = 10x + 120,000;$
$R(x) = 60x$

12. $C(x) = 30x + 49,500;$
$R(x) = 85x$

13. $C(x) = 40x + 22,500;$
$R(x) = 85x$

14. $C(x) = 20x + 10,000;$
$R(x) = 100x$

15. $C(x) = 22x + 16,000;$
$R(x) = 40x$

16. $C(x) = 15x + 75,000;$
$R(x) = 55x$

Aha! **17.** $C(x) = 75x + 100,000;$
$R(x) = 125x$

18. $C(x) = 20x + 120,000;$
$R(x) = 50x$

Find the equilibrium point for each of the following pairs of demand and supply functions.

19. $D(p) = 1000 - 10p,$
$S(p) = 230 + p$

20. $D(p) = 2000 - 60p,$
$S(p) = 460 + 94p$

21. $D(p) = 760 - 13p,$
$S(p) = 430 + 2p$

22. $D(p) = 800 - 43p,$
$S(p) = 210 + 16p$

23. $D(p) = 7500 - 25p,$
$S(p) = 6000 + 5p$

24. $D(p) = 8800 - 30p,$
$S(p) = 7000 + 15p$

25. $D(p) = 1600 - 53p,$
$S(p) = 320 + 75p$

26. $D(p) = 5500 - 40p,$
$S(p) = 1000 + 85p$

Solve.

27. *Computer manufacturing.* Biz.com Electronics is planning to introduce a new line of computers. The fixed costs for production are $125,300. The variable costs for producing each computer are $450. The revenue from each computer is $800. Find the following.

a) The total cost $C(x)$ of producing x computers

b) The total revenue $R(x)$ from the sale of x computers

c) The total profit $P(x)$ from the production and sale of x computers

d) The profit or loss from the production and sale of 100 computers; of 400 computers

e) The break-even point

28. *Manufacturing CD players.* SoundGen, Inc., is planning to manufacture a new type of CD player. The fixed costs for production are $22,500. The variable costs for producing each CD player are estimated to be $40. The revenue from each CD player is to be $85. Find the following.

a) The total cost $C(x)$ of producing x CD players

b) The total revenue $R(x)$ from the sale of x CD players

c) The total profit $P(x)$ from the production and sale of x CD players

d) The profit or loss from the production and sale of 3000 CD players; of 400 CD players

e) The break-even point

29. *Manufacturing caps.* Martina's Custom Printing is planning on adding painter's caps to its product line. For the first year, the fixed costs for setting up production are $16,404. The variable costs for producing a dozen caps are $6.00. The revenue on each dozen caps will be $18.00. Find the following.

a) The total cost $C(x)$ of producing x dozen caps

b) The total revenue $R(x)$ from the sale of x dozen caps

c) The total profit $P(x)$ from the production and sale of x dozen caps

d) The profit or loss from the production and sale of 3000 dozen caps; of 1000 dozen caps

e) The break-even point

30. *Sport coat production.* Sarducci's is planning a new line of sport coats. For the first year, the fixed costs for setting up production are $10,000. The variable costs for producing each coat are $30. The revenue from each coat is to be $80. Find the following.

a) The total cost $C(x)$ of producing x coats

b) The total revenue $R(x)$ from the sale of x coats

c) The total profit $P(x)$ from the production and sale of x coats

d) The profit or loss from the production and sale of 2000 coats; of 50 coats

e) The break-even point

31. In Example 1, the slope of the line representing Revenue is the sum of the slopes of the other two lines. This is not a coincidence. Explain why.

32. Variable costs and fixed costs are often compared to the slope and the *y*-intercept, respectively, of an equation for a line. Explain why you feel this analogy is or is not valid.

SKILL MAINTENANCE

Solve. [1.3]

33. $3x - 9 = 27$

34. $4x - 7 = 53$

35. $4x - 5 = 7x - 13$

36. $2x + 9 = 8x - 15$

37. $7 - 2(x - 8) = 14$

38. $6 - 4(3x - 2) = 10$

SYNTHESIS

39. Ian claims that since his fixed costs are $1000, he need sell only 20 birdbaths at $50 each in order to break even. Does this sound plausible? Why or why not?

40. In this section, we examined supply and demand functions for coffee. Does it seem realistic to you for the graph of *D* to have a constant slope? Why or why not?

41. *Yo-yo production.*   Bing Boing Hobbies is willing to produce 100 yo-yo's at $2.00 each and 500 yo-yo's at $8.00 each. Research indicates that the public will buy 500 yo-yo's at $1.00 each and 100 yo-yo's at $9.00 each. Find the equilibrium point.

42. *Loudspeaker production.*   Fidelity Speakers, Inc., has fixed costs of $15,400 and variable costs of $100 for each pair of speakers produced. If the speakers sell for $250 a pair, how many pairs of speakers must be produced (and sold) in order to have enough profit to cover the fixed costs of two additional facilities? Assume that all fixed costs are identical.

Use a graphing calculator to solve.

43. *Dog food production.*   Puppy Love, Inc., will soon begin producing a new line of puppy food. The marketing department predicts that the demand function will be $D(p) = -14.97p + 987.35$ and the supply function will be $S(p) = 98.55p - 5.13$.

  a) To the nearest cent, what price per unit should be charged in order to have equilibrium between supply and demand?

  b) The production of the puppy food involves $87,985 in fixed costs and $5.15 per unit in variable costs. If the price per unit is the value you found in part (a), how many units must be sold in order to break even?

44. *Computer production.*   Number Cruncher Computers, Inc., is planning a new line of computers, each of which will sell for $970. The fixed costs in setting up production are $1,235,580 and the variable costs for each computer are $697.

  a) What is the break-even point? (Round to the nearest whole number.)

  b) The marketing department at Number Cruncher is not sure that $970 is the best price. Their demand function for the new computers is given by $D(p) = -304.5p + 374,580$ and their supply function is given by $S(p) = 788.7p - 576,504$. To the nearest dollar, what price *p* would result in equilibrium between supply and demand?

3 Study Summary

Because so many real-world problems translate into two or more equations in two or more variables, **systems of equations** are studied in great detail (p. 151). Most of the systems studied in this chapter are **consistent**, meaning that they have at least one solution, although we also studied some **inconsistent** systems, for which there is no solution (p. 155). The equations in the systems we solved were **independent**, except for those cases in which one equation could be written as a multiple and/or sum of the other equation(s) (such equations are called **dependent**) (p. 155).

Three methods—graphing, substitution, and elimination—can be used to solve systems of equations (pp. 153, 159, 161). Of the three methods, graphing is the easiest to visualize.

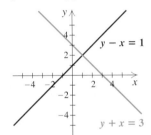

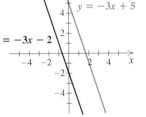

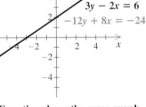

Graphs intersect at one point.
The system is *consistent* and has one solution. Since neither equation is a multiple of the other, they are *independent*.

Graphs are parallel.
The system is *inconsistent* because there is no solution. Since the equations are not equivalent, they are *independent*.

Equations have the same graph.
The system is *consistent* and has an infinite number of solutions. The equations are *dependent* since they are equivalent.

Graphing is especially useful when working with **revenue, cost,** and **profit** functions to determine a **break-even point** (p. 207). It is also used when working with **supply** and **demand** functions to determine an **equilibrium point** (p. 210).

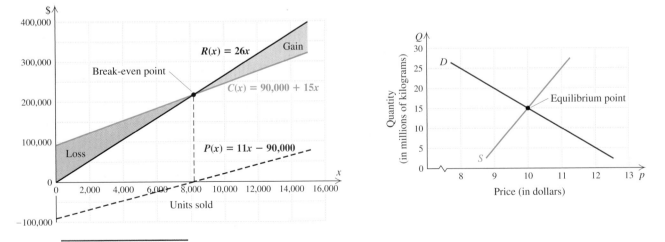

The substitution and elimination methods are the most commonly used methods for solving **total-value**, **mixture**, and **motion problems**.

Total Value

King Street Printing recently charged 1.9¢ per sheet of paper, but 2.4¢ per sheet for paper made of recycled fibers. Darren's bill for 150 sheets of paper was $3.41. How many sheets of each type were used? (Exercise 15, p. 177)

Mixture

Etch Clean Graphics uses one cleanser that is 25% acid and a second that is 50% acid. How many liters of each should be mixed to get 30 L of a solution that is 40% acid? (Exercise 24, p. 177)

Motion

Mia's motorboat took 3 hr to make a trip downstream with a 6-mph current. The return trip against the same current took 5 hr. Find the speed of the boat in still water. (Exercise 37, p. 179)

The elimination method can be extended to systems of three or more equations in three or more variables. One way in which this is accomplished is through the use of **matrices** (singular: **matrix**) (p. 196). A matrix is an array of numbers that are displayed in **rows** and **columns** (p. 196). The individual numbers are called **entries** or **elements** (p. 196).

By using **row-equivalent** operations, we can use matrices to solve systems of equations without needing to continually rewrite all of the variables (p. 199). The solution of systems is streamlined even further when **determinants** are used as part of **Cramer's rule** (pp. 201–204).

Determinant of a 2 × 2 Matrix

$$\begin{vmatrix} a & c \\ b & d \end{vmatrix} = ad - bc$$

Determinant of a 3 × 3 Matrix

$$\begin{vmatrix} a_1 & b_1 & c_1 \\ a_2 & b_2 & c_2 \\ a_3 & b_3 & c_3 \end{vmatrix} = a_1 \begin{vmatrix} b_2 & c_2 \\ b_3 & c_3 \end{vmatrix} - a_2 \begin{vmatrix} b_1 & c_1 \\ b_3 & c_3 \end{vmatrix} + a_3 \begin{vmatrix} b_1 & c_1 \\ b_2 & c_2 \end{vmatrix}$$

Cramer's Rule: 2 × 2 Systems

The solution of the system

$$a_1 x + b_1 y = c_1,$$
$$a_2 x + b_2 y = c_2,$$

if it is unique, is given by

$$x = \frac{\begin{vmatrix} c_1 & b_1 \\ c_2 & b_2 \end{vmatrix}}{\begin{vmatrix} a_1 & b_1 \\ a_2 & b_2 \end{vmatrix}}, \qquad y = \frac{\begin{vmatrix} a_1 & c_1 \\ a_2 & c_2 \end{vmatrix}}{\begin{vmatrix} a_1 & b_1 \\ a_2 & b_2 \end{vmatrix}}.$$

Cramer's Rule: 3 × 3 Systems

The solution of the system

$$a_1 x + b_1 y + c_1 z = d_1,$$
$$a_2 x + b_2 y + c_2 z = d_2,$$
$$a_3 x + b_3 y + c_3 z = d_3,$$

if it is unique, is given by

$$x = \frac{\begin{vmatrix} d_1 & b_1 & c_1 \\ d_2 & b_2 & c_2 \\ d_3 & b_3 & c_3 \end{vmatrix}}{\begin{vmatrix} a_1 & b_1 & c_1 \\ a_2 & b_2 & c_2 \\ a_3 & b_3 & c_3 \end{vmatrix}}, \qquad y = \frac{\begin{vmatrix} a_1 & d_1 & c_1 \\ a_2 & d_2 & c_2 \\ a_3 & d_3 & c_3 \end{vmatrix}}{\begin{vmatrix} a_1 & b_1 & c_1 \\ a_2 & b_2 & c_2 \\ a_3 & b_3 & c_3 \end{vmatrix}}, \qquad z = \frac{\begin{vmatrix} a_1 & b_1 & d_1 \\ a_2 & b_2 & d_2 \\ a_3 & b_3 & d_3 \end{vmatrix}}{\begin{vmatrix} a_1 & b_1 & c_1 \\ a_2 & b_2 & c_2 \\ a_3 & b_3 & c_3 \end{vmatrix}}.$$

3 Review Exercises

 Concept Reinforcement *Complete each of the following sentences.*

1. The system
$$5x + 3y = 7,$$
$$y = 2x + 1$$
is most easily solved using the _____ method. [3.2]

2. The system
$$-2x + 3y = 8,$$
$$2x + 2y = 7$$
is most easily solved using the _____ method. [3.2]

3. A weakness in using graphs to solve a system is that when solutions involve fractions or decimals, the graph may yield only a(n) _____ solution. [3.2]

4. When one equation in a system is a multiple of another equation in that system, the equations are said to be _____. [3.1]

5. A system for which there is no solution is said to be _____. [3.1]

6. When using elimination to solve a system of two equations, if an identity is obtained, we know that there is a(n) _____ number of solutions. [3.2]

7. When we are graphing to solve a system of two equations, if there is no solution, the lines will be _____. [3.1]

8. When a matrix has the same number of rows and columns, it is said to be _____. [3.7]

9. Cramer's rule is a formula in which the numerator and the denominator of each fraction is a(n) _____. [3.7]

10. At the break-even point, the value of the profit function is _____. [3.8]

For Exercises 11–19, if a system has an infinite number of solutions, use set-builder notation to write the solution set. If a system has no solution, state this.

Solve graphically. [3.1]

11. $3x + 2y = -4,$
$y = 3x + 7$

12. $2x + 3y = 12,$
$4x - y = 10$

Solve using the substitution method. [3.2]

13. $9x - 6y = 2,$
$x = 4y + 5$

14. $y = x + 2,$
$y - x = 8$

15. $x - 3y = -2,$
$7y - 4x = 6$

Solve using the elimination method. [3.2]

16. $8x - 2y = 10,$
$-4y - 3x = -17$

17. $4x - 7y = 18,$
$9x + 14y = 40$

18. $3x - 5y = -4,$
$5x - 3y = 4$

19. $1.5x - 3 = -2y,$
$3x + 4y = 6$

Solve. [3.3]

20. Luther bought two DVD's and one videocassette for $48. If he had purchased one DVD and two videocassettes, he would have spent $3 less. What is the price of a DVD? What is the price of a videocassette?

21. A freight train leaves Houston at midnight traveling north at a speed of 44 mph. One hour later, a passenger train, going 55 mph, travels north from Houston on a parallel track. How many hours will the passenger train travel before it overtakes the freight train?

22. Yolanda wants 14 L of fruit punch that is 10% juice. At the store, she finds punch that is 15% juice and punch that is 8% juice. How much of each should she purchase?

Solve. If a system's equations are dependent or if there is no solution, state this.

23. $x + 4y + 3z = 2,$
$2x + y + z = 10,$
$-x + y + 2z = 8$
[3.4]

24. $4x + 2y - 6z = 34,$
$2x + y + 3z = 3,$
$6x + 3y - 3z = 37$
[3.4]

25. $2x - 5y - 2z = -4,$
$7x + 2y - 5z = -6,$
$-2x + 3y + 2z = 4$
[3.4]

26. $-5x + 5y = -6,$
$2x - 2y = 4$
[3.2]

27. $3x + y = 2,$
$x + 3y + z = 0,$
$x + z = 2$ [3.4]

Solve.

28. In triangle ABC, the measure of angle A is four times the measure of angle C, and the measure of angle B is 45° more than the measure of angle C. What are the measures of the angles of the triangle? [3.5]

29. *Nontoxic floor wax.*   A nontoxic floor wax can be made from lemon juice and food-grade linseed oil. The amount of oil should be twice the amount of lemon juice. How much of each ingredient is needed to make 32 oz of floor wax? (The mix should be spread with a rag and buffed when dry.) [3.3]

30. *Lumber production.*   Denison Lumber can convert logs into either lumber or plywood. In a given day, the mill turns out 42 pallets of plywood and lumber. It makes a profit of $75 on a pallet of lumber and $120 on a pallet of plywood. How many pallets of each type must be produced and sold in order to make a profit of $3735? [3.3]

Solve using matrices. Show your work. [3.6]

31. $3x + 4y = -13,$
$5x + 6y = 8$

32. $3x - y + z = -1,$
$2x + 3y + z = 4,$
$5x + 4y + 2z = 5$

Evaluate. [3.7]

33. $\begin{vmatrix} -2 & 4 \\ -3 & 5 \end{vmatrix}$

34. $\begin{vmatrix} 2 & 3 & 0 \\ 1 & 4 & -2 \\ 2 & -1 & 5 \end{vmatrix}$

Solve using Cramer's rule. Show your work. [3.7]

35. $2x + 3y = 6,$
$x - 4y = 14$

36. $2x + y + z = -2,$
$2x - y + 3z = 6,$
$3x - 5y + 4z = 7$

37. Find the equilibrium point for the demand and supply functions

$$S(p) = 60 + 7p$$

and

$$D(p) = 120 - 13p. \ [3.8]$$

38. Auriel is beginning to produce organic honey. For the first year, the fixed costs for setting up production are $9000. The variable costs for producing each pint of honey are $0.75. The revenue from each pint of honey is $5.25. Find the following. [3.8]

a) The total cost $C(x)$ of producing x pints of honey

b) The total revenue $R(x)$ from the sale of x pints of honey

c) The total profit $P(x)$ from the production and sale of x pints of honey

d) The profit or loss from the production and sale of 1500 pints of honey; of 5000 pints of honey

e) The break-even point

SYNTHESIS

39. How would you go about solving a problem that involves four variables? [3.5]

40. Explain how a system of equations can be both dependent and inconsistent. [3.4]

41. Auriel is quitting a job that pays $27,000 a year to make honey (see Exercise 38). How many pints of honey must she produce and sell in order to make the same amount that she made in the job she left? [3.8]

42. Solve graphically:

$$y = x + 2,$$
$$y = x^2 + 2. \ [3.1]$$

43. The graph of $f(x) = ax^2 + bx + c$ contains the points $(-2, 3)$, $(1, 1)$, and $(0, 3)$. Find a, b, and c and give a formula for the function. [3.5]

3 Chapter Test

1. Solve graphically:

$$2x + y = 8,$$
$$y - x = 2.$$

Solve, if possible, using the substitution method.

2. $x + 3y = -8,$
$4x - 3y = 23$

3. $2x + 4y = -6,$
$y = 3x - 9$

Solve, if possible, using the elimination method.

4. $4x - 6y = 3,$
$6x - 4y = -3$

5. $4y + 2x = 18,$
$3x + 6y = 26$

6. The perimeter of a rectangle is 96. The length of the rectangle is 6 less than twice the width. Find the dimensions of the rectangle.

7. Pepperidge Farm® Goldfish is a snack food for which 40% of its calories come from fat. Rold Gold® Pretzels receive 9% of their calories from fat. How many grams of each would be needed to make 620 g of a snack mix for which 15% of the calories are from fat?

Solve. If a system's equations are dependent or if there is no solution, state this.

8. $-3x + y - 2z = 8,$
$-x + 2y - z = 5,$
$2x + y + z = -3$

9. $6x + 2y - 4z = 15,$
$-3x - 4y + 2z = -6,$
$4x - 6y + 3z = 8$

10. $2x + 2y = 0,$
$4x + 4z = 4,$
$2x + y + z = 2$

11. $3x + 3z = 0,$
$2x + 2y = 2,$
$3y + 3z = 3$

Solve using matrices.

12. $7x - 8y = 10,$
$9x + 5y = -2$

13. $x + 3y - 3z = 12,$
$3x - y + 4z = 0,$
$-x + 2y - z = 1$

Evaluate.

14. $\begin{vmatrix} 4 & -2 \\ 3 & 7 \end{vmatrix}$

15. $\begin{vmatrix} 3 & 4 & 2 \\ 2 & -5 & 4 \\ 4 & 5 & -3 \end{vmatrix}$

16. Solve using Cramer's rule:

$$8x - 3y = 5,$$
$$2x + 6y = 3.$$

17. An electrician, a carpenter, and a plumber are hired to work on a house. The electrician earns $21 per hour, the carpenter $19.50 per hour, and the plumber $24 per hour. The first day on the job, they worked a total of 21.5 hr and earned a total of $469.50. If the plumber worked 2 more hours than the carpenter did, how many hours did each work?

18. Find the equilibrium point for the demand and supply functions

$$D(p) = 79 - 8p \quad \text{and} \quad S(p) = 37 + 6p.$$

19. Kick Back, Inc., is producing a new hammock. For the first year, the fixed costs for setting up production are $40,000. The variable costs for producing each hammock are $25. The revenue from each hammock is $70. Find the following.

a) The total cost $C(x)$ of producing x hammocks

b) The total revenue $R(x)$ from the sale of x hammocks

c) The total profit $P(x)$ from the production and sale of x hammocks

d) The profit or loss from the production and sale of 300 hammocks; of 900 hammocks

e) The break-even point

SYNTHESIS

20. The graph of the function $f(x) = mx + b$ contains the points $(-1, 3)$ and $(-2, -4)$. Find m and b.

21. At a county fair, an adult's ticket sold for $5.50, a senior citizen's ticket for $4.00, and a child's ticket for $1.50. On opening day, the number of adults' and senior citizens' tickets sold was 30 more than the number of children's tickets sold. The number of adults' tickets sold was 6 more than four times the number of senior citizens' tickets sold. Total receipts from the ticket sales were $11,219.50. How many of each type of ticket were sold?

Cumulative Review

Solve. [1.3]

1. $-14.3 + 29.17 = x$

2. $x + 9.4 = -12.6$

3. $3.9(-11) = x$

4. $-2.4x = -48$

5. $\frac{3}{8}x + 7 = -14$

6. $-3 + 5x = 2x + 15$

7. $3n - (4n - 2) = 7$

8. $6y - 5(3y - 4) = 10$

9. $14 + 2c = -3(c + 4) - 6$

10. $5x - [4 - 2(6x - 1)] = 12$

Simplify. Do not leave negative exponents in your answers.

11. $x^4 \cdot x^{-6} \cdot x^{13}$ [1.6]

12. $(4x^{-3}y^2)(-10x^4y^{-7})$ [1.6]

13. $(6x^2y^3)^2(-2x^0y^4)^3$ [1.6]

14. $\dfrac{y^4}{y^{-6}}$ [1.6]

15. $\dfrac{-10a^7b^{-11}}{25a^{-4}b^{22}}$ [1.6]

16. $\left(\dfrac{3x^4y^{-2}}{4x^{-5}}\right)^4$ [1.6]

17. $(1.95 \times 10^{-3})(5.73 \times 10^8)$ [1.7]

18. $\dfrac{2.42 \times 10^5}{6.05 \times 10^{-2}}$ [1.7]

19. Solve $A = \frac{1}{2}h(b + t)$ for b. [1.5]

20. Determine whether $(-3, 4)$ is a solution of $5a - 2b = -23$. [2.1]

Graph.

21. $f(x) = -2x + 8$ [2.3]

22. $y = x^2 - 1$ [2.1]

23. $4x + 16 = 0$ [2.4]

24. $-3x + 2y = 6$ [2.3]

25. Find the slope and the y-intercept of the line with equation $-4y + 9x = 12$. [2.3]

26. Find the slope, if it exists, of the line containing the points $(2, 7)$ and $(-1, 3)$. [2.3]

27. Find an equation of the line with slope -3 and containing the point $(2, -11)$. [2.5]

28. Find an equation of the line containing the points $(-6, 3)$ and $(4, 2)$. [2.5]

29. Determine whether the lines are parallel, perpendicular, or neither:

$$2x = 4y + 7,$$
$$x - 2y = 5. \ [2.5]$$

30. Find an equation of the line containing the point $(2, 1)$ and perpendicular to the line $x - 2y = 5$. [2.5]

31. For the graph of f shown, determine the domain, the range, $f(-3)$, and any value of x for which $f(x) = 5$. [2.2]

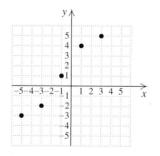

32. Determine the domain of the function given by

$$f(x) = \frac{7}{2x - 1}. \ [2.2]$$

Given $g(x) = 4x - 3$ and $h(x) = -2x^2 + 1$, find the following function values.

33. $h(4)$ [2.2]

34. $-g(0)$ [2.2]

35. $(g \cdot h)(-1)$ [2.6]

36. $g(a) - h(2a)$ [2.6]

Solve.

37. $3x + y = 4,$
$6x - y = 5$ [3.2]

38. $4x + 4y = 4,$
$5x - 3y = -19$ [3.2]

39. $6x - 10y = -22,$
$-11x - 15y = 27$ [3.2]

40. $x + y + z = -5,$
$2x + 3y - 2z = 8,$
$x - y + 4z = -21$ [3.4]

41. $2x + 5y - 3z = -11,$
$-5x + 3y - 2z = -7,$
$3x - 2y + 5z = 12$ [3.4]

Evaluate. [3.7]

42. $\begin{vmatrix} 2 & -3 \\ 4 & 1 \end{vmatrix}$

43. $\begin{vmatrix} 1 & 0 & 1 \\ -1 & 2 & 1 \\ 2 & 1 & 3 \end{vmatrix}$

44. The sum of two numbers is 26. Three times the smaller plus twice the larger is 60. Find the numbers. [3.3]

45. At the end of 2000, there were 1015 endangered or threatened species of U.S. animals and plants that had recovery plans. By the end of 2003, there were 1114 species with recovery plans. Find the rate at which recovery plans were being formed. [2.3]
Source: U.S. Fish and Wildlife Service, Department of the Interior

46. The number of U.S. domestic trips, in millions, t years after 1994 can be estimated by $P(t) = 8.8t + 954$. What do the numbers 8.8 and 954 signify? [2.3]
Source: Travel Industry Association of America

47. In 1994, there were 7.5 million U.S. aircraft departures, and in 2002, there were 9.2 million departures. Let $A(t)$ represent the number of departures, in millions, t years after 1994. [2.5]
Source: Bureau of Transportation Statistics
  a) Find an equation for a linear function that fits the data.
  b) Use the function of part (a) to predict the number of departures in 2010.

48. "Soakem" is 34% salt and the rest water. "Rinsem" is 61% salt and the rest water. How many ounces of each would be needed to obtain 120 oz of a mixture that is 50% salt? [3.3]

49. Find three consecutive odd numbers such that the sum of four times the first number and five times the third number is 47. [1.4]

50. Belinda's scores on four tests are 83, 92, 100, and 85. What must the score be on the fifth test so that the average will be 90? [1.4]

51. The perimeter of a rectangle is 32 cm. If five times the width equals three times the length, what are the dimensions of the rectangle? [3.3]

52. There are 4 more nickels than dimes in a bank. The total amount of money in the bank is $2.45. How many of each type of coin are in the bank? [3.3]

53. One month Lori and Jon spent $680 for electricity, rent, and telephone. The electric bill was $\frac{1}{4}$ of the rent and the rent was $400 more than the phone bill. How much was the electric bill? [3.5]

54. A hockey team played 64 games one season. It won 15 more games than it tied and lost 10 more games than it won. How many games did it win? lose? tie? [3.5]

55. Reggie, Jenna, and Achmed are counting calories. For lunch one day, Reggie ate two cookies and a banana, for a total of 260 calories. Jenna had a cup of yogurt and a banana, for a total of 245 calories. Achmed ate a cookie, a cup of yogurt, and two bananas, for a total of 415 calories. How many calories are in each item? [3.5]

SYNTHESIS

56. Simplify: $(6x^{a+2}y^{b+2})(-2x^{a-2}y^{y+1})$. [1.6]

57. Chaney Chevrolet discovers that when $1000 is spent on radio advertising, weekly sales increase by $101,000. When $1250 is spent on radio advertising, weekly sales increase by $126,000. Assuming that sales increase according to a linear equation, by what amount would sales increase when $1500 is spent on radio advertising? [2.5]

58. Given that $f(x) = mx + b$ and that $f(5) = -3$ when $f(-4) = 2$, find m and b. [2.5], [3.3]

4

Inequalities and Problem Solving

AN APPLICATION

The yearly U.S. production of crude oil $C(t)$, in millions of barrels, t years after 1990, can be approximated by the equation

$$C(t) = -53.5t + 2683$$

(*Source*: Based on data from the *Statistical Abstract of the United States* 2003). Determine (using an inequality) those years for which domestic production will be less than 1750 million barrels.

This problem appears as Example 7 in Section 4.1.

Melissa Leadley
OILFIELD SERVICES ENGINEER
Bakersfield, California

When trying to find oil and gas, we are always solving for the unknown. We use math to calculate important information like oil saturation and rock permeability, which can influence multimillion dollar decisions on whether to produce an oil well, stimulate it, or abandon it.

Inequalities are mathematical sentences containing symbols such as < (is less than). Principles similar to those used for solving equations enable us to solve inequalities and the problems that translate to inequalities. In this chapter, we develop and use these principles to solve a variety of inequalities, systems of inequalities, and real-world applications.

4.1 Inequalities and Applications

Solving Inequalities • Interval Notation • The Addition Principle for Inequalities • The Multiplication Principle for Inequalities • Using the Principles Together • Problem Solving

Solving Inequalities

We now extend our equation-solving skills to the solving of *inequalities*. An **inequality** is any sentence containing $<$, $>$, $\leq$, $\geq$, or $\neq$ (see Section 1.2)—for example,

$$-2 < a, \qquad x > 4, \qquad x + 3 \leq 6, \qquad 6 - 7y \geq 10y - 4, \quad \text{and} \quad 5x \neq 10.$$

Any value for the variable that makes an inequality true is called a **solution**. The set of all solutions is called the **solution set**. When all solutions of an inequality are found, we say that we have **solved** the inequality.

EXAMPLE 1 Determine whether the given number is a solution of the inequality.

a) $x + 3 < 6$; 5

b) $-3 > -5 - 2x$; 1

Solution

a) We substitute to get $5 + 3 < 6$, or $8 < 6$, a false sentence. Thus, 5 *is not* a solution.

b) We substitute to get $-3 > -5 - 2 \cdot 1$, or $-3 > -7$, a true sentence. Thus, 1 *is* a solution.

Study Skills

Create Your Own Glossary

Understanding mathematical terminology is essential for success in any math course. Consider writing your own glossary of important words toward the back of your notebook. Often, just the act of writing out a word's definition can help you remember what the word means.

The *graph* of an inequality is a visual representation of the inequality's solution set. An inequality in one variable can be graphed on a number line. Inequalities in two variables are graphed on a coordinate plane, and appear later in this chapter.

EXAMPLE 2 Graph $x < 4$ on a number line.

Solution The solutions are all real numbers less than 4, so we shade all numbers less than 4. Since 4 is not a solution, we use an open dot at 4.

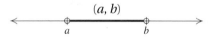

To write the solution set, we use *set-builder notation* (see Section 1.1):

$\{x \mid x < 4\}$.

This is read

"The set of all x such that x is less than 4."

Interval Notation

Another way to write solutions of an inequality in one variable is to use **interval notation**. Interval notation uses parentheses, ( ), and brackets, [ ].

If a and b are real numbers such that $a < b$, we define the **open interval** $(\boldsymbol{a}, \boldsymbol{b})$ as the set of all numbers x for which $a < x < b$. Thus,

$(a, b) = \{x \mid a < x < b\}$.   Parentheses are used to exclude endpoints.

Its graph excludes the endpoints:

(a, b)

```
        ⊕─────────⊕
   ←────┼─────────┼────→
        a         b
```

> *Caution!* Do not confuse the *interval* (a, b) with the *ordered pair* (a, b). The context in which the notation appears usually makes the meaning clear.

The **closed interval** $[\boldsymbol{a}, \boldsymbol{b}]$ is defined as the set of all numbers x for which $a \le x \le b$. Thus,

$[a, b] = \{x \mid a \le x \le b\}$.   Brackets are used to include endpoints.

Its graph includes the endpoints, as indicated by solid dots*:

$[a, b]$

```
        ●─────────●
   ←────┼─────────┼────→
        a         b
```

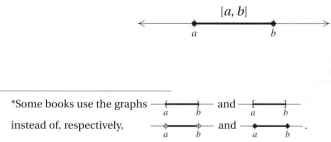

*Some books use the graphs ─(────────)─ and ─[────────]─ instead of, respectively, ─⊕─●──────●─⊕─ and ─●────────●─.

There are two kinds of **half-open intervals**, defined as follows:

1. $(a, b] = \{x \mid a < x \leq b\}$. This is open on the left. Its graph is as follows:

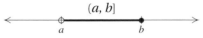

2. $[a, b) = \{x \mid a \leq x < b\}$. This is open on the right. Its graph is as follows:

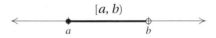

We use the symbols ∞ and $-\infty$ to represent positive and negative infinity, respectively. Thus the notation (a, ∞) represents the set of all real numbers greater than a, and $(-\infty, a)$ represents the set of all real numbers less than a.

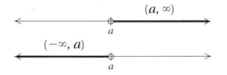

The notations $[a, \infty)$ and $(-\infty, a]$ are used when we want to include the endpoint a.

EXAMPLE 3 Graph $y \geq -2$ on a number line and write the solution set using both set-builder and interval notations.

Solution Using set-builder notation, we write the solution set as $\{y \mid y \geq -2\}$; using interval notation, we write $[-2, \infty)$.* To graph the solution, we shade all numbers to the right of -2 and use a solid dot to indicate that -2 is also a solution.

The Addition Principle for Inequalities

Two inequalities are *equivalent* if they have the same solution set. For example, the inequalities $x > 4$ and $4 < x$ are equivalent. Just as the addition principle for equations produces equivalent equations, the addition principle for inequalities produces equivalent inequalities.

---

*Any letter can be used in place of y. For example, $\{t \mid t \geq -2\}$ and $\{x \mid x \geq -2\}$ are both equal to $\{y \mid y \geq -2\}$. In $\{y \mid y \geq -2\}$, y is a *dummy variable*.

> **The Addition Principle for Inequalities**
>
> For any real numbers a, b, and c:
>
> $$a < b \text{ is equivalent to } a + c < b + c;$$
> $$a > b \text{ is equivalent to } a + c > b + c.$$
>
> Similar statements hold for $\leq$ and $\geq$.

As with equations, we try to get the variable alone on one side in order to determine solutions easily.

EXAMPLE 4   Solve and graph: **(a)** $x + 5 > 1$; **(b)** $4x - 1 \geq 5x - 2$.

Solution

a) $x + 5 > 1$

$x + 5 - 5 > 1 - 5$ Using the addition principle to add -5 to both sides

$x > -4$

When an inequality—like this last one—has an infinite number of solutions, we cannot possibly check them all. Instead, we can perform a partial check by substituting one member of the solution set (here we use -2) into the original inequality: $x + 5 = -2 + 5 = 3$ and $3 > 1$, so -2 is a solution and we have our check. The solution set is $\{x \mid x > -4\}$, or $(-4, \infty)$. The graph is as follows:

b) $4x - 1 \geq 5x - 2$

$4x - 1 + 2 \geq 5x - 2 + 2$ Adding 2 to both sides

$4x + 1 \geq 5x$ Simplifying

$4x + 1 - 4x \geq 5x - 4x$ Adding $-4x$ to both sides

$1 \geq x$ Simplifying

We know that $1 \geq x$ has the same meaning as $x \leq 1$. You can check that any number less than or equal to 1 is a solution. The solution set is $\{x \mid 1 \geq x\}$ or, more commonly, $\{x \mid x \leq 1\}$. Using interval notation, we write that the solution set is $(-\infty, 1]$. The graph is as follows:

The Multiplication Principle for Inequalities

The multiplication principle for inequalities differs from the multiplication principle for equations.

Consider this true inequality: $4 < 9$. If we multiply both sides of $4 < 9$ by 2, we get another true inequality:

$$4 \cdot 2 < 9 \cdot 2, \quad \text{or} \quad 8 < 18.$$

If we multiply both sides of $4 < 9$ by -2, we get a false inequality:

$$\text{FALSE} \longrightarrow 4(-2) < 9(-2), \quad \text{or} \quad -8 < -18. \longleftarrow \text{FALSE}$$

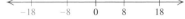

This is because multiplication (or division) by a negative number changes the sign of the number being multiplied (or divided). When the signs of both numbers in an inequality are changed, the position of the numbers with respect to each other is reversed.

$$-8 > -18. \longleftarrow \text{TRUE}$$

The $<$ symbol has been reversed!

The Multiplication Principle for Inequalities

For any real numbers a and b, and for any *positive* number c,

$$a < b \text{ is equivalent to } ac < bc;$$
$$a > b \text{ is equivalent to } ac > bc.$$

For any real numbers a and b, and for any *negative* number c,

$$a < b \text{ is equivalent to } ac > bc;$$
$$a > b \text{ is equivalent to } ac < bc.$$

Similar statements hold for $\leq$ and $\geq$.

Since division by c is the same as multiplication by $1/c$, there is no need for a separate division principle.

Caution!   Remember that whenever we multiply or divide both sides of an inequality by a negative number, we must reverse the inequality symbol.

EXAMPLE 5   Solve and graph: **(a)** $3y < \frac{3}{4}$; **(b)** $-5x \geq -80$.

Solution

a) $3y < \frac{3}{4}$

The symbol stays the same.

$\frac{1}{3} \cdot 3y < \frac{1}{3} \cdot \frac{3}{4}$ Multiplying both sides by $\frac{1}{3}$ or dividing both sides by 3

$y < \frac{1}{4}$

Any number less than $\frac{1}{4}$ is a solution. The solution set is $\{y \mid y < \frac{1}{4}\}$, or $\left(-\infty, \frac{1}{4}\right)$. The graph is as follows:

Student Notes _____

Try to remember to reverse the inequality symbol as soon as both sides are multiplied or divided by a negative number. Don't wait until after the multiplication or division has been carried out to reverse the symbol.

b) $-5x \geq -80$

The symbol must be reversed.

$$\frac{-5x}{-5} \leq \frac{-80}{-5}$$ Dividing both sides by -5 or multiplying both sides by $-\frac{1}{5}$

$$x \leq 16$$

The solution set is $\{x \,|\, x \leq 16\}$, or $(-\infty, 16]$. The graph is as follows:

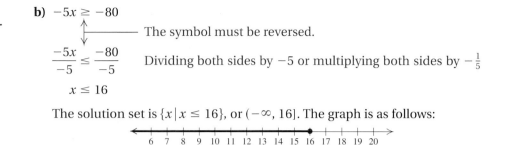

Using the Principles Together

We use the addition and multiplication principles together in solving inequalities in much the same way as in solving equations.

EXAMPLE 6

Solve: **(a)** $16 - 7y \geq 10y - 4$; **(b)** $-3(x + 8) - 5x > 4x - 9$.

Solution

a)
$$16 - 7y \geq 10y - 4$$
$$-16 + 16 - 7y \geq -16 + 10y - 4 \qquad \text{Adding } -16 \text{ to both sides}$$
$$-7y \geq 10y - 20$$
$$-10y + (-7y) \geq -10y + 10y - 20 \qquad \text{Adding } -10y \text{ to both sides}$$
$$-17y \geq -20$$

The symbol must be reversed.

$$-\tfrac{1}{17} \cdot (-17y) \leq -\tfrac{1}{17} \cdot (-20) \qquad \begin{array}{l}\text{Multiplying both sides by } -\frac{1}{17} \text{ or}\\ \text{dividing both sides by } -17\end{array}$$
$$y \leq \tfrac{20}{17}$$

The solution set is $\left\{y \,\middle|\, y \leq \tfrac{20}{17}\right\}$, or $\left(-\infty, \tfrac{20}{17}\right]$.

b)
$$-3(x + 8) - 5x > 4x - 9$$
$$-3x - 24 - 5x > 4x - 9 \qquad \text{Using the distributive law}$$
$$-24 - 8x > 4x - 9$$
$$-24 - 8x + 8x > 4x - 9 + 8x \qquad \text{Adding } 8x \text{ to both sides}$$
$$-24 > 12x - 9$$
$$-24 + 9 > 12x - 9 + 9 \qquad \text{Adding } 9 \text{ to both sides}$$
$$-15 > 12x$$

The symbol stays the same.

$$-\tfrac{5}{4} > x \qquad \text{Dividing by 12 and simplifying}$$

The solution set is $\left\{x \,\middle|\, -\tfrac{5}{4} > x\right\}$, or $\left\{x \,\middle|\, x < -\tfrac{5}{4}\right\}$, or $\left(-\infty, -\tfrac{5}{4}\right)$.

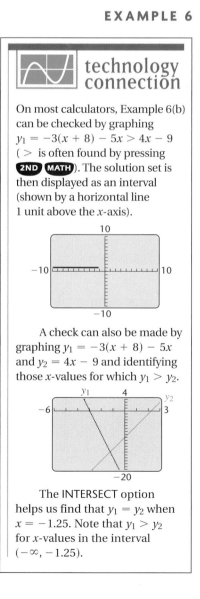

technology connection

On most calculators, Example 6(b) can be checked by graphing $y_1 = -3(x + 8) - 5x > 4x - 9$ ($>$ is often found by pressing **2ND MATH**). The solution set is then displayed as an interval (shown by a horizontal line 1 unit above the x-axis).

A check can also be made by graphing $y_1 = -3(x + 8) - 5x$ and $y_2 = 4x - 9$ and identifying those x-values for which $y_1 > y_2$.

The INTERSECT option helps us find that $y_1 = y_2$ when $x = -1.25$. Note that $y_1 > y_2$ for x-values in the interval $(-\infty, -1.25)$.

Problem Solving

Many problem-solving situations translate to inequalities. In addition to "is less than" and "is more than," other phrases are commonly used.

Important Words	Sample Sentence	Translation
is at least	Max is at least 5 years old.	$m \geq 5$
are at most	There are at most 6 people in the car.	$n \leq 6$
cannot exceed	Total weight in the elevator cannot exceed 2000 pounds.	$w \leq 2000$
must exceed	The speed must exceed 15 mph.	$s > 15$
is between	Heather's income is between $23,000 and $35,000.	$23{,}000 < h < 35{,}000$
no more than	Bing weighs no more than 90 pounds.	$w \leq 90$
no less than	Saul would accept no less than $5000 for his used car.	$t \geq 5000$

EXAMPLE 7   Domestic oil production.   The yearly U.S. production of crude oil $C(t)$, in millions of barrels, t years after 1990, can be approximated by the equation

$$C(t) = -53.5t + 2683$$

(*Source*: Based on data from the *Statistical Abstract of the United States* 2003). Determine (using an inequality) those years for which domestic production will be less than 1750 million barrels.

Solution

1. **Familiarize.**   We already have a formula. To become more familiar with it, we might make a substitution for t. Suppose we want to predict production after 20 years, in 2010. We substitute 20 for t:

   $$C(20) = -53.5 \cdot 20 + 2683 = 1613.$$

   We see that by 2010, production will be less than 1750 million barrels. To predict the exact years in which fewer than 1750 million barrels will be produced, we could check other substitutions. Instead, we proceed to the next step.

2. **Translate.**   We are asked to find the years for which U.S. oil production $C(t)$ will be *less than* 1750 million barrels. Thus we have

   $$C(t) < 1750.$$

   We replace $C(t)$ with $-53.5t + 2683$ to find the times t that solve the inequality:

   $$-53.5t + 2683 < 1750. \qquad \text{Substituting}$$

3. Carry out. We solve the inequality:

$$-53.5t + 2683 < 1750$$

$$-53.5t < -933 \qquad \text{Adding } -2683 \text{ to both sides}$$

$$t > 17.44. \qquad \text{Dividing both sides by } -53.5, \text{ reversing the symbol, and rounding}$$

4. Check. A partial check is to substitute a value for t greater than 17.44. We did that in the *Familiarize* step.

5. State. U.S. oil production will fall below 1750 million barrels about 17.4 years after 1990, or in 2007, and will remain below 1750 million for all years after that.

EXAMPLE 8

Job offers. After graduation, Rose had two job offers in sales:

Uptown Fashions: A salary of $600 per month, plus a commission of 4% of sales;

Ergo Designs: A salary of $800 per month, plus a commission of 6% of sales in excess of $10,000.

If sales always exceed $10,000, for what amount of sales would Uptown Fashions provide higher pay?

Solution

1. Familiarize. Listing the given information in a table will be helpful.

Uptown Fashions Monthly Income	Ergo Designs Monthly Income
$600 salary 4% of sales *Total*: $600 + 4% of sales	$800 salary 6% of sales over $10,000 *Total*: $800 + 6% of sales over $10,000

Next, suppose that Rose sold a certain amount—say, $12,000—in one month. Which plan would be better? Working for Uptown, she would earn $600 plus 4% of $12,000, or

$$600 + 0.04(12,000) = \$1080.$$

Since with Ergo Designs commissions are paid only on sales in excess of $10,000, Rose would earn $800 plus 6% of ($12,000 − $10,000), or

$$800 + 0.06(2000) = \$920.$$

This shows that for monthly sales of $12,000, Uptown pays better. Similar calculations will show that for sales of $30,000 a month, Ergo pays better. To determine *all* values for which Uptown pays more money, we must solve an inequality that is based on the calculations above.

2. **Translate.**   We let $S =$ the amount of monthly sales, in dollars, and will assume $S > 10,000$ so that both plans will pay a commission. Examining the calculations in the *Familiarize* step, we see that monthly income from Uptown is $600 + 0.04S$ and from Ergo is $800 + 0.06(S - 10,000)$. We want to find all values of S for which

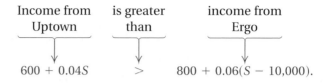

$$600 + 0.04S \quad > \quad 800 + 0.06(S - 10,000).$$

3. **Carry out.**   We solve the inequality:

$600 + 0.04S > 800 + 0.06(S - 10,000)$	
$600 + 0.04S > 800 + 0.06S - 600$	Using the distributive law
$600 + 0.04S > 200 + 0.06S$	Combining like terms
$400 > 0.02S$	Subtracting 200 and $0.04S$ from both sides
$20,000 > S$, or $S < 20,000$.	Dividing both sides by 0.02

4. **Check.**   The above steps indicate that income from Uptown Fashions is higher than income from Ergo Designs for sales less than $20,000. In the *Familiarize* step, we saw that for sales of $12,000, Uptown pays more. Since $12,000 < 20,000$, this is a partial check.

5. **State.**   When monthly sales are less than $20,000, Uptown Fashions provides the higher pay.

Exercise Set

4.1

FOR EXTRA HELP

Student's Solutions Manual | Digital Video Tutor CD 3 Videotape 7 | AW Math Tutor Center | MathXL Tutorials on CD | MathXL | MyMathLab

❧ *Concept Reinforcement*   Classify each of the following as equivalent inequalities, equivalent equations, equivalent expressions, or not equivalent.

1. $x - 7 > -2$, $x > 5$

2. $t + 3 < 1$, $t < 2$

3. $5x + 7 = 6 - 3x$, $8x + 7 = 6$

4. $2(4x + 1)$, $8x + 2$

5. $-4t \le 12$, $t \le -3$

6. $\frac{3}{5}a + \frac{1}{5} = 2$, $3a + 1 = 10$

7. $6a + 9$, $3(2a + 3)$

8. $-4x \ge -8$, $x \ge 2$

9. $-\frac{1}{2}x < 7$, $x > 14$

10. $-\frac{1}{3}t \le -5$, $t \ge 15$

Determine whether the given numbers are solutions of the inequality.

11. $x - 3 \ge 5$; $-4, 0, 8, 13$

12. $3x + 5 \le -10$; $-5, -10, 0, 27$

13. $t - 6 > 2t - 1$; $0, -8, -9, -3$

14. $5y - 9 < 3 - y$; $2, -3, 0, 3$

Graph each inequality, and write the solution set using both interval and set-builder notations.

15. $y < 6$

16. $x > 4$

17. $x \geq -4$

18. $t \leq 6$

19. $t > -3$

20. $y < -3$

21. $x \leq -7$

22. $x \geq -6$

Solve. Then graph.

23. $x + 8 > 2$

24. $x + 5 > 2$

25. $a + 7 \leq -13$

26. $a + 9 \leq -12$

27. $x - 8 \leq 9$

28. $t + 14 \geq 9$

29. $y - 9 > -18$

30. $y - 8 > -14$

31. $y - 20 \leq -6$

32. $x - 11 \leq -2$

33. $9t < -81$

34. $8x \geq 24$

35. $0.3x < -18$

36. $0.5x < 25$

37. $-9x \geq -8.1$

38. $-8y \leq 3.2$

39. $-\frac{3}{4}x \geq -\frac{5}{8}$

40. $-\frac{5}{6}y \leq -\frac{3}{4}$

41. $\dfrac{2x + 7}{5} < -9$

42. $\dfrac{5y + 13}{4} > -2$

43. $\dfrac{3t - 7}{-4} \leq 5$

44. $\dfrac{2t - 9}{-3} \geq 7$

45. Let $f(x) = 2x + 1$ and $g(x) = x + 7$. Find all values of x for which $f(x) \geq g(x)$.

46. Let $f(x) = 5 - x$ and $g(x) = 4x - 5$. Find all values of x for which $f(x) \geq g(x)$.

47. Let $f(x) = 7 - 3x$ and $g(x) = 2x - 3$. Find all values of x for which $f(x) \leq g(x)$.

48. Let $f(x) = 8x - 9$ and $g(x) = 3x - 11$. Find all values of x for which $f(x) \leq g(x)$.

49. Let $f(x) = 2x - 7$ and $g(x) = 5x - 9$. Find all values of x for which $f(x) < g(x)$.

50. Let $f(x) = 0.4x + 5$ and $g(x) = 1.2x - 4$. Find all values of x for which $g(x) \geq f(x)$.

51. Let $f(x) = \frac{3}{8} + 2x$ and $g(x) = 3x - \frac{1}{8}$. Find all values of x for which $g(x) \geq f(x)$.

52. Let $f(x) = 2x + 1$ and $g(x) = -\frac{1}{2}x + 6$. Find all values of x for which $f(x) < g(x)$.

Solve.

53. $4(3y - 2) \geq 9(2y + 5)$

54. $4m + 7 \geq 14(m - 3)$

55. $5(t - 3) + 4t < 2(7 + 2t)$

56. $2(4 + 2x) > 2x + 3(2 - 5x)$

57. $5[3m - (m + 4)] > -2(m - 4)$

58. $8x - 3(3x + 2) - 5 \geq 3(x + 4) - 2x$

59. $19 - (2x + 3) \leq 2(x + 3) + x$

60. $13 - (2c + 2) \geq 2(c + 2) + 3c$

61. $\frac{1}{4}(8y + 4) - 17 < -\frac{1}{2}(4y - 8)$

62. $\frac{1}{3}(6x + 24) - 20 > -\frac{1}{4}(12x - 72)$

63. $2[8 - 4(3 - x)] - 2 \geq 8[2(4x - 3) + 7] - 50$

64. $5[3(7 - t) - 4(8 + 2t)] - 20 \leq -6[2(6 + 3t) - 4]$

Solve.

65. *Truck rentals.* Campus Entertainment rents a truck for $45 plus 20¢ per mile. A budget of $75 has been set for the rental. For what mileages will they not exceed the budget?

66. *Truck rentals.* Metro Concerts can rent a truck for either $55 with unlimited mileage or $29 plus 40¢ per mile. For what mileages would the unlimited mileage plan save money?

67. *Insurance claims.* After a serious automobile accident, most insurance companies will replace the damaged car with a new one if repair costs exceed 80% of the NADA, or "blue-book," value of the car. Miguel's car recently sustained $9200 worth of damage but was not replaced. What was the blue-book value of his car?

68. *Phone rates.* A long-distance telephone call using Down East Calling costs 10 cents for the first minute and 8 cents for each additional minute. The same call, placed on Long Call Systems, cost 15 cents for the first minute and 6 cents for each additional minute. For what length phone calls is Down East Calling less expensive?

Phone rates.   *In Vermont, Verizon recently charged customers $13.55 for monthly service plus 2.2¢ per minute for local phone calls between 9 A.M. and 9 P.M. weekdays. The charge for off-peak local calls was 0.5¢ per minute. Calls were free after the total monthly charges reached $39.40.*

69. Assume that only peak local calls were made. For how long must a customer speak on the phone if the $39.40 maximum charge is to apply?

70. Assume that only off-peak calls were made. For how long must a customer speak on the phone if the $39.40 maximum charge is to apply?

71. *Checking-account rates.*   The Hudson Bank offers two checking-account plans. Their Anywhere plan charges 20¢ per check whereas their Acu-checking plan costs $2 per month plus 12¢ per check. For what numbers of checks per month will the Acu-checking plan cost less?

72. *Moving costs.*   Musclebound Movers charges $85 plus $40 an hour to move households across town. Champion Moving charges $60 an hour for cross-town moves. For what lengths of time is Champion more expensive?

73. *Wages.*   Toni can be paid in one of two ways:

> *Plan A*: A salary of $400 per month, plus a commission of 8% of gross sales;

> *Plan B*: A salary of $610 per month, plus a commission of 5% of gross sales.

For what amount of gross sales should Toni select plan A?

74. *Wages.*   Branford can be paid for his masonry work in one of two ways:

> *Plan A*: $300 plus $9.00 per hour;

> *Plan B*: Straight $12.50 per hour.

Suppose that the job takes n hours. For what values of n is plan B better for Branford?

75. *Wedding costs.*   The Arnold Inn offers two plans for wedding parties. Under plan A, the inn charges $30 for each person in attendance. Under plan B, the inn charges $1300 plus $20 for each person in excess of the first 25 who attend. For what size parties will plan B cost less? (Assume that more than 25 guests will attend.)

76. *Insurance benefits.*   Bayside Insurance offers two plans. Under plan A, Giselle would pay the first $50 of her medical bills and 20% of all bills after that. Under plan B, Giselle would pay the first $250 of bills, but only 10% of the rest. For what amount of medical bills will plan B save Giselle money? (Assume that her bills will exceed $250.)

77. *Show business.*   Slobberbone receives $750 plus 15% of receipts over $750 for playing a club date. If a club charges a $6 cover charge, how many people must attend in order for the band to receive at least $1200?

78. *Temperature conversion.*   The function
$$C(F) = \tfrac{5}{9}(F - 32)$$
can be used to find the Celsius temperature $C(F)$ that corresponds to $F°$ Fahrenheit.

a) Gold is solid at Celsius temperatures less than 1063°C. Find the Fahrenheit temperatures for which gold is solid.

b) Silver is solid at Celsius temperatures less than 960.8°C. Find the Fahrenheit temperatures for which silver is solid.

79. *Manufacturing.*   Ergs, Inc., is planning to make a new kind of lamp. Fixed costs will be $90,000, and variable costs will be $15 for the production of each lamp. The total-cost function for x lamps is
$$C(x) = 90{,}000 + 15x.$$
The company makes $26 in revenue for each lamp sold. The total-revenue function for x lamps is
$$R(x) = 26x.$$
(See Section 3.8.)

a) When $R(x) < C(x)$, the company loses money. Find the values of x for which the company loses money.

b) When $R(x) > C(x)$, the company makes a profit. Find the values of x for which the company makes a profit.

80. *Publishing.*   The demand and supply functions for a locally produced poetry book are approximated by
$$D(p) = 2000 - 60p \quad \text{and}$$
$$S(p) = 460 + 94p,$$
where p is the price in dollars (see Section 3.8).

a) Find those values of p for which demand exceeds supply.

b) Find those values of p for which demand is less than supply.

81. Explain in your own words why the inequality symbol must be reversed when both sides of an inequality are multiplied by a negative number.

82. Why isn't roster notation used to write solutions of inequalities?

SKILL MAINTENANCE

Find the domain of f. [2.2]

83. $f(x) = \dfrac{3}{x - 2}$

84. $f(x) = \dfrac{x - 5}{4x + 12}$

85. $f(x) = \dfrac{5x}{7 - 2x}$

86. $f(x) = \dfrac{x + 3}{9 - 4x}$

Simplify. [1.2]

87. $9x - 2(x - 5)$

88. $8x + 7(2x - 1)$

SYNTHESIS

89. A Presto photocopier costs $510 and an Exact Image photocopier costs $590. Write a problem that involves the cost of the copiers, the cost per page of photocopies, and the number of copies for which the Presto machine is the more expensive machine to own.

90. Explain how the addition principle can be used to avoid ever needing to multiply or divide both sides of an inequality by a negative number.

Solve for x and y. Assume that a, b, c, d, and m are positive constants.

91. $3ax + 2x \geq 5ax - 4$; assume $a > 1$

92. $6by - 4y \leq 7by + 10$

93. $a(by - 2) \geq b(2y + 5)$; assume $a > 2$

94. $c(6x - 4) < d(3 + 2x)$; assume $3c > d$

95. $c(2 - 5x) + dx > m(4 + 2x)$; assume $5c + 2m < d$

96. $a(3 - 4x) + cx < d(5x + 2)$; assume $c > 4a + 5d$

Determine whether each statement is true or false. If false, give an example that shows this.

97. For any real numbers a, b, c, and d, if $a < b$ and $c < d$, then $a - c < b - d$.

98. For all real numbers x and y, if $x < y$, then $x^2 < y^2$.

99. Are the inequalities
$$x < 3 \quad \text{and} \quad x + \frac{1}{x} < 3 + \frac{1}{x}$$
equivalent? Why or why not?

100. Are the inequalities
$$x < 3 \quad \text{and} \quad 0 \cdot x < 0 \cdot 3$$
equivalent? Why or why not?

Solve. Then graph.

101. $x + 5 \leq 5 + x$

102. $x + 8 < 3 + x$

103. $x^2 > 0$

104. Assume that the graphs of $y_1 = -\frac{1}{2}x + 5$, $y_2 = x - 1$, and $y_3 = 2x - 3$ are as shown below. Solve each inequality, referring only to the figure.

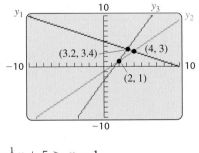

a) $-\frac{1}{2}x + 5 > x - 1$

b) $x - 1 \leq 2x - 3$

c) $2x - 3 \geq -\frac{1}{2}x + 5$

105. Using an approach similar to that in the Technology Connection on p. 227, use a graphing calculator to check your answers to Exercises 23, 47, 63, and 71.

CORNER

Reduce, Reuse, and Recycle

Focus: Inequalities and problem solving
Time: 15–20 minutes
Group size: 2

In the United States, the amount of solid waste (rubbish) being recycled is slowly catching up to the amount being generated. In 1991, each person generated, on average, 4.3 lb of solid waste every day, of which 0.8 lb was recycled. In 2001, each person generated, on average, 4.4 lb of solid waste, of which 1.3 lb was recycled. (*Sources*: U.S. Bureau of the Census, *Statistical Abstract of the United States* 2003, and EPA Municipal Solid Waste Factbook)

ACTIVITY

Assume that the amount of solid waste being generated and the amount recycled are both increasing linearly. One group member should find a linear function w for which $w(t)$ represents the number of pounds of waste generated per person per day t years after 1991. The other group member should find a linear function r for which $r(t)$ represents the number of pounds recycled per person per day t years after 1991. Finally, working together, the group should determine those years for which the amount recycled will meet or exceed the amount generated.

4.2 Intersections, Unions, and Compound Inequalities

> Intersections of Sets and Conjunctions of Sentences •
> Unions of Sets and Disjunctions of Sentences • Interval
> Notation and Domains

Two inequalities joined by the word "and" or the word "or" are called **compound inequalities**. Thus, "$2x - 7 < 3$ *or* $x - 1 > 4$" and "$7x < 9$ *and* $x - 1 > -5$" are two examples of compound inequalities. In order to discuss how to solve compound inequalities, we must first study ways in which sets can be combined.

Intersections of Sets and Conjunctions of Sentences

The **intersection** of two sets A and B is the set of all elements that are common to both A and B. We denote the intersection of sets A and B as

$$A \cap B.$$

$A \cap B$

The intersection of two sets is represented by the purple region shown in the figure at left. For example, if $A = \{$all students who are more than 5′4″ tall$\}$ and $B = \{$all students who weigh more than 120 lb$\}$, then $A \cap B = \{$all students who are more than 5′4″ tall and weigh more than 120 lb$\}$.

EXAMPLE 1   Find the intersection: $\{1, 2, 3, 4, 5\} \cap \{-2, -1, 0, 1, 2, 3\}$.

Solution   The numbers 1, 2, and 3 are common to both sets, so the intersection is $\{1, 2, 3\}$.

Study Skills

Two Books Are Better Than One

Many students find it helpful to use a second book as a reference when studying. Perhaps you or a friend own a text from a previous math course that can serve as a resource. Often professors have older texts that they will happily give away. Library book sales and thrift shops can also be excellent sources for extra books. Saving your text when you finish a math course can provide you with an excellent aid for your next course.

When two or more sentences are joined by the word *and* to make a compound sentence, the new sentence is called a **conjunction** of the sentences. The following is a conjunction of inequalities:

$$-2 < x \quad and \quad x < 1.$$

A number is a solution of a conjunction if it is a solution of *both* of the separate parts. For example, -1 is a solution because it is a solution of $-2 < x$ as well as $x < 1$.

Below we show the graph of $-2 < x$, followed by the graph of $x < 1$, and finally the graph of the conjunction $-2 < x$ and $x < 1$. *Note that the solution set of a conjunction is the intersection of the solution sets of the individual sentences.*

$\{x \mid -2 < x\}$ $(-2, \infty)$

$\{x \mid x < 1\}$ $(-\infty, 1)$

$\{x \mid -2 < x\} \cap \{x \mid x < 1\}$
$= \{x \mid -2 < x \ and \ x < 1\}$ $(-2, 1)$

Because there are numbers that are both greater than -2 and less than 1, the conjunction $-2 < x$ *and* $x < 1$ can be abbreviated by $-2 < x < 1$. Thus the interval $(-2, 1)$ can be represented as $\{x \mid -2 < x < 1\}$, the set of all numbers that are *simultaneously* greater than -2 *and* less than 1. Note that for $a < b$,

$$\boldsymbol{a < x \quad and \quad x < b \quad \text{can be abbreviated} \quad a < x < b;}$$

and, equivalently,

$$\boldsymbol{b > x \quad and \quad x > a \quad \text{can be abbreviated} \quad b > x > a.}$$

EXAMPLE 2   Solve and graph: $-1 \le 2x + 5 < 13$.

Solution   This inequality is an abbreviation for the conjunction

$$-1 \le 2x + 5 \quad and \quad 2x + 5 < 13.$$

The word *and* corresponds to set *intersection*. To solve the conjunction, we solve each of the two inequalities separately and then find the intersection of the solution sets:

$$-1 \le 2x + 5 \quad and \quad 2x + 5 < 13$$

$$-6 \le 2x \qquad and \qquad 2x < 8 \qquad \text{Subtracting 5 from both sides of each inequality}$$

$$-3 \le x \qquad and \qquad x < 4. \qquad \text{Dividing both sides of each inequality by 2}$$

We now abbreviate the answer:

$$-3 \leq x < 4.$$

The solution set is $\{x \mid -3 \leq x < 4\}$, or, in interval notation, $[-3, 4)$. The graph is the intersection of the two separate solution sets.

The steps in Example 2 are often combined as follows:

$$-1 \leq 2x + 5 < 13$$
$$-1 - 5 \leq 2x + 5 - 5 < 13 - 5 \qquad \text{Subtracting 5 from all three regions}$$
$$-6 \leq 2x < 8$$
$$-3 \leq x < 4.$$

Such an approach saves some writing and will prove useful in Section 4.3.

> **Caution!** The abbreviated form of a conjunction, like $-3 \leq x < 4$, can be written only if both inequality symbols point in the same direction. It is *not acceptable* to write a sentence like $-1 > x < 5$ since doing so does not indicate if *both* $-1 > x$ and $x < 5$ must be true or if it is enough for one of the separate inequalities to be true.

EXAMPLE 3 Solve and graph: $2x - 5 \geq -3 \ and \ 5x + 2 \geq 17$.

Solution   We first solve each inequality separately, retaining the word *and*:

$$2x - 5 \geq -3 \quad and \quad 5x + 2 \geq 17$$
$$2x \geq 2 \quad and \quad 5x \geq 15$$
$$x \geq 1 \quad and \quad x \geq 3.$$

Next, we find the intersection of the two separate solution sets.

The numbers common to both sets are those greater than or equal to 3. Thus the solution set is $\{x \mid x \geq 3\}$, or, in interval notation, $[3, \infty)$. You should check that any number in $[3, \infty)$ satisfies the conjunction whereas numbers outside $[3, \infty)$ do not.

Mathematical Use of the Word "and"

The word "and" corresponds to "intersection" and to the symbol " $\cap$ ". Any solution of a conjunction must make each part of the conjunction true.

Sometimes there is no way to solve both parts of a conjunction at once.

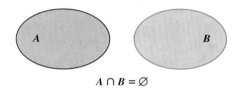

When $A \cap B = \varnothing$,
A and B are said to
be *disjoint*.

$A \cap B = \varnothing$

EXAMPLE 4   Solve and graph: $2x - 3 > 1 \ and \ 3x - 1 < 2$.

Solution   We solve each inequality separately:

$$2x - 3 > 1 \quad and \quad 3x - 1 < 2$$
$$2x > 4 \quad and \quad 3x < 3$$
$$x > 2 \quad and \quad x < 1.$$

The solution set is the intersection of the individual inequalities.

$\{x \mid x > 2\}$ $(2, \infty)$

$\{x \mid x < 1\}$ $(-\infty, 1)$

$\{x \mid x > 2\} \cap \{x \mid x < 1\}$
$= \{x \mid x > 2 \ and \ x < 1\} = \varnothing$ $\varnothing$

Since no number is both greater than 2 and less than 1, the solution set is the empty set, $\varnothing$.

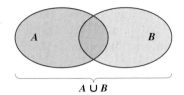

$A \cup B$

Unions of Sets and Disjunctions of Sentences

The **union** of two sets A and B is the collection of elements belonging to A and/or B. We denote the union of A and B by

$$A \cup B.$$

The union of two sets is often pictured as shown at left. For example, if $A = \{$all parents$\}$ and $B = \{$all people who are at least 30 yr old$\}$, then $A \cup B = \{$all people who are parents *or* who are at least 30 yr old$\}$. Note that this set includes people who are parents *and* at least 30 yr old.

EXAMPLE 5   Find the union: $\{2, 3, 4\} \cup \{3, 5, 7\}$.

Solution   The numbers in either or both sets are 2, 3, 4, 5, and 7, so the union is $\{2, 3, 4, 5, 7\}$.

Student Notes _____

Remember that the union or intersection of two sets is itself a set and should be written with set braces.

When two or more sentences are joined by the word *or* to make a compound sentence, the new sentence is called a **disjunction** of the sentences. Here is an example:

$$x < -3 \quad or \quad x > 3.$$

A number is a solution of a disjunction if it is a solution of at least one of the separate parts. For example, -5 is a solution of this disjunction since -5 is a solution of $x < -3$. Below we show the graph of $x < -3$, followed by the graph of $x > 3$, and finally the graph of the disjunction $x < -3 \ or \ x > 3$. *Note that the solution set of a disjunction is the union of the solution sets of the individual sentences.*

$\{x \mid x < -3\}$ $(-\infty, -3)$

$\{x \mid x > 3\}$ $(3, \infty)$

$\{x \mid x < -3\} \cup \{x \mid x > 3\}$
$= \{x \mid x < -3 \ or \ x > 3\}$ $(-\infty, -3) \cup (3, \infty)$

The solution set of $x < -3 \ or \ x > 3$ is $\{x \mid x < -3 \ or \ x > 3\}$, or, in interval notation, $(-\infty, -3) \cup (3, \infty)$. There is no simpler way to write the solution.

> **Mathematical Use of the Word "or"**
>
> The word "or" corresponds to "union" and to the symbol " $\cup$ ". For a number to be a solution of a disjunction, it must be in *at least one* of the solution sets of the individual sentences.

EXAMPLE 6   Solve and graph: $7 + 2x < -1$ *or* $13 - 5x \le 3$.

Solution   We solve each inequality separately, retaining the word *or*:

$$7 + 2x < -1 \quad or \quad 13 - 5x \le 3$$
$$2x < -8 \quad or \quad -5x \le -10$$

Dividing by a negative and reversing the symbol

$$x < -4 \quad or \quad x \ge 2.$$

To find the solution set of the disjunction, we consider the individual graphs. We graph $x < -4$ and then $x \ge 2$. Then we take the union of the graphs.

$\{x \mid x < -4\}$ $(-\infty, -4)$

$\{x \mid x \ge 2\}$ $[2, \infty)$

$\{x \mid x < -4\} \cup \{x \mid x \ge 2\}$ $(-\infty, -4) \cup [2, \infty)$
$= \{x \mid x < -4 \text{ or } x \ge 2\}$

The solution set is $\{x \mid x < -4 \text{ or } x \ge 2\}$, or $(-\infty, -4) \cup [2, \infty)$.

> *Caution!*   A compound inequality like
>
> $$x < -4 \quad or \quad x \ge 2,$$
>
> as in Example 6, *cannot* be expressed as $2 \le x < -4$ because to do so would be to say that x is *simultaneously* less than -4 and greater than or equal to 2. No number is both less than -4 *and* greater than 2, but many are less than -4 *or* greater than 2.

EXAMPLE 7   Solve: $-2x - 5 < -2$ *or* $x - 3 < -10$.

Solution   We solve the individual inequalities separately, retaining the word *or*:

$$-2x - 5 < -2 \quad or \quad x - 3 < -10$$
$$-2x < 3 \quad or \quad x < -7$$

Dividing by a negative and reversing the symbol   Keep the word "or."

$$x > -\frac{3}{2} \quad or \quad x < -7.$$

The solution set is $\left\{x \mid x < -7 \text{ or } x > -\frac{3}{2}\right\}$, or $(-\infty, -7) \cup \left(-\frac{3}{2}, \infty\right)$.

EXAMPLE 8   Solve: $3x - 11 < 4$ *or* $4x + 9 \geq 1$.

Solution   We solve the individual inequalities separately, retaining the word *or*:

$$3x - 11 < 4 \quad or \quad 4x + 9 \geq 1$$
$$3x < 15 \quad or \qquad 4x \geq -8$$
$$x < 5 \quad or \qquad x \geq -2.$$

Keep the word "or."

To find the solution set, we first look at the individual graphs.

$\{x \mid x < 5\}$

$(-\infty, 5)$

$\{x \mid x \geq -2\}$

$[-2, \infty)$

$\{x \mid x < 5\} \cup \{x \mid x \geq -2\}$
$= \{x \mid x < 5 \text{ or } x \geq -2\}$

$(-\infty, \infty) = \mathbb{R}$

Since *all* numbers are less than 5 or greater than or equal to -2, the two sets fill the entire number line. Thus the solution set is $\mathbb{R}$, the set of all real numbers.

Interval Notation and Domains

In Section 2.2, we saw that if $g(x) = \dfrac{5x - 2}{3x - 7}$, then the domain of $g = \{x \mid x$ is a real number *and* $x \neq \frac{7}{3}\}$. We can now represent such a set using interval notation:

$$\{x \mid x \text{ is a real number } and\ x \neq \tfrac{7}{3}\} = \left(-\infty, \tfrac{7}{3}\right) \cup \left(\tfrac{7}{3}, \infty\right).$$

$\left(-\infty, \frac{7}{3}\right) \cup \left(\frac{7}{3}, \infty\right)$

$\frac{7}{3}$

EXAMPLE 9   Use interval notation to write the domain of f if $f(x) = \sqrt{x + 2}$.

Solution   The expression $\sqrt{x + 2}$ is not a real number when $x + 2$ is negative. Thus the domain of f is the set of all x-values for which $x + 2 \geq 0$:

$$x + 2 \geq 0 \qquad x + 2 \text{ cannot be negative.}$$
$$x \geq -2. \qquad \text{Adding } -2 \text{ to both sides}$$

$[-2, \infty)$

We have the domain of $f = \{x \mid x \geq -2\} = [-2, \infty)$.

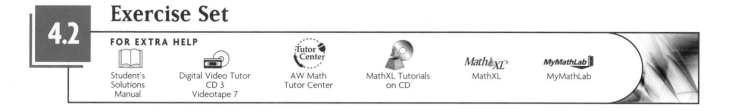

Exercise Set

FOR EXTRA HELP

Student's Solutions Manual   Digital Video Tutor CD 3 Videotape 7   AW Math Tutor Center   MathXL Tutorials on CD   MathXL   MyMathLab

⤶ *Concept Reinforcement   In each of Exercises 1–10, match the set with the most appropriate choice from the column on the right.*

1. ____ $\{x \mid x < -2 \; or \; x > 2\}$

2. ____ $\{x \mid x < -2 \; and \; x > 2\}$

3. ____ $\{x \mid x > -2\} \cap \{x \mid x < 2\}$

4. ____ $\{x \mid x \leq -2\} \cup \{x \mid x \geq 2\}$

5. ____ $\{x \mid x \leq -2\} \cup \{x \mid x \leq 2\}$

6. ____ $\{x \mid x \leq -2\} \cap \{x \mid x \leq 2\}$

7. ____ $\{x \mid x \geq -2\} \cap \{x \mid x \geq 2\}$

8. ____ $\{x \mid x \geq -2\} \cup \{x \mid x \geq 2\}$

9. ____ $\{x \mid x \leq 2\} \; and \; \{x \mid x \geq -2\}$

10. ____ $\{x \mid x \leq 2\} \; or \; \{x \mid x \geq -2\}$

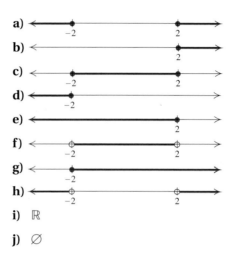

i) $\mathbb{R}$

j) $\varnothing$

Find each indicated intersection or union.

11. $\{5, 9, 11\} \cap \{9, 11, 18\}$

12. $\{2, 4, 8\} \cup \{8, 9, 10\}$

13. $\{0, 5, 10, 15\} \cup \{5, 15, 20\}$

14. $\{2, 5, 9, 13\} \cap \{5, 8, 10\}$

15. $\{a, b, c, d, e, f\} \cap \{b, d, f\}$

16. $\{a, b, c\} \cup \{a, c\}$

17. $\{r, s, t\} \cup \{r, u, t, s, v\}$

18. $\{m, n, o, p\} \cap \{m, o, p\}$

19. $\{3, 6, 9, 12\} \cap \{5, 10, 15\}$

20. $\{1, 5, 9\} \cup \{4, 6, 8\}$

21. $\{3, 5, 7\} \cup \varnothing$

22. $\{3, 5, 7\} \cap \varnothing$

Graph and write interval notation for each compound inequality.

23. $3 < x < 7$

24. $0 \leq y \leq 4$

25. $-6 \leq y \leq -2$

26. $-9 \leq x < -5$

27. $x < -1 \; or \; x > 4$

28. $x < -5 \; or \; x > 1$

29. $x \leq -2 \; or \; x > 1$

30. $x \leq -5 \; or \; x > 2$

31. $-4 \leq -x < 2$

32. $x > -7 \; and \; x < -2$

33. $x > -2 \; and \; x < 4$

34. $3 > -x \geq -1$

35. $5 > a \; or \; a > 7$

36. $t \geq 2 \; or \; -3 > t$

37. $x \geq 5 \; or \; -x \geq 4$

38. $-x < 3 \; or \; x < -6$

39. $7 > y \; and \; y \geq -3$

40. $6 > -x \geq 0$

41. $x < 7 \; and \; x \geq 3$

42. $x \geq -3 \; and \; x < 3$

Aha! 43. $t < 2 \; or \; t < 5$

44. $t > 4 \; or \; t > -1$

Solve and graph each solution set.

45. $-2 < t + 1 < 8$

46. $-3 < t + 1 \leq 5$

47. $2 < x + 3 \; and \; x + 1 \leq 5$

48. $-1 < x + 2 \; and \; x - 4 < 3$

49. $-7 \le 2a - 3 \; and \; 3a + 1 < 7$

50. $-4 \le 3n + 5 \; and \; 2n - 3 \le 7$

Aha!   **51.** $x + 7 \le -2 \; or \; x + 7 \ge -3$

52. $x + 5 < -3 \; or \; x + 5 \ge 4$

53. $5 > \dfrac{x - 3}{4} > 1$

54. $3 \ge \dfrac{x - 1}{2} \ge -4$

55. $-7 \le 4x + 5 \le 13$

56. $-4 \le 2x + 3 \le 15$

57. $2 \le f(x) \le 8$, where $f(x) = 3x - 1$

58. $7 \ge g(x) \ge -2$, where $g(x) = 3x - 5$

59. $-21 \le f(x) < 0$, where $f(x) = -2x - 7$

60. $4 > g(t) \ge 2$, where $g(t) = -3t - 8$

61. $f(x) \le 2 \; or \; f(x) \ge 8$, where $f(x) = 3x - 1$

62. $g(x) \le -2 \; or \; g(x) \ge 10$, where $g(x) = 3x - 5$

63. $f(x) < -3 \; or \; f(x) > 5$, where $f(x) = 2x - 7$

64. $g(x) < -7 \; or \; g(x) > 7$, where $g(x) = 3x + 5$

65. $6 > 2a - 1 \; or \; -4 \le -3a + 2$

66. $3a - 7 > -10 \; or \; 5a + 2 \le 22$

67. $a + 3 < -2 \; and \; 3a - 4 < 8$

68. $1 - a < -2 \; and \; 2a + 1 > 9$

69. $3x + 2 < 2 \; or \; 4 - 2x < 14$

70. $2x - 1 > 5 \; or \; 3 - 2x \ge 7$

71. $2t - 7 \le 5 \; or \; 5 - 2t > 3$

72. $5 - 3a \le 8 \; or \; 2a + 1 > 7$

For $f(x)$ as given, use interval notation to write the domain of f.

73. $f(x) = \dfrac{9}{x + 8}$

74. $f(x) = \dfrac{2}{x + 3}$

75. $f(x) = \sqrt{x - 6}$

76. $f(x) = \sqrt{x - 2}$

77. $f(x) = \dfrac{x + 3}{2x - 8}$

78. $f(x) = \dfrac{x - 1}{3x + 6}$

79. $f(x) = \sqrt{2x + 7}$

80. $f(x) = \sqrt{8 - 5x}$

81. $f(x) = \sqrt{8 - 2x}$

82. $f(x) = \sqrt{10 - 2x}$

83. Why can the conjunction $2 < x \; and \; x < 5$ be rewritten as $2 < x < 5$, but the disjunction $2 < x \; or \; x < 5$ cannot be rewritten as $2 < x < 5$?

84. Can the solution set of a disjunction be empty? Why or why not?

SKILL MAINTENANCE

Graph.

85. $y = 5$ [2.4]

86. $y = -2$ [2.4]

87. $f(x) = |x|$ [2.2]

88. $g(x) = x - 1$ [2.3]

Solve each system graphically. [3.1]

89. $y = x - 3$,
$y = 5$

90. $y = x + 2$,
$y = -3$

SYNTHESIS

91. What can you conclude about a, b, c, and d, if $[a, b] \cup [c, d] = [a, d]$? Why?

92. What can you conclude about a, b, c, and d, if $[a, b] \cap [c, d] = [a, b]$? Why?

93. Use the accompanying graph of $f(x) = 2x - 5$ to solve $-7 < 2x - 5 < 7$.

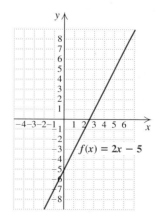

94. Use the accompanying graph of $g(x) = 4 - x$ to solve $4 - x < -2$ or $4 - x > 7$.

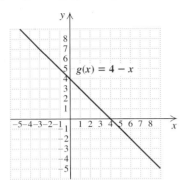

95. *Childless women.*   On the basis of trends from the late 1900s, the function given by

$$P(t) = 0.44t + 10.2$$

can be used to estimate the percentage of U.S. women age 40–44, $P(t)$, t years after 1980, who have not given birth. For what years will the percentage of childless 40–44-year-old women be between 19 and 30 percent?
Sources: Based on data from the U.S. Bureau of the Census; *Deseret Morning News*, 11/03

96. *Pressure at sea depth.*   The function given by

$$P(d) = 1 + \frac{d}{33}$$

gives the pressure, in atmospheres (atm), at a depth of d feet in the sea. For what depths d is the pressure at least 1 atm and at most 7 atm?

97. *Converting dress sizes.*   The function given by

$$f(x) = 2(x + 10)$$

can be used to convert dress sizes x in the United States to dress sizes $f(x)$ in Italy. For what dress sizes in the United States will dress sizes in Italy be between 32 and 46?

98. *Solid-waste generation.*   The function given by

$$w(t) = 0.01t + 4.3$$

can be used to estimate the number of pounds of solid waste, $w(t)$, produced daily, on average, by each person in the United States, t years after 1991. For what years will waste production range from 4.5 to 4.75 lb per person per day?

99. *Records in the women's 100-m dash.*   Florence Griffith Joyner set a world record of 10.49 sec in the women's 100-m dash in 1988. The function given by

$$R(t) = -0.0433t + 10.49$$

can be used to predict the world record in the women's 100-m dash t years after 1988. Predict (using an inequality) those years for which the world record was between 11.5 and 10.8 sec. (Measure from the middle of 1988.)
Sources: *Guinness Book of World Records* 2004; www.Runnersworld.com

100. *Temperatures of liquids.*   The formula

$$C = \tfrac{5}{9}(F - 32)$$

can be used to convert Fahrenheit temperatures F to Celsius temperatures C.

a) Gold is liquid for Celsius temperatures C such that $1063° \le C < 2660°$. Find a comparable inequality for Fahrenheit temperatures.

b) Silver is liquid for Celsius temperatures C such that $960.8° \le C < 2180°$. Find a comparable inequality for Fahrenheit temperatures.

101. *Minimizing tolls.*   A $3.00 toll is charged to cross the bridge from Sanibel Island to mainland Florida. A six-month pass, costing $15.00, reduces the toll to $0.50. A one-year pass, costing $150, allows for free crossings. How many crossings per year does it take, on average, for two consecutive six-month passes to be the most economical choice? Assume a constant number of trips per month.
Source: www.leewayinfo.com

Solve and graph.

102. $4a - 2 \le a + 1 \le 3a + 4$

103. $4m - 8 > 6m + 5$ or $5m - 8 < -2$

104. $x - 10 < 5x + 6 \le x + 10$

105. $3x < 4 - 5x < 5 + 3x$

Determine whether each sentence is true or false for all real numbers a, b, and c.

106. If $-b < -a$, then $a < b$.

107. If $a \leq c$ and $c \leq b$, then $b > a$.

108. If $a < c$ and $b < c$, then $a < b$.

109. If $-a < c$ and $-c > b$, then $a > b$.

For f(x) as given, use interval notation to write the domain of f.

110. $f(x) = \dfrac{\sqrt{5 + 2x}}{x - 1}$

111. $f(x) = \dfrac{\sqrt{3 - 4x}}{x + 7}$

112. Let $y_1 = -1$, $y_2 = 2x + 5$, and $y_3 = 13$. Then use the graphs of y_1, y_2, and y_3 to check the solution to Example 2.

113. Let $y_1 = -2x - 5$, $y_2 = -2$, $y_3 = x - 3$, and $y_4 = -10$. Then use the graphs of y_1, y_2, y_3, and y_4 to check the solution to Example 7.

114. Use a graphing calculator to check your answers to Exercises 43–46 and Exercises 63–66.

115. On many graphing calculators, the TEST key provides access to inequality symbols, while the LOGIC option of that same key accesses the conjunction *and* and the disjunction *or*. Thus, if $y_1 = x > -2$ and $y_2 = x < 4$, Exercise 33 can be checked by forming the expression $y_3 = y_1$ and y_2. The interval(s) in the solution set appears as a horizontal line 1 unit above the *x*-axis. (Be careful to "deselect" y_1 and y_2 so that only y_3 is drawn.)

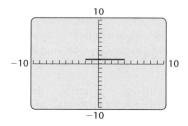

CORNER

Saving on Shipping Costs

COLLABORATIVE

Focus: Compound inequalities and solution sets

Time: 20–30 minutes

Group size: 2–3

For Priority Mail, the United States Postal Service charges (at present, in 2004) $3.85 for the first pound plus $0.51 for each pound or part thereof after the first. For UPS Ground Residential shipping, the first pound costs $6.25 and each additional pound or part thereof costs $0.17.*

*This activity is based on an article by Michael Contino in *Mathematics Teacher*, May 1995.

ACTIVITY

1. One group member should determine the function p, where $p(x)$ represents the cost, in dollars, of mailing x pounds as Priority Mail.
2. One member should determine the function r, where $r(x)$ represents the cost, in dollars, of shipping x pounds using UPS Ground.
3. A third member should graph p and r on the same set of axes.
4. Finally, working together, use the graph to determine those weights for which Priority Mail is less expensive than UPS Ground Residential shipping. Express your answer in both set-builder and interval notations.

Absolute-Value Equations and Inequalities

4.3

Equations with Absolute Value   •   Inequalities with Absolute Value

Equations with Absolute Value

Recall from Section 1.2 the definition of absolute value.

> **Absolute Value**
>
> The absolute value of x, denoted $|x|$, is defined as
>
> $$|x| = \begin{cases} x, & \text{if } x \geq 0, \\ -x, & \text{if } x < 0. \end{cases}$$
>
> (When x is nonnegative, the absolute value of x is x. When x is negative, the absolute value of x is the opposite of x.)

To better understand this definition, suppose x is -5. Then $|x| = |-5| = 5$, and 5 is the opposite of -5. This shows that when x represents a negative number, the absolute value of x is the opposite of x (which is positive).

Since distance is always nonnegative, we can think of a number's absolute value as its distance from zero on a number line.

EXAMPLE 1

Find the solution set: **(a)** $|x| = 4$; **(b)** $|x| = 0$; **(c)** $|x| = -7$.

Solution

a) We interpret $|x| = 4$ to mean that the number x is 4 units from zero on a number line. There are two such numbers, 4 and -4. Thus the solution set is $\{-4, 4\}$.

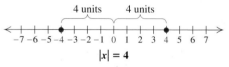

A second way to visualize this problem is to graph $f(x) = |x|$ (see Section 2.1). We also graph $g(x) = 4$. The x-values of the points of intersection are the solutions of $|x| = 4$.

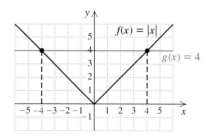

b) We interpret $|x| = 0$ to mean that x is 0 units from zero on a number line. The only number that satisfies this is 0 itself. Thus the solution set is $\{0\}$.

c) Since distance is always nonnegative, it doesn't make sense to talk about a number that is -7 units from zero. Remember: The absolute value of a number is never negative. Thus, $|x| = -7$ has no solution; the solution set is $\varnothing$.

Example 1 leads us to the following principle for solving equations.

The Absolute-Value Principle for Equations

For any positive number p and any algebraic expression X:

a) The solutions of $|X| = p$ are those numbers that satisfy $X = -p$ or $X = p$.
b) The equation $|X| = 0$ is equivalent to the equation $X = 0$.
c) The equation $|X| = -p$ has no solution.

EXAMPLE 2

Find the solution set: **(a)** $|2x + 5| = 13$; **(b)** $|4 - 7x| = -8$.

Solution

a) We use the absolute-value principle, knowing that $2x + 5$ must be either 13 or -13:

$$|X| = p$$
$$|2x + 5| = 13 \qquad \text{Substituting}$$
$$2x + 5 = -13 \quad or \quad 2x + 5 = 13$$
$$2x = -18 \quad or \quad 2x = 8$$
$$x = -9 \quad or \quad x = 4.$$

Check: For -9:

$$\begin{array}{c|c} |2x + 5| = 13 \\ \hline |2(-9) + 5| & 13 \\ |-18 + 5| & \\ |-13| & \\ & 13 \stackrel{?}{=} 13 \quad \text{TRUE} \end{array}$$

For 4:

$$\begin{array}{c|c} |2x + 5| = 13 \\ \hline |2 \cdot 4 + 5| & 13 \\ |8 + 5| & \\ |13| & \\ & 13 \stackrel{?}{=} 13 \quad \text{TRUE} \end{array}$$

The number $2x + 5$ is 13 units from zero if x is replaced with -9 or 4. The solution set is $\{-9, 4\}$.

b) The absolute-value principle reminds us that absolute value is always nonnegative. The equation $|4 - 7x| = -8$ has no solution. The solution set is $\varnothing$.

To use the absolute-value principle, we must be sure that the absolute-value expression is alone on one side of the equation.

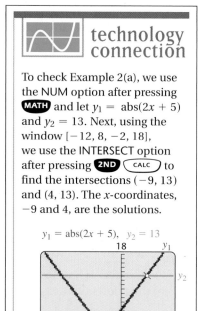

technology connection

To check Example 2(a), we use the NUM option after pressing **MATH** and let $y_1 = \text{abs}(2x + 5)$ and $y_2 = 13$. Next, using the window $[-12, 8, -2, 18]$, we use the INTERSECT option after pressing **2ND** **CALC** to find the intersections $(-9, 13)$ and $(4, 13)$. The x-coordinates, -9 and 4, are the solutions.

$y_1 = \text{abs}(2x + 5), \quad y_2 = 13$

1. Use a graphing calculator to show that Example 2(b) has no solution.

EXAMPLE 3

Given that $f(x) = 2|x + 3| + 1$, find all x for which $f(x) = 15$.

Solution Since we are looking for $f(x) = 15$, we substitute:

$$f(x) = 15$$

$2|x + 3| + 1 = 15$ Replacing $f(x)$ with $2|x + 3| + 1$

$2|x + 3| = 14$ Subtracting 1 from both sides

$|x + 3| = 7$ Dividing both sides by 2

$x + 3 = -7$ *or* $x + 3 = 7$ Using the absolute-value principle
for equations

$x = -10$ *or* $x = 4.$

We leave it to the student to check that $f(-10) = f(4) = 15$. The solution set is $\{-10, 4\}$.

EXAMPLE 4

Solve: $|x - 2| = 3$.

Solution Because this equation is of the form $|a - b| = c$, it can be solved in two different ways.

Method 1. We interpret $|x - 2| = 3$ as stating that the number $x - 2$ is 3 units from zero. Using the absolute-value principle, we replace X with $x - 2$ and p with 3:

$$|X| = p$$
$$|x - 2| = 3$$
$x - 2 = -3$ *or* $x - 2 = 3$ Using the absolute-value principle
$x = -1$ *or* $x = 5.$

Method 2. This approach is helpful in calculus. The expressions $|a - b|$ and $|b - a|$ can be used to represent the *distance between a and b* on the number line. For example, the distance between 7 and 8 is given by $|8 - 7|$ or $|7 - 8|$. From this viewpoint, the equation $|x - 2| = 3$ states that the distance between x and 2 is 3 units. We draw a number line and locate all numbers that are 3 units from 2.

The solutions of $|x - 2| = 3$ are -1 and 5.

Check: The check consists of observing that both methods give the same solutions. The solution set is $\{-1, 5\}$.

Sometimes an equation has two absolute-value expressions. Consider $|a| = |b|$. This means that a and b are the same distance from zero.

If a and b are the same distance from zero, then either they are the same number or they are opposites.

EXAMPLE 5   Solve: $|2x - 3| = |x + 5|$.

Solution   The given equation tells us that $2x - 3$ and $x + 5$ are the same distance from 0. This means that they are either the same number or opposites:

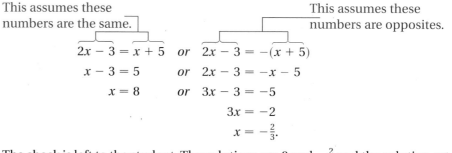

This assumes these numbers are the same.   This assumes these numbers are opposites.

$$2x - 3 = x + 5 \quad or \quad 2x - 3 = -(x + 5)$$
$$x - 3 = 5 \quad or \quad 2x - 3 = -x - 5$$
$$x = 8 \quad or \quad 3x - 3 = -5$$
$$3x = -2$$
$$x = -\tfrac{2}{3}.$$

The check is left to the student. The solutions are 8 and $-\tfrac{2}{3}$ and the solution set is $\left\{ -\tfrac{2}{3}, 8 \right\}$.

Inequalities with Absolute Value

Our methods for solving equations with absolute value can be adapted for solving inequalities. Inequalities of this sort arise regularly in more advanced courses.

EXAMPLE 6   Solve $|x| < 4$. Then graph.

Solution   The solutions of $|x| < 4$ are all numbers whose *distance from zero is less than* 4. By substituting or by looking at the number line, we can see that numbers like $-3, -2, -1, -\tfrac{1}{2}, -\tfrac{1}{4}, 0, \tfrac{1}{4}, \tfrac{1}{2}, 1, 2$, and 3 are all solutions. In fact, the solutions are all the numbers between -4 and 4. The solution set is $\{x \mid -4 < x < 4\}$. In interval notation, the solution set is $(-4, 4)$. The graph is as follows:

$$\xleftarrow{\qquad} \underset{-7\ -6\ -5\ -4\ -3\ -2\ -1\ \ 0\ \ 1\ \ 2\ \ 3\ \ 4\ \ 5\ \ 6\ \ 7}{\circ\text{———}\circ} \xrightarrow{\qquad}$$

$$|x| < 4$$

We can also visualize Example 6 by graphing $f(x) = |x|$ and $g(x) = 4$, as in Example 1. The solution set consists of all x-values for which $(x, f(x))$ is below the horizontal line $g(x) = 4$. These x-values comprise the interval $(-4, 4)$.

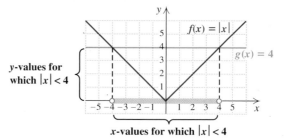

y-values for which $|x| < 4$

x-values for which $|x| < 4$

EXAMPLE 7   Solve $|x| \geq 4$. Then graph.

Solution   The solutions of $|x| \geq 4$ are all numbers that are at least 4 units from zero—in other words, those numbers x for which $x \leq -4$ *or* $4 \leq x$. The solution set is $\{x \mid x \leq -4 \ or \ x \geq 4\}$. In interval notation, the solution set is $(-\infty, -4] \cup [4, \infty)$. We can check mentally with numbers like $-4.1, -5, 4.1$, and 5. The graph is as follows:

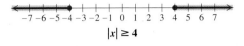

$|x| \geq 4$

As with Examples 1 and 6, Example 7 can be visualized by graphing $f(x) = |x|$ and $g(x) = 4$. The solution set of $|x| \geq 4$ consists of all x-values for which $(x, f(x))$ is on or above the horizontal line $g(x) = 4$. These x-values comprise $(-\infty, -4] \cup [4, \infty)$.

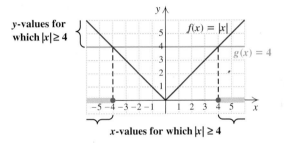

Examples 1, 6, and 7 illustrate three types of problems in which absolute-value symbols appear. The following is a general principle for solving such problems.

Principles for Solving Absolute-Value Problems

For any positive number p and any expression X:

a) The solutions of $|X| = p$ are those numbers that satisfy $X = -p \ or \ X = p$.

b) The solutions of $|X| < p$ are those numbers that satisfy $-p < X < p$.

c) The solutions of $|X| > p$ are those numbers that satisfy $X < -p \ or \ p < X$.

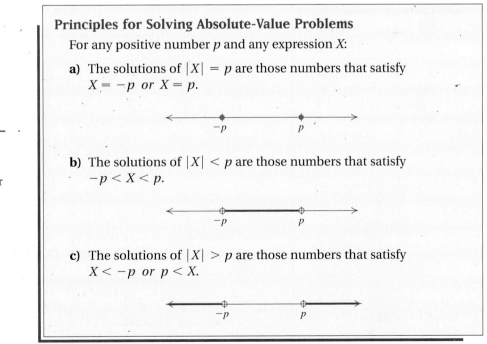

Of course, if p is negative, any value of X will satisfy the inequality $|X| > p$ because absolute value is never negative. (The solution set is $\mathbb{R}$.) By the same reasoning, $|X| < p$ has no solution when p is not positive. (The solution set is $\varnothing$). Thus, $|2x - 7| > -3$ is true for any real number x, and $|2x - 7| < -3$ has no solution.

Note that an inequality of the form $|X| < p$ corresponds to a *con*junction, whereas an inequality of the form $|X| > p$ corresponds to a *dis*junction.

EXAMPLE 8   Solve $|3x - 2| < 4$. Then graph.

Solution   The number $3x - 2$ must be less than 4 units from 0. This is of the form $|X| < p$, so part (b) of the principles listed above applies:

$$|X| < p$$

$$|3x - 2| < 4 \qquad \text{Replacing } X \text{ with } 3x - 2 \text{ and } p \text{ with } 4$$

$$-4 < 3x - 2 < 4 \qquad \text{The number } 3x - 2 \text{ must be within 4 units}$$
$$\text{of zero.}$$

$$-2 < \quad 3x \quad < 6 \qquad \text{Adding 2}$$

$$-\tfrac{2}{3} < \quad x \quad < 2. \qquad \text{Multiplying by } \tfrac{1}{3}$$

The solution set is $\left\{x \mid -\tfrac{2}{3} < x < 2\right\}$. In interval notation, the solution set is $\left(-\tfrac{2}{3}, 2\right)$. The graph is as follows:

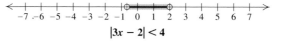

$$|3x - 2| < 4$$

EXAMPLE 9   Given that $f(x) = |4x + 2|$, find all x for which $f(x) \geq 6$.

Solution   We have

$$f(x) \geq 6,$$

or $\qquad |4x + 2| \geq 6. \qquad$ Substituting

To solve, we use part (c) of the principles listed above. In this case, X is $4x + 2$ and p is 6:

$$|X| \geq p$$

$$|4x + 2| \geq 6 \qquad \text{Replacing } X \text{ with } 4x + 2 \text{ and } p \text{ with } 6$$

$$4x + 2 \leq -6 \quad or \quad 6 \leq 4x + 2 \qquad \text{The number } 4x + 2 \text{ must be at}$$
$$\text{least 6 units from zero.}$$

$$4x \leq -8 \quad or \quad 4 \leq 4x \qquad \text{Adding } -2$$

$$x \leq -2 \quad or \quad 1 \leq x. \qquad \text{Multiplying by } \tfrac{1}{4}$$

The solution set is $\{x \mid x \leq -2 \text{ or } x \geq 1\}$. In interval notation, the solution is $(-\infty, -2] \cup [1, \infty)$. The graph is as follows:

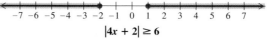

$$|4x + 2| \geq 6$$

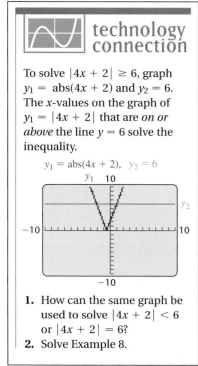

technology connection

To solve $|4x + 2| \geq 6$, graph $y_1 = \text{abs}(4x + 2)$ and $y_2 = 6$. The x-values on the graph of $y_1 = |4x + 2|$ that are *on or above* the line $y = 6$ solve the inequality.

$y_1 = \text{abs}(4x + 2)$, $y_2 = 6$

1. How can the same graph be used to solve $|4x + 2| < 6$ or $|4x + 2| = 6$?
2. Solve Example 8.

Exercise Set

4.3

↘ *Concept Reinforcement*   *Classify each of the following as either true or false.*

1. If x is negative, then $|x| = -x$.

2. $|x|$ is never negative.

3. $|x|$ is always positive.

4. The distance between a and b can be expressed as $|a - b|$.

5. The number a is $|a|$ units from 0.

6. There are two solutions of $|3x - 8| = 17$.

7. There is no solution of $|4x + 9| > -5$.

8. All real numbers are solutions of $|2x - 7| < -3$.

Solve.

9. $|x| = 7$

10. $|x| = 9$

Aha! **11.** $|x| = -6$

12. $|x| = -3$

13. $|p| = 0$

14. $|y| = 7.3$

15. $|t| = 5.5$

16. $|m| = 0$

17. $|2x - 3| = 4$

18. $|5x + 2| = 7$

19. $|3x - 5| = -8$

20. $|7x - 2| = -9$

21. $|x - 2| = 6$

22. $|x - 3| = 8$

23. $|x - 5| = 3$

24. $|x - 6| = 1$

25. $|x - 7| = 9$

26. $|x - 4| = 5$

27. $|5x| - 3 = 37$

28. $|2y| - 5 = 13$

29. $7|q| - 2 = 9$

30. $7|z| + 2 = 16$

31. $\left| \dfrac{2x - 1}{3} \right| = 5$

32. $\left| \dfrac{4 - 5x}{6} \right| = 3$

33. $|m + 5| + 9 = 16$

34. $|t - 7| + 1 = 4$

35. $5 - 2|3x - 4| = -5$

36. $3|2x - 5| - 7 = -1$

37. Let $f(x) = |2x + 6|$. Find all x for which $f(x) = 8$.

38. Let $f(x) = |2x + 4|$. Find all x for which $f(x) = 10$.

39. Let $f(x) = |x| - 3$. Find all x for which $f(x) = 5.7$.

40. Let $f(x) = |x| + 7$. Find all x for which $f(x) = 18$.

41. Let $f(x) = \left| \dfrac{3x - 2}{5} \right|$. Find all x for which $f(x) = 2$.

42. Let $f(x) = \left| \dfrac{1 - 2x}{3} \right|$. Find all x for which $f(x) = 1$.

Solve.

43. $|x + 4| = |2x - 7|$

44. $|3x + 5| = |x - 6|$

45. $|x + 4| = |x - 3|$

46. $|x - 9| = |x + 6|$

47. $|3a - 1| = |2a + 4|$

48. $|5t + 7| = |4t + 3|$

Aha! **49.** $|n - 3| = |3 - n|$

50. $|y - 2| = |2 - y|$

51. $|7 - a| = |a + 5|$

52. $|6 - t| = |t + 7|$

53. $\left| \frac{1}{2}x - 5 \right| = \left| \frac{1}{4}x + 3 \right|$

54. $\left| 2 - \frac{2}{3}x \right| = \left| 4 + \frac{7}{8}x \right|$

Solve and graph.

55. $|a| \leq 9$

56. $|x| < 2$

57. $|x| > 8$

58. $|a| \geq 3$

59. $|t| > 0$

60. $|t| \geq 1.7$

61. $|x - 1| < 4$

62. $|x - 1| < 3$

63. $|x + 2| \leq 6$

64. $|x + 4| \leq 1$

65. $|x - 3| + 2 > 7$

66. $|x - 4| + 5 > 2$

Aha! **67.** $|2y - 9| > -5$

68. $|3y - 4| > 8$

69. $|3a - 4| + 2 \geq 8$

70. $|2a - 5| + 1 \geq 9$

71. $|y - 3| < 12$

72. $|p - 2| < 3$

73. $9 - |x + 4| \leq 5$

74. $12 - |x - 5| \leq 9$

75. $|4 - 3y| > 8$

76. $|7 - 2y| < -6$

Aha!　**77.** $|5 - 4x| < -6$

78. $7 + |4a - 5| \leq 26$

79. $\left|\dfrac{2 - 5x}{4}\right| \geq \dfrac{2}{3}$

80. $\left|\dfrac{1 + 3x}{5}\right| > \dfrac{7}{8}$

81. $|m + 3| + 8 \leq 14$

82. $|t - 7| + 3 \geq 4$

83. $25 - 2|a + 3| > 19$

84. $30 - 4|a + 2| > 12$

85. Let $f(x) = |2x - 3|$. Find all x for which $f(x) \leq 4$.

86. Let $f(x) = |5x + 2|$. Find all x for which $f(x) \leq 3$.

87. Let $f(x) = 5 + |3x - 4|$. Find all x for which $f(x) \geq 16$.

88. Let $f(x) = |2 - 9x|$. Find all x for which $f(x) \geq 25$.

89. Let $f(x) = 7 + |2x - 1|$. Find all x for which $f(x) < 16$.

90. Let $f(x) = 5 + |3x + 2|$. Find all x for which $f(x) < 19$.

91. Explain in your own words why -7 is not a solution of $|x| < 5$.

92. Explain in your own words why $[6, \infty)$ is only part of the solution of $|x| \geq 6$.

SKILL MAINTENANCE

Solve using substitution or elimination. [3.2]

93. $2x - 3y = 7,$
$3x + 2y = -10$

94. $3x - 5y = 9,$
$4x - 3y = 1$

95. $x = -2 + 3y,$
$x - 2y = 2$

96. $y = 3 - 4x,$
$2x - y = -9$

Solve graphically. [3.1]

97. $x + 2y = 9,$
$3x - y = -1$

98. $2x + y = 7,$
$-3x - 2y = 10$

SYNTHESIS

99. Is it possible for an equation in x of the form $|ax + b| = c$ to have exactly one solution? Why or why not?

100. Explain why the inequality $|x + 5| \geq 2$ can be interpreted as "the number x is at least 2 units from -5."

101. From the definition of absolute value, $|x| = x$ only when $x \geq 0$. Solve $|3t - 5| = 3t - 5$ using this same reasoning.

Solve.

102. $|3x - 5| = x$

103. $|x + 2| > x$

104. $2 \leq |x - 1| \leq 5$

105. $|5t - 3| = 2t + 4$

106. $t - 2 \leq |t - 3|$

Find an equivalent inequality with absolute value.

107. $-3 < x < 3$

108. $-5 \leq y \leq 5$

109. $x \leq -6 \ or \ 6 \leq x$

110. $x < -4 \ or \ 4 < x$

111. $x < -8 \ or \ 2 < x$

112. $-5 < x < 1$

113. x is less than 2 units from 7.

114. x is less than 1 unit from 5.

Write an absolute-value inequality for which the interval shown is the solution.

115.
```
←+++++++●+++++++→
 -7-6-5-4-3-2-1 0 1 2 3 4 5 6 7
```

116.
```
←+○++++++++++++○+→
 -5-4-3-2-1 0 1 2 3 4 5 6 7 8 9
```

117.
```
←○++++++○+++++++→
 -7-6-5-4-3-2-1 0 1 2 3 4 5 6 7
```

118.
```
←++●+++++++++++●++→
  0 1 2 3 4 5 6 7 8 9 10 11 12 13 14
```

119. *Bungee jumping.* A bungee jumper is bouncing up and down so that her distance d above a river satisfies the inequality $|d - 60 \text{ ft}| \leq 10 \text{ ft}$ (see the figure below). If the bridge from which she jumped is 150 ft above the river, how far is the bungee jumper from the bridge at any given time?

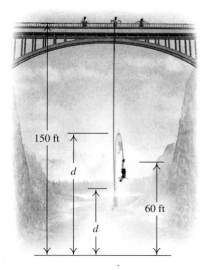

120. *Water level.*   Depending on how dry or wet the weather has been, water in a well will rise and fall. The distance d that a well's water level is below the ground satisfies the inequality $|d - 15| \le 2.5$ (see the figure below).

a) Solve for d.

b) How tall a column of water is in the well at any given time?

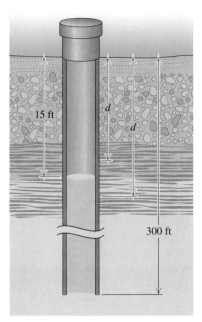

15 ft

d

d

300 ft

121. Use the accompanying graph of $f(x) = |2x - 6|$ to solve $|2x - 6| \le 4$.

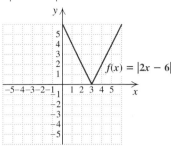

$f(x) = |2x - 6|$

122. Describe a procedure that could be used to solve any equation of the form $g(x) < c$ graphically.

123. Use a graphing calculator to check the solutions to Examples 3 and 5.

124. Use a graphing calculator to check your answers to Exercises 9, 17, 23, 49, 61, 71, 79, and 103.

125. Isabel is using the following graph to solve $|x + 3| < 4$. How can you tell that a mistake has been made in entering $y = \text{abs}(x + 3)$?

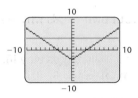

4.4   Inequalities in Two Variables

Graphs of Linear Inequalities • Systems of Linear Inequalities

In Section 4.1, we graphed inequalities in one variable on a number line. Now we graph inequalities in two variables on a plane.

Graphs of Linear Inequalities

When the equals sign in a linear equation is replaced with an inequality sign, a **linear inequality** is formed. Solutions of linear inequalities are ordered pairs.

EXAMPLE 1

Determine whether $(-3, 2)$ and $(6, -7)$ are solutions of the inequality $5x - 4y > 13$.

Solution   Below, on the left, we replace x with -3 and y with 2. On the right, we replace x with 6 and y with -7.

$$\frac{5x - 4y > 13}{5(-3) - 4 \cdot 2 \,\big|\, 13}$$
$$-15 - 8 \,\big|$$
$$-23 \overset{?}{>} 13 \quad \text{FALSE}$$

Since $-23 > 13$ is false, $(-3, 2)$ is not a solution.

$$\frac{5x - 4y > 13}{5(6) - 4(-7) \,\big|\, 13}$$
$$30 + 28 \,\big|$$
$$58 \overset{?}{>} 13 \quad \text{TRUE}$$

Since $58 > 13$ is true, $(6, -7)$ is a solution.

The graph of a linear equation is a straight line. The graph of a linear inequality is a **half-plane**, with a **boundary** that is a straight line. To find the equation of the boundary line, we simply replace the inequality sign with an equals sign.

EXAMPLE 2

Graph: $y \le x$.

Solution   We first graph the equation of the boundary, $y = x$. Every solution of $y = x$ is an ordered pair, like $(3, 3)$, in which both coordinates are the same. The graph of $y = x$ is shown on the left below. Since the inequality symbol is $\le$, the line is drawn solid and is part of the graph of $y \le x$.

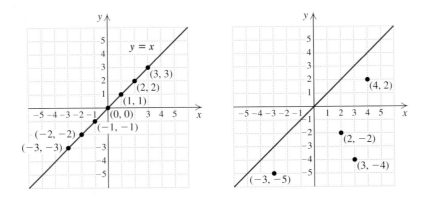

Note that in the graph on the right each ordered pair on the half-plane below $y = x$ contains a y-coordinate that is less than the x-coordinate. All these pairs represent solutions of $y \le x$. We check one pair, $(4, 2)$, as follows:

$$\frac{y \le x}{2 \,\big|\, 4} \quad \text{TRUE}$$

It turns out that *any* point on the same side of $y = x$ as $(4, 2)$ is also a solution. Thus, if one point in a half-plane is a solution, then *all* points in that half-plane are solutions.

We finish drawing the solution set by shading the half-plane below $y = x$. The complete solution set consists of the shaded half-plane as well as the boundary line itself.

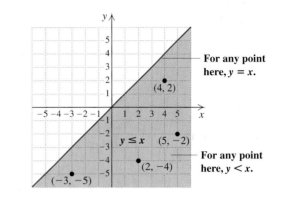

For any point here, $y = x$.

$y \leq x$ $(5, -2)$

For any point here, $y < x$.

$(4, 2)$

$(2, -4)$

$(-3, -5)$

From Example 2, we see that for any inequality of the form $y \leq f(x)$ or $y < f(x)$, we shade *below* the graph of $y = f(x)$.

EXAMPLE 3   Graph: $8x + 3y > 24$.

Solution   First, we sketch the graph of $8x + 3y = 24$. Since the inequality sign is $>$, points on this line do not represent solutions of the inequality, so the line is drawn dashed. Points representing solutions of $8x + 3y > 24$ are in either the half-plane above the line or the half-plane below the line. To determine which, we select a point that is not on the line and determine whether it is a solution of $8x + 3y > 24$. Let's use $(1, 1)$ as this *test point*:

$$\begin{array}{c|c} 8x + 3y > 24 \\ \hline 8(1) + 3(1) & 24 \\ 8 + 3 & \\ 11 \overset{?}{>} 24 & \text{FALSE} \end{array}$$

Since $11 > 24$ is *false*, $(1, 1)$ is not a solution. Thus no point in the half-plane containing $(1, 1)$ is a solution. The points in the other half-plane *are* solutions, so we shade that half-plane and obtain the graph shown at right.

This point is not a solution.

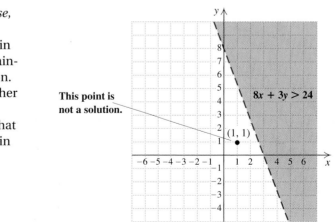

$8x + 3y > 24$

$(1, 1)$

Steps for Graphing Linear Inequalities

1. Replace the inequality sign with an equals sign and graph this line as the boundary. If the inequality symbol is $<$ or $>$, draw the line dashed. If the inequality symbol is $\leq$ or $\geq$, draw the line solid.
2. The graph of the inequality consists of a half-plane on one side of the line and, if the line is solid, the line as well.

   a) If the inequality is of the form $y < mx + b$ or $y \leq mx + b$, shade *below* the line.
   If the inequality is of the form $y > mx + b$ or $y \geq mx + b$, shade *above* the line.
   b) If y is not isolated, either solve for y and graph as in part (a) or simply graph the boundary and use a test point (as in Example 3). If the test point *is* a solution, shade the half-plane containing the point. If it is not a solution, shade the other half-plane.

EXAMPLE 4   Graph: $6x - 2y < 12$.

Solution   We could graph $6x - 2y = 12$ and use a test point, as in Example 3. Instead, let's solve $6x - 2y < 12$ for y:

$$6x - 2y < 12$$
$$-2y < -6x + 12 \qquad \text{Adding } -6x \text{ to both sides}$$
$$y > 3x - 6. \qquad \text{Dividing both sides by } -2 \text{ and reversing the } < \text{ symbol}$$

The graph consists of the half plane above the dashed boundary line $y = 3x - 6$ (see the graph below).

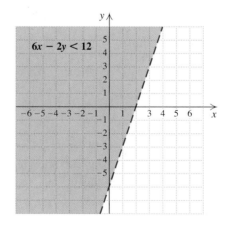

EXAMPLE 5   Graph $x > -3$ on a plane.

Solution   There is a missing variable in this inequality. If we graph the inequality on a line, its graph is as follows:

However, we can also write this inequality as $x + 0y > -3$ and graph it on a plane. We can use the same technique as in the examples above. First, we graph the boundary $x = -3$ in the plane, using a dashed line. Then we test some point, say, $(2, 5)$:

$$\frac{x + 0y > -3}{2 + 0 \cdot 5 \mid -3}$$
$$2 \overset{?}{>} -3 \quad \text{TRUE}$$

Since $(2, 5)$ is a solution, all points in the half-plane containing $(2, 5)$ are solutions. We shade that half-plane. Another approach is to simply note that the solutions of $x > -3$ are all pairs with first coordinates greater than -3.

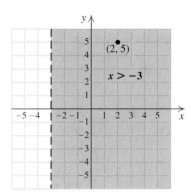

![technology connection]

On most graphing calculators, an inequality like $y < \frac{6}{5}x + 3.49$ can be drawn by entering $(6/5)x + 3.49$ as y_1, moving the cursor to the GraphStyle icon just to the left of y_1, pressing **ENTER** until ◣ appears, and then pressing **GRAPH**.

Many newer calculators have an INEQUALZ program that is accessed using the **APPS** key. Running this program allows us to write inequalities at the **Y=** screen by pressing **ALPHA** and then one of the five keys just below the screen.

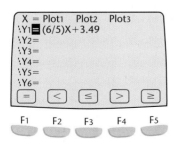

Although the graphs should be identical regardless of the method used, on the newer calculators the boundary line appears dashed when $<$ or $>$ is selected.

$$y_1 < (6/5)x + 3.49, \text{ or}$$
$$◣ \, y_1 = (6/5)x + 3.49$$

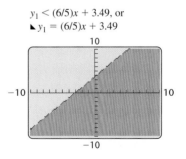

Inequalities containing $\leq$ or $\geq$ are handled in a similar manner.

Graph each of the following. Solve for y first if necessary.

1. $y > x + 3.5$
2. $7y \leq 2x + 5$
3. $8x - 2y < 11$
4. $11x + 13y + 4 \geq 0$

EXAMPLE 6   Graph $y \le 4$ on a plane.

Solution   The inequality is of the form $y \le mx + b$ (with $m = 0$), so we shade below the solid horizontal line representing $y = 4$.

      This inequality can also be graphed by drawing $y = 4$ and testing a point above or below the line. The student should check that this results in a graph identical to the one at right.

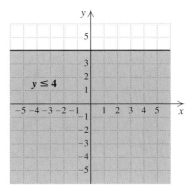

Systems of Linear Inequalities

To graph a system of equations, we graph the individual equations and then find the intersection of the individual graphs. We do the same thing for a system of inequalities, that is, we graph each inequality and find the intersection of the individual graphs.

EXAMPLE 7   Graph the system

$$x + y \le 4,$$
$$x - y < 4.$$

Solution   To graph $x + y \le 4$, we graph $x + y = 4$ using a solid line. Since the test point $(0, 0)$ *is* a solution and $(0, 0)$ is below the line, we shade the half-plane below the graph red. The arrows near the ends of the line are another way of indicating the half-plane containing solutions.

      Next, we graph $x - y < 4$. We graph $x - y = 4$ using a dashed line and consider $(0, 0)$ as a test point. Again, $(0, 0)$ is a solution, so we shade that side of the line blue. The solution set of the system is the region that is shaded purple (both red and blue) and part of the line $x + y = 4$.

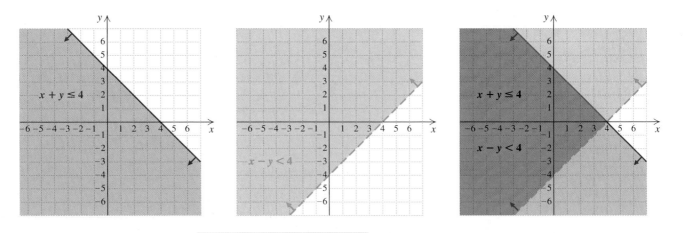

EXAMPLE 8

Student Notes ———

If you don't use differently colored pencils or pens to shade different regions, consider using a pencil to make slashes that tilt in different directions in each region. You may also find it useful to attach arrows to the lines, as in the examples shown.

Graph: $-2 < x \le 3$.

Solution   This is a system of inequalities:

$$-2 < x,$$
$$x \le 3.$$

We graph the equation $-2 = x$, and see that the graph of the first inequality is the half-plane to the right of the boundary $-2 = x$. It is shaded red.

We graph the second inequality, starting with the boundary $x = 3$, and find that its graph is the line and also the half-plane to its left. It is shaded blue.

The solution set of the system is the region that is the intersection of the individual graphs. Since it is shaded both blue and red, it appears to be purple. All points in this region have x-coordinates that are greater than -2 but do not exceed 3.

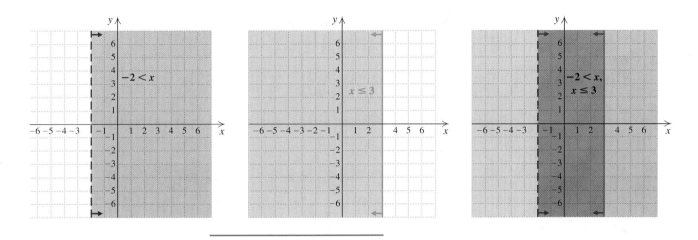

A system of inequalities may have a graph that consists of a polygon and its interior. In Section 4.5, we will have use for the corners, or *vertices* (singular, *vertex*), of such a graph.

EXAMPLE 9

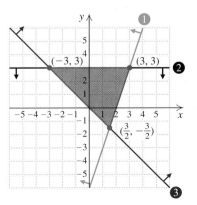

Graph the system of inequalities. Find the coordinates of any vertices formed.

$$6x - 2y \le 12, \qquad (1)$$
$$y - 3 \le 0, \qquad (2)$$
$$x + y \ge 0 \qquad (3)$$

Solution   We graph the boundaries

$$6x - 2y = 12,$$
$$y - 3 = 0,$$
and $$x + y = 0$$

using solid lines. The regions for each inequality are indicated by the arrows near the ends of the lines. We note where the regions overlap and shade the region of solutions purple.

To find the vertices, we solve three different systems of two equations. The system of boundary equations from inequalities (1) and (2) is

$$6x - 2y = 12,$$
$$y - 3 = 0.$$

Solving, we obtain the vertex $(3, 3)$.

     The system of boundary equations from inequalities (1) and (3) is

$$6x - 2y = 12,$$
$$x + y = 0.$$

Solving, we obtain the vertex $\left(\frac{3}{2}, -\frac{3}{2}\right)$.

     The system of boundary equations from inequalities (2) and (3) is

$$y - 3 = 0,$$
$$x + y = 0.$$

Solving, we obtain the vertex $(-3, 3)$.

 **technology connection**

Systems of inequalities can be graphed by solving for y and then graphing each inequality as in the Technology Connection on p. 257. To graph systems directly using the INEQUALZ application, enter the correct inequalities, press (GRAPH), and then press **ALPHA** and Shades ((F1) or (F2)). At the SHADES menu, select Ineq Intersection to see the final graph. To find the vertices, or points of intersection, select PoI-Trace from the graph menu.

$$y_1 \geq 3x - 6, \quad y_2 \leq 3, \quad y_3 \geq -x$$

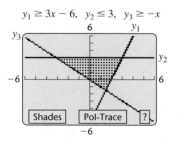

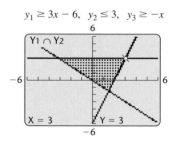

1. Use a graphing calculator to check the solution of Example 7.

CONNECTING THE CONCEPTS

We have now solved a variety of equations, inequalities, systems of equations, and systems of inequalities. In each case, there are different ways to represent the solution. Below is a list of the different types of problems we have solved, along with illustrations of each type.

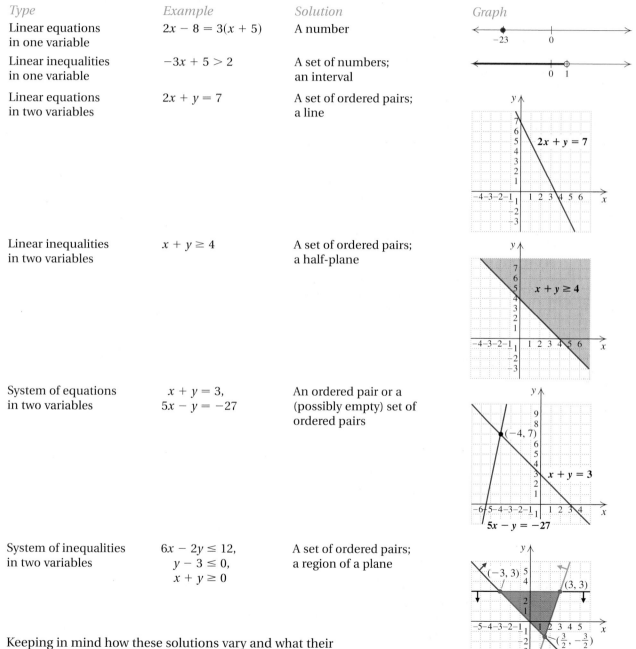

Type	*Example*	*Solution*	*Graph*
Linear equations in one variable	$2x - 8 = 3(x + 5)$	A number	
Linear inequalities in one variable	$-3x + 5 > 2$	A set of numbers; an interval	
Linear equations in two variables	$2x + y = 7$	A set of ordered pairs; a line	
Linear inequalities in two variables	$x + y \geq 4$	A set of ordered pairs; a half-plane	
System of equations in two variables	$x + y = 3,$ $5x - y = -27$	An ordered pair or a (possibly empty) set of ordered pairs	
System of inequalities in two variables	$6x - 2y \leq 12,$ $y - 3 \leq 0,$ $x + y \geq 0$	A set of ordered pairs; a region of a plane	

Keeping in mind how these solutions vary and what their graphs look like will help you as you progress further in this book and in mathematics in general.

Exercise Set

4.4

FOR EXTRA HELP

Student's Solutions Manual | Digital Video Tutor CD 3 Videotape 8 | AW Math Tutor Center | MathXL Tutorials on CD | MathXL | MyMathLab

✎ *Concept Reinforcement   In each of Exercises 1–6, match the phrase with the most appropriate choice from the column on the right.*

1. ____ A solution of a linear inequality

2. ____ The graph of a linear inequality

3. ____ The graph of a system of linear inequalities

4. ____ Often a convenient test point

5. ____ The name for the corners of a graph of a system of linear inequalities

6. ____ A dashed line

a) $(0, 0)$

b) Vertices

c) A half-plane

d) The intersection of two or more half-planes

e) An ordered pair that satisfies the inequality

f) Indicates the line is not part of the solution

Determine whether each ordered pair is a solution of the given inequality.

7. $(-4, 2)$; $2x + 3y < -1$

8. $(3, -6)$; $4x + 2y \leq -2$

9. $(8, 14)$; $2y - 3x \geq 9$

10. $(7, 20)$; $3x - y > -1$

Graph on a plane.

11. $y > \frac{1}{2}x$

12. $y > 2x$

13. $y \geq x - 3$

14. $y < x + 3$

15. $y \leq x + 5$

16. $y > x - 2$

17. $x - y \leq 4$

18. $x + y < 4$

19. $2x + 3y < 6$

20. $3x + 4y \leq 12$

21. $2y - x \leq 4$

22. $2y - 3x > 6$

23. $2x - 2y \geq 8 + 2y$

24. $3x - 2 \leq 5x + y$

25. $y \geq 3$

26. $x < -5$

27. $x \leq 6$

28. $y > -3$

29. $-2 < y < 7$

30. $-4 < y < -1$

31. $-4 \leq x \leq 2$

32. $-3 \leq y \leq 4$

33. $0 \leq y \leq 3$

34. $0 \leq x \leq 6$

Graph each system.

35. $y > -x$, $y < x + 2$

36. $y < x$, $y > -x + 1$

37. $y \geq x$, $y \geq 2x - 4$

38. $y \geq x$, $y \leq -x + 4$

39. $y \leq -3$, $x \geq -1$

40. $y \geq -3$, $x \geq 1$

41. $x > -4$, $y < -2x + 3$

42. $x < 3$, $y > -3x + 2$

43. $y \leq 5$, $y \geq -x + 4$

44. $y \geq -2$, $y \geq x + 3$

45. $x + y \leq 6$, $x - y \leq 4$

46. $x + y < 1$, $x - y < 2$

47. $y + 3x > 0$, $y + 3x < 2$

48. $y - 2x \geq 1$, $y - 2x \leq 3$

Graph each system of inequalities. Find the coordinates of any vertices formed.

49. $y \leq 2x - 3$, $y \geq -2x + 1$, $x \leq 5$

50. $2y - x \leq 2$, $y - 3x \geq -4$, $y \geq -1$

51. $x + 2y \leq 12$, $2x + y \leq 12$, $x \geq 0$, $y \geq 0$

52. $x - y \leq 2$, $x + 2y \geq 8$, $y \leq 4$

53. $8x + 5y \leq 40$, $x + 2y \leq 8$, $x \geq 0$, $y \geq 0$

54. $4y - 3x \geq -12$, $4y + 3x \geq -36$, $y \leq 0$, $x \leq 0$

55. $y - x \geq 2$, $y - x \leq 4$, $2 \leq x \leq 5$

56. $3x + 4y \geq 12$, $5x + 6y \leq 30$, $1 \leq x \leq 3$

📝 57. In Example 7, is the point $(4, 0)$ part of the solution set? Why or why not?

58. When graphing linear inequalities, Ron makes a habit of always shading above the line when the symbol ≥ is used. Is this wise? Why or why not?

SKILL MAINTENANCE

Solve.

59. *Catering.*   Sandy's Catering needs to provide 10 lb of mixed nuts for a wedding reception. Peanuts cost $2.50 per pound and fancy nuts cost $7 per pound. If $40 has been allocated for nuts, how many pounds of each type should be mixed? [3.3]

60. *Household waste.*   The Hendersons generate two and a half times as much trash as their neighbors, the Savickis. Together, the two households produce 14 bags of trash each month. How much trash does each household produce? [3.3]

61. *Paid admissions.*   There were 203 tickets sold for a volleyball game. For activity-card holders the price was $2.50, and for noncard holders the price was $4. The total amount of money collected was $620. How many of each type of ticket were sold? [3.3]

62. *Paid admissions.*   There were 200 tickets sold for a women's basketball game. Tickets for students were $4 each and for adults were $6 each. The total amount collected was $1060. How many of each type of ticket were sold? [3.3]

63. *Landscaping.*   Grass seed is being spread on a triangular traffic island. If the grass seed can cover an area of 200 ft^2 and the island's base is 16 ft long, how tall a triangle can the seed fill? [1.4]

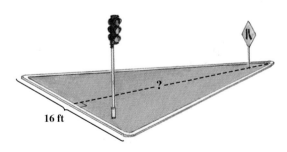

64. *Interest rate.*   What rate of interest is required in order for a principal of $1280 to earn $17.60 in half a year? [1.4]

SYNTHESIS

65. Explain how a system of linear inequalities could have a solution set containing exactly one pair.

66. Do all systems of linear inequalities have solutions? Why or why not?

Graph.

67. $x + y > 8,$
$\quad x + y \le -2$

68. $x + y \ge 1,$
$\quad -x + y \ge 2,$
$\qquad\quad x \ge -2,$
$\qquad\quad y \ge 2,$
$\qquad\quad y \le 4,$
$\qquad\quad x \le 2$

69. $\quad x - 2y \le 0,$
$\;-2x + \;y \le 2,$
$\qquad\quad x \le 2,$
$\qquad\quad y \le 2,$
$\;\;x + \;\;y \le 4$

70. Write four systems of four inequalities that describe a 2-unit by 2-unit square that has (0, 0) as one of the vertices.

71. *Luggage size.*   Unless an additional fee is paid, most major airlines will not check any luggage for which the sum of the item's length, width, and height exceeds 62 in. The U.S. Postal Service will ship a package only if the sum of the package's length and girth (distance around its midsection) does not exceed 130 in. Video Promotions is ordering several 30-in. long cases that will be both mailed and checked as luggage. Using w and h for width and height (in inches), respectively, write and graph an inequality that represents all acceptable combinations of width and height.
Sources: U.S. Postal Service; www.case2go.com

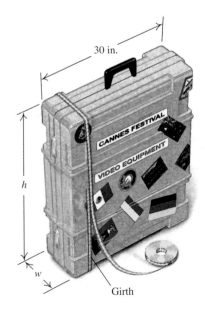

72. *Hockey wins and losses.* The Skating Stars figure that they need at least 60 points for the season in order to make the playoffs. A win is worth 2 points and a tie is worth 1 point. Graph a system of inequalities that describes the situation. (*Hint*: Let w = the number of wins and t = the number of ties.)

73. *Elevators.* Many elevators have a capacity of 1 metric ton (1000 kg). Suppose that c children, each weighing 35 kg, and a adults, each 75 kg, are on an elevator. Graph a system of inequalities that indicates when the elevator is overloaded.

74. *Widths of a basketball floor.* Sizes of basketball floors vary due to building sizes and other constraints such as cost. The length L is to be at most 94 ft and the width W is to be at most 50 ft. Graph a system of inequalities that describes the possible dimensions of a basketball floor.

75. Use a graphing calculator to graph each inequality.
   a) $3x + 6y > 2$
   b) $x - 5y \leq 10$
   c) $13x - 25y + 10 \leq 0$
   d) $2x + 5y > 0$

76. Use a graphing calculator to check your answers to Exercises 35–48. Then use INTERSECT to determine any point(s) of intersection.

COLLABORATIVE CORNER

How Old Is Old Enough?

Focus: Linear inequalities

Time: 15–25 minutes

Group size: 2

It is not unusual for the ages of a bride and groom to differ significantly. Yet is it possible for the difference in age to be too great? In answer to this question, the following rule of thumb has emerged: *The younger spouse's age should be at least seven more than half the age of the older spouse.* (*Source*: http://home.earthlink.net/~mybrainhurts/2002_06_01_archive.html)

ACTIVITY

1. Let b = the age of the bride, in years, and g = the age of the groom, in years. One group member should write an equation for calculating the bride's minimum age if the groom's age is known. The other group member should write an equation for finding the groom's minimum age if the bride's age is known. The equations should look similar.

2. Convert each equation into an inequality by selecting the appropriate symbol from $<$, $>$, $\leq$, and $\geq$. Be sure to reflect the rule of thumb stated above.

3. Graph both inequalities from step 2 as a system of linear inequalities. What does the solution set represent?

4. If your group feels that a minimum or maximum age for marriage should exist, adjust your graph accordingly.

5. Compare your finished graph with those of other groups.

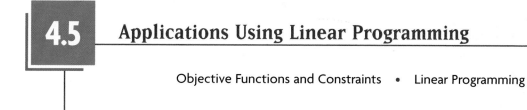

4.5   Applications Using Linear Programming

Objective Functions and Constraints   •   Linear Programming

There are many real-world situations in which we need to find a greatest value (a maximum) or a least value (a minimum). For example, most businesses like to know how to make the *most* profit with the *least* expense possible. Some such problems can be solved using systems of inequalities.

Objective Functions and Constraints

Often a quantity we wish to maximize depends on two or more other quantities. For example, a gardener's profits P might depend on the number of shrubs s and the number of trees t that are planted. If the gardener makes a $10 profit from each shrub and an $18 profit from each tree, the total profit, in dollars, is given by the **objective function**

$$P = 10s + 18t.$$

Thus the gardener might be tempted to simply plant lots of trees since they yield the greater profit. This would be a good idea were it not for the fact that the number of trees and shrubs planted—and thus the total profit—is subject to the demands, or **constraints**, of the situation. For example, to improve drainage, the gardener might be required to plant at least 3 shrubs. Thus the objective function would be subject to the *constraint*

$$s \geq 3.$$

Because of the limited space, the gardener might also be required to plant no more than 10 plants. This would subject the objective function to a *second* constraint:

$$s + t \leq 10.$$

Finally, the gardener might be told to spend no more than $700 on the plants. If the shrubs cost $40 each and the trees cost $100 each, the objective function is subject to a *third* constraint:

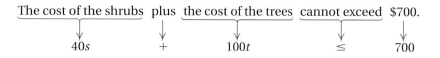

The cost of the shrubs	plus	the cost of the trees	cannot exceed	$700.
40s	+	100t	≤	700

In short, the gardener wishes to maximize the objective function

$$P = 10s + 18t,$$

subject to the constraints

$$s \geq 3,$$
$$s + t \leq 10,$$
$$40s + 100t \leq 700,$$
$$\left.\begin{array}{r} s \geq 0, \\ t \geq 0. \end{array}\right\}$$ Because the number of trees and shrubs cannot be negative

These constraints form a system of linear inequalities that can be graphed.

Linear Programming

The gardener's problem is "How many shrubs and trees should be planted, subject to the constraints listed, in order to maximize profit?" To solve such a problem, we use an important result from a branch of mathematics known as **linear programming**.

The Corner Principle

Suppose that an objective function $F = ax + by + c$ depends on x and y (with a, b, and c constant). Suppose also that F is subject to constraints on x and y, which form a system of linear inequalities. If F has a minimum or a maximum value, then it can be found as follows:

1. Graph the system of inequalities and find the vertices.
2. Find the value of the objective function at each vertex. The greatest and the least of those values are the maximum and the minimum of the function, respectively.
3. The ordered pair at which the maximum or minimum occurs indicates the choice of (x, y) for which that maximum or minimum occurs.

   This result was proven during World War II, when linear programming was developed to help with shipping troops and supplies from North America to Europe.

EXAMPLE 1   Solve the gardener's problem discussed above.

Solution   We are asked to maximize $P = 10s + 18t$, subject to the constraints

$$s \geq 3,$$
$$s + t \leq 10,$$
$$40s + 100t \leq 700,$$
$$s \geq 0,$$
$$t \geq 0.$$

We graph the system, using the techniques of Section 4.4. The portion of the graph that is shaded represents all pairs that satisfy the constraints. It is sometimes called the *feasible region*.

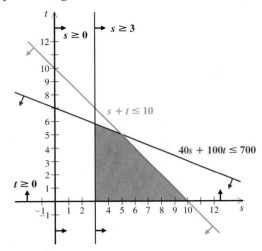

According to the corner principle, *P* is maximized at one of the vertices of the shaded region. To determine the coordinates of the vertices, we solve the following systems:

$$\left.\begin{array}{r} 40s + 100t = 700, \\ s = 3; \end{array}\right\}$$
The student can verify that the solution of this system is (3, 5.8).

$$\left.\begin{array}{r} s + t = 10, \\ 40s + 100t = 700; \end{array}\right\}$$
The student can verify that the solution of this system is (5, 5).

$$\left.\begin{array}{r} s + t = 10, \\ t = 0; \end{array}\right\}$$
The solution of this system is (10, 0).

$$\left.\begin{array}{r} t = 0, \\ s = 3. \end{array}\right\}$$
The solution of this system is (3, 0).

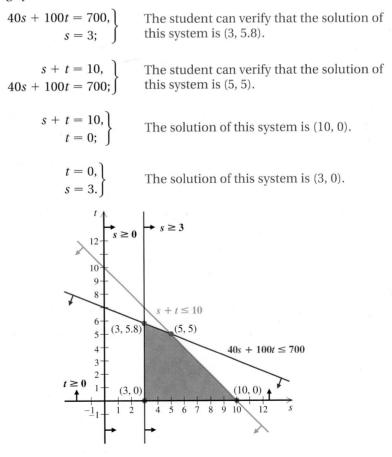

We now find the value of P at each vertex.

Vertex (s, t)	Profit $P = 10s + 18t$	
(3, 5.8)	$10(3) + 18(5.8) = 134.4$	
(5, 5)	$10(5) + 18(5) = 140$	← Maximum
(10, 0)	$10(10) + 18(0) = 100$	
(3, 0)	$10(3) + 18(0) = 30$	← Minimum

The greatest value of P occurs at (5, 5). Thus profit is maximized at $140 if the gardener plants 5 shrubs and 5 trees. Incidentally, we have also shown that profit is minimized at $30 if 3 shrubs and 0 trees are planted.

EXAMPLE 2   *Test scores.*   Corinna is taking a test in which multiple-choice questions are worth 10 points each and short-answer questions are worth 15 points each. It takes her 3 min to answer each multiple-choice question and 6 min to answer each short-answer question. The total time allowed is 60 min, and no more than 16 questions can be answered. Assuming that all her answers are correct, how many items of each type should Corinna answer in order to get the best score?

Solution

1. Familiarize.   Tabulating information will help us to see the picture.

Type	Number of Points for Each	Time Required for Each	Number Answered
Multiple-choice	10	3 min	x
Short-answer	15	6 min	y
Total time: 60 min			
Total number of items: 16 or fewer			

Note that we use x to represent the number of multiple-choice questions and y to represent the number of short-answer questions that are answered.

2. Translate.   In this case, it helps to extend the table.

Type	Number of Points for Each	Time Required for Each	Number Answered	Total Time for Each Type	Total Points for Each Type
Multiple-choice	10	3 min	x	$3x$	$10x$
Short-answer	15	6 min	y	$6y$	$15y$
Total			$x + y \leq 16$	$3x + 6y \leq 60$	$10x + 15y$

Because no more than 16 items may be answered

Because the time cannot exceed 60 min

This is what we want to maximize: the total score on the test.

Student Notes _____

It is very important that you clearly label what each variable represents. It is also important to clearly label what the function is that is being maximized or minimized and how that function is evaluated.

Let T represent the total score. We express this as a function of the same variables used in the constraints—in this case, x and y:

$$T = 10x + 15y.$$

We wish to maximize T subject to the constraints listed above:

$$x + y \leq 16,$$
$$3x + 6y \leq 60,$$
$$\left.\begin{array}{l} x \geq 0, \\ y \geq 0. \end{array}\right\}$$ We include this because the number of questions answered cannot be negative.

3. **Carry out.**   The mathematical manipulation consists of graphing the system and evaluating T at each vertex. The graph is as follows:

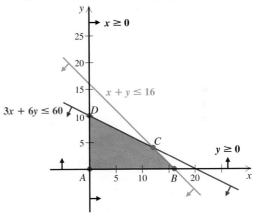

We find the coordinates of each vertex by solving a system of two linear equations. The coordinates of point A are obviously $(0, 0)$. To find the coordinates of point C, we solve the system

$$x + y = 16, \qquad (1)$$
$$3x + 6y = 60, \qquad (2)$$

as follows:

$$-3x - 3y = -48 \qquad \text{Multiplying both sides of equation (1) by } -3$$
$$\underline{3x + 6y = 60}$$
$$ 3y = 12 \qquad \text{Adding}$$
$$y = 4.$$

Then we find that $x = 12$. Thus the coordinates of vertex C are $(12, 4)$. Point B is the x-intercept of the line given by $x + y = 16$, so B is $(16, 0)$. Point D is the y-intercept of $3x + 6y = 60$, so D is $(0, 10)$. Computing the test score for each ordered pair, we obtain the table at left.

The greatest score in the table is 180, obtained when 12 multiple-choice and 4 short-answer questions are answered.

4. **Check.**   We can check that $T \leq 180$ for any other pair in the shaded region. This is left to the student.

5. **State.**   In order to maximize her score, Corinna should answer 12 multiple-choice questions and 4 short-answer questions.

Vertex (x, y)	Score $T = 10x + 15y$
A $(0, 0)$	0
B $(16, 0)$	160
C $(12, 4)$	180
D $(0, 10)$	150

Exercise Set

4.5

FOR EXTRA HELP

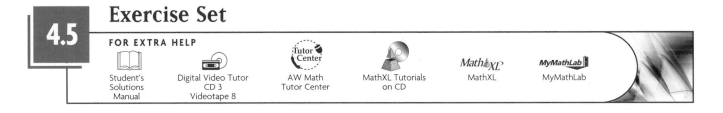

Student's Solutions Manual | Digital Video Tutor CD 3 Videotape 8 | AW Math Tutor Center | MathXL Tutorials on CD | MathXL | MyMathLab

---

👈 *Concept Reinforcement*   *Complete each of the following sentences.*

1. In linear programing, a function is maximized or _____ subject to the demands of a given situation.

2. In linear programming, the function on which the demands are placed is known as the _____ function.

3. In linear programming, the demands arising from the given situation are known as _____ .

4. To solve a linear programming problem, we make use of the _____ principle.

5. The shaded portion of a graph that represents all points that satisfy a problem's constraints is known as the _____ region.

6. In linear programming, the corners of the shaded portion of the graph are referred to as _____ .

Find the maximum and the minimum values of each objective function and the values of x and y at which they occur.

7. $F = 2x + 14y$,
   subject to
   $5x + 3y \le 34$,
   $3x + 5y \le 30$,
   $x \ge 0$,
   $y \ge 0$

8. $G = 7x + 8y$,
   subject to
   $3x + 2y \le 12$,
   $2y - x \le 4$,
   $x \ge 0$,
   $y \ge 0$

9. $P = 8x - y + 20$,
   subject to
   $6x + 8y \le 48$,
   $0 \le y \le 4$,
   $0 \le x \le 7$

10. $Q = 24x - 3y + 52$,
    subject to
    $5x + 4y \le 20$,
    $0 \le y \le 4$,
    $0 \le x \le 3$

11. $F = 2y - 3x$,
    subject to
    $y \le 2x + 1$,
    $y \ge -2x + 3$,
    $x \le 3$

12. $G = 5x + 2y + 4$,
    subject to
    $y \le 2x + 1$,
    $y \ge -x + 3$,
    $x \le 5$

13. *Gas mileage.* Roschelle owns a car and a moped. She has at most 12 gal of gasoline to be used between the car and the moped. The car's tank holds at most 18 gal and the moped's 3 gal. The mileage for the car is 20 mpg and for the moped is 100 mpg. How many gallons of gasoline should each vehicle use if Roschelle wants to travel as far as possible? What is the maximum number of miles?

14. *Lunch-time profits.* Elrod's lunch cart sells burritos and chili. To stay in business, Elrod must sell at least 10 orders of chili and 30 burritos each day. Because of limited space, no more than 40 orders of chili or 70 burritos can be made. The total number of orders cannot exceed 90. If profit is $1.65 per chili order and $1.05 per burrito, how many of each item should Elrod sell in order to maximize profit?

15. *Milling.* Johnson Lumber can convert logs into either lumber or plywood. In a given week, the mill can turn out 400 units of production, of which 100 units of lumber and 150 units of plywood are required by regular customers. The profit on a unit of lumber is $20 and on a unit of plywood is $30. How many units of each type should the mill produce in order to maximize profit?

16. *Cycle production.* Yawaka manufactures motor-cycles and bicycles. In order to stay in business, the company determines that the number of bicycles made cannot exceed 3 times the number of motorcycles made. Yawaka lacks the facilities to produce more than 60 motorcycles or more than 120 bicycles. The total production of motorcycles and bicycles cannot exceed 160. The profit on a motorcycle is $1340 and on a bicycle is $200. Find the number of each that should be manufactured in order to maximize profit.

Aha! **17.** *Investing.*   Rosa is planning to invest up to $40,000 in corporate or municipal bonds, or both. She must invest from $6000 to $22,000 in corporate bonds, and she does not want to invest more than $30,000 in municipal bonds. The interest on corporate bonds is 8% and on municipal bonds is $7\frac{1}{2}$%. This is simple interest for one year. How much should Rosa invest in each type of bond in order to earn the most interest? What is the maximum interest?

18. *Investing.*   Jamaal is planning to invest up to $22,000 in City Bank or the Southwick Credit Union, or both. He wants to invest at least $2000 but no more than $14,000 in City Bank. Because of insurance limitations, he will invest no more than $15,000 in the Southwick Credit Union. The interest in City Bank is 6% and in the credit union is $6\frac{1}{2}$%. This is simple interest for one year. How much should Jamaal invest in each bank in order to earn the most interest? What is the maximum interest?

19. *Test scores.*   Phil is about to take a test that contains matching questions worth 10 points each and essay questions worth 25 points each. He must do at least 3 matching questions, but time restricts doing more than 12. Phil must do at least 4 essays, but time restricts doing more than 15. If no more than 20 questions can be answered, how many of each type should Phil do in order to maximize his score? What is this maximum score?

20. *Test scores.*   Edy is about to take a test that contains short-answer questions worth 4 points each and word problems worth 7 points each. Edy must do at least 5 short-answer questions, but time restricts doing more than 10. She must do at least 3 word problems, but time restricts doing more than 10. Edy can do no more than 18 questions in total. How many of each type of question must Edy do in order to maximize her score? What is this maximum score?

21. *Grape growing.*   Auggie's vineyard consists of 240 acres upon which he wishes to plant Merlot and Cabernet grapes. Profit per acre of Merlot is $400 and profit per acre of Cabernet is $300. Furthermore, the total number of hours of labor available during the harvest season is 3200. Each acre of Merlot requires 20 hr of labor and each acre of Cabernet requires 10 hr of labor. Determine how the land should be divided between Merlot and Cabernet in order to maximize profit.

22. *Coffee blending.*   The Coffee Peddler has 1440 lb of Sumatran coffee and 700 lb of Kona coffee. A batch of Hawaiian Blend requires 8 lb of Kona and 12 lb of Sumatran, and yields a profit of $90. A batch of Classic Blend requires 4 lb of Kona and 16 lb of Sumatran, and yields a $55 profit. How many batches of each kind should be made in order to maximize profit? What is the maximum profit? (*Hint*: Organize the information in a table.)

23. *Textile production.*   It takes Cosmic Stitching 2 hr of cutting and 4 hr of sewing to make a knit suit. To make a worsted suit, it takes 4 hr of cutting and 2 hr of sewing. At most 20 hr per day are available for cutting and at most 16 hr per day are available for sewing. The profit on a knit suit is $68 and on a worsted suit is $62. How many of each kind of suit should be made in order to maximize profit?

24. *Biscuit production.*   The Broad St. Biscuit Factory makes two types of biscuits, Biscuit Jumbos and Mitimite Biscuits. The oven can cook at most 200 biscuits per hour. Jumbos each require 2 oz of flour, Mitimites require 1 oz of flour, and there is at most 1440 oz of flour available. The income from Jumbos is $1.00 and from Mitimites is $0.80. How many of each type of biscuit should be made in order to maximize income? What is the maximum income?

25. Before a student begins work in this section, what three sections of the text would you suggest he or she study? Why?

26. What does the use of the word "constraint" in this section have in common with the use of the word in everyday speech?

SKILL MAINTENANCE

Evaluate.

27. $5x^3 - 4x^2 - 7x + 2$, for $x = -2$ [1.1], [1.2]

28. $6t^3 - 3t^2 + 5t$, for $t = 2$ [1.1]

Simplify. [1.2]

29. $3(2x - 5) + 4(x + 5)$

30. $4(5t - 7) + 6(t + 8)$

31. $6x - 3(x + 2)$

32. $8t - 2(3t - 1)$

SYNTHESIS

33. Explain how Exercises 17 and 18 can be answered by logical reasoning without linear programming.

34. Write a linear programming problem for a classmate to solve. Devise the problem so that profit must be maximized subject to at least two (nontrivial) constraints.

35. *Airplane production.* Alpha Tours has two types of airplanes, the T3 and the S5, and contracts requiring accommodations for a minimum of 2000 first-class, 1500 tourist-class, and 2400 economy-class passengers. The T3 costs $30 per mile to operate and can accommodate 40 first-class, 40 tourist-class, and 120 economy-class passengers, whereas the S5 costs $25 per mile to operate and can accommodate 80 first-class, 30 tourist-class, and 40 economy-class passengers. How many of each type of airplane should be used in order to minimize the operating cost?

36. *Airplane production.* A new airplane, the T4, is now available, having an operating cost of $37.50 per mile and accommodating 40 first-class, 40 tourist-class, and 80 economy-class passengers. If the T3 of Exercise 35 were replaced with the T4, how many S5's and how many T4's would be needed in order to minimize the operating cost?

37. *Furniture production.* P. J. Edward Furniture Design produces chairs and sofas. The chairs require 20 ft of wood, 1 lb of foam rubber, and 2 sq yd of fabric. The sofas require 100 ft of wood, 50 lb of foam rubber, and 20 sq yd of fabric. The company has 1900 ft of wood, 500 lb of foam rubber, and 240 sq yd of fabric. The chairs can be sold for $80 each and the sofas for $1200 each. How many of each should be produced in order to maximize income?

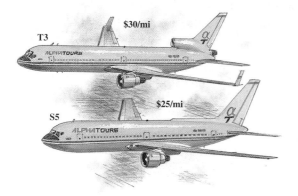

STUDY SUMMARY: CHAPTER 4

4 Study Summary

Our focus in this chapter has been on **inequalities** and their **solutions** (p. 222). Because most inequalities have an infinite number of solutions, **set-builder notation** and **interval notation** are used to write **solution sets** (pp. 222–223). To solve an inequality, principles similar to those used for solving equations are used (pp. 225–226).

The Addition Principle for Inequalities

For any real numbers a, b, and c:

$a < b$ is equivalent to $a + c < b + c$; $a > b$ is equivalent to $a + c > b + c$.

The Multiplication Principle for Inequalities

For any real numbers a and b, and for any positive number c,

$a < b$ is equivalent to $ac < bc$; $a > b$ is equivalent to $ac > bc$.

For any real numbers a and b, and for any *negative* number c,

$a < b$ is equivalent to $ac > bc$; $a > b$ is equivalent to $ac < bc$.

Similar statements hold for $\leq$ and $\geq$.

These principles can be used, for example, to show that the solution set of $-2x + 7 > 1$ is $\{x \mid x < 3\}$ or $(-\infty, 3)$.

The solution of a **conjunction** like $x + 5 \leq 7$ *and* $2x \geq -3$ is an **intersection** of sets (pp. 234–235). The solution of a **disjunction** like $3x < 12$ *or* $x - 9 > 5$ is a **union** of sets (pp. 237–238). Both conjunctions and disjunctions are **compound inequalities** (p. 234). We use compound inequalities to solve **absolute-value** problems like $|5x + 2| < 9$ (pp. 245–249).

The Absolute-Value Principles for Equations and Inequalities

For any positive number p and any algebraic expression X:

a) The solutions of $|X| = p$ are those numbers that satisfy $X = -p$ *or* $X = p$.
b) The solutions of $|X| < p$ are those numbers that satisfy $-p < X < p$.
c) The solutions of $|X| > p$ are those numbers that satisfy $X < -p$ *or* $p < X$.

If $|X| = 0$, then $X = 0$. If p is negative, then $|X| = p$ and $|X| < p$ have no solution, and any value of X will satisfy $|X| > p$.

Linear inequalities like $2x + 3y \leq 10$ are solved by graphing the **boundary** $2x + 3y = 10$ (pp. 253–254). To determine which **half-plane** is part of the solution set, we use a convenient **test point** (pp. 254, 255). When more than one linear inequality is to be solved at a time, we have a **system of linear inequalities** (p. 258). The solution of such a system requires finding the intersection of two or more half-planes.

Maximization or minimization problems in **linear programming** require us to identify a **feasible region** by solving a system of linear inequalities that represent the **constraints** of the problem (pp. 265–267). The corners, or **vertices**, of the region are then tested in the problem's **objective function** (pp. 259, 265). The **corner principle** states that if a maximum or minimum exists, then it will occur at one of the vertices (p. 266).

4 Review Exercises

🔖 *Concept Reinforcement*   *Classify each of the following as either true or false.*

1. The addition and multiplication principles for inequalities are used to write equivalent inequalities. [4.1]

2. It is always true that if $a > b$, then $ac > bc$. [4.1]

3. The solution of $|3x - 5| \leq 8$ is a closed interval. [4.3]

4. The inequality $2 < 5x + 1 < 9$ is equivalent to $2 < 5x + 1 \ or \ 5x + 1 < 9$. [4.2]

5. The solution set of a disjunction is the union of two solution sets. [4.2]

6. The equation $|x| = -p$ has no solution when p is positive. [4.3]

7. $|f(x)| > 3$ is equivalent to $f(x) < -3 \ or \ f(x) > 3$. [4.3]

8. A test point is used to determine whether the line in a linear inequality is drawn solid or dashed. [4.4]

9. The graph of a system of linear inequalities is always a half-plane. [4.4]

10. The corner principle states that every objective function has a maximum or minimum value. [4.5]

Graph each inequality and write the solution set using both set-builder and interval notations. [4.1]

11. $x \leq -2$

12. $a + 7 \leq -14$

13. $y - 5 \geq -12$

14. $4y > -15$

15. $-0.3y < 9$

16. $-6x - 5 < 4$

17. $-\frac{1}{2}x - \frac{1}{4} > \frac{1}{2} - \frac{1}{4}x$

18. $0.3y - 7 < 2.6y + 15$

19. $-2(x - 5) \geq 6(x + 7) - 12$

20. Let $f(x) = 3x - 5$ and $g(x) = 11 - x$. Find all values of x for which $f(x) \leq g(x)$. [4.1]

Solve. [4.1]

21. Rose can choose between two summer jobs. She can work as a checker in a discount store for $8.40 an hour, or she can mow lawns for $12.00 an hour. In order to mow lawns, she must buy a $450 lawnmower. For how many hours must Rose work in order for the mowing to be more profitable than checking?

22. Clay is going to invest $9000, part at 3% and the rest at 3.5%. What is the most he can invest at 3% and still be guaranteed $300 in interest each year?

23. Find the intersection:

$\{1, 2, 5, 6, 9\} \cap \{1, 3, 5, 9\}$. [4.2]

24. Find the union:

$\{1, 2, 5, 6, 9\} \cup \{1, 3, 5, 9\}$. [4.2]

Graph and write interval notation. [4.2]

25. $x \leq 3$ *and* $x > -5$

26. $x \leq 3$ *or* $x > -5$

Solve and graph each solution set. [4.2]

27. $-4 < x + 8 \leq 5$

28. $-15 < -4x - 5 < 0$

29. $3x < -9$ *or* $-5x < -5$

30. $2x + 5 < -17$ *or* $-4x + 10 \leq 34$

31. $2x + 7 \leq -5$ *or* $x + 7 \geq 15$

32. $f(x) < -5$ *or* $f(x) > 5$, where $f(x) = 3 - 5x$

For $f(x)$ as given, use interval notation to write the domain of f. [4.2]

33. $f(x) = \dfrac{2x}{x - 8}$

34. $f(x) = \sqrt{x + 5}$

35. $f(x) = \sqrt{8 - 3x}$

Solve. [4.3]

36. $|x| = 5$

37. $|t| \geq 3.5$

38. $|x - 3| = 7$

39. $|2x + 5| < 12$

40. $|3x - 4| \geq 15$

41. $|2x + 5| = |x - 9|$

42. $|5n + 6| = -8$

43. $\left|\dfrac{x + 4}{6}\right| \leq 2$

44. $2|x - 5| - 7 > 3$

45. Let $f(x) = |3x - 5|$. Find all x for which $f(x) < 0$. [4.3]

46. Graph $x - 2y \geq 6$ on a plane. [4.4]

Graph each system of inequalities. Find the coordinates of any vertices formed. [4.4]

47. $x + 3y > -1$,
$x + 3y < 4$

48. $x - 3y \leq 3$,
$x + 3y \geq 9$,
$y \leq 6$

49. Find the maximum and the minimum values of

$F = 3x + y + 4$

subject to

$y \leq 2x + 1,$
$x \leq 7,$
$y \geq 3.$ [4.5]

50. Custom Computers has two manufacturing plants. The Oregon plant cannot produce more than 60 computers a week, while the Ohio plant cannot produce more than 120 computers a week. The Electronics Outpost sells at least 160 Custom computers each week. It costs \$40 to ship a computer to The Electronics Outpost from the Oregon plant and \$25 to ship from the Ohio plant. How many computers should be shipped from each plant in order to minimize cost? [4.5]

SYNTHESIS

51. Explain in your own words why $|X| = p$ has two solutions when p is positive and no solution when p is negative. [4.3]

52. Explain why the graph of the solution of a system of linear inequalities is the intersection, not the union, of the individual graphs. [4.4]

53. Solve: $|2x + 5| \leq |x + 3|$. [4.3]

54. Classify as true or false: If $x < 3$, then $x^2 < 9$. If false, give an example showing why. [4.1]

55. Just-For-Fun manufactures marbles with a 1.1-cm diameter and a ± 0.03-cm manufacturing tolerance, or allowable variation in diameter. Write the tolerance as an inequality with absolute value. [4.3]

4 Chapter Test

Graph each inequality and write the solution set using both set-builder and interval notations.

1. $x - 2 < 10$

2. $-0.6y < 30$

3. $-4y - 3 \geq 5$

4. $3a - 5 \leq -2a + 6$

5. $3(7 - x) < 2x + 5$

6. $-8(2x + 3) + 6(4 - 5x) \geq 2(1 - 7x) - 4(4 + 6x)$

7. Let $f(x) = -5x - 1$ and $g(x) = -9x + 3$. Find all values of x for which $f(x) > g(x)$.

8. Lia can rent a van for either $40 with unlimited mileage or $30 with 100 free miles and an extra charge of 15¢ for each mile over 100. For what numbers of miles traveled would the unlimited mileage plan save Lia money?

9. A refrigeration repair company charges $80 for the first half-hour of work and $60 for each additional hour. Blue Mountain Camp has budgeted $200 to repair its walk-in cooler. For what lengths of a service call will the budget not be exceeded?

10. Find the intersection:
$$\{1, 3, 5, 7, 9\} \cap \{3, 5, 11, 13\}.$$

11. Find the union:
$$\{1, 3, 5, 7, 9\} \cup \{3, 5, 11, 13\}.$$

12. Write the domain of f using interval notation if $f(x) = \sqrt{8 - 2x}$.

Solve and graph each solution set.

13. $-2 < x - 3 < 5$

14. $-11 \leq -5t - 2 < 0$

15. $3x - 2 < 7 \ or \ x - 2 > 4$

16. $-3x > 12 \ or \ 4x > -10$

17. $-\frac{1}{3} \leq \frac{1}{6}x - 1 < \frac{1}{4}$

18. $|x| = 13$

19. $|a| > 7$

20. $|3x - 1| < 7$

21. $|-5t - 3| \geq 10$

22. $|2 - 5x| = -12$

23. $g(x) < -3 \ or \ g(x) > 3$, where $g(x) = 4 - 2x$

24. Let $f(x) = |x + 10|$ and $g(x) = |x - 12|$. Find all values of x for which $f(x) = g(x)$.

Graph the system of inequalities. Find the coordinates of any vertices formed.

25. $x + y \geq 3,$
$x - y \geq 5$

26. $2y - x \geq -7,$
$2y + 3x \leq 15,$
$y \leq 0,$
$x \leq 0$

27. Find the maximum and the minimum values of
$$F = 5x + 3y$$
subject to
$$x + y \leq 15,$$
$$1 \leq x \leq 6,$$
$$0 \leq y \leq 12.$$

28. Sassy Salon makes $12 on each manicure and $18 on each haircut. A manicure takes 30 minutes and a haircut takes 50 minutes, and there are 5 stylists who each work 6 hours a day. If the salon can schedule 50 appointments a day, how many should be manicures and how many haircuts in order to maximize profit? What is the maximum profit?

SYNTHESIS

Solve. Write the solution set using interval notation.

29. $|2x - 5| \leq 7 \ and \ |x - 2| \geq 2$

30. $7x < 8 - 3x < 6 + 7x$

31. Write an absolute-value inequality for which the interval shown is the solution.

5

Polynomials and Polynomial Functions

AN APPLICATION

During the 2004 Winter X-Games, Simon Dumont launched himself a record 22 ft above the edge of a halfpipe. His speed $v(t)$, in miles per hour, upon re-entering the pipe can be approximated by $v(t) = 10.9t$, where t is the number of seconds for which he was airborne. (*Source*: Based on data from EXPN.com) If Dumont was airborne for 2.34 sec, what was his speed when he re-entered the halfpipe?

This problem appears as Exercise 35 in Section 5.1.

Chris Gunnarson
SKI ENGINEER
Truckee, California

Knowing how many cubic feet of snow is needed is essential when designing a good terrain park. We constantly measure such elements as snow depth, run width, and steepness of a run's pitch.

e have already used many polynomials, like $2x + 5$ or $x^2 - 3$, in this text. In many ways, polynomials are central to the study of algebra. Here in Chapter 5 we will clearly define what polynomials are and how to manipulate them. We will then use polynomials and polynomial functions in problem solving.

5.1 Introduction to Polynomials and Polynomial Functions

Algebraic Expressions and Polynomials • Polynomial Functions • Adding Polynomials • Opposites and Subtraction

CONNECTING THE CONCEPTS

Let's briefly summarize our work up to this point in the text: In Chapter 1, we learned several ways to write equivalent expressions as well as methods for writing equivalent equations and for problem solving. In Chapter 2, we studied equations in two variables for which graphs represent the solution sets. Graphs also enabled us to visualize the solutions of systems of equations in Chapter 3. In Chapter 4, we continued finding solutions, but this time our work included absolute-value functions and inequalities.

Now, in Chapter 5, we will first focus on writing equivalent expressions for polynomials. Our work with equivalent expressions will then allow us to solve a new type of equation at the end of this chapter.

Chapters 6 and 7 will follow a similar pattern: After learning new ways of writing equivalent expressions, we will learn to solve new types of equations toward the end of each chapter.

In this section, we introduce a type of algebraic expression known as a *polynomial*. After developing some vocabulary, we study addition and subtraction of polynomials, and evaluate *polynomial functions*.

Algebraic Expressions and Polynomials

In Chapter 1, we introduced algebraic expressions like

$$5x^2, \quad \frac{3}{x^2 + 5}, \quad 9a^3b^4, \quad 3x^{-2}, \quad 6x^2 + 3x + 1, \quad -9, \quad \text{and} \quad 5 - 2x.$$

Of the expressions listed, $5x^2, 9a^3b^4, 3x^{-2}$, and -9 are examples of *terms*. A **term** is simply a number or a variable raised to a power or a product of numbers and/or variables raised to powers. The power used may be 1 and thus not written.

When all variables in a term are raised to whole-number powers, the term is a **monomial**. Of the terms listed above, $5x^2$, $9a^3b^4$, and -9 are monomials. Since -2 is not a whole number, $3x^{-2}$ is *not* a monomial. The **degree** of a monomial is the sum of the exponents of the variables. Thus, $5x^2$ has degree 2 and $9a^3b^4$ has degree 7. Nonzero constant terms, like -9, can be written $-9x^0$ and therefore have degree 0. The term 0 itself is said to have no degree.

The number 5 is said to be the **coefficient** of $5x^2$. Thus the coefficient of $9a^3b^4$ is 9 and the coefficient of $-2x$ is -2. The coefficient of a constant term is just that constant.

A **polynomial** is a monomial or a sum of monomials. Of the expressions listed, $5x^2$, $9a^3b^4$, $6x^2 + 3x + 1$, -9, and $5 - 2x$ are polynomials. In fact, with the exception of $9a^3b^4$, these are all polynomials *in one variable*. The expression $9a^3b^4$ is a *polynomial in two variables*. Note that $5 - 2x$ is the sum of 5 and $-2x$. Thus, 5 and $-2x$ are the terms in the polynomial $5 - 2x$.

The **leading term** of a polynomial is the term of highest degree. Its coefficient is called the **leading coefficient**. The **degree of a polynomial** is the same as the degree of its leading term.

EXAMPLE 1   For each polynomial given, find the degree of each term, the degree of the polynomial, the leading term, and the leading coefficient.

a) $2x^3 + 8x^2 - 17x - 3$

b) $6x^2 + 8x^2y^3 - 17xy - 24xy^2z^4 + 2y + 3$

Solution

a) $2x^3 + 8x^2 - 17x - 3$ **b)** $6x^2 + 8x^2y^3 - 17xy - 24xy^2z^4 + 2y + 3$

Term	$2x^3$	$8x^2$	$-17x$	-3	$6x^2$	$8x^2y^3$	$-17xy$	$-24xy^2z^4$	$2y$	3
Degree	3	2	1	0	2	5	2	7	1	0
Leading Term	$2x^3$				$-24xy^2z^4$					
Leading Coefficient	2				-24					
Degree of Polynomial	3				7					

A polynomial of degree 0 or 1 is called **linear**. A polynomial in one variable is said to be **quadratic** if it is of degree 2, **cubic** if it is of degree 3, and **quartic** if it is of degree 4.

The following are some names for certain kinds of polynomials.

Type	Definition	Examples				
Monomial	A polynomial of one term	4	$-3p$	$5x^2$	$-7a^2b^3$	0	xyz
Binomial	A polynomial of two terms	$2x + 7$		$a - 3b$		$5x^2 + 7y^3$	
Trinomial	A polynomial of three terms	$x^2 - 7x + 12$		$4a^2 + 2ab + b^2$			

We generally arrange polynomials in one variable so that the exponents *decrease* from left to right. This is called **descending order**. Some polynomials may be written with exponents *increasing* from left to right, which is **ascending order**. Generally, if an exercise is written in one kind of order, the answer is written in that same order.

EXAMPLE 2   Arrange in ascending order: $12 + 2x^3 - 7x + x^2$.

Solution

$$12 + 2x^3 - 7x + x^2 = 12 - 7x + x^2 + 2x^3$$

Polynomials in several variables can be arranged with respect to the powers of one of the variables.

EXAMPLE 3   Arrange in descending powers of x: $y^4 + 2 - 5x^2 + 3x^3y + 7xy^2$.

Solution

$$y^4 + 2 - 5x^2 + 3x^3y + 7xy^2 = 3x^3y - 5x^2 + 7xy^2 + y^4 + 2$$

Polynomial Functions

A *polynomial function* is a function in which ordered pairs are determined by evaluating a polynomial. For example, the function P given by

$$P(x) = 5x^7 + 3x^5 - 4x^2 - 5$$

is an example of a polynomial function. To evaluate a polynomial function, we substitute a number for the variable just as in Chapter 2. In this text, we limit ourselves to polynomial functions in one variable.

EXAMPLE 4   For the polynomial function $P(x) = -x^2 + 4x - 1$, find the following: **(a)** $P(2)$; **(b)** $P(10)$; **(c)** $P(-10)$.

Solution

a) $P(2) = -2^2 + 4(2) - 1$ We square the input before taking its opposite.

$\quad = -4 + 8 - 1 = 3$

b) $P(10) = -10^2 + 4(10) - 1$

$\quad\quad = -100 + 40 - 1 = -61$

c) $P(-10) = -(-10)^2 + 4(-10) - 1$

$\quad\quad = -100 - 40 - 1 = -141$

EXAMPLE 5   Medicine.   Ibuprofen is a medication used to relieve pain. The polynomial function

$$M(t) = 0.5t^4 + 3.45t^3 - 96.65t^2 + 347.7t, \quad 0 \le t \le 6$$

can be used to estimate the number of milligrams of ibuprofen in the bloodstream t hours after 400 mg of the medication has been swallowed (*Source:* Based on data from Dr. P. Carey, Burlington, VT).

a) How many milligrams of ibuprofen are in the bloodstream 2 hr after 400 mg has been swallowed?

b) Use the graph below to estimate $M(4)$.

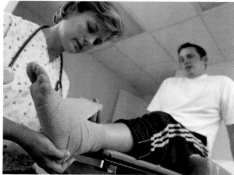

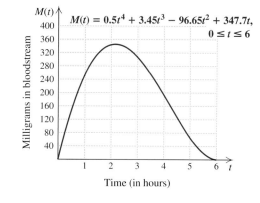

technology connection

One way to evaluate a function is to enter and graph it as y_1 and then select TRACE. We can then enter any x-value that appears in that window and the corresponding y-value will appear. We use this approach below to check Example 5(a).

$y_1 = 0.5t^4 + 3.45t^3 - 96.65t^2 + 347.7t$

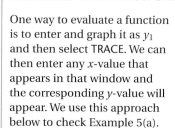

$y_1 = 0.5t^4 + 3.45t^3 - 96.65t^2 + 347.7t$

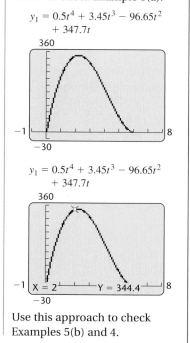

Use this approach to check Examples 5(b) and 4.

Solution

a) We evaluate the function for $t = 2$:

$$\begin{aligned}
M(2) &= 0.5(2)^4 + 3.45(2)^3 - 96.65(2)^2 + 347.7(2) \\
&= 0.5(16) + 3.45(8) - 96.65(4) + 695.4 \\
&= 8 + 27.6 - 386.6 + 695.4 \\
&= 344.4.
\end{aligned}$$

We carry out the calculation using the rules for order of operations.

Approximately 344 mg of ibuprofen is in the bloodstream 2 hr after 400 mg has been swallowed.

b) To estimate $M(4)$, the amount in the bloodstream after 4 hr, we locate 4 on the horizontal axis. From there we move vertically to the graph of the function and then horizontally to the $M(t)$-axis, as shown below. This locates a value of about $M(4) \approx 190$.

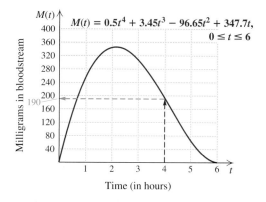

After 4 hr, approximately 190 mg of ibuprofen is still in a person's bloodstream, assuming an original dosage of 400 mg.

Adding Polynomials

Recall from Section 1.3 that when two terms have the same variable(s) raised to the same power(s), they are **similar**, or **like, terms** and can be "combined" or "collected."

EXAMPLE 6   Combine like terms.

a) $3x^2 - 4y + 2x^2$ **b)** $9x^3 + 5x - 4x^2 - 2x^3 + 5x^2$

c) $3x^2y + 5xy^2 - 3x^2y - xy^2$

Solution

a) $3x^2 - 4y + 2x^2 = 3x^2 + 2x^2 - 4y$ Rearranging terms using the commutative law for addition

$\qquad\qquad\qquad\qquad = (3 + 2)x^2 - 4y$ Using the distributive law

$\qquad\qquad\qquad\qquad = 5x^2 - 4y$

b) $9x^3 + 5x - 4x^2 - 2x^3 + 5x^2 = 7x^3 + x^2 + 5x$ We usually perform the middle steps mentally and write just the answer.

c) $3x^2y + 5xy^2 - 3x^2y - xy^2 = 4xy^2$

The sum of two polynomials can be found by writing a plus sign between them and then combining like terms.

EXAMPLE 7   Add: $(-3x^3 + 2x - 4) + (4x^3 + 3x^2 + 2)$.

Solution

$$(-3x^3 + 2x - 4) + (4x^3 + 3x^2 + 2) = x^3 + 3x^2 + 2x - 2$$

Using columns is sometimes helpful. To do so, we write the polynomials one under the other, listing like terms under one another and leaving spaces for any missing terms.

EXAMPLE 8   Add: $4ax^2 + 4bx - 5$ and $-6ax^2 + 8$.

Solution

$$
\begin{array}{l}
4ax^2 + 4bx - 5 \\
\underline{-6ax^2 \qquad\quad + 8} \qquad \text{Leaving space for the missing } bx\text{-term} \\
-2ax^2 + 4bx + 3 \qquad \text{Combining like terms}
\end{array}
$$

EXAMPLE 9   Add: $13x^3y + 3x^2y - 5y$ and $x^3y + 4x^2y - 3xy$.

Solution

$$(13x^3y + 3x^2y - 5y) + (x^3y + 4x^2y - 3xy) = 14x^3y + 7x^2y - 3xy - 5y$$

technology connection

By pressing **2ND** **TBLSET** and selecting AUTO, we can use a Table to check that polynomials in one variable have been added or subtracted correctly. To check Example 7, we enter $y_1 = (-3x^3 + 2x - 4) + (4x^3 + 3x^2 + 2)$ and $y_2 = x^3 + 3x^2 + 2x - 2$. If the addition is correct, the values of y_1 and y_2 will match, regardless of the x-values used.

We can also check by using graphs, as discussed on p. 296.

X	Y₁	Y₂
-2	-2	-2
-1	-2	-2
0	-2	-2
1	4	4
2	22	22
3	58	58
4	118	118

X = -2

Use a table to determine whether the sum or difference is correct.

1. $(x^3 - 2x^2 + 3x - 7) + (3x^2 - 4x + 5) = x^3 + x^2 - x - 2$
2. $(2x^2 + 3x - 6) + (5x^2 - 7x + 4) = 7x^2 + 4x - 2$
3. $(4x^3 + 3x^2 + 2) + (-3x^3 + 2x - 4) = x^3 + 3x^2 + 2x - 2$
4. $(3x^4 - 2x^2 - 1) - (2x^4 - 3x^2 - 4) = x^4 + x^2 - 5$

Opposites and Subtraction

If the sum of two polynomials is 0, the polynomials are *opposites*, or *additive inverses*, of each other. For example,

$$(3x^2 - 5x + 2) + (-3x^2 + 5x - 2) = 0,$$

so the opposite of $(3x^2 - 5x + 2)$ must be $(-3x^2 + 5x - 2)$. We can say the same thing using algebraic symbolism, as follows:

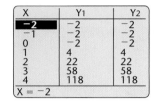

The opposite of $(3x^2 - 5x + 2)$ is $(-3x^2 + 5x - 2)$.

$$-\quad(3x^2 - 5x + 2)\quad=\quad-3x^2 + 5x - 2$$

To form the opposite of a polynomial, we can think of distributing the "−" sign, or multiplying each term of the polynomial by −1, and removing the parentheses. The effect is to change the sign of each term in the polynomial.

The Opposite of a Polynomial

The *opposite* of a polynomial P can be written as $-P$ or, equivalently, by replacing each term with its opposite.

EXAMPLE 10   Write two equivalent expressions for the opposite of

$$7xy^2 - 6xy - 4y + 3.$$

Solution

a) The opposite of $7xy^2 - 6xy - 4y + 3$ can be written as

$$-(7xy^2 - 6xy - 4y + 3).$$ Writing the opposite
of P as $-P$

b) The opposite of $7xy^2 - 6xy - 4y + 3$ can also be written as

$$-7xy^2 + 6xy + 4y - 3.$$ Multiplying each term
by -1

To subtract a polynomial, we add its opposite.

EXAMPLE 11   Subtract: $(-5x^2 + 4) - (2x^2 + 3x - 1)$.

Solution

$$(-5x^2 + 4) - (2x^2 + 3x - 1)$$
$$= (-5x^2 + 4) + (-2x^2 - 3x + 1)$$ Adding the opposite
$$= -5x^2 + 4 - 2x^2 - 3x + 1$$ Try to go directly to
this step.
$$= -7x^2 - 3x + 5$$ Combining like
terms

With practice, you will find that you can skip more steps, by mentally taking the opposite of each term and then combining like terms.

To use columns for subtraction, we mentally change the signs of the terms being subtracted.

EXAMPLE 12   Subtract:

$$(4x^2y - 6x^3y^2 + x^2y^2) - (4x^2y + x^3y^2 + 3x^2y^3 - 8x^2y^2).$$

Solution

Write: (Subtract)   *Think*: (Add)

$$\begin{array}{l} 4x^2y - 6x^3y^2 \qquad\qquad + \ x^2y^2 \\ -(4x^2y + \ x^3y^2 + 3x^2y^3 - 8x^2y^2) \end{array}$$

$$\begin{array}{l} 4x^2y - 6x^3y^2 \qquad\qquad + \ x^2y^2 \\ \underline{-4x^2y - \ x^3y^2 - 3x^2y^3 + 8x^2y^2} \\ \qquad\quad -7x^3y^2 - 3x^2y^3 + 9x^2y^2 \end{array}$$

Take the opposite of each term mentally and add.

Exercise Set

5.1

➥ *Concept Reinforcement   In each of Exercises 1–10, match the item with the best example of that item from the column on the right.*

1. _____ A binomial

2. _____ A trinomial

3. _____ A monomial

4. _____ A third-degree polynomial

5. _____ A polynomial written in ascending powers of t

6. _____ A term that is not a monomial

7. _____ A polynomial with a leading term of degree 5

8. _____ A polynomial with a leading coefficient of 5

9. _____ A quadratic polynomial

10. _____ A polynomial containing similar terms

a) $9a^7$

b) $6s^2 - 2t + 4st^2 - st^3$

c) $4t^{-2}$

d) $t^4 - st + s^3$

e) $7t^3 - 13 + 5t^4 - 2t$

f) $4t^2 + 12t + 9$

g) $5 + a$

h) $4t^3 + 7t - 8t^2 + 5$

i) $8st^3 - 6s^2t + 4st^3 - 2$

j) $7s^3t^2 - 4s^2t + 3st^2 + 1$

Determine the degree of each term and the degree of the polynomial.

11. $-6x^5 - 8x^3 + x^2 + 3x - 4$

12. $t^3 - 5t^2 + 2t + 7$

13. $y^3 + 2y^7 + x^2y^4 - 8$

14. $-2u^2 + 3v^5 - u^3v^4 - 7$

15. $a^5 + 4a^2b^4 + 6ab + 4a - 3$

16. $8p^6 + 2p^4t^4 - 7p^3t + 5p^2 - 14$

Arrange in descending order. Then find the leading term and the leading coefficient.

17. $3 - 5y + 6y^2 + 11y^3 - 18y^4$

18. $9 - 3x - 10x^4 + 7x^2$

19. $a + 5a^3 - a^7 - 19a^2 + 8a^5$

20. $a^3 - 7 + 11a^4 + a^9 - 5a^2$

Arrange in ascending powers of x.

21. $6x - 9 + 3x^4 - 5x^2$

22. $-3x^4 + 4x - x^3 + 9$

23. $5x^2y^2 - 9xy + 8x^3y^2 - 5x^4$

24. $4ax - 7ab + 4x^6 - 7ax^2$

Evaluate each polynomial for $x = 4$.

25. $-7x + 5$

26. $4x - 13$

27. $x^3 - 5x^2 + x$

28. $7 - x + 3x^2$

Evaluate each polynomial function for $x = -1$.

29. $f(x) = -5x^3 + 3x^2 - 4x - 3$

30. $g(x) = -4x^3 + 2x^2 + 5x - 7$

Find the specified function values.

31. Find $F(2)$ and $F(5)$: $F(x) = 2x^2 - 6x - 9$.

32. Find $P(4)$ and $P(0)$: $P(x) = 3x^2 - 2x + 7$.

33. Find $Q(-3)$ and $Q(0)$:
$$Q(y) = -8y^3 + 7y^2 - 4y - 9.$$

34. Find $G(3)$ and $G(-1)$:
$$G(x) = -6x^2 - 5x + x^3 - 1.$$

35. *Freestyle skiing.*   During the 2004 Winter X-Games, Simon Dumont established a world record by launching himself 22 ft above the edge of a halfpipe. His speed $v(t)$, in miles per hour, upon re-entering the pipe can be approximated by $v(t) = 10.9t$, where t is the number of seconds for which he was airborne. Dumont was airborne for approximately 2.34 sec. How fast was he going when he re-entered the halfpipe?
Source: Based on data from EXPN.com

36. *Cliff diving.*   The speed $v(t)$, in miles per hour, at which a swimmer enters the water can be approximated by $v(t) = 21.82t$, where t is the number of seconds the diver is falling. Cliff divers in Acapulco, Mexico, often are in the air for a total of 5 sec. Estimate the speed of such a cliff diver when he or she enters the water.
Source: Based on information from www.guinnessworldrecords.com

37. *Skateboarding.*   The distance $s(t)$, in feet, traveled by a body falling freely from rest in t seconds is approximated by
$$s(t) = 16t^2.$$
In 2002, Adil Dyani set a record for the highest skateboard drop into a quarterpipe. Dyani was airborne for about 1.36 sec. How far did he drop?
Source: Based on data from *Guinness Book of World Records* 2004

38. A paintbrush falls from a scaffold and takes 3 sec to hit the ground (see Exercise 37). How high is the scaffold?

Electing officers.   *For a club consisting of p people, the number of ways N in which a president, vice president, and treasurer can be elected can be determined using the function given by*
$$N(p) = p^3 - 3p^2 + 2p.$$

39. The Southside Rugby Club has 20 members. In how many ways can they elect a president, vice president, and treasurer?

40. The Stage Right drama club has 12 members. In how many ways can a president, vice president, and treasurer be elected?

NASCAR attendance.   *Attendance at NASCAR auto races has grown rapidly over a 10-yr span. Attendance A, in millions, can be approximated by the polynomial function given by*
$$A(x) = 0.0024x^3 - 0.005x^2 + 0.31x + 3,$$
where x is the number of years since 1989. Use the following graph for Exercises 41–44.

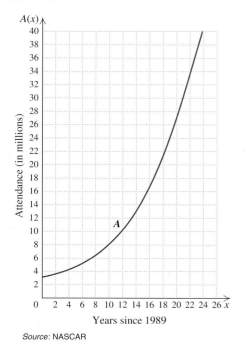

Source: NASCAR

41. Estimate the attendance at NASCAR races in 2009.

42. Estimate the attendance at NASCAR races in 2006.

43. Approximate $A(8)$.

44. Approximate $A(12)$.

45. *Stacking spheres.* In 2004, the journal *Annals of Mathematics* accepted a proof of the so-called Kepler Conjecture: that the most efficient way to pack spheres is in the shape of a square pyramid.

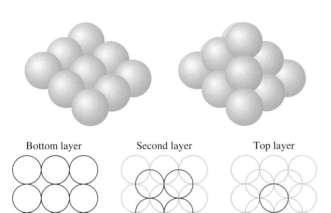

| Bottom layer | Second layer | Top layer |

The number N of balls in the stack is given by the polynomial function

$$N(x) = \tfrac{1}{3}x^3 + \tfrac{1}{2}x^2 + \tfrac{1}{6}x,$$

where x is the number of layers. Use both the function and the figure to find $N(3)$. Then calculate the number of oranges in a pyramid with 5 layers.
Source: *The New York Times* 4/6/04

46. *Stacking cannonballs.* The function in Exercise 45 was discovered by Thomas Harriot, assistant to Sir Walter Raleigh, when preparing for an expedition at sea. How many cannonballs did they pack if there were 10 layers to their pyramid?
Source: *The New York Times* 4/7/04

Veterinary science. *Gentamicin is an antibiotic frequently used by veterinarians. The concentration, in micrograms per milliliter (mcg/mL), of Gentamicin in a horse's bloodstream t hours after injection can be approximated by the polynomial function*

$$C(t) = -0.005t^4 + 0.003t^3 + 0.35t^2 + 0.5t.$$

Use the following graph for Exercises 47–50.
Source: Michele Tulis, DVM, telephone interview

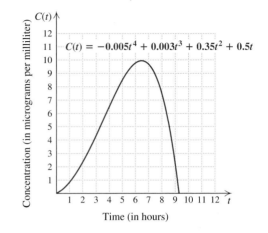

47. Estimate the concentration, in mcg/mL, of Gentamicin in the bloodstream 2 hr after injection.

48. Estimate the concentration, in mcg/mL, of Gentamicin in the bloodstream 4 hr after injection.

49. Approximate $C(2)$.

50. Approximate $C(4)$.

Surface area of a right circular cylinder. *The surface area of a right circular cylinder is given by the polynomial*

$$2\pi rh + 2\pi r^2,$$

where h is the height, r is the radius of the base, and h and r are given in the same units.

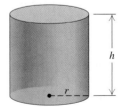

51. A 16-oz beverage can has height 6.3 in. and radius 1.2 in. Find the surface area of the can. (Use a calculator with a π key or use 3.141592654 for π.)

52. A 12-oz beverage can has height 4.7 in. and radius 1.2 in. Find the surface area of the can. (Use a calculator with a π key or use 3.141592654 for π.)

Total revenue. An electronics firm is marketing a new kind of DVD player. The firm determines that when it sells x DVD players, its total revenue is

$$R(x) = 280x - 0.4x^2 \text{ dollars.}$$

53. What is the total revenue from the sale of 75 DVD players?

54. What is the total revenue from the sale of 100 DVD players?

Total cost. The electronics firm determines that the total cost, in dollars, of producing x DVD players is given by

$$C(x) = 5000 + 0.6x^2.$$

55. What is the total cost of producing 75 DVD players?

56. What is the total cost of producing 100 DVD players?

Combine like terms to write an equivalent expression.

57. $5a + 6 - 4 + 2a^3 - 6a + 2$

58. $6x + 13 - 8 - 7x + 5x^2 + 10$

59. $3a^2b + 4b^2 - 9a^2b - 7b^2$

60. $5x^2y^2 + 4x^3 - 8x^2y^2 - 12x^3$

61. $9x^2 - 3xy + 12y^2 + x^2 - y^2 + 5xy + 4y^2$

62. $a^2 - 2ab + b^2 + 9a^2 + 5ab - 4b^2 + a^2$

Add.

63. $(7x - 5y + 3z) + (9x + 12y - 8z)$

64. $(a^2 - 3b^2 + 4c^2) + (-5a^2 + 2b^2 - c^2)$

65. $(x^2 + 2x - 3xy - 7) + (-3x^2 - x + 2xy + 6)$

66. $(3a^2 - 2b + ab + 6) + (-a^2 + 5b - 5ab - 2)$

67. $(8x^2y - 3xy^2 + 4xy) + (-2x^2y - xy^2 + xy)$

68. $(9ab - 3ac + 5bc) + (13ab - 15ac - 8bc)$

69. $(2r^2 + 12r - 11) + (6r^2 - 2r + 4) + (r^2 - r - 2)$

70. $(5x^2 + 19x - 23) + (7x^2 - 2x + 1) + (-x^2 - 9x + 8)$

71. $\left(\frac{1}{8}xy - \frac{3}{5}x^3y^2 + 4.3y^3\right) + \left(-\frac{1}{3}xy - \frac{3}{4}x^3y^2 - 2.9y^3\right)$

72. $\left(\frac{2}{3}xy + \frac{5}{6}xy^2 + 5.1x^2y\right) + \left(-\frac{4}{5}xy + \frac{3}{4}xy^2 - 3.4x^2y\right)$

Write two expressions, one with parentheses and one without, for the opposite of each polynomial.

73. $5x^3 - 7x^2 + 3x - 9$

74. $-8y^4 - 18y^3 + 4y - 7$

75. $-12y^5 + 4ay^4 - 7by^2$

76. $7ax^3y^2 - 8by^4 - 7abx - 12ay$

Subtract.

77. $(7x - 5) - (-3x + 4)$

78. $(8y + 2) - (-6y - 5)$

79. $(-3x^2 + 2x + 9) - (x^2 + 5x - 4)$

80. $(-7y^2 + 5y + 6) - (4y^2 + 3y - 2)$

81. $(8a - 3b + c) - (2a + 3b - 4c)$

82. $(9r - 5s - t) - (7r - 5s + 3t)$

83. $(6a^2 + 5ab - 4b^2) - (8a^2 - 7ab + 3b^2)$

84. $(4y^2 - 13yz - 9z^2) - (9y^2 - 6yz + 3z^2)$

85. $(6ab - 4a^2b + 6ab^2) - (3ab^2 - 10ab - 12a^2b)$

86. $(10xy - 4x^2y^2 - 3y^3) - (-9x^2y^2 + 4y^3 - 7xy)$

87. $\left(\frac{5}{8}x^4 - \frac{1}{4}x^2 - \frac{1}{2}\right) - \left(-\frac{3}{8}x^4 + \frac{3}{4}x^2 + \frac{1}{2}\right)$

88. $\left(\frac{5}{6}y^4 - \frac{1}{2}y^2 - 7.8y\right) - \left(-\frac{3}{8}y^4 + \frac{3}{4}y^2 + 3.4y\right)$

Perform the indicated operations.

89. $(5x^2 + 7) - (2x^2 + 1) + (x^2 + 2x)$

90. $(3t^2 + 5) - (t^2 + 2) + (4t^2 + 3t)$

91. $(8r^2 - 6r) - (2r - 6) + (5r^2 - 7)$

92. $(7s^2 - 5s) - (4s - 1) + (3s^2 - 5)$

Aha! **93.** $(x^2 - 4x + 7) + (3x^2 - 9) - (x^2 - 4x + 7)$

94. $(t^2 - 5t + 6) + (5t - 8) - (t^2 + 3t - 4)$

Total profit. Total profit is defined as total revenue minus total cost. In Exercises 95 and 96, $R(x)$ and $C(x)$ are the revenue and the cost, respectively, from the sale of x futons.

95. If $R(x) = 280x - 0.4x^2$ and $C(x) = 5000 + 0.6x^2$, find the profit from the sale of 70 futons.

96. If $R(x) = 280x - 0.7x^2$ and $C(x) = 8000 + 0.5x^2$, find the profit from the sale of 100 futons.

97. Is the sum of two binomials always a binomial? Why or why not?

98. Ani claims that she can add any two polynomials but finds subtraction difficult. What advice would you offer her?

SKILL MAINTENANCE

Simplify. [1.6]

99. $(x^2)^5$

100. $(t^3)^4$

101. $(5x^3)^2$

102. $(7t^4)^2$

103. $x^5 \cdot x^4$

104. $a^2 \cdot a^6$

SYNTHESIS

105. Write a problem in which revenue and cost functions are given and a profit function, $P(x)$, is required. Devise the problem so that $P(0) < 0$ and $P(100) > 0$.

106. Write a problem in which revenue and cost functions are given and a profit function, $P(x)$, is required. Devise the problem so that $P(10) < 0$ and $P(50) > 0$.

For P(x) and Q(x) as given, find the following.

$P(x) = 13x^5 - 22x^4 - 36x^3 + 40x^2 - 16x + 75,$
$Q(x) = 42x^5 - 37x^4 + 50x^3 - 28x^2 + 34x + 100$

107. $2[P(x)] + Q(x)$

108. $3[P(x)] - Q(x)$

109. $2[Q(x)] - 3[P(x)]$

110. $4[P(x)] + 3[Q(x)]$

111. *Volume of a display.*   The number of spheres in a triangular pyramid with x layers is given by the function

$$N(x) = \tfrac{1}{6}x^3 + \tfrac{1}{2}x^2 + \tfrac{1}{3}x.$$

The volume of a sphere of radius r is given by the function

$$V(r) = \tfrac{4}{3}\pi r^3,$$

where π can be approximated as 3.14.

Chocolate Heaven has a window display of truffles piled in a triangular pyramid formation 5 layers deep. If the diameter of each truffle is 3 cm, find the volume of chocolate in the display.

112. If one large truffle were to have the same volume as the display of truffles in Exercise 111, what would be its diameter?

113. Find a polynomial function that gives the outside surface area of a box like this one, with an open top and dimensions as shown.

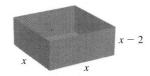

114. Develop a formula for the surface area of a right circular cylinder in which h is the height, in *centimeters*, and r is the radius, in *meters*. (See Exercises 51 and 52.)

Perform the indicated operation. Assume that the exponents are natural numbers.

115. $(2x^{2a} + 4x^a + 3) + (6x^{2a} + 3x^a + 4)$

116. $(3x^{6a} - 5x^{5a} + 4x^{3a} + 8) - (2x^{6a} + 4x^{4a} + 3x^{3a} + 2x^{2a})$

117. $(2x^{5b} + 4x^{4b} + 3x^{3b} + 8) - (x^{5b} + 2x^{3b} + 6x^{2b} + 9x^b + 8)$

118. Use a graphing calculator to check your answers to Exercises 29, 39, and 57.

119. Use a graphing calculator to check your answers to Exercises 32, 40, and 58.

120. A student who is trying to graph

$$p(x) = 0.05x^4 - x^2 + 5$$

gets the following screen. How can the student tell at a glance that a mistake has been made?

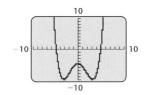

CORNER

How Many Handshakes?

Focus: Polynomial functions

Time: 20 minutes

Group size: 5

ACTIVITY

1. All group members should shake hands with each other. Without "double counting," determine how many handshakes occurred.
2. Complete the table in the next column.
3. Join another group to determine the number of handshakes for a group of size 10.
4. Try to find a function of the form $H(n) = an^2 + bn$, for which $H(n)$ is the number of different handshakes that are possible in a group of n people. Make sure

that $H(n)$ produces all of the values in the table above. (*Hint*: Use the table to twice select n and $H(n)$. Then solve the resulting system of equations for a and b.)

Group Size	Number of Handshakes
1	
2	
3	
4	
5	

5.2   Multiplication of Polynomials

Multiplying Monomials   •   Multiplying Monomials and Binomials   •   Multiplying Any Two Polynomials   •   The Product of Two Binomials: FOIL   •   Squares of Binomials   • Products of Sums and Differences   •   Function Notation

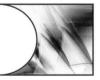

Just like numbers, polynomials can be multiplied. The product of two polynomials $P(x)$ and $Q(x)$ is a polynomial $R(x)$ that gives the same value as $P(x) \cdot Q(x)$ for any replacement of x.

Multiplying Monomials

To multiply monomials, we first multiply their coefficients. Then we multiply the variables using the rules for exponents and the commutative and associative laws. With practice, we can work mentally, writing only the answer.

EXAMPLE 1   Multiply and simplify: **(a)** $(-8x^4y^7)(5x^3y^2)$; **(b)** $(-2x^2yz^5)(-6x^5y^{10}z^2)$.

Solution

a) $(-8x^4y^7)(5x^3y^2) = -8 \cdot 5 \cdot x^4 \cdot x^3 \cdot y^7 \cdot y^2$ Using the associative and commutative laws

$$= -40x^{4+3}y^{7+2}$$ Multiplying coefficients; adding exponents

$$= -40x^7y^9$$

b) $(-2x^2yz^5)(-6x^5y^{10}z^2) = (-2)(-6) \cdot x^2 \cdot x^5 \cdot y \cdot y^{10} \cdot z^5 \cdot z^2$

$$= 12x^7y^{11}z^7$$ Multiplying coefficients; adding exponents

Multiplying Monomials and Binomials

The distributive law is the basis for multiplying polynomials other than monomials. We first multiply a monomial and a binomial.

EXAMPLE 2   Multiply: **(a)** $2x(3x - 5)$; **(b)** $3a^2b(a^2 - b^2)$.

Solution

a) $2x(3x - 5) = 2x \cdot 3x - 2x \cdot 5$ Using the distributive law

$$= 6x^2 - 10x$$ Multiplying monomials

b) $3a^2b(a^2 - b^2) = 3a^2b \cdot a^2 - 3a^2b \cdot b^2$ Using the distributive law

$$= 3a^4b - 3a^2b^3$$

The distributive law is also used for multiplying two binomials. In this case, however, we begin by distributing a *binomial* rather than a monomial. With practice, some of the following steps can be combined.

EXAMPLE 3   Multiply: $(y^3 - 5)(2y^3 + 4)$.

Solution

$$(y^3 - 5)(2y^3 + 4) = (y^3 - 5)2y^3 + (y^3 - 5)4$$ "Distributing" the $y^3 - 5$

$$= 2y^3(y^3 - 5) + 4(y^3 - 5)$$ Using the commutative law for multiplication. Try to do this step mentally.

$$= 2y^3 \cdot y^3 - 2y^3 \cdot 5 + 4 \cdot y^3 - 4 \cdot 5$$ Using the distributive law (twice)

$$= 2y^6 - 10y^3 + 4y^3 - 20$$ Multiplying the monomials

$$= 2y^6 - 6y^3 - 20$$ Combining like terms

Multiplying Any Two Polynomials

Repeated use of the distributive law enables us to multiply *any* two polynomials, regardless of how many terms are in each.

EXAMPLE 4   Multiply: $(p + 2)(p^4 - 2p^3 + 3)$.

Solution   We can use the distributive law from right to left if we wish:

$$(p + 2)(p^4 - 2p^3 + 3) = p(p^4 - 2p^3 + 3) + 2(p^4 - 2p^3 + 3)$$
$$= p \cdot p^4 - p \cdot 2p^3 + p \cdot 3 + 2 \cdot p^4 - 2 \cdot 2p^3 + 2 \cdot 3$$
$$= p^5 - 2p^4 + 3p + 2p^4 - 4p^3 + 6$$
$$= p^5 - 4p^3 + 3p + 6. \qquad \text{Combining like terms}$$

> *The Product of Two Polynomials*
>
> The *product* of two polynomials $P(x)$ and $Q(x)$ is found by multiplying each term of $P(x)$ by every term of $Q(x)$ and then combining like terms.

It is also possible to stack the polynomials, multiplying each term at the top by every term below, keeping like terms in columns, and leaving spaces for missing terms. Then we add just as we do in long multiplication with numbers.

EXAMPLE 5   Multiply: $(5x^3 + x - 4)(-2x^2 + 3x + 6)$.

Solution

$$
\begin{array}{r}
5x^3 + \quad x - \ 4 \\
-2x^2 + \ 3x + \ 6 \\
\hline
30x^3 \qquad\quad + \ 6x - 24 \qquad \text{Multiplying by } 6 \\
15x^4 \qquad\quad + \ 3x^2 - 12x \qquad \text{Multiplying by } 3x \\
-10x^5 \qquad\quad - \ 2x^3 + \ 8x^2 \qquad\qquad \text{Multiplying by } -2x^2 \\
\hline
-10x^5 + 15x^4 + 28x^3 + 11x^2 - \ 6x - 24 \qquad \text{Adding}
\end{array}
$$

The Product of Two Binomials: FOIL

We now consider what are called *special products*. These products of polynomials occur often and can be simplified using shortcuts that we now develop.

To find a faster special-product rule for the product of two binomials, consider $(x + 7)(x + 4)$. We multiply each term of $(x + 7)$ by each term of $(x + 4)$.

$$(x + 7)(x + 4) = x \cdot x + x \cdot 4 + 7 \cdot x + 7 \cdot 4.$$

A visualization of
$(x + 7)(x + 4)$ using areas

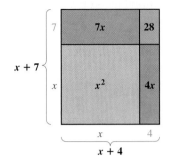

This multiplication illustrates a pattern that occurs whenever two binomials are multiplied:

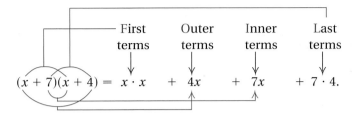

We use the mnemonic device FOIL to remember this method for multiplying.

The FOIL Method

To multiply two binomials $A + B$ and $C + D$, multiply the First terms AC, the Outer terms AD, the Inner terms BC, and then the Last terms BD. Then combine like terms, if possible.

$$(A + B)(C + D) = AC + AD + BC + BD$$

1. Multiply First terms: AC.
2. Multiply Outer terms: AD.
3. Multiply Inner terms: BC.
4. Multiply Last terms: BD.
      ↓
    FOIL

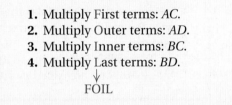

EXAMPLE 6   Multiply.

a) $(x + 5)(x - 8)$

b) $(2x + 3y)(x - 4y)$

c) $(t + 2)(t - 4)(t + 5)$

Solution

$$ \overset{\text{F}}{} \quad \overset{\text{O}}{} \quad \overset{\text{I}}{} \quad \overset{\text{L}}{}$$

a) $(x + 5)(x - 8) = x^2 - 8x + 5x - 40$

$ = x^2 - 3x - 40 \qquad$ Combining like terms

b) $(2x + 3y)(x - 4y) = 2x^2 - 8xy + 3xy - 12y^2 \qquad$ Using FOIL

$ = 2x^2 - 5xy - 12y^2 \qquad$ Combining like terms

c) $(t + 2)(t - 4)(t + 5) = (t^2 - 4t + 2t - 8)(t + 5) \qquad$ Using FOIL

$ = (t^2 - 2t - 8)(t + 5)$

$ = (t^2 - 2t - 8) \cdot t + (t^2 - 2t - 8) \cdot 5 \qquad$ Using the distributive law

$ = t^3 - 2t^2 - 8t + 5t^2 - 10t - 40$

$ = t^3 + 3t^2 - 18t - 40 \qquad$ Combining like terms

A visualization of
$(A + B)^2$ using areas

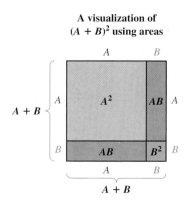

Squares of Binomials

A fast method for squaring binomials can be developed using FOIL:

$$(A + B)^2 = (A + B)(A + B)$$
$$= A^2 + AB + AB + B^2 \qquad \text{Note that } AB \text{ occurs twice.}$$
$$= A^2 + 2AB + B^2;$$

$$(A - B)^2 = (A - B)(A - B)$$
$$= A^2 - AB - AB + B^2 \qquad \text{Note that } -AB \text{ occurs twice.}$$
$$= A^2 - 2AB + B^2.$$

Squaring a Binomial

$$(A + B)^2 = A^2 + 2AB + B^2;$$
$$(A - B)^2 = A^2 - 2AB + B^2$$

The square of a binomial is the square of the first term, plus twice the product of the two terms, plus the square of the last term.
   Trinomials that can be written in the form $A^2 + 2AB + B^2$ or $A^2 - 2AB + B^2$ are called *perfect-square trinomials*.

It can help to remember the words of the rules and say them while multiplying.

EXAMPLE 7   Multiply: **(a)** $(y - 5)^2$; **(b)** $(2x + 3y)^2$; **(c)** $\left(\frac{1}{2}x - 3y^4\right)^2$.

Solution

$$(A - B)^2 = A^2 - 2 \cdot A \cdot B + B^2$$

a) $(y - 5)^2 = y^2 - 2 \cdot y \cdot 5 + 5^2 \qquad$ Note that $-2 \cdot y \cdot 5$ is twice the product of y and -5.

$$= y^2 - 10y + 25$$

b) $(2x + 3y)^2 = (2x)^2 + 2 \cdot 2x \cdot 3y + (3y)^2$
$$= 4x^2 + 12xy + 9y^2 \qquad \text{Raising a product to a power}$$

c) $\left(\frac{1}{2}x - 3y^4\right)^2 = \left(\frac{1}{2}x\right)^2 - 2 \cdot \frac{1}{2}x \cdot 3y^4 + (3y^4)^2 \qquad \begin{array}{l} 2 \cdot \frac{1}{2}x \cdot (-3y^4) = \\ -2 \cdot \frac{1}{2}x \cdot 3y^4 \end{array}$

$$= \tfrac{1}{4}x^2 - 3xy^4 + 9y^8 \qquad \begin{array}{l}\text{Raising a product to a power;} \\ \text{multiplying exponents}\end{array}$$

> *Caution!*   Note that $(y - 5)^2 \neq y^2 - 5^2$. (To see this, replace y with 6 and note that $(6 - 5)^2 = 1^2 = 1$ and $6^2 - 5^2 = 36 - 25 = 11$.) More generally,
>
> $$(A + B)^2 \neq A^2 + B^2 \quad \text{and} \quad (A - B)^2 \neq A^2 - B^2.$$

Student Notes _____

Being able to quickly multiply products of the form $(A + B)(A - B)$ or $(A - B)(A + B)$ is a skill heavily relied on in Section 5.5.

Products of Sums and Differences

Another pattern emerges when we are multiplying a sum and difference of the same two terms. Note the following:

$$
\begin{array}{cccc}
\text{F} & \text{O} & \text{I} & \text{L} \\
\downarrow & \downarrow & \downarrow & \downarrow
\end{array}
$$

$$(A + B)(A - B) = A^2 - AB + AB - B^2$$
$$= A^2 - B^2. \qquad -AB + AB = 0$$

The Product of a Sum and a Difference

$$(A + B)(A - B) = A^2 - B^2 \qquad \text{This is called a } \textit{difference of two squares.}$$

The product of the sum and difference of the same two terms is the square of the first term minus the square of the second term.

EXAMPLE 8   Multiply.

a) $(t + 5)(t - 5)$

b) $(2xy^2 + 3x)(2xy^2 - 3x)$

c) $(0.2t - 1.4m)(0.2t + 1.4m)$

d) $\left(\frac{2}{3}n - m^3\right)\left(\frac{2}{3}n + m^3\right)$

Solution

$$
\begin{array}{cccccc}
(A & + & B) & (A & - & B) = A^2 - B^2 \\
\downarrow & \downarrow & \downarrow & \downarrow & \downarrow & \downarrow
\end{array}
$$

a) $(t + 5)(t - 5) = t^2 - 5^2$ Replacing A with t and B with 5
$$\qquad\qquad = t^2 - 25$$ Try to do problems like this mentally.

b) $(2xy^2 + 3x)(2xy^2 - 3x) = (2xy^2)^2 - (3x)^2$
$$\qquad\qquad = 4x^2y^4 - 9x^2$$ Raising a product to a power

c) $(0.2t - 1.4m)(0.2t + 1.4m) = (0.2t)^2 - (1.4m)^2$
$$\qquad\qquad = 0.04t^2 - 1.96m^2$$

d) $\left(\frac{2}{3}n - m^3\right)\left(\frac{2}{3}n + m^3\right) = \left(\frac{2}{3}n\right)^2 - (m^3)^2$
$$\qquad\qquad = \frac{4}{9}n^2 - m^6$$

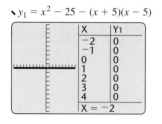

technology connection

One way to check problems like Example 8(a) is to use a Table, much as in the Technology Connection on p. 283. Another check is to note that if the multiplication is correct, then $(t + 5)(t - 5) = t^2 - 25$ is an identity and $t^2 - 25 - (t + 5)(t - 5)$ must be 0. In the window below, we set the MODE to G-T so that we can view both a graph and a table. We use a heavy line to distinguish the graph from the x-axis.

$$y_1 = x^2 - 25 - (x + 5)(x - 5)$$

X	Y₁
−2	0
−1	0
0	0
1	0
2	0
3	0
4	0

X = −2

Had we found $y_1 \neq 0$, we would have known that a mistake had been made.

1. Use this procedure to show that $(x - 3)(x + 3) = x^2 - 9$.
2. Use this procedure to show that $(t - 4)^2 = t^2 - 8t + 16$.
3. Show that the graphs of $y_1 = x^2 - 4$ and $y_2 = (x + 2)(x - 2)$ coincide, using the Sequential MODE with a heavier-weight line for y_2. Then, using the Y-VARS option of the VARS key, let $y_3 = y_2 - y_1$. What do you expect the graph of y_3 to look like?

EXAMPLE 9   Multiply.

a) $(5y + 4 + 3x)(5y + 4 - 3x)$ **b)** $(3xy^2 + 4y)(-3xy^2 + 4y)$
c) $(2t + 3)^2 - (t - 1)(t + 1)$

Solution

a) By far the easiest way to multiply $(5y + 4 + 3x)(5y + 4 - 3x)$ is to note that it is in the form $(A + B)(A - B)$:

$$(5y + 4 + 3x)(5y + 4 - 3x) = (5y + 4)^2 - (3x)^2 \qquad \text{Try to be alert for situations like this.}$$

$$= 25y^2 + 40y + 16 - 9x^2$$

We can also multiply $(5y + 4 + 3x)(5y + 4 - 3x)$ using columns, but not as quickly.

b) $(3xy^2 + 4y)(-3xy^2 + 4y) = (4y + 3xy^2)(4y - 3xy^2)$ Rewriting
$$= (4y)^2 - (3xy^2)^2$$
$$= 16y^2 - 9x^2y^4$$

c) $(2t + 3)^2 - (t - 1)(t + 1) = 4t^2 + 12t + 9 - (t^2 - 1)$ Squaring a binomial; simplifying $(t - 1)(t + 1)$
$$= 4t^2 + 12t + 9 - t^2 + 1 \qquad \text{Subtracting}$$
$$= 3t^2 + 12t + 10 \qquad \text{Combining like terms}$$

Function Notation

Let's stop for a moment and look back at what we have done in this section. We have shown, for example, that

$$(x - 2)(x + 2) = x^2 - 4,$$

that is, $x^2 - 4$ and $(x - 2)(x + 2)$ are equivalent expressions.

From the viewpoint of functions, if

$$f(x) = x^2 - 4$$

and

$$g(x) = (x - 2)(x + 2),$$

then for any given input x, the outputs $f(x)$ and $g(x)$ are identical. Thus the graphs of these functions are identical and we say that f and g represent the same function. Functions like these are graphed in detail in Chapter 8.

x	$f(x)$	$g(x)$
3	5	5
2	0	0
1	−3	−3
0	−4	−4
−1	−3	−3
−2	0	0
−3	5	5

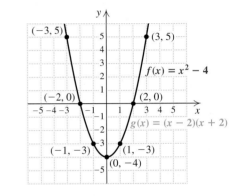

Our work with multiplying can be used when evaluating functions.

EXAMPLE 10   Given $f(x) = x^2 - 4x + 5$, find and simplify each of the following.

a) $f(a) + 3$ **b)** $f(a + 3)$ **c)** $f(a + h) - f(a)$

Solution

a) To find $f(a) + 3$, we replace x with a to find $f(a)$. Then we add 3 to the result:

$$f(a) + 3 = a^2 - 4a + 5 + 3 \qquad \text{Evaluating } f(a)$$
$$= a^2 - 4a + 8.$$

b) To find $f(a + 3)$, we replace x with $a + 3$. Then we simplify:

$$f(a + 3) = (a + 3)^2 - 4(a + 3) + 5$$
$$= a^2 + 6a + 9 - 4a - 12 + 5$$
$$= a^2 + 2a + 2.$$

Caution!   Note from parts (a) and (b) that, in general,

$$f(a + 3) \neq f(a) + 3.$$

---

c) To find $f(a + h)$ and $f(a)$, we replace x with $a + h$ and a, respectively:

$$\begin{aligned}
f(a + h) - f(a) &= [(a + h)^2 - 4(a + h) + 5] - [a^2 - 4a + 5] \\
&= [a^2 + 2ah + h^2 - 4a - 4h + 5] - [a^2 - 4a + 5] \\
&= a^2 + 2ah + h^2 - 4a - 4h + 5 - a^2 + 4a - 5 \\
&= 2ah + h^2 - 4h.
\end{aligned}$$

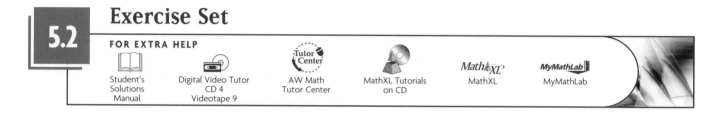

Exercise Set
5.2

FOR EXTRA HELP

Student's Solutions Manual | Digital Video Tutor CD 4 Videotape 9 | AW Math Tutor Center | MathXL Tutorials on CD | MathXL | MyMathLab

Concept Reinforcement   *Classify each of the following as either true or false.*

1. The product of two monomials is a monomial.

2. The product of $5a^2$ and $3a^4$ is $15a^8$.

3. The product of a monomial and a binomial is found using the distributive law.

4. To simplify the product of two binomials, we often need to combine like terms.

5. FOIL can be used whenever a monomial and a binomial are multiplied.

6. The square of a binomial is a difference of two squares.

7. The product of two polynomials of the form $(A - B)(A + B)$ is a difference of two squares.

8. In general, $f(a + 5) \neq f(a) + 5$.

Multiply.

9. $8a^2 \cdot 5a$

10. $-5x^3 \cdot 2x$

11. $5x(-4x^2 y)$

12. $-3ab^2(2a^2 b^2)$

13. $(4x^3 y^2)(-5x^2 y^4)$

14. $(7a^2 bc^4)(-8ab^3 c^2)$

15. $7x(3 - x)$

16. $3a(a^2 - 4a)$

17. $5cd(4c^2 d - 5cd^2)$

18. $a^2(2a^2 - 5a^3)$

19. $(x + 3)(x + 5)$

20. $(t + 1)(t + 4)$

21. $(t - 2)(t + 7)$

22. $(x - 5)(x + 6)$

23. $(2a + 3)(4a + 1)$

24. $(3r + 4)(2r + 1)$

25. $(5x - 4y)(2x + y)$

26. $(4t + 3)(2t - 7)$

27. $(x + 2)(x^2 - 3x + 1)$

28. $(a + 3)(a^2 - 4a + 2)$

29. $(t - 5)(t^2 + 2t - 3)$

30. $(x - 4)(x^2 + x - 7)$

31. $(a^2 + a - 1)(a^2 + 4a - 5)$

32. $(x^2 - 2x + 1)(x^2 + x + 2)$

33. $(x + 3)(x^2 - 3x + 9)$

34. $(y + 4)(y^2 - 4y + 16)$

35. $(a - b)(a^2 + ab + b^2)$

36. $(x - y)(x^2 + xy + y^2)$

37. $\left(t - \frac{1}{3}\right)\left(t - \frac{1}{4}\right)$

38. $\left(x - \frac{1}{2}\right)\left(x - \frac{1}{5}\right)$

39. $(r + 3)(r + 2)(r - 1)$

40. $(t + 4)(t + 1)(t - 2)$

41. $(x + 5)^2$

42. $(t + 6)^2$

43. $(1.2t + 3s)(2.5t - 5s)$

44. $(30a - 0.24b)(0.2a + 10b)$

45. $(2y - 7)^2$

46. $(3x - 4)^2$

47. $(5a - 3b)^2$

48. $(7x - 4y)^2$

49. $(2a^3 - 3b^2)^2$

50. $(3s^2 + 4t^3)^2$

51. $(x^3y^4 + 5)^2$

52. $(a^4b^2 + 3)^2$

53. Let $P(x) = 3x^2 - 5$ and $Q(x) = 4x^2 - 7x + 1$. Find $P(x) \cdot Q(x)$.

54. Let $P(x) = x^2 - x + 1$ and $Q(x) = x^3 + x^2 + 5$. Find $P(x) \cdot Q(x)$.

55. Let $P(x) = 5x - 2$. Find $P(x) \cdot P(x)$.

56. Let $Q(x) = 3x^2 + 1$. Find $Q(x) \cdot Q(x)$.

57. Let $F(x) = 2x - \frac{1}{3}$. Find $[F(x)]^2$.

58. Let $G(x) = 5x - \frac{1}{2}$. Find $[G(x)]^2$.

Multiply.

59. $(c + 7)(c - 7)$

60. $(x - 3)(x + 3)$

61. $(4x + 1)(4x - 1)$

62. $(3 - 2x)(3 + 2x)$

63. $(3m - 2n)(3m + 2n)$

64. $(3x + 5y)(3x - 5y)$

65. $(x^3 + yz)(x^3 - yz)$

66. $(4a^3 + 5ab)(4a^3 - 5ab)$

67. $(-mn + m^2)(mn + m^2)$

68. $(-3b + a^2)(3b + a^2)$

69. $(x + 7)^2 - (x + 3)(x - 3)$

70. $(t + 5)^2 - (t - 4)(t + 4)$

71. $(2m - n)(2m + n) - (m - 2n)^2$

72. $(3x + y)(3x - y) - (2x + y)^2$

Aha! **73.** $(a + b + 1)(a + b - 1)$

74. $(m + n + 2)(m + n - 2)$

75. $(2x + 3y + 4)(2x + 3y - 4)$

76. $(3a - 2b + c)(3a - 2b - c)$

77. *Compounding interest.* Suppose that P dollars is invested in a savings account at interest rate i, compounded annually, for 2 yr. The amount A in the account after 2 yr is given by
$$A = P(1 + i)^2.$$
Find an equivalent expression for A.

78. *Compounding interest.* Suppose that P dollars is invested in a savings account at interest rate i, compounded semiannually, for 1 yr. The amount A in the account after 1 yr is given by
$$A = P\left(1 + \frac{i}{2}\right)^2.$$
Find an equivalent expression for A.

79. Given $f(x) = x^2 + 5$, find and simplify.
   a) $f(t - 1)$
   b) $f(a + h) - f(a)$
   c) $f(a) - f(a - h)$

80. Given $f(x) = x^2 + 7$, find and simplify.
   a) $f(p + 1)$
   b) $f(a + h) - f(a)$
   c) $f(a) - f(a - h)$

81. Find two binomials whose product is $x^2 - 25$ and explain how you decided on those two binomials.

82. Find two binomials whose product is $x^2 - 6x + 9$ and explain how you decided on those two binomials.

SKILL MAINTENANCE

Solve. [1.5]

83. $ab + ac = d$, for a

84. $xy + yz = w$, for y

85. $mn + m = p$, for m

86. $rs + s = t$, for s

87. *Value of coins.*   There are 50 dimes in a roll of dimes, 40 nickels in a roll of nickels, and 40 quarters in a roll of quarters. Kacie has 13 rolls of coins, which have a total value of $89. There are three more rolls of dimes than nickels. How many of each type of roll does she have? [3.5]

88. *Wages.*   Takako worked a total of 17 days last month at her father's restaurant. She earned $50 a day during the week and $60 a day during the weekend. Last month Takako earned $940. How many weekdays did she work? [3.3]

SYNTHESIS

89. We have seen that $(a - b)(a + b) = a^2 - b^2$. Explain how this result can be used to develop a fast way of multiplying $95 \cdot 105$.

90. A student incorrectly claims that since $2x^2 \cdot 2x^2 = 4x^4$, it follows that $5x^5 \cdot 5x^5 = 25x^{25}$. What mistake is the student making?

Multiply. Assume that variables in exponents represent natural numbers.

91. $(x^2 + y^n)(x^2 - y^n)$

92. $(a^n + b^n)^2$

93. $x^2y^3(5x^n + 4y^n)$

94. $a^n b^m(7a^2 - 3b^3)$

95. $(t^n + 3)(t^{2n} - 2t^n + 1)$

96. $(x^n - 4)(x^{2n} + 3x^n - 2)$

Aha! **97.** $(a - b + c - d)(a + b + c + d)$

98. $[(a + b)(a - b)][5 - (a + b)][5 + (a + b)]$

99. $\left(\frac{2}{3}x + \frac{1}{3}y + 1\right)\left(\frac{2}{3}x - \frac{1}{3}y - 1\right)$

100. $(x^2 - 3x + 5)(x^2 + 3x + 5)$

101. $(x^a + y^b)(x^a - y^b)(x^{2a} + y^{2b})$

102. $(x - 1)(x^2 + x + 1)(x^3 + 1)$

103. $(x^{a-b})^{a+b}$

104. $(M^{x+y})^{x+y}$

Aha! **105.** $(x - a)(x - b)(x - c) \cdots (x - z)$

106. Given $f(x) = x^2 + 7$, find and simplify
$$\frac{f(a + h) - f(a)}{h}.$$

107. Given $g(x) = x^2 - 9$, find and simplify
$$\frac{g(a + h) - g(a)}{h}.$$

108. Draw rectangles similar to those on p. 293 to show that $(x + 2)(x + 5) = x^2 + 7x + 10$.

109. Use a graphing calculator to determine whether each of the following is an identity.
a) $(x - 1)^2 = x^2 - 1$
b) $(x - 2)(x + 3) = x^2 + x - 6$
c) $(x - 1)^3 = x^3 - 3x^2 + 3x - 1$
d) $(x + 1)^4 = x^4 + 1$
e) $(x + 1)^4 = x^4 + 4x^3 + 8x^2 + 4x + 1$

110. Use a graphing calculator to check your answers to Exercises 15, 33, and 57.

CORNER

Algebra and Number Tricks

Focus: Polynomial multiplication
Time: 15–20 minutes
Group size: 2

Consider the following dialogue:

Jinny: Cal, let me do a number trick with you. Think of a number between 1 and 7. I'll have you perform some manipulations to this number, you'll tell me the result, and I'll tell me your number.

Cal: Okay. I've thought of a number.

Jinny: Good. Write it down so I can't see it, double it, and then subtract x from the result.

Cal: Hey, this is algebra!

Jinny: I know. Now square your binomial and subtract x^2.

Cal: How did you know I had an x^2? I *thought* this was rigged!

Jinny: It is. Now, divide by 4 and tell me either your constant term or your x-term. I'll tell you the other term and the number you chose.

Cal: Okay. The constant term is 16.

Jinny: Then the other term is $-4x$ and the number you chose is 4.

Cal: You're right! How did you do it?

ACTIVITY

1. Each group member should follow Jinny's instructions. Then determine how Jinny determined Cal's number and the other term.
2. Suppose that, at the end, Cal told Jinny the x-term. How would Jinny have determined Cal's number and the other term?
3. Would Jinny's "trick" work with *any* real number? Why do you think she specified numbers between 1 and 7?
4. Each group member should create a new number "trick" and perform it on the other group member. Be sure to include a variable so that both members can gain practice with polynomials.

5.3   Common Factors and Factoring by Grouping

Terms with Common Factors   •   Factoring by Grouping

Factoring is the reverse of multiplication. To **factor** an expression means to write an equivalent expression that is a product. Skill at factoring will assist us when working with polynomial functions and solving polynomial equations later in this chapter.

CONNECTING THE CONCEPTS

Despite all the equals signs in Sections 5.1 and 5.2, we have not solved any equations since Chapter 4. We have concentrated instead on writing equivalent expressions by adding, subtracting, and multiplying polynomials. We found that we could not have multiplied polynomials had we not first learned how to add or subtract them. In a similar manner, we will find that our work with factoring polynomials in Sections 5.3–5.7 relies heavily on

our ability to multiply polynomials. Not until Section 5.8 will we return to solving equations. At that point, however, we will study a new type of equation that cannot be solved without under-standing how to factor. As we have seen before, mathematics consistently builds on learned concepts. The more mastery you develop in factoring, the better prepared you will be to solve the equations of Section 5.8.

Terms with Common Factors

When factoring a polynomial, we look for factors common to every term and then use the distributive law.

EXAMPLE 1   Factor out a common factor: $6y^2 - 18$.

Solution   We have

$$6y^2 - 18 = 6 \cdot y^2 - 6 \cdot 3 \qquad \text{Noting that 6 is a common factor}$$
$$= 6(y^2 - 3). \qquad \text{Using the distributive law}$$

Check: $6(y^2 - 3) = 6y^2 - 18$.

Suppose in Example 1 that the common factor 2 were used:

$$6y^2 - 18 = 2 \cdot 3y^2 - 2 \cdot 9 \qquad \text{2 is a common factor.}$$
$$= 2(3y^2 - 9). \qquad \text{Using the distributive law}$$

Note that $3y^2 - 9$ itself has a common factor, 3. It is standard practice to factor out the *largest*, or *greatest, common factor*, so that the polynomial factor cannot be factored any further. Thus, by now factoring out 3, we can complete the factorization:

$$6y^2 - 18 = 2(3y^2 - 9)$$
$$= 2 \cdot 3(y^2 - 3) = 6(y^2 - 3). \qquad \begin{array}{l}\text{Remember to multiply the}\\\text{two common factors:}\\ 2 \cdot 3 = 6.\end{array}$$

To find the greatest common factor of a polynomial, we multiply the greatest common factor of the coefficients by the greatest common factor of the variable factors appearing in every term. Thus, to find the greatest common factor of $30x^4 + 20x^5$, we multiply the greatest common factor of 30 and 20, which is 10, by the greatest common factor of x^4 and x^5, which is x^4:

$$30x^4 + 20x^5 = 10 \cdot 3 \cdot x^4 + 10 \cdot 2 \cdot x^4 \cdot x$$
$$= 10x^4(3 + 2x). \qquad \text{The greatest common factor is } 10x^4.$$

EXAMPLE 2   Write an expression equivalent to $8p^6q^2 - 4p^5q^3 + 10p^4q^4$ by factoring out the greatest common factor.

Solution   First, we look for the greatest positive common factor in the coefficients:

$$8, \ -4, \ 10 \ \longrightarrow \text{Greatest common factor} = 2.$$

Second, we look for the greatest common factor in the powers of p:

$$p^6, \ p^5, \ p^4 \longrightarrow \text{Greatest common factor} = p^4.$$

Third, we look for the greatest common factor in the powers of q:

$$q^2, \ q^3, \ q^4 \longrightarrow \text{Greatest common factor} = q^2.$$

Thus, $2p^4q^2$ is the greatest common factor of the given polynomial. Then

$$8p^6q^2 - 4p^5q^3 + 10p^4q^4 = 2p^4q^2 \cdot 4p^2 - 2p^4q^2 \cdot 2pq + 2p^4q^2 \cdot 5q^2$$
$$= 2p^4q^2(4p^2 - 2pq + 5q^2).$$

As a final check, note that

$$2p^4q^2(4p^2 - 2pq + 5q^2) = 2p^4q^2 \cdot 4p^2 - 2p^4q^2 \cdot 2pq + 2p^4q^2 \cdot 5q^2$$
$$= 8p^6q^2 - 4p^5q^3 + 10p^4q^4,$$

which is the original polynomial. Since $4p^2 - 2pq + 5q^2$ has no common factor, we know that $2p^4q^2$ is the greatest common factor.

The polynomials in Examples 1 and 2 have been **factored completely**. They cannot be factored further. The factors in the resulting factorizations are said to be **prime polynomials**.

When the leading coefficient is a negative number, we generally factor out a common factor with a negative coefficient.

EXAMPLE 3   Write an equivalent expression by factoring out a common factor with a negative coefficient.

a) $-4x - 24$ **b)** $-2x^3 + 6x^2 - 2x$

Solution

a) $-4x - 24 = -4(x + 6)$

b) $-2x^3 + 6x^2 - 2x = -2x(x^2 - 3x + 1)$ The 1 is essential—without it, the factorization does not check.

EXAMPLE 4

Height of a thrown object.   Suppose that a baseball is thrown upward with an initial velocity of 64 ft/sec. Its height in feet, $h(t)$, after t seconds is given by

$$h(t) = -16t^2 + 64t.$$

Find an equivalent expression for $h(t)$ by factoring out a common factor.

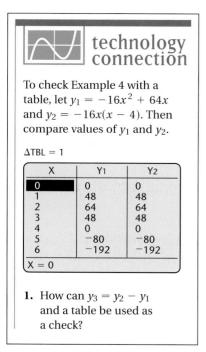

technology connection

To check Example 4 with a table, let $y_1 = -16x^2 + 64x$ and $y_2 = -16x(x - 4)$. Then compare values of y_1 and y_2.

ΔTBL = 1

X	Y1	Y2
0	0	0
1	48	48
2	64	64
3	48	48
4	0	0
5	−80	−80
6	−192	−192

X = 0

1. How can $y_3 = y_2 - y_1$ and a table be used as a check?

$h(t) = -16t^2 + 64t$

Solution   We factor out $-16t$ as follows:

$$h(t) = -16t^2 + 64t = -16t(t - 4).$$ *Check*: $-16t \cdot t = -16t^2$ and $-16t(-4) = 64t.$

Note that we can obtain function values using either expression for $h(t)$, since factoring forms equivalent expressions. For example,

$$h(1) = -16 \cdot 1^2 + 64 \cdot 1 = 48$$

and

$$h(1) = -16 \cdot 1(1 - 4) = 48.$$ Using the factorization

In Example 4, we could have evaluated $-16t^2 + 64t$ and $-16t(t - 4)$ using any value for t. The results should always match. Thus a quick partial check of any factorization is to evaluate the factorization and the original polynomial for one or two convenient replacements. The check in Example 4 becomes foolproof if three replacements are used. In general, an nth-degree factorization is correct if it checks for $n + 1$ different replacements. The proof of this useful result is beyond the scope of this text.

Factoring by Grouping

The largest common factor is sometimes a binomial.

EXAMPLE 5

Write an equivalent expression by factoring:

$$(a - b)(x + 5) + (a - b)(x - y^2).$$

Solution   Here the largest common factor is the binomial $a - b$:

$$(a - b)(x + 5) + (a - b)(x - y^2) = (a - b)[(x + 5) + (x - y^2)]$$
$$= (a - b)[2x + 5 - y^2].$$

Often, in order to identify a common binomial factor, we must regroup into two groups of two terms each.

EXAMPLE 6   Write an equivalent expression by factoring.

a) $y^3 + 3y^2 + 4y + 12$ **b)** $4x^3 - 15 + 20x^2 - 3x$

Solution

a) $y^3 + 3y^2 + 4y + 12 = (y^3 + 3y^2) + (4y + 12)$ Each grouping has a common factor.

$$= y^2(y + 3) + 4(y + 3)$$ Factoring out a common factor from each binomial

$$= (y + 3)(y^2 + 4)$$ Factoring out $y + 3$

b) When we try grouping $4x^3 - 15 + 20x^2 - 3x$ as

$$(4x^3 - 15) + (20x^2 - 3x),$$

we are unable to factor $4x^3 - 15$. When this happens, we can rearrange the polynomial and try a different grouping:

$$4x^3 - 15 + 20x^2 - 3x = 4x^3 + 20x^2 - 3x - 15$$ Using the commutative law to rearrange the terms

$$= 4x^2(x + 5) - 3(x + 5)$$ By factoring out -3 instead of 3,

$$= (x + 5)(4x^2 - 3).$$ we see that $x + 5$ is a common factor.

In Example 8 of Section 1.3 (see p. 27), we saw that

$$b - a, \quad -a + b, \quad -(a - b), \quad \text{and} \quad -1(a - b)$$

are equivalent. Remembering this can help whenever we wish to reverse the order in subtraction (see the third step below).

EXAMPLE 7   Write an equivalent expression by factoring: $ax - bx + by - ay$.

Solution   We have

$$ax - bx + by - ay = (ax - bx) + (by - ay)$$ Grouping

$$= x(a - b) + y(b - a)$$ Factoring each binomial

$$= x(a - b) + y(-1)(a - b)$$ Factoring out -1 to reverse $b - a$

$$= x(a - b) - y(a - b)$$ Simplifying

$$= (a - b)(x - y).$$ Factoring out $a - b$

Check: To check, note that $a - b$ and $x - y$ are both prime and

$$(a - b)(x - y) = ax - ay - bx + by = ax - bx + by - ay.$$

Student Notes

In Example 7, make certain that you understand why -1 or $-y$ must be factored from $by - ay$.

Some polynomials with four terms, like $x^3 + x^2 + 3x - 3$, are prime. Not only is there no common monomial factor, but no matter how we group terms, there is no common binomial factor:

$$x^3 + x^2 + 3x - 3 = x^2(x + 1) + 3(x - 1); \quad \text{No common factor}$$
$$x^3 + 3x + x^2 - 3 = x(x^2 + 3) + (x^2 - 3); \quad \text{No common factor}$$
$$x^3 - 3 + x^2 + 3x = (x^3 - 3) + x(x + 3). \quad \text{No common factor}$$

Exercise Set

5.3

FOR EXTRA HELP

Student's Solutions Manual | Digital Video Tutor CD 4 Videotape 9 | AW Math Tutor Center | MathXL Tutorials on CD | MathXL MathXL | MyMathLab MyMathLab

Concept Reinforcement   Classify each statement as either true or false.

1. The largest common factor of $10x^4 + 15x^2$ is $5x$.

2. The largest common factor of a polynomial always has the same degree as the polynomial itself.

3. It is possible for a polynomial to contain several different common factors.

4. When the leading coefficient of a polynomial is negative, we generally factor out a common factor with a negative coefficient.

5. A polynomial is not prime if it contains a common factor other than 1 or −1.

6. It is possible for a polynomial with four terms to factor into a product of two binomials.

7. The expressions $b - a$, $-(a - b)$, and $-1(a - b)$ are all equivalent.

8. The complete factorization of $12x^3 - 20x^2$ is $4x(3x^2 - 5x)$.

Write an equivalent expression by factoring out the greatest common factor.

9. $3y^2 + 6y$

10. $2t^2 + 8t$

11. $x^2 + 9x$

12. $y^2 - 5y$

13. $x^3 + 8x^2$

14. $y^3 + 9y^2$

15. $8y^2 + 4y^4$

16. $15x^2 - 5x^4$

17. $5x^2y^3 + 15x^3y^2$

18. $4x^2y - 12xy^2$

19. $5x^2 - 5x + 15$

20. $3y^2 - 3y - 9$

21. $8xy + 10xz - 14xw$

22. $6ab - 4ad + 12ac$

23. $9x^3y^6z^2 - 12x^4y^4z^4 + 15x^2y^5z^3$

24. $14a^4b^3c^5 + 21a^3b^5c^4 - 35a^4b^4c^3$

Write an equivalent expression by factoring out a factor with a negative coefficient.

25. $-5x - 40$

26. $-5x + 35$

27. $-8t + 72$

28. $-6y - 72$

29. $-2x^2 + 12x + 40$

30. $-2x^2 + 4x - 12$

31. $7x - 56y$

32. $3y - 24x$

33. $5r - 10s$

34. $7s - 14t$

35. $-p^3 - 4p^2 + 11$

36. $-x^2 + 5x - 9$

37. $-m^3 - m^2 + m - 2$

38. $-a^4 + 2a^3 - 13a$

Write an equivalent expression by factoring.

39. $a(b - 5) + c(b - 5)$

40. $r(t - 3) - s(t - 3)$

41. $(x + 7)(x - 1) + (x + 7)(x - 2)$

42. $(a + 5)(a - 2) + (a + 5)(a + 1)$

43. $a^2(x - y) + 5(y - x)$

44. $5x^2(x - 6) + 2(6 - x)$

45. $xy + xz + wy + wz$

46. $ac + ad + bc + bd$

47. $y^3 - y^2 + 3y - 3$

48. $b^3 - b^2 + 2b - 2$

49. $t^3 + 6t^2 - 2t - 12$

50. $a^3 - 3a^2 + 6 - 2a$

51. $12a^4 - 21a^3 - 9a^2$

52. $72x^3 - 36x^2 + 24x$

53. $y^4 - y^3 - y + y^2$

54. $x^6 - x^5 - x^3 + x^4$

55. $2xy - x^2y - 6 + 3x$

56. $2y^4 + 6y^2 + 5y^2 + 15$

57. *Height of a baseball.*   A baseball is popped up with an upward velocity of 72 ft/sec. Its height in feet, $h(t)$, after t seconds is given by

$$h(t) = -16t^2 + 72t.$$

   a) Find an equivalent expression for $h(t)$ by factoring out a common factor with a negative coefficient.

   b) Perform a partial check of part (a) by evaluating both expressions for $h(t)$ at $t = 1$.

58. *Height of a rocket.*   A water rocket is launched upward with an initial velocity of 96 ft/sec. Its height in feet, $h(t)$, after t seconds is given by

$$h(t) = -16t^2 + 96t.$$

   a) Find an equivalent expression for $h(t)$ by factoring out a common factor with a negative coefficient.

   b) Check your factoring by evaluating both expressions for $h(t)$ at $t = 1$.

59. *Airline routes.*   When an airline links n cities so that from any one city it is possible to fly directly to each of the other cities, the total number of direct routes is given by

$$R(n) = n^2 - n.$$

   Find an equivalent expression for $R(n)$ by factoring out a common factor.

60. *Surface area of a silo.*   A silo is a structure that is shaped like a right circular cylinder with a half sphere on top. The surface area of a silo of height h and radius r (including the area of the base) is

given by the polynomial $2\pi rh + \pi r^2$. Find an equivalent expression by factoring out a common factor.

61. *Total profit.*   After t weeks of production, Claw Foot, Inc., is making a profit of $P(t) = t^2 - 5t$ from sales of their surfboards. Find an equivalent expression by factoring out a common factor.

62. *Total profit.*   When x hundred CD players are sold, Rolics Electronics collects a profit of $P(x)$, where

$$P(x) = x^2 - 3x,$$

   and $P(x)$ is in thousands of dollars. Find an equivalent expression by factoring out a common factor.

63. *Total revenue.*   Urban Sounds is marketing a new MP3 player. The firm determines that when it sells x units, the total revenue R is given by the polynomial function

$$R(x) = 280x - 0.4x^2 \text{ dollars.}$$

   Find an equivalent expression for $R(x)$ by factoring out $0.4x$.

64. *Total cost.*   Urban Sounds determines that the total cost C of producing x MP3 players is given by the polynomial function

$$C(x) = 0.18x + 0.6x^2.$$

   Find an equivalent expression for $C(x)$ by factoring out $0.6x$.

65. *Number of diagonals.*   The number of diagonals of a polygon having n sides is given by the polynomial function

$$P(n) = \tfrac{1}{2}n^2 - \tfrac{3}{2}n.$$

   Find an equivalent expression for $P(n)$ by factoring out $\tfrac{1}{2}$.

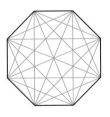

66. *Number of games in a league.*   If there are n teams in a league and each team plays every other team once, we can find the total number of games played by using the polynomial function $f(n) = \frac{1}{2}n^2 - \frac{1}{2}n$. Find an equivalent expression by factoring out $\frac{1}{2}$.

67. *High-fives.*   When a team of n players all give each other high-fives, a total of $H(n)$ hand slaps occurs, where

$$H(n) = \frac{1}{2}n^2 - \frac{1}{2}n.$$

Find an equivalent expression by factoring out $\frac{1}{2}n$.

68. *Counting spheres in a pile.*   The number N of spheres in a triangular pile like the one shown here is a polynomial function given by

$$N(x) = \frac{1}{6}x^3 + \frac{1}{2}x^2 + \frac{1}{3}x,$$

where x is the number of layers and $N(x)$ is the number of spheres. Find an equivalent expression for $N(x)$ by factoring out $\frac{1}{6}$.

69. Under what conditions would it be easier to evaluate a polynomial *after* it has been factored?

70. Explain in your own words why $-(a - b) = b - a$.

SKILL MAINTENANCE

Simplify. [1.2]

71. $2(-3) + 4(-5)$ **72.** $-7(-2) + 5(-3)$

73. $4(-6) - 3(2)$ **74.** $5(-3) + 2(-2)$

75. *Geometry.*   The perimeter of a triangle is 174. The lengths of the three sides are consecutive even numbers. What are the lengths of the sides of the triangle? [1.4]

76. *Manufacturing.*   In a factory, there are three machines A, B, and C. When all three are running, they produce 222 suitcases per day. If A and B work but C does not, they produce 159 suitcases per day. If B and C work but A does not, they produce 147 suitcases. What is the daily production of each machine? [3.5]

SYNTHESIS

77. Is it true that if a polynomial's coefficients and exponents are all prime numbers, then the polynomial itself is prime? Why or why not?

78. Following Example 4, we stated that checking the factorization of a second-degree polynomial by making a single replacement is only a *partial* check. Write an *incorrect* factorization and explain how evaluating both the polynomial and the factorization might not catch the mistake.

Complete each of the following.

79. $x^5 y^4 + \underline{\hspace{1cm}} = x^3 y\,(\underline{\hspace{1cm}} + xy^5)$

80. $a^3 b^7 - \underline{\hspace{1cm}} = \underline{\hspace{1cm}}(ab^4 - c^2)$

Write an equivalent expression by factoring.

81. $rx^2 - rx + 5r + sx^2 - sx + 5s$

82. $3a^2 + 6a + 30 + 7a^2 b + 14ab + 70b$

83. $a^4 x^4 + a^4 x^2 + 5a^4 + a^2 x^4 + a^2 x^2 + 5a^2 + 5x^4 + 5x^2 + 25$
(*Hint*: Use three groups of three.)

Write an equivalent expression by factoring out the smallest power of x in each of the following.

84. $x^{-8} + x^{-4} + x^{-6}$ **85.** $x^{-6} + x^{-9} + x^{-3}$

86. $x^{3/4} + x^{1/2} - x^{1/4}$ **87.** $x^{1/3} - 5x^{1/2} + 3x^{3/4}$

88. $x^{-3/2} + x^{-1/2}$ **89.** $x^{-5/2} + x^{-3/2}$

90. $x^{-3/4} - x^{-5/4} + x^{-1/2}$ **91.** $x^{-4/5} - x^{-7/5} + x^{-1/3}$

Write an equivalent expression by factoring. Assume that all exponents are natural numbers.

92. $2x^{3a} + 8x^a + 4x^{2a}$

93. $3a^{n+1} + 6a^n - 15a^{n+2}$

94. $4x^{a+b} + 7x^{a-b}$

95. $7y^{2a+b} - 5y^{a+b} + 3y^{a+2b}$

96. Use the TABLE feature of a graphing calculator to check your answers to Exercises 25, 35, and 41.

97. Use a graphing calculator to show that
$$(x^2 - 3x + 2)^4 = x^8 + 81x^4 + 16$$
is *not* an identity.

5.4 Factoring Trinomials

Factoring Trinomials of the Type $x^2 + bx + c$ •
Factoring Trinomials of the Type $ax^2 + bx + c, a \neq 1$

Our study of the factoring of trinomials begins with trinomials of the type $x^2 + bx + c$. We then move on to the type $ax^2 + bx + c$, where $a \neq 1$.

Factoring Trinomials of the Type $x^2 + bx + c$

When trying to factor trinomials of the type $x^2 + bx + c$, we can use a trial-and-error procedure.

Constant Term Positive

Recall the FOIL method of multiplying two binomials:

$$\begin{array}{cccc} & F & O \quad I & L \\ (x+3)(x+5) = & x^2 + & \underbrace{5x + 3x} & + 15 \\ & & \downarrow \qquad \downarrow & \downarrow \\ = & x^2 + & 8x & + 15. \end{array}$$

Because the leading coefficient in each binomial is 1, the leading coefficient in the product is also 1. To factor $x^2 + 8x + 15$, we think of FOIL: The first term, x^2, is the product of the First terms of two binomial factors, so the first term in each binomial must be x. The challenge is to find two numbers p and q such that

$$x^2 + 8x + 15 = (x + p)(x + q)$$
$$= x^2 + qx + px + pq.$$

Note that the Outer and Inner products, qx and px, can be written as $(p + q)x$. The Last product, pq, will be a constant. Thus the numbers p and q must be selected so that their product is 15 and their sum is 8. In this case, we know from above that these numbers are 3 and 5. The factorization is

$$(x + 3)(x + 5), \quad \text{or} \quad (x + 5)(x + 3). \qquad \text{Using a commutative law}$$

In general, to factor $x^2 + (p + q)x + pq$, we use FOIL in reverse:

$$x^2 + (p + q)x + pq = (x + p)(x + q).$$

EXAMPLE 1 Write an equivalent expression by factoring: $x^2 + 9x + 8$.

Solution We think of FOIL in reverse. The first term of each factor is x. We are looking for numbers p and q such that:

$$x^2 + 9x + 8 = (x + p)(x + q) = x^2 + (p + q)x + pq.$$

Thus we search for factors of 8 whose sum is 9.

Pair of Factors	Sum of Factors
2, 4	6
1, 8	9 ←

The numbers we need are 1 and 8.

The factorization is thus $(x + 1)(x + 8)$. The student should check by multiplying to confirm that the product is the original trinomial.

When factoring trinomials with a leading coefficient of 1, it suffices to consider all pairs of factors along with their sums, as we did above. At times, however, you may be tempted to form factors without calculating any sums. It is essential that you check any attempt made in this manner! For example, if we attempt the factorization

$$x^2 + 9x + 8 \stackrel{?}{=} (x + 2)(x + 4),$$

a check reveals that

$$(x + 2)(x + 4) = x^2 + 6x + 8 \neq x^2 + 9x + 8.$$

This type of trial-and-error procedure becomes easier to use with time. As you gain experience, you will find that many trials can be performed mentally.

When the constant term of a trinomial is positive, the constant terms in the binomial factors must either both be positive or both be negative. This ensures a positive product. The sign used is that of the trinomial's middle term.

EXAMPLE 2   Factor: $y^2 - 9y + 20$.

Solution   Since the constant term is positive and the coefficient of the middle term is negative, we look for a factorization of 20 in which both factors are negative. Their sum must be -9.

Pair of Factors	Sum of Factors
−1, −20	−21
−2, −10	−12
−4, −5	−9 ←

The numbers we need are −4 and −5.

Check: $(y - 4)(y - 5) = y^2 - 5y - 4y + 20 = y^2 - 9y + 20.$

The factorization is $(y - 4)(y - 5)$.

Constant Term Negative

When the constant term of a trinomial is negative, we look for one negative factor and one positive factor. The sum of the factors must still be the coefficient of the middle term.

EXAMPLE 3

Factor: $x^3 - x^2 - 30x$.

Solution   *Always* look first for a common factor! This time there is one, x. We factor it out:

$$x^3 - x^2 - 30x = x(x^2 - x - 30).$$

Now we consider $x^2 - x - 30$. We need a factorization of -30 in which one factor is positive, the other factor is negative, and the sum of the factors is -1. Since the sum is to be negative, the negative factor must have the greater absolute value. Thus we need only consider the following pairs of factors.

Pair of Factors	Sum of Factors
1, −30	−29
3, −10	−7
5, −6	−1 ←

The numbers we need are 5 and −6.

The factorization of $x^2 - x - 30$ is $(x + 5)(x - 6)$. *Don't forget to include the factor that was factored out earlier!*

Check: $x(x + 5)(x - 6) = x[x^2 - 6x + 5x - 30]$
$$= x[x^2 - x - 30]$$
$$= x^3 - x^2 - 30x.$$

The factorization is $x(x + 5)(x - 6)$.

EXAMPLE 4

technology connection

The method described in the Technology Connection on p. 296 can be used to check Example 4: Let

$y_1 = 2x^2 + 34x - 220$,
$y_2 = 2(x - 5)(x + 22)$, and
$y_3 = y_2 - y_1$.

1. How should the graphs of y_1 and y_2 compare?
2. What should the graph of y_3 look like?
3. Check Example 3 with a graphing calculator.
4. Use graphs to show that $(2x + 5)(x - 3)$ is *not* a factorization of $2x^2 + x - 15$.

Factor: $2x^2 + 34x - 220$.

Solution   *Always* look first for a common factor! This time we can factor out 2:

$$2x^2 + 34x - 220 = 2(x^2 + 17x - 110).$$

We next look for a factorization of -110 in which one factor is positive, the other factor is negative, and the sum of the factors is 17. Since the sum is to be positive, we examine only pairs of factors in which the positive term has the larger absolute value.

Pair of Factors	Sum of Factors
−1, 110	109
−2, 55	53
−5, 22	17 ←

The numbers we need are −5 and 22.

Thus, $x^2 + 17x - 110 = (x - 5)(x + 22)$. The factorization of the original trinomial, $2x^2 + 34x - 220$, is $2(x - 5)(x + 22)$. The check is left to the student.

Some polynomials are not factorable using integers.

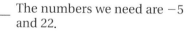

EXAMPLE 5   Factor: $x^2 - x - 7$.

Solution   There are no factors of -7 whose sum is -1. This trinomial is *not* factorable into binomials with integer coefficients. Although $x^2 - x - 7$ can be factored using more advanced techniques, for our purposes the polynomial is *prime*.

Tips for Factoring $x^2 + bx + c$

1. If necessary, rewrite the trinomial in descending order.
2. Find a pair of factors that have c as their product and b as their sum. Remember the following:

   - If c is positive, its factors will have the same sign as b.
   - If c is negative, one factor will be positive and the other will be negative. Select the factors such that the factor with the larger absolute value is the factor with the same sign as b.
   - If the sum of the two factors is the opposite of b, changing the signs of both factors will give the desired factors whose sum is b (see Example 7).

3. Check the result by multiplying the binomials.

These tips still apply when a trinomial has more than one variable.

EXAMPLE 6   Factor: $x^2 - 2xy - 48y^2$.

Solution   We look for numbers p and q such that

$$x^2 - 2xy - 48y^2 = (x + py)(x + qy).$$ The x's and y's can be written in the binomials in advance.

Our thinking is much the same as if we were factoring $x^2 - 2x - 48$. We look for factors of -48 whose sum is -2. Those factors are 6 and -8. Thus,

$$x^2 - 2xy - 48y^2 = (x + 6y)(x - 8y).$$

The check is left to the student.   ↶

Factoring Trinomials of the Type $ax^2 + bx + c$, $a \neq 1$

Now we look at trinomials in which the leading coefficient is not 1. We consider two methods. Use what works best for you or what your instructor chooses for you. Regardless of the method used, it is always wise to check your work.

Method 1: Reversing FOIL
We first consider the **FOIL method** for factoring trinomials of the type

$$ax^2 + bx + c, \quad \text{where } a \neq 1.$$

Consider the following multiplication.

$$
\begin{array}{cccc}
\text{F} & \text{O} & \text{I} & \text{L} \\
\downarrow & \downarrow & \downarrow & \downarrow
\end{array}
$$

$$(3x + 2)(4x + 5) = 12x^2 + 15x + 8x + 10$$

$$= 12x^2 + 23x + 10$$

To factor $12x^2 + 23x + 10$, we must reverse what we just did. We look for two binomials whose product is this trinomial. The product of the First terms must be $12x^2$. The product of the Outer terms plus the product of the Inner terms must be $23x$. The product of the Last terms must be 10. We know from the preceding discussion that the answer is

$$(3x + 2)(4x + 5).$$

In general, however, finding such an answer involves trial and error. We use the following method.

To Factor $ax^2 + bx + c$ by Reversing FOIL

1. Factor out the largest common factor, if one exists. Here we assume none does.

2. Find two **First** terms whose product is ax^2:

$$(\,\blacksquare x + \quad)(\,\blacksquare + \quad) = ax^2 + bx + c.$$
$$\underbrace{\qquad\qquad\qquad}_{\text{FOIL}}$$

3. Find two **Last** terms whose product is c:

$$(\quad x + \blacksquare)(\quad x + \blacksquare) = ax^2 + bx + c.$$
$$\underbrace{\qquad\qquad\qquad}_{\text{FOIL}}$$

4. Repeat steps (2) and (3) until a combination is found for which the sum of the **O**uter and **I**nner products is bx:

$$(\,\blacksquare x + \blacksquare)(\,\blacksquare x + \blacksquare) = ax^2 + bx + c.$$

EXAMPLE 7   Factor: $3x^2 + 10x - 8$.

Solution

1. First, observe that there is no common factor (other than 1 or -1).

2. Next, factor the first term, $3x^2$. The only possibility for factors is $3x \cdot x$. Thus, if a factorization exists, it must be of the form

$$(3x + \blacksquare)(x + \blacksquare).$$

We need to find the right numbers for the blanks.

3. The constant term, -8, can be factored as $(-8)(1)$, $8(-1)$, $(-2)4$, and $2(-4)$, as well as $(1)(-8)$, $(-1)8$, $4(-2)$, and $(-4)2$.

4. Find a pair of factors for which the sum of the Outer and Inner products is the middle term, $10x$. Each possibility should be checked by multiplying:

$$(3x - 8)(x + 1) = 3x^2 - 5x - 8. \qquad O + I = 3x + (-8x) = -5x$$

This gives a middle term with a negative coefficient. Since a positive coefficient is needed, a second possibility must be tried:

$$(3x + 8)(x - 1) = 3x^2 + 5x - 8. \qquad O + I = -3x + 8x = 5x$$

Note that changing the signs of the two constant terms changes only the sign of the middle term. We try again:

$$(3x - 2)(x + 4) = 3x^2 + 10x - 8. \qquad \text{This is what we wanted.}$$

Thus the desired factorization is $(3x - 2)(x + 4)$.

EXAMPLE 8

Factor: $6x^6 - 19x^5 + 10x^4$.

Solution

1. First, factor out the greatest common factor x^4:

$$x^4(6x^2 - 19x + 10).$$

2. Note that $6x^2 = 6x \cdot x$ and $6x^2 = 3x \cdot 2x$. Thus, $6x^2 - 19x + 10$ may factor into

$$(3x + \blacksquare)(2x + \blacksquare) \quad \text{or} \quad (6x + \blacksquare)(x + \blacksquare).$$

3. We factor the last term, 10. The possibilities are $10 \cdot 1$, $(-10)(-1)$, $5 \cdot 2$, and $(-5)(-2)$, as well as $1 \cdot 10$, $(-1)(-10)$, $2 \cdot 5$, and $(-2)(-5)$.

4. There are 8 possibilities for *each* factorization in step (2). We need factors for which the sum of the products (the "outer" and "inner" parts of FOIL) is the middle term, $-19x$. Since the x-coefficient is negative, we consider pairs of negative factors. Each possible factorization must be checked by multiplying:

$$(3x - 10)(2x - 1) = 6x^2 - 23x + 10.$$

We try again:

$$(3x - 5)(2x - 2) = 6x^2 - 16x + 10.$$

Actually this last attempt could have been rejected by simply noting that $2x - 2$ has a common factor, 2. Since the *largest* common factor was removed in step (1), no other common factors can exist. We try again, reversing the -5 and -2:

$$(3x - 2)(2x - 5) = 6x^2 - 19x + 10. \qquad \text{This is what we wanted.}$$

The factorization of $6x^2 - 19x + 10$ is $(3x - 2)(2x - 5)$. *But do not forget the common factor!* We must include it to get the complete factorization of the original trinomial:

$$6x^6 - 19x^5 + 10x^4 = x^4(3x - 2)(2x - 5).$$

Student Notes _____

Keep your work organized so that you can see what you have already considered. When factoring $6x^2 - 19x + 10$, we can list all possibilities and cross out those in which a common factor appears:

$(3x - 10)(2x - 1)$,

~~$(3x - 1)(2x - 10)$~~,

~~$(3x - 5)(2x - 2)$~~,

$(3x - 2)(2x - 5)$,

~~$(6x - 10)(x - 1)$~~,

$(6x - 1)(x - 10)$,

$(6x - 5)(x - 2)$,

~~$(6x - 2)(x - 5)$~~.

By being organized and not erasing, we can see that there are only four possible factorizations.

Tips for Factoring with FOIL

1. If the largest common factor has been factored out of the original trinomial, then no binomial factor can have a common factor (other than 1 or -1).
2. If a and c are both positive, then the signs in the factors will be the same as the sign of b.
3. When a possible factoring produces the opposite of the desired middle term, reverse the signs of the constants in the factors.
4. Be systematic about your trials. Keep track of those possibilities that you have tried and those that you have not.

Keep in mind that this method of factoring involves trial and error. With practice, you will find yourself making fewer and better guesses.

Method 2: The Grouping Method

The second method for factoring trinomials of the type $ax^2 + bx + c, a \neq 1$, is known as the *grouping method*. It involves not only trial and error and FOIL but also factoring by grouping. We know that

$$x^2 + 7x + 10 = x^2 + 2x + 5x + 10$$
$$= x(x + 2) + 5(x + 2)$$
$$= (x + 2)(x + 5),$$

but what if the leading coefficient is not 1? Consider $6x^2 + 23x + 20$. The method is similar to what we just did with $x^2 + 7x + 10$, but we need two more steps.* First, multiply the leading coefficient, 6, and the constant, 20, to get 120. Then find a factorization of 120 in which the sum of the factors is the coefficient of the middle term: 23. The middle term is then split into a sum or difference using these factors.

$$6x^2 + 23x + 20$$

(1) Multiply 6 and 20: $6 \cdot 20 = 120$.

(2) Factor 120: $120 = 8 \cdot 15$, and $8 + 15 = 23$.

(3) Split the middle term: $23x = 8x + 15x$.

(4) Factor by grouping.

We factor by grouping as follows:

$$6x^2 + 23x + 20 = 6x^2 + 8x + 15x + 20$$
$$= 2x(3x + 4) + 5(3x + 4)$$
$$= (3x + 4)(2x + 5).$$

Factoring by grouping

---

*The rationale behind these steps is outlined in Exercise 111.

> *To Factor $ax^2 + bx + c$ Using Grouping*
> 1. Make sure that any common factors have been factored out.
> 2. Multiply the leading coefficient a and the constant c.
> 3. Find a pair of factors, p and q, so that $pq = ac$ and $p + q = b$.
> 4. Rewrite the trinomial's middle term, bx, as $px + qx$.
> 5. Factor by grouping.

EXAMPLE 9

Factor: $3x^2 + 10x - 8$.

Solution

1. First, look for a common factor. There is none (other than 1 or -1).
2. Multiply the leading coefficient and the constant, 3 and -8:

$$3(-8) = -24.$$

3. Try to factor -24 so that the sum of the factors is 10:

$$-24 = 12(-2) \quad \text{and} \quad 12 + (-2) = 10.$$

4. Split $10x$ using the results of step (3):

$$10x = 12x - 2x.$$

5. Finally, factor by grouping:

$$3x^2 + 10x - 8 = 3x^2 + 12x - 2x - 8 \qquad \text{Substituting } 12x - 2x \text{ for } 10x$$

$$\left. \begin{aligned} &= 3x(x + 4) - 2(x + 4) \\ &= (x + 4)(3x - 2). \end{aligned} \right\} \quad \text{Factoring by grouping}$$

The check is left to the student.

EXAMPLE 10

Factor: $6x^4 - 116x^3 - 80x^2$.

Solution

1. First, factor out the greatest common factor, if any. The expression $2x^2$ is common to all three terms: $2x^2(3x^2 - 58x - 40)$.
2. To factor $3x^2 - 58x - 40$, we first multiply the leading coefficient, 3, and the constant, -40: $3(-40) = -120$.
3. Next, try to factor -120 so that the sum of the factors is -58. Since -58 is negative, the negative factor of -120 must have the larger absolute value. We see from the table at left that the factors we need are 2 and -60.
4. Split the middle term, $-58x$, using the results of step (3): $-58x = 2x - 60x$.
5. Factor by grouping:

Pairs of Factors	Sums of Factors
1, -120	-119
2, -60	-58
3, -40	-37
4, -30	-26
5, -24	-19
6, -20	-14
8, -15	-7
10, -12	-2

$$3x^2 - 58x - 40 = 3x^2 + 2x - 60x - 40 \qquad \text{Substituting } 2x - 60x \text{ for } -58x$$

$$\left. \begin{aligned} &= x(3x + 2) - 20(3x + 2) \\ &= (3x + 2)(x - 20). \end{aligned} \right\} \quad \text{Factoring by grouping}$$

The factorization of $3x^2 - 58x - 40$ is $(3x + 2)(x - 20)$. *But don't forget the common factor!* We must include it to factor the original trinomial:

$$6x^4 - 116x^3 - 80x^2 = 2x^2(3x + 2)(x - 20).$$

Check: $2x^2(3x + 2)(x - 20) = 2x^2(3x^2 - 58x - 40) = 6x^4 - 116x^3 - 80x^2.$

The complete factorization is $2x^2(3x + 2)(x - 20)$.

Exercise Set

5.4

FOR EXTRA HELP

Student's Solutions Manual   Digital Video Tutor CD 4 Videotape 9   AW Math Tutor Center   MathXL Tutorials on CD   MathXL   MyMathLab

🍂 *Concept Reinforcement*   *Classify each of the following as either true or false.*

1. Before factoring any polynomial, it is always best to look for a common factor.

2. Whenever the sum of a negative number and a positive number is negative, the negative number has the greater absolute value.

3. Whenever the product of a pair of factors is negative, the factors have the same sign.

4. If $p + q = -17$, then $-p + (-q) = 17$.

5. If a trinomial has no common factor, then none of its binomial factors can have a common factor.

6. If $(3x - 1)(2x - 5)$ is not a correct factorization of a trinomial, then $(3x - 5)(2x - 1)$ cannot possibly be a correct factorization.

7. If b and c are positive and $x^2 + bx + c$ can be factored as $(x + p)(x + q)$, then it follows that p and q are both negative.

8. If b is negative and c is positive and $x^2 + bx + c$ can be factored as $(x + p)(x + q)$, then it follows that p and q are both positive.

Factor. If a polynomial is prime, state this.

9. $x^2 + 6x + 5$

10. $x^2 + 8x + 12$

11. $y^2 + 12y + 27$

12. $t^2 + 8t + 15$

13. $t^2 - 15 - 2t$

14. $x^2 - 27 - 6x$

15. $2a^2 - 16a + 32$

16. $2n^2 - 20n + 50$

17. $x^3 + 3x^2 - 54x$

18. $a^3 - a^2 - 72a$

19. $12y + y^2 + 32$

20. $14x + x^2 + 45$

21. $p^2 - 3p - 40$

22. $y^2 + 2y - 63$

23. $a^2 - 11a + 28$

24. $t^2 - 14t + 45$

25. $x + x^2 - 6$

26. $3x + x^2 - 10$

27. $5y^2 + 40y + 35$

28. $3x^2 + 15x + 18$

29. $32 + 4y - y^2$

30. $56 + x - x^2$

31. $56x + x^2 - x^3$

32. $32y + 4y^2 - y^3$

33. $y^4 + 5y^3 - 84y^2$

34. $x^4 + 11x^3 - 80x^2$

35. $x^2 - 3x + 7$

36. $x^2 + 12x + 13$

37. $x^2 + 12xy + 27y^2$

38. $p^2 - 5pq - 24q^2$

39. $x^2 - 14xy + 49y^2$

40. $y^2 + 8yz + 16z^2$

41. $x^4 - 50x^3 + 49x^2$

42. $p^4 - 80p^3 + 79p^2$

43. $x^6 + 2x^5 - 63x^4$

44. $x^6 + 7x^5 - 18x^4$

45. $3x^2 - 16x - 12$

46. $6x^2 - 5x - 25$

47. $6x^3 - 15x - x^2$

48. $10y^3 - 12y - 7y^2$

49. $3a^2 - 10a + 8$

50. $24a^2 - 14a + 2$

51. $9a^2 + 18a + 8$

52. $35y^2 + 34y + 8$

53. $8x + 30x^2 - 6$

54. $4t + 10t^2 - 6$

55. $18x^2 - 24 - 6x$

56. $8x^2 - 16 - 28x$

57. $t^8 + 5t^7 - 14t^6$

58. $a^6 + a^5 - 6a^4$

59. $70x^4 - 68x^3 + 16x^2$

60. $14x^4 - 19x^3 - 3x^2$

61. $12a^2 - 14a - 20$

62. $12a^2 - 4a - 16$

63. $9x^2 + 15x + 4$

64. $6y^2 + 7y + 2$

Aha! **65.** $4x^2 + 15x + 9$
(*Hint:* See Exercise 63.)

66. $2y^2 + 7y + 6$

67. $-8t^2 - 8t + 30$

68. $-36a^2 + 21a - 3$

69. $18xy^3 + 3xy^2 - 10xy$

70. $3x^3y^2 - 5x^2y^2 - 2xy^2$

71. $24x^2 - 2 - 47x$

72. $15y^2 - 10 - 47y$

73. $63x^3 + 111x^2 + 36x$

74. $50y^3 + 115y^2 + 60y$

75. $48x^4 + 4x^3 - 30x^2$

76. $40y^4 + 4y^3 - 12y^2$

77. $12a^2 - 17ab + 6b^2$

78. $20p^2 - 23pq + 6q^2$

79. $2x^2 + xy - 6y^2$

80. $8m^2 - 6mn - 9n^2$

81. $6x^2 - 29xy + 28y^2$

82. $10p^2 + 7pq - 12q^2$

83. $9x^2 - 30xy + 25y^2$

84. $4p^2 + 12pq + 9q^2$

85. $9x^2y^2 + 5xy - 4$

86. $7a^2b^2 + 13ab + 6$

87. How can one conclude that $x^2 + 5x + 200$ is a prime polynomial without performing any trials?

88. How can one conclude that $x^2 - 59x + 6$ is a prime polynomial without performing any trials?

SKILL MAINTENANCE

Simplify. [1.6]

89. $(3t)^3$

90. $(5a^4)^3$

91. $(4s^3)^3$

92. $(-2x^2)^3$

93. If $g(x) = -5x^2 - 7x$, find $g(-3)$. [2.2]

94. *Height of a rocket.* A water rocket is launched upward with an initial velocity of 96 ft/sec from a height of 880 ft. Its height in feet, $h(t)$, after t seconds is given by

$$h(t) = -16t^2 + 96t + 880.$$

What is its height after 0 sec, 1 sec, 3 sec, 8 sec, and 10 sec? [2.2]

SYNTHESIS

95. Describe in your own words an approach that can be used to factor any trinomial of the form $ax^2 + bx + c$ that is not prime.

96. Suppose $(rx + p)(sx - q) = ax^2 - bx + c$ is true. Explain how this can be used to factor $ax^2 + bx + c$.

Factor. Assume that variables in exponents represent positive integers.

97. $60x^8y^6 + 35x^4y^3 + 5$

98. $x^2 - \frac{4}{25} + \frac{3}{5}x$

99. $y^2 - \frac{8}{49} + \frac{2}{7}y$

100. $y^2 + 0.4y - 0.05$

101. $4x^{2a} - 4x^a - 3$

102. $20a^3b^6 - 3a^2b^4 - 2ab^2$

103. $x^{2a} + 5x^a - 24$

104. $bdx^2 + adx + bcx + ac$

105. $2ar^2 + 4asr + as^2 - asr$

106. $a^2p^{2a} + a^2p^a - 2a^2$

Aha! **107.** $(x + 3)^2 - 2(x + 3) - 35$

108. $6(x - 7)^2 + 13(x - 7) - 5$

109. Find all integers m for which $x^2 + mx + 75$ can be factored.

110. Find all integers q for which $x^2 + qx - 32$ can be factored.

111. To better understand factoring $ax^2 + bx + c$ by grouping, suppose that

$$ax^2 + bx + c = (mx + r)(nx + s).$$

Show that if $P = ms$ and $Q = rn$, then $P + Q = b$ and $PQ = ac$.

112. One factor of $x^2 - 345x - 7300$ is $x + 20$. Find the other factor.

113. Use the TABLE feature to check your answers to Exercises 15, 57, and 99.

114. Let $y_1 = 3x^2 + 10x - 8$, $y_2 = (x + 4)(3x - 2)$, and $y_3 = y_2 - y_1$ to check Example 9 graphically.

115. Explain how the following graph of

$$y = x^2 + 3x - 2 - (x - 2)(x + 1)$$

can be used to show that

$$x^2 + 3x - 2 \neq (x - 2)(x + 1).$$

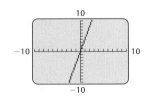

Factoring Perfect-Square Trinomials and Differences of Squares

5.5

Perfect-Square Trinomials • Differences of Squares •
More Factoring by Grouping

We now introduce a faster way to factor trinomials that are squares of binomials. A method for factoring differences of squares is also developed.

Perfect-Square Trinomials

Student Notes————

If you're not already quick to recognize that $1^2 = 1$, $2^2 = 4$, $3^2 = 9$, $4^2 = 16$, $5^2 = 25$, $6^2 = 36$, $7^2 = 49$, $8^2 = 64$, $9^2 = 81$, $10^2 = 100$, $11^2 = 121$, and $12^2 = 144$, this is a good time to familiarize yourself with these numbers.

Consider the trinomial

$$x^2 + 6x + 9.$$

To factor it, we can proceed as in Section 5.4 and look for factors of 9 that add to 6. These factors are 3 and 3 and the factorization is

$$x^2 + 6x + 9 = (x + 3)(x + 3) = (x + 3)^2.$$

Note that the result is the square of a binomial. Because of this, we call $x^2 + 6x + 9$ a **perfect-square trinomial**. Although trial and error can be used to factor a perfect-square trinomial, once recognized, a perfect-square trinomial can be quickly factored.

> **To Recognize a Perfect-Square Trinomial**
> * Two terms must be squares, such as A^2 and B^2.
> * There must be no minus sign before A^2 or B^2.
> * The remaining term must be $2AB$ or its opposite, $-2AB$.

EXAMPLE 1   Determine whether each polynomial is a perfect-square trinomial.

a) $x^2 + 10x + 25$

b) $4x + 16 + 3x^2$

c) $100y^2 + 81 - 180y$

Solution

a) • Two of the terms in $x^2 + 10x + 25$ are squares: x^2 and 25.
  • There is no minus sign before x^2 or 25.
  • The remaining term, $10x$, is twice the product of the square roots, x and 5.

   Thus, $x^2 + 10x + 25$ *is* a perfect square.

b) In $4x + 16 + 3x^2$, only one term, 16, is a square ($3x^2$ is not a square because 3 is not a perfect square; $4x$ is not a square because x is not a square).

   Thus, $4x + 16 + 3x^2$ *is not* a perfect square.

c) It can help to first write the polynomial in descending order:

$$100y^2 - 180y + 81.$$

- Two of the terms, $100y^2$ and 81, are squares.
- There is no minus sign before either $100y^2$ or 81.
- If the product of the square roots, $10y$ and 9, is doubled, we get the opposite of the remaining term: $2(10y)(9) = 180y$ (the opposite of $-180y$).

Thus, $100y^2 + 81 - 180y$ *is* a perfect-square trinomial.

To factor a perfect-square trinomial, we reuse the patterns that we learned in Section 5.2.

Factoring a Perfect-Square Trinomial
$$A^2 + 2AB + B^2 = (A + B)^2;$$
$$A^2 - 2AB + B^2 = (A - B)^2$$

EXAMPLE 2   Factor.

a) $x^2 - 10x + 25$

b) $16y^2 + 49 + 56y$

c) $-20xy + 4y^2 + 25x^2$

Solution

a) $x^2 - 10x + 25 = (x - 5)^2$ We find the square terms and write the square roots with a minus sign between them.

Note the sign!

As always, any factorization can be checked by multiplying:

$$(x - 5)^2 = (x - 5)(x - 5) = x^2 - 5x - 5x + 25 = x^2 - 10x + 25.$$

b) $16y^2 + 49 + 56y = 16y^2 + 56y + 49$ Using a commutative law

$$= (4y + 7)^2$$

We find the square terms and write the square roots with a plus sign between them.

The check is left to the student.

c) $-20xy + 4y^2 + 25x^2 = 4y^2 - 20xy + 25x^2$ Writing descending order with respect to y

$$= (2y - 5x)^2$$

This square can also be expressed as

$$25x^2 - 20xy + 4y^2 = (5x - 2y)^2.$$

The student should confirm that both factorizations check.

When factoring, always look first for a factor common to all the terms.

EXAMPLE 3   Factor: **(a)** $2x^2 - 12xy + 18y^2$; **(b)** $-4y^2 - 144y^8 + 48y^5$.

Solution

a) We first look for a common factor. This time, there is a common factor, 2.

$$2x^2 - 12xy + 18y^2 = 2(x^2 - 6xy + 9y^2) \qquad \text{Factoring out the 2}$$
$$= 2(x - 3y)^2 \qquad \text{Factoring the perfect-square trinomial}$$

The check is left to the student.

b) $-4y^2 - 144y^8 + 48y^5 = -4y^2(1 + 36y^6 - 12y^3) \qquad \text{Factoring out the common factor}$

$$= -4y^2(36y^6 - 12y^3 + 1) \qquad \begin{array}{l}\text{Changing order.}\\ \text{Note that}\\ (y^3)^2 = y^6.\end{array}$$

$$= -4y^2(6y^3 - 1)^2 \qquad \text{Factoring the perfect-square trinomial}$$

Check: $-4y^2(6y^3 - 1)^2 = -4y^2(6y^3 - 1)(6y^3 - 1)$
$$= -4y^2(36y^6 - 12y^3 + 1)$$
$$= -144y^8 + 48y^5 - 4y^2$$
$$= -4y^2 - 144y^8 + 48y^5$$

The factorization $-4y^2(6y^3 - 1)^2$ checks.

Differences of Squares

When an expression like $x^2 - 9$ is recognized as a difference of two squares, we can reverse another pattern first seen in Section 5.2.

> *Factoring a Difference of Two Squares*
> $$A^2 - B^2 = (A + B)(A - B)$$
>
> To factor a difference of two squares, write the product of the sum and the difference of the quantities being squared.

EXAMPLE 4   Factor: **(a)** $x^2 - 9$; **(b)** $25y^6 - 49x^2$.

Solution

a) $x^2 - 9 = x^2 - 3^2 = (x + 3)(x - 3)$

$$A^2 \quad - \quad B^2 \quad = (A \quad + \quad B)(A \quad - \quad B)$$

b) $25y^6 - 49x^2 = (5y^3)^2 - (7x)^2 = (5y^3 + 7x)(5y^3 - 7x)$

As always, the first step in factoring is to look for common factors.

EXAMPLE 5   Factor: **(a)** $5 - 5x^2y^6$; **(b)** $16x^4y - 81y$.

Solution

a) $5 - 5x^2y^6 = 5(1 - x^2y^6)$ Factoring out the common factor

$\qquad\qquad = 5[1^2 - (xy^3)^2]$ Rewriting x^2y^6 as a quantity squared

$\qquad\qquad = 5(1 + xy^3)(1 - xy^3)$ Factoring the difference of squares

Check: $\quad 5(1 + xy^3)(1 - xy^3) = 5(1 - xy^3 + xy^3 - x^2y^6)$

$\qquad\qquad\qquad\qquad\qquad\quad = 5(1 - x^2y^6)$

$\qquad\qquad\qquad\qquad\qquad\quad = 5 - 5x^2y^6$

The factorization $5(1 + xy^3)(1 - xy^3)$ checks.

b) $16x^4y - 81y = y(16x^4 - 81)$ Factoring out the common factor

$\qquad\qquad = y[(4x^2)^2 - 9^2]$

$\qquad\qquad = y(4x^2 + 9)(4x^2 - 9)$ Factoring the difference of squares

$\qquad\qquad = y(4x^2 + 9)(2x + 3)(2x - 3)$ Factoring $4x^2 - 9$, which is itself a difference of squares

The check is left to the student.

Note in Example 5(b) that $4x^2 - 9$ *could* be factored further. Whenever a factor itself can be factored, do so. We say that we have factored completely when none of the factors can be factored further.

More Factoring by Grouping

Sometimes, when factoring a polynomial with four terms, we may be able to factor further.

EXAMPLE 6   Factor: $x^3 + 3x^2 - 4x - 12$.

Solution

$x^3 + 3x^2 - 4x - 12 = x^2(x + 3) - 4(x + 3)$ Factoring by grouping

$\qquad\qquad\qquad\qquad = (x + 3)(x^2 - 4)$ Factoring out $x + 3$

$\qquad\qquad\qquad\qquad = (x + 3)(x + 2)(x - 2)$ Factoring $x^2 - 4$

A difference of squares can have four or more terms. For example, one of the squares may be a trinomial. In this case, a type of grouping can be used.

EXAMPLE 7   Factor: **(a)** $x^2 + 6x + 9 - y^2$; **(b)** $a^2 - b^2 + 8b - 16$.

Solution

a) $x^2 + 6x + 9 - y^2 = (x^2 + 6x + 9) - y^2$ Grouping as a perfect-square trinomial minus y^2 to show a difference of squares

$$= (x + 3)^2 - y^2$$
$$= (x + 3 + y)(x + 3 - y)$$

b) Grouping $a^2 - b^2 + 8b - 16$ into two groups of two terms does not yield a common binomial factor, so we look for a perfect-square trinomial. In this case, the perfect-square trinomial is being subtracted from a^2:

$$a^2 - b^2 + 8b - 16 = a^2 - (b^2 - 8b + 16)$$ Factoring out -1 and rewriting as subtraction

$$= a^2 - (b - 4)^2$$ Factoring the perfect-square trinomial

$$= (a + (b - 4))(a - (b - 4))$$ Factoring a difference of squares

$$= (a + b - 4)(a - b + 4).$$ Removing parentheses

Exercise Set

5.5

FOR EXTRA HELP

Student's Solutions Manual | Digital Video Tutor CD 4 Videotape 10 | AW Math Tutor Center | MathXL Tutorials on CD | MathXL | MyMathLab

---

✎ *Concept Reinforcement   Classify each of the following as a perfect-square trinomial, a difference of two squares, a polynomial having a common factor, or none of these.*

1. $9t^2 - 49$

2. $25x^2 - 20x + 4$

3. $36x^2 - 12x + 1$

4. $36a^2 - 25$

5. $4r^2 + 8r + 9$

6. $9x^2 - 12$

7. $4x^2 + 8x + 10$

8. $t^2 - 6t + 8$

9. $4t^2 + 9s^2 + 12st$

10. $9rt^2 - 5rt + 6r$

Factor completely.

11. $t^2 + 6t + 9$

12. $x^2 - 8x + 16$

13. $a^2 - 14a + 49$

14. $a^2 + 16a + 64$

15. $4a^2 - 16a + 16$

16. $2a^2 + 8a + 8$

17. $y^2 + 36 + 12y$

18. $y^2 + 36 - 12y$

19. $-18y^2 + y^3 + 81y$

20. $24a^2 + a^3 + 144a$

21. $2x^2 - 40x + 200$

22. $32x^2 + 48x + 18$

23. $1 - 8d + 16d^2$

24. $64 + 25y^2 - 80y$

25. $y^3 + 8y^2 + 16y$

26. $a^3 - 10a^2 + 25a$

27. $0.25x^2 + 0.30x + 0.09$

28. $0.04x^2 - 0.28x + 0.49$

29. $p^2 - 2pq + q^2$

30. $m^2 + 2mn + n^2$

31. $25a^2 + 30ab + 9b^2$

32. $49p^2 - 84pq + 36q^2$

33. $5a^2 - 10ab + 5b^2$

34. $4t^2 - 8tr + 4r^2$

35. $y^2 - 100$

36. $x^2 - 16$

37. $m^2 - 64$

38. $p^2 - 49$

39. $p^2q^2 - 25$

40. $a^2b^2 - 81$

41. $8x^2 - 8y^2$

42. $6x^2 - 6y^2$

43. $7xy^4 - 7xz^4$

44. $25ab^4 - 25az^4$

45. $4a^3 - 49a$

46. $9x^4 - 25x^2$

47. $3x^8 - 3y^8$

48. $9a^4 - a^2b^2$

49. $9a^4 - 25a^2b^4$

50. $16x^6 - 121x^2y^4$

51. $\frac{1}{49} - x^2$

52. $\frac{1}{16} - y^2$

53. $(a + b)^2 - 9$

54. $(p + q)^2 - 25$

55. $x^2 - 6x + 9 - y^2$

56. $a^2 - 8a + 16 - b^2$

57. $t^3 + 8t^2 - t - 8$

58. $x^3 - 7x^2 - 4x + 28$

59. $r^3 - 3r^2 - 9r + 27$

60. $t^3 + 2t^2 - 4t - 8$

61. $m^2 - 2mn + n^2 - 25$

62. $x^2 + 2xy + y^2 - 9$

63. $36 - (x + y)^2$

64. $49 - (a + b)^2$

65. $r^2 - 2r + 1 - 4s^2$

66. $c^2 + 4cd + 4d^2 - 9p^2$

Aha! **67.** $16 - a^2 - 2ab - b^2$

68. $9 - x^2 - 2xy - y^2$

69. $x^3 + 5x^2 - 4x - 20$

70. $t^3 + 6t^2 - 9t - 54$

71. $a^3 - ab^2 - 2a^2 + 2b^2$

72. $p^2q - 25q + 3p^2 - 75$

73. Are the product and power rules for exponents (see Section 1.6) important when factoring differences of squares? Why or why not?

74. Describe a procedure that could be used to find a polynomial with four terms that can be factored as a difference of two squares.

SKILL MAINTENANCE

Simplify.

75. $(2a^4b^5)^3$ [1.6]

76. $(5x^2y^4)^3$ [1.6]

77. $(x + y)^3$ [5.2]

78. $(a + 1)^3$ [5.2]

Solve.

79. $\begin{aligned} x - y + z &= 6, \\ 2x + y - z &= 0, \\ x + 2y + z &= 3 \end{aligned}$ [3.4]

80. $|5 - 7x| \geq 9$ [4.3]

81. $|5 - 7x| \leq 9$ [4.3]

82. $5 - 7x > -9 + 12x$ [4.1]

SYNTHESIS

83. Without finding the entire factorization, determine the number of factors of $x^{256} - 1$. Explain how you arrived at your answer.

84. Under what conditions can a sum of two squares be factored?

Factor completely. Assume that variables in exponents represent positive integers.

85. $-\frac{8}{27}r^2 - \frac{10}{9}rs - \frac{1}{6}s^2 + \frac{2}{3}rs$

86. $\frac{1}{36}x^8 + \frac{2}{9}x^4 + \frac{4}{9}$

87. $0.09x^8 + 0.48x^4 + 0.64$

88. $a^2 + 2ab + b^2 - c^2 + 6c - 9$

89. $r^2 - 8r - 25 - s^2 - 10s + 16$

90. $x^{2a} - y^2$

91. $x^{4a} - y^{2b}$

92. $4y^{4a} + 20y^{2a} + 20y^{2a} + 100$

93. $25y^{2a} - (x^{2b} - 2x^b + 1)$

94. $(t + 1)^2 - 4(t + 1) + 3$

95. $(a - 3)^2 - 8(a - 3) + 16$

96. $3(x + 1)^2 + 12(x + 1) + 12$

97. $m^2 + 4mn + 4n^2 + 5m + 10n$

98. $s^2 - 4st + 4t^2 + 4s - 8t + 4$

99. $5c^{100} - 80d^{100}$

100. $9x^{2n} - 6x^n + 1$

101. $c^{2w+1} + 2c^{w+1} + c$

102. If $P(x) = x^2$, use factoring to simplify

$P(a + h) - P(a)$.

103. If $P(x) = x^4$, use factoring to simplify

$P(a + h) - P(a)$.

104. *Volume of carpeting.* The volume of a carpet that is rolled up can be estimated by the polynomial $\pi R^2 h - \pi r^2 h$.

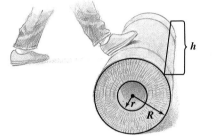

a) Factor the polynomial.
b) Use both the original and the factored forms to find the volume of a roll for which $R = 50$ cm, $r = 10$ cm, and $h = 4$ m. Use 3.14 for π.

105. Use a graphing calculator to check your answers to Exercises 11, 35, and 51 graphically by examining $y_1 = $ the original polynomial, $y_2 = $ the factored polynomial, and $y_3 = y_2 - y_1$.

106. Check your answers to Exercises 11, 35, and 51 by using tables of values (see Exercise 105).

5.6   Factoring Sums or Differences of Cubes

Formulas for Factoring Sums or Differences of Cubes •
Using the Formulas

Formulas for Factoring Sums or Differences of Cubes

We have seen that a difference of two squares can always be factored, but a *sum* of two squares is usually prime. The situation is different with cubes: The difference *or* sum of two cubes can always be factored. To see this, consider the following products:

$$(A + B)(A^2 - AB + B^2) = A(A^2 - AB + B^2) + B(A^2 - AB + B^2)$$
$$= A^3 - A^2B + AB^2 + A^2B - AB^2 + B^3$$
$$= A^3 + B^3 \qquad \text{Combining like terms}$$

and

$$(A - B)(A^2 + AB + B^2) = A(A^2 + AB + B^2) - B(A^2 + AB + B^2)$$
$$= A^3 + A^2B + AB^2 - A^2B - AB^2 - B^3$$
$$= A^3 - B^3. \qquad \text{Combining like terms}$$

These products allow us to factor a sum or difference of two cubes. Observe how the location of the $+$ and $-$ signs changes. Both formulas are worth remembering.

Factoring a Sum or Difference of Two Cubes

$A^3 + B^3 = (A + B)(A^2 - AB + B^2);$

$A^3 - B^3 = (A - B)(A^2 + AB + B^2)$

Using the Formulas

When factoring a sum or difference of cubes, it can be helpful to remember that $2^3 = 8, 3^3 = 27, 4^3 = 64, 5^3 = 125, 6^3 = 216,$ and so on. We say that 2 is the *cube root* of 8, that 3 is the cube root of 27, and so on.

EXAMPLE 1   Write an equivalent expression by factoring: $x^3 - 27.$

Solution   We first observe that

$$x^3 - 27 = x^3 - 3^3.$$

Next, in one set of parentheses, we write the first cube root, x, minus the second cube root, 3:

$$(x - 3)(\qquad).$$

To get the other factor, we think of $x - 3$ and do the following:

Square the first term: $x^2.$
Multiply the terms and then change the sign: $3x.$
Square the second term: $(-3)^2,$ or 9.

$$(x - 3)(x^2 + 3x + 9).$$

Check: $(x - 3)(x^2 + 3x + 9) = x^3 + 3x^2 + 9x - 3x^2 - 9x - 27$
$$= x^3 - 27. \qquad \text{Combining like terms}$$

Thus, $x^3 - 27 = (x - 3)(x^2 + 3x + 9).$

In Example 2, you will see that the pattern used to write the trinomial factor in Example 1 can be used when factoring a *sum* of two cubes as well.

EXAMPLE 2   Factor.

a) $125x^3 + y^3$ **b)** $m^6 + 64$

c) $128y^7 - 250x^6y$ **d)** $r^6 - s^6$

Solution

a) We have

$$125x^3 + y^3 = (5x)^3 + y^3.$$

In one set of parentheses, we write the cube root of the first term, $5x$, plus the cube root of the second term, y:

$$(5x + y)(\qquad).$$

Student Notes _____

If you think of $A^3 - B^3$ as $A^3 + (-B)^3$, it is then sufficient to remember only the pattern for factoring a sum of two cubes. Be sure to simplify your result if you do this.

To get the other factor, we think of $5x + y$ and do the following:

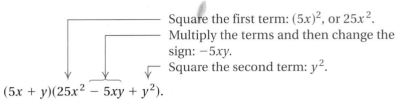

Square the first term: $(5x)^2$, or $25x^2$.
Multiply the terms and then change the sign: $-5xy$.
Square the second term: y^2.

$$(5x + y)(25x^2 - 5xy + y^2).$$

Check:

$$(5x + y)(25x^2 - 5xy + y^2) = 125x^3 - 25x^2y + 5xy^2 + 25x^2y - 5xy^2 + y^3$$

$$= 125x^3 + y^3 \qquad \text{Combining like terms}$$

Thus, $125x^3 + y^3 = (5x + y)(25x^2 - 5xy + y^2)$.

b) We have

$$m^6 + 64 = (m^2)^3 + 4^3. \qquad \text{Rewriting as quantities cubed}$$

Next, we reuse the pattern used in part (a) above:

$$A^3 + B^3 = (A + B)(A^2 - A \cdot B + B^2)$$

$$(m^2)^3 + 4^3 = (m^2 + 4)((m^2)^2 - m^2 \cdot 4 + 4^2)$$

$$= (m^2 + 4)(m^4 - 4m^2 + 16). \qquad \text{The check is left to the student.}$$

c) We have

$$128y^7 - 250x^6y = 2y(64y^6 - 125x^6) \qquad \text{Remember: } \textit{Always} \text{ look for a common factor.}$$

$$= 2y[(4y^2)^3 - (5x^2)^3]. \qquad \text{Rewriting as quantities cubed}$$

To factor $(4y^2)^3 - (5x^2)^3$, we reuse the pattern in Example 1:

$$A^3 - B^3 = (A - B)(A^2 + A \cdot B + B^2)$$

$$(4y^2)^3 - (5x^2)^3 = (4y^2 - 5x^2)((4y^2)^2 + 4y^2 \cdot 5x^2 + (5x^2)^2)$$

$$= (4y^2 - 5x^2)(16y^4 + 20x^2y^2 + 25x^4).$$

The check is left to the student. We have

$$128y^7 - 250x^6y = 2y(4y^2 - 5x^2)(16y^4 + 20x^2y^2 + 25x^4).$$

d) We have

$$r^6 - s^6 = (r^3)^2 - (s^3)^2$$

$$= (r^3 + s^3)(r^3 - s^3) \qquad \text{Factoring a difference of two } \textit{squares}$$

$$= (r + s)(r^2 - rs + s^2)(r - s)(r^2 + rs + s^2). \qquad \text{Factoring the sum and difference of two cubes}$$

To check, simply read the steps in reverse order and inspect the multiplication.

In Example 2(d), suppose we first factored $r^6 - s^6$ as a difference of two cubes:

$$(r^2)^3 - (s^2)^3 = (r^2 - s^2)(r^4 + r^2s^2 + s^4)$$
$$= (r + s)(r - s)(r^4 + r^2s^2 + s^4).$$

In this case, we might have missed some factors; $r^4 + r^2s^2 + s^4$ can be factored as $(r^2 - rs + s^2)(r^2 + rs + s^2)$, but we probably would never have suspected that such a factorization exists. Given a choice, it is generally better to factor as a difference of squares before factoring as a sum or difference of cubes.

take to last

Useful Factoring Facts

Sum of cubes: $A^3 + B^3 = (A + B)(A^2 - AB + B^2)$;

Difference of cubes: $A^3 - B^3 = (A - B)(A^2 + AB + B^2)$;

Difference of squares: $A^2 - B^2 = (A + B)(A - B)$;

There is no formula for factoring a sum of two squares.

Exercise Set

5.6

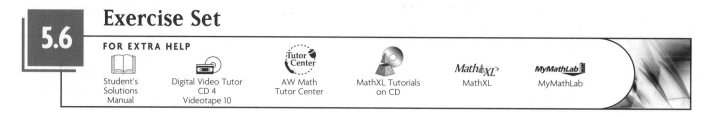

FOR EXTRA HELP

Student's Solutions Manual | Digital Video Tutor CD 4 Videotape 10 | AW Math Tutor Center | MathXL Tutorials on CD | MathXL | MyMathLab

🖐 *Concept Reinforcement   Classify each binomial as either a sum of two cubes, a difference of two cubes, a difference of two squares, both a difference of two cubes and a difference of two squares, a prime polynomial, or none of these.*

1. $8x^3 - 27$

2. $64t^3 + 27$

3. $9x^4 - 25$

4. $9x^2 + 25$

5. $1000t^3 + 1$

6. $m^3 - 8n^3$

7. $25x^2 + 8x$

8. $64t^6 - 1$

9. $s^{12} - t^6$

10. $14x^3 - 2x$

Factor completely.

11. $x^3 + 64$

12. $t^3 + 27$

13. $z^3 - 1$

14. $x^3 - 8$

15. $x^3 - 27$

16. $m^3 - 64$

17. $27x^3 + 1$

18. $8a^3 + 1$

19. $64 - 125x^3$

20. $27 - 8t^3$

21. $27y^3 + 64$

22. $8x^3 + 27$

23. $x^3 - y^3$

24. $y^3 - z^3$

25. $a^3 + \frac{1}{8}$

26. $x^3 + \frac{1}{27}$

27. $8t^3 - 8$

28. $2y^3 - 128$

29. $54x^3 + 2$

30. $8a^3 + 1000$

31. $ab^3 + 125a$

32. $rs^3 + 64r$

33. $5x^3 - 40z^3$

34. $2y^3 - 54z^3$

35. $x^3 + 0.001$

36. $y^3 + 0.125$

37. $64x^6 - 8t^6$

38. $125c^6 - 8d^6$

39. $2y^4 - 128y$

40. $3z^5 - 3z^2$

41. $z^6 - 1$

42. $t^6 + 1$

43. $t^6 + 64y^6$

44. $p^6 - q^6$

45. $x^{12} - y^3z^{12}$

46. $a^9 + b^{12}c^{15}$

47. How could you use factoring to convince someone that $x^3 + y^3 \neq (x + y)^3$?

48. Is the following statement true or false and why? If A^3 and B^3 have a common factor, then A and B have a common factor.

SKILL MAINTENANCE

49. The width of a rectangle is 7 ft less than its length. If the width is increased by 2 ft, the perimeter is then 66 ft. What is the area of the original rectangle? [3.3]

50. If $f(x) = 7 - x^2$, find $f(-3)$. [2.2]

51. Find the slope and the y-intercept of the line given by $4x - 3y = 8$. [2.3]

52. *Height of a baseball.*   A baseball is thrown upward with an initial velocity of 80 ft/sec from a 224-ft-high cliff. Its height in feet, $h(t)$, after t seconds is given by

$$h(t) = -16t^2 + 80t + 224.$$

What is the height of the ball after 0 sec, 1 sec, 3 sec, 4 sec, and 6 sec? [2.2]

Solve. [1.3]

53. $3x - 5 = 0$

54. $2x + 7 = 0$

SYNTHESIS

55. Explain how the geometric model below can be used to verify the formula for factoring $a^3 - b^3$.

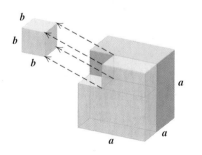

56. Explain how someone could construct a binomial that is both a difference of two cubes and a difference of two squares.

Factor.

57. $x^{6a} - y^{3b}$

58. $2x^{3a} + 16y^{3b}$

Aha! **59.** $(x + 5)^3 + (x - 5)^3$

60. $\frac{1}{16}x^{3a} + \frac{1}{2}y^{6a}z^{9b}$

61. $5x^3y^6 - \frac{5}{8}$

62. $x^3 - (x + y)^3$

63. $x^{6a} - (x^{2a} + 1)^3$

64. $(x^{2a} - 1)^3 - x^{6a}$

65. $t^4 - 8t^3 - t + 8$

66. If $P(x) = x^3$, use factoring to simplify

$$P(a + h) - P(a).$$

67. If $Q(x) = x^6$, use factoring to simplify

$$Q(a + h) - Q(a).$$

68. Using one set of axes, graph the following.
   a) $f(x) = x^3$
   b) $g(x) = x^3 - 8$
   c) $h(x) = (x - 2)^3$

69. Use a graphing calculator to check Example 1: Let $y_1 = x^3 - 27$, $y_2 = (x - 3)(x^2 + 3x + 9)$, and $y_3 = y_1 - y_2$.

70. Use the approach of Exercise 69 to check your answers to Exercises 15, 25, and 39.

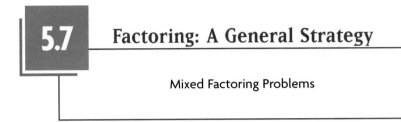

5.7 Factoring: A General Strategy

Mixed Factoring Problems

Factoring is an important algebraic skill. Once you recognize the kind of expression you have, factoring can be done without too much difficulty.

A Strategy for Factoring

A. Always factor out the greatest common factor, if possible.

B. Once the greatest common factor has been factored out, *count the number of terms* in the other factor:

Two terms: Try factoring as a difference of squares first. Next, try factoring as a sum or a difference of cubes.

Three terms: Try factoring as a perfect-square trinomial. Next, try trial and error, either reversing the FOIL method or using the grouping method.

Four or more terms: Try factoring by grouping and factoring out a common binomial factor. Next, try grouping into a difference of squares, one of which is a trinomial.

C. Always *factor completely*. If a factor with more than one term can itself be factored further, do so.

D. *Check* the factorization by multiplying.

EXAMPLE 1   Write an equivalent expression by factoring: $10a^2x - 40b^2x$.

Solution

A. Look first for a common factor:

$$10a^2x - 40b^2x = 10x(a^2 - 4b^2).$$ Factoring out the largest common factor

B. The factor $a^2 - 4b^2$ has two terms. It is a difference of squares. We factor it, keeping the common factor:

$$10a^2x - 40b^2x = 10x(a + 2b)(a - 2b).$$

C. Have we factored completely? Yes, because no factor with more than one term can be factored further.

D. *Check:* $10x(a + 2b)(a - 2b) = 10x(a^2 - 4b^2) = 10a^2x - 40b^2x$.

EXAMPLE 2   Factor: $x^6 - 64$.

Solution

A. Look for a common factor. There is none (other than 1 or -1).

B. There are two terms, a difference of squares: $(x^3)^2 - (8)^2$. We factor it:

$$x^6 - 64 = (x^3 + 8)(x^3 - 8). \qquad \text{Note that } x^6 = (x^3)^2.$$

C. One factor is a sum of two cubes, and the other factor is a difference of two cubes. We factor both:

$$x^6 - 64 = (x + 2)(x^2 - 2x + 4)(x - 2)(x^2 + 2x + 4).$$

The factorization is complete because no factor can be factored further.

D. The check is left to the student.

EXAMPLE 3   Factor: $7x^6 + 35y^2$.

Solution

A. Factor out the largest common factor:

$$7x^6 + 35y^2 = 7(x^6 + 5y^2).$$

B. The binomial $x^6 + 5y^2$ cannot be factored.

C. We cannot factor further.

D. *Check:* $7(x^6 + 5y^2) = 7x^6 + 35y^2$.

EXAMPLE 4   Factor: $2x^2 + 50a^2 - 20ax$.

Solution

A. Factor out the largest common factor: $2(x^2 + 25a^2 - 10ax)$.

B. Next, we rearrange the trinomial in descending powers of x:
$2(x^2 - 10ax + 25a^2)$. The trinomial is a perfect-square trinomial:

$$2x^2 + 50a^2 - 20ax = 2(x^2 - 10ax + 25a^2) = 2(x - 5a)^2.$$

C. No factor with more than one term can be factored further. Had we used descending powers of a, we would have discovered an equivalent factorization, $2(5a - x)^2$.

D. *Check:* $2(x - 5a)^2 = 2(x^2 - 10ax + 25a^2)$

$$= 2x^2 - 20ax + 50a^2$$
$$= 2x^2 + 50a^2 - 20ax.$$

EXAMPLE 5   Factor: $12x^2 - 40x - 32$.

Solution

A. Factor out the largest common factor: $4(3x^2 - 10x - 8)$.

B. The trinomial factor is not a square. We factor using trial and error:

$$12x^2 - 40x - 32 = 4(x - 4)(3x + 2).$$

C. We cannot factor further.

D. *Check:* $4(x - 4)(3x + 2) = 4(3x^2 + 2x - 12x - 8)$

$$= 4(3x^2 - 10x - 8)$$

$$= 12x^2 - 40x - 32.$$

EXAMPLE 6   Factor: $3x + 12 + ax^2 + 4ax$.

Solution

A. There is no common factor (other than 1 or -1).
B. There are four terms. We try grouping to find a common binomial factor:

$$3x + 12 + ax^2 + 4ax = 3(x + 4) + ax(x + 4) \quad \text{Factoring two grouped binomials}$$

$$= (x + 4)(3 + ax). \quad \text{Removing the common binomial factor}$$

C. We cannot factor further.
D. *Check:* $(x + 4)(3 + ax) = 3x + ax^2 + 12 + 4ax = 3x + 12 + ax^2 + 4ax$.

EXAMPLE 7   Factor: $y^2 - 9a^2 + 12y + 36$.

Solution

A. There is no common factor (other than 1 or -1).
B. There are four terms. We try grouping to remove a common binomial factor, but find none. Next, we try grouping as a difference of squares:

$$(y^2 + 12y + 36) - 9a^2 \quad \text{Grouping}$$

$$= (y + 6)^2 - (3a)^2 \quad \text{Rewriting as a difference of squares}$$

$$= (y + 6 + 3a)(y + 6 - 3a). \quad \text{Factoring the difference of squares}$$

C. No factor with more than one term can be factored further.
D. The check is left to the student.

EXAMPLE 8   Factor: $x^3 - xy^2 + x^2y - y^3$.

Solution

A. There is no common factor (other than 1 or -1).
B. There are four terms. We try grouping to remove a common binomial factor:

$$x^3 - xy^2 + x^2y - y^3$$

$$= x(x^2 - y^2) + y(x^2 - y^2) \quad \text{Factoring two grouped binomials}$$

$$= (x^2 - y^2)(x + y). \quad \text{Removing the common binomial factor}$$

C. The factor $x^2 - y^2$ can be factored further:

$$x^3 - xy^2 + x^2y - y^3 = (x + y)(x - y)(x + y), \text{ or } (x + y)^2(x - y).$$

No factor can be factored further, so we have factored completely.
D. The check is left to the student.

Exercise Set
5.7

FOR EXTRA HELP

Student's Solutions Manual | Digital Video Tutor CD 4 Videotape 10 | AW Math Tutor Center | MathXL Tutorials on CD | MathXL | MyMathLab

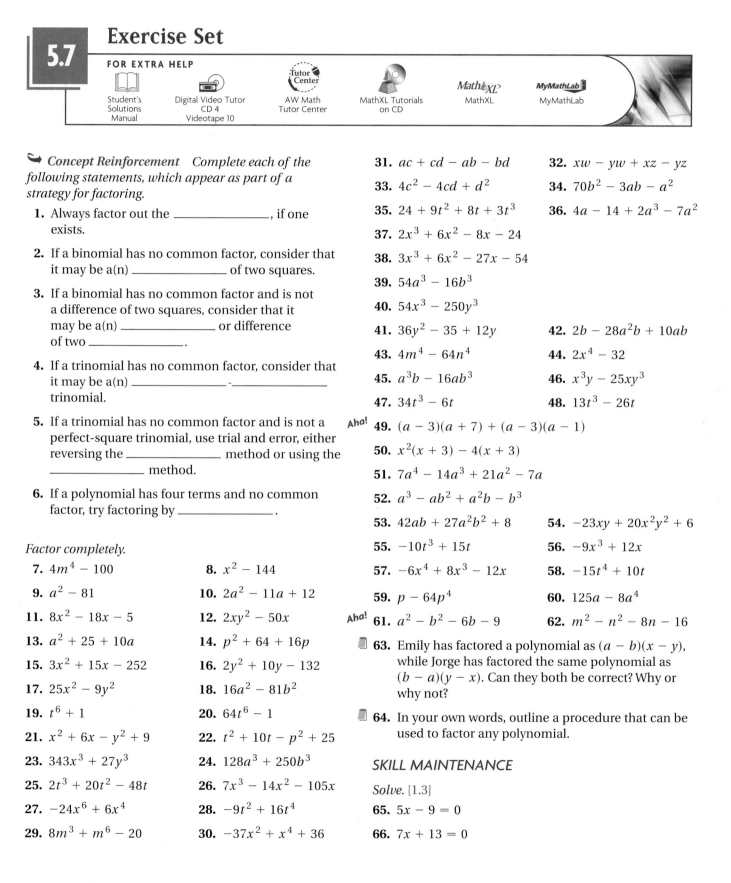

↪ *Concept Reinforcement   Complete each of the following statements, which appear as part of a strategy for factoring.*

1. Always factor out the _____, if one exists.

2. If a binomial has no common factor, consider that it may be a(n) _____ of two squares.

3. If a binomial has no common factor and is not a difference of two squares, consider that it may be a(n) _____ or difference of two _____.

4. If a trinomial has no common factor, consider that it may be a(n) _____-_____ trinomial.

5. If a trinomial has no common factor and is not a perfect-square trinomial, use trial and error, either reversing the _____ method or using the _____ method.

6. If a polynomial has four terms and no common factor, try factoring by _____.

Factor completely.

7. $4m^4 - 100$

8. $x^2 - 144$

9. $a^2 - 81$

10. $2a^2 - 11a + 12$

11. $8x^2 - 18x - 5$

12. $2xy^2 - 50x$

13. $a^2 + 25 + 10a$

14. $p^2 + 64 + 16p$

15. $3x^2 + 15x - 252$

16. $2y^2 + 10y - 132$

17. $25x^2 - 9y^2$

18. $16a^2 - 81b^2$

19. $t^6 + 1$

20. $64t^6 - 1$

21. $x^2 + 6x - y^2 + 9$

22. $t^2 + 10t - p^2 + 25$

23. $343x^3 + 27y^3$

24. $128a^3 + 250b^3$

25. $2t^3 + 20t^2 - 48t$

26. $7x^3 - 14x^2 - 105x$

27. $-24x^6 + 6x^4$

28. $-9t^2 + 16t^4$

29. $8m^3 + m^6 - 20$

30. $-37x^2 + x^4 + 36$

31. $ac + cd - ab - bd$

32. $xw - yw + xz - yz$

33. $4c^2 - 4cd + d^2$

34. $70b^2 - 3ab - a^2$

35. $24 + 9t^2 + 8t + 3t^3$

36. $4a - 14 + 2a^3 - 7a^2$

37. $2x^3 + 6x^2 - 8x - 24$

38. $3x^3 + 6x^2 - 27x - 54$

39. $54a^3 - 16b^3$

40. $54x^3 - 250y^3$

41. $36y^2 - 35 + 12y$

42. $2b - 28a^2b + 10ab$

43. $4m^4 - 64n^4$

44. $2x^4 - 32$

45. $a^3b - 16ab^3$

46. $x^3y - 25xy^3$

47. $34t^3 - 6t$

48. $13t^3 - 26t$

Aha! 49. $(a - 3)(a + 7) + (a - 3)(a - 1)$

50. $x^2(x + 3) - 4(x + 3)$

51. $7a^4 - 14a^3 + 21a^2 - 7a$

52. $a^3 - ab^2 + a^2b - b^3$

53. $42ab + 27a^2b^2 + 8$

54. $-23xy + 20x^2y^2 + 6$

55. $-10t^3 + 15t$

56. $-9x^3 + 12x$

57. $-6x^4 + 8x^3 - 12x$

58. $-15t^4 + 10t$

59. $p - 64p^4$

60. $125a - 8a^4$

Aha! 61. $a^2 - b^2 - 6b - 9$

62. $m^2 - n^2 - 8n - 16$

63. Emily has factored a polynomial as $(a - b)(x - y)$, while Jorge has factored the same polynomial as $(b - a)(y - x)$. Can they both be correct? Why or why not?

64. In your own words, outline a procedure that can be used to factor any polynomial.

SKILL MAINTENANCE

Solve. [1.3]

65. $5x - 9 = 0$

66. $7x + 13 = 0$

Graph. [2.3]

67. $g(x) = 3x - 7$

68. $f(x) = -2x + 8$

69. *Exam scores.* There are 75 questions on a college entrance examination. Two points are awarded for each correct answer, and one half point is deducted for each incorrect answer. Ralph scored 100 on the exam. How many correct and how many incorrect answers did Ralph have if all questions were answered? [3.3]

70. *Perimeter.* A pentagon with all five sides the same size has the same perimeter as an octagon in which all eight sides are the same size. One side of the pentagon is 2 less than three times the length of one side of the octagon. Find the perimeters. [3.3]

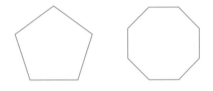

SYNTHESIS

71. Explain how one could construct a polynomial that is a difference of squares that contains a sum of two cubes and a difference of two cubes as factors.

72. Explain how one could construct a polynomial with four terms that can be factored by grouping three terms together.

Factor completely.

73. $28a^3 - 25a^2bc + 3ab^2c^2$

74. $-16 + 17(5 - y^2) - (5 - y^2)^2$

Aha! **75.** $(x - p)^2 - p^2$

76. $a^4 - 50a^2b^2 + 49b^4$

77. $(y - 1)^4 - (y - 1)^2$

78. $27x^{6s} + 64y^{3t}$

79. $x^6 - 2x^5 + x^4 - x^2 + 2x - 1$

80. $4x^2 + 4xy + y^2 - r^2 + 6rs - 9s^2$

81. $(1 - x)^3 - (x - 1)^6$

82. $24t^{2a} - 6$

83. $a^{2w+1} + 2a^{w+1} + a$

84. $\dfrac{x^{27}}{1000} - 1$

85. $a - by^8 + b - ay^8$

86. $3(x + 1)^2 - 9(x + 1) - 12$

87. $3a^2 + 3b^2 - 3c^2 - 3d^2 + 6ab - 6cd$

88. $3(a + 2)^2 + 30(a + 2) + 75$

89. $(m - 1)^3 - (m + 1)^3$

90. If $\left(x + \dfrac{2}{x}\right)^2 = 6$, find $x^3 + \dfrac{8}{x^3}$.

5.8 Applications of Polynomial Equations

The Principle of Zero Products • Polynomial Functions
and Graphs • Problem Solving

We now turn our focus to solving a new type of equation in which factoring plays an important role.

Whenever two polynomials are set equal to each other, we have a **polynomial equation**. Some examples of polynomial equations are

$$4x^3 + x^2 + 5x = 6x - 3, \qquad x^2 - x = 6, \quad \text{and} \quad 3y^4 + 2y^2 + 2 = 0.$$

The *degree of a polynomial equation* is the same as the highest degree of any term in the equation. Thus, from left to right, the degree of each equation listed above is 3, 2, and 4. A second-degree polynomial equation in one variable is usually called a **quadratic equation**. Of the equations listed above, only $x^2 - x = 6$ is a quadratic equation.

Polynomial equations, and quadratic equations in particular, occur frequently in applications, so the ability to solve them is an important skill. One way of solving certain polynomial equations involves factoring.

The Principle of Zero Products

When we multiply two or more numbers, the product is 0 if any one of those numbers (factors) is 0. Conversely, if a product is 0, then at least one of the factors must be 0. This property of 0 gives us a new principle for solving equations.

> **The Principle of Zero Products**
>
> For any real numbers a and b:
>
> If $ab = 0$, then $a = 0$ or $b = 0$. If $a = 0$ or $b = 0$, then $ab = 0$.

Thus, if $(t - 7)(2t + 5) = 0$, then $t - 7 = 0$ or $2t + 5 = 0$. To solve a quadratic equation using the principle of zero products, we first write it in *standard form*: with 0 on one side of the equation and the leading coefficient positive. We then factor and determine when each factor is 0.

EXAMPLE 1   Solve: $x^2 - x = 6$.

Solution   To apply the principle of zero products, we need 0 on one side of the equation. Thus we subtract 6 from both sides:

$$x^2 - x - 6 = 0. \qquad \text{Getting 0 on one side}$$

To express the polynomial as a product, we factor:

$$(x - 3)(x + 2) = 0. \qquad \text{Factoring}$$

The principle of zero products says that since $(x - 3)(x + 2)$ is 0, then

$$x - 3 = 0 \quad or \quad x + 2 = 0. \qquad \text{Using the principle of zero products}$$

Each of these linear equations is then solved separately:

$$x = 3 \quad or \quad x = -2.$$

We check as follows:

Check:

$$\frac{x^2 - x = 6}{\begin{array}{c|c} 3^2 - 3 & 6 \\ 9 - 3 & \\ & 6 \overset{?}{=} 6 \quad \text{TRUE} \end{array}} \qquad \frac{x^2 - x = 6}{\begin{array}{c|c} (-2)^2 - (-2) & 6 \\ 4 + 2 & \\ & 6 \overset{?}{=} 6 \quad \text{TRUE} \end{array}}$$

Both 3 and -2 are solutions. The solution set is $\{3, -2\}$.

To Use the Principle of Zero Products

1. Write an equivalent equation with 0 on one side, using the addition principle.
2. Factor the nonzero side of the equation.
3. Set each factor that is not a constant equal to 0.
4. Solve the resulting equations.

Caution! When using the principle of zero products, we must make sure that there is a 0 on one side of the equation. If neither side of the equation is 0, the procedure will not work.

To see this, consider $x^2 - x = 6$ in Example 1 as

$$x(x - 1) = 6.$$

Knowing that the product of two numbers is 6 tells us nothing about either number. The numbers may be $2 \cdot 3$ or $6 \cdot 1$ or $12 \cdot \frac{1}{2}$ or $-\frac{3}{5} \cdot (-10)$ and so on.

Suppose we *incorrectly* set each factor equal to 6:

$$x = 6 \quad or \quad x - 1 = 6 \longleftarrow \text{This is wrong!}$$
$$x = 7.$$

Neither 6 nor 7 checks, as shown below:

$x^2 - x = 6$		$x^2 - x = 6$	
$6^2 - 6$	6	$7^2 - 7$	6
$36 - 6$		$49 - 7$	
$30 \overset{?}{=} 6$	FALSE	$42 \overset{?}{=} 6$	FALSE

EXAMPLE 2   Solve: **(a)** $5b^2 = 10b$; **(b)** $x^2 - 6x + 9 = 0$.

Solution

a) We have

$$5b^2 = 10b$$
$$5b^2 - 10b = 0 \qquad \text{Getting 0 on one side}$$
$$5b(b - 2) = 0 \qquad \text{Factoring}$$
$$5b = 0 \quad or \quad b - 2 = 0 \qquad \text{Using the principle of zero products}$$
$$b = 0 \quad or \qquad b = 2. \qquad \text{The checks are left to the student.}$$

The solutions are 0 and 2. The solution set is $\{0, 2\}$.

b) We have

$$x^2 - 6x + 9 = 0$$
$$(x - 3)(x - 3) = 0 \qquad \text{Factoring}$$
$$x - 3 = 0 \quad or \quad x - 3 = 0 \qquad \text{Using the principle of zero products}$$
$$x = 3 \quad or \qquad x = 3. \qquad \textit{Check:}$$
$$3^2 - 6 \cdot 3 + 9 = 9 - 18 + 9 = 0.$$

There is only one solution, 3. The solution set is $\{3\}$.

EXAMPLE 3   Given that $f(x) = 3x^2 - 4x$, find all values of a for which $f(a) = 4$.

Solution   We want all numbers a for which $f(a) = 4$. Since $f(a) = 3a^2 - 4a$, we must have

$$3a^2 - 4a = 4 \qquad \text{Setting } f(a) \text{ equal to 4}$$
$$3a^2 - 4a - 4 = 0 \qquad \text{Getting 0 on one side}$$
$$(3a + 2)(a - 2) = 0 \qquad \text{Factoring}$$
$$3a + 2 = 0 \quad \text{ or } \quad a - 2 = 0$$
$$a = -\tfrac{2}{3} \quad \text{ or } \qquad a = 2.$$

Check: $f\left(-\tfrac{2}{3}\right) = 3\left(-\tfrac{2}{3}\right)^2 - 4\left(-\tfrac{2}{3}\right) = 3 \cdot \tfrac{4}{9} + \tfrac{8}{3} = \tfrac{4}{3} + \tfrac{8}{3} = \tfrac{12}{3} = 4.$
   $f(2) = 3(2)^2 - 4(2) = 3 \cdot 4 - 8 = 12 - 8 = 4.$

To have $f(a) = 4$, we must have $a = -\tfrac{2}{3}$ or $a = 2$.

EXAMPLE 4   Let $f(x) = 3x^3 - 30x$ and $g(x) = 9x^2$. Find all x-values for which $f(x) = g(x)$.

Solution   We substitute the polynomial expressions for $f(x)$ and $g(x)$ and solve the resulting equation:

$$f(x) = g(x)$$
$$3x^3 - 30x = 9x^2 \qquad \text{Substituting}$$
$$3x^3 - 9x^2 - 30x = 0 \qquad \begin{array}{l}\text{Getting 0 on one side and}\\ \text{writing in descending order}\end{array}$$
$$3x(x^2 - 3x - 10) = 0 \qquad \begin{array}{l}\text{Factoring out a common}\\ \text{factor}\end{array}$$
$$3x(x + 2)(x - 5) = 0 \qquad \text{Factoring the trinomial}$$
$$3x = 0 \quad \text{ or } \quad x + 2 = 0 \quad \text{ or } \quad x - 5 = 0 \qquad \begin{array}{l}\text{Using the principle of}\\ \text{zero products}\end{array}$$
$$x = 0 \quad \text{ or } \qquad x = -2 \quad \text{ or } \qquad x = 5.$$

Check: $f(0) = 3 \cdot 0^3 - 30 \cdot 0 = 3 \cdot 0 - 0 = 0$, and
   $g(0) = 9 \cdot 0^2 = 9 \cdot 0 = 0$;

   $f(-2) = 3(-2)^3 - 30(-2) = 3(-8) - (-60) = -24 + 60 = 36$, and
   $g(-2) = 9(-2)^2 = 9 \cdot 4 = 36$;

   $f(5) = 3 \cdot 5^3 - 30 \cdot 5 = 3 \cdot 125 - 150 = 375 - 150 = 225$, and
   $g(5) = 9 \cdot 5^2 = 9 \cdot 25 = 225$.

For $x = 0, -2$, and 5, we have $f(x) = g(x)$.

EXAMPLE 5   Find the domain of F if $F(x) = \dfrac{x - 2}{x^2 + 2x - 15}$.

Solution   The domain of F is the set of all values for which $F(x)$ is a real number. Since division by 0 is undefined, $F(x)$ cannot be calculated for any x-value for which the denominator, $x^2 + 2x - 15$, is 0. To make sure these values are *excluded*, we solve:

$$x^2 + 2x - 15 = 0 \qquad \text{Setting the denominator equal to 0}$$
$$(x - 3)(x + 5) = 0 \qquad \text{Factoring}$$
$$x - 3 = 0 \quad or \quad x + 5 = 0$$
$$x = 3 \quad or \qquad x = -5. \qquad \text{These are the values to } \textit{exclude.}$$

The domain of F is $\{x \mid x$ is a real number *and* $x \neq -5$ *and* $x \neq 3\}$.

technology connection

To use the INTERSECT option (see p. 119 in Section 2.4) to check Example 1, let $y_1 = x^2 - x$ and $y_2 = 6$. One intersection occurs at $(-2, 6)$. You should confirm that the other intersection occurs at $(3, 6)$.

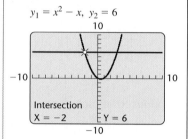

Another approach is to find where the graph of $y_3 = x^2 - x - 6$ crosses the x-axis, using the ZERO option of the CALC menu (see p. 116 in Section 2.4).

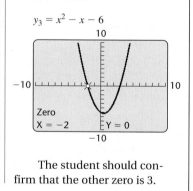

The student should confirm that the other zero is 3.

Polynomial Functions and Graphs

Let's return for a moment to the equation in Example 1, $x^2 - x = 6$. One way to begin solving this equation is to either graph by hand or use a graphing calculator to draw the graph of the function given by $f(x) = x^2 - x$. We then look for any x-value that is paired with 6, as shown on the left below. In Chapter 8, we will develop methods for quickly generating these graphs by hand.

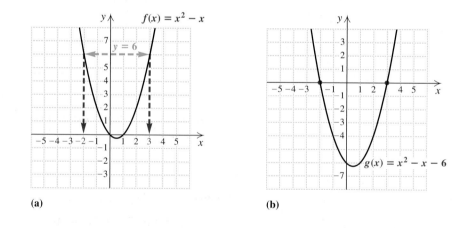

(a)   (b)

Equivalently, we could graph the function given by $g(x) = x^2 - x - 6$ and look for values of x for which $g(x) = 0$. See the figure on the right above. Here you can visualize what we call the *roots*, or *zeros*, of a polynomial function.

It appears from the graph that $f(x) = 6$ and $g(x) = 0$ when $x \approx -2$ or $x \approx 3$. Although making a graph is not the fastest or most precise method of solving this equation, it gives us a visualization and is useful with problems that are more difficult to solve algebraically. In some cases, the x-intercepts of a graph can be used to help find a factorization.

Problem Solving

Some problems can be translated to quadratic equations, which we can now solve. The problem-solving process is the same as for other kinds of problems.

EXAMPLE 6

Prize tee shirts.   During intermission at sporting events, it has become common for team mascots to use a powerful slingshot to launch tightly rolled tee shirts into the stands. The height $h(t)$, in feet, of an airborne tee shirt t seconds after being launched can be approximated by

$$h(t) = -15t^2 + 75t + 10.$$

After peaking, a rolled-up tee shirt is caught by a fan 70 ft above ground level. How long was the tee shirt in the air?

Solution

1. Familiarize.   We make a drawing and label it, using the information provided (see the figure). If we wanted to, we could evaluate $h(t)$ for a few values of t. Note that t cannot be negative, since it represents time from launch.

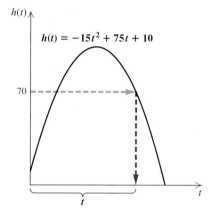

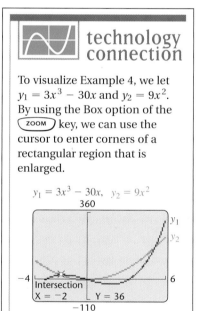

technology connection

To visualize Example 4, we let $y_1 = 3x^3 - 30x$ and $y_2 = 9x^2$. By using the Box option of the ⟨ ZOOM ⟩ key, we can use the cursor to enter corners of a rectangular region that is enlarged.

1. Use the INTERSECT option of the CALC menu (see p. 119) to confirm all three solutions of Example 4.
2. As a second check, show that $y_3 = 3x^3 - 9x^2 - 30x$ has zeros at 0, −2, and 5.

2. Translate.   The relevant function has been provided. Since we are asked to determine how long it will take for the shirt to reach someone 70 ft above ground level, we are interested in the value of t for which $h(t) = 70$:

$$-15t^2 + 75t + 10 = 70.$$

3. Carry out.   We solve the quadratic equation:

$$-15t^2 + 75t + 10 = 70$$
$$-15t^2 + 75t - 60 = 0 \qquad \text{Subtracting 70 from both sides}$$
$$\left. \begin{array}{l} -15(t^2 - 5t + 4) = 0 \\ -15(t - 4)(t - 1) = 0 \end{array} \right\} \quad \text{Factoring}$$
$$t - 4 = 0 \quad or \quad t - 1 = 0$$
$$t = 4 \quad or \qquad t = 1.$$

The solutions appear to be 4 and 1.

4. Check.   We have

$$h(4) = -15 \cdot 4^2 + 75 \cdot 4 + 10 = -240 + 300 + 10 = 70 \text{ ft};$$
$$h(1) = -15 \cdot 1^2 + 75 \cdot 1 + 10 = -15 + 75 + 10 = 70 \text{ ft}.$$

Both 4 and 1 check. However, the problem states that the tee shirt is caught after peaking. Thus we reject 1 since that would indicate when the height of the tee shirt was 70 ft on the way *up*.

5. State.   The tee shirt was in the air for 4 sec before being caught 70 ft above ground level.

The following problem involves the **Pythagorean theorem**, which relates the lengths of the sides of a right triangle. A **right triangle** has a 90°, or right, angle, which is denoted in the triangle by the symbol ⌐ or ⌐. The longest side, opposite the 90° angle, is called the **hypotenuse**. The other sides, called **legs**, form the two sides of the right angle.

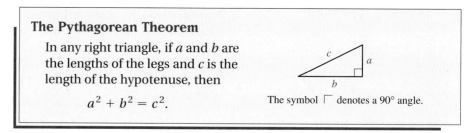

The Pythagorean Theorem

In any right triangle, if a and b are the lengths of the legs and c is the length of the hypotenuse, then

$$a^2 + b^2 = c^2.$$

The symbol ⌐ denotes a 90° angle.

EXAMPLE 7   Carpentry.   In order to build a deck at a right angle to their house, Lucinda and Felipe decide to plant a stake in the ground a precise distance from the back wall of their house. This stake will combine with two marks on the house to form a right triangle. From a course in geometry, Lucinda remembers that there are three consecutive integers that can work as sides of a right triangle. Find the measurements of that triangle.

Solution

1. Familiarize.   Recall that x, $x + 1$, and $x + 2$ can be used to represent three unknown consecutive integers. Since $x + 2$ is the largest number, it must represent the hypotenuse. The legs serve as the sides of the right angle, so one leg must be formed by the marks on the house. We make a drawing in which

$x =$ the distance between the marks on the house,

$x + 1 =$ the length of the other leg,

and

$x + 2 =$ the length of the hypotenuse.

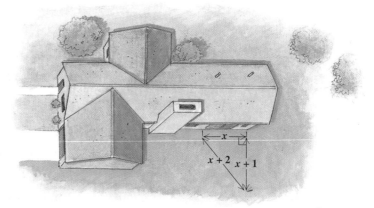

2. **Translate.**   Applying the Pythagorean theorem, we translate as follows:

$$a^2 + b^2 = c^2$$
$$x^2 + (x + 1)^2 = (x + 2)^2.$$

3. **Carry out.**   We solve the equation as follows:

$$x^2 + (x^2 + 2x + 1) = x^2 + 4x + 4 \qquad \text{Squaring the binomials}$$
$$2x^2 + 2x + 1 = x^2 + 4x + 4 \qquad \text{Combining like terms}$$
$$x^2 - 2x - 3 = 0 \qquad \text{Subtracting } x^2 + 4x + 4 \text{ from both sides}$$
$$(x - 3)(x + 1) = 0 \qquad \text{Factoring}$$
$$x - 3 = 0 \quad or \quad x + 1 = 0 \qquad \text{Using the principle of zero products}$$
$$x = 3 \quad or \qquad x = -1.$$

4. **Check.**   The integer -1 cannot be a length of a side because it is negative. For $x = 3$, we have $x + 1 = 4$, and $x + 2 = 5$. Since $3^2 + 4^2 = 5^2$, the lengths 3, 4, and 5 determine a right triangle. Thus, 3, 4, and 5 check.

5. **State.**   Lucinda and Felipe should use a triangle with sides having a ratio of $3:4:5$. Thus, if the marks on the house are 3 yd apart, they should locate the stake at the point in the yard that is precisely 4 yd from one mark and 5 yd from the other mark.

EXAMPLE 8   Display of a sports card.   A valuable sports card is 4 cm wide and 5 cm long. The card is to be sandwiched by two pieces of Lucite, each of which is $5\frac{1}{2}$ times the area of the card. Determine the dimensions of the Lucite that will ensure a uniform border around the card.

Solution

1. **Familiarize.**   We make a drawing and label it, using x to represent the width of the border, in centimeters. Since the border extends uniformly around the entire card, the length of the Lucite must be $5 + 2x$ and the width must be $4 + 2x$.

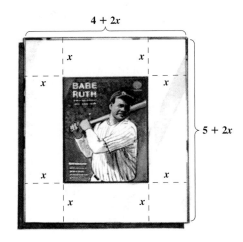

2. Translate. We rephrase the information given and translate as follows:

$$\underbrace{\text{Area of Lucite}}_{(5 + 2x)(4 + 2x)} \quad \text{is} \quad \underbrace{5\tfrac{1}{2} \text{ times}}_{= \quad 5\tfrac{1}{2}} \cdot \quad \underbrace{\text{area of card.}}_{5 \cdot 4}$$

3. Carry out. We solve the equation:

$$(5 + 2x)(4 + 2x) = 5\tfrac{1}{2} \cdot 5 \cdot 4$$

$$20 + 10x + 8x + 4x^2 = 110 \qquad \text{Multiplying}$$

$$4x^2 + 18x - 90 = 0 \qquad \text{Finding standard form}$$

$$\left.\begin{array}{l} 2(2x^2 + 9x - 45) = 0 \\ 2(2x + 15)(x - 3) = 0 \end{array}\right\} \qquad \text{Factoring}$$

$$2x + 15 = 0 \quad \text{or} \quad x - 3 = 0 \qquad \text{Principle of zero products}$$

$$x = -7\tfrac{1}{2} \quad \text{or} \qquad x = 3.$$

4. Check. We check 3 in the original problem. (Note that $-7\tfrac{1}{2}$ is not a solution because measurements cannot be negative.) If the border is 3 cm wide, the Lucite will have a length of $5 + 2 \cdot 3$, or 11 cm, and a width of $4 + 2 \cdot 3$, or 10 cm. The area of the Lucite is thus $11 \cdot 10$, or 110 cm^2. Since the area of the card is 20 cm^2 and 110 cm^2 is $5\tfrac{1}{2}$ times 20 cm^2, the number 3 checks.

5. State. Each piece of Lucite should be 11 cm long and 10 cm wide.

Exercise Set

5.8

FOR EXTRA HELP

Student's Solutions Manual | Digital Video Tutor CD 4 Videotape 10 | AW Math Tutor Center | MathXL Tutorials on CD | MathXL | MyMathLab

✎ *Concept Reinforcement* *Classify each statement as either true or false.*

1. The equations $3x^2 - 5x + 2 = 0$, $4t^2 = t + 2$, and $9n^2 = 4$ are all examples of quadratic equations.

2. Not every polynomial equation is quadratic.

3. Every quadratic equation has two solutions.

4. It is impossible for a quadratic equation to have only one solution.

5. To use the principle of zero products, we must have an equation with 0 on one side.

6. If $a \cdot b = 0$, then we know that $a = 0$ or $b = 0$.

Solve.

7. $(t - 3)(t + 4) = 0$

8. $(t + 5)(t - 1) = 0$

9. $t^2 + t - 6 = 0$

10. $t^2 + 2t - 8 = 0$

11. $5x(2x - 3) = 0$

12. $7x(4x - 5) = 0$

13. $15x^2 - 21x = 0$

14. $6x^2 - 14x = 0$

15. $(2t + 5)(t - 7) = 0$

16. $(2t - 7)(t + 2) = 0$

17. $x^2 - 7x + 10 = 0$

18. $x^2 - 6x + 8 = 0$

19. $t^2 - 10t = 0$

20. $t^2 - 8t = 0$

21. $(3x - 1)(4x - 5) = 0$

22. $(5x + 3)(2x - 7) = 0$

23. $4a^2 = 10a$

24. $6a^2 = 8a$

25. $t^2 - 6t - 16 = 0$

26. $t^2 - 3t - 18 = 0$

27. $t^2 - 3t = 28$

28. $x^2 - 4x = 45$

29. $r^2 + 16 = 8r$

30. $a^2 + 1 = 2a$

31. $y^2 + 16y + 64 = 0$

32. $x^2 + 12x + 36 = 0$

33. $8y + y^2 + 15 = 0$

34. $9x + x^2 + 20 = 0$

Aha! **35.** $r^2 - 9 = 0$

36. $t^2 - 16 = 0$

37. $x^3 - 2x^2 = 63x$

38. $a^3 - 3a^2 = 40a$

39. $r^2 = 49$

40. $t^2 = 100$

41. $(a - 4)(a + 4) = 20$

42. $(t - 6)(t + 6) = 45$

43. $-9x^2 + 15x - 4 = 0$

44. $3x^2 - 8x + 4 = 0$

45. $-8y^3 - 10y^2 - 3y = 0$

46. $-4t^3 - 11t^2 - 6t = 0$

47. $(z + 4)(z - 2) = -5$

48. $(y - 3)(y + 2) = 14$

49. $x(5 + 12x) = 28$

50. $a(1 + 21a) = 10$

51. $a^2 - \frac{1}{64} = 0$

52. $x^2 - \frac{1}{25} = 0$

53. $t^4 - 26t^2 + 25 = 0$

54. $t^4 - 13t^2 + 36 = 0$

55. Let $f(x) = x^2 + 12x + 40$. Find a such that $f(a) = 8$.

56. Let $f(x) = x^2 + 14x + 50$. Find a such that $f(a) = 5$.

57. Let $g(x) = 2x^2 + 5x$. Find a such that $g(a) = 12$.

58. Let $g(x) = 2x^2 - 15x$. Find a such that $g(a) = -7$.

59. Let $h(x) = 12x + x^2$. Find a such that $h(a) = -27$.

60. Let $h(x) = 4x - x^2$. Find a such that $h(a) = -32$.

61. If $f(x) = 12x^2 - 15x$ and $g(x) = 8x - 5$, find all x-values for which $f(x) = g(x)$.

62. If $f(x) = 10x^2 + 20$ and $g(x) = 43x - 8$, find all x-values for which $f(x) = g(x)$.

63. If $f(x) = 2x^3 - 5x$ and $g(x) = 10x - 7x^2$, find all x-values for which $f(x) = g(x)$.

64. If $f(x) = 3x^3 - 4x$ and $g(x) = 8x^2 + 12x$, find all x-values for which $f(x) = g(x)$.

Find the domain of the function f given by each of the following.

65. $f(x) = \dfrac{3}{x^2 - 4x - 5}$

66. $f(x) = \dfrac{2}{x^2 - 7x + 6}$

67. $f(x) = \dfrac{x}{6x^2 - 54}$

68. $f(x) = \dfrac{2x}{5x^2 - 20}$

69. $f(x) = \dfrac{x - 5}{9x - 18x^2}$

70. $f(x) = \dfrac{1 + x}{3x - 15x^2}$

71. $f(x) = \dfrac{7}{5x^3 - 35x^2 + 50x}$

72. $f(x) = \dfrac{3}{2x^3 - 2x^2 - 12x}$

Solve.

73. A photo is 5 cm longer than it is wide. Find the length and the width if the area is 84 cm^2.

74. An envelope is 4 cm longer than it is wide. The area is 96 cm^2. Find the length and the width.

75. *Geometry.* If each of the sides of a square is lengthened by 4 m, the area becomes 49 m^2. Find the length of a side of the original square.

76. *Geometry.* If each of the sides of a square is lengthened by 6 cm, the area becomes 144 cm^2. Find the length of a side of the original square.

77. *Framing a picture.* A picture frame measures 12 cm by 20 cm, and 84 cm^2 of picture shows. Find the width of the frame.

78. *Framing a picture.* A picture frame measures 14 cm by 20 cm, and 160 cm^2 of picture shows. Find the width of the frame.

79. *Landscaping.* A rectangular lawn measures 60 ft by 80 ft. Part of the lawn is torn up to install a sidewalk of uniform width around it. The area of the new lawn is 2400 ft^2. How wide is the sidewalk?

80. *Landscaping.* A rectangular garden is 30 ft by 40 ft. Part of the garden is removed in order to install a walkway of uniform width around it. The area of the new garden is one-half the area of the old garden. How wide is the walkway?

81. Three consecutive even integers are such that the square of the third is 76 more than the square of the second. Find the three integers.

82. Three consecutive even integers are such that the square of the first plus the square of the third is 136. Find the three integers.

83. *Tent design.* The triangular entrance to a tent is 2 ft taller than it is wide. The area of the entrance is 12 ft². Find the height and the base.

Area = 12 ft²

84. *Antenna wires.* A wire is stretched from the ground to the top of an antenna tower, as shown. The wire is 20 ft long. The height of the tower is 4 ft greater than the distance d from the tower's base to the bottom of the wire. Find the distance d and the height of the tower.

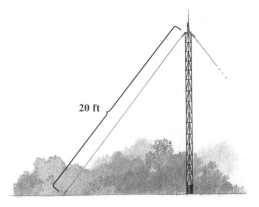

20 ft

85. *Sailing.* A triangular sail is 9 m taller than it is wide. The area is 56 m². Find the height and the base of the sail.

Area = 56 m²

86. *Ladder location.* The foot of an extension ladder is 9 ft from a wall. The height that the ladder reaches on the wall and the length

of the ladder are consecutive integers. How long is the ladder?

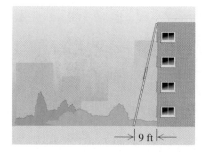

9 ft

87. *Ladder location.* The foot of an extension ladder is 10 ft from a wall. The ladder is 2 ft longer than the height that it reaches on the wall. How far up the wall does the ladder reach?

88. *Garden design.* Ignacio is planning a garden that is 25 m longer than it is wide. The garden will have an area of 7500 m². What will its dimensions be?

89. *Garden design.* A flower bed is to be 3 m longer than it is wide. The flower bed will have an area of 108 m². What will its dimensions be?

90. *Cabinet making.* Dovetail Woodworking determines that the revenue R, in thousands of dollars, from the sale of x sets of cabinets is given by $R(x) = 2x^2 + x$. If the cost C, in thousands of dollars, of producing x sets of cabinets is given by $C(x) = x^2 - 2x + 10$, how many sets must be produced and sold in order for the company to break even?

91. *Camcorder production.* Suppose that the cost of making x video cameras is $C(x) = \frac{1}{9}x^2 + 2x + 1$, where $C(x)$ is in thousands of dollars. If the revenue from the sale of x video cameras is given by $R(x) = \frac{5}{36}x^2 + 2x$, where $R(x)$ is in thousands of dollars, how many cameras must be sold in order for the firm to break even?

92. *Prize tee shirts.* Using the model in Example 6, determine how long a tee shirt has been airborne if it is caught on the way *up* by a fan 100 ft above ground level.

93. *Prize tee shirts.* Using the model in Example 6, determine how long a tee shirt has been airborne if it is caught on the way *down* by a fan 10 ft above ground level.

94. *Fireworks displays.* Fireworks are typically launched from a mortar with an upward velocity (initial speed) of about 64 ft/sec. The height $h(t)$, in feet, of a "weeping willow" display, t seconds after having been launched from an 80-ft-high rooftop, is given by

$$h(t) = -16t^2 + 64t + 80.$$

After how long will the cardboard shell from the fireworks reach the ground?

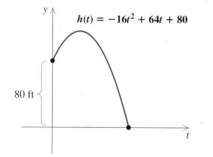

$h(t) = -16t^2 + 64t + 80$

80 ft

95. *Home-healthcare.* The number of home-healthcare patients in the United States per 10,000 people, t years after 1992, can be approximated by $N(t) = -2t^2 + 15t + 51$. How many years after 1992 were there about 69 home-healthcare patients per 10,000 people?
Source: Based on data from Centers for Disease Control, 2003

96. *Safety flares.* Suppose that a flare is launched upward with an initial velocity of 80 ft/sec from a height of 224 ft. Its height in feet, $h(t)$, after t seconds is given by

$$h(t) = -16t^2 + 80t + 224.$$

After how long will the flare reach the ground?

97. Suppose that you are given a detailed graph of $y = p(x)$, where $p(x)$ is some polynomial in x. How could the graph be used to help solve the equation $p(x) = 0$?

98. Can the number of solutions of a quadratic equation exceed two? Why or why not?

SKILL MAINTENANCE

Simplify. [1.2]

99. $\dfrac{5 - 10 \cdot 3}{-4 + 11 \cdot 4}$

100. $\dfrac{2 \cdot 3 - 5 \cdot 2}{7 - 3^2}$

101. *Driving.* At noon, two cars start from the same location going in opposite directions at different speeds. After 7 hr, they are 651 mi apart. If one car is traveling 15 mph slower than the other car, what are their respective speeds? [3.3]

102. *Television sales.* At the beginning of the month, J.C.'s Appliances had 150 televisions in stock. During the month, they sold 45% of their conventional televisions and 60% of their surround-sound televisions. If they sold a total of 78 televisions, how many of each type did they sell? [3.3]

Solve.

103. $2x - 14 + 9x > -8x + 16 + 10x$ [4.1]

104. $x + y = 0,$
$z - y = -2,$
$x - z = 6$ [3.4]

SYNTHESIS

105. Explain how one could write a quadratic equation that has -3 and 5 as solutions.

106. If the graph of $f(x) = ax^2 + bx + c$ has no x-intercepts, what can you conclude about the equation $ax^2 + bx + c = 0$?

Solve.

107. $(8x + 11)(12x^2 - 5x - 2) = 0$

108. $(x + 1)^3 = (x - 1)^3 + 26$

109. Use the following graph of $g(x) = -x^2 - 2x + 3$ to solve $-x^2 - 2x + 3 = 0$ and to solve $-x^2 - 2x + 3 \geq -5$.

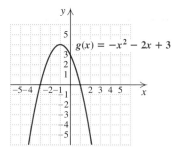

110. Find a polynomial function f for which $f(2) = 0, f(-1) = 0, f(3) = 0$, and $f(0) = 30$.

111. Find a polynomial function g for which $g(-3) = 0, g(1) = 0, g(5) = 0$, and $g(0) = 45$.

112. Use the following graph of $f(x) = x^2 - 2x - 3$ to solve $x^2 - 2x - 3 = 0$ and to solve $x^2 - 2x - 3 < 5$.

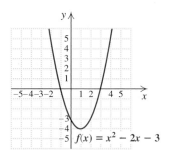

113. *Box construction.*   A rectangular piece of tin is twice as long as it is wide. Squares 2 cm on a side are cut out of each corner, and the ends are turned up to make a box whose volume is 480 cm^3. What are the dimensions of the piece of tin?

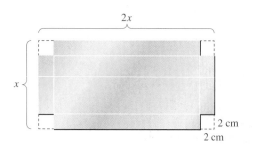

114. *Navigation.*   A tugboat and a freighter leave the same port at the same time at right angles. The freighter travels 7 km/h slower than the tugboat. After 4 hr, they are 68 km apart. Find the speed of each boat.

115. *Skydiving.*   During the first 13 sec of a jump, a skydiver falls approximately $11.12t^2$ feet in t seconds. A small heavy object (with less wind resistance) falls about $15.4t^2$ feet in t seconds. Suppose that a skydiver jumps from 30,000 ft, and 1 sec later a camera falls out of the airplane. How long will it take the camera to catch up to the skydiver?

116. Use the TABLE feature of a graphing calculator to check that -5 and 3 are not in the domain of F as shown in Example 5.

117. Use the TABLE feature of a graphing calculator to check your answers to Exercises 67, 69, and 71.

In Exercises 118–121, use a graphing calculator to find any real-number solutions that exist accurate to two decimal places.

118. $x^2 - 2x - 8 = 0$ (Check by factoring.)

119. $-x^2 + 13.80x = 47.61$

120. $-x^2 + 3.63x + 34.34 = x^2$

121. $x^3 - 3.48x^2 + x = 3.48$

122. Mary Louise is attempting to solve $x^3 + 20x^2 + 4x + 80 = 0$ with a graphing calculator. Unfortunately, when she graphs $y_1 = x^3 + 20x^2 + 4x + 80$ in a standard $[-10, 10, -10, 10]$ window, she sees no graph at all, let alone any x-intercept. Can this problem be solved graphically? If so, how? If not, why?

5 Study Summary

In many ways, **polynomials**, like $x^2 + 5x$ or $3a^4b^3 - 7ab + 9$, are to algebra as whole numbers are to arithmetic (p. 278). Polynomials can be *added, subtracted, multiplied,* and—as we will see in Chapter 6—*divided* (pp. 282–284, 290–296). Like whole numbers, many polynomials can be expressed as products. The process of writing a polynomial as a product is called **factoring**, and the polynomials used in the product are called **factors** (p. 301). In some instances, all terms in the polynomial being factored share a **common factor** that can be factored out by using the distributive law (p. 302). Other times, the polynomial being factored is a **perfect-square trinomial**, a **difference of two squares**, or a **sum or difference of two cubes** and can be factored using a recognizable pattern (pp. 319–321, 326). Often trinomials can be factored either by **trial and error**, by reversing **FOIL**, or by **grouping** (pp. 313, 316). A polynomial that cannot be factored is said to be **prime** (p. 303). In order for a polynomial to be **factored completely**, each factor in the factorization must be prime and accomplishing this often requires more than one step (p. 303). (For a general strategy for factoring, see p. 330.)

$15x^3 + 10x^2 = 5x^2(3x + 2)$ ⟵———————— $5x^2$ is the greatest common factor.

$$
\begin{aligned}
ad + ac + bc + bd &= a(d + c) + b(c + d) \longleftarrow \text{Factoring by grouping}\\
&= (d + c)(a + b)
\end{aligned}
$$

$4t^2 + 20t + 25 = (2t + 5)^2$ ⟵———————— Factoring a perfect-square trinomial

$16x^2 - 9y^2 = (4x - 3y)(4x + 3y)$ ⟵———————— Factoring a difference of two squares

$27a^3 - 8 = (3a - 2)(9a^2 + 6a + 4)$ ⟵———————— Factoring a difference of two cubes

$x^3 + 64 = (x + 4)(x^2 - 4x + 16)$ ⟵———————— Factoring a sum of two cubes

$4t^2 - 21t - 18 = (4t + 3)(t - 6)$ ⟵———————— Using trial and error (FOIL) or grouping

Certain equations can be solved by factoring and using the **principle of zero products** (p. 335). Often **quadratic equations** can be solved in this manner (p. 335):

$$
\begin{aligned}
3x^2 - 4x - 4 &= 0\\
(3x + 2)(x - 2) &= 0\\
3x + 2 = 0 \quad &or \quad x - 2 = 0\\
3x = -2 \quad &or \quad x = 2.\\
x = -\tfrac{2}{3} \quad &
\end{aligned}
$$

The solution set is $\left\{-\tfrac{2}{3}, 2\right\}$.

Quadratic equations sometimes arise from use of the **Pythagorean theorem** (p. 340):

$$a^2 + b^2 = c^2.$$

This indicates 90°.

5 Review Exercises

↘ *Concept Reinforcement   In each of Exercises 1–10, match the item with the most appropriate choice from the column on the right.*

1. ____ A polynomial with four terms [5.1]

2. ____ A difference of two squares [5.5]

3. ____ A perfect-square trinomial [5.5]

4. ____ A difference of two cubes [5.6]

5. ____ The principle of zero products [5.8]

6. ____ A quadratic equation [5.8]

7. ____ A term that is not a monomial [5.1]

8. ____ The longest side in any right triangle [5.8]

9. ____ A polynomial written in ascending order [5.1]

10. ____ A polynomial that cannot be factored [5.7]

a) $5x + 2x^2 - 4x^3$

b) $3x^{-1}$

c) If $a \cdot b = 0$, then $a = 0$ or $b = 0$.

d) Prime

e) $9 - t^2$

f) Hypotenuse

g) $8x^3 - 4x^2 + 12x + 14$

h) $27 - t^3$

i) $2x^2 - 4x = 7$

j) $4a^2 - 12a + 9$

11. Given the polynomial
$$2xy^6 - 7x^8y^3 + 2x^3 + 9,$$
determine the degree of each term and the degree of the polynomial. [5.1]

12. Given the polynomial
$$3x - 5x^3 + 2x^2 + 9,$$
arrange in descending order and determine the leading term and the leading coefficient. [5.1]

13. Arrange in ascending powers of x:
$$8x^6y - 7x^8y^3 + 2x^3 - 3x^2.\ [5.1]$$

14. Find $P(0)$ and $P(-1)$:
$$P(x) = x^3 - x^2 + 4x.\ [5.1]$$

15. Evaluate the polynomial function for $x = -2$:
$$P(x) = 4 - 2x - x^2.\ [5.1]$$

Combine like terms. [5.1]

16. $6 - 4a + a^2 - 2a^3 - 10 + a$

17. $4x^2y - 3xy^2 - 5x^2y + xy^2$

Add. [5.1]

18. $(-7x^3 - 4x^2 + 3x + 2) + (5x^3 + 2x + 6x^2 + 1)$

19. $(3x^4 + 3x^3 - 8x + 9) + (-6x^4 + 4x + 7 + 3x)$

20. $(-9xy^2 - xy + 6x^2y) + (-5x^2y - xy + 4xy^2)$

Subtract. [5.1]

21. $(8x - 5) - (-6x + 2)$

22. $(4a - b + 3c) - (6a - 7b - 4c)$

23. $(8x^2 - 4xy + y^2) - (2x^2 + 3xy - 2y^2)$

Simplify as indicated. [5.2]

24. $(3x^2y)(-6xy^3)$

25. $(x^4 - 2x^2 + 3)(x^4 + x^2 - 1)$

26. $(4ab + 3c)(2ab - c)$

27. $(2x + 5y)(2x - 5y)$

28. $(3x - 4y)^2$

29. $(x + 3)(2x - 1)$

30. $(x^2 + 4y^3)^2$

31. $(3t - 5)^2 - (2t + 3)^2$

32. $\left(x - \frac{1}{3}\right)\left(x - \frac{1}{6}\right)$

Factor.

33. $7x^2 + 6x$ [5.3]

34. $9y^4 - 3y^2$ [5.3]

35. $15x^4 - 18x^3 + 21x^2 - 9x$ [5.3]

36. $a^2 - 12a + 27$ [5.4]

37. $3m^2 + 14m + 8$ [5.4]

38. $25x^2 + 20x + 4$ [5.5]

39. $4y^2 - 16$ [5.5]

40. $5x^2 + x^3 - 14x$ [5.4]

41. $ax + 2bx - ay - 2by$ [5.3]

42. $3y^3 + 6y^2 - 5y - 10$ [5.3]

43. $a^4 - 81$ [5.5]

44. $4x^4 + 4x^2 + 20$ [5.3]

45. $27x^3 - 8$ [5.6]

46. $0.064b^3 - 0.125c^3$ [5.6]

47. $y^5 + y$ [5.3]

48. $2z^8 - 16z^6$ [5.3]

49. $54x^6y - 2y$ [5.6]

50. $36x^2 - 120x + 100$ [5.5]

51. $6t^2 + 17pt + 5p^2$ [5.4]

52. $x^3 + 2x^2 - 9x - 18$ [5.5]

53. $a^2 - 2ab + b^2 - 4t^2$ [5.5]

Solve. [5.8]

54. $x^2 - 16x = -64$

55. $6b^2 - 13b + 6 = 0$

56. $8y^2 = 14y$

57. $r^2 = 16$

58. $a^3 = 4a^2 + 21a$

59. $(y - 1)(y - 4) = 10$

60. Let $f(x) = x^2 - 7x - 40$. Find a such that $f(a) = 4$. [5.8]

61. Find the domain of the function f given by

$$f(x) = \frac{x - 5}{3x^2 + 19x - 14}. \ [5.8]$$

62. The area of a square is 5 more than four times the length of a side. What is the length of a side of the square? [5.8]

63. The sum of the squares of three consecutive odd numbers is 83. Find the numbers. [5.8]

64. A photograph is 3 in. longer than it is wide. When a 2-in. border is placed around the photograph, the total area of the photograph and the border is 108 in². Find the dimensions of the photograph. [5.8]

65. Tim is designing a rectangular garden with a width of 8 ft. The path that leads diagonally across the garden is 2 ft longer than the length of the garden. How long is the path? [5.8]

8 ft

SYNTHESIS

66. Explain how to find the roots of a polynomial function from its graph. [5.8]

67. Explain in your own words why there must be a 0 on one side of an equation before you can use the principle of zero products. [5.8]

Factor. [5.6]

68. $128x^6 - 2y^6$

69. $(x - 1)^3 - (x + 1)^3$

Multiply.

70. $[a - (b - 1)][(b - 1)^2 + a(b - 1) + a^2]$ [5.2], [5.6]

71. $(z^{n^2})^{n^3}(z^{4n^3})^{n^2}$ [5.2]

72. Solve: $(x + 1)^3 = x^2(x + 1)$. [5.8]

5 Chapter Test

Given the polynomial $8xy^3 - 14x^2y + 5x^5y^4 - 9x^4y$.

1. Determine the degree of the polynomial.

2. Arrange in descending powers of x.

3. Determine the leading term of the polynomial $7a - 12 + a^2 - 5a^3$.

4. Given $P(x) = 2x^3 + 3x^2 - x + 4$, find $P(0)$ and $P(-2)$.

5. Given $P(x) = x^2 - 3x$, find and simplify
$$P(a + h) - P(a).$$

6. Combine like terms:
$$6xy - 2xy^2 - 2xy + 5xy^2.$$

Add.

7. $(-6x^3 + 3x^2 - 4y) + (3x^3 - 2y - 7y^2)$

8. $(5m^3 - 4m^2n - 6mn^2 - 3n^3) + (9mn^2 - 4n^3 + 2m^3 + 6m^2n)$

Subtract.

9. $(8a - 4b) - (3a + 4b)$

10. $(9y^2 - 2y - 5y^3) - (4y^2 - 7y - 6y^3)$

Multiply.

11. $(-4x^2y)(-16xy^2)$

12. $(6a - 5b)(2a + b)$

13. $(x - y)(x^2 - xy - y^2)$

14. $(2x^3 - 4)^2$

15. $(5t - 9)^2$

16. $(x - 2y)(x + 2y)$

Factor.

17. $15x^2 - 5x^4$

18. $y^3 + 5y^2 - 4y - 20$

19. $p^2 - 12p - 28$

20. $12m^2 + 20m + 3$

21. $9y^2 - 25$

22. $3r^3 - 3$

23. $9x^2 + 25 - 30x$

24. $x^8 - y^8$

25. $y^2 + 8y + 16 - 100t^2$

26. $20a^2 - 5b^2$

27. $24x^2 - 46x + 10$

28. $16a^7b + 54ab^7$

29. $4y^4x + 36yx^2 + 8y^2x^3 - 16xy$

Solve.

30. $x^2 - 30 = 7x$

31. $5y^2 = 125$

32. $2x^3 + 21x = -17x^2$

33. $9x^2 + 3x = 0$

34. Let $f(x) = 3x^2 - 15x + 11$. Find a such that $f(a) = 11$.

35. Find the domain of the function f given by
$$f(x) = \frac{8 - x}{x^2 + 2x + 1}.$$

36. A photograph is 3 cm longer than it is wide. Its area is 40 cm^2. Find its length and its width.

37. To celebrate July 4th, fireworks are launched over a lake from a dam 36 ft above the water. The height of a display, t seconds after it has been launched, is given by
$$h(t) = -16t^2 + 64t + 36.$$

How long will it take for a "dud" firework to reach the water?

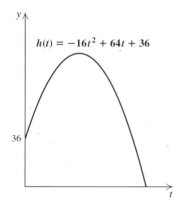

SYNTHESIS

38. a) Multiply: $(x^2 + x + 1)(x^3 - x^2 + 1)$.
   b) Factor: $x^5 + x + 1$.

39. Factor: $6x^{2n} - 7x^n - 20$.

6

Rational Expressions, Equations, and Functions

AN APPLICATION

The ultraviolet, or UV, index is a measure issued daily by the National Weather Service that indicates the strength of the sun's rays in a particular locale. For those people whose skin is quite sensitive, a UV rating of 6 will cause sunburn after 10 min (*Source: The Electronic Textbook of Dermatology* found at www.telemedicine.org, January 2004). Given that the number of minutes it takes to burn, *t*, varies inversely as the UV rating *u*, how long will it take a highly sensitive person to burn on a day with a UV rating of 4?

This problem appears as Example 7 in Section 6.8.

Amanda Rainwater
DERMATOLOGIST
Scottsdale, Arizona

Math is an integral part of my practice. I use it to calculate children's prescriptions according to their weight and to formulate solutions for injections. I also need math, of course, to figure costs.

A rational expression is an expression, similar to the fractions in arithmetic, that indicates division. In this chapter, we add, subtract, multiply, and divide rational expressions, and use them in equations and functions. We then use rational expressions to solve problems that we could not have solved before.

6.1 Rational Expressions and Functions: Multiplying and Dividing

Rational Functions • Multiplying • Simplifying Rational Expressions and Functions • Dividing and Simplifying

An expression that consists of a polynomial divided by a nonzero polynomial is called a **rational expression**. The following are examples of rational expressions:

$$\frac{3}{4}, \quad \frac{x}{y}, \quad \frac{9}{a+b}, \quad \frac{x^2 + 7xy - 4}{x^3 - y^3}, \quad \frac{1 + z^3}{1 - z^6}.$$

Rational Functions

Like polynomials, certain rational expressions are used to describe functions. Such functions are called **rational functions**.

EXAMPLE 1   The function given by

$$H(t) = \frac{t^2 + 5t}{2t + 5}$$

gives the time, in hours, for two machines, working together, to complete a job that the first machine could do alone in t hours and the other machine could do in $t + 5$ hours. How long will the two machines, working together, require for the job if the first machine alone would take **(a)** 1 hour? **(b)** 6 hours?

Solution

a) $H(1) = \dfrac{1^2 + 5 \cdot 1}{2 \cdot 1 + 5} = \dfrac{1 + 5}{2 + 5} = \dfrac{6}{7} \text{hr}$

b) $H(6) = \dfrac{6^2 + 5 \cdot 6}{2 \cdot 6 + 5} = \dfrac{36 + 30}{12 + 5} = \dfrac{66}{17} \text{ or } 3\dfrac{15}{17} \text{hr}$

Study Skills _____

Try an Exercise Break

Often the best way to regain your energy or focus is to take a break from your studies in order to exercise. Jogging, biking, or walking briskly are just a few of the ways in which you can improve your concentration for when you return to your studies.

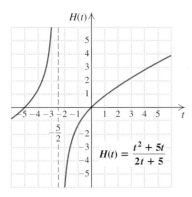

$H(t) = \dfrac{t^2 + 5t}{2t + 5}$

In Section 5.8, we found that the domain of a rational function must exclude any numbers for which the denominator is 0. For a function like H above, the denominator is 0 when t is $-\frac{5}{2}$, so the domain of H is $\left(-\infty, -\frac{5}{2}\right) \cup \left(-\frac{5}{2}, \infty\right)$.

Although graphing rational functions is beyond the scope of this course, it is educational to examine a computer-generated graph of the above function. Note that the graph consists of two unconnected "branches." Since $-\frac{5}{2}$ is not in the domain of H, a vertical line drawn at $-\frac{5}{2}$ does not touch the graph of H.

Multiplying

The calculations that are performed with rational expressions resemble those performed in arithmetic.

Products of Rational Expressions

To multiply two rational expressions, multiply numerators and multiply denominators:

$$\frac{A}{B} \cdot \frac{C}{D} = \frac{AC}{BD}, \quad \text{where } B \neq 0, D \neq 0.$$

EXAMPLE 2   Multiply: $\dfrac{x + 1}{y - 3} \cdot \dfrac{x^2}{y + 1}$.

Solution

$$\frac{x + 1}{y - 3} \cdot \frac{x^2}{y + 1} = \frac{(x + 1)x^2}{(y - 3)(y + 1)} \qquad \text{Multiplying the numerators and multiplying the denominators}$$

Recall from arithmetic that multiplication by 1 can be used to find equivalent expressions:

$$\frac{3}{5} = \frac{3}{5} \cdot \frac{2}{2} \qquad \text{Multiplying by } \tfrac{2}{2}, \text{ which is } 1$$

$$= \frac{6}{10}. \qquad \tfrac{3}{5} \text{ and } \tfrac{6}{10} \text{ represent the same number.}$$

Similarly, multiplication by 1 can be used to find equivalent rational expressions:

$$\left.\begin{array}{l} \dfrac{x - 5}{x + 2} = \dfrac{x - 5}{x + 2} \cdot \dfrac{x + 3}{x + 3} \\[2mm] \qquad = \dfrac{(x - 5)(x + 3)}{(x + 2)(x + 3)}. \end{array}\right\} \quad \begin{array}{l} \text{Multiplying by } \dfrac{x + 3}{x + 3}, \text{ which, provided} \\[2mm] x \neq -3, \text{ is } 1 \end{array}$$

The expressions

$$\frac{x - 5}{x + 2} \quad \text{and} \quad \frac{(x - 5)(x + 3)}{(x + 2)(x + 3)}$$

are equivalent: So long as x is replaced with a number other than -2 or -3, both expressions represent the same number. For example, if $x = 4$, then

$$\frac{x - 5}{x + 2} = \frac{4 - 5}{4 + 2} = \frac{-1}{6}$$

and

$$\frac{(x - 5)(x + 3)}{(x + 2)(x + 3)} = \frac{(4 - 5)(4 + 3)}{(4 + 2)(4 + 3)} = \frac{-1 \cdot 7}{6 \cdot 7} = \frac{-7}{42} = \frac{-1}{6}.$$

Simplifying Rational Expressions and Functions

As in arithmetic, rational expressions are *simplified* by "removing" a factor equal to 1. This reverses the process shown above:

$$\frac{6}{10} = \frac{3 \cdot 2}{5 \cdot 2} = \frac{3}{5} \cdot \frac{2}{2} = \frac{3}{5}. \qquad \text{We "removed" the factor that equals 1:}\ \frac{2}{2} = 1.$$

Similarly,

$$\frac{(x - 5)(x + 3)}{(x + 2)(x + 3)} = \frac{x - 5}{x + 2} \cdot \frac{x + 3}{x + 3} = \frac{x - 5}{x + 2}. \qquad \text{We "removed" the factor that equals 1:}\ \frac{x + 3}{x + 3} = 1.$$

Because rational expressions often appear when we are writing functions, it is important that the function's domain not be changed as a result of simplifying. For example, the domain of the function given by

$$F(x) = \frac{(x - 5)(x + 3)}{(x + 2)(x + 3)}$$

is assumed to be all real numbers for which the denominator is nonzero. Thus,

Domain of $F = \{x \,|\, x \neq -2, x \neq -3\}$.

Above, we wrote $\dfrac{(x - 5)(x + 3)}{(x + 2)(x + 3)}$ in simplified form as $\dfrac{x - 5}{x + 2}$. There is a serious problem with stating that these are equivalent. The difficulty arises from the fact that, unless we specify otherwise, the domain of the function given by

$$G(x) = \frac{x - 5}{x + 2} \qquad \text{is assumed to be}\ \{x \,|\, x \neq -2\}.$$

Thus, as presently written, the domain of G includes -3, but the domain of F does not. This problem is easily addressed by specifying

$$\frac{(x - 5)(x + 3)}{(x + 2)(x + 3)} = \frac{x - 5}{x + 2} \qquad \text{with } x \neq -3.$$

EXAMPLE 3 Write the function given by

$$f(t) = \frac{7t^2 + 21t}{14t}$$

in simplified form.

Solution We first factor the numerator and the denominator, looking for the largest factor common to both. Once the greatest common factor is found, we use it to write 1 and simplify:

$$f(t) = \frac{7t^2 + 21t}{14t}$$ Note that the domain of $f = \{t \mid t \neq 0\}$.

$$= \frac{7t(t + 3)}{7 \cdot 2 \cdot t}$$ Factoring. The greatest common factor is $7t$.

$$= \frac{7t}{7t} \cdot \frac{t + 3}{2}$$ Rewriting as a product of two rational expressions

$$= 1 \cdot \frac{t + 3}{2}$$ For $t \neq 0$, we have $\frac{7t}{7t} = 1$; try to do this step mentally.

$$= \frac{t + 3}{2}, \ t \neq 0.$$ Removing the factor 1. To keep the same domain, we specify that $t \neq 0$.

Thus simplified form is $f(t) = \dfrac{t + 3}{2}$, *with $t \neq 0$.*

A rational expression is said to be **simplified** when no factors equal to 1 can be removed. Often, this requires two or more steps. For example, suppose we remove $7/7$ instead of $(7t)/(7t)$ in Example 3. We would then have

$$\left. \begin{aligned} \frac{7t^2 + 21t}{14t} &= \frac{7(t^2 + 3t)}{7 \cdot 2t} \\ &= \frac{t^2 + 3t}{2t}. \end{aligned} \right\}$$ Removing a factor equal to 1: $\frac{7}{7} = 1$. Note that $t \neq 0$.

Here, since another common factor remains, we need to simplify further:

$$\left. \begin{aligned} \frac{t^2 + 3t}{2t} &= \frac{t(t + 3)}{t \cdot 2} \\ &= \frac{t + 3}{2}, \ t \neq 0. \end{aligned} \right\}$$ Removing another factor equal to 1: $t/t = 1$. The rational expression is now simplified. We must still specify $t \neq 0$.

EXAMPLE 4 Write the function given by

$$g(x) = \frac{x^2 + 3x - 10}{2x^2 - 3x - 2}$$

in simplified form.

Solution We have

$$g(x) = \frac{x^2 + 3x - 10}{2x^2 - 3x - 2}$$

$$= \frac{(x - 2)(x + 5)}{(2x + 1)(x - 2)} \qquad \text{Factoring the numerator and the denominator. Note that } x \neq -\tfrac{1}{2} \text{ and } x \neq 2.$$

$$= \frac{x - 2}{x - 2} \cdot \frac{x + 5}{2x + 1} \qquad \text{Rewriting as a product of two rational expressions}$$

$$= \frac{x + 5}{2x + 1}, \quad x \neq -\frac{1}{2}, 2. \qquad \text{Removing a factor equal to 1:} \\ \frac{x - 2}{x - 2} = 1$$

Thus, $g(x) = \dfrac{x + 5}{2x + 1}$, $x \neq -\dfrac{1}{2}, 2.$

EXAMPLE 5 Simplify: **(a)** $\dfrac{9x^2 + 6xy - 3y^2}{12x^2 - 12y^2}$; **(b)** $\dfrac{4 - t}{3t - 12}$.

Solution

a) $\dfrac{9x^2 + 6xy - 3y^2}{12x^2 - 12y^2} = \dfrac{3(x + y)(3x - y)}{12(x + y)(x - y)}$ Factoring the numerator and the denominator

$$= \frac{3(x + y)}{3(x + y)} \cdot \frac{3x - y}{4(x - y)} \qquad \text{Rewriting as a product of two rational expressions}$$

$$= \frac{3x - y}{4(x - y)} \qquad \text{Removing a factor equal to 1:} \\ \frac{3(x + y)}{3(x + y)} = 1$$

For purposes of later work, we usually do not multiply out the numerator and the denominator of the simplified expression.

b) $\dfrac{4 - t}{3t - 12} = \dfrac{4 - t}{3(t - 4)}$ Factoring

Since $4 - t$ is the opposite of $t - 4$, we factor out -1 to reverse the subtraction:

$$\frac{4 - t}{3t - 12} = \frac{-1(t - 4)}{3(t - 4)} \qquad 4 - t = -1(-4 + t) = -1(t - 4)$$

$$= \frac{-1}{3} \cdot \frac{t - 4}{t - 4}$$

$$= -\frac{1}{3}. \qquad \text{Removing a factor equal to 1:} \\ (t - 4)/(t - 4) = 1$$

technology connection

To check that a simplified expression is equivalent to the original expression, we can let y_1 = the original expression, y_2 = the simplified expression, and $y_3 = y_1 - y_2$ (or $y_2 - y_1$). If y_1 and y_2 are indeed equivalent, ⬭TABLE or ⬭TRACE can be used to show that, except when y_1 or y_2 is undefined, we have $y_1 = y_2$ and $y_3 = 0$.

1. Use a graphing calculator to check Example 3. Be sure to use parentheses as needed.
2. Use a graphing calculator to show that $\dfrac{x+3}{x} \neq 3$.
   (See the Caution! box concerning factoring.)

Canceling

"Canceling" is a shortcut often used for removing a factor equal to 1 when working with fractions. With some misgivings, we mention it as a possible way to speed up your work. Canceling removes factors equal to 1 in products. It *cannot* be done in sums or when adding expressions together. If your instructor permits canceling (not all do), it must be done with care and understanding. Example 5(a) might have been done faster as follows:

$$\frac{9x^2 + 6xy - 3y^2}{12x^2 - 12y^2} = \frac{3\cancel{(x+y)}(3x-y)}{3 \cdot 4\cancel{(x+y)}(x-y)}$$ When a factor that equals 1 is found, it is "canceled" as shown.

$$= \frac{3x-y}{4(x-y)}.$$ Removing a factor equal to 1: $\dfrac{3(x+y)}{3(x+y)} = 1$

Caution!   Canceling is often performed incorrectly:

$$\frac{\cancel{x}+3}{\cancel{x}} = 3, \qquad \frac{\cancel{4}x+3}{2} = 2x+3, \qquad \frac{\cancel{3}}{\cancel{3}+x} = \frac{1}{x}$$ To check that these are not equivalent, substitute a number for x.

Incorrect!   Incorrect!   Incorrect!

In each situation, the expressions canceled are *not* both factors. Factors are parts of products. For example, 4 is *not* a factor of the numerator $4x + 3$. If you can't factor, you can't cancel! When in doubt, don't cancel!

If, after multiplying two rational expressions, it is possible to simplify, it is common practice to do so.

EXAMPLE 6   Multiply. Then simplify by removing a factor equal to 1.

a) $\dfrac{x+2}{x-3} \cdot \dfrac{x^2-4}{x^2+x-2}$

b) $\dfrac{1-a^3}{a^2} \cdot \dfrac{a^5}{a^2-1}$

Solution

a) $\dfrac{x+2}{x-3} \cdot \dfrac{x^2-4}{x^2+x-2} = \dfrac{(x+2)(x^2-4)}{(x-3)(x^2+x-2)}$ Multiplying the numerators and also the denominators

$$= \frac{(x+2)(x-2)(x+2)}{(x-3)(x+2)(x-1)}$$ Factoring the numerator and the denominator and finding common factors

$$= \frac{(x+2)\cancel{(x+2)}(x-2)}{(x-3)\cancel{(x+2)}(x-1)}$$ Removing a factor equal to 1: $\dfrac{x+2}{x+2} = 1$

$$= \frac{(x+2)(x-2)}{(x-3)(x-1)}$$ Simplifying

b) $\dfrac{1 - a^3}{a^2} \cdot \dfrac{a^5}{a^2 - 1} = \dfrac{(1 - a^3)a^5}{a^2(a^2 - 1)}$

$\qquad\qquad\qquad = \dfrac{(1 - a)(1 + a + a^2)a^5}{a^2(a - 1)(a + 1)}$ Factoring a difference of cubes and a difference of squares

$\qquad\qquad\qquad = \dfrac{-1(a - 1)(1 + a + a^2)a^5}{a^2(a - 1)(a + 1)}$ *Important!* Factoring out -1 reverses the subtraction.

$\qquad\qquad\qquad = \dfrac{\cancel{(a - 1)}a^2 \cdot a^3(-1)(1 + a + a^2)}{\cancel{(a - 1)}a^2(a + 1)}$ Rewriting a^5 as $a^2 \cdot a^3$; removing a factor equal to 1: $\dfrac{(a - 1)a^2}{(a - 1)a^2} = 1$

$\qquad\qquad\qquad = \dfrac{-a^3(1 + a + a^2)}{a + 1}$ Simplifying

As in Example 5, there is no need for us to multiply out the numerator or the denominator of the final result.

Dividing and Simplifying

Two expressions are reciprocals of each other if their product is 1. As in arithmetic, to find the reciprocal of a rational expression, we interchange numerator and denominator.

The reciprocal of $\dfrac{x}{x^2 + 3}$ is $\dfrac{x^2 + 3}{x}$.

The reciprocal of $y - 8$ is $\dfrac{1}{y - 8}$.

Student Notes _____

The procedures covered in this chapter are by their nature rather long. As is often the case in mathematics, it usually helps to write out each step as you do the problems. Use plenty of paper and, if you are using lined paper, consider using two spaces at a time, writing the fraction bar on a line of the paper.

> **Quotients of Rational Expressions**
>
> For any rational expressions A/B and C/D, with $B, C, D \neq 0$,
>
> $$\dfrac{A}{B} \div \dfrac{C}{D} = \dfrac{A}{B} \cdot \dfrac{D}{C}.$$
>
> (To divide two rational expressions, multiply by the reciprocal of the divisor. We often say that we "*invert* and multiply.")

EXAMPLE 7   Divide. Simplify by removing a factor equal to 1 if possible.

$$\dfrac{x - 2}{x + 1} \div \dfrac{x + 5}{x - 3}$$

Solution

$$\frac{x-2}{x+1} \div \frac{x+5}{x-3} = \frac{x-2}{x+1} \cdot \frac{x-3}{x+5}$$ Multiplying by the reciprocal of the divisor

$$= \frac{(x-2)(x-3)}{(x+1)(x+5)}$$ Multiplying the numerators and the denominators

EXAMPLE 8 Write the function given by

$$g(a) = \frac{a^2 - 2a + 1}{a+1} \div \frac{a^3 - a}{a-2}$$

in simplified form, and list all restrictions on the domain.

Solution The denominators of the rational expressions are $a + 1$ and $a - 2$, so the domain of g cannot contain -1 or 2. Also, division by $(a^3 - a)/(a - 2)$ is defined only when $(a^3 - a)/(a - 2)$ is nonzero. This expression will be zero when the numerator is zero, so we solve $a^3 - a = 0$:

$$a^3 - a = 0$$
$$a(a^2 - 1) = 0$$
$$a(a + 1)(a - 1) = 0$$
$$a = 0 \quad or \quad a + 1 = 0 \quad or \quad a - 1 = 0$$
$$a = 0 \quad or \quad a = -1 \quad or \quad a = 1.$$

Thus the domain of g cannot contain 0, -1, or 1.

$$g(a) = \frac{a^2 - 2a + 1}{a+1} \div \frac{a^3 - a}{a-2} = \frac{a^2 - 2a + 1}{a+1} \cdot \frac{a-2}{a^3 - a}$$ Multiplying by the reciprocal of the divisor. Note that $a \neq -1, 0, 1, 2$.

$$= \frac{(a^2 - 2a + 1)(a - 2)}{(a + 1)(a^3 - a)}$$ Multiplying the numerators and the denominators

$$= \frac{(a - 1)(a - 1)(a - 2)}{(a + 1)a(a + 1)(a - 1)}$$ Factoring the numerator and the denominator

$$= \frac{(a - 1)\cancel{(a - 1)}(a - 2)}{(a + 1)a(a + 1)\cancel{(a - 1)}}$$ Removing a factor equal to 1: $\frac{a-1}{a-1} = 1$

$$= \frac{(a - 1)(a - 2)}{a(a + 1)(a + 1)}, \quad a \neq -1, 0, 1, 2$$ Simplifying

↪ *Concept Reinforcement* In each of Exercises 1–10, *match the function described with the appropriate domain from the column on the right.*

1. ____ $f(x) = \dfrac{2 - x}{x - 5}$

2. ____ $g(x) = \dfrac{x + 2}{x + 5}$

3. ____ $h(x) = \dfrac{x - 5}{x - 2}$

4. ____ $f(x) = \dfrac{x + 5}{x + 2}$

5. ____ $g(x) = \dfrac{x - 3}{(x - 2)(x - 5)}$

6. ____ $h(x) = \dfrac{x - 3}{(x + 2)(x + 5)}$

7. ____ $f(x) = \dfrac{x + 3}{(x - 2)(x + 5)}$

8. ____ $g(x) = \dfrac{x + 3}{(x + 2)(x - 5)}$

9. ____ $h(x) = \dfrac{(x - 2)(x - 3)}{x + 3}$

10. ____ $f(x) = \dfrac{(x + 2)(x + 3)}{x - 3}$

a) $\{x \,|\, x \neq -5, x \neq 2\}$

b) $\{x \,|\, x \neq 3\}$

c) $\{x \,|\, x \neq -2\}$

d) $\{x \,|\, x \neq -3\}$

e) $\{x \,|\, x \neq 5\}$

f) $\{x \,|\, x \neq -2, x \neq 5\}$

g) $\{x \,|\, x \neq 2\}$

h) $\{x \,|\, x \neq -2, x \neq -5\}$

i) $\{x \,|\, x \neq 2, x \neq 5\}$

j) $\{x \,|\, x \neq -5\}$

Photo developing. *Rik usually takes 3 hr more than Pearl does to process a day's orders at Liberty Place Photo. If Pearl takes t hr to process a day's orders, the function given by*

$$H(t) = \frac{t^2 + 3t}{2t + 3}$$

can be used to determine how long it would take if they worked together.

11. How long will it take them, working together, to complete a day's orders if Pearl can process the orders alone in 5 hr?

12. How long will it take them, working together, to complete a day's orders if Pearl can process the orders alone in 7 hr?

For each rational function, find the function values indicated, provided the value exists.

13. $v(t) = \dfrac{4t^2 - 5t + 2}{t + 3}$; $v(0)$, $v(-2)$, $v(7)$

14. $f(x) = \dfrac{5x^2 + 4x - 12}{6 - x}$; $f(0)$, $f(-1)$, $f(3)$

15. $g(x) = \dfrac{2x^3 - 9}{x^2 - 4x + 4}$; $g(0), g(2), g(-1)$

16. $r(t) = \dfrac{t^2 - 5t + 4}{t^2 - 9}$; $r(1), r(2), r(-3)$

Multiply to obtain equivalent expressions. Do not simplify. Assume that all denominators are nonzero.

17. $\dfrac{4x}{4x} \cdot \dfrac{x - 3}{x + 2}$

18. $\dfrac{3 - a^2}{a - 7} \cdot \dfrac{-1}{-1}$

19. $\dfrac{t - 2}{t + 3} \cdot \dfrac{-1}{-1}$

20. $\dfrac{x - 4}{x + 5} \cdot \dfrac{x - 5}{x - 5}$

Simplify by removing a factor equal to 1.

21. $\dfrac{15x}{5x^2}$

22. $\dfrac{7a^3}{21a}$

23. $\dfrac{18t^3 s^2}{27t^7 s^9}$

24. $\dfrac{8y^5 z^3}{4y^9 z^4}$

25. $\dfrac{2a - 10}{2}$

26. $\dfrac{3a + 12}{3}$

27. $\dfrac{15}{25a - 30}$

28. $\dfrac{21}{6x - 9}$

29. $\dfrac{3x - 12}{3x + 15}$

30. $\dfrac{4y - 20}{4y + 12}$

Write simplified form for each of the following. Be sure to list all restrictions on the domain, as in Example 4.

31. $f(x) = \dfrac{3x + 21}{x^2 + 7x}$

32. $f(x) = \dfrac{5x + 20}{x^2 + 4x}$

33. $g(x) = \dfrac{x^2 - 9}{5x + 15}$

34. $g(x) = \dfrac{8x - 16}{x^2 - 4}$

35. $h(x) = \dfrac{4 - x}{5x - 20}$

36. $h(x) = \dfrac{7 - x}{3x - 21}$

37. $f(t) = \dfrac{t^2 - 16}{t^2 - 8t + 16}$

38. $f(t) = \dfrac{t^2 - 25}{t^2 + 10t + 25}$

39. $g(t) = \dfrac{21 - 7t}{3t - 9}$

40. $g(t) = \dfrac{12 - 6t}{5t - 10}$

41. $h(t) = \dfrac{t^2 + 5t + 4}{t^2 - 8t - 9}$

42. $h(t) = \dfrac{t^2 - 3t - 4}{t^2 + 9t + 8}$

43. $f(x) = \dfrac{9x^2 - 4}{3x - 2}$

44. $f(x) = \dfrac{4x^2 - 1}{2x - 1}$

45. $g(t) = \dfrac{16 - t^2}{t^2 - 8t + 16}$

46. $g(p) = \dfrac{25 - p^2}{p^2 + 10p + 25}$

Multiply and, if possible, simplify.

47. $\dfrac{5a^3}{3b} \cdot \dfrac{7b^3}{10a^7}$

48. $\dfrac{25a}{9b^8} \cdot \dfrac{3b^5}{5a^2}$

49. $\dfrac{8x - 16}{5x} \cdot \dfrac{x^3}{5x - 10}$

50. $\dfrac{5t^3}{4t - 8} \cdot \dfrac{6t - 12}{10t}$

51. $\dfrac{x^2 - 16}{x^2} \cdot \dfrac{x^2 - 4x}{x^2 - x - 12}$

52. $\dfrac{y^2 + 10y + 25}{y^2 - 9} \cdot \dfrac{y^2 + 3y}{y + 5}$

53. $\dfrac{7a - 14}{4 - a^2} \cdot \dfrac{5a^2 + 6a + 1}{35a + 7}$

54. $\dfrac{a^2 - 1}{2 - 5a} \cdot \dfrac{15a - 6}{a^2 + 5a - 6}$

Aha! **55.** $\dfrac{t^3 - 4t}{t - t^4} \cdot \dfrac{t^4 - t}{4t - t^3}$

56. $\dfrac{x^2 - 6x + 9}{12 - 4x} \cdot \dfrac{x^6 - 9x^4}{x^3 - 3x^2}$

57. $\dfrac{c^3 + 8}{c^5 - 4c^3} \cdot \dfrac{c^6 - 4c^5 + 4c^4}{c^2 - 2c + 4}$

58. $\dfrac{x^3 - 27}{x^4 - 9x^2} \cdot \dfrac{x^5 - 6x^4 + 9x^3}{x^2 + 3x + 9}$

59. $\dfrac{a^3 - b^3}{3a^2 + 9ab + 6b^2} \cdot \dfrac{a^2 + 2ab + b^2}{a^2 - b^2}$

60. $\dfrac{x^3 + y^3}{x^2 + 2xy - 3y^2} \cdot \dfrac{x^2 - y^2}{3x^2 + 6xy + 3y^2}$

Divide and, if possible, simplify.

61. $\dfrac{9x^5}{8y^2} \div \dfrac{3x}{16y^9}$

62. $\dfrac{16a^7}{3b^5} \div \dfrac{8a^3}{6b}$

63. $\dfrac{5x + 10}{x^8} \div \dfrac{x + 2}{x^3}$

64. $\dfrac{3y + 15}{y^7} \div \dfrac{y + 5}{y^2}$

65. $\dfrac{25x^2 - 4}{x^2 - 9} \div \dfrac{2 - 5x}{x + 3}$

66. $\dfrac{4a^2 - 1}{a^2 - 4} \div \dfrac{2a - 1}{2 - a}$

67. $\dfrac{5y - 5x}{15y^3} \div \dfrac{x^2 - y^2}{3x + 3y}$

68. $\dfrac{x^2 - y^2}{4x + 4y} \div \dfrac{3y - 3x}{12x^2}$

69. $\dfrac{x^2 - 16}{x^2 - 10x + 25} \div \dfrac{3x - 12}{x^2 - 3x - 10}$

70. $\dfrac{y^2 - 36}{y^2 - 8y + 16} \div \dfrac{3y - 18}{y^2 - y - 12}$

71. $\dfrac{x^3 - 64}{x^3 + 64} \div \dfrac{x^2 - 16}{x^2 - 4x + 16}$

72. $\dfrac{8y^3 - 27}{64y^3 - 1} \div \dfrac{4y^2 - 9}{16y^2 + 4y + 1}$

Write simplified form for each of the following. Be sure to list all restrictions on the domain.

73. $f(t) = \dfrac{t^2 - 16}{4t + 12} \cdot \dfrac{t + 3}{t - 4}$

74. $g(m) = \dfrac{m^2 - 9}{3m + 3} \cdot \dfrac{m + 3}{m - 3}$

75. $g(x) = \dfrac{x^2 - 2x - 35}{2x^3 - 3x^2} \cdot \dfrac{4x^3 - 9x}{7x - 49}$

76. $h(t) = \dfrac{t^2 - 10t + 9}{t^2 - 1} \cdot \dfrac{1 - t^2}{t^2 - 5t - 36}$

77. $f(x) = \dfrac{x^2 - 4}{x^3} \div \dfrac{x^5 - 2x^4}{x + 4}$

78. $g(x) = \dfrac{x^2 - 9}{x^2} \div \dfrac{x^5 + 3x^4}{x + 2}$

79. $h(n) = \dfrac{n^3 + 3n}{n^2 - 9} \div \dfrac{n^2 + 5n - 14}{n^2 + 4n - 21}$

80. $f(x) = \dfrac{x^3 + 4x}{x^2 - 16} \div \dfrac{x^2 + 8x + 15}{x^2 + x - 20}$

Perform the indicated operations and, if possible, simplify.

81. $\dfrac{4x^2 - 9y^2}{8x^3 - 27y^3} \div \dfrac{4x + 6y}{3x - 9y} \cdot \dfrac{4x^2 + 6xy + 9y^2}{4x^2 - 8xy + 3y^2}$

82. $\dfrac{5x^2 - 5y^2}{27x^3 + 8y^3} \div \dfrac{x^2 - 2xy + y^2}{9x^2 - 6xy + 4y^2} \cdot \dfrac{6x + 4y}{10x - 15y}$

83. $\dfrac{a^3 - ab^2}{2a^2 + 3ab + b^2} \cdot \dfrac{4a^2 - b^2}{a^2 - 2ab + b^2} \div \dfrac{a^2 + a}{a - 1}$

84. $\dfrac{2x + 4y}{2x^2 + 5xy + 2y^2} \cdot \dfrac{4x^2 - y^2}{8x^2 - 8} \div \dfrac{x^2 + 4xy + 4y^2}{x^2 - 6xy + 9y^2}$

85. Is it possible to understand how to simplify rational expressions without first understanding how to multiply rational expressions? Why or why not?

86. Nancy *incorrectly* simplifies $\dfrac{x + 2}{x}$ as

$$\frac{x + 2}{x} = \frac{\cancel{x} + 2}{\cancel{x}} = 1 + 2 = 3.$$

She insists this is correct because it checks when x is replaced with 1. Explain her misconception.

SKILL MAINTENANCE

Simplify.

87. $\dfrac{3}{10} - \dfrac{8}{15}$ [1.2]

88. $\dfrac{3}{8} - \dfrac{7}{10}$ [1.2]

89. $\dfrac{2}{3} \cdot \dfrac{5}{7} - \dfrac{5}{7} \cdot \dfrac{1}{6}$ [1.2]

90. $\dfrac{4}{7} \cdot \dfrac{1}{5} - \dfrac{3}{10} \cdot \dfrac{2}{7}$ [1.2]

91. $(8x^3 - 5x^2 + 6x + 2) - (4x^3 + 2x^2 - 3x + 7)$ [5.1]

92. $(6t^4 + 9t^3 - t^2 + 4t) - (8t^4 - 2t^3 - 6t + 3)$ [5.1]

SYNTHESIS

93. Explain why the graphs of $f(x) = 5x$ and $g(x) = \dfrac{5x^2}{x}$ differ.

94. Tony *incorrectly* argues that since

$$\frac{a^2 - 4}{a - 2} = \frac{a^2}{a} + \frac{-4}{-2} = a + 2$$

is correct, it follows that

$$\frac{x^2 + 9}{x + 1} = \frac{x^2}{x} + \frac{9}{1} = x + 9.$$

Explain his misconception.

Recall that a fraction bar is a grouping symbol. The next example shows that when a numerator is subtracted, care must be taken to subtract, or change the sign of, *each* term in that polynomial.

EXAMPLE 3

If

$$f(x) = \frac{4x + 5}{x + 3} - \frac{x - 2}{x + 3},$$

find a simplified form for $f(x)$ and list all restrictions on the domain.

Solution

$$f(x) = \frac{4x + 5}{x + 3} - \frac{x - 2}{x + 3} \qquad \text{Note that } x \neq -3.$$

$$= \frac{4x + 5 - (x - 2)}{x + 3} \qquad \begin{array}{l}\text{The parentheses remind us}\\ \text{to subtract } both \text{ terms.}\end{array}$$

$$= \frac{4x + 5 - x + 2}{x + 3}$$

$$= \frac{3x + 7}{x + 3}, \ x \neq -3$$

technology connection

Example 3 can be checked by comparing the graphs of

$$y_1 = \frac{4x + 5}{x + 3} - \frac{x - 2}{x + 3}$$

and

$$y_2 = \frac{3x + 7}{x + 3}$$

on the same set of axes. Since the equations are equivalent, one curve (it has two branches) should appear. Equivalently, you can show that $y_3 = y_2 - y_1$ is 0 for all x not equal to -3. The TABLE or TRACE feature can assist in either type of check.

When Denominators Are Different

In order to add rational expressions such as

$$\frac{7}{12xy^2} + \frac{8}{15x^3y} \quad \text{or} \quad \frac{x}{x^2 - y^2} + \frac{y}{x^2 - 4xy + 3y^2},$$

we must first find common denominators. As in arithmetic, our work is easier when we use the *least common multiple* (LCM) of the denominators.

> **Least Common Multiple**
>
> To find the least common multiple (LCM) of two or more expressions, find the prime factorization of each expression and form a product that contains each factor the greatest number of times that it occurs in any one prime factorization.

EXAMPLE 4

Find the least common multiple of each pair of polynomials.

a) $21x$ and $3x^2$ 　　　　　　　　　　　**b)** $x^2 + x - 12$ and $x^2 - 16$

Solution

a) We write the prime factorizations of $21x$ and $3x^2$:

$$21x = 3 \cdot 7 \cdot x \quad \text{and} \quad 3x^2 = 3 \cdot x \cdot x.$$

The factors 3, 7, and x must appear in the LCM if $21x$ is to be a factor of the LCM. The other polynomial, $3x^2$, is not a factor of $3 \cdot 7 \cdot x$ because the prime factors of $3x^2$—namely, 3, x, and x—do not all appear in $3 \cdot 7 \cdot x$. However, if $3 \cdot 7 \cdot x$ is multiplied by another factor of x, a product is formed that contains both $21x$ and $3x^2$ as factors:

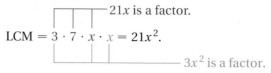

Note that each factor (3, 7, and x) is used the greatest number of times that it occurs as a factor of either $21x$ or $3x^2$. The LCM is $3 \cdot 7 \cdot x \cdot x$, or $21x^2$.

b) We factor both expressions:

$$x^2 + x - 12 = (x - 3)(x + 4),$$
$$x^2 - 16 = (x + 4)(x - 4).$$

The LCM must contain each polynomial as a factor. By multiplying the factors of $x^2 + x - 12$ by $x - 4$, we form a product that contains both $x^2 + x - 12$ and $x^2 - 16$ as factors:

$$\text{LCM} = (x - 3)(x + 4)(x - 4).$$

— $x^2 + x - 12$ is a factor.

There is no need to multiply this out.

— $x^2 - 16$ is a factor.

> **To Add or Subtract Rational Expressions**
>
> 1. Determine the *least common denominator* (LCD) by finding the least common multiple of the denominators.
> 2. Rewrite each of the original rational expressions, as needed, in an equivalent form that has the LCD.
> 3. Add or subtract the resulting rational expressions, as indicated.
> 4. Simplify the result, if possible, and list any restrictions on the domain of functions.

EXAMPLE 5   Add: $\dfrac{2}{21x} + \dfrac{5}{3x^2}$.

Solution   In Example 4(a), we found that the LCD is $3 \cdot 7 \cdot x \cdot x$, or $21x^2$. We now multiply each rational expression by 1, using expressions for 1 that give us the LCD in each expression. To determine what to use, ask "$21x$ times what is $21x^2$?" and "$3x^2$ times what is $21x^2$?" The answers are x and 7, respectively, so we multiply by x/x and $7/7$:

$$\frac{2}{21x} \cdot \frac{x}{x} + \frac{5}{3x^2} \cdot \frac{7}{7} = \frac{2x}{21x^2} + \frac{35}{21x^2} \qquad \text{We now have a common denominator.}$$

$$= \frac{2x + 35}{21x^2}. \qquad \text{This expression cannot be simplified.}$$

EXAMPLE 6   Add: $\dfrac{x^2}{x^2 + 2xy + y^2} + \dfrac{2x - 2y}{x^2 - y^2}$.

Solution   To find the LCD, we first factor the denominators:

$$\frac{x^2}{x^2 + 2xy + y^2} + \frac{2x - 2y}{x^2 - y^2} = \frac{x^2}{(x + y)(x + y)} + \frac{2x - 2y}{(x + y)(x - y)}.$$

Although the numerators need not always be factored, we do so if it enables us to simplify. In this case, the rightmost rational expression can be simplified:

$$\frac{x^2}{x^2 + 2xy + y^2} + \frac{2x - 2y}{x^2 - y^2} = \frac{x^2}{(x + y)(x + y)} + \frac{2(x - y)}{(x + y)(x - y)}$$ Factoring

$$= \frac{x^2}{(x + y)(x + y)} + \frac{2}{x + y}.$$ Removing a factor equal to 1: $\dfrac{x - y}{x - y} = 1$

Note that the LCM of $(x + y)(x + y)$ and $(x + y)$ is $(x + y)(x + y)$. To get the LCD in the second expression, we multiply by 1, using $(x + y)/(x + y)$. Then we add and, if possible, simplify.

$$\frac{x^2}{(x + y)(x + y)} + \frac{2}{x + y} = \frac{x^2}{(x + y)(x + y)} + \frac{2}{x + y} \cdot \frac{x + y}{x + y}$$

$$= \frac{x^2}{(x + y)(x + y)} + \frac{2x + 2y}{(x + y)(x + y)}$$ We now have the LCD.

$$= \frac{x^2 + 2x + 2y}{(x + y)(x + y)}$$ Since the numerator cannot be factored, we cannot simplify further.

EXAMPLE 7   Subtract: $\dfrac{2y + 1}{y^2 - 7y + 6} - \dfrac{y + 3}{y^2 - 5y - 6}.$

Solution

$$\frac{2y + 1}{y^2 - 7y + 6} - \frac{y + 3}{y^2 - 5y - 6}$$

$$= \frac{2y + 1}{(y - 6)(y - 1)} - \frac{y + 3}{(y - 6)(y + 1)}$$ The LCD is $(y - 6)(y - 1)(y + 1)$.

$$= \frac{2y + 1}{(y - 6)(y - 1)} \cdot \frac{y + 1}{y + 1} - \frac{y + 3}{(y - 6)(y + 1)} \cdot \frac{y - 1}{y - 1}$$

⎣Multiplying by 1 to get⎦
the LCD in each expression

$$= \frac{(2y + 1)(y + 1) - (y + 3)(y - 1)}{(y - 6)(y - 1)(y + 1)}$$

$$= \frac{2y^2 + 3y + 1 - (y^2 + 2y - 3)}{(y - 6)(y - 1)(y + 1)}$$ The parentheses are important.

$$= \frac{2y^2 + 3y + 1 - y^2 - 2y + 3}{(y - 6)(y - 1)(y + 1)}$$

$$= \frac{y^2 + y + 4}{(y - 6)(y - 1)(y + 1)}$$ We leave the denominator in factored form.

EXAMPLE 8 Add: $\dfrac{3}{8a} + \dfrac{1}{-8a}$.

Solution

$$\dfrac{3}{8a} + \dfrac{1}{-8a} = \dfrac{3}{8a} + \dfrac{-1}{-1} \cdot \dfrac{1}{-8a}$$

> When denominators are opposites, we multiply one rational expression by $-1/-1$ to get the LCD.

$$= \dfrac{3}{8a} + \dfrac{-1}{8a} = \dfrac{2}{8a}$$

$$= \dfrac{2 \cdot 1}{2 \cdot 4a} = \dfrac{1}{4a} \qquad \text{Simplifying by removing a factor}$$

equal to 1: $\dfrac{2}{2} = 1$

EXAMPLE 9 Subtract: $\dfrac{5x}{x - 2y} - \dfrac{3y - 7}{2y - x}$.

Solution

$$\dfrac{5x}{x - 2y} - \dfrac{3y - 7}{2y - x} = \dfrac{5x}{x - 2y} - \dfrac{-1}{-1} \cdot \dfrac{3y - 7}{2y - x} \qquad \text{Note that } x - 2y \text{ and} \\ 2y - x \text{ are opposites.}$$

$$= \dfrac{5x}{x - 2y} - \dfrac{7 - 3y}{x - 2y} \qquad \begin{array}{l}\text{Performing the multiplication.} \\ Note: -1(2y - x) = -2y + x \\ \qquad\qquad\qquad = x - 2y.\end{array}$$

$$= \dfrac{5x - (7 - 3y)}{x - 2y}$$

$$= \dfrac{5x - 7 + 3y}{x - 2y} \qquad \begin{array}{l}\text{Subtracting. The parentheses} \\ \text{are important.}\end{array}$$

In Example 9, you may have noticed that when $3y - 7$ is multiplied by -1 and subtracted, the result is $-7 + 3y$, which is equivalent to the original $3y - 7$. Thus, instead of multiplying the numerator by -1 and then subtracting, we could have simply *added* $3y - 7$ to $5x$, as in the following:

$$\dfrac{5x}{x - 2y} - \dfrac{3y - 7}{2y - x} = \dfrac{5x}{x - 2y} + (-1) \cdot \dfrac{3y - 7}{2y - x} \qquad \begin{array}{l}\text{Rewriting subtraction} \\ \text{as addition}\end{array}$$

$$= \dfrac{5x}{x - 2y} + \dfrac{1}{-1} \cdot \dfrac{3y - 7}{2y - x} \qquad \text{Writing } -1 \text{ as } \dfrac{1}{-1}$$

$$= \dfrac{5x}{x - 2y} + \dfrac{3y - 7}{x - 2y} \qquad \begin{array}{l}\text{The opposite of } 2y - x \\ \text{is } x - 2y.\end{array}$$

$$= \dfrac{5x + 3y - 7}{x - 2y}. \qquad \begin{array}{l}\text{This checks with} \\ \text{the answer to} \\ \text{Example 9.}\end{array}$$

EXAMPLE 10   Find simplified form for the function given by

$$f(x) = \frac{2x}{x^2 - 4} + \frac{5}{2 - x} - \frac{1}{2 + x}$$

and list all restrictions on the domain.

Solution   We have

$$\frac{2x}{x^2 - 4} + \frac{5}{2 - x} - \frac{1}{2 + x} = \frac{2x}{(x - 2)(x + 2)} + \frac{5}{2 - x} - \frac{1}{2 + x} \qquad \begin{array}{l}\text{Factoring.}\\ \text{Note that}\\ x \neq -2, 2.\end{array}$$

$$= \frac{2x}{(x - 2)(x + 2)} + \frac{-1}{-1} \cdot \frac{5}{(2 - x)} - \frac{1}{x + 2} \qquad \begin{array}{l}\text{Multiplying by } -1/-1 \text{ since}\\ 2 - x \text{ is the opposite of}\\ x - 2\end{array}$$

$$= \frac{2x}{(x - 2)(x + 2)} + \frac{-5}{x - 2} - \frac{1}{x + 2} \qquad \text{The LCD is } (x - 2)(x + 2).$$

$$= \frac{2x}{(x - 2)(x + 2)} + \frac{-5}{x - 2} \cdot \frac{x + 2}{x + 2} - \frac{1}{x + 2} \cdot \frac{x - 2}{x - 2} \qquad \begin{array}{l}\text{Multiplying by}\\ \text{1 to get the}\\ \text{LCD}\end{array}$$

$$= \frac{2x - 5(x + 2) - (x - 2)}{(x - 2)(x + 2)} = \frac{2x - 5x - 10 - x + 2}{(x - 2)(x + 2)}$$

$$= \frac{-4x - 8}{(x - 2)(x + 2)} = \frac{-4(x + 2)}{(x - 2)(x + 2)}$$

$$= \frac{-4\cancel{(x + 2)}}{(x - 2)\cancel{(x + 2)}} \qquad \begin{array}{l}\text{Removing a factor equal to 1:}\\ \dfrac{x + 2}{x + 2} = 1, x \neq -2\end{array}$$

$$= \frac{-4}{x - 2}, \text{ or } -\frac{4}{x - 2}, \ x \neq \pm 2.$$

An equivalent answer is $\dfrac{4}{2 - x}$. It is found by writing $-\dfrac{4}{x - 2}$ as $\dfrac{4}{-(x - 2)}$ and then using the distributive law to remove parentheses.

Student Notes _____

Adding or subtracting rational expressions can involve many steps. Therefore, it is important to double-check each step of your work as you proceed. Waiting until the end of the problem to check your work is generally a less efficient use of your time.

Our work in Example 10 indicates that if

$$f(x) = \frac{2x}{x^2 - 4} + \frac{5}{2 - x} - \frac{1}{2 + x} \quad \text{and} \quad g(x) = \frac{-4}{x - 2},$$

then, for $x \neq -2$ and $x \neq 2$, we have $f = g$. Note that whereas the domain of f includes all real numbers except -2 or 2, the domain of g excludes only 2. This is illustrated in the graphs at the top of the following page. Methods for drawing such graphs by hand are discussed in more advanced courses. The graphs are for visualization only.

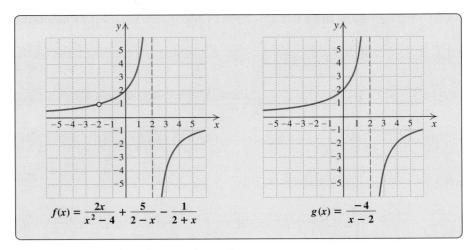

$$f(x) = \frac{2x}{x^2 - 4} + \frac{5}{2 - x} - \frac{1}{2 + x}$$

$$g(x) = \frac{-4}{x - 2}$$

A computer-generated visualization of Example 10

A quick, partial check of any simplification is to evaluate both the original and the simplified expressions for a convenient choice of x. For instance, to check Example 10, if $x = 1$, we have

$$f(1) = \frac{2 \cdot 1}{1^2 - 4} + \frac{5}{2 - 1} - \frac{1}{2 + 1}$$

$$= \frac{2}{-3} + \frac{5}{1} - \frac{1}{3} = 5 - \frac{3}{3} = 4$$

and

$$g(1) = \frac{-4}{1 - 2} = \frac{-4}{-1} = 4.$$

Since both functions include the pair $(1, 4)$, our algebra was *probably* correct. Although this is only a partial check (on rare occasions, an incorrect answer might "check"), because it is so easy to perform, it is quite useful. Further evaluation provides a more definitive check.

Exercise Set

6.2

FOR EXTRA HELP

Student's Solutions Manual | Digital Video Tutor CD 4 Videotape 11 | AW Math Tutor Center | MathXL Tutorials on CD | MathXL MathXL | MyMathLab MyMathLab

🪱 *Concept Reinforcement* *Classify each of the following statements as either true or false.*

1. To add or subtract rational expressions, a common denominator is required.

2. To find the least common denominator, we find the least common multiple of the denominators.

3. It is rarely necessary to factor in order to find a common denominator.

4. The sum of two rational expressions is the sum of the numerators over the sum of the denominators.

5. The least common multiple of two expressions always contains each of those expressions as a factor.

6. The least common multiple of two expressions is always the product of those two expressions.

7. To add two rational expressions, it is often necessary to multiply at least one of those expressions by a form of 1.

8. After two rational expressions are added, it is unnecessary to simplify the result.

Perform the indicated operations. Simplify when possible.

9. $\dfrac{3}{2y} + \dfrac{5}{2y}$

10. $\dfrac{5}{3a} + \dfrac{7}{3a}$

11. $\dfrac{5}{3m^2n^2} - \dfrac{4}{3m^2n^2}$

12. $\dfrac{1}{4a^2b} - \dfrac{5}{4a^2b}$

13. $\dfrac{x - 3y}{x + y} + \dfrac{x + 5y}{x + y}$

14. $\dfrac{a - 5b}{a + b} + \dfrac{a + 7b}{a + b}$

15. $\dfrac{3t + 2}{t - 4} - \dfrac{t - 2}{t - 4}$

16. $\dfrac{4y + 2}{y - 2} - \dfrac{y - 3}{y - 2}$

Find simplified form for $f(x)$ and list all restrictions on the domain.

17. $f(x) = \dfrac{9x}{x + 3} + \dfrac{4}{5x}$

18. $f(x) = \dfrac{7x}{x + 2} + \dfrac{3}{2x}$

19. $f(x) = \dfrac{5}{x^2 - 9} - \dfrac{2}{x^2 - 9}$

20. $f(x) = \dfrac{8}{x^2 - 4} - \dfrac{1}{x^2 - 4}$

21. $f(x) = \dfrac{7}{3x^2} + \dfrac{2}{x^2 - 1}$

22. $f(x) = \dfrac{9}{4x^3} + \dfrac{3}{x^2 - 4}$

23. $f(x) = \dfrac{2}{x - 1} + \dfrac{3}{1 - x}$

24. $f(x) = \dfrac{4}{x - 7} + \dfrac{2}{7 - x}$

Perform the indicated operations. Simplify when possible.

25. $\dfrac{5 - 4x}{x^2 - 6x - 7} + \dfrac{5x - 4}{x^2 - 6x - 7}$

26. $\dfrac{3 - 2x}{x^2 - 5x + 4} + \dfrac{3x - 4}{x^2 - 5x + 4}$

27. $\dfrac{2a - 5}{a^2 - 9} - \dfrac{3a - 8}{a^2 - 9}$

28. $\dfrac{3a - 2}{a^2 - 25} - \dfrac{4a - 7}{a^2 - 25}$

29. $\dfrac{s^2}{r - s} + \dfrac{r^2}{s - r}$

30. $\dfrac{a^2}{a - b} + \dfrac{b^2}{b - a}$

31. $\dfrac{2}{a} - \dfrac{5}{-a}$

32. $\dfrac{7}{x} - \dfrac{8}{-x}$

33. $\dfrac{y - 4}{y^2 - 25} - \dfrac{9 - 2y}{25 - y^2}$

34. $\dfrac{x - 7}{x^2 - 16} - \dfrac{x - 1}{16 - x^2}$

35. $\dfrac{y^2 - 5}{y^4 - 81} + \dfrac{4}{81 - y^4}$

36. $\dfrac{t^2 + 3}{t^4 - 16} + \dfrac{7}{16 - t^4}$

37. $\dfrac{r - 6s}{r^3 - s^3} - \dfrac{5s}{s^3 - r^3}$

38. $\dfrac{m - 3n}{m^3 - n^3} - \dfrac{2n}{n^3 - m^3}$

39. $\dfrac{a + 2}{a - 4} + \dfrac{a - 2}{a + 3}$

40. $\dfrac{a + 3}{a - 5} + \dfrac{a - 2}{a + 4}$

41. $4 + \dfrac{x - 3}{x + 1}$

42. $3 + \dfrac{y + 2}{y - 5}$

43. $\dfrac{4xy}{x^2 - y^2} + \dfrac{x - y}{x + y}$

44. $\dfrac{5ab}{a^2 - b^2} + \dfrac{a + b}{a - b}$

45. $\dfrac{8}{2x^2 - 7x + 5} + \dfrac{3x + 2}{2x^2 - x - 10}$

46. $\dfrac{7}{3y^2 + y - 4} + \dfrac{9y + 2}{3y^2 - 2y - 8}$

47. $\dfrac{4}{x + 1} + \dfrac{x + 2}{x^2 - 1} + \dfrac{3}{x - 1}$

48. $\dfrac{-2}{y + 2} + \dfrac{5}{y - 2} + \dfrac{y + 3}{y^2 - 4}$

49. $\dfrac{x + 6}{5x + 10} - \dfrac{x - 2}{4x + 8}$

50. $\dfrac{a + 3}{5a + 25} - \dfrac{a - 1}{3a + 15}$

51. $\dfrac{5ab}{a^2 - b^2} - \dfrac{a - b}{a + b}$

52. $\dfrac{6xy}{x^2 - y^2} - \dfrac{x + y}{x - y}$

53. $\dfrac{x}{x^2 + 9x + 20} - \dfrac{4}{x^2 + 7x + 12}$

54. $\dfrac{x}{x^2 + 11x + 30} - \dfrac{5}{x^2 + 9x + 20}$

55. $\dfrac{3y}{y^2 - 7y + 10} - \dfrac{2y}{y^2 - 8y + 15}$

56. $\dfrac{5x}{x^2 - 6x + 8} - \dfrac{3x}{x^2 - x - 12}$

57. $\dfrac{2x + 1}{x - y} + \dfrac{5x^2 - 5xy}{x^2 - 2xy + y^2}$

58. $\dfrac{2 - 3a}{a - b} + \dfrac{3a^2 + 3ab}{a^2 - b^2}$

59. $\dfrac{3y + 2}{y^2 + 5y - 24} + \dfrac{7}{y^2 + 4y - 32}$

60. $\dfrac{3x + 2}{x^2 - 7x + 10} + \dfrac{2x}{x^2 - 8x + 15}$

61. $\dfrac{2y - 6}{y^2 - 9} - \dfrac{y}{y - 1} + \dfrac{y^2 + 2}{y^2 + 2y - 3}$

62. $\dfrac{x - 1}{x^2 - 1} - \dfrac{x}{x - 2} + \dfrac{x^2 + 2}{x^2 - x - 2}$

Aha! **63.** $\dfrac{5y}{1 - 4y^2} - \dfrac{2y}{2y + 1} + \dfrac{5y}{4y^2 - 1}$

64. $\dfrac{4x}{x^2 - 1} + \dfrac{3x}{1 - x} - \dfrac{4}{x - 1}$

Find simplified form for f(x) and list all restrictions on the domain.

65. $f(x) = 2 + \dfrac{x}{x - 3} - \dfrac{18}{x^2 - 9}$

66. $f(x) = 5 + \dfrac{x}{x + 2} - \dfrac{8}{x^2 - 4}$

67. $f(x) = \dfrac{3x - 1}{x^2 + 2x - 3} - \dfrac{x + 4}{x^2 - 16}$

68. $f(x) = \dfrac{3x - 2}{x^2 + 2x - 24} - \dfrac{x - 3}{x^2 - 9}$

69. $f(x) = \dfrac{1}{x^2 + 5x + 6} - \dfrac{2}{x^2 + 3x + 2} - \dfrac{1}{x^2 + 5x + 6}$

70. $f(x) = \dfrac{2}{x^2 - 5x + 6} - \dfrac{4}{x^2 - 2x - 3} + \dfrac{2}{x^2 + 4x + 3}$

71. Janine found that the sum of two rational expressions was $(3 - x)/(x - 5)$. The answer given at the back of the book is $(x - 3)/(5 - x)$. Is Janine's answer incorrect? Why or why not?

72. When two rational expressions are added or subtracted, should the numerator of the result be factored? Why or why not?

SKILL MAINTENANCE

Simplify. Use only positive exponents in your answer. [1.6]

73. $\dfrac{15x^{-7}y^{12}z^4}{35x^{-2}y^6z^{-3}}$

74. $\dfrac{21a^{-4}b^6c^8}{27a^{-2}b^{-5}c}$

75. $\dfrac{34s^9t^{-40}r^{30}}{10s^{-3}t^{20}r^{-10}}$

76. Find an equation for the line that passes through the point $(-2, 3)$ and is perpendicular to the line $f(x) = -\frac{4}{5}x + 7$. [2.5]

77. *Value of coins.* There are 50 dimes in a roll of dimes, 40 nickels in a roll of nickels, and 40 quarters in a roll of quarters. Robert has a total of 12 rolls of coins with a total value of $70.00. If he has 3 more rolls of nickels than dimes, how many of each roll of coins does he have? [3.5]

78. *Audiotapes.*   Anna wants to buy tapes for her work at the campus radio station. She needs some 30-min tapes and some 60-min tapes. If she buys 12 tapes with a total recording time of 10 hr, how many tapes of each length did she buy? [3.3]

SYNTHESIS

79. Many students make the mistake of always multiplying denominators when looking for a common denominator. Use Example 7 to explain why this approach can yield results that are more difficult to simplify.

80. Is the sum of two rational expressions always a rational expression? Why or why not?

81. *Prescription drugs.*   After visiting her doctor, Corinna went to the pharmacy for a two-week supply of Zyrtec®, a 20-day supply of Albuterol, and a 30-day supply of Pepcid®. Corinna refills each prescription as soon as her supply runs out. How long will it be until she can refill all three prescriptions on the same day?

82. *Astronomy.*   The earth, Jupiter, Saturn, and Uranus all revolve around the sun. The earth takes 1 yr, Jupiter 12 yr, Saturn 30 yr, and Uranus 84 yr. How frequently do these four planets line up with each other?

83. *Music.*   To duplicate a common African polyrhythm, a drummer needs to play sextuplets (6 beats per measure) on a tom-tom while simultaneously playing quarter notes (4 beats per measure) on a bass drum. Into how many equally sized parts must a measure be divided, in order to precisely execute this rhythm?

84. *Home appliances.*   Refrigerators last an average of 20 yr, clothes washers about 14 yr, and dishwashers about 10 yr. In 1980, Westgate College bought new refrigerators for its dormitories. In 1990, the college bought new clothes washers and dishwashers. Predict the year in which the college will need to replace all three types of appliances at once.

Source: U.S. Department of Energy

Find the LCM.

85. $x^8 - x^4$, $x^5 - x^2$, $x^5 - x^3$, $x^5 + x^2$

86. $2a^3 + 2a^2b + 2ab^2$, $a^6 - b^6$, $2b^2 + ab - 3a^2$, $2a^2b + 4ab^2 + 2b^3$

87. The LCM of two expressions is $8a^4b^7$. One of the expressions is $2a^3b^7$. List all the possibilities for the other expression.

88. Determine the domain and the range of the function graphed below.

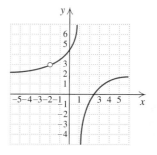

If

$$f(x) = \frac{x^3}{x^2 - 4} \quad and \quad g(x) = \frac{x^2}{x^2 + 3x - 10},$$

find each of the following.

89. $(f + g)(x)$

90. $(f - g)(x)$

91. $(f \cdot g)(x)$

92. $(f/g)(x)$

Perform the indicated operations and simplify.

93. $5(x - 3)^{-1} + 4(x + 3)^{-1} - 2(x + 3)^{-2}$

94. $4(y - 1)(2y - 5)^{-1} + 5(2y + 3)(5 - 2y)^{-1} + (y - 4)(2y - 5)^{-1}$

95. $\dfrac{x + 4}{6x^2 - 20x} \cdot \left(\dfrac{x}{x^2 - x - 20} + \dfrac{2}{x + 4} \right)$

96. $\dfrac{x^2 - 7x + 12}{x^2 - x - 29/3} \cdot \left(\dfrac{3x + 2}{x^2 + 5x - 24} + \dfrac{7}{x^2 + 4x - 32} \right)$

97. $\dfrac{8t^5}{2t^2 - 10t + 12} \div \left(\dfrac{2t}{t^2 - 8t + 15} - \dfrac{3t}{t^2 - 7t + 10} \right)$

98. $\dfrac{9t^3}{3t^3 - 12t^2 + 9t} \div \left(\dfrac{t + 4}{t^2 - 9} - \dfrac{3t - 1}{t^2 + 2t - 3} \right)$

99. Use a graphing calculator to check your answers to Exercises 17, 33, and 61. Use the method discussed on p. 365.

100. Let

$$f(x) = 2 + \dfrac{x - 3}{x + 1}.$$

Use algebra, together with a graphing calculator, to determine the domain and the range of f.

6.3 Complex Rational Expressions

Multiplying by 1 • Dividing Two Rational Expressions

A **complex rational expression** is a rational expression that contains rational expressions within its numerator and/or its denominator. Here are some examples:

$$\dfrac{x + \dfrac{5}{x}}{4x}, \quad \dfrac{\dfrac{x - y}{x + y}}{\dfrac{2x - y}{3x + y}}, \quad \dfrac{\dfrac{7x}{3} - \dfrac{4}{x}}{\dfrac{5x}{6} + \dfrac{8}{3}}, \quad \dfrac{\dfrac{r}{6} + \dfrac{r^2}{144}}{\dfrac{r}{12}}.$$

The rational expressions within each complex rational expression are red.

Complex rational expressions arise in a variety of real-world applications. For example, the last complex rational expression in the list above is used when calculating the size of certain loan payments.

Two methods are used to simplify complex rational expressions. Determining restrictions on variables may now require the solution of equations not yet studied. *Thus for this section we will not state restrictions on variables.*

Method 1: Multiplying by 1

One method of simplifying a complex rational expression is to multiply the entire expression by 1. To write 1, we use the LCD of the rational expressions within the complex rational expression.

EXAMPLE 1 Simplify:

$$\dfrac{\dfrac{1}{a^3b} + \dfrac{1}{b}}{\dfrac{1}{a^2b^2} - \dfrac{1}{b^2}}.$$

Student Notes _____

In Example 1, we found the LCD for four rational expressions. To find this LCD, you may prefer to (1) find one LCD for the two expressions in the numerator, (2) find a second LCD for the expressions in the denominator, and (3) find the LCM of those two LCDs.

Solution The denominators within the complex rational expression are $a^3b, b, a^2b^2,$ and b^2. Thus the LCD is a^3b^2. We multiply by 1, using $(a^3b^2)/(a^3b^2)$:

$$\frac{\dfrac{1}{a^3b} + \dfrac{1}{b}}{\dfrac{1}{a^2b^2} - \dfrac{1}{b^2}} = \frac{\dfrac{1}{a^3b} + \dfrac{1}{b}}{\dfrac{1}{a^2b^2} - \dfrac{1}{b^2}} \cdot \frac{a^3b^2}{a^3b^2}$$

Multiplying by 1, using the LCD

$$= \frac{\left(\dfrac{1}{a^3b} + \dfrac{1}{b}\right)a^3b^2}{\left(\dfrac{1}{a^2b^2} - \dfrac{1}{b^2}\right)a^3b^2}$$

Multiplying the numerator and the denominator. Remember to use parentheses.

$$= \frac{\dfrac{1}{a^3b} \cdot a^3b^2 + \dfrac{1}{b} \cdot a^3b^2}{\dfrac{1}{a^2b^2} \cdot a^3b^2 - \dfrac{1}{b^2} \cdot a^3b^2}$$

Using the distributive law to carry out the multiplications

$$= \frac{\dfrac{a^3b}{a^3b} \cdot b + \dfrac{b}{b} \cdot a^3b}{\dfrac{a^2b^2}{a^2b^2} \cdot a - \dfrac{b^2}{b^2} \cdot a^3}$$

Removing factors that equal 1. Study this carefully.

$$= \frac{b + a^3b}{a - a^3}$$

Simplifying

$$= \frac{b(1 + a^3)}{a(1 - a^2)}$$

Factoring

$$= \frac{b(1 + a)(1 - a + a^2)}{a(1 + a)(1 - a)}$$

Factoring further and identifying a factor that equals 1

$$= \frac{b(1 - a + a^2)}{a(1 - a)}.$$

Simplifying

> ### Using Multiplication by 1 to Simplify a Complex Rational Expression
>
> 1. Find the LCD of all rational expressions *within* the complex rational expression.
> 2. Multiply the complex rational expression by 1, writing 1 as the LCD divided by itself.
> 3. Distribute and simplify so that the numerator and the denominator of the complex rational expression are polynomials.
> 4. Factor and, if possible, simplify.

Note that use of the LCD, when multiplying by a form of 1, clears the numerator and the denominator of the complex rational expression of all rational expressions.

EXAMPLE 2   Simplify:

$$\frac{\dfrac{3}{2x-2}-\dfrac{1}{x+1}}{\dfrac{1}{x-1}+\dfrac{x}{x^2-1}}.$$

Solution   In this case, to find the LCD, we have to factor first:

$$\frac{\dfrac{3}{2x-2}-\dfrac{1}{x+1}}{\dfrac{1}{x-1}+\dfrac{x}{x^2-1}}=\frac{\dfrac{3}{2(x-1)}-\dfrac{1}{x+1}}{\dfrac{1}{x-1}+\dfrac{x}{(x-1)(x+1)}}\qquad\text{The LCD is } 2(x-1)(x+1).$$

$$=\frac{\dfrac{3}{2(x-1)}-\dfrac{1}{x+1}}{\dfrac{1}{x-1}+\dfrac{x}{(x-1)(x+1)}}\cdot\frac{2(x-1)(x+1)}{2(x-1)(x+1)}\qquad\begin{array}{l}\text{Multiplying by 1,}\\\text{using the LCD}\end{array}$$

$$=\frac{\dfrac{3}{2(x-1)}\cdot 2(x-1)(x+1)-\dfrac{1}{x+1}\cdot 2(x-1)(x+1)}{\dfrac{1}{x-1}\cdot 2(x-1)(x+1)+\dfrac{x}{(x-1)(x+1)}\cdot 2(x-1)(x+1)}$$

Using the distributive law

$$=\frac{\dfrac{2(x-1)}{2(x-1)}\cdot 3(x+1)-\dfrac{x+1}{x+1}\cdot 2(x-1)}{\dfrac{x-1}{x-1}\cdot 2(x+1)+\dfrac{(x-1)(x+1)}{(x-1)(x+1)}\cdot 2x}\qquad\begin{array}{l}\text{Removing factors}\\\text{that equal 1}\end{array}$$

$$=\frac{3(x+1)-2(x-1)}{2(x+1)+2x}\qquad\text{Simplifying}$$

$$=\frac{3x+3-2x+2}{2x+2+2x}\qquad\text{Using the distributive law}$$

$$=\frac{x+5}{4x+2}.\qquad\text{Combining like terms}$$

Method 2: Dividing Two Rational Expressions

Another method for simplifying complex rational expressions involves first adding or subtracting, as necessary, to obtain one rational expression in the numerator and one rational expression in the denominator. The problem is then regarded as one rational expression divided by another.

EXAMPLE 3   Simplify:

$$\frac{\dfrac{3}{x} - \dfrac{2}{x^2}}{\dfrac{3}{x-2} + \dfrac{1}{x^2}}.$$

Solution

$$\frac{\dfrac{3}{x} - \dfrac{2}{x^2}}{\dfrac{3}{x-2} + \dfrac{1}{x^2}} = \frac{\dfrac{3}{x}\cdot\dfrac{x}{x} - \dfrac{2}{x^2}}{\dfrac{3}{x-2}\cdot\dfrac{x^2}{x^2} + \dfrac{1}{x^2}\cdot\dfrac{x-2}{x-2}}$$

Multiplying $3/x$ by 1 to obtain x^2 as a common denominator

Multiplying by 1, twice, to obtain $x^2(x-2)$ as a common denominator

$$= \frac{\dfrac{3x}{x^2} - \dfrac{2}{x^2}}{\dfrac{3x^2}{(x-2)x^2} + \dfrac{x-2}{x^2(x-2)}}$$

There is now a common denominator in the numerator and a common denominator in the denominator of the complex rational expression.

$$= \frac{\dfrac{3x-2}{x^2}}{\dfrac{3x^2+x-2}{(x-2)x^2}}$$

Subtracting in the numerator and adding in the denominator. We now have one rational expression divided by another rational expression.

$$= \frac{3x-2}{x^2} \div \frac{3x^2+x-2}{(x-2)x^2}$$

Rewriting with a division symbol

$$= \frac{3x-2}{x^2} \cdot \frac{(x-2)x^2}{3x^2+x-2}$$

To divide, multiply by the reciprocal of the divisor.

$$= \frac{(3x-2)(x-2)x^2}{x^2(3x-2)(x+1)}$$

Factoring and removing a factor equal to 1: $\dfrac{x^2(3x-2)}{x^2(3x-2)} = 1$

$$= \frac{x-2}{x+1}$$

Using Division to Simplify a Complex Rational Expression

1. Add or subtract, as necessary, to get one rational expression in the numerator.
2. Add or subtract, as necessary, to get one rational expression in the denominator.
3. Perform the indicated division (invert the divisor and multiply).
4. Simplify, if possible, by removing any factors that equal 1.

EXAMPLE 4   Simplify:

$$\frac{1 + \dfrac{2}{x}}{1 - \dfrac{4}{x^2}}.$$

Solution   We have

$$\frac{1 + \dfrac{2}{x}}{1 - \dfrac{4}{x^2}} = \frac{\left.\dfrac{x}{x} + \dfrac{2}{x}\right\}}{\left.\dfrac{x^2}{x^2} - \dfrac{4}{x^2}\right\}} \qquad \begin{array}{l}\text{Finding a common denominator}\\[1.5em]\text{Finding a common denominator}\end{array}$$

$$= \frac{\dfrac{x + 2}{x}}{\dfrac{x^2 - 4}{x^2}} \qquad \begin{array}{l}\text{Adding in the numerator}\\[1.5em]\text{Subtracting in the denominator}\end{array}$$

$$= \frac{x + 2}{x} \cdot \frac{x^2}{x^2 - 4} \qquad \begin{array}{l}\text{Multiplying by the reciprocal of}\\\text{the divisor}\end{array}$$

$$= \frac{(x + 2) \cdot x^2}{x(x + 2)(x - 2)} \qquad \begin{array}{l}\text{Factoring. Remember to simplify}\\\text{when possible.}\end{array}$$

$$= \frac{\cancel{(x + 2)}\cancel{x} \cdot x}{\cancel{x}\cancel{(x + 2)}(x - 2)} \qquad \begin{array}{l}\text{Removing a factor equal to 1:}\\ \dfrac{(x + 2)x}{(x + 2)x} = 1\end{array}$$

$$= \frac{x}{x - 2}. \qquad \text{Simplifying}$$

As a quick partial check, we select a convenient value for x—say, 1:

$$\frac{1 + \dfrac{2}{1}}{1 - \dfrac{4}{1^2}} = \frac{1 + 2}{1 - 4} = \frac{3}{-3} = -1 \qquad \begin{array}{l}\text{We evaluated the original expression}\\\text{for } x = 1.\end{array}$$

and

$$\frac{1}{1 - 2} = \frac{1}{-1} = -1. \qquad \begin{array}{l}\text{We evaluated the simplified}\\\text{expression for } x = 1.\end{array}$$

Since both expressions yield the same result, our simplification is probably correct. More evaluation would provide a more definitive check.

If negative exponents occur, we first find an equivalent expression using positive exponents and then simplify as in the preceding examples.

EXAMPLE 5   Simplify:

$$\frac{a^{-1} + b^{-1}}{a^{-3} + b^{-3}}.$$

**technology
connection**

To check Example 4, we can show that the graphs of

$$y_1 = \frac{1 + \dfrac{2}{x}}{1 - \dfrac{4}{x^2}}$$

and

$$y_2 = \frac{x}{x - 2}$$

coincide, or we can show that (except for $x = -2$ or 0) their tables of values are identical. We could also check by showing that (except for $x = -2$ or 0 or 2) $y_2 - y_1 = 0$.

1. Use a graphing calculator to check Example 3. What values, if any, can x *not* equal?

Solution

$$\frac{a^{-1} + b^{-1}}{a^{-3} + b^{-3}} = \frac{\dfrac{1}{a} + \dfrac{1}{b}}{\dfrac{1}{a^3} + \dfrac{1}{b^3}}$$

Rewriting with positive exponents. We continue, using method 2.

$$= \frac{\dfrac{1}{a} \cdot \dfrac{b}{b} + \dfrac{1}{b} \cdot \dfrac{a}{a}}{\dfrac{1}{a^3} \cdot \dfrac{b^3}{b^3} + \dfrac{1}{b^3} \cdot \dfrac{a^3}{a^3}}$$

Finding a common denominator

Finding a common denominator

$$= \frac{\dfrac{b}{ab} + \dfrac{a}{ab}}{\dfrac{b^3}{a^3 b^3} + \dfrac{a^3}{a^3 b^3}}$$

$$= \frac{\dfrac{b + a}{ab}}{\dfrac{b^3 + a^3}{a^3 b^3}}$$

Adding in the numerator

Adding in the denominator

$$= \frac{b + a}{ab} \cdot \frac{a^3 b^3}{b^3 + a^3}$$

Multiplying by the reciprocal of the divisor

$$= \frac{(b + a) \cdot ab \cdot a^2 b^2}{ab(b + a)(b^2 - ab + a^2)}$$

Factoring and looking for common factors

$$= \frac{\cancel{(b + a)} \cdot \cancel{ab} \cdot a^2 b^2}{\cancel{ab}\cancel{(b + a)}(b^2 - ab + a^2)}$$

Removing a factor equal to 1: $\dfrac{(b + a)ab}{ab(b + a)} = 1$

$$= \frac{a^2 b^2}{b^2 - ab + a^2}$$

There is no one method that is better to use. For expressions like

$$\frac{\dfrac{3x + 1}{x - 5}}{\dfrac{2 - x}{x + 3}} \quad \text{or} \quad \frac{\dfrac{3}{x} - \dfrac{2}{x}}{\dfrac{1}{x + 1} + \dfrac{5}{x + 1}},$$

the second method is probably easier to use since it is little or no work to write the expression as a quotient of two rational expressions.

On the other hand, expressions like

$$\frac{\dfrac{3}{a^2 b} - \dfrac{4}{bc^3}}{\dfrac{1}{b^3 c} + \dfrac{2}{ac^4}} \quad \text{or} \quad \frac{\dfrac{5}{a^2 - b^2} + \dfrac{2}{a^2 + 2ab + b^2}}{\dfrac{1}{a - b} + \dfrac{4}{a + b}}$$

require fewer steps if we use the first method. Either method can be used with any complex rational expression.

Exercise Set

6.3

FOR EXTRA HELP

Student's Solutions Manual   Digital Video Tutor CD 5 Videotape 11   AW Math Tutor Center   MathXL Tutorials on CD   MathXL   MyMathLab

🍂 *Concept Reinforcement*   *In each of Exercises 1–6, match the complex rational expression with an equivalent expression from the column on the right.*

1. ___ $\dfrac{\dfrac{5}{x} + \dfrac{3}{y}}{\dfrac{3}{x} - \dfrac{2}{y}}$

2. ___ $\dfrac{\dfrac{5}{x+y}}{\dfrac{3}{x-y}}$

3. ___ $\dfrac{\dfrac{3}{x} + \dfrac{2}{y}}{\dfrac{5}{x} - \dfrac{3}{y}}$

4. ___ $\dfrac{\dfrac{3}{x} - \dfrac{2}{y}}{\dfrac{5}{x} + \dfrac{3}{y}}$

5. ___ $\dfrac{\dfrac{5}{x} - \dfrac{3}{y}}{\dfrac{3}{x} + \dfrac{2}{y}}$

6. ___ $\dfrac{\dfrac{3}{x-y}}{\dfrac{5}{x+y}}$

a) $\dfrac{3}{x-y} \div \dfrac{5}{x+y}$

b) $\dfrac{\dfrac{5}{x} \cdot \dfrac{y}{y} - \dfrac{3}{y} \cdot \dfrac{x}{x}}{\dfrac{3}{x} \cdot \dfrac{y}{y} + \dfrac{2}{y} \cdot \dfrac{x}{x}}$

c) $\dfrac{\dfrac{3}{x} - \dfrac{2}{y}}{\dfrac{5}{x} + \dfrac{3}{y}} \cdot \dfrac{xy}{xy}$

d) $\dfrac{5}{x+y} \div \dfrac{3}{x-y}$

e) $\dfrac{\dfrac{5}{x} \cdot \dfrac{y}{y} + \dfrac{3}{y} \cdot \dfrac{x}{x}}{\dfrac{3}{x} \cdot \dfrac{y}{y} - \dfrac{2}{y} \cdot \dfrac{x}{x}}$

f) $\dfrac{\dfrac{3}{x} + \dfrac{2}{y}}{\dfrac{5}{x} - \dfrac{3}{y}} \cdot \dfrac{xy}{xy}$

9. $\dfrac{\dfrac{5}{a} - \dfrac{4}{b}}{\dfrac{2}{a} + \dfrac{3}{b}}$

10. $\dfrac{\dfrac{2}{r} - \dfrac{3}{t}}{\dfrac{4}{r} + \dfrac{5}{t}}$

11. $\dfrac{\dfrac{3}{z} + \dfrac{2}{y}}{\dfrac{4}{z} - \dfrac{1}{y}}$

12. $\dfrac{\dfrac{6}{x} + \dfrac{7}{y}}{\dfrac{7}{x} - \dfrac{6}{y}}$

13. $\dfrac{\dfrac{a^2 - b^2}{ab}}{\dfrac{a-b}{b}}$

14. $\dfrac{\dfrac{x^2 - y^2}{xy}}{\dfrac{x-y}{y}}$

15. $\dfrac{1 - \dfrac{2}{3x}}{x - \dfrac{4}{9x}}$

16. $\dfrac{\dfrac{3x}{y} - x}{2y - \dfrac{y}{x}}$

17. $\dfrac{\dfrac{x^{-1} + y^{-1}}{x^2 - y^2}}{xy}$

18. $\dfrac{\dfrac{a^{-1} + b^{-1}}{a^2 - b^2}}{ab}$

19. $\dfrac{\dfrac{1}{a-h} - \dfrac{1}{a}}{h}$

20. $\dfrac{\dfrac{1}{x+h} - \dfrac{1}{x}}{h}$

21. $\dfrac{\dfrac{a^2 - 4}{a^2 + 3a + 2}}{\dfrac{a^2 - 5a - 6}{a^2 - 6a - 7}}$

22. $\dfrac{\dfrac{x^2 - x - 12}{x^2 - 2x - 15}}{\dfrac{x^2 + 8x + 12}{x^2 - 5x - 14}}$

23. $\dfrac{\dfrac{x}{x^2 + 3x - 4} - \dfrac{1}{x^2 + 3x - 4}}{\dfrac{x}{x^2 + 6x + 8} + \dfrac{3}{x^2 + 6x + 8}}$

Simplify. If possible, use a second method or evaluation as a check.

7. $\dfrac{\dfrac{x+2}{x-1}}{\dfrac{x+4}{x-3}}$

8. $\dfrac{\dfrac{x-1}{x+3}}{\dfrac{x-6}{x+2}}$

24. $\dfrac{\dfrac{x}{x^2 + 5x - 6} + \dfrac{6}{x^2 + 5x - 6}}{\dfrac{x}{x^2 - 5x + 4} - \dfrac{2}{x^2 - 5x + 4}}$

25. $\dfrac{\dfrac{1}{y} + 2}{\dfrac{1}{y} - 3}$

26. $\dfrac{7 + \dfrac{1}{a}}{\dfrac{1}{a} - 3}$

27. $\dfrac{y + y^{-1}}{y - y^{-1}}$

28. $\dfrac{x - x^{-1}}{x + x^{-1}}$

29. $\dfrac{\dfrac{1}{x - 2} + \dfrac{3}{x - 1}}{\dfrac{2}{x - 1} + \dfrac{5}{x - 2}}$

30. $\dfrac{\dfrac{2}{y - 3} + \dfrac{1}{y + 1}}{\dfrac{3}{y + 1} + \dfrac{4}{y - 3}}$

31. $\dfrac{a(a + 3)^{-1} - 2(a - 1)^{-1}}{a(a + 3)^{-1} - (a - 1)^{-1}}$

32. $\dfrac{a(a + 2)^{-1} - 3(a - 3)^{-1}}{a(a + 2)^{-1} - (a - 3)^{-1}}$

33. $\dfrac{\dfrac{2}{a^2 - 1} + \dfrac{1}{a + 1}}{\dfrac{3}{a^2 - 1} + \dfrac{2}{a - 1}}$

34. $\dfrac{\dfrac{3}{a^2 - 9} + \dfrac{2}{a + 3}}{\dfrac{4}{a^2 - 9} + \dfrac{1}{a + 3}}$

35. $\dfrac{\dfrac{5}{x^2 - 4} - \dfrac{3}{x - 2}}{\dfrac{4}{x^2 - 4} - \dfrac{2}{x + 2}}$

36. $\dfrac{\dfrac{4}{x^2 - 1} - \dfrac{3}{x + 1}}{\dfrac{5}{x^2 - 1} - \dfrac{2}{x - 1}}$

37. $\dfrac{\dfrac{y}{y^2 - 4} + \dfrac{5}{4 - y^2}}{\dfrac{y^2}{y^2 - 4} + \dfrac{25}{4 - y^2}}$

38. $\dfrac{\dfrac{y}{y^2 - 1} + \dfrac{3}{1 - y^2}}{\dfrac{y^2}{y^2 - 1} + \dfrac{9}{1 - y^2}}$

39. $\dfrac{\dfrac{y^2}{y^2 - 25} - \dfrac{y}{y - 5}}{\dfrac{y}{y^2 - 25} - \dfrac{1}{y + 5}}$

40. $\dfrac{\dfrac{y^2}{y^2 - 9} - \dfrac{y}{y + 3}}{\dfrac{y}{y^2 - 9} - \dfrac{1}{y - 3}}$

41. $\dfrac{\dfrac{a}{a + 2} + \dfrac{5}{a}}{\dfrac{a}{2a + 4} + \dfrac{1}{3a}}$

42. $\dfrac{\dfrac{a}{a + 3} + \dfrac{4}{5a}}{\dfrac{a}{2a + 6} + \dfrac{3}{a}}$

43. $\dfrac{\dfrac{1}{x^2 - 3x + 2} + \dfrac{1}{x^2 - 4}}{\dfrac{1}{x^2 + 4x + 4} + \dfrac{1}{x^2 - 4}}$

44. $\dfrac{\dfrac{1}{x^2 + 3x + 2} + \dfrac{1}{x^2 - 1}}{\dfrac{1}{x^2 - 1} + \dfrac{1}{x^2 - 4x + 3}}$

45. $\dfrac{\dfrac{3}{a^2 - 4a + 3} + \dfrac{3}{a^2 - 5a + 6}}{\dfrac{3}{a^2 - 3a + 2} + \dfrac{3}{a^2 + 3a - 10}}$

46. $\dfrac{\dfrac{1}{a^2 + 7a + 10} - \dfrac{2}{a^2 - 7a + 12}}{\dfrac{2}{a^2 - a - 6} - \dfrac{1}{a^2 + a - 20}}$

Aha! **47.** $\dfrac{\dfrac{y}{y^2 - 4} - \dfrac{2y}{y^2 + y - 6}}{\dfrac{2y}{y^2 + y - 6} - \dfrac{y}{y^2 - 4}}$

48. $\dfrac{\dfrac{y}{y^2 - 1} - \dfrac{3y}{y^2 + 5y + 4}}{\dfrac{3y}{y^2 - 1} - \dfrac{y}{y^2 - 4y + 3}}$

49. $\dfrac{\dfrac{3}{x^2 + 2x - 3} - \dfrac{1}{x^2 - 3x - 10}}{\dfrac{3}{x^2 - 6x + 5} - \dfrac{1}{x^2 + 5x + 6}}$

50. $\dfrac{\dfrac{1}{a^2 + 7a + 12} + \dfrac{1}{a^2 + a - 6}}{\dfrac{1}{a^2 + 2a - 8} + \dfrac{1}{a^2 + 5a + 4}}$

51. Michael *incorrectly* simplifies

$$\frac{a + b^{-1}}{a + c^{-1}} \quad \text{as} \quad \frac{a + c}{a + b}.$$

What mistake is he making and how could you convince him that this is incorrect?

52. To simplify a complex rational expression in which the sum of two fractions is divided by the difference of the same two fractions, which method is easier? Why?

SKILL MAINTENANCE

Solve. [1.3]

53. $2(3x - 1) + 5(4x - 3) = 3(2x + 1)$

54. $5(2x + 3) - 3(4x + 1) = 2(3x - 5)$

55. Solve for y: $\dfrac{t}{s + y} = r$. [1.5]

56. Solve: $|2x - 3| = 7$. [4.3]

57. *Framing.* Andrea has two rectangular frames. The first frame is 3 cm shorter, and 4 cm narrower, than the second frame. If the perimeter of the second frame is 1 cm less than twice the perimeter of the first, what is the perimeter of each frame? [3.3]

58. *Earnings.* Antonio received $28 in tips on Monday, $22 in tips on Tuesday, and $36 in tips on Wednesday. How much will Antonio need to receive in tips on Thursday if his average for the four days is to be $30? [1.4]

SYNTHESIS

59. In arithmetic, we are taught that

$$\frac{a}{b} \div \frac{c}{d} = \frac{a}{b} \cdot \frac{d}{c}$$

(to divide by a fraction, we invert and multiply). Use method 1 to explain *why* we do this.

60. Lisa claims that she can simplify complex rational expressions without knowing how to add rational expressions. Is this possible? Why or why not?

Simplify.

61. $\dfrac{5x^{-2} + 10x^{-1}y^{-1} + 5y^{-2}}{3x^{-2} - 3y^{-2}}$

62. $(a^2 - ab + b^2)^{-1}(a^2b^{-1} + b^2a^{-1}) \times (a^{-2} - b^{-2})(a^{-2} + 2a^{-1}b^{-1} + b^{-2})^{-1}$

63. *Astronomy.* When two galaxies are moving in opposite directions at velocities v_1 and v_2, an observer in one of the galaxies would see the other galaxy receding at speed

$$\frac{v_1 + v_2}{1 + \dfrac{v_1 v_2}{c^2}},$$

where c is the speed of light. Determine the observed speed if v_1 and v_2 are both one-fourth the speed of light.

Find and simplify

$$\frac{f(x + h) - f(x)}{h}$$

for each rational function f in Exercises 64–67.

64. $f(x) = \dfrac{2}{x^2}$

65. $f(x) = \dfrac{3}{x}$

66. $f(x) = \dfrac{x}{1 - x}$

67. $f(x) = \dfrac{2x}{1 + x}$

68. If

$$F(x) = \frac{3 + \dfrac{1}{x}}{2 - \dfrac{8}{x^2}},$$

find the domain of F.

69. If

$$G(x) = \frac{x - \dfrac{1}{x^2 - 1}}{\dfrac{1}{9} - \dfrac{1}{x^2 - 16}},$$

find the domain of G.

70. Find the reciprocal of y if

$$y = x^2 + x + 1 + \frac{1}{x} + \frac{1}{x^2}.$$

71. For $f(x) = \dfrac{2}{2 + x}$, find $f(f(a))$.

72. For $g(x) = \dfrac{x + 3}{x - 1}$, find $g(g(a))$.

73. Let

$$f(x) = \left[\frac{\dfrac{x + 3}{x - 3} + 1}{\dfrac{x + 3}{x - 3} - 1} \right]^4.$$

Find a simplified form of $f(x)$ and specify the domain of f.

74. Use a graphing calculator to check your answers to Exercises 27, 29, 41, and 68.

75. Use a graphing calculator to check your answers to Exercises 15, 25, 45, and 69.

76. Use algebra to determine the domain of the function given by

$$f(x) = \dfrac{\dfrac{1}{x-2}}{\dfrac{x}{x-2} - \dfrac{5}{x-2}}.$$

Then explain how a graphing calculator could be used to check your answer.

77. *Financial planning.*   Alexis wishes to invest a portion of each month's pay in an account that pays 7.5% interest. If he wants to have $30,000 in the account after 10 yr, the amount invested each month is given by

$$\dfrac{30{,}000 \cdot \dfrac{0.075}{12}}{\left(1 + \dfrac{0.075}{12}\right)^{120} - 1}.$$

Find the amount of Alexis' monthly investment.

CORNER

Which Method Is Better?

COLLABORATIVE

Focus: Complex rational expressions

Time: 10–15 minutes

Group size: 3–4

ACTIVITY

Consider the steps in Examples 2 and 3 for simplifying a complex rational expression by each of the two methods. Then, work as a group to simplify

$$\dfrac{\dfrac{5}{x+1} - \dfrac{1}{x}}{\dfrac{2}{x^2} + \dfrac{4}{x}}$$

subject to the following conditions.

1. The group should predict which method will more easily simplify this expression.

2. Using the method selected in part (1), one group member should perform the first step in the simplification and then pass the problem on to another member of the group. That person then checks the work, performs the next step, and passes the problem on to another group member. If a mistake is found, the problem should be passed to the person who made the mistake for repair. This process continues until, eventually, the simplification is complete.

3. At the same time that part (2) is being performed, another group member should perform the first step of the solution using the method not selected in part (1). He or she should then pass the problem to another group member and so on, just as in part (2).

4. What method *was* easier? Why? Compare your responses with those of other groups.

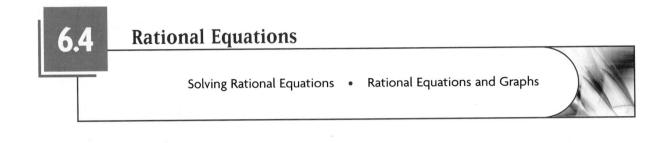

6.4   Rational Equations

Solving Rational Equations   •   Rational Equations and Graphs

As we mentioned in the Connecting the Concepts feature of Chapter 5, it is not unusual to learn a skill that enables us to write equivalent expressions and then to put that skill to work solving a new type of equation. That is precisely what we are now doing: Sections 6.1–6.3 have been devoted to writing equivalent expressions in which rational expressions appeared. There we used least common denominators to add or subtract. Here in Section 6.4, we return to the task of solving equations, but this time the equations contain rational expressions and the LCD is used as part of the multiplication principle.

Solving Rational Equations

In Sections 6.1–6.3, we learned how to *simplify expressions*. We now learn to *solve* a new type of *equation*. A **rational equation** is an equation that contains one or more rational expressions. Here are some examples:

$$\frac{2}{3} - \frac{5}{6} = \frac{1}{t}, \qquad \frac{a-1}{a-5} = \frac{4}{a^2-25}, \qquad x^3 + \frac{6}{x} = 5.$$

As you will see in Section 6.5, equations of this type occur frequently in applications. To solve rational equations, recall that one way to *clear fractions* from an equation is to multiply both sides of the equation by the LCD.

> *To Solve a Rational Equation*
>
> Multiply both sides of the equation by the LCD. This is called *clearing fractions* and produces an equation similar to those we have already solved.

Recall that division by 0 is undefined. Note, too, that variables usually appear in at least one denominator of a rational equation. Thus certain numbers can often be ruled out as possible solutions before we even attempt to solve a given rational equation.

EXAMPLE 1   Solve: $\dfrac{x+4}{3x} + \dfrac{x+8}{5x} = 2$.

Solution   Because the left side of this equation is undefined when x is 0, we state at the outset that $x \neq 0$. Next, we multiply both sides of the equation by the LCD, $3 \cdot 5 \cdot x$, or $15x$:

$$15x\left(\frac{x+4}{3x} + \frac{x+8}{5x}\right) = 15x \cdot 2 \qquad \text{Multiplying by the LCD to clear fractions}$$

$$15x \cdot \frac{x+4}{3x} + 15x \cdot \frac{x+8}{5x} = 15x \cdot 2 \qquad \text{Using the distributive law}$$

$$\frac{5 \cdot 3x \cdot (x+4)}{3x} + \frac{3 \cdot 5x \cdot (x+8)}{5x} = 30x \qquad \text{Locating factors equal to 1}$$

$$5(x+4) + 3(x+8) = 30x \qquad \text{Removing factors equal to 1:}$$
$$\frac{3x}{3x} = 1; \frac{5x}{5x} = 1$$

We have solved equations like this before.

$$5x + 20 + 3x + 24 = 30x \qquad \text{Using the distributive law}$$

$$8x + 44 = 30x$$

$$44 = 22x$$

$$2 = x. \qquad \text{This should check since } x \neq 0.$$

Check:
$$\frac{x+4}{3x} + \frac{x+8}{5x} = 2$$

$$\frac{2+4}{3 \cdot 2} + \frac{2+8}{5 \cdot 2} \;\Big|\; 2$$

$$\frac{6}{6} + \frac{10}{10} \;\Big|\;$$

$$2 \overset{?}{=} 2 \quad \text{TRUE}$$

The number 2 is the solution.

_____

It is important that when we clear fractions, all denominators are eliminated. This produces an equation without rational expressions, which is easier to solve.

EXAMPLE 2   Solve: $\dfrac{x-1}{x-5} = \dfrac{4}{x-5}$.

Solution   To ensure that neither denominator is 0, we state at the outset the restriction that $x \neq 5$. Then we proceed as before, multiplying both sides by the LCD, $x - 5$:

$$(x-5) \cdot \frac{x-1}{x-5} = (x-5) \cdot \frac{4}{x-5} \qquad \text{Using the multiplication principle}$$

$$x - 1 = 4$$

$$x = 5. \qquad \text{But recall that } x \neq 5.$$

In this case, it is important to remember that, because of the restriction above, 5 cannot be a solution. A check confirms the necessity of that restriction.

Check:

$$\frac{x-1}{x-5} = \frac{4}{x-5}$$

$$\frac{\dfrac{5-1}{5-5}}{} \Bigg| \frac{4}{5-5}$$

$$\frac{4}{0} \overset{?}{=} \frac{4}{0} \qquad \text{Division by 0 is undefined.}$$

This equation has no solution.

___

To see why 5 is not a solution of Example 2, note that the multiplication principle for equations requires that we multiply both sides by a *nonzero* number. When both sides of an equation are multiplied by an expression containing variables, it is possible that certain replacements will make that expression equal to 0. Thus it is safe to say that *if* a solution of

$$\frac{x-1}{x-5} = \frac{4}{x-5}$$

exists, then it is also a solution of $x - 1 = 4$. We *cannot* conclude that every solution of $x - 1 = 4$ is a solution of the original equation.

> *Caution!* When solving rational equations, be sure to list any restrictions as part of the first step. Refer to the restriction(s) as you proceed.

EXAMPLE 3 Solve: $\dfrac{x^2}{x-3} = \dfrac{9}{x-3}$.

Solution Note that $x \neq 3$. Since the LCD is $x - 3$, we multiply both sides by $x - 3$:

$$(x - 3) \cdot \frac{x^2}{x - 3} = (x - 3) \cdot \frac{9}{x - 3}$$

$$x^2 = 9 \qquad \text{Simplifying}$$

$$x^2 - 9 = 0 \qquad \text{Getting 0 on one side}$$

$$(x - 3)(x + 3) = 0 \qquad \text{Factoring}$$

$$x = 3 \quad or \quad x = -3. \qquad \text{Using the principle of zero products}$$

Although 3 is a solution of $x^2 = 9$, it must be rejected as a solution of the rational equation because of the restriction stated in red above. You should check that -3 *is* a solution despite the fact that 3 is not.

EXAMPLE 4 Solve: $\dfrac{2}{x+5} + \dfrac{1}{x-5} = \dfrac{16}{x^2 - 25}$.

Solution   To find all restrictions and to assist in finding the LCD, we factor:

$$\frac{2}{x+5} + \frac{1}{x-5} = \frac{16}{(x+5)(x-5)}. \qquad \text{Factoring } x^2 - 25$$

Note that $x \neq -5$ and $x \neq 5$. We multiply by the LCD, $(x+5)(x-5)$, and then use the distributive law:

$$(x+5)(x-5)\left(\frac{2}{x+5} + \frac{1}{x-5}\right) = (x+5)(x-5) \cdot \frac{16}{(x+5)(x-5)}$$

$$(x+5)(x-5)\frac{2}{x+5} + (x+5)(x-5)\frac{1}{x-5} = \frac{(x+5)(x-5)16}{(x+5)(x-5)}$$

$$2(x-5) + (x+5) = 16$$

$$2x - 10 + x + 5 = 16$$

$$3x - 5 = 16$$

$$3x = 21$$

$$x = 7.$$

A check will confirm that the solution is 7.

Rational equations often appear when we are working with functions.

EXAMPLE 5   Let $f(x) = x + \dfrac{6}{x}$. Find all values of a for which $f(a) = 5$.

Solution   Since $f(a) = a + \dfrac{6}{a}$, the problem asks that we find all values of a for which

$$a + \frac{6}{a} = 5.$$

First note that $a \neq 0$. To solve for a, we multiply both sides of the equation by the LCD, a:

$$a\left(a + \frac{6}{a}\right) = 5 \cdot a \qquad \text{Multiplying both sides by } a.$$
$$\qquad\qquad\qquad\qquad \text{Parentheses are important.}$$

$$a \cdot a + a \cdot \frac{6}{a} = 5a \qquad \text{Using the distributive law}$$

$$a^2 + 6 = 5a \qquad \text{Simplifying}$$

$$a^2 - 5a + 6 = 0 \qquad \text{Getting 0 on one side}$$

$$(a-3)(a-2) = 0 \qquad \text{Factoring}$$

$$a = 3 \quad or \quad a = 2. \qquad \text{Using the principle of zero products}$$

Check: $f(3) = 3 + \dfrac{6}{3} = 3 + 2 = 5;$

$\qquad\qquad f(2) = 2 + \dfrac{6}{2} = 2 + 3 = 5.$

The solutions are 2 and 3. For $a = 2$ or $a = 3$, we have $f(a) = 5$.

technology connection

There are several ways in which Example 5 can be checked. One way is to confirm that the graphs of $y_1 = x + 6/x$ and $y_2 = 5$ intersect at $x = 2$ and $x = 3$. You can also use a table to check that $y_1 = y_2$ when x is 2 and again when x is 3.

Use a graphing calculator to check Examples 1–3.

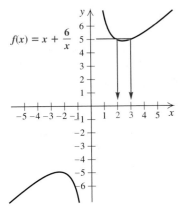

$f(x) = x + \dfrac{6}{x}$

A computer-generated visualization of Example 5

Rational Equations and Graphs

One way to visualize the solution to Example 5 is to make a graph. This can be done by graphing

$$f(x) = x + \frac{6}{x}$$

with a calculator, with a computer, or by hand. We then inspect the graph for any x-values that are paired with 5. (Note that no y-value is paired with 0, since 0 is not in the domain of f.) It appears from the graph that $f(x) = 5$ when $x \approx 2$ or $x \approx 3$. Although making a graph is not the fastest or most precise method of solving a rational equation, it provides visualization and is useful when problems are too difficult to solve algebraically.

Exercise Set

6.4

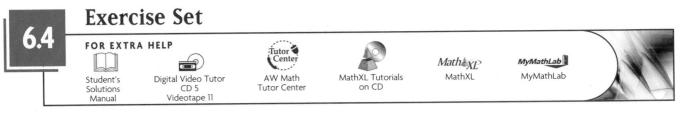

FOR EXTRA HELP

Student's Solutions Manual   Digital Video Tutor CD 5 Videotape 11   AW Math Tutor Center   MathXL Tutorials on CD   MathXL   MyMathLab

↪ *Concept Reinforcement   Classify each of the following as either an expression or an equation.*

1. $\dfrac{5x}{x + 2} - \dfrac{3}{x} = 7$

2. $\dfrac{3}{x - 4} + \dfrac{2}{x + 4}$

3. $\dfrac{4}{t^2 - 1} + \dfrac{3}{t + 1}$

4. $\dfrac{2}{t^2 - 1} + \dfrac{3}{t + 1} = 5$

5. $\dfrac{2}{x + 7} + \dfrac{6}{5x} = 4$

6. $\dfrac{5}{2x} - \dfrac{3}{x^2} = 7$

7. $\dfrac{t + 3}{t - 4} = \dfrac{t - 5}{t - 7}$

8. $\dfrac{7t}{2t - 3} \div \dfrac{3t}{2t + 3}$

9. $\dfrac{5x}{x^2 - 4} \cdot \dfrac{7}{x^2 - 5x + 4}$

10. $\dfrac{7x}{2 - x} = \dfrac{3}{4 - x}$

Solve. If no solution exists, state this.

11. $\dfrac{t}{2} + \dfrac{t}{3} = 7$

12. $\dfrac{y}{4} + \dfrac{y}{3} = 8$

13. $\dfrac{5}{8} - \dfrac{1}{x} = \dfrac{3}{4}$

14. $\dfrac{1}{3} - \dfrac{1}{t} = \dfrac{5}{6}$

15. $\dfrac{2x + 1}{3} + \dfrac{x + 2}{5} = \dfrac{4x}{15}$

16. $\dfrac{3x + 2}{2} + \dfrac{x + 4}{8} = \dfrac{2x}{5}$

17. $\dfrac{2}{3} - \dfrac{1}{t} = \dfrac{7}{3t}$

18. $\dfrac{1}{2} - \dfrac{2}{t} = \dfrac{3}{2t}$

19. $\dfrac{4}{x - 1} + \dfrac{5}{6} = \dfrac{2}{3x - 3}$

20. $\dfrac{3}{2x + 10} + \dfrac{5}{4} = \dfrac{7}{x + 5}$

Aha! **21.** $\dfrac{2}{6} + \dfrac{1}{2x} = \dfrac{1}{3}$

22. $\dfrac{12}{15} - \dfrac{1}{3x} = \dfrac{4}{5}$

23. $y + \dfrac{4}{y} = -5$

24. $t + \dfrac{6}{t} = -5$

25. $x - \dfrac{12}{x} = 4$

26. $y - \dfrac{14}{y} = 5$

27. $\dfrac{t-1}{t-3} = \dfrac{2}{t-3}$

28. $\dfrac{x-2}{x-4} = \dfrac{2}{x-4}$

29. $\dfrac{x}{x-5} = \dfrac{25}{x^2-5x}$

30. $\dfrac{t}{t-6} = \dfrac{36}{t^2-6t}$

31. $\dfrac{5}{4t} = \dfrac{7}{5t-2}$

32. $\dfrac{3}{x-2} = \dfrac{5}{x+4}$

Aha! **33.** $\dfrac{x^2+4}{x-1} = \dfrac{5}{x-1}$

34. $\dfrac{x^2-1}{x+2} = \dfrac{3}{x+2}$

35. $\dfrac{6}{a+1} = \dfrac{a}{a-1}$

36. $\dfrac{4}{a-7} = \dfrac{-2a}{a+3}$

37. $\dfrac{60}{t-5} - \dfrac{18}{t} = \dfrac{40}{t}$

38. $\dfrac{50}{t-2} - \dfrac{16}{t} = \dfrac{30}{t}$

39. $\dfrac{3}{x-3} + \dfrac{5}{x+2} = \dfrac{5x}{x^2-x-6}$

40. $\dfrac{2}{x-2} + \dfrac{1}{x+4} = \dfrac{x}{x^2+2x-8}$

41. $\dfrac{3}{x} + \dfrac{x}{x+2} = \dfrac{4}{x^2+2x}$

42. $\dfrac{x}{x+1} + \dfrac{5}{x} = \dfrac{1}{x^2+x}$

43. $\dfrac{5}{x+2} - \dfrac{3}{x-2} = \dfrac{2x}{4-x^2}$

44. $\dfrac{y+3}{y+2} - \dfrac{y}{y^2-4} = \dfrac{y}{y-2}$

45. $\dfrac{3}{x^2-6x+9} + \dfrac{x-2}{3x-9} = \dfrac{x}{2x-6}$

46. $\dfrac{3-2y}{y+1} - \dfrac{10}{y^2-1} = \dfrac{2y+3}{1-y}$

In Exercises 47–52, a rational function f is given. Find all values of a for which f(a) is the indicated value.

47. $f(x) = 2x - \dfrac{15}{x}$; $f(a) = 7$

48. $f(x) = 2x - \dfrac{6}{x}$; $f(a) = 1$

49. $f(x) = \dfrac{x-5}{x+1}$; $f(a) = \dfrac{3}{5}$

50. $f(x) = \dfrac{x-3}{x+2}$; $f(a) = \dfrac{1}{5}$

51. $f(x) = \dfrac{12}{x} - \dfrac{12}{2x}$; $f(a) = 8$

52. $f(x) = \dfrac{6}{x} - \dfrac{6}{2x}$; $f(a) = 5$

For each pair of functions f and g, find all values of a for which f(a) = g(a).

53. $f(x) = \dfrac{3x-1}{x^2-7x+10}$,

$g(x) = \dfrac{x-1}{x^2-4} + \dfrac{2x+1}{x^2-3x-10}$

54. $f(x) = \dfrac{2x+5}{x^2+4x+3}$, $g(x) = \dfrac{x+2}{x^2-9} + \dfrac{x-1}{x^2-2x-3}$

55. $f(x) = \dfrac{2}{x^2-8x+7}$, $g(x) = \dfrac{3}{x^2-2x-3} - \dfrac{1}{x^2-1}$

56. $f(x) = \dfrac{4}{x^2+3x-10}$,

$g(x) = \dfrac{3}{x^2-x-12} + \dfrac{1}{x^2+x-6}$

57. Explain how one can easily produce rational equations for which no solution exists. (*Hint*: Examine Example 2.)

58. Below are unlabeled graphs of $f(x) = x + 2$ and $g(x) = (x^2 - 4)/(x - 2)$. How could you determine which graph represents *f* and which graph represents *g*?

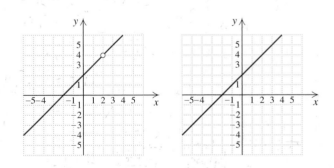

SKILL MAINTENANCE

59. *Test questions.*   There are 70 questions on a test. The questions are either multiple-choice, true–false, or fill-in. There are twice as many true–false as fill-in and 5 fewer multiple-choice than true–false. How many of each type of question are there on the test? [3.5]

60. *Dinner prices.* The Danville Volunteer Fire Department served 250 dinners. A child's dinner cost $3 and an adult's dinner cost $8. The total amount of money collected was $1410. How many of each type of dinner was served? [3.3]

61. *Agriculture.* The perimeter of a rectangular corn field is 628 m. The length of the field is 6 m greater than the width. Find the area of the field. [1.4]

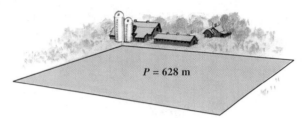

$P = 628$ m

62. Find two consecutive positive even numbers whose product is 288. [5.8]

63. Determine whether each of the following systems is consistent or inconsistent. [3.1]
  a) $2x - 3y = 4$,
       $4x - 6y = 7$
  b) $x + 3y = 2$,
       $2x - 3y = 1$

64. Solve: $|x - 2| > 3$. [4.3]

SYNTHESIS

65. Is the following statement true or false: "For any real numbers a, b, and c, if $ac = bc$, then $a = b$"? Explain why you answered as you did.

66. When checking a possible solution of a rational equation, is it sufficient to check that the "solution" does not make any denominator equal to 0? Why or why not?

For each pair of functions f and g, find all values of a for which $f(a) = g(a)$.

67. $f(x) = \dfrac{x - \dfrac{2}{3}}{x + \dfrac{1}{2}}$, $g(x) = \dfrac{x + \dfrac{2}{3}}{x - \dfrac{3}{2}}$

68. $f(x) = \dfrac{2 - \dfrac{x}{4}}{2}$, $g(x) = \dfrac{\dfrac{x}{4} - 2}{\dfrac{x}{2} + 2}$

69. $f(x) = \dfrac{x + 3}{x + 2} - \dfrac{x + 4}{x + 3}$, $g(x) = \dfrac{x + 5}{x + 4} - \dfrac{x + 6}{x + 5}$

70. $f(x) = \dfrac{1}{1 + x} + \dfrac{x}{1 - x}$, $g(x) = \dfrac{1}{1 - x} - \dfrac{x}{1 + x}$

71. $f(x) = \dfrac{0.793}{x} + 18.15$, $g(x) = \dfrac{6.034}{x} - 43.17$

72. $f(x) = \dfrac{2.315}{x} - \dfrac{12.6}{17.4}$, $g(x) = \dfrac{6.71}{x} + 0.763$

Recall that identities are true for any possible replacement of the variable(s). Determine whether each of the following equations is an identity.

73. $\dfrac{x^2 + 6x - 16}{x - 2} = x + 8$, $x \ne 2$

74. $\dfrac{x^3 + 8}{x^2 - 4} = \dfrac{x^2 - 2x + 4}{x - 2}$, $x \ne -2, x \ne 2$

75. Use a graphing calculator to check your answers to Exercises 11, 45, and 49.

76. Use a graphing calculator to check your answers to Exercises 12, 34, and 50.

77. Use a graphing calculator with a TABLE feature to show that 2 is not in the domain of g, if $g(x) = (x^2 - 4)/(x - 2)$. (See Exercise 58.)

78. Can Exercise 77 be answered on a graphing calculator using only graphs? Why or why not?

6.5 Solving Applications Using Rational Equations

Problems Involving Work • Problems Involving Motion

Now that we are able to solve rational equations, it is possible to solve problems that we could not have handled before. The five problem-solving steps remain the same.

Problems Involving Work

EXAMPLE 1

The roof of Grady and Cleo's townhouse needs to be reshingled. Grady can do the job alone in 8 hr and Cleo can do the job alone in 10 hr. How long will it take the two of them, working together, to reshingle the roof?

Solution

1. **Familiarize.** We familiarize ourselves with the problem by considering two *incorrect* ways of translating the problem to mathematical language.

   a) One *incorrect* way to translate the problem is to add the two times:

   $$8 \text{ hr} + 10 \text{ hr} = 18 \text{ hr}.$$

   Think about this. Grady can do the job *alone* in 8 hr. If Grady and Cleo work together, whatever time it takes them must be *less* than 8 hr.

   b) Another *incorrect* approach is to assume that each person reshingles half of the roof. Were this the case,

   Grady would reshingle $\frac{1}{2}$ the roof in $\frac{1}{2}(8 \text{ hr})$, or 4 hr,

   and

   Cleo would reshingle $\frac{1}{2}$ the roof in $\frac{1}{2}(10 \text{ hr})$, or 5 hr.

   But time would be wasted since Grady would finish 1 hr before Cleo. Were Grady to help Cleo after completing his half, the entire job would take between 4 and 5 hr. This information provides a partial check on any answer we get—the answer should be between 4 and 5 hr.

   Let's consider how much of the job each person completes in 1 hr, 2 hr, 3 hr, and so on. Since Grady takes 8 hr to reshingle the entire roof, in 1 hr he reshingles $\frac{1}{8}$ of the roof. Since Cleo takes 10 hr to reshingle the entire roof, in 1 hr she reshingles $\frac{1}{10}$ of the roof. Thus Grady works at a rate of $\frac{1}{8}$ roof per hour, and Cleo works at a rate of $\frac{1}{10}$ roof per hour.

   Working together, Grady and Cleo reshingle $\frac{1}{8} + \frac{1}{10}$ roof in 1 hr, so their rate—as a team—is $\frac{1}{8} + \frac{1}{10} = \frac{5}{40} + \frac{4}{40} = \frac{9}{40}$ roof per hour.

In 2 hr, Grady reshingles $\frac{1}{8} \cdot 2$ of the roof and Cleo reshingles $\frac{1}{10} \cdot 2$ of the roof. Working together, they reshingle

$$\frac{1}{8} \cdot 2 + \frac{1}{10} \cdot 2, \text{ or } \frac{9}{20} \text{ of the roof in 2 hr.} \qquad \text{Note that } \frac{9}{40} \cdot 2 = \frac{9}{20}.$$

We are looking for the time required to reshingle 1 entire roof, not just a part of it. To find this time, we form a table and extend the pattern developed above.

Time	Fraction of the Roof Reshingled		
	By Grady	**By Cleo**	**Together**
1 hr	$\frac{1}{8}$	$\frac{1}{10}$	$\frac{1}{8} + \frac{1}{10}$, or $\frac{9}{40}$
2 hr	$\frac{1}{8} \cdot 2$	$\frac{1}{10} \cdot 2$	$\left(\frac{1}{8} + \frac{1}{10}\right)2$, or $\frac{9}{40} \cdot 2$, or $\frac{9}{20}$
3 hr	$\frac{1}{8} \cdot 3$	$\frac{1}{10} \cdot 3$	$\left(\frac{1}{8} + \frac{1}{10}\right)3$, or $\frac{9}{40} \cdot 3$, or $\frac{27}{40}$
t hr	$\frac{1}{8} \cdot t$	$\frac{1}{10} \cdot t$	$\left(\frac{1}{8} + \frac{1}{10}\right)t$, or $\frac{9}{40} \cdot t$

2. Translate. From the table, we see that t must be some number for which

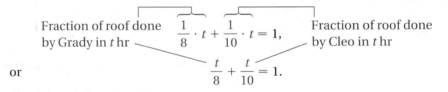

Fraction of roof done by Grady in t hr $\quad \dfrac{1}{8} \cdot t + \dfrac{1}{10} \cdot t = 1,\quad$ Fraction of roof done by Cleo in t hr

or

$$\frac{t}{8} + \frac{t}{10} = 1.$$

3. Carry out. We solve the equation:

$$\frac{t}{8} + \frac{t}{10} = 1$$

$$40\left(\frac{t}{8} + \frac{t}{10}\right) = 40 \cdot 1 \qquad \text{Multiplying by the LCD}$$

$$\frac{40t}{8} + \frac{40t}{10} = 40 \qquad \text{Distributing the 40}$$

$$5t + 4t = 40 \qquad \text{Simplifying}$$

$$9t = 40$$

$$t = \frac{40}{9}, \text{ or } 4\frac{4}{9}.$$

4. Check. In $\frac{40}{9}$ hr, Grady reshingles $\frac{1}{8} \cdot \frac{40}{9}$, or $\frac{5}{9}$, of the roof and Cleo reshingles $\frac{1}{10} \cdot \frac{40}{9}$, or $\frac{4}{9}$, of the roof. Together, they reshingle $\frac{5}{9} + \frac{4}{9}$, or 1 roof. The fact that our solution is between 4 and 5 hr (see step 1 above) is also a check.

5. State. It will take $4\frac{4}{9}$ hr for Grady and Cleo, working together, to reshingle the roof.

EXAMPLE 2   It takes Pepe 9 hr longer than Wendy to rebuild an engine. Working together, they can do the job in 20 hr. How long would it take each, working alone, to rebuild an engine?

Solution

1. **Familiarize.**   Unlike Example 1, this problem does not provide us with the times required by the individuals to do the job alone. Let's have w = the number of hours it would take Wendy working alone and $w + 9$ = the number of hours it would take Pepe working alone.

2. **Translate.**   Using the same reasoning as in Example 1, we see that Wendy completes $\dfrac{1}{w}$ of the job in 1 hr and Pepe completes $\dfrac{1}{w + 9}$ of the job in 1 hr.

   In 2 hr, Wendy completes $\dfrac{1}{w} \cdot 2$ of the job and Pepe completes $\dfrac{1}{w + 9} \cdot 2$ of the job. We are told that, working together, Wendy and Pepe can complete the entire job in 20 hr. This gives the following:

   Fraction of job done by Wendy in 20 hr $\dfrac{1}{w} \cdot 20 + \dfrac{1}{w + 9} \cdot 20 = 1,$ Fraction of job done by Pepe in 20 hr

   or $\dfrac{20}{w} + \dfrac{20}{w + 9} = 1.$

3. **Carry out.**   We solve the equation:

$$\frac{20}{w} + \frac{20}{w + 9} = 1$$

$$w(w + 9)\left(\frac{20}{w} + \frac{20}{w + 9}\right) = w(w + 9)1 \qquad \text{Multiplying by the LCD}$$

$$(w + 9)20 + w \cdot 20 = w(w + 9) \qquad \text{Distributing and simplifying}$$

$$40w + 180 = w^2 + 9w$$

$$0 = w^2 - 31w - 180 \qquad \text{Getting 0 on one side}$$

$$0 = (w - 36)(w + 5) \qquad \text{Factoring}$$

$$w - 36 = 0 \quad or \quad w + 5 = 0 \qquad \text{Principle of zero products}$$

$$w = 36 \quad or \qquad w = -5.$$

4. **Check.**   Since negative time has no meaning in the problem, -5 is not a solution to the original problem. The number 36 checks since, if Wendy takes 36 hr alone and Pepe takes $36 + 9 = 45$ hr alone, in 20 hr they would have rebuilt

$$\frac{20}{36} + \frac{20}{45} = \frac{5}{9} + \frac{4}{9} = 1 \text{ complete engine.}$$

5. **State.**   It would take Wendy 36 hr to rebuild an engine alone, and Pepe 45 hr.

The equations used in Examples 1 and 2 can be generalized as follows.

Modeling Work Problems

If

a = the time needed for A to complete the work alone,

b = the time needed for B to complete the work alone, and

t = the time needed for A and B to complete the work together,

then

$$\frac{t}{a} + \frac{t}{b} = 1.$$

The following are equivalent equations that can also be used:

$$\frac{1}{a} \cdot t + \frac{1}{b} \cdot t = 1 \quad \text{and} \quad \frac{1}{a} + \frac{1}{b} = \frac{1}{t}.$$

Problems Involving Motion

Problems dealing with distance, rate (or speed), and time are called **motion problems.** To translate them, we use either the basic motion formula, $d = rt$, or the formulas $r = d/t$ or $t = d/r$, which can be derived from $d = rt$.

EXAMPLE 3   On her road bike, Ignacia bikes 15 km/h faster than Franklin does on his mountain bike. In the time it takes Ignacia to travel 80 km, Franklin travels 50 km. Find the speed of each bicyclist.

Solution

1. **Familiarize.**   Let's guess that Franklin is going 10 km/h. Ignacia would then be traveling 10 + 15, or 25 km/h. At 25 km/h, she would travel 80 km in $\frac{80}{25} = 3.2$ hr. Going 10 km/h, Franklin would cover 50 km in $\frac{50}{10} = 5$ hr. Since $3.2 \neq 5$, our guess was wrong, but we can see that if r = the rate, in kilometers per hour, of Franklin's bike, then the rate of Ignacia's bike = $r + 15$.
   Making a drawing and constructing a table can be helpful.

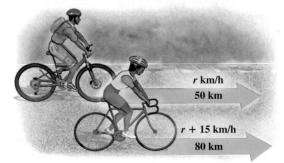

r km/h
50 km

r + 15 km/h
80 km

	Distance	Speed	Time
Franklin's Mountain Bike	50	r	t
Ignacia's Road Bike	80	$r + 15$	t

2. Translate. By looking at how we checked our guess, we see that in the **Time** column of the table, the t's can be replaced, using the formula *Time = Distance/Rate*, as follows.

	Distance	Speed	Time
Franklin's Mountain Bike	50	r	$50/r$
Ignacia's Road Bike	80	$r + 15$	$80/(r + 15)$

Since we are told that the times must be the same, we can write an equation:

$$\frac{50}{r} = \frac{80}{r + 15}.$$

3. Carry out. We solve the equation:

$$\frac{50}{r} = \frac{80}{r + 15}$$

$$r(r + 15)\frac{50}{r} = r(r + 15)\frac{80}{r + 15} \qquad \text{Multiplying by the LCD}$$

$$50r + 750 = 80r \qquad\qquad \text{Simplifying}$$

$$750 = 30r$$

$$25 = r.$$

4. Check. If our answer checks, Franklin's mountain bike is going 25 km/h and Ignacia's road bike is going $25 + 15 = 40$ km/h.

Traveling 80 km at 40 km/h, Ignacia is riding for $\frac{80}{40} = 2$ hr. Traveling 50 km at 25 km/h, Franklin is riding for $\frac{50}{25} = 2$ hr. Our answer checks since the two times are the same.

5. State. Ignacia's speed is 40 km/h, and Franklin's speed is 25 km/h.

In the following example, although the distance is the same in both directions, the key to the translation lies in an additional piece of given information.

EXAMPLE 4   A Hudson River tugboat goes 10 mph in still water. It travels 24 mi upstream and 24 mi back in a total time of 5 hr. What is the speed of the current? (*Sources*: Based on information from the Department of the Interior, U.S. Geological Survey, and *The Tugboat Captain*, Montgomery County Community College)

Solution

1. Familiarize. Let's guess that the speed of the current is 4 mph. The tugboat would then be moving $10 - 4 = 6$ mph upstream and $10 + 4 = 14$ mph downstream. The tugboat would require $\frac{24}{6} = 4$ hr to travel 24 mi upstream and $\frac{24}{14} = 1\frac{5}{7}$ hr to travel 24 mi downstream. Since the total time, $4 + 1\frac{5}{7} = 5\frac{5}{7}$ hr, is not the 5 hr mentioned in the problem, we know that our guess is wrong.

Suppose that the current's speed $= c$ mph. The tugboat's speed would then be $10 - c$ mph going upstream and $10 + c$ mph going downstream. A sketch and table can help display the information.

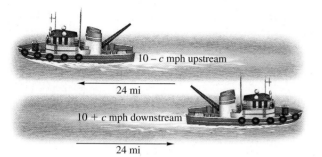

$10 - c$ mph upstream

24 mi

$10 + c$ mph downstream

24 mi

	Distance	Speed	Time
Upstream	24	$10 - c$	t_1
Downstream	24	$10 + c$	t_2

2. Translate. From examining our guess, we see that the time traveled can be represented using the formula *Time = Distance/Rate*:

	Distance	Speed	Time
Upstream	24	$10 - c$	$24/(10 - c)$
Downstream	24	$10 + c$	$24/(10 + c)$

Since the total time upstream and back is 5 hr, we use the last column of the table to form an equation:

$$\frac{24}{10 - c} + \frac{24}{10 + c} = 5.$$

3. Carry out. We solve the equation:

$$\frac{24}{10 - c} + \frac{24}{10 + c} = 5$$

$$(10 - c)(10 + c)\left[\frac{24}{10 - c} + \frac{24}{10 + c}\right] = (10 - c)(10 + c)5 \qquad \text{Multiplying by the LCD}$$

$$24(10 + c) + 24(10 - c) = (100 - c^2)5$$

$$480 = 500 - 5c^2 \qquad \text{Simplifying}$$

$$5c^2 - 20 = 0$$

$$5(c^2 - 4) = 0$$

$$5(c - 2)(c + 2) = 0$$

$$c = 2 \quad or \quad c = -2.$$

4. Check. Since speed cannot be negative in this problem, -2 cannot be a solution. You should confirm that 2 checks in the original problem.

5. State. The speed of the current is 2 mph.

6.5 Exercise Set

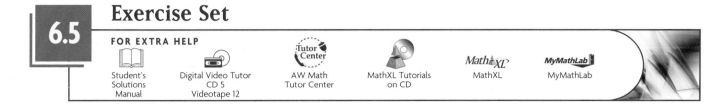
Solve.

1. The reciprocal of 3, plus the reciprocal of 6, is the reciprocal of what number?

2. The reciprocal of 5, plus the reciprocal of 7, is the reciprocal of what number?

3. The sum of a number and 6 times its reciprocal is −5. Find the number.

4. The sum of a number and 21 times its reciprocal is −10. Find the number.

5. The reciprocal of the product of two consecutive integers is $\frac{1}{42}$. Find the two integers.

6. The reciprocal of the product of two consecutive integers is $\frac{1}{72}$. Find the two integers.

7. *Photo processing.* Nadine can develop a day's accumulation of film orders in 5 hr. Willy, a new employee, needs 9 hr to complete the same job. Working together, how long will it take them to do the job?

8. *Home restoration.* Cedric can refinish the floor of an apartment in 8 hr. Carolyn can refinish the floor in 6 hr. How long will it take them, working together, to refinish the floor?

9. *Filling a pool.* The San Paulo community swimming pool can be filled in 12 hr if water enters through a pipe alone or in 30 hr if water enters through a hose alone. If water is entering through both the pipe and the hose, how long will it take to fill the pool?

10. *Filling a tank.* A community water tank can be filled in 18 hr by the town office well alone and in 22 hr by the high school well alone. How long will it take to fill the tank if both wells are working?

11. *Pumping water.* A $\frac{1}{3}$ HP Craftsman Thermoplastic sump pump can remove water from Helen's flooded basement in 44 min. The $\frac{1}{2}$ HP Simer Thermoplastic sump pump can complete the same job in 36 min. How long would it take the two pumps together to pump out the basement?
Sources: Based on data from Sears, Northern Tool & Electric Co., and Home Depot

12. *Hotel management.* The Honeywell HQ17 air cleaner can clean the air in a 12-ft by 14-ft conference room in 10 min. The HQ174 can clean the air in a room of the same size in 6 min. How long would it take the two machines together to clean the air in such a room?

13. *Photocopiers.* The HP Officejet 6110 takes twice the time required by the Canon Imageclass D680 to photocopy brochures for the New Bretton Arts Council year-end concert. If working together the two machines can complete the job in 24 min, how long would it take each machine, working alone, to copy the brochures?
Source: Manufacturers' marketing brochures

14. *Computer printers.* The HP Laser Jet 9000 works twice as fast as the Laser Jet 2300. If the machines work together, a university can produce all its staff manuals in 15 hr. Find the time it would take each machine, working alone, to complete the same job.
Source: www.hewlettpackard.com

15. *Hotel management.* The Blueair 402 can purify the air in a conference hall in 10 fewer minutes than it takes the Panasonic F-P20HU1 to do the same job. Together the two machines can purify the air in the conference hall in $\frac{120}{7}$, or $17\frac{1}{7}$ min. How long would it take each machine, working alone, to purify the air in the room?
Source: Based on information from manufacturers' websites

16. *Cutting firewood.* Jake can cut and split a cord of firewood in 6 fewer hr than Skyler can. When they work together, it takes them 4 hr. How long would it take each of them to do the job alone?

17. *Forest fires.* The Erickson Air-Crane helicopter can scoop water and douse a certain forest fire four times as fast as an S-58T helicopter. Working together, the two helicopters can douse the fire in 8 hr. How long would it take each helicopter, working alone, to douse the fire?
Sources: Based on information from www.emergency.com and www.arishelicopters.com

18. *Waxing a car.* Rosita can wax her car in 2 hr. When she works together with Helga, they can wax the car in 45 min. How long would it take Helga, working by herself, to wax the car?

19. *Newspaper delivery.* Zsuzanna can deliver papers three times as fast as Stan can. If they work together, it takes them 1 hr. How long would it take each to deliver the papers alone?

20. *Sorting recyclables.* Together, it takes John and Deb 2 hr 55 min to sort recyclables. Alone, John would require 2 more hr than Deb. How long would it take Deb to do the job alone? (*Hint*: Convert minutes to hours or hours to minutes.)

21. *Paving.* Together, Larry and Mo require 4 hr 48 min to pave a driveway. Alone, Larry would require 4 hr more than Mo. How long would it take Mo to do the job alone? (*Hint*: Convert minutes to hours.)

22. *Painting.* Sara takes 3 hr longer to paint a floor than it takes Kate. When they work together, it takes them 2 hr. How long would each take to do the job alone?

23. *Kayaking.* The speed of the current in Catamount Creek is 3 mph. Zeno can kayak 4 mi upstream in the same time it takes him to kayak 10 mi downstream. What is the speed of Zeno's kayak in still water?

24. *Boating.* The current in the Lazy River moves at a rate of 4 mph. Monica's dinghy motors 6 mi upstream in the same time it takes to motor 12 mi downstream. What is the speed of the dinghy in still water?

25. *Moving sidewalks.* Newark Airport's moving sidewalk moves at a speed of 1.7 ft/sec. Walking on the moving sidewalk, Benny can travel 120 ft forward in the same time it takes to travel 52 ft in the opposite direction. How fast would Benny be walking on a nonmoving sidewalk?

26. *Moving sidewalks.* The moving sidewalk at O'Hare Airport in Chicago moves 1.8 ft/sec. Walking on the moving sidewalk, Camille travels 105 ft forward in the time it takes to travel 51 ft in the opposite direction. How fast would Camille be walking on a nonmoving sidewalk?

27. *Train speed.* The speed of the A&M freight train is 14 mph less than the speed of the A&M passenger train. The passenger train travels 400 mi in the same time that the freight train travels 330 mi. Find the speed of each train.

28. *Walking.* Rosanna walks 2 mph slower than Simone. In the time it takes Simone to walk 8 mi, Rosanna walks 5 mi. Find the speed of each person.

Aha! **29.** *Bus travel.* A local bus travels 7 mph slower than the express. The express travels 45 mi in the time it takes the local to travel 38 mi. Find the speed of each bus.

30. *Train speed.* The A train goes 12 mph slower than the E train. The A train travels 230 mi in the same time that the E train travels 290 mi. Find the speed of each train.

31. *Boating.* Laverne's Mercruiser travels 15 km/h in still water. She motors 140 km downstream in the same time it takes to travel 35 km upstream. What is the speed of the river?

32. *Boating.* Audrey's paddleboat travels 2 km/h in still water. The boat is paddled 4 km downstream in the same time it takes to go 1 km upstream. What is the speed of the river?

33. *Shipping.* A barge moves 7 km/h in still water. It travels 45 km upriver and 45 km downriver in a total time of 14 hr. What is the speed of the current?

34. *Moped speed.*   Jaime's moped travels 8 km/h faster than Mara's. Jaime travels 69 km in the same time that Mara travels 45 km. Find the speed of each person's moped.

35. *Aviation.*   A Citation II Jet travels 350 mph in still air and flies 487.5 mi into the wind and 487.5 mi with the wind in a total of 2.8 hr. Find the wind speed.
Source: Eastern Air Charter

36. *Canoeing.*   Al paddles 55 m per minute in still water. He paddles 150 m upstream and 150 m downstream in a total time of 5.5 min. What is the speed of the current?

37. *Train travel.*   A freight train covered 120 mi at a certain speed. Had the train been able to travel 10 mph faster, the trip would have been 2 hr shorter. How fast did the train go?

38. *Boating.*   Julia's Boston Whaler cruised 45 mi upstream and 45 mi back in a total of 8 hr. The speed of the river is 3 mph. Find the speed of the boat in still water.

39. Two steamrollers are paving a parking lot. Working together, will the two steamrollers take less than half as long as the slower steamroller would working alone? Why or why not?

40. Two fuel lines are filling a freighter with oil. Will the faster fuel line take more or less than twice as long to fill the freighter by itself? Why?

SKILL MAINTENANCE

Simplify.

41. $\dfrac{35a^6b^8}{7a^2b^2}$ [1.6]

42. $\dfrac{20x^9y^6}{4x^3y^2}$ [1.6]

43. $\dfrac{36s^{15}t^{10}}{9s^5t^2}$ [1.6]

44. $6x^4 - 3x^2 + 9x - (8x^4 + 4x^2 - 2x)$ [5.1]

45. $2(x^3 + 4x^2 - 5x + 7) - 5(2x^3 - 4x^2 + 3x - 1)$ [5.1]

46. $9x^4 + 7x^3 + x^2 - 8 - (-2x^4 + 3x^2 + 4x + 2)$ [5.1]

SYNTHESIS

47. Write a work problem for a classmate to solve. Devise the problem so that the solution is "Liane and Michele will take 4 hr to complete the job, working together."

48. Write a work problem for a classmate to solve. Devise the problem so that the solution is "Jen takes 5 hr and Pablo takes 6 hr to complete the job alone."

49. *Filling a bog.*   The Norwich cranberry bog can be filled in 9 hr and drained in 11 hr. How long will it take to fill the bog if the drainage gate is left open?

50. *Filling a tub.*   Justine's hot tub can be filled in 10 min and drained in 8 min. How long will it take to empty a full tub if the water is left on?

51. Refer to Exercise 24. How long will it take Monica to motor 3 mi downstream?

52. Refer to Exercise 23. How long will it take Zeno to kayak 5 mi downstream?

53. *Escalators.*   Together, a 100-cm wide escalator and a 60-cm wide escalator can empty a 1575-person auditorium in 14 min. The wider escalator moves twice as many people as the narrower one. How many people per hour does the 60-cm wide escalator move?
Source: *McGraw-Hill Encyclopedia of Science and Technology*

54. *Aviation.*   A Coast Guard plane has enough fuel to fly for 6 hr, and its speed in still air is 240 mph. The plane departs with a 40-mph tailwind and returns to the same airport flying into the same wind. How far can the plane travel under these conditions?

55. *Boating.*   Shoreline Travel operates a 3-hr paddle-boat cruise on the Missouri River. If the speed of the boat in still water is 12 mph, how far upriver can the pilot travel against a 5-mph current before it is time to turn around?

56. *Travel by car.*   Melissa drives to work at 50 mph and arrives 1 min late. She drives to work at 60 mph and arrives 5 min early. How far does Melissa live from work?

57. *Photocopying.*   The printer in an admissions office can print a 500-page document in 50 min, while the printer in the business office can print the same document in 40 min. If the two printers work together to print the document, with the faster machine starting on page 1 and the slower machine working backwards from page 500, at what page will the two machines meet to complete the job?

58. At what time after 4:00 will the minute hand and the hour hand of a clock first be in the same position?

59. At what time after 10:30 will the hands of a clock first be perpendicular?

Average speed is defined as total distance divided by total time.

60. Lenore drove 200 km. For the first 100 km of the trip, she drove at a speed of 40 km/h. For the second half of the trip, she traveled at a speed of 60 km/h. What was the average speed of the entire trip? (It was *not* 50 km/h.)

61. For the first 50 mi of a 100-mi trip, Chip drove 40 mph. What speed would he have to travel for the last half of the trip so that the average speed for the entire trip would be 45 mph?

CORNER

Does the Model Hold Water?

COLLABORATIVE

Focus: Testing a mathematical model

Time: 20–30 minutes

Group size: 2–3

Materials: An empty 1-gal plastic jug, a kitchen or laboratory sink, a stopwatch or a watch capable of measuring seconds, an inexpensive pen or pair of scissors or a nail or knife for poking holes in plastic

Problems like Exercises 49 and 50 can be solved algebraically and then checked at home or in a laboratory.

ACTIVITY

1. While one group member fills the empty jug with water, the other group member(s) should record how many seconds this takes.
2. After carefully poking a few holes in the bottom of the jug, record how many seconds it takes the full jug to empty.
3. Using the information found in parts (1) and (2) above, use algebra to predict how long it will take to fill the punctured jug.
4. Test your prediction by timing how long it takes for the pierced jug to be filled. Be sure to run the water at the same rate as in part (1).
5. How accurate was your prediction? How might your prediction have been made more accurate?

6.6 Division of Polynomials

Dividing by a Monomial • Dividing by a Polynomial

A rational expression indicates division. Division of polynomials, like division of real numbers, relies on our multiplication and subtraction skills.

Dividing by a Monomial

To divide a monomial by a monomial, we divide coefficients and, if the bases are the same, subtract exponents (see Section 1.6).

$$\text{Dividend} \longrightarrow \frac{45x^{10}}{3x^4} = 15x^{10-4} = 15x^6, \qquad \frac{8a^2b^5}{-2ab^2} = -4a^{2-1}b^{5-2} = -4ab^3.$$
$$\text{Divisor} \longrightarrow$$

Quotient

To divide a polynomial by a monomial, we regard the division as a sum of quotients of monomials. This uses the fact that since

$$\frac{A}{C} + \frac{B}{C} = \frac{A+B}{C}, \quad \text{we know that} \quad \frac{A+B}{C} = \frac{A}{C} + \frac{B}{C}.$$

EXAMPLE 1   Divide $12x^3 + 8x^2 + x + 4$ by $4x$.

Solution

$$(12x^3 + 8x^2 + x + 4) \div (4x) = \frac{12x^3 + 8x^2 + x + 4}{4x} \qquad \text{Writing a rational expression}$$

$$= \frac{12x^3}{4x} + \frac{8x^2}{4x} + \frac{x}{4x} + \frac{4}{4x} \qquad \text{Writing as a sum of quotients}$$

$$= 3x^2 + 2x + \frac{1}{4} + \frac{1}{x} \qquad \text{Performing the four indicated divisions}$$

EXAMPLE 2   Divide: $(8x^4y^5 - 3x^3y^4 + 5x^2y^3) \div (-x^2y^3)$.

Solution

$$\frac{8x^4y^5 - 3x^3y^4 + 5x^2y^3}{-x^2y^3} = \frac{8x^4y^5}{-x^2y^3} - \frac{3x^3y^4}{-x^2y^3} + \frac{5x^2y^3}{-x^2y^3} \qquad \text{Try to perform this step mentally.}$$

$$= -8x^2y^2 + 3xy - 5$$

> **Division by a Monomial**
>
> To divide a polynomial by a monomial, divide each term of the polynomial by the monomial.

Dividing by a Polynomial

When the divisor has more than one term, we use a procedure very similar to long division in arithmetic.

EXAMPLE 3   Divide $2x^2 - 7x - 15$ by $x - 5$.

Solution   We have

$$
\begin{array}{r}
2x \\
x - 5\overline{)2x^2 - 7x - 15} \\
-(2x^2 - 10x) \\
\hline
3x - 15
\end{array}
$$

Divide $2x^2$ by x: $2x^2/x = 2x$.

Multiply $x - 5$ by $2x$.

Subtract by mentally changing signs and adding: $-7x + 10x = 3x$.

We next divide the leading term of this remainder, $3x$, by the leading term of the divisor, x.

$$
\begin{array}{r}
2x + 3 \\
x - 5\overline{)2x^2 - 7x - 15} \\
2x^2 - 10x \\
\hline
3x - 15 \\
-(3x - 15) \\
\hline
0
\end{array}
$$

Divide $3x$ by x: $3x/x = 3$.

Multiply $x - 5$ by 3.

Subtract. Our remainder is now 0.

Check: $(x - 5)(2x + 3) = 2x^2 - 7x - 15$. The answer checks.

The quotient is $2x + 3$.

To understand why we perform long division as we do, note that Example 3 amounts to "filling in" an unknown polynomial:

$$(x - 5)(\,?\,) = 2x^2 - 7x - 15.$$

We see that $2x$ must be in the unknown polynomial if we are to get the first term, $2x^2$, from the multiplication. To see what else is needed, note that

$$(x - 5)(2x\quad) = 2x^2 - 10x \neq 2x^2 - 7x - 15.$$

The $2x$ can be regarded as a (poor) approximation of the quotient that we are seeking. To see how far off the approximation is, we subtract:

$$
\left.
\begin{array}{r}
2x^2 - 7x - 15 \\
-(2x^2 - 10x) \\
\hline
3x - 15
\end{array}
\right\}
$$

Note where this appeared in the long division above.

$\leftarrow$ This is the first remainder.

To get the needed terms, $3x - 15$, we need another term in the unknown polynomial. We use 3 because $(x - 5) \cdot 3$ is $3x - 15$:

$$(x - 5)(2x + 3) = 2x^2 - 10x + 3x - 15$$
$$= 2x^2 - 7x - 15.$$

Now when we subtract the product $(x - 5)(2x + 3)$ from $2x^2 - 7x - 15$, the remainder is 0.

If a nonzero remainder occurs, when do we stop dividing? We continue until the degree of the remainder is less than the degree of the divisor.

EXAMPLE 4   Divide $x^2 + 5x + 8$ by $x + 3$.

Solution   We have

$$
\begin{array}{r}
x \phantom{{}+ 5x + 8} \\
x + 3 \overline{) x^2 + 5x + 8} \\
\underline{x^2 + 3x} \\
2x + 8
\end{array}
$$

Divide the first term of the dividend by the first term of the divisor: $x^2/x = x$.
Multiply x above by $x + 3$.
Subtract: $5x - 3x = 2x$.

Note that

$$x^2 + 5x + 8 - (x^2 + 3x) = x^2 + 5x + 8 - x^2 - 3x.$$

Remember: To subtract, add the opposite (change the sign of every term, then add).

We now focus on the current remainder, $2x + 8$, and repeat the process:

$$
\begin{array}{r}
x + 2 \\
x + 3 \overline{) x^2 + 5x + 8} \\
\underline{x^2 + 3x} \\
2x + 8 \\
\underline{2x + 6} \\
2
\end{array}
$$

Divide the first term by the first term: $2x/x = 2$.
$2x + 8$ is the first remainder.
Multiply 2 by $x + 3$.
Subtract: $(2x + 8) - (2x + 6)$.

The quotient is $x + 2$, with remainder 2. Note that the degree of the remainder is 0 and the degree of the divisor, $x + 3$, is 1. Since $0 < 1$, the process stops.

Check: $(x + 3)(x + 2) + 2 = x^2 + 5x + 6 + 2$ Add the remainder to the product.

$$= x^2 + 5x + 8$$

We write our answer as $x + 2$, R 2, or as

$$\text{Quotient} + \frac{\text{Remainder}}{\text{Divisor}}$$

$$x + 2 + \frac{2}{x + 3}.$$

This is how answers are listed at the back of the book.

Student Notes _____

See if, after studying Examples 3 and 4, you can explain why long division of numbers is performed as it is.

The last answer in Example 4 can also be checked by multiplying:

$$(x + 3)\left[(x + 2) + \frac{2}{x + 3}\right] = (x + 3)(x + 2) + (x + 3)\frac{2}{x + 3}$$

Using the distributive law

$$= x^2 + 5x + 6 + 2$$
$$= x^2 + 5x + 8. \qquad \text{This was the dividend in Example 4.}$$

As shown in Example 4, the quicker check is to multiply the divisor by the quotient and then add the remainder.

You may have noticed that it is helpful to have all polynomials written in descending order.

Tips for Dividing Polynomials

1. Arrange polynomials in descending order.
2. If there are missing terms in the dividend, either write them with 0 coefficients or leave space for them.
3. Continue the long division process until the degree of the remainder is less than the degree of the divisor.

EXAMPLE 5

Divide: $(9a^2 + a^3 - 5) \div (a^2 - 1)$.

Solution We rewrite the problem in descending order:

$$(a^3 + 9a^2 - 5) \div (a^2 - 1).$$

Thus,

$$
\begin{array}{r}
a + 9 \\
a^2 - 1 \overline{\smash{\big)}\, a^3 + 9a^2 + 0a - 5} \\
\underline{a^3 \qquad\quad - a} \\
9a^2 + a - 5 \\
\underline{9a^2 \qquad - 9} \\
a + 4
\end{array}
$$

When there is a missing term in the dividend, we can write it in, as shown here, or leave space, as in Example 6 below.

Subtracting: $0a - (-a) = a$.

The degree of the remainder is less than the degree of the divisor, so we are finished.

The answer is $a + 9 + \dfrac{a + 4}{a^2 - 1}$.

EXAMPLE 6

Let $f(x) = 125x^3 - 8$ and $g(x) = 5x - 2$. If $F(x) = (f/g)(x)$, find a simplified expression for $F(x)$ and list all restrictions on the domain.

Solution Recall that $(f/g)(x) = f(x)/g(x)$. Thus,

$$F(x) = \frac{125x^3 - 8}{5x - 2}$$

and

$$5x - 2 \overline{)\begin{array}{l} 25x^2 + 10x + 4 \\ 125x^3 \qquad\qquad - 8 \end{array}}$$

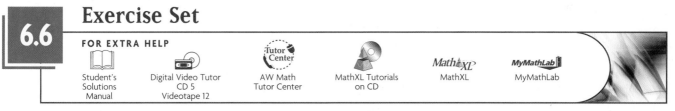

	Leaving space for the missing terms
$\dfrac{125x^3 - 50x^2}{50x^2\qquad - 8}$	Subtracting: $125x^3 - (125x^3 - 50x^2) = 50x^2$
$\dfrac{50x^2 - 20x}{20x - 8}$	
$\dfrac{20x - 8}{0}$	Subtracting

Note that, because $F(x) = f(x)/g(x)$, $g(x)$ cannot be 0. Since $g(x)$ is 0 for $x = \frac{2}{5}$ (check this), we have

$$F(x) = 25x^2 + 10x + 4, \quad \text{provided } x \neq \tfrac{2}{5}.$$

6.6 Exercise Set

FOR EXTRA HELP

Student's Solutions Manual   Digital Video Tutor CD 5 Videotape 12   Tutor Center / AW Math Tutor Center   MathXL Tutorials on CD   *MathXL* MathXL   *MyMathLab* MyMathLab

↘ *Concept Reinforcement   Classify each statement as either true or false.*

1. To divide a polynomial by a monomial, we divide each term of the polynomial by the monomial.

2. To divide a monomial by a monomial, we can subtract coefficients and divide exponents.

3. When the divisor has more than one term, polynomial division is very similar to long division in arithmetic.

4. In long division, we subtract to find how far off our approximation of the quotient is.

5. When we are dividing by $2x + 3$, the remainder in our answer may include a variable.

6. When we are dividing by $x^2 - 5$, the remainder in our answer may include a variable.

Divide and check.

7. $\dfrac{32x^6 + 18x^5 - 27x^2}{6x^2}$

8. $\dfrac{30y^8 - 15y^6 + 40y^4}{5y^4}$

9. $\dfrac{21a^3 + 7a^2 - 3a - 14}{7a}$

10. $\dfrac{-25x^3 + 20x^2 - 3x + 7}{5x}$

11. $\dfrac{18t^4 - 15t + 21}{-3t}$

12. $\dfrac{14a^5 - 21a + 7}{-7a}$

13. $\dfrac{16y^4z^2 - 8y^6z^4 + 12y^8z^3}{-4y^4z}$

14. $\dfrac{6p^2q^2 - 9p^2q + 12pq^2}{-3pq}$

15. $(16y^3 - 9y^2 - 8y) \div (2y^2)$

16. $(6a^4 + 9a^2 - 8) \div (2a)$

17. $(15x^7 - 21x^4 - 3x^2) \div (-3x^2)$

18. $(36y^6 - 18y^4 - 12y^2) \div (-6y)$

19. $(a^2b - a^3b^3 - a^5b^5) \div (a^2b)$

20. $(x^3y^2 - x^3y^3 - x^4y^2) \div (x^2y^2)$

Aha! **21.** $(x^2 + 10x + 21) \div (x + 7)$

22. $(y^2 - 8y + 16) \div (y - 4)$

23. $(a^2 - 8a - 16) \div (a + 4)$

24. $(y^2 - 10y - 25) \div (y - 5)$

25. $(x^2 - 9x + 21) \div (x - 4)$

26. $(x^2 - 11x + 23) \div (x - 6)$

27. $(y^2 - 25) \div (y + 5)$

28. $(a^2 - 81) \div (a - 9)$

29. $(y^3 - 4y^2 + 3y - 6) \div (y - 2)$

30. $(x^3 - 5x^2 + 4x - 7) \div (x - 3)$

31. $(2x^3 + 3x^2 - x - 3) \div (x + 2)$

32. $(3x^3 - 5x^2 - 3x - 2) \div (x - 2)$

33. $(a^3 - a + 10) \div (a - 4)$

34. $(x^3 - x + 6) \div (x + 2)$

35. $(10y^3 + 6y^2 - 9y + 10) \div (5y - 2)$

36. $(6x^3 - 11x^2 + 11x - 2) \div (2x - 3)$

37. $(2x^4 - x^3 - 5x^2 + x - 6) \div (x^2 + 2)$

38. $(3x^4 + 2x^3 - 11x^2 - 2x + 5) \div (x^2 - 2)$

For Exercises 39–46, $f(x)$ and $g(x)$ are as given. Find a simplified expression for $F(x)$ if $F(x) = (f/g)(x)$. (See Example 6.) Be sure to list all restrictions on the domain of $F(x)$.

39. $f(x) = 8x^3 - 27, \ g(x) = 2x - 3$

40. $f(x) = 64x^3 + 8, \ g(x) = 4x + 2$

41. $f(x) = 6x^2 - 11x - 10, \ g(x) = 3x + 2$

42. $f(x) = 8x^2 - 22x - 21, \ g(x) = 2x - 7$

43. $f(x) = x^4 - 24x^2 - 25, \ g(x) = x^2 - 25$

44. $f(x) = x^4 - 3x^2 - 54, \ g(x) = x^2 - 9$

45. $f(x) = 8x^2 - 3x^4 - 2x^3 + 2x^5 - 5, \ g(x) = x^2 - 1$

46. $f(x) = 4x - x^3 - 10x^2 + 3x^4 - 8, \ g(x) = x^2 - 4$

47. Explain how factoring could be used to solve Example 6.

48. Explain how to construct a polynomial of degree 4 that has a remainder of 3 when divided by $x + 1$.

SKILL MAINTENANCE

Solve. [1.5]

49. $ab - cd = k$, for c **50.** $xy - wz = t$, for z

51. Find three consecutive positive integers such that the product of the first and second integers is 26 less than the product of the second and third integers. [5.8]

52. If $f(x) = 2x^3$, find $f(-3a)$. [2.2]

Solve. [4.3]

53. $|2x - 3| > 7$ **54.** $|3x - 1| < 8$

SYNTHESIS

55. Explain how to construct a polynomial of degree 4 that has a remainder of 2 when divided by $x + c$.

56. Do addition, subtraction, and multiplication of polynomials always result in a polynomial? Does division? Why or why not?

Divide.

57. $(4a^3b + 5a^2b^2 + a^4 + 2ab^3) \div (a^2 + 2b^2 + 3ab)$

58. $(x^4 - x^3y + x^2y^2 + 2x^2y - 2xy^2 + 2y^3) \div (x^2 - xy + y^2)$

59. $(a^7 + b^7) \div (a + b)$

60. Find k such that when $x^3 - kx^2 + 3x + 7k$ is divided by $x + 2$, the remainder is 0.

61. When $x^2 - 3x + 2k$ is divided by $x + 2$, the remainder is 7. Find k.

62. Let

$$f(x) = \frac{3x + 7}{x + 2}.$$

a) Use division to find an expression equivalent to $f(x)$. Then graph f.

b) On the same set of axes, sketch both $g(x) = 1/(x + 2)$ and $h(x) = 1/x$.

c) How do the graphs of f, g, and h compare?

63. Jamaladeen incorrectly states that

$$(x^3 + 9x^2 - 6) \div (x^2 - 1) = x + 9 + \frac{x + 4}{x^2 - 1}.$$

Without performing any long division, how could you show Jamaladeen that his division cannot possibly be correct?

64. Use a graphing calculator to check Example 3 by setting $y_1 = (2x^2 - 7x - 15)/(x - 5)$ and $y_2 = 2x + 3$. Then press either (TRACE) (after selecting the ZOOM Z INTEGER option) or (TABLE) (with TblMin $= 0$ and ΔTbl $= 1$) to show that $y_1 \neq y_2$ for $x = 5$.

65. Use a graphing calculator to check Example 5. Perform the check using
$y_1 = (9x^2 + x^3 - 5)/(x^2 - 1),$
$y_2 = x + 9 + (x + 4)/(x^2 - 1),$ and $y_3 = y_2 - y_1.$

6.7   Synthetic Division

Streamlining Long Division   •   The Remainder Theorem

Streamlining Long Division

To divide a polynomial by a binomial of the type $x - a$, we can streamline the usual procedure to develop a process called *synthetic division*.

Compare the following. In each stage, we attempt to write a bit less than in the previous stage, while retaining enough essentials to solve the problem. At the end, we will return to the usual polynomial notation.

Stage 1

When a polynomial is written in descending order, the coefficients provide the essential information:

$$
\begin{array}{r}
4x^2 + 5x + 11 \\
x - 2\overline{)4x^3 - 3x^2 + x + 7} \\
\underline{4x^3 - 8x^2} \\
5x^2 + x \\
\underline{5x^2 - 10x} \\
11x + 7 \\
\underline{11x - 22} \\
29
\end{array}
\qquad
\begin{array}{r}
4 + 5 + 11 \\
1 - 2\overline{)4 - 3 + 1 + 7} \\
\underline{4 - 8} \\
5 + 1 \\
\underline{5 - 10} \\
11 + 7 \\
\underline{11 - 22} \\
29
\end{array}
$$

Because the leading coefficient in the divisor is 1, each time we multiply the divisor by a term in the answer, the leading coefficient of that product duplicates a coefficient in the answer. In the next stage, we don't bother to duplicate these numbers. We also show where -2 is used and drop the 1 from the divisor.

Stage 2

$$
\begin{array}{r}
4x^2 + 5x\ +\ 11 \\
x-2\overline{)4x^3 - 3x^2 +\ \ \ x +\ 7} \\
\underline{4x^3 - 8x^2} \\
5x^2 +\ \ \ x \\
\underline{5x^2 - 10x} \\
11x +\ 7 \\
\underline{11x - 22} \\
29
\end{array}
$$

$$
\begin{array}{r}
4 + 5 + 11 \\
-2\overline{)4 - 3 +\ 1 +\ 7} \\
\underline{-\ 8} \\
5 +\ 1 \\
\underline{-\ 10} \\
11 +\ 7 \\
-22 \\
29
\end{array}
$$

Multiply: $-2 \cdot 4 = -8$.
Subtract: $-3 - (-8) = 5$.
Multiply: $-2 \cdot 5 = -10$.
Subtract: $1 - (-10) = 11$.
Multiply: $-2 \cdot 11 = -22$.
Subtract: $7 - (-22) = 29$.

To simplify further, we now reverse the sign of the -2 in the divisor and, in exchange, *add* at each step in the long division.

Stage 3

$$
\begin{array}{r}
4x^2 + 5x\ +\ 11 \\
x-2\overline{)4x^3 - 3x^2 +\ \ \ x +\ 7} \\
\underline{4x^3 - 8x^2} \\
5x^2 +\ \ \ x \\
\underline{5x^2 - 10x} \\
11x +\ 7 \\
\underline{11x - 22} \\
29
\end{array}
$$

$$
\begin{array}{r}
4 + 5 + 11 \\
2\overline{)4 - 3 +\ 1 +\ 7} \\
8 \\
5 +\ 1 \\
10 \\
11 +\ 7 \\
22 \\
29
\end{array}
$$

Replace the -2 with 2.
Multiply: $2 \cdot 4 = 8$.
Add: $-3 + 8 = 5$.
Multiply: $2 \cdot 5 = 10$.
Add: $1 + 10 = 11$.
Multiply: $2 \cdot 11 = 22$.
Add: $7 + 22 = 29$.

The blue numbers can be eliminated if we look at the red numbers instead.

Stage 4

$$
\begin{array}{r}
4x^2 + 5x\ +\ 11 \\
x-2\overline{)4x^3 - 3x^2 +\ \ \ x +\ 7} \\
\underline{4x^3 - 8x^2} \\
5x^2 +\ \ \ x \\
\underline{5x^2 - 10x} \\
11x +\ 7 \\
\underline{11x - 22} \\
29
\end{array}
$$

$$
\begin{array}{r}
4\ \ \ \ 5\ \ 11 \\
2\overline{)4\ \ -3\ \ \ 1\ \ \ 7} \\
\underline{8\ \ \ 10\ \ 22} \\
5\ \ \ 11\ \ 29
\end{array}
$$

Student Notes

You will not need to write out all five stages when performing synthetic division on your own. We show the steps to help you understand the reasoning behind the method.

Don't lose sight of how the products 8, 10, and 22 are found. Also, note that the 5 and 11 preceding the remainder 29 coincide with the 5 and 11 following the 4 on the top line. By writing a 4 to the left of 5 on the bottom line, we can eliminate the top line in stage 4 and read our answer from the bottom line. This final stage is commonly called **synthetic division**.

Stage 5

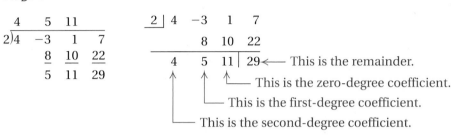

The quotient is $4x^2 + 5x + 11$. The remainder is 29.

> Remember that in order for this method to work, the divisor must be of the form $x - a$, that is, a variable minus a constant. Both the coefficient and the exponent of the variable must be 1.

EXAMPLE 1   Use synthetic division to divide: $(x^3 + 6x^2 - x - 30) \div (x - 2)$.

Solution

$$
\begin{array}{r|rrrr}
2 & 1 & 6 & -1 & -30 \\
\hline
 & 1 & & &
\end{array}
$$
Write the 2 of $x - 2$ and the coefficients of the dividend.
Bring down the first coefficient.

$$
\begin{array}{r|rrrr}
2 & 1 & 6 & -1 & -30 \\
 & & 2 & & \\
\hline
 & 1 & 8 & &
\end{array}
$$
Multiply 1 by 2 to get 2.
Add 6 and 2.

$$
\begin{array}{r|rrrr}
2 & 1 & 6 & -1 & -30 \\
 & & 2 & 16 & \\
\hline
 & 1 & 8 & 15 &
\end{array}
$$
Multiply 8 by 2.
Add -1 and 16.

$$
\begin{array}{r|rrrr}
2 & 1 & 6 & -1 & -30 \\
 & & 2 & 16 & 30 \\
\hline
 & 1 & 8 & 15 & 0
\end{array}
$$
Multiply 15 by 2 and add.

The answer is $x^2 + 8x + 15$ with R 0, or just $x^2 + 8x + 15$.

EXAMPLE 2   Use synthetic division to divide.

a) $(2x^3 + 7x^2 - 5) \div (x + 3)$

b) $(10x^2 - 13x + 3x^3 - 20) \div (4 + x)$

Solution

a) $(2x^3 + 7x^2 - 5) \div (x + 3)$

The dividend has no x-term, so we need to write 0 for its coefficient of x. Note that $x + 3 = x - (-3)$, so we write -3 inside the __ .

$$
\begin{array}{r|rrrr}
-3 & 2 & 7 & 0 & -5 \\
   &   & -6 & -3 & 9 \\
\hline
   & 2 & 1 & -3 & 4
\end{array}
$$

The answer is $2x^2 + x - 3$, with R 4, or $2x^2 + x - 3 + \dfrac{4}{x + 3}$.

b) We first rewrite $(10x^2 - 13x + 3x^3 - 20) \div (4 + x)$ in descending order:

$$(3x^3 + 10x^2 - 13x - 20) \div (x + 4).$$

Next, we use synthetic division. Note that $x + 4 = x - (-4)$.

$$
\begin{array}{r|rrrr}
-4 & 3 & 10 & -13 & -20 \\
   &   & -12 & 8 & 20 \\
\hline
   & 3 & -2 & -5 & 0
\end{array}
$$

The answer is $3x^2 - 2x - 5$.

technology connection

In Example 1, the division by $x - 2$ gave a remainder of 0. The remainder theorem tells us that this means that when $x = 2$, the value of $x^3 + 6x^2 - x - 30$ is 0. Check this both graphically and algebraically (by substitution). Then perform a similar check for Example 2(b).

The Remainder Theorem

Because the remainder is 0, Example 1 shows that $x - 2$ is a factor of $x^3 + 6x^2 - x - 30$ and that we can write $x^3 + 6x^2 - x - 30$ as $(x - 2)(x^2 + 8x + 15)$. Using this result and the principle of zero products, we know that if $f(x) = x^3 + 6x^2 - x - 30$, then $f(2) = 0$ (since $x - 2$ is a factor of $f(x)$). Similarly, from Example 2(b), we know that $x + 4$ is a factor of $g(x) = 10x^2 - 13x + 3x^3 - 20$. This tells us that $g(-4) = 0$. In both examples, the remainder from the division, 0, can serve as a function value. Remarkably, this pattern extends to nonzero remainders. To see this, note that the remainder in Example 2(a) is 4, and if $f(x) = 2x^3 + 7x^2 - 5$, then $f(-3)$ is also 4 (you should check this). The fact that the remainder and the function value coincide is predicted by the remainder theorem.

> **The Remainder Theorem**
>
> The remainder obtained by dividing $P(x)$ by $x - r$ is $P(r)$.

A proof of this result is outlined in Exercise 39.

EXAMPLE 3   Let $f(x) = 8x^5 - 6x^3 + x - 8$. Use synthetic division to find $f(2)$.

Solution   The remainder theorem tells us that $f(2)$ is the remainder when $f(x)$ is divided by $x - 2$. We use synthetic division to find that remainder:

$$\begin{array}{r|rrrrrr} 2 & 8 & 0 & -6 & 0 & 1 & -8 \\ & & 16 & 32 & 52 & 104 & 210 \\ \hline & 8 & 16 & 26 & 52 & 105 & 202 \end{array}$$

Although the bottom line can be used to find the quotient for the division $(8x^5 - 6x^3 + x - 8) \div (x - 2)$, what we are really interested in is the remainder. It tells us that $f(2) = 202$.

In practice, the remainder theorem is often used to check division. Thus Example 2(a) can be checked by evaluating $P(-3)$, with $P(x) = 2x^3 + 7x^2 - 5$. Since $P(-3) = 4$ (check this) and the remainder in Example 2(a) is also 4, our division was probably correct.

Exercise Set

6.7

FOR EXTRA HELP

Student's Solutions Manual   Digital Video Tutor CD 5 Videotape 12   AW Math Tutor Center   MathXL Tutorials on CD   Math XL MathXL   MyMathLab MyMathLab

🔖 *Concept Reinforcement*   *Classify each statement as either true or false.*

1. If $x - 2$ is a factor of some polynomial $P(x)$, then $P(2) = 0$.

2. If $p(3) = 0$ for some polynomial $p(x)$, then $x - 3$ is a factor of $p(x)$.

3. If $P(-5) = 39$ and $P(x) = x^3 + 7x^2 + 3x + 4$, then

$$\begin{array}{r|rrrr} -5 & 1 & 7 & 3 & 4 \\ & & -5 & -10 & 35 \\ \hline & 1 & 2 & -7 & 39. \end{array}$$

4. In order for $f(x)/g(x)$ to exist, $g(x)$ must be 0.

5. In order to use synthetic division, we must be sure that the divisor is of the form $x - a$.

6. Synthetic division can be used in problems in which long division could not be used.

7. Both $3(-4)^3 + 10(-4)^2 - 13(-4) - 20 = 0$ and

$$\begin{array}{r|rrrr} -4 & 3 & 10 & -13 & -20 \\ & & -12 & 8 & 20 \\ \hline & 3 & -2 & -5 & 0 \end{array}$$

indicate that for $p(x) = 3x^3 + 10x^2 - 13x - 20$, $p(-4) = 0$.

8. Synthetic division can be used to show that for any polynomial function P, $P(a)$ and $P(-a)$ are opposites.

Use synthetic division to divide.

9. $(x^3 - 2x^2 + 2x - 7) \div (x - 1)$

10. $(x^3 - 2x^2 + 2x - 7) \div (x + 1)$

11. $(a^2 + 8a + 11) \div (a + 3)$

12. $(a^2 + 8a + 11) \div (a + 5)$

13. $(x^3 - 7x^2 - 13x + 3) \div (x + 2)$

14. $(x^3 - 7x^2 - 13x + 3) \div (x - 2)$

15. $(3x^3 + 7x^2 - 4x + 3) \div (x + 3)$

16. $(3x^3 + 7x^2 - 4x + 3) \div (x - 3)$

17. $(y^3 - 3y + 10) \div (y - 2)$

18. $(x^3 - 2x^2 + 8) \div (x + 2)$

19. $(x^5 - 32) \div (x - 2)$

20. $(y^5 - 1) \div (y - 1)$

21. $(3x^3 + 1 - x + 7x^2) \div \left(x + \frac{1}{3}\right)$

22. $(8x^3 - 1 + 7x - 6x^2) \div \left(x - \frac{1}{2}\right)$

Use synthetic division to find the indicated function value.

23. $f(x) = 5x^4 + 12x^3 + 28x + 9;\ f(-3)$

24. $g(x) = 3x^4 - 25x^2 - 18;\ g(3)$

25. $P(x) = 6x^4 - x^3 - 7x^2 + x + 2;\ P(-1)$

26. $F(x) = 3x^4 + 8x^3 + 2x^2 - 7x - 4;\ F(-2)$

27. $f(x) = x^4 - x^3 - 19x^2 + 49x - 30;\ f(4)$

28. $p(x) = x^4 + 7x^3 + 11x^2 - 7x - 12;\ p(2)$

29. Why is it that we *add* when performing synthetic division, but *subtract* when performing long division?

30. Explain how synthetic division could be useful when attempting to factor a polynomial.

SKILL MAINTENANCE

Solve. [1.5]

31. $9 + cb = a - b$, for b

32. $8 + ac = bd + ab$, for a

Find the domain of f. [5.8]

33. $f(x) = \dfrac{5}{3x^2 - 75}$

34. $f(x) = \dfrac{7}{2x^2 + 7x - 9}$

Graph.

35. $y - 2 = \frac{3}{4}(x + 1)$ [2.5]

36. $y = -\frac{4}{3}x + 2$ [2.3]

SYNTHESIS

37. Let $Q(x)$ be a polynomial function with $p(x)$ a factor of $Q(x)$. If $p(3) = 0$, does it follow that $Q(3) = 0$? Why or why not? If $Q(3) = 0$, does it follow that $p(3) = 0$? Why or why not?

38. What adjustments must be made if synthetic division is to be used to divide a polynomial by a binomial of the form $ax + b$, with $a > 1$?

39. To prove the remainder theorem, note that any polynomial $P(x)$ can be rewritten as $(x - r) \cdot Q(x) + R$, where $Q(x)$ is the quotient polynomial that arises when $P(x)$ is divided by $x - r$, and R is some constant (the remainder).
   a) How do we know that R must be a constant?
   b) Show that $P(r) = R$ (this says that $P(r)$ is the remainder when $P(x)$ is divided by $x - r$).

40. Let $f(x) = 6x^3 - 13x^2 - 79x + 140$. Find $f(4)$ and then solve the equation $f(x) = 0$.

41. Let $f(x) = 4x^3 + 16x^2 - 3x - 45$. Find $f(-3)$ and then solve the equation $f(x) = 0$.

42. Use the TRACE feature on a graphing calculator to check your answer to Exercise 40.

43. Use the TRACE feature on a graphing calculator to check your answer to Exercise 41.

Nested evaluation.   One way to evaluate a polynomial function like $P(x) = 3x^4 - 5x^3 + 4x^2 - 1$ is to successively factor out x as shown:
$$P(x) = x(x(x(3x - 5) + 4) + 0) - 1.$$
Computations are then performed using this "nested" form of $P(x)$.

44. Use nested evaluation to find $f(4)$ in Exercise 40. Note the similarities to the calculations performed with synthetic division.

45. Use nested evaluation to find $f(-3)$ in Exercise 41. Note the similarities to the calculations performed with synthetic division.

6.8   Formulas, Applications, and Variation

Formulas • Direct Variation • Inverse Variation •
Joint and Combined Variation

Formulas

Formulas occur frequently as mathematical models. Many formulas contain rational expressions, and to solve such formulas for a specified letter, we proceed as when solving rational equations.

EXAMPLE 1   Electronics.   The formula

$$\frac{1}{R} = \frac{1}{r_1} + \frac{1}{r_2}$$

is used by electricians to determine the resistance R of two resistors r_1 and r_2 connected in parallel.* Solve for r_1.

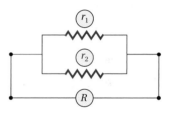

Solution   We use the same approach as in Section 6.4:

$$Rr_1r_2 \cdot \frac{1}{R} = Rr_1r_2 \cdot \left(\frac{1}{r_1} + \frac{1}{r_2} \right) \qquad \text{Multiplying both sides by the LCD}$$

$$Rr_1r_2 \cdot \frac{1}{R} = Rr_1r_2 \cdot \frac{1}{r_1} + Rr_1r_2 \cdot \frac{1}{r_2} \qquad \text{Multiplying to remove parentheses}$$

$$r_1r_2 = Rr_2 + Rr_1. \qquad \text{Simplifying by removing factors equal to 1: } \frac{R}{R} = 1; \frac{r_1}{r_1} = 1; \frac{r_2}{r_2} = 1$$

At this point it is tempting to multiply by $1/r_2$ to get r_1 alone on the left, *but* note that there is an r_1 on the right. We must get all the terms involving r_1 on the *same side* of the equation.

$$r_1r_2 - Rr_1 = Rr_2 \qquad \text{Subtracting } Rr_1 \text{ from both sides}$$

$$r_1(r_2 - R) = Rr_2 \qquad \text{Factoring out } r_1 \text{ in order to combine like terms}$$

$$r_1 = \frac{Rr_2}{r_2 - R} \qquad \text{Dividing both sides by } r_2 - R \text{ to get } r_1 \text{ alone}$$

This formula can be used to calculate r_1 whenever R and r_2 are known.

*Recall that the subscripts 1 and 2 merely indicate that r_1 and r_2 are different variables representing similar quantities.

EXAMPLE 2

Astronomy. The formula

$$\frac{V^2}{R^2} = \frac{2g}{R + h}$$

is used to find a satellite's *escape velocity* V, where R is a planet's radius, h is the satellite's height above the planet, and g is the planet's gravitational constant. Solve for h.

Solution We first clear fractions by multiplying by the LCD, which is $R^2(R + h)$:

$$\frac{V^2}{R^2} = \frac{2g}{R + h}$$

$$R^2(R + h)\frac{V^2}{R^2} = R^2(R + h)\frac{2g}{R + h}$$

$$\frac{R^2(R + h)V^2}{R^2} = \frac{R^2(R + h)2g}{R + h}$$

$(R + h)V^2 = R^2 \cdot 2g.$ Removing factors equal to 1:
$$\frac{R^2}{R^2} = 1 \text{ and } \frac{R + h}{R + h} = 1$$

Remember: We are solving for h. Although we *could* distribute V^2, since h appears only within the factor $R + h$, it is easier to divide both sides by V^2:

$$\frac{(R + h)V^2}{V^2} = \frac{2R^2g}{V^2}$$ Dividing both sides by V^2

$$R + h = \frac{2R^2g}{V^2}$$ Removing a factor equal to 1: $\frac{V^2}{V^2} = 1$

$$h = \frac{2R^2g}{V^2} - R.$$ Subtracting R from both sides

The last equation can be used to determine the height of a satellite above a planet when the planet's radius and gravitational constant, along with the satellite's escape velocity, are known.

EXAMPLE 3

Acoustics (the Doppler Effect). The formula

$$f = \frac{sg}{s + v}$$

is used to determine the frequency f of a sound that is moving at velocity v toward a listener who hears the sound as frequency g. Here s is the speed of sound in a particular medium. Solve for s.

Student Notes _____

The steps used to solve equations are precisely the same steps used to solve formulas. If you feel "rusty" in this regard, study the earlier section in which this type of equation first appeared. Then make sure that you can consistently solve those equations before returning to the work with formulas.

Solution   We first clear fractions by multiplying by the LCD, $s + v$:

$$f \cdot (s + v) = \frac{sg}{s + v}(s + v)$$

$$fs + fv = sg. \qquad \text{The variable for which we are solving, } s,$$
$$\text{appears on both sides, forcing us to}$$
$$\text{distribute on the left side.}$$

Next, we must get all terms containing s on one side:

$$fv = sg - fs \qquad \text{Subtracting } fs \text{ from both sides}$$
$$fv = s(g - f) \qquad \text{Factoring out } s. \text{ This is like combining like terms.}$$
$$\frac{fv}{g - f} = s. \qquad \text{Dividing both sides by } g - f$$

Since s is isolated on one side, we have solved for s. This last equation can be used to determine the speed of sound whenever f, v, and g are known.

> **To Solve a Rational Equation for a Specified Variable**
>
> 1. If necessary, multiply both sides by the LCD to clear fractions.
> 2. Multiply, as needed, to remove parentheses.
> 3. Get all terms with the specified variable alone on one side.
> 4. Factor out the specified variable if it is in more than one term.
> 5. Multiply or divide on both sides to isolate the specified variable.

Variation

To extend our study of formulas and functions, we now examine three real-world situations: direct variation, inverse variation, and combined variation.

Direct Variation

A registered nurse earns $22 per hour. In 1 hr, $22 is earned. In 2 hr, $44 is earned. In 3 hr, $66 is earned, and so on. This gives rise to a set of ordered pairs:

(1, 22), (2, 44), (3, 66), (4, 88), and so on.

Note that the ratio of earnings E to time t is $\frac{22}{1}$ in every case.

If a situation gives rise to pairs of numbers in which the ratio is constant, we say that there is **direct variation**. Here earnings *vary directly* as the time:

We have $\dfrac{E}{t} = 22$, so $E = 22t$ or, using function notation, $E(t) = 22t$.

Direct Variation

When a situation gives rise to a linear function of the form $f(x) = kx$, or $y = kx$, where k is a nonzero constant, we say that there is *direct variation*, that *y varies directly* as *x*, or that *y is proportional to x*. The number k is called the *variation constant*, or *constant of proportionality*.

Note that for $k > 0$, any equation of the form $y = kx$ indicates that as x increases, y increases as well.

EXAMPLE 4   Find the variation constant and an equation of variation if y varies directly as x, and $y = 32$ when $x = 2$.

Solution   We know that $(2, 32)$ is a solution of $y = kx$. Therefore,

$$32 = k \cdot 2 \qquad \text{Substituting}$$

$$\frac{32}{2} = k, \quad \text{or} \quad k = 16. \qquad \text{Solving for } k$$

The variation constant is 16. The equation of variation is $y = 16x$. The notation $y(x) = 16x$ or $f(x) = 16x$ is also used.

EXAMPLE 5   Water from melting snow.   The number of centimeters W of water produced from melting snow varies directly as the number of centimeters S of snow. Meteorologists know that under certain conditions, 150 cm of snow will melt to 16.8 cm of water. The average annual snowfall in Alta, Utah, is 500 in. Assuming the above conditions, how much water will replace the 500 in. of snow?

Alta, Utah

Solution

1. **Familiarize.**   Because of the phrase "*W* … varies directly as … *S*," we express the amount of water as a function of the amount of snow. Thus, $W(S) = kS$, where k is the variation constant. Knowing that 150 cm of snow becomes 16.8 cm of water, we have $W(150) = 16.8$. Because we are using ratios, it does not matter whether we work in inches or centimeters, provided the same units are used for W and S.

2. **Translate.**   We find the variation constant using the data and then use it to write the equation of variation:

$$W(S) = kS$$
$$W(150) = k \cdot 150 \qquad \text{Replacing } S \text{ with } 150$$
$$16.8 = k \cdot 150 \qquad \text{Replacing } W(150) \text{ with } 16.8$$
$$\frac{16.8}{150} = k \qquad \text{Solving for } k$$
$$0.112 = k. \qquad \text{This is the variation constant.}$$

The equation of variation is $W(S) = 0.112S$. This is the translation.

3. **Carry out.**   To find how much water 500 in. of snow will become, we compute $W(500)$:

$$W(S) = 0.112S$$
$$W(500) = 0.112(500) \qquad \text{Substituting 500 for } S$$
$$W = 56.$$

4. **Check.**   To check, we could reexamine all our calculations. Note that our answer seems reasonable since 500/56 and 150/16.8 are equal.

5. **State.**   Alta's 500 in. of snow will become 56 in. of water.

Inverse Variation

To see what we mean by inverse variation, suppose a bus is traveling 20 mi. At 20 mph, the trip will take 1 hr. At 40 mph, it will take $\frac{1}{2}$ hr. At 60 mph, it will take $\frac{1}{3}$ hr, and so on. This gives rise to pairs of numbers, all having the same product:

$$(20, 1), \left(40, \tfrac{1}{2}\right), \left(60, \tfrac{1}{3}\right), \left(80, \tfrac{1}{4}\right), \quad \text{and so on.}$$

Note that the product of each pair of numbers is 20. Whenever a situation gives rise to pairs of numbers for which the product is constant, we say that there is **inverse variation**. Since $r \cdot t = 20$, the time t, in hours, required for the bus to travel 20 mi at r mph is given by

$$t = \frac{20}{r} \quad \text{or, using function notation,} \quad t(r) = \frac{20}{r}.$$

> **Inverse Variation**
>
> When a situation gives rise to a rational function of the form $f(x) = k/x$, or $y = k/x$, where k is a nonzero constant, we say that there is *inverse variation*, that *y varies inversely as x*, or that *y is inversely proportional to x*. The number k is called the *variation constant*, or *constant of proportionality*.

Note that for $k > 0$, any equation of the form $y = k/x$ indicates that as x increases, y decreases.

EXAMPLE 6   Find the variation constant and an equation of variation if y varies inversely as x, and $y = 32$ when $x = 0.2$.

Solution   We know that (0.2, 32) is a solution of

$$y = \frac{k}{x}.$$

Therefore,

$$32 = \frac{k}{0.2} \qquad \text{Substituting}$$
$$(0.2)32 = k$$
$$6.4 = k. \qquad \text{Solving for } k$$

The variation constant is 6.4. The equation of variation is

$$y = \frac{6.4}{x}.$$

There are many real-life problems that translate to an equation of inverse variation.

EXAMPLE 7   Ultraviolet index.   The ultraviolet, or UV, index is a measure issued daily by the National Weather Service that indicates the strength of the sun's rays in a particular locale. For those people whose skin is quite sensitive, a UV rating of 6 will cause sunburn after 10 min (*Source: The Electronic Textbook of Dermatology* found at www.telemedicine.org, January 2004). Given that the number of minutes it takes to burn, t, varies inversely as the UV rating u, how long will it take a highly sensitive person to burn on a day with a UV rating of 4?

Solution

1. **Familiarize.**   Because of the phrase "… varies inversely as the UV index," we express the amount of time needed to burn as a function of the UV rating: $t(u) = k/u$.

2. **Translate.**   We use the given information to solve for k. Then we use that result to write the equation of variation.

$$t(u) = \frac{k}{u} \qquad \text{Using function notation}$$

$$t(6) = \frac{k}{6} \qquad \text{Replacing } u \text{ with } 6$$

$$10 = \frac{k}{6} \qquad \text{Replacing } t(6) \text{ with } 10$$

$$60 = k \qquad \text{Solving for } k, \text{ the variation constant}$$

The equation of variation is $t(u) = 60/u$. This is the translation.

3. **Carry out.**   To find how long it would take a highly sensitive person to burn on a day with a UV index of 4, we calculate $t(4)$:

$$t(4) = \frac{60}{4} = 15. \qquad t = 15 \text{ when } u = 4$$

4. **Check.**   We could now recheck each step. Note that, as expected, as the UV rating goes *down*, the time it takes to burn goes *up*.

5. **State.**   On a day with a UV rating of 4, a highly sensitive person will begin to burn after 15 min of exposure.

Joint and Combined Variation

When a variable varies directly with more than one other variable, we say that there is *joint variation*. For example, in the formula for the volume of a right circular cylinder, $V = \pi r^2 h$, we say that V varies *jointly* as h and the square of r.

> **Joint Variation**
>
> y varies *jointly* as x and z if, for some nonzero constant k, $y = kxz$.

EXAMPLE 8   Find an equation of variation if y varies jointly as x and z, and $y = 30$ when $x = 2$ and $z = 3$.

Solution   We have

$$y = kxz,$$

so

$$30 = k \cdot 2 \cdot 3$$

$$k = 5. \qquad \text{The variation constant is } 5.$$

The equation of variation is $y = 5xz$.

Joint variation is one form of *combined variation*. In general, when a variable varies directly and/or inversely, at the same time, with more than one other variable, there is **combined variation**. Examples 8 and 9 are both examples of combined variation.

EXAMPLE 9   Find an equation of variation if y varies jointly as x and z and inversely as the square of w, and $y = 105$ when $x = 3$, $z = 20$, and $w = 2$.

Solution   The equation of variation is of the form

$$y = k \cdot \frac{xz}{w^2},$$

so, substituting, we have

$$105 = k \cdot \frac{3 \cdot 20}{2^2}$$

$$105 = k \cdot 15$$

$$k = 7.$$

Thus,

$$y = 7 \cdot \frac{xz}{w^2}.$$

Exercise Set

6.8

FOR EXTRA HELP

Student's Solutions Manual   Digital Video Tutor CD 5 Videotape 12   AW Math Tutor Center   MathXL Tutorials on CD   MathXL   MyMathLab

↜ *Concept Reinforcement*   *Complete each of the following statements.*

1. To solve a formula for a variable appearing in a denominator, we multiply both sides of the equation by the _____.

2. All terms containing the variable being solved for must ultimately be on the _____ side of the equation.

3. If the variable being solved for appears in more than one term on the same side of the equation, it is usually necessary to _____.

4. If y varies directly as x, then $y = kx$ and k is called the _____ of _____.

↜ *Concept Reinforcement*   *Determine whether each situation represents direct or inverse variation.*

5. Two painters can scrape a house in 9 hr, whereas three painters can scrape the house in 6 hr.

6. Fanny planted 5 bulbs in 20 min and 7 bulbs in 28 min.

7. Glen swam 2 laps in 7 min and 6 laps in 21 min.

8. It took 2 band members 80 min to set up for a show; with 4 members working, it took 40 min.

9. It took 3 hr for 4 volunteers to wrap the campus' collection of Toys for Tots, but only 1.5 hr with 8 volunteers working.

10. Fatuma's air conditioner cooled off 1000 ft^3 in 10 min and 3000 ft^3 in 30 min.

Solve each formula for the specified variable.

11. $f = \dfrac{L}{d}$; d

12. $\dfrac{W_1}{W_2} = \dfrac{d_1}{d_2}$; W_1

13. $s = \dfrac{(v_1 + v_2)t}{2}$; v_1

14. $s = \dfrac{(v_1 + v_2)t}{2}$; t

15. $\dfrac{t}{a} + \dfrac{t}{b} = 1$; b

16. $\dfrac{1}{R} = \dfrac{1}{r_1} + \dfrac{1}{r_2}$; R

17. $I = \dfrac{2V}{R + 2r}$; R

18. $I = \dfrac{2V}{R + 2r}$; r

19. $R = \dfrac{gs}{g + s}$; g

20. $K = \dfrac{rt}{r - t}$; t

21. $I = \dfrac{nE}{R + nr}$; n

22. $I = \dfrac{nE}{R + nr}$; r

23. $\dfrac{1}{p} + \dfrac{1}{q} = \dfrac{1}{f}$; q

24. $\dfrac{1}{p} + \dfrac{1}{q} = \dfrac{1}{f}$; p

25. $S = \dfrac{H}{m(t_1 - t_2)}$; t_1

26. $S = \dfrac{H}{m(t_1 - t_2)}$; H

27. $\dfrac{E}{e} = \dfrac{R + r}{r}$; r

28. $\dfrac{E}{e} = \dfrac{R + r}{R}$; R

29. $S = \dfrac{a}{1 - r}$; r

30. $S = \dfrac{a - ar^n}{1 - r}$; a

Aha! **31.** $c = \dfrac{f}{(a + b)c}$; $a + b$

32. $d = \dfrac{g}{d(c + f)}$; $c + f$

33. *Interest.*   The formula

$$P = \dfrac{A}{1 + r}$$

is used to determine what principal P should be invested for one year at $(100 \cdot r)$% simple interest in order to have A dollars after a year. Solve for r.

34. *Taxable interest.*   The formula

$$I_t = \dfrac{I_f}{1 - T}$$

gives the *taxable interest rate* I_t equivalent to the *tax-free interest rate* I_f for a person in the $(100 \cdot T)$% tax bracket. Solve for T.

35. *Average speed.*   The formula

$$v = \dfrac{d_2 - d_1}{t_2 - t_1}$$

gives an object's average speed v when that object has traveled d_1 miles in t_1 hours and d_2 miles in t_2 hours. Solve for t_2.

36. *Average acceleration.*   The formula

$$a = \dfrac{v_2 - v_1}{t_2 - t_1}$$

gives a vehicle's *average acceleration* when its velocity changes from v_1 at time t_1 to v_2 at time t_2. Solve for t_1.

37. *Planetary orbits.*   The formula

$$\dfrac{x^2}{a^2} + \dfrac{y^2}{b^2} = 1$$

can be used to plot a planet's elliptical orbit of width $2a$ and length $2b$ (see p. 671 in Section 10.2). Solve for b^2.

38. *Work rate.*   The formula

$$\dfrac{1}{t} = \dfrac{1}{a} + \dfrac{1}{b}$$

gives the total time t required for two workers to complete a job, if the workers' individual times are a and b. Solve for t.

39. *Semester average.*   The formula

$$A = \dfrac{2Tt + Qq}{2T + Q}$$

gives a student's average A after T tests and Q quizzes, where each test counts as 2 quizzes, t is the test average, and q is the quiz average. Solve for Q.

40. *Astronomy.*   The formula

$$L = \dfrac{dR}{D - d},$$

where D is the diameter of the sun, d is the diameter of the earth, R is the earth's distance from the sun, and L is some fixed distance, is used in calculating when lunar eclipses occur. Solve for D.

Find the variation constant and an equation of variation if y varies directly as x and the following conditions apply.

41. $y = 28$ when $x = 4$

42. $y = 5$ when $x = 12$

43. $y = 3.4$ when $x = 2$

44. $y = 2$ when $x = 5$

45. $y = 2$ when $x = \frac{1}{3}$

46. $y = 0.9$ when $x = 0.5$

Find the variation constant and an equation of variation in which y varies inversely as x, and the following conditions exist.

47. $y = 3$ when $x = 20$

48. $y = 16$ when $x = 4$

49. $y = 28$ when $x = 4$

50. $y = 9$ when $x = 5$

51. $y = 27$ when $x = \frac{1}{3}$

52. $y = 81$ when $x = \frac{1}{9}$

53. *Use of aluminum cans.* The number N of aluminum cans used each year varies directly as the number of people using the cans. If 250 people use 60,000 cans in one year, how many cans are used each year in Dallas, which has a population of 1,008,000?

54. *Hooke's law.* Hooke's law states that the distance d that a spring is stretched by a hanging object varies directly as the mass m of the object. If the distance is 20 cm when the mass is 3 kg, what is the distance when the mass is 5 kg?

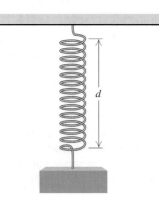

55. *Ohm's law.* The electric current I, in amperes, in a circuit varies directly as the voltage V. When 15 volts are applied, the current is 5 amperes. What is the current when 18 volts are applied?

56. *Pumping rate.* The time t required to empty a tank varies inversely as the rate r of pumping. If a Briggs and Stratton pump can empty a tank in 45 min at the rate of 600 kL/min, how long will it take the pump to empty the tank at 1000 kL/min?

57. *Work rate.* The time T required to do a job varies inversely as the number of people P working. It takes 5 hr for 7 volunteers to pick up rubbish from 1 mi of roadway. How long would it take 10 volunteers to complete the job?

58. *Weekly allowance.* According to Fidelity Investments *Investment Vision Magazine*, the average weekly allowance A of children varies directly as their grade level, G. In a recent year, the average allowance of a 9th-grade student was $13.50 per week. What was the average weekly allowance of a 4th-grade student?
Source: Based on data from www.kidsmoney.org

Aha! **59.** *Mass of water in a human.* The number of kilograms W of water in a human body varies directly as the mass of the body. A 96-kg person contains 64 kg of water. How many kilograms of water are in a 48-kg person?

60. *Weight on Mars.* The weight M of an object on Mars varies directly as its weight E on Earth. A person who weighs 95 lb on Earth weighs 38 lb on Mars. How much would a 100-lb person weigh on Mars?

61. *Bicycling.* The number of calories burned by a bicyclist is directly proportional to the time spent bicycling. At 10 mph, it takes 30 min to burn 150 calories. How long would it take to burn 250 calories when biking 10 mph?
Source: Physical Activity and Health: A Report of the Surgeon General

62. *Wavelength and frequency.* The wavelength W of a radio wave varies inversely as its frequency F. A wave with a frequency of 1200 kilohertz has a length of 300 meters. What is the length of a wave with a frequency of 800 kilohertz?

63. *Ultraviolet index.*   At an ultraviolet, or UV, rating of 4, those people who are less sensitive to the sun will burn in 75 min. Given that the number of minutes it takes to burn, *t*, varies inversely with the UV rating, *u*, how long will it take less sensitive people to burn when the UV rating is 14?
Source: *The Electronic Textbook of Dermatology* at www.telemedicine.org

64. *Current and resistance.*   The current *I* in an electrical conductor varies inversely as the resistance *R* of the conductor. If the current is $\frac{1}{2}$ ampere when the resistance is 240 ohms, what is the current when the resistance is 540 ohms?

65. *Air pollution.*   The average U.S. household of 2.6 people released 1.1 tons of carbon monoxide into the environment in a recent year. How many tons were released nationally? Use 289,000,000 as the U.S. population.
Sources: Based on data from the U.S. Environmental Protection Agency and the U.S. Bureau of the Census

66. *Relative aperture.*   The relative aperture, or f-stop, of a 23.5-mm lens is directly proportional to the focal length *F* of the lens. If a lens with a 150-mm focal length has an f-stop of 6.3, find the f-stop of a 23.5-mm lens with a focal length of 80 mm.

Find an equation of variation in which:

67. *y* varies directly as the square of *x*, and *y* = 6 when *x* = 3.

68. *y* varies directly as the square of *x*, and *y* = 0.15 when *x* = 0.1.

69. *y* varies inversely as the square of *x*, and *y* = 6 when *x* = 3.

70. *y* varies inversely as the square of *x*, and *y* = 0.15 when *x* = 0.1.

71. *y* varies jointly as *x* and the square of *z*, and *y* = 105 when *x* = 14 and *z* = 5.

72. *y* varies jointly as *x* and *z* and inversely as *w*, and *y* = $\frac{3}{2}$ when *x* = 2, *z* = 3, and *w* = 4.

73. *y* varies jointly as *w* and the square of *x* and inversely as *z*, and *y* = 49 when *w* = 3, *x* = 7, and *z* = 12.

74. *y* varies directly as *x* and inversely as *w* and the square of *z*, and *y* = 4.5 when *x* = 15, *w* = 5, and *z* = 2.

75. *Electrical safety.*   The amount of time *t* needed for an electrical shock to stop a 150-lb person's heart from beating varies inversely as the square of the current flowing through the body. It is known that a 0.089-amp current is deadly to a 150-lb person after 3.4 sec. How long would it take a 0.096-amp current to be deadly?
Source: Safety Consulting Services

76. *Stopping distance of a car.*   The stopping distance *d* of a car after the brakes have been applied varies directly as the square of the speed *r*. Once the brakes are applied, a car traveling 60 mph can stop in 138 ft. What stopping distance corresponds to a speed of 40 mph?
Source: Based on data from Edmunds.com

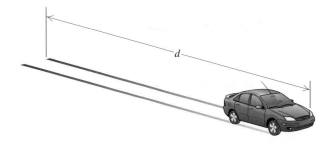

77. *Volume of a gas.*   The volume *V* of a given mass of a gas varies directly as the temperature *T* and inversely as the pressure *P*. If *V* = 231 cm³ when *T* = 300°K (Kelvin) and *P* = 20 lb/cm², what is the volume when *T* = 320°K and *P* = 16 lb/cm²?

78. *Intensity of a signal.* The intensity I of a television signal varies inversely as the square of the distance d from the transmitter. If the intensity is 25 W/m^2 at a distance of 2 km, what is the intensity 6.25 km from the transmitter?

79. *Atmospheric drag.* Wind resistance, or atmospheric drag, tends to slow down moving objects. Atmospheric drag W varies jointly as an object's surface area A and velocity v. If a car traveling at a speed of 40 mph with a surface area of 37.8 ft^2 experiences a drag of 222 N (Newtons), how fast must a car with 51 ft^2 of surface area travel in order to experience a drag force of 430 N?

80. *Drag force.* The drag force F on a boat varies jointly as the wetted surface area A and the square of the velocity of the boat. If a boat traveling 6.5 mph experiences a drag force of 86 N when the wetted surface area is 41.2 ft^2, find the wetted surface area of a boat traveling 8.2 mph with a drag force of 94 N.

81. Which exercise did you find easier to work: Exercise 13 or Exercise 19? Why?

82. If y varies directly as x, does doubling x cause y to be doubled as well? Why or why not?

SKILL MAINTENANCE

Find the domain of f. [2.2]

83. $f(x) = \dfrac{2x - 1}{x^2 + 1}$

84. $f(x) = |2x - 1|$

85. Graph on a plane: $6x - y < 6$. [4.4]

86. If $f(x) = x^3 - x$, find $f(2a)$. [2.2]

87. Factor: $t^3 + 8b^3$. [5.6]

88. Solve: $6x^2 = 11x + 35$. [5.8]

SYNTHESIS

89. Suppose that the number of customer complaints is inversely proportional to the number of employees hired. Will a firm reduce the number of complaints more by expanding from 5 to 10 employees, or from 20 to 25? Explain. Consider using a graph to help justify your answer.

90. Why do you think subscripts are used in Exercises 13 and 25 but not in Exercises 27 and 28?

91. *Escape velocity.* A satellite's escape velocity is 6.5 mi/sec, the radius of the earth is 3960 mi, and the earth's gravitational constant is 32.2 ft/sec^2. How far is the satellite from the surface of the earth? (See Example 2.)

92. The *harmonic mean* of two numbers a and b is a number M such that the reciprocal of M is the average of the reciprocals of a and b. Find a formula for the harmonic mean.

93. *Health-care.* Young's rule for determining the size of a particular child's medicine dosage c is

$$c = \frac{a}{a + 12} \cdot d,$$

where a is the child's age and d is the typical adult dosage. If a child's age is doubled, the dosage increases. Find the ratio of the larger dosage to the smaller dosage. By what percent does the dosage increase?
Source: Olsen, June Looby, Leon J. Ablon, and Anthony Patrick Giangrasso, *Medical Dosage Calculations,* 6th ed.

94. Solve for x:
$$x^2\left(1 - \frac{2pq}{x}\right) = \frac{2p^2q^3 - pq^2x}{-q}.$$

95. *Average acceleration.* The formula

$$a = \frac{\dfrac{d_4 - d_3}{t_4 - t_3} - \dfrac{d_2 - d_1}{t_2 - t_1}}{t_4 - t_2}$$

can be used to approximate average acceleration, where the d's are distances and the t's are the corresponding times. Solve for t_1.

96. If y varies inversely as the cube of x and x is multiplied by 0.5, what is the effect on y?

97. *Intensity of light.* The intensity I of light from a bulb varies directly as the wattage of the bulb and inversely as the square of the distance d from the bulb. If the wattage of a light source and its distance from reading matter are both doubled, how does the intensity change?

98. Describe in words the variation represented by $W = \dfrac{km_1M_1}{d^2}$. Assume k is a constant.

99. *Tension of a musical string.* The tension T on a string in a musical instrument varies jointly as the string's mass per unit length m, the square of its length l, and the square of its fundamental frequency f. A 2-m long string of mass 5 gm/m with a fundamental frequency of 80 has a tension of 100 N. How long should the same string be if its tension is going to be changed to 72 N?

100. *Volume and cost.* A peanut butter jar in the shape of a right circular cylinder is 4 in. high and 3 in. in diameter and sells for $1.20. If we assume that cost is proportional to volume, how much should a jar 6 in. high and 6 in. in diameter cost?

101. *Golf distance finder.* A device used in golf to estimate the distance d to a hole measures the size s that the 7-ft pin *appears* to be in a viewfinder. The viewfinder uses the principle, diagrammed here, that s gets bigger when d gets smaller. If $s = 0.56$ in. when $d = 50$ yd, find an equation of variation that expresses d as a function of s. What is d when $s = 0.40$ in.?

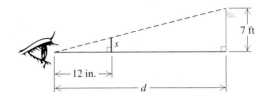

HOW IT WORKS:

Just sight the flagstick through the viewfinder…
fit flag between top dashed line and the solid line below…
…read the distance, 50 – 220 yards.

| 50 | 70 | 90 | 110 | 130 | 150 | 170 | 190 | 210 |

RANGE YARDS

Nothing to focus.

Gives you exact distance that your ball lies from the flagstick.

Choose proper club on every approach shot.

Figure new pin placement instantly.

Train your naked eye for formal and tournament play.

Eliminate the need to remember every stake, tree, and bush on the course.

CORNER

How Many Is a Million?

COLLABORATIVE

Focus: Direct variation and estimation

Time: 15 minutes

Group size: 2 or 3 and entire class

The National Park Service's estimates of crowd sizes for static (stationary) mass demonstrations vary directly as the area covered by the crowd. Park Service officials have found that at basic "shoulder-to-shoulder" demonstrations, 1 acre of land (about 45,000 ft^2) holds about 9000 people. Using aerial photographs, officials impose a grid to estimate the total area covered by the demonstrators. Once this has been accomplished, estimates of crowd size can be prepared.

ACTIVITY

1. In the grid imposed on the photograph below, each square represents 10,000 ft^2.

Estimate the size of the crowd photographed. Then compare your group's estimate with those of other groups. What might explain discrepancies between estimates? List ways in which your group's estimate could be made more accurate.

2. Park Service officials use an "acceptable margin of error" of no more than 20%. Using all estimates from part (1) above and allowing for error, find a range of values within which you feel certain that the actual crowd size lies.

3. The Million Man March of 1995 was not a static demonstration because of a periodic turnover of people in attendance (many people stayed for only part of the day's festivities). How might you change your methodology to compensate for this complication?

6 Study Summary

The focus of Chapter 6 is *rational expressions* similar to

$$\frac{x^2 + 6x}{x - 7} \quad \text{or} \quad \frac{5rt - s}{r^2 - rt + 7} \quad \text{(p. 352)}.$$

Much like fractions, rational expressions can be **multiplied, simplified, divided, added,** and **subtracted** (pp. 353, 354, 358, 364). For addition or subtraction, it is essential that a common denominator be used. If A, B, C, and D are rational expressions, with B, C, and D nonzero, then

$$\frac{A}{B} \cdot \frac{C}{D} = \frac{AC}{BD}; \qquad \frac{A}{B} \div \frac{C}{D} = \frac{A}{B} \cdot \frac{D}{C} = \frac{AD}{BC};$$

$$\frac{A}{B} + \frac{C}{B} = \frac{A + C}{B}; \qquad \frac{A}{B} - \frac{C}{B} = \frac{A - C}{B}.$$

To add or subtract two rational expressions that do not have a common denominator, it is necessary to multiply at least one of the expressions by a form of 1. To write this form of 1, we use the factors of the **least common denominator** that are absent from the original denominator (p. 366):

$$\frac{3}{(x - 2)(x + 5)} + \frac{4}{(x - 2)(x + 1)} = \frac{3}{(x - 2)(x + 5)} \cdot \frac{x + 1}{x + 1} + \frac{4}{(x - 2)(x + 1)} \cdot \frac{x + 5}{x + 5}$$

$$= \frac{3x + 3 + 4x + 20}{(x - 2)(x + 1)(x + 5)} \qquad \begin{array}{l} \text{The LCD is} \\ (x - 2)(x + 1)(x + 5). \end{array}$$

$$= \frac{7x + 23}{(x - 2)(x + 1)(x + 5)}.$$

Because rational expressions often appear in **rational functions**, it is important to specify restrictions on the domain of the function (p. 352). For example, if

$$f(x) = \frac{x^2 - 9}{(x - 3)(4x - 1)},$$

it is assumed that $x \neq 3$ and $x \neq \frac{1}{4}$, since either of those values would force us to divide by 0. However, when simplification occurs, we must clearly state the restriction $x \neq 3$:

$$f(x) = \frac{x^2 - 9}{(x - 3)(4x - 1)} = \frac{(x - 3)(x + 3)}{(x - 3)(4x - 1)} = \frac{x + 3}{4x - 1}, \quad x \neq 3 \quad \text{(p. 354)}.$$

When a rational expression itself appears within a rational expression, we have a **complex rational expression**, such as

$$\frac{\dfrac{x}{x-2}+3x}{\dfrac{x-5}{x+1}} \quad \text{or} \quad \frac{2+\dfrac{1}{r}+\dfrac{2}{s}}{\dfrac{r}{s}-\dfrac{2}{r}} \quad \text{(p. 374)}.$$

Two methods exist for simplifying complex rational expressions—multiplying by a form of 1 to clear fractions and dividing the numerator of the complex rational expression by its denominator (pp. 375, 377).

Sometimes the degree of the numerator of a rational expression exceeds the degree of the denominator, as in

$$\frac{5x^3+2x^2-3x+7}{x-4}.$$

When this occurs, **long division** or **synthetic division** can be used to write the expression as a quotient plus a remainder (pp. 403, 409). The **remainder theorem** states that when a polynomial like $5x^3+2x^2-3x+7$ is divided by $x-4$, the remainder is $5\cdot4^3+2\cdot4^2-3\cdot4+7$:

$$
\begin{array}{r}
5x^2+\ 22x+\ 85+\dfrac{347}{x-4} \\
x-4\overline{)5x^3+\ 2x^2-\ 3x+\ \ \ 7} \\
\underline{5x^3-20x^2} \\
22x^2-\ 3x+\ \ \ 7 \\
\underline{22x^2-88x} \\
85x+\ \ \ 7 \\
\underline{85x-\ 340} \\
347
\end{array}
\quad \text{or}
$$

$$
\begin{array}{r|rrrr}
4 & 5 & 2 & -3 & 7 \\
  &   & 20 & 88 & 340 \\
\hline
  & 5 & 22 & 85\ | & 347 \end{array} \leftarrow \text{Remainder} \leftarrow
$$

$$
\begin{aligned}
5\cdot4^3+2\cdot4^2-3\cdot4+7 &= 5\cdot64+2\cdot16-12+7 \\
&= 320+32-12+7 \\
&= 347
\end{aligned}
$$

When a rational expression appears in an equation, we have a **rational equation** (p. 384). To solve a rational equation, we multiply both sides of the equation by the least common denominator to clear fractions. Because the LCD is often a variable expression, it is important to check all potential solutions in the original equation:

$$\frac{2x+1}{x+3}+1 = \frac{2x+1}{x+3} \qquad \text{The LCD is } x+3. \text{ Note that } x\neq-3.$$

$$(x+3)\cdot\frac{2x+1}{x+3}+(x+3)\cdot1 = \frac{2x+1}{x+3}\cdot(x+3)$$

$$2x+1+x+3 = 2x+1$$

$$3x+4 = 2x+1$$

$$x = -3. \qquad \begin{array}{l}\text{Subtracting } 2x \text{ and subtracting } 4 \\ \text{on both sides}\end{array}$$

However, since $x\neq-3$, this equation has no solution.

Rational equations arise in a variety of real-world applications. Among these are **work problems** and **problems involving motion** or **speed** (pp. 391, 394).

When a rational equation contains more than one variable, it is a formula (p. 413). To solve such a formula for a specified variable, we use the same approach as for solving a rational equation.

Some rational equations arise as a result of developing equations of variation. **Direct variation** produces equations of the form $y = kx$, and **inverse variation** produces equations of the form $y = k/x$ (pp. 415, 417). In both cases, k is the **constant of proportionality** (pp. 416, 418).

Direct variation: $y = kx$;   for $k > 0$, as x increases, y increases.

Inverse variation: $y = k/x$;   for $k > 0$, as x increases, y decreases.

In **combined** and **joint variation**, one variable depends on at least two other variables (p. 419).

6 Review Exercises

🖙 *Concept Reinforcement   Classify each of the following statements as either true or false.*

1. If $f(x) = \dfrac{x - 3}{x^2 - 4}$, the domain of f is assumed to be $\{x \,|\, x \neq -2, x \neq 2\}$. [6.1]

2. We can simplify a rational expression whenever a term that appears in the numerator also appears in the denominator. [6.1]

3. To add rational expressions in which the denominators are opposites, we use the product of the denominators as the least common denominator. [6.2]

4. A complex rational expression can always be simplified by multiplying its numerator and its denominator by the LCD of those expressions. [6.3]

5. Checking the solution of a rational equation is no more important than checking the solution of any other equation. [6.4]

6. If Hannah can do a job alone in t_1 hr and Jacob can do the same job in t_2 hr, then working together it will take them $(t_1 + t_2)/2$ hr. [6.5]

7. If Carlie swims 5 km/h in still water and heads into a current of 2 km/h, her speed will change to 3 km/h. [6.5]

8. To divide a polynomial by a monomial, we divide the monomial by each term in the polynomial. [6.6]

9. To divide two polynomials using synthetic division, we must make sure that the divisor is of the form $x - a$. [6.7]

10. If x varies inversely as y, then there exists some constant k for which $x = k/y$. [6.8]

11. If

$$f(t) = \frac{t^2 - 3t + 2}{t^2 - 9},$$

find the following function values. [6.1]

a) $f(0)$

b) $f(-1)$

c) $f(2)$

Find the LCD. [6.2]

12. $\dfrac{5}{6x^3}, \dfrac{15}{16x^2}$

13. $\dfrac{x-2}{x^2+x-20}, \dfrac{x-3}{x^2+3x-10}$

Perform the indicated operations and, if possible, simplify.

14. $\dfrac{x^2}{x-3} - \dfrac{9}{x-3}$ [6.2]

15. $\dfrac{3a^2b^3}{5c^3d^2} \cdot \dfrac{15c^9d^4}{9a^7b}$ [6.1]

16. $\dfrac{5}{6m^2n^3p} + \dfrac{7}{9mn^4p^2}$ [6.2]

17. $\dfrac{x^3-8}{x^2-25} \cdot \dfrac{x^2+10x+25}{x^2+2x+4}$ [6.1]

18. $\dfrac{x^2-3x-18}{x^2+5x+6} \div \dfrac{x^2-16}{x^3-64}$ [6.1]

19. $\dfrac{x}{x^2+5x+6} - \dfrac{2}{x^2+3x+2}$ [6.2]

20. $\dfrac{-4xy}{x^2-y^2} + \dfrac{x+y}{x-y}$ [6.2]

21. $\dfrac{2x^2}{x-y} + \dfrac{2y^2}{y-x}$ [6.2]

22. $\dfrac{3}{y+4} - \dfrac{y}{y-1} + \dfrac{y^2+3}{y^2+3y-4}$ [6.2]

Find simplified form for f(x) and list all restrictions on the domain.

23. $f(x) = \dfrac{4x-2}{x^2-5x+4} - \dfrac{3x+2}{x^2-5x+4}$ [6.2]

24. $f(x) = \dfrac{x+8}{x+5} \cdot \dfrac{2x+10}{x^2-64}$ [6.1]

25. $f(x) = \dfrac{9x^2-1}{x^2-9} \div \dfrac{3x+1}{x+3}$ [6.1]

Simplify. [6.3]

26. $\dfrac{\dfrac{4}{x} - 4}{\dfrac{9}{x} - 9}$

27. $\dfrac{\dfrac{3}{a} + \dfrac{3}{b}}{\dfrac{6}{a^3} + \dfrac{6}{b^3}}$

28. $\dfrac{\dfrac{y^2+4y-77}{y^2-10y+25}}{\dfrac{y^2-5y-14}{y^2-25}}$

29. $\dfrac{\dfrac{5}{x^2-9} - \dfrac{3}{x+3}}{\dfrac{4}{x^2+6x+9} + \dfrac{2}{x-3}}$

Solve. [6.4]

30. $\dfrac{3}{x} + \dfrac{7}{x} = 5$

31. $\dfrac{5}{3x+2} = \dfrac{3}{2x}$

32. $\dfrac{4x}{x+1} + \dfrac{4}{x} + 9 = \dfrac{4}{x^2+x}$

33. $\dfrac{x+6}{x^2+x-6} + \dfrac{x}{x^2+4x+3} = \dfrac{x+2}{x^2-x-2}$

34. If

$$f(x) = \dfrac{2}{x-1} + \dfrac{2}{x+2},$$

find all a for which $f(a) = 1$. [6.4]

Solve. [6.5]

35. Kim can arrange the books for a book sale in 9 hr. Kelly can set up for the same book sale in 12 hr. How long would it take them, working together, to set up for the book sale?

36. A research company uses personal computers to process data while the owner is not using the computer. A Pentium 4 3.0-gigahertz processor can process a megabyte of data in 15 sec less time than a Celeron 2.53-gigahertz processor. Working together, the computers can process a megabyte of data in 18 sec. How long does it take each computer to process one megabyte of data?

37. The Black River's current is 6 mph. A boat travels 50 mi downstream in the same time that it takes to travel 30 mi upstream. What is the speed of the boat in still water?

38. A car and a motorcycle leave a rest area at the same time, with the car traveling 8 mph faster than the motorcycle. The car then travels 105 mi in the time it takes the motorcycle to travel 93 mi. Find the speed of each vehicle.

Divide. [6.6]

39. $(30r^2s^3 + 25r^2s^2 - 20r^3s^3) \div (5r^2s)$

40. $(y^3 + 125) \div (y + 5)$

41. $(4x^3 + 3x^2 - 5x - 2) \div (x^2 + 1)$

42. Divide using synthetic division: [6.7]

$(x^3 + 3x^2 + 2x - 6) \div (x - 3)$.

43. If $f(x) = 4x^3 - 6x^2 - 9$, use synthetic division to find $f(5)$. [6.7]

Solve. [6.8]

44. $R = \dfrac{gs}{g + s}$, for s

45. $S = \dfrac{H}{m(t_1 - t_2)}$, for m

46. $\dfrac{1}{ac} = \dfrac{2}{ab} - \dfrac{3}{bc}$, for c

47. $T = \dfrac{A}{v(t_2 - t_1)}$, for t_1

48. The amount of waste generated by a family varies directly as the number of people in the family. The average U.S. family has 3.14 people and generates 13.8 lb of waste daily. How many pounds of waste would be generated daily by a family of 5?
Sources: Based on data from the U.S. Bureau of the Census and the U.S. Statistical Abstract 2003

49. A warning dye is used by people in lifeboats to aid search planes. The volume V of the dye used varies directly as the square of the diameter d of the circular patch of water formed by the dye. If 4 L of dye is required for a 10-m wide circle, how much dye is needed for a 40-m wide circle?

50. Find an equation of variation in which y varies inversely as x, and $y = 3$ when $x = \frac{1}{4}$.

SYNTHESIS

51. Discuss at least three different uses of the LCD studied in this chapter. [6.2], [6.3], [6.4]

52. Explain the difference between a rational expression and a rational equation. [6.1]

Solve.

53. $\dfrac{5}{x - 13} - \dfrac{5}{x} = \dfrac{65}{x^2 - 13x}$ [6.4]

54. $\dfrac{\dfrac{x}{x^2 - 25} + \dfrac{2}{x - 5}}{\dfrac{3}{x - 5} - \dfrac{4}{x^2 - 10x + 25}} = 1$ [6.3], [6.4]

55. A Xeon 3.6-gigahertz processor can process a megabyte of data in 20 sec. How long would it take a Xeon working together with the Pentium 4 and Celeron processors (see Exercise 36) to process a megabyte of data? [6.5]

6 Chapter Test

Simplify.

1. $\dfrac{t + 1}{t + 3} \cdot \dfrac{3t + 9}{4t^2 - 4}$

2. $\dfrac{x^3 + 27}{x^2 - 16} \div \dfrac{x^2 + 8x + 15}{x^2 + x - 20}$

3. Find the LCD:

$$\dfrac{3x}{x^2 + 5x - 24}, \quad \dfrac{x + 1}{x^2 - 12x + 27}.$$

Perform the indicated operation and simplify when possible.

4. $\dfrac{25x}{x + 5} + \dfrac{x^3}{x + 5}$

5. $\dfrac{3a^2}{a - b} - \dfrac{3b^2 - 6ab}{b - a}$

6. $\dfrac{4ab}{a^2 - b^2} + \dfrac{a^2 + b^2}{a + b}$

7. $\dfrac{6}{x^3 - 64} - \dfrac{4}{x^2 - 16}$

Find simplified form for f(x) and list all restrictions on the domain.

8. $f(x) = \dfrac{4}{x+3} - \dfrac{x}{x-2} + \dfrac{x^2+4}{x^2+x-6}$

9. $f(x) = \dfrac{x^2-1}{x+2} \div \dfrac{x^2-2x}{x^2+x-2}$

Simplify.

10. $\dfrac{\dfrac{2}{a} + \dfrac{3}{b}}{\dfrac{5}{ab} + \dfrac{1}{a^2}}$

11. $\dfrac{\dfrac{x^2-5x-36}{x^2-36}}{\dfrac{x^2+x-12}{x^2-12x+36}}$

12. $\dfrac{\dfrac{2}{x+3} - \dfrac{1}{x^2-3x+2}}{\dfrac{3}{x-2} + \dfrac{4}{x^2+2x-3}}$

Solve.

13. $\dfrac{4}{2x-5} = \dfrac{6}{5x+3}$

14. $\dfrac{t+11}{t^2-t-12} + \dfrac{1}{t-4} = \dfrac{4}{t+3}$

For Exercises 15 and 16, let $f(x) = \dfrac{x+3}{x-1}$.

15. Find $f(2)$ and $f(-3)$.

16. Find all a for which $f(a) = 7$.

17. Kyla can install a vinyl kitchen floor in 3.5 hr. Brock can perform the same job in 4.5 hr. How long will it take them, working together, to install the vinyl?

Divide.

18. $(16ab^3c - 10ab^2c^2 + 12a^2b^2c) \div (4a^2b)$

19. $(y^2 - 20y + 64) \div (y - 6)$

20. $(6x^4 + 3x^2 + 5x + 4) \div (x^2 + 2)$

21. Divide using synthetic division:
$(x^3 + 5x^2 + 4x - 7) \div (x - 4)$.

22. If $f(x) = 3x^4 - 5x^3 + 2x - 7$, use synthetic division to find $f(4)$.

23. Solve $A = \dfrac{h(b_1 + b_2)}{2}$ for b_1.

24. The product of the reciprocals of two consecutive integers is $\frac{1}{30}$. Find the integers.

25. Emma bicycles 12 mph with no wind. Against the wind, she bikes 8 mi in the same time that it takes to bike 14 mi with the wind. What is the speed of the wind?

26. The number of workers n needed to clean a stadium after a game varies inversely as the amount of time t allowed for the cleanup. If it takes 25 workers to clean the stadium when there are 6 hr allowed for the job, how many workers are needed if the stadium must be cleaned in 5 hr?

27. The surface area of a balloon varies directly as the square of its radius. The area is 325 in^2 when the radius is 5 in. What is the area when the radius is 7 in.?

SYNTHESIS

28. Let
$$f(x) = \dfrac{1}{x+3} + \dfrac{5}{x-2}.$$
Find all a for which $f(a) = f(a+5)$.

29. Solve: $\dfrac{6}{x-15} - \dfrac{6}{x} = \dfrac{90}{x^2-15x}$.

30. Find the x- and y-intercepts for the function given by
$$f(x) = \dfrac{\dfrac{5}{x+4} - \dfrac{3}{x-2}}{\dfrac{2}{x-3} + \dfrac{1}{x+4}}.$$

31. One summer, Hans mowed 4 lawns for every 3 lawns mowed by his brother Franz. Together, they mowed 98 lawns. How many lawns did each mow?

Cumulative Review

1. Evaluate
$$\frac{2x - y^2}{x + y}$$
for $x = 3$ and $y = -4$. [1.1], [1.2]

2. Convert to scientific notation: 391,000,000. [1.7]

3. Determine the slope and the y-intercept for the line given by $7x - 4y = 12$. [2.3]

4. Find an equation for the line that passes through the points $(-1, 7)$ and $(2, -3)$. [2.5]

5. Solve the system
$$5x - 2y = -23,$$
$$3x + 4y = 7. \ [3.2]$$

6. Solve the system
$$-3x + 4y + \ z = -5,$$
$$x - 3y - \ z = 6,$$
$$2x + 3y + 5z = -8. \ [3.4]$$

7. Folsom Elementary School sold 45 pizzas for a fundraiser. Small pizzas sold for $7.00 each and large pizzas for $10.00 each. The total amount of funds received from the sale was $402. How many of each size pizza were sold? [3.3]

8. The sum of three numbers is 20. The first number is 3 less than twice the third number. The second number minus the third number is -7. What are the numbers? [3.5]

9. Trex Company makes decking material from waste wood fibers and reclaimed polyethylene. Its sales rose from $49.2 million in 1998 to $191 million in 2003. Calculate the rate at which sales were rising. [2.3]
Source: U.S. Securities and Exchange Commission

10. In 1989, the average length of a visit to a physician in an HMO was 15.4 min; and in 1998, it was 17.9 min. Let V represent the average length of a visit t years after 1989. [2.5]
Sources: Rutgers University Study and National Center for Health Statistics

   a) Find a linear function $V(t)$ that fits the data.
   b) Use the function of part (a) to predict the average length of a visit in 2005.

11. If
$$f(x) = \frac{x - 2}{x - 5},$$
find (a) $f(3)$ and (b) the domain of f. [2.2], [6.1]

Solve.

12. $8x = 1 + 16x^2$ [5.8]

13. $144 = 49y^2$ [5.8]

14. $20 > 2 - 6x$ [4.1]

15. $\frac{1}{3}x - \frac{1}{5} \geq \frac{1}{5}x - \frac{1}{3}$ [4.1]

16. $-8 < x + 2 < 15$ [4.2]

17. $3x - 2 < -6 \ or \ x + 3 > 9$ [4.2]

18. $|x| > 6.4$ [4.3]

19. $|3x - 2| \leq 14$ [4.3]

20. $\frac{8}{n} - \frac{2}{n} = 3$ [6.4]

21. $\frac{6}{x - 5} = \frac{2}{2x}$ [6.4]

22. $\frac{3x}{x - 2} - \frac{6}{x + 2} = \frac{24}{x^2 - 4}$ [6.4]

23. $\frac{3x^2}{x + 2} + \frac{5x - 22}{x - 2} = \frac{-48}{x^2 - 4}$ [6.4]

24. Let $f(x) = |3x - 5|$. Find all values of x for which $f(x) = 2$. [4.3]

25. Write the domain of f using interval notation if $f(x) = \sqrt{x - 9}$. [4.2]

Solve.

26. $5m - 3n = 4m + 12$, for n [1.5]

27. $P = \dfrac{4a}{a + b}$, for a [6.8]

Graph on a plane.

28. $4x \geq 5y + 12$ [4.4]

29. $y = \frac{1}{3}x - 2$ [2.3]

Perform the indicated operations and simplify.

30. $(2x^2 - 8x + 7) + (6x - 3x^3 + 9x^2 - 4)$ [5.1]

31. $(8x^3y^2)(-3xy^2)$ [5.2]

32. $(3a + b - 2c) - (-4b + 3c - 2a)$ [5.1]

33. $(5x^2 - 2x + 1)(3x^2 + x - 2)$ [5.2]

34. $(3x^2 + y)^2$ [5.2]

35. $(2x^2 - y)(2x^2 + y)$ [5.2]

36. $(-5m^3n^2 - 3mn^3) + (-4m^2n^2 + 4m^3n^2) - (2mn^3 - 3m^2n^2)$ [5.1]

37. $\dfrac{y^2 - 36}{2y + 8} \cdot \dfrac{y + 4}{y + 6}$ [6.1]

38. $\dfrac{x^4 - 1}{x^2 - x - 2} \div \dfrac{x^2 + 1}{x - 2}$ [6.1]

39. $\dfrac{5ab}{a^2 - b^2} + \dfrac{a + b}{a - b}$ [6.2]

40. $\dfrac{2}{m + 1} + \dfrac{3}{m - 5} - \dfrac{m^2 - 1}{m^2 - 4m - 5}$ [6.2]

41. $y - \dfrac{2}{3y}$ [6.2]

42. Simplify: $\dfrac{\dfrac{1}{x} - \dfrac{1}{y}}{x + y}$. [6.3]

43. Divide: $(9x^3 + 5x^2 + 2) \div (x + 2)$. [6.6]

Factor.

44. $4x^3 - 14x^2$ [5.3]

45. $x^2 + 8x - 84$ [5.4]

46. $16y^2 - 81$ [5.5]

47. $64x^3 + 8$ [5.6]

48. $t^2 - 16t + 64$ [5.5]

49. $x^6 - x^2$ [5.5]

50. $0.027b^3 - 0.008c^3$ [5.6]

51. $20x^2 - 7x - 3$ [5.4]

52. $3t^2 + 17t - 28$ [5.4]

53. $x^5 - x^3y + x^2y - y^2$ [5.3]

54. If $f(x) = x^2 - 4$ and $g(x) = x^2 - 7x + 10$, find the domain of f/g. [2.6], [5.8]

55. A digital data circuit can transmit a particular set of data in 4 sec. An analog phone circuit can transmit the same data in 20 sec. How long would it take, working together, for both circuits to transmit the data? [6.5]

56. The floor area of a rental trailer is rectangular. The length is 3 ft more than the width. A rug of area 54 ft^2 exactly fills the floor of the trailer. Find the perimeter of the trailer. [5.8]

57. The sum of the squares of three consecutive even integers is equal to 8 more than three times the square of the second number. Find the integers. [5.8]

58. *Logging.* The volume of wood V in a tree trunk varies jointly as the height h and the square of the girth g (girth is distance around). If the volume is 35 ft^3 when the height is 20 ft and the girth is 5 ft, what is the height when the volume is 85.75 ft^3 and the girth is 7 ft? [6.8]

SYNTHESIS

59. Multiply: $(x - 4)^3$. [5.2]

60. Find all roots for $f(x) = x^4 - 34x^2 + 225$. [5.8]

Solve.

61. $4 \leq |3 - x| \leq 6$ [4.2], [4.3]

62. $\dfrac{18}{x - 9} + \dfrac{10}{x + 5} = \dfrac{28x}{x^2 - 4x - 45}$ [6.4]

63. $16x^3 = x$ [5.8]

7

Exponents and Radicals

AN APPLICATION

The function given by $f(x) = 262 \cdot 2^{x/12}$ can be used to determine the frequency, in cycles per second, of a musical note that is x half-steps above a piano's middle C. Show that the G that is 7 half-steps (a perfect 5th) above middle C has a frequency that is about 1.5 times that of middle C.

This problem appears as Exercise 113 in Section 7.2.

R. J. Malloy
GUITAR MAKER
Englewood, Florida

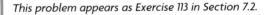

Music is mathematics with a soul. From determining what thickness and length of a string will produce the vibrations of a pure C note to determining how a new bracing system will affect the tone of a guitar, I use math constantly. Even the shape of a violin is determined by algebraic formulas that are 500 years old and based on the speed of sound through solid materials.

*I*n this chapter, we learn about square roots, cube roots, fourth roots, and so on. These roots are studied in connection with the manipulation of radical expressions and the solution of real-world applications. Exponents that are fractions are also studied and used to ease some of our work with radicals. The chapter closes with an examination of the complex-number system.

7.1 Radical Expressions and Functions

Square Roots and Square-Root Functions • Expressions of the Form $\sqrt{a^2}$ • Cube Roots • Odd and Even *n*th Roots

In this section, we consider roots, such as square roots and cube roots. We look at the symbolism that is used and ways in which symbols can be manipulated to get equivalent expressions. All of this will be important in problem solving.

Square Roots and Square-Root Functions

When a number is multiplied by itself, we say that the number is squared. Often we need to know what number was squared in order to produce some value a. If such a number can be found, we call that number a *square root* of a.

> **Square Root**
> The number c is a *square root* of a if $c^2 = a$.

For example,

9 has -3 and 3 as square roots because $(-3)^2 = 9$ and $3^2 = 9$.

25 has -5 and 5 as square roots because $(-5)^2 = 25$ and $5^2 = 25$.

-4 does not have a real-number square root because there is no real number c for which $c^2 = -4$.

Note that every positive number has two square roots, whereas 0 has only itself as a square root. Negative numbers do not have real-number square roots, although later in this chapter we introduce the *complex-number* system in which such square roots do exist.

EXAMPLE 1 Find the two square roots of 64.

Solution The square roots are 8 and -8, because $8^2 = 64$ and $(-8)^2 = 64$.

Whenever we refer to *the* square root of a number, we mean the nonnegative square root of that number. This is often referred to as the *principal square root* of the number.

Student Notes _____

It is important to remember the difference between *the* square root of 9 and *a* square root of 9. *A* square root of 9 means either 3 or -3, whereas *the* square root of 9, denoted $\sqrt{9}$, means the principal square root of 9, or 3.

> ### Principal Square Root
>
> The *principal square root* of a nonnegative number is its nonnegative square root. The symbol $\sqrt{}$ is called a *radical sign* and is used to indicate the principal square root of the number over which it appears.

EXAMPLE 2   Simplify each of the following.

a) $\sqrt{25}$

b) $\sqrt{\dfrac{25}{64}}$

c) $-\sqrt{64}$

d) $\sqrt{0.0049}$

Solution

a) $\sqrt{25} = 5$ $\sqrt{}$ indicates the principal square root. Note that $\sqrt{25} \neq -5$.

b) $\sqrt{\dfrac{25}{64}} = \dfrac{5}{8}$ Since $\left(\dfrac{5}{8}\right)^2 = \dfrac{25}{64}$

c) $-\sqrt{64} = -8$ Since $\sqrt{64} = 8$, $-\sqrt{64} = -8$.

d) $\sqrt{0.0049} = 0.07$ $(0.07)(0.07) = 0.0049$. Note too that $0.0049 = \dfrac{49}{10,000}$ and $\sqrt{49/10,000} = \dfrac{7}{100}$.

In addition to being read as "the principal square root of *a*," $\sqrt{a}$ is also read as "the square root of *a*," "root *a*," or "radical *a*." Any expression in which a radical sign appears is called a *radical expression*. The following are radical expressions:

$$\sqrt{5}, \qquad \sqrt{a}, \qquad -\sqrt{3x}, \qquad \sqrt{\dfrac{y^2 + 7}{y}}, \qquad \sqrt{x} + 8.$$

The expression under the radical sign is called the **radicand**. In the expressions above, the radicands are 5, *a*, 3*x*, $(y^2 + 7)/y$, and *x*.

All but the most basic calculators give values for square roots. These values are, for the most part, approximations. For example, on many calculators, if you enter 5 and then press $\boxed{\checkmark}$, a number like

2.23606798

appears, depending on how the calculator rounds. (On some calculators, the $\boxed{\checkmark}$ key is pressed first.) The exact value of $\sqrt{5}$, for example, is not given by any repeating or terminating decimal. In general, for any whole number *a* that is not a perfect square, $\sqrt{a}$ is a nonterminating, nonrepeating decimal. We discussed such *irrational numbers* in Chapter 1.

The square-root function, given by

$$f(x) = \sqrt{x},$$

has the interval $[0, \infty)$ as its domain. We can draw its graph by selecting convenient values for x and calculating the corresponding outputs. Once these ordered pairs have been graphed, a smooth curve can be drawn.

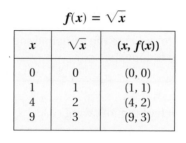

$$f(x) = \sqrt{x}$$

x	$\sqrt{x}$	$(x, f(x))$
0	0	$(0, 0)$
1	1	$(1, 1)$
4	2	$(4, 2)$
9	3	$(9, 3)$

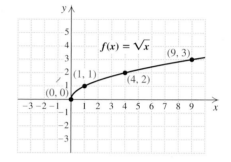

EXAMPLE 3   For each function, find the indicated function value.

a) $f(x) = \sqrt{3x - 2}$; $f(1)$ **b)** $g(z) = -\sqrt{6z + 4}$; $g(3)$

Solution

a) $f(1) = \sqrt{3 \cdot 1 - 2}$ Substituting

$\quad\quad\ = \sqrt{1} = 1$ Simplifying

b) $g(3) = -\sqrt{6 \cdot 3 + 4}$ Substituting

$\quad\quad\ = -\sqrt{22}$ Simplifying. This is the most exact way to write the answer.

$\quad\quad\ \approx -4.69041576$ Using a calculator to write an approximate answer

Expressions of the Form $\sqrt{a^2}$

It is tempting to write $\sqrt{a^2} = a$, but the next example shows that, as a rule, this is untrue.

EXAMPLE 4   Evaluate $\sqrt{a^2}$ for the following values: **(a)** 5; **(b)** 0; **(c)** -5.

Solution

a) $\sqrt{5^2} = \sqrt{25} = 5$

$\quad\quad\quad\quad\quad\quad$ Same

b) $\sqrt{0^2} = \sqrt{0} = 0$

$\quad\quad\quad\quad\quad\quad$ Same

c) $\sqrt{(-5)^2} = \sqrt{25} = 5$

$\quad\quad\quad\quad\quad\quad$ Opposites   Note that $\sqrt{(-5)^2} \neq -5$.

You may have noticed that evaluating $\sqrt{a^2}$ is just like evaluating $|a|$.

> **Simplifying $\sqrt{a^2}$**
>
> For any real number a,
>
> $$\sqrt{a^2} = |a|.$$
>
> (The principal square root of a^2 is the absolute value of a.)

When a radicand is the square of a variable expression, like $(x + 5)^2$ or $36t^2$, absolute-value signs are needed when simplifying. We use absolute-value signs unless we know that the expression being squared is nonnegative. This assures that our result is never negative.

EXAMPLE 5   Simplify each expression. Assume that the variable can represent any real number.

a) $\sqrt{(x + 1)^2}$ **b)** $\sqrt{x^2 - 8x + 16}$

c) $\sqrt{a^8}$ **d)** $\sqrt{t^6}$

Solution

a) $\sqrt{(x + 1)^2} = |x + 1|$ Since $x + 1$ might be negative (for example, if $x = -3$), absolute-value notation is necessary.

b) $\sqrt{x^2 - 8x + 16} = \sqrt{(x - 4)^2} = |x - 4|$ Since $x - 4$ might be negative, absolute-value notation is necessary.

c) Note that $(a^4)^2 = a^8$ and that a^4 is never negative. Thus,

$$\sqrt{a^8} = a^4.$$ Absolute-value notation is unnecessary here.

d) Note that $(t^3)^2 = t^6$. Thus,

$$\sqrt{t^6} = |t^3|.$$ Since t^3 might be negative, absolute-value notation is necessary.

---

technology connection

To see the necessity of the absolute-value signs, let y_1 represent the left side and y_2 the right side of each of the following equations. Then use a graph or table to determine whether these equations are true.

1. $\sqrt{x^2} \overset{?}{=} x$

2. $\sqrt{x^2} \overset{?}{=} |x|$

3. $x \overset{?}{=} |x|$

EXAMPLE 6   Simplify each expression. Assume that no radicands were formed by raising negative quantities to even powers.

a) $\sqrt{y^2}$ **b)** $\sqrt{a^{10}}$ **c)** $\sqrt{9x^2 - 6x + 1}$

Solution

a) $\sqrt{y^2} = y$ We are assuming that y is nonnegative, so no absolute-value notation is necessary. When y *is* negative, $\sqrt{y^2} \neq y$.

b) $\sqrt{a^{10}} = a^5$ Assuming that a^5 is nonnegative. Note that $(a^5)^2 = a^{10}$.

c) $\sqrt{9x^2 - 6x + 1} = \sqrt{(3x - 1)^2} = 3x - 1$ Assuming that $3x - 1$ is nonnegative

Cube Roots

We often need to know what number was cubed in order to produce a certain value. When such a number is found, we say that we have found a *cube root*. For example,

2 is the cube root of 8 because $2^3 = 2 \cdot 2 \cdot 2 = 8$;

-4 is the cube root of -64 because $(-4)^3 = (-4)(-4)(-4) = -64$.

Cube Root

The number c is the *cube root* of a if $c^3 = a$. In symbols, we write $\sqrt[3]{a}$ to denote the cube root of a.

Because each real number has only one cube root, a function can be formed. The cube-root function, given by

$$f(x) = \sqrt[3]{x},$$

has $\mathbb{R}$ as its domain and $\mathbb{R}$ as its range. To draw its graph, we select convenient values for x and calculate the corresponding outputs. Once these ordered pairs have been graphed, a smooth curve is drawn. Note that the cube root of a positive number is positive, and the cube root of a negative number is negative.

$$f(x) = \sqrt[3]{x}$$

x	$\sqrt[3]{x}$	$(x, f(x))$
0	0	(0, 0)
1	1	(1, 1)
8	2	(8, 2)
-1	-1	$(-1, -1)$
-8	-2	$(-8, -2)$

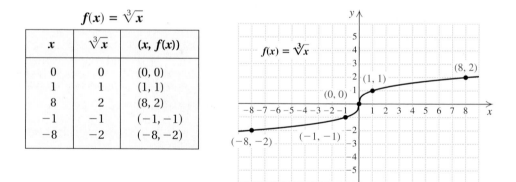

EXAMPLE 7   For each function, find the indicated function value.

a) $f(y) = \sqrt[3]{y}$; $f(125)$

b) $g(x) = \sqrt[3]{x - 1}$; $g(-26)$

Solution

a) $f(125) = \sqrt[3]{125} = 5$ Since $5 \cdot 5 \cdot 5 = 125$

b) $g(-26) = \sqrt[3]{-26 - 1}$

$\qquad\quad = \sqrt[3]{-27}$

$\qquad\quad = -3$ Since $(-3)(-3)(-3) = -27$

EXAMPLE 8   Simplify: $\sqrt[3]{-8y^3}$.

Solution

$$\sqrt[3]{-8y^3} = -2y \qquad \text{Since } (-2y)(-2y)(-2y) = -8y^3$$

Odd and Even *n*th Roots

The fourth root of a number a is the number c for which $c^4 = a$. There are also 5th roots, 6th roots, and so on. We write $\sqrt[n]{a}$ for the nth root. The number n is called the *index* (plural, *indices*). When the index is 2, we do not write it.

EXAMPLE 9   Find each of the following.

a) $\sqrt[5]{32}$

b) $\sqrt[5]{-32}$

c) $-\sqrt[5]{32}$

d) $-\sqrt[5]{-32}$

Solution

a) $\sqrt[5]{32} = 2$ Since $2^5 = 32$

b) $\sqrt[5]{-32} = -2$ Since $(-2)^5 = -32$

c) $-\sqrt[5]{32} = -2$ Taking the opposite of $\sqrt[5]{32}$

d) $-\sqrt[5]{-32} = -(-2) = 2$ Taking the opposite of $\sqrt[5]{-32}$

Note that every number has exactly one real root when n is odd. Odd roots of positive numbers are positive and odd roots of negative numbers are negative. Absolute-value signs are not used when finding odd roots.

EXAMPLE 10   Find each of the following.

a) $\sqrt[7]{x^7}$

b) $\sqrt[9]{(t-1)^9}$

Solution

a) $\sqrt[7]{x^7} = x$

b) $\sqrt[9]{(t-1)^9} = t - 1$

When the index n is even, we say that we are taking an *even root*. Every positive real number has two real nth roots when n is even—one positive and one negative. For example, the fourth roots of 16 are -2 and 2. Negative numbers do not have real nth roots when n is even.

When n is even, the notation $\sqrt[n]{a}$ indicates the nonnegative nth root. Thus, to write even nth roots, absolute-value signs are often required.

EXAMPLE 11   Simplify each expression, if possible. Assume that variables can represent any real number.

a) $\sqrt[4]{16}$ **b)** $-\sqrt[4]{16}$

c) $\sqrt[4]{-16}$ **d)** $\sqrt[4]{81x^4}$

e) $\sqrt[6]{(y+7)^6}$

Solution

a) $\sqrt[4]{16} = 2$ Since $2^4 = 16$

b) $-\sqrt[4]{16} = -2$ Taking the opposite of $\sqrt[4]{16}$

c) $\sqrt[4]{-16}$ cannot be simplified. $\sqrt[4]{-16}$ is not a real number.

d) $\sqrt[4]{81x^4} = |3x|$, or $3|x|$ Use absolute-value notation since x could represent a negative number.

e) $\sqrt[6]{(y+7)^6} = |y+7|$ Use absolute-value notation since $y+7$ is negative for $y < -7$.

We summarize as follows.

Simplifying *n*th Roots

n	a		$\sqrt[n]{a}$	$\sqrt[n]{a^n}$
Even	Positive	Positive		a
	Negative	Not a real number		$-a$
Odd	Positive	Positive		a
	Negative	Negative		a

EXAMPLE 12   Determine the domain of $g(x) = \sqrt[6]{7 - 3x}$.

Solution   Since the index is even, the radicand, $7 - 3x$, must be nonnegative. We solve the inequality:

$$7 - 3x \geq 0 \qquad \text{We cannot find the 6th root of a negative number.}$$

$$-3x \geq -7$$

$$x \leq \tfrac{7}{3}. \qquad \text{Multiplying both sides by } -\tfrac{1}{3} \text{ and reversing the inequality}$$

Thus,

$$\text{Domain of } g = \left\{x \mid x \leq \tfrac{7}{3}\right\}$$

$$= \left(-\infty, \tfrac{7}{3}\right].$$

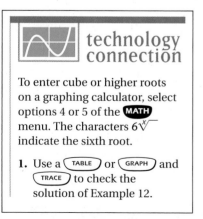

technology connection

To enter cube or higher roots on a graphing calculator, select options 4 or 5 of the **MATH** menu. The characters $6\sqrt[x]{}$ indicate the sixth root.

1. Use a **TABLE** or **GRAPH** and **TRACE** to check the solution of Example 12.

Exercise Set

7.1

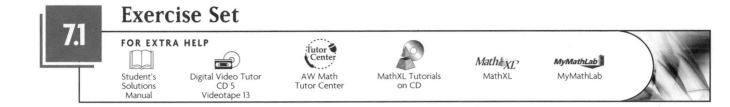

↪ *Concept Reinforcement Select the appropriate word to complete each of the following.*

1. Every positive number has _____ square root(s). one/two

2. The principal square root is never _____. negative/positive

3. For any _____ number a, we have negative/positive
   $\sqrt{a^2} = a$.

4. For any _____ number a, we have negative/positive
   $\sqrt{a^2} = -a$.

5. If a is a whole number that is not a perfect square, then $\sqrt{a}$ is a(n) _____ number. irrational/rational

6. The domain of the function f given by $f(x) = \sqrt[3]{x}$ is all _____ numbers. whole/real/positive

7. If $\sqrt[4]{x}$ is a real number, then x must be _____. negative/positive/nonnegative

8. If $\sqrt[3]{x}$ is negative, then x must be _____. negative/positive

For each number, find all of its square roots.

9. 49

10. 81

11. 144

12. 9

13. 400

14. 2500

15. 900

16. 225

Simplify.

17. $-\sqrt{\dfrac{36}{49}}$

18. $-\sqrt{\dfrac{361}{9}}$

19. $\sqrt{441}$

20. $\sqrt{196}$

21. $-\sqrt{\dfrac{16}{81}}$

22. $-\sqrt{\dfrac{81}{144}}$

23. $\sqrt{0.04}$

24. $\sqrt{0.36}$

25. $-\sqrt{0.0025}$

26. $\sqrt{0.0144}$

Identify the radicand and the index for each expression.

27. $5\sqrt{p^2 + 4}$

28. $-7\sqrt{y^2 - 8}$

29. $x^2 y^3 \sqrt[3]{\dfrac{x}{y + 4}}$

30. $a^2 b^3 \sqrt[3]{\dfrac{a}{a^2 - b}}$

For each function, find the specified function value, if it exists.

31. $f(t) = \sqrt{5t - 10}$; $f(6), f(2), f(1), f(-1)$

32. $g(x) = \sqrt{x^2 - 25}$; $g(-6), g(3), g(6), g(13)$

33. $t(x) = -\sqrt{2x + 1}$; $t(4), t(0), t(-1), t\left(-\tfrac{1}{2}\right)$

34. $p(z) = \sqrt{2z^2 - 20}$; $p(4), p(3), p(-5), p(0)$

35. $f(t) = \sqrt{t^2 + 1}$; $f(0), f(-1), f(-10)$

36. $g(x) = -\sqrt{(x + 1)^2}$; $g(-3), g(4), g(-5)$

37. $g(x) = \sqrt{x^3 + 9}$; $g(-2), g(-3), g(3)$

38. $f(t) = \sqrt{t^3 - 10}$; $f(2), f(3), f(4)$

Simplify. Remember to use absolute-value notation when necessary. If a root cannot be simplified, state this.

39. $\sqrt{36x^2}$

40. $\sqrt{25t^2}$

41. $\sqrt{(-6b)^2}$

42. $\sqrt{(-7c)^2}$

43. $\sqrt{(8 - t)^2}$

44. $\sqrt{(a + 3)^2}$

45. $\sqrt{y^2 + 16y + 64}$

46. $\sqrt{x^2 - 4x + 4}$

47. $\sqrt{4x^2 + 28x + 49}$

48. $\sqrt{9x^2 - 30x + 25}$

49. $\sqrt[4]{256}$

50. $-\sqrt[4]{625}$

51. $\sqrt[5]{-1}$

52. $-\sqrt[5]{3^5}$

53. $\sqrt[5]{-\dfrac{32}{243}}$

54. $\sqrt[5]{-\dfrac{1}{32}}$

55. $\sqrt[6]{x^6}$

56. $\sqrt[8]{y^8}$

57. $\sqrt[4]{(6a)^4}$

58. $\sqrt[4]{(7b)^4}$

59. $\sqrt[10]{(-6)^{10}}$

60. $\sqrt[12]{(-10)^{12}}$

61. $\sqrt[414]{(a + b)^{414}}$

62. $\sqrt[1976]{(2a + b)^{1976}}$

63. $\sqrt{a^{22}}$

64. $\sqrt{x^{10}}$

65. $\sqrt{-25}$

66. $\sqrt{-16}$

Simplify. Assume that no radicands were formed by raising negative quantities to even powers.

67. $\sqrt{16x^2}$

68. $\sqrt{25t^2}$

69. $\sqrt{(3t)^2}$

70. $\sqrt{(7c)^2}$

71. $\sqrt{(a+1)^2}$

72. $\sqrt{(5+b)^2}$

73. $\sqrt{4x^2+8x+4}$

74. $\sqrt{9x^2+36x+36}$

75. $\sqrt{9t^2-12t+4}$

76. $\sqrt{25t^2-20t+4}$

77. $\sqrt[3]{27}$

78. $-\sqrt[3]{64}$

79. $\sqrt[4]{16x^4}$

80. $\sqrt[4]{81x^4}$

81. $\sqrt[3]{-216}$

82. $-\sqrt[5]{-100,000}$

83. $-\sqrt[3]{-125y^3}$

84. $-\sqrt[3]{-64x^3}$

85. $\sqrt{t^{18}}$

86. $\sqrt{a^{14}}$

87. $\sqrt{(x-2)^8}$

88. $\sqrt{(x+3)^{10}}$

For each function, find the specified function value, if it exists.

89. $f(x) = \sqrt[3]{x+1}$; $f(7), f(26), f(-9), f(-65)$

90. $g(x) = -\sqrt[3]{2x-1}$; $g(0), g(-62), g(-13), g(63)$

91. $g(t) = \sqrt[4]{t-3}$; $g(19), g(-13), g(1), g(84)$

92. $f(t) = \sqrt[4]{t+1}$; $f(0), f(15), f(-82), f(80)$

Determine the domain of each function described.

93. $f(x) = \sqrt{x-6}$

94. $g(x) = \sqrt{x+8}$

95. $g(t) = \sqrt[4]{t+8}$

96. $f(x) = \sqrt[4]{x-9}$

97. $g(x) = \sqrt[4]{2x-10}$

98. $g(t) = \sqrt[3]{2t-6}$

99. $f(t) = \sqrt[5]{8-3t}$

100. $f(t) = \sqrt[6]{4-3t}$

101. $h(z) = -\sqrt[6]{5z+2}$

102. $d(x) = -\sqrt[4]{7x-5}$

Aha! **103.** $f(t) = 7 + \sqrt[8]{t^8}$

104. $g(t) = 9 + \sqrt[6]{t^6}$

105. Explain how to write the negative square root of a number using radical notation.

106. Does the square root of a number's absolute value always exist? Why or why not?

SKILL MAINTENANCE

Simplify. Do not use negative exponents in your answer. [1.6]

107. $(a^3b^2c^5)^3$

108. $(5a^7b^8)(2a^3b)$

109. $(2a^{-2}b^3c^{-4})^{-3}$

110. $(5x^{-3}y^{-1}z^2)^{-2}$

111. $\dfrac{8x^{-2}y^5}{4x^{-6}z^{-2}}$

112. $\dfrac{10a^{-6}b^{-7}}{2a^{-2}c^{-3}}$

SYNTHESIS

113. Under what conditions does the *n*th root of x^3 exist? Explain your reasoning.

114. Under what conditions does the *n*th root of x^2 exist? Explain your reasoning.

115. *Dairy farming.*   As a calf grows, it needs more milk for nourishment. The number of pounds of milk, M, required by a calf weighing x pounds can be estimated using the formula

$$M = -5 + \sqrt{6.7x - 444}.$$

Estimate the number of pounds of milk required by a calf of the given weight: **(a)** 300 lb; **(b)** 100 lb; **(c)** 200 lb; **(d)** 400 lb.
Source: www.ext.vt.edu

116. *Spaces in a parking lot.*   A parking lot has attendants to park the cars. The number N of stalls needed for waiting cars before attendants can get to them is given by the formula $N = 2.5\sqrt{A}$, where A is the number of arrivals in peak hours. Find the number of spaces needed for the given number of arrivals in peak hours: **(a)** 25; **(b)** 36; **(c)** 49; **(d)** 64.

Determine the domain of each function described. Then draw the graph of each function.

117. $f(x) = \sqrt{x+5}$

118. $g(x) = \sqrt{x} + 5$

119. $g(x) = \sqrt{x} - 2$

120. $f(x) = \sqrt{x-2}$

121. Find the domain of f if

$$f(x) = \frac{\sqrt{x+3}}{\sqrt[4]{2-x}}.$$

122. Find the domain of g if

$$g(x) = \frac{\sqrt[4]{5-x}}{\sqrt[6]{x+4}}.$$

123. Find the domain of F if $F(x) = \dfrac{x}{\sqrt{x^2-5x-6}}$.

124. Use a graphing calculator to check your answers to Exercises 39, 45, and 57. On some graphing calculators, a MATH key is needed to enter higher roots.

125. Use a graphing calculator to check your answers to Exercises 117 and 118. (See Exercise 124.)

7.2 Rational Numbers as Exponents

Rational Exponents   •   Negative Rational Exponents   •
Laws of Exponents   •   Simplifying Radical Expressions

In Section 1.1, we considered the natural numbers as exponents. Our discussion of exponents was expanded in Section 1.6 to include all integers. We now expand the study still further—to include all rational numbers. This will give meaning to expressions like $a^{1/3}$, $7^{-1/2}$, and $(3x)^{4/5}$. Such notation will help us simplify certain radical expressions.

Study Skills

Avoid Overconfidence

Sometimes a topic that seems familiar—perhaps the material in this section—reappears and students take a vacation from their studies, thinking they "know it already." Try to avoid this tendency. Often a new wrinkle is included that will catch the unsuspecting student by surprise.

Rational Exponents

Consider $a^{1/2} \cdot a^{1/2}$. If we still want to add exponents when multiplying, it must follow that $a^{1/2} \cdot a^{1/2} = a^{1/2+1/2}$, or a^1. This suggests that $a^{1/2}$ is a square root of a. Similarly, $a^{1/3} \cdot a^{1/3} \cdot a^{1/3} = a^{1/3+1/3+1/3}$, or a^1, so $a^{1/3}$ should mean $\sqrt[3]{a}$.

$$a^{1/n} = \sqrt[n]{a}$$

$a^{1/n}$ means $\sqrt[n]{a}$. When a is nonnegative, n can be any natural number greater than 1. When a is negative, n must be odd.

Note that the denominator of the exponent becomes the index and the base becomes the radicand.

EXAMPLE 1   Write an equivalent expression using radical notation.

a) $x^{1/2}$

b) $(-8)^{1/3}$

c) $(abc)^{1/5}$

d) $(25x^{16})^{1/2}$

Solution

a) $x^{1/2} = \sqrt{x}$ }

b) $(-8)^{1/3} = \sqrt[3]{-8} = -2$ }   The denominator of the exponent becomes the index. The base becomes the radicand.

c) $(abc)^{1/5} = \sqrt[5]{abc}$ }   Recall that for square roots, the index 2 is understood without being written.

d) $(25x^{16})^{1/2} = 25^{1/2}x^8 = \sqrt{25} \cdot x^8 = 5x^8$

EXAMPLE 2   Write an equivalent expression using exponential notation.

a) $\sqrt[5]{9xy}$

b) $\sqrt[7]{\dfrac{x^3y}{4}}$

c) $\sqrt{5x}$

Solution   Parentheses are required to indicate the base.

a) $\sqrt[5]{9xy} = (9xy)^{1/5}$

b) $\sqrt[7]{\dfrac{x^3 y}{4}} = \left(\dfrac{x^3 y}{4}\right)^{1/7}$ The index becomes the denominator of the exponent. The radicand becomes the base.

c) $\sqrt{5x} = (5x)^{1/2}$ The index 2 is understood without being written. We assume $x \geq 0$.

How shall we define $a^{2/3}$? If the property for multiplying exponents is to hold, we must have $a^{2/3} = (a^{1/3})^2$ and $a^{2/3} = (a^2)^{1/3}$. This would suggest that $a^{2/3} = (\sqrt[3]{a})^2$ and $a^{2/3} = \sqrt[3]{a^2}$. We make our definition accordingly.

Positive Rational Exponents

For any natural numbers m and n ($n \neq 1$) and any real number a for which $\sqrt[n]{a}$ exists,

$$a^{m/n} \quad \text{means} \quad (\sqrt[n]{a})^m, \quad \text{or} \quad \sqrt[n]{a^m}.$$

EXAMPLE 3   Write an equivalent expression using radical notation and simplify.

a) $27^{2/3}$

b) $25^{3/2}$

Solution

a) $27^{2/3}$ means $(\sqrt[3]{27})^2$ or, equivalently, $\sqrt[3]{27^2}$. Let's see which is easier to simplify:

$$(\sqrt[3]{27})^2 = 3^2 \qquad \qquad \sqrt[3]{27^2} = \sqrt[3]{729}$$
$$= 9; \qquad \qquad \qquad = 9.$$

The simplification on the left is probably easier for most people.

b) $25^{3/2}$ means $(\sqrt[2]{25})^3$ or, equivalently, $\sqrt[2]{25^3}$ (the index 2 is normally omitted). Since $\sqrt{25}$ is more commonly known than $\sqrt{25^3}$, we use that form:

$$25^{3/2} = (\sqrt{25})^3 = 5^3 = 125.$$

Student Notes _____

It is important to remember both meanings of $a^{m/n}$. When the root of the base a is known, $(\sqrt[n]{a})^m$ is generally easier to work with. When it is not known, $\sqrt[n]{a^m}$ is often more convenient.

EXAMPLE 4   Write an equivalent expression using exponential notation.

a) $\sqrt[3]{9^4}$

b) $(\sqrt[4]{7xy})^5$

Solution

a) $\sqrt[3]{9^4} = 9^{4/3}$

b) $(\sqrt[4]{7xy})^5 = (7xy)^{5/4}$ The index becomes the denominator of the fraction that is the exponent.

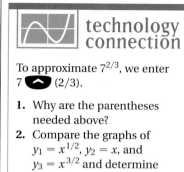

Negative Rational Exponents

Recall from Section 1.6 that $x^{-2} = 1/x^2$. Negative rational exponents behave similarly.

Negative Rational Exponents

For any rational number m/n and any nonzero real number a for which $a^{m/n}$ exists,

$$a^{-m/n} \text{ means } \frac{1}{a^{m/n}}.$$

Caution!   A negative exponent does not indicate that the expression in which it appears is negative: $a^{-1} \neq -a$.

EXAMPLE 5   Write an equivalent expression with positive exponents and, if possible, simplify.

a) $9^{-1/2}$
b) $(5xy)^{-4/5}$
c) $64^{-2/3}$
d) $4x^{-2/3}y^{1/5}$
e) $\left(\dfrac{3r}{7s}\right)^{-5/2}$

Solution

a) $9^{-1/2} = \dfrac{1}{9^{1/2}}$ $9^{-1/2}$ is the reciprocal of $9^{1/2}$.

Since $9^{1/2} = \sqrt{9} = 3$, the answer simplifies to $\dfrac{1}{3}$.

b) $(5xy)^{-4/5} = \dfrac{1}{(5xy)^{4/5}}$ $(5xy)^{-4/5}$ is the reciprocal of $(5xy)^{4/5}$.

c) $64^{-2/3} = \dfrac{1}{64^{2/3}}$ $64^{-2/3}$ is the reciprocal of $64^{2/3}$.

Since $64^{2/3} = \left(\sqrt[3]{64}\right)^2 = 4^2 = 16$, the answer simplifies to $\dfrac{1}{16}$.

d) $4x^{-2/3}y^{1/5} = 4 \cdot \dfrac{1}{x^{2/3}} \cdot y^{1/5} = \dfrac{4y^{1/5}}{x^{2/3}}$

e) In Section 1.6, we found that $(a/b)^{-n} = (b/a)^n$. This property holds for *any* negative exponent:

$$\left(\frac{3r}{7s}\right)^{-5/2} = \left(\frac{7s}{3r}\right)^{5/2}.$$ Writing the reciprocal of the base and changing the sign of the exponent

Laws of Exponents

The same laws hold for rational exponents as for integer exponents.

Laws of Exponents

For any real numbers a and b and any rational exponents m and n for which a^m, a^n, and b^m are defined:

1. $a^m \cdot a^n = a^{m+n}$ In multiplying, add exponents if the bases are the same.

2. $\dfrac{a^m}{a^n} = a^{m-n}$ In dividing, subtract exponents if the bases are the same. (Assume $a \neq 0$.)

3. $(a^m)^n = a^{m \cdot n}$ To raise a power to a power, multiply the exponents.

4. $(ab)^m = a^m b^m$ To raise a product to a power, raise each factor to the power and multiply.

EXAMPLE 6

Use the laws of exponents to simplify.

a) $3^{1/5} \cdot 3^{3/5}$

b) $\dfrac{a^{1/4}}{a^{1/2}}$

c) $(7.2^{2/3})^{3/4}$

d) $(a^{-1/3}b^{2/5})^{1/2}$

Solution

a) $3^{1/5} \cdot 3^{3/5} = 3^{1/5+3/5} = 3^{4/5}$ Adding exponents

b) $\dfrac{a^{1/4}}{a^{1/2}} = a^{1/4-1/2} = a^{1/4-2/4}$ Subtracting exponents after finding a common denominator

$= a^{-1/4}$, or $\dfrac{1}{a^{1/4}}$ $a^{-1/4}$ is the reciprocal of $a^{1/4}$.

c) $(7.2^{2/3})^{3/4} = 7.2^{(2/3)(3/4)} = 7.2^{6/12}$ Multiplying exponents

$= 7.2^{1/2}$ Using arithmetic to simplify the exponent

d) $(a^{-1/3}b^{2/5})^{1/2} = a^{(-1/3)(1/2)} \cdot b^{(2/5)(1/2)}$ Raising a product to a power and multiplying exponents

$= a^{-1/6}b^{1/5}$, or $\dfrac{b^{1/5}}{a^{1/6}}$

Simplifying Radical Expressions

Many radical expressions contain radicands or factors of radicands that are powers. When these powers and the index share a common factor, rational exponents can be used to simplify the expression.

> **To Simplify Radical Expressions**
> 1. Convert radical expressions to exponential expressions.
> 2. Use arithmetic and the laws of exponents to simplify.
> 3. Convert back to radical notation as needed.

EXAMPLE 7

Use rational exponents to simplify. Do not use exponents that are fractions in the final answer.

a) $\sqrt[6]{(5x)^3}$ **b)** $\sqrt[5]{t^{20}}$

c) $\left(\sqrt[3]{ab^2c}\right)^{12}$ **d)** $\sqrt{\sqrt[3]{x}}$

Solution

a) $\sqrt[6]{(5x)^3} = (5x)^{3/6}$ Converting to exponential notation

$\qquad\qquad = (5x)^{1/2}$ Simplifying the exponent

$\qquad\qquad = \sqrt{5x}$ Returning to radical notation

b) $\sqrt[5]{t^{20}} = t^{20/5}$ Converting to exponential notation

$\qquad\quad = t^4$ Simplifying the exponent

c) $\left(\sqrt[3]{ab^2c}\right)^{12} = (ab^2c)^{12/3}$ Converting to exponential notation

$\qquad\qquad\quad = (ab^2c)^4$ Simplifying the exponent

$\qquad\qquad\quad = a^4b^8c^4$ Using the laws of exponents

d) $\sqrt{\sqrt[3]{x}} = \sqrt{x^{1/3}}$ Converting the radicand to exponential notation

$\qquad\quad = (x^{1/3})^{1/2}$ Try to go directly to this step.

$\qquad\quad = x^{1/6}$ Using the laws of exponents

$\qquad\quad = \sqrt[6]{x}$ Returning to radical notation

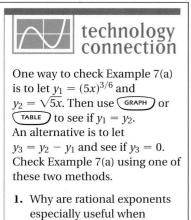

technology connection

One way to check Example 7(a) is to let $y_1 = (5x)^{3/6}$ and $y_2 = \sqrt{5x}$. Then use [GRAPH] or [TABLE] to see if $y_1 = y_2$. An alternative is to let $y_3 = y_2 - y_1$ and see if $y_3 = 0$. Check Example 7(a) using one of these two methods.

1. Why are rational exponents especially useful when working on a graphing calculator?

7.2 # Exercise Set

FOR EXTRA HELP

Student's Solutions Manual | Digital Video Tutor CD 5 Videotape 13 | AW Math Tutor Center | MathXL Tutorials on CD | MathXL MathXL | MyMathLab MyMathLab

↪ *Concept Reinforcement* *In each of Exercises 1–8, match the expression with the equivalent expression from the column on the right.*

1. ____ $x^{2/5}$ **2.** ____ $x^{5/2}$ **a)** $x^{3/5}$ **b)** $\left(\sqrt[5]{x}\right)^4$ **c)** $\sqrt{x^5}$

3. ____ $x^{-5/2}$ **4.** ____ $x^{-2/5}$ **d)** $x^{1/2}$ **e)** $\dfrac{1}{\left(\sqrt{x}\right)^5}$ **f)** $\sqrt[4]{x^5}$

5. ____ $x^{1/5} \cdot x^{2/5}$ **6.** ____ $(x^{1/5})^{5/2}$

7. ____ $\sqrt[5]{x^4}$ **8.** ____ $\left(\sqrt[4]{x}\right)^5$ **g)** $\sqrt[5]{x^2}$ **h)** $\dfrac{1}{\left(\sqrt[5]{x}\right)^2}$

Note: Assume for all exercises that even roots are of nonnegative quantities and that all denominators are nonzero.

Write an equivalent expression using radical notation and, if possible, simplify.

9. $x^{1/6}$ **10.** $y^{1/5}$

11. $16^{1/2}$ **12.** $8^{1/3}$

13. $81^{1/4}$ **14.** $64^{1/6}$

15. $9^{1/2}$ **16.** $25^{1/2}$

17. $(xyz)^{1/3}$ **18.** $(ab)^{1/4}$

19. $(a^2b^2)^{1/5}$ **20.** $(x^3y^3)^{1/4}$

21. $t^{2/5}$ **22.** $b^{3/2}$

23. $16^{3/4}$ **24.** $4^{7/2}$

25. $27^{4/3}$ **26.** $9^{5/2}$

27. $(81x)^{3/4}$ **28.** $(125a)^{2/3}$

29. $(25x^4)^{3/2}$ **30.** $(9y^6)^{3/2}$

Write an equivalent expression using exponential notation.

31. $\sqrt[3]{20}$ **32.** $\sqrt[3]{19}$

33. $\sqrt{17}$ **34.** $\sqrt{6}$

35. $\sqrt{x^3}$ **36.** $\sqrt{a^5}$

37. $\sqrt[5]{m^2}$ **38.** $\sqrt[5]{n^4}$

39. $\sqrt[4]{cd}$ **40.** $\sqrt[5]{xy}$

41. $\sqrt[5]{xy^2z}$ **42.** $\sqrt[7]{x^3y^2z^2}$

43. $(\sqrt{3mn})^3$ **44.** $(\sqrt[3]{7xy})^4$

45. $(\sqrt[7]{8x^2y})^5$ **46.** $(\sqrt[6]{2a^5b})^7$

47. $\dfrac{2x}{\sqrt[3]{z^2}}$ **48.** $\dfrac{3a}{\sqrt[5]{c^2}}$

Write an equivalent expression with positive exponents and, if possible, simplify.

49. $x^{-1/3}$ **50.** $y^{-1/4}$

51. $(2rs)^{-3/4}$ **52.** $(5xy)^{-5/6}$

53. $\left(\dfrac{1}{16}\right)^{-3/4}$ **54.** $\left(\dfrac{1}{8}\right)^{-2/3}$

55. $\dfrac{2c}{a^{-3/5}}$ **56.** $\dfrac{3b}{a^{-5/7}}$

57. $5x^{-2/3}y^{4/5}z$ **58.** $2a^{3/4}b^{-1/2}c^{2/3}$

59. $3^{-5/2}a^3b^{-7/3}$ **60.** $2^{-1/3}x^4y^{-2/7}$

61. $\left(\dfrac{2ab}{3c}\right)^{-5/6}$ **62.** $\left(\dfrac{7x}{8yz}\right)^{-3/5}$

63. $\dfrac{6a}{\sqrt[4]{b}}$ **64.** $\dfrac{7x}{\sqrt[3]{z}}$

Use the laws of exponents to simplify. Do not use negative exponents in any answers.

65. $7^{3/4} \cdot 7^{1/8}$ **66.** $11^{2/3} \cdot 11^{1/2}$

67. $\dfrac{3^{5/8}}{3^{-1/8}}$ **68.** $\dfrac{8^{7/11}}{8^{-2/11}}$

69. $\dfrac{5.2^{-1/6}}{5.2^{-2/3}}$ **70.** $\dfrac{2.3^{-3/10}}{2.3^{-1/5}}$

71. $(10^{3/5})^{2/5}$ **72.** $(5^{5/4})^{3/7}$

73. $a^{2/3} \cdot a^{5/4}$ **74.** $x^{3/4} \cdot x^{1/3}$

Aha!   **75.** $(64^{3/4})^{4/3}$ **76.** $(27^{-2/3})^{3/2}$

77. $(m^{2/3}n^{-1/4})^{1/2}$ **78.** $(x^{-1/3}y^{2/5})^{1/4}$

Use rational exponents to simplify. Do not use fraction exponents in the final answer.

79. $\sqrt[6]{x^4}$ **80.** $\sqrt[6]{a^2}$

81. $\sqrt[4]{a^{12}}$ **82.** $\sqrt[3]{x^{15}}$

83. $\sqrt[5]{a^{10}}$ **84.** $\sqrt[6]{x^{18}}$

85. $(\sqrt[7]{xy})^{14}$ **86.** $(\sqrt[3]{ab})^{15}$

87. $\sqrt[4]{(7a)^2}$ **88.** $\sqrt[8]{(3x)^2}$

89. $(\sqrt[8]{2x})^6$ **90.** $(\sqrt[10]{3a})^5$

91. $\sqrt[3]{\sqrt[6]{a}}$ **92.** $\sqrt[4]{\sqrt{x}}$

93. $\sqrt[4]{(xy)^{12}}$ **94.** $\sqrt{(ab)^6}$

95. $(\sqrt[5]{a^2b^4})^{15}$ **96.** $(\sqrt[3]{x^2y^5})^{12}$

97. $\sqrt[3]{\sqrt[4]{xy}}$ **98.** $\sqrt[5]{\sqrt{2a}}$

99. If $f(x) = (x + 5)^{1/2}(x + 7)^{-1/2}$, find the domain of f. Explain how you found your answer.

100. Explain why $\sqrt[3]{x^6} = x^2$ for any value of x, whereas $\sqrt[2]{x^6} = x^3$ only when $x \geq 0$.

SKILL MAINTENANCE

Simplify. [5.2]

101. $3x(x^3 - 2x^2) + 4x^2(2x^2 + 5x)$

102. $5t^3(2t^2 - 4t) - 3t^4(t^2 - 6t)$

103. $(3a - 4b)(5a + 3b)$

104. $(7x - y)^2$

105. *Real estate taxes.*   For homes under $100,000, the property transfer tax in Vermont is 0.5% of the selling price. Find the selling price of a home that had a transfer tax of $467.50. [1.4]

106. What numbers are their own squares? [5.8]

SYNTHESIS

107. Let $f(x) = 5x^{-1/3}$. Under what condition will we have $f(x) > 0$? Why?

108. If $g(x) = x^{3/n}$, in what way does the domain of g depend on whether n is odd or even?

Use rational exponents to simplify.

109. $\sqrt{x\sqrt[3]{x^2}}$ **110.** $\sqrt[4]{\sqrt[3]{8x^3y^6}}$

111. $\sqrt[12]{p^2 + 2pq + q^2}$

Music.   The function given by $f(x) = k2^{x/12}$ can be used to determine the frequency, in cycles per second, of a musical note that is x half-steps above a note with frequency k.[*]

112. The frequency of concert A for a trumpet is 440 cycles per second. Find the frequency of the A that is two octaves (24 half-steps) above concert A (few trumpeters can reach this note.)

113. Show that the G that is 7 half-steps (a "perfect fifth") above middle C (262 cycles per second) has a frequency that is about 1.5 times that of middle C.

---
[*]This application was inspired by information provided by Dr. Homer B. Tilton of Pima Community College East.

114. Show that the C sharp that is 4 half-steps (a "major third") above concert A (see Exercise 112) has a frequency that is about 25% greater than that of concert A.

115. *Baseball.*   The statistician Bill James has found that a baseball team's winning percentage P can be approximated by

$$P = \frac{r^{1.83}}{r^{1.83} + \sigma^{1.83}},$$

where r is the total number of runs scored by that team and σ is the total number of runs scored by their opponents. During a recent season, the San Francisco Giants scored 799 runs and their opponents scored 749 runs. Use James's formula to predict the Giants' winning percentage (the team actually won 55.6% of their games).

Source: M. Bittinger, *One Man's Journey Through Mathematics.* Boston: Addison-Wesley, 2004

116. *Road pavement messages.*   In a psychological study, it was determined that the proper length L of the letters of a word printed on pavement is given by

$$L = \frac{0.000169d^{2.27}}{h},$$

where d is the distance of a car from the lettering and h is the height of the eye above the surface of the road. All units are in meters. This formula says that if a person is h meters above the surface of the road and is to be able to recognize a message d meters away, that message will be the most recognizable if the length of the letters is L. Find L to the nearest tenth of a meter, given d and h.

a) $h = 1$ m, $d = 60$ m
b) $h = 0.9906$ m, $d = 75$ m
c) $h = 2.4$ m, $d = 80$ m
d) $h = 1.1$ m, $d = 100$ m

117. *Dating fossils.* The function $r(t) = 10^{-12}2^{-t/5700}$ expresses the ratio of carbon isotopes to carbon atoms in a fossil that is t years old. What ratio of carbon isotopes to carbon atoms would be present in a 1900-year-old bone?

118. *Physics.* The equation $m = m_0(1 - v^2c^{-2})^{-1/2}$, developed by Albert Einstein, is used to determine the mass m of an object that is moving v meters per second and has mass m_0 before the motion begins. The constant c is the speed of light, approximately 3×10^8 m/sec. Suppose that a particle with mass 8 mg is accelerated to a speed of $\frac{9}{5} \times 10^8$ m/sec. Without using a calculator, find the new mass of the particle.

119. Using a graphing calculator, select the **MODE** SIMUL and the FORMAT ExprOff. Then graph
$$y_1 = x^{1/2}, \qquad y_2 = 3x^{2/5},$$
$$y_3 = x^{4/7}, \quad \text{and} \quad y_4 = \tfrac{1}{5}x^{3/4}.$$

Looking only at coordinates, match each graph with its equation.

COLLABORATIVE CORNER

Are Equivalent Fractions Equivalent Exponents?

Focus: Functions and rational exponents

Time: 10–20 minutes

Group size: 3

Materials: Graph paper

In arithmetic, we have seen that $\frac{1}{3}, \frac{1}{6} \cdot 2$, and $2 \cdot \frac{1}{6}$ all represent the same number. Interestingly,
$$f(x) = x^{1/3},$$
$$g(x) = (x^{1/6})^2, \quad \text{and}$$
$$h(x) = (x^2)^{1/6}$$

represent three *different* functions.

ACTIVITY

1. Selecting a variety of values for x and using the definition of positive rational exponents, one group member should graph f, a second group member should graph g, and a third group member should graph h. Be sure to check whether negative x-values are in the domain of the function.

2. Compare the three graphs and check each other's work. How and why do the graphs differ?

3. Decide as a group which graph, if any, would best represent the graph of $k(x) = x^{2/6}$. Then be prepared to explain your reasoning to the entire class. (*Hint*: Study the definition of $a^{m/n}$ on p. 446 carefully.)

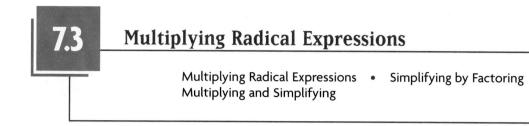

7.3   Multiplying Radical Expressions

Multiplying Radical Expressions • Simplifying by Factoring •
Multiplying and Simplifying

Multiplying Radical Expressions

Note that $\sqrt{4}\,\sqrt{25} = 2 \cdot 5 = 10$. Also $\sqrt{4 \cdot 25} = \sqrt{100} = 10$. Likewise,

$$\sqrt[3]{27}\,\sqrt[3]{8} = 3 \cdot 2 = 6 \quad \text{and} \quad \sqrt[3]{27 \cdot 8} = \sqrt[3]{216} = 6.$$

These examples suggest the following.

The Product Rule for Radicals

For any real numbers $\sqrt[n]{a}$ and $\sqrt[n]{b}$,

$$\sqrt[n]{a} \cdot \sqrt[n]{b} = \sqrt[n]{a \cdot b}.$$

(The product of two *n*th roots is the *n*th root of the product of the
two radicands.)

Rational exponents can be used to derive this rule:

$$\sqrt[n]{a} \cdot \sqrt[n]{b} = a^{1/n} \cdot b^{1/n} = (a \cdot b)^{1/n} = \sqrt[n]{a \cdot b}.$$

EXAMPLE 1   Multiply.

a) $\sqrt{3} \cdot \sqrt{5}$ **b)** $\sqrt{x+3}\,\sqrt{x-3}$

c) $\sqrt[3]{4} \cdot \sqrt[3]{5}$ **d)** $\sqrt[4]{\dfrac{y}{5}} \cdot \sqrt[4]{\dfrac{7}{x}}$

Solution

a) When no index is written, roots are understood to be square roots with an
unwritten index of two. We apply the product rule:

$$\sqrt{3} \cdot \sqrt{5} = \sqrt{3 \cdot 5}$$
$$= \sqrt{15}.$$

b) $\sqrt{x+3}\,\sqrt{x-3} = \sqrt{(x+3)(x-3)}$ The product of two square roots is
$$= \sqrt{x^2 - 9}$$ the square root of the product.

Caution! $\sqrt{x^2 - 9} \neq \sqrt{x^2} - \sqrt{9}.$

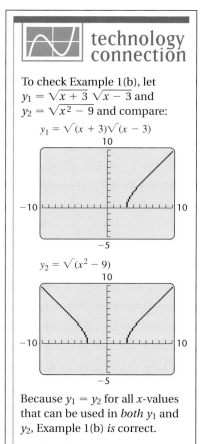

To check Example 1(b), let $y_1 = \sqrt{x+3}\,\sqrt{x-3}$ and $y_2 = \sqrt{x^2 - 9}$ and compare:

$$y_1 = \sqrt{(x+3)}\sqrt{(x-3)}$$

$$y_2 = \sqrt{(x^2 - 9)}$$

Because $y_1 = y_2$ for all x-values that can be used in *both* y_1 and y_2, Example 1(b) *is* correct.

1. Why do the graphs above differ in appearance? (*Hint:* What are the domains of the two related functions?)

---

c) Both $\sqrt[3]{4}$ and $\sqrt[3]{5}$ have indices of three, so to multiply we can use the product rule:

$$\sqrt[3]{4} \cdot \sqrt[3]{5} = \sqrt[3]{4 \cdot 5} = \sqrt[3]{20}.$$

d) $\sqrt[4]{\dfrac{y}{5}} \cdot \sqrt[4]{\dfrac{7}{x}} = \sqrt[4]{\dfrac{y}{5} \cdot \dfrac{7}{x}} = \sqrt[4]{\dfrac{7y}{5x}}$ In Section 7.4, we discuss other ways to write answers like this.

> *Caution!*   The product rule for radicals applies only when radicals have the same index:
>
> $$\sqrt[n]{a} \cdot \sqrt[m]{b} \neq \sqrt[nm]{a \cdot b}.$$

Simplifying by Factoring

The number p is a *perfect square* if there exists a rational number q for which $q^2 = p$. We say that p is a *perfect cube* if $q^3 = p$ for some rational number q. In general, p is a *perfect nth power* if $q^n = p$ for some rational number q. Thus, 16 and $\frac{1}{10,000}$ are both perfect 4th powers since $2^4 = 16$ and $\left(\frac{1}{10}\right)^4 = \frac{1}{10,000}$.

The product rule allows us to simplify $\sqrt[n]{ab}$ whenever ab contains a factor that is a perfect nth power.

> **Using the Product Rule to Simplify**
> $$\sqrt[n]{ab} = \sqrt[n]{a} \cdot \sqrt[n]{b}.$$
> ($\sqrt[n]{a}$ and $\sqrt[n]{b}$ must both be real numbers.)

To illustrate, suppose we wish to simplify $\sqrt{20}$. Since this is a *square* root, we check to see if there is a factor of 20 that is a perfect square. There is one, 4, so we express 20 as $4 \cdot 5$ and use the product rule:

$$\sqrt{20} = \sqrt{4 \cdot 5} \qquad \text{Factoring the radicand (4 is a perfect square)}$$
$$= \sqrt{4} \cdot \sqrt{5} \qquad \text{Factoring into two radicals}$$
$$= 2\sqrt{5}. \qquad \text{Finding the square root of 4}$$

> *To Simplify a Radical Expression with Index n by Factoring*
> **1.** Express the radicand as a product in which one factor is the largest perfect nth power possible.
> **2.** Take the nth root of each factor.
> **3.** Simplification is complete when no radicand has a factor that is a perfect nth power.

It is often safe to assume that a radicand does not represent a negative number raised to an even power. We will henceforth make this assumption—unless functions are involved—and discontinue use of absolute-value notation when taking even roots.

EXAMPLE 2   Simplify by factoring: **(a)** $\sqrt{200}$; **(b)** $\sqrt{18x^2y}$; **(c)** $\sqrt[3]{72}$; **(d)** $\sqrt[4]{162x^6}$.

Solution

a) $\sqrt{200} = \sqrt{100 \cdot 2}$ 100 is the largest perfect-square factor of 200.

$= \sqrt{100} \cdot \sqrt{2} = 10\sqrt{2}$

b) $\sqrt{18x^2y} = \sqrt{9 \cdot 2 \cdot x^2 \cdot y}$ $9x^2$ is the largest perfect-square factor of $18x^2y$.

$= \sqrt{9x^2} \cdot \sqrt{2y}$ Factoring into two radicals

$= 3x\sqrt{2y}$ Taking the square root of $9x^2$

c) $\sqrt[3]{72} = \sqrt[3]{8 \cdot 9}$ 8 is the largest perfect-cube (third-power) factor of 72.

$= \sqrt[3]{8} \cdot \sqrt[3]{9} = 2\sqrt[3]{9}$

Let's look at this example another way. We write a complete factorization and look for triples of factors. Each triple of factors makes a cube:

$$\sqrt[3]{72} = \sqrt[3]{\underline{2 \cdot 2 \cdot 2} \cdot 3 \cdot 3}$$ Each triple of factors is a cube.

$$= 2\sqrt[3]{3 \cdot 3} = 2\sqrt[3]{9}.$$

d) $\sqrt[4]{162x^6} = \sqrt[4]{81 \cdot 2 \cdot x^4 \cdot x^2}$ $81 \cdot x^4$ is the largest perfect fourth-power factor of $162x^6$.

$= \sqrt[4]{81x^4} \cdot \sqrt[4]{2x^2}$ Factoring into two radicals

$= 3x\sqrt[4]{2x^2}$ Taking fourth roots

Let's look at this example another way. We write a complete factorization and look for quadruples of factors. Each quadruple makes a perfect fourth power:

$$\sqrt[4]{162x^6} = \sqrt[4]{\underline{3 \cdot 3 \cdot 3 \cdot 3} \cdot 2 \cdot \underline{x \cdot x \cdot x \cdot x} \cdot x \cdot x}$$ Each quadruple of factors is a power of 4.

$$= 3 \cdot x \cdot \sqrt[4]{2 \cdot x \cdot x}$$

$$= 3x\sqrt[4]{2x^2}.$$

EXAMPLE 3   If $f(x) = \sqrt{3x^2 - 6x + 3}$, find a simplified form for $f(x)$. Assume that x can be any real number.

Solution

$$f(x) = \sqrt{3x^2 - 6x + 3}$$

$$= \sqrt{3(x^2 - 2x + 1)} \Big\}$$ Factoring the radicand; $x^2 - 2x + 1$ is a perfect square.

$$= \sqrt{(x-1)^2 \cdot 3}$$

$$= \sqrt{(x-1)^2} \cdot \sqrt{3}$$ Factoring into two radicals

$$= |x-1|\sqrt{3}$$ Finding the square root of $(x-1)^2$

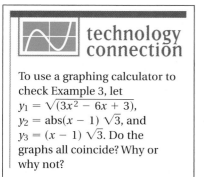

technology connection

To use a graphing calculator to check Example 3, let
$y_1 = \sqrt{(3x^2 - 6x + 3)}$,
$y_2 = \text{abs}(x-1)\sqrt{3}$, and
$y_3 = (x-1)\sqrt{3}$. Do the graphs all coincide? Why or why not?

EXAMPLE 4   Simplify: **(a)** $\sqrt{x^7 y^{11} z^9}$; **(b)** $\sqrt[3]{16 a^7 b^{14}}$.

Solution

a) There are many ways to factor $x^7 y^{11} z^9$. Because of the square root (index of 2), we identify the largest exponents that are multiples of 2:

$$\sqrt{x^7 y^{11} z^9} = \sqrt{x^6 \cdot x \cdot y^{10} \cdot y \cdot z^8 \cdot z} \qquad \text{Using the largest even powers of } x, y, \text{ and } z$$

$$= \sqrt{x^6}\, \sqrt{y^{10}}\, \sqrt{z^8}\, \sqrt{xyz} \qquad \text{Factoring into several radicals}$$

$$= x^{6/2}\, y^{10/2}\, z^{8/2} \sqrt{xyz} \qquad \text{Converting to rational exponents}$$

$$= x^3 y^5 z^4 \sqrt{xyz}.$$

Check: $\left(x^3 y^5 z^4 \sqrt{xyz}\right)^2 = (x^3)^2 (y^5)^2 (z^4)^2 \left(\sqrt{xyz}\right)^2$
$$= x^6 \cdot y^{10} \cdot z^8 \cdot xyz = x^7 y^{11} z^9$$

Our check shows that $x^3 y^5 z^4 \sqrt{xyz}$ is the square root of $x^7 y^{11} z^9$.

b) There are many ways to factor $16 a^7 b^{14}$. Because of the cube root (index of 3), we identify factors with the largest exponents that are multiples of 3:

$$\sqrt[3]{16 a^7 b^{14}} = \sqrt[3]{8 \cdot 2 \cdot a^6 \cdot a \cdot b^{12} \cdot b^2} \qquad \text{Using the largest perfect-cube factors; 6 and 12 are multiples of 3}$$

$$= \sqrt[3]{8}\, \sqrt[3]{a^6}\, \sqrt[3]{b^{12}}\, \sqrt[3]{2 a b^2} \qquad \text{Factoring into several radicals}$$

$$= 2\ a^{6/3}\ b^{12/3} \sqrt[3]{2 a b^2} \qquad \text{Converting to rational exponents}$$

$$= 2 a^2 b^4 \sqrt[3]{2 a b^2}$$

As a check, let's redo the problem using a complete factorization of the radicand:

$$\sqrt[3]{16 a^7 b^{14}} = \sqrt[3]{\underline{2 \cdot 2 \cdot 2} \cdot 2 \cdot \underline{a \cdot a \cdot a} \cdot a \cdot \underline{b \cdot b \cdot b} \cdot \underline{b \cdot b \cdot b} \cdot \underline{b \cdot b \cdot b} \cdot \underline{b \cdot b \cdot b} \cdot b \cdot b}$$

Each triple of factors makes a cube.

$$= 2 \cdot a \cdot a \cdot b \cdot b \cdot b \cdot b \cdot \sqrt[3]{2 \cdot a \cdot b \cdot b}$$
$$= 2 a^2 b^4 \sqrt[3]{2 a b^2}. \qquad \text{Our answer checks.}$$

> *Remember*: To simplify an nth root, identify factors in the radicand with exponents that are multiples of n.

Multiplying and Simplifying

We have used the product rule for radicals to find products and also to simplify radical expressions. For some radical expressions, it is possible to do both: First find a product and then simplify.

EXAMPLE 5   Multiply and simplify.

a) $\sqrt{15}\,\sqrt{6}$
b) $3\sqrt[3]{25} \cdot 2\sqrt[3]{5}$
c) $\sqrt[4]{8x^3y^5}\,\sqrt[4]{4x^2y^3}$

Solution

a) $\sqrt{15}\,\sqrt{6} = \sqrt{15 \cdot 6}$ Multiplying radicands
$= \sqrt{90} = \sqrt{9 \cdot 10}$ 9 is a perfect square.
$= 3\sqrt{10}$

b) $3\sqrt[3]{25} \cdot 2\sqrt[3]{5} = 3 \cdot 2 \cdot \sqrt[3]{25 \cdot 5}$ Using a commutative law; multiplying radicands
$= 6 \cdot \sqrt[3]{125}$ 125 is a perfect cube.
$= 6 \cdot 5,\text{ or } 30$

c) $\sqrt[4]{8x^3y^5}\,\sqrt[4]{4x^2y^3} = \sqrt[4]{32x^5y^8}$ Multiplying radicands
$= \sqrt[4]{16x^4y^8 \cdot 2x}$ Identifying perfect fourth-power factors
$= \sqrt[4]{16}\,\sqrt[4]{x^4}\,\sqrt[4]{y^8}\,\sqrt[4]{2x}$ Factoring into radicals
$= 2xy^2\sqrt[4]{2x}$ Finding the fourth roots; assume $x \geq 0$.

The checks are left to the student.

Student Notes

To multiply $\sqrt{x} \cdot \sqrt{x}$, remember what $\sqrt{x}$ represents and go directly to the product, x. Too often students unnecessarily write $\sqrt{x} \cdot \sqrt{x} = \sqrt{x^2} = x$.

Exercise Set

7.3

FOR EXTRA HELP

Student's Solutions Manual   Digital Video Tutor CD 5 Videotape 13   AW Math Tutor Center   MathXL Tutorials on CD   MathXL   MyMathLab

✎ *Concept Reinforcement*   *Classify each of the following statements as either true or false.*

1. For any real numbers $\sqrt[n]{a}$ and $\sqrt[n]{b}$, $\sqrt[n]{a} \cdot \sqrt[n]{b} = \sqrt[n]{ab}$.

2. For any real numbers $\sqrt[n]{a}$ and $\sqrt[n]{b}$, $\sqrt[n]{a} + \sqrt[n]{b} = \sqrt[n]{a + b}$.

3. For any real numbers $\sqrt[n]{a}$ and $\sqrt[m]{b}$, $\sqrt[n]{a} \cdot \sqrt[m]{b} = \sqrt[nm]{ab}$.

4. For $x > 0$, $\sqrt{x^2 - 9} = x - 3$.

5. The expression $\sqrt[3]{X}$ is not simplified if X contains a factor that is a perfect cube.

6. It is often possible to simplify $\sqrt{A \cdot B}$ even though $\sqrt{A}$ and $\sqrt{B}$ cannot be simplified.

Multiply.

7. $\sqrt{5}\,\sqrt{7}$

8. $\sqrt{10}\,\sqrt{7}$

9. $\sqrt[3]{7}\,\sqrt[3]{2}$

10. $\sqrt[3]{2}\,\sqrt[3]{5}$

11. $\sqrt[4]{6}\,\sqrt[4]{3}$

12. $\sqrt[4]{8}\,\sqrt[4]{9}$

13. $\sqrt{2x}\,\sqrt{13y}$

14. $\sqrt{5a}\,\sqrt{6b}$

15. $\sqrt[3]{8y^3}\,\sqrt[3]{10y}$

16. $\sqrt[5]{9t^2}\,\sqrt[5]{2t}$

17. $\sqrt{y - b}\,\sqrt{y + b}$

18. $\sqrt{x - a}\,\sqrt{x + a}$

19. $\sqrt[3]{0.7y}\,\sqrt[3]{0.3y}$

20. $\sqrt[3]{0.5x}\,\sqrt[3]{0.2x}$

21. $\sqrt[5]{x - 2}\,\sqrt[5]{(x - 2)^2}$

22. $\sqrt[4]{x - 1}\,\sqrt[4]{x^2 + x + 1}$

23. $\sqrt{\dfrac{7}{t}}\sqrt{\dfrac{s}{11}}$ **24.** $\sqrt{\dfrac{x}{6}}\sqrt{\dfrac{7}{y}}$

25. $\sqrt[7]{\dfrac{x-3}{4}}\sqrt[7]{\dfrac{5}{x+2}}$

26. $\sqrt[6]{\dfrac{a}{b-2}}\sqrt[6]{\dfrac{3}{b+2}}$

Simplify by factoring.

27. $\sqrt{18}$ **28.** $\sqrt{50}$ **29.** $\sqrt{27}$

30. $\sqrt{45}$ **31.** $\sqrt{8}$ **32.** $\sqrt{75}$

33. $\sqrt{198}$ **34.** $\sqrt{325}$ **35.** $\sqrt{36a^4b}$

36. $\sqrt{175y^8}$ **37.** $\sqrt[3]{8x^3y^2}$ **38.** $\sqrt[3]{27ab^6}$

39. $\sqrt[3]{-16x^6}$ **40.** $\sqrt[3]{-32a^6}$

Find a simplified form of $f(x)$. Assume that x can be any real number.

41. $f(x) = \sqrt[3]{125x^5}$

42. $f(x) = \sqrt[3]{16x^6}$

43. $f(x) = \sqrt{49(x-3)^2}$

44. $f(x) = \sqrt{81(x-1)^2}$

45. $f(x) = \sqrt{5x^2 - 10x + 5}$

46. $f(x) = \sqrt{2x^2 + 8x + 8}$

Simplify. Assume that no radicands were formed by raising negative numbers to even powers.

47. $\sqrt{a^6b^7}$ **48.** $\sqrt{x^6y^9}$

49. $\sqrt[3]{x^5y^6z^{10}}$ **50.** $\sqrt[3]{a^6b^7c^{13}}$

51. $\sqrt[5]{-32a^7b^{11}}$ **52.** $\sqrt[4]{16x^5y^{11}}$

53. $\sqrt[5]{x^{13}y^8z^{17}}$ **54.** $\sqrt[5]{a^6b^8c^9}$

55. $\sqrt[3]{-80a^{14}}$ **56.** $\sqrt[4]{810x^9}$

Multiply and simplify.

57. $\sqrt{6}\sqrt{3}$ **58.** $\sqrt{15}\sqrt{5}$

59. $\sqrt{15}\sqrt{21}$ **60.** $\sqrt{10}\sqrt{14}$

61. $\sqrt[3]{9}\sqrt[3]{3}$ **62.** $\sqrt[3]{2}\sqrt[3]{4}$

Aha! **63.** $\sqrt{18a^3}\sqrt{18a^3}$ **64.** $\sqrt{75x^7}\sqrt{75x^7}$

65. $\sqrt[3]{5a^2}\sqrt[3]{2a}$ **66.** $\sqrt[3]{7x}\sqrt[3]{3x^2}$

67. $\sqrt{2x^5}\sqrt{10x^2}$ **68.** $\sqrt{5a^7}\sqrt{15a^3}$

69. $\sqrt[3]{s^2t^4}\sqrt[3]{s^4t^6}$ **70.** $\sqrt[3]{x^2y^4}\sqrt[3]{x^2y^6}$

71. $\sqrt[3]{(x+5)^2}\sqrt[3]{(x+5)^4}$

72. $\sqrt[3]{(a-b)^5}\sqrt[3]{(a-b)^7}$

73. $\sqrt[4]{20a^3b^7}\sqrt[4]{4a^2b^5}$

74. $\sqrt[4]{9x^7y^2}\sqrt[4]{9x^2y^9}$

75. $\sqrt[5]{x^3(y+z)^6}\sqrt[5]{x^3(y+z)^4}$

76. $\sqrt[5]{a^3(b-c)^4}\sqrt[5]{a^7(b-c)^4}$

77. Why do we need to know how to multiply radical expressions before learning how to simplify radical expressions?

78. Why is it incorrect to say that, in general, $\sqrt{x^2} = x$?

SKILL MAINTENANCE

Perform the indicated operation and, if possible, simplify. [6.2]

79. $\dfrac{3x}{16y} + \dfrac{5y}{64x}$

80. $\dfrac{2}{a^3b^4} + \dfrac{6}{a^4b}$

81. $\dfrac{4}{x^2-9} - \dfrac{7}{2x-6}$

82. $\dfrac{8}{x^2-25} - \dfrac{3}{2x-10}$

Simplify. [1.6]

83. $\dfrac{9a^4b^7}{3a^2b^5}$ **84.** $\dfrac{12a^2b^7}{4ab^2}$

SYNTHESIS

85. Explain why it is true that $\sqrt[n]{ab} = \sqrt[n]{a} \cdot \sqrt[n]{b}$.

86. Is the equation $\sqrt{(2x+3)^8} = (2x+3)^4$ always, sometimes, or never true? Why?

87. *Radar range.* The function given by
$$R(x) = \frac{1}{2}\sqrt[4]{\frac{x \cdot 3.0 \times 10^6}{\pi^2}}$$
can be used to determine the maximum range $R(x)$, in miles, of an ARSR-3 surveillance radar with a peak power of x watts. Determine the maximum radar range when the peak power is 5×10^4 watts.
Source: Introduction to RADAR Techniques, Federal Aviation Administration, 1988

88. *Speed of a skidding car.*   Police can estimate the speed at which a car was traveling by measuring its skid marks. The function given by

$$r(L) = 2\sqrt{5L}$$

can be used, where L is the length of a skid mark, in feet, and $r(L)$ is the speed, in miles per hour. Find the exact speed and an estimate (to the nearest tenth mile per hour) for the speed of a car that left skid marks **(a)** 20 ft long; **(b)** 70 ft long; **(c)** 90 ft long. See also Exercise 102.

89. *Wind chill temperature.*   When the temperature is T degrees Celsius and the wind speed is v meters per second, the *wind chill temperature*, T_w, is the temperature (with no wind) that it feels like. Here is a formula for finding wind chill temperature:

$$T_w = 33 - \frac{\left(10.45 + 10\sqrt{v} - v\right)\left(33 - T\right)}{22}.$$

Estimate the wind chill temperature (to the nearest tenth of a degree) for the given actual temperatures and wind speeds.

a) $T = 7°C, \ v = 8 \text{ m/sec}$
b) $T = 0°C, \ v = 12 \text{ m/sec}$
c) $T = -5°C, \ v = 14 \text{ m/sec}$
d) $T = -23°C, \ v = 15 \text{ m/sec}$

Simplify. Assume that all variables are nonnegative.

90. $\left(\sqrt{r^3 t}\right)^7$

91. $\left(\sqrt[3]{25x^4}\right)^4$

92. $\left(\sqrt[3]{a^2 b^4}\right)^5$

93. $\left(\sqrt{a^3 b^5}\right)^7$

Draw and compare the graphs of each group of equations.

94. $f(x) = \sqrt{x^2 - 2x + 1}$,
$g(x) = x - 1$,
$h(x) = |x - 1|$

95. $f(x) = \sqrt{x^2 + 2x + 1}$,
$g(x) = x + 1$,
$h(x) = |x + 1|$

96. If $f(t) = \sqrt{t^2 - 3t - 4}$, what is the domain of f?

97. What is the domain of g, if $g(x) = \sqrt{x^2 - 6x + 8}$?

Solve.

98. $\sqrt[3]{5x^{k+1}} \ \sqrt[3]{25x^k} = 5x^7$, for k

99. $\sqrt[5]{4a^{3k+2}} \ \sqrt[5]{8a^{6-k}} = 2a^4$, for k

100. Use a graphing calculator to check your answers to Exercises 21 and 41.

101. Rony is puzzled. When he uses a graphing calculator to graph $y = \sqrt{x} \cdot \sqrt{x}$, he gets the following screen. Explain why Rony did not get the complete line $y = x$.

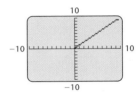

102. Does a car traveling twice as fast as another car leave a skid mark that is twice as long? (See Exercise 88.) Why or why not?

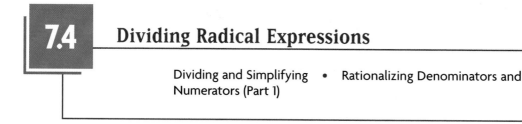

7.4 Dividing Radical Expressions

Dividing and Simplifying • Rationalizing Denominators and Numerators (Part 1)

Dividing and Simplifying

Just as the root of a product can be expressed as the product of two roots, the root of a quotient can be expressed as the quotient of two roots. For example,

$$\sqrt[3]{\frac{27}{8}} = \frac{3}{2} \quad \text{and} \quad \frac{\sqrt[3]{27}}{\sqrt[3]{8}} = \frac{3}{2}.$$

This example suggests the following.

The Quotient Rule for Radicals

For any real numbers $\sqrt[n]{a}$ and $\sqrt[n]{b}$, $b \neq 0$,

$$\sqrt[n]{\frac{a}{b}} = \frac{\sqrt[n]{a}}{\sqrt[n]{b}}.$$

Remember that an nth root is simplified when its radicand has no factors that are perfect nth powers. Recall too that we assume that no radicands represent negative quantities raised to an even power.

EXAMPLE 1 Simplify by taking the roots of the numerator and the denominator.

a) $\sqrt[3]{\frac{27}{125}}$

b) $\sqrt{\frac{25}{y^2}}$

Solution

a) $\sqrt[3]{\frac{27}{125}} = \frac{\sqrt[3]{27}}{\sqrt[3]{125}} = \frac{3}{5}$ Taking the cube roots of the numerator and the denominator

b) $\sqrt{\frac{25}{y^2}} = \frac{\sqrt{25}}{\sqrt{y^2}} = \frac{5}{y}$ Taking the square roots of the numerator and the denominator. Assume $y > 0$.

As in Section 7.3, any radical expressions appearing in the answers should be simplified as much as possible.

EXAMPLE 2 Simplify: **(a)** $\sqrt{\dfrac{16x^3}{y^8}}$; **(b)** $\sqrt[3]{\dfrac{27y^{14}}{8x^3}}$.

Solution

a) $\sqrt{\dfrac{16x^3}{y^8}} = \dfrac{\sqrt{16x^3}}{\sqrt{y^8}}$

$= \dfrac{\sqrt{16x^2 \cdot x}}{\sqrt{y^8}}$

$= \dfrac{4x\sqrt{x}}{y^4}$ Simplifying the numerator and the denominator

b) $\sqrt[3]{\dfrac{27y^{14}}{8x^3}} = \dfrac{\sqrt[3]{27y^{14}}}{\sqrt[3]{8x^3}}$

$= \dfrac{\sqrt[3]{27y^{12}y^2}}{\sqrt[3]{8x^3}}$ y^{12} is the largest perfect-cube factor of y^{14}.

$= \dfrac{\sqrt[3]{27y^{12}}\,\sqrt[3]{y^2}}{\sqrt[3]{8x^3}}$

$= \dfrac{3y^4\sqrt[3]{y^2}}{2x}$ Simplifying the numerator and the denominator

If we read from right to left, the quotient rule tells us that to divide two radical expressions that have the same index, we can divide the radicands.

EXAMPLE 3 Divide and, if possible, simplify.

Student Notes _____

When writing radical signs, pay careful attention to what is included as the radicand. Each of the following represents a *different* number:

$$\sqrt{\dfrac{5 \cdot 2}{3}}, \quad \dfrac{\sqrt{5 \cdot 2}}{3}, \quad \dfrac{\sqrt{5} \cdot 2}{3}.$$

a) $\dfrac{\sqrt{80}}{\sqrt{5}}$

b) $\dfrac{5\sqrt[3]{32}}{\sqrt[3]{2}}$

c) $\dfrac{\sqrt{72xy}}{2\sqrt{2}}$

d) $\dfrac{\sqrt[4]{18a^9b^5}}{\sqrt[4]{3b}}$

Solution

a) $\dfrac{\sqrt{80}}{\sqrt{5}} = \sqrt{\dfrac{80}{5}} = \sqrt{16} = 4$

> Because the indices match, we can divide the radicands.

b) $\dfrac{5\sqrt[3]{32}}{\sqrt[3]{2}} = 5\sqrt[3]{\dfrac{32}{2}} = 5\sqrt[3]{16}$

$= 5\sqrt[3]{8 \cdot 2}$ 8 is the largest perfect-cube factor of 16.

$= 5\sqrt[3]{8}\,\sqrt[3]{2} = 5 \cdot 2\sqrt[3]{2}$

$= 10\sqrt[3]{2}$

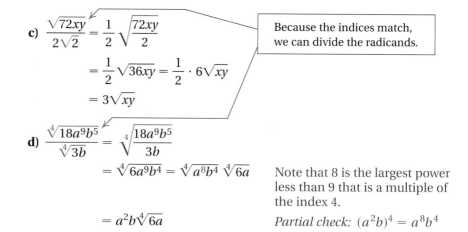

c) $\dfrac{\sqrt{72xy}}{2\sqrt{2}} = \dfrac{1}{2}\sqrt{\dfrac{72xy}{2}}$

> Because the indices match, we can divide the radicands.

$= \dfrac{1}{2}\sqrt{36xy} = \dfrac{1}{2} \cdot 6\sqrt{xy}$

$= 3\sqrt{xy}$

d) $\dfrac{\sqrt[4]{18a^9b^5}}{\sqrt[4]{3b}} = \sqrt[4]{\dfrac{18a^9b^5}{3b}}$

$= \sqrt[4]{6a^9b^4} = \sqrt[4]{a^8b^4}\,\sqrt[4]{6a}$

$= a^2b\sqrt[4]{6a}$

Note that 8 is the largest power less than 9 that is a multiple of the index 4.

Partial check: $(a^2b)^4 = a^8b^4$

Rationalizing Denominators and Numerators (Part 1)*

When a radical expression appears in a denominator, it can be useful to find an equivalent expression in which the denominator no longer contains a radical.† The procedure for finding such an expression is called **rationalizing the denominator**. We carry this out by multiplying by 1 in either of two ways.

One way is to multiply by 1 *under* the radical to make the denominator of the radicand a perfect power.

EXAMPLE 4   Rationalize each denominator.

a) $\sqrt{\dfrac{7}{3}}$

b) $\sqrt[3]{\dfrac{5}{16}}$

Solution

a) We multiply by 1 under the radical, using $\frac{3}{3}$. We do this so that the denominator of the radicand will be a perfect square:

$\sqrt{\dfrac{7}{3}} = \sqrt{\dfrac{7}{3} \cdot \dfrac{3}{3}}$ Multiplying by 1 under the radical

$= \sqrt{\dfrac{21}{9}}$ The denominator, 9, is now a perfect square.

$= \dfrac{\sqrt{21}}{\sqrt{9}}$ Using the quotient rule for radicals

$= \dfrac{\sqrt{21}}{3}.$

---

*Denominators and numerators with two terms are rationalized in Section 7.5.
†See Exercise 73 on p. 466.

b) Note that $16 = 4^2$. Thus, to make the denominator a perfect cube, we multiply under the radical by $\frac{4}{4}$:

$$\sqrt[3]{\frac{5}{16}} = \sqrt[3]{\frac{5}{4 \cdot 4} \cdot \frac{4}{4}} \qquad \text{Since the index is 3, we need 3 identical factors in the denominator.}$$

$$= \sqrt[3]{\frac{20}{4^3}} \qquad \text{The denominator is now a perfect cube.}$$

$$= \frac{\sqrt[3]{20}}{\sqrt[3]{4^3}}$$

$$= \frac{\sqrt[3]{20}}{4}.$$

Another way to rationalize a denominator is to multiply by 1 *outside* the radical.

EXAMPLE 5 Rationalize each denominator.

a) $\sqrt{\dfrac{4}{5b}}$

b) $\dfrac{\sqrt[3]{a}}{\sqrt[3]{9x}}$

c) $\dfrac{3x}{\sqrt[5]{2x^2y^3}}$

Solution

a) We rewrite the expression as a quotient of two radicals. Then we simplify and multiply by 1:

$$\sqrt{\frac{4}{5b}} = \frac{\sqrt{4}}{\sqrt{5b}} = \frac{2}{\sqrt{5b}} \qquad \text{We assume } b > 0.$$

$$= \frac{2}{\sqrt{5b}} \cdot \frac{\sqrt{5b}}{\sqrt{5b}} \qquad \text{Multiplying by 1}$$

$$= \frac{2\sqrt{5b}}{(\sqrt{5b})^2} \qquad \text{Try to do this step mentally.}$$

$$= \frac{2\sqrt{5b}}{5b}.$$

b) To rationalize the denominator $\sqrt[3]{9x}$, note that $9x$ is $3 \cdot 3 \cdot x$. In order for this radicand to be a cube, we need another factor of 3 and two more factors of x. Thus we multiply by 1, using $\sqrt[3]{3x^2}/\sqrt[3]{3x^2}$:

$$\frac{\sqrt[3]{a}}{\sqrt[3]{9x}} = \frac{\sqrt[3]{a}}{\sqrt[3]{9x}} \cdot \frac{\sqrt[3]{3x^2}}{\sqrt[3]{3x^2}} \qquad \text{Multiplying by 1}$$

$$= \frac{\sqrt[3]{3ax^2}}{\sqrt[3]{27x^3}} \longleftarrow \text{This radicand is now a perfect cube.}$$

$$= \frac{\sqrt[3]{3ax^2}}{3x}.$$

c) To change the radicand $2x^2y^3$ into a perfect fifth power, we need four more factors of 2, three more factors of x, and two more factors of y. Thus we multiply by 1, using $\sqrt[5]{2^4x^3y^2}/\sqrt[5]{2^4x^3y^2}$, or $\sqrt[5]{16x^3y^2}/\sqrt[5]{16x^3y^2}$:

$$\frac{3x}{\sqrt[5]{2x^2y^3}} = \frac{3x}{\sqrt[5]{2x^2y^3}} \cdot \frac{\sqrt[5]{16x^3y^2}}{\sqrt[5]{16x^3y^2}} \qquad \text{Multiplying by 1}$$

$$= \frac{3x\sqrt[5]{16x^3y^2}}{\sqrt[5]{32x^5y^5}} \longleftarrow \qquad \begin{array}{l}\text{This radicand is now a} \\ \text{perfect fifth power.}\end{array}$$

$$= \frac{3x\sqrt[5]{16x^3y^2}}{2xy} = \frac{3\sqrt[5]{16x^3y^2}}{2y}. \qquad \text{Always simplify if possible.}$$

Sometimes in calculus it is necessary to rationalize a numerator. To do so, we multiply by 1 to make the radicand in the *numerator* a perfect power.

EXAMPLE 6   Rationalize each numerator: **(a)** $\sqrt{\dfrac{7}{5}}$; **(b)** $\dfrac{\sqrt[3]{4a^2}}{\sqrt[3]{5b}}$.

Solution

a) $\sqrt{\dfrac{7}{5}} = \sqrt{\dfrac{7}{5} \cdot \dfrac{7}{7}}$ Multiplying by 1 under the radical. We also could have multiplied by $\sqrt{7}/\sqrt{7}$ outside the radical.

$$= \sqrt{\frac{49}{35}} \qquad \text{The numerator is now a perfect square.}$$

$$= \frac{\sqrt{49}}{\sqrt{35}} \qquad \text{Using the quotient rule for radicals}$$

$$= \frac{7}{\sqrt{35}}$$

b) $\dfrac{\sqrt[3]{4a^2}}{\sqrt[3]{5b}} = \dfrac{\sqrt[3]{4a^2}}{\sqrt[3]{5b}} \cdot \dfrac{\sqrt[3]{2a}}{\sqrt[3]{2a}}$ Multiplying by 1

$$= \frac{\sqrt[3]{8a^3}}{\sqrt[3]{10ba}} \longleftarrow \quad \text{This radicand is now a perfect cube.}$$

$$= \frac{2a}{\sqrt[3]{10ab}}$$

In Section 7.5, we will discuss rationalizing denominators and numerators in which two terms appear.

90. Show that $\dfrac{\sqrt[n]{a}}{\sqrt[n]{b}}$ is the nth root of $\dfrac{a}{b}$ by raising it to the nth power and simplifying.

91. Let $f(x) = \sqrt{18x^3}$ and $g(x) = \sqrt{2x}$. Find $(f/g)(x)$ and specify the domain of f/g.

92. Let $f(t) = \sqrt{2t}$ and $g(t) = \sqrt{50t^3}$. Find $(f/g)(t)$ and specify the domain of f/g.

93. Let $f(x) = \sqrt{x^2 - 9}$ and $g(x) = \sqrt{x - 3}$. Find $(f/g)(x)$ and specify the domain of f/g.

7.5 Expressions Containing Several Radical Terms

Adding and Subtracting Radical Expressions • Products and Quotients of Two or More Radical Terms • Rationalizing Denominators and Numerators (Part 2) • Terms with Differing Indices

Radical expressions like $6\sqrt{7} + 4\sqrt{7}$ or $\left(\sqrt{a} + \sqrt{b}\right)\left(\sqrt{a} - \sqrt{b}\right)$ contain more than one *radical term* and can sometimes be simplified.

Adding and Subtracting Radical Expressions

When two radical expressions have the same indices and radicands, they are said to be **like radicals**. Like radicals can be combined (added or subtracted) in much the same way that we combined like terms earlier in this text.

EXAMPLE 1   Simplify by combining like radical terms.

a) $6\sqrt{7} + 4\sqrt{7}$

b) $\sqrt[3]{2} - 7x\sqrt[3]{2} + 5\sqrt[3]{2}$

c) $6\sqrt[5]{4x} + 3\sqrt[5]{4x} - \sqrt[3]{4x}$

Solution

a) $6\sqrt{7} + 4\sqrt{7} = (6 + 4)\sqrt{7}$ Using the distributive law (factoring out $\sqrt{7}$)

$\qquad\qquad\qquad\;\; = 10\sqrt{7}$ You can think: 6 square roots of 7 plus 4 square roots of 7 results in 10 square roots of 7.

b) $\sqrt[3]{2} - 7x\sqrt[3]{2} + 5\sqrt[3]{2} = (1 - 7x + 5)\sqrt[3]{2}$ Factoring out $\sqrt[3]{2}$

$\qquad\qquad\qquad\qquad\qquad\; = (6 - 7x)\sqrt[3]{2}$ These parentheses are important!

c) $6\sqrt[5]{4x} + 3\sqrt[5]{4x} - \sqrt[3]{4x} = (6 + 3)\sqrt[5]{4x} - \sqrt[3]{4x}$ Try to do this step mentally.

$\qquad\qquad\qquad\qquad\qquad\quad = 9\sqrt[5]{4x} - \sqrt[3]{4x}$ Because the indices differ, we are done.

Study Skills

Review Material on Your Own

Never hesitate to review earlier material in which you feel a lack of confidence. For example, if you feel unsure about how to multiply with fraction notation, be sure to review that material before studying any new material that involves multiplication of fractions. Doing (and checking) some practice problems from that section is also a good way to sharpen any skills that may not have been used for a while.

Our ability to simplify radical expressions can help us to find like radicals even when, at first, it may appear that none exists.

EXAMPLE 2   Simplify by combining like radical terms, if possible.

a) $3\sqrt{8} - 5\sqrt{2}$

b) $9\sqrt{5} - 4\sqrt{3}$

c) $\sqrt[3]{2x^6y^4} + 7\sqrt[3]{2y}$

Solution

a) $3\sqrt{8} - 5\sqrt{2} = 3\sqrt{4 \cdot 2} - 5\sqrt{2}$

$\qquad\qquad\qquad = 3\sqrt{4} \cdot \sqrt{2} - 5\sqrt{2}$ $\left.\rule{0cm}{1.2cm}\right\}$ Simplifying $\sqrt{8}$

$\qquad\qquad\qquad = 3 \cdot 2 \cdot \sqrt{2} - 5\sqrt{2}$

$\qquad\qquad\qquad = 6\sqrt{2} - 5\sqrt{2}$

$\qquad\qquad\qquad = \sqrt{2}$ $\qquad$ Combining like radicals

b) $9\sqrt{5} - 4\sqrt{3}$ cannot be simplified.

c) $\sqrt[3]{2x^6y^4} + 7\sqrt[3]{2y} = \sqrt[3]{x^6y^3 \cdot 2y} + 7\sqrt[3]{2y}$

$\qquad\qquad\qquad\qquad = \sqrt[3]{x^6y^3} \cdot \sqrt[3]{2y} + 7\sqrt[3]{2y}$ $\left.\rule{0cm}{1.2cm}\right\}$ Simplifying $\sqrt[3]{2x^6y^4}$

$\qquad\qquad\qquad\qquad = x^2y \cdot \sqrt[3]{2y} + 7\sqrt[3]{2y}$

$\qquad\qquad\qquad\qquad = (x^2y + 7)\sqrt[3]{2y}$ $\qquad$ Factoring to combine like radical terms

Products and Quotients of Two or More Radical Terms

Radical expressions often contain factors that have more than one term. Multiplying such expressions is similar to finding products of polynomials. Some products will yield like radical terms, which we can now combine.

EXAMPLE 3   Multiply.

a) $\sqrt{3}(x - \sqrt{5})$ $\qquad\qquad\qquad$ **b)** $\sqrt[3]{y}(\sqrt[3]{y^2} + \sqrt[3]{2})$

c) $(4\sqrt{3} + \sqrt{2})(\sqrt{3} - 5\sqrt{2})$ $\qquad$ **d)** $(\sqrt{a} + \sqrt{b})(\sqrt{a} - \sqrt{b})$

Solution

a) $\sqrt{3}(x - \sqrt{5}) = \sqrt{3} \cdot x - \sqrt{3} \cdot \sqrt{5}$ $\qquad$ Using the distributive law

$\qquad\qquad\qquad = x\sqrt{3} - \sqrt{15}$ $\qquad\qquad$ Multiplying radicals

b) $\sqrt[3]{y}(\sqrt[3]{y^2} + \sqrt[3]{2}) = \sqrt[3]{y} \cdot \sqrt[3]{y^2} + \sqrt[3]{y} \cdot \sqrt[3]{2}$ $\qquad$ Using the distributive law

$\qquad\qquad\qquad\qquad = \sqrt[3]{y^3} + \sqrt[3]{2y}$ $\qquad\qquad$ Multiplying radicals

$\qquad\qquad\qquad\qquad = y + \sqrt[3]{2y}$ $\qquad\qquad$ Simplifying $\sqrt[3]{y^3}$

$\qquad\qquad\qquad\qquad\qquad$ F $\qquad$ O $\qquad$ I $\qquad$ L

c) $(4\sqrt{3} + \sqrt{2})(\sqrt{3} - 5\sqrt{2}) = 4(\sqrt{3})^2 - 20\sqrt{3} \cdot \sqrt{2} + \sqrt{2} \cdot \sqrt{3} - 5(\sqrt{2})^2$

$\qquad\qquad\qquad\qquad\qquad = 4 \cdot 3 - 20\sqrt{6} + \sqrt{6} - 5 \cdot 2$ $\qquad$ Multiplying radicals

$\qquad\qquad\qquad\qquad\qquad = 12 - 20\sqrt{6} + \sqrt{6} - 10$

$\qquad\qquad\qquad\qquad\qquad = 2 - 19\sqrt{6}$ $\qquad$ Combining like terms

d) $(\sqrt{a} + \sqrt{b})(\sqrt{a} - \sqrt{b}) = (\sqrt{a})^2 - \sqrt{a}\,\sqrt{b} + \sqrt{a}\,\sqrt{b} - (\sqrt{b})^2$ Using FOIL

$$= a - b \qquad \text{Combining like terms}$$

In Example 3(d) above, you may have noticed that since the outer and inner products in FOIL are opposites, the result, $a - b$, is not itself a radical expression. Pairs of radical terms, like $\sqrt{a} + \sqrt{b}$ and $\sqrt{a} - \sqrt{b}$, are called **conjugates**.

Rationalizing Denominators and Numerators (Part 2)

The use of conjugates allows us to rationalize denominators or numerators with two terms.

EXAMPLE 4   Rationalize each denominator: **(a)** $\dfrac{4}{\sqrt{3} + x}$; **(b)** $\dfrac{4 + \sqrt{2}}{\sqrt{5} - \sqrt{2}}$.

Solution

a) $\dfrac{4}{\sqrt{3} + x} = \dfrac{4}{\sqrt{3} + x} \cdot \dfrac{\sqrt{3} - x}{\sqrt{3} - x}$ Multiplying by 1, using the conjugate of $\sqrt{3} + x$, which is $\sqrt{3} - x$

$$= \dfrac{4(\sqrt{3} - x)}{(\sqrt{3} + x)(\sqrt{3} - x)} \qquad \text{Multiplying numerators and denominators}$$

$$= \dfrac{4(\sqrt{3} - x)}{(\sqrt{3})^2 - x^2} \qquad \text{Using FOIL in the denominator}$$

$$= \dfrac{4\sqrt{3} - 4x}{3 - x^2} \qquad \text{Simplifying. No radicals remain in the denominator.}$$

b) $\dfrac{4 + \sqrt{2}}{\sqrt{5} - \sqrt{2}} = \dfrac{4 + \sqrt{2}}{\sqrt{5} - \sqrt{2}} \cdot \dfrac{\sqrt{5} + \sqrt{2}}{\sqrt{5} + \sqrt{2}}$ Multiplying by 1, using the conjugate of $\sqrt{5} - \sqrt{2}$, which is $\sqrt{5} + \sqrt{2}$

$$= \dfrac{(4 + \sqrt{2})(\sqrt{5} + \sqrt{2})}{(\sqrt{5} - \sqrt{2})(\sqrt{5} + \sqrt{2})} \qquad \text{Multiplying numerators and denominators}$$

$$= \dfrac{4\sqrt{5} + 4\sqrt{2} + \sqrt{2}\,\sqrt{5} + (\sqrt{2})^2}{(\sqrt{5})^2 - (\sqrt{2})^2} \qquad \text{Using FOIL}$$

$$= \dfrac{4\sqrt{5} + 4\sqrt{2} + \sqrt{10} + 2}{5 - 2} \qquad \text{Squaring in the denominator and the numerator}$$

$$= \dfrac{4\sqrt{5} + 4\sqrt{2} + \sqrt{10} + 2}{3} \qquad \text{No radicals remain in the denominator.}$$

To rationalize a numerator with more than one term, we use the conjugate of the numerator.

EXAMPLE 5   Rationalize the numerator: $\dfrac{4 + \sqrt{2}}{\sqrt{5} - \sqrt{2}}$.

Solution

$$\frac{4 + \sqrt{2}}{\sqrt{5} - \sqrt{2}} = \frac{4 + \sqrt{2}}{\sqrt{5} - \sqrt{2}} \cdot \frac{4 - \sqrt{2}}{4 - \sqrt{2}} \qquad \begin{array}{l}\text{Multiplying by 1, using} \\ \text{the conjugate of } 4 + \sqrt{2}, \\ \text{which is } 4 - \sqrt{2}\end{array}$$

$$= \frac{16 - \left(\sqrt{2}\right)^2}{4\sqrt{5} - \sqrt{5}\,\sqrt{2} - 4\sqrt{2} + \left(\sqrt{2}\right)^2}$$

$$= \frac{14}{4\sqrt{5} - \sqrt{10} - 4\sqrt{2} + 2}$$

Terms with Differing Indices

To multiply or divide radical terms with different indices, we can convert to exponential notation, use the rules for exponents, and then convert back to radical notation.

EXAMPLE 6   Divide and, if possible, simplify: $\dfrac{\sqrt[4]{(x + y)^3}}{\sqrt{x + y}}$.

Solution

$$\frac{\sqrt[4]{(x + y)^3}}{\sqrt{x + y}} = \frac{(x + y)^{3/4}}{(x + y)^{1/2}} \qquad \text{Converting to exponential notation}$$

$$= (x + y)^{3/4 - 1/2} \qquad \begin{array}{l}\text{Since the bases are identical, we can} \\ \text{subtract exponents:} \\ \frac{3}{4} - \frac{1}{2} = \frac{3}{4} - \frac{2}{4} = \frac{1}{4}.\end{array}$$

$$\left.\begin{array}{l}= (x + y)^{1/4} \\ = \sqrt[4]{x + y}\end{array}\right\} \qquad \text{Converting back to radical notation}$$

Student Notes

Expressions similar to the one in Example 6 are most easily simplified by rewriting the expression using exponents in place of radicals. After simplifying, remember to write your final result in radical notation. In general, if a problem is presented in one form, it is expected that the final result be presented in the same form.

The steps used in Example 6 can be used in a variety of situations.

To Simplify Products or Quotients with Differing Indices

1. Convert all radical expressions to exponential notation.
2. When the bases are identical, subtract exponents to divide and add exponents to multiply. This may require finding a common denominator.
3. Convert back to radical notation and, if possible, simplify.

EXAMPLE 7   Multiply and simplify: $\sqrt{x^3}\,\sqrt[3]{x}$.

Solution

$$\sqrt{x^3}\,\sqrt[3]{x} = x^{3/2}\cdot x^{1/3} \qquad \text{Converting to exponential notation}$$

$$= x^{11/6} \qquad\qquad \text{Adding exponents: } \tfrac{3}{2}+\tfrac{1}{3}=\tfrac{9}{6}+\tfrac{2}{6}$$

$$= \sqrt[6]{x^{11}} \qquad\qquad \text{Converting back to radical notation}$$

$$\left.\begin{aligned} &= \sqrt[6]{x^6}\,\sqrt[6]{x^5} \\ &= x\sqrt[6]{x^5} \end{aligned}\right\} \quad \text{Simplifying}$$

EXAMPLE 8   If $f(x) = \sqrt[3]{x^2}$ and $g(x) = \sqrt{x} + \sqrt[4]{x}$, find $(f\cdot g)(x)$.

Solution   Recall from Section 2.6 that $(f\cdot g)(x) = f(x)\cdot g(x)$. Thus,

$$(f\cdot g)(x) = \sqrt[3]{x^2}\big(\sqrt{x} + \sqrt[4]{x}\big) \qquad \begin{aligned} &x \text{ is assumed to be} \\ &\text{nonnegative.} \end{aligned}$$

$$= x^{2/3}(x^{1/2} + x^{1/4}) \qquad \begin{aligned}&\text{Converting to exponential} \\ &\text{notation}\end{aligned}$$

$$= x^{2/3}\cdot x^{1/2} + x^{2/3}\cdot x^{1/4} \qquad \text{Using the distributive law}$$

$$= x^{2/3+1/2} + x^{2/3+1/4} \qquad \text{Adding exponents:}$$

$$= x^{7/6} + x^{11/12} \qquad \tfrac{2}{3}+\tfrac{1}{2}=\tfrac{4}{6}+\tfrac{3}{6}; \tfrac{2}{3}+\tfrac{1}{4}=\tfrac{8}{12}+\tfrac{3}{12}$$

$$= \sqrt[6]{x^7} + \sqrt[12]{x^{11}} \qquad \begin{aligned}&\text{Converting back to radical} \\ &\text{notation}\end{aligned}$$

$$\left.\begin{aligned} &= \sqrt[6]{x^6}\,\sqrt[6]{x} + \sqrt[12]{x^{11}} \\ &= x\sqrt[6]{x} + \sqrt[12]{x^{11}} \end{aligned}\right\} \quad \text{Simplifying}$$

If factors are raised to powers that share a common denominator, we can write the final result as a single radical expression.

EXAMPLE 9   Divide and, if possible, simplify: $\dfrac{\sqrt[3]{a^2b^4}}{\sqrt{ab}}$.

Solution

$$\frac{\sqrt[3]{a^2b^4}}{\sqrt{ab}} = \frac{(a^2b^4)^{1/3}}{(ab)^{1/2}} \qquad \text{Converting to exponential notation}$$

$$= \frac{a^{2/3}b^{4/3}}{a^{1/2}b^{1/2}} \qquad \text{Using the product and power rules}$$

$$= a^{2/3-1/2}b^{4/3-1/2} \qquad \text{Subtracting exponents}$$

$$= a^{1/6}b^{5/6}$$

$$= \sqrt[6]{a}\,\sqrt[6]{b^5} \qquad \text{Converting to radical notation}$$

$$= \sqrt[6]{ab^5} \qquad \text{Using the product rule for radicals}$$

Exercise Set

7.5

↪ *Concept Reinforcement* *For each of Exercises 1–6, fill in the blanks by selecting from the following words (which may be used more than once):*

radicand(s), indices, conjugate(s), base(s), denominator(s), numerator(s).

1. To add radical expressions, the _____ and the _____ must be the same.

2. To multiply radical expressions, the _____ must be the same.

3. To find a product by adding exponents, the _____ must be the same.

4. To add rational expressions, the _____ must be the same.

5. To rationalize the _____ of $\dfrac{\sqrt{a+0.1}-\sqrt{a}}{0.1}$, we multiply by a form of 1, using the _____ of $\sqrt{a+0.1}-\sqrt{a}$, or $\sqrt{a+0.1}+\sqrt{a}$, to write 1.

6. To find a quotient by subtracting exponents, the _____ must be the same.

Add or subtract. Simplify by combining like radical terms, if possible. Assume that all variables and radicands represent positive real numbers.

7. $2\sqrt{5} + 7\sqrt{5}$

8. $4\sqrt{7} + 2\sqrt{7}$

9. $7\sqrt[3]{4} - 5\sqrt[3]{4}$

10. $14\sqrt[5]{2} - 6\sqrt[5]{2}$

11. $\sqrt[3]{y} + 9\sqrt[3]{y}$

12. $9\sqrt[4]{t} - 3\sqrt[4]{t}$

13. $8\sqrt{2} - 6\sqrt{2} + 5\sqrt{2}$

14. $\sqrt{6} + 8\sqrt{6} - 3\sqrt{6}$

15. $9\sqrt[3]{7} - \sqrt{3} + 4\sqrt[3]{7} + 2\sqrt{3}$

16. $5\sqrt{7} - 8\sqrt[4]{11} + \sqrt{7} + 9\sqrt[4]{11}$

17. $4\sqrt{27} - 3\sqrt{3}$

18. $9\sqrt{50} - 4\sqrt{2}$

19. $3\sqrt{45} + 7\sqrt{20}$

20. $5\sqrt{12} + 16\sqrt{27}$

21. $3\sqrt[3]{16} + \sqrt[3]{54}$

22. $\sqrt[3]{27} - 5\sqrt[3]{8}$

23. $\sqrt{5a} + 2\sqrt{45a^3}$

24. $4\sqrt{3x^3} - \sqrt{12x}$

25. $\sqrt[3]{6x^4} + \sqrt[3]{48x}$

26. $\sqrt[3]{54x} - \sqrt[3]{2x^4}$

27. $\sqrt{4a-4} + \sqrt{a-1}$

28. $\sqrt{9y+27} + \sqrt{y+3}$

29. $\sqrt{x^3-x^2} + \sqrt{9x-9}$

30. $\sqrt{4x-4} - \sqrt{x^3-x^2}$

Multiply. Assume all variables represent nonnegative real numbers.

31. $\sqrt{3}(4 + \sqrt{3})$

32. $\sqrt{7}(3 - \sqrt{7})$

33. $3\sqrt{5}(\sqrt{5} - \sqrt{2})$

34. $4\sqrt{2}(\sqrt{3} - \sqrt{5})$

35. $\sqrt{2}(3\sqrt{10} - 2\sqrt{2})$

36. $\sqrt{3}(2\sqrt{5} - 3\sqrt{4})$

37. $\sqrt[3]{3}(\sqrt[3]{9} - 4\sqrt[3]{21})$

38. $\sqrt[3]{2}(\sqrt[3]{4} - 2\sqrt[3]{32})$

39. $\sqrt[3]{a}(\sqrt[3]{a^2} + \sqrt[3]{24a^2})$

40. $\sqrt[3]{x}(\sqrt[3]{3x^2} - \sqrt[3]{81x^2})$

41. $(2 + \sqrt{6})(5 - \sqrt{6})$

42. $(4 - \sqrt{5})(2 + \sqrt{5})$

43. $(\sqrt{2} + \sqrt{7})(\sqrt{3} - \sqrt{7})$

44. $(\sqrt{7} - \sqrt{2})(\sqrt{5} + \sqrt{2})$

45. $(3 - \sqrt{5})(3 + \sqrt{5})$

46. $(6 - \sqrt{7})(6 + \sqrt{7})$

47. $(\sqrt{6} + \sqrt{8})(\sqrt{6} - \sqrt{8})$

48. $(\sqrt{5} + \sqrt{3})(\sqrt{5} - \sqrt{3})$

49. $\left(3\sqrt{7} + 2\sqrt{5}\right)\left(2\sqrt{7} - 4\sqrt{5}\right)$

50. $\left(4\sqrt{5} - 3\sqrt{2}\right)\left(2\sqrt{5} + 4\sqrt{2}\right)$

51. $\left(2 + \sqrt{3}\right)^2$

52. $\left(3 + \sqrt{7}\right)^2$

53. $\left(\sqrt{3} - \sqrt{2}\right)^2$

54. $\left(\sqrt{5} - \sqrt{3}\right)^2$

55. $\left(\sqrt{2t} + \sqrt{5}\right)^2$

56. $\left(\sqrt{3x} - \sqrt{2}\right)^2$

57. $\left(3 - \sqrt{x+5}\right)^2$

58. $\left(4 + \sqrt{x-3}\right)^2$

59. $\left(2\sqrt[4]{7} - \sqrt[4]{6}\right)\left(3\sqrt[4]{9} + 2\sqrt[4]{5}\right)$

60. $\left(4\sqrt[3]{3} + \sqrt[3]{10}\right)\left(2\sqrt[3]{7} + 5\sqrt[3]{6}\right)$

Rationalize each denominator.

61. $\dfrac{5}{4 - \sqrt{3}}$

62. $\dfrac{3}{4 - \sqrt{7}}$

63. $\dfrac{2 + \sqrt{5}}{6 + \sqrt{3}}$

64. $\dfrac{1 + \sqrt{2}}{3 + \sqrt{5}}$

65. $\dfrac{\sqrt{a}}{\sqrt{a} + \sqrt{b}}$

66. $\dfrac{\sqrt{z}}{\sqrt{x} - \sqrt{z}}$

Aha! 67. $\dfrac{\sqrt{7} - \sqrt{3}}{\sqrt{3} - \sqrt{7}}$

68. $\dfrac{\sqrt{7} + \sqrt{5}}{\sqrt{5} + \sqrt{2}}$

69. $\dfrac{3\sqrt{2} - \sqrt{7}}{4\sqrt{2} + 2\sqrt{5}}$

70. $\dfrac{5\sqrt{3} - \sqrt{11}}{2\sqrt{3} - 5\sqrt{2}}$

Rationalize each numerator. If possible, simplify your result.

71. $\dfrac{\sqrt{7} + 2}{5}$

72. $\dfrac{\sqrt{3} + 1}{4}$

73. $\dfrac{\sqrt{6} - 2}{\sqrt{3} + 7}$

74. $\dfrac{\sqrt{10} + 4}{\sqrt{2} - 3}$

75. $\dfrac{\sqrt{x} - \sqrt{y}}{\sqrt{x} + \sqrt{y}}$

76. $\dfrac{\sqrt{a} + \sqrt{b}}{\sqrt{a} - \sqrt{b}}$

77. $\dfrac{\sqrt{a+h} - \sqrt{a}}{h}$

78. $\dfrac{\sqrt{x-h} - \sqrt{x}}{h}$

Perform the indicated operation and simplify. Assume all variables represent positive real numbers.

79. $\sqrt{a}\,\sqrt[4]{a^3}$

80. $\sqrt[3]{x^2}\,\sqrt[6]{x^5}$

81. $\sqrt[5]{b^2}\,\sqrt{b^3}$

82. $\sqrt[4]{a^3}\,\sqrt[3]{a^2}$

83. $\sqrt{xy^3}\,\sqrt[3]{x^2y}$

84. $\sqrt[5]{a^3b}\,\sqrt{ab}$

85. $\sqrt[4]{9ab^3}\,\sqrt{3a^4b}$

86. $\sqrt{2x^3y^3}\,\sqrt[3]{4xy^2}$

87. $\sqrt{a^4b^3c^4}\,\sqrt[3]{ab^2c}$

88. $\sqrt[3]{xy^2z}\,\sqrt{x^3yz^2}$

89. $\dfrac{\sqrt[3]{a^2}}{\sqrt[4]{a}}$

90. $\dfrac{\sqrt[3]{x^2}}{\sqrt[5]{x}}$

91. $\dfrac{\sqrt[4]{x^2y^3}}{\sqrt[3]{xy}}$

92. $\dfrac{\sqrt[5]{a^4b}}{\sqrt[3]{ab}}$

93. $\dfrac{\sqrt{ab^3}}{\sqrt[5]{a^2b^3}}$

94. $\dfrac{\sqrt[5]{x^3y^4}}{\sqrt{xy}}$

95. $\dfrac{\sqrt[4]{(3x-1)^3}}{\sqrt[5]{(3x-1)^3}}$

96. $\dfrac{\sqrt[3]{(2+5x)^2}}{\sqrt[4]{2+5x}}$

97. $\dfrac{\sqrt[3]{(2x+1)^2}}{\sqrt[5]{(2x+1)^2}}$

98. $\dfrac{\sqrt[4]{(5+3x)^3}}{\sqrt[3]{(5+3x)^2}}$

99. $\sqrt[3]{x^2y}\left(\sqrt{xy} - \sqrt[5]{xy^3}\right)$

100. $\sqrt[4]{a^2b}\left(\sqrt[3]{a^2b} - \sqrt[5]{a^2b^2}\right)$

101. $\left(m + \sqrt[3]{n^2}\right)\left(2m + \sqrt[4]{n}\right)$

102. $\left(r - \sqrt[4]{s^3}\right)\left(3r - \sqrt[5]{s}\right)$

In Exercises 103–106, f(x) and g(x) are as given. Find $(f \cdot g)(x)$. Assume all variables represent nonnegative real numbers.

103. $f(x) = \sqrt[4]{x}$, $g(x) = \sqrt[4]{2x} - \sqrt[4]{x^{11}}$

104. $f(x) = \sqrt[4]{x^7} + \sqrt[4]{3x^2}$, $g(x) = \sqrt[4]{x}$

105. $f(x) = x + \sqrt{7}$, $g(x) = x - \sqrt{7}$

106. $f(x) = x - \sqrt{2}$, $g(x) = x + \sqrt{6}$

Let $f(x) = x^2$. Find each of the following.

107. $f\left(5 + \sqrt{2}\right)$

108. $f\left(7 + \sqrt{3}\right)$

109. $f\left(\sqrt{3} - \sqrt{5}\right)$

110. $f\left(\sqrt{6} - \sqrt{3}\right)$

111. In what way(s) is combining like radical terms similar to combining like terms that are monomials?

112. Why do we need to know how to multiply radical expressions before learning how to add them?

SKILL MAINTENANCE

Solve.

113. $\dfrac{12x}{x-4} - \dfrac{3x^2}{x+4} = \dfrac{384}{x^2-16}$ [6.4]

114. $\dfrac{2}{3} + \dfrac{1}{t} = \dfrac{4}{5}$ [6.4]

115. $5x^2 - 6x + 1 = 0$ [5.8]

116. $7t^2 - 8t + 1 = 0$ [5.8]

117. The sum of a number and its square is 20. Find the number. [5.8]

118. The width of a rectangle is one-fourth the length. The area is twice the perimeter. Find the dimensions of the rectangle. [5.8]

SYNTHESIS

119. Ramon *incorrectly* writes
$$\sqrt[5]{x^2} \cdot \sqrt{x^3} = x^{2/5} \cdot x^{3/2} = \sqrt[5]{x^3}.$$
What mistake do you suspect he is making?

120. After examining the expression $\sqrt[4]{25xy^3}\sqrt{5x^4y}$, Dyan (correctly) concludes that x and y are both nonnegative. Explain how she could reach this conclusion.

Find a simplified form for $f(x)$. Assume $x \ge 0$.

121. $f(x) = \sqrt{20x^2 + 4x^3} - 3x\sqrt{45 + 9x} + \sqrt{5x^2 + x^3}$

122. $f(x) = \sqrt{x^3 - x^2} + \sqrt{9x^3 - 9x^2} - \sqrt{4x^3 - 4x^2}$

123. $f(x) = \sqrt[4]{x^5 - x^4} + 3\sqrt[4]{x^9 - x^8}$

124. $f(x) = \sqrt[4]{16x^4 + 16x^5} - 2\sqrt[4]{x^8 + x^9}$

Simplify.

125. $\frac{1}{2}\sqrt{36a^5bc^4} - \frac{1}{2}\sqrt[3]{64a^4bc^6} + \frac{1}{6}\sqrt{144a^3bc^6}$

126. $7x\sqrt{(x+y)^3} - 5xy\sqrt{x+y} - 2y\sqrt{(x+y)^3}$

127. $\sqrt{27a^5(b+1)}\,\sqrt[3]{81a(b+1)^4}$

128. $\sqrt{8x(y+z)^5}\,\sqrt[3]{4x^2(y+z)^2}$

129. $\dfrac{\dfrac{1}{\sqrt{w}} - \sqrt{w}}{\dfrac{\sqrt{w}+1}{\sqrt{w}}}$

130. $\dfrac{1}{4+\sqrt{3}} + \dfrac{1}{\sqrt{3}} + \dfrac{1}{\sqrt{3}-4}$

Express each of the following as the product of two radical expressions.

131. $x - 5$

132. $y - 7$

133. $x - a$

Multiply.

134. $\sqrt{9 + 3\sqrt{5}}\,\sqrt{9 - 3\sqrt{5}}$

135. $\left(\sqrt{x+2} - \sqrt{x-2}\right)^2$

136. Use a graphing calculator to check your answers to Exercises 25, 39, and 81.

7.6

Solving Radical Equations

The Principle of Powers • Equations with Two Radical Terms

CONNECTING THE CONCEPTS

In Sections 7.1–7.5, we learned how to manipulate radical expressions as well as expressions containing rational exponents. We performed this work to find *equivalent expressions*.

Now that we know how to work with radicals and rational exponents, we can learn how to solve a new type of equation. As in our earlier work with

equations, finding *equivalent equations* will be part of our strategy. What is different, however, is that now we will use an equation-solving step that does not always produce equivalent equations. Checking solutions will therefore be more important than ever.

The Principle of Powers

A **radical equation** is an equation in which the variable appears in a radicand. Examples are

$$\sqrt[3]{2x} + 1 = 5, \quad \sqrt{a} + \sqrt{a - 2} = 7, \quad \text{and} \quad 4 - \sqrt{3x + 1} = \sqrt{6 - x}.$$

To solve such equations, we need a new principle. Suppose $a = b$ is true. If we square both sides, we get another true equation: $a^2 = b^2$. This can be generalized.

> **The Principle of Powers**
>
> If $a = b$, then $a^n = b^n$ for any exponent n.

Note that the principle of powers is an "if–then" statement. The statement obtained by interchanging the two parts of the sentence—"if $a^n = b^n$ for some exponent n, then $a = b$"—*is not always true*. For example, "if $x = 3$, then $x^2 = 9$" is true, but the statement "if $x^2 = 9$, then $x = 3$" is *not* true when x is replaced with -3. For this reason, when both sides of an equation are raised to an even exponent, it is essential to check the answer(s) in the *original* equation.

EXAMPLE 1   Solve: $\sqrt{x} - 3 = 4$.

Solution   Before using the principle of powers, we need to isolate the radical term:

$$
\begin{aligned}
\sqrt{x} - 3 &= 4 \\
\sqrt{x} &= 7 \qquad &&\text{Isolating the radical by adding 3 to both sides} \\
(\sqrt{x})^2 &= 7^2 \qquad &&\text{Using the principle of powers} \\
x &= 49.
\end{aligned}
$$

Check: $$\dfrac{\sqrt{x} - 3 = 4}{\sqrt{49} - 3 \mid 4}$$
$$7 - 3 \mid$$
$$4 \overset{?}{=} 4 \quad \text{TRUE}$$

The solution is 49.

EXAMPLE 2   Solve: $\sqrt{x} - 5 = -7$.

Solution

$$\sqrt{x} - 5 = -7$$
$$\sqrt{x} = -2 \qquad \text{Isolating the radical by adding 5 to both sides}$$

> The equation $\sqrt{x} = -2$ has no solution because the principal square root of a number is never negative. We continue as in Example 1 for comparison.

$$\left(\sqrt{x}\right)^2 = (-2)^2 \qquad \text{Using the principle of powers}$$
$$x = 4$$

Check: $$\dfrac{\sqrt{x} - 5 = -7}{\sqrt{4} - 5 \mid -7}$$
$$2 - 5 \mid$$
$$-3 \overset{?}{=} -7 \quad \text{FALSE}$$

The number 4 does not check. Thus $\sqrt{x} - 5 = -7$ has no solution.

> **Caution!**   Raising both sides of an equation to an even power may not produce an equivalent equation. In this case, a check is essential.

Note in Example 2 that $x = 4$ has solution 4, but that $\sqrt{x} - 5 = -7$ has *no* solution. Thus the equations $x = 4$ and $\sqrt{x} - 5 = -7$ are *not* equivalent.

> *To Solve an Equation with a Radical Term*
>
> **1.** Isolate the radical term on one side of the equation.
> **2.** Use the principle of powers and solve the resulting equation.
> **3.** Check any possible solution in the original equation.

EXAMPLE 3   Solve: $x = \sqrt{x + 7} + 5$.

Solution

$$x = \sqrt{x + 7} + 5$$
$$x - 5 = \sqrt{x + 7}$$ Isolating the radical by subtracting 5 from both sides

$$\left.\begin{array}{l}(x - 5)^2 = \left(\sqrt{x + 7}\right)^2 \\ x^2 - 10x + 25 = x + 7\end{array}\right\}$$ Using the principle of powers; squaring both sides

$$x^2 - 11x + 18 = 0$$ Adding $-x - 7$ to both sides to write the quadratic equation in standard form

$$(x - 9)(x - 2) = 0$$ Factoring

$$x = 9 \quad or \quad x = 2$$ Using the principle of zero products

The possible solutions are 9 and 2. Let's check.

Check:

For 9:

$$\begin{array}{c|c} x = \sqrt{x + 7} + 5 \\ \hline 9 & \sqrt{9 + 7} + 5 \\ 9 \overset{?}{=} 9 \end{array}$$ TRUE

For 2:

$$\begin{array}{c|c} x = \sqrt{x + 7} + 5 \\ \hline 2 & \sqrt{2 + 7} + 5 \\ 2 \overset{?}{=} 8 \end{array}$$ FALSE

Since 9 checks but 2 does not, the solution is 9.

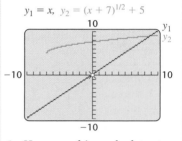

It is important to isolate a radical term before using the principle of powers. Suppose in Example 3 that both sides of the equation were squared *before* isolating the radical. We then would have had the expression $\left(\sqrt{x + 7} + 5\right)^2$ or $x + 7 + 10\sqrt{x + 7} + 25$ on the right side, and the radical would have remained in the problem.

EXAMPLE 4   Solve: $(2x + 1)^{1/3} + 5 = 0$.

Solution   We need not use radical notation to solve:

$$(2x + 1)^{1/3} + 5 = 0$$
$$(2x + 1)^{1/3} = -5$$ Subtracting 5 from both sides
$$[(2x + 1)^{1/3}]^3 = (-5)^3$$ Cubing both sides
$$(2x + 1)^1 = (-5)^3$$ Multiplying exponents. Try to do this mentally.
$$2x + 1 = -125$$
$$2x = -126$$ Subtracting 1 from both sides
$$x = -63.$$

Because both sides were raised to an *odd* power, it is not *essential* that we check the answer. Students can confirm that -63 checks and is the solution.

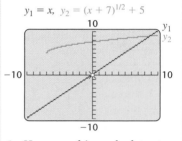

technology connection

To solve Example 3, we can graph $y_1 = x$ and $y_2 = (x + 7)^{1/2} + 5$ and then use the INTERSECT option of the CALC menu to find the point of intersection. The intersection occurs at $x = 9$. Note that there is no intersection when $x = 2$, as predicted in the check of Example 3.

$y_1 = x, \ y_2 = (x + 7)^{1/2} + 5$

1. Use a graphing calculator to solve Examples 1, 2, 4, 5, and 6. Compare your answers with those found using the algebraic methods shown.

Equations with Two Radical Terms

A strategy for solving equations with two or more radical terms is as follows.

To Solve an Equation with Two or More Radical Terms

1. Isolate one of the radical terms.
2. Use the principle of powers.
3. If a radical remains, perform steps (1) and (2) again.
4. Solve the resulting equation.
5. Check possible solutions in the original equation.

EXAMPLE 5   Solve: $\sqrt{2x - 5} = 1 + \sqrt{x - 3}$.

Solution

$$\sqrt{2x - 5} = 1 + \sqrt{x - 3}$$

$$\left(\sqrt{2x - 5}\right)^2 = \left(1 + \sqrt{x - 3}\right)^2 \quad \text{One radical is already isolated.}$$
$$\text{We square both sides.}$$

> This is like squaring a binomial. We square 1, then find twice the product of 1 and $\sqrt{x - 3}$, and then the square of $\sqrt{x - 3}$. Study this carefully.

$$2x - 5 = 1 + 2\sqrt{x - 3} + \left(\sqrt{x - 3}\right)^2$$

$$2x - 5 = 1 + 2\sqrt{x - 3} + (x - 3)$$

$$x - 3 = 2\sqrt{x - 3} \qquad \text{Isolating the remaining radical term}$$

$$(x - 3)^2 = \left(2\sqrt{x - 3}\right)^2 \qquad \text{Squaring both sides}$$

$$x^2 - 6x + 9 = 4(x - 3) \qquad \text{Remember to square both the 2 and the } \sqrt{x - 3} \text{ on the right side.}$$

$$x^2 - 6x + 9 = 4x - 12$$

$$x^2 - 10x + 21 = 0$$

$$(x - 7)(x - 3) = 0 \qquad \text{Factoring}$$

$$x = 7 \quad or \quad x = 3 \qquad \text{Using the principle of zero products}$$

We leave it to the student to show that 7 and 3 both check and are the solutions.

Caution!   A common error in solving equations like

$$\sqrt{2x - 5} = 1 + \sqrt{x - 3}$$

is to obtain $1 + (x - 3)$ as the square of the right side. This is wrong because $(A + B)^2 \neq A^2 + B^2$. For example,

$$\left.\begin{array}{c} (1 + 2)^2 \neq 1^2 + 2^2 \\ 3^2 \neq 1 + 4 \\ 9 \neq 5. \end{array}\right\} \quad \text{See Example 5 for the correct expansion of } \left(1 + \sqrt{x - 3}\right)^2.$$

EXAMPLE 6   Let $f(x) = \sqrt{x + 5} - \sqrt{x - 7}$. Find all x-values for which $f(x) = 2$.

Solution   We must have $f(x) = 2$, or

$$\sqrt{x + 5} - \sqrt{x - 7} = 2. \qquad \text{Substituting for } f(x)$$

To solve, we isolate one radical term and square both sides:

$$\sqrt{x + 5} = 2 + \sqrt{x - 7} \qquad \begin{array}{l}\text{Adding } \sqrt{x - 7} \text{ to both sides.}\\ \text{This isolates one of the}\\ \text{radical terms.}\end{array}$$

$$(\sqrt{x + 5})^2 = (2 + \sqrt{x - 7})^2 \qquad \begin{array}{l}\text{Using the principle of powers}\\ \text{(squaring both sides)}\end{array}$$

$$x + 5 = 4 + 4\sqrt{x - 7} + (x - 7) \qquad \begin{array}{l}\text{Using}\\ (A + B)^2 = A^2 + 2AB + B^2\end{array}$$

$$5 = 4\sqrt{x - 7} - 3 \qquad \begin{array}{l}\text{Adding } -x \text{ to both sides and}\\ \text{combining like terms}\end{array}$$

$$8 = 4\sqrt{x - 7} \qquad \begin{array}{l}\text{Isolating the remaining}\\ \text{radical term}\end{array}$$

$$2 = \sqrt{x - 7}$$
$$2^2 = (\sqrt{x - 7})^2 \qquad \text{Squaring both sides}$$
$$4 = x - 7$$
$$11 = x.$$

Student Notes

Be careful when checking answers. You don't want to discard a correct answer because of a careless mistake.

Check: $f(11) = \sqrt{11 + 5} - \sqrt{11 - 7}$
$$= \sqrt{16} - \sqrt{4}$$
$$= 4 - 2 = 2.$$

We have $f(x) = 2$ when $x = 11$.

Exercise Set

7.6

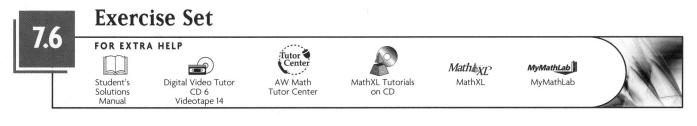

FOR EXTRA HELP

Student's Solutions Manual   Digital Video Tutor CD 6 Videotape 14   AW Math Tutor Center   MathXL Tutorials on CD   MathXL   MyMathLab

↪ *Concept Reinforcement*   *Classify each of the following as either true or false.*

1. If $x^2 = 25$, then $x = 5$.

2. If $t = 7$, then $t^2 = 49$.

3. If $\sqrt{x} = 3$, then $(\sqrt{x})^2 = 3^2$.

4. If $x^2 = 36$, then $x = 6$.

5. $\sqrt{x} - 8 = 7$ is equivalent to $\sqrt{x} = 15$.

6. $\sqrt{t} + 5 = 8$ is equivalent to $\sqrt{t} = 3$.

Solve.

7. $\sqrt{5x - 2} = 7$

8. $\sqrt{3x - 2} = 6$

9. $\sqrt{3x} + 1 = 6$

10. $\sqrt{2x} - 1 = 2$

11. $\sqrt{y + 1} - 5 = 8$

12. $\sqrt{x - 2} - 7 = -4$

13. $\sqrt{x - 7} + 3 = 10$

14. $\sqrt{y + 4} + 6 = 7$

15. $\sqrt[3]{x + 5} = 2$

16. $\sqrt[3]{x - 2} = 3$

17. $\sqrt[4]{y - 1} = 3$

18. $\sqrt[4]{x + 3} = 2$

19. $3\sqrt{x} = x$

20. $8\sqrt{y} = y$

21. $2y^{1/2} - 7 = 9$

22. $3x^{1/2} + 12 = 9$

23. $\sqrt[3]{x} = -3$

24. $\sqrt[3]{y} = -4$

25. $t^{1/3} - 2 = 3$

26. $x^{1/4} - 2 = 1$

Aha! **27.** $(y - 3)^{1/2} = -2$

28. $(x + 2)^{1/2} = -4$

29. $\sqrt[4]{3x + 1} - 4 = -1$

30. $\sqrt[4]{2x + 3} - 5 = -2$

31. $(x + 7)^{1/3} = 4$

32. $(y - 7)^{1/4} = 3$

33. $\sqrt[3]{3y + 6} + 7 = 8$

34. $\sqrt[3]{6x + 9} + 5 = 2$

35. $\sqrt{3t + 4} = \sqrt{4t + 3}$

36. $\sqrt{2t - 7} = \sqrt{3t - 12}$

37. $3(4 - t)^{1/4} = 6^{1/4}$

38. $2(1 - x)^{1/3} = 4^{1/3}$

39. $3 + \sqrt{5 - x} = x$

40. $x = \sqrt{x - 1} + 3$

41. $\sqrt{4x - 3} = 2 + \sqrt{2x - 5}$

42. $3 + \sqrt{z - 6} = \sqrt{z + 9}$

43. $\sqrt{20 - x} + 8 = \sqrt{9 - x} + 11$

44. $4 + \sqrt{10 - x} = 6 + \sqrt{4 - x}$

45. $\sqrt{x + 2} + \sqrt{3x + 4} = 2$

46. $\sqrt{6x + 7} - \sqrt{3x + 3} = 1$

47. If $f(x) = \sqrt{x} + \sqrt{x - 9}$, find any x for which $f(x) = 1$.

48. If $g(x) = \sqrt{x} + \sqrt{x - 5}$, find any x for which $g(x) = 5$.

49. If $f(t) = \sqrt{t - 2} - \sqrt{4t + 1}$, find any t for which $f(t) = -3$.

50. If $g(t) = \sqrt{2t + 7} - \sqrt{t + 15}$, find any t for which $g(t) = -1$.

51. If $f(x) = \sqrt{2x - 3}$ and $g(x) = \sqrt{x + 7} - 2$, find any x for which $f(x) = g(x)$.

52. If $f(x) = 2\sqrt{3x + 6}$ and $g(x) = 5 + \sqrt{4x + 9}$, find any x for which $f(x) = g(x)$.

53. If $f(t) = 4 - \sqrt{t - 3}$ and $g(t) = (t + 5)^{1/2}$, find any t for which $f(t) = g(t)$.

54. If $f(t) = 7 + \sqrt{2t - 5}$ and $g(t) = 3(t + 1)^{1/2}$, find any t for which $f(t) = g(t)$.

55. Explain in your own words why it is important to check your answers when using the principle of powers.

56. The principle of powers is an "if–then" statement that becomes false when the sentence parts are interchanged. Give an example of another such if–then statement from everyday life (answers will vary).

SKILL MAINTENANCE

57. The base of a triangle is 2 in. longer than the height. The area is $31\frac{1}{2}$ in^2. Find the height and the base. [5.8]

58. During a one-hour television show, there were 12 commercials. Some of the commercials were 30 sec long and the others were 60 sec long. If the number of 30-sec commercials was 6 less than the total number of minutes of commercial time during the show, how many 60-sec commercials were used? [3.3]

Graph.

59. $f(x) = \frac{2}{3}x - 5$ [2.3]

60. $g(x) = 3x + 4$ [2.3]

61. $F(x) < -2x + 4$ [4.4]

62. $G(x) > -3x + 2$ [4.4]

SYNTHESIS

63. Describe a procedure that could be used to create radical equations that have no solution.

64. Is checking essential when the principle of powers is used with an odd power n? Why or why not?

Steel manufacturing. *In the production of steel and other metals, the temperature of the molten metal is so great that conventional thermometers melt. Instead, sound is transmitted across the surface of the metal to a receiver on the far side and the speed of the sound is measured. The formula*

$$S(t) = 1087.7\sqrt{\frac{9t + 2617}{2457}}$$

gives the speed of sound $S(t)$, in feet per second, at a temperature of t degrees Celsius.

Sighting to the horizon.   *The function $D(h) = 1.2\sqrt{h}$ can be used to approximate the distance D, in miles, that a person can see to the horizon from a height h, in feet.*

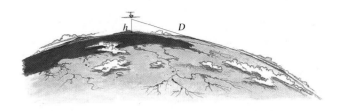

72. How far above sea level must a pilot fly in order to see a horizon that is 180 mi away?

73. How high above sea level must a sailor climb in order to see 10.2 mi out to sea?

Solve.

74. $\left(\dfrac{z}{4} - 5\right)^{2/3} = \dfrac{1}{25}$

75. $\dfrac{x + \sqrt{x + 1}}{x - \sqrt{x + 1}} = \dfrac{5}{11}$

76. $\sqrt{\sqrt{y} + 49} = 7$

77. $(z^2 + 17)^{3/4} = 27$

78. $x^2 - 5x - \sqrt{x^2 - 5x - 2} = 4$
(*Hint*: Let $u = x^2 - 5x - 2$.)

79. $\sqrt{8 - b} = b\sqrt{8 - b}$

Without graphing, determine the x-intercepts of the graphs given by each of the following.

80. $f(x) = \sqrt{x - 2} - \sqrt{x + 2} + 2$

81. $g(x) = 6x^{1/2} + 6x^{-1/2} - 37$

82. $f(x) = (x^2 + 30x)^{1/2} - x - (5x)^{1/2}$

83. Use a graphing calculator to check your answers to Exercises 9, 15, and 31.

84. Saul is trying to solve Exercise 73 using a graphing calculator. Without resorting to trial and error, how can he determine a suitable viewing window for finding the solution?

85. Use a graphing calculator to check your answers to Exercises 27, 35, and 41.

65. Find the temperature of a blast furnace where sound travels 1880 ft/sec.

66. Find the temperature of a blast furnace where sound travels 1502.3 ft/sec.

67. Solve the above equation for *t*.

Automotive repair.   *For an engine with a displacement of 2.8 L, the function given by*

$$d(n) = 0.75\sqrt{2.8n}$$

can be used to determine the diameter size of the carburetor's opening, in millimeters. Here n is the number of rpm's at which the engine achieves peak performance.
Source: macdizzy.com

68. If a carburetor's opening is 81 mm, for what number of rpm's will the engine produce peak power?

69. If a carburetor's opening is 84 mm, for what number of rpm's will the engine produce peak power?

Escape velocity.   *A formula for the escape velocity v of a satellite is*

$$v = \sqrt{2gr}\,\sqrt{\dfrac{h}{r + h}},$$

where g is the force of gravity, r is the planet or star's radius, and h is the height of the satellite above the planet or star's surface.

70. Solve for *h*.

71. Solve for *r*.

CORNER

Tailgater Alert

Focus: Radical equations and problem solving

Time: 15–25 minutes

Group size: 2–3

Materials: Calculators or square-root tables

The faster a car is traveling, the more distance it needs to stop. Thus it is important for drivers to allow sufficient space between their vehicle and the vehicle in front of them. Police recommend that for each 10 mph of speed, a driver allow 1 car length. Thus a driver going 30 mph should have at least 3 car lengths between his or her vehicle and the one in front.

In Exercise Set 7.3, the function $r(L) = 2\sqrt{5L}$ was used to find the speed, in miles per hour, that a car was traveling when it left skid marks L feet long.

ACTIVITY

1. Each group member should estimate the length of a car in which he or she frequently travels. (Each should use a different length, if possible.)
2. Using a calculator as needed, each group member should complete the table below.

Column 1 gives a car's speed s, and column 2 lists the minimum amount of space between cars traveling s miles per hour, as recommended by police. Column 3 is the speed that a vehicle *could* travel were it forced to stop in the distance listed in column 2, using the above function.

Column 1 s (in miles per hour)	Column 2 $L(s)$ (in feet)	Column 3 $r(L)$ (in miles per hour)
20		
30		
40		
50		
60		
70		

3. Determine whether there are any speeds at which the "1 car length per 10 mph" guideline might not suffice. On what reasoning do you base your answer? Compare tables to determine how car length affects the results. What recommendations would your group make to a new driver?

7.7 Geometric Applications

Using the Pythagorean Theorem • Two Special Triangles

Using the Pythagorean Theorem

There are many kinds of problems that involve powers and roots. Many also involve right triangles and the Pythagorean theorem, which we studied in Section 5.8 and restate here.

Study Skills _____

Making Sketches

One need not be an artist to make highly useful mathematical sketches. That said, it is important to make sure that your sketches are drawn accurately enough to represent the relative sizes within each shape. For example, if one side of a triangle is clearly the longest, make sure your drawing reflects this.

The Pythagorean Theorem*

In any right triangle, if a and b are the lengths of the legs and c is the length of the hypotenuse, then

$$a^2 + b^2 = c^2.$$

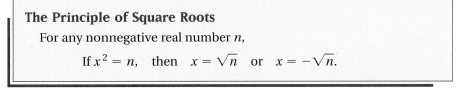

In using the Pythagorean theorem, we often make use of the following principle.

The Principle of Square Roots

For any nonnegative real number n,

$$\text{If } x^2 = n, \quad \text{then} \quad x = \sqrt{n} \quad \text{or} \quad x = -\sqrt{n}.$$

For most real-world applications involving length or distance, $-\sqrt{n}$ is not needed.

EXAMPLE 1

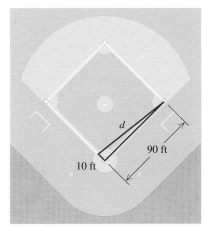

Baseball. A baseball diamond is actually a square 90 ft on a side. Suppose a catcher fields a ball while standing on the third-base line 10 ft from home plate. How far is the catcher's throw to first base? Give an exact answer and an approximation to three decimal places.

Solution We make a drawing and let d = the distance, in feet, to first base. Note that a right triangle is formed in which the leg from home plate to first base measures 90 ft and the leg from home plate to where the catcher fields the ball measures 10 ft.

We substitute these values into the Pythagorean theorem to find d:

$$d^2 = 90^2 + 10^2$$
$$d^2 = 8100 + 100$$
$$d^2 = 8200.$$

We now use the principle of square roots: If $d^2 = 8200$, then $d = \sqrt{8200}$ or $d = -\sqrt{8200}$. Since d represents a length, it follows that d is the positive square root of 8200:

$$d = \sqrt{8200} \text{ ft} \qquad \text{This is an exact answer.}$$
$$d \approx 90.554 \text{ ft.} \qquad \text{Using a calculator for an approximation}$$

---

*The converse of the Pythagorean theorem also holds. That is, if a, b, and c are the lengths of the sides of a triangle and $a^2 + b^2 = c^2$, then the triangle is a right triangle.

EXAMPLE 2   Guy wires.   The base of a 40-ft-long guy wire is located 15 ft from the telephone pole that it is anchoring. How high up the pole does the guy wire reach? Give an exact answer and an approximation to three decimal places.

Solution   We make a drawing and let h = the height, in feet, to which the guy wire reaches. A right triangle is formed in which one leg measures 15 ft and the hypotenuse measures 40 ft. Using the Pythagorean theorem, we have

$$h^2 + 15^2 = 40^2$$
$$h^2 + 225 = 1600$$
$$h^2 = 1375$$
$$h = \sqrt{1375}.$$

Exact answer:
$$h = \sqrt{1375} \text{ ft}$$

Approximation:

$$h \approx 37.081 \text{ ft}$$ Using a calculator

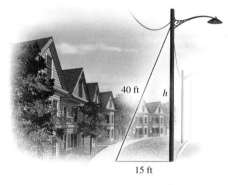

Two Special Triangles

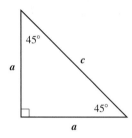

When both legs of a right triangle are the same size, we call the triangle an *isosceles right triangle,* as shown at left. If one leg of an isosceles right triangle has length a, we can find a formula for the length of the hypotenuse as follows:

$$c^2 = a^2 + b^2$$
$$c^2 = a^2 + a^2$$ Because the triangle is isosceles, both legs are the same size: $a = b$.
$$c^2 = 2a^2.$$ Combining like terms

Next, we use the principle of square roots. Because a, b, and c are lengths, there is no need to consider negative square roots or absolute values. Thus,

$$c = \sqrt{2a^2}$$ Using the principle of square roots
$$c = \sqrt{a^2 \cdot 2} = a\sqrt{2}.$$

EXAMPLE 3   One leg of an isosceles right triangle measures 7 cm. Find the length of the hypotenuse. Give an exact answer and an approximation to three decimal places.

Solution   We substitute:

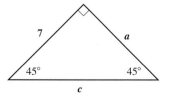

$$c = a\sqrt{2}$$ This equation is worth memorizing.
$$c = 7\sqrt{2}.$$

Exact answer: $c = 7\sqrt{2}$ cm
Approximation: $c \approx 9.899$ cm   Using a calculator

When the hypotenuse of an isosceles right triangle is known, the lengths of the legs can be found.

EXAMPLE 4   The hypotenuse of an isosceles right triangle is 5 ft long. Find the length of a leg. Give an exact answer and an approximation to three decimal places.

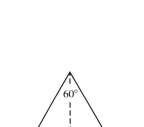

Solution   We replace c with 5 and solve for a:

$$5 = a\sqrt{2} \qquad \text{Substituting 5 for } c \text{ in } c = a\sqrt{2}$$

$$\frac{5}{\sqrt{2}} = a \qquad \text{Dividing both sides by } \sqrt{2}$$

$$\frac{5\sqrt{2}}{2} = a. \qquad \text{Rationalize the denominator if desired.}$$

Exact answer: $a = \dfrac{5}{\sqrt{2}}$ ft, or $\dfrac{5\sqrt{2}}{2}$ ft

Approximation: $a \approx 3.536$ ft   Using a calculator

A second special triangle is known as a 30°–60°–90° right triangle, so named because of the measures of its angles. Note that in an equilateral triangle, all sides have the same length and all angles are 60°. An altitude, drawn dashed in the figure, bisects, or splits in half, one angle and one side. Two 30°–60°–90° right triangles are thus formed.

Because of the way in which the altitude is drawn, if a represents the length of the shorter leg in a 30°–60°–90° right triangle, then $2a$ represents the length of the hypotenuse. We have

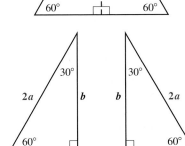

$$a^2 + b^2 = (2a)^2 \qquad \text{Using the Pythagorean theorem}$$
$$a^2 + b^2 = 4a^2$$
$$b^2 = 3a^2 \qquad \text{Subtracting } a^2 \text{ from both sides}$$
$$b = \sqrt{3a^2}$$
$$b = \sqrt{a^2 \cdot 3}$$
$$b = a\sqrt{3}.$$

EXAMPLE 5   The shorter leg of a 30°–60°–90° right triangle measures 8 in. Find the lengths of the other sides. Give exact answers and, where appropriate, an approximation to three decimal places.

Solution   The hypotenuse is twice as long as the shorter leg, so we have

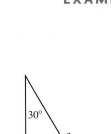

$$c = 2a \qquad\qquad \text{This relationship is worth memorizing.}$$
$$= 2 \cdot 8 = 16 \text{ in.} \qquad \text{This is the length of the hypotenuse.}$$

The length of the longer leg is the length of the shorter leg times $\sqrt{3}$. This gives us

$$b = a\sqrt{3} \qquad\qquad \text{This is also worth memorizing.}$$
$$= 8\sqrt{3} \text{ in.} \qquad \text{This is the length of the longer leg.}$$

Exact answer: $c = 16$ in., $b = 8\sqrt{3}$ in.

Approximation: $b \approx 13.856$ in.

EXAMPLE 6   The length of the longer leg of a 30°–60°–90° right triangle is 14 cm. Find the length of the hypotenuse. Give an exact answer and an approximation to three decimal places.

Solution   The length of the hypotenuse is twice the length of the shorter leg. We first find a, the length of the shorter leg, by using the length of the longer leg:

$$14 = a\sqrt{3} \qquad \text{Substituting 14 for } b \text{ in } b = a\sqrt{3}$$

$$\frac{14}{\sqrt{3}} = a. \qquad \text{Dividing by } \sqrt{3}$$

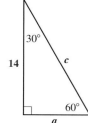

Since the hypotenuse is twice as long as the shorter leg, we have

$$c = 2a$$

$$= 2 \cdot \frac{14}{\sqrt{3}} \qquad \text{Substituting}$$

$$= \frac{28}{\sqrt{3}} \text{ cm.}$$

Exact answer: $c = \dfrac{28}{\sqrt{3}}$ cm, or $\dfrac{28\sqrt{3}}{3}$ cm if the denominator is rationalized.

Approximation: $c \approx 16.166$ cm

Student Notes

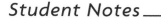

Perhaps the easiest way to remember the important results listed in the adjacent box is to write out, on your own, the derivations shown on pp. 484 and 485.

Lengths Within Isosceles and 30°–60°–90° Right Triangles

The length of the hypotenuse in an isosceles right triangle is the length of a leg times $\sqrt{2}$.

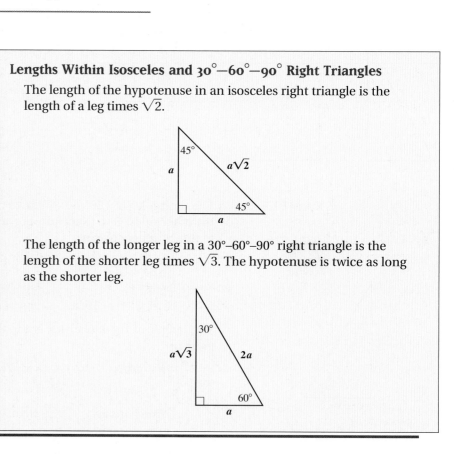

The length of the longer leg in a 30°–60°–90° right triangle is the length of the shorter leg times $\sqrt{3}$. The hypotenuse is twice as long as the shorter leg.

Exercise Set

7.7

FOR EXTRA HELP

Student's Solutions Manual | Digital Video Tutor CD 6 Videotape 14 | AW Math Tutor Center | MathXL Tutorials on CD | MathXL MathXL | MyMathLab MyMathLab

↪ *Concept Reinforcement*   *Complete each of the following sentences.*

1. In any _____ triangle, the square of the length of the _____ is the sum of the squares of the lengths of the legs.

2. The shortest side of a right triangle is always one of the two _____.

3. The principle of _____ _____ states that if $x^2 = n$, then $x = \sqrt{n}$ or $x = -\sqrt{n}$.

4. In a(n) _____ right triangle, both legs have the same length.

5. In a(n) ____–____–____ right triangle, the hypotenuse is twice as long as the shorter _____.

6. If both legs have measure a, then the hypotenuse measures _____.

In a right triangle, find the length of the side not given. Give an exact answer and, where appropriate, an approximation to three decimal places.

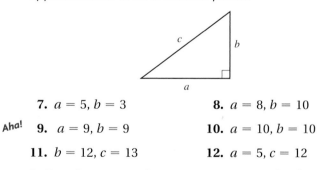

7. $a = 5, b = 3$

8. $a = 8, b = 10$

Aha! 9. $a = 9, b = 9$

10. $a = 10, b = 10$

11. $b = 12, c = 13$

12. $a = 5, c = 12$

In Exercises 13–18, give an exact answer and, where appropriate, an approximation to three decimal places.

13. A right triangle's hypotenuse is 8 m and one leg is $4\sqrt{3}$ m. Find the length of the other leg.

14. A right triangle's hypotenuse is 6 cm and one leg is $\sqrt{5}$ cm. Find the length of the other leg.

15. The hypotenuse of a right triangle is $\sqrt{20}$ in. and one leg measures 1 in. Find the length of the other leg.

16. The hypotenuse of a right triangle is $\sqrt{15}$ ft and one leg measures 2 ft. Find the length of the other leg.

Aha! 17. One leg in a right triangle is 1 m and the hypotenuse measures $\sqrt{2}$ m. Find the length of the other leg.

18. One leg of a right triangle is 1 yd and the hypotenuse measures 2 yd. Find the length of the other leg.

In Exercises 19–28, give an exact answer and, where appropriate, an approximation to three decimal places.

19. *Bicycling.*   Clare routinely bicycles across a rectangular parking lot on her way to class. If the lot is 200 ft long and 150 ft wide, how far does Clare travel when she rides across the lot diagonally?

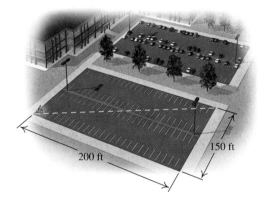

150 ft

200 ft

20. *Guy wire.*   How long is a guy wire if it reaches from the top of a 15-ft pole to a point on the ground 10 ft from the pole?

21. *Softball.*   A slow-pitch softball diamond is actually a square 65 ft on a side. How far is it from home plate to second base?

22. *Baseball.*   Suppose the catcher in Example 1 makes a throw to second base from the same location. How far is that throw?

23. *Television sets.* What does it mean to refer to a 20-in. TV set or a 25-in. TV set? Such units refer to the diagonal of the screen. A 20-in. TV set has a width of 16 in. What is its height?

24. *Television sets.* A 25-in. TV set has a screen with a height of 15 in. What is its width? (See Exercise 23.)

25. *Speaker placement.* A stereo receiver is in a corner of a 12-ft by 14-ft room. Speaker wire will run under a rug, diagonally, to a speaker in the far corner. If 4 ft of slack is required on each end, how long a piece of wire should be purchased?

26. *Distance over water.* To determine the width of a pond, a surveyor locates two stakes at either end of the pond and uses instrumentation to place a third stake so that the distance across the pond is the length of a hypotenuse. If the third stake is 90 m from one stake and 70 m from the other, what is the distance across the pond?

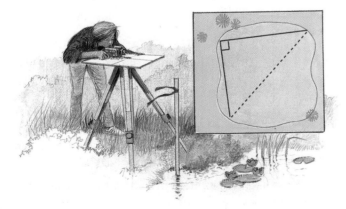

27. *Walking.* Students at Compkin Community College have worn a path that cuts diagonally across the campus "quad." If the quad is actually a rectangle that Marissa measured to be 70 paces

long and 40 paces wide, how many paces will Marissa save by using the diagonal path?

28. *Crosswalks.* The diagonal crosswalk at the intersection of State St. and Main St. is the hypotenuse of a triangle in which the crosswalks across State St. and Main St. are the legs. If State St. is 28 ft wide and Main St. is 40 ft wide, how much shorter is the distance traveled by pedestrians using the diagonal crosswalk?

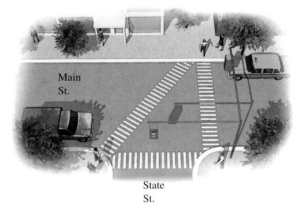

For each triangle, find the missing length(s). Give an exact answer and, where appropriate, an approximation to three decimal places.

29.

45°
5
?
?

30.

?
?
45°
14

31.

14
30°
?
?

32.

?
60°
18
?

33.

?
60°
?
15

34.

?
?
45°
8

35.

?
?
45°
13

36.

7
30°
?
?

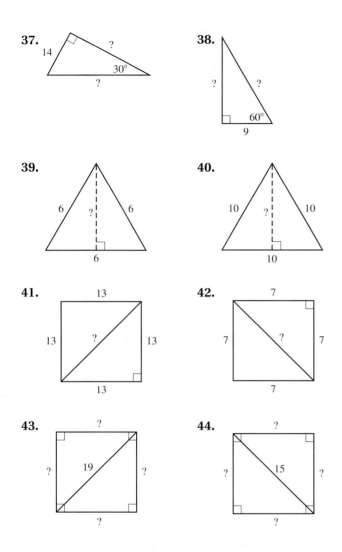

37.

38.

39.

40.

41.

42.

43.

44.

In Exercises 45–50, give an exact answer and, where appropriate, an approximation to three decimal places.

45. *Bridge expansion.*   During the summer heat, a 2-mi bridge expands 2 ft in length. If we assume that the bulge occurs straight up the middle, how high is the bulge? (The answer may surprise you. Most bridges have expansion spaces to avoid such buckling.)

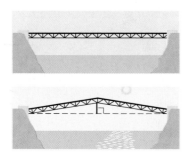

46. Triangle *ABC* has sides of lengths 25 ft, 25 ft, and 30 ft. Triangle *PQR* has sides of lengths 25 ft, 25 ft, and 40 ft. Which triangle, if either, has the greater area and by how much?

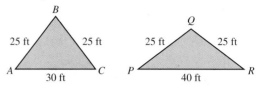

47. *Camping tent.*   The entrance to a pup tent is the shape of an equilateral triangle. If the base of the tent is 4 ft wide, how tall is the tent?

48. The length and the width of a rectangle are given by consecutive integers. The area of the rectangle is 90 cm^2. Find the length of a diagonal of the rectangle.

49. Find all points on the *y*-axis of a Cartesian coordinate system that are 5 units from the point (3, 0).

50. Find all points on the *x*-axis of a Cartesian coordinate system that are 5 units from the point (0, 4).

51. Write a problem for a classmate to solve in which the solution is: "The height of the tepee is $5\sqrt{3}$ yd."

52. Write a problem for a classmate to solve in which the solution is: "The height of the window is $15\sqrt{3}$ ft."

SKILL MAINTENANCE

Simplify. [1.6]

53. $47(-1)^{19}$

54. $(-5)(-1)^{13}$

Factor. [5.5]

55. $x^3 - 9x$

56. $7a^3 - 28a$

Solve. [4.3]

57. $|3x - 5| = 7$

58. $|2x - 3| = |x + 7|$

SYNTHESIS

59. Are there any right triangles, other than those with sides measuring 3, 4, and 5, that have consecutive numbers for the lengths of the sides? Why or why not?

60. If a 30°–60°–90° triangle and an isosceles right triangle have the same perimeter, which will have the greater area? Why?

61. The perimeter of a regular hexagon is 72 cm. Determine the area of the shaded region shown.

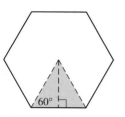

62. If the perimeter of a regular hexagon is 120 ft, what is its area? (*Hint*: See Exercise 61.)

63. Each side of a regular octagon has length s. Find a formula for the distance d between the parallel sides of the octagon.

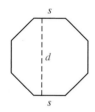

64. *Roofing.* Kit's home, which is 24 ft wide and 32 ft long, needs a new roof. By counting clapboards that are 4 in. apart, Kit determines that the peak of the roof is 6 ft higher than the sides. If one packet of shingles covers 100 square feet, how many packets will the job require?

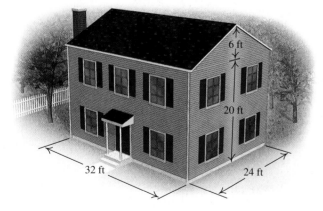

65. *Painting.* (Refer to Exercise 64.) A gallon of Benjamin Moore® exterior acrylic paint covers 450–500 square feet. If Kit's house has dimensions as shown above, how many gallons of paint should be bought to paint the house? What assumption(s) is made in your answer?

66. *Contracting.* Oxford Builders has an extension cord on their generator that permits them to work, with electricity, anywhere in a circular area of 3850 ft². Find the dimensions of the largest square room they could work on without having to relocate the generator to reach each corner of the floor plan.

67. *Contracting.* Cleary Construction has a hose attached to their insulation blower that permits them to work, with electricity, anywhere in a circular area of 6160 ft². Find the dimensions of the largest square room with 12-ft ceilings in which they could reach all corners with the hose while leaving the blower centrally located. Assume that the blower sits on the floor.

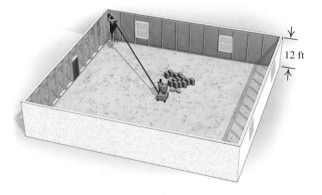

68. A cube measures 5 cm on each side. How long is the diagonal that connects two opposite corners of the cube? Give an exact answer.

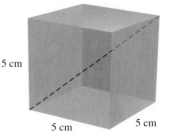

7.8   The Complex Numbers

Imaginary and Complex Numbers • Addition and Subtraction • Multiplication • Conjugates and Division • Powers of i

Study Skills

Studying Together by Phone

Working with a classmate over the telephone can be a very effective way to receive or give help. The fact that you cannot point to figures on paper forces you to verbalize the mathematics and the act of speaking mathematics will improve your understanding of the material. In some cases it may be more effective to study with a classmate over the phone than in person.

Imaginary and Complex Numbers

Negative numbers do not have square roots in the real-number system. However, a larger number system that contains the real-number system is designed so that negative numbers *do* have square roots. That system is called the **complex-number system**, and it will allow us to solve equations like $x^2 + 1 = 0$. The complex-number system makes use of i, a number that is, by definition, a square root of -1.

> **The Number i**
>
> i is the unique number for which $i = \sqrt{-1}$ and $i^2 = -1$.

We can now define the square root of a negative number as follows:

$$\sqrt{-p} = \sqrt{-1}\,\sqrt{p} = i\sqrt{p} \text{ or } \sqrt{p}i, \text{ for any positive number } p.$$

EXAMPLE 1   Express in terms of i: (a) $\sqrt{-7}$;   (b) $\sqrt{-16}$;   (c) $-\sqrt{-13}$;   (d) $-\sqrt{-50}$.

Solution

a) $\sqrt{-7} = \sqrt{-1 \cdot 7} = \sqrt{-1} \cdot \sqrt{7} = i\sqrt{7}$, or $\sqrt{7}i$ $\boxed{i \text{ is } not \text{ under the radical.}}$

b) $\sqrt{-16} = \sqrt{-1 \cdot 16} = \sqrt{-1} \cdot \sqrt{16} = i \cdot 4 = 4i$

c) $-\sqrt{-13} = -\sqrt{-1 \cdot 13} = -\sqrt{-1} \cdot \sqrt{13} = -i\sqrt{13}$, or $-\sqrt{13}i$

d) $-\sqrt{-50} = -\sqrt{-1} \cdot \sqrt{25} \cdot \sqrt{2} = -i \cdot 5 \cdot \sqrt{2} = -5i\sqrt{2}$, or $-5\sqrt{2}i$

> **Imaginary Numbers**
>
> An *imaginary number* is a number that can be written in the form $a + bi$, where a and b are real numbers and $b \neq 0$.

Don't let the name "imaginary" fool you. Imaginary numbers appear in fields such as engineering and the physical sciences. The following are examples of imaginary numbers:

$5 + 4i$,   Here $a = 5$, $b = 4$.

$\sqrt{3} - \pi i$,   Here $a = \sqrt{3}$, $b = -\pi$.

$\sqrt{7}i$ Here $a = 0$, $b = \sqrt{7}$.

The union of the set of all imaginary numbers and the set of all real numbers is the set of all **complex numbers**.

Complex Numbers

A *complex number* is any number that can be written in the form $a + bi$, where a and b are real numbers. (Note that a and b both can be 0.)

The following are examples of complex numbers:

$7 + 3i$ (here $a \neq 0, b \neq 0$); $4i$ (here $a = 0, b \neq 0$);

8 (here $a \neq 0, b = 0$); 0 (here $a = 0, b = 0$).

Complex numbers like $17i$ or $4i$, in which $a = 0$ and $b \neq 0$, are imaginary numbers with no real part. Such numbers are called *pure imaginary numbers*.

Note that when $b = 0$, we have $a + 0i = a$, so every real number is a complex number. The relationships among various real and complex numbers are shown below.

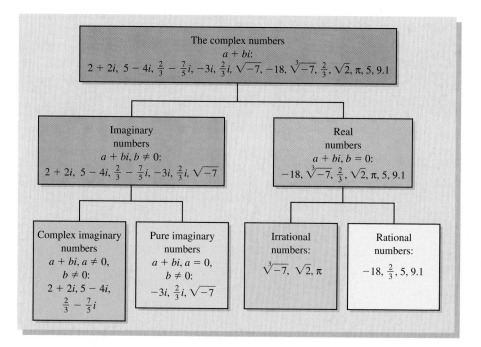

Note that although $\sqrt{-7}$ and $\sqrt[3]{-7}$ are both complex numbers, $\sqrt{-7}$ is imaginary whereas $\sqrt[3]{-7}$ is real.

Addition and Subtraction

The complex numbers obey the commutative, associative, and distributive laws. Thus we can add and subtract them as we do binomials.

EXAMPLE 2   Add or subtract and simplify.

a) $(8 + 6i) + (3 + 2i)$ **b)** $(4 + 5i) - (6 - 3i)$

Solution

a) $(8 + 6i) + (3 + 2i) = (8 + 3) + (6i + 2i)$ Combining the real parts
                                                                                 and the imaginary parts

$$= 11 + (6 + 2)i = 11 + 8i$$

b) $(4 + 5i) - (6 - 3i) = (4 - 6) + [5i - (-3i)]$ Note that the 6 and the
                                                                                                $-3i$ are *both* being
                                                                                                subtracted.

$$= -2 + 8i$$

Student Notes

The rule developed in Section 7.3, $\sqrt[n]{a} \cdot \sqrt[n]{b} = \sqrt[n]{a \cdot b}$, does *not* apply when n is 2 and either a or b is negative. Indeed this condition is stated on p. 453 when it is specified that $\sqrt[n]{a}$ and $\sqrt[n]{b}$ are both *real* numbers.

Multiplication

To multiply square roots of negative real numbers, we first express them in terms of i. For example,

$$\sqrt{-2} \cdot \sqrt{-5} = \sqrt{-1} \cdot \sqrt{2} \cdot \sqrt{-1} \cdot \sqrt{5}$$
$$= i \cdot \sqrt{2} \cdot i \cdot \sqrt{5}$$
$$= i^2 \cdot \sqrt{10}$$
$$= -1\sqrt{10} = -\sqrt{10} \text{ is correct!}$$

Caution!   With complex numbers, simply multiplying radicands is *incorrect* when both radicands are negative: $\sqrt{-2} \cdot \sqrt{-5} \neq \sqrt{10}$.

With this in mind, we can now multiply complex numbers.

EXAMPLE 3   Multiply and simplify. When possible, write answers in the form $a + bi$.

a) $\sqrt{-16} \cdot \sqrt{-25}$ **b)** $\sqrt{-5} \cdot \sqrt{-7}$ **c)** $-3i \cdot 8i$

d) $-4i(3 - 5i)$ **e)** $(1 + 2i)(1 + 3i)$

Solution

a) $\sqrt{-16} \cdot \sqrt{-25} = \sqrt{-1} \cdot \sqrt{16} \cdot \sqrt{-1} \cdot \sqrt{25}$

$$= i \cdot 4 \cdot i \cdot 5$$
$$= i^2 \cdot 20$$
$$= -1 \cdot 20 \qquad i^2 = -1$$
$$= -20$$

b) $\sqrt{-5} \cdot \sqrt{-7} = \sqrt{-1} \cdot \sqrt{5} \cdot \sqrt{-1} \cdot \sqrt{7}$ Try to do this step mentally.

$$= i \cdot \sqrt{5} \cdot i \cdot \sqrt{7}$$
$$= i^2 \cdot \sqrt{35}$$
$$= -1 \cdot \sqrt{35} \qquad i^2 = -1$$
$$= -\sqrt{35}$$

c) $-3i \cdot 8i = -24 \cdot i^2$

$\qquad\qquad = -24 \cdot (-1) \qquad i^2 = -1$

$\qquad\qquad = 24$

d) $-4i(3 - 5i) = -4i \cdot 3 + (-4i)(-5i)$ Using the distributive law

$\qquad\qquad\qquad = -12i + 20i^2$

$\qquad\qquad\qquad = -12i - 20 \qquad\qquad\qquad\quad i^2 = -1$

$\qquad\qquad\qquad = -20 - 12i \qquad\qquad\qquad$ Writing in the form $a + bi$

e) $(1 + 2i)(1 + 3i) = 1 + 3i + 2i + 6i^2$ Multiplying each term of $1 + 3i$ by each term of $1 + 2i$ (FOIL)

$\qquad\qquad\qquad\qquad = 1 + 3i + 2i - 6 \qquad i^2 = -1$

$\qquad\qquad\qquad\qquad = -5 + 5i \qquad\qquad$ Combining like terms

Conjugates and Division

Recall that the conjugate of $4 + \sqrt{2}$ is $4 - \sqrt{2}$.

    Conjugates of complex numbers are defined in a similar manner.

> **Conjugate of a Complex Number**
>
> The *conjugate* of a complex number $a + bi$ is $a - bi$, and the *conjugate* of $a - bi$ is $a + bi$.

EXAMPLE 4   Find the conjugate of each number.

a) $-3 + 7i$ **b)** $14 - 5i$ **c)** $4i$

Solution

a) $-3 + 7i$ The conjugate is $-3 - 7i$.

b) $14 - 5i$ The conjugate is $14 + 5i$.

c) $4i$ The conjugate is $-4i$. Note that $4i = 0 + 4i$.

The product of a complex number and its conjugate is a real number.

EXAMPLE 5   Multiply: $(5 + 7i)(5 - 7i)$.

Solution

$\qquad (5 + 7i)(5 - 7i) = 5^2 - (7i)^2$ Using $(A + B)(A - B) = A^2 - B^2$

$\qquad\qquad\qquad\qquad = 25 - 49i^2$

$\qquad\qquad\qquad\qquad = 25 - 49(-1) \qquad i^2 = -1$

$\qquad\qquad\qquad\qquad = 25 + 49 = 74$

Conjugates are used when dividing complex numbers. The procedure is much like that used to rationalize denominators in Section 7.5.

EXAMPLE 6   Divide and simplify to the form $a + bi$.

a) $\dfrac{-5 + 9i}{1 - 2i}$

b) $\dfrac{7 + 3i}{5i}$

Solution

a) To divide and simplify $(-5 + 9i)/(1 - 2i)$, we multiply by 1, using the conjugate of the denominator to form 1:

$$\frac{-5 + 9i}{1 - 2i} = \frac{-5 + 9i}{1 - 2i} \cdot \frac{1 + 2i}{1 + 2i} \qquad \begin{array}{l}\text{Multiplying by 1 using the} \\ \text{conjugate of the denominator} \\ \text{in the symbol for 1}\end{array}$$

$$= \frac{(-5 + 9i)(1 + 2i)}{(1 - 2i)(1 + 2i)} \qquad \begin{array}{l}\text{Multiplying numerators;} \\ \text{multiplying denominators}\end{array}$$

$$= \frac{-5 - 10i + 9i + 18i^2}{1^2 - 4i^2} \qquad \text{Using FOIL}$$

$$= \frac{-5 - i - 18}{1 - 4(-1)} \qquad i^2 = -1$$

$$\left.\begin{array}{l}= \dfrac{-23 - i}{5} \\[2mm] = -\dfrac{23}{5} - \dfrac{1}{5}i\end{array}\right\} \qquad \begin{array}{l}\text{Writing in the form } a + bi; \\ \text{note that } \dfrac{X + Y}{Z} = \dfrac{X}{Z} + \dfrac{Y}{Z}\end{array}$$

b) The conjugate of $5i$ is $-5i$, so we *could* multiply by $-5i/(-5i)$. However, when the denominator is a pure imaginary number, it is easiest if we multiply by i/i:

$$\frac{7 + 3i}{5i} = \frac{7 + 3i}{5i} \cdot \frac{i}{i} \qquad \begin{array}{l}\text{Multiplying by 1 using } i/i. \text{ We can also use} \\ \text{the conjugate of } 5i \text{ to write } -5i/(-5i).\end{array}$$

$$= \frac{7i + 3i^2}{5i^2} \qquad \text{Multiplying}$$

$$= \frac{7i + 3(-1)}{5(-1)} \qquad i^2 = -1$$

$$= \frac{7i - 3}{-5}$$

$$= \frac{-3}{-5} + \frac{7}{-5}i, \text{ or } \frac{3}{5} - \frac{7}{5}i. \qquad \text{Writing in the form } a + bi$$

Powers of i

Answers to problems involving complex numbers are generally written in the form $a + bi$. In the following discussion, we show why there is no need to use powers of i (other than 1) when writing answers.

Recall that -1 raised to an *even* power is 1, and -1 raised to an *odd* power is -1. Simplifying powers of i can then be done by using the fact that $i^2 = -1$ and expressing the given power of i in terms of i^2. Consider the following:

$$i^2 = -1, \longleftarrow$$
$$i^3 = i^2 \cdot i = (-1)i = -i,$$
$$i^4 = (i^2)^2 = (-1)^2 = 1,$$
$$i^5 = i^4 \cdot i = (i^2)^2 \cdot i = (-1)^2 \cdot i = i,$$
$$i^6 = (i^2)^3 = (-1)^3 = -1. \longleftarrow \text{The pattern is now repeating.}$$

The powers of i cycle themselves through the values i, -1, $-i$, and 1. Even powers of i are -1 or 1 whereas odd powers of i are i or $-i$.

EXAMPLE 7   Simplify: **(a)** i^{18}; **(b)** i^{24}.

Solution

a) $i^{18} = (i^2)^9$ Using the power rule

$\qquad = (-1)^9 = -1$ Raising -1 to a power

b) $i^{24} = (i^2)^{12}$

$\qquad = (-1)^{12} = 1$

To simplify i^n when n is odd, we rewrite i^n as $i^{n-1} \cdot i$.

EXAMPLE 8   Simplify: **(a)** i^{29}; **(b)** i^{75}.

Solution

a) $i^{29} = i^{28}i^1$ Using the product rule. This is a key step
                                        when i is raised to an odd power.

$\qquad = (i^2)^{14}i$ Using the power rule

$\qquad = (-1)^{14}i$

$\qquad = 1 \cdot i = i$

b) $i^{75} = i^{74}i^1$ Using the product rule

$\qquad = (i^2)^{37}i$ Using the power rule

$\qquad = (-1)^{37}i$

$\qquad = -1 \cdot i = -i$

Concept Reinforcement   *Classify each statement as either true or false.*

1. Imaginary numbers are so named because they have no real-world applications.

2. Every real number is imaginary, but not every imaginary number is real.

3. Every imaginary number is a complex number, but not every complex number is imaginary.

4. Every real number is a complex number, but not every complex number is real.

5. Addition and subtraction of complex numbers has much in common with addition and subtraction of polynomials.

6. The product of a complex number and its conjugate is always a real number.

7. The square of a complex number is always a real number.

8. The quotient of two complex numbers is always a complex number.

Express in terms of i.

9. $\sqrt{-36}$

10. $\sqrt{-25}$

11. $\sqrt{-13}$

12. $\sqrt{-19}$

13. $\sqrt{-18}$

14. $\sqrt{-98}$

15. $\sqrt{-3}$

16. $\sqrt{-4}$

17. $\sqrt{-81}$

18. $\sqrt{-27}$

19. $-\sqrt{-300}$

20. $-\sqrt{-75}$

21. $6 - \sqrt{-84}$

22. $4 - \sqrt{-60}$

23. $-\sqrt{-76} + \sqrt{-125}$

24. $\sqrt{-4} + \sqrt{-12}$

25. $\sqrt{-18} - \sqrt{-100}$

26. $\sqrt{-72} - \sqrt{-25}$

Perform the indicated operation and simplify. Write each answer in the form a + bi.

27. $(6 + 7i) + (5 + 3i)$

28. $(4 - 5i) + (3 + 9i)$

29. $(9 + 8i) - (5 + 3i)$

30. $(9 + 7i) - (2 + 4i)$

31. $(7 - 4i) - (5 - 3i)$

32. $(5 - 3i) - (9 + 2i)$

33. $(-5 - i) - (7 + 4i)$

34. $(-2 + 6i) - (-7 + i)$

35. $7i \cdot 6i$

36. $6i \cdot 9i$

37. $(-4i)(-6i)$

38. $7i \cdot (-8i)$

39. $\sqrt{-36}\,\sqrt{-9}$

40. $\sqrt{-49}\,\sqrt{-25}$

41. $\sqrt{-5}\,\sqrt{-2}$

42. $\sqrt{-6}\,\sqrt{-7}$

43. $\sqrt{-6}\,\sqrt{-21}$

44. $\sqrt{-15}\,\sqrt{-10}$

45. $5i(2 + 6i)$

46. $2i(7 + 3i)$

47. $-7i(3 - 4i)$

48. $-4i(6 - 5i)$

49. $(1 + i)(3 + 2i)$

50. $(1 + 5i)(4 + 3i)$

51. $(6 - 5i)(3 + 4i)$

52. $(5 - 6i)(2 + 5i)$

53. $(7 - 2i)(2 - 6i)$

54. $(-4 + 5i)(3 - 4i)$

55. $(-2 + 3i)(-2 + 5i)$

56. $(-3 + 6i)(-3 + 4i)$

57. $(-5 - 4i)(3 + 7i)$

58. $(2 + 9i)(-3 - 5i)$

59. $(4 - 2i)^2$

60. $(1 - 2i)^2$

61. $(2 + 3i)^2$

62. $(3 + 2i)^2$

63. $(-2 + 3i)^2$

64. $(-5 - 2i)^2$

65. $\dfrac{7}{4 + i}$

66. $\dfrac{3}{8 + i}$

67. $\dfrac{2}{3 - 2i}$

68. $\dfrac{4}{2 - 3i}$

69. $\dfrac{2i}{5 + 3i}$

70. $\dfrac{3i}{4 + 2i}$

71. $\dfrac{5}{6i}$

72. $\dfrac{4}{7i}$

73. $\dfrac{5 - 3i}{4i}$

74. $\dfrac{2 + 7i}{5i}$

Aha! 75. $\dfrac{7i + 14}{7i}$

76. $\dfrac{6i + 3}{3i}$

77. $\dfrac{4 + 5i}{3 - 7i}$

78. $\dfrac{5 + 3i}{7 - 4i}$

79. $\dfrac{2 + 3i}{2 + 5i}$

80. $\dfrac{3 + 2i}{4 + 3i}$

81. $\dfrac{3 - 2i}{4 + 3i}$

82. $\dfrac{5 - 2i}{3 + 6i}$

Simplify.

83. i^7

84. i^{11}

85. i^{24}

86. i^{35}

87. i^{42}

88. i^{64}

89. i^9

90. $(-i)^{71}$

91. $(-i)^6$

92. $(-i)^4$

93. $(5i)^3$

94. $(-3i)^5$

95. $i^2 + i^4$

96. $5i^5 + 4i^3$

97. Is the product of two imaginary numbers always an imaginary number? Why or why not?

98. In what way(s) are conjugates of complex numbers similar to the conjugates used in Section 7.5?

SKILL MAINTENANCE

Graph.

99. $f(x) = 3x - 5$ [2.3]

100. $g(x) = -2x + 6$ [2.3]

101. $F(x) = x^2$ [2.2]

102. $G(x) = |x|$ [2.2]

Solve.

103. $28 = 3x^2 - 17x$ [5.8]

104. $|3x + 7| < 22$ [4.3]

SYNTHESIS

105. Is the set of real numbers a subset of the set of complex numbers? Why or why not?

106. Is the union of the set of imaginary numbers and the set of real numbers the set of complex numbers? Why or why not?

A function g is given by

$$g(z) = \dfrac{z^4 - z^2}{z - 1}.$$

107. Find $g(3i)$.

108. Find $g(1 + i)$.

109. Find $g(5i - 1)$.

110. Find $g(2 - 3i)$.

111. Evaluate

$$\dfrac{1}{w - w^2} \quad \text{for} \quad w = \dfrac{1 - i}{10}.$$

Simplify.

112. $\dfrac{i^5 + i^6 + i^7 + i^8}{(1 - i)^4}$

113. $(1 - i)^3(1 + i)^3$

114. $\dfrac{5 - \sqrt{5}i}{\sqrt{5}i}$

115. $\dfrac{6}{1 + \dfrac{3}{i}}$

116. $\left(\dfrac{1}{2} - \dfrac{1}{3}i\right)^2 - \left(\dfrac{1}{2} + \dfrac{1}{3}i\right)^2$

117. $\dfrac{i - i^{38}}{1 + i}$

7 Study Summary

Radical expressions involve **square roots**, **cube roots**, and, more generally, **nth roots** (pp. 436, 440, and 441):

c is a square root of a if $c^2 = a$;

c is a cube root of a if $c^3 = a$.

The notation $\sqrt{a}$ indicates the **principal** square root of a, the notation $\sqrt[3]{a}$ indicates the cube root of a, and $\sqrt[n]{a}$ indicates the nth root of a (p. 441). For these and other **radical expressions**, we use the following vocabulary (pp. 437, 441):

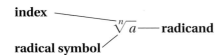

index ——
$\sqrt[n]{a}$ —— radicand
radical symbol ——

Radical notation can also be written using **rational exponents** (p. 445):

$a^{1/n}$ means $\sqrt[n]{a}$. When a is nonnegative, n can be any natural number greater than 1. When a is negative, n must be odd.

For any natural numbers m and n ($n \neq 1$), and any real number a,

$a^{m/n}$ means $(\sqrt[n]{a})^m$ or $\sqrt[n]{a^m}$.

When a is negative, n must be odd.

Radical expressions can be added, subtracted, multiplied, and divided. For multiplication and division, the product and quotient rules, respectively, are used (pp. 453, 460):

The Product Rule for Radicals

For any real numbers $\sqrt[n]{a}$ and $\sqrt[n]{b}$, $\sqrt[n]{a}\,\sqrt[n]{b} = \sqrt[n]{a \cdot b}$.

The Quotient Rule for Radicals

For any real numbers $\sqrt[n]{a}$ and $\sqrt[n]{b}$, $b \neq 0$, $\sqrt[n]{\dfrac{a}{b}} = \dfrac{\sqrt[n]{a}}{\sqrt[n]{b}}$.

When working with radical expressions, it is understood that all results should be written in simplified form. Rational exponents can be especially useful when simplifying certain products or quotients:

Some Ways to Simplify Radical Expressions

1. *Simplifying by factoring.* Factor the radicand and look for factors raised to powers that are divisible by the index.

   Example: $\sqrt[3]{a^6 b} = \sqrt[3]{a^6}\,\sqrt[3]{b} = a^2 \sqrt[3]{b}$

2. *Using rational exponents to simplify.* Convert to exponential notation and then use arithmetic and the laws of exponents to simplify the exponents. Then convert back to radical notation as needed.

   Example: $\sqrt[3]{p} \cdot \sqrt[4]{q^3} = p^{1/3} \cdot q^{3/4}$
   $$= p^{4/12} \cdot q^{9/12}$$
   $$= \sqrt[12]{p^4 q^9}$$

3. *Combining like radical terms.*

   Example: $\sqrt{8} + 3\sqrt{2} = \sqrt{4} \cdot \sqrt{2} + 3\sqrt{2}$
   $$= 2\sqrt{2} + 3\sqrt{2} = 5\sqrt{2}$$

There sometimes arises a need to **rationalize** a numerator or a denominator (pp. 462, 464). This is accomplished by multiplying the original expression by a form of 1 that is written using the **conjugate** of the portion of the fraction being rationalized (p. 469):

$$\frac{3}{2 + \sqrt{5}} = \frac{3}{2 + \sqrt{5}} \cdot \frac{2 - \sqrt{5}}{2 - \sqrt{5}} \longleftarrow \quad \text{This is the conjugate of } 2 + \sqrt{5}.$$
$$= \frac{6 - 3\sqrt{5}}{4 - 5} = \frac{6 - 3\sqrt{5}}{-1} = -6 + 3\sqrt{5}.$$

Radical equations are solved using the **principle of powers** (p. 475): If $a = b$, then $a^n = b^n$ for any exponent n. Because the principle of powers does not always produce equivalent equations, it is important that the solutions be checked in the original equation.

To solve an equation with a radical term:

1. Isolate the radical term on one side of the equation.
2. Use the principle of powers and solve the resulting equation.
3. Check any possible solution in the original equation.

To solve an equation with two or more radical terms:

1. Isolate one of the radical terms.
2. Use the principle of powers.
3. If a radical remains, repeat steps (1) and (2).
4. Solve the resulting equation.
5. Check possible solutions in the original equation.

Radical notation arises frequently when we are using the **Pythagorean theorem** and in work with **isosceles** or **30°–60°–90° right triangles** (pp. 483, 486):

The Pythagorean Theorem

For any right triangle with legs of lengths a and b and hypotenuse of length c,

$$a^2 + b^2 = c^2.$$

Special Triangles

The length of the hypotenuse in an isosceles right triangle is the length of a leg times $\sqrt{2}$.

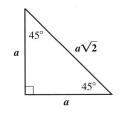

The length of the longer leg in a 30°–60°–90° right triangle is the length of the shorter leg times $\sqrt{3}$. The hypotenuse is twice as long as the shorter leg.

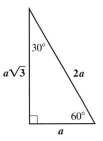

Real numbers and **imaginary** numbers make up the **complex** numbers (pp. 491, 492). Addition, subtraction, multiplication, and division of complex numbers make use of the facts that $i = \sqrt{-1}$ and the product of a complex number and its **conjugate** is a real number (p. 494):

$$(3 + 2i) + (4 - 7i) = 7 - 5i;$$

$$(8 + 6i) - (5 + 2i) = 3 + 4i;$$

$$(2 + 3i)(4 - i) = 8 - 2i + 12i - 3i^2$$
$$= 8 + 10i - 3(-1) = 11 + 10i;$$

$$\frac{1 - 4i}{3 - 2i} = \frac{1 - 4i}{3 - 2i} \cdot \frac{3 + 2i}{3 + 2i} \quad \longleftarrow \quad \text{The conjugate of } 3 - 2i$$

$$= \frac{3 + 2i - 12i - 8i^2}{9 + 6i - 6i - 4i^2}$$

$$= \frac{3 - 10i - 8(-1)}{9 - 4(-1)} = \frac{11 - 10i}{13} = \frac{11}{13} - \frac{10}{13}i.$$

7 Review Exercises

↩ *Concept Reinforcement*   *Classify each of the following as either true or false.*

1. $\sqrt{ab} = \sqrt{a} \cdot \sqrt{b}$ for any real numbers $\sqrt{a}$ and $\sqrt{b}$. [7.3]

2. $\sqrt{a^2} = a$, for any real number a. [7.1]

3. $\sqrt[3]{a^3} = a$, for any real number a. [7.1]

4. $x^{2/5}$ means $\sqrt[5]{x^2}$ and $\left(\sqrt[5]{x}\right)^2$. [7.2]

5. A hypotenuse is never shorter than either leg. [7.7]

6. $i^{13} = (i^2)^6 i = i$. [7.8]

7. Some radical equations have no solution. [7.6]

8. If $f(x) = \sqrt{x-5}$, then the domain of f is the set of all nonnegative real numbers. [7.1]

Simplify. [7.1]

9. $\sqrt{\dfrac{49}{9}}$

10. $-\sqrt{0.25}$

Let $f(x) = \sqrt{2x - 7}$. *Find the following.* [7.1]

11. $f(16)$

12. The domain of f

Simplify. Assume that each variable can represent any real number.

13. $\sqrt{25t^2}$ [7.1]

14. $\sqrt{(c + 8)^2}$ [7.1]

15. $\sqrt{x^2 - 6x + 9}$ [7.1]

16. $\sqrt{4x^2 + 4x + 1}$ [7.1]

17. $\sqrt[5]{-32}$ [7.1]

18. $\sqrt[3]{-\dfrac{64x^6}{27}}$ [7.4]

19. $\sqrt[4]{x^{12}y^8}$ [7.3]

20. $\sqrt[6]{64x^{12}}$ [7.3]

21. Write an equivalent expression using exponential notation: $\left(\sqrt[3]{5ab}\right)^4$. [7.2]

22. Write an equivalent expression using radical notation: $(16a^6)^{3/4}$. [7.2]

Use rational exponents to simplify. Assume $x, y \geq 0$. [7.2]

23. $\sqrt{x^6 y^{10}}$

24. $\left(\sqrt[6]{x^2 y}\right)^2$

Simplify. Do not use negative exponents in the answers. [7.2]

25. $(x^{-2/3})^{3/5}$

26. $\dfrac{7^{-1/3}}{7^{-1/2}}$

27. If $f(x) = \sqrt{25(x-6)^2}$, find a simplified form for $f(x)$. [7.3]

Perform the indicated operation and, if possible, simplify. Write all answers using radical notation.

28. $\sqrt{2x}\sqrt{3y}$ [7.3]

29. $\sqrt[3]{a^5 b}\sqrt[3]{27b}$ [7.3]

30. $\sqrt[3]{-24x^{10}y^8}\sqrt[3]{18x^7y^4}$ [7.3]

31. $\dfrac{\sqrt[3]{60xy^3}}{\sqrt[3]{10x}}$ [7.4]

32. $\dfrac{\sqrt{75x}}{2\sqrt{3}}$ [7.4]

33. $\sqrt[4]{\dfrac{48a^{11}}{c^8}}$ [7.4]

34. $5\sqrt[3]{x} + 2\sqrt[3]{x}$ [7.5]

35. $2\sqrt{75} - 9\sqrt{3}$ [7.5]

36. $\sqrt[3]{8x^4} + \sqrt[3]{xy^6}$ [7.5]

37. $\sqrt{50} + 2\sqrt{18} + \sqrt{32}$ [7.5]

38. $\left(\sqrt{3} - 3\sqrt{8}\right)\left(\sqrt{5} + 2\sqrt{8}\right)$ [7.5]

39. $\sqrt[4]{x}\sqrt{x}$ [7.5]

40. $\dfrac{\sqrt[3]{x^2}}{\sqrt[4]{x}}$ [7.5]

41. If $f(x) = x^2$, find $f\left(a - \sqrt{2}\right)$. [7.5]

42. Rationalize the denominator:
$$\dfrac{4\sqrt{5}}{\sqrt{2} + \sqrt{3}}.\ [7.5]$$

43. Rationalize the numerator of the expression in Exercise 42. [7.5]

Solve. [7.6]

44. $\sqrt{y + 6} - 2 = 3$

45. $(x + 1)^{1/3} = -5$

46. $1 + \sqrt{x} = \sqrt{3x - 3}$

47. If $f(x) = \sqrt[4]{x + 2}$, find a such that $f(a) = 2$. [7.6]

Solve. Give an exact answer and, where appropriate, an approximation to three decimal places. [7.7]

48. The diagonal of a square has length 10 cm. Find the length of a side of the square.

49. A skate-park jump has a ramp that is 6 ft long and is 2 ft high. How long is its base?

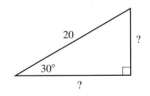

6 ft   2 ft
?

50. Find the missing lengths. Give exact answers and, where appropriate, an approximation to three decimal places.

20   ?
30°
?

51. Express in terms of i and simplify: $-\sqrt{-8}$. [7.8]

52. Add: $(-4 + 3i) + (2 - 12i)$. [7.8]

53. Subtract: $(9 - 7i) - (3 - 8i)$. [7.8]

Simplify. [7.8]

54. $(2 + 5i)(2 - 5i)$ **55.** i^{18}

56. Solve:

$$\sqrt{11x + \sqrt{6 + x}} = 6. \text{ [7.6]}$$

57. Simplify:

$$\frac{2}{1 - 3i} - \frac{3}{4 + 2i}. \text{ [7.8]}$$

58. Simplify: $(6 - 3i)(2 - i)$. [7.8]

59. Divide and simplify to the form $a + bi$:

$$\frac{7 - 2i}{3 + 4i}. \text{ [7.8]}$$

60. What makes some complex numbers real and others imaginary? [7.8]

SYNTHESIS

61. Explain why $\sqrt[n]{x^n} = |x|$ when n is even, but $\sqrt[n]{x^n} = x$ when n is odd. [7.1]

62. Write a quotient of two imaginary numbers that is a real number (answers may vary). [7.8]

7 Chapter Test

Simplify. Assume that variables can represent any real number.

1. $\sqrt{50}$

2. $\sqrt[3]{-\dfrac{8}{x^6}}$

3. $\sqrt{81a^2}$

4. $\sqrt{x^2 - 8x + 16}$

5. $\sqrt[5]{x^{12}y^8}$

6. $\sqrt{\dfrac{25x^2}{36y^4}}$

7. $\sqrt[3]{3z}\,\sqrt[3]{5y^2}$

8. $\dfrac{\sqrt[5]{x^3y^4}}{\sqrt[5]{xy^2}}$

9. $\sqrt[4]{x^3y^2}\,\sqrt[4]{xy}$

10. $\dfrac{\sqrt[5]{a^2}}{\sqrt[4]{a}}$

11. $8\sqrt{2} - 2\sqrt{2}$

12. $\sqrt{x^4y} + \sqrt{9y^3}$

13. $(7 + \sqrt{x})(2 - 3\sqrt{x})$

14. Write an equivalent expression using exponential notation: $\sqrt{7xy}$.

15. Write an equivalent expression using radical notation: $(4a^3b)^{5/6}$.

16. If $f(x) = \sqrt{2x - 10}$, determine the domain of f.

17. If $f(x) = x^2$, find $f(5 + \sqrt{2})$.

18. Rationalize the denominator:
$$\dfrac{\sqrt{3}}{5 + \sqrt{2}}.$$

Solve.

19. $x = \sqrt{2x - 5} + 4$

20. $\sqrt{x} = \sqrt{x + 1} - 5$

Solve. Give exact answers and approximations to three decimal places.

21. One leg of a 30°–60°–90° right triangle is 7 cm long. Find the possible lengths of the other leg.

22. A referee jogs diagonally from one corner of a 50-ft by 90-ft basketball court to the far corner. How far does she jog?

23. Express in terms of i and simplify: $\sqrt{-50}$.

24. Subtract: $(9 + 8i) - (-3 + 6i)$.

25. Multiply: $\sqrt{-16}\,\sqrt{-36}$.

26. Multiply. Write the answer in the form $a + bi$.
$$(4 - i)^2$$

27. Divide and simplify to the form $a + bi$:
$$\dfrac{-2 + i}{3 - 5i}.$$

28. Simplify: i^{37}.

SYNTHESIS

29. Solve:
$$\sqrt{2x - 2} + \sqrt{7x + 4} = \sqrt{13x + 10}.$$

30. Simplify:
$$\dfrac{1 - 4i}{4i(1 + 4i)^{-1}}.$$

31. Drake's Discount Shoe Center has two locations. The sign at the original location is shaped like an isosceles right triangle. The sign at the newer location is shaped like a 30°–60°–90° triangle. The hypotenuse of each sign measures 6 ft. Which sign has the greater area and by how much? (Round to three decimal places.)

8

Quadratic Functions and Equations

AN APPLICATION

An athlete's hang time T, in seconds, is related to vertical leap V, in inches, by the formula $V = 48T^2$. The NBA's Steve Francis reportedly has a vertical leap of 45 in. What is his hang time?

This problem appears as Exercise 37 in Section 8.3.

Elizabeth Bluebird
PHYSICAL THERAPIST
Oklahoma City, Oklahoma

A treadmill test helps me to set up an exercise program for people with heart disease. I use an algebraic formula to figure a single-stage treadmill test using variables such as age, heart rate and speed. I also use math every day when measuring a person's range of motion—the arc or angle through which a body part moves. Knowing this angle, I can help make a diagnosis and show progress in my patient's therapy.

In translating problem situations to mathematics, we often obtain a function or equation containing a second-degree polynomial in one variable. Such functions or equations are said to be quadratic. *In this chapter, we examine a variety of equations, inequalities, and applications for which we will need to solve quadratic equations or graph quadratic functions.*

8.1 Quadratic Equations

The Principle of Square Roots • Completing the Square • Problem Solving

ALGEBRAIC—GRAPHICAL CONNECTION

Let's reexamine the graphical connections to the algebraic equation-solving concepts we have studied before.

In Chapter 2, we introduced the graph of a quadratic function given by

$$f(x) = ax^2 + bx + c, \quad a \neq 0.$$

For example, the graphs of $f(x) = x^2 + 6x + 8$ and $g(x) = 0$ are shown below.

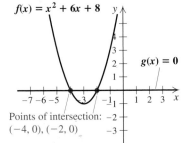

$f(x) = x^2 + 6x + 8$

$g(x) = 0$

Points of intersection: $(-4, 0), (-2, 0)$

Note that $(-4, 0)$ and $(-2, 0)$ are the points of intersection of the graphs of $f(x) = x^2 + 6x + 8$ and $g(x) = 0$ (the x-axis). In Sections 8.6 and 8.7, we will develop efficient ways to graph quadratic functions. For now, the graphs simply aid in visualizing solutions.

In Chapter 5, we solved equations like $x^2 + 6x + 8 = 0$ by factoring:

$$x^2 + 6x + 8 = 0$$

$$(x + 4)(x + 2) = 0 \qquad \text{Factoring}$$

$$x + 4 = 0 \quad \text{or} \quad x + 2 = 0 \qquad \text{Using the principle of zero products}$$

$$x = -4 \quad \text{or} \qquad x = -2.$$

Note that -4 and -2 are the first coordinates of the points of intersection (or the x-intercepts) above.

In this section and the next, we develop algebraic methods for solving *any* quadratic equation, whether it is factorable or not.

EXAMPLE 1   Solve: $x^2 = 25$.

Solution   We have

$$x^2 = 25$$

$$x^2 - 25 = 0 \qquad \text{Writing in standard form}$$

$$(x - 5)(x + 5) = 0 \qquad \text{Factoring}$$

$$x - 5 = 0 \quad or \quad x + 5 = 0 \qquad \text{Using the principle of zero products}$$

$$x = 5 \quad or \qquad x = -5.$$

The solutions are 5 and -5. A graph in which $f(x) = x^2$ represents the left side of the original equation and $g(x) = 25$ represents the right side provides a check (see the figure at left). Of course, we can also check by substituting 5 and -5 into the original equation.

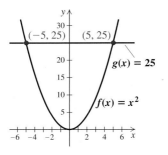

A visualization of Example 1

The Principle of Square Roots

Let's reconsider $x^2 = 25$. We know from Chapter 7 that the number 25 has two real-number square roots, 5 and -5, the solutions of the equation in Example 1. Thus we see that square roots can provide quick solutions for equations of the type $x^2 = k$.

> **The Principle of Square Roots**
>
> For any real number k, if $x^2 = k$, then $x = \sqrt{k}$ or $x = -\sqrt{k}$.

EXAMPLE 2   Solve: $3x^2 = 6$. Give exact solutions and approximations to three decimal places.

Solution   We have

$$3x^2 = 6$$

$$x^2 = 2 \qquad \text{Isolating } x^2$$

$$x = \sqrt{2} \quad or \quad x = -\sqrt{2}. \qquad \text{Using the principle of square roots}$$

We often use the symbol $\pm\sqrt{2}$ to represent both of the solutions.

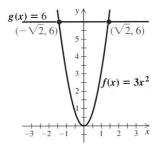

A visualization of Example 2

Check: For $\sqrt{2}$:

$$\frac{3x^2 = 6}{3(\sqrt{2})^2 \;\big|\; 6}$$
$$3 \cdot 2 \;\big|$$
$$6 \stackrel{?}{=} 6 \quad \text{TRUE}$$

For $-\sqrt{2}$:

$$\frac{3x^2 = 6}{3(-\sqrt{2})^2 \;\big|\; 6}$$
$$3 \cdot 2 \;\big|$$
$$6 \stackrel{?}{=} 6 \quad \text{TRUE}$$

The solutions are $\sqrt{2}$ and $-\sqrt{2}$, or $\pm\sqrt{2}$, which round to 1.414 and -1.414.

EXAMPLE 3   Solve: $-5x^2 + 2 = 0$.

Solution   We have

$$-5x^2 + 2 = 0$$

$$x^2 = \frac{2}{5} \qquad \text{Isolating } x^2$$

$$x = \sqrt{\frac{2}{5}} \quad or \quad x = -\sqrt{\frac{2}{5}}. \qquad \text{Using the principle of square roots}$$

The solutions are $\sqrt{\dfrac{2}{5}}$ and $-\sqrt{\dfrac{2}{5}}$. This can also be written as $\pm\sqrt{\dfrac{2}{5}}$. In the days before calculators, it was standard practice to rationalize denominators when simplifying answers. If we rationalize the denominator, the solutions are $\pm\dfrac{\sqrt{10}}{5}$. The checks are left to the student.

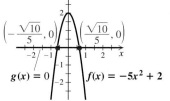

A visualization of Example 3

Sometimes we get solutions that are imaginary numbers.

EXAMPLE 4   Solve: $4x^2 + 9 = 0$.

Solution   We have

$$4x^2 + 9 = 0$$

$$x^2 = -\frac{9}{4} \qquad \text{Isolating } x^2$$

$$x = \sqrt{-\frac{9}{4}} \qquad or \qquad x = -\sqrt{-\frac{9}{4}} \qquad \text{Using the principle of square roots}$$

$$x = \sqrt{\tfrac{9}{4}}\,\sqrt{-1} \quad or \quad x = -\sqrt{\tfrac{9}{4}}\,\sqrt{-1}$$

$$x = \tfrac{3}{2}i \qquad\quad or \quad x = -\tfrac{3}{2}i. \qquad \text{Recall that } \sqrt{-1} = i.$$

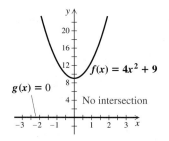

A visualization of Example 4

Check: For $\frac{3}{2}i$:

$$\begin{array}{c|c}
4x^2 + 9 = 0 \\ \hline
4\left(\tfrac{3}{2}i\right)^2 + 9 & 0 \\
4 \cdot \tfrac{9}{4} \cdot i^2 + 9 & \\
9(-1) + 9 & \\
& 0 \overset{?}{=} 0 \quad \text{TRUE}
\end{array}$$

For $-\frac{3}{2}i$:

$$\begin{array}{c|c}
4x^2 + 9 = 0 \\ \hline
4\left(-\tfrac{3}{2}i\right)^2 + 9 & 0 \\
4 \cdot \tfrac{9}{4} \cdot i^2 + 9 & \\
9(-1) + 9 & \\
& 0 \overset{?}{=} 0 \quad \text{TRUE}
\end{array}$$

The solutions are $\frac{3}{2}i$ and $-\frac{3}{2}i$, or $\pm\frac{3}{2}i$. The graph at left confirms that there are no real-number solutions.

The principle of square roots can be restated in a more general form that pertains to more complicated algebraic expressions than just x.

> **The Principle of Square Roots (Generalized Form)**
> For any real number k and any algebraic expression X,
> If $X^2 = k$, then $X = \sqrt{k}$ or $X = -\sqrt{k}$.

EXAMPLE 5 Let $f(x) = (x - 2)^2$. Find all x-values for which $f(x) = 7$.

Solution We are asked to find all x-values for which

$$f(x) = 7,$$

or

$$(x - 2)^2 = 7. \qquad \text{Substituting } (x - 2)^2 \text{ for } f(x)$$

The generalized principle of square roots gives us

$$x - 2 = \sqrt{7} \qquad or \quad x - 2 = -\sqrt{7} \qquad \text{Using the principle of}$$
$$\text{square roots}$$

$$x = 2 + \sqrt{7} \quad or \qquad x = 2 - \sqrt{7}.$$

Check: $f\left(2 + \sqrt{7}\right) = \left(2 + \sqrt{7} - 2\right)^2 = \left(\sqrt{7}\right)^2 = 7.$

Similarly,

$$f\left(2 - \sqrt{7}\right) = \left(2 - \sqrt{7} - 2\right)^2 = \left(-\sqrt{7}\right)^2 = 7.$$

The solutions are $2 + \sqrt{7}$ and $2 - \sqrt{7}$, or simply $2 \pm \sqrt{7}$.

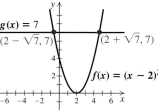

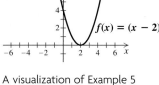

A visualization of Example 5

Example 5 is of the form $(x - a)^2 = c$, where a and c are constants. Sometimes we must factor in order to obtain this form.

EXAMPLE 6 Solve: $x^2 + 6x + 9 = 2$.

Solution We have

$$x^2 + 6x + 9 = 2 \qquad \text{The left side is the square of a binomial.}$$
$$(x + 3)^2 = 2 \qquad \text{Factoring}$$
$$x + 3 = \sqrt{2} \qquad or \quad x + 3 = -\sqrt{2} \qquad \text{Using the principle of square roots}$$
$$x = -3 + \sqrt{2} \quad or \qquad x = -3 - \sqrt{2}. \qquad \text{Adding } -3 \text{ to both sides}$$

The solutions are $-3 + \sqrt{2}$ and $-3 - \sqrt{2}$, or $-3 \pm \sqrt{2}$. The checks are left to the student.

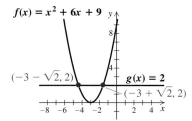

A visualization of Example 6

Completing the Square

Not all quadratic equations can be solved as we did Examples 1–6. By using a method called *completing the square*, we can use the principle of square roots to solve *any* quadratic equation.

EXAMPLE 7   Solve: $x^2 + 6x + 4 = 0$.

Solution   We have

$$x^2 + 6x + 4 = 0$$
$$x^2 + 6x = -4 \qquad\qquad \text{Subtracting 4 from both sides}$$
$$x^2 + 6x + 9 = -4 + 9 \qquad\qquad \text{Adding 9 to both sides. We explain this}$$
shortly.
$$(x + 3)^2 = 5 \qquad\qquad \text{Factoring the perfect-square trinomial}$$
$$x + 3 = \pm\sqrt{5} \qquad\qquad \text{Using the principle of square roots.}$$
Remember that $\pm\sqrt{5}$ represents two numbers.
$$x = -3 \pm \sqrt{5}. \qquad\qquad \text{Adding } -3 \text{ to both sides}$$

Check: For $-3 + \sqrt{5}$:

$$\begin{array}{c|c} x^2 + 6x + 4 = 0 & \\ \hline (-3 + \sqrt{5})^2 + 6(-3 + \sqrt{5}) + 4 & 0 \\ 9 - 6\sqrt{5} + 5 - 18 + 6\sqrt{5} + 4 & \\ 9 + 5 - 18 + 4 - 6\sqrt{5} + 6\sqrt{5} & \\ & 0 \overset{?}{=} 0 \quad \text{TRUE} \end{array}$$

For $-3 - \sqrt{5}$:

$$\begin{array}{c|c} x^2 + 6x + 4 = 0 & \\ \hline (-3 - \sqrt{5})^2 + 6(-3 - \sqrt{5}) + 4 & 0 \\ 9 + 6\sqrt{5} + 5 - 18 - 6\sqrt{5} + 4 & \\ 9 + 5 - 18 + 4 + 6\sqrt{5} - 6\sqrt{5} & \\ & 0 \overset{?}{=} 0 \quad \text{TRUE} \end{array}$$

The solutions are $-3 + \sqrt{5}$ and $-3 - \sqrt{5}$, or $-3 \pm \sqrt{5}$.

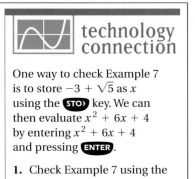

technology connection

One way to check Example 7 is to store $-3 + \sqrt{5}$ as x using the **STO** key. We can then evaluate $x^2 + 6x + 4$ by entering $x^2 + 6x + 4$ and pressing **ENTER**.

1. Check Example 7 using the method described above.

In Example 7, we chose to add 9 to both sides because it creates a perfect-square trinomial on the left side. The 9 was determined by taking half of the coefficient of x and squaring it—that is,

$$\left(\tfrac{1}{2} \cdot 6\right)^2 = 3^2, \quad \text{or} \quad 9.$$

To help see why this procedure works, examine the following drawings.

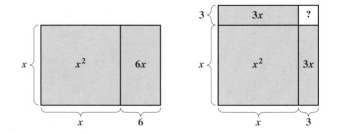

Note that the shaded areas in both figures represent the same area, $x^2 + 6x$. However, only the figure on the right, in which the $6x$ is halved, can be converted into a square with the addition of a constant term. The constant 9 is the "missing" piece that *completes* the square.

To complete the square for $x^2 + bx$, we add $(b/2)^2$.

Example 8, which follows, provides practice in finding numbers that complete the square. We will then use this skill to solve equations.

EXAMPLE 8   Replace the blanks in each equation with constants to form a true equation.

a) $x^2 + 14x + \underline{\quad} = (x + \underline{\quad})^2$

b) $x^2 - 5x + \underline{\quad} = (x - \underline{\quad})^2$

c) $x^2 + \frac{3}{4}x + \underline{\quad} = (x + \underline{\quad})^2$

Solution   We take half of the coefficient of x and square it.

a) Half of 14 is 7, and $7^2 = 49$. Thus, $x^2 + 14x + 49$ is a perfect-square trinomial and is equivalent to $(x + 7)^2$. We have

$$x^2 + 14x + 49 = (x + 7)^2.$$

Student Notes

In problems like Examples 8(b) and (c), it is best to avoid decimal notation. Most students have an easier time recognizing $\frac{9}{64}$ as $\left(\frac{3}{8}\right)^2$ than regarding 0.140625 as 0.375^2.

b) Half of -5 is $-\frac{5}{2}$, and $\left(-\frac{5}{2}\right)^2 = \frac{25}{4}$. Thus, $x^2 - 5x + \frac{25}{4}$ is a perfect-square trinomial and is equivalent to $\left(x - \frac{5}{2}\right)^2$. We have

$$x^2 - 5x + \frac{25}{4} = \left(x - \frac{5}{2}\right)^2.$$

c) Half of $\frac{3}{4}$ is $\frac{3}{8}$, and $\left(\frac{3}{8}\right)^2 = \frac{9}{64}$. Thus, $x^2 + \frac{3}{4}x + \frac{9}{64}$ is a perfect-square trinomial and is equivalent to $\left(x + \frac{3}{8}\right)^2$. We have

$$x^2 + \frac{3}{4}x + \frac{9}{64} = \left(x + \frac{3}{8}\right)^2.$$

We can now use the method of completing the square to solve equations similar to Example 7.

EXAMPLE 9   Solve: **(a)** $x^2 - 8x - 7 = 0$; **(b)** $x^2 + 5x - 3 = 0$.

Solution

a)

$$x^2 - 8x - 7 = 0$$

$$x^2 - 8x \quad\quad = 7 \qquad\qquad \text{Adding 7 to both sides. We can now complete the square on the left side.}$$

$$x^2 - 8x + 16 = 7 + 16 \qquad \text{Adding 16 to both sides to complete the square: } \tfrac{1}{2}(-8) = -4, \text{ and } (-4)^2 = 16$$

$$(x - 4)^2 = 23 \qquad\qquad \text{Factoring and simplifying}$$

$$x - 4 = \pm\sqrt{23} \qquad\quad \text{Using the principle of square roots}$$

$$x = 4 \pm \sqrt{23} \qquad\quad \text{Adding 4 to both sides}$$

The solutions are $4 - \sqrt{23}$ and $4 + \sqrt{23}$, or $4 \pm \sqrt{23}$. The checks are left to the student.

b)

$$x^2 + 5x - 3 = 0$$

$$x^2 + 5x = 3 \qquad \text{Adding 3 to both sides}$$

$$x^2 + 5x + \frac{25}{4} = 3 + \frac{25}{4} \qquad \text{Completing the square: } \frac{1}{2} \cdot 5 = \frac{5}{2}, \text{ and } \left(\frac{5}{2}\right)^2 = \frac{25}{4}$$

$$\left(x + \frac{5}{2}\right)^2 = \frac{37}{4} \qquad \text{Factoring and simplifying}$$

$$x + \frac{5}{2} = \pm\frac{\sqrt{37}}{2} \qquad \begin{array}{l}\text{Using the principle of square roots and} \\ \text{the quotient rule for radicals}\end{array}$$

$$x = -\frac{5}{2} \pm \frac{\sqrt{37}}{2}, \quad \text{or} \quad \frac{-5 \pm \sqrt{37}}{2} \qquad \text{Adding } -\frac{5}{2} \text{ to both sides}$$

The checks are left to the student. The solutions are $-\dfrac{5}{2} \pm \dfrac{\sqrt{37}}{2}$, or $\dfrac{-5 \pm \sqrt{37}}{2}$. This can be written as $-\dfrac{5}{2} - \dfrac{\sqrt{37}}{2}$ and $-\dfrac{5}{2} + \dfrac{\sqrt{37}}{2}$, or $\dfrac{-5 - \sqrt{37}}{2}$ and $\dfrac{-5 + \sqrt{37}}{2}$.

Before we complete the square, the x^2-coefficient must be 1. When it is not 1, we divide both sides of the equation by whatever that coefficient may be.

EXAMPLE 10   Find the x-intercepts of the function given by $f(x) = 3x^2 + 7x - 2$.

Solution   The value of $f(x)$ must be 0 at any x-intercepts. Thus,

$$f(x) = 0 \qquad \text{We set } f(x) \text{ equal to 0.}$$

$$3x^2 + 7x - 2 = 0 \qquad \text{Substituting}$$

$$3x^2 + 7x = 2 \qquad \text{Adding 2 to both sides}$$

$$x^2 + \frac{7}{3}x = \frac{2}{3} \qquad \text{Dividing both sides by 3}$$

$$x^2 + \frac{7}{3}x + \frac{49}{36} = \frac{2}{3} + \frac{49}{36} \qquad \text{Completing the square: } \left(\frac{1}{2} \cdot \frac{7}{3}\right)^2 = \frac{49}{36}$$

$$\left(x + \frac{7}{6}\right)^2 = \frac{73}{36} \qquad \text{Factoring and simplifying}$$

$$x + \frac{7}{6} = \pm\frac{\sqrt{73}}{6} \qquad \begin{array}{l}\text{Using the principle of square roots} \\ \text{and the quotient rule for radicals}\end{array}$$

$$x = -\frac{7}{6} \pm \frac{\sqrt{73}}{6}, \quad \text{or} \quad \frac{-7 \pm \sqrt{73}}{6}. \qquad \text{Adding } -\frac{7}{6} \text{ to both sides}$$

The x-intercepts are

$$\left(-\frac{7}{6} - \frac{\sqrt{73}}{6}, 0\right) \quad \text{and} \quad \left(-\frac{7}{6} + \frac{\sqrt{73}}{6}, 0\right), \quad \text{or}$$

$$\left(\frac{-7 - \sqrt{73}}{6}, 0\right) \quad \text{and} \quad \left(\frac{-7 + \sqrt{73}}{6}, 0\right).$$

The checks are left to the student.

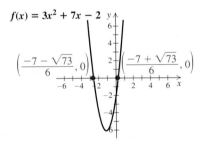

$f(x) = 3x^2 + 7x - 2$

$\left(\dfrac{-7 - \sqrt{73}}{6}, 0\right)$ $\left(\dfrac{-7 + \sqrt{73}}{6}, 0\right)$

A visualization of Example 10

The procedure used in Example 10 is important because it can be used to solve *any* quadratic equation.

To Solve a Quadratic Equation in x by Completing the Square

1. Isolate the terms with variables on one side of the equation, and arrange them in descending order.
2. Divide both sides by the coefficient of x^2 if that coefficient is not 1.
3. Complete the square by taking half of the coefficient of x and adding its square to both sides.
4. Express the trinomial as the square of a binomial (factor the trinomial) and simplify the other side.
5. Use the principle of square roots (find the square roots of both sides).
6. Solve for x by adding or subtracting on both sides.

Problem Solving

After one year, an amount of money P, invested at 4% per year, is worth 104% of P, or $P(1.04)$. If that amount continues to earn 4% interest per year, after the second year the investment will be worth 104% of $P(1.04)$, or $P(1.04)^2$. This is called **compounding interest** since after the first time period, interest is earned on both the initial investment *and* the interest from the first time period. Continuing the above pattern, we see that after the third year, the investment will be worth 104% of $P(1.04)^2$. Generalizing, we have the following.

The Compound-Interest Formula

If an amount of money P is invested at interest rate r, compounded annually, then in t years, it will grow to the amount A given by

$$A = P(1 + r)^t. \qquad (r \text{ is written in decimal notation.})$$

We can use quadratic equations to solve certain interest problems.

EXAMPLE 11   *Investment growth.*   Rosa invested $4000 at interest rate r, compounded annually. In 2 yr, it grew to $4410. What was the interest rate?

Solution

1. **Familiarize.**   We are already familiar with the compound-interest formula. If we were not, we would need to consult an outside source.
2. **Translate.**   The translation consists of substituting into the formula:

$$A = P(1 + r)^t$$
$$4410 = 4000(1 + r)^2. \qquad \text{Substituting}$$

3. Carry out.   We solve for r:

$$4410 = 4000(1 + r)^2$$

$$\frac{4410}{4000} = (1 + r)^2 \qquad \text{Dividing both sides by 4000}$$

$$\frac{441}{400} = (1 + r)^2 \qquad \text{Simplifying}$$

$$\pm\sqrt{\frac{441}{400}} = 1 + r \qquad \text{Using the principle of square roots}$$

$$\pm\frac{21}{20} = 1 + r \qquad \text{Simplifying}$$

$$-\frac{20}{20} \pm \frac{21}{20} = r \qquad \text{Adding } -1, \text{ or } -\frac{20}{20}, \text{ to both sides}$$

$$\frac{1}{20} = r \quad or \quad -\frac{41}{20} = r.$$

4. Check.   Since the interest rate cannot be negative, we need only check $\frac{1}{20}$, or 5%. If \$4000 were invested at 5% interest, compounded annually, then in 2 yr it would grow to $4000(1.05)^2$, or \$4410. The number 5% checks.

5. State.   The interest rate was 5%.

EXAMPLE 12   *Free-falling objects.*   The formula $s = 16t^2$ is used to approximate the distance s, in feet, that an object falls freely from rest in t seconds. Ireland's Cliffs of Moher are 702 ft tall (*Source*: Based on data from 4windstravel.com). How long will it take a stone to fall from the top? Round to the nearest tenth of a second.

Solution

1. Familiarize.   We make a drawing to help visualize the problem and agree to disregard air resistance.

2. Translate.   We substitute into the formula:

$$s = 16t^2$$
$$702 = 16t^2.$$

3. Carry out.   We solve for t:

$$702 = 16t^2$$
$$\frac{702}{16} = t^2$$
$$43.875 = t^2$$
$$\sqrt{43.875} = t \qquad \text{Using the principle of square roots;}$$
$$\text{rejecting the negative square root since } t$$
$$\text{cannot be negative in this problem}$$
$$6.6 \approx t. \qquad \text{Using a calculator and rounding to}$$
$$\text{the nearest tenth}$$

4. Check.   Since $16(6.6)^2 \approx 697 \approx 702$, our answer checks.

5. State.   It takes about 6.6 sec for a stone to fall freely from the Cliffs of Moher.

technology connection

As we saw in Section 5.8, a graphing calculator can be used to find approximate solutions of any quadratic equation that has real-number solutions.

To check Example 9(a), we graph $y = x^2 - 8x - 7$ and use the ZERO or ROOT option of the CALC menu. When asked for a Left and Right Bound, we enter cursor positions to the left of and to the right of the root. A Guess between the bounds is entered and a value for the root then appears.

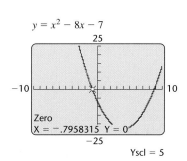

$y = x^2 - 8x - 7$

Zero
X = −.7958315 Y = 0

Yscl = 5

1. Use a graphing calculator to check the second solution of Example 9(a).
2. Use a graphing calculator to confirm the solutions in Examples 7 and 9(b).
3. Can a graphing calculator be used to find *exact* solutions in Example 10? Why or why not?
4. Use a graphing calculator to confirm that there are no real-number solutions of $x^2 - 6x + 11 = 0$.

8.1 Exercise Set

FOR EXTRA HELP

Student's Solutions Manual   Digital Video Tutor CD 6 Videotape 15   Tutor Center AW Math Tutor Center   MathXL Tutorials on CD   MathXL MathXL   MyMathLab MyMathLab

Concept Reinforcement   Complete each of the following to form true statements.

1. The principle of square roots states that if $x^2 = k$, then $x = $ ____ or $x = $ ____.

2. If $(x + 5)^2 = 49$, then $x + 5 = $ ____ or $x + 5 = $ ____.

3. If $t^2 + 6t + 9 = 17$, then $($ ____ $)^2 = 17$ and ____ $= \pm\sqrt{17}$.

4. The equations $x^2 + 8x + $ ____ $= 23$ and $x^2 + 8x = 7$ are equivalent.

5. The expressions $t^2 + 10t + $ ____ and $(t + $ ____ $)^2$ are equivalent.

6. The expressions $x^2 - 6x + $ ____ and $(x - $ ____ $)^2$ are equivalent.

Solve.

7. $4x^2 = 20$

8. $7x^2 = 21$

9. $9x^2 + 16 = 0$

10. $25x^2 + 4 = 0$

11. $5t^2 - 7 = 0$

12. $3t^2 - 2 = 0$

13. $(x - 1)^2 = 49$

14. $(x + 2)^2 = 25$

15. $(a - 13)^2 = 18$

16. $(a + 5)^2 = 8$

17. $(x + 1)^2 = -9$

18. $(x - 1)^2 = -49$

19. $\left(y + \frac{3}{4}\right)^2 = \frac{17}{16}$

20. $\left(t + \frac{3}{2}\right)^2 = \frac{7}{2}$

21. $x^2 - 10x + 25 = 64$

22. $x^2 - 6x + 9 = 100$

23. Let $f(x) = (x - 5)^2$. Find x such that $f(x) = 16$.

24. Let $g(x) = (x - 2)^2$. Find x such that $g(x) = 25$.

25. Let $F(t) = (t + 4)^2$. Find t such that $F(t) = 13$.

26. Let $f(t) = (t + 6)^2$. Find t such that $f(t) = 15$.

Aha! 27. Let $g(x) = x^2 + 14x + 49$. Find x such that $g(x) = 49$.

28. Let $F(x) = x^2 + 8x + 16$. Find x such that $F(x) = 9$.

Replace the blanks in each equation with constants to complete the square and form a true equation.

29. $x^2 + 16x +$ ___ $= (x +$ ___ $)^2$

30. $x^2 + 8x +$ ___ $= (x +$ ___ $)^2$

31. $t^2 - 10t +$ ___ $= (t -$ ___ $)^2$

32. $t^2 - 6t +$ ___ $= (t -$ ___ $)^2$

33. $x^2 + 3x +$ ___ $= (x +$ ___ $)^2$

34. $x^2 + 7x +$ ___ $= (x +$ ___ $)^2$

35. $t^2 - 9t +$ ___ $= (t -$ ___ $)^2$

36. $t^2 - 3t +$ ___ $= (t -$ ___ $)^2$

37. $x^2 + \frac{2}{5}x +$ ___ $= (x +$ ___ $)^2$

38. $x^2 + \frac{2}{3}x +$ ___ $= (x +$ ___ $)^2$

39. $t^2 - \frac{5}{6}t +$ ___ $= (t -$ ___ $)^2$

40. $t^2 - \frac{5}{3}t +$ ___ $= (t -$ ___ $)^2$

Solve by completing the square. Show your work.

41. $x^2 + 6x = 7$

42. $x^2 + 8x = 9$

43. $t^2 - 10t = -24$

44. $t^2 - 10t = -21$

45. $x^2 + 10x + 9 = 0$

46. $x^2 + 8x + 7 = 0$

47. $t^2 + 8t - 3 = 0$

48. $t^2 + 6t - 5 = 0$

Complete the square to find the x-intercepts of each function given by the equation listed.

49. $f(x) = x^2 + 6x + 7$

50. $f(x) = x^2 + 5x + 3$

51. $g(x) = x^2 + 12x + 25$

52. $g(x) = x^2 + 4x + 2$

53. $f(x) = x^2 - 10x - 22$

54. $f(x) = x^2 - 8x - 10$

Solve by completing the square. Remember to first divide, as in Example 10, to make sure that the coefficient of x^2 is 1.

55. $9x^2 + 18x = -8$

56. $4x^2 + 8x = -3$

57. $3x^2 - 5x - 2 = 0$

58. $2x^2 - 5x - 3 = 0$

59. $5x^2 + 4x - 3 = 0$

60. $4x^2 + 3x - 5 = 0$

61. Find the x-intercepts of the function given by $f(x) = 4x^2 + 2x - 3$.

62. Find the x-intercepts of the function given by $f(x) = 3x^2 + x - 5$.

63. Find the x-intercepts of the function given by $g(x) = 2x^2 - 3x - 1$.

64. Find the x-intercepts of the function given by $g(x) = 3x^2 - 5x - 1$.

Interest.   Use $A = P(1 + r)^t$ to find the interest rate in Exercises 65–70. Refer to Example 11.

65. $2000 grows to $2420 in 2 yr

66. $2560 grows to $2890 in 2 yr

67. $1280 grows to $1805 in 2 yr

68. $1000 grows to $1440 in 2 yr

69. $6250 grows to $6760 in 2 yr

70. $6250 grows to $7290 in 2 yr

Free-falling objects.   Use $s = 16t^2$ for Exercises 71–74. Refer to Example 12 and neglect air resistance.

71. Suspended 1053 ft above the water, the bridge over Colorado's Royal Gorge is the world's highest bridge. How long would it take an object to fall freely from the bridge?
Source: *The Guinness Book of Records*

72. The CN Tower in Toronto, at 1815 ft, is the world's tallest self-supporting tower (no guy wires). How long would it take an object to fall freely from the top?
Source: *The Guinness Book of Records*

73. The Sears Tower in Chicago is 1454 ft tall. How long would it take an object to fall freely from the top?

74. The Gateway Arch in St. Louis is 630 ft high. How long would it take an object to fall freely from the top?
Source: www.icivilengineer.com

75. Explain in your own words a sequence of steps that can be used to solve any quadratic equation in the quickest way.

76. Write an interest-rate problem for a classmate to solve. Devise the problem so that the solution is "The loan was made at 7% interest."

SKILL MAINTENANCE

Evaluate. [1.2]

77. $at^2 - bt$, for $a = 3$, $b = 5$, and $t = 4$

78. $mn^2 - mp$, for $m = -2$, $n = 7$, and $p = 3$

Simplify. [7.3]

79. $\sqrt[3]{270}$

80. $\sqrt{80}$

Let $f(x) = \sqrt{3x - 5}$. [7.1]

81. Find $f(10)$.

82. Find $f(18)$.

SYNTHESIS

83. What would be better: to receive 3% interest every 6 months, or to receive 6% interest every 12 months? Why?

84. Write a problem involving a free-falling object for a classmate to solve (see Example 12). Devise the problem so that the solution is "The object takes about 4.5 sec to fall freely from the top of the structure."

Find b such that each trinomial is a square.

85. $x^2 + bx + 81$

86. $x^2 + bx + 49$

87. If $f(x) = 2x^5 - 9x^4 - 66x^3 + 45x^2 + 280x$ and $x^2 - 5$ is a factor of $f(x)$, find all a for which $f(a) = 0$.

88. If $f(x) = \left(x - \frac{1}{3}\right)(x^2 + 6)$ and $g(x) = \left(x - \frac{1}{3}\right)\left(x^2 - \frac{2}{3}\right)$, find all a for which $(f + g)(a) = 0$.

89. *Boating.* A barge and a fishing boat leave a dock at the same time, traveling at a right angle to each other. The barge travels 7 km/h slower than the fishing boat. After 4 hr, the boats are 68 km apart. Find the speed of each boat.

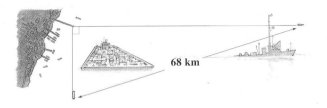

68 km

90. Find three consecutive integers such that the square of the first plus the product of the other two is 67.

91. Exercises 15, 23, and 43 can be solved on a graphing calculator without first rewriting in standard form. Simply let y_1 represent the left side of the equation and y_2 the right side. Then use a graphing calculator to determine the x-coordinate of any point of intersection. Use a graphing calculator to solve Exercises 15, 23, and 43 in this manner.

92. Use a graphing calculator to check your answers to Exercises 5, 11, 61, and 63.

93. Example 11 can be solved with a graphing calculator by graphing each side of

$$4410 = 4000(1 + r)^2.$$

How could you determine, from a reading of the problem, a suitable viewing window? What might that window be?

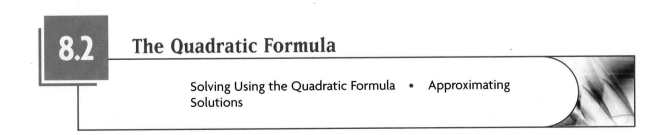

8.2   The Quadratic Formula

Solving Using the Quadratic Formula • Approximating Solutions

There are at least two reasons for learning to complete the square. One is to help graph certain equations that appear in this and other chapters. Another is to develop a general formula for solving quadratic equations.

Solving Using the Quadratic Formula

Each time we solve by completing the square, the procedure is the same. In mathematics, when a procedure is repeated many times, a formula can often be developed to speed up our work.

We begin with a quadratic equation in standard form,

$$ax^2 + bx + c = 0,$$

with $a > 0$. For $a < 0$, a slightly different derivation is needed (see Exercise 58), but the result is the same. Let's solve by completing the square. As the steps are performed, compare them with Example 10 on p. 512.

$$ax^2 + bx = -c \qquad \text{Adding } -c \text{ to both sides}$$

$$x^2 + \frac{b}{a}x = -\frac{c}{a} \qquad \text{Dividing both sides by } a$$

Half of $\dfrac{b}{a}$ is $\dfrac{b}{2a}$ and $\left(\dfrac{b}{2a}\right)^2$ is $\dfrac{b^2}{4a^2}$. We add $\dfrac{b^2}{4a^2}$ to both sides:

$$x^2 + \frac{b}{a}x + \frac{b^2}{4a^2} = -\frac{c}{a} + \frac{b^2}{4a^2} \qquad \text{Adding } \dfrac{b^2}{4a^2} \text{ to complete the square}$$

$$\left(x + \frac{b}{2a}\right)^2 = -\frac{4ac}{4a^2} + \frac{b^2}{4a^2} \qquad \text{Factoring on the left side; finding a common denominator on the right side}$$

$$\left(x + \frac{b}{2a}\right)^2 = \frac{b^2 - 4ac}{4a^2}$$

$$x + \frac{b}{2a} = \pm\frac{\sqrt{b^2 - 4ac}}{2a} \qquad \text{Using the principle of square roots and the quotient rule for radicals; since } a > 0, \sqrt{4a^2} = 2a$$

$$x = \frac{-b \pm \sqrt{b^2 - 4ac}}{2a}. \qquad \text{Adding } -\dfrac{b}{2a} \text{ to both sides}$$

It is important to remember the quadratic formula and know how to use it.

The Quadratic Formula

The solutions of $ax^2 + bx + c = 0$, $a \neq 0$, are given by

$$x = \frac{-b \pm \sqrt{b^2 - 4ac}}{2a}.$$

EXAMPLE 1   Solve $5x^2 + 8x = -3$ using the quadratic formula.

Solution   We first find standard form and determine a, b, and c:

$$5x^2 + 8x + 3 = 0; \qquad \text{Adding 3 to both sides to get 0 on one side}$$

$$a = 5, \quad b = 8, \quad c = 3.$$

Next, we use the quadratic formula:

$$x = \frac{-b \pm \sqrt{b^2 - 4ac}}{2a}$$

$$x = \frac{-8 \pm \sqrt{8^2 - 4 \cdot 5 \cdot 3}}{2 \cdot 5}$$ Substituting

$$x = \frac{-8 \pm \sqrt{64 - 60}}{10}$$

Be sure to write the fraction bar all the way across.

$$x = \frac{-8 \pm \sqrt{4}}{10} = \frac{-8 \pm 2}{10}$$

$$x = \frac{-8 + 2}{10} \quad or \quad x = \frac{-8 - 2}{10}$$

$$x = \frac{-6}{10} \quad or \quad x = \frac{-10}{10}$$

$$x = -\frac{3}{5} \quad or \quad x = -1.$$

The solutions are $-\frac{3}{5}$ and -1. The checks are left to the student.

Because $5x^2 + 8x + 3$ can be factored, the quadratic formula may not have been the fastest way of solving Example 1. However, because the quadratic formula works for *any* quadratic equation, we need not spend too much time struggling to solve a quadratic equation by factoring.

Study Skills

Know It "By Heart"

When memorizing something like the quadratic formula, try to first understand and write out the derivation. Doing this two or three times will help you remember the formula.

To Solve a Quadratic Equation

1. If the equation can be easily written in the form $ax^2 = p$ or $(x + k)^2 = d$, use the principle of square roots as in Section 8.1.
2. If step (1) does not apply, write the equation in the form $ax^2 + bx + c = 0$.
3. Try factoring and using the principle of zero products.
4. If factoring seems difficult or impossible, use the quadratic formula. Completing the square can also be used, but is slower.

The solutions of a quadratic equation can always be found using the quadratic formula. They cannot always be found by factoring.

Recall that a second-degree polynomial in one variable is said to be quadratic. Similarly, a second-degree polynomial function in one variable is said to be a **quadratic function.**

EXAMPLE 2   For the quadratic function given by $f(x) = 3x^2 - 6x - 4$, find all x for which $f(x) = 0$.

Solution   We substitute and solve for x:

$$f(x) = 0$$

$$3x^2 - 6x - 4 = 0 \qquad \text{Substituting. You can try to solve this by factoring.}$$

$$a = 3; \quad b = -6; \quad c = -4.$$

We then substitute into the quadratic formula:

$$x = \frac{-(-6) \pm \sqrt{(-6)^2 - 4 \cdot 3 \cdot (-4)}}{2 \cdot 3}$$

$$= \frac{6 \pm \sqrt{36 + 48}}{6}$$

$$= \frac{6 \pm \sqrt{84}}{6} \qquad \text{Note that 4 is a perfect-square factor of 84.}$$

$$= \frac{6}{6} \pm \frac{\sqrt{84}}{6}$$

$$= 1 \pm \frac{\sqrt{4}\sqrt{21}}{6} \qquad 84 = 4 \cdot 21$$

$$\left.\begin{array}{l} = 1 \pm \dfrac{2\sqrt{21}}{6} \\[2mm] = 1 \pm \dfrac{\sqrt{21}}{3}. \end{array}\right\} \qquad \text{Simplifying}$$

The solutions are $1 - \dfrac{\sqrt{21}}{3}$ and $1 + \dfrac{\sqrt{21}}{3}$. The checks are left to the student.

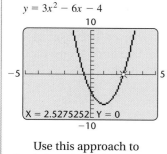

technology connection

To check Example 2 by graphing $y_1 = 3x^2 - 6x - 4$, press [TRACE] and enter $1 + \sqrt{21}/3$. A rational approximation and the y-value 0 should appear.

$y = 3x^2 - 6x - 4$

X = 2.5275252 Y = 0

Use this approach to check the other solution of Example 2.

Some quadratic equations have solutions that are imaginary numbers.

EXAMPLE 3   Solve: $x^2 + 2 = -x$.

Solution   We first find standard form:

$$x^2 + x + 2 = 0. \qquad \text{Adding } x \text{ to both sides}$$

Since we cannot factor $x^2 + x + 2$, we use the quadratic formula with $a = 1$, $b = 1$, and $c = 2$:

$$x = \frac{-1 \pm \sqrt{1^2 - 4 \cdot 1 \cdot 2}}{2 \cdot 1} \qquad \text{Substituting}$$

$$= \frac{-1 \pm \sqrt{1 - 8}}{2}$$

$$= \frac{-1 \pm \sqrt{-7}}{2}$$

$$= \frac{-1 \pm i\sqrt{7}}{2}, \text{ or } -\frac{1}{2} \pm \frac{\sqrt{7}}{2}i.$$

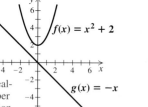

$f(x) = x^2 + 2$

$g(x) = -x$

No real-number solution exists

A visualization of Example 3

The solutions are $-\dfrac{1}{2} - \dfrac{\sqrt{7}}{2}i$ and $-\dfrac{1}{2} + \dfrac{\sqrt{7}}{2}i$. The checks are left to the student.

The quadratic formula can be used to solve certain rational equations.

EXAMPLE 4   If $f(x) = 2 + \dfrac{7}{x}$ and $g(x) = \dfrac{4}{x^2}$, find all x for which $f(x) = g(x)$.

Solution   We set $f(x)$ equal to $g(x)$ and solve:

$$f(x) = g(x)$$

$$2 + \frac{7}{x} = \frac{4}{x^2}.\qquad \text{Substituting. Note that } x \neq 0.$$

This is a rational equation similar to those in Section 6.4. To solve, we multiply both sides by the LCD, x^2:

$$x^2 \left(2 + \frac{7}{x} \right) = x^2 \cdot \frac{4}{x^2}$$

$$2x^2 + 7x = 4 \qquad \text{Simplifying}$$

$$2x^2 + 7x - 4 = 0. \qquad \text{Subtracting 4 from both sides}$$

We have

$$a = 2, \quad b = 7, \quad \text{and} \quad c = -4.$$

Substituting then gives us

$$x = \frac{-7 \pm \sqrt{7^2 - 4 \cdot 2 \cdot (-4)}}{2 \cdot 2}$$

$$= \frac{-7 \pm \sqrt{49 + 32}}{4}$$

$$= \frac{-7 \pm \sqrt{81}}{4}$$

$$= \frac{-7 \pm 9}{4}$$

$$x = \frac{2}{4} = \frac{1}{2} \text{ or } x = \frac{-16}{4} = -4. \qquad \begin{array}{l}\text{Both answers should check}\\ \text{since } x \neq 0.\end{array}$$

You can confirm that $f\left(\tfrac{1}{2}\right) = g\left(\tfrac{1}{2}\right)$ and $f(-4) = g(-4)$. The solutions are $\tfrac{1}{2}$ and -4.

technology connection

We saw in Sections 5.8 and 8.1 how graphing calculators can solve quadratic equations. To determine whether quadratic equations are solved more quickly on a graphing calculator or by using the quadratic formula, solve Examples 2 and 4 both ways. Which method is faster? Which method is more precise? Why?

Approximating Solutions

When the solution of an equation is irrational, a rational-number approximation is often useful. This is often the case in real-world applications similar to those found in Section 8.3.

EXAMPLE 5   Use a calculator to approximate the solutions of Example 2.

Student Notes

It is important that you understand both the rules for order of operations *and* the manner in which your calculator applies those rules.

Solution   On most calculators, one of the following sequences of keystrokes can be used to approximate $1 + \sqrt{21}/3$:

Similar keystrokes can be used to approximate $1 - \sqrt{21}/3$.
The solutions are approximately 2.527525232 and -0.5275252317.

<table>
<tr><td>

Exercise Set

8.2
</td></tr>
</table>

FOR EXTRA HELP

Student's Solutions Manual   Digital Video Tutor CD 6 Videotape 15   AW Math Tutor Center   MathXL Tutorials on CD   MathXL   MyMathLab

---

🐟 *Concept Reinforcement*   *Classify each of the following as either true or false.*

1. The quadratic formula can be used to solve *any* quadratic equation.

2. The quadratic formula does not work if solutions are imaginary numbers.

3. Solving by factoring is always slower than using the quadratic formula.

4. The steps used to derive the quadratic formula are the same as those used when solving by completing the square.

5. A quadratic equation can have as many as four solutions.

6. It is possible for a quadratic equation to have no real-number solutions.

Solve.

7. $x^2 + 7x - 3 = 0$

8. $x^2 - 7x + 4 = 0$

9. $3p^2 = 18p - 6$

10. $3u^2 = 8u - 5$

11. $x^2 + x + 1 = 0$

12. $x^2 + x + 2 = 0$

13. $x^2 + 13 = 4x$

14. $x^2 + 13 = 6x$

15. $h^2 + 4 = 6h$

16. $r^2 + 3r = 8$

17. $\dfrac{1}{x^2} - 3 = \dfrac{8}{x}$

18. $\dfrac{9}{x} - 2 = \dfrac{5}{x^2}$

19. $3x + x(x - 2) = 4$

20. $4x + x(x - 3) = 5$

21. $12t^2 + 9t = 1$

22. $15t^2 + 7t = 2$

23. $25x^2 - 20x + 4 = 0$

24. $36x^2 + 84x + 49 = 0$

25. $7x(x + 2) + 5 = 3x(x + 1)$

26. $5x(x - 1) - 7 = 4x(x - 2)$

27. $14(x - 4) - (x + 2) = (x + 2)(x - 4)$

28. $11(x - 2) + (x - 5) = (x + 2)(x - 6)$

29. $5x^2 = 13x + 17$

30. $25x = 3x^2 + 28$

31. $x^2 + 9 = 4x$

32. $x^2 + 7 = 3x$

33. $x^3 - 8 = 0$ (*Hint*: Factor the difference of cubes. Then use the quadratic formula.)

34. $x^3 + 1 = 0$

35. Let $f(x) = 3x^2 - 5x + 2$. Find x such that $f(x) = 0$.

36. Let $g(x) = 4x^2 - 2x - 3$. Find x such that $g(x) = 0$.

37. Let
$$f(x) = \frac{7}{x} + \frac{7}{x + 4}.$$
Find all x for which $f(x) = 1$.

38. Let
$$g(x) = \frac{2}{x} + \frac{2}{x + 3}.$$
Find all x for which $g(x) = 1$.

39. Let
$$F(x) = \frac{x + 3}{x} \quad \text{and} \quad G(x) = \frac{x - 4}{3}.$$
Find all x for which $F(x) = G(x)$.

40. Let
$$f(x) = \frac{3 - x}{4} \quad \text{and} \quad g(x) = \frac{1}{4x}.$$
Find all x for which $f(x) = g(x)$.

41. Let
$$f(x) = \frac{15 - 2x}{6} \quad \text{and} \quad g(x) = \frac{3}{x}.$$
Find all x for which $f(x) = g(x)$.

42. Let
$$f(x) = x + 5 \quad \text{and} \quad g(x) = \frac{3}{x - 5}.$$
Find all x for which $f(x) = g(x)$.

Solve. Use a calculator to approximate, as precisely as possible, the solutions as rational numbers.

43. $x^2 + 4x - 7 = 0$

44. $x^2 + 6x + 4 = 0$

45. $x^2 - 6x + 4 = 0$

46. $x^2 - 4x + 1 = 0$

47. $2x^2 - 3x - 7 = 0$

48. $3x^2 - 3x - 2 = 0$

49. Are there any equations that can be solved by the quadratic formula but not by completing the square? Why or why not?

50. If you had to choose between remembering the method of completing the square and remembering the quadratic formula, which would you choose? Why?

SKILL MAINTENANCE

51. *Coffee beans.* Twin Cities Roasters has Kenyan coffee for which they pay $6.75 a pound and Kona coffee for which they pay $11.25 a pound. How much of each kind should be mixed in order to obtain a 50-lb mixture that costs them $8.55 a pound? [3.3]

52. *Donuts.* South Street Bakers charges $1.10 for a cream-filled donut and 85¢ for a glazed donut. On a recent Sunday, a total of 90 glazed and cream-filled donuts were sold for $88.00. How many of each type were sold? [3.3]

Simplify.

53. $\sqrt{27a^2b^5} \cdot \sqrt{6a^3b}$ [7.3]

54. $\sqrt{8a^3b} \cdot \sqrt{12ab^5}$ [7.3]

55. $\dfrac{\dfrac{3}{x - 1}}{\dfrac{1}{x + 1} + \dfrac{2}{x - 1}}$ [6.3]

56. $\dfrac{\dfrac{4}{a^2b}}{\dfrac{3}{a} - \dfrac{4}{b^2}}$ [6.3]

SYNTHESIS

57. Suppose you had a large number of quadratic equations to solve and none of the equations had a constant term. Would you use factoring or the quadratic formula to solve these equations? Why?

58. If $a < 0$ and $ax^2 + bx + c = 0$, then $-a$ is positive and the equivalent equation, $-ax^2 - bx - c = 0$, can be solved using the quadratic formula.

a) Find this solution, replacing a, b, and c in the formula with $-a$, $-b$, and $-c$ from the equation.

b) How does the result of part (a) indicate that the quadratic formula "works" regardless of the sign of a?

For Exercises 59–61, let

$$f(x) = \frac{x^2}{x-2} + 1 \quad and \quad g(x) = \frac{4x-2}{x-2} + \frac{x+4}{2}.$$

59. Find the x-intercepts of the graph of f.

60. Find the x-intercepts of the graph of g.

61. Find all x for which $f(x) = g(x)$.

Solve.

62. $x^2 - 0.75x - 0.5 = 0$

63. $z^2 + 0.84z - 0.4 = 0$

64. $\left(1 + \sqrt{3}\right)x^2 - \left(3 + 2\sqrt{3}\right)x + 3 = 0$

65. $\sqrt{2}x^2 + 5x + \sqrt{2} = 0$

66. $ix^2 - 2x + 1 = 0$

67. One solution of $kx^2 + 3x - k = 0$ is -2. Find the other.

68. Use a graphing calculator to solve Exercises 9, 23, and 39.

69. Use a graphing calculator to solve Exercises 15, 31, and 37. Use the method of graphing each side of the equation.

70. Can a graphing calculator be used to solve *any* quadratic equation? Why or why not?

8.3 Applications Involving Quadratic Equations

Solving Problems • Solving Formulas

Solving Problems

As we found in Section 6.5, some problems translate to rational equations. The solution of such rational equations can involve quadratic equations.

EXAMPLE 1 Motorcycle travel. Makita rode her motorcycle 300 mi at a certain average speed. Had she averaged 10 mph more, the trip would have taken 1 hr less. Find the average speed of the motorcycle.

Solution

1. **Familiarize.** We make a drawing, labeling it with the information provided. As in Section 6.5, we can create a table. We let r represent the rate, in miles per hour, and t the time, in hours, for Makita's trip.

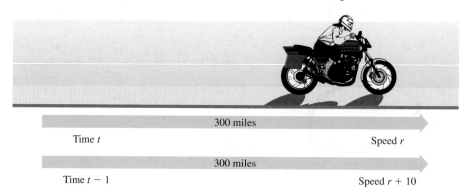

Time t 300 miles   Speed r

Time $t - 1$ 300 miles   Speed $r + 10$

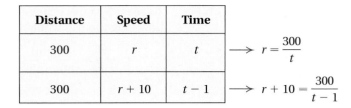

Distance	Speed	Time
300	r	t
300	$r + 10$	$t - 1$

Recall that the definition of speed, $r = d/t$, relates the three quantities.

2. **Translate.**   From the table, we obtain

$$r = \frac{300}{t} \quad \text{and} \quad r + 10 = \frac{300}{t - 1}.$$

3. **Carry out.**   A system of equations has been formed. We substitute for r from the first equation into the second and solve the resulting equation:

$$\frac{300}{t} + 10 = \frac{300}{t - 1} \qquad \text{Substituting } 300/t \text{ for } r$$

$$t(t - 1) \cdot \left[\frac{300}{t} + 10\right] = t(t - 1) \cdot \frac{300}{t - 1} \qquad \text{Multiplying by the LCD}$$

$$\cancel{t}(t - 1) \cdot \frac{300}{\cancel{t}} + t(t - 1) \cdot 10 = t\cancel{(t - 1)} \cdot \frac{300}{\cancel{t - 1}} \qquad \begin{array}{l}\text{Using the} \\ \text{distributive law} \\ \text{and removing} \\ \text{factors that} \\ \text{equal 1: } \dfrac{t}{t} = 1; \\ \dfrac{t - 1}{t - 1} = 1\end{array}$$

$$\left.\begin{array}{l}300(t - 1) + 10(t^2 - t) = 300t \\ 300t - 300 + 10t^2 - 10t = 300t \\ 10t^2 - 10t - 300 = 0\end{array}\right\} \qquad \begin{array}{l}\text{Rewriting in} \\ \text{standard form}\end{array}$$

$$t^2 - t - 30 = 0 \qquad \begin{array}{l}\text{Multiplying by } \frac{1}{10} \\ \text{or dividing by 10}\end{array}$$

$$(t - 6)(t + 5) = 0 \qquad \text{Factoring}$$

$$t = 6 \quad \text{or} \quad t = -5. \qquad \begin{array}{l}\text{Principle of zero} \\ \text{products}\end{array}$$

4. **Check.**   Note that we have solved for t, not r as required. Since negative time has no meaning here, we disregard the -5 and use 6 hr to find r:

$$r = \frac{300 \text{ mi}}{6 \text{ hr}} = 50 \text{ mph}.$$

> *Caution!*   Always make sure that you find the quantity asked for in the problem.

To see if 50 mph checks, we increase the speed 10 mph to 60 mph and see how long the trip would have taken at that speed:

$$t = \frac{d}{r} = \frac{300 \text{ mi}}{60 \text{ mph}} = 5 \text{ hr.}$$ Note that mi/mph = mi $\div \dfrac{\text{mi}}{\text{hr}}$ =

$$\cancel{\text{mi}} \cdot \frac{\text{hr}}{\cancel{\text{mi}}} = \text{hr.}$$

This is 1 hr less than the trip actually took, so the answer checks.

5. State.   Makita's motorcycle traveled at an average speed of 50 mph.

Solving Formulas

Recall that to solve a formula for a certain letter, we use the principles for solving equations to get that letter alone on one side.

EXAMPLE 2   *Period of a pendulum.*   The time T required for a pendulum of length l to swing back and forth (complete one period) is given by the formula $T = 2\pi\sqrt{l/g}$, where g is the earth's gravitational constant. Solve for l.

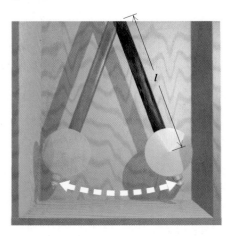

Solution   We have

$$T = 2\pi\sqrt{\frac{l}{g}}$$ This is a radical equation (see Section 7.6).

$$T^2 = \left(2\pi\sqrt{\frac{l}{g}}\right)^2$$ Principle of powers (squaring both sides)

$$T^2 = 2^2\pi^2\frac{l}{g}$$

$$gT^2 = 4\pi^2 l$$ Multiplying both sides by g to clear fractions

$$\frac{gT^2}{4\pi^2} = l.$$ Dividing both sides by $4\pi^2$

We now have l alone on one side and l does not appear on the other side, so the formula is solved for l.

In formulas for which variables represent only nonnegative numbers, there is no need for absolute-value signs when taking square roots.

EXAMPLE 3   Hang time.*   An athlete's *hang time* is the amount of time that the athlete can remain airborne when jumping. A formula relating an athlete's vertical leap V, in inches, to hang time T, in seconds, is $V = 48T^2$. Solve for T.

Solution   We have

$$48T^2 = V$$

$$T^2 = \frac{V}{48} \qquad \text{Dividing by 48 to isolate } T^2$$

$$T = \frac{\sqrt{V}}{\sqrt{48}} \qquad \begin{array}{l}\text{Using the principle of square roots} \\ \text{and the quotient rule for radicals.} \\ \text{We assume } V, T \geq 0.\end{array}$$

$$\left. \begin{array}{l} = \dfrac{\sqrt{V}}{\sqrt{16}\,\sqrt{3}} = \dfrac{\sqrt{V}}{4\sqrt{3}} \\[2mm] = \dfrac{\sqrt{V}}{4\sqrt{3}} \cdot \dfrac{\sqrt{3}}{\sqrt{3}} = \dfrac{\sqrt{3V}}{12}. \end{array} \right\} \quad \text{Rationalizing the denominator}$$

EXAMPLE 4   Falling distance.   An object tossed downward with an initial speed (velocity) of v_0 will travel a distance of s meters, where $s = 4.9t^2 + v_0 t$ and t is measured in seconds. Solve for t.

Solution   Since t is squared in one term and raised to the first power in the other term, the equation is quadratic in t.

$$4.9t^2 + v_0 t = s$$

$$4.9t^2 + v_0 t - s = 0 \qquad \text{Writing standard form}$$

$$a = 4.9, \quad b = v_0, \quad c = -s$$

$$t = \frac{-v_0 \pm \sqrt{v_0^2 - 4(4.9)(-s)}}{2(4.9)} \qquad \text{Using the quadratic formula}$$

*This formula is taken from an article by Peter Brancazio, "The Mechanics of a Slam Dunk," *Popular Mechanics*, November 1991. Courtesy of Professor Peter Brancazio, Brooklyn College.

Since the negative square root would yield a negative value for t, we use only the positive root:

$$t = \frac{-v_0 + \sqrt{v_0^2 + 19.6s}}{9.8}.$$

The following list of steps should help you when solving formulas for a given letter. Try to remember that when solving a formula, you use the same approach that you would to solve an equation.

Student Notes

After identifying which numbers to use as a, b, and c, be careful to replace only the *letters* in the quadratic formula.

To Solve a Formula for a Letter—Say, b

1. Clear fractions and use the principle of powers, as needed. Perform these steps until radicals containing b are gone and b is not in any denominator.
2. Combine all like terms.
3. If the only power of b is b^1, the equation can be solved as in Sections 1.5 and 6.8.
4. If b^2 appears but b does not, solve for b^2 and use the principle of square roots to solve for b.
5. If there are terms containing both b and b^2, put the equation in standard form and use the quadratic formula.

Exercise Set

8.3

FOR EXTRA HELP

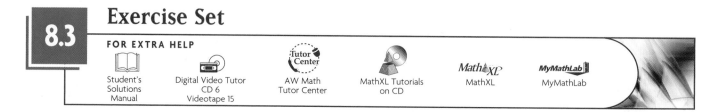

Student's Solutions Manual | Digital Video Tutor CD 6 Videotape 15 | AW Math Tutor Center | MathXL Tutorials on CD | MathXL | MyMathLab

Solve.

1. *Car trips.* During the first part of a trip, Trudy's Honda traveled 120 mi at a certain speed. Trudy then drove another 100 mi at a speed that was 10 mph slower. If the total time of Trudy's trip was 4 hr, what was her speed on each part of the trip?

2. *Canoeing.* During the first part of a canoe trip, Tim covered 60 km at a certain speed. He then traveled 24 km at a speed that was 4 km/h slower. If the total time for the trip was 8 hr, what was the speed on each part of the trip?

3. *Car trips.* Petra's Plymouth travels 200 mi averaging a certain speed. If the car had gone 10 mph faster, the trip would have taken 1 hr less. Find Petra's average speed.

4. *Car trips.* Sandi's Subaru travels 280 mi averaging a certain speed. If the car had gone 5 mph faster, the trip would have taken 1 hr less. Find Sandi's average speed.

5. *Air travel.* A Cessna flies 600 mi at a certain speed. A Beechcraft flies 1000 mi at a speed that is 50 mph faster, but takes 1 hr longer. Find the speed of each plane.

6. *Air travel.*   A turbo-jet flies 50 mph faster than a super-prop plane. If a turbo-jet goes 2000 mi in 3 hr less time than it takes the super-prop to go 2800 mi, find the speed of each plane.

7. *Bicycling.*   Naoki bikes the 40 mi to Hillsboro averaging a certain speed. The return trip is made at a speed that is 6 mph slower. Total time for the round trip is 14 hr. Find Naoki's average speed on each part of the trip.

8. *Car speed.*   On a sales trip, Gail drives the 600 mi to Richmond averaging a certain speed. The return trip is made at an average speed that is 10 mph slower. Total time for the round trip is 22 hr. Find Gail's average speed on each part of the trip.

9. *Navigation.*   The Hudson River flows at a rate of 3 mph. A patrol boat travels 60 mi upriver and returns in a total time of 9 hr. What is the speed of the boat in still water?

10. *Navigation.*   The current in a typical Mississippi River shipping route flows at a rate of 4 mph. In order for a barge to travel 24 mi upriver and then return in a total of 5 hr, approximately how fast must the barge be able to travel in still water?

11. *Filling a pool.*   A well and a spring are filling a swimming pool. Together, they can fill the pool in 4 hr. The well, working alone, can fill the pool in 6 hr less time than the spring. How long would the spring take, working alone, to fill the pool?

12. *Filling a tank.*   Two pipes are connected to the same tank. Working together, they can fill the tank in 2 hr. The larger pipe, working alone, can fill the tank in 3 hr less time than the smaller one. How long would the smaller one take, working alone, to fill the tank?

13. *Paddleboats.*   Ellen paddles 1 mi upstream and 1 mi back in a total time of 1 hr. The speed of the river is 2 mph. Find the speed of Ellen's paddleboat in still water.

14. *Rowing.*   Dan rows 10 km upstream and 10 km back in a total time of 3 hr. The speed of the river is 5 km/h. Find Dan's speed in still water.

Solve each formula for the indicated letter. Assume that all variables represent nonnegative numbers.

15. $A = 4\pi r^2$, for r
(Surface area of a sphere of radius r)

16. $A = 6s^2$, for s
(Surface area of a cube with sides of length s)

17. $A = 2\pi r^2 + 2\pi rh$, for r
(Surface area of a right cylindrical solid with radius r and height h)

18. $F = \dfrac{Gm_1 m_2}{r^2}$, for r
(Law of gravity)

19. $N = \dfrac{kQ_1 Q_2}{s^2}$, for s
(Number of phone calls between two cities)

20. $A = \pi r^2$, for r
(Area of a circle)

21. $T = 2\pi \sqrt{\dfrac{l}{g}}$, for g
(A pendulum formula)

22. $a^2 + b^2 = c^2$, for b
(Pythagorean formula in two dimensions)

23. $a^2 + b^2 + c^2 = d^2$, for c
(Pythagorean formula in three dimensions)

24. $N = \dfrac{k^2 - 3k}{2}$, for k
(Number of diagonals of a polygon with k sides)

25. $s = v_0 t + \dfrac{gt^2}{2}$, for t
(A motion formula)

26. $A = \pi r^2 + \pi rs$, for r
(Surface area of a cone)

27. $N = \frac{1}{2}(n^2 - n)$, for n
(Number of games if n teams play each other once)

28. $A = A_0 (1 - r)^2$, for r
(A business formula)

29. $V = 3.5\sqrt{h}$, for h
(Distance to horizon from a height)

30. $W = \sqrt{\dfrac{1}{LC}}$, for L
(An electricity formula)

Aha!   **31.** $at^2 + bt + c = 0$, for t
(An algebraic formula)

32. $A = P_1(1 + r)^2 + P_2(1 + r)$, for r
(Amount in an account when P_1 is invested for 2 yr and P_2 for 1 yr at interest rate r)

Solve.

33. *Falling distance.* (Use $4.9t^2 + v_0t = s$.)

   a) A bolt falls off an airplane at an altitude of 500 m. Approximately how long does it take the bolt to reach the ground?

   b) A ball is thrown downward at a speed of 30 m/sec from an altitude of 500 m. Approximately how long does it take the ball to reach the ground?

   c) Approximately how far will an object fall in 5 sec, when thrown downward at an initial velocity of 30 m/sec from a plane?

34. *Falling distance.* (Use $4.9t^2 + v_0t = s$.)

   a) A ring is dropped from a helicopter at an altitude of 75 m. Approximately how long does it take the ring to reach the ground?

   b) A coin is tossed downward with an initial velocity of 30 m/sec from an altitude of 75 m. Approximately how long does it take the coin to reach the ground?

   c) Approximately how far will an object fall in 2 sec, if thrown downward at an initial velocity of 20 m/sec from a helicopter?

35. *Bungee jumping.* Jesse is tied to one end of a 40-m elasticized (bungee) cord. The other end of the cord is tied to the middle of a bridge. If Jesse jumps off the bridge, for how long will he fall before the cord begins to stretch? (Use $4.9t^2 = s$.)

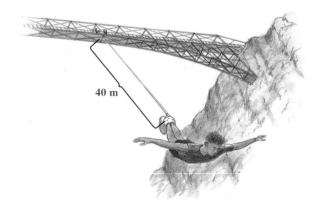

40 m

36. *Bungee jumping.* Sheila is tied to a bungee cord (see Exercise 35) and falls for 2.5 sec before her cord begins to stretch. How long is the bungee cord?

37. *Hang time.* The NBA's Steve Francis reportedly has a vertical leap of 45 in. What is his hang time? (Use $V = 48T^2$.)
Source: www.maximonline.com

38. *League schedules.* In a bowling league, each team plays each of the other teams once. If a total of 66 games is played, how many teams are in the league? (See Exercise 27.)

For Exercises 39 and 40, use $4.9t^2 + v_0t = s$.

39. *Downward speed.* An object thrown downward from a 100-m cliff travels 51.6 m in 3 sec. What was the initial velocity of the object?

40. *Downward speed.* An object thrown downward from a 200-m cliff travels 91.2 m in 4 sec. What was the initial velocity of the object?

For Exercises 41 and 42, use $A = P_1(1 + r)^2 + P_2(1 + r)$. (See Exercise 32.)

41. *Compound interest.* A firm invests $3000 in a savings account for 2 yr. At the beginning of the second year, an additional $1700 is invested. If a total of $5253.70 is in the account at the end of the second year, what is the annual interest rate?

42. *Compound interest.* A business invests $10,000 in a savings account for 2 yr. At the beginning of the second year, an additional $3500 is invested. If a total of $15,569.75 is in the account at the end of the second year, what is the annual interest rate?

43. Marti is tied to a bungee cord that is twice as long as the cord tied to Pedro. Will Marti's fall take twice as long as Pedro's before their cords begin to stretch? Why or why not? (See Exercises 35 and 36.)

44. Under what circumstances would a negative value for t, time, have meaning?

SKILL MAINTENANCE

Evaluate.

45. $b^2 - 4ac$, for $a = 5$, $b = 6$, and $c = 7$ [1.2]

46. $\sqrt{b^2 - 4ac}$, for $a = 3$, $b = 4$, and $c = 5$ [7.8]

Simplify.

47. $\dfrac{x^2 + xy}{2x}$ [6.1]

48. $\dfrac{a^3 - ab^2}{ab}$ [6.1]

49. $\dfrac{3 + \sqrt{45}}{6}$ [7.3]

50. $\dfrac{2 - \sqrt{28}}{10}$ [7.3]

SYNTHESIS

51. Write a problem for a classmate to solve. Devise the problem so that **(a)** the solution is found after solving a rational equation and **(b)** the solution is "The express train travels 90 mph."

52. In what ways do the motion problems of this section (like Example 1) differ from the motion problems in Chapter 6 (see p. 394)?

53. *Biochemistry.* The equation
$$A = 6.5 - \frac{20.4t}{t^2 + 36}$$
is used to calculate the acid level A in a person's blood t minutes after sugar is consumed. Solve for t.

54. *Special relativity.* Einstein found that an object with initial mass m_0 and traveling velocity v has mass
$$m = \frac{m_0}{\sqrt{1 - \dfrac{v^2}{c^2}}},$$
where c is the speed of light. Solve the formula for c.

55. Find a number for which the reciprocal of 1 less than the number is the same as 1 more than the number.

56. *Purchasing.* A discount store bought a quantity of beach towels for $250 and sold all but 15 at a profit of $3.50 per towel. With the total amount received, the manager could buy 4 more than twice as many as were bought before. Find the cost per towel.

57. *Art and aesthetics.* For over 2000 yr, artists, sculptors, and architects have regarded the proportions of a "golden" rectangle as visually appealing. A rectangle of width w and length l is considered "golden" if
$$\frac{w}{l} = \frac{l}{w + l}.$$
Solve for l.

58. *Diagonal of a cube.* Find a formula that expresses the length of the three-dimensional diagonal of a cube as a function of the cube's surface area.

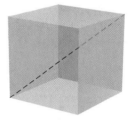

59. Solve for n:
$$mn^4 - r^2pm^3 - r^2n^2 + p = 0.$$

60. *Surface area.* Find a formula that expresses the diameter of a right cylindrical solid as a function of its surface area and its height. (See Exercise 17.)

61. A sphere is inscribed in a cube as shown in the figure below. Express the surface area of the sphere as a function of the surface area S of the cube. (See Exercise 15.)

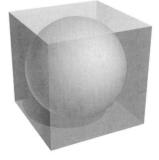

Studying Solutions of Quadratic Equations

The Discriminant   •   Writing Equations from Solutions

Study Skills

When Something Seems Easy

Every so often, you may encounter a lesson that you remember from a previous math course. When this occurs, make sure that *all* of that lesson is understood. If anything is amiss, be sure to review all tricky spots.

The Discriminant

It is sometimes enough to know what *type* of number a solution will be, without actually solving the equation. Suppose we want to know if $4x^2 - 5x - 2 = 0$ has rational solutions (and thus can be solved by factoring). Using the quadratic formula, we would have

$$x = \frac{-b \pm \sqrt{b^2 - 4ac}}{2a}$$

$$= \frac{-(-5) \pm \sqrt{(-5)^2 - 4 \cdot 4(-2)}}{2 \cdot 4}.$$

Note that the radicand, $(-5)^2 - 4 \cdot 4 \cdot (-2)$, determines what type of number the solutions will be. Since $(-5)^2 - 4 \cdot 4 \cdot (-2) = 25 - 16(-2) = 25 + 32 = 57$ and since 57 is not a perfect square, we know that the solutions of the equation are not rational numbers. This means that $4x^2 - 5x - 2 = 0$ *cannot* be solved by factoring.

It is $b^2 - 4ac$, known as the **discriminant**, that determines what type of number the solutions of a quadratic equation are. If a, b, and c are rational, then:

- When $b^2 - 4ac$ simplifies to 0, it doesn't matter if we use $+\sqrt{b^2 - 4ac}$ or $-\sqrt{b^2 - 4ac}$; we get the same solution twice. Thus, when the discriminant is 0, there is one *repeated* solution and it is rational.

  Example: $9x^2 + 6x + 1 = 0 \rightarrow b^2 - 4ac = 6^2 - 4 \cdot 9 \cdot 1 = 0$.

- When $b^2 - 4ac$ is positive, there are two different real-number solutions: If $b^2 - 4ac$ is a perfect square, these solutions are rational numbers.

  Example: $6x^2 + 5x + 1 = 0 \rightarrow b^2 - 4ac = 5^2 - 4 \cdot 6 \cdot 1 = 1$.

- When $b^2 - 4ac$ is positive but not a perfect square, there are two irrational solutions and they are conjugates of each other (see p. 469).

  Example: $2x^2 + 4x + 1 = 0 \rightarrow b^2 - 4ac = 4^2 - 4 \cdot 2 \cdot 1 = 8$.

- When the discriminant is negative, there are two imaginary-number solutions and they are complex conjugates of each other.

  Example: $3x^2 + 2x + 1 = 0 \rightarrow b^2 - 4ac = 2^2 - 4 \cdot 3 \cdot 1 = -8$.

Note that any equation for which $b^2 - 4ac$ is a perfect square can be solved by factoring.

Discriminant $b^2 - 4ac$	Nature of Solutions
0	One solution; a rational number

Positive
  Perfect square
  Not a perfect square
	Two different real-number solutions
	  Solutions are rational.
	  Solutions are irrational conjugates.

Negative	Two different imaginary-number solutions (complex conjugates)

EXAMPLE 1   For each equation, determine what type of number the solutions are and how many solutions exist.

a) $9x^2 - 12x + 4 = 0$

b) $x^2 + 5x + 8 = 0$

c) $2x^2 + 7x - 3 = 0$

Solution

a) For $9x^2 - 12x + 4 = 0$, we have

$$a = 9, \quad b = -12, \quad c = 4.$$

We substitute and compute the discriminant:

$$b^2 - 4ac = (-12)^2 - 4 \cdot 9 \cdot 4$$
$$= 144 - 144 = 0.$$

There is exactly one solution, and it is rational. This indicates that $9x^2 - 12x + 4 = 0$ can be solved by factoring.

b) For $x^2 + 5x + 8 = 0$, we have

$$a = 1, \quad b = 5, \quad c = 8.$$

We substitute and compute the discriminant:

$$b^2 - 4ac = 5^2 - 4 \cdot 1 \cdot 8$$
$$= 25 - 32 = -7.$$

Since the discriminant is negative, there are two imaginary-number solutions that are complex conjugates of each other.

c) For $2x^2 + 7x - 3 = 0$, we have

$$a = 2, \quad b = 7, \quad c = -3;$$
$$b^2 - 4ac = 7^2 - 4 \cdot 2(-3)$$
$$= 49 - (-24) = 73.$$

The discriminant is a positive number that is not a perfect square. Thus there are two irrational solutions that are conjugates of each other.

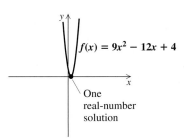

A visualization of part (a)

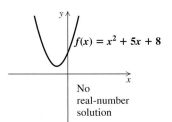

A visualization of part (b)

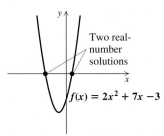

A visualization of part (c)

Discriminants can also be used to determine the number of real-number solutions of $ax^2 + bx + c = 0$. This can be used as an aid in graphing.

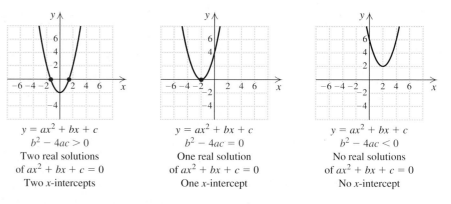

$y = ax^2 + bx + c$
$b^2 - 4ac > 0$
Two real solutions
of $ax^2 + bx + c = 0$
Two x-intercepts

$y = ax^2 + bx + c$
$b^2 - 4ac = 0$
One real solution
of $ax^2 + bx + c = 0$
One x-intercept

$y = ax^2 + bx + c$
$b^2 - 4ac < 0$
No real solutions
of $ax^2 + bx + c = 0$
No x-intercept

Writing Equations from Solutions

We know by the principle of zero products that $(x - 2)(x + 3) = 0$ has solutions 2 and -3. If we know the solutions of an equation, we can write an equation, using the principle in reverse.

EXAMPLE 2   Find an equation for which the given numbers are solutions.

a) 3 and $-\frac{2}{5}$

b) $2i$ and $-2i$

c) $5\sqrt{7}$ and $-5\sqrt{7}$

d) $-4, 0$, and 1

Solution

a)
$$x = 3 \quad or \quad x = -\tfrac{2}{5}$$
$$x - 3 = 0 \quad or \quad x + \tfrac{2}{5} = 0 \qquad \text{Getting 0's on one side}$$
$$(x - 3)\left(x + \tfrac{2}{5}\right) = 0 \qquad \text{Using the principle of zero products (multiplying)}$$
$$x^2 + \tfrac{2}{5}x - 3x - 3 \cdot \tfrac{2}{5} = 0 \qquad \text{Multiplying}$$
$$x^2 - \tfrac{13}{5}x - \tfrac{6}{5} = 0 \qquad \text{Combining like terms}$$
$$5x^2 - 13x - 6 = 0 \qquad \text{Multiplying both sides by 5 to clear fractions}$$

Note that multiplying both sides by the LCD, 5, clears the equation of fractions. Had we preferred, we could have multiplied $x + \frac{2}{5} = 0$ by 5, thus clearing fractions *before* using the principle of zero products.

b)
$$x = 2i \quad or \quad x = -2i$$
$$x - 2i = 0 \quad or \quad x + 2i = 0 \qquad \text{Getting 0's on one side}$$
$$(x - 2i)(x + 2i) = 0 \qquad \text{Using the principle of zero products (multiplying)}$$
$$x^2 - (2i)^2 = 0 \qquad \text{Finding the product of a sum and difference}$$
$$x^2 - 4i^2 = 0$$
$$x^2 + 4 = 0 \qquad i^2 = -1$$

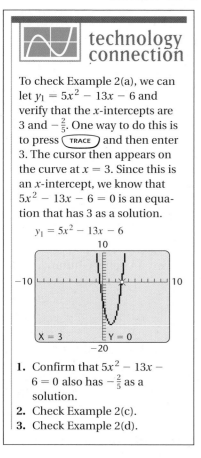

c)
$$x = 5\sqrt{7} \quad or \quad x = -5\sqrt{7}$$
$$x - 5\sqrt{7} = 0 \quad or \quad x + 5\sqrt{7} = 0 \qquad \text{Getting 0's on one side}$$
$$(x - 5\sqrt{7})(x + 5\sqrt{7}) = 0 \qquad \text{Using the principle of zero products}$$
$$x^2 - (5\sqrt{7})^2 = 0 \qquad \text{Finding the product of a sum and difference}$$
$$x^2 - 25 \cdot 7 = 0$$
$$x^2 - 175 = 0$$

d)
$$x = -4 \quad or \quad x = 0 \quad or \quad x = 1$$
$$x + 4 = 0 \quad or \quad x = 0 \quad or \quad x - 1 = 0 \qquad \text{Getting 0's on one side}$$
$$(x + 4)x(x - 1) = 0 \qquad \text{Using the principle of zero products}$$
$$x(x^2 + 3x - 4) = 0 \qquad \text{Multiplying}$$
$$x^3 + 3x^2 - 4x = 0$$

To check any of these equations, we can simply substitute one or more of the given solutions. For example, in Example 2(d) above,
$$(-4)^3 + 3(-4)^2 - 4(-4) = -64 + 3 \cdot 16 + 16$$
$$= -64 + 48 + 16 = 0.$$
The other checks are left to the student.

Exercise Set

8.4

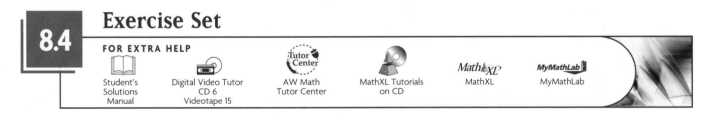

FOR EXTRA HELP

Student's Solutions Manual   Digital Video Tutor CD 6 Videotape 15   AW Math Tutor Center   MathXL Tutorials on CD   MathXL   MyMathLab

🖋 *Concept Reinforcement   Complete each of the following.*

1. In the quadratic formula, the expression $b^2 - 4ac$ is called the _____.

2. When $b^2 - 4ac$ is 0, there is/are _____ solution(s).

3. When $b^2 - 4ac$ is positive, there is/are _____ solution(s).

4. When $b^2 - 4ac$ is negative, there is/are _____ solution(s).

5. When $b^2 - 4ac$ is a perfect square, the answers are _____ numbers.

6. When $b^2 - 4ac$ is negative, the answer(s) is/are _____ numbers.

For each equation, determine what type of number the solutions are and how many solutions exist.

7. $x^2 - 7x + 5 = 0$

8. $x^2 - 5x + 3 = 0$

9. $x^2 + 3 = 0$

10. $x^2 + 5 = 0$

11. $x^2 - 5 = 0$

12. $x^2 - 3 = 0$

13. $4x^2 + 8x - 5 = 0$

14. $4x^2 - 12x + 9 = 0$

15. $x^2 + 4x + 6 = 0$

16. $x^2 - 2x + 4 = 0$

17. $9t^2 - 48t + 64 = 0$

18. $6t^2 - 19t - 20 = 0$

19. $10x^2 - x - 2 = 0$

20. $6x^2 + 5x - 4 = 0$

Aha! 21. $9t^2 - 3t = 0$

22. $4m^2 + 7m = 0$

23. $x^2 + 4x = 8$

24. $x^2 + 5x = 9$

25. $2a^2 - 3a = -5$

26. $3a^2 + 5 = 7a$

27. $y^2 + \frac{9}{4} = 4y$

28. $x^2 = \frac{1}{2}x - \frac{3}{5}$

Write a quadratic equation having the given numbers as solutions.

29. $-7, 3$

30. $-6, 4$

31. 3, only solution (*Hint*: It must be a repeated solution.)

32. -5, only solution

33. $-1, -3$

34. $-2, -5$

35. $5, \frac{3}{4}$

36. $4, \frac{2}{3}$

37. $-\frac{1}{4}, -\frac{1}{2}$

38. $\frac{1}{2}, \frac{1}{3}$

39. $2.4, -0.4$

40. $-0.6, 1.4$

41. $-\sqrt{3}, \sqrt{3}$

42. $-\sqrt{7}, \sqrt{7}$

43. $2\sqrt{5}, -2\sqrt{5}$

44. $3\sqrt{2}, -3\sqrt{2}$

45. $4i, -4i$

46. $3i, -3i$

47. $2 - 7i, 2 + 7i$

48. $5 - 2i, 5 + 2i$

49. $3 - \sqrt{14}, 3 + \sqrt{14}$

50. $2 - \sqrt{10}, 2 + \sqrt{10}$

51. $1 - \dfrac{\sqrt{21}}{3}, 1 + \dfrac{\sqrt{21}}{3}$

52. $\dfrac{5}{4} - \dfrac{\sqrt{33}}{4}, \dfrac{5}{4} + \dfrac{\sqrt{33}}{4}$

Write a third-degree equation having the given numbers as solutions.

53. $-2, 1, 5$

54. $-5, 0, 2$

55. $-1, 0, 3$

56. $-2, 2, 3$

57. Under what condition(s) is the discriminant *not* the fastest way to determine how many and what type of solutions exist?

58. Describe a procedure that could be used to write an equation having the first 7 natural numbers as solutions.

SKILL MAINTENANCE

Simplify. [1.6]

59. $(3a^2)^4$

60. $(4x^3)^2$

Find the x-intercepts of the graph of f. [5.8]

61. $f(x) = x^2 - 7x - 8$

62. $f(x) = x^2 - 6x + 8$

63. During a one-hour television show, there were 12 commercials. Some of the commercials were 30 sec long and the others were 60 sec long. The amount of time for 30-sec commercials was 6 min less than the total number of minutes of commercial time during the show. How many 30-sec commercials were used? [3.3]

64. Graph: $y = -\frac{3}{7}x + 4$. [2.3]

SYNTHESIS

65. If we assume that a quadratic equation has integers for coefficients, will the product of the solutions always be a real number? Why or why not?

66. Can a fourth-degree equation have exactly three irrational solutions? Why or why not?

67. The graph of an equation of the form
$$y = ax^2 + bx + c$$
is a curve similar to the one shown below. Determine *a*, *b*, and *c* from the information given.

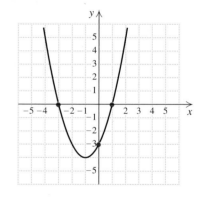

68. Show that the product of the solutions of $ax^2 + bx + c = 0$ is c/a.

For each equation under the given condition, **(a)** *find k and* **(b)** *find the other solution.*

69. $kx^2 - 2x + k = 0$; one solution is -3

70. $x^2 - kx + 2 = 0$; one solution is $1 + i$

71. $x^2 - (6 + 3i)x + k = 0$; one solution is 3

72. Show that the sum of the solutions of $ax^2 + bx + c = 0$ is $-b/a$.

73. Show that whenever there is just one solution of $ax^2 + bx + c = 0$, that solution is of the form $-b/2a$.

74. Find h and k, where $3x^2 - hx + 4k = 0$, the sum of the solutions is -12, and the product of the solutions is 20. (*Hint*: See Exercises 68 and 72.)

75. Suppose that $f(x) = ax^2 + bx + c$, with $f(-3) = 0$, $f\left(\frac{1}{2}\right) = 0$, and $f(0) = -12$. Find a, b, and c.

76. Find an equation for which $2 - \sqrt{3}$, $2 + \sqrt{3}$, $5 - 2i$, and $5 + 2i$ are solutions.

Aha! **77.** Find an equation with integer coefficients for which $1 - \sqrt{5}$ and $3 + 2i$ are two of the solutions.

78. A discriminant that is a perfect square indicates that factoring can be used to solve the quadratic equation. Why?

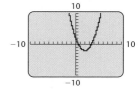

79. While solving a quadratic equation of the form $ax^2 + bx + c = 0$ with a graphing calculator, Shawn-Marie gets the following screen. How could the sign of the discriminant help her check the graph?

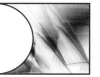

8.5

Equations Reducible to Quadratic

Recognizing Equations in Quadratic Form • Radical and Rational Equations

*Study Skills*_____

Use Your Words

Although you may at first feel a bit hesitant to do so, when a new word or phrase (such as *reducible to quadratic*) arises, try to use it in conversation with classmates, your instructor, or a study partner. Doing so will help you in your reading, deepen your level of understanding, and increase your confidence.

Recognizing Equations in Quadratic Form

Certain equations that are not really quadratic can be thought of in such a way that they can be solved as quadratic. For example, because the square of x^2 is x^4, the equation $x^4 - 9x^2 + 8 = 0$ is said to be "quadratic in x^2":

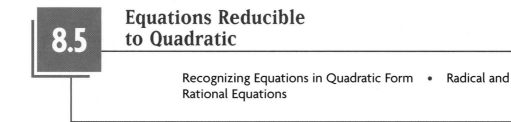

$$x^4 - 9x^2 + 8 = 0$$

$$(x^2)^2 - 9(x^2) + 8 = 0 \qquad \text{Thinking of } x^4 \text{ as } (x^2)^2$$

$$u^2 - 9u + 8 = 0. \qquad \text{To make this clearer, write } u \text{ instead of } x^2.$$

The equation $u^2 - 9u + 8 = 0$ can be solved by factoring or by the quadratic formula. Then, remembering that $u = x^2$, we can solve for x. Equations that can be solved like this are *reducible to quadratic*, or *in quadratic form*.

EXAMPLE 1   Solve: $x^4 - 9x^2 + 8 = 0$.

Solution   Let $u = x^2$. Then we solve by substituting u for x^2 and u^2 for x^4:

$$u^2 - 9u + 8 = 0$$

$$(u - 8)(u - 1) = 0 \qquad \text{Factoring}$$

$$u - 8 = 0 \quad or \quad u - 1 = 0 \qquad \text{Principle of zero products}$$

$$u = 8 \quad or \qquad u = 1.$$

Student Notes _____

To recognize an equation in quadratic form, note that there will be two terms with variables. The exponent of the first term is twice the exponent of the second term. It is the variable expression in the *second* term that is written as u.

We replace u with x^2 and solve these equations:

$$x^2 = 8 \qquad or \quad x^2 = 1$$
$$x = \pm\sqrt{8} \quad or \quad x = \pm 1$$
$$x = \pm 2\sqrt{2} \quad or \quad x = \pm 1.$$

To check, note that for both $x = 2\sqrt{2}$ and $-2\sqrt{2}$, we have $x^2 = 8$ and $x^4 = 64$. Similarly, for both $x = 1$ and -1, we have $x^2 = 1$ and $x^4 = 1$. Thus instead of making four checks, we need make only two.

Check:

For $\pm 2\sqrt{2}$:

$$\begin{array}{c|c} x^4 - 9x^2 + 8 = 0 & \\ \hline (\pm 2\sqrt{2})^4 - 9(\pm 2\sqrt{2})^2 + 8 & 0 \\ 64 - 9 \cdot 8 + 8 & \\ & 0 \overset{?}{=} 0 \quad \text{TRUE} \end{array}$$

For ± 1:

$$\begin{array}{c|c} x^4 - 9x^2 + 8 = 0 & \\ \hline (\pm 1)^4 - 9(\pm 1)^2 + 8 & 0 \\ 1 - 9 + 8 & \\ & 0 \overset{?}{=} 0 \quad \text{TRUE} \end{array}$$

The solutions are 1, -1, $2\sqrt{2}$, and $-2\sqrt{2}$.

> **Caution!** A common error on problems like Example 1 is to solve for u but forget to solve for x. Remember to solve for the *original* variable!

Example 1 can be solved directly by factoring:

$$x^4 - 9x^2 + 8 = 0$$
$$(x^2 - 1)(x^2 - 8) = 0$$
$$x^2 - 1 = 0 \quad or \quad x^2 - 8 = 0$$
$$x^2 = 1 \quad or \qquad x^2 = 8$$
$$x = \pm 1 \quad or \qquad x = \pm 2\sqrt{2}.$$

There is nothing wrong with this approach. However, in the examples that follow, you will note that it becomes increasingly difficult to solve the equation without first making a substitution.

Radical and Rational Equations

Sometimes rational equations, radical equations, or equations containing exponents that are fractions are reducible to quadratic. It is especially important that answers to these equations be checked in the original equation.

EXAMPLE 2 Solve: $x - 3\sqrt{x} - 4 = 0$.

Solution This radical equation could be solved using the method discussed in Section 7.6. However, if we note that the square of $\sqrt{x}$ is x, we can regard the equation as "quadratic in $\sqrt{x}$."

We let $u = \sqrt{x}$ and consequently $u^2 = x$:

$$x - 3\sqrt{x} - 4 = 0$$
$$u^2 - 3u - 4 = 0 \qquad \text{Substituting}$$
$$(u - 4)(u + 1) = 0$$
$$u = 4 \quad or \quad u = -1. \qquad \text{Using the principle of zero products}$$

Next, we replace u with $\sqrt{x}$ and solve these equations:

$$\sqrt{x} = 4 \quad or \quad \sqrt{x} = -1.$$

Squaring gives us $x = 16$ or $x = 1$ and also makes checking essential.

Check: For 16:   For 1:

$$\begin{array}{c|c} x - 3\sqrt{x} - 4 = 0 & \\ \hline 16 - 3\sqrt{16} - 4 & 0 \\ 16 - 3 \cdot 4 - 4 & \\ & 0 \overset{?}{=} 0 \quad \text{TRUE} \end{array}$$

$$\begin{array}{c|c} x - 3\sqrt{x} - 4 = 0 & \\ \hline 1 - 3\sqrt{1} - 4 & 0 \\ 1 - 3 \cdot 1 - 4 & \\ & -6 \overset{?}{=} 0 \quad \text{FALSE} \end{array}$$

The number 16 checks, but 1 does not. Had we noticed that $\sqrt{x} = -1$ has no solution (since principal roots are never negative), we could have solved only the equation $\sqrt{x} = 4$. The solution is 16.

EXAMPLE 3   Find the x-intercepts of the graph of $f(x) = (x^2 - 1)^2 - (x^2 - 1) - 2$.

technology connection

Check Example 3 with a graphing calculator. Use the ZERO, ROOT, or INTERSECT option, if possible.

Solution   The x-intercepts occur where $f(x) = 0$ so we must have

$$(x^2 - 1)^2 - (x^2 - 1) - 2 = 0. \qquad \text{Setting } f(x) \text{ equal to } 0$$

This equation is quadratic in $x^2 - 1$, so we let $u = x^2 - 1$ and $u^2 = (x^2 - 1)^2$:

$$u^2 - u - 2 = 0 \qquad \begin{array}{l}\text{Substituting in} \\ (x^2 - 1)^2 - (x^2 - 1) - 2 = 0\end{array}$$

$$(u - 2)(u + 1) = 0$$

$$u = 2 \quad or \quad u = -1. \qquad \text{Using the principle of zero products}$$

Next, we replace u with $x^2 - 1$ and solve these equations:

$$x^2 - 1 = 2 \qquad or \quad x^2 - 1 = -1$$
$$x^2 = 3 \qquad or \qquad x^2 = 0 \qquad \text{Adding 1 to both sides}$$
$$x = \pm\sqrt{3} \quad or \qquad x = 0. \qquad \begin{array}{l}\text{Using the principle of} \\ \text{square roots}\end{array}$$

The x-intercepts occur at $\left(-\sqrt{3}, 0\right)$, $(0, 0)$, and $\left(\sqrt{3}, 0\right)$.

Sometimes great care must be taken in deciding what substitution to make.

EXAMPLE 4   Solve: $m^{-2} - 6m^{-1} + 4 = 0$.

Solution   Note that the square of m^{-1} is $(m^{-1})^2$, or m^{-2}. This allows us to regard the equation as quadratic in m^{-1}.

We let $u = m^{-1}$ and $u^2 = m^{-2}$:

$$u^2 - 6u + 4 = 0 \qquad \text{Substituting}$$

$$u = \frac{-(-6) \pm \sqrt{(-6)^2 - 4 \cdot 1 \cdot 4}}{2 \cdot 1} \qquad \begin{array}{l}\text{Using the quadratic}\\ \text{formula}\end{array}$$

$$\left. \begin{array}{l} u = \dfrac{6 \pm \sqrt{20}}{2} = \dfrac{2 \cdot 3 \pm 2\sqrt{5}}{2} \\[2mm] u = 3 \pm \sqrt{5}. \end{array} \right\} \quad \text{Simplifying}$$

Next, we replace u with m^{-1} and solve:

$$m^{-1} = 3 \pm \sqrt{5}$$

$$\frac{1}{m} = 3 \pm \sqrt{5} \qquad \text{Recall that } m^{-1} = \frac{1}{m}.$$

$$1 = m(3 \pm \sqrt{5}) \qquad \text{Multiplying both sides by } m$$

$$\frac{1}{3 \pm \sqrt{5}} = m. \qquad \text{Dividing both sides by } 3 \pm \sqrt{5}$$

We can check both solutions as follows.

Check:

For $1/(3 - \sqrt{5})$:

$$\begin{array}{c|c} m^{-2} - 6m^{-1} + 4 = 0 & \\ \hline \left(\dfrac{1}{3 - \sqrt{5}}\right)^{-2} - 6\left(\dfrac{1}{3 - \sqrt{5}}\right)^{-1} + 4 & 0 \\ (3 - \sqrt{5})^2 - 6(3 - \sqrt{5}) + 4 & \\ 9 - 6\sqrt{5} + 5 - 18 + 6\sqrt{5} + 4 & \\ & 0 \overset{?}{=} 0 \quad \text{TRUE} \end{array}$$

For $1/(3 + \sqrt{5})$:

$$\begin{array}{c|c} m^{-2} - 6m^{-1} + 4 = 0 & \\ \hline \left(\dfrac{1}{3 + \sqrt{5}}\right)^{-2} - 6\left(\dfrac{1}{3 + \sqrt{5}}\right)^{-1} + 4 & 0 \\ (3 + \sqrt{5})^2 - 6(3 + \sqrt{5}) + 4 & \\ 9 + 6\sqrt{5} + 5 - 18 - 6\sqrt{5} + 4 & \\ & 0 \overset{?}{=} 0 \quad \text{TRUE} \end{array}$$

Both numbers check. The solutions are $1/(3 - \sqrt{5})$ and $1/(3 + \sqrt{5})$, or approximately 1.309016994 and 0.1909830056.

EXAMPLE 5   Solve: $t^{2/5} - t^{1/5} - 2 = 0$.

Solution   Note that the square of $t^{1/5}$ is $(t^{1/5})^2$, or $t^{2/5}$. The equation is therefore quadratic in $t^{1/5}$, so we let $u = t^{1/5}$ and $u^2 = t^{2/5}$:

$$u^2 - u - 2 = 0 \qquad \text{Substituting}$$

$$(u - 2)(u + 1) = 0$$

$$u = 2 \quad or \quad u = -1. \qquad \text{Using the principle of zero products}$$

Now we replace u with $t^{1/5}$ and solve:

$$t^{1/5} = 2 \quad or \quad t^{1/5} = -1$$

$$t = 32 \quad or \quad t = -1. \qquad \begin{array}{l}\text{Principle of powers; raising to the}\\ \text{5th power}\end{array}$$

Check:

For 32:

$$t^{2/5} - t^{1/5} - 2 = 0$$

$$\begin{array}{c|c} 32^{2/5} - 32^{1/5} - 2 & 0 \\ (32^{1/5})^2 - 32^{1/5} - 2 & \\ 2^2 - 2 - 2 & \end{array}$$

$$0 \overset{?}{=} 0 \quad \text{TRUE}$$

For −1:

$$t^{2/5} - t^{1/5} - 2 = 0$$

$$\begin{array}{c|c} (-1)^{2/5} - (-1)^{1/5} - 2 & 0 \\ [(-1)^{1/5}]^2 - (-1)^{1/5} - 2 & \\ (-1)^2 - (-1) - 2 & \end{array}$$

$$0 \overset{?}{=} 0 \quad \text{TRUE}$$

Both numbers check. The solutions are 32 and −1.

The following tips may prove useful.

To Solve an Equation That Is Reducible to Quadratic

1. The equation is quadratic in form if the variable factor in one term is the square of the variable factor in the other variable term.
2. Write down any substitutions that you are making.
3. Whenever you make a substitution, remember to solve for the variable that is used in the original equation.
4. Check possible answers in the original equation.

Exercise Set

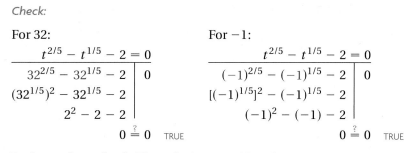

8.5

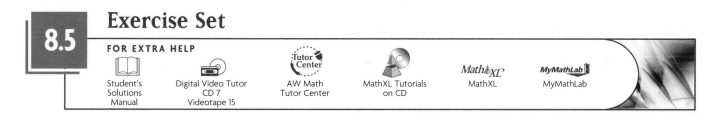

FOR EXTRA HELP

Student's Solutions Manual   Digital Video Tutor CD 7 Videotape 15   AW Math Tutor Center   MathXL Tutorials on CD   Math XL MathXL   MyMathLab MyMathLab

🖎 *Concept Reinforcement   In each of Exercises 1–8, match the equation with an appropriate substitution from the column on the right that could be used to reduce the equation to quadratic form.*

1. _____ $4x^6 - 2x^3 + 1 = 0$

2. _____ $3x^4 + 4x^2 - 7 = 0$

3. _____ $5x^8 + 2x^4 - 3 = 0$

4. _____ $2x^{2/3} - 5x^{1/3} + 4 = 0$

5. _____ $3x^{4/3} + 4x^{2/3} - 7 = 0$

6. _____ $2x^{-2/3} + x^{-1/3} + 6 = 0$

7. _____ $4x^{-4/3} - 2x^{-2/3} + 3 = 0$

8. _____ $3x^{-4} + 4x^{-2} - 2 = 0$

a) $u = x^{-1/3}$

b) $u = x^{1/3}$

c) $u = x^{-2}$

d) $u = x^2$

e) $u = x^{-2/3}$

f) $u = x^3$

g) $u = x^{2/3}$

h) $u = x^4$

Solve.

9. $x^4 - 5x^2 + 4 = 0$

10. $x^4 - 10x^2 + 9 = 0$

11. $x^4 - 9x^2 + 20 = 0$

12. $x^4 - 12x^2 + 27 = 0$

13. $4t^4 - 19t^2 + 12 = 0$

14. $9t^4 - 14t^2 + 5 = 0$

15. $r - 2\sqrt{r} - 6 = 0$

16. $s - 4\sqrt{s} - 1 = 0$

17. $(x^2 - 7)^2 - 3(x^2 - 7) + 2 = 0$

18. $(x^2 - 1)^2 - 5(x^2 - 1) + 6 = 0$

19. $\left(1 + \sqrt{x}\right)^2 + 5\left(1 + \sqrt{x}\right) + 6 = 0$

20. $\left(3 + \sqrt{x}\right)^2 + 3\left(3 + \sqrt{x}\right) - 10 = 0$

21. $x^{-2} - x^{-1} - 6 = 0$

22. $2x^{-2} - x^{-1} - 1 = 0$

23. $4x^{-2} + x^{-1} - 5 = 0$

24. $m^{-2} + 9m^{-1} - 10 = 0$

25. $t^{2/3} + t^{1/3} - 6 = 0$

26. $w^{2/3} - 2w^{1/3} - 8 = 0$

27. $y^{1/3} - y^{1/6} - 6 = 0$

28. $t^{1/2} + 3t^{1/4} + 2 = 0$

29. $t^{1/3} + 2t^{1/6} = 3$

30. $m^{1/2} + 6 = 5m^{1/4}$

31. $\left(3 - \sqrt{x}\right)^2 - 10\left(3 - \sqrt{x}\right) + 23 = 0$

32. $\left(5 + \sqrt{x}\right)^2 - 12\left(5 + \sqrt{x}\right) + 33 = 0$

33. $16\left(\dfrac{x - 1}{x - 8}\right)^2 + 8\left(\dfrac{x - 1}{x - 8}\right) + 1 = 0$

34. $9\left(\dfrac{x + 2}{x + 3}\right)^2 - 6\left(\dfrac{x + 2}{x + 3}\right) + 1 = 0$

Find all x-intercepts of the given function f. If none exist, state this.

35. $f(x) = 5x + 13\sqrt{x} - 6$

36. $f(x) = 3x + 10\sqrt{x} - 8$

37. $f(x) = (x^2 - 3x)^2 - 10(x^2 - 3x) + 24$

38. $f(x) = (x^2 - 6x)^2 - 2(x^2 - 6x) - 35$

39. $f(x) = x^{2/5} + x^{1/5} - 6$

40. $f(x) = x^{1/2} - x^{1/4} - 6$

Aha! **41.** $f(x) = \left(\dfrac{x^2 + 2}{x}\right)^4 + 7\left(\dfrac{x^2 + 2}{x}\right)^2 + 5$

42. $f(x) = \left(\dfrac{x^2 + 1}{x}\right)^4 + 4\left(\dfrac{x^2 + 1}{x}\right)^2 + 12$

43. To solve $25x^6 - 10x^3 + 1 = 0$, Don lets $u = 5x^3$ and Robin lets $u = x^3$. Can they both be correct? Why or why not?

44. Can the examples and exercises of this section be understood without knowing the rules for exponents? Why or why not?

SKILL MAINTENANCE

Graph.

45. $f(x) = \frac{3}{2}x$ [2.3]

46. $f(x) = -\frac{2}{3}x$ [2.3]

47. $f(x) = \dfrac{2}{x}$ [2.1]

48. $f(x) = \dfrac{3}{x}$ [2.1]

49. Hiker's Mix is 18% peanuts and Trail Snax is 45% peanuts. How much of each should be mixed together in order to get 12 lb of a mixture that is 36% peanuts? [3.3]

50. If $g(x) = x^2 - x$, find $g(a + 1)$. [5.2]

SYNTHESIS

51. Describe a procedure that could be used to solve any equation of the form $ax^4 + bx^2 + c = 0$.

52. Describe a procedure that could be used to write an equation that is quadratic in $3x^2 - 1$. Then explain how the procedure could be adjusted to write equations that are quadratic in $3x^2 - 1$ and have no real-number solution.

Solve.

53. $5x^4 - 7x^2 + 1 = 0$

54. $3x^4 + 5x^2 - 1 = 0$

55. $(x^2 - 4x - 2)^2 - 13(x^2 - 4x - 2) + 30 = 0$

56. $(x^2 - 5x - 1)^2 - 18(x^2 - 5x - 1) + 65 = 0$

57. $\dfrac{x}{x - 1} - 6\sqrt{\dfrac{x}{x - 1}} - 40 = 0$

58. $\left(\sqrt{\dfrac{x}{x - 3}}\right)^2 - 24 = 10\sqrt{\dfrac{x}{x - 3}}$

59. $a^5(a^2 - 25) + 13a^3(25 - a^2) + 36a(a^2 - 25) = 0$

60. $a^3 - 26a^{3/2} - 27 = 0$

61. $x^6 - 28x^3 + 27 = 0$

62. $x^6 + 7x^3 - 8 = 0$

 63. Use a graphing calculator to check your answers to Exercises 9, 11, 37, and 55.

 64. Use a graphing calculator to solve
$$x^4 - x^3 - 13x^2 + x + 12 = 0.$$

65. While trying to solve $0.05x^4 - 0.8 = 0$ with a graphing calculator, Murray gets the following screen. Can Murray solve this equation with a graphing calculator? Why or why not?

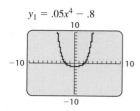

$y_1 = .05x^4 - .8$

8.6 Quadratic Functions and Their Graphs

The Graph of $f(x) = ax^2$ •   The Graph of $f(x) = a(x - h)^2$ •
The Graph of $f(x) = a(x - h)^2 + k$

We have already used quadratic functions when we solved equations earlier in this chapter. In this section and the next, we learn to graph such functions.

The Graph of $f(x) = ax^2$

The most basic quadratic function is $f(x) = x^2$.

EXAMPLE 1   Graph: $f(x) = x^2$.

Study Skills—————

Take a Peek Ahead

Try to at least glance at the next section of material that will be covered in class. This will make it easier to concentrate on your instructor's lecture instead of trying to write everything down.

Solution   We choose some values for x and compute $f(x)$ for each. Then we plot the ordered pairs and connect them with a smooth curve.

x	$f(x) = x^2$	$(x, f(x))$
-3	9	$(-3, 9)$
-2	4	$(-2, 4)$
-1	1	$(-1, 1)$
0	0	$(0, 0)$
1	1	$(1, 1)$
2	4	$(2, 4)$
3	9	$(3, 9)$

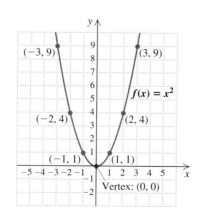

Student Notes _____

By paying attention to the symmetry of each parabola and the location of the vertex, you save yourself considerable work. Note too that if the x^2-coefficient is a, the x-values 1 unit to the right or left of the vertex are paired with the y-value a units above the vertex. Thus the graph of $y = \frac{3}{2}x^2$ includes the points $\left(-1, \frac{3}{2}\right)$ and $\left(1, \frac{3}{2}\right)$.

All quadratic functions have graphs similar to the one in Example 1. Such curves are called *parabolas*. They are U-shaped and symmetric with respect to a vertical line known as the parabola's *axis of symmetry*. For the graph of $f(x) = x^2$, the y-axis (the vertical line $x = 0$) is the axis of symmetry. Were the paper folded on this line, the two halves of the curve would match. The point $(0, 0)$ is known as the *vertex* of this parabola.

By plotting points, we can compare the graphs of $g(x) = \frac{1}{2}x^2$ and $h(x) = 2x^2$ with the graph of $f(x) = x^2$.

x	$g(x) = \frac{1}{2}x^2$
-3	$\frac{9}{2}$
-2	2
-1	$\frac{1}{2}$
0	0
1	$\frac{1}{2}$
2	2
3	$\frac{9}{2}$

x	$h(x) = 2x^2$
-3	18
-2	8
-1	2
0	0
1	2
2	8
3	18

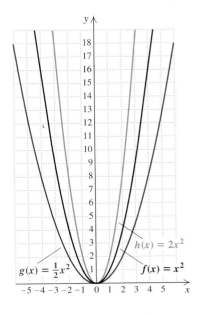

technology connection

To explore the effect of a on the graph of $y = ax^2$, let $y_1 = x^2$, $y_2 = 3x^2$, and $y_3 = \frac{1}{3}x^2$. Graph the equations and use ⟨TRACE⟩ to see how the y-values compare, using ⌃ or ⌄ to hop the cursor from one curve to the next.

On many graphing calculators, the ⟨APPS⟩ key will access the Transfrm application. If you run that application and let $y_1 = Ax^2$, the graph becomes interactive and allows for A to be changed while viewing the graph. Selecting the SETTINGS option, after pressing ⟨WINDOW⟩, permits us to adjust the step by which A can be changed.

1. Compare the graphs of $y_1 = \frac{1}{5}x^2$, $y_2 = x^2$, $y_3 = \frac{5}{2}x^2$, $y_4 = -\frac{1}{5}x^2$, $y_5 = -x^2$, and $y_6 = -\frac{5}{2}x^2$.
2. Describe the effect that A has on each graph.

Note that the graph of $g(x) = \frac{1}{2}x^2$ is "wider" than the graph of $f(x) = x^2$, and the graph of $h(x) = 2x^2$ is "narrower." The vertex and the axis of symmetry, however, remain $(0, 0)$ and the line $x = 0$, respectively.

When we consider the graph of $k(x) = -\frac{1}{2}x^2$, we see that the parabola is the same shape as the graph of $g(x) = \frac{1}{2}x^2$, but opens downward. We say that the graphs of k and g are *reflections* of each other across the x-axis.

x	$k(x) = -\frac{1}{2}x^2$
-3	$-\frac{9}{2}$
-2	-2
-1	$-\frac{1}{2}$
0	0
1	$-\frac{1}{2}$
2	-2
3	$-\frac{9}{2}$

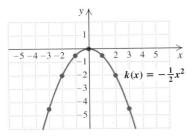

> *Graphing f(x) = ax²*
>
> The graph of $f(x) = ax^2$ is a parabola with $x = 0$ as its axis of symmetry. Its vertex is the origin.
>
> For $a > 0$, the parabola opens upward. For $a < 0$, the parabola opens downward.
>
> If $|a|$ is greater than 1, the parabola is narrower than $y = x^2$.
>
> If $|a|$ is between 0 and 1, the parabola is wider than $y = x^2$.

The Graph of $f(x) = a(x - h)^2$

We *could* next consider graphs of

$$f(x) = ax^2 + bx + c,$$

where b and c are not both 0. In effect, we will do that, but in a disguised form. It turns out to be convenient to first graph $f(x) = a(x - h)^2$, where h is some constant. This allows us to observe similarities to the graphs drawn above.

EXAMPLE 2

Graph: $f(x) = (x - 3)^2$.

Solution   We choose some values for x and compute $f(x)$. It is important to note that when an input here is 3 more than an input for Example 1, the outputs match. We plot the points and draw the curve.

x	$f(x) = (x - 3)^2$	
-1	16	
0	9	
1	4	
2	1	
3	0	← Vertex
4	1	
5	4	
6	9	

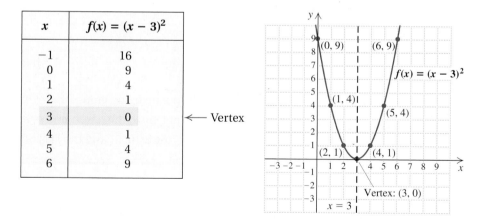

Note that $f(x)$ is smallest when $x - 3$ is 0, that is, for $x = 3$. Thus the line $x = 3$ is now the axis of symmetry and the point $(3, 0)$ is the vertex. Had we recognized earlier that $x = 3$ is the axis of symmetry, we could have computed some values on one side, such as $(4, 1)$, $(5, 4)$, and $(6, 9)$, and then used symmetry to get their mirror images $(2, 1)$, $(1, 4)$, and $(0, 9)$ without further computation.

EXAMPLE 3   Graph: $g(x) = -2(x + 4)^2$.

Solution   We choose some values for x and compute $g(x)$. Note that $g(x)$ is greatest when $x + 4$ is 0, that is, for $x = -4$. Thus the line given by $x = -4$ is the axis of symmetry and the point $(-4, 0)$ is the vertex. We plot some points and draw the curve.

technology connection

To explore the effect of h on the graph of $f(x) = a(x - h)^2$, let $y_1 = 7x^2$ and $y_2 = 7(x - 1)^2$. Graph both y_1 and y_2 and compare y-values, beginning at $x = 1$ and increasing x by one unit at a time. The G-T or HORIZ **MODE** can be used to view a split screen showing both the graph and a table.

Next, let $y_3 = 7(x - 2)^2$ and compare its graph and y-values with those of y_1 and y_2. Then let $y_4 = 7(x + 1)^2$ and $y_5 = 7(x + 2)^2$.

1. Compare graphs and y-values and describe the effect of h on the graph of $f(x) = a(x - h)^2$.
2. If the Transfrm application is available, let $y_1 = A(x - B)^2$ and describe the effect that A and B have on each graph.

x	$g(x) = -2(x + 4)^2$
-6	-8
-5	-2
-4	0
-3	-2
-2	-8

← Vertex

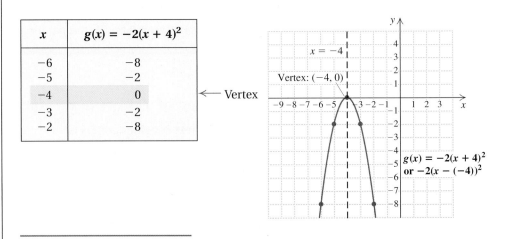

In Example 2, the graph of $f(x) = (x - 3)^2$ looks just like the graph of $y = x^2$, except that it is moved, or *translated*, 3 units to the right. In Example 3, the graph of $g(x) = -2(x + 4)^2$ looks like the graph of $y = -2x^2$, except that it is shifted 4 units to the left. These results are generalized as follows.

> *Graphing* $f(x) = a(x - h)^2$
>
> The graph of $f(x) = a(x - h)^2$ has the same shape as the graph of $y = ax^2$.
>
> If h is positive, the graph of $y = ax^2$ is shifted h units to the right.
>
> If h is negative, the graph of $y = ax^2$ is shifted $|h|$ units to the left.
>
> The vertex is $(h, 0)$ and the axis of symmetry is $x = h$.

The Graph of $f(x) = a(x - h)^2 + k$

Given a graph of $f(x) = a(x - h)^2$, what happens if we add a constant k? Suppose that we add 2. This increases $f(x)$ by 2, so the curve is moved up. If k is negative, the curve is moved down. The axis of symmetry for the parabola remains $x = h$, but the vertex will be at (h, k), or, equivalently, $(h, f(h))$.

Note that if a parabola opens upward ($a > 0$), the function value, or y-value, at the vertex is a least, or *minimum*, value. That is, it is less than the y-value at any other point on the graph. If the parabola opens downward ($a < 0$), the function value at the vertex is a greatest, or *maximum*, value.

Graphs of $f(x) = a(x - h)^2 + k$

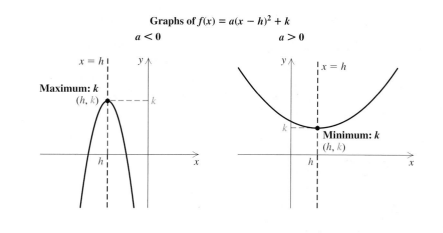

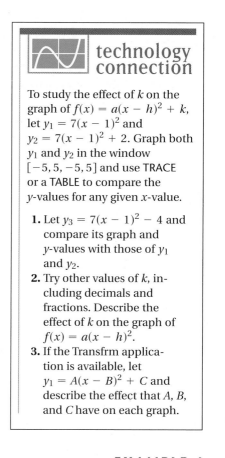

To study the effect of k on the graph of $f(x) = a(x - h)^2 + k$, let $y_1 = 7(x - 1)^2$ and $y_2 = 7(x - 1)^2 + 2$. Graph both y_1 and y_2 in the window $[-5, 5, -5, 5]$ and use TRACE or a TABLE to compare the y-values for any given x-value.

1. Let $y_3 = 7(x - 1)^2 - 4$ and compare its graph and y-values with those of y_1 and y_2.
2. Try other values of k, including decimals and fractions. Describe the effect of k on the graph of $f(x) = a(x - h)^2$.
3. If the Transfrm application is available, let $y_1 = A(x - B)^2 + C$ and describe the effect that A, B, and C have on each graph.

Graphing $f(x) = a(x - h)^2 + k$

The graph of $f(x) = a(x - h)^2 + k$ has the same shape as the graph of $y = a(x - h)^2$.

If k is positive, the graph of $y = a(x - h)^2$ is shifted k units up.

If k is negative, the graph of $y = a(x - h)^2$ is shifted $|k|$ units down.

The vertex is (h, k), and the axis of symmetry is $x = h$.

For $a > 0$, k is the minimum function value. For $a < 0$, k is the maximum function value.

EXAMPLE 4

Graph $g(x) = (x - 3)^2 - 5$, and find the minimum function value.

Solution   The graph will look like that of $f(x) = (x - 3)^2$ (see Example 2) but shifted 5 units down. You can confirm this by plotting some points. For instance, $g(4) = (4 - 3)^2 - 5 = -4$, whereas in Example 2, $f(4) = (4 - 3)^2 = 1$.

The vertex is now $(3, -5)$, and the minimum function value is -5.

x	$g(x) = (x - 3)^2 - 5$	
0	4	
1	-1	
2	-4	
3	-5	← Vertex
4	-4	
5	-1	
6	4	

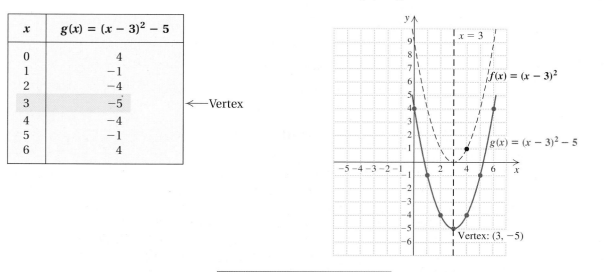

EXAMPLE 5   Graph $h(x) = \frac{1}{2}(x - 3)^2 + 6$, and find the minimum function value.

Solution   The graph looks just like that of $f(x) = \frac{1}{2}x^2$ but moved 3 units to the right and 6 units up. The vertex is $(3, 6)$, and the axis of symmetry is $x = 3$. We draw $f(x) = \frac{1}{2}x^2$ and then shift the curve over and up. The minimum function value is 6. By plotting some points, we have a check.

x	$h(x) = \frac{1}{2}(x - 3)^2 + 6$
0	$10\frac{1}{2}$
1	8
3	6
5	8
6	$10\frac{1}{2}$

←—Vertex

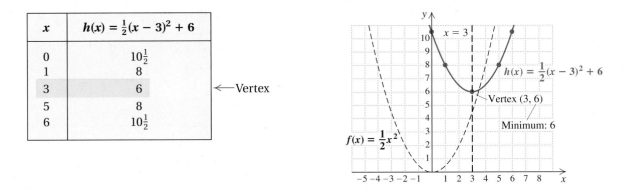

EXAMPLE 6   Graph $y = -2(x + 3)^2 + 5$. Find the vertex, the axis of symmetry, and the maximum or minimum value.

Solution   We first express the equation in the equivalent form

$$y = -2[x - (-3)]^2 + 5.$$

The graph looks like that of $y = -2x^2$ translated 3 units to the left and 5 units up. The vertex is $(-3, 5)$, and the axis of symmetry is $x = -3$. Since -2 is negative, we know that 5, the second coordinate of the vertex, is the maximum y-value.

We compute a few points as needed, selecting convenient x-values on either side of the vertex. The graph is shown here.

x	$y = -2(x + 3)^2 + 5$
-4	3
-3	5
-2	3

←—Vertex

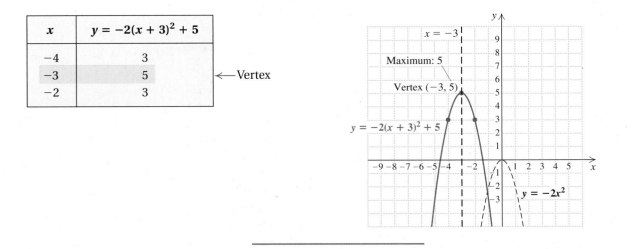

CONNECTING THE CONCEPTS

The ability to graph a function is an important skill. Later in this chapter, as well as in future courses, you will find that graphs of polynomial functions can be used as a tool for solving equations, inequalities, and real-world applications. In the process of learning how to graph quadratic functions, we have developed the ability to reflect or shift (translate) a graph. This skill will prove useful not only in future courses, but in Chapters 9 and 10 as well.

Exercise Set

8.6

FOR EXTRA HELP

Student's Solutions Manual   Digital Video Tutor CD 7 Videotape 16   AW Math Tutor Center   MathXL Tutorials on CD   MathXL   MyMathLab

Concept Reinforcement In each of Exercises 1–8, match the equation with the corresponding graph from those shown.

1. _____ $f(x) = 2(x - 1)^2 + 3$

2. _____ $f(x) = -2(x - 1)^2 + 3$

3. _____ $f(x) = 2(x + 1)^2 + 3$

4. _____ $f(x) = 2(x - 1)^2 - 3$

5. _____ $f(x) = -2(x + 1)^2 + 3$

6. _____ $f(x) = -2(x + 1)^2 - 3$

7. _____ $f(x) = 2(x + 1)^2 - 3$

8. _____ $f(x) = -2(x - 1)^2 - 3$

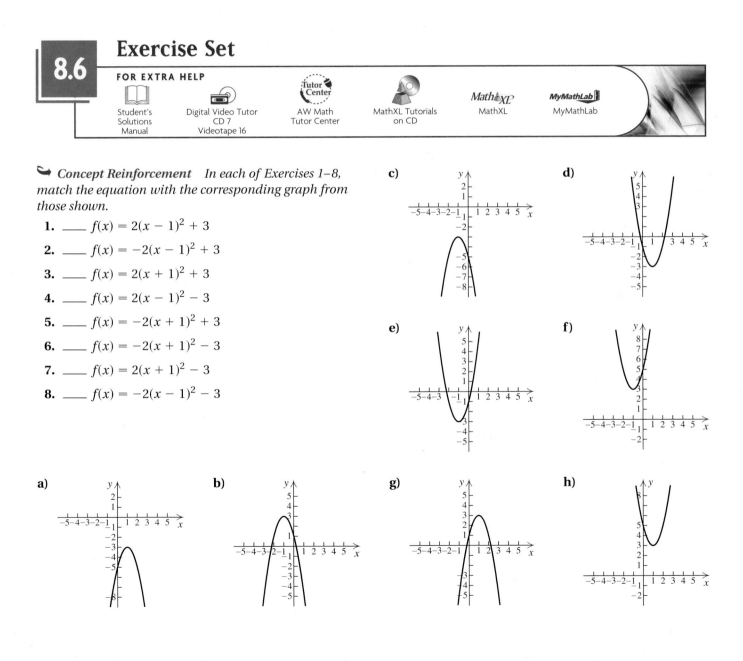

Graph.

9. $f(x) = x^2$

10. $f(x) = -x^2$

11. $f(x) = -2x^2$

12. $f(x) = -3x^2$

13. $g(x) = \frac{1}{3}x^2$

14. $g(x) = \frac{1}{4}x^2$

Aha! **15.** $h(x) = -\frac{1}{3}x^2$

16. $h(x) = -\frac{1}{4}x^2$

17. $f(x) = \frac{5}{2}x^2$

18. $f(x) = \frac{3}{2}x^2$

For each of the following, graph the function, label the vertex, and draw the axis of symmetry.

19. $g(x) = (x + 1)^2$

20. $g(x) = (x + 4)^2$

21. $f(x) = (x - 2)^2$

22. $f(x) = (x - 1)^2$

23. $h(x) = (x - 3)^2$

24. $h(x) = (x - 4)^2$

25. $f(x) = -(x + 1)^2$

26. $f(x) = -(x - 1)^2$

27. $g(x) = -(x - 2)^2$

28. $g(x) = -(x + 4)^2$

29. $f(x) = 2(x + 1)^2$

30. $f(x) = 2(x + 4)^2$

31. $h(x) = -\frac{1}{2}(x - 4)^2$

32. $h(x) = -\frac{3}{2}(x - 2)^2$

33. $f(x) = \frac{1}{2}(x - 1)^2$

34. $f(x) = \frac{1}{3}(x + 2)^2$

35. $f(x) = -2(x + 5)^2$

36. $f(x) = 2(x + 7)^2$

37. $h(x) = -3\left(x - \frac{1}{2}\right)^2$

38. $h(x) = -2\left(x + \frac{1}{2}\right)^2$

For each of the following, graph the function and find the vertex, the axis of symmetry, and the maximum value or the minimum value.

39. $f(x) = (x - 5)^2 + 2$

40. $f(x) = (x + 3)^2 - 2$

41. $f(x) = (x + 1)^2 - 3$

42. $f(x) = (x - 1)^2 + 2$

43. $g(x) = (x + 4)^2 + 1$

44. $g(x) = -(x - 2)^2 - 4$

45. $h(x) = -2(x - 1)^2 - 3$

46. $h(x) = -2(x + 1)^2 + 4$

47. $f(x) = 2(x + 4)^2 + 1$

48. $f(x) = 2(x - 5)^2 - 3$

49. $g(x) = -\frac{3}{2}(x - 1)^2 + 4$

50. $g(x) = \frac{3}{2}(x + 2)^2 - 3$

Without graphing, find the vertex, the axis of symmetry, and the maximum value or the minimum value.

51. $f(x) = 6(x - 8)^2 + 7$

52. $f(x) = 4(x + 5)^2 - 6$

53. $h(x) = -\frac{2}{7}(x + 6)^2 + 11$

54. $h(x) = -\frac{3}{11}(x - 7)^2 - 9$

55. $f(x) = 7\left(x + \frac{1}{4}\right)^2 - 13$

56. $f(x) = 6\left(x - \frac{1}{4}\right)^2 + 15$

57. $f(x) = \sqrt{2}(x + 4.58)^2 + 65\pi$

58. $f(x) = 4\pi(x - 38.2)^2 - \sqrt{34}$

59. Explain, without plotting points, why the graph of $y = x^2 - 4$ looks like the graph of $y = x^2$ translated 4 units down.

60. Explain, without plotting points, why the graph of $y = (x + 2)^2$ looks like the graph of $y = x^2$ translated 2 units to the left.

SKILL MAINTENANCE

Graph using intercepts. [2.4]

61. $2x - 7y = 28$

62. $6x - 3y = 36$

Solve each system. [3.2]

63. $3x + 4y = -19,$
$\quad 7x - 6y = -29$

64. $5x + 7y = 9,$
$\quad 3x - 4y = -11$

Replace the blanks with constants to form a true equation. [8.1]

65. $x^2 + 5x + \underline{\quad} = \left(x + \underline{\quad}\right)^2$

66. $x^2 - 9x + \underline{\quad} = \left(x - \underline{\quad}\right)^2$

SYNTHESIS

67. Before graphing a quadratic function, Sophie always plots five points. First, she calculates and plots the coordinates of the vertex. Then she plots *four* more points after calculating *two* more ordered pairs. How is this possible?

68. If the graphs of $f(x) = a_1(x - h_1)^2 + k_1$ and $g(x) = a_2(x - h_2)^2 + k_2$ have the same shape, what, if anything, can you conclude about the a's, the h's, and the k's? Why?

Write an equation for a function having a graph with the same shape as the graph of $f(x) = \frac{3}{5}x^2$, but with the given point as the vertex.

69. $(4, 1)$

70. $(2, 6)$

71. $(3, -1)$

72. $(5, -6)$

73. $(-2, -5)$

74. $(-4, -2)$

For each of the following, write the equation of the parabola that has the shape of $f(x) = 2x^2$ or $g(x) = -2x^2$ and has a maximum or minimum value at the specified point.

75. Minimum: $(2, 0)$

76. Minimum: $(-4, 0)$

77. Maximum: $(0, 3)$

78. Maximum: $(3, 8)$

Find an equation for a quadratic function F that satisfies the following conditions.

79. The graph of F is the same shape as the graph of f, where $f(x) = 3(x + 2)^2 + 7$, and $F(x)$ is a minimum at the same point that $g(x) = -2(x - 5)^2 + 1$ is a maximum.

80. The graph of F is the same shape as the graph of f, where $f(x) = -\frac{1}{3}(x - 2)^2 + 7$, and $F(x)$ is a maximum at the same point that $g(x) = 2(x + 4)^2 - 6$ is a minimum.

Functions other than parabolas can be translated. When calculating $f(x)$, if we replace x with $x - h$, where h is a constant, the graph will be moved horizontally. If we replace $f(x)$ with $f(x) + k$, the graph will be moved vertically. Use the graph below for Exercises 81–86.

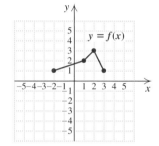

Draw a graph of each of the following.

81. $y = f(x - 1)$

82. $y = f(x + 2)$

83. $y = f(x) + 2$

84. $y = f(x) - 3$

85. $y = f(x + 3) - 2$

86. $y = f(x - 3) + 1$

87. Use the TRACE and/or TABLE features of a graphing calculator to confirm the maximum and minimum values given as answers to Exercises 51, 53, and 55. Be sure to adjust the window appropriately. On many graphing calculators, a maximum or minimum option may be available by using a CALC key.

88. Use a graphing calculator to check your graphs for Exercises 18, 28, and 48.

89. While trying to graph $y = -\frac{1}{2}x^2 + 3x + 1$, Omar gets the following screen. How can Omar tell at a glance that a mistake has been made?

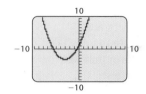

CORNER

Match the Graph

Focus: Graphing quadratic functions

Time: 15–20 minutes

Group size: 6

Materials: Index cards

ACTIVITY

1. On each of six index cards, write one of the following equations:

$y = \frac{1}{2}(x - 3)^2 + 1$; $y = \frac{1}{2}(x - 1)^2 + 3$;
$y = \frac{1}{2}(x + 1)^2 - 3$; $y = \frac{1}{2}(x + 3)^2 + 1$;
$y = \frac{1}{2}(x + 3)^2 - 1$; $y = \frac{1}{2}(x + 1)^2 + 3$.

2. Fold each index card and mix up the six cards in a hat or bag. Then, one by one, each group member should select one of the equations. Do not let anyone see your equation.

3. Each group member should carefully graph the equation selected. Make the graph large enough so that when it is finished, it can be easily viewed by the rest of the group. Be sure to scale the axes and label the vertex, but **do not label the graph with the equation used.**

4. When all group members have drawn a graph, place the graphs in a pile. The group should then match and agree on the correct equation for each graph *with no help from the person who drew the graph.* If a mistake has been made and a graph has no match, determine what its equation *should* be.

5. Compare your group's labeled graphs with those of other groups to reach consensus within the class on the correct label for each graph.

COLLABORATIVE

More About Graphing
8.7 Quadratic Functions

Completing the Square   •   Finding Intercepts

Completing the Square

By *completing the square* (see Section 8.1), we can rewrite any polynomial $ax^2 + bx + c$ in the form $a(x - h)^2 + k$. Once that has been done, the procedures discussed in Section 8.6 will enable us to graph any quadratic function.

EXAMPLE 1   Graph: $g(x) = x^2 - 6x + 4$.

Solution   We have

$$g(x) = x^2 - 6x + 4$$
$$= (x^2 - 6x) + 4.$$

To complete the square inside the parentheses, we take half the x-coefficient, $\frac{1}{2} \cdot (-6) = -3$, and square it to get $(-3)^2 = 9$. Then we add $9 - 9$ inside the parentheses:

$$g(x) = (x^2 - 6x + 9 - 9) + 4 \qquad \text{The effect is of adding 0.}$$
$$= (x^2 - 6x + 9) + (-9 + 4) \qquad \text{Using the associative law of addition to regroup}$$
$$= (x - 3)^2 - 5. \qquad \text{Factoring and simplifying}$$

This equation appeared as Example 4 of Section 8.6. The graph is that of $f(x) = x^2$ translated right 3 units and down 5 units. The vertex is $(3, -5)$.

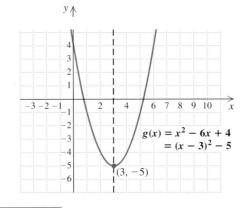

When the leading coefficient is not 1, we factor out that number from the first two terms. Then we complete the square and use the distributive law.

EXAMPLE 2   Graph: $f(x) = 3x^2 + 12x + 13$.

Solution   Since the coefficient of x^2 is not 1, we need to factor out that number—in this case, 3—from the first two terms. Remember that we want the form $f(x) = a(x - h)^2 + k$:

$$f(x) = 3x^2 + 12x + 13$$
$$= 3(x^2 + 4x) + 13.$$

Now we complete the square as before. We take half of the x-coefficient, $\frac{1}{2} \cdot 4 = 2$, and square it: $2^2 = 4$. Then we add $4 - 4$ inside the parentheses:

$$f(x) = 3(x^2 + 4x + 4 - 4) + 13. \qquad \text{Adding } 4 - 4, \text{ or 0, inside the parentheses}$$

The distributive law allows us to separate the -4 from the perfect-square trinomial so long as it is multiplied by 3. *This step is critical*:

$$f(x) = 3(x^2 + 4x + 4) + 3(-4) + 13 \qquad \text{This leaves a perfect-square trinomial inside the parentheses.}$$
$$= 3(x + 2)^2 + 1. \qquad \text{Factoring and simplifying}$$

The vertex is $(-2, 1)$, and the axis of symmetry is $x = -2$. The coefficient of x^2 is 3, so the graph is narrow and opens upward. We choose a few x-values on either side of the vertex, compute y-values, and then graph the parabola.

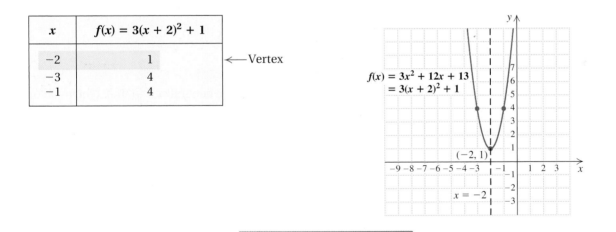

x	$f(x) = 3(x + 2)^2 + 1$	
-2	1	←—Vertex
-3	4	
-1	4	

EXAMPLE 3   Graph: $f(x) = -2x^2 + 10x - 7$.

Solution   We first find the vertex by completing the square. To do so, we factor out -2 from the first two terms of the expression. This makes the coefficient of x^2 inside the parentheses 1:

$$f(x) = -2x^2 + 10x - 7$$
$$= -2(x^2 - 5x) - 7.$$

Now we complete the square as before. We take half of the x-coefficient and square it to get $\frac{25}{4}$. Then we add $\frac{25}{4} - \frac{25}{4}$ inside the parentheses:

$$f(x) = -2\left(x^2 - 5x + \frac{25}{4} - \frac{25}{4}\right) - 7$$

$$= -2\left(x^2 - 5x + \frac{25}{4}\right) + (-2)\left(-\frac{25}{4}\right) - 7 \qquad \begin{array}{l}\text{Multiplying by } -2,\\ \text{using the distributive}\\ \text{law, and regrouping}\end{array}$$

$$= -2\left(x - \frac{5}{2}\right)^2 + \frac{11}{2}. \qquad \begin{array}{l}\text{Factoring and}\\ \text{simplifying}\end{array}$$

The vertex is $\left(\frac{5}{2}, \frac{11}{2}\right)$, and the axis of symmetry is $x = \frac{5}{2}$. The coefficient of x^2, -2, is negative, so the graph opens downward. We plot a few points on either side of the vertex, including the y-intercept, $f(0)$, and graph the parabola.

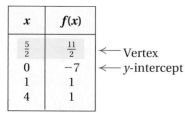

x	$f(x)$	
$\frac{5}{2}$	$\frac{11}{2}$	← Vertex
0	-7	← y-intercept
1	1	
4	1	

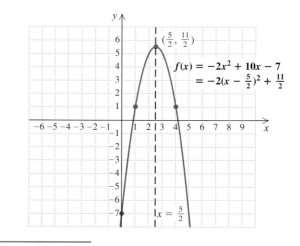

The method used in Examples 1–3 can be generalized to find a formula for locating the vertex. We complete the square as follows:

$$f(x) = ax^2 + bx + c$$

$$= a\left(x^2 + \frac{b}{a}x\right) + c. \qquad \begin{array}{l}\text{Factoring } a \text{ out of the first two terms.}\\ \text{Check by multiplying.}\end{array}$$

Half of the x-coefficient, $\frac{b}{a}$, is $\frac{b}{2a}$. We square it to get $\frac{b^2}{4a^2}$ and add $\frac{b^2}{4a^2} - \frac{b^2}{4a^2}$ inside the parentheses. Then we distribute the a and regroup terms:

$$f(x) = a\left(x^2 + \frac{b}{a}x + \frac{b^2}{4a^2} - \frac{b^2}{4a^2}\right) + c$$

$$= a\left(x^2 + \frac{b}{a}x + \frac{b^2}{4a^2}\right) + a\left(-\frac{b^2}{4a^2}\right) + c \qquad \begin{array}{l}\text{Using the distribu-}\\ \text{tive law}\end{array}$$

$$= a\left(x + \frac{b}{2a}\right)^2 + \frac{-b^2}{4a} + \frac{4ac}{4a} \qquad \begin{array}{l}\text{Factoring and find-}\\ \text{ing a common}\\ \text{denominator}\end{array}$$

$$= a\left[x - \left(-\frac{b}{2a}\right)\right]^2 + \frac{4ac - b^2}{4a}.$$

Student Notes _____

The easiest way to remember a formula is to understand its derivation. Check with your instructor to determine what, if any, formulas you will be expected to remember.

Thus we have the following.

The Vertex of a Parabola

The vertex of the parabola given by $f(x) = ax^2 + bx + c$ is

$$\left(-\frac{b}{2a}, f\left(-\frac{b}{2a}\right)\right) \quad \text{or} \quad \left(-\frac{b}{2a}, \frac{4ac - b^2}{4a}\right).$$

The x-coordinate of the vertex is $-b/(2a)$. The axis of symmetry is $x = -b/(2a)$. The second coordinate of the vertex is most commonly found by computing $f\left(-\dfrac{b}{2a}\right)$.

Let's reexamine Example 3 to see how we could have found the vertex directly. From the formula above,

$$\text{the } x\text{-coordinate of the vertex is } -\frac{b}{2a} = -\frac{10}{2(-2)} = \frac{5}{2}.$$

Substituting $\frac{5}{2}$ into $f(x) = -2x^2 + 10x - 7$, we find the second coordinate of the vertex:

$$f\left(\tfrac{5}{2}\right) = -2\left(\tfrac{5}{2}\right)^2 + 10\left(\tfrac{5}{2}\right) - 7$$
$$= -2\left(\tfrac{25}{4}\right) + 25 - 7$$
$$= -\tfrac{25}{2} + 18$$
$$= -\tfrac{25}{2} + \tfrac{36}{2} = \tfrac{11}{2}.$$

The vertex is $\left(\tfrac{5}{2}, \tfrac{11}{2}\right)$. The axis of symmetry is $x = \tfrac{5}{2}$.

We have actually developed two methods for finding the vertex. One is by completing the square and the other is by using a formula. You should check to see if your instructor prefers one method over the other or wants you to use both.

Finding Intercepts

For any function f, the y-intercept occurs at $f(0)$. Thus, for $f(x) = ax^2 + bx + c$, the y-intercept is simply $(0, c)$. To find x-intercepts, we look for points where $y = 0$ or $f(x) = 0$. Thus, for $f(x) = ax^2 + bx + c$, the x-intercepts occur at those x-values for which

$$ax^2 + bx + c = 0.$$

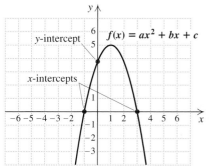

EXAMPLE 4   Find the x- and y-intercepts of the graph of $f(x) = x^2 - 2x - 2$.

Solution   The y-intercept is simply $(0, f(0))$, or $(0, -2)$. To find the x-intercepts, we solve the equation

$$0 = x^2 - 2x - 2.$$

We are unable to factor $x^2 - 2x - 2$, so we use the quadratic formula and get $x = 1 \pm \sqrt{3}$. Thus the x-intercepts are $\left(1 - \sqrt{3}, 0\right)$ and $\left(1 + \sqrt{3}, 0\right)$. If graphing, we would approximate, to get $(-0.7, 0)$ and $(2.7, 0)$.

Exercise Set

8.7

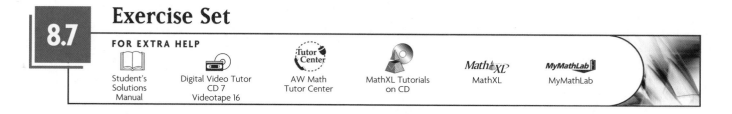

FOR EXTRA HELP

Student's Solutions Manual   Digital Video Tutor CD 7 Videotape 16   AW Math Tutor Center   MathXL Tutorials on CD   MathXL   MyMathLab

---

🔖 *Concept Reinforcement   Complete each of the following.*

1. The expressions $x^2 + 6x + 5$ and $(x^2 + 6x + 9) - \underline{\quad} + 5$ are equivalent.

2. The expressions $x^2 + 8x + 3$ and $(x^2 + 8x + \underline{\quad}) - 16 + 3$ are equivalent.

3. The functions given by $f(x) = 2x^2 + 12x - 7$ and $f(x) = 2(x^2 + 6x + \underline{\quad}) - 18 - 7$ are equivalent.

4. The functions given by $g(x) = 3x^2 - 6x - 5$ and $g(x) = 3(x^2 - 2x + \underline{\quad}) - 3 - 5$ are equivalent.

5. The graph of $f(x) = -2(x - \underline{\quad})^2 + 7$ has its vertex at $(3, 7)$.

6. The graph of $g(x) = -3(x - 4)^2 + \underline{\quad}$ has its vertex at $(4, 1)$.

7. The graph of $f(x) = 2(x - \underline{\quad})^2 + \underline{\quad}$ has its axis of symmetry at $x = \frac{5}{2}$ and has its vertex at $\left(\frac{5}{2}, -4\right)$.

8. The graph of $g(x) = \frac{1}{2}(x - \underline{\quad})^2 + \frac{7}{2}$ has $x = -2$ as its axis of symmetry and $\underline{\quad}$ as its vertex.

*For each quadratic function, **(a)** find the vertex and the axis of symmetry and **(b)** graph the function.*

9. $f(x) = x^2 + 4x + 5$

10. $f(x) = x^2 + 2x - 5$

11. $g(x) = x^2 - 6x + 13$

12. $g(x) = x^2 - 4x + 5$

13. $f(x) = x^2 + 8x + 20$

14. $f(x) = x^2 - 10x + 21$

15. $h(x) = 2x^2 - 16x + 25$

16. $h(x) = 2x^2 + 16x + 23$

17. $f(x) = -x^2 + 2x + 5$

18. $f(x) = -x^2 - 2x + 7$

19. $g(x) = x^2 + 3x - 10$

20. $g(x) = x^2 + 5x + 4$

21. $f(x) = 3x^2 - 24x + 50$

22. $f(x) = 4x^2 + 8x - 3$

23. $h(x) = x^2 + 7x$

24. $h(x) = x^2 - 5x$

25. $f(x) = -2x^2 - 4x - 6$

26. $f(x) = -3x^2 + 6x + 2$

27. $g(x) = 2x^2 - 8x + 3$

28. $g(x) = 2x^2 + 5x - 1$

29. $f(x) = -3x^2 + 5x - 2$

30. $f(x) = -3x^2 - 7x + 2$

31. $h(x) = \frac{1}{2}x^2 + 4x + \frac{19}{3}$

32. $h(x) = \frac{1}{2}x^2 - 3x + 2$

Find the x- and y-intercepts. If no x-intercepts exist, state this.

33. $f(x) = x^2 - 6x + 3$ **34.** $f(x) = x^2 + 5x + 2$

35. $g(x) = -x^2 + 2x + 3$ **36.** $g(x) = x^2 - 6x + 9$

Aha! **37.** $f(x) = x^2 - 9x$ **38.** $f(x) = x^2 - 7x$

39. $h(x) = -x^2 + 4x - 4$ **40.** $h(x) = 4x^2 - 12x + 3$

41. $f(x) = 2x^2 - 4x + 6$ **42.** $f(x) = x^2 - x + 2$

43. Does the graph of every quadratic function have a y-intercept? Why or why not?

44. Is it possible for the graph of a quadratic function to have only one x-intercept if the vertex is off the x-axis? Why or why not?

SKILL MAINTENANCE

Solve each system.

45. $5x - 3y = 16,$
$4x + 2y = 4$ [3.2]

46. $2x - 5y = 9,$
$5x - 15y = 20$ [3.2]

47. $4a - 5b + \ c = 3,$
$3a - 4b + 2c = 3,$
$a + \ b - 7c = -2$
[3.4]

48. $2a - 7b + \ c = 25,$
$a + 5b - 2c = -18,$
$3a - \ b + 4c = 14$
[3.4]

Solve. [7.6]

49. $\sqrt{4x - 4} = \sqrt{x + 4} + 1$

50. $\sqrt{5x - 4} + \sqrt{13 - x} = 7$

SYNTHESIS

51. If the graphs of two quadratic functions have the same x-intercepts, will they also have the same vertex? Why or why not?

52. Suppose that the graph of $f(x) = ax^2 + bx + c$ has $(x_1, 0)$ and $(x_2, 0)$ as x-intercepts. Explain why the graph of $g(x) = -ax^2 - bx - c$ will also have $(x_1, 0)$ and $(x_2, 0)$ as x-intercepts.

For each quadratic function, find **(a)** *the maximum or minimum value and* **(b)** *the x- and y-intercepts.*

53. $f(x) = 2.31x^2 - 3.135x - 5.89$

54. $f(x) = -18.8x^2 + 7.92x + 6.18$

55. Graph the function
$$f(x) = x^2 - x - 6.$$
Then use the graph to approximate solutions to each of the following equations.
a) $x^2 - x - 6 = 2$
b) $x^2 - x - 6 = -3$

56. Graph the function
$$f(x) = \frac{x^2}{2} + x - \frac{3}{2}.$$
Then use the graph to approximate solutions to each of the following equations.
a) $\dfrac{x^2}{2} + x - \dfrac{3}{2} = 0$

b) $\dfrac{x^2}{2} + x - \dfrac{3}{2} = 1$

c) $\dfrac{x^2}{2} + x - \dfrac{3}{2} = 2$

Find an equivalent equation of the type
$$f(x) = a(x - h)^2 + k.$$
57. $f(x) = mx^2 - nx + p$

58. $f(x) = 3x^2 + mx + m^2$

59. A quadratic function has $(-1, 0)$ as one of its intercepts and $(3, -5)$ as its vertex. Find an equation for the function.

60. A quadratic function has $(4, 0)$ as one of its intercepts and $(-1, 7)$ as its vertex. Find an equation for the function.

Graph.

61. $f(x) = |x^2 - 1|$

62. $f(x) = |x^2 - 3x - 4|$

63. $f(x) = |2(x - 3)^2 - 5|$

64. Use a graphing calculator to check your answers to Exercises 17, 31, 41, 53, and 55.

8.8 Problem Solving and Quadratic Functions

Maximum and Minimum Problems • Fitting Quadratic Functions to Data

Let's look now at some of the many situations in which quadratic functions are used for problem solving.

Maximum and Minimum Problems

We have seen that for any quadratic function f, the value of $f(x)$ at the vertex is either a maximum or a minimum. Thus problems in which a quantity must be maximized or minimized can be solved by finding the coordinates of a vertex, assuming the problem can be modeled with a quadratic function.

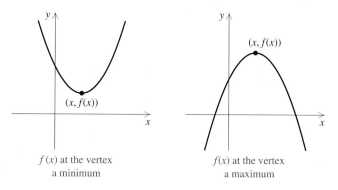

$f(x)$ at the vertex a minimum

$f(x)$ at the vertex a maximum

EXAMPLE 1

Newborn calves. The number of pounds of milk per day recommended for a calf that is x weeks old can be approximated by $p(x)$, where $p(x) = -0.2x^2 + 1.3x + 6.2$ (*Source*: C. Chaloux, University of Vermont, 1998). When is a calf's milk consumption greatest and how much milk does it consume at that time?

Solution

1., 2. Familiarize and **Translate.** We are given the function for milk consumption by a calf. Note that it is a quadratic function of x, the calf's age in weeks. Since the coefficient of x^2 is negative, it appears that milk consumption increases and then decreases. The calculator-generated graph at left confirms this.

3. Carry out. We can either complete the square,

$$\begin{aligned}
p(x) &= -0.2x^2 + 1.3x + 6.2 \\
&= -0.2(x^2 - 6.5x) + 6.2 \\
&= -0.2(x^2 - 6.5x + 3.25^2 - 3.25^2) + 6.2 \qquad \text{Completing the} \\
&\qquad\qquad\qquad\qquad\qquad\qquad\qquad\qquad\qquad\quad \text{square;} \\
&\qquad\qquad\qquad\qquad\qquad\qquad\qquad\qquad\qquad\quad 6.5/2 = 3.25 \\
&= -0.2(x^2 - 6.5x + 3.25^2) + (-0.2)(-3.25^2) + 6.2 \\
&= -0.2(x - 3.25)^2 + 8.3125, \qquad \text{Factoring and simplifying}
\end{aligned}$$

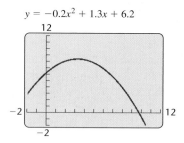

$y = -0.2x^2 + 1.3x + 6.2$

A visualization for Example 1

or we can use $-b/(2a) = -1.3/(-0.4) = 3.25$. Using a calculator, we find that

$$p(3.25) = -0.2(3.25)^2 + 1.3(3.25) + 6.2 = 8.3125.$$

4. **Check.**   Both of the approaches in step (3) indicate that a maximum occurs when $x = 3.25$, or $3\frac{1}{4}$. The graph also serves as a check.

5. **State.**   A calf's milk consumption is greatest when the calf is $3\frac{1}{4}$ weeks old. At that time, it drinks about 8.3 lb of milk per day.

EXAMPLE 2

Swimming area.   A lifeguard has 100 m of roped-together flotation devices with which to cordon off a rectangular swimming area at Lakeside Beach. If the shoreline forms one side of the rectangle, what dimensions will maximize the size of the area for swimming?

Solution

1. **Familiarize.**   We make a drawing and label it, letting $w =$ the width of the rectangle, in meters, and $l =$ the length of the rectangle, in meters.

    Recall that Area $= l \cdot w$ and Perimeter $= 2w + 2l$. Since the beach forms one length of the rectangle, the flotation devices comprise three sides. Thus

    $$2w + l = 100.$$

    To get a better feel for the problem, we can look at some possible dimensions for a rectangular area that can be enclosed with 100 m of flotation devices. All possibilities are chosen so that $2w + l = 100$.

l	w	**Rope Length**	**Area**
40 m	30 m	100 m	1200 m^2
30 m	35 m	100 m	1050 m^2
20 m	40 m	100 m	800 m^2
⋮	⋮	⋮	⋮

What choice of l and w will maximize A?

2. **Translate.**   We have two equations: One guarantees that all 100 m of flotation devices are used; the other expresses area in terms of length and width.

    $$2w + l = 100,$$
    $$A = l \cdot w$$

3. **Carry out.**   We need to express A as a function of l or w but not both. To do so, we solve for l in the first equation to obtain $l = 100 - 2w$. Substituting for l in the second equation, we get a quadratic function:

    $$A = (100 - 2w)w \qquad \text{Substituting for } l$$
    $$= 100w - 2w^2. \qquad \text{This represents a parabola opening downward, so a maximum exists.}$$

Factoring and completing the square, we get

$$A = -2(w^2 - 50w + 625 - 625)$$ We could also use the vertex formula.

$$= -2(w - 25)^2 + 1250.$$ This suggests a maximum of 1250 m^2 when $w = 25$ m.

The maximum area, 1250 m^2, occurs when $w = 25$ m and $l = 100 - 2(25)$, or 50 m.

4. **Check.** Note that 1250 m^2 is greater than any of the values for A found in the *Familiarize* step. To be more certain, we could check values other than those used in that step. For example, if $w = 26$ m, then $l = 100 - 2 \cdot 26 = 48$ m, and $A = 26 \cdot 48 = 1248 \text{ m}^2$. Since 1250 m^2 is greater than 1248 m^2, it appears that we have a maximum.

5. **State.** The largest rectangular area for swimming that can be enclosed is 25 m by 50 m.

Fitting Quadratic Functions to Data

Whenever a certain quadratic function fits a situation, that function can be determined if three inputs and their outputs are known. Each of the given ordered pairs is called a *data point*.

EXAMPLE 3   The decline of teen smoking.   Deadly and less fashionable than ever, teen smoking has declined since 1997. According to the Centers for Disease Control and Prevention, the percentage of high-school students who reported having smoked a cigarette in the preceding 30 days increased from 30.5% in 1993 to 36.4% in 1997, but has declined since then to 21.9% in 2003.

Years After 1993	Percentage of High-School Students Who Smoked a Cigarette in the Preceding 30 Days
0	30.5
4	36.4
10	21.9

Source: Centers for Disease Control and Prevention, *Morbidity and Mortality Weekly Report* 6/18/04

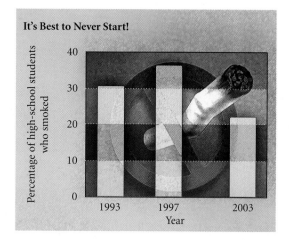

The rise and fall in the percentage of teen smokers suggests that the situation can be modeled by a quadratic function.

a) Use the data points $(0, 30.5)$, $(4, 36.4)$, and $(10, 21.9)$ to find a quadratic function that fits the data.

b) Use the function from part (a) to estimate the percentage of high-school students in 2005 who smoked a cigarette in the preceding 30 days.

Solution

a) We are looking for a function of the form $T(x) = ax^2 + bx + c$, given that $T(0) = 30.5$, $T(4) = 36.4$, and $T(10) = 21.9$. Thus,

$$30.5 = a \cdot 0^2 + b \cdot 0 + c, \qquad \text{Using the data point } (0, 30.5)$$
$$36.4 = a \cdot 4^2 + b \cdot 4 + c, \qquad \text{Using the data point } (4, 36.4)$$
$$21.9 = a \cdot 10^2 + b \cdot 10 + c. \qquad \text{Using the data point } (10, 21.9)$$

After simplifying, we see that we need to solve the system

$$30.5 = c, \qquad\qquad\qquad \textbf{(1)}$$
$$36.4 = 16a + 4b + c, \qquad \textbf{(2)}$$
$$21.9 = 100a + 10b + c. \quad \textbf{(3)}$$

We know from equation (1) that $c = 30.5$. Substituting that value into equations (2) and (3), we have

$$36.4 = 16a + 4b + 30.5,$$
$$21.9 = 100a + 10b + 30.5.$$

Subtracting 30.5 from both sides of each equation, we have

$$5.9 = 16a + 4b,$$
$$-8.6 = 100a + 10b$$

or

$$59 = 160a + 40b, \qquad \textbf{(4)} \ \Big\} \quad \text{Multiplying both sides}$$
$$-86 = 1000a + 100b. \quad \textbf{(5)} \ \Big\} \quad \text{by 10 to clear decimals}$$

To solve, we multiply equation (4) by 5 and equation (5) by -2. We then add to eliminate b:

$$295 = 800a + 200b$$
$$\underline{172 = -2000a - 200b}$$
$$467 = -1200a$$

$$-\frac{467}{1200} = a, \quad \text{or} \quad a \approx -0.39. \qquad \text{Converting to decimal notation}$$

Next, we solve for b, using equation (5) above:

$$-86 = 1000\left(-\frac{467}{1200}\right) + 100b$$

$$-86 = -\frac{2335}{6} + 100b$$

$$\frac{1819}{6} = 100b \qquad\qquad \text{Adding } \frac{2335}{6} \text{ to both sides and simplifying}$$

$$\frac{1819}{600} = b, \quad \text{or} \quad b \approx 3.03. \qquad \text{Dividing both sides by 100 and converting to decimal notation}$$

We can now write $T(x) = ax^2 + bx + c$ as

$$T(x) = -\frac{467}{1200}x^2 + \frac{1819}{600}x + 30.5 \quad \text{or} \quad T(x) = -0.39x^2 + 3.03x + 30.5.$$

Student Notes

Try to keep the "big picture" in mind on problems like Example 3. Solving a system of three equations is but one part of the solution.

b) To find the percentage of high-school students who will have smoked at least once during 30 days in 2005, we evaluate the function. Note that 2005 is 12 yr after 1993. Thus,

$$T(12) = -\frac{467}{1200} \cdot 12^2 + \frac{1819}{600} \cdot 12 + 30.5$$

$$= 10.84.$$

In 2005, an estimated 10.8% of all high-schoolers smoked at least 1 cigarette during the 30 days preceding the survey in 2005.

---

technology connection

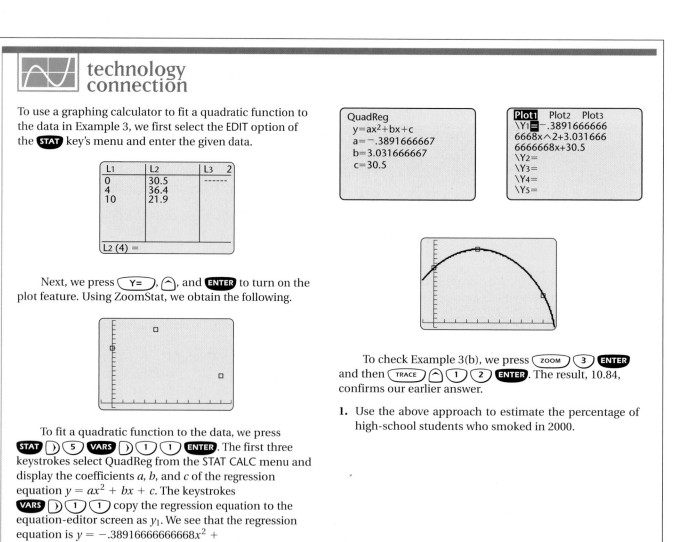

To use a graphing calculator to fit a quadratic function to the data in Example 3, we first select the EDIT option of the **STAT** key's menu and enter the given data.

L1	L2	L3 2
0	30.5	------
4	36.4	
10	21.9	

L2 (4) =

Next, we press **Y=**, ⌃, and **ENTER** to turn on the plot feature. Using ZoomStat, we obtain the following.

To fit a quadratic function to the data, we press **STAT** 〉 **5** **VARS** 〉 **1** **1** **ENTER**. The first three keystrokes select QuadReg from the STAT CALC menu and display the coefficients a, b, and c of the regression equation $y = ax^2 + bx + c$. The keystrokes **VARS** 〉 **1** **1** copy the regression equation to the equation-editor screen as y_1. We see that the regression equation is $y = -.38916666666668x^2 + 3.0316666666668x + 30.5$. Pressing **ZOOM** **9**, we see the regression equation graphed with the data points.

QuadReg
y=ax²+bx+c
a=-.3891666667
b=3.031666667
c=30.5

Plot1 Plot2 Plot3
\Y1■-.3891666666
6668x∧2+3.031666
6666668x+30.5
\Y2=
\Y3=
\Y4=
\Y5=

To check Example 3(b), we press **ZOOM** **3** **ENTER** and then **TRACE** ⌃ **1** **2** **ENTER**. The result, 10.84, confirms our earlier answer.

1. Use the above approach to estimate the percentage of high-school students who smoked in 2000.

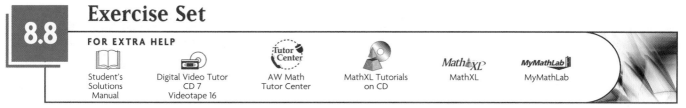

Exercise Set

8.8

FOR EXTRA HELP

Student's Solutions Manual | Digital Video Tutor CD 7 Videotape 16 | Tutor Center AW Math Tutor Center | MathXL Tutorials on CD | Math XL MathXL | MyMathLab MyMathLab

Concept Reinforcement *In each of Exercises 1–6, match the description with the graph that displays that characteristic.*

1. ____ A minimum value of $f(x)$ exists.

2. ____ A maximum value of $f(x)$ exists.

3. ____ No maximum or minimum value of $f(x)$ exists.

4. ____ The data points appear to suggest a linear model.

5. ____ The data points appear to suggest a quadratic model with a maximum.

6. ____ The data points appear to suggest a quadratic model with a minimum.

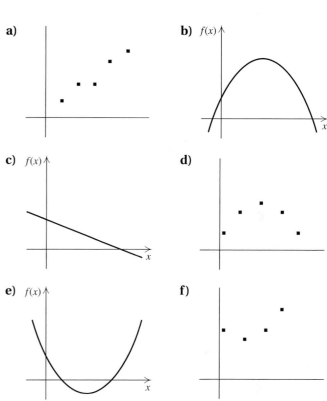

Solve.

7. *Ticket sales.* The number of tickets sold each day for an upcoming Los Lobos show can be approximated by

$$N(x) = -0.4x^2 + 9x + 11,$$

where x is the number of days since the concert was first announced. When will daily ticket sales peak and how many tickets will be sold that day?

8. *Stock prices.* The value of a share of I. J. Solar can be represented by $V(x) = x^2 - 6x + 13$, where x is the number of months after January 2004. What is the lowest value $V(x)$ will reach, and when did that occur?

9. *Minimizing cost.* Sweet Harmony Crafts has determined that when x hundred Dobros are built, the average cost per Dobro can be estimated by

$$C(x) = 0.1x^2 - 0.7x + 2.425,$$

where $C(x)$ is in hundreds of dollars. What is the minimum average cost per Dobro and how many Dobros should be built to achieve that minimum?

10. *Maximizing profit.* Recall that total profit P is the difference between total revenue R and total cost C. Given $R(x) = 1000x - x^2$ and $C(x) = 3000 + 20x$, find the total profit, the maximum value of the total profit, and the value of x at which it occurs.

11. *Furniture design.* A furniture builder is designing a rectangular end table with a perimeter of 128 in. What dimensions will yield the maximum area?

12. *Architecture.* An architect is designing an atrium for a hotel. The atrium is to be rectangular with a perimeter of 720 ft of brass piping. What dimensions will maximize the area of the atrium?

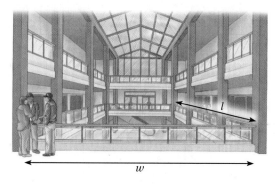

13. *Patio design.* A stone mason has enough stones to enclose a rectangular patio with 60 ft of perimeter, assuming that the attached house forms one side of the rectangle. What is the maximum area that the mason can enclose? What should the dimensions of the patio be in order to yield this area?

14. *Garden design.* Ginger is fencing in a rectangular garden, using the side of her house as one side of the rectangle. What is the maximum area that she can enclose with 40 ft of fence? What should the dimensions of the garden be in order to yield this area?

15. *Molding plastics.* Economite Plastics plans to produce a one-compartment vertical file by bending the long side of an 8-in. by 14-in. sheet of plastic along two lines to form a U shape. How tall should the file be in order to maximize the volume that the file can hold?

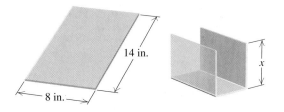

16. *Composting.* A rectangular compost container is to be formed in a corner of a fenced yard, with 8 ft of chicken wire completing the other two sides of the rectangle. If the chicken wire is 3 ft high, what dimensions of the base will maximize the container's volume?

17. What is the maximum product of two numbers that add to 18? What numbers yield this product?

18. What is the maximum product of two numbers that add to 26? What numbers yield this product?

19. What is the minimum product of two numbers that differ by 8? What are the numbers?

20. What is the minimum product of two numbers that differ by 7? What are the numbers?

Aha! **21.** What is the maximum product of two numbers that add to −10? What numbers yield this product?

22. What is the maximum product of two numbers that add to −12? What numbers yield this product?

Choosing models.   *For the scatterplots and graphs in Exercises 23–34, determine which, if any, of the following functions might be used as a model for the data: Linear, with $f(x) = mx + b$; quadratic, with $f(x) = ax^2 + bx + c, a > 0$; quadratic, with $f(x) = ax^2 + bx + c, a < 0$; neither quadratic nor linear.*

23.

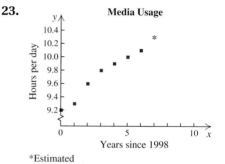

Media Usage

*Estimated
Source: Statistical Abstract of the United States, 2003*

24.

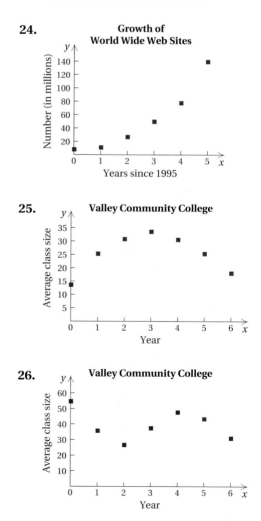

27.

Driver Fatalities by Age

Source: National Highway Traffic Administration

28.

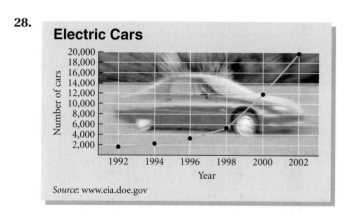

Electric Cars

Source: www.eia.doe.gov

29.

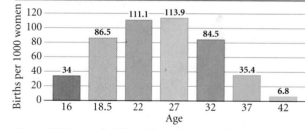

Average Number of Live Births per 1000 Women

Source: U.S. Centers for Disease Control

30.

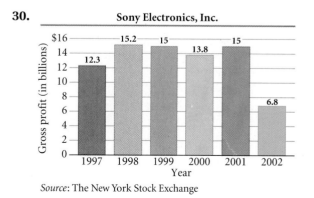

Sony Electronics, Inc.

Source: The New York Stock Exchange

31.

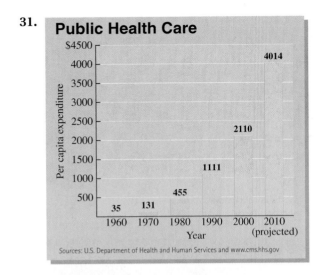

Public Health Care

Sources: U.S. Department of Health and Human Services and www.cms.hhs.gov

32.

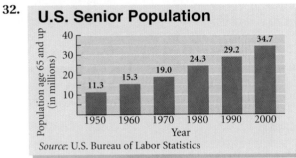

U.S. Senior Population

Source: U.S. Bureau of Labor Statistics

33.

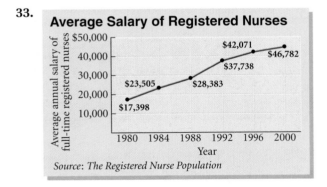

Average Salary of Registered Nurses

Source: The Registered Nurse Population

34.

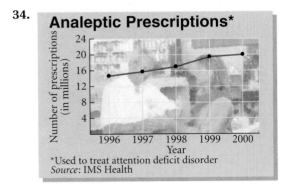

Analeptic Prescriptions*

*Used to treat attention deficit disorder
Source: IMS Health

Find a quadratic function that fits the set of data points.

35. $(1, 4), (-1, -2), (2, 13)$ **36.** $(1, 4), (-1, 6), (-2, 16)$

37. $(2, 0), (4, 3), (12, -5)$ **38.** $(-3, -30), (3, 0), (6, 6)$

39. a) Find a quadratic function that fits the following data.

Travel Speed (in kilometers per hour)	Number of Nighttime Accidents (for every 200 million kilometers driven)
60	400
80	250
100	250

b) Use the function to estimate the number of nighttime accidents that occur at 50 km/h.

40. a) Find a quadratic function that fits the following data.

Travel Speed (in kilometers per hour)	Number of Daytime Accidents (for every 200 million kilometers driven)
60	100
80	130
100	200

b) Use the function to estimate the number of daytime accidents that occur at 50 km/h.

41. *Archery.* The Olympic flame tower at the 1992 Summer Olympics was lit at a height of about 27 m by a flaming arrow that was launched about 63 m from the base of the tower. If the arrow landed about 63 m beyond the tower, find a quadratic function that expresses the height h of the arrow as a function of the distance d that it traveled horizontally.

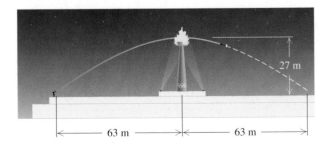

27 m

63 m   63 m

42. *Pizza prices.* Pizza Unlimited has the following prices for pizzas.

Diameter	Price
8 in.	$ 6.00
12 in.	$ 8.50
16 in.	$11.50

Is price a quadratic function of diameter? It probably should be, because the price should be proportional to the area, and the area is a quadratic function of the diameter. (The area of a circular region is given by $A = \pi r^2$ or $(\pi/4) \cdot d^2$.)

a) Express price as a quadratic function of diameter using the data points (8, 6), (12, 8.50), and (16, 11.50).

b) Use the function to find the price of a 14-in. pizza.

43. Does every nonlinear function have a minimum or maximum value? Why or why not?

44. Explain how the leading coefficient of a quadratic function can be used to determine if a maximum or a minimum function value exists.

SKILL MAINTENANCE

Simplify.

45. $\dfrac{x}{x^2 + 17x + 72} - \dfrac{8}{x^2 + 15x + 56}$ [6.2]

46. $\dfrac{x^2 - 9}{x^2 - 8x + 7} \div \dfrac{x^2 + 6x + 9}{x^2 - 1}$ [6.1]

47. $\dfrac{t^2 - 4}{t^2 - 7t - 8} \cdot \dfrac{t^2 - 64}{t^2 - 5t + 6}$ [6.1]

48. $\dfrac{t}{t^2 - 10t + 21} + \dfrac{t}{t^2 - 49}$ [6.2]

Solve. [4.1]

49. $5x - 9 < 31$

50. $3x - 8 \geq 22$

SYNTHESIS

51. Write a problem for a classmate to solve. Design the problem so that its solution requires finding the minimum or maximum value.

52. Explain what restrictions should be placed on the quadratic functions developed in Exercises 39 and 42 and why such restrictions are needed.

53. *Bridge design.* The cables supporting a straight-line suspension bridge are nearly parabolic in shape. Suppose that a suspension bridge is being designed with concrete supports 160 ft apart and with vertical cables 30 ft above road level at the midpoint of the bridge and 80 ft above road level at a point 50 ft from the midpoint of the bridge. How long are the longest vertical cables?

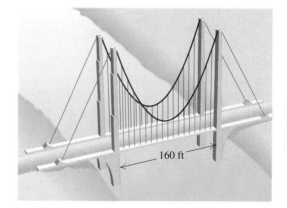

54. *Trajectory of a launched object.* The height above the ground of a launched object is a quadratic function of the time that it is in the air. Suppose that a flare is launched from a cliff 64 ft above sea level. If 3 sec after being launched the flare is again level with the cliff, and if 2 sec after that it lands in the sea, what is the maximum height that the flare will reach?

55. *Cover charges.* When the owner of Sweet Sounds charges a $10 cover charge, an average of 80 people will attend a show. For each 25¢ increase in admission price, the average number attending decreases by 1. What should the owner charge in order to make the most money?

56. *Crop yield.* An orange grower finds that she gets an average yield of 40 bushels (bu) per tree when she plants 20 trees on an acre of ground. Each time she adds a tree to an acre, the yield per tree decreases by 1 bu, due to congestion. How many trees per acre should she plant for maximum yield?

57. *Norman window.* A *Norman window* is a rectangle with a semicircle on top. Big Sky Windows is designing a Norman window that will require 24 ft of trim. What dimensions will allow the maximum amount of light to enter a house?

58. *Minimizing area.* A 36-in. piece of string is cut into two pieces. One piece is used to form a circle while the other is used to form a square. How should the string be cut so that the sum of the areas is a minimum?

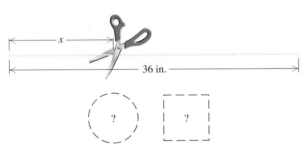

Regression can be used to find the "best"-fitting quadratic function when more than three data points are provided. In Exercises 59 and 60, six data points are given, but the approach used in the Technology Connection on p. 562 still applies.

59. *Alternative fueled vehicles.* The number of cars fueled by electricity in the United States during several years is shown in the table below.

Year	Number of Cars
1992	1,607
1994	2,224
1996	3,280
1998	5,243
2000	11,834
2002	19,755

Source: www.eia.doe.gov

a) Use regression to find a quadratic function that can be used to estimate the number of cars *c* that are fueled by electricity *x* years after 1992.

b) Use the function found in part (a) to predict the number of cars fueled by electricity in 2008.

60. *Hydrology.* The drawing below shows the cross section of a river. Typically rivers are deepest in the middle, with the depth decreasing to 0 at the edges. A hydrologist measures the depths *D*, in feet, of a river at distances *x*, in feet, from one bank. The results are listed in the table below.

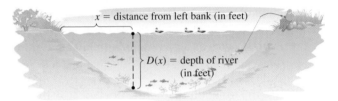

Distance *x*, from the Left Bank (in feet)	Depth, *D*, of the River (in feet)
0	0
15	10.2
25	17
50	20
90	7.2
100	0

a) Use regression to find a quadratic function that fits the data.

b) Use the function to estimate the depth of the river 70 ft from the left bank.

CORNER

Quadratic Counter Settings

Focus: Modeling quadratic functions

Time: 20–30 minutes

Group size: 3 or 4

Materials: Graphing calculators are optional.

The Panasonic Portable Stereo System RX-DT680® has a counter for finding locations on an audio cassette. When a fully wound cassette with 45 min of music on a side begins to play, the counter is at 0. After 15 min of music has played, the counter reads 250 and after 35 min, it reads 487. When the 45-min side is finished playing, the counter reads 590.

ACTIVITY

1. The paragraph above describes four ordered pairs of the form (counter number, minutes played). Three pairs are enough to find a function of the form

$$T(n) = an^2 + bn + c,$$

where $T(n)$ represents the time, in minutes, that the tape has run at counter reading n hundred. Each group member should select a different set of three points from the four given and then fit a quadratic function to the data.

2. Of the 3 or 4 functions found in part (1) above, which fits the data "best"? One way to answer this is to see how well each function predicts other pairs. The same counter used above reads 432 after a 45-min tape has played for 30 min. Which function comes closest to predicting this?

3. If a graphing calculator is available with a QUADREG option (see Exercise 59), what function does it fit to the four pairs originally listed?

4. If a class member has access to a Panasonic System RX-DT680, see how well the functions developed above predict the counter readings for a tape that has played for 5 or 10 min.

8.9 Polynomial and Rational Inequalities

Quadratic and Other Polynomial Inequalities •
Rational Inequalities

Quadratic and Other Polynomial Inequalities

Inequalities like the following are called *polynomial inequalities*:

$$x^3 - 5x > x^2 + 7, \qquad 4x - 3 < 9, \qquad 5x^2 - 3x + 2 \geq 0.$$

Second-degree polynomial inequalities in one variable are called *quadratic inequalities*. To solve polynomial inequalities, we often focus attention on where the outputs of a polynomial function are positive and where they are negative.

EXAMPLE 1

Solve: $x^2 + 3x - 10 > 0$.

Solution   Consider the "related" function $f(x) = x^2 + 3x - 10$ and its graph. Its graph opens upward since the leading coefficient is positive. Thus y-values are positive outside the interval formed by the x-intercepts. To find the intercepts, we set the polynomial equal to 0 and solve:

$$x^2 + 3x - 10 = 0$$
$$(x + 5)(x - 2) = 0$$
$$x + 5 = 0 \quad or \quad x - 2 = 0$$
$$x = -5 \quad or \quad x = 2.$$

Test values can be used to confirm that $f(x)$ is positive outside the interval $[-5, 2]$:

$$f(3) = 3^2 + 3 \cdot 3 - 10 = 9 + 9 - 10 = 8; \leftarrow \text{---Positive}$$
$$f(-6) = (-6)^2 + 3(-6) - 10 = 36 - 18 - 10 = 8.$$

Thus the solution set of the inequality is

$$\{x \,|\, x < -5 \ or \ x > 2\}, \quad or \quad (-\infty, -5) \cup (2, \infty).$$

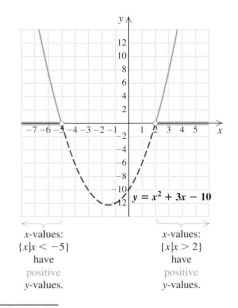

x-values: $\{x|x < -5\}$ have positive y-values.

x-values: $\{x|x > 2\}$ have positive y-values.

Any inequality with 0 on one side can be solved by considering a graph of the related function and finding intercepts as in Example 1. Sometimes the quadratic formula is needed to find the intercepts.

EXAMPLE 2

Solve: $x^2 - 2x \leq 2$.

Solution   We first write the quadratic inequality in standard form, with 0 on one side:

$$x^2 - 2x - 2 \leq 0. \qquad \text{This is equivalent to the original inequality.}$$

The graph of $f(x) = x^2 - 2x - 2$ is a parabola opening upward. Values of $f(x)$ are negative for x-values between the x-intercepts. We find the x-intercepts by solving $f(x) = 0$:

$$x = \frac{-b \pm \sqrt{b^2 - 4ac}}{2a}$$

$$= \frac{-(-2) \pm \sqrt{(-2)^2 - 4 \cdot 1(-2)}}{2 \cdot 1}$$

$$= \frac{2 \pm \sqrt{12}}{2} = \frac{2}{2} \pm \frac{2\sqrt{3}}{2} = 1 \pm \sqrt{3}.$$

Inputs in this interval have negative or 0 outputs.

At the x-intercepts, $1 - \sqrt{3}$ and $1 + \sqrt{3}$, the value of $f(x)$ is 0. Thus the solution set of the inequality is

$$\left[1 - \sqrt{3}, 1 + \sqrt{3}\right], \quad \text{or} \quad \left\{x \mid 1 - \sqrt{3} \le x \le 1 + \sqrt{3}\right\}.$$

In Example 2, it was not essential to draw the graph. The important information came from finding the x-intercepts and the sign of $f(x)$ on each side of those intercepts. We now solve a third-degree polynomial inequality, without graphing, by locating the x-intercepts, or **zeros**, of f and then using *test points* to determine the sign of $f(x)$ over each interval of the x-axis.

EXAMPLE 3 For $f(x) = 5x^3 + 10x^2 - 15x$, find all x-values for which $f(x) > 0$.

Solution We first solve the related equation:

$$f(x) = 0$$
$$5x^3 + 10x^2 - 15x = 0 \qquad \text{Substituting}$$
$$5x(x^2 + 2x - 3) = 0$$
$$5x(x + 3)(x - 1) = 0$$
$$5x = 0 \quad or \quad x + 3 = 0 \quad or \quad x - 1 = 0$$
$$x = 0 \quad or \qquad x = -3 \quad or \qquad x = 1.$$

The zeros of f are -3, 0, and 1. These zeros divide the number line, or x-axis, into four intervals: A, B, C, and D.

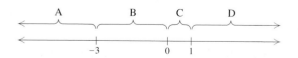

Next, selecting one convenient test value from each interval, we determine the sign of $f(x)$ for that interval. We know that, within each interval, the sign of $f(x)$ cannot change. If it did, there would need to be another zero in that interval.

Student Notes _____

When we are evaluating test values, there is often no need to do lengthy computations since all we need to determine is the sign of the result.

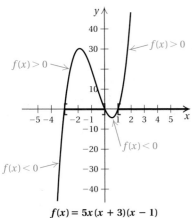

$$f(x) = 5x(x + 3)(x - 1)$$

A visualization of Example 3

Using the factored form of $f(x)$ eases the computations:

$$f(x) = 5x(x + 3)(x - 1).$$

For interval A,

$$f(-4) = 5(-4)((-4) + 3)((-4) - 1)$$ -4 is a convenient value in interval A.

$$= -20(-1)(-5)$$

$$= -100.$$ $f(-4)$ is negative.

For interval B,

$$f(-1) = 5(-1)((-1) + 3)((-1) - 1)$$ -1 is a convenient value in interval B.

$$= -5(2)(-2)$$

$$= 20.$$ $f(-1)$ is positive.

For interval C,

$$f\left(\tfrac{1}{2}\right) = 5 \cdot \tfrac{1}{2} \quad \cdot \left(\tfrac{1}{2} + 3\right) \cdot \quad \left(\tfrac{1}{2} - 1\right).$$ $\tfrac{1}{2}$ is a convenient value in interval C.

$$\underbrace{\text{Positive}\quad\text{Positive}\quad\text{Negative}}_{\text{Negative}}$$

Only the sign is important. The product is negative, so $f\left(\tfrac{1}{2}\right)$ is negative.

For interval D,

$$f(2) = \underbrace{5 \cdot 2}_{\text{Positive}} \cdot \underbrace{(2 + 3)}_{\text{Positive}} \cdot \underbrace{(2 - 1)}_{\text{Positive}}.$$ 2 is a convenient value in interval D.

$f(2)$ is positive.

Recall that we are looking for all x for which $5x^3 + 10x^2 - 15x > 0$. The calculations above indicate that $f(x)$ is positive for any number in intervals B and D. The solution set of the original inequality is

$$(-3, 0) \cup (1, \infty), \quad \text{or} \quad \{x \mid -3 < x < 0 \text{ or } x > 1\}.$$

The calculations in Example 3 were made simpler by using a factored form of the polynomial. This process was simplified further when, for intervals C and D, we concentrated on only the *sign* of $f(x)$. In the next example, we determine the sign of a polynomial function over each interval by tracking the sign of each factor. By looking at how many positive or negative factors are being multiplied, we will be able to determine the sign of the polynomial function.

EXAMPLE 4   For $f(x) = 4x^3 - 4x$, find all x-values for which $f(x) \leq 0$.

Solution   We first solve the related equation:

$$f(x) = 0$$
$$4x^3 - 4x = 0$$
$$4x(x^2 - 1) = 0$$
$$4x(x + 1)(x - 1) = 0$$
$$4x = 0 \quad \text{or} \quad x + 1 = 0 \quad \text{or} \quad x - 1 = 0$$
$$x = 0 \quad \text{or} \quad x = -1 \quad \text{or} \quad x = 1.$$

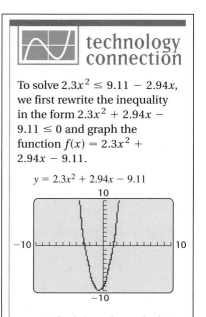

To solve $2.3x^2 \leq 9.11 - 2.94x$, we first rewrite the inequality in the form $2.3x^2 + 2.94x - 9.11 \leq 0$ and graph the function $f(x) = 2.3x^2 + 2.94x - 9.11$.

$$y = 2.3x^2 + 2.94x - 9.11$$

To find the values of x for which $f(x) \leq 0$, we focus on the region in which the graph lies *on or below* the x-axis. It appears that this region begins somewhere between -3 and -2, and continues to somewhere between 1 and 2. Using the ZERO or ROOT option of CALC, we can find the endpoints of this region. To two decimal places, the endpoints are -2.73 and 1.45. The solution set is approximately $\{x \mid -2.73 \leq x \leq 1.45\}$.

Had the inequality been $2.3x^2 > 9.11 - 2.94x$, we would look for portions of the graph that lie *above* the x-axis. An approximate solution set of such an inequality would be $\{x \mid x < -2.73 \text{ or } x > 1.45\}$.

Use a graphing calculator to solve each inequality. Round the values of the endpoints to the nearest hundredth.

1. $4.32x^2 - 3.54x - 5.34 \leq 0$
2. $7.34x^2 - 16.55x - 3.89 \geq 0$
3. $10.85x^2 + 4.28x + 4.44 > 7.91x^2 + 7.43x + 13.03$
4. $5.79x^3 - 5.68x^2 + 10.68x > 2.11x^3 + 16.90x - 11.69$

The function f has zeros at $-1, 0$, and 1. Rather than use test values, as in Example 3, let's use the factorization $f(x) = 4x(x + 1)(x - 1)$. The product $4x(x + 1)(x - 1)$ is positive or negative, depending on the signs of $4x$, $x + 1$, and $x - 1$. This is easily determined using a chart.

Interval:	$(-\infty, -1)$	$(-1, 0)$	$(0, 1)$	$(1, \infty)$
Sign of $4x$:	$-$	$-$	$+$	$+$
Sign of $x + 1$:	$-$	$+$	$+$	$+$
Sign of $x - 1$:	$-$	$-$	$-$	$+$
		-1	0	1
Sign of product $4x(x + 1)(x - 1)$:	$-$	$+$	$-$	$+$

A product is negative when it has an odd number of negative factors. Since the $\leq$ sign allows for equality, the endpoints $-1, 0$, and 1 are solutions. From the chart, we see that the solution set is

$$(-\infty, -1] \cup [0, 1], \text{ or } \{x \mid x \leq -1 \text{ or } 0 \leq x \leq 1\}.$$

To Solve a Polynomial Inequality Using Factors

1. Add or subtract to get 0 on one side and solve the related polynomial equation by factoring.
2. Use the numbers found in step (1) to divide the number line into intervals.
3. Using a test value from each interval, determine the sign of each factor over that interval.
4. Determine the sign of the product of the factors over each interval. Remember that the product of an odd number of negative numbers is negative.
5. Select the interval(s) for which the inequality is satisfied and write set-builder notation or interval notation for the solution set. Include the endpoints of the intervals when $\leq$ or $\geq$ is used.

Rational Inequalities

Inequalities involving rational expressions are called **rational inequalities**. Like polynomial inequalities, rational inequalities can be solved using test values. Unlike polynomials, however, rational expressions often have values for which the expression is undefined.

EXAMPLE 5   Solve: $\dfrac{x-3}{x+4} \geq 2$.

Solution   We write the related equation by changing the $\geq$ symbol to $=$:

$$\frac{x-3}{x+4} = 2. \qquad \text{Note that } x \neq -4.$$

Next, we solve this related equation:

$$(x+4) \cdot \frac{x-3}{x+4} = (x+4) \cdot 2 \qquad \begin{array}{l}\text{Multiplying both sides} \\ \text{by the LCD, } x+4\end{array}$$

$$x - 3 = 2x + 8$$

$$-11 = x. \qquad\qquad \text{Solving for } x$$

In the case of rational inequalities, we must always find any values that make the denominator 0. As noted at the beginning of this example, $x \neq -4$.

Now we use -11 and -4 to divide the number line into intervals:

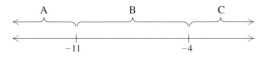

We test a number in each interval to see where the original inequality is satisfied:

$$\frac{x-3}{x+4} \geq 2.$$

A: Test -15, $\dfrac{-15-3}{-15+4} = \dfrac{-18}{-11}$

$$= \frac{18}{11} \not\geq 2 \qquad \begin{array}{l}-15 \text{ } \textit{is not} \text{ a solution, so interval A is} \\ \text{not part of the solution set.}\end{array}$$

B: Test -8, $\dfrac{-8-3}{-8+4} = \dfrac{-11}{-4}$

$$= \frac{11}{4} \geq 2 \qquad \begin{array}{l}-8 \text{ } \textit{is} \text{ a solution, so interval B is part of} \\ \text{the solution set.}\end{array}$$

C: Test 1, $\dfrac{1-3}{1+4} = \dfrac{-2}{5}$

$$= -\frac{2}{5} \not\geq 2 \qquad \begin{array}{l}1 \text{ } \textit{is not} \text{ a solution, so interval C is not} \\ \text{part of the solution set.}\end{array}$$

The solution set includes interval B. The endpoint -11 is included because the inequality symbol is $\geq$ and -11 is a solution of the related equation. The number -4 is *not* included because $(x-3)/(x+4)$ is undefined for $x = -4$. Thus the solution set of the original inequality is

$$[-11, -4), \quad \text{or} \quad \{x \,|\, -11 \leq x < -4\}.$$

ALGEBRAIC—GRAPHICAL CONNECTION

Let's compare the algebraic solution of Example 5 to a graphical solution. By graphing $f(x) = (x - 3)/(x + 4)$, we can find the solutions of $(x - 3)/(x + 4) \geq 2$ by sketching the line $y = 2$ and locating all x-values for which $f(x) \geq 2$.

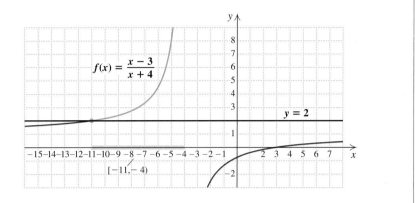

Because graphing rational functions can be very time-consuming, we generally just use test values.

To Solve a Rational Inequality

1. Change the inequality symbol to an equals sign and solve the related equation.
2. Find any replacements for which the rational expression is undefined.
3. Use the numbers found in steps (1) and (2) to divide the number line into intervals.
4. Substitute a test value from each interval into the inequality. If the number is a solution, then the interval to which it belongs is part of the solution set.
5. Select the interval(s) and any endpoints for which the inequality is satisfied and write set-builder or interval notation for the solution set. If the inequality symbol is $\leq$ or $\geq$, then the solutions from step (1) are also included in the solution set. Those numbers found in step (2) should be excluded from the solution set, even if they are solutions from step (1).

Exercise Set

8.9

⬐ *Concept Reinforcement*   *Classify each of the following as either true or false.*

1. The solution of $(x - 3)(x + 2) \le 0$ is $[-2, 3]$.

2. The solution of $(x + 5)(x - 4) \ge 0$ is $[-5, 4]$.

3. The solution of $(x - 1)(x - 6) > 0$ is $\{x \mid x < 1 \text{ or } x > 6\}$.

4. The solution of $(x + 4)(x + 2) < 0$ is $(-4, -2)$.

5. To solve $\dfrac{x - 5}{x + 4} \ge 0$ using intervals, we divide the number line into the intervals $(-\infty, -4)$, $(-4, 5)$, and $(5, \infty)$.

6. To solve $\dfrac{x + 2}{x - 3} < 0$ using intervals, we divide the number line into the intervals $(-\infty, -2)$, $(-2, 3)$, and $(3, \infty)$.

7. The solution of $\dfrac{3}{x - 5} \le 0$ is $[5, \infty)$.

8. The solution of $\dfrac{-2}{x + 3} \ge 0$ is $(-\infty, -3]$.

Solve.

9. $(x + 4)(x - 3) < 0$

10. $(x - 5)(x + 2) > 0$

11. $(x + 7)(x - 2) \ge 0$

12. $(x - 1)(x + 4) \le 0$

13. $x^2 - x - 2 > 0$

14. $x^2 + x - 2 < 0$

Aha! **15.** $x^2 + 4x + 4 < 0$

16. $x^2 + 6x + 9 < 0$

17. $x^2 - 4x < 12$

18. $x^2 + 6x > -8$

19. $3x(x + 2)(x - 2) < 0$

20. $5x(x + 1)(x - 1) > 0$

21. $(x - 1)(x + 2)(x - 4) \ge 0$

22. $(x + 3)(x + 2)(x - 1) < 0$

23. For $f(x) = x^2 - 1$, find all x-values for which $f(x) \le 3$.

24. For $f(x) = x^2 - 20$, find all x-values for which $f(x) > 5$.

25. For $g(x) = (x - 2)(x - 3)(x + 1)$, find all x-values for which $g(x) > 0$.

26. For $g(x) = (x + 3)(x - 2)(x + 1)$, find all x-values for which $g(x) < 0$.

27. For $F(x) = x^3 - 7x^2 + 10x$, find all x-values for which $F(x) \le 0$.

28. For $G(x) = x^3 - 8x^2 + 12x$, find all x-values for which $G(x) \ge 0$.

Solve.

29. $\dfrac{1}{x + 5} < 0$

30. $\dfrac{1}{x + 4} > 0$

31. $\dfrac{x + 1}{x - 3} \ge 0$

32. $\dfrac{x - 2}{x + 4} \le 0$

33. $\dfrac{x + 1}{x + 6} \ge 1$

34. $\dfrac{x - 1}{x - 2} \le 1$

35. $\dfrac{(x - 2)(x + 1)}{x - 5} \le 0$

36. $\dfrac{(x + 4)(x - 1)}{x + 3} \ge 0$

37. $\dfrac{x}{x + 3} \ge 0$

38. $\dfrac{x - 2}{x} \le 0$

39. $\dfrac{x - 5}{x} < 1$

40. $\dfrac{x}{x - 1} > 2$

41. $\dfrac{x - 1}{(x - 3)(x + 4)} \le 0$

42. $\dfrac{x + 2}{(x - 2)(x + 7)} \ge 0$

43. For $f(x) = \dfrac{5 - 2x}{4x + 3}$, find all x-values for which $f(x) \ge 0$.

44. For $g(x) = \dfrac{2 + 3x}{2x - 4}$, find all x-values for which $g(x) \geq 0$.

45. For $G(x) = \dfrac{1}{x - 2}$, find all x-values for which $G(x) \leq 1$.

46. For $F(x) = \dfrac{1}{x - 3}$, find all x-values for which $F(x) \leq 2$.

47. Explain how any quadratic inequality can be solved by examining a parabola.

48. Describe a method for creating a quadratic inequality for which there is no solution.

SKILL MAINTENANCE

Simplify. [1.6]

49. $(2a^3b^2c^4)^3$

50. $(5a^4b^7)^2$

51. 2^{-5}

52. 3^{-4}

53. If $f(x) = 3x^2$, find $f(a + 1)$. [5.2]

54. If $g(x) = 5x - 3$, find $g(a + 2)$. [2.2]

SYNTHESIS

55. Step (5) on p. 575 states that even when the inequality symbol is $\leq$ or $\geq$, the solutions from step (1) are not always part of the solution set. Why?

56. Describe a method that could be used to create quadratic inequalities that have $(-\infty, a] \cup [b, \infty)$ as the solution set. Assume $a < b$.

Find each solution set.

57. $x^2 + 2x < 5$

58. $x^4 + 2x^2 \geq 0$

59. $x^4 + 3x^2 \leq 0$

60. $\left| \dfrac{x + 2}{x - 1} \right| \leq 3$

61. *Total profit.* Derex, Inc., determines that its total-profit function is given by
$$P(x) = -3x^2 + 630x - 6000.$$
   a) Find all values of x for which Derex makes a profit.
   b) Find all values of x for which Derex loses money.

62. *Height of a thrown object.* The function
$$S(t) = -16t^2 + 32t + 1920$$
gives the height S, in feet, of an object thrown from a cliff that is 1920 ft high. Here t is the time, in seconds, that the object is in the air.
   a) For what times does the height exceed 1920 ft?
   b) For what times is the height less than 640 ft?

63. *Number of handshakes.* There are n people in a room. The number N of possible handshakes by the people is given by the function
$$N(n) = \dfrac{n(n - 1)}{2}.$$
For what number of people n is $66 \leq N \leq 300$?

64. *Number of diagonals.* A polygon with n sides has D diagonals, where D is given by the function
$$D(n) = \dfrac{n(n - 3)}{2}.$$
Find the number of sides n if
$$27 \leq D \leq 230.$$

Use a graphing calculator to graph each function and find solutions of $f(x) = 0$. Then solve the inequalities $f(x) < 0$ and $f(x) > 0$.

65. $f(x) = x^3 - 2x^2 - 5x + 6$

66. $f(x) = \dfrac{1}{3}x^3 - x + \dfrac{2}{3}$

67. $f(x) = x + \dfrac{1}{x}$

68. $f(x) = x - \sqrt{x}, x \geq 0$

69. $f(x) = \dfrac{x^3 - x^2 - 2x}{x^2 + x - 6}$

70. $f(x) = x^4 - 4x^3 - x^2 + 16x - 12$

71. Use a graphing calculator to solve Exercises 39 and 45 by drawing two curves, one for each side of the inequality.

8 Study Summary

Any equation that can be written in the form $ax^2 + bx + c = 0$, with a, b, and c constant, is said to be **quadratic** (p. 506). There are several methods for solving quadratic equations:

Factoring (p. 506)

Easiest method to use *if* you can factor the polynomial.

$$x^2 - 3x - 10 = 0$$
$$(x + 2)(x - 5) = 0$$
$$x + 2 = 0 \quad or \quad x - 5 = 0$$
$$x = -2 \quad or \quad x = 5$$

Principle of Square Roots (p. 507)

Works only if a perfect-square trinomial is on one side and a constant is on the other side.

$$x^2 - 8x + 16 = 25$$
$$(x - 4)^2 = 25$$
$$x - 4 = -5 \quad or \quad x - 4 = 5$$
$$x = -1 \quad or \quad x = 9$$

Completing the Square (p. 509)

Can be used to solve *any* quadratic equation, but calculations can be lengthy.

$$x^2 + 6x = 1$$
$$x^2 + 6x + \left(\tfrac{6}{2}\right)^2 = 1 + \left(\tfrac{6}{2}\right)^2$$
$$x^2 + 6x + 9 = 1 + 9$$
$$(x + 3)^2 = 10$$
$$x + 3 = \pm\sqrt{10}$$
$$x = -3 \pm \sqrt{10}$$

Quadratic Formula (p. 518)

> If $ax^2 + bx + c = 0$,
> then $x = \dfrac{-b \pm \sqrt{b^2 - 4ac}}{2a}$.

Is based on completing the square but is quicker to use. Can be used to solve *any* quadratic equation.

$$3x^2 - 2x - 5 = 0$$
$$x = \frac{-(-2) \pm \sqrt{(-2)^2 - 4 \cdot 3(-5)}}{2 \cdot 3}$$
$$x = \frac{2 \pm \sqrt{4 + 60}}{6}$$
$$x = \frac{2 \pm \sqrt{64}}{6}$$
$$x = \frac{2 \pm 8}{6}$$
$$x = \frac{10}{6} = \frac{5}{3} \quad or \quad x = \frac{-6}{6} = -1$$

The **discriminant** of the quadratic formula can be used to find the nature of the solution(s) of a quadratic equation (p. 532):

$$b^2 - 4ac = 0 \;\rightarrow\; \text{One solution; a rational number.}$$

$$b^2 - 4ac > 0 \;\rightarrow\; \text{Two real solutions, both rational if } b^2 - 4ac \text{ is a perfect square.}$$

$$b^2 - 4ac < 0 \;\rightarrow\; \text{Two imaginary-number solutions.}$$

The methods for solving quadratic equations can be applied to equations that are not quadratic but are in **quadratic form** (p. 537):

$$x^4 - 10x^2 + 9 = 0 \qquad \text{Think of } x^2 \text{ as } u \text{ and } x^4 \text{ as } u^2.$$

$$u^2 - 10u + 9 = 0$$

$$(u - 9)(u - 1) = 0$$

$$u - 9 = 0 \quad or \quad u - 1 = 0$$

$$u = 9 \quad or \quad u = 1$$

$$x^2 = 9 \quad or \quad x^2 = 1$$

$$x = \pm 3 \quad or \quad x = \pm 1.$$

The graph of a quadratic function is a **parabola** (p. 544). The graph of $f(x) = ax^2 + bx + c$ opens upward for $a > 0$ and downward for $a < 0$. The graph of an equation of the form $f(x) = a(x - h)^2 + k$ looks like the graph of $y = ax^2$ **translated** $|h|$ units to the right (for $h > 0$) or left (for $h < 0$) and $|k|$ units up (for $k > 0$) or down (for $k < 0$).

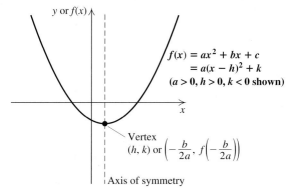

Every parabola has an **axis of symmetry** and **vertex** as shown above (p. 544). For a quadratic function, the second coordinate of the vertex is a **minimum** (for $a > 0$) or **maximum** (for $a < 0$) function value (p. 546).

The x-intercepts, or **zeros**, of a function are used to divide the x-axis into intervals when solving **polynomial** or **rational inequalities** (pp. 569, 573). Rational inequalities also require forming intervals on either side of any x-value(s) for which the denominator is 0. We then use test values from each interval to identify the solution set.

8 Review Exercises

Classify each statement as either true or false.

1. Every quadratic equation has two different solutions. [8.4]

2. Every quadratic equation has at least one solution. [8.4]

3. If an equation cannot be solved by completing the square, it cannot be solved by the quadratic formula. [8.2]

4. A negative discriminant indicates two imaginary-number solutions of a quadratic equation. [8.4]

5. Certain radical or rational equations can be written in quadratic form. [8.5]

6. The graph of $f(x) = 2(x + 3)^2 - 4$ has its vertex at $(3, -4)$. [8.6]

7. The graph of $g(x) = 5x^2$ has $x = 0$ as its axis of symmetry. [8.6]

8. The graph of $f(x) = -2x^2 + 1$ has no minimum value. [8.6]

9. The zeros of $g(x) = x^2 - 9$ are -3 and 3. [8.6]

10. To solve a polynomial inequality, we often must solve a polynomial equation. [8.9]

Solve.

11. $4x^2 - 9 = 0$ [8.1]   12. $8x^2 + 6x = 0$ [8.1]

13. $x^2 - 12x + 36 = 9$ [8.1]   14. $x^2 - 4x + 8 = 0$ [8.2]

15. $x(3x + 4) = 4x(x - 1) + 15$ [8.2]

16. $x^2 + 9x = 1$ [8.2]

17. $x^2 - 5x - 2 = 0$. Use a calculator to approximate the solutions with rational numbers. [8.2]

18. Let $f(x) = 4x^2 - 3x - 1$. Find x such that $f(x) = 0$. [8.2]

Replace the blanks with constants to form a true equation. [8.1]

19. $x^2 - 12x +$ ___ $= (x -$ ___ $)^2$

20. $x^2 + \frac{3}{5}x +$ ___ $= \left(x +$ ___ $\right)^2$

21. Solve by completing the square. Show your work.
$x^2 - 6x + 1 = 0$ [8.1]

22. \$2500 grows to \$3025 in 2 yr. Use the formula $A = P(1 + r)^t$ to find the interest rate. [8.1]

23. The Peachtree Center Plaza in Atlanta, Georgia, is 723 ft tall. Use $s = 16t^2$ to approximate how long it would take an object to fall from the top. [8.1]

Solve. [8.3]

24. A corporate pilot must fly from company headquarters to a manufacturing plant and back in 4 hr. The distance between headquarters and the plant is 300 mi. If there is a 20-mph headwind going and a 20-mph tailwind returning, how fast must the plane be able to travel in still air?

25. Working together, Erica and Shawna can answer a day's worth of customer-service e-mails in 4 hr. Working alone, Erica takes 6 hr longer than Shawna. How long would it take Shawna to answer the e-mails alone?

For each equation, determine whether the solutions are real or imaginary. If they are real, specify whether they are rational or irrational. [8.4]

26. $x^2 + 3x - 6 = 0$ 27. $x^2 + 2x + 5 = 0$

28. Write a quadratic equation having the solutions $\sqrt{5}$ and $-\sqrt{5}$. [8.4]

29. Write a quadratic equation having -4 as its only solution. [8.4]

30. Find all x-intercepts of the graph of $f(x) = x^4 - 13x^2 + 36$. [8.5]

Solve. [8.5]

31. $15x^{-2} - 2x^{-1} - 1 = 0$

32. $(x^2 - 4)^2 - (x^2 - 4) - 6 = 0$

33. **a)** Graph: $f(x) = -3(x + 2)^2 + 4$. [8.6]
     b) Label the vertex.
     c) Draw the axis of symmetry.
     d) Find the maximum or the minimum value.

34. For the function given by $f(x) = 2x^2 - 12x + 23$: [8.7]

    a) find the vertex and the axis of symmetry;

    b) graph the function.

35. Find the x- and y-intercepts of

$$f(x) = x^2 - 9x + 14. \quad [8.7]$$

36. Solve $N = 3\pi\sqrt{1/p}$ for p. [8.3]

37. Solve $2A + T = 3T^2$ for T. [8.3]

State whether each graph appears to represent a quadratic or linear function. [8.8]

38.

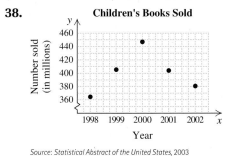

Source: Statistical Abstract of the United States, 2003

39.

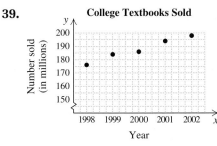

Source: Statistical Abstract of the United States, 2003

40. Eastgate Consignments wants to build a rectangular area in a corner for children to play in while their parents shop. They have 30 ft of low fencing. What is the maximum area they can enclose? What dimensions will yield this area? [8.8]

41. The following table lists the number of children's books sold (in millions) x years after 1998. (See Exercise 38.) [8.8]

Years Since 1998	Number of Children's Books Sold (in millions)
1	406
2	447
3	404

    a) Find the quadratic function that fits the data.

    b) Use the function to estimate the number of children's books sold in 2002.

Solve. [8.9]

42. $x^3 - 3x > 2x^2$

43. $\dfrac{x - 5}{x + 3} \le 0$

SYNTHESIS

44. Explain how the x-intercepts of a quadratic function can be used to help find the maximum or minimum value of the function. [8.7], [8.8]

45. Suppose that the quadratic formula is used to solve a quadratic equation. If the discriminant is a perfect square, could factoring have been used to solve the equation? Why or why not? [8.2], [8.4]

46. What is the greatest number of solutions that an equation of the form $ax^4 + bx^2 + c = 0$ can have? Why? [8.5]

47. Discuss two ways in which completing the square was used in this chapter. [8.1], [8.2], [8.7]

48. A quadratic function has x-intercepts at -3 and 5. If the y-intercept is at -7, find an equation for the function. [8.7]

49. Find h and k if, for $3x^2 - hx + 4k = 0$, the sum of the solutions is 20 and the product of the solutions is 80. [8.4]

50. The average of two positive integers is 171. One of the numbers is the square root of the other. Find the integers. [8.5]

8 Chapter Test

Solve.

1. $4x^2 - 11 = 0$

2. $4x(x - 2) - 3x(x + 1) = -18$

3. $x^2 + 2x + 3 = 0$

4. $2x + 5 = x^2$ **5.** $x^{-2} - x^{-1} = \frac{3}{4}$

6. $x^2 + 3x = 5$. Use a calculator to approximate the solutions with rational numbers.

7. Let $f(x) = 12x^2 - 19x - 21$. Find x such that $f(x) = 0$.

Replace the blanks with constants to form a true equation.

8. $x^2 - 16x + \underline{\quad} = (x - \underline{\quad})^2$

9. $x^2 + \frac{2}{7}x + \underline{\quad} = \left(x + \underline{\quad}\right)^2$

10. Solve by completing the square. Show your work.
$$x^2 + 10x + 15 = 0$$

Solve.

11. The Connecticut River flows at a rate of 4 km/h for the length of a popular scenic route. In order for a cruiser to travel 60 km upriver and then return in a total of 8 hr, how fast must the boat be able to travel in still water?

12. Brock and Ian can assemble a swing set in $1\frac{1}{2}$ hr. Working alone, it takes Ian 4 hr longer than Brock to assemble the swing set. How long would it take Brock, working alone, to assemble the swing set?

13. Determine the type of number that the solutions of $x^2 + 5x + 13 = 0$ will be.

14. Write a quadratic equation having solutions -2 and $\frac{1}{3}$.

15. Find all x-intercepts of the graph of
$$f(x) = x^4 - 15x^2 - 16.$$

16. a) Graph: $f(x) = 4(x - 3)^2 + 5$.
   b) Label the vertex.
   c) Draw the axis of symmetry.
   d) Find the maximum or the minimum function value.

17. For the function $f(x) = 2x^2 + 4x - 6$:
   a) find the vertex and the axis of symmetry;
   b) graph the function.

18. Find the x- and y-intercepts of
$$f(x) = x^2 - x - 6.$$

19. Solve $V = \frac{1}{3}\pi(R^2 + r^2)$ for r. Assume all variables are positive.

20. State whether the graph appears to represent a linear function, a quadratic function, or neither.

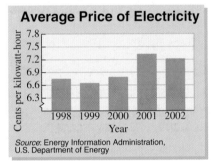

Average Price of Electricity

Source: Energy Information Administration, U.S. Department of Energy

21. Jay's Custom Pickups has determined that when x hundred truck caps are built, the average cost per cap is given by
$$C(x) = 0.2x^2 - 1.3x + 3.4025,$$
where $C(x)$ is in hundreds of dollars. What is the minimum cost per truck cap and how many caps should be built to achieve that minimum?

22. Find the quadratic function that fits the data points $(0, 0)$, $(3, 0)$, and $(5, 2)$.

Solve.

23. $x^2 + 5x < 6$

24. $x - \dfrac{1}{x} \geq 0$

SYNTHESIS

25. One solution of $kx^2 + 3x - k = 0$ is -2. Find the other solution.

26. Find a fourth-degree polynomial equation, with integer coefficients, for which $2 - \sqrt{3}$ and $5 - i$ are solutions.

27. Find a polynomial equation, with integer coefficients, for which 5 is a repeated root and $\sqrt{2}$ and $\sqrt{3}$ are solutions.

9

Exponential and Logarithmic Functions

AN APPLICATION

As more Americans make cell phones their *only* phones, the percentage of phone lines that are land lines is shrinking and can be estimated by

$$P(t) = 63.03(0.95)^t,$$

where *t* is the number of years since 2000. In what year will the percentage of phones that are land lines drop below 25%?

This problem appears as Exercise 7 in Section 9.7.

Teresa Matos
ENGINEER
Falls Church, Virginia

We use math with everything. Without math, we could not have technology. We use math to quantify cost as well. With knowledge of math, people can do anything.

he functions that we consider in this chapter are interesting not only from a purely intellectual point of view, but also for their rich applications to many fields. We will look at applications such as spread of a disease and population growth, to name just two.

The theory centers on functions with variable exponents (exponential func-tions). Results follow from those functions, their properties, and the inverses of those functions.

9.1 Composite and Inverse Functions

Composite Functions • Inverses and One-to-One Functions • Finding Formulas for Inverses • Graphing Functions and Their Inverses • Inverse Functions and Composition

Study Skills

Divide and Conquer

On longer sections of reading, there are almost always subsections. Rather than feel obliged to read the entire assignment at once, make use of the subsections as natural resting points. Reading two or three subsections and then taking a break can increase your comprehension and is thus an efficient use of your time.

Composite Functions

In the real world, functions frequently occur in which some quantity depends on a variable that, in turn, depends on another variable. For instance, a firm's profits may depend on the number of items the firm produces, which may in turn depend on the number of employees hired. Functions like this are called **composite functions**.

For example, the function g that gives a correspondence between women's shoe sizes in the United States and those in Italy is given by $g(x) = 2x + 24$, where x is the U.S. size and $g(x)$ is the Italian size. Thus a U.S. size 4 corresponds to a shoe size of $g(4) = 2 \cdot 4 + 24$, or 32, in Italy.

A different function gives a correspondence between women's shoe sizes in Italy and those in Britain. This particular function is given by $f(x) = \frac{1}{2}x - 14$, where x is the Italian size and $f(x)$ is the corresponding British size. Thus an Italian size 32 corresponds to a British size $f(32) = \frac{1}{2} \cdot 32 - 14$, or 2.

It seems reasonable to conclude that a U.S. size 4 corresponds to a British size 2 and that some function h describes this correspondence. Can we find a formula for h? If we look at the following tables, we might guess that such a formula is $h(x) = x - 2$, and that is indeed correct. But, for more complicated formulas, we would need to use algebra.

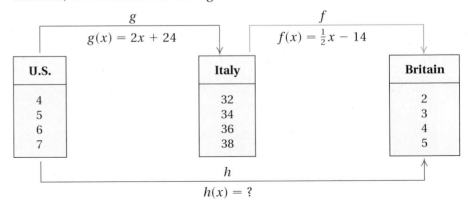

Size x shoes in the United States correspond to size $g(x)$ shoes in Italy, where

$$g(x) = 2x + 24.$$

Size n shoes in Italy correspond to size $f(n)$ shoes in Britain. Similarly, size $g(x)$ shoes in Italy correspond to size $f(g(x))$ shoes in Britain. Since the x in the expression $f(g(x))$ represents a U.S. shoe size, we can find the British shoe size that corresponds to a U.S. size x as follows:

$$f(g(x)) = f(2x + 24) = \tfrac{1}{2} \cdot (2x + 24) - 14 \qquad \text{Using } g(x) \text{ as an input}$$
$$= x + 12 - 14 = x - 2.$$

This gives a formula for h: $h(x) = x - 2$. Thus U.S. size 4 corresponds to British size $h(4) = 4 - 2$, or 2. The function h is the *composition* of f and g and is denoted $f \circ g$ (read "the composition of f and g," "f composed with g," or "f circle g").

Composition of Functions

The *composite function* $f \circ g$, the *composition* of f and g, is defined as

$$(f \circ g)(x) = f(g(x)).$$

We can visualize the composition of functions as follows.

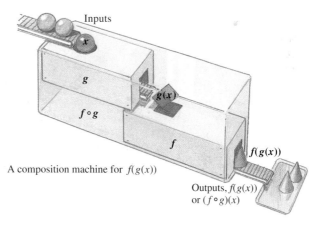

A composition machine for $f(g(x))$

Outputs, $f(g(x))$ or $(f \circ g)(x)$

EXAMPLE 1   Given $f(x) = 3x$ and $g(x) = 1 + x^2$:

a) Find $(f \circ g)(5)$ and $(g \circ f)(5)$.   **b)** Find $(f \circ g)(x)$ and $(g \circ f)(x)$.

Solution   Consider each function separately:

$$f(x) = 3x \qquad \text{This function multiplies each input by 3.}$$

and

$$g(x) = 1 + x^2. \qquad \text{This function adds 1 to the square of each input.}$$

technology connection

In Example 3, we see that if $g(x) = 7x + 3$ and $f(x) = x^2$, then $f(g(x)) = (7x + 3)^2$. One way to show this is to let $y_1 = 7x + 3$ and $y_2 = (y_1)^2$ (this is accomplished by using the Y-VARS option of the VARS key's menu). We then let $y_3 = (7x + 3)^2$ and use graphs or a table to show that $y_2 = y_3$.

Another approach is to let $y_1 = 7x + 3$ and $y_2 = x^2$ and have $y_4 = y_2(y_1)$. If we again let $y_3 = (7x + 3)^2$, we complete the check by showing that $y_3 = y_4$.

1. Check Example 2 by using one of the above approaches.

---

a) To find $(f \circ g)(5)$, we find $g(5)$ and then use that as an input for f:

$$(f \circ g)(5) = f(g(5)) = f(1 + 5^2) \qquad \text{Using } g(x) = 1 + x^2$$
$$= f(26) = 3 \cdot 26 = 78. \qquad \text{Using } f(x) = 3x$$

To find $(g \circ f)(5)$, we find $f(5)$ and then use that as an input for g:

$$(g \circ f)(5) = g(f(5)) = g(3 \cdot 5) \qquad \text{Note that } f(5) = 3 \cdot 5 = 15.$$
$$= g(15) = 1 + 15^2 = 1 + 225 = 226.$$

b) We find $(f \circ g)(x)$ by substituting $g(x)$ for x in the equation for $f(x)$:

$$(f \circ g)(x) = f(g(x)) = f(1 + x^2) \qquad \text{Using } g(x) = 1 + x^2$$
$$= 3 \cdot (1 + x^2) = 3 + 3x^2. \qquad \text{Using } f(x) = 3x$$

To find $(g \circ f)(x)$, we substitute $f(x)$ for x in the equation for $g(x)$:

$$(g \circ f)(x) = g(f(x)) = g(3x) \qquad \text{Substituting } 3x \text{ for } f(x)$$
$$= 1 + (3x)^2 = 1 + 9x^2.$$

As a check, note that $(g \circ f)(5) = 1 + 9 \cdot 5^2 = 1 + 9 \cdot 25 = 226$, as expected from part (a) above.

Example 1 shows that, in general, $(f \circ g)(5) \neq (g \circ f)(5)$ and $(f \circ g)(x) \neq (g \circ f)(x)$.

EXAMPLE 2   Given $f(x) = \sqrt{x}$ and $g(x) = x - 1$, find $(f \circ g)(x)$ and $(g \circ f)(x)$.

Solution

$$(f \circ g)(x) = f(g(x)) = f(x - 1) = \sqrt{x - 1} \qquad \text{Using } g(x) = x - 1$$
$$(g \circ f)(x) = g(f(x)) = g(\sqrt{x}) = \sqrt{x} - 1 \qquad \text{Using } f(x) = \sqrt{x}$$

In fields ranging from chemistry to geology and economics, one needs to recognize how a function can be regarded as the composition of two "simpler" functions. This is sometimes called *de*composition.

EXAMPLE 3   If $h(x) = (7x + 3)^2$, find f and g such that $h(x) = (f \circ g)(x)$.

Solution   We can think of $h(x)$ as the result of first finding $7x + 3$ and then squaring that. This suggests that $g(x) = 7x + 3$ and $f(x) = x^2$. We check by forming the composition:

$$(f \circ g)(x) = f(g(x))$$
$$= f(7x + 3) = (7x + 3)^2 = h(x), \text{ as desired.}$$

This is probably the most "obvious" answer to the question. There are other less obvious answers. For example, if

$$f(x) = (x - 1)^2 \quad \text{and} \quad g(x) = 7x + 4,$$

Student Notes

Throughout this chapter, keep in mind that functions are defined using dummy variables that represent an arbitrary member of the domain. Thus, $f(x) = x^2 - 3x$ and $f(t) = t^2 - 3t$ describe the same function f.

then

$$(f \circ g)(x) = f(g(x)) = f(7x + 4)$$
$$= (7x + 4 - 1)^2 = (7x + 3)^2 = h(x).$$

Inverses and One-to-One Functions

Let's view the following two functions as relations, or correspondences.

Selected Professions and Their Median Yearly Salary in 2003*

Domain (Set of Inputs)	Range (Set of Outputs)
Registered nurse	→ $49,550
Film and video editor	→ $40,600
Firefighter	→ $37,060
Computer programmer	→ $61,340
Secondary-school teacher	→ $44,580
Architect	→ $57,950

U.S. Senators and Their States

Domain (Set of Inputs)	Range (Set of Outputs)
Feinstein	California
Boxer	
Obama	Illinois
Durban	
Clinton	New York
Schumer	

Suppose we reverse the arrows. We obtain what is called the **inverse relation**. Are these inverse relations functions?

Selected Professions and Their Median Yearly Salary in 2003

Range (Set of Outputs)	Domain (Set of Inputs)
Registered nurse ←	$49,550
Film and video editor ←	$40,600
Firefighter ←	$37,060
Computer programmer ←	$61,340
Secondary-school teacher ←	$44,580
Architect ←	$57,950

U.S. Senators and Their States

Range (Set of Outputs)	Domain (Set of Inputs)
Feinstein ←	California
Boxer ←	
Obama ←	Illinois
Durban ←	
Clinton ←	New York
Schumer ←	

Recall that for each input, a function provides exactly one output. However, a function can have the same output for two or more different inputs. Thus it is possible for different inputs to correspond to the same output. Only when this possibility is *excluded* will the inverse be a function. For the functions listed above, this means the inverse of the "Selected Professions" correspondence is a function, but the inverse of the "U.S. Senator" correspondence is not.

**Source:* U.S. Bureau of Labor Statistics, May 2003, Occupational Employment and Wage Estimates

In the Selected Professions function, different inputs have different outputs, so it is a **one-to-one function**. In the U.S. Senator function, *Clinton* and *Schumer* are both paired with *New York*. Thus the U.S. Senator function is not one-to-one.

One-To-One Function

A function *f* is *one-to-one* if different inputs have different outputs. That is, if for *a* and *b* in the domain of *f* with $a \neq b$, we have $f(a) \neq f(b)$, then the function *f* is one-to-one. If a function is one-to-one, then its inverse correspondence is also a function.

How can we tell graphically whether a function is one-to-one?

EXAMPLE 4   At left is the graph of a function similar to those we will study in Section 9.2. Determine whether the function is one-to-one and thus has an inverse that is a function.

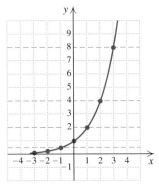

Solution   A function is one-to-one if different inputs have different outputs—that is, if no two *x*-values have the same *y*-value. For this function, we cannot find two *x*-values that have the same *y*-value. Note that this means that no horizontal line can be drawn so that it crosses the graph more than once. The function is one-to-one so its inverse is a function.

The graph of every function must pass the vertical-line test. In order for a function to have an inverse that is a function, it must pass the *horizontal-line test* as well.

The Horizontal-Line Test

If it is impossible to draw a horizontal line that intersects a function's graph more than once, then the function is one-to-one. For every one-to-one function, an inverse function exists.

EXAMPLE 5   Determine whether the function $f(x) = x^2$ is one-to-one and thus has an inverse that is a function.

Solution   The graph of $f(x) = x^2$ is shown here. Many horizontal lines cross the graph more than once. For example, the line $y = 4$ crosses where the first coordinates are -2 and 2. Although these are different inputs, they have the same output. That is, $-2 \neq 2$, but

$$f(-2) = (-2)^2 = 4 = 2^2 = f(2).$$

Thus the function is not one-to-one and no inverse function exists.

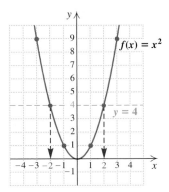

Finding Formulas for Inverses

When the inverse of f is also a function, it is denoted f^{-1} (read "f-inverse").

> *Caution!*   The -1 in f^{-1} is *not* an exponent!

Suppose a function is described by a formula. If its inverse is a function, how do we find a formula for that inverse? For any equation in two variables, if we interchange the variables, we form an equation of the inverse correspondence. If it is a function, we proceed as follows to find a formula for f^{-1}.

> *To Find a Formula for f^{-1}*
>
> First make sure that f is one-to-one. Then:
>
> 1. Replace $f(x)$ with y.
> 2. Interchange x and y. (This gives the inverse function.)
> 3. Solve for y.
> 4. Replace y with $f^{-1}(x)$. (This is inverse function notation.)

EXAMPLE 6   Determine whether each function is one-to-one and if it is, find a formula for $f^{-1}(x)$.

a) $f(x) = x + 2$ b) $f(x) = 2x - 3$

Solution

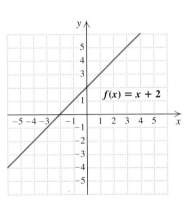

a) The graph of $f(x) = x + 2$ is shown at left. It passes the horizontal-line test, so it is one-to-one. Thus its inverse is a function.

    1. Replace $f(x)$ with y: $y = x + 2.$

    2. Interchange x and y: $x = y + 2.$ This gives the inverse function.

    3. Solve for y: $x - 2 = y.$

    4. Replace y with $f^{-1}(x)$: $f^{-1}(x) = x - 2.$ We also "reversed" the equation.

In this case, the function f adds 2 to all inputs. Thus, to "undo" f, the function f^{-1} must subtract 2 from its inputs.

b) The function $f(x) = 2x - 3$ is also linear. Any linear function that is not constant will pass the horizontal-line test. Thus, f is one-to-one.

    1. Replace $f(x)$ with y: $y = 2x - 3.$

    2. Interchange x and y: $x = 2y - 3.$

    3. Solve for y: $x + 3 = 2y$

$$\frac{x + 3}{2} = y.$$

    4. Replace y with $f^{-1}(x)$: $f^{-1}(x) = \dfrac{x + 3}{2}.$

In this case, the function f doubles all inputs and then subtracts 3. Thus, to "undo" f, the function f^{-1} adds 3 to each input and then divides by 2.

Graphing Functions and Their Inverses

How do the graphs of a function and its inverse compare?

EXAMPLE 7   Graph $f(x) = 2x - 3$ and $f^{-1}(x) = (x + 3)/2$ on the same set of axes. Then compare.

Solution   The graph of each function follows. Note that the graph of f^{-1} can be drawn by reflecting the graph of f across the line $y = x$. That is, if we graph $f(x) = 2x - 3$ in wet ink and fold the paper along the line $y = x$, the graph of $f^{-1}(x) = (x + 3)/2$ will appear as the impression made by f.

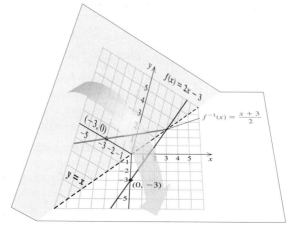

When x and y are interchanged to find a formula for the inverse, we are, in effect, reflecting or flipping the graph of $f(x) = 2x - 3$ across the line $y = x$. For example, when $(0, -3)$, the coordinates of the y-intercept of the graph of f, are reversed, we get $(-3, 0)$, the x-intercept of the graph of f^{-1}.

Visualizing Inverses

The graph of f^{-1} is a reflection of the graph of f across the line $y = x$.

EXAMPLE 8   Consider $g(x) = x^3 + 2$.

a) Determine whether the function is one-to-one.

b) If it is one-to-one, find a formula for its inverse.

c) Graph the inverse, if it exists.

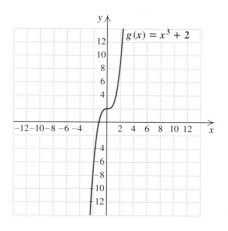

Solution

a) The graph of $g(x) = x^3 + 2$ is shown at right. It passes the horizontal-line test and thus has an inverse.

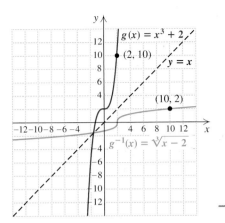

b) 1. Replace $g(x)$ with y: $y = x^3 + 2.$ Using $g(x) = x^3 + 2$
2. Interchange x and y: $x = y^3 + 2.$
3. Solve for y: $x - 2 = y^3$
$$\sqrt[3]{x - 2} = y.$$ Each real number has only one cube root, so we can solve for y.

4. Replace y with $g^{-1}(x)$: $g^{-1}(x) = \sqrt[3]{x - 2}.$

c) To find the graph, we reflect the graph of $g(x) = x^3 + 2$ across the line $y = x$, as we did in Example 7. We can also substitute into $g^{-1}(x) = \sqrt[3]{x - 2}$ and plot points. Note that $(2, 10)$ is on the graph of g, whereas $(10, 2)$ is on the graph of g^{-1}. The graphs of g and g^{-1} are shown together at left.

Inverse Functions and Composition

Let's consider inverses of functions in terms of function machines. Suppose that a one-to-one function f is programmed into a machine. If the machine has a reverse switch, when the switch is thrown, the machine performs the inverse function f^{-1}. Inputs then enter at the opposite end, and the entire process is reversed.

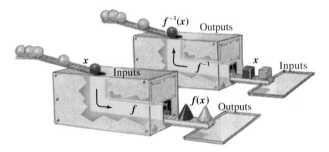

Consider $f(x) = x^3 + 2$ and $f^{-1}(x) = \sqrt[3]{x - 2}$ from Example 8. For the input 3,

$$f(3) = 3^3 + 2 = 27 + 2 = 29.$$

The output is 29. Now we use 29 for the input in the inverse:

$$f^{-1}(29) = \sqrt[3]{29 - 2} = \sqrt[3]{27} = 3.$$

The function f takes 3 to 29. The inverse function f^{-1} takes the number 29 back to 3.

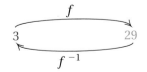

In general, for any output $f(x)$, the function f^{-1} takes that output back to x. Similarly, for any output $f^{-1}(x)$, the function f takes that output back to x.

Composition and Inverses

If a function f is one-to-one, then f^{-1} is the unique function for which

$$(f^{-1} \circ f)(x) = f^{-1}(f(x)) = x \quad \text{and} \quad (f \circ f^{-1})(x) = f(f^{-1}(x)) = x.$$

EXAMPLE 9 Let $f(x) = 2x + 1$. Show that

$$f^{-1}(x) = \frac{x-1}{2}.$$

Solution We find $(f^{-1} \circ f)(x)$ and $(f \circ f^{-1})(x)$ and check to see that each is x.

$$(f^{-1} \circ f)(x) = f^{-1}(f(x)) = f^{-1}(2x + 1)$$

$$= \frac{(2x + 1) - 1}{2}$$

$$= \frac{2x}{2} = x$$

$$(f \circ f^{-1})(x) = f(f^{-1}(x)) = f\left(\frac{x-1}{2}\right)$$

$$= 2 \cdot \frac{x-1}{2} + 1$$

$$= x - 1 + 1 = x$$

technology connection

To determine whether $y_1 = 2x + 6$ and $y_2 = \frac{1}{2}x - 3$ are inverses of each other, we can graph both functions, along with the line $y = x$, on a "squared" set of axes. It *appears* that y_1 and y_2 are inverses of each other. A more precise check is achieved by selecting the DRAWINV option of the (DRAW) menu. The resulting graph of the inverse of y_1 should coincide with y_2.

$$y_1 = 2x + 6, \quad y_2 = \frac{1}{2}x - 3$$

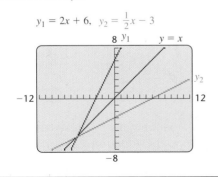

For a more dependable check, examine a TABLE in which $y_1 = 2x + 6$ and $y_2 = \frac{1}{2} \cdot y_1 - 3$. Note that y_2 "undoes" what y_1 does.

TBL MIN $= -3$ ΔTBL $= 1$ $y_2 = \frac{1}{2}y_1 - 3$

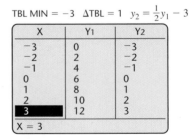

X	Y₁	Y₂
−3	0	−3
−2	2	−2
−1	4	−1
0	6	0
1	8	1
2	10	2
3	12	3

X = 3

1. Use a graphing calculator to check Examples 7, 8, and 9.
2. Will DRAWINV work for *any* choice of y_1? Why or why not?

Exercise Set

9.1

🖢 *Concept Reinforcement*   *Classify each statement as either true or false.*

1. The composition of two functions f and g is written $f \circ g$.

2. The notation $(f \circ g)(x)$ means $f(g(x))$.

3. If $f(x) = x^2$ and $g(x) = x + 3$, then $(g \circ f)(x) = (x + 3)^2$.

4. If $f(2) = 15$ and $g(15) = 25$, then $(g \circ f)(2) = 25$.

5. The function f is one-to-one if $f(1) = 1$.

6. For f^{-1} to be a function, f cannot be one-to-one.

7. The function f is the inverse of f^{-1}.

8. If g and h are inverses of each other, then $(g \circ h)(x) = x$.

Find $(f \circ g)(1)$, $(g \circ f)(1)$, $(f \circ g)(x)$, and $(g \circ f)(x)$.

9. $f(x) = x^2 + 1$; $g(x) = 2x - 3$

10. $f(x) = 2x + 1$; $g(x) = x^2 - 5$

11. $f(x) = x - 3$; $g(x) = 2x^2 - 7$

12. $f(x) = 3x^2 + 4$; $g(x) = 4x - 1$

13. $f(x) = x + 7$; $g(x) = 1/x^2$

14. $f(x) = 1/x^2$; $g(x) = x + 2$

15. $f(x) = \sqrt{x}$; $g(x) = x + 3$

16. $f(x) = 10 - x$; $g(x) = \sqrt{x}$

17. $f(x) = \sqrt{4x}$; $g(x) = 1/x$

18. $f(x) = \sqrt{x + 3}$; $g(x) = 13/x$

19. $f(x) = x^2 + 4$; $g(x) = \sqrt{x - 1}$

20. $f(x) = x^2 + 8$; $g(x) = \sqrt{x + 17}$

Find $f(x)$ and $g(x)$ such that $h(x) = (f \circ g)(x)$. Answers may vary.

21. $h(x) = (7 + 5x)^2$

22. $h(x) = (3x - 1)^2$

23. $h(x) = \sqrt{2x + 7}$

24. $h(x) = \sqrt{5x + 2}$

25. $h(x) = \dfrac{2}{x - 3}$

26. $h(x) = \dfrac{3}{x} + 4$

Determine whether each function is one-to-one.

27. $f(x) = x - 5$

28. $f(x) = 5 - 2x$

Aha! 29. $f(x) = x^2 + 1$

30. $f(x) = 1 - x^2$

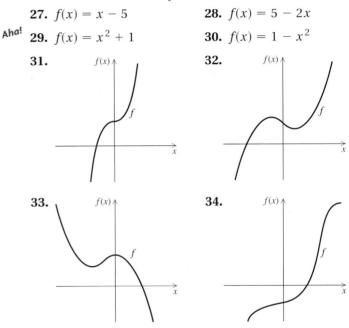

31.

32.

33.

34.

For each function, (a) determine whether it is one-to-one; (b) if it is one-to-one, find a formula for the inverse.

35. $f(x) = x + 4$

36. $f(x) = x + 2$

37. $f(x) = 2x$

38. $f(x) = 3x$

39. $g(x) = 3x - 1$

40. $g(x) = 2x - 3$

41. $f(x) = \dfrac{1}{2}x + 1$

42. $f(x) = \dfrac{1}{3}x + 2$

43. $g(x) = x^2 + 5$

44. $g(x) = x^2 - 4$

45. $h(x) = -2x + 4$

46. $h(x) = -3x + 1$

Aha! 47. $f(x) = \dfrac{1}{x}$

48. $f(x) = \dfrac{3}{x}$

49. $G(x) = 4$

50. $H(x) = 2$

51. $f(x) = \dfrac{2x + 1}{3}$

52. $f(x) = \dfrac{3x + 2}{5}$

53. $f(x) = x^3 - 5$

54. $f(x) = x^3 + 2$

55. $g(x) = (x - 2)^3$

56. $g(x) = (x + 7)^3$

57. $f(x) = \sqrt{x}$

58. $f(x) = \sqrt{x - 1}$

Graph each function and its inverse using the same set of axes.

59. $f(x) = \frac{2}{3}x + 4$

60. $g(x) = \frac{1}{4}x + 2$

61. $f(x) = x^3 + 1$

62. $f(x) = x^3 - 1$

63. $g(x) = \frac{1}{2}x^3$

64. $g(x) = \frac{1}{3}x^3$

65. $F(x) = -\sqrt{x}$

66. $f(x) = \sqrt{x}$

67. $f(x) = -x^2, x \geq 0$

68. $f(x) = x^2 - 1, x \leq 0$

69. Let $f(x) = \sqrt[3]{x - 4}$. Show that
$$f^{-1}(x) = x^3 + 4.$$

70. Let $f(x) = 3/(x + 2)$. Show that
$$f^{-1}(x) = \frac{3}{x} - 2.$$

71. Let $f(x) = (1 - x)/x$. Show that
$$f^{-1}(x) = \frac{1}{x + 1}.$$

72. Let $f(x) = x^3 - 5$. Show that
$$f^{-1}(x) = \sqrt[3]{x + 5}.$$

73. *Dress sizes in the United States and Italy.*   A size-6 dress in the United States is size 36 in Italy. A function that converts dress sizes in the United States to those in Italy is
$$f(x) = 2(x + 12).$$
  a) Find the dress sizes in Italy that correspond to sizes 8, 10, 14, and 18 in the United States.
  b) Determine whether this function has an inverse that is a function. If so, find a formula for the inverse.
  c) Use the inverse function to find dress sizes in the United States that correspond to sizes 40, 44, 52, and 60 in Italy.

74. *Dress sizes in the United States and France.*   A size-6 dress in the United States is size 38 in France. A function that converts dress sizes in the United States to those in France is
$$f(x) = x + 32.$$
  a) Find the dress sizes in France that correspond to sizes 8, 10, 14, and 18 in the United States.
  b) Determine whether this function has an inverse that is a function. If so, find a formula for the inverse.
  c) Use the inverse function to find dress sizes in the United States that correspond to sizes 40, 42, 46, and 50 in France.

75. Is there a one-to-one relationship between the letters and the numbers on the keypad of a telephone? Why or why not?

76. Mathematicians usually try to select "logical" words when forming definitions. Does the term "one-to-one" seem logical? Why or why not?

SKILL MAINTENANCE

Simplify.

77. $(a^5b^4)^2(a^3b^5)$ [1.6]

78. $(x^3y^5)^2(x^4y^2)$ [1.6]

79. $27^{4/3}$ [7.2]

80. $25^{3/2}$ [7.2]

Solve. [1.5]

81. $x = \frac{2}{3}y - 7$, for y

82. $x = 10 - 3y$, for y

SYNTHESIS

83. The function $V(t) = 750(1.2)^t$ is used to predict the value, $V(t)$, of a certain rare stamp t years from 2001. Do not calculate $V^{-1}(t)$, but explain how V^{-1} could be used.

84. An organization determines that the cost per person of chartering a bus is given by the function
$$C(x) = \frac{100 + 5x}{x},$$
where x is the number of people in the group and $C(x)$ is in dollars. Determine $C^{-1}(x)$ and explain how this inverse function could be used.

For Exercises 85 and 86, graph the inverse of f.

85.

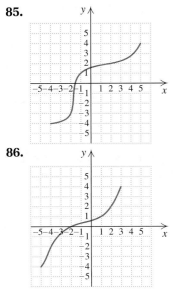

86.

87. *Dress sizes in France and Italy.*   Use the information in Exercises 73 and 74 to find a function for the French dress size that corresponds to a size x dress in Italy.

88. *Dress sizes in Italy and France.*   Use the information in Exercises 73 and 74 to find a function for the Italian dress size that corresponds to a size x dress in France.

89. What relationship exists between the answers to Exercises 87 and 88? Explain how you determined this.

90. Show that function composition is associative by showing that $((f \circ g) \circ h)(x) = (f \circ (g \circ h))(x)$.

91. Show that if $h(x) = (f \circ g)(x)$, then $h^{-1}(x) = (g^{-1} \circ f^{-1})(x)$. (*Hint*: Use Exercise 90.)

Determine whether or not the given pairs of functions are inverses of each other.

92. $f(x) = 0.75x^2 + 2$; $g(x) = \sqrt{\dfrac{4(x-2)}{3}}$

93. $f(x) = 1.4x^3 + 3.2$; $g(x) = \sqrt[3]{\dfrac{x - 3.2}{1.4}}$

94. $f(x) = \sqrt{2.5x + 9.25}$; $g(x) = 0.4x^2 - 3.7, x \geq 0$

95. $f(x) = 0.8x^{1/2} + 5.23$; $g(x) = 1.25(x^2 - 5.23), x \geq 0$

96. $f(x) = 2.5(x^3 - 7.1)$; $g(x) = \sqrt[3]{0.4x + 7.1}$

97. Match each function in Column A with its inverse from Column B.

Column A

(1) $y = 5x^3 + 10$

(2) $y = (5x + 10)^3$

(3) $y = 5(x + 10)^3$

(4) $y = (5x)^3 + 10$

Column B

A. $y = \dfrac{\sqrt[3]{x} - 10}{5}$

B. $y = \sqrt[3]{\dfrac{x}{5}} - 10$

C. $y = \sqrt[3]{\dfrac{x - 10}{5}}$

D. $y = \dfrac{\sqrt[3]{x - 10}}{5}$

98. Examine the following table. Is it possible that f and g are inverses of each other? Why or why not?

x	$f(x)$	$g(x)$
6	6	6
7	6.5	8
8	7	10
9	7.5	12
10	8	14
11	8.5	16
12	9	18

99. The following window appears on a graphing calculator.

X	Y1	Y2
0	1	−2
1	1.5	0
2	2	2
3	2.5	4
4	3	6
5	3.5	8
6	4	10

X = 0

a) What evidence is there that the functions Y1 and Y2 are inverses of each other?

b) Find equations for Y1 and Y2, assuming that both are linear functions.

c) On the basis of your answer to part (b), are Y1 and Y2 inverses of each other?

9.2 Exponential Functions

Graphing Exponential Functions • Equations with *x* and *y*
Interchanged • Applications of Exponential Functions

CONNECTING THE CONCEPTS

Composite and inverse functions, as shown in Section 9.1, are very useful in and of themselves. The reason they are included in this chapter, however, is that they are needed in order to understand the logarithmic functions that appear in Section 9.3. Here in Section 9.2, we make no reference to composite or inverse functions. Instead, we introduce a new type of function, the *exponential function*, so that we can study both it and its inverse in Sections 9.3–9.7.

Study Skills

Know Your Machine

Whether you use a scientific or a graphing calculator, it is a wise investment of time to study the user's manual. If you cannot find a paper manual to consult, an electronic version can usually be found, online, at the manufacturer's website. Experimenting by pressing various combinations of keystrokes is also useful.

Consider the graph below. The rapidly rising curve approximates the graph of an *exponential function*. We now consider such functions and some of their applications.

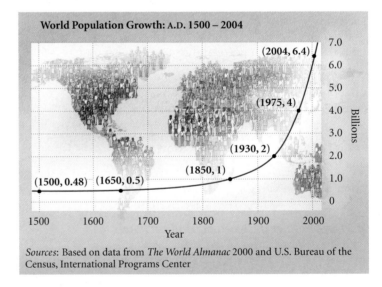

Sources: Based on data from *The World Almanac* 2000 and U.S. Bureau of the Census, International Programs Center

Graphing Exponential Functions

In Chapter 7, we studied exponential expressions with rational-number exponents, such as

$$5^{1/4}, \qquad 3^{-3/4}, \qquad 7^{2.34}, \qquad 5^{1.73}.$$

For example, $5^{1.73}$, or $5^{173/100}$, represents the 100th root of 5 raised to the 173rd power. What about expressions with irrational exponents, such as $5^{\sqrt{3}}$ or $7^{-\pi}$?

To attach meaning to $5^{\sqrt{3}}$, consider a rational approximation, r, of $\sqrt{3}$. As r gets closer to $\sqrt{3}$, the value of 5^r gets closer to some real number p.

r closes in on $\sqrt{3}$.	5^r closes in on some real number p.
$1.7 < r < 1.8$	$15.426 \approx 5^{1.7} < p < 5^{1.8} \approx 18.119$
$1.73 < r < 1.74$	$16.189 \approx 5^{1.73} < p < 5^{1.74} \approx 16.452$
$1.732 < r < 1.733$	$16.241 \approx 5^{1.732} < p < 5^{1.733} \approx 16.267$

We define $5^{\sqrt{3}}$ to be the number p. To eight decimal places,

$$5^{\sqrt{3}} \approx 16.24245082.$$

Any positive irrational exponent can be interpreted in a similar way. Negative irrational exponents are then defined using reciprocals. Thus, so long as a is positive, a^x has meaning for *any* real number x. All of the laws of exponents still hold, but we will not prove that here. We now define an *exponential function*.

Exponential Function

The function $f(x) = a^x$, where a is a positive constant, $a \neq 1$, is called the *exponential function*, base a.

We require the base a to be positive to avoid imaginary numbers that would result from taking even roots of negative numbers. The restriction $a \neq 1$ is made to exclude the constant function $f(x) = 1^x$, or $f(x) = 1$.

The following are examples of exponential functions:

$$f(x) = 2^x, \qquad f(x) = \left(\tfrac{1}{3}\right)^x, \qquad f(x) = 5^{-3x}. \qquad \begin{array}{l}\text{Note that} \\ 5^{-3x} = (5^{-3})^x.\end{array}$$

Like polynomial functions, the domain of an exponential function is the set of all real numbers. Unlike polynomial functions, exponential functions have a variable exponent. Because of this, graphs of exponential functions either rise or fall dramatically.

EXAMPLE 1 Graph the exponential function given by $y = f(x) = 2^x$.

Solution We compute some function values, thinking of y as $f(x)$, and list the results in a table. It is a good idea to start by letting $x = 0$.

$$f(0) = 2^0 = 1; \qquad\qquad f(-1) = 2^{-1} = \frac{1}{2^1} = \frac{1}{2};$$
$$f(1) = 2^1 = 2;$$
$$f(2) = 2^2 = 4; \qquad\qquad f(-2) = 2^{-2} = \frac{1}{2^2} = \frac{1}{4};$$
$$f(3) = 2^3 = 8;$$
$$\qquad\qquad f(-3) = 2^{-3} = \frac{1}{2^3} = \frac{1}{8}$$

Next, we plot these points and connect them with a smooth curve.

x	y, or f(x)
0	1
1	2
2	4
3	8
−1	$\frac{1}{2}$
−2	$\frac{1}{4}$
−3	$\frac{1}{8}$

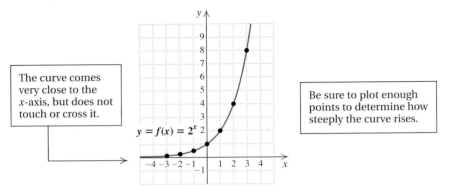

The curve comes very close to the x-axis, but does not touch or cross it.

Be sure to plot enough points to determine how steeply the curve rises.

$y = f(x) = 2^x$

Note that as x increases, the function values increase without bound. As x decreases, the function values decrease, getting very close to 0. The x-axis, or the line $y = 0$, is a horizontal *asymptote*, meaning that the curve gets closer and closer to this line the further we move to the left.

EXAMPLE 2

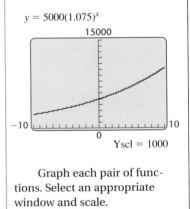

Graphing calculators are helpful when graphing equations like $y = 5000(1.075)^x$. To choose a window, we note that y-values are positive and increase rapidly. One suitable window is $[-10, 10, 0, 15000]$, with a y-scale of 1000.

$y = 5000(1.075)^x$

15000

−10   10

0

Yscl = 1000

Graph each pair of functions. Select an appropriate window and scale.

1. $y_1 = \left(\frac{5}{2}\right)^x$ and $y_2 = \left(\frac{2}{5}\right)^x$
2. $y_1 = 3.2^x$ and $y_2 = 3.2^{-x}$
3. $y_1 = \left(\frac{3}{7}\right)^x$ and $y_2 = \left(\frac{7}{3}\right)^x$
4. $y_1 = 5000(1.08)^x$ and $y_2 = 5000(1.08)^{x-3}$

Graph: $y = f(x) = \left(\frac{1}{2}\right)^x$.

Solution   We compute some function values, thinking of y as $f(x)$, and list the results in a table. Before we do this, note that

$$y = f(x) = \left(\tfrac{1}{2}\right)^x = (2^{-1})^x = 2^{-x}.$$

Then we have

$f(0) = 2^{-0} = 1;$

$f(1) = 2^{-1} = \dfrac{1}{2^1} = \dfrac{1}{2};$

$f(2) = 2^{-2} = \dfrac{1}{2^2} = \dfrac{1}{4};$

$f(3) = 2^{-3} = \dfrac{1}{2^3} = \dfrac{1}{8};$

$f(-1) = 2^{-(-1)} = 2^1 = 2;$

$f(-2) = 2^{-(-2)} = 2^2 = 4;$

$f(-3) = 2^{-(-3)} = 2^3 = 8.$

x	y, or f(x)
0	1
1	$\frac{1}{2}$
2	$\frac{1}{4}$
3	$\frac{1}{8}$
−1	2
−2	4
−3	8

Next, we plot these points and connect them with a smooth curve. This curve is a mirror image, or *reflection*, of the graph of $y = 2^x$ (see Example 1) across the y-axis. The line $y = 0$ is again the asymptote.

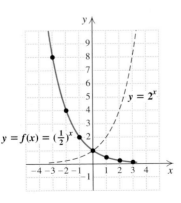

$y = 2^x$

$y = f(x) = \left(\frac{1}{2}\right)^x$

From Examples 1 and 2, we can make the following observations.

A. For $a > 1$, the graph of $f(x) = a^x$ increases from left to right. The greater the value of a, the steeper the curve. (See the figure on the left below.)

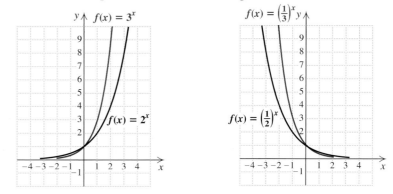

B. For $0 < a < 1$, the graph of $f(x) = a^x$ decreases from left to right. For smaller values of a, the curve becomes steeper. (See the figure on the right above.)
C. All graphs of $f(x) = a^x$ go through the y-intercept $(0, 1)$.
D. All graphs of $f(x) = a^x$ have the x-axis as the asymptote.
E. If $f(x) = a^x$, with $a > 0$, $a \neq 1$, the domain of f is all real numbers, and the range of f is all positive real numbers.
F. For $a > 0$, $a \neq 1$, the function given by $f(x) = a^x$ is one-to-one. Its graph passes the horizontal-line test.

EXAMPLE 3

Graph: $y = f(x) = 2^{x-2}$.

Solution   We construct a table of values. Then we plot the points and connect them with a smooth curve. Here $x - 2$ is the *exponent*.

$$f(0) = 2^{0-2} = 2^{-2} = \frac{1}{4};$$ $$f(-1) = 2^{-1-2} = 2^{-3} = \frac{1}{8};$$

$$f(1) = 2^{1-2} = 2^{-1} = \frac{1}{2};$$ $$f(-2) = 2^{-2-2} = 2^{-4} = \frac{1}{16}$$

$$f(2) = 2^{2-2} = 2^0 = 1;$$

$$f(3) = 2^{3-2} = 2^1 = 2;$$

$$f(4) = 2^{4-2} = 2^2 = 4;$$

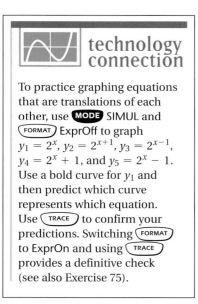
x	y, or f(x)
0	$\frac{1}{4}$
1	$\frac{1}{2}$
2	1
3	2
4	4
-1	$\frac{1}{8}$
-2	$\frac{1}{16}$

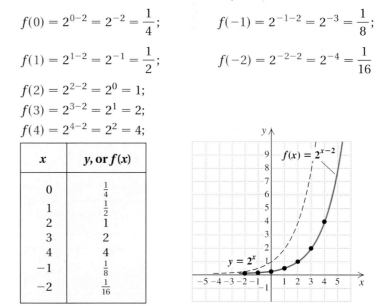

The graph looks just like the graph of $y = 2^x$, but it is translated 2 units to the right. The y-intercept of $y = 2^x$ is $(0, 1)$. The y-intercept of $y = 2^{x-2}$ is $\left(0, \frac{1}{4}\right)$. The line $y = 0$ is again the asymptote.

Equations with x and y Interchanged

It will be helpful in later work to be able to graph an equation in which the x and the y in $y = a^x$ are interchanged.

EXAMPLE 4   Graph: $x = 2^y$.

Solution   Note that x is alone on one side of the equation. To find ordered pairs that are solutions, we choose values for y and then compute values for x:

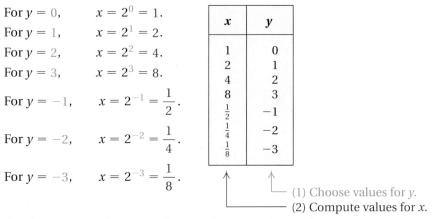

For $y = 0$, $x = 2^0 = 1$.
For $y = 1$, $x = 2^1 = 2$.
For $y = 2$, $x = 2^2 = 4$.
For $y = 3$, $x = 2^3 = 8$.
For $y = -1$, $x = 2^{-1} = \dfrac{1}{2}$.
For $y = -2$, $x = 2^{-2} = \dfrac{1}{4}$.
For $y = -3$, $x = 2^{-3} = \dfrac{1}{8}$.

x	y
1	0
2	1
4	2
8	3
$\frac{1}{2}$	-1
$\frac{1}{4}$	-2
$\frac{1}{8}$	-3

(1) Choose values for y.
(2) Compute values for x.

We plot the points and connect them with a smooth curve.

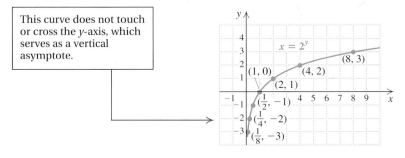

This curve does not touch or cross the y-axis, which serves as a vertical asymptote.

Note too that this curve looks just like the graph of $y = 2^x$, except that it is reflected across the line $y = x$, as shown here.

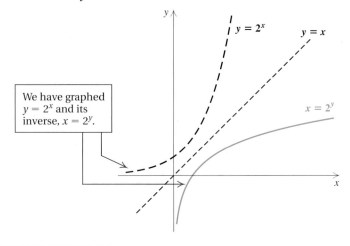

We have graphed $y = 2^x$ and its inverse, $x = 2^y$.

Applications of Exponential Functions

EXAMPLE 5   Interest compounded annually.   The amount of money A that a principal P will be worth after t years at interest rate i, compounded annually, is given by the formula

$$A = P(1 + i)^t.$$ You might review Example 11 in Section 8.1.

Suppose that $100,000 is invested at 8% interest, compounded annually.

a) Find a function for the amount in the account after t years.

b) Find the amount of money in the account at $t = 0$, $t = 4$, $t = 8$, and $t = 10$.

c) Graph the function.

Solution

a) If $P = \$100{,}000$ and $i = 8\% = 0.08$, we can substitute these values and form the following function:

$$A(t) = \$100{,}000(1 + 0.08)^t \qquad \text{Using } A = P(1 + i)^t$$
$$= \$100{,}000(1.08)^t.$$

b) To find the function values, a calculator with a power key is helpful.

$$A(0) = \$100{,}000(1.08)^0 \qquad\qquad A(8) = \$100{,}000(1.08)^8$$
$$= \$100{,}000(1) \qquad\qquad\qquad\;\; \approx \$100{,}000(1.85093021)$$
$$= \$100{,}000 \qquad\qquad\qquad\qquad \approx \$185{,}093.02$$

$$A(4) = \$100{,}000(1.08)^4 \qquad\quad A(10) = \$100{,}000(1.08)^{10}$$
$$= \$100{,}000(1.36048896) \qquad\;\; \approx \$100{,}000(2.158924997)$$
$$\approx \$136{,}048.90 \qquad\qquad\qquad\;\; \approx \$215{,}892.50$$

 **technology connection**

Graphing calculators can quickly find many function values at the touch of a few keys. To see this, let $y_1 = 100{,}000(1.08)^x$. Then use the TABLE feature to check Example 5(b).

c) We use the function values computed in part (b), and others if we wish, to draw the graph as follows. Note that the axes are scaled differently because of the large numbers.

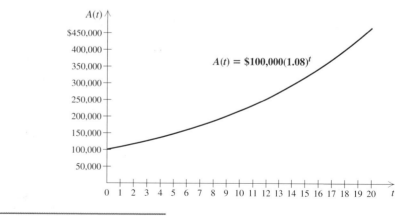

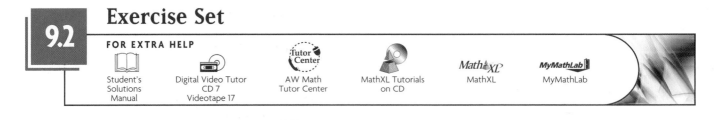

Exercise Set

9.2

FOR EXTRA HELP

Student's Solutions Manual | Digital Video Tutor CD 7 Videotape 17 | Tutor Center AW Math Tutor Center | MathXL Tutorials on CD | MathXL MathXL | MyMathLab MyMathLab

☞ *Concept Reinforcement   Classify each of the following as either true or false.*

1. The graph of $f(x) = a^x$ always passes through the point $(0, 1)$.

2. The graph of $g(x) = \left(\frac{1}{2}\right)^x$ gets closer and closer to the x-axis as x gets larger and larger.

3. The graph of $f(x) = 2^{x-3}$ looks just like the graph of $y = 2^x$, but it is translated 3 units to the right.

4. The graph of $g(x) = 2^x - 3$ looks just like the graph of $y = 2^x$, but it is translated 3 units up.

5. The graph of $y = 3^x$ gets close to, but never touches, the y-axis.

6. The graph of $x = 3^y$ gets close to, but never touches, the y-axis.

Graph.

7. $y = f(x) = 3^x$

8. $y = f(x) = 2^x$

9. $y = 6^x$

10. $y = 5^x$

11. $y = 2^x + 1$

12. $y = 2^x + 3$

13. $y = 3^x - 2$

14. $y = 3^x - 1$

15. $y = 2^x - 5$

16. $y = 2^x - 4$

17. $y = 2^{x-2}$

18. $y = 2^{x-1}$

19. $y = 2^{x+1}$

20. $y = 2^{x+3}$

21. $y = \left(\frac{1}{4}\right)^x$

22. $y = \left(\frac{1}{5}\right)^x$

23. $y = \left(\frac{1}{3}\right)^x$

24. $y = \left(\frac{1}{2}\right)^x$

25. $y = 2^{x+1} - 3$

26. $y = 2^{x-3} - 1$

27. $x = 6^y$

28. $x = 3^y$

29. $x = 3^{-y}$

30. $x = 2^{-y}$

31. $x = 4^y$

32. $x = 5^y$

33. $x = \left(\frac{4}{3}\right)^y$

34. $x = \left(\frac{3}{2}\right)^y$

Graph each pair of equations on the same set of axes.

35. $y = 3^x,\ x = 3^y$

36. $y = 2^x,\ x = 2^y$

37. $y = \left(\frac{1}{2}\right)^x,\ x = \left(\frac{1}{2}\right)^y$

38. $y = \left(\frac{1}{4}\right)^x,\ x = \left(\frac{1}{4}\right)^y$

Solve.

39. *Population growth.* The world population $P(t)$, in billions, t years after 1980, can be approximated by
$$P(t) = 4.495(1.015)^t.$$

Source: Based on data from U.S. Bureau of the Census, International Data Base

a) Predict the world population in 2008, 2012, and 2016.

b) Graph the function.

40. *Growth of bacteria.* The bacteria *Escherichia coli* are commonly found in the human bladder. Suppose that 3000 of the bacteria are present at time $t = 0$. Then t minutes later, the number of bacteria present can be approximated by
$$N(t) = 3000(2)^{t/20}.$$

a) How many bacteria will be present after 10 min? 20 min? 30 min? 40 min? 60 min?

b) Graph the function.

41. *Smoking cessation.* The percentage of smokers P who receive telephone counseling to quit smoking and are still successful t months later can be approximated by
$$P(t) = 21.4(0.914)^t.$$

Sources: *New England Journal of Medicine*; data from California's Smokers' Hotline

a) Estimate the percentage of smokers receiving telephone counseling who are successful in quitting for 1 month, 3 months, and 1 year.

b) Graph the function.

42. *Smoking cessation.* The percentage of smokers P who, without telephone counseling, have successfully quit smoking for t months (see Exercise 41) can be approximated by
$$P(t) = 9.02(0.93)^t.$$

Sources: *New England Journal of Medicine*; data from California's Smokers' Hotline

a) Estimate the percentage of smokers not receiving telephone counseling who are successful in quitting for 1 month, 3 months, and 1 year.

b) Graph the function.

43. *Marine biology.*   Due to excessive whaling prior to the mid 1970s, the humpback whale is considered an endangered species. The worldwide population of humpbacks, $P(t)$, in thousands, t years after 1900 ($t < 70$) can be approximated by*
$$P(t) = 150(0.960)^t.$$
a) How many humpback whales were alive in 1930? in 1960?
b) Graph the function.

44. *Salvage value.*   A photocopier is purchased for $5200. Its value each year is about 80% of the value of the preceding year. Its value, in dollars, after t years is given by the exponential function
$$V(t) = 5200(0.8)^t.$$
a) Find the value of the machine after 0 yr, 1 yr, 2 yr, 5 yr, and 10 yr.
b) Graph the function.

45. *Marine biology.*   As a result of preservation efforts in most countries in which whaling was common, the humpback whale population has grown since the 1970s. The worldwide population of humpbacks, $P(t)$, in thousands, t years after 1982 can be approximated by*
$$P(t) = 5.5(1.047)^t.$$
a) How many humpback whales were alive in 1992? in 2004?
b) Graph the function.

46. *Recycling aluminum cans.*   It is estimated that $\frac{1}{2}$ of all aluminum cans distributed will be recycled each year. A beverage company distributes 250,000 cans. The number still in use after time t, in years, is given by the exponential function
$$N(t) = 250{,}000\left(\tfrac{1}{2}\right)^t.$$
Source: The Aluminum Association, Inc., May 2004
a) How many cans are still in use after 0 yr? 1 yr? 4 yr? 10 yr?
b) Graph the function.

47. *Spread of zebra mussels.*   Beginning in 1988, infestations of zebra mussels started spreading throughout North American waters.† These mussels spread with such speed that water treatment facilities, power plants, and entire

---
*Based on information from the American Cetacean Society, 2001, and the ASK Archive, 1998.
†Many thanks to Dr. Gerald Mackie of the Department of Zoology at the University of Guelph in Ontario for the background information for this exercise.

ecosystems can become threatened. The function
$$A(t) = 10 \cdot 34^t$$
can be used to estimate the number of square centimeters of lake bottom that will be covered with mussels t years after an infestation covering 10 cm^2 first occurs.
a) How many square centimeters of lake bottom will be covered with mussels 5 years after an infestation covering 10 cm^2 first appears? 7 years after the infestation first appears?
b) Graph the function.

48. *Cell phones.*   The number of cell phones in use in the United States is increasing exponentially. The number N, in millions, in use can be estimated by
$$N(t) = 7.12(1.3)^t,$$
where t is the number of years after 1990.
Source: Cellular Telecommunications and Internet Association
a) Estimate the number of cell phones in use in 1995, 2005, and 2010.
b) Graph the function.

49. Without using a calculator, explain why 2^π must be greater than 8 but less than 16.

50. Suppose that $1000 is invested for 5 yr at 7% interest, compounded annually. In what year will the most interest be earned? Why?

SKILL MAINTENANCE

Simplify.

51. 5^{-2} [1.6]

52. 2^{-5} [1.6]

53. $1000^{2/3}$ [7.2]

54. $25^{-3/2}$ [7.2]

55. $\dfrac{10a^8b^7}{2a^2b^4}$ [1.6]

56. $\dfrac{24x^6y^4}{4x^2y^3}$ [1.6]

SYNTHESIS

57. Examine Exercise 48. Do you believe that the equation for the number of cell phones in use in the United States will be accurate 20 yr from now? Why or why not?

58. Why was it necessary to discuss irrational exponents before graphing exponential functions?

Determine which of the two numbers is larger. Do not use a calculator.

59. $\pi^{1.3}$ or $\pi^{2.4}$

60. $\sqrt{8^3}$ or $8^{\sqrt{3}}$

Graph.

61. $f(x) = 3.8^x$

62. $f(x) = 2.3^x$

63. $y = 2^x + 2^{-x}$

64. $y = \left| \left(\frac{1}{2} \right)^x - 1 \right|$

65. $y = |2^x - 2|$

66. $y = 2^{-(x-1)^2}$

67. $y = |2^{x^2} - 1|$

68. $y = 3^x + 3^{-x}$

Graph both equations using the same set of axes.

69. $y = 3^{-(x-1)}, \ x = 3^{-(y-1)}$ **70.** $y = 1^x, \ x = 1^y$

71. *Sales of DVD players.* As prices of DVD players continue to drop, sales have grown from $171 million in 1997 to $1099 million in 1999 and $2697 million in 2001. After pressing **STAT**, use the ExpReg option in the CALC menu to find an exponential function that models the total sales of DVD players t years after 1997. Then use that function to predict the total sales in 2008.
Source: *Statistical Abstract of the United States*, 2003

72. *Keyboarding speed.* Ali is studying keyboarding. After he has studied for t hours, Ali's speed, in words per minute, is given by the exponential function

$$S(t) = 200[1 - (0.99)^t].$$

Use a graph and/or table of values to predict Ali's speed after studying for 10 hr, 40 hr, and 80 hr.

73. *Spread of AIDS.* In 2000, a total of 40,282 cases of AIDS was reported in the United States; in 2001, a total of 41,450 cases; and in 2002, a total of 42,745 cases.

  a) Graph the data points, letting t represent the number of years since 2000.
  b) Which function best fits the data: linear, exponential, or quadratic? Why?

74. Consider any exponential function of the form $f(x) = a^x$ with $a > 1$. Will it always follow that $f(3) - f(2) > f(2) - f(1)$, and, in general, $f(n + 2) - f(n + 1) > f(n + 1) - f(n)$? Why or why not? (*Hint*: Think graphically.)

75. On many graphing calculators, it is possible to enter and graph $y_1 = A \wedge (X - B) + C$ after first pressing **APPS** Transfrm. Use this application to graph $f(x) = 2.5^{x-3} + 2$, $g(x) = 2.5^{x+3} + 2$, $h(x) = 2.5^{x-3} - 2$, and $k(x) = 2.5^{x+3} - 2$.

CORNER

The True Cost of a New Car

COLLABORATIVE

Focus: Car loans and exponential functions
Time: 30 minutes
Group size: 2
Materials: Calculators with exponentiation keys

The formula

$$M = \frac{Pr}{1 - (1 + r)^{-n}}$$

is used to determine the payment size, M, when a loan of P dollars is to be repaid in n equally sized monthly payments. Here r represents the monthly interest rate. Loans repaid in this fashion are said to be *amortized* (spread out equally) over a period of n months.

ACTIVITY

1. Suppose one group member is selling the other a car for $2600, financed at 1% interest per month for 24 months. What should be the size of each monthly payment?

2. Suppose both group members are shopping for the same model new car. To save time, each group member visits a different dealer. One dealer offers the car for $13,000 at 10.5% interest (0.00875 monthly interest) for 60 months (no down payment). The other dealer offers the same car for $12,000, but at 12% interest (0.01 monthly interest) for 48 months (no down payment).

  a) Determine the monthly payment size for each offer. Then determine the total amount paid for the car under each offer. How much of each total is interest?
  b) Work together to find the annual interest rate for which the total cost of 60 monthly payments for the $13,000 car would equal the total amount paid for the $12,000 car (as found in part a above).

9.3 | Logarithmic Functions

Graphs of Logarithmic Functions   •   Equivalent Equations   •
Solving Certain Logarithmic Equations

We are now ready to study inverses of exponential functions. These functions
have many applications and are called *logarithm,* or *logarithmic, functions.*

Graphs of Logarithmic Functions

Consider the exponential function $f(x) = 2^x$. Like all exponential functions,
f is one-to-one. Can a formula for f^{-1} be found? To answer this, we use the
method of Section 9.1:

1. Replace $f(x)$ with y: $y = 2^x$.
2. Interchange x and y: $x = 2^y$.
3. Solve for y: $y =$ the exponent to which we raise 2 to
                                       get x.
4. Replace y with $f^{-1}(x)$: $f^{-1}(x) =$ the exponent to which we raise 2 to
                                       get x.

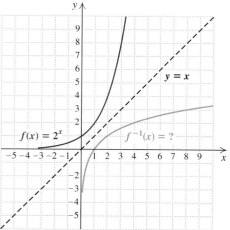

We now define a new symbol to replace the words "the exponent to which
we raise 2 to get x":

> **$\log_2 x$, read "the logarithm, base 2, of x," or "log, base 2, of x," means
> "the exponent to which we raise 2 to get x."**

Thus if $f(x) = 2^x$, then $f^{-1}(x) = \log_2 x$. Note that $f^{-1}(8) = \log_2 8 = 3$,
because 3 is *the exponent to which we raise* 2 *to get* 8.

EXAMPLE 1   Simplify: **(a)** $\log_2 32$; **(b)** $\log_2 1$; **(c)** $\log_2 \frac{1}{8}$.

Solution

a) Think of $\log_2 32$ as the exponent to which we raise 2 to get 32. That exponent is 5. Therefore, $\log_2 32 = 5$.

b) We ask ourselves: "To what exponent do we raise 2 in order to get 1?" That exponent is 0 (recall that $2^0 = 1$). Thus, $\log_2 1 = 0$.

c) To what exponent do we raise 2 in order to get $\frac{1}{8}$? Since $2^{-3} = \frac{1}{8}$, we have $\log_2 \frac{1}{8} = -3$.

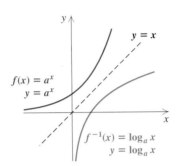

$f(x) = a^x$
$y = a^x$

$f^{-1}(x) = \log_a x$
$y = \log_a x$

Although numbers like $\log_2 13$ can only be approximated, we must remember that $\log_2 13$ represents *the exponent to which we raise 2 to get 13.* That is, $2^{\log_2 13} = 13$. A calculator can be used to show that $\log_2 13 \approx 3.7$ and $2^{3.7} \approx 13$. Later in this chapter, we will use a calculator to find such approximations.

For any exponential function $f(x) = a^x$, the inverse is called a **logarithmic function, base a.** The graph of the inverse can be drawn by reflecting the graph of $f(x) = a^x$ across the line $y = x$. It will be helpful to remember that the inverse of $f(x) = a^x$ is given by $f^{-1}(x) = \log_a x$.

Student Notes────────

As an aid in remembering what $\log_a x$ means, note that a is called the *base*, just as it is the base in $a^y = x$.

The Meaning of $\log_a x$

For $x > 0$ and a a positive constant other than 1, $\log_a x$ is the exponent to which a must be raised in order to get x. Thus,

$$\log_a x = m \text{ means } a^m = x$$

or equivalently,

$$\log_a x \text{ is that unique exponent for which } a^{\log_a x} = x.$$

It is important to remember that *a logarithm is an exponent.* It might help to repeat several times: "The logarithm, base a, of a number x is the exponent to which a must be raised in order to get x."

EXAMPLE 2   Simplify: $7^{\log_7 85}$.

Solution   Remember that $\log_7 85$ is the exponent to which 7 is raised to get 85. Raising 7 to that exponent, we have

$$7^{\log_7 85} = 85.$$

Because logarithmic and exponential functions are inverses of each other, the result in Example 2 should come as no surprise: If $f(x) = \log_7 x$, then

for $f(x) = \log_7 x$, we have $f^{-1}(x) = 7^x$

and $f^{-1}(f(x)) = f^{-1}(\log_7 x) = 7^{\log_7 x} = x.$

Thus, $f^{-1}(f(85)) = 7^{\log_7 85} = 85.$

The following is a comparison of exponential and logarithmic functions.

Exponential Function	Logarithmic Function
$y = a^x$	$x = a^y$
$f(x) = a^x$	$g(x) = \log_a x$
$a > 0, a \neq 1$	$a > 0, a \neq 1$
The domain is $\mathbb{R}$.	The range is $\mathbb{R}$.
$y > 0$ (Outputs are positive.)	$x > 0$ (Inputs are positive.)
$f^{-1}(x) = \log_a x$	$g^{-1}(x) = a^x$

EXAMPLE 3

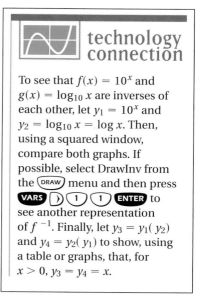

technology connection

To see that $f(x) = 10^x$ and $g(x) = \log_{10} x$ are inverses of each other, let $y_1 = 10^x$ and $y_2 = \log_{10} x = \log x$. Then, using a squared window, compare both graphs. If possible, select DrawInv from the (DRAW) menu and then press (VARS) (▷) (1) (1) (ENTER) to see another representation of f^{-1}. Finally, let $y_3 = y_1(y_2)$ and $y_4 = y_2(y_1)$ to show, using a table or graphs, that, for $x > 0$, $y_3 = y_4 = x$.

Graph: $y = f(x) = \log_5 x$.

Solution   If $y = \log_5 x$, then $5^y = x$. We can find ordered pairs that are solutions by choosing values for y and computing the x-values.

For $y = 0$, $x = 5^0 = 1$.
For $y = 1$, $x = 5^1 = 5$.
For $y = 2$, $x = 5^2 = 25$.
For $y = -1$, $x = 5^{-1} = \frac{1}{5}$.
For $y = -2$, $x = 5^{-2} = \frac{1}{25}$.

(1) Select y.
(2) Compute x.

This table shows the following:

$\log_5 1 = 0;$
$\log_5 5 = 1;$
$\log_5 25 = 2;$
$\log_5 \frac{1}{5} = -1;$
$\log_5 \frac{1}{25} = -2.$

These can all be checked using the equations above.

x, or 5^y	y
1	0
5	1
25	2
$\frac{1}{5}$	-1
$\frac{1}{25}$	-2

We plot the set of ordered pairs and connect the points with a smooth curve. The graphs of $y = 5^x$ and $y = x$ are shown only for reference.

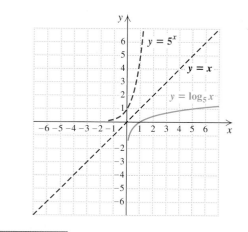

Equivalent Equations

We use the definition of logarithm to rewrite a *logarithmic equation* as an equivalent *exponential equation* or the other way around:

$$m = \log_a x \quad \text{is equivalent to} \quad a^m = x.$$

> *Caution!*   **Do not forget this relationship!** It is probably the most important definition in the chapter. Many times this definition will be used to justify a property we are considering.

EXAMPLE 4   Rewrite each as an equivalent exponential equation: **(a)** $y = \log_3 5$; **(b)** $-2 = \log_a 7$; **(c)** $a = \log_b d$.

Solution

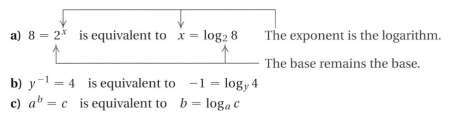

a) $y = \log_3 5$ is equivalent to $3^y = 5$ The logarithm is the exponent.

  The base remains the base.

b) $-2 = \log_a 7$ is equivalent to $a^{-2} = 7$

c) $a = \log_b d$ is equivalent to $b^a = d$

We also use the definition of logarithm to rewrite an exponential equation as an equivalent logarithmic equation.

EXAMPLE 5   Rewrite each as an equivalent logarithmic equation: **(a)** $8 = 2^x$; **(b)** $y^{-1} = 4$; **(c)** $a^b = c$.

Solution

a) $8 = 2^x$ is equivalent to $x = \log_2 8$ The exponent is the logarithm.

  The base remains the base.

b) $y^{-1} = 4$ is equivalent to $-1 = \log_y 4$

c) $a^b = c$ is equivalent to $b = \log_a c$

Solving Certain Logarithmic Equations

Logarithmic equations are often solved by rewriting them as equivalent exponential equations.

EXAMPLE 6   Solve: **(a)** $\log_2 x = -3$; **(b)** $\log_x 16 = 2$.

Solution

a) $\log_2 x = -3$

$\quad\quad 2^{-3} = x$ Rewriting as an exponential equation

$\quad\quad \frac{1}{8} = x$ Computing 2^{-3}

Check: $\log_2 \frac{1}{8}$ is the exponent to which 2 is raised to get $\frac{1}{8}$. Since that exponent is -3, we have a check. The solution is $\frac{1}{8}$.

b) $\log_x 16 = 2$

$$x^2 = 16 \qquad \text{Rewriting as an exponential equation}$$

$$x = 4 \quad or \quad x = -4 \qquad \text{Principle of square roots}$$

Check: $\log_4 16 = 2$ because $4^2 = 16$. Thus, 4 is a solution of $\log_x 16 = 2$. Because all logarithmic bases must be positive, -4 cannot be a solution. Logarithmic bases must be positive because logarithms are defined using exponential functions that require positive bases. The solution is 4.

One method for solving certain logarithmic and exponential equations relies on the following property, which results from the fact that exponential functions are one-to-one.

The Principle of Exponential Equality

For any real number b, where $b \neq -1, 0,$ or $1,$

$$b^{x_1} = b^{x_2} \quad \text{is equivalent to} \quad x_1 = x_2.$$

(Powers of the same base are equal if and only if the exponents are equal.)

EXAMPLE 7   Solve: **(a)** $\log_{10} 1000 = x$;   **(b)** $\log_4 1 = t$.

Solution

a) We rewrite $\log_{10} 1000 = x$ in exponential form and solve:

$$10^x = 1000 \qquad \text{Rewriting as an exponential equation}$$

$$10^x = 10^3 \qquad \text{Writing 1000 as a power of 10}$$

$$x = 3. \qquad \text{Equating exponents}$$

Check: This equation can also be solved directly by determining the exponent to which we raise 10 in order to get 1000. In both cases we find that $\log_{10} 1000 = 3$, so we have a check. The solution is 3.

b) We rewrite $\log_4 1 = t$ in exponential form and solve:

$$4^t = 1 \qquad \text{Rewriting as an exponential equation}$$

$$4^t = 4^0 \qquad \text{Writing 1 as a power of 4. This can be done mentally.}$$

$$t = 0. \qquad \text{Equating exponents}$$

Check: As in part (a), this equation can be solved directly by determining the exponent to which we raise 4 in order to get 1. In both cases we find that $\log_4 1 = 0$, so we have a check. The solution is 0.

Example 7 illustrates an important property of logarithms.

$\log_a 1$

The logarithm, base a, of 1 is always 0: $\log_a 1 = 0$.

This follows from the fact that $a^0 = 1$ is equivalent to the logarithmic equation $\log_a 1 = 0$. Thus, $\log_{10} 1 = 0$, $\log_7 1 = 0$, and so on.

Another property results from the fact that $a^1 = a$. This is equivalent to the equation $\log_a a = 1$.

$\log_a a$

The logarithm, base a, of a is always 1: $\log_a a = 1$.

Thus, $\log_{10} 10 = 1$, $\log_8 8 = 1$, and so on.

Exercise Set

9.3

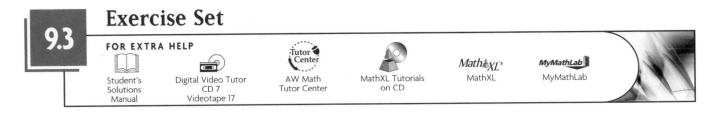

FOR EXTRA HELP

Student's Solutions Manual | Digital Video Tutor CD 7 Videotape 17 | AW Math Tutor Center | MathXL Tutorials on CD | MathXL | MyMathLab

↪ *Concept Reinforcement*   *In each of Exercises 1–8, match the expression or equation with an equivalent expression or equation from the column on the right.*

1. _____ $\log_5 25$

2. _____ $2^5 = x$

3. _____ $\log_5 5$

4. _____ $\log_2 1$

5. _____ $\log_5 5^x$

6. _____ $\log_x 27 = 5$

7. _____ $8 = 2^x$

8. _____ $x^{-2} = 5$

a) 1

b) x

c) $x^5 = 27$

d) $\log_2 x = 5$

e) $\log_2 8 = x$

f) $\log_x 5 = -2$

g) 2

h) 0

Simplify.

9. $\log_{10} 1000$

10. $\log_{10} 100$

11. $\log_2 16$

12. $\log_2 8$

13. $\log_3 81$ 4

14. $\log_3 27$

15. $\log_4 \frac{1}{16}$

16. $\log_4 \frac{1}{4}$

17. $\log_7 \frac{1}{7}$

18. $\log_7 \frac{1}{49}$

19. $\log_5 625$

20. $\log_5 125$

21. $\log_8 8$

22. $\log_7 1$

23. $\log_8 1$

24. $\log_8 8$

Aha! 25. $\log_9 9^5$

26. $\log_9 9^{10}$

27. $\log_{10} 0.01$

28. $\log_{10} 0.1$

29. $\log_9 3$

30. $\log_{16} 4$

31. $\log_9 27$

32. $\log_{16} 64$

33. $\log_{1000} 100$

34. $\log_{27} 9$

35. $5^{\log_5 7}$

36. $6^{\log_6 13}$

Graph.

37. $y = \log_{10} x$

38. $y = \log_2 x$

39. $y = \log_3 x$

40. $y = \log_7 x$

41. $f(x) = \log_6 x$

42. $f(x) = \log_4 x$

43. $f(x) = \log_{2.5} x$

44. $f(x) = \log_{1/2} x$

Graph both functions using the same set of axes.

45. $f(x) = 3^x$, $f^{-1}(x) = \log_3 x$

46. $f(x) = 4^x$, $f^{-1}(x) = \log_4 x$

Rewrite each of the following as an equivalent exponential equation. Do not solve.

47. $t = \log_5 9$

48. $h = \log_7 10$

49. $\log_5 25 = 2$

50. $\log_6 6 = 1$

51. $\log_{10} 0.1 = -1$

52. $\log_{10} 0.01 = -2$

53. $\log_{10} 7 = 0.845$

54. $\log_{10} 3 = 0.4771$

55. $\log_c m = 8$

56. $\log_b n = 23$

57. $\log_t Q = r$

58. $\log_m P = a$

59. $\log_e 0.25 = -1.3863$

60. $\log_e 0.989 = -0.0111$

61. $\log_r T = -x$

62. $\log_c M = -w$

Rewrite each of the following as an equivalent logarithmic equation. Do not solve.

63. $10^2 = 100$

64. $10^4 = 10,000$

65. $4^{-5} = \frac{1}{1024}$

66. $5^{-3} = \frac{1}{125}$

67. $16^{3/4} = 8$

68. $8^{1/3} = 2$

69. $10^{0.4771} = 3$

70. $10^{0.3010} = 2$

71. $z^m = 6$

72. $m^n = r$

73. $p^m = V$

74. $Q^t = x$

75. $e^3 = 20.0855$

76. $e^2 = 7.3891$

77. $e^{-4} = 0.0183$

78. $e^{-2} = 0.1353$

Solve.

79. $\log_3 x = 2$

80. $\log_4 x = 3$

81. $\log_5 125 = x$

82. $\log_4 64 = x$

83. $\log_2 16 = x$

84. $\log_3 27 = x$

85. $\log_x 7 = 1$

86. $\log_x 8 = 1$

87. $\log_3 x = -2$

88. $\log_2 x = -1$

89. $\log_{32} x = \frac{2}{5}$

90. $\log_8 x = \frac{2}{3}$

91. Express in words what number is represented by $\log_b c$.

92. Is it true that $2 = b^{\log_b 2}$? Why or why not?

SKILL MAINTENANCE

Simplify.

93. $\dfrac{x^{12}}{x^4}$ [1.6]

94. $\dfrac{a^{15}}{a^3}$ [1.6]

95. $(a^4 b^6)(a^3 b^2)$ [1.6]

96. $(x^3 y^5)(x^2 y^7)$ [1.6]

97. $\dfrac{\dfrac{3}{x} - \dfrac{2}{xy}}{\dfrac{2}{x^2} + \dfrac{1}{xy}}$ [6.3]

98. $\dfrac{\dfrac{4 + x}{x^2 + 2x + 1}}{\dfrac{3}{x + 1} - \dfrac{2}{x + 2}}$ [6.3]

SYNTHESIS

99. Would a manufacturer be pleased or unhappy if sales of a product grew logarithmically? Why?

100. Explain why the number $\log_2 13$ must be between 3 and 4.

101. Graph both equations using the same set of axes:
$$y = \left(\tfrac{3}{2}\right)^x, \qquad y = \log_{3/2} x.$$

Graph.

102. $y = \log_2(x - 1)$

103. $y = \log_3 |x + 1|$

Solve.

104. $|\log_3 x| = 2$

105. $\log_4 (3x - 2) = 2$

106. $\log_8 (2x + 1) = -1$

107. $\log_{10} (x^2 + 21x) = 2$

Simplify.

108. $\log_{1/4} \frac{1}{64}$

109. $\log_{1/5} 25$

110. $\log_{81} 3 \cdot \log_3 81$

111. $\log_{10} (\log_4 (\log_3 81))$

112. $\log_2 (\log_2 (\log_4 256))$

113. Show that $b^{x_1} = b^{x_2}$ is *not* equivalent to $x_1 = x_2$ for $b = 0$ or $b = 1$.

114. If $\log_b a = x$, does it follow that $\log_a b = 1/x$? Why or why not?

Properties of Logarithmic Functions

Logarithms of Products • Logarithms of Powers •
Logarithms of Quotients • Using the Properties Together

Logarithmic functions are important in many applications and in more advanced mathematics. We now establish some basic properties that are useful in manipulating expressions involving logarithms. As their proofs reveal, the properties of logarithms are related to the properties of exponents.

Logarithms of Products

The first property we discuss is related to the product rule for exponents: $a^m \cdot a^n = a^{m+n}$. Its proof appears immediately after Example 2.

> **The Product Rule for Logarithms**
>
> For any positive numbers M, N, and a ($a \neq 1$),
>
> $$\log_a (MN) = \log_a M + \log_a N.$$
>
> (The logarithm of a product is the sum of the logarithms of the factors.)

EXAMPLE 1

Express as an equivalent expression that is a sum of logarithms: $\log_2 (4 \cdot 16)$.

Solution We have

$$\log_2 (4 \cdot 16) = \log_2 4 + \log_2 16. \qquad \text{Using the product rule for logarithms}$$

As a check, note that

$$\log_2 (4 \cdot 16) = \log_2 64 = 6 \qquad 2^6 = 64$$

and that

$$\log_2 4 + \log_2 16 = 2 + 4 = 6. \qquad 2^2 = 4 \text{ and } 2^4 = 16$$

EXAMPLE 2

Express as an equivalent expression that is a single logarithm: $\log_b 7 + \log_b 5$.

Solution We have

$$\log_b 7 + \log_b 5 = \log_b (7 \cdot 5) \qquad \text{Using the product rule for logarithms}$$

$$= \log_b 35.$$

A Proof of the Product Rule.   Let $\log_a M = x$ and $\log_a N = y$. Converting to exponential equations, we have $a^x = M$ and $a^y = N$.

Now we multiply the left side of the first exponential equation by the left side of the second equation and similarly multiply the right sides to obtain

$$MN = a^x \cdot a^y, \text{ or } MN = a^{x+y}.$$

Converting back to a logarithmic equation, we get

$$\log_a (MN) = x + y.$$

Recalling what x and y represent, we have

$$\log_a (MN) = \log_a M + \log_a N.$$

Logarithms of Powers

The second basic property is related to the power rule for exponents: $(a^m)^n = a^{mn}$. Its proof follows Example 3.

The Power Rule for Logarithms

For any positive numbers M and a ($a \neq 1$), and any real number p,

$$\log_a M^p = p \cdot \log_a M.$$

(The logarithm of a power of M is the exponent times the logarithm of M.)

To better understand the power rule, note that

$$\log_a M^3 = \log_a (M \cdot M \cdot M) = \log_a M + \log_a M + \log_a M = 3 \log_a M.$$

EXAMPLE 3   Use the power rule for logarithms to write an equivalent expression that is a product: **(a)** $\log_a 9^{-5}$; **(b)** $\log_7 \sqrt[3]{x}$.

Solution

a) $\log_a 9^{-5} = -5 \log_a 9$ Using the power rule for logarithms

b) $\log_7 \sqrt[3]{x} = \log_7 x^{1/3}$ Writing exponential notation

$\qquad\qquad = \frac{1}{3} \log_7 x$ Using the power rule for logarithms

---

A Proof of the Power Rule.   Let $x = \log_a M$. We then write the equivalent exponential equation, $a^x = M$. Raising both sides to the pth power, we get

$$(a^x)^p = M^p, \text{ or } a^{xp} = M^p.   \text{ Multiplying exponents}$$

Converting back to a logarithmic equation gives us

$$\log_a M^p = xp.$$

But $x = \log_a M$, so substituting, we have

$$\log_a M^p = (\log_a M)p = p \cdot \log_a M.$$

Student Notes

Without understanding and *remembering* the rules of this section, it will be extremely difficult to solve the equations of Section 9.6.

Logarithms of Quotients

The third property that we study is similar to the quotient rule for exponents: $a^m/a^n = a^{m-n}$.

The Quotient Rule for Logarithms

For any positive numbers M, N, and a ($a \neq 1$),

$$\log_a \frac{M}{N} = \log_a M - \log_a N.$$

(The logarithm of a quotient is the logarithm of the dividend minus the logarithm of the divisor.)

To better understand the quotient rule, note that

$$\log_a \left(\frac{b^5}{b^3} \right) = \log_a b^2 = 2 \log_a b = 5 \log_a b - 3 \log_a b$$

$$= \log_a b^5 - \log_a b^3.$$

EXAMPLE 4   Express as an equivalent expression that is a difference of logarithms: $\log_t (6/U)$.

Solution

$$\log_t \frac{6}{U} = \log_t 6 - \log_t U \qquad \text{Using the quotient rule for logarithms}$$

EXAMPLE 5   Express as an equivalent expression that is a single logarithm: $\log_b 17 - \log_b 27$.

Solution

$$\log_b 17 - \log_b 27 = \log_b \frac{17}{27} \qquad \begin{array}{l} \text{Using the quotient rule for} \\ \text{logarithms "in reverse"} \end{array}$$

A Proof of the Quotient Rule.   Our proof uses both the product and power rules:

$$\log_a \frac{M}{N} = \log_a MN^{-1} \qquad \text{Rewriting } \frac{M}{N} \text{ as } MN^{-1}$$

$$= \log_a M + \log_a N^{-1} \qquad \begin{array}{l} \text{Using the product rule for} \\ \text{logarithms} \end{array}$$

$$= \log_a M + (-1)\log_a N \qquad \begin{array}{l} \text{Using the power rule for} \\ \text{logarithms} \end{array}$$

$$= \log_a M - \log_a N.$$

Using the Properties Together

EXAMPLE 6   Express as an equivalent expression, using the individual logarithms of x, y, and z.

a) $\log_b \dfrac{x^3}{yz}$

b) $\log_a \sqrt[4]{\dfrac{xy}{z^3}}$

Solution

a) $\log_b \dfrac{x^3}{yz} = \log_b x^3 - \log_b yz$ Using the quotient rule for logarithms

$\qquad = 3 \log_b x - \log_b yz$ Using the power rule for logarithms

$\qquad = 3 \log_b x - (\log_b y + \log_b z)$ Using the product rule for logarithms. Because of the subtraction, parentheses are essential.

$\qquad = 3 \log_b x - \log_b y - \log_b z$ Using the distributive law

b) $\log_a \sqrt[4]{\dfrac{xy}{z^3}} = \log_a \left(\dfrac{xy}{z^3}\right)^{1/4}$ Writing exponential notation

$\qquad = \dfrac{1}{4} \cdot \log_a \dfrac{xy}{z^3}$ Using the power rule for logarithms

$\qquad = \dfrac{1}{4} (\log_a xy - \log_a z^3)$ Using the quotient rule for logarithms. Parentheses are important.

$\qquad = \dfrac{1}{4} (\log_a x + \log_a y - 3 \log_a z)$ Using the product and power rules for logarithms

Caution!   Because the product and quotient rules replace one term with two, it is often best to use the rules within parentheses, as in Example 6.

EXAMPLE 7   Express as an equivalent expression that is a single logarithm.

a) $\dfrac{1}{2} \log_a x - 7 \log_a y + \log_a z$

b) $\log_a \dfrac{b}{\sqrt{x}} + \log_a \sqrt{bx}$

Solution

a) $\dfrac{1}{2} \log_a x - 7 \log_a y + \log_a z$

$\qquad = \log_a x^{1/2} - \log_a y^7 + \log_a z$ Using the power rule for logarithms

$\qquad = \left(\log_a \sqrt{x} - \log_a y^7\right) + \log_a z$ Using parentheses to emphasize the order of operations; $x^{1/2} = \sqrt{x}$

$\qquad = \log_a \dfrac{\sqrt{x}}{y^7} + \log_a z$ Using the quotient rule for logarithms

$\qquad = \log_a \dfrac{z\sqrt{x}}{y^7}$ Using the product rule for logarithms

b) $\log_a \dfrac{b}{\sqrt{x}} + \log_a \sqrt{bx} = \log_a \dfrac{b \cdot \sqrt{bx}}{\sqrt{x}}$ Using the product rule for logarithms

$\qquad\qquad\qquad\qquad = \log_a b\sqrt{b}$ Removing a factor equal to 1: $\dfrac{\sqrt{x}}{\sqrt{x}} = 1$

$\qquad\qquad\qquad\qquad = \log_a b^{3/2}$, or $\dfrac{3}{2}\log_a b$ Since $b\sqrt{b} = b^1 \cdot b^{1/2}$

If we know the logarithms of two different numbers (to the same base), the properties allow us to calculate other logarithms.

EXAMPLE 8   Given $\log_a 2 = 0.431$ and $\log_a 3 = 0.683$, calculate a numerical value for each of the following.

a) $\log_a 6$ **b)** $\log_a \frac{2}{3}$ **c)** $\log_a 81$

d) $\log_a \frac{1}{3}$ **e)** $\log_a 2a$ **f)** $\log_a 5$

Solution

a) $\log_a 6 = \log_a(2 \cdot 3) = \log_a 2 + \log_a 3$ Using the product rule for logarithms

$\qquad\qquad = 0.431 + 0.683 = 1.114$

> *Check:* $a^{1.114} = a^{0.431} \cdot a^{0.683} = 2 \cdot 3 = 6$

b) $\log_a \frac{2}{3} = \log_a 2 - \log_a 3$ Using the quotient rule for logarithms

$\qquad\qquad = 0.431 - 0.683 = -0.252$

c) $\log_a 81 = \log_a 3^4 = 4 \log_a 3$ Using the power rule for logarithms

$\qquad\qquad = 4(0.683) = 2.732$

d) $\log_a \frac{1}{3} = \log_a 1 - \log_a 3$ Using the quotient rule for logarithms

$\qquad\qquad = 0 - 0.683 = -0.683$

e) $\log_a 2a = \log_a 2 + \log_a a$ Using the product rule for logarithms

$\qquad\qquad = 0.431 + 1 = 1.431$

f) $\log_a 5$ *cannot be found using these properties.* ($\log_a 5 \neq \log_a 2 + \log_a 3$)

A final property follows from the product rule: Since $\log_a a^k = k \log_a a$, and $\log_a a = 1$, we have $\log_a a^k = k$.

> **The Logarithm of the Base to an Exponent**
>
> For any base a,
>
> $$\log_a a^k = k.$$
>
> (The logarithm, base a, of a to an exponent is the exponent.

This property also follows from the definition of logarithm: k is the exponent to which you raise a in order to get a^k.

EXAMPLE 9   Simplify: **(a)** $\log_3 3^7$; **(b)** $\log_{10} 10^{-5.2}$.

Solution

a) $\log_3 3^7 = 7$ 7 is the exponent to which you raise 3 in order to get 3^7.

b) $\log_{10} 10^{-5.2} = -5.2$

We summarize the properties of logarithms as follows.

> For any positive numbers M, N, and a ($a \neq 1$):
>
> $$\log_a MN = \log_a M + \log_a N; \qquad \log_a M^p = p \cdot \log_a M;$$
>
> $$\log_a \frac{M}{N} = \log_a M - \log_a N; \qquad \log_a a^k = k.$$

> *Caution!*   Keep in mind that, in general,
>
> $$\log_a (M + N) \neq \log_a M + \log_a N, \qquad \log_a MN \neq (\log_a M)(\log_a N),$$
>
> $$\log_a (M - N) \neq \log_a M - \log_a N, \qquad \log_a \frac{M}{N} \neq \frac{\log_a M}{\log_a N}.$$

Exercise Set
9.4

FOR EXTRA HELP

Student's Solutions Manual | Digital Video Tutor CD 7 Videotape 17 | AW Math Tutor Center | MathXL Tutorials on CD | MathXL | MyMathLab

↪ *Concept Reinforcement*   *In each of Exercises 1–6, match the expression with an equivalent expression from the column on the right.*

1. ____ $\log_7 20$

2. ____ $\log_7 5^4$

3. ____ $\log_7 \frac{5}{4}$

4. ____ $\log_7 7$

5. ____ $\log_7 1$

6. ____ $\log_7 5 + \log_7 6$

a) $\log_7 5 - \log_7 4$

b) 1

c) 0

d) $\log_7 30$

e) $\log_7 5 + \log_7 4$

f) $4 \log_7 5$

Express as an equivalent expression that is a sum of logarithms.

7. $\log_3 (81 \cdot 27)$

8. $\log_2 (16 \cdot 32)$

9. $\log_4 (64 \cdot 16)$

10. $\log_5 (25 \cdot 125)$

11. $\log_c (rst)$

12. $\log_t (3ab)$

Express as an equivalent expression that is a single logarithm.

13. $\log_a 5 + \log_a 14$

14. $\log_b 65 + \log_b 2$

15. $\log_c t + \log_c y$

16. $\log_t H + \log_t M$

Express as an equivalent expression that is a product.

17. $\log_a r^8$

18. $\log_b t^5$

19. $\log_c y^6$

20. $\log_{10} y^7$

21. $\log_b C^{-3}$

22. $\log_c M^{-5}$

Express as an equivalent expression that is a difference of two logarithms.

23. $\log_2 \frac{25}{13}$

24. $\log_3 \frac{23}{9}$

25. $\log_b \frac{m}{n}$

26. $\log_a \frac{y}{x}$

Express as an equivalent expression that is a single logarithm.

27. $\log_a 17 - \log_a 6$

28. $\log_b 32 - \log_b 7$

29. $\log_b 36 - \log_b 4$

30. $\log_a 26 - \log_a 2$

31. $\log_a 7 - \log_a 18$

32. $\log_b 5 - \log_b 13$

Express as an equivalent expression, using the individual logarithms of w, x, y, and z.

33. $\log_a (xyz)$

34. $\log_a (wxy)$

35. $\log_a (x^3 z^4)$

36. $\log_a (x^2 y^5)$

37. $\log_a (x^2 y^{-2} z)$

38. $\log_a (xy^2 z^{-3})$

39. $\log_a \frac{x^4}{y^3 z}$

40. $\log_a \frac{x^4}{yz^2}$

41. $\log_b \frac{xy^2}{wz^3}$

42. $\log_b \frac{w^2 x}{y^3 z}$

43. $\log_a \sqrt{\frac{x^7}{y^5 z^8}}$

44. $\log_c \sqrt[3]{\frac{x^4}{y^3 z^2}}$

45. $\log_a \sqrt[3]{\frac{x^6 y^3}{a^2 z^7}}$

46. $\log_a \sqrt[4]{\frac{x^8 y^{12}}{a^3 z^5}}$

Express as an equivalent expression that is a single logarithm and, if possible, simplify.

47. $8 \log_a x + 3 \log_a z$

48. $2 \log_b m + \frac{1}{2} \log_b n$

49. $\log_a x^2 - 2 \log_a \sqrt{x}$

50. $\log_a \frac{a}{\sqrt{x}} - \log_a \sqrt{ax}$

51. $\frac{1}{2} \log_a x + 5 \log_a y - 2 \log_a x$

52. $\log_a 2x + 3(\log_a x - \log_a y)$

53. $\log_a (x^2 - 4) - \log_a (x + 2)$

54. $\log_a (2x + 10) - \log_a (x^2 - 25)$

Given $\log_b 3 = 0.792$ and $\log_b 5 = 1.161$. If possible, calculate numerical values for each of the following.

55. $\log_b 15$

56. $\log_b \frac{5}{3}$

57. $\log_b \frac{3}{5}$

58. $\log_b \frac{1}{3}$

59. $\log_b \frac{1}{5}$

60. $\log_b \sqrt{b}$

61. $\log_b \sqrt{b^3}$

62. $\log_b 3b$

63. $\log_b 8$

64. $\log_b 45$

Simplify.

Aha! **65.** $\log_t t^7$

66. $\log_p p^4$

67. $\log_e e^m$

68. $\log_Q Q^{-2}$

69. A student *incorrectly* reasons that

$$\log_b \frac{1}{x} = \log_b \frac{x}{xx}$$

$$= \log_b x - \log_b x + \log_b x = \log_b x.$$

What mistake has the student made?

70. How could you convince someone that

$$\log_a c \neq \log_c a?$$

SKILL MAINTENANCE

Graph. [7.1]

71. $f(x) = \sqrt{x} - 3$

72. $g(x) = \sqrt{x} + 2$

73. $g(x) = \sqrt[3]{x} + 1$

74. $f(x) = \sqrt[3]{x} - 1$

Simplify. [1.6]

75. $(a^3 b^2)^5 (a^2 b^7)$

76. $(x^5 y^3 z^2)(x^2 yz^2)^3$

SYNTHESIS

77. Is it possible to express $\log_b \frac{x}{5}$ as an equivalent expression that is a difference of two logarithms without using the quotient rule? Why or why not?

78. Is it true that $\log_a x + \log_b x = \log_{ab} x$? Why or why not?

Express as an equivalent expression that is a single logarithm and, if possible, simplify.

79. $\log_a (x^8 - y^8) - \log_a (x^2 + y^2)$

80. $\log_a (x + y) + \log_a (x^2 - xy + y^2)$

Express as an equivalent expression that is a sum or difference of logarithms and, if possible, simplify.

81. $\log_a \sqrt{1 - s^2}$

82. $\log_a \dfrac{c - d}{\sqrt{c^2 - d^2}}$

83. If $\log_a x = 2$, $\log_a y = 3$, and $\log_a z = 4$, what is
$$\log_a \dfrac{\sqrt[3]{x^2 z}}{\sqrt[3]{y^2 z^{-2}}}?$$

84. If $\log_a x = 2$, what is $\log_a (1/x)$?

85. If $\log_a x = 2$, what is $\log_{1/a} x$?

Classify each of the following as true or false. Assume a, x, P, and Q > 0, $a \neq 1$.

86. $\log_a \left(\dfrac{P}{Q} \right)^x = x \log_a P - \log_a Q$

87. $\log_a (Q + Q^2) = \log_a Q + \log_a (Q + 1)$

 **88.** Use graphs to show that
$$\log x^2 \neq \log x \cdot \log x.$$
(*Note*: log means $\log_{10}$.)

9.5   Common and Natural Logarithms

Common Logarithms on a Calculator   •   The Base *e* and Natural Logarithms on a Calculator   •   Changing Logarithmic Bases   •   Graphs of Exponential and Logarithmic Functions, Base *e*

Any positive number other than 1 can serve as the base of a logarithmic function. However, some numbers are easier to use than others, and there are logarithmic bases that fit into certain applications more naturally than others.

Base-10 logarithms, called **common logarithms**, are useful because they have the same base as our "commonly" used decimal system. Before calculators became widely available, common logarithms helped with tedious calculations. In fact, that is why logarithms were devised.

The logarithmic base most widely used today is an irrational number named *e*. We will consider *e* and base *e*, or *natural*, logarithms later in this section. First we examine common logarithms.

Common Logarithms on a Calculator

Before the advent of scientific calculators, tables were developed to list common logarithms. Today we find common logarithms using calculators.

Here, and in most books, the abbreviation **log**, with no base written, is understood to mean logarithm base 10, or a common logarithm. Thus,

log 17   means $\log_{10} 17$.   It is important to remember this abbreviation.

On most calculators, the key for common logarithms is marked **LOG**. To find the common logarithm of a number, we key in that number and press **LOG**. On most graphing calculators, we press **LOG**, the number, and then **ENTER**.

EXAMPLE 1

Use a scientific calculator to approximate each number to four decimal places.

a) $\log 53{,}128$

b) $\dfrac{\log 6500}{\log 0.007}$

Solution

a) We enter 53,128 and then press **LOG**. We find that

$$\log 53{,}128 \approx 4.7253. \qquad \text{Rounded to four decimal places}$$

b) We enter 6500 and then press **LOG**. Next, we press ÷ , enter 0.007, and then press **LOG** = . Be careful not to round until the end:

$$\frac{\log 6500}{\log 0.007} \approx -1.7694. \qquad \text{Rounded to four decimal places}$$

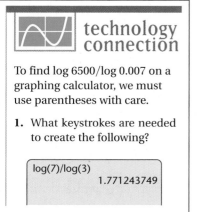

technology connection

To find log 6500/log 0.007 on a graphing calculator, we must use parentheses with care.

1. What keystrokes are needed to create the following?

log(7)/log(3)
                1.771243749

The inverse of a logarithmic function is an exponential function. Because of this, on many calculators the **LOG** key doubles as the **10ˣ** key after a **2ND** or SHIFT key is pressed. Calculators lacking a **10ˣ** key may have a key labeled x^y , a^x , or ⌃ . Such a key can raise any positive real number to any real-numbered exponent.

EXAMPLE 2

Use a calculator to approximate $10^{3.417}$ to four decimal places.

Solution We enter 3.417 and then press **10ˣ**. On most graphing calculators, **10ˣ** is pressed first, followed by 3.417 and **ENTER**. Rounding to four decimal places, we have

$$10^{3.417} \approx 2612.1614.$$

The Base *e* and Natural Logarithms on a Calculator

When interest is compounded *n* times a year, the compound interest formula is

$$A = P\left(1 + \frac{r}{n}\right)^{nt},$$

where *A* is the amount that an initial investment *P* will be worth after *t* years at interest rate *r*. Suppose that $1 is invested at 100% interest for 1 year (no bank would pay this). The preceding formula becomes a function *A* defined in terms of the number of compounding periods *n*:

$$A(n) = \left(1 + \frac{1}{n}\right)^{n}.$$

Let's find some function values. We round to six decimal places, using a calculator.

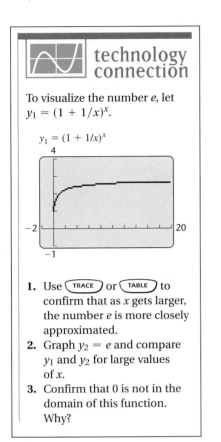

n	$A(n) = \left(1 + \dfrac{1}{n}\right)^n$
1 (compounded annually)	$2.00
2 (compounded semiannually)	$2.25
3	$2.370370
4 (compounded quarterly)	$2.441406
5	$2.488320
100	$2.704814
365 (compounded daily)	$2.714567
8760 (compounded hourly)	$2.718127

The numbers in this table approach a very important number in mathematics, called e. Because e is irrational, its decimal representation does not terminate or repeat.

> **The Number e**
>
> $e \approx 2.7182818284\ldots$

Logarithms base e are called **natural logarithms**, or **Napierian logarithms**, in honor of John Napier (1550–1617), who first "discovered" logarithms.

The abbreviation "ln" is generally used with natural logarithms. Thus,

$$\ln 53 \quad \text{means} \quad \log_e 53. \qquad \text{It is important to remember this abbreviation.}$$

On most scientific calculators, to find the natural logarithm of a number, we enter that number and press **LN**. On most graphing calculators, we press **LN**, the number, and then **ENTER**.

EXAMPLE 3   Use a scientific calculator to approximate $\ln 4568$ to four decimal places.

Solution   We enter 4568 and then press **LN**. We find that

$$\ln 4568 \approx 8.4268. \qquad \text{Rounded to four decimal places}$$

On many calculators, the **LN** key doubles as the e^x key after a **2ND** or SHIFT key has been pressed.

EXAMPLE 4   Use a calculator to approximate $e^{-1.524}$ to four decimal places.

Solution   We enter -1.524 and then press e^x. On most graphing calculators, e^x is pressed first, followed by -1.524 and **ENTER**. Since $e^{-1.524}$ is irrational, our answer is approximate:

$$e^{-1.524} \approx 0.2178. \qquad \text{Rounded to four decimal places}$$

Changing Logarithmic Bases

Most calculators can find both common logarithms and natural logarithms. To find a logarithm with some other base, a conversion formula is usually needed.

The Change-of-Base Formula

For any logarithmic bases a and b, and any positive number M,

$$\log_b M = \frac{\log_a M}{\log_a b}.$$

(To find the log, base b, of M, we typically compute $\log M/\log b$ or $\ln M/\ln b$.)

Proof. Let $x = \log_b M$. Then,

$b^x = M$	$\log_b M = x$ is equivalent to $b^x = M$.
$\log_a b^x = \log_a M$	Taking the logarithm, base a, on both sides
$x \log_a b = \log_a M$	Using the power rule for logarithms
$x = \dfrac{\log_a M}{\log_a b}.$	Dividing both sides by $\log_a b$

But at the outset we stated that $x = \log_b M$. Thus, by substitution, we have

$$\log_b M = \frac{\log_a M}{\log_a b}. \qquad \text{This is the change-of-base formula.}$$

EXAMPLE 5   Find $\log_5 8$ using the change-of-base formula.

Solution   We use the change-of-base formula with $a = 10$, $b = 5$, and $M = 8$:

$$\log_5 8 = \frac{\log_{10} 8}{\log_{10} 5} \qquad \text{Substituting into } \log_b M = \frac{\log_a M}{\log_a b}$$

$$\approx \frac{0.903089987}{0.6989700043} \qquad \text{Using } \boxed{\text{LOG}} \text{ twice}$$

$$\approx 1.2920. \qquad \text{When using a calculator, it is best not to round before dividing.}$$

To check, note that $\ln 8/\ln 5 \approx 1.2920$. We can also use a calculator to verify that $5^{1.2920} \approx 8$.

EXAMPLE 6   Find $\log_4 31$.

Solution   As shown in the check of Example 5, base e can also be used.

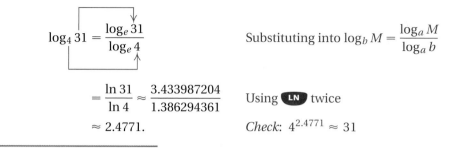

$$\log_4 31 = \frac{\log_e 31}{\log_e 4} \qquad \text{Substituting into } \log_b M = \frac{\log_a M}{\log_a b}$$

$$= \frac{\ln 31}{\ln 4} \approx \frac{3.433987204}{1.386294361} \qquad \text{Using } \boxed{\text{LN}} \text{ twice}$$

$$\approx 2.4771. \qquad \textit{Check: } 4^{2.4771} \approx 31$$

Graphs of Exponential and Logarithmic Functions, Base e

EXAMPLE 7   Graph $f(x) = e^x$ and $g(x) = e^{-x}$ and state the domain and the range of f and g.

Solution   We use a calculator with an $\boxed{e^x}$ key to find approximate values of e^x and e^{-x}. Using these values, we can graph the functions.

x	e^x	e^{-x}
0	1	1
1	2.7	0.4
2	7.4	0.1
−1	0.4	2.7
−2	0.1	7.4

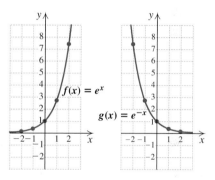

The domain of each function is $\mathbb{R}$ and the range of each function is $(0, \infty)$.

EXAMPLE 8   Graph $f(x) = e^{-x} + 2$ and state the domain and the range of f.

Solution   We find some solutions with a calculator, plot them, and then draw the graph. For example, $f(2) = e^{-2} + 2 \approx 0.1 + 2 \approx 2.1$. The graph is exactly like the graph of $g(x) = e^{-x}$, but is translated up 2 units.

x	$e^{-x} + 2$
0	3
1	2.4
2	2.1
−1	4.7
−2	9.4

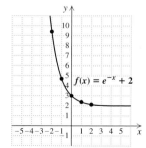

The domain of f is $\mathbb{R}$ and the range is $(2, \infty)$.

EXAMPLE 9   Graph and state the domain and the range of each function.

a) $g(x) = \ln x$ **b)** $f(x) = \ln (x + 3)$

Solution

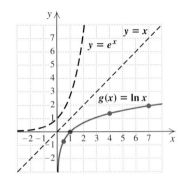

a) We find some solutions with a calculator and then draw the graph. As expected, the graph is a reflection across the line $y = x$ of the graph of $y = e^x$.

x	$\ln x$
1	0
4	1.4
7	1.9
0.5	−0.7

The domain of g is $(0, \infty)$ and the range is $\mathbb{R}$.

b) We find some solutions with a calculator, plot them, and draw the graph.

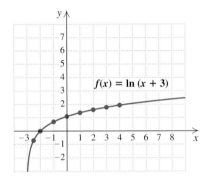

x	$\ln (x + 3)$
0	1.1
1	1.4
2	1.6
3	1.8
4	1.9
−1	0.7
−2	0
−2.5	−0.7

The graph of $y = \ln (x + 3)$ is the graph of $y = \ln x$ translated 3 units to the left. Since $x + 3$ must be positive, the domain is $(-3, \infty)$ and the range is $\mathbb{R}$.

technology connection

Logarithmic functions with bases other than 10 or e can be easily drawn using the change-of-base formula. For example, $y = \log_5 x$ can be written $y = \ln x/\ln 5$.

1. Graph $y = \log_7 x$.
2. Graph $y = \log_5(x + 2)$.
3. Graph $y = \log_7 x + 2$.

Exercise Set

9.5

FOR EXTRA HELP

Student's Solutions Manual   Digital Video Tutor CD 8 Videotape 18   Tutor Center AW Math Tutor Center   MathXL Tutorials on CD   Math XL MathXL   MyMathLab MyMathLab

↰ *Concept Reinforcement* *Classify each of the following as either true or false.*

1. The expression $\log 23$ means $\log_{10} 23$.

2. The expression $\ln 7$ means $\log_e 7$.

3. The number e is approximately 2.7.

4. The expressions $\log 9$ and $\log 18/\log 2$ are equivalent.

5. The expressions $\log_2 9$ and $\ln 9/\ln 2$ are equivalent.

6. The domain of the function given by $f(x) = \ln{(x + 2)}$ is $(-2, \infty)$.

7. The range of the function given by $g(x) = e^x$ is $(0, \infty)$.

8. The range of the function given by $f(x) = \ln{x}$ is $(-\infty, \infty)$.

Use a calculator to find each of the following to the nearest ten-thousandth.

9. $\log 6$

10. $\log 5$

11. $\log 72.8$

12. $\log 73.9$

Aha! **13.** $\log 1000$

14. $\log 100$

15. $\log 0.527$

16. $\log 0.493$

17. $\dfrac{\log 8200}{\log 150}$

18. $\dfrac{\log 5700}{\log 90}$

19. $10^{2.3}$

20. $10^{3.4}$

21. $10^{0.173}$

22. $10^{0.247}$

23. $10^{-2.9523}$

24. $10^{-3.2046}$

25. $\ln 5$

26. $\ln 2$

27. $\ln 57$

28. $\ln 30$

29. $\ln 0.0062$

30. $\ln 0.00073$

31. $\dfrac{\ln 2300}{0.08}$

32. $\dfrac{\ln 1900}{0.07}$

33. $e^{2.71}$

34. $e^{3.06}$

35. $e^{-3.49}$

36. $e^{-2.64}$

37. $e^{4.7}$

38. $e^{1.23}$

Find each of the following logarithms using the change-of-base formula. Round answers to the nearest ten-thousandth.

39. $\log_6 92$

40. $\log_3 78$

41. $\log_2 100$

42. $\log_7 100$

43. $\log_7 65$

44. $\log_5 42$

45. $\log_{0.5} 5$

46. $\log_{0.1} 3$

47. $\log_2 0.2$

48. $\log_2 0.08$

49. $\log_\pi 58$

50. $\log_\pi 200$

Graph and state the domain and the range of each function.

51. $f(x) = e^x$

52. $f(x) = e^{-x}$

53. $f(x) = e^x + 3$

54. $f(x) = e^x + 2$

55. $f(x) = e^x - 2$

56. $f(x) = e^x - 3$

57. $f(x) = 0.5e^x$

58. $f(x) = 2e^x$

59. $f(x) = 0.5e^{2x}$

60. $f(x) = 2e^{-0.5x}$

61. $f(x) = e^{x-3}$

62. $f(x) = e^{x-2}$

63. $f(x) = e^{x+2}$

64. $f(x) = e^{x+3}$

65. $f(x) = -e^x$

66. $f(x) = -e^{-x}$

67. $g(x) = \ln x + 1$

68. $g(x) = \ln x + 3$

69. $g(x) = \ln x - 2$

70. $g(x) = \ln x - 1$

71. $g(x) = 2 \ln x$

72. $g(x) = 3 \ln x$

73. $g(x) = -2 \ln x$

74. $g(x) = -\ln x$

75. $g(x) = \ln{(x + 2)}$

76. $g(x) = \ln{(x + 1)}$

77. $g(x) = \ln{(x - 1)}$

78. $g(x) = \ln{(x - 3)}$

79. Using a calculator, Zeno *incorrectly* says that $\log 79$ is between 4 and 5. How could you convince him, without using a calculator, that he is mistaken?

80. Examine Exercise 79. What mistake do you believe Zeno made?

SKILL MAINTENANCE

Solve.

81. $4x^2 - 25 = 0$ [5.8]

82. $5x^2 - 7x = 0$ [5.8]

83. $17x - 15 = 0$ [1.3]

84. $9 - 13x = 0$ [1.3]

85. $x^{1/2} - 6x^{1/4} + 8 = 0$ [8.5]

86. $2y - 7\sqrt{y} + 3 = 0$ [8.5]

SYNTHESIS

87. Explain how the graph of $f(x) = e^x$ could be used to graph the function given by $g(x) = 1 + \ln x$.

88. Explain how the graph of $f(x) = \ln x$ could be used to graph the function given by $g(x) = e^{x-1}$.

Knowing only that $\log 2 \approx 0.301$ and $\log 3 \approx 0.477$, find each of the following.

89. $\log_6 81$

90. $\log_9 16$

91. $\log_{12} 36$

92. Find a formula for converting common logarithms to natural logarithms.

93. Find a formula for converting natural logarithms to common logarithms.

Solve for x.

94. $\log{(275x^2)} = 38$

95. $\log{(492x)} = 5.728$

96. $\dfrac{3.01}{\ln x} = \dfrac{28}{4.31}$

97. $\log 692 + \log x = \log 3450$

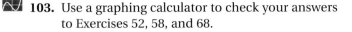 *For each function given below,* **(a)** *determine the domain and the range,* **(b)** *set an appropriate window, and* **(c)** *draw the graph. Graphs may vary, depending on the scale used.*

98. $f(x) = 7.4e^x \ln x$

99. $f(x) = 3.4 \ln x - 0.25e^x$

100. $f(x) = x \ln (x - 2.1)$

101. $f(x) = 2x^3 \ln x$

102. Use a graphing calculator to check your answers to Exercises 53, 61, and 75.

103. Use a graphing calculator to check your answers to Exercises 52, 58, and 68.

104. In an attempt to solve $\ln x = 1.5$, Emma gets the following graph. How can Emma tell at a glance that she has made a mistake?

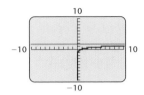

9.6 Solving Exponential and Logarithmic Equations

Solving Exponential Equations •
Solving Logarithmic Equations

Solving Exponential Equations

Equations with variables in exponents, such as $5^x = 12$ and $2^{7x} = 64$, are called **exponential equations**. In Section 9.3, we solved certain exponential equations by using the principle of exponential equality. We restate that principle below.

The Principle of Exponential Equality

For any real number b, where $b \neq -1, 0$, or 1,

$$b^x = b^y \quad \text{is equivalent to} \quad x = y.$$

(Powers of the same base are equal if and only if the exponents are equal.)

EXAMPLE 1 Solve: $4^x = 16$.

Solution Note that $16 = 4^2$. Thus we can write each side as a power of the same base:

$$4^x = 4^2.$$

Since the base is the same, 4, the exponents must be equal. Thus, x must be 2. The solution is 2.

Example 1 is not difficult to solve, but what if the right side of the equation were not a power of 4?

When it seems impossible to write both sides of an equation as powers of the same base, we use the following principle and write an equivalent logarithmic equation.

The Principle of Logarithmic Equality

For any logarithmic base a, and for $x, y > 0$,

$$x = y \quad \text{is equivalent to} \quad \log_a x = \log_a y.$$

(Two expressions are equal if and only if the logarithms of those expressions are equal.)

The principle of logarithmic equality, used together with the power rule for logarithms, allows us to solve equations in which the variable is an exponent.

EXAMPLE 2   Solve: $7^{x-2} = 60$.

Solution   We have

$$7^{x-2} = 60$$

$$\log 7^{x-2} = \log 60$$ Using the principle of logarithmic equality to take the common logarithm on both sides. Natural logarithms also would work.

$$(x - 2) \log 7 = \log 60$$ Using the power rule for logarithms

$$x - 2 = \frac{\log 60}{\log 7} \longleftarrow \boxed{\text{\emph{Caution!}   This is \emph{not} } \log 60 - \log 7.}$$

$$x = \frac{\log 60}{\log 7} + 2$$ Adding 2 to both sides

$$x \approx 4.1041.$$ Using a calculator and rounding to four decimal places

Since $7^{4.1041-2} \approx 60.0027$, we have a check. We can also note that since $7^{4-2} = 49$, we expect a solution near 4. The solution is $\dfrac{\log 60}{\log 7} + 2$, or approximately 4.1041.

ALGEBRAIC–GRAPHICAL CONNECTION

We can obtain a visual check of the solution of an exponential equation by graphing. For example, the solution of Example 1 can be visualized by graphing $y = 4^x$ and $y = 16$ on the same set of axes. The y-values for both equations are the same where the graphs intersect. The x-value at that point is the solution of the equation. That x-value appears to be 2. Since $4^2 = 16$, we see that this is indeed the case.

Similarly, the solution of Example 2 can be visualized by graphing $y = 7^{x-2}$ and $y = 60$ and identifying the x-value at the point of intersection. As expected, this value appears to be approximately 4.1.

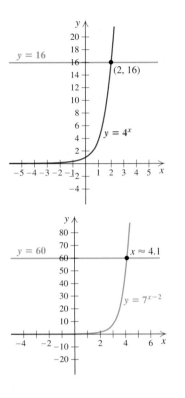

EXAMPLE 3 Solve: $e^{0.06t} = 1500$.

Solution Since one side is a power of e, it is easiest to take the *natural logarithm* on both sides:

$\ln e^{0.06t} = \ln 1500$ Taking the natural logarithm on both sides

$0.06t = \ln 1500$ Finding the logarithm of the base to a power: $\log_a a^k = k$. Logarithmic and exponential functions are inverses of each other.

$t = \dfrac{\ln 1500}{0.06}$ Dividing both sides by 0.06

$\approx 121.887.$ Using a calculator and rounding to three decimal places

---

To Solve an Equation of the Form $a^t = b$ for t

1. Take the logarithm (either natural or common) of both sides.

2. Use the power rule for exponents so that the variable is no longer written as an exponent.

3. Divide both sides by the coefficient of the variable to isolate the variable.

4. If appropriate, use a calculator to find an approximate solution in decimal form.

Solving Logarithmic Equations

Equations containing logarithmic expressions are called **logarithmic equations**. We saw in Section 9.3 that certain logarithmic equations can be solved by writing an equivalent exponential equation.

EXAMPLE 4   Solve: **(a)** $\log_4 (8x - 6) = 3$; **(b)** $\ln 5x = 27$.

Solution

a) $\log_4 (8x - 6) = 3$

$$4^3 = 8x - 6 \qquad \text{Writing the equivalent exponential equation}$$

$$64 = 8x - 6$$

$$70 = 8x \qquad \text{Adding 6 to both sides}$$

$$x = \tfrac{70}{8}, \text{ or } \tfrac{35}{4}.$$

Check:

$$
\begin{array}{c|c}
\log_4 (8x - 6) = 3 & \\
\hline
\log_4 \left(8 \cdot \tfrac{35}{4} - 6\right) & 3 \\
\log_4 (2 \cdot 35 - 6) & \\
\log_4 64 & \\
3 \stackrel{?}{=} 3 & \text{TRUE}
\end{array}
$$

The solution is $\tfrac{35}{4}$.

b) $\ln 5x = 27$ Remember: $\ln 5x$ means $\log_e 5x$.

$$e^{27} = 5x \qquad \text{Writing the equivalent exponential equation}$$

$$\frac{e^{27}}{5} = x \qquad \text{This is a very large number.}$$

The solution is $\dfrac{e^{27}}{5}$. The check is left to the student.

Often the properties for logarithms are needed. The goal is to first write an equivalent equation in which the variable appears in just one logarithmic expression. We then isolate that expression and solve as in Example 4.

EXAMPLE 5   Solve.

a) $\log x + \log (x - 3) = 1$

b) $\log_2 (x + 7) - \log_2 (x - 7) = 3$

c) $\log_7 (x + 1) + \log_7 (x - 1) = \log_7 8$

Solution

a) To increase understanding, we write in the base, 10.

$$\log_{10} x + \log_{10} (x - 3) = 1$$

$$\log_{10} [x(x - 3)] = 1 \qquad \text{Using the product rule for logarithms to obtain a single logarithm}$$

$$x(x - 3) = 10^1 \qquad \text{Writing an equivalent exponential equation}$$

$$x^2 - 3x = 10$$

$$x^2 - 3x - 10 = 0$$

$$(x + 2)(x - 5) = 0 \qquad \text{Factoring}$$

$$x + 2 = 0 \quad or \quad x - 5 = 0 \qquad \text{Using the principle of zero products}$$

$$x = -2 \quad or \qquad x = 5$$

Student Notes ⎯⎯⎯⎯

It is essential that you remember the properties of logarithms from Section 9.4. Consider reviewing the properties before attempting to solve equations similar to those in Example 5.

Check:

For -2:

$$\log x + \log (x - 3) = 1$$

$$\log (-2) + \log (-2 - 3) \overset{?}{=} 1 \quad \text{FALSE}$$

For 5:

$$\log x + \log (x - 3) = 1$$

$$\log 5 + \log (5 - 3) \;\bigg|\; 1$$

$$\log 5 + \log 2 \;\bigg|\;$$

$$\log 10 \;\bigg|\;$$

$$1 \overset{?}{=} 1 \quad \text{TRUE}$$

The number -2 *does not check* because the logarithm of a negative number is undefined. The solution is 5.

b) We have

$$\log_2 (x + 7) - \log_2 (x - 7) = 3$$

$$\log_2 \frac{x + 7}{x - 7} = 3 \qquad \text{Using the quotient rule for logarithms to obtain a single logarithm}$$

$$\frac{x + 7}{x - 7} = 2^3 \qquad \text{Writing an equivalent exponential equation}$$

$$\frac{x + 7}{x - 7} = 8$$

$$x + 7 = 8(x - 7) \qquad \text{Multiplying by the LCD, } x - 7$$

$$x + 7 = 8x - 56 \qquad \text{Using the distributive law}$$

$$63 = 7x$$

$$9 = x. \qquad \text{Dividing by 7}$$

Check:

$$\log_2 (x + 7) - \log_2 (x - 7) = 3$$

$$\begin{array}{c|c} \log_2 (9 + 7) - \log_2 (9 - 7) & 3 \\ \log_2 16 - \log_2 2 & \\ 4 - 1 & \\ 3 \overset{?}{=} 3 & \text{TRUE} \end{array}$$

The solution is 9.

c) We have

$$\begin{aligned} \log_7 (x + 1) + \log_7 (x - 1) &= \log_7 8 \\ \log_7 [(x + 1)(x - 1)] &= \log_7 8 \qquad &&\text{Using the product rule} \\ &&&\text{for logarithms} \\ \log_7 (x^2 - 1) &= \log_7 8 \qquad &&\text{Multiplying. Note that both} \\ &&&\text{sides are base-7 logarithms.} \\ x^2 - 1 &= 8 \qquad &&\text{Using the principle of} \\ &&&\text{logarithmic equality. Study} \\ &&&\text{this step carefully.} \\ x^2 - 9 &= 0 \\ (x - 3)(x + 3) &= 0 \qquad &&\text{Solving the quadratic} \\ &&&\text{equation} \\ x = 3 \quad or \quad x &= -3. \end{aligned}$$

We leave it to the student to show that 3 checks but -3 does not. The solution is 3.

---

 technology connection

To solve exponential and logarithmic equations, we can use INTERSECT to determine the x-coordinate at each intersection.

    For example, to solve $e^{0.5x} - 7 = 2x + 6$, we graph $y_1 = e^{0.5x} - 7$ and $y_2 = 2x + 6$ as shown. We then use the INTERSECT option of the ⬚ CALC ⬚ menu. The x-coordinates at the intersections are approximately -6.48 and 6.52.

    Use a graphing calculator to solve each equation to the nearest hundredth.

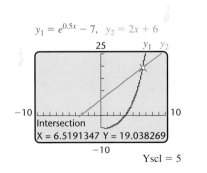

$y_1 = e^{0.5x} - 7, \quad y_2 = 2x + 6$

Intersection
X = 6.5191347 Y = 19.038269
Yscl = 5

1. $e^{7x} = 14$

2. $8e^{0.5x} = 3$

3. $xe^{3x-1} = 5$

4. $4 \ln (x + 3.4) = 2.5$

5. $\ln 3x = 0.5x - 1$

6. $\ln x^2 = -x^2$

Exercise Set

9.6

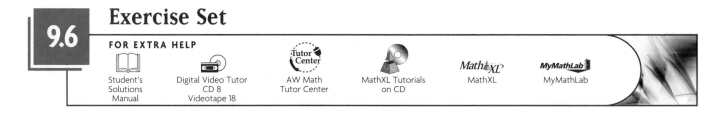

🍂 *Concept Reinforcement* *In each of Exercises 1–8, match the equation with an equivalent equation from the column on the right that could be the next step in the solution process.*

1. ____ $5^x = 3$

2. ____ $e^{5x} = 3$

3. ____ $\ln x = 3$

4. ____ $\log_x 5 = 3$

5. ____ $\log_5 x + \log_5(x - 2) = 3$

6. ____ $\log_5 x - \log_5(x - 2) = 3$

7. ____ $\ln x - \ln(x - 2) = 3$

8. ____ $\log x + \log(x - 2) = 3$

a) $\ln e^{5x} = \ln 3$

b) $\log_5 (x^2 - 2x) = 3$

c) $\log (x^2 - 2x) = 3$

d) $\log_5 \dfrac{x}{x - 2} = 3$

e) $\log 5^x = \log 3$

f) $e^3 = x$

g) $\ln \dfrac{x}{x - 2} = 3$

h) $x^3 = 5$

Solve. Where appropriate, include approximations to the nearest thousandth.

9. $2^x = 19$

10. $2^x = 15$

11. $8^{x-1} = 17$

12. $4^{x+1} = 13$

13. $e^t = 1000$

14. $e^t = 100$

15. $e^{0.03t} + 2 = 7$

16. $e^{-0.07t} + 3 = 3.08$

17. $5 = 3^{x+1}$

18. $7 = 3^{x-1}$

Aha! 19. $2^{x+3} = 16$

20. $4^{x+1} = 64$

21. $4.9^x - 87 = 0$

22. $7.2^x - 65 = 0$

23. $19 = 2e^{4x}$

24. $29 = 3e^{2x}$

25. $7 + 3e^{5x} = 13$

26. $4 + 5e^{4x} = 9$ **Aha!**

27. $\log_3 x = 4$

28. $\log_2 x = 6$

29. $\log_2 x = -3$

30. $\log_5 x = 3$

31. $\ln x = 5$

32. $\ln x = 4$

33. $\log_8 x = \frac{1}{3}$

34. $\log_4 x = \frac{1}{2}$

35. $\ln 4x = 3$

36. $\ln 3x = 2$

37. $\log x = 2.5$

38. $\log x = 0.5$

39. $\ln (2x + 1) = 4$

40. $\ln (4x - 2) = 3$

Aha! 41. $\ln x = 1$

42. $\log x = 1$

43. $5 \ln x = -15$

44. $3 \ln x = -3$

45. $\log_2 (8 - 6x) = 5$

46. $\log_5 (2x - 7) = 3$

47. $\log (x - 9) + \log x = 1$

48. $\log (x + 9) + \log x = 1$

49. $\log x - \log (x + 3) = 1$

50. $\log x - \log (x + 7) = -1$

51. $\log_4 (x + 3) = 2 + \log_4 (x - 5)$

52. $\log_2 (x + 3) = 4 + \log_2 (x - 3)$

53. $\log_7 (x + 1) + \log_7 (x + 2) = \log_7 6$

54. $\log_6 (x + 3) + \log_6 (x + 2) = \log_6 20$

55. $\log_5 (x + 4) + \log_5 (x - 4) = \log_5 20$

56. $\log_4 (x + 2) + \log_4 (x - 7) = \log_4 10$

57. $\ln (x + 5) + \ln (x + 1) = \ln 12$

58. $\ln (x - 6) + \ln (x + 3) = \ln 22$

59. $\log_2 (x - 3) + \log_2 (x + 3) = 4$

60. $\log_3 (x - 4) + \log_3 (x + 4) = 2$

61. $\log_{12} (x + 5) - \log_{12} (x - 4) = \log_{12} 3$

62. $\log_6 (x + 7) - \log_6 (x - 2) = \log_6 5$

63. $\log_2 (x - 2) + \log_2 x = 3$

64. $\log_4 (x + 6) - \log_4 x = 2$

65. Could Example 2 have been solved by taking the natural logarithm on both sides? Why or why not?

66. Christina finds that the solution of $\log_3 (x + 4) = 1$ is -1, but rejects -1 as an answer. What mistake is she making?

SKILL MAINTENANCE

67. Find an equation of variation if y varies directly as x, and $y = 7.2$ when $x = 0.8$. [6.8]

68. Find an equation of variation if y varies inversely as x, and $y = 3.5$ when $x = 6.1$. [6.8]

Solve. [8.3]

69. $T = 2\pi\sqrt{L/32}$, for L

70. $E = mc^2$, for c
(Assume $E, m, c > 0$.)

71. Joni can key in a musical score in 2 hr. Miles takes 3 hr to key in the same score. How long would it take them, working together, to key in the score? [6.5]

72. The side exit at the Flynn Theater can empty a capacity crowd in 25 min. The main exit can empty a capacity crowd in 15 min. How long will it take to empty a capacity crowd when both exits are in use? [6.5]

SYNTHESIS

73. Can the principle of logarithmic equality be expanded to include all functions? That is, is the statement "$m = n$ is equivalent to $f(m) = f(n)$" true for any function f? Why or why not?

74. Explain how Exercises 37 and 38 could be solved using the graph of $f(x) = \log x$.

Solve. If no solution exists, state this.

75. $27^x = 81^{2x-3}$

76. $8^x = 16^{3x+9}$

77. $\log_x (\log_3 27) = 3$

78. $\log_6 (\log_2 x) = 0$

79. $x \log \frac{1}{8} = \log 8$

80. $\log_5 \sqrt{x^2 - 9} = 1$

81. $2^{x^2+4x} = \frac{1}{8}$

82. $\log (\log x) = 5$

83. $\log_5 |x| = 4$

84. $\log x^2 = (\log x)^2$

85. $\log \sqrt{2x} = \sqrt{\log 2x}$

86. $1000^{2x+1} = 100^{3x}$

87. $3^{x^2} \cdot 3^{4x} = \frac{1}{27}$

88. $3^{3x} \cdot 3^{x^2} = 81$

89. $\log x^{\log x} = 25$

90. $3^{2x} - 8 \cdot 3^x + 15 = 0$

91. $(81^{x-2})(27^{x+1}) = 9^{2x-3}$

92. $3^{2x} - 3^{2x-1} = 18$

93. Given that $2^y = 16^{x-3}$ and $3^{y+2} = 27^x$, find the value of $x + y$.

94. If $x = (\log_{125} 5)^{\log_5 125}$, what is the value of $\log_3 x$?

95. Find the value of x for which the natural logarithm is the same as the common logarithm.

96. Use a graphing calculator to check your answers to Exercises 3, 29, 41, and 57.

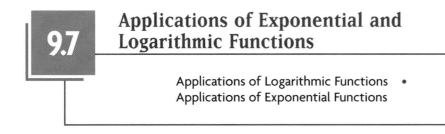

9.7 # Applications of Exponential and Logarithmic Functions

Applications of Logarithmic Functions •
Applications of Exponential Functions

We now consider applications of exponential and logarithmic functions.

Applications of Logarithmic Functions

EXAMPLE 1   *Sound levels.*   To measure the volume, or "loudness," of a sound, the *decibel* scale is used. The loudness L, in decibels (dB), of a sound is given by

$$L = 10 \cdot \log \frac{I}{I_0},$$

where I is the intensity of the sound, in watts per square meter (W/m^2), and $I_0 = 10^{-12}$ W/m^2. (I_0 is approximately the intensity of the softest sound that can be heard by the human ear.)

a) It is common for the intensity of sound at live performances of rock music to reach 10^{-1} W/m^2 (even higher close to the stage). How loud, in decibels, is the sound level?

b) The Occupational Safety and Health Administration (OSHA) considers sound levels of 85 dB and above unsafe. What is the intensity of such sounds?

Solution

a) To find the loudness, in decibels, we use the above formula:

$$
\begin{aligned}
L &= 10 \cdot \log \frac{I}{I_0} \\
&= 10 \cdot \log \frac{10^{-1}}{10^{-12}} \qquad \text{Substituting} \\
&= 10 \cdot \log 10^{11} \qquad \text{Subtracting exponents} \\
&= 10 \cdot 11 \qquad\qquad \log 10^a = a \\
&= 110.
\end{aligned}
$$

The volume of the music is 110 decibels.

Sarah Fisher puts in her earplugs in the pits at the Indianapolis Motor Speedway on May 19, 2002. She is the youngest woman to qualify for the Indy 500.

b) We substitute and solve for I:

$$L = 10 \cdot \log \frac{I}{I_0}$$

$$85 = 10 \cdot \log \frac{I}{10^{-12}} \qquad \text{Substituting}$$

$$8.5 = \log \frac{I}{10^{-12}} \qquad \text{Dividing both sides by 10}$$

$$8.5 = \log I - \log 10^{-12} \qquad \text{Using the quotient rule for logarithms}$$

$$8.5 = \log I - (-12) \qquad \log 10^a = a$$

$$-3.5 = \log I \qquad \text{Adding } -12 \text{ to both sides}$$

$$10^{-3.5} = I. \qquad \text{Converting to an exponential equation}$$

Earplugs would be recommended for sounds with intensities exceeding $10^{-3.5}$ W/m^2.

EXAMPLE 2

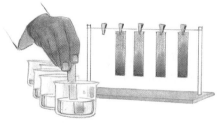

Chemistry: pH of liquids.   In chemistry, the pH of a liquid is a measure of its acidity. We calculate pH as follows:

$$\text{pH} = -\log[\text{H}^+],$$

where $[\text{H}^+]$ is the hydrogen ion concentration in moles per liter.

a) The hydrogen ion concentration of human blood is normally about 3.98×10^{-8} moles per liter. Find the pH.

b) The pH of seawater is about 8.3. Find the hydrogen ion concentration.

Solution

a) To find the pH of blood, we use the above formula:

$$\text{pH} = -\log[\text{H}^+]$$

$$= -\log[3.98 \times 10^{-8}]$$

$$\approx -(-7.400117) \qquad \text{Using a calculator}$$

$$\approx 7.4.$$

The pH of human blood is normally about 7.4.

b) We substitute and solve for $[\text{H}^+]$:

$$8.3 = -\log[\text{H}^+] \qquad \text{Using pH} = -\log[\text{H}^+]$$

$$-8.3 = \log[\text{H}^+] \qquad \text{Dividing both sides by } -1$$

$$10^{-8.3} = [\text{H}^+] \qquad \text{Converting to an exponential equation}$$

$$5.01 \times 10^{-9} \approx [\text{H}^+]. \qquad \text{Using a calculator; writing scientific notation}$$

The hydrogen ion concentration of seawater is about 5.01×10^{-9} moles per liter.

Applications of Exponential Functions

EXAMPLE 3   Interest compounded annually.   Suppose that $25,000 is invested at 4% interest, compounded annually. In t years, it will grow to the amount A given by the function

$$A(t) = 25,000(1.04)^t.$$

(See Example 5 in Section 9.2.)

a) How long will it take to accumulate $80,000 in the account?

b) Find the amount of time it takes for the $25,000 to double itself.

Solution

a) We set $A(t) = 80,000$ and solve for t:

$$80,000 = 25,000(1.04)^t$$

$$\frac{80,000}{25,000} = 1.04^t \qquad \text{Dividing both sides by 25,000}$$

$$3.2 = 1.04^t$$

$$\log 3.2 = \log 1.04^t \qquad \begin{array}{l}\text{Taking the common} \\ \text{logarithm on both sides}\end{array}$$

$$\log 3.2 = t \log 1.04 \qquad \text{Using the power rule for logarithms}$$

$$\frac{\log 3.2}{\log 1.04} = t \qquad \text{Dividing both sides by } \log 1.04$$

$$29.7 \approx t. \qquad \text{Using a calculator}$$

Remember that when doing a calculation like this on a calculator, it is best to wait until the end to round off. At an interest rate of 4% per year, it will take about 29.7 yr for $25,000 to grow to $80,000.

Student Notes _____

Study the different steps in the solution of Example 3(b). Note that if 50,000 and 25,000 are replaced with 8000 and 4000, the doubling time is unchanged.

b) To find the *doubling time*, we replace $A(t)$ with 50,000 and solve for t:

$$50,000 = 25,000(1.04)^t$$

$$2 = (1.04)^t \qquad \text{Dividing both sides by 25,000}$$

$$\log 2 = \log (1.04)^t \qquad \begin{array}{l}\text{Taking the common logarithm} \\ \text{on both sides}\end{array}$$

$$\log 2 = t \log 1.04 \qquad \begin{array}{l}\text{Using the power rule for} \\ \text{logarithms}\end{array}$$

$$t = \frac{\log 2}{\log 1.04} \approx 17.7. \qquad \begin{array}{l}\text{Dividing both sides by } \log 1.04 \text{ and} \\ \text{using a calculator}\end{array}$$

At an interest rate of 4% per year, the doubling time is about 17.7 yr.

Like investments, populations often grow exponentially.

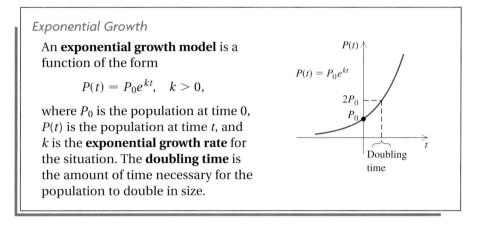

Exponential Growth

An **exponential growth model** is a function of the form

$$P(t) = P_0 e^{kt}, \quad k > 0,$$

where P_0 is the population at time 0, $P(t)$ is the population at time t, and k is the **exponential growth rate** for the situation. The **doubling time** is the amount of time necessary for the population to double in size.

The exponential growth rate is the rate of growth of a population at any *instant* in time. Since the population is continually growing, the percent of total growth after one year will exceed the exponential growth rate.

EXAMPLE 4 Growth of zebra mussel populations. Zebra mussels, inadvertently imported from Europe, began fouling North American waters in 1988. These mussels are so prolific that lake and river bottoms, as well as water intake pipes, can become blanketed with them, altering an entire ecosystem. In 2000, a portion of the Hudson River contained an average of 10 zebra mussels per square mile. The exponential growth rate was 340% per year.

a) Find the exponential growth function that models the data.

b) Predict the number of mussels per square mile in 2007.

Solution

a) In 2000, at $t = 0$, the population was $10/\text{mi}^2$. We substitute 10 for P_0 and 340%, or 3.4, for k. This gives the exponential growth function

$$P(t) = 10e^{3.4t}.$$

b) In 2007, we have $t = 7$ (since 7 yr have passed since 2000). To find the population in 2007, we compute $P(7)$:

$$P(7) = 10e^{3.4(7)} \qquad \text{Using } P(t) = 10e^{3.4t} \text{ from part (a)}$$
$$= 10e^{23.8}$$
$$\approx 217{,}000{,}000{,}000. \qquad \text{Using a calculator}$$

The population of zebra mussels in the specified portion of the Hudson River will reach approximately 217,000,000,000 per square mile in 2007.

EXAMPLE 5   *Spread of a computer virus.* The number of computers infected by a virus t hours after it first appears usually increases exponentially. In 2004, the "MyDoom" worm spread from 100 computers to about 100,000 computers in 24 hr. (*Source*: Based on data from IDG News Service)

a) Find the exponential growth rate and the exponential growth function.

b) Assuming exponential growth, estimate how long it took the MyDoom worm to infect 9000 computers.

Solution

a) We use $N(t) = N_0e^{kt}$, where t is the number of hours since the first 100 computers were infected. Substituting 100 for N_0 gives

$$N(t) = 100e^{kt}.$$

To find the exponential growth rate, k, note that after 24 hr, 100,000 computers had been infected:

$$\left. \begin{array}{l} N(24) = 100e^{k \cdot 24} \\ 100{,}000 = 100e^{24k} \end{array} \right\} \quad \text{Substituting}$$

$$1000 = e^{24k} \qquad \text{Dividing both sides by 100}$$
$$\ln 1000 = \ln e^{24k} \qquad \text{Taking the natural logarithm on both sides}$$
$$\ln 1000 = 24k \qquad \ln e^a = a$$
$$\frac{\ln 1000}{24} = k \qquad \text{Dividing both sides by 24}$$
$$0.288 \approx k. \qquad \text{Using a calculator and rounding}$$

The exponential growth rate is 28.8% and the exponential growth function is given by $N(t) = 100e^{0.288t}$.

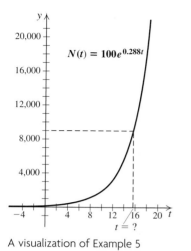

A visualization of Example 5

b) To estimate how long it took for 9000 computers to be infected, we replace $N(t)$ with 9000 and solve for t:

$$9000 = 100e^{0.288t}$$

$$90 = e^{0.288t} \qquad \text{Dividing both sides by 100}$$

$$\ln 90 = \ln e^{0.288t} \qquad \text{Taking the natural logarithm on both sides}$$

$$\ln 90 = 0.288t \qquad \ln e^a = a$$

$$\frac{\ln 90}{0.288} = t \qquad \text{Dividing both sides by 0.288}$$

$$15.6 \approx t. \qquad \text{Using a calculator}$$

Rounding up to 16, we see that, according to this model, it took about 16 hr for 9000 computers to be infected.

EXAMPLE 6   Interest compounded continuously.   When an amount of money P_0 is invested at interest rate k, compounded *continuously*, interest is computed every "instant" and added to the original amount. The balance $P(t)$, after t years, is given by the exponential growth model

$$P(t) = P_0 e^{kt}.$$

a) Suppose that \$30,000 is invested and grows to \$44,754.75 in 5 yr. Find the exponential growth function.

b) What is the doubling time?

Solution

a) We have $P(0) = 30{,}000$. Thus the exponential growth function is

$$P(t) = 30{,}000e^{kt}, \quad \text{where } k \text{ must still be determined.}$$

Knowing that for $t = 5$ we have $P(5) = 44{,}754.75$, it is possible to solve for k:

$$44{,}754.75 = 30{,}000e^{k(5)} = 30{,}000e^{5k}$$

$$\frac{44{,}754.75}{30{,}000} = e^{5k} \qquad \text{Dividing both sides by 30,000}$$

$$1.491825 = e^{5k}$$

$$\ln 1.491825 = \ln e^{5k} \qquad \text{Taking the natural logarithm on both sides}$$

$$\ln 1.491825 = 5k \qquad \ln e^a = a$$

$$\frac{\ln 1.491825}{5} = k \qquad \text{Dividing both sides by 5}$$

$$0.08 \approx k. \qquad \text{Using a calculator and rounding}$$

The interest rate is about 0.08, or 8%, compounded continuously. Because interest is being compounded continuously, the yearly interest rate is a bit more than 8%. The exponential growth function is

$$P(t) = 30{,}000e^{0.08t}.$$

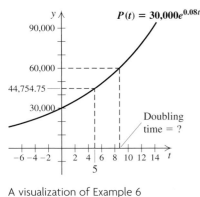

A visualization of Example 6

b) To find the doubling time T, we replace $P(T)$ with 60,000 and solve for T:

$$60,000 = 30,000e^{0.08T}$$

$2 = e^{0.08T}$	Dividing both sides by 30,000
$\ln 2 = \ln e^{0.08T}$	Taking the natural logarithm on both sides
$\ln 2 = 0.08T$	$\ln e^a = a$
$\dfrac{\ln 2}{0.08} = T$	Dividing both sides by 0.08
$8.7 \approx T.$	Using a calculator and rounding

Thus the original investment of $30,000 will double in about 8.7 yr.

For any specified interest rate, continuous compounding gives the highest yield and the shortest doubling time.

In some real-life situations, a quantity or population is *decreasing* or *decaying* exponentially.

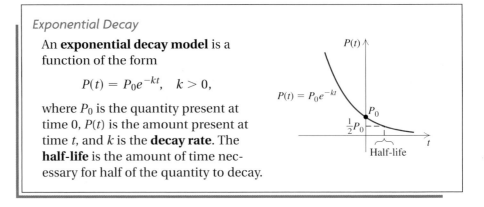

Exponential Decay

An **exponential decay model** is a function of the form

$$P(t) = P_0e^{-kt}, \quad k > 0,$$

where P_0 is the quantity present at time 0, $P(t)$ is the amount present at time t, and k is the **decay rate**. The **half-life** is the amount of time necessary for half of the quantity to decay.

EXAMPLE 7

Carbon dating. The radioactive element carbon-14 has a half-life of 5750 yr. The percentage of carbon-14 present in the remains of organic matter can be used to determine the age of that organic matter. Recently, while digging in Chaco Canyon, New Mexico, archaeologists found corn pollen that had lost 38.1% of its carbon-14. The age of this corn pollen was evidence that Indians had been cultivating crops in the Southwest centuries earlier than scientists had thought. What was the age of the pollen? (*Source: American Anthropologist*)

Solution We first find k. To do so, we use the concept of half-life. When $t = 5750$ (the half-life), $P(t)$ will be half of P_0. Then

$0.5P_0 = P_0e^{-k(5750)}$	Substituting in $P(t) = P_0e^{-kt}$
$0.5 = e^{-5750k}$	Dividing both sides by P_0
$\ln 0.5 = \ln e^{-5750k}$	Taking the natural logarithm on both sides
$\ln 0.5 = -5750k$	$\ln e^a = a$
$\dfrac{\ln 0.5}{-5750} = k$	Dividing
$0.00012 \approx k.$	Using a calculator and rounding

Chaco Canyon, New Mexico

Now we have a function for the decay of carbon-14:

$$P(t) = P_0 e^{-0.00012t}. \longleftarrow \text{This completes the first part of our solution.}$$

(*Note*: This equation can be used for any subsequent carbon-dating problem.) If the corn pollen has lost 38.1% of its carbon-14 from an initial amount P_0, then $100\% - 38.1\%$, or 61.9%, of P_0 is still present. To find the age t of the pollen, we solve this equation for t:

$0.619 P_0 = P_0 e^{-0.00012t}$	We want to find t for which $P(t) = 0.619 P_0$.
$0.619 = e^{-0.00012t}$	Dividing both sides by P_0
$\ln 0.619 = \ln e^{-0.00012t}$	Taking the natural logarithm on both sides
$\ln 0.619 = -0.00012t$	$\ln e^a = a$
$\dfrac{\ln 0.619}{-0.00012} = t$	Dividing
$4000 \approx t.$	Using a calculator

The pollen is about 4000 yr old.

Exercise Set

9.7

FOR EXTRA HELP

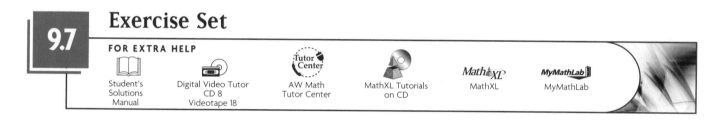

Student's Solutions Manual   Digital Video Tutor CD 8 Videotape 18   Tutor Center AW Math Tutor Center   MathXL Tutorials on CD   Math XL MathXL   MyMathLab MyMathLab

Solve.

1. *DVD players.*   Yearly sales of DVD players $S(t)$, in millions of dollars, t years after 1997 can be estimated by

$$S(t) = 200 \cdot 2^t.$$

Source: *Statistical Abstract of the United States*, 2003

a) Determine the year in which sales of DVD players first reached $2800 million.

b) What is the doubling time for the yearly sales of DVD players?

2. *Cell phones.*   The number of cell phones in use in the United States, in millions, t years after 1990 can be estimated by

$$N(t) = 7.12(1.3)^t.$$

Source: Cellular Telecommunications and Internet Association

a) In what year did the number of cell phones in use first reach 200 million?

b) What is the doubling time for the number of cell phones in use?

3. *Skateboarding.*   The number of skateboarders of age x, in thousands, can be approximated by

$$N(x) = 1337(0.9)^x, \quad 7 \le x \le 60.$$

Sources: Based on figures from the National Sporting Goods Association and *Statistical Abstract of the United States*, 2003

a) Estimate the number of 21-year-old skateboarders.

b) At what age are there only 6300 skateboarders?

4. *Recycling aluminum cans.* Approximately one-half of all aluminum cans distributed will be recycled each year. A beverage company distributes 250,000 cans. The number still in use after t years is given by the function

$$N(t) = 250,000\left(\tfrac{1}{2}\right)^t.$$

Source: The Aluminum Association, Inc., May 2004

a) After how many years will 60,000 cans still be in use?

b) After what amount of time will only 1000 cans still be in use?

5. *Student loan repayment.* A college loan of $29,000 is made at 3% interest, compounded annually. After t years, the amount due, A, is given by the function

$$A(t) = 29,000(1.03)^t.$$

a) After what amount of time will the amount due reach $35,000?

b) Find the doubling time.

6. *Spread of a rumor.* The number of people who have heard a rumor increases exponentially. If all who hear a rumor repeat it to two people a day, and if 20 people start the rumor, the number of people N who have heard the rumor after t days is given by

$$N(t) = 20(3)^t.$$

a) After what amount of time will 1000 people have heard the rumor?

b) What is the doubling time for the number of people who have heard the rumor?

7. *Telephone lines.* As more Americans make cell phones their *only* phones, the percentage of phone lines that are land lines has been shrinking.

The percentage of U.S. phone lines that are land lines $P(t)$, in use t years after 2000, can be estimated by

$$P(t) = 63.03(0.95)^t.$$

Sources: Based on data from Federal Communications Commission; Cellular Telecommunications and Internet Association

a) In what year did/will the percentage of phones that are land lines drop below 50%?

b) In what year will the percentage of phones that are land lines drop below 25%?

8. *Smoking.* The percentage of smokers who received telephone counseling and had successfully quit smoking for t months is given by

$$P(t) = 21.4(0.914)^t.$$

Sources: *New England Journal of Medicine*; data from California's Smoker's Hotline

a) In what month will 15% of those who quit and used telephone counseling still be smoke-free?

b) In what month will 5% of those who quit and used phone counseling still be smoke-free?

9. *Marine biology.* As a result of preservation efforts in countries in which whaling was once common, the humpback whale population has grown since the 1970s. The worldwide population $P(t)$, in thousands, t years after 1982 can be estimated by

$$P(t) = 5.5(1.047)^t.$$

a) In what year will the humpback whale population reach 30,000?

b) Find the doubling time.

10. *World population.* The world population $P(t)$, in billions, t years after 1980 can be approximated by

$$P(t) = 4.495(1.015)^t.$$

Sources: Based on data from U.S. Bureau of the Census; International Data Base

a) In what year will the world population reach 7 billion?

b) Find the doubling time.

Use the pH formula given in Example 2 for Exercises 11–14.

11. *Chemistry.* The hydrogen ion concentration of fresh-brewed coffee is about 1.3×10^{-5} moles per liter. Find the pH.

12. *Chemistry.* The hydrogen ion concentration of milk is about 1.6×10^{-7} moles per liter. Find the pH.

13. *Medicine.* When the pH of a patient's blood drops below 7.4, a condition called *acidosis* sets in. Acidosis can be deadly when the patient's pH reaches 7.0. What would the hydrogen ion concentration of the patient's blood be at that point?

14. *Medicine.* When the pH of a patient's blood rises above 7.4, a condition called *alkalosis* sets in. Alkalosis can be deadly when the patient's pH reaches 7.8. What would the hydrogen ion concentration of the patient's blood be at that point?

Use the decibel formula given in Example 1 for Exercises 15–18.

15. *Audiology.* The intensity of sound in normal conversation is about $3.2 \times 10^{-6} \, \text{W/m}^2$. How loud in decibels is this sound level?

16. *Audiology.* The intensity of a riveter at work is about $3.2 \times 10^{-3} \, \text{W/m}^2$. How loud in decibels is this sound level?

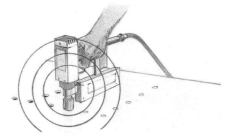

17. *Music.* The band U2 recently performed and sound measurements of 105 dB were recorded. What is the intensity of such sounds?

18. *Music.* The band Strange Folk performed in Burlington, VT, and reached sound levels of 111 dB. What is the intensity of such sounds?
Source: Melissa Garrido, *Burlington Free Press*

Use the compound-interest formula in Example 6 for Exercises 19 and 20.

19. *Interest compounded continuously.* Suppose that P_0 is invested in a savings account where interest is compounded continuously at 2.5% per year.
a) Express $P(t)$ in terms of P_0 and 0.025.
b) Suppose that $5000 is invested. What is the balance after 1 yr? after 2 yr?
c) When will an investment of $5000 double itself?

20. *Interest compounded continuously.* Suppose that P_0 is invested in a savings account where interest is compounded continuously at 3.1% per year.
a) Express $P(t)$ in terms of P_0 and 0.031.
b) Suppose that $1000 is invested. What is the balance after 1 yr? after 2 yr?
c) When will an investment of $1000 double itself?

21. *Population growth.* In 2004, the population of the United States was 292.80 million and the exponential growth rate was 0.9% per year.
Source: U.S. Bureau of the Census
a) Find the exponential growth function.
b) Predict the U.S. population in 2005.
c) When will the U.S. population reach 325 million?

22. *World population growth.* In 2004, the world population was 6.3 billion and the exponential growth rate was 1.1% per year.
Source: U.S. Bureau of the Census
a) Find the exponential growth function.
b) Predict the world population in 2009.
c) When will the world population be 8.0 billion?

23. *iPod sales.* The number of iPods sold since January 1, 2003, has grown at an exponential growth rate of 10.3% per month. What is the doubling time for iPod sales?
Source: Based on data from iPodlounge.com

24. *Population growth.* The exponential growth rate of the population of Saudi Arabia is 3.3% per year (one of the highest in the world). What is the doubling time?
Sources: Based on data from U.S. Bureau of the Census; International Data Base 2002

25. *World population.* The function
$$Y(x) = 67.17 \ln \frac{x}{4.5}$$
can be used to estimate the number of years $Y(x)$ after 1980 required for the world population to reach x billion people.
Sources: Based on data from U.S. Bureau of the Census; International Data Base
a) In what year will the world population reach 7 billion?
b) In what year will the world population reach 8 billion?
c) Graph the function.

26. *Marine biology.* The function

$$Y(x) = 21.77 \ln \frac{x}{5.5}$$

can be used to estimate the number of years $Y(x)$ after 1982 required for the world's humpback whale population to reach x thousand whales.
a) In what year will the whale population reach 15,000?
b) In what year will the whale population reach 25,000?
c) Graph the function.

27. *Forgetting.* Students in an English class took a final exam. They took equivalent forms of the exam at monthly intervals thereafter. The average score $S(t)$, in percent, after t months was found to be given by

$$S(t) = 68 - 20 \log (t + 1), \quad t \geq 0.$$

a) What was the average score when they initially took the test, $t = 0$?
b) What was the average score after 4 months? after 24 months?
c) Graph the function.
d) After what time t was the average score 50%?

28. *Advertising.* A model for advertising response is given by

$$N(a) = 2000 + 500 \log a, \quad a \geq 1,$$

where $N(a)$ is the number of units sold and a is the amount spent on advertising, in thousands of dollars.
a) How many units were sold after spending $1000 ($a = 1$) on advertising?
b) How many units were sold after spending $8000?
c) Graph the function.
d) How much would have to be spent in order to sell 5000 units?

29. *Sexually transmitted disease.* Suppose that in 2000 an outbreak of Herpes infected 17 people at a large university, and that by 2001 the number of those infected had grown to 29.
a) Find an exponential growth function that fits the data.
b) Predict the number of people who will be infected in 2006.

30. *iPod sales.* Sales of iPods have grown exponentially since January 1, 2003, at which time 656,000 units had been sold. Approximately 16 months later, in May 2004, the 3-millionth unit was sold.
Source: Based on data from iPodlounge.com
a) Find an exponential growth function that fits the data.
b) Predict the number of units sold as of April 2005.

31. *Decline in farmland.* The number of acres of farmland in the United States has decreased from 987 million acres in 1990 to 941 million acres in 2002. Assume the number of acres of farmland is decreasing exponentially.
Source: Statistical Abstract of the United States, 2003
a) Find the value k, and write an equation for an exponential function that can predict the number of acres of U.S. farmland t years after 1990.
b) Predict the number of acres of farmland in 2008.
c) In what year (theoretically) will there be only 800 million acres of U.S. farmland remaining?

32. *Decline in cases of mumps.* The number of cases of mumps has dropped exponentially from 900 in 1995 to 300 in 2001.
Source: U.S. Centers for Disease Control and Prevention
a) Find the value k, and write an exponential function that can be used to estimate the number of cases t years after 1995.
b) Estimate the number of cases of mumps in 2008.
c) In what year (theoretically) will there be only 1 case of mumps?

33. *Archaeology.* When archaeologists found the Dead Sea Scrolls, they determined that the linen wrapping had lost 22.3% of its carbon-14. How old is the linen wrapping? (See Example 7.)

34. *Archaeology.* In 1996, researchers found an ivory tusk that had lost 18% of its carbon-14. How old was the tusk? (See Example 7.)

35. *Chemistry.* The exponential decay rate of iodine-131 is 9.6% per day. What is its half-life?

36. *Chemistry.* The decay rate of krypton-85 is 6.3% per year. What is its half-life?

37. *Home construction.* The chemical urea formaldehyde was found in some insulation used in houses built during the mid to late 1960s. Unknown at the time was the fact that urea formaldehyde emitted toxic fumes as it decayed. The half-life of urea formaldehyde is 1 yr. What is its decay rate?

38. *Plumbing.* Lead pipes and solder are often found in older buildings. Unfortunately, as lead decays, toxic chemicals can get in the water resting in the pipes. The half-life of lead is 22 yr. What is its decay rate?

39. *Value of a sports card.* Legend has it that because he objected to smoking, and because his first baseball card was issued in cigarette packs, the great shortstop Honus Wagner halted production of his card before many were produced. One of these cards was purchased in 1991 by hockey great Wayne Gretzky (and a partner) for $451,000. The same card was sold in 2000 for $1.1 million. For the following questions, assume that the card's value increases exponentially, as it has for many years.

WAGNER. PITTSBURG

a) Find the exponential growth rate k, and determine an exponential function V that can be used to estimate the dollar value, $V(t)$, of the card t years after 1991.
b) Predict the value of the card in 2006.
c) What is the doubling time for the value of the card?
d) In what year will the value of the card first exceed $3,000,000?

40. *Art masterpieces.* As of August 2004, the most ever paid for a painting is $104,168,000, paid in 2004 for Pablo Picasso's "Garçon à la Pipe." The same painting sold for $30,000 in 1950.
Source: BBC News, 5/6/04

a) Find the exponential growth rate k, and determine the exponential growth function V, for which $V(t)$ is the painting's value, in millions of dollars, t years after 1950.
b) Estimate the value of the painting in 2009.
c) What is the doubling time for the value of the painting?
d) How long after 1950 will the value of the painting be $1 billion?

41. Write a problem for a classmate to solve in which information is provided and the classmate is asked to find an exponential growth function. Make the problem as realistic as possible.

42. Examine the restriction on t in Exercise 27.
a) What upper limit might be placed on t?
b) In practice, would this upper limit ever be enforced? Why or why not?

SKILL MAINTENANCE

Graph. [8.7]

43. $y = x^2 - 8x$ **44.** $y = x^2 - 5x - 6$

45. $f(x) = 3x^2 - 5x - 1$ **46.** $g(x) = 2x^2 - 6x + 3$

Solve by completing the square. [8.1]

47. $x^2 - 8x = 7$ **48.** $x^2 + 10x = 6$

SYNTHESIS

49. Will the model used to predict the number of DVD players in Exercise 1 still be realistic in 2020? Why or why not?

50. *Atmospheric pressure.* Atmospheric pressure P at altitude a is given by

$$P = P_0 e^{-0.00005a},$$

where P_0 is the pressure at sea level $\approx 14.7 \ \text{lb/in}^2$ (pounds per square inch). Explain how a barometer, or some other device for measuring atmospheric pressure, can be used to find the height of a skyscraper.

51. *Sports salaries.* As of November 2004, Alex Rodriguez of the New York Yankees has the largest contract in sports history. As part of the 10-year $252-million deal, he will receive $24 million in 2010 (part from the Yankees and part from his former team, the Texas Rangers). How much money would need to be invested in 2004, at 4% interest compounded continuously, in order to have $24 million for Rodriguez in 2010? (This is much like finding what $24 million in 2010 is worth in 2004 dollars.)
Source: *The San Francisco Chronicle*

52. *Supply and demand.* The supply and demand for the sale of stereos by Sound Ideas are given by

$$S(x) = e^x \quad \text{and} \quad D(x) = 162{,}755e^{-x},$$

where $S(x)$ is the price at which the company is willing to supply x stereos and $D(x)$ is the demand price for a quantity of x stereos. Find the equilibrium point. (For reference, see Section 3.8.)

53. Use Exercise 7 to form a model for the percentage of U.S. phone lines that are cellular t years after 2000.

54. Use the model developed in Exercise 53 to predict the percentage of U.S. phone lines that will be cellular in 2020. Does your prediction seem plausible? Why or why not?

55. *Nuclear energy.* Plutonium-239 (Pu-239) is used in nuclear energy plants. The half-life of Pu-239 is 24,360 yr. How long will it take for a fuel rod of Pu-239 to lose 90% of its radioactivity?
Source: *Microsoft Encarta 97 Encyclopedia*

56. *Growth of bacteria.* The bacteria *Escherichia coli* (*E. coli*) are commonly found in the human bladder. Suppose that 3000 of the bacteria are present at time $t = 0$. Then t minutes later, the number of bacteria present is

$$N(t) = 3000(2)^{t/20}.$$

If 100,000,000 bacteria accumulate, a bladder infection can occur. If, at 11:00 A.M., a patient's bladder contains 25,000 *E. coli* bacteria, at what time can infection occur?

57. Show that for exponential growth at rate k, the doubling time T is given by $T = \dfrac{\ln 2}{k}$.

58. Show that for exponential decay at rate k, the half-life T is given by $T = \dfrac{\ln 2}{k}$.

59. *Heart transplants.* In 1967, Dr. Christiaan Barnard of South Africa stunned the world by performing the first heart transplant. Since that time, the operation's popularity has both grown and declined, as shown in the table below.

Year	Number of Heart Transplants Worldwide*
1982	189
1983	318
1984	669
1985	1189
1986	2167
1987	2720
1988	3157
1989	3378
1990	4016
1991	4186
1992	4199
1993	4346
1994	4402
1995	4314
1996	4128
1997	4039
1998	3744
1999	3419
2000	3246
2001	3122
2002	3265
2003	3020

Source: International Society for Heart & Lung Transplantation

a) Using 1982 as $t = 0$, graph the data.
b) Does it appear that an exponential function might have ever served as an appropriate model for these data? If so, for what years would this have been the case?
c) Considering *all* of the data points on your graph, which would be the most appropriate model: a linear, a quadratic, or an exponential function? Why?

CORNER

Investments in Collectibles

Focus: Exponential-growth models

Time: 30 minutes

Group size: 7

Collecting comic books has long been a popular hobby. It can also be quite profitable since many comic books that originally sold for less than $1 are now worth hundreds or, in some cases, thousands of dollars.

Collectors often estimate the future value of their collections by examining the value's growth in the past. As with many collectibles, the value of comic books often grows exponentially.

ACTIVITY

1. Suppose that in 2005, each group member invests in the comic books listed in the following table. Looking only at the approximate value in 2005, each student should select $1200 worth of comic books. More than one comic of each type can be selected. The comics chosen will become that student's portfolio.

2. Each group member should select a different one of the comic books. That person should then form an exponential growth model for the value of that comic book using the year of issue and the original cost of the comic book.

3. Using the models developed above, each group member should predict the value of his or her portfolio in 2009. Compare the values. Why, when buying a collector's item, is it important to consider its previous worth?

4. Look at the predicted value of each comic book in 2009. Does an exponential growth model seem appropriate? If possible, find the current value of some of the comic books and compare those values with the predicted values.

Comic Book	Approximate Value in 2005	Year of Issue	Original Cost
Bugs Bunny #28	$ 70	1952	15¢
Josie and the Pussycats #45	140	1969	15¢
Amazing Spider-Man #121	250	1973	25¢
Men In Black #1	45	1990	$2.25
Sonic the Hedgehog #1	35	1993	1.25
G. I. Joe #155	35	1994	1.75
Ultimate Spider-Man #1	90	2000	2.99

Source: Prices based on data kindly provided by Tim Reynolds at Comic Carnival, Indianapolis, IN.

9 Study Summary

Chapter 9 focuses on two special types of functions: **exponential** and **logarithmic** (pp. 596, 605). To understand how these functions relate to each other, it is important to first understand what **composite functions** are and how function **composition** is performed (pp. 584, 585).

The composition of f and g, or $f \circ g$: $(f \circ g)(x) = f(g(x))$

Thus

> if $f(x) = 2x + 1$ and $g(x) = x^2$,
> then $(f \circ g)(3) = f(g(3)) = f(9) = 2 \cdot 9 + 1 = 19.$

Composition of functions is developed so we can discuss **inverse functions** (p. 587). To find a function f^{-1}, which is the inverse of f, f must be **one-to-one**, meaning that no two members of f's domain are paired with the same member of the range (p. 588). One quick test for a one-to-one function is the **horizontal-line test**:

A function f is one-to-one if it is not possible to draw a horizontal line that crosses the graph of f at more than one point (p. 588):

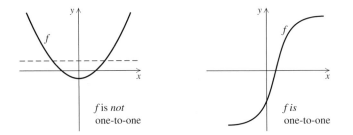

If f is one-to-one, it is then possible to find its inverse:

1. Replace $f(x)$ with y.
2. Interchange x and y.
3. Solve for y.
4. Replace y with $f^{-1}(x)$.

For any two inverse functions f and f^{-1},

$$(f^{-1} \circ f)(x) = (f \circ f^{-1})(x) = x,$$

or, equivalently,

$$f^{-1}(f(x)) = x \quad \text{and} \quad f(f^{-1}(x)) = x.$$

Probably the most important pair of inverse functions in mathematics are the exponential and logarithmic functions:

f is an exponential function if $f(x) = a^x$ for $a > 0$, $a \neq 1$.

g is a logarithmic function if $g(x) = \log_a x$ for $a > 0$, $a \neq 1$.

Note that $y = \log_a x$ means $a^y = x$.

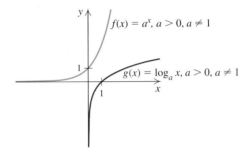

Common (base-10) and **natural** (base-e) logarithms are the most frequently used types of logarithms (pp. 619, 620). Note that $e \approx 2.7$.

Exponential and logarithmic expressions frequently appear in equations. To solve such **exponential** or **logarithmic equations**, we must use certain principles and properties:

Properties of Logarithms

$\log_a (MN) = \log_a M + \log_a N$: $\log_7 10 = \log_7 5 + \log_7 2$;

$\log_a \dfrac{M}{N} = \log_a M - \log_a N$: $\log_5 \dfrac{14}{3} = \log_5 14 - \log_5 3$;

$\log_a M^p = p \cdot \log_a M$: $\log_8 5^{12} = 12 \log_8 5$;

$\log_a 1 = 0$: $\log_9 1 = 0$;

$\log_a a = 1$: $\log_4 4 = 1$;

$\log_a a^k = k$: $\log_3 3^8 = 8$;

$\log M = \log_{10} M$: $\log 43 = \log_{10} 43$;

$\ln M = \log_e M$: $\ln 37 = \log_e 37$;

$\log_b M = \dfrac{\log_a M}{\log_a b}$: $\log_6 31 = \dfrac{\log 31}{\log 6}$

The Principle of Exponential Equality

For any real number b, $b \neq -1, 0$, or 1: $b^x = b^y$ is equivalent to $x = y$.

The Principle of Logarithmic Equality

For any logarithmic base a, and for $x, y > 0$: $x = y$ is equivalent to $\log_a x = \log_a y$.

Thus,

$$25 = 5^x \quad \text{is equivalent to} \quad 5^2 = 5^x, \text{ or } 2 = x,$$

and

$$7^x = 83 \quad \text{is equivalent to} \quad \log 7^x = \log 83, \text{ or } x \log 7 = \log 83, \text{ or } x = \frac{\log 83}{\log 7}.$$

To solve an equation of the form $a^t = b$ for t:

1. Take the logarithm (either natural or common) of both sides.
2. Use the power rule for exponents so that the variable is no longer written as an exponent.
3. Divide both sides by the coefficient of the variable to isolate the variable.
4. If appropriate, use a calculator to find an approximate solution in decimal form.

Exponential and logarithmic equations arise in a wide variety of applications.

Exponential Growth

An **exponential growth model** is a function of the form

$$P(t) = P_0 e^{kt}, \quad k > 0,$$

where P_0 is the population at time 0, $P(t)$ is the population at time t, and k is the **exponential growth rate** for the situation. The **doubling time** is the amount of time necessary for the population to double in size.

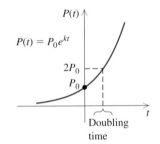

Exponential Decay

An **exponential decay model** is a function of the form

$$P(t) = P_0 e^{-kt}, \quad k > 0,$$

where P_0 is the quantity present at time 0, $P(t)$ is the amount present at time t, and k is the **decay rate**. The **half-life** is the amount of time necessary for half of the quantity to decay.

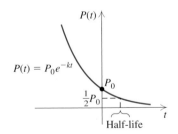

9 Review Exercises

↩ *Concept Reinforcement* *In each of Exercises 1–10, classify the statement as either true or false.*

1. The functions given by $f(x) = e^x$ and $g(x) = \ln x$ are inverses of each other. [9.3]

2. A function's doubling time is the amount of time t for which $f(t) = 2f(0)$. [9.7]

3. A radioactive isotope's half-life is the amount of time t for which $f(t) = \frac{1}{2} f(0)$. [9.7]

4. $\ln(ab) = \ln a - \ln b$ [9.4]

5. $\log x^a = x \ln a$ [9.4]

6. $\log_a \dfrac{m}{n} = \log_a m - \log_a n$ [9.4]

7. For $f(x) = 3^x$, the domain of f is $[0, \infty)$. [9.2]

8. For $g(x) = \log_2 x$, the domain of g is $[0, \infty)$. [9.3]

9. The function F is not one-to-one if $F(-2) = F(5)$. [9.1]

10. The function g is one-to-one if it passes the vertical-line test. [9.1]

11. Find $(f \circ g)(x)$ and $(g \circ f)(x)$ if $f(x) = x^2 + 1$ and $g(x) = 2x - 3$. [9.1]

12. If $h(x) = \sqrt{3 - x}$, find $f(x)$ and $g(x)$ such that $h(x) = (f \circ g)(x)$. Answers may vary. [9.1]

13. Determine whether $f(x) = 4 - x^2$ is one-to-one. [9.1]

Find a formula for the inverse of each function. [9.1]

14. $f(x) = x - 8$

15. $g(x) = \dfrac{3x + 1}{2}$

16. $f(x) = 27x^3$

Graph.

17. $f(x) = 3^x + 1$ [9.2]

18. $x = \left(\frac{1}{4}\right)^y$ [9.2]

19. $y = \log_5 x$ [9.3]

Simplify. [9.3]

20. $\log_3 9$

21. $\log_{10} \frac{1}{100}$

22. $\log_5 5^7$

23. $\log_9 3$

Rewrite as an equivalent logarithmic equation. [9.3]

24. $10^{-2} = \frac{1}{100}$

25. $25^{1/2} = 5$

Rewrite as an equivalent exponential equation. [9.3]

26. $\log_4 16 = x$

27. $\log_8 1 = 0$

Express as an equivalent expression using the individual logarithms of x, y, and z. [9.4]

28. $\log_a x^4 y^2 z^3$

29. $\log_a \dfrac{x^5}{yz^2}$

30. $\log \sqrt[4]{\dfrac{z^2}{x^3 y}}$

Express as an equivalent expression that is a single logarithm and, if possible, simplify. [9.4]

31. $\log_a 7 + \log_a 8$

32. $\log_a 72 - \log_a 12$

33. $\frac{1}{2} \log a - \log b - 2 \log c$

34. $\frac{1}{3}[\log_a x - 2 \log_a y]$

Simplify. [9.4]

35. $\log_m m$

36. $\log_m 1$

37. $\log_m m^{17}$

Given $\log_a 2 = 1.8301$ and $\log_a 7 = 5.0999$, find each of the following. [9.4]

38. $\log_a 14$

39. $\log_a \frac{2}{7}$

40. $\log_a 28$

41. $\log_a 3.5$

42. $\log_a \sqrt{7}$

43. $\log_a \frac{1}{4}$

▦ *Use a calculator to find each of the following to the nearest ten-thousandth.* [9.5]

44. $\log 75$

45. $10^{1.789}$

46. $\ln 0.05$

47. $e^{-0.98}$

▦ *Find each of the following logarithms using the change-of-base formula. Round answers to the nearest ten-thousandth.* [9.5]

48. $\log_5 2$

49. $\log_{12} 70$

Graph and state the domain and the range of each function. [9.5]

50. $f(x) = e^x - 1$

51. $g(x) = 0.6 \ln x$

Solve. Where appropriate, include approximations to the nearest ten-thousandth. [9.6]

52. $2^x = 32$

53. $3^x = \frac{1}{9}$

54. $\log_3 x = -4$

55. $\log_x 16 = 4$

56. $\log x = -3$

57. $3 \ln x = -6$

58. $4^{2x-5} = 19$

59. $2^{x^2} \cdot 2^{4x} = 32$

60. $4^x = 8.3$

61. $e^{-0.1t} = 0.03$

62. $2 \ln x = -6$

63. $\log_3 (2x - 5) = 1$

64. $\log_4 x + \log_4 (x - 6) = 2$

65. $\log x + \log (x - 15) = 2$

66. $\log_3 (x - 4) = 3 - \log_3 (x + 4)$

67. In a business class, students were tested at the end of the course with a final exam. They were then tested again 6 months later. The forgetting formula was determined to be

$$S(t) = 82 - 18 \log (t + 1),$$

where t is the time, in months, after taking the final exam. [9.7]

a) Determine the average score when they first took the exam (when $t = 0$).

▦ **b)** What was the average score after 6 months?

▦ **c)** After what time was the average score 54?

▦ **68.** A color photocopier is purchased for $5200. Its value each year is about 80% of its value in the preceding year. Its value in dollars after t years is given by the exponential function

$$V(t) = 5200(0.8)^t. \ [9.7]$$

a) After what amount of time will the copier's value be $1200?

b) After what amount of time will the copier's value be half the original value?

▦ **69.** In 1999, Lucille invested in a building lot that cost $10,000. By 2004, fair-market value for the lot was $19,000. Assume that the value of the lot is increasing exponentially. [9.7]

a) Find the value k, and write an exponential function that describes the value of Lucille's lot t years after 1999.

b) Predict the value of the lot in 2009.

c) In what year will the value of the lot first reach $35,000?

70. The value of Jose's stock market portfolio doubled in 3 yr. What was the exponential growth rate? [9.7]

▦ **71.** How long will it take $7600 to double itself if it is invested at 4.2%, compounded continuously? [9.7]

▦ **72.** How old is a skull that has lost 34% of its carbon-14? (Use $P(t) = P_0 e^{-0.00012t}$.) [9.7]

73. What is the pH of a substance if its hydrogen ion concentration is 2.3×10^{-7} moles per liter? (Use pH $= -\log [H^+]$.) [9.7]

74. The intensity of the sound of water at the foot of the Niagara Falls is about 10^{-3} W/m². * How loud in decibels is this sound level? [9.7]

$$\left(\text{Use } L = 10 \cdot \log \frac{I}{10^{-12}}. \right)$$

SYNTHESIS

▯ **75.** Explain why negative numbers do not have logarithms. [9.3]

▯ **76.** Explain why taking the natural or common logarithm on each side of an equation produces an equivalent equation. [9.6]

Solve. [9.6]

77. $\ln (\ln x) = 3$

78. $2^{x^2+4x} = \frac{1}{8}$

79. Solve the system:
$$5^{x+y} = 25,$$
$$2^{2x-y} = 64. \ [9.6]$$

Sound and Hearing, Life Science Library. (New York: Time Incorporated, 1965), p. 173.

9 Chapter Test

1. Find $(f \circ g)(x)$ and $(g \circ f)(x)$ if $f(x) = x + x^2$ and $g(x) = 2x + 1$.

2. If
$$h(x) = \frac{1}{2x^2 + 1},$$
find $f(x)$ and $g(x)$ such that $h(x) = (f \circ g)(x)$. Answers may vary.

3. Determine whether $f(x) = |x - 3|$ is one-to-one.

Find a formula for the inverse of each function.

4. $f(x) = 3x + 4$

5. $g(x) = (x + 1)^3$

Graph.

6. $f(x) = 2^x - 3$

7. $g(x) = \log_7 x$

Simplify.

8. $\log_5 125$

9. $\log_{100} 10$

10. $3^{\log_3 18}$

Rewrite as an equivalent logarithmic equation.

11. $4^{-3} = \frac{1}{64}$

12. $256^{1/2} = 16$

Rewrite as an equivalent exponential equation.

13. $m = \log_7 49$

14. $\log_3 81 = 4$

15. Express as an equivalent expression using the individual logarithms of a, b, and c:
$$\log \frac{a^3 b^{1/2}}{c^2}.$$

16. Express as an equivalent expression that is a single logarithm:
$$\tfrac{1}{3} \log_a x + 2 \log_a z.$$

Simplify.

17. $\log_p p$

18. $\log_t t^{23}$

19. $\log_c 1$

Given $\log_a 2 = 0.301$, $\log_a 6 = 0.778$, and $\log_a 7 = 0.845$, find each of the following.

20. $\log_a 14$

21. $\log_a 3$

22. $\log_a 16$

Use a calculator to find each of the following to the nearest ten-thousandth.

23. $\log 12.3$

24. $10^{-0.8}$

25. $\ln 0.035$

26. $e^{4.8}$

27. Find $\log_3 14$ using the change-of-base formula. Round to the nearest ten-thousandth.

Graph and state the domain and the range of each function.

28. $f(x) = e^x + 3$

29. $g(x) = \ln (x - 4)$

Solve. Where appropriate, include approximations to the nearest ten-thousandth.

30. $2^x = \frac{1}{32}$

31. $\log_x 25 = 2$

32. $\log_4 x = \frac{1}{2}$

33. $\log x = 4$

34. $5^{4-3x} = 87$

35. $7^x = 1.2$

36. $\ln x = \frac{1}{4}$

37. $\log (x - 3) + \log (x + 1) = \log 5$

38. The average walking speed R of people living in a city of population P, in thousands, is given by $R = 0.37 \ln P + 0.05$, where R is in feet per second.

   a) The population of Tulsa, Oklahoma, is 878,000. Find the average walking speed.
   b) Baton Rouge, Louisiana, has an average walking speed of about 2.48 ft/sec. Find the population.

39. The population of Nigeria was about 130.5 million in 2002, and the exponential growth rate was 2.4% per year.

   a) Write an exponential function describing the population of Nigeria.
   b) What will the population be in 2007? in 2012?
   c) When will the population be 175 million?
   d) What is the doubling time?

40. The average cost of a year at a private four-year college grew exponentially from $19,070 in 1997 to $22,968 in 2003.

Source: National Center for Education Statistics, *Digest of Education Statistics, 2002*

   a) Find the value k, and write an exponential function that approximates the cost of a year of college t years after 1997.

   b) Predict the cost of a year of college in 2010.

   c) In what year will the average cost of college be $50,000?

41. An investment with interest compounded continuously doubled itself in 15 yr. What is the interest rate?

42. How old is an animal bone that has lost 43% of its carbon-14? (Use $P(t) = P_0 e^{-0.00012t}$.)

43. The sound of traffic at a busy intersection averages 75 dB. What is the intensity of such a sound?

$$\left(\text{Use } L = 10 \cdot \log \frac{I}{I_0}. \right)$$

44. The hydrogen ion concentration of water is 1.0×10^{-7} moles per liter. What is the pH? (Use $\text{pH} = -\log [\text{H}^+]$.)

SYNTHESIS

45. Solve: $\log_5 |2x - 7| = 4$.

46. If $\log_a x = 2$, $\log_a y = 3$, and $\log_a z = 4$, find

$$\log_a \frac{\sqrt[3]{x^2 z}}{\sqrt[3]{y^2 z^{-1}}}.$$

1–9 Cumulative Review

1. Evaluate $\dfrac{x^0 + y}{-z}$ for $x = 6$, $y = 9$, and $z = -5$.
[1.1], [1.6]

Simplify.

2. $\left| -\frac{5}{2} + \left(-\frac{7}{2} \right) \right|$ [1.2]

3. $(-2x^2 y^{-3})^{-4}$ [1.6]

4. $(-5x^4 y^{-3} z^2)(-4x^2 y^2)$ [1.6]

5. $\dfrac{3x^4 y^6 z^{-2}}{-9x^4 y^2 z^3}$ [1.6]

6. $4x - 3 - 2[5 - 3(2 - x)]$ [1.3]

7. $3^3 + 2^2 - (32 \div 4 - 16 \div 8)$ [1.1]

Solve.

8. $5(2x - 3) = 9 - 5(2 - x)$ [1.3]

9. $4x - 3y = 15,$
$3x + 5y = 4$ [3.2]

10. $\quad x + y - 3z = -1,$
$\quad 2x - y + z = 4,$
$\quad -x - y + z = 1$ [3.4]

11. $x(x - 3) = 10$ [5.8]

12. $\dfrac{7}{x^2 - 5x} - \dfrac{2}{x - 5} = \dfrac{4}{x}$ [6.4]

13. $\dfrac{8}{x + 1} + \dfrac{11}{x^2 - x + 1} = \dfrac{24}{x^3 + 1}$ [6.4], [8.2]

14. $\sqrt{4 - 5x} = 2x - 1$ [7.6]

15. $\sqrt[3]{2x} = 1$ [7.6]

16. $3x^2 + 75 = 0$ [8.1]

17. $x - 8\sqrt{x} + 15 = 0$ [8.5]

18. $x^4 - 13x^2 + 36 = 0$ [8.5]

19. $\log_7 x = 1$ [9.3]

20. $\log_x 36 = 2$ [9.3]

21. $9^x = 27$ [9.6]

22. $3^{5x} = 7$ [9.6]

23. $\ln x - \ln (x - 8) = 1$ [9.6]

24. $x^2 + 4x > 5$ [8.9]

25. If $f(x) = x^2 + 6x$, find a such that $f(a) = 11$. [8.2]

26. If $f(x) = |2x - 3|$, find all x for which $f(x) \geq 7$. [4.3]

Solve.

27. $D = \dfrac{ab}{b + a}$, for a [6.8]

28. $\dfrac{1}{p} + \dfrac{1}{q} = \dfrac{1}{f}$, for q [6.8]

29. $M = \dfrac{2}{3}(A + B)$, for B [1.5]

Evaluate. [3.7]

30. $\begin{vmatrix} 6 & -5 \\ 4 & -3 \end{vmatrix}$

31. $\begin{vmatrix} 7 & -6 & 0 \\ -2 & 1 & 2 \\ -1 & 1 & -1 \end{vmatrix}$

32. Find the domain of the function f given by
$$f(x) = \frac{x + 4}{3x^2 - 5x - 2}.\ [5.8]$$

Solve.

33. *Gasoline consumption.* The number of barrels of gasoline consumed per day in the United States has increased from 7.2 million in 1990 to 8.9 million in 2003.
Source: U.S. Department of Energy, Energy Information Administration

   a) At what rate did gasoline consumption increase from 1990 to 2003? [2.3]

   b) Find a linear function g that fits the data. Let t represent the number of years since 1990. [2.5]

   c) Find an exponential function G that fits the data. [9.7]

34. The perimeter of a rectangular garden is 112 m. The length is 16 m more than the width. Find the length and the width. [1.4]

35. In triangle ABC, the measure of angle B is three times the measure of angle A. The measure of angle C is 105° greater than the measure of angle A. Find the angle measures. [1.4]

36. Good's Candies of Indiana makes all their chocolates by hand. It takes Anne 10 min to coat a tray of candies in chocolate. It takes Clay 12 min to coat a tray of candies. How long would it take Anne and Clay, working together, to coat the candies? [6.5]

37. Joe's Thick and Tasty salad dressing gets 45% of its calories from fat. The Light and Lean dressing gets 20% of its calories from fat. How many ounces of each should be mixed in order to get 15 oz of dressing that gets 30% of its calories from fat? [3.3]

38. A fishing boat with a trolling motor can move at a speed of 5 km/h in still water. The boat travels 42 km downriver in the same time that it takes to travel 12 km upriver. What is the speed of the river? [6.5]

39. What is the minimum product of two numbers whose difference is 14? What are the numbers that yield this product? [8.8]

Students in a biology class just took a final exam. A formula for determining what the average exam grade on a similar test will be t months later is
$$S(t) = 78 - 15 \log (t + 1).$$

40. The average score when the students first took the exam occurs when $t = 0$. Find the students' average score on the final exam. [9.7]

41. What would the average score be on a retest after 4 months? [9.7]

The population of Kenya was 33.7 million in 2002, and the exponential growth rate was 1.0% per year. [9.7]

42. Write an exponential function describing the growth of the population of Kenya.

43. Predict what the population will be in 2008 and in 2014.

44. What is the doubling time of the population?

45. y varies directly as the square of x and inversely as z, and $y = 2$ when $x = 5$ and $z = 100$. What is y when $x = 3$ and $z = 4$? [6.8]

Perform the indicated operations and simplify.

46. $(5p^2q^3 + 6pq - p^2 + p) + (2p^2q^3 + p^2 - 5pq - 9)$ [5.1]

47. $(11x^2 - 6x - 3) - (3x^2 + 5x - 2)$ [5.1]

48. $(3x^2 - 2y)^2$ [5.2]

49. $(5a + 3b)(2a - 3b)$ [5.2]

50. $\dfrac{x^2 + 8x + 16}{2x + 6} \div \dfrac{x^2 + 3x - 4}{x^2 - 9}$ [6.1]

51. $\dfrac{1 + \dfrac{3}{x}}{x - 1 - \dfrac{12}{x}}$ [6.3]

52. $\dfrac{a^2 - a - 6}{a^3 - 27} \cdot \dfrac{a^2 + 3a + 9}{6}$ [6.1]

53. $\dfrac{3}{x + 6} - \dfrac{2}{x^2 - 36} + \dfrac{4}{x - 6}$ [6.2]

Factor.

54. $xy + 2xz - xw$ [5.3]

55. $8 - 125x^3$ [5.6]

56. $6x^2 + 8xy - 8y^2$ [5.4]

57. $x^4 - 4x^3 + 7x - 28$ [5.3]

58. $2m^2 + 12mn + 18n^2$ [5.5]

59. $x^4 - 16y^4$ [5.5]

60. For the function described by
$$h(x) = -3x^2 + 4x + 8,$$
find $h(-2)$. [2.2]

61. Divide: $(x^4 - 5x^3 + 2x^2 - 6) \div (x - 3)$. [6.6]

62. Multiply $(5.2 \times 10^4)(3.5 \times 10^{-6})$. Write scientific notation for the answer. [1.7]

For the radical expressions that follow, assume that all variables represent positive numbers.

63. Divide and simplify:
$$\frac{\sqrt[3]{40xy^8}}{\sqrt[3]{5xy}}. \text{ [7.4]}$$

64. Multiply and simplify: $\sqrt{7xy^3} \cdot \sqrt{28x^2y}$. [7.3]

65. Write as an equivalent expression without rational exponents: $(27a^6b)^{4/3}$. [7.2]

66. Rationalize the denominator:
$$\frac{3 - \sqrt{y}}{2 - \sqrt{y}}. \text{ [7.5]}$$

67. Divide and simplify:
$$\frac{\sqrt{x + 5}}{\sqrt[5]{x + 5}}. \text{ [7.5]}$$

68. Multiply these complex numbers:
$$\left(2 - i\sqrt{3}\right)\left(6 + 2i\sqrt{3}\right). \text{ [7.8]}$$

69. Add: $(8 + 2i) + (5 - 3i)$. [7.8]

70. Find the inverse of f if $f(x) = 9 - 2x$. [9.1]

71. Find a linear function with a graph that contains the points $(0, -8)$ and $(-1, 2)$. [2.5]

72. Find an equation of the line whose graph has a y-intercept of $(0, 7)$ and is perpendicular to the line given by $2x + y = 6$. [2.5]

Graph.

73. $5x = 15 + 3y$ [2.4]

74. $y = 2x^2 - 4x - 1$ [8.7]

75. $y = \log_3 x$ [9.3]

76. $y = 3^x$ [9.2]

77. $-2x - 3y \leq 12$ [4.4]

78. Graph: $f(x) = 2(x + 3)^2 + 1$. [8.7]
   a) Label the vertex.
   b) Draw the axis of symmetry.
   c) Find the maximum or minimum value.

79. Graph $f(x) = 2e^x$ and determine the domain and the range. [9.5]

80. Express in terms of logarithms of a, b, and c:
$$\log\left(\frac{a^2c^3}{b}\right). \text{ [9.4]}$$

81. Express as a single logarithm:
$$3 \log x - \tfrac{1}{2} \log y - 2 \log z. \text{ [9.4]}$$

82. Convert to an exponential equation: $\log_a 5 = x$. [9.3]

83. Convert to a logarithmic equation: $x^3 = t$. [9.3]

Find each of the following using a calculator. Round to the nearest ten-thousandth. [9.5]

84. $\log 0.05566$ **85.** $10^{2.89}$

86. $\ln 12.78$ **87.** $e^{-1.4}$

SYNTHESIS

Solve.

88. $\dfrac{5}{3x - 3} + \dfrac{10}{3x + 6} = \dfrac{5x}{x^2 + x - 2}$ [6.4]

89. $\log \sqrt{3x} = \sqrt{\log 3x}$ [9.6]

90. A train travels 280 mi at a certain speed. If the speed had been increased by 5 mph, the trip could have been made in 1 hr less time. Find the actual speed. [8.3]

10 Conic Sections

AN APPLICATION

The spotlight on a violin soloist casts an ellipse of light on the floor below her that is 6 ft wide and 10 ft long. Find an equation of that ellipse if the performer is in its center, x is the distance from the performer to the side of the ellipse, and y is the distance from the performer to the top of the ellipse.

This problem appears as Exercise 49 in Section 10.2.

Tony Penna
LIGHTING DESIGNER
Clemson, South Carolina

As a lighting designer, I use math in almost every aspect of my work. Before the lighting instruments are loaded into the theatre, I must determine the angle at which each instrument will be focused at the stage, which requires a great deal of geometry in three dimensions. For each light I use, I must also choose which type of lighting instrument will produce the appropriate sized beam of light.

*T*he ellipse described in the chapter opener is one example of a conic section, meaning that it can be regarded as a cross section of a cone. This chapter presents a variety of applications and equations with graphs that are conic sections. We have already worked with two conic sections, lines *and* parabolas, *in* Chapters 2 and 8.

10.1   Conic Sections: Parabolas and Circles

Parabolas • The Distance and Midpoint Formulas •
Circles

This section and the next two examine curves formed by cross sections of cones. These curves are all graphs of $Ax^2 + By^2 + Cxy + Dx + Ey + F = 0$. The constants A, B, C, D, E, and F determine which of the following shapes will serve as the graph.

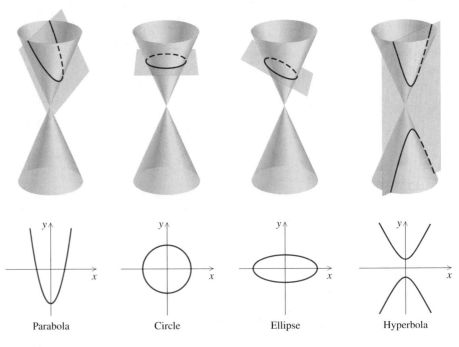

Parabola   Circle   Ellipse   Hyperbola

Parabolas

When a cone is cut as shown in the first figure above, the conic section formed is a **parabola**. Parabolas have many applications in electricity, mechanics, and optics. A cross section of a contact lens or satellite dish is a parabola, and arches that support certain bridges are parabolas.

> **Equation of a Parabola**
>
> A parabola with a vertical axis of symmetry opens upward or downward and has an equation that can be written in the form
>
> $$y = ax^2 + bx + c.$$
>
> A parabola with a horizontal axis of symmetry opens to the right or left and has an equation that can be written in the form
>
> $$x = ay^2 + by + c.$$

Parabolas with equations of the form $f(x) = ax^2 + bx + c$ were graphed in Chapter 8.

EXAMPLE 1   Graph: $y = x^2 - 4x + 9$.

Solution   To locate the vertex, we can use either of two approaches. One way is to complete the square:

$$y = (x^2 - 4x) + 9 \qquad$$ Note that half of -4 is -2, and $(-2)^2 = 4$.

$$= (x^2 - 4x + 4 - 4) + 9 \qquad$$ Adding and subtracting 4

$$= (x^2 - 4x + 4) + (-4 + 9) \qquad$$ Regrouping

$$= (x - 2)^2 + 5. \qquad$$ Factoring and simplifying

The vertex is $(2, 5)$.

A second way to find the vertex is to recall that the x-coordinate of the vertex of the parabola given by $y = ax^2 + bx + c$ is $-b/(2a)$:

$$x = -\frac{b}{2a} = -\frac{-4}{2(1)} = 2.$$

To find the y-coordinate of the vertex, we substitute 2 for x:

$$y = x^2 - 4x + 9 = 2^2 - 4(2) + 9 = 5.$$

Either way, the vertex is $(2, 5)$. Next, we calculate and plot some points on each side of the vertex. As expected for a positive coefficient of x^2, the graph opens upward.

x	y	
2	5	← Vertex
0	9	← y-intercept
1	6	
3	6	
4	9	

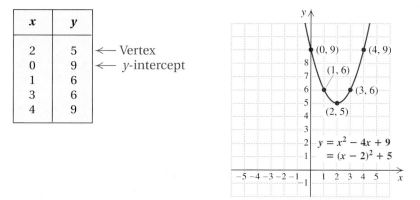

To Graph an Equation of the Form $y = ax^2 + bx + c$

1. Find the vertex (h, k) either by completing the square to find an equivalent equation

$$y = a(x - h)^2 + k,$$

or by using $-b/(2a)$ to find the x-coordinate and substituting to find the y-coordinate.
2. Choose other values for x on each side of the vertex, and compute the corresponding y-values.
3. The graph opens upward for $a > 0$ and downward for $a < 0$.

Equations of the form $x = ay^2 + by + c$ represent horizontal parabolas. These parabolas open to the right for $a > 0$, open to the left for $a < 0$, and have axes of symmetry parallel to the x-axis.

EXAMPLE 2   Graph: $x = y^2 - 4y + 9$.

Solution   This equation is like that in Example 1 but with x and y interchanged. The vertex is $(5, 2)$ instead of $(2, 5)$. To find ordered pairs, we choose values for y on each side of the vertex. Then we compute values for x. Note that the x- and y-values of the table in Example 1 are now switched. You should confirm that, by completing the square, we get $x = (y - 2)^2 + 5$.

x	y	
5	2	← Vertex
9	0	← x-intercept
6	1	
6	3	
9	4	

(1) Choose these values for y.
(2) Compute these values for x.

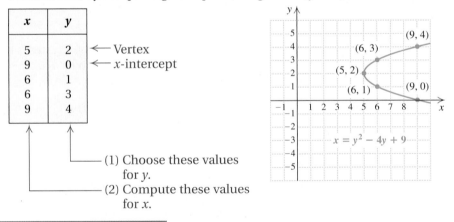

To Graph an Equation of the Form $x = ay^2 + by + c$

1. Find the vertex (h, k) either by completing the square to find an equivalent equation

$$x = a(y - k)^2 + h,$$

or by using $-b/(2a)$ to find the y-coordinate and substituting to find the x-coordinate.
2. Choose other values for y that are above and below the vertex, and compute the corresponding x-values.
3. The graph opens to the right if $a > 0$ and to the left if $a < 0$.

EXAMPLE 3   Graph: $x = -2y^2 + 10y - 7$.

Solution   We find the vertex by completing the square:

$$x = -2y^2 + 10y - 7$$
$$= -2(y^2 - 5y\qquad) - 7$$
$$= -2\left(y^2 - 5y + \tfrac{25}{4}\right) - 7 - (-2)\tfrac{25}{4}$$ $\tfrac{1}{2}(-5) = \tfrac{-5}{2}; \left(\tfrac{-5}{2}\right)^2 = \tfrac{25}{4};$ we
add and subtract $(-2)\tfrac{25}{4}$.
$$= -2\left(y - \tfrac{5}{2}\right)^2 + \tfrac{11}{2}.$$ Factoring and simplifying

The vertex is $\left(\tfrac{11}{2}, \tfrac{5}{2}\right)$.

For practice, we also find the vertex by first computing its y-coordinate, $-b/(2a)$, and then substituting to find the x-coordinate:

$$y = -\frac{b}{2a} = -\frac{10}{2(-2)} = \frac{5}{2}$$
$$x = -2y^2 + 10y - 7 = -2\left(\tfrac{5}{2}\right)^2 + 10\left(\tfrac{5}{2}\right) - 7$$
$$= \tfrac{11}{2}.$$

To find ordered pairs, we choose values for y on each side of the vertex and then compute values for x. A table is shown below, together with the graph. The graph opens to the left because the y^2-coefficient, -2, is negative.

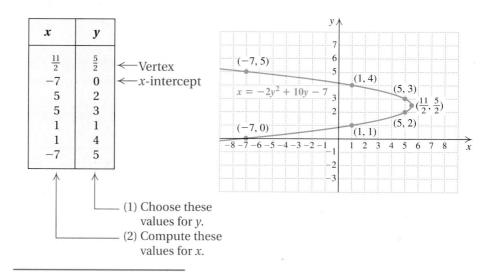

x	y	
$\tfrac{11}{2}$	$\tfrac{5}{2}$	←Vertex
-7	0	←x-intercept
5	2	
5	3	
1	1	
1	4	
-7	5	

(1) Choose these values for y.
(2) Compute these values for x.

The Distance and Midpoint Formulas

If two points are on a horizontal line, they have the same second coordinate. We can find the distance between them by subtracting their first coordinates. This difference may be negative, depending on the order in which we subtract. So, to make sure we get a positive number, we take the absolute value of this difference. The distance between the points (x_1, y_1) and (x_2, y_1) on a horizontal line is thus $|x_2 - x_1|$. Similarly, the distance between the points (x_2, y_1) and (x_2, y_2) on a vertical line is $|y_2 - y_1|$.

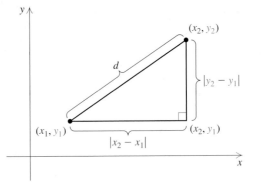

Now consider *any* two points (x_1, y_1) and (x_2, y_2). If $x_1 \neq x_2$ and $y_1 \neq y_2$, these points, along with the point (x_2, y_1), describe a right triangle. The lengths of the legs are $|x_2 - x_1|$ and $|y_2 - y_1|$. We find d, the length of the hypotenuse, by using the Pythagorean theorem:

$$d^2 = |x_2 - x_1|^2 + |y_2 - y_1|^2.$$

Since the square of a number is the same as the square of its opposite, we can replace the absolute-value signs with parentheses:

$$d^2 = (x_2 - x_1)^2 + (y_2 - y_1)^2.$$

Taking the principal square root, we have a formula for distance.

The Distance Formula

The distance d between any two points (x_1, y_1) and (x_2, y_2) is given by

$$d = \sqrt{(x_2 - x_1)^2 + (y_2 - y_1)^2}.$$

EXAMPLE 4   Find the distance between $(5, -1)$ and $(-4, 6)$. Find an exact answer and an approximation to three decimal places.

Solution   We substitute into the distance formula:

$$d = \sqrt{(-4 - 5)^2 + [6 - (-1)]^2} \qquad \text{Substituting}$$
$$= \sqrt{(-9)^2 + 7^2}$$
$$= \sqrt{130} \qquad \text{This is exact.}$$
$$\approx 11.402. \qquad \text{Using a calculator for an approximation}$$

The distance formula is needed to develop the formula for a circle, which follows, and to verify certain properties of conic sections. It is also needed to verify a formula for the coordinates of the *midpoint* of a segment connecting two points. We state the midpoint formula and leave its proof to the exercises.

Student Notes

To help remember the formulas correctly, note that the distance formula (a variation on the Pythagorean theorem) involves both subtraction and addition, whereas the midpoint formula does not include any subtraction.

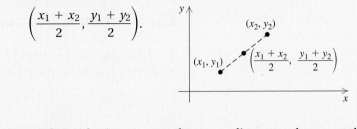

The Midpoint Formula

If the endpoints of a segment are (x_1, y_1) and (x_2, y_2), then the coordinates of the midpoint are

$$\left(\frac{x_1 + x_2}{2}, \frac{y_1 + y_2}{2} \right).$$

(To locate the midpoint, average the x-coordinates and average the y-coordinates.)

EXAMPLE 5   Find the midpoint of the segment with endpoints $(-2, 3)$ and $(4, -6)$.

Solution   Using the midpoint formula, we obtain

$$\left(\frac{-2 + 4}{2}, \frac{3 + (-6)}{2} \right), \quad \text{or} \quad \left(\frac{2}{2}, \frac{-3}{2} \right), \quad \text{or} \quad \left(1, -\frac{3}{2} \right).$$

Circles

One conic section, the **circle**, is a set of points in a plane that are a fixed distance r, called the **radius** (plural, **radii**), from a fixed point (h, k), called the **center**. Note that the word radius can mean either any segment connecting a point on a circle to the center or the length of such a segment. If (x, y) is on the circle, then by the definition of a circle and the distance formula, it follows that

$$r = \sqrt{(x - h)^2 + (y - k)^2}.$$

Squaring both sides gives the equation of a circle in standard form.

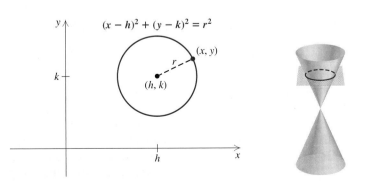

> *Equation of a Circle (Standard Form)*
>
> The equation of a circle, centered at (h, k), with radius r, is given by
>
> $$(x - h)^2 + (y - k)^2 = r^2.$$

Note that for $h = 0$ and $k = 0$, the circle is centered at the origin. Otherwise, the circle is translated $|h|$ units horizontally and $|k|$ units vertically.

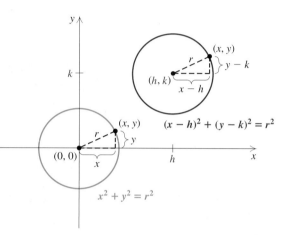

EXAMPLE 6   Find an equation of the circle having center $(4, 5)$ and radius 6.

Solution   Using the standard form, we obtain

$$(x - 4)^2 + (y - 5)^2 = 6^2, \qquad \text{Using } (x - h)^2 + (y - k)^2 = r^2$$

or

$$(x - 4)^2 + (y - 5)^2 = 36.$$

EXAMPLE 7   Find the center and the radius and then graph each circle.

a) $(x - 2)^2 + (y + 3)^2 = 4^2$

b) $x^2 + y^2 + 8x - 2y + 15 = 0$

Solution

a) We write standard form:

$$(x - 2)^2 + [y - (-3)]^2 = 4^2.$$

The center is $(2, -3)$ and the radius is 4. To graph, we plot the points $(2, 1)$, $(2, -7)$, $(-2, -3)$, and $(6, -3)$, which are, respectively, 4 units above, below, left, and right of $(2, -3)$. We then either sketch a circle by hand or use a compass.

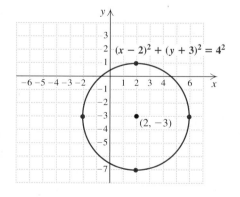

b) To write the equation $x^2 + y^2 + 8x - 2y + 15 = 0$ in standard form, we complete the square twice, once with $x^2 + 8x$ and once with $y^2 - 2y$:

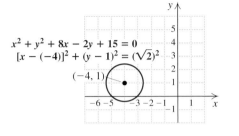

$x^2 + y^2 + 8x - 2y + 15 = 0$
$[x - (-4)]^2 + (y - 1)^2 = (\sqrt{2})^2$
$(-4, 1)$

$$x^2 + y^2 + 8x - 2y + 15 = 0$$

$$x^2 + 8x \quad\quad + y^2 - 2y \quad\quad = -15 \qquad \text{Grouping the } x\text{-terms and the } y\text{-terms; adding } -15 \text{ to both sides}$$

$$x^2 + 8x + 16 + y^2 - 2y + 1 = -15 + 16 + 1 \qquad \text{Adding } \left(\tfrac{8}{2}\right)^2 \text{, or } 16, \text{ and } \left(-\tfrac{2}{2}\right)^2 \text{, or } 1, \text{ to both sides to get standard form}$$

$$(x + 4)^2 + (y - 1)^2 = 2 \qquad \text{Factoring}$$
$$[x - (-4)]^2 + (y - 1)^2 = \left(\sqrt{2}\right)^2. \qquad \text{Writing standard form}$$

The center is $(-4, 1)$ and the radius is $\sqrt{2}$.

 **technology connection**

Most graphing calculators graph only functions, so graphing the equation of a circle usually requires two steps:

1. Solve the equation for y. The result will include a $\pm$ sign in front of a radical.
2. Graph two functions, one for the $+$ sign and the other for the $-$ sign, on the same set of axes.

For example, to graph $(x - 3)^2 + (y + 1)^2 = 16$, solve for $y + 1$ and then y:

$$(y + 1)^2 = 16 - (x - 3)^2$$
$$y + 1 = \pm\sqrt{16 - (x - 3)^2}$$
$$y = -1 \pm \sqrt{16 - (x - 3)^2},$$

or $\quad\quad y_1 = -1 + \sqrt{16 - (x - 3)^2}$

and $\quad\quad y_2 = -1 - \sqrt{16 - (x - 3)^2}.$

When both functions are graphed (in a "squared" window to eliminate distortion), the result is as follows.

$y_1 = -1 + \sqrt{16 - (x - 3)^2},$
$y_2 = -1 - \sqrt{16 - (x - 3)^2}$

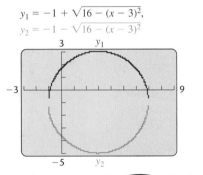

On many calculators, pressing **APPS** and selecting Conics and then Circle accesses a program in which equations in standard form can be graphed directly and then Traced.

Graph each of the following equations.

1. $x^2 + y^2 - 16 = 0$
2. $(x - 1)^2 + (y - 2)^2 = 25$
3. $(x + 3)^2 + (y - 5)^2 = 16$
4. $(x - 5)^2 + (y + 6)^2 = 49$

Exercise Set

10.1

☙ *Concept Reinforcement* *In each of Exercises 1–8, match the equation with the graph of that equation from those shown.*

1. ___ $(x - 2)^2 + (y + 5)^2 = 9$

2. ___ $(x + 2)^2 + (y - 5)^2 = 9$

3. ___ $(x - 5)^2 + (y + 2)^2 = 9$

4. ___ $(x + 5)^2 + (y - 2)^2 = 9$

5. ___ $y = (x - 2)^2 - 5$

6. ___ $y = (x - 5)^2 - 2$

7. ___ $x = (y - 2)^2 - 5$

8. ___ $x = (y - 5)^2 - 2$

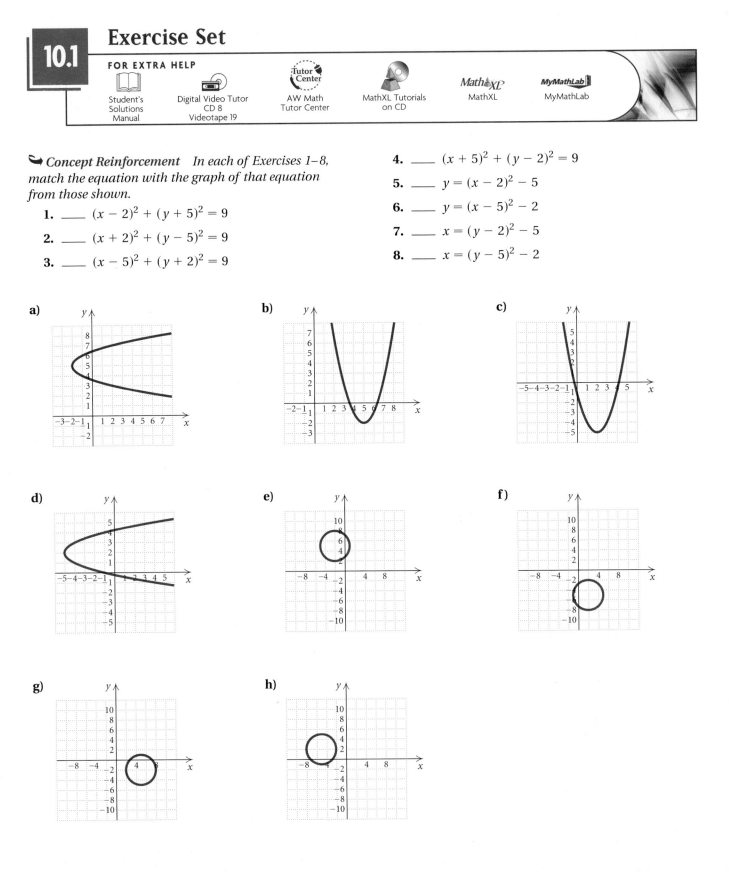

a)

b)

c)

d)

e)

f)

g)

h)

Graph. Be sure to label each vertex.

9. $y = -x^2$

10. $y = 2x^2$

11. $y = -x^2 + 4x - 5$

12. $x = 4 - 3y - y^2$

13. $x = y^2 - 4y + 2$

14. $y = x^2 + 2x + 3$

15. $x = y^2 + 3$

16. $x = 2y^2$

17. $x = -\frac{1}{2}y^2$

18. $x = y^2 - 1$

19. $x = -y^2 - 4y$

20. $x = y^2 + y - 6$

21. $x = 4 - y - y^2$

22. $y = x^2 + 2x + 1$

23. $y = x^2 - 2x + 1$

24. $y = -\frac{1}{2}x^2$

25. $x = -y^2 + 2y - 1$

26. $x = -y^2 - 2y + 3$

27. $x = -2y^2 - 4y + 1$

28. $x = 2y^2 + 4y - 1$

Find the distance between each pair of points. Where appropriate, find an approximation to three decimal places.

29. $(1, 6)$ and $(5, 9)$

30. $(1, 10)$ and $(7, 2)$

31. $(0, -7)$ and $(3, -4)$

32. $(6, 2)$ and $(6, -8)$

33. $(-4, 4)$ and $(6, -6)$

34. $(5, 21)$ and $(-3, 1)$

Aha!   **35.** $(8.6, -3.4)$ and $(-9.2, -3.4)$

36. $(5.9, 2)$ and $(3.7, -7.7)$

37. $\left(\frac{5}{7}, \frac{1}{14}\right)$ and $\left(\frac{1}{7}, \frac{11}{14}\right)$

38. $\left(0, \sqrt{7}\right)$ and $\left(\sqrt{6}, 0\right)$

39. $\left(-\sqrt{6}, \sqrt{2}\right)$ and $(0, 0)$

40. $\left(\sqrt{5}, -\sqrt{3}\right)$ and $(0, 0)$

41. $(-4, -2)$ and $(-7, -11)$

42. $(-3, -7)$ and $(-1, -5)$

Find the midpoint of each segment with the given endpoints.

43. $(-7, 6)$ and $(9, 2)$

44. $(6, 7)$ and $(7, -9)$

45. $(2, -1)$ and $(5, 8)$

46. $(-1, 2)$ and $(1, -3)$

47. $(-8, -5)$ and $(6, -1)$

48. $(8, -2)$ and $(-3, 4)$

49. $(-3.4, 8.1)$ and $(2.9, -8.7)$

50. $(4.1, 6.9)$ and $(5.2, -6.9)$

51. $\left(\frac{1}{6}, -\frac{3}{4}\right)$ and $\left(-\frac{1}{3}, \frac{5}{6}\right)$

52. $\left(-\frac{4}{5}, -\frac{2}{3}\right)$ and $\left(\frac{1}{8}, \frac{3}{4}\right)$

53. $\left(\sqrt{2}, -1\right)$ and $\left(\sqrt{3}, 4\right)$

54. $\left(9, 2\sqrt{3}\right)$ and $\left(-4, 5\sqrt{3}\right)$

Find an equation of the circle satisfying the given conditions.

55. Center $(0, 0)$, radius 6

56. Center $(0, 0)$, radius 5

57. Center $(7, 3)$, radius $\sqrt{5}$

58. Center $(5, 6)$, radius $\sqrt{2}$

59. Center $(-4, 3)$, radius $4\sqrt{3}$

60. Center $(-2, 7)$, radius $2\sqrt{5}$

61. Center $(-7, -2)$, radius $5\sqrt{2}$

62. Center $(-5, -8)$, radius $3\sqrt{2}$

63. Center $(0, 0)$, passing through $(-3, 4)$

64. Center $(3, -2)$, passing through $(11, -2)$

65. Center $(-4, 1)$, passing through $(-2, 5)$

66. Center $(-1, -3)$, passing through $(-4, 2)$

Find the center and the radius of each circle. Then graph the circle.

67. $x^2 + y^2 = 64$

68. $x^2 + y^2 = 36$

69. $(x + 1)^2 + (y + 3)^2 = 36$

70. $(x - 2)^2 + (y + 3)^2 = 4$

71. $(x - 4)^2 + (y + 3)^2 = 10$

72. $(x + 5)^2 + (y - 1)^2 = 15$

73. $x^2 + y^2 = 10$

74. $x^2 + y^2 = 7$

75. $(x - 5)^2 + y^2 = \frac{1}{4}$

76. $x^2 + (y - 1)^2 = \frac{1}{25}$

77. $x^2 + y^2 + 8x - 6y - 15 = 0$

78. $x^2 + y^2 + 6x - 4y - 15 = 0$

79. $x^2 + y^2 - 8x + 2y + 13 = 0$

80. $x^2 + y^2 + 6x + 4y + 12 = 0$

81. $x^2 + y^2 + 10y - 75 = 0$

82. $x^2 + y^2 - 8x - 84 = 0$

83. $x^2 + y^2 + 7x - 3y - 10 = 0$

84. $x^2 + y^2 - 21x - 33y + 17 = 0$

85. $36x^2 + 36y^2 = 1$

86. $4x^2 + 4y^2 = 1$

87. Describe a procedure that would use the distance formula to determine whether three points, (x_1, y_1), (x_2, y_2), and (x_3, y_3), are vertices of a right triangle.

88. Does the graph of an equation of a circle include the point that is the center? Why or why not?

SKILL MAINTENANCE

Solve. [6.4]

89. $\dfrac{x}{4} + \dfrac{5}{6} = \dfrac{2}{3}$

90. $\dfrac{t}{6} - \dfrac{1}{9} = \dfrac{7}{12}$

91. A rectangle 10 in. long and 6 in. wide is bordered by a strip of uniform width. If the perimeter of the larger rectangle is twice that of the smaller rectangle, what is the width of the border? [1.4]

92. One airplane flies 60 mph faster than another. To fly a certain distance, the faster plane takes 4 hr and the slower plane takes 4 hr and 24 min. What is the distance? [3.3]

Solve each system. [3.2]

93. $3x - 8y = 5,$
     $2x + 6y = 5$

94. $4x - 5y = 9,$
     $12x - 10y = 18$

SYNTHESIS

95. Outline a procedure that would use the distance formula to determine whether three points, (x_1, y_1), (x_2, y_2), and (x_3, y_3), are collinear (lie on the same line).

96. Why does the discussion of the distance formula precede the discussion of circles?

Find an equation of a circle satisfying the given conditions.

97. Center $(3, -5)$ and tangent to (touching at one point) the y-axis

98. Center $(-7, -4)$ and tangent to the x-axis

99. The endpoints of a diameter are $(7, 3)$ and $(-1, -3)$.

100. Center $(-3, 5)$ with a circumference of 8π units

101. Find the point on the y-axis that is equidistant from $(2, 10)$ and $(6, 2)$.

102. Find the point on the x-axis that is equidistant from $(-1, 3)$ and $(-8, -4)$.

103. *Wrestling.*   The equation $x^2 + y^2 = \frac{81}{4}$, where x and y represent the number of meters from the center, can be used to draw the outer circle on a wrestling mat used in International, Olympic, and World Championship wrestling. The equation $x^2 + y^2 = 16$ can be used to draw the inner edge of the red zone. Find the area of the red zone.

Source: Based on data from the Government of Western Australia

104. *Snowboarding.*   Each side edge of the Salomon Freestyle 500 Pro snowboard is an arc of a circle

with a "running length" of 1180 mm and a "side-cut depth" of 21.5 mm (see the figure below).

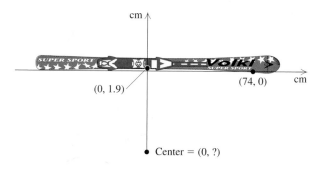

a) Using the coordinates shown, locate the center of the circle. (*Hint*: Equate distances.)
b) What radius is used for the edge of the board?

⊞ **105.** *Snowboarding.* The Elan Jason Evans 155 snowboard has a running length of 1170 mm and a sidecut depth of 23 mm (see Exercise 104). What radius is used for the edge of this snowboard?

⊞ **106.** *Skiing.* The Völkl Supersport 5 Star ski, when lying flat and viewed from above, has edges that are arcs of a circle. (Actually, each edge is made of two arcs of slightly different radii. The arc for the rear half of the ski edge has a slightly larger radius.)

a) Using the coordinates shown, locate the center of the circle. (*Hint*: Equate distances.)

b) What radius is used for the arc passing through (0, 1.9) and (74, 0)?

107. *Doorway construction.* Ace Carpentry needs to cut an arch for the top of an entranceway. The arch needs to be 8 ft wide and 2 ft high. To draw the arch, the carpenters will use a stretched string with chalk attached at an end as a compass.

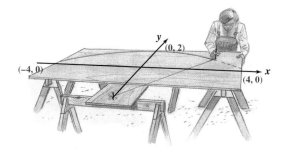

a) Using a coordinate system, locate the center of the circle.
b) What radius should the carpenters use to draw the arch?

108. *Archaeology.* During an archaeological dig, Martina finds the bowl fragment shown below. What was the original diameter of the bowl?

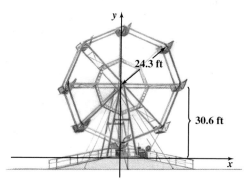

109. *Ferris wheel design.* A ferris wheel has a radius of 24.3 ft. Assuming that the center is 30.6 ft off the ground and that the origin is below the center, as in the following figure, find an equation of the circle.

∿ **110.** Use a graph of the equation $x = y^2 - y - 6$ to approximate to the nearest tenth the solutions of each of the following equations.
a) $y^2 - y - 6 = 2$ (*Hint*: Graph $x = 2$ on the same set of axes as the graph of $x = y^2 - y - 6$.)
b) $y^2 - y - 6 = -3$

111. *Power of a motor.*   The horsepower of a certain kind of engine is given by the formula

$$H = \frac{D^2 N}{2.5},$$

where N is the number of cylinders and D is the diameter, in inches, of each piston. Graph this equation, assuming that $N = 6$ (a six-cylinder engine). Let D run from 2.5 to 8.

112. Prove the midpoint formula by showing that

i) the distance from (x_1, y_1) to

$$\left(\frac{x_1 + x_2}{2}, \frac{y_1 + y_2}{2} \right)$$

equals the distance from (x_2, y_2) to

$$\left(\frac{x_1 + x_2}{2}, \frac{y_1 + y_2}{2} \right);$$

and

ii) the points

$$(x_1, y_1), \left(\frac{x_1 + x_2}{2}, \frac{y_1 + y_2}{2} \right),$$

and

$$(x_2, y_2)$$

lie on the same line (see Exercise 95).

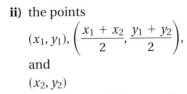 **113.** If the equation $x^2 + y^2 - 6x + 2y - 6 = 0$ is written as $y^2 + 2y + (x^2 - 6x - 6) = 0$, it can be regarded as quadratic in y.
  a) Use the quadratic formula to solve for y.
  b) Show that the graph of your answer to part (a) coincides with the graph in the Technology Connection on p. 665.

114. How could a graphing calculator best be used to help you sketch the graph of an equation of the form $x = ay^2 + by + c$?

115. Why should a graphing calculator's window be "squared" before graphing a circle?

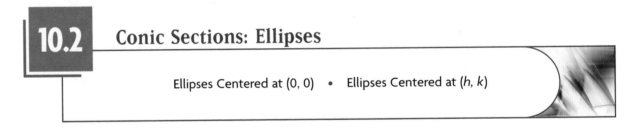

10.2   Conic Sections: Ellipses

Ellipses Centered at (0, 0)   •   Ellipses Centered at (h, k)

When a cone is cut at an angle, as shown below, the conic section formed is an *ellipse*. To draw an ellipse, stick two tacks in a piece of cardboard. Then tie a loose string to the tacks, place a pencil as shown, and draw an oval by moving the pencil while stretching the string tight.

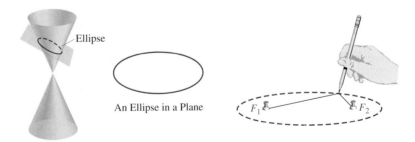

Ellipse

An Ellipse in a Plane

Ellipses Centered at (0, 0)

An **ellipse** is defined as the set of all points in a plane for which the sum of the distances from two fixed points F_1 and F_2 is constant. The points F_1 and F_2 are called **foci** (pronounced fō-sī), the plural of focus. In the figure above, the tacks are at the foci and the length of the string is the constant sum of the distances. The midpoint of the segment F_1F_2 is the **center**. The equation of an ellipse is as follows. Its derivation is left to the exercises.

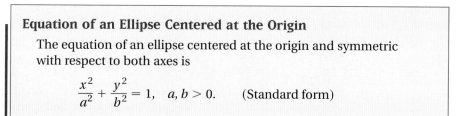

> **Equation of an Ellipse Centered at the Origin**
>
> The equation of an ellipse centered at the origin and symmetric with respect to both axes is
>
> $$\frac{x^2}{a^2} + \frac{y^2}{b^2} = 1, \quad a, b > 0. \qquad \text{(Standard form)}$$

To graph an ellipse centered at the origin, it helps to first find the intercepts. If we replace x with 0, we can find the y-intercepts:

$$\frac{0^2}{a^2} + \frac{y^2}{b^2} = 1$$

$$\frac{y^2}{b^2} = 1$$

$$y^2 = b^2 \quad \text{or} \quad y = \pm b.$$

Thus the y-intercepts are $(0, b)$ and $(0, -b)$. Similarly, the x-intercepts are $(a, 0)$ and $(-a, 0)$. If $a > b$, the ellipse is said to be horizontal and $(-a, 0)$ and $(a, 0)$ are referred to as the **vertices** (singular, **vertex**). If $b > a$, the ellipse is said to be vertical and $(0, -b)$ and $(0, b)$ are then the vertices.

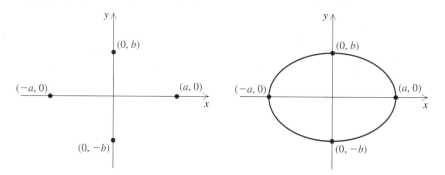

Plotting these four points and drawing an oval-shaped curve, we graph the ellipse. If a more precise graph is desired, we can plot more points.

> **Using a and b to Graph an Ellipse**
>
> For the ellipse
>
> $$\frac{x^2}{a^2} + \frac{y^2}{b^2} = 1,$$
>
> the x-intercepts are $(-a, 0)$ and $(a, 0)$. The y-intercepts are $(0, -b)$ and $(0, b)$. For $a^2 > b^2$, the ellipse is horizontal. For $b^2 > a^2$, the ellipse is vertical.

EXAMPLE 1   Graph the ellipse

$$\frac{x^2}{4} + \frac{y^2}{9} = 1.$$

Solution   Note that

$$\frac{x^2}{4} + \frac{y^2}{9} = \frac{x^2}{2^2} + \frac{y^2}{3^2}.$$ Identifying a and b. Since $b > a$, the ellipse is vertical.

Thus the x-intercepts are $(-2, 0)$ and $(2, 0)$, and the y-intercepts are $(0, -3)$ and $(0, 3)$. We plot these points and connect them with an oval-shaped curve. To plot two other points, we let $x = 1$ and solve for y:

$$\frac{1^2}{4} + \frac{y^2}{9} = 1$$

$$36\left(\frac{1}{4} + \frac{y^2}{9}\right) = 36 \cdot 1$$

$$36 \cdot \frac{1}{4} + 36 \cdot \frac{y^2}{9} = 36$$

$$9 + 4y^2 = 36$$

$$4y^2 = 27$$

$$y^2 = \frac{27}{4}$$

$$y = \pm\sqrt{\frac{27}{4}}$$

$$y \approx \pm 2.6.$$

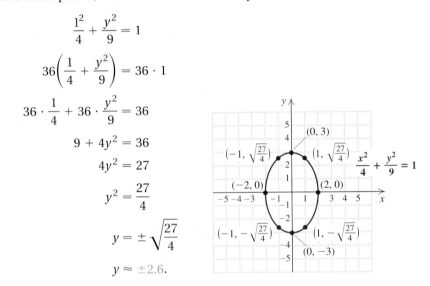

Thus, $(1, 2.6)$ and $(1, -2.6)$ can also be used to draw the graph. Similarly, the points $(-1, 2.6)$ and $(-1, -2.6)$ should appear on the graph.

EXAMPLE 2   Graph: $4x^2 + 25y^2 = 100$.

Solution   To write the equation in standard form, we divide both sides by 100 to get 1 on the right side:

$$\frac{4x^2 + 25y^2}{100} = \frac{100}{100}$$ Dividing by 100 to get 1 on the right side

$$\left.\begin{array}{l} \dfrac{4x^2}{100} + \dfrac{25y^2}{100} = 1 \\[2mm] \dfrac{x^2}{25} + \dfrac{y^2}{4} = 1 \end{array}\right\}$$ Simplifying

$$\frac{x^2}{5^2} + \frac{y^2}{2^2} = 1.$$ $a = 5, b = 2$

Student Notes _____

Note that any equation of the form $Ax^2 + By^2 = C$ can be rewritten as an equivalent equation in standard form. The graph is an ellipse.

The x-intercepts are $(-5, 0)$ and $(5, 0)$, and the y-intercepts are $(0, -2)$ and $(0, 2)$. We plot the intercepts and connect them with an oval-shaped curve. Other points can also be computed and plotted.

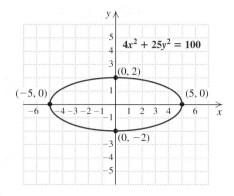

Ellipses Centered at (h, k)

Horizontal and vertical translations, similar to those used in Chapter 8, can be used to graph ellipses that are not centered at the origin.

Equation of an Ellipse Centered at (h, k)

The standard form of a horizontal or vertical ellipse centered at (h, k) is

$$\frac{(x - h)^2}{a^2} + \frac{(y - k)^2}{b^2} = 1.$$

The vertices are $(h + a, k)$ and $(h - a, k)$ if horizontal; $(h, k + b)$ and $(h, k - b)$ if vertical.

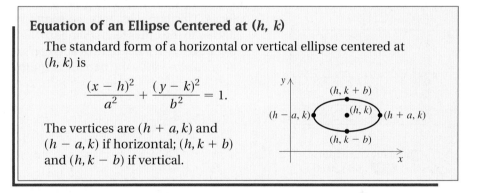

EXAMPLE 3 Graph the ellipse

$$\frac{(x - 1)^2}{4} + \frac{(y + 5)^2}{9} = 1.$$

Solution Note that

$$\frac{(x - 1)^2}{4} + \frac{(y + 5)^2}{9} = \frac{(x - 1)^2}{2^2} + \frac{(y + 5)^2}{3^2}.$$

Thus, $a = 2$ and $b = 3$. To determine the center of the ellipse, (h, k), note that

$$\frac{(x - 1)^2}{2^2} + \frac{(y + 5)^2}{3^2} = \frac{(x - 1)^2}{2^2} + \frac{(y - (-5))^2}{3^2}.$$

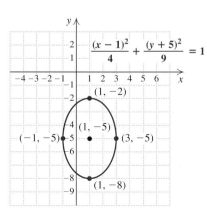

Thus the center is $(1, -5)$. We plot the points 2 units to the left and right of center, as well as the points 3 units above and below center. These are the points $(3, -5)$, $(-1, -5)$, $(1, -2)$, and $(1, -8)$. The graph of the ellipse is shown at left.

Note that this ellipse is the same as the ellipse in Example 1 but translated 1 unit to the right and 5 units down.

Graphing an ellipse on a graphing calculator is much like graphing a circle: We graph it in two pieces after solving for y. To illustrate, let's check Example 2:

$$4x^2 + 25y^2 = 100$$
$$25y^2 = 100 - 4x^2$$
$$y^2 = 4 - \frac{4}{25}x^2$$
$$y = \pm\sqrt{4 - \frac{4}{25}x^2}.$$

Using a squared window, we have our check:

$$y_1 = -\sqrt{4 - \frac{4}{25}x^2}, \quad y_2 = \sqrt{4 - \frac{4}{25}x^2}$$

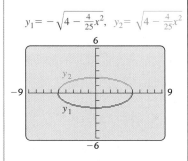

On many calculators, pressing **APPS** and selecting Conics and then Ellipse accesses a program in which equations in Standard Form can be graphed directly.

Ellipses have many applications. Communications satellites move in elliptical orbits with the earth as a focus while the earth itself follows an elliptical path around the sun. A medical instrument, the lithotripter, uses shock waves originating at one focus to crush a kidney stone located at the other focus.

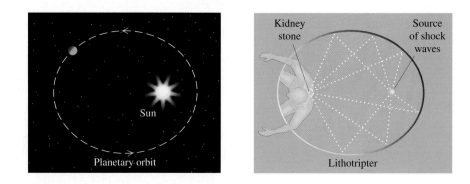

In some buildings, an ellipsoidal ceiling creates a "whispering gallery" in which a person at one focus can whisper and still be heard clearly at the other focus. This happens because sound waves coming from one focus are all reflected to the other focus. Similarly, light waves bouncing off an ellipsoidal mirror are used in a dentist's or surgeon's reflector light. The light source is located at one focus while the patient's mouth or surgical field is at the other.

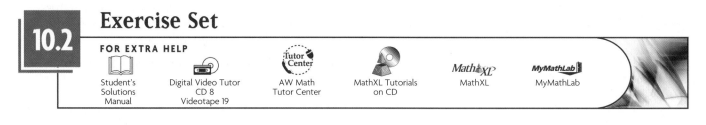

Exercise Set

10.2

FOR EXTRA HELP

Student's Solutions Manual | Digital Video Tutor CD 8 Videotape 19 | AW Math Tutor Center | MathXL Tutorials on CD | MathXL | MyMathLab

☙ *Concept Reinforcement   Classify each of the following as either true or false.*

1. The graph of $\frac{x^2}{9} + \frac{y^2}{25} = 1$ includes the points $(-3, 0)$ and $(3, 0)$.

2. The graph of $\frac{x^2}{36} + \frac{y^2}{25} = 1$ includes the points $(0, -5)$ and $(0, 5)$.

3. The graph of $\frac{x^2}{28} + \frac{y^2}{48} = 1$ is a vertical ellipse.

4. The graph of $\dfrac{x^2}{30} + \dfrac{y^2}{20} = 1$ is a vertical ellipse.

5. The graph of $\dfrac{x^2}{25} - \dfrac{y^2}{9} = 1$ is a horizontal ellipse.

6. The graph of $\dfrac{-x^2}{20} + \dfrac{y^2}{16} = 1$ is a horizontal ellipse.

7. The graph of $\dfrac{(x+3)^2}{25} + \dfrac{(y-2)^2}{36} = 1$ is an ellipse centered at $(-3, 2)$.

8. The graph of $\dfrac{(x-2)^2}{49} + \dfrac{(y+5)^2}{9} = 1$ is an ellipse centered at $(2, -5)$.

Graph each of the following equations.

9. $\dfrac{x^2}{1} + \dfrac{y^2}{9} = 1$

10. $\dfrac{x^2}{9} + \dfrac{y^2}{1} = 1$

11. $\dfrac{x^2}{25} + \dfrac{y^2}{9} = 1$

12. $\dfrac{x^2}{16} + \dfrac{y^2}{25} = 1$

13. $4x^2 + 9y^2 = 36$

14. $9x^2 + 4y^2 = 36$

15. $16x^2 + 9y^2 = 144$

16. $9x^2 + 16y^2 = 144$

17. $2x^2 + 3y^2 = 6$

18. $5x^2 + 7y^2 = 35$

Aha! **19.** $5x^2 + 5y^2 = 125$

20. $8x^2 + 5y^2 = 80$

21. $3x^2 + 7y^2 - 63 = 0$

22. $3x^2 + 8y^2 - 72 = 0$

23. $8x^2 = 96 - 3y^2$

24. $6y^2 = 24 - 8x^2$

25. $16x^2 + 25y^2 = 1$

26. $9x^2 + 4y^2 = 1$

27. $\dfrac{(x-3)^2}{9} + \dfrac{(y-2)^2}{25} = 1$

28. $\dfrac{(x-2)^2}{25} + \dfrac{(y-4)^2}{9} = 1$

29. $\dfrac{(x+4)^2}{16} + \dfrac{(y-3)^2}{49} = 1$

30. $\dfrac{(x+5)^2}{4} + \dfrac{(y-2)^2}{36} = 1$

31. $12(x-1)^2 + 3(y+4)^2 = 48$
(*Hint*: Divide both sides by 48.)

32. $4(x-6)^2 + 9(y+2)^2 = 36$

Aha! **33.** $4(x+3)^2 + 4(y+1)^2 - 10 = 90$

34. $9(x+6)^2 + (y+2)^2 - 20 = 61$

35. Is the center of an ellipse part of the ellipse itself? Why or why not?

36. Can an ellipse ever be the graph of a function? Why or why not?

SKILL MAINTENANCE

Solve.

37. $\dfrac{3}{x-2} - \dfrac{5}{x-2} = 9$ [6.4]

38. $\dfrac{7}{x+3} - \dfrac{2}{x+3} = 8$ [6.4]

39. $\dfrac{x}{x-4} - \dfrac{3}{x-5} = \dfrac{2}{x-4}$ [8.2]

40. $\dfrac{7}{x-3} - \dfrac{x}{x-2} = \dfrac{4}{x-2}$ [8.2]

41. $9 - \sqrt{2x+1} = 7$ [7.6]

42. $5 - \sqrt{x+3} = 9$ [7.6]

SYNTHESIS

43. An eccentric person builds a pool table in the shape of an ellipse with a hole at one focus and a tiny dot at the other. Guests are amazed at how many bank shots the owner of the pool table makes. Explain why this occurs.

44. Can a circle be considered a special type of ellipse? Why or why not?

Find an equation of an ellipse that contains the following points.

45. $(-9, 0), (9, 0), (0, -11),$ and $(0, 11)$

46. $(-7, 0), (7, 0), (0, -5),$ and $(0, 5)$

47. $(-2, -1), (6, -1), (2, -4),$ and $(2, 2)$

48. $(4, 3), (-6, 3), (-1, -1)$ and $(-1, 7)$

49. *Theatrical lighting.* The spotlight on a violin soloist casts an ellipse of light on the floor below her that is 6 ft wide and 10 ft long. Find an equation of that ellipse if the performer is in its center, x is the distance from the performer to the side of the ellipse, and y is the distance from the performer to the top of the ellipse.

50. *Astronomy.* The maximum distance of the planet Mars from the sun is 2.48×10^8 mi. The minimum distance is 3.46×10^7 mi. The sun is at one focus of the elliptical orbit. Find the distance from the sun to the other focus.

51. Let $(-c, 0)$ and $(c, 0)$ be the foci of an ellipse. Any point $P(x, y)$ is on the ellipse if the sum of the distances from the foci to P is some constant. Use $2a$ to represent this constant.
   a) Show that an equation for the ellipse is given by
   $$\frac{x^2}{a^2} + \frac{y^2}{a^2 - c^2} = 1.$$
   b) Substitute b^2 for $a^2 - c^2$ to get standard form.

52. *President's office.* The Oval Office of the President of the United States is an ellipse 31 ft wide and 38 ft long. Show in a sketch precisely where the President and an adviser could sit to best hear each other using the room's acoustics. (*Hint*: See Exercise 51(b) and the discussion following Example 3.)

53. *Dentistry.* The light source in a dental lamp shines against a reflector that is shaped like a portion of an ellipse in which the light source is one focus of the ellipse. Reflected light enters a patient's mouth at the other focus of the ellipse. If the ellipse from which the reflector was formed is 2 ft wide and 6 ft long, how far should the patient's mouth be from the light source? (*Hint*: See Exercise 51(b).)

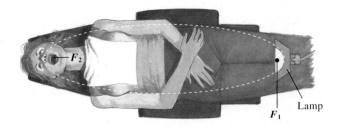

54. *Firefighting.* The size and shape of certain forest fires can be approximated as the union of two "half-ellipses." For the blaze modeled below, the equation of the smaller ellipse—the part of the fire moving *into* the wind—is
$$\frac{x^2}{40,000} + \frac{y^2}{10,000} = 1.$$

The equation of the other ellipse—the part moving *with* the wind—is
$$\frac{x^2}{250,000} + \frac{y^2}{10,000} = 1.$$

Determine the width and the length of the fire.
Source for figure: "Predicting Wind-Driven Wild Land Fire Size and Shape," Hal E. Anderson, Research Paper INT-305, U.S. Department of Agriculture, Forest Service, February 1983

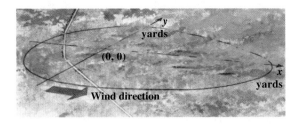

For each of the following equations, complete the square as needed and find an equivalent equation in standard form. Then graph the ellipse.

55. $x^2 - 4x + 4y^2 + 8y - 8 = 0$

56. $4x^2 + 24x + y^2 - 2y - 63 = 0$

57. Use a graphing calculator to check your answers to Exercises 11, 25, 29, and 33.

CORNER

COLLABORATIVE

A Cosmic Path

Focus: Ellipses

Time: 20–30 minutes

Group size: 2

Materials: Scientific calculators

In March 1996, the comet Hyakutake came within 21 million mi of the sun, and closer to Earth than any comet in over 500 yr (*Source*: Associated Press newspaper story, 3/20/96). Hyakutake is traveling in an elliptical orbit with the sun at one focus. The comet's average speed is about 100,000 mph (it actually goes much faster near its foci and slower as it gets further from the foci) and one orbit takes about 15,000 yr. (Astronomers estimate the time at 10,000–20,000 yr.)

ACTIVITY

1. The elliptical orbit of Hyakutake is so elongated that the distance traveled in one orbit can be estimated by $4a$ (see the following figure). Use the information above to estimate the distance, in millions of miles, traveled in one orbit. Then determine a.

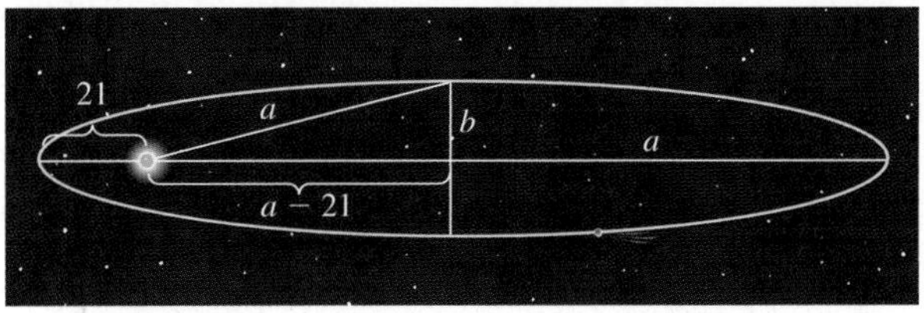

Units are millions of miles.

2. Using the figure above, express b^2 as a function of a. Then solve for b using the value found for a in part (1).

3. Approximately how far will Hyakutake be from the sun at the most distant part of its orbit?

4. Repeat parts (1)–(3), with one group member using the lower estimate of orbit time (10,000 yr) and the other using the upper estimate of orbit time (20,000 yr). By how much do the three answers to part (3) vary?

10.3 Conic Sections: Hyperbolas

Hyperbolas • Hyperbolas (Nonstandard Form) • Classifying Graphs of Equations

Hyperbolas

A **hyperbola** looks like a pair of parabolas, but the shapes are actually different. A hyperbola has two **vertices** and the line through the vertices is known as the **axis**. The point halfway between the vertices is called the **center**. The two curves that comprise a hyperbola are called **branches**.

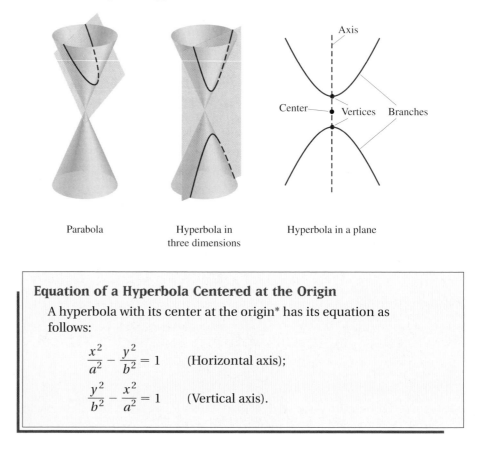

Parabola   Hyperbola in three dimensions   Hyperbola in a plane

---

Equation of a Hyperbola Centered at the Origin

A hyperbola with its center at the origin* has its equation as follows:

$$\frac{x^2}{a^2} - \frac{y^2}{b^2} = 1 \qquad \text{(Horizontal axis);}$$

$$\frac{y^2}{b^2} - \frac{x^2}{a^2} = 1 \qquad \text{(Vertical axis).}$$

Note that both equations have a 1 on the right-hand side and a subtraction symbol between the terms. For the discussion that follows, we assume $a, b > 0$.

---

*Hyperbolas with horizontal or vertical axes and centers *not* at the origin are discussed in Exercises 59–64.

To graph a hyperbola, it helps to begin by graphing two lines called **asymptotes**. Although the asymptotes themselves are not part of the graph, they serve as guidelines for an accurate sketch.

As a hyperbola gets farther away from the origin, it gets closer and closer to its asymptotes. The larger $|x|$ gets, the closer the graph gets to an asymptote. The asymptotes act to "constrain" the graph of a hyperbola. Parabolas are *not* constrained by any asymptotes.

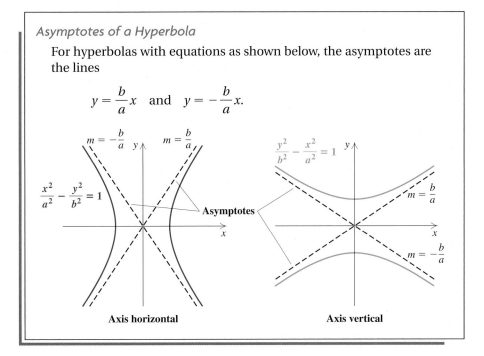

Asymptotes of a Hyperbola

For hyperbolas with equations as shown below, the asymptotes are the lines

$$y = \frac{b}{a}x \quad \text{and} \quad y = -\frac{b}{a}x.$$

Axis horizontal   **Axis vertical**

In Section 10.2, we used a and b to determine the width and the length of an ellipse. For hyperbolas, a and b are used to determine the base and the height of a rectangle that can be used as an aid in sketching asymptotes and locating vertices. This is illustrated in the following example.

EXAMPLE 1   Graph: $\dfrac{x^2}{4} - \dfrac{y^2}{9} = 1.$

Solution   Note that

$$\frac{x^2}{4} - \frac{y^2}{9} = \frac{x^2}{2^2} - \frac{y^2}{3^2}, \qquad \text{Identifying } a \text{ and } b$$

so $a = 2$ and $b = 3$. The asymptotes are thus

$$y = \frac{3}{2}x \quad \text{and} \quad y = -\frac{3}{2}x.$$

To help us sketch asymptotes and locate vertices, we use a and b—in this case, 2 and 3—to form the pairs $(-2, 3)$, $(2, 3)$, $(2, -3)$, and $(-2, -3)$. We plot these pairs and lightly sketch a rectangle. The asymptotes pass through the corners and, since this is a horizontal hyperbola, the vertices are where the rectangle intersects the x-axis. Finally, we draw the hyperbola, as shown below.

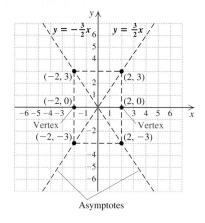

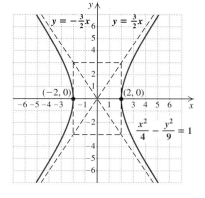

Asymptotes

EXAMPLE 2   Graph: $\dfrac{y^2}{36} - \dfrac{x^2}{4} = 1$.

Student Notes _____

Regarding the orientation of a hyperbola, you may find it helpful to think as follows: "The axis is parallel to the x-axis if $\dfrac{x^2}{a^2}$ is the positive term. The axis is parallel to the y-axis if $\dfrac{y^2}{b^2}$ is the positive term."

Solution   Note that

$$\frac{y^2}{36} - \frac{x^2}{4} = \frac{y^2}{6^2} - \frac{x^2}{2^2} = 1.$$

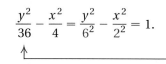

> Whether the hyperbola is horizontal or vertical is determined by the nonnegative term. Here there is a y in this term, so the hyperbola is vertical.

Using ± 2 as x-coordinates and ± 6 as y-coordinates, we plot $(2, 6)$, $(2, -6)$, $(-2, 6)$, and $(-2, -6)$, and lightly sketch a rectangle through them. The asymptotes pass through the corners (see the figure on the left below). Since the hyperbola is vertical, its vertices are $(0, 6)$ and $(0, -6)$. Finally, we draw curves through the vertices toward the asymptotes, as shown below.

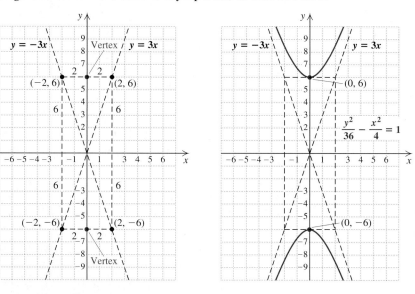

Hyperbolas (Nonstandard Form)

The equations for hyperbolas just examined are the standard ones, but there are other hyperbolas. We consider some of them.

Equation of a Hyperbola in Nonstandard Form

Hyperbolas having the x- and y-axes as asymptotes have equations as follows:

$$xy = c, \quad \text{where } c \text{ is a nonzero constant.}$$

EXAMPLE 3   Graph: $xy = -8$.

Solution   We first solve for y:

$$y = -\frac{8}{x}. \qquad \text{Dividing both sides by } x. \text{ Note that } x \neq 0.$$

Next, we find some solutions, keeping the results in a table. Note that x cannot be 0 and that for large values of $|x|$, y will be close to 0. Thus the x- and y-axes serve as asymptotes. We plot the points and draw two curves.

x	y
2	−4
−2	4
4	−2
−4	2
1	−8
−1	8
8	−1
−8	1

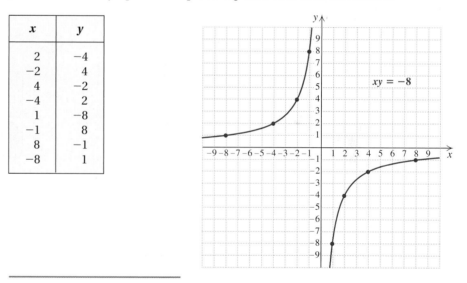

Hyperbolas have many applications. A jet breaking the sound barrier creates a sonic boom with a wave front the shape of a cone. The intersection of the cone with the ground is one branch of a hyperbola. Some comets travel in hyperbolic orbits, and a cross section of many lenses is hyperbolic in shape.

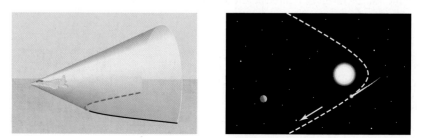

CONNECTING THE CONCEPTS

Recall that the vertical-line test tells us that circles, ellipses, and hyperbolas in standard form do not represent functions. Of the graphs examined in this chapter, only vertical parabolas and hyperbolas similar to the one in Example 3 can represent functions. Because functions are so important, circles and ellipses generally appear in applications of a purely geometric nature, whereas vertical parabolas and nonstandard hyperbolas can be used in applications involving geometry or functions.

In Section 10.4, we return to the challenge of solving real-world problems that translate to a system of equations. There we will find that knowing the general shape of the graph of an equation can help us determine how many solutions, if any, exist.

technology connection

The procedure used to graph a hyperbola in standard form is similar to that used to draw a circle or an ellipse. Consider the graph of

$$\frac{x^2}{25} - \frac{y^2}{49} = 1.$$

The student should confirm that solving for y yields

$$y_1 = \frac{\sqrt{49x^2 - 1225}}{5}$$

$$= \frac{7}{5}\sqrt{x^2 - 25}$$

and

$$y_2 = \frac{-\sqrt{49x^2 - 1225}}{5}$$

$$= -\frac{7}{5}\sqrt{x^2 - 25},$$

or $y_2 = -y_1.$

When the two pieces are drawn on the same squared window, the result is as shown. The gaps occur where the graph is nearly vertical.

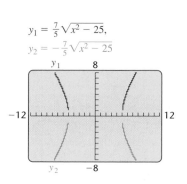

$$y_1 = \frac{7}{5}\sqrt{x^2 - 25},$$
$$y_2 = -\frac{7}{5}\sqrt{x^2 - 25}$$

On many calculators, pressing **APPS** and selecting Conics and then Hyperbola accesses a program in which hyperbolas in standard form can be graphed directly.

Graph.

1. $\dfrac{x^2}{16} - \dfrac{y^2}{60} = 1$

2. $16x^2 - 3y^2 = 64$

3. $\dfrac{y^2}{20} - \dfrac{x^2}{64} = 1$

4. $45y^2 - 9x^2 = 441$

Classifying Graphs of Equations

CONNECTING THE CONCEPTS

We summarize the equations and the graphs of the conic sections studied. The examples resume on p. 685.

PARABOLA

$y = ax^2 + bx + c,\ a > 0$
$\quad = a(x - h)^2 + k$

$y = ax^2 + bx + c,\ a < 0$
$\quad = a(x - h)^2 + k$

$x = ay^2 + by + c,\ a > 0$
$\quad = a(y - k)^2 + h$

$x = ay^2 + by + c,\ a < 0$
$\quad = a(y - k)^2 + h$

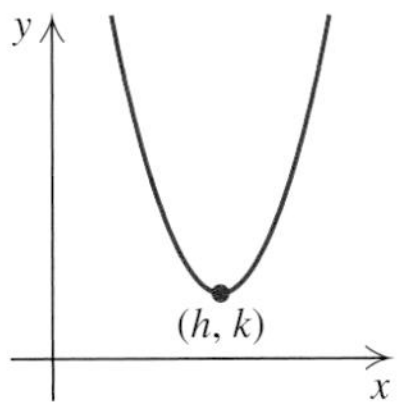

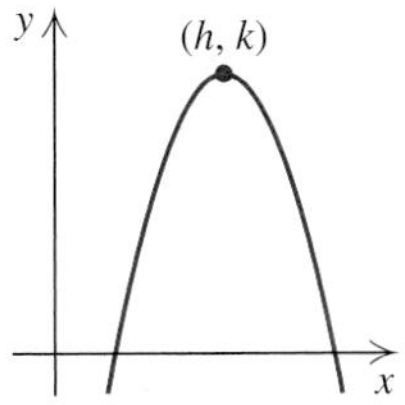

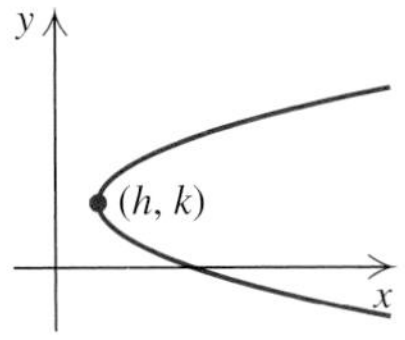

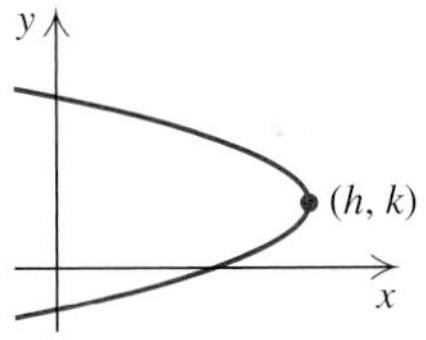

CIRCLE

Center at the origin:
$x^2 + y^2 = r^2$

Center at (h, k):
$(x - h)^2 + (y - k)^2 = r^2$

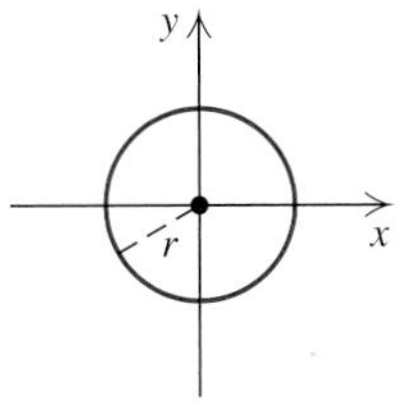

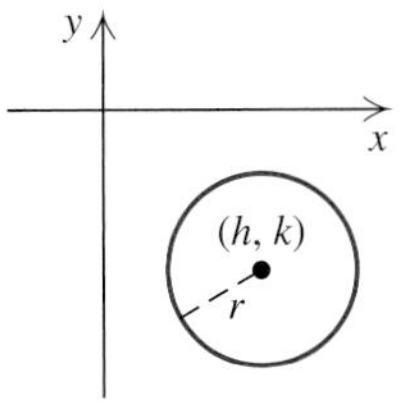

(*continued*)

CONNECTING THE CONCEPTS

HYPERBOLA

Center at the origin:

$$\frac{x^2}{a^2} - \frac{y^2}{b^2} = 1$$

$$\frac{y^2}{b^2} - \frac{x^2}{a^2} = 1$$

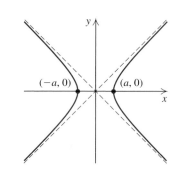

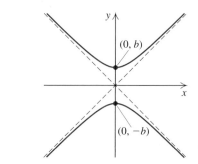

$xy = c, \ c > 0$

$xy = c, \ c < 0$

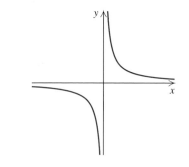

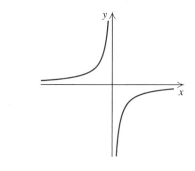

Center at (h, k)*:

$$\frac{(x - h)^2}{a^2} - \frac{(y - k)^2}{b^2} = 1$$

$$\frac{(y - k)^2}{b^2} - \frac{(x - h)^2}{a^2} = 1$$

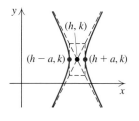

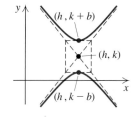

(*continued*)

*See Exercises 59–64.

ELLIPSE

Center at the origin:

$$\frac{x^2}{a^2} + \frac{y^2}{b^2} = 1$$

Center at (h, k):

$$\frac{(x - h)^2}{a^2} + \frac{(y - k)^2}{b^2} = 1$$

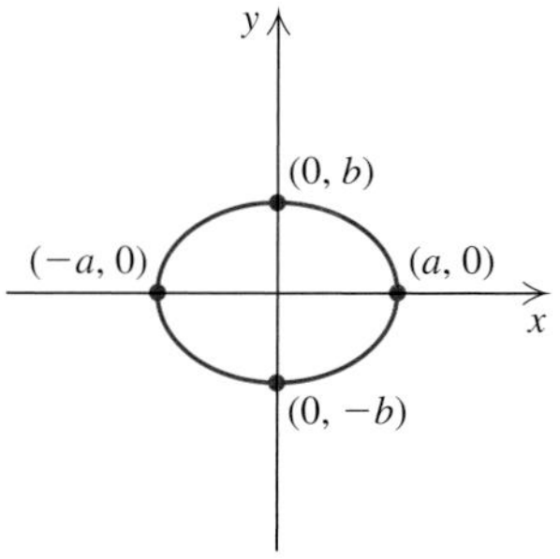

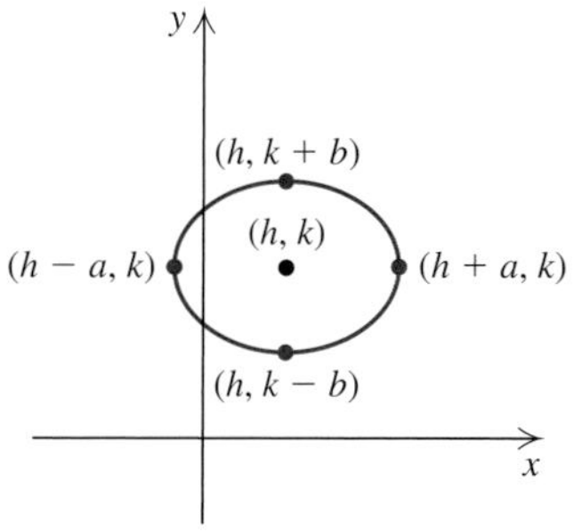

At the beginning of this chapter, we stated that the conic sections represent graphs of $Ax^2 + By^2 + Cxy + Dx + Ey + F = 0$ (we have assumed $C = 0^*$):

If A or B (but not both) is 0, the equation can be written in the form $y = ax^2 + bx + c$ or $x = ay^2 + by + c$, and represents a parabola.

If $A = B$, the equation can be written in the form $x^2 + y^2 = c$ or $(x - h)^2 + (y - k)^2 = r^2$, and represents a circle.

If $A \neq B$, but both A and B have the same sign, the equation can be written in the form $b^2x^2 + a^2y^2 = c$ or $b^2(x - h)^2 + a^2(y - k)^2 = c$, and represents an ellipse.

If A and B have opposite signs, the equation can be written in the form $b^2x^2 - a^2y^2 = c$ or $b^2(x - h)^2 - a^2(y - k)^2 = c$, and represents a hyperbola.

Algebraic manipulations may be needed to express an equation in one of the preceding forms.

EXAMPLE 4   Classify the graph of each equation as a circle, an ellipse, a parabola, or a hyperbola.

a) $5x^2 = 20 - 5y^2$

b) $x + 3 + 8y = y^2$

c) $x^2 = y^2 + 4$

d) $x^2 = 16 - 4y^2$

---

*For $C \neq 0$, the graphs are not symmetric with respect to both axes.

Solution

a) We get the terms with variables on one side by adding $5y^2$:

$$5x^2 + 5y^2 = 20.$$

Since x and y are *both* squared, we do not have a parabola. The fact that the squared terms are *added* tells us that we do not have a hyperbola. Do we have a circle? To find out, we need to get $x^2 + y^2$ by itself. We can do that by factoring the 5 out of both terms on the left and then dividing by 5:

$$5(x^2 + y^2) = 20 \qquad \text{Factoring out 5}$$
$$x^2 + y^2 = 4 \qquad \text{Dividing both sides by 5}$$
$$x^2 + y^2 = 2^2. \qquad \text{This is an equation for a circle.}$$

We can see that the graph is a circle with center at the origin and radius 2.

b) The equation $x + 3 + 8y = y^2$ has only one variable squared, so we solve for the other variable:

$$x = y^2 - 8y - 3. \qquad \text{This is an equation for a parabola.}$$

The graph is a horizontal parabola that opens to the right.

c) In $x^2 = y^2 + 4$, both variables are squared, so the graph is not a parabola. We subtract y^2 on both sides and divide by 4 to obtain

$$\frac{x^2}{2^2} - \frac{y^2}{2^2} = 1. \qquad \text{This is an equation for a hyperbola.}$$

The minus sign here indicates that the graph of this equation is a hyperbola. Because it is the x^2-term that is nonnegative, the hyperbola is horizontal.

d) In $x^2 = 16 - 4y^2$, both variables are squared, so the graph cannot be a parabola. We obtain the following equivalent equation:

$$x^2 + 4y^2 = 16.$$

If the coefficients of the terms were the same, we would have the graph of a circle, as in part (a), but they are not. Dividing both sides by 16 yields

$$\frac{x^2}{16} + \frac{y^2}{4} = 1. \qquad \text{This is an equation for an ellipse.}$$

The graph of this equation is a horizontal ellipse.

Exercise Set

10.3

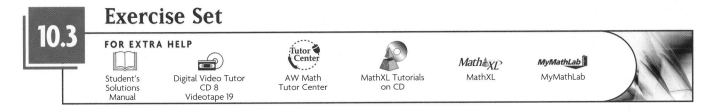

FOR EXTRA HELP

Student's Solutions Manual | Digital Video Tutor CD 8 Videotape 19 | AW Math Tutor Center | MathXL Tutorials on CD | MathXL | MyMathLab

🖐 *Concept Reinforcement* *In each of Exercises 1–8, match the conic section with the equation in the column on the right that represents that type of conic section.*

1. ___ A hyperbola with a horizontal axis

2. ___ A hyperbola with a vertical axis

3. ___ An ellipse with its center not at the origin

4. ___ An ellipse with its center at the origin

5. ___ A circle with its center at the origin

6. ___ A circle with its center not at the origin

7. ___ A parabola opening upward or downward

8. ___ A parabola opening to the right or left

a) $\dfrac{x^2}{10} + \dfrac{y^2}{12} = 1$

b) $(x + 1)^2 + (y - 3)^2 = 30$

c) $y - x^2 = 5$

d) $\dfrac{x^2}{9} - \dfrac{y^2}{10} = 1$

e) $x - 2y^2 = 3$

f) $\dfrac{y^2}{20} - \dfrac{x^2}{35} = 1$

g) $3x^2 + 3y^2 = 75$

h) $\dfrac{(x - 1)^2}{10} + \dfrac{(y - 4)^2}{8} = 1$

Graph each hyperbola. Label all vertices and sketch all asymptotes.

9. $\dfrac{y^2}{16} - \dfrac{x^2}{16} = 1$

10. $\dfrac{x^2}{9} - \dfrac{y^2}{9} = 1$

11. $\dfrac{x^2}{4} - \dfrac{y^2}{25} = 1$

12. $\dfrac{y^2}{16} - \dfrac{x^2}{9} = 1$

13. $\dfrac{y^2}{36} - \dfrac{x^2}{9} = 1$

14. $\dfrac{x^2}{25} - \dfrac{y^2}{36} = 1$

15. $y^2 - x^2 = 25$

16. $x^2 - y^2 = 4$

17. $25x^2 - 16y^2 = 400$

18. $4y^2 - 9x^2 = 36$

Graph.

19. $xy = -5$

20. $xy = 5$

21. $xy = 4$

22. $xy = -9$

23. $xy = -2$

24. $xy = -1$

25. $xy = 1$

26. $xy = 2$

Classify each of the following as the equation of a circle, an ellipse, a parabola, or a hyperbola.

27. $x^2 + y^2 - 6x + 4y - 30 = 0$

28. $y + 9 = 3x^2$

29. $9x^2 + 4y^2 - 36 = 0$

30. $x + 3y = 2y^2 - 1$

31. $4x^2 - 9y^2 - 72 = 0$

32. $y^2 + x^2 = 8$

33. $x^2 + y^2 = 2x + 4y + 4$

34. $2y + 13 + x^2 = 8x - y^2$

35. $4x^2 = 64 - y^2$

36. $y = \dfrac{7}{x}$

37. $x - \dfrac{8}{y} = 0$

38. $x - 4 = y^2 - 3y$

39. $y + 6x = x^2 + 5$

40. $x^2 = 16 + y^2$

41. $9y^2 = 36 + 4x^2$

42. $3x^2 + 5y^2 + x^2 = y^2 + 49$

43. $3x^2 + y^2 - x = 2x^2 - 9x + 10y + 40$

44. $4y^2 + 20x^2 + 1 = 8y - 5x^2$

45. $16x^2 + 5y^2 - 12x^2 + 8y^2 - 3x + 4y = 568$

46. $56x^2 - 17y^2 = 234 - 13x^2 - 38y^2$

47. What does graphing hyperbolas have in common with graphing ellipses?

48. Is it possible for a hyperbola to represent the graph of a function? Why or why not?

SKILL MAINTENANCE

Solve each system. [3.2]

49. $5x + 6y = -12,$
$3x + 9y = 15$

50. $2x + 6y = -6,$
$3x + 5y = 7$

Solve. [5.8]

51. $y^2 - 3 = 6$

52. $x^2 + 3 = 4$

53. The price of a lawn chair, including 5% sales tax, is $36.75. Find the price of the chair before the tax was added. [1.4]

54. A basketball team increases its score by 7 points in each of the two consecutive games after the home opener. If the team scored a total of 228 points in all three games, what was its score in the home opener? [1.4]

SYNTHESIS

55. What is it in the equation of a hyperbola that controls how wide open the branches are? Explain your reasoning.

56. If, in
$$\frac{x^2}{a^2} - \frac{y^2}{b^2} = 1,$$
$a = b$, what are the asymptotes of the graph? Why?

Find an equation of a hyperbola satisfying the given conditions.

57. Having intercepts $(0, 6)$ and $(0, -6)$ and asymptotes $y = 3x$ and $y = -3x$

58. Having intercepts $(8, 0)$ and $(-8, 0)$ and asymptotes $y = 4x$ and $y = -4x$

The standard equations for horizontal or vertical hyperbolas centered at (h, k) are as follows:

$$\frac{(x - h)^2}{a^2} - \frac{(y - k)^2}{b^2} = 1$$

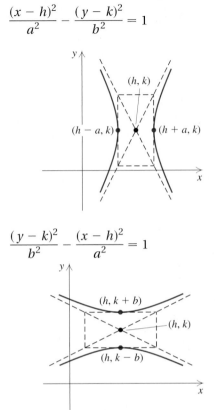

$$\frac{(y - k)^2}{b^2} - \frac{(x - h)^2}{a^2} = 1$$

The vertices are as labeled and the asymptotes are
$$y - k = \frac{b}{a}(x - h) \quad and \quad y - k = -\frac{b}{a}(x - h).$$

For each of the following equations of hyperbolas, complete the square, if necessary, and write in standard form. Find the center, the vertices, and the asymptotes. Then graph the hyperbola.

59. $\dfrac{(x - 5)^2}{36} - \dfrac{(y - 2)^2}{25} = 1$

60. $\dfrac{(x - 2)^2}{9} - \dfrac{(y - 1)^2}{4} = 1$

61. $8(y + 3)^2 - 2(x - 4)^2 = 32$

62. $25(x - 4)^2 - 4(y + 5)^2 = 100$

63. $4x^2 - y^2 + 24x + 4y + 28 = 0$

64. $4y^2 - 25x^2 - 8y - 100x - 196 = 0$

65. Use a graphing calculator to check your answers to Exercises 13, 25, 31, and 59.

10.4 | Nonlinear Systems of Equations

Systems Involving One Nonlinear Equation   •   Systems of Two Nonlinear Equations   •   Problem Solving

The equations appearing in systems of two equations have thus far in our discussion always been linear. We now consider systems of two equations in which at least one equation is nonlinear.

Systems Involving One Nonlinear Equation

Suppose that a system consists of an equation of a circle and an equation of a line. In what ways can the circle and the line intersect? The figures below represent three ways in which the situation can occur. We see that such a system will have 0, 1, or 2 real solutions.

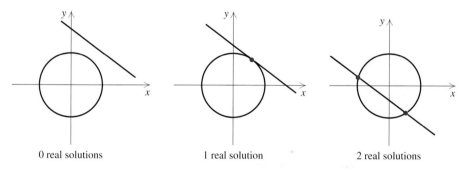

0 real solutions   1 real solution   2 real solutions

Recall that graphing, *elimination*, and *substitution* were all used to solve systems of linear equations. To solve systems in which one equation is of first degree and one is of second degree, it is preferable to use the *substitution* method.

EXAMPLE 1   Solve the system

$$x^2 + y^2 = 25, \quad (1) \quad \text{(The graph is a circle.)}$$
$$3x - 4y = 0. \quad (2) \quad \text{(The graph is a line.)}$$

Solution   First, we solve the linear equation, (2), for x:

$$x = \tfrac{4}{3}y. \quad (3) \qquad \text{We could have solved for } y \text{ instead.}$$

Then we substitute $\tfrac{4}{3}y$ for x in equation (1) and solve for y:

$$\left(\tfrac{4}{3}y\right)^2 + y^2 = 25$$
$$\tfrac{16}{9}y^2 + y^2 = 25$$
$$\tfrac{25}{9}y^2 = 25$$
$$y^2 = 9 \qquad \text{Multiplying both sides by } \tfrac{9}{25}$$
$$y = \pm 3. \qquad \text{Using the principle of square roots}$$

*Student Notes*_____

Be sure to either list each solution of a system as an ordered pair or separately state the value of each variable.

Now we substitute these numbers for y in equation (3) and solve for x:

$$\text{for } y = 3, \quad x = \tfrac{4}{3}(3) = 4;$$
$$\text{for } y = -3, \quad x = \tfrac{4}{3}(-3) = -4.$$

Check: For (4, 3):

$$
\begin{array}{c|c}
x^2 + y^2 = 25 & \\
\hline
4^2 + 3^2 & 25 \\
16 + 9 & \\
25 \overset{?}{=} 25 & \text{TRUE}
\end{array}
\qquad
\begin{array}{c|c}
3x - 4y = 0 & \\
\hline
3(4) - 4(3) & 0 \\
12 - 12 & \\
0 \overset{?}{=} 0 & \text{TRUE}
\end{array}
$$

It is left to the student to confirm that $(-4, -3)$ also checks in both equations.
    The pairs (4, 3) and $(-4, -3)$ check, so they are solutions. We can see the solutions in the graph. Intersections occur at (4, 3) and $(-4, -3)$.

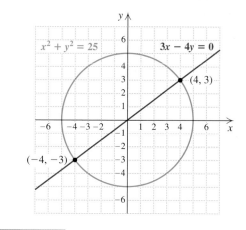

Even if we do not know what the graph of each equation in a system looks like, the algebraic approach of Example 1 can still be used.

EXAMPLE 2   Solve the system

$$
\begin{aligned}
y + 3 &= 2x, & (1)\\
x^2 + 2xy &= -1. & (2)
\end{aligned}
$$

Solution   First, we solve the linear equation (1) for y:

$$y = 2x - 3. \qquad (3)$$

Then we substitute $2x - 3$ for y in equation (2) and solve for x:

$$
\begin{aligned}
x^2 + 2x(2x - 3) &= -1 \\
x^2 + 4x^2 - 6x &= -1 \\
5x^2 - 6x + 1 &= 0 \\
(5x - 1)(x - 1) &= 0 && \text{Factoring} \\
5x - 1 = 0 \quad &or \quad x - 1 = 0 && \text{Using the principle of} \\
&&& \text{zero products} \\
x = \tfrac{1}{5} \quad &or \qquad\quad x = 1.
\end{aligned}
$$

Now we substitute these numbers for x in equation (3) and solve for y:

$$\text{for } x = \tfrac{1}{5}, \quad y = 2\left(\tfrac{1}{5}\right) - 3 = -\tfrac{13}{5};$$
$$\text{for } x = 1, \quad y = 2(1) - 3 = -1.$$

You can confirm that $\left(\tfrac{1}{5}, -\tfrac{13}{5}\right)$ and $(1, -1)$ check, so they are both solutions.

EXAMPLE 3

Solve the system

$$x + y = 5, \qquad (1) \qquad \text{(The graph is a line.)}$$
$$y = 3 - x^2. \qquad (2) \qquad \text{(The graph is a parabola.)}$$

Solution   We substitute $3 - x^2$ for y in the first equation:

$$x + 3 - x^2 = 5$$
$$-x^2 + x - 2 = 0 \qquad \text{Adding } -5 \text{ to both sides and rearranging}$$
$$x^2 - x + 2 = 0. \qquad \text{Multiplying both sides by } -1$$

Since $x^2 - x + 2$ does not factor, we need the quadratic formula:

$$x = \frac{-b \pm \sqrt{b^2 - 4ac}}{2a}$$
$$= \frac{-(-1) \pm \sqrt{(-1)^2 - 4 \cdot 1 \cdot 2}}{2(1)} \qquad \text{Substituting}$$
$$= \frac{1 \pm \sqrt{1 - 8}}{2} = \frac{1 \pm \sqrt{-7}}{2} = \frac{1}{2} \pm \frac{\sqrt{7}}{2}i.$$

Solving equation (1) for y gives us $y = 5 - x$. Substituting values for x gives

$$y = 5 - \left(\frac{1}{2} + \frac{\sqrt{7}}{2}i\right) = \frac{9}{2} - \frac{\sqrt{7}}{2}i \quad \text{and}$$
$$y = 5 - \left(\frac{1}{2} - \frac{\sqrt{7}}{2}i\right) = \frac{9}{2} + \frac{\sqrt{7}}{2}i.$$

The solutions are

$$\left(\frac{1}{2} + \frac{\sqrt{7}}{2}i, \frac{9}{2} - \frac{\sqrt{7}}{2}i\right) \quad \text{and} \quad \left(\frac{1}{2} - \frac{\sqrt{7}}{2}i, \frac{9}{2} + \frac{\sqrt{7}}{2}i\right).$$

There are no real-number solutions. Note in the figure at right that the graphs do not intersect. Getting only nonreal solutions tells us that the graphs do not intersect.

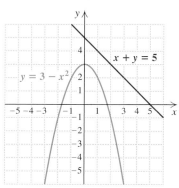

technology connection

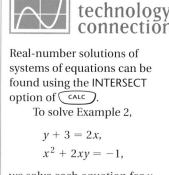

Real-number solutions of systems of equations can be found using the INTERSECT option of ⬭CALC⬭.

To solve Example 2,

$$y + 3 = 2x,$$
$$x^2 + 2xy = -1,$$

we solve each equation for y and then graph:

$$\left. \begin{array}{l} y_1 = 2x - 3, \\ y_2 = \dfrac{-1 - x^2}{2x}. \end{array} \right\} \quad \begin{array}{l} \text{Note that} \\ x, y \neq 0. \end{array}$$

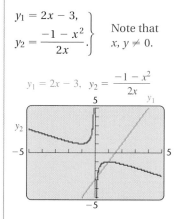

Using INTERSECT, we find the solutions to be $(0.2, -2.6)$ and $(1, -1)$.

Solve each system. Round all values to two decimal places.

1. $4xy - 7 = 0,$
     $x - 3y - 2 = 0$
2. $x^2 + y^2 = 14,$
     $16x + 7y^2 = 0$

Systems of Two Nonlinear Equations

We now consider systems of two second-degree equations. Graphs of such systems can involve any two conic sections. The following figure shows some ways in which a circle and a hyperbola can intersect.

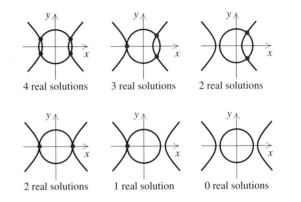

4 real solutions   3 real solutions   2 real solutions

2 real solutions   1 real solution   0 real solutions

To solve systems of two second-degree equations, we either substitute or eliminate. The elimination method is generally better when both equations are of the form $Ax^2 + By^2 = C$. Then we can eliminate an x^2- or y^2-term in a manner similar to the procedure used in Chapter 3.

EXAMPLE 4   Solve the system

$$2x^2 + 5y^2 = 22, \quad (1)$$
$$3x^2 - y^2 = -1. \quad (2)$$

Solution   Here we multiply equation (2) by 5 and then add:

$$
\begin{array}{ll}
2x^2 + 5y^2 = \ 22 & \\
\underline{15x^2 - 5y^2 = -5} & \text{Multiplying both sides of equation (2) by 5} \\
17x^2 \qquad\ \ = 17 & \text{Adding} \\
\qquad x^2 = \ 1 & \\
\qquad\ x = \pm 1. &
\end{array}
$$

There is no x-term, and whether x is -1 or 1, we have $x^2 = 1$. Thus we can simultaneously substitute 1 and -1 for x in equation (2):

$$
\left.
\begin{array}{l}
3 \cdot (\pm 1)^2 - y^2 = -1 \\
\qquad\quad 3 - y^2 = -1 \\
\qquad\qquad -y^2 = -4
\end{array}
\right\}
\quad
\begin{array}{l}
\text{Since } (-1)^2 = 1^2, \text{ we can evaluate for} \\
x = -1 \text{ and } x = 1 \text{ simultaneously.}
\end{array}
$$

$$y^2 = 4 \quad \text{or} \quad y = \pm 2.$$

Thus, if $x = 1$, then $y = 2$ or $y = -2$; and if $x = -1$, then $y = 2$ or $y = -2$. The four possible solutions are $(1, 2)$, $(1, -2)$, $(-1, 2)$, and $(-1, -2)$.

Check: Since $(2)^2 = (-2)^2$ and $(1)^2 = (-1)^2$, we can check all four pairs at once.

$$\frac{2x^2 + 5y^2 = 22}{2(\pm 1)^2 + 5(\pm 2)^2 \mid 22}$$
$$2 + 20$$
$$22 \overset{?}{=} 22 \quad \text{TRUE}$$

$$\frac{3x^2 - y^2 = -1}{3(\pm 1)^2 - (\pm 2)^2 \mid -1}$$
$$3 - 4$$
$$-1 \overset{?}{=} -1 \quad \text{TRUE}$$

The solutions are $(1, 2)$, $(1, -2)$, $(-1, 2)$, and $(-1, -2)$.

When a product of variables is in one equation and the other equation is of the form $Ax^2 + By^2 = C$, we often solve for a variable in the equation with the product and then use substitution.

EXAMPLE 5 Solve the system

$$x^2 + 4y^2 = 20, \quad (1)$$
$$xy = 4. \quad (2)$$

Solution First, we solve equation (2) for y:

$$y = \frac{4}{x}. \quad \text{Dividing both sides by } x. \text{ Note that } x \neq 0.$$

Then we substitute $4/x$ for y in equation (1) and solve for x:

$$x^2 + 4\left(\frac{4}{x}\right)^2 = 20$$
$$x^2 + \frac{64}{x^2} = 20$$
$$x^4 + 64 = 20x^2 \quad \text{Multiplying by } x^2$$
$$x^4 - 20x^2 + 64 = 0 \quad \text{Obtaining standard form. This equation is reducible to quadratic.}$$
$$(x^2 - 4)(x^2 - 16) = 0 \quad \text{Factoring. If you prefer, let } u = x^2 \text{ and substitute.}$$
$$(x - 2)(x + 2)(x - 4)(x + 4) = 0 \quad \text{Factoring again}$$
$$x = 2 \quad or \quad x = -2 \quad or \quad x = 4 \quad or \quad x = -4. \quad \text{Using the principle of zero products}$$

Since $y = 4/x$, for $x = 2$, we have $y = 4/2$, or 2. Thus, $(2, 2)$ is a solution. Similarly, $(-2, -2)$, $(4, 1)$, and $(-4, -1)$ are solutions. You can show that all four pairs check.

---

technology connection

Before Example 4 can be checked by graphing, each equation must be solved for y. For equation (1), this yields $y_1 = \sqrt{(22 - 2x^2)/5}$ and $y_2 = -\sqrt{(22 - 2x^2)/5}$. For equation (2), this yields $y_3 = \sqrt{3x^2 + 1}$ and $y_4 = -\sqrt{3x^2 + 1}$.

1. Perform a visual check for Example 5.

Problem Solving

We now consider applications that can be modeled by a system of equations in which at least one equation is not linear.

EXAMPLE 6   Architecture.   For a college gymnasium, an architect wants to lay out a rectangular piece of land that has a perimeter of 204 m and an area of 2565 m^2. Find the dimensions of the piece of land.

Solution

1. **Familiarize.**   We draw and label a sketch, letting l = the length and w = the width, both in meters.

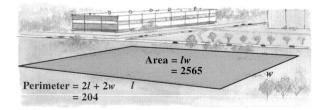

Area = lw
= 2565

Perimeter = $2l + 2w$ l
= 204

2. **Translate.**   We then have the following translation:

$$\text{Perimeter:} \quad 2w + 2l = 204;$$
$$\text{Area:} \quad lw = 2565.$$

3. **Carry out.**   We solve the system

$$2w + 2l = 204,$$
$$lw = 2565.$$

Solving the second equation for l gives us $l = 2565/w$. Then we substitute $2565/w$ for l in the first equation and solve for w:

$$2w + 2\left(\frac{2565}{w}\right) = 204$$

$$2w^2 + 2(2565) = 204w \qquad \text{Multiplying both sides by } w$$

$$2w^2 - 204w + 2(2565) = 0 \qquad \text{Standard form}$$

$$w^2 - 102w + 2565 = 0 \qquad \text{Multiplying by } \tfrac{1}{2}$$

> Factoring could be used instead of the quadratic formula, but the numbers are quite large.

$$w = \frac{-(-102) \pm \sqrt{(-102)^2 - 4 \cdot 1 \cdot 2565}}{2 \cdot 1}$$

$$w = \frac{102 \pm \sqrt{144}}{2} = \frac{102 \pm 12}{2}$$

$$w = 57 \quad or \quad w = 45.$$

If $w = 57$, then $l = 2565/w = 2565/57 = 45$. If $w = 45$, then $l = 2565/w = 2565/45 = 57$. Since length is usually considered to be longer than width, we have the solution $l = 57$ and $w = 45$, or $(57, 45)$.

4. **Check.**   If $l = 57$ and $w = 45$, the perimeter is $2 \cdot 57 + 2 \cdot 45$, or 204. The area is $57 \cdot 45$, or 2565. The numbers check.

5. **State.**   The length is 57 m and the width is 45 m.

EXAMPLE 7   HDTV dimensions.   High-definition television (HDTV) offers greater clarity than conventional television. The Kaplans' new HDTV screen has an area of 1296 in^2 and has a $\sqrt{3033}$-in. (about 55-in.) diagonal screen. Find the width and the length of the screen.

Solution

1. **Familiarize.**   We make a drawing and label it. Note the right triangle in the figure. We let $l =$ the length and $w =$ the width, both in inches.

55 in.

2. **Translate.**   We translate to a system of equations:

$$l^2 + w^2 = \sqrt{3033}^2, \qquad \text{Using the Pythagorean theorem}$$
$$lw = 1296. \qquad\qquad \text{Using the formula for the area of a rectangle}$$

3. **Carry out.**   We solve the system

$$\left.\begin{array}{l} l^2 + w^2 = 3033, \\ lw = 1296 \end{array}\right\} \quad \text{You should complete the solution of this system.}$$

to get $(48, 27)$, $(27, 48)$, $(-48, -27)$, and $(-27, -48)$.

4. **Check.**   Measurements must be positive and length is usually greater than width, so we check only $(48, 27)$. In the right triangle, $48^2 + 27^2 = 2304 + 729 = 3033$ or $\sqrt{3033}^2$. The area is $48 \cdot 27 = 1296$, so our answer checks.

5. **State.**   The length is 48 in. and the width is 27 in.

Exercise Set

10.4

FOR EXTRA HELP

Student's Solutions Manual

Digital Video Tutor CD 8 Videotape 19

Tutor Center AW Math Tutor Center

MathXL Tutorials on CD

Math XL MathXL

MyMathLab MyMathLab

⤳ *Concept Reinforcement   Classify each statement as either true or false.*

1. A system of equations that represent a line and an ellipse can have 0, 1, or 2 solutions.

2. A system of equations that represent a parabola and a circle can have up to 4 solutions.

3. A system of equations representing a hyperbola and a circle can have no fewer than 2 solutions.

4. A system of equations representing an ellipse and a line has either 0 or 2 solutions.

5. Systems containing one first-degree equation and one second-degree equation are most easily solved using the substitution method.

6. Systems containing two second-degree equations of the form $Ax^2 + By^2 = C$ are most easily solved using the elimination method.

Solve. Remember that graphs can be used to confirm all real solutions.

7. $x^2 + y^2 = 25,$
$y - x = 1$

8. $x^2 + y^2 = 100,$
$y - x = 2$

9. $4x^2 + 9y^2 = 36,$
$3y + 2x = 6$

10. $9x^2 + 4y^2 = 36,$
$3x + 2y = 6$

11. $y^2 = x + 3,$
$2y = x + 4$

12. $y = x^2,$
$3x = y + 2$

13. $x^2 - xy + 3y^2 = 27,$
$x - y = 2$

14. $2y^2 + xy + x^2 = 7,$
$x - 2y = 5$

15. $x^2 + 4y^2 = 25,$
$x + 2y = 7$

16. $x^2 - y^2 = 16,$
$x - 2y = 1$

17. $x^2 - xy + 3y^2 = 5,$
$x - y = 2$

18. $m^2 + 3n^2 = 10,$
$m - n = 2$

19. $3x + y = 7,$
$4x^2 + 5y = 24$

20. $2y^2 + xy = 5,$
$4y + x = 7$

21. $a + b = 7,$
$ab = 4$

22. $p + q = -6,$
$pq = -7$

23. $2a + b = 1,$
$b = 4 - a^2$

24. $4x^2 + 9y^2 = 36,$
$x + 3y = 3$

25. $a^2 + b^2 = 89,$
$a - b = 3$

26. $xy = 4,$
$x + y = 5$

Aha! **27.** $y = x^2,$
$x = y^2$

28. $x^2 + y^2 = 25,$
$y^2 = x + 5$

29. $x^2 + y^2 = 9,$
$x^2 - y^2 = 9$

30. $y^2 - 4x^2 = 4,$
$4x^2 + y^2 = 4$

31. $x^2 + y^2 = 25,$
$xy = 12$

32. $x^2 - y^2 = 16,$
$x + y^2 = 4$

33. $x^2 + y^2 = 9,$
$25x^2 + 16y^2 = 400$

34. $x^2 + y^2 = 4,$
$9x^2 + 16y^2 = 144$

35. $x^2 + y^2 = 14,$
$x^2 - y^2 = 4$

36. $x^2 + y^2 = 16,$
$y^2 - 2x^2 = 10$

37. $x^2 + y^2 = 20,$
$xy = 8$

38. $x^2 + y^2 = 5,$
$xy = 2$

39. $x^2 + 4y^2 = 20,$
$xy = 4$

40. $x^2 + y^2 = 13,$
$xy = 6$

41. $2xy + 3y^2 = 7,$
$3xy - 2y^2 = 4$

42. $3xy + x^2 = 34,$
$2xy - 3x^2 = 8$

43. $4a^2 - 25b^2 = 0,$
$2a^2 - 10b^2 = 3b + 4$

44. $xy - y^2 = 2,$
$2xy - 3y^2 = 0$

45. $ab - b^2 = -4,$
$ab - 2b^2 = -6$

46. $x^2 - y = 5,$
$x^2 + y^2 = 25$

Solve.

47. *Computer parts.* Dataport Electronics needs a rectangular memory board that has a perimeter of 28 cm and a diagonal of length 10 cm. What should the dimensions of the board be?

48. *Geometry.* A rectangle has an area of 2 yd^2 and a perimeter of 6 yd. Find its dimensions.

49. *Geometry.* A rectangle has an area of 20 in^2 and a perimeter of 18 in. Find its dimensions.

50. *Tile design.*   The New World tile company wants to make a new rectangular tile that has a perimeter of 6 in. and a diagonal of length $\sqrt{5}$ in. What should the dimensions of the tile be?

51. *Design of a van.*   The cargo area of a delivery van must be 60 ft^2, and the length of a diagonal must accommodate a 13-ft board. Find the dimensions of the cargo area.

52. *Dimensions of a rug.*   The diagonal of a Persian rug is 25 ft. The area of the rug is 300 ft^2. Find the length and the width of the rug.

53. The product of two numbers is 60. The sum of their squares is 136. Find the numbers.

54. *Investments.*   A certain amount of money saved for 1 yr at a certain interest rate yielded $225 in interest. If $750 more had been invested and the rate had been 1% less, the interest would have been the same. Find the principal and the rate.

55. *Garden design.*   A garden contains two square peanut beds. Find the length of each bed if the sum of their areas is 832 ft^2 and the difference of their areas is 320 ft^2.

56. The area of a rectangle is $\sqrt{3}$ m^2, and the length of a diagonal is 2 m. Find the dimensions.

57. The product of the lengths of the legs of a right triangle is 156. The hypotenuse has length $\sqrt{313}$. Find the lengths of the legs.

58. The area of a rectangle is $\sqrt{2}$ m^2, and the length of a diagonal is $\sqrt{3}$ m. Find the dimensions.

59. How can an understanding of conic sections be helpful when a system of nonlinear equations is being solved algebraically?

60. Suppose a system of equations is comprised of one linear equation and one nonlinear equation. Is it possible for such a system to have three solutions? Why or why not?

SKILL MAINTENANCE

Simplify. [1.2]

61. $(-1)^9(-2)^4$

62. $(-1)^{10}(-2)^5$

Evaluate each of the following. [1.2]

63. $\dfrac{(-1)^k}{k-5}$, for $k = 6$

64. $\dfrac{(-1)^k}{k-5}$, for $k = 9$

65. $\dfrac{n}{2}(3 + n)$, for $n = 8$

66. $\dfrac{7(1 - r^2)}{1 - r}$, for $r = 3$

SYNTHESIS

67. Write a problem that translates to a system of two equations. Design the problem so that at least one equation is nonlinear and so that no real solution exists.

68. Write a problem for a classmate to solve. Devise the problem so that a system of two nonlinear equations with exactly one real solution is solved.

69. Find the equation of a circle that passes through $(-2, 3)$ and $(-4, 1)$ and whose center is on the line $5x + 8y = -2$.

70. Find the equation of an ellipse centered at the origin that passes through the points $(2, -3)$ and $\left(1, \sqrt{13}\right)$.

Solve.

71. $p^2 + q^2 = 13,$

$\dfrac{1}{pq} = -\dfrac{1}{6}$

72. $a + b = \dfrac{5}{6},$

$\dfrac{a}{b} + \dfrac{b}{a} = \dfrac{13}{6}$

73. *Fence design.*   A roll of chain-link fencing contains 100 ft of fence. The fencing is bent at a 90° angle to enclose a rectangular work area of 2475 ft², as shown. Determine the length and the width of the rectangle.

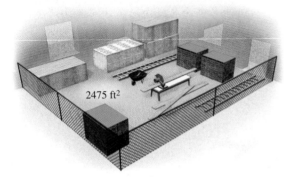

2475 ft²

74. A piece of wire 100 cm long is to be cut into two pieces and those pieces are each to be bent to make a square. The area of one square is to be 144 cm² greater than that of the other. How should the wire be cut?

75. *Box design.*   Four squares with sides 5 in. long are cut from the corners of a rectangular metal sheet that has an area of 340 in². The edges are bent up to form an open box with a volume of 350 in³. Find the dimensions of the box.

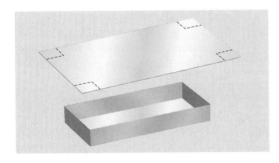

76. *Computer screens.*   The ratio of the length to the height of the screen on a computer monitor is 4 to 3. A Dell Inspiron notebook has a 15-in. diagonal screen. Find the dimensions of the screen.

15 in.

77. *HDTV screens.*   The ratio of the length to the height of an HDTV screen (see Example 7) is 16 to 9. The Remton Lounge has an HDTV screen with a $\sqrt{4901}$-in. (about 70-in.) diagonal screen. Find the dimensions of the screen.

78. *Railing sales.*   Fireside Castings finds that the total revenue R from the sale of x units of railing is given by

$$R = 100x + x^2.$$

Fireside also finds that the total cost C of producing x units of the same product is given by

$$C = 80x + 1500.$$

A break-even point is a value of x for which total revenue is the same as total cost; that is, $R = C$. How many units must be sold to break even?

79. Use a graphing calculator to check your answers to Exercises 13, 25, and 47.

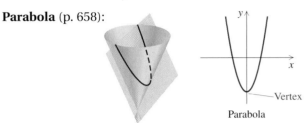

Study Summary

The curves formed by cross sections of cones are **conic sections** (p. 658).

Parabola (p. 658):

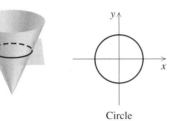

Parabola

$y = ax^2 + bx + c$ (opens upward or downward)

$= a(x - h)^2 + k$ (vertex at (h, k))

$x = ay^2 + by + c$ (opens right or left)

$= a(y - k)^2 + h$ (vertex at (h, k))

Circle (p. 663):

Circle

$x^2 + y^2 = r^2$ (center at $(0, 0)$)

$(x - h)^2 + (y - k)^2 = r^2$ (center at (h, k))

Ellipse (p. 670):

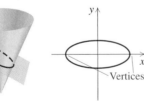

Ellipse

$\dfrac{x^2}{a^2} + \dfrac{y^2}{b^2} = 1$ (center at $(0, 0)$)

$\dfrac{(x - h)^2}{a^2} + \dfrac{(y - k)^2}{b^2} = 1$ (center at (h, k))

Hyperbola (p. 678):

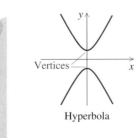

Hyperbola

$\dfrac{x^2}{a^2} - \dfrac{y^2}{b^2} = 1$ (opens right and left)

$\dfrac{y^2}{b^2} - \dfrac{x^2}{a^2} = 1$ (opens upward and downward)

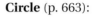

To discuss the conic sections in detail, it is necessary to be able to calculate the **distance** d between any two points (x_1, y_1) and (x_2, y_2):

$$d = \sqrt{(x_2 - x_1)^2 + (y_2 - y_1)^2} \quad \text{(p. 662)}.$$

Using the distance formula, it is possible to develop the formula for the coordinates of the **midpoint** of the segment connecting any two points (x_1, y_1) and (x_2, y_2):

Midpoint: $\left(\dfrac{x_1 + x_2}{2}, \dfrac{y_1 + y_2}{2} \right)$ (p. 663).

10 Review Exercises

↪ *Concept Reinforcement   Classify each statement as either true or false.*

1. To use the distance formula, one must have an understanding of radical notation. [10.1]

2. The midpoint of the segment connecting (x_1, y_1) and (x_2, y_2) is $(x_1 + x_2, y_1 + y_2)$. [10.1]

3. The center of a circle is part of the circle itself. [10.1]

4. The foci of an ellipse always share the same second coordinate. [10.2]

5. Every parabola that opens upward or downward can represent the graph of a function. [10.1]

6. It is possible for a hyperbola to represent the graph of a function. [10.3]

7. Every system of nonlinear equations has at least one real solution. [10.4]

8. Both substitution and elimination can be used as methods for solving a system of nonlinear equations. [10.4]

Find the distance between each pair of points. Where appropriate, find an approximation to three decimal places. [10.1]

9. $(3, 6)$ and $(7, 6)$

10. $(-1, 1)$ and $(-5, 4)$

11. $(1.4, 3.6)$ and $(4.7, -5.3)$

12. $(2, 3a)$ and $(-1, a)$

Find the midpoint of the segment with the given endpoints. [10.1]

13. $(2, -1)$ and $(7, -1)$

14. $(-1, 10)$ and $(-5, 4)$

15. $\left(1, \sqrt{3}\right)$ and $\left(\frac{1}{2}, -\sqrt{2}\right)$

16. $(2, 3a)$ and $(-1, a)$

Find the center and the radius of each circle. [10.1]

17. $(x + 3)^2 + (y - 2)^2 = 7$

18. $(x - 5)^2 + y^2 = 49$

19. $x^2 + y^2 - 6x - 2y + 1 = 0$

20. $x^2 + y^2 + 8x - 6y = 10$

21. Find an equation of the circle with center $(-4, 3)$ and radius $4\sqrt{3}$. [10.1]

22. Find an equation of the circle with center $(7, -2)$ and radius $2\sqrt{5}$. [10.1]

Classify each equation as a circle, an ellipse, a parabola, or a hyperbola. Then graph.

23. $5x^2 + 5y^2 = 80$ [10.1], [10.3]

24. $9x^2 + 2y^2 = 18$ [10.2], [10.3]

25. $y = -x^2 + 2x - 3$ [10.1], [10.3]

26. $\dfrac{y^2}{9} - \dfrac{x^2}{4} = 1$ [10.3]

27. $xy = 9$ [10.3]

28. $x = y^2 + 2y - 2$ [10.1], [10.3]

29. $\dfrac{(x+1)^2}{3} + (y-3)^2 = 1$ [10.2], [10.3]

30. $x^2 + y^2 + 6x - 8y - 39 = 0$ [10.1], [10.3]

Solve. [10.4]

31. $x^2 - y^2 = 33$,
   $x + y = 11$

32. $x^2 - 2x + 2y^2 = 8$,
   $2x + y = 6$

33. $x^2 - y = 3$,
   $2x - y = 3$

34. $x^2 + y^2 = 25$,
   $x^2 - y^2 = 7$

35. $x^2 - y^2 = 3$,
   $y = x^2 - 3$

36. $x^2 + y^2 = 18$,
   $2x + y = 3$

37. $x^2 + y^2 = 100$,
   $2x^2 - 3y^2 = -120$

38. $x^2 + 2y^2 = 12$,
   $xy = 4$

39. A rectangular bandstand has a perimeter of 38 m and an area of 84 m². What are the dimensions of the bandstand? [10.4]

40. One type of carton used by tableproducts.com exactly fits both a rectangular napkin of area 108 in² and a candle of length 15 in., laid diagonally on top of the napkin. Find the length and the width of the carton. [10.4]

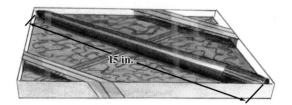

15 in.

41. The perimeter of a square mirror is 12 cm more than the perimeter of another square mirror. Its area exceeds the area of the other by 39 cm². Find the perimeter of each mirror. [10.4]

42. The sum of the areas of two circles is 130π ft². The difference of the circumferences is 16π ft. Find the radius of each circle. [10.4]

SYNTHESIS

43. How does the graph of a hyperbola differ from the graph of a parabola? [10.1], [10.3]

44. Explain why function notation rarely appears in this chapter, and list the graphs discussed for which function notation could be used. [10.1], [10.2], [10.3]

45. Solve: [10.4]
$$4x^2 - x - 3y^2 = 9,$$
$$-x^2 + x + y^2 = 2.$$

46. Find the points whose distance from $(8, 0)$ and from $(-8, 0)$ is 10. [10.1]

47. Find an equation of the circle that passes through $(-2, -4)$, $(5, -5)$, and $(6, 2)$. [10.1], [10.4]

48. Find an equation of the ellipse with the following intercepts: $(-9, 0)$, $(9, 0)$, $(0, -5)$, and $(0, 5)$. [10.2]

49. Find the point on the x-axis that is equidistant from $(-3, 4)$ and $(5, 6)$. [10.1]

10 Chapter Test

Find the distance between each pair of points. Where appropriate, find an approximation to three decimal places.

1. $(5, -1)$ and $(-4, 8)$

2. $(3, -a)$ and $(-3, a)$

Find the midpoint of the segment with the given endpoints.

3. $(4, -1)$ and $(-5, 8)$

4. $(3, -a)$ and $(-3, a)$

Find the center and the radius of each circle.

5. $(x + 5)^2 + (y - 1)^2 = 81$

6. $x^2 + y^2 + 4x - 6y + 4 = 0$

Classify the equation as a circle, an ellipse, a parabola, or a hyperbola. Then graph.

7. $y = x^2 - 4x - 1$

8. $x^2 + y^2 + 2x + 6y + 6 = 0$

9. $\dfrac{x^2}{16} - \dfrac{y^2}{9} = 1$

10. $16x^2 + 4y^2 = 64$

11. $xy = -5$

12. $x = -y^2 + 4y$

Solve.

13. $\dfrac{x^2}{4} + \dfrac{y^2}{9} = 1,$
$3x + 4y = 12$

14. $x^2 + y^2 = 16,$
$\dfrac{x^2}{16} - \dfrac{y^2}{9} = 1$

15. $x^2 - 2y^2 = 1,$
$xy = 6$

16. $x^2 + y^2 = 10,$
$x^2 = y^2 + 2$

17. A rectangular bookmark with diagonal of length $5\sqrt{5}$ has an area of 22. Find the dimensions of the bookmark.

18. Two squares are such that the sum of their areas is $8\ \text{m}^2$ and the difference of their areas is $2\ \text{m}^2$. Find the length of a side of each square.

19. A rectangular dance floor has a diagonal of length 40 ft and a perimeter of 112 ft. Find the dimensions of the dance floor.

20. Nikki invested a certain amount of money for 1 yr and earned $72 in interest. Erin invested $240 more than Nikki at an interest rate that was $\frac{5}{6}$ of the rate given to Nikki, but she earned the same amount of interest. Find the principal and the interest rate for Nikki's investment.

SYNTHESIS

21. Find an equation of the ellipse passing through $(6, 0)$ and $(6, 6)$ with vertices at $(1, 3)$ and $(11, 3)$.

22. Find the point on the y-axis that is equidistant from $(-3, -5)$ and $(4, -7)$.

23. The sum of two numbers is 36, and the product is 4. Find the sum of the reciprocals of the numbers.

24. *Theatrical production.* An E.T.C. spotlight for a college's production of *Hamlet* projects an ellipse of light on a stage that is 8 ft wide and 14 ft long. Find an equation of that ellipse if an actor is in its center and x represents the number of feet, horizontally, from the actor to the edge of the ellipse and y represents the number of feet, vertically, from the actor to the edge of the ellipse.

11

Sequences, Series, and the Binomial Theorem

AN APPLICATION

Approximately 534,000 new apartments and houses were built in the United States in 1991. Since then, the number has grown by about 5.35% per year. (*Sources*: Based on data from the U.S. Bureau of the Census and the U.S. Department of Housing and Urban Development) How many new apartments and houses were built in the United States from 1991 through 2004?

This problem appears as Exercise 65 in Section 11.3.

Beverley Dockeray-Ojo
CITY PLANNING DIRECTOR
Atlanta, Georgia

Mathematics is applied in the urban planning field intensively. Specifically used in statistical analysis and demographic projections including population, employment, and income, it is also applied in basic land, floor area, and density calculations. Like many other professions, the numbers tell the story, and provide the basis for problem identification and possible solutions.

*T*he first three sections of this chapter are devoted to sequences *and* series. A sequence is simply an ordered list. For example, when a baseball coach writes a batting order, a sequence is being formed. When the members of a sequence are numbers, they can be added. Such a sum is called a series.

Section 11.4 presents the binomial theorem, *which is used to expand* expressions of the form $(a + b)^n$. Such an expansion is itself a series.

11.1   Sequences and Series

Sequences • Finding the General Term •
Sums and Series • Sigma Notation

Sequences

Suppose that $10,000 is borrowed at 5%, compounded annually. The value of the loan at the start of years 1, 2, 3, 4, and so on, is

$10,000, $10,500, $11,025, $11,576.25, ...

We can regard this as a function that pairs 1 with $10,000, 2 with $10,500, 3 with $11,025, and so on. A **sequence** (or **progression**) is thus a function, where the domain is a set of consecutive positive integers beginning with 1, and the range varies from sequence to sequence.

If we continue computing the amounts in the account forever, we obtain an **infinite sequence**, with function values

$10,000, $10,500, $11,025, $11,576.25, $12,155.06, ...

The three dots at the end indicate that the sequence goes on without stopping. If we stop after a certain number of years, we obtain a **finite sequence**:

$10,000, $10,500, $11,025, $11,576.25.

Sequences

An *infinite sequence* is a function having for its domain the set of natural numbers: $\{1, 2, 3, 4, 5, \ldots\}$.

A *finite sequence* is a function having for its domain a set of natural numbers: $\{1, 2, 3, 4, 5, \ldots, n\}$, for some natural number n.

As another example, consider the sequence given by

$$a(n) = 2^n, \quad \text{or} \quad a_n = 2^n.$$

The notation a_n means the same as $a(n)$ but is used more commonly with sequences. Some function values (also called *terms* of the sequence) follow:

$$a_1 = 2^1 = 2,$$
$$a_2 = 2^2 = 4,$$
$$a_3 = 2^3 = 8,$$
$$a_6 = 2^6 = 64.$$

The first term of the sequence is a_1, the fifth term is a_5, and the nth term, or **general term**, is a_n. This sequence can also be denoted in the following ways:

$$2, 4, 8, \ldots;$$

or $2, 4, 8, \ldots, 2^n, \ldots$.   The 2^n emphasizes that the nth term of this sequence is found by raising 2 to the nth power.

EXAMPLE 1   Find the first four terms and the 57th term of the sequence for which the general term is given by $a_n = (-1)^n/(n + 1)$.

Solution   We have

$$a_1 = \frac{(-1)^1}{1 + 1} = -\frac{1}{2},$$

$$a_2 = \frac{(-1)^2}{2 + 1} = \frac{1}{3},$$

$$a_3 = \frac{(-1)^3}{3 + 1} = -\frac{1}{4},$$

$$a_4 = \frac{(-1)^4}{4 + 1} = \frac{1}{5},$$

$$a_{57} = \frac{(-1)^{57}}{57 + 1} = -\frac{1}{58}.$$

Note that the expression $(-1)^n$ causes the signs of the terms to alternate between positive and negative, depending on whether n is even or odd.

---

technology connection

Sequences are entered and graphed much like functions. The difference is that the SEQUENCE MODE must be selected. You can then enter U_n or V_n using n as the variable. Use this approach to check Example 1 with a table of values for the sequence.

Finding the General Term

When only the first few terms of a sequence are known, it is impossible to be certain what the general term is. Still, a prediction can be made by looking for a pattern.

EXAMPLE 2   For each sequence, predict the general term.

a) $1, 4, 9, 16, 25, \ldots$ **b)** $-1, 2, -4, 8, -16, \ldots$

c) $2, 4, 8, \ldots$

Solution

a) $1, 4, 9, 16, 25, \ldots$

These are squares of consecutive positive integers, so the general term could be n^2.

b) $-1, 2, -4, 8, -16, \ldots$

These are powers of 2 with alternating signs, so the general term may be $(-1)^n[2^{n-1}]$. To check, note that 8 is the fourth term, and

$$(-1)^4[2^{4-1}] = 1 \cdot 2^3$$
$$= 8.$$

c) $2, 4, 8, \ldots$

We regard the pattern as powers of 2, in which case 16 would be the next term and 2^n the general term. The sequence could then be written with more terms as

$$2, 4, 8, 16, 32, 64, 128, \ldots .$$

In part (c) above, suppose that the second term is found by adding 2, the third term by adding 4, the next term by adding 6, and so on. In this case, 14 would be the next term and the sequence would be

$$2, 4, 8, 14, 22, 32, 44, 58, \ldots .$$

This illustrates that the fewer terms we are given, the greater the uncertainty about the nth term.

Sums and Series

Series

Given the infinite sequence

$$a_1, \ a_2, \ a_3, \ a_4, \ \ldots, \ a_n, \ldots,$$

the sum of the terms

$$a_1 + a_2 + a_3 + \cdots + a_n + \cdots$$

is called an *infinite series* and is denoted S_∞. A *partial sum* is the sum of the first n terms:

$$a_1 + a_2 + a_3 + \cdots + a_n.$$

A partial sum is also called a *finite series* and is denoted S_n.

EXAMPLE 3   For the sequence $-2, 4, -6, 8, -10, 12, -14$, find: **(a)** S_2; **(b)** S_3; **(c)** S_7.

Solution

a) $S_2 = -2 + 4 = 2$ This is the sum of the first 2 terms.

b) $S_3 = -2 + 4 + (-6) = -4$ This is the sum of the first 3 terms.

c) $S_7 = -2 + 4 + (-6) + 8 + (-10) + 12 + (-14) = -8$ This is the sum of the first 7 terms.

Sigma Notation

When the general term of a sequence is known, the Greek letter Σ (capital sigma) can be used to write a series. For example, the sum of the first four terms of the sequence 3, 5, 7, 9, 11,..., $2k + 1$,... can be named as follows, using *sigma notation*, or *summation notation*:

$$\sum_{k=1}^{4} (2k + 1).$$

This represents
$(2 \cdot 1 + 1) + (2 \cdot 2 + 1) + (2 \cdot 3 + 1) + (2 \cdot 4 + 1).$

This is read "the sum as k goes from 1 to 4 of $(2k + 1)$." The letter k is called the *index of summation*. The index of summation need not always start at 1.

EXAMPLE 4   Write out and evaluate each sum.

a) $\displaystyle\sum_{k=1}^{5} k^2$ 　　　　　　　**b)** $\displaystyle\sum_{k=4}^{6} (-1)^k(2k)$ 　　　　　**c)** $\displaystyle\sum_{k=0}^{3} (2^k + 5)$

Solution

a) $\displaystyle\sum_{k=1}^{5} k^2 = 1^2 + 2^2 + 3^2 + 4^2 + 5^2 = 1 + 4 + 9 + 16 + 25 = 55$

Evaluate k^2 for all integers from 1 through 5. Then add.

b) $\displaystyle\sum_{k=4}^{6} (-1)^k(2k) = (-1)^4(2 \cdot 4) + (-1)^5(2 \cdot 5) + (-1)^6(2 \cdot 6)$

$$= 8 - 10 + 12 = 10$$

c) $\displaystyle\sum_{k=0}^{3} (2^k + 5) = (2^0 + 5) + (2^1 + 5) + (2^2 + 5) + (2^3 + 5)$

$$= 6 + 7 + 9 + 13 = 35$$

Student Notes

A great deal of information is condensed into sigma notation. Be careful to pay attention to what values the index of summation will take on. Evaluate the expression following sigma, the general term, for each value and then add the results.

EXAMPLE 5   Write sigma notation for each sum.

a) $1 + 4 + 9 + 16 + 25$

b) $-1 + 3 - 5 + 7$

c) $3 + 9 + 27 + 81 + \cdots$

Solution

a) $1 + 4 + 9 + 16 + 25$

Note that this is a sum of squares, $1^2 + 2^2 + 3^2 + 4^2 + 5^2$, so the general term is k^2. Sigma notation is

$$\sum_{k=1}^{5} k^2.$$ 　　The sum starts with 1^2 and ends with 5^2.

Answers can vary here. For example, another—perhaps less obvious—way of writing $1 + 4 + 9 + 16 + 25$ is

$$\sum_{k=2}^{6} (k - 1)^2.$$

b) $-1 + 3 - 5 + 7$

Except for the alternating signs, this is the sum of the first four positive odd numbers. It is useful to note that $2k - 1$ is a formula for the kth positive odd number. It is also important to note that since $(-1)^k = 1$ when k is even and $(-1)^k = -1$ when k is odd, the factor $(-1)^k$ can be used to create the alternating signs. The general term is thus $(-1)^k(2k - 1)$, beginning with $k = 1$. Sigma notation is

$$\sum_{k=1}^{4} (-1)^k(2k - 1).$$

To check, we can evaluate $(-1)^k(2k - 1)$ using 1, 2, 3, and 4. Then we can write the sum of the four terms. We leave this to the student.

c) $3 + 9 + 27 + 81 + \cdots$

This is a sum of powers of 3, and it is also an infinite series. We use the symbol ∞ for infinity and write the series using sigma notation:

$$\sum_{k=1}^{\infty} 3^k.$$

Exercise Set

11.1

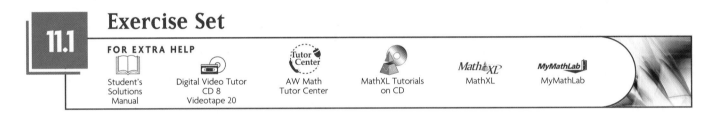

FOR EXTRA HELP

Student's Solutions Manual | Digital Video Tutor CD 8 Videotape 20 | Tutor Center AW Math Tutor Center | MathXL Tutorials on CD | MathXL MathXL | MyMathLab MyMathLab

↝ *Concept Reinforcement   In each of Exercises 1–6, match the expression with the most appropriate expression from the column on the right.*

1. ___ $\displaystyle\sum_{k=1}^{4} k^2$

2. ___ $\displaystyle\sum_{k=3}^{6} (-1)^k$

3. ___ $5 + 10 + 15 + 20$

4. ___ $a_n = 5^n$

5. ___ $a_n = 3n + 2$

6. ___ $a_1 + a_2 + a_3$

a) $-1 + 1 + (-1) + 1$

b) $a_2 = 25$

c) $a_2 = 8$

d) $\displaystyle\sum_{k=1}^{4} 5k$

e) S_3

f) $1 + 4 + 9 + 16$

In each of the following, the nth term of a sequence is given. In each case, find the first 4 terms; the 10th term, a_{10}; and the 15th term, a_{15}.

7. $a_n = 2n + 3$

8. $a_n = 5n - 2$

9. $a_n = n^2 + 2$

10. $a_n = \dfrac{n}{n + 1}$

11. $a_n = \dfrac{n^2 - 1}{n^2 + 1}$

12. $a_n = n^2 - 2n$

13. $a_n = \left(-\dfrac{1}{2}\right)^{n-1}$

14. $a_n = n + \dfrac{1}{n}$

15. $a_n = (-1)^n(n + 3)$

16. $a_n = (-1)^n n^2$

17. $a_n = (-1)^n(n^3 - 1)$

18. $a_n = (-1)^{n+1}(3n - 5)$

Find the indicated term of each sequence.

19. $a_n = 2n - 3;\ a_8$

20. $a_n = 3n + 2;\ a_8$

21. $a_n = (3n + 1)(2n - 5);\ a_9$

22. $a_n = (3n + 2)^2$; a_6

23. $a_n = (-1)^{n-1}(3.4n - 17.3)$; a_{12}

24. $a_n = (-2)^{n-2}(45.68 - 1.2n)$; a_{23}

25. $a_n = 3n^2(9n - 100)$; a_{11}

26. $a_n = 4n^2(2n - 39)$; a_{22}

27. $a_n = \left(1 + \dfrac{1}{n}\right)^2$; a_{20}

28. $a_n = \left(1 - \dfrac{1}{n}\right)^3$; a_{15}

Look for a pattern and then predict the general term, or nth term, a_n, of each sequence. Answers may vary.

29. $2, 4, 6, 8, 10, \ldots$

30. $1, 3, 5, 7, \ldots$

31. $1, -1, 1, -1, \ldots$

32. $-1, 1, -1, 1, \ldots$

33. $-1, 2, -3, 4, \ldots$

34. $1, -2, 3, -4, \ldots$

35. $3, 5, 7, 9, \ldots$

36. $4, 6, 8, 10, \ldots$

37. $-2, 6, -18, 54, \ldots$

38. $-2, 3, 8, 13, 18, \ldots$

39. $\frac{1}{2}, \frac{2}{3}, \frac{3}{4}, \frac{4}{5}, \frac{5}{6}, \ldots$

40. $1 \cdot 2, 2 \cdot 3, 3 \cdot 4, 4 \cdot 5, \ldots$

41. $5, 25, 125, 625, \ldots$

42. $4, 16, 64, 256, \ldots$

43. $-1, 4, -9, 16, \ldots$

44. $1, -4, 9, -16, \ldots$

Find the indicated partial sum for each sequence.

45. $1, -2, 3, -4, 5, -6, \ldots$; S_7

46. $1, -3, 5, -7, 9, -11, \ldots$; S_8

47. $2, 4, 6, 8, \ldots$; S_5

48. $1, \frac{1}{4}, \frac{1}{9}, \frac{1}{16}, \frac{1}{25}, \ldots$; S_5

Write out and evaluate each sum.

49. $\displaystyle\sum_{k=1}^{5} \dfrac{1}{2k}$

50. $\displaystyle\sum_{k=1}^{6} \dfrac{1}{2k - 1}$

51. $\displaystyle\sum_{k=0}^{4} 3^k$

52. $\displaystyle\sum_{k=4}^{7} \sqrt{2k + 1}$

53. $\displaystyle\sum_{k=2}^{8} \dfrac{k}{k - 1}$

54. $\displaystyle\sum_{k=2}^{5} \dfrac{k - 2}{k + 3}$

55. $\displaystyle\sum_{k=1}^{8} (-1)^{k+1} 2^k$

56. $\displaystyle\sum_{k=1}^{7} (-1)^k 4^{k+1}$

57. $\displaystyle\sum_{k=0}^{5} (k^2 - 2k + 3)$

58. $\displaystyle\sum_{k=0}^{5} (k^2 - 3k + 4)$

59. $\displaystyle\sum_{k=3}^{5} \dfrac{(-1)^k}{k(k + 1)}$

60. $\displaystyle\sum_{k=3}^{7} \dfrac{k}{2^k}$

Rewrite each sum using sigma notation. Answers may vary.

61. $\dfrac{2}{3} + \dfrac{3}{4} + \dfrac{4}{5} + \dfrac{5}{6} + \dfrac{6}{7}$

62. $3 + 6 + 9 + 12 + 15$

63. $1 + 4 + 9 + 16 + 25 + 36$

64. $\dfrac{1}{1^2} + \dfrac{1}{2^2} + \dfrac{1}{3^2} + \dfrac{1}{4^2} + \dfrac{1}{5^2}$

65. $4 - 9 + 16 - 25 + \cdots + (-1)^n n^2$

66. $9 - 16 + 25 + \cdots + (-1)^{n+1} n^2$

67. $5 + 10 + 15 + 20 + 25 + \cdots$

68. $7 + 14 + 21 + 28 + 35 + \cdots$

69. $\dfrac{1}{1 \cdot 2} + \dfrac{1}{2 \cdot 3} + \dfrac{1}{3 \cdot 4} + \dfrac{1}{4 \cdot 5} + \cdots$

70. $\dfrac{1}{1 \cdot 2^2} + \dfrac{1}{2 \cdot 3^2} + \dfrac{1}{3 \cdot 4^2} + \dfrac{1}{4 \cdot 5^2} + \cdots$

71. The sequence $1, 4, 9, 16, \ldots$ can be written as $f(x) = x^2$ with the domain the set of all positive integers. Explain how the graph of f would compare with the graph of $y = x^2$.

72. Eric says he expects he will prefer sequences to functions because he dislikes fractions. Will his expectations prove correct? Why or why not?

SKILL MAINTENANCE

Evaluate. [1.1]

73. $\frac{7}{2}(a_1 + a_7)$, for $a_1 = 8$ and $a_7 = 14$

74. $a_1 + (n - 1)d$, for $a_1 = 3$, $n = 6$, and $d = 4$

Multiply. [5.2]

75. $(x + y)^3$

76. $(a - b)^3$

77. $(2a - b)^3$

78. $(2x + y)^3$

SYNTHESIS

79. Explain why the equation

$$\sum_{k=1}^{n}(a_k + b_k) = \sum_{k=1}^{n} a_k + \sum_{k=1}^{n} b_k$$

is true for any positive integer n. What laws are used to justify this result?

80. Consider the sums

$$\sum_{k=1}^{5} 3k^2 \quad \text{and} \quad 3\sum_{k=1}^{5} k^2.$$

a) Which is easier to evaluate and why?
b) Is it true that

$$\sum_{k=1}^{n} ca_k = c\sum_{k=1}^{n} a_k?$$

Why or why not?

Some sequences are given by a recursive definition. The value of the first term, a_1, is given, and then we are told how to find any subsequent term from the term preceding it. Find the first six terms of each of the following recursively defined sequences.

81. $a_1 = 1$, $a_{n+1} = 5a_n - 2$

82. $a_1 = 0$, $a_{n+1} = a_n^2 + 3$

83. *Value of a copier.*   The value of a color photo-copier is $5200. Its scrap value each year is 75% of its value the year before. Give a sequence that lists the scrap value of the machine at the start of each year for a 10-yr period.

84. *Cell biology.*   A single cell of bacterium divides into two every 15 min. Suppose that the same rate of division is maintained for 4 hr. Give a sequence that lists the number of cells after successive 15-min periods.

85. Find S_{100} and S_{101} for the sequence in which $a_n = (-1)^n$.

Find the first five terms of each sequence; then find S_5.

86. $a_n = \frac{1}{2^n} \log 1000^n$

87. $a_n = i^n$, $i = \sqrt{-1}$

88. Find all values for x that solve the following:

$$\sum_{k=1}^{x} i^k = -1.$$

89. The nth term of a sequence is given by

$$a_n = n^5 - 14n^4 + 6n^3 + 416n^2 - 655n - 1050.$$

Use a graphing calculator with a TABLE feature to determine what term in the sequence is 6144.

90. To define a sequence recursively on a graphing calculator (see Exercises 81 and 82), the SEQ MODE is used. The general term U_n or V_n can often be expressed in terms of U_{n-1} or V_{n-1} by pressing **2ND** **7** or **2ND** **8**. The starting values of U_n, V_n, and n are set as one of the WINDOW variables.

Use recursion to determine how many handshakes will occur if a group of 50 people shake hands with one another. To develop the recursion formula, begin with a group of 2 and determine how many additional handshakes occur with the arrival of each new group member. (See the Collaborative Corner following Exercise Set 5.1 on p. 290.)

11.2 Arithmetic Sequences and Series

Arithmetic Sequences • Sum of the First *n* Terms of an
Arithmetic Sequence • Problem Solving

In this section, we concentrate on sequences and series that are said to be
arithmetic (pronounced ar-ith-MET-ik).

Arithmetic Sequences

In an **arithmetic sequence** (or **progression**), any term (other than the first) can
be found by adding the same number to its preceding term. For example, the
sequence 2, 5, 8, 11, 14, 17,... is arithmetic because adding 3 to any term produces the next term.

Arithmetic Sequence

A sequence is *arithmetic* if there exists a number d, called the
common difference, such that $a_{n+1} = a_n + d$ for any integer $n \geq 1$.

EXAMPLE 1   For each arithmetic sequence, identify the first term, a_1, and the common
difference, d.

a) 4, 9, 14, 19, 24,...   **b)** 27, 20, 13, 6, −1, −8,...

Solution   To find a_1, we simply use the first term listed. To find d, we choose
any term other than a_1 and subtract the preceding term from it.

Sequence	First Term, a_1	Common Difference, d
a) 4, 9, 14, 19, 24,...	4	$5 \leftarrow 9 - 4 = 5$
b) 27, 20, 13, 6, −1, −8,...	27	$-7 \leftarrow 20 - 27 = -7$

To find the common difference, we subtracted a_1 from a_2. Had we subtracted
a_2 from a_3 or a_3 from a_4, we would have found the same values for d.

Check: As a check, note that when d is added to each term, the result is the
next term in the sequence.

a) $4 + 5 = 9$, $9 + 5 = 14$, $14 + 5 = 19$, $19 + 5 = 24$

b) $27 + (-7) = 20$, $20 + (-7) = 13$, $13 + (-7) = 6$, $6 + (-7) = -1$,
    $-1 + (-7) = -8$

To develop a formula for the general, nth, term of any arithmetic sequence, we denote the common difference by d and write out the first few terms:

$a_1,$

$a_2 = a_1 + d,$

$a_3 = a_2 + d = (a_1 + d) + d = a_1 + 2d,$ Substituting $a_1 + d$ for a_2

$a_4 = a_3 + d = (a_1 + 2d) + d = a_1 + 3d.$ Substituting $a_1 + 2d$ for a_3

Note that the coefficient of d in each case is 1 less than the subscript.

Generalizing, we obtain the following formula.

To Find a_n for an Arithmetic Sequence

The nth term of an arithmetic sequence with common difference d is

$$a_n = a_1 + (n - 1)d, \quad \text{for any integer } n \geq 1.$$

EXAMPLE 2   Find the 14th term of the arithmetic sequence $6, 9, 12, 15, \ldots$.

Solution   First we note that $a_1 = 6$, $d = 3$, and $n = 14$. Using the formula for the nth term of an arithmetic sequence, we have

$$a_n = a_1 + (n - 1)d$$
$$a_{14} = 6 + (14 - 1) \cdot 3 = 6 + 13 \cdot 3 = 6 + 39 = 45.$$

The 14th term is 45.

EXAMPLE 3   For the sequence in Example 2, which term is 300? That is, find n if $a_n = 300$.

Solution   We substitute into the formula for the nth term of an arithmetic sequence and solve for n:

$$a_n = a_1 + (n - 1)d$$
$$300 = 6 + (n - 1) \cdot 3$$
$$300 = 6 + 3n - 3$$
$$297 = 3n$$
$$99 = n.$$

The term 300 is the 99th term of the sequence.

Given two terms and their places in an arithmetic sequence, we can construct the sequence.

EXAMPLE 4   The 3rd term of an arithmetic sequence is 14, and the 16th term is 79. Find a_1 and d and construct the sequence.

Solution   We know that $a_3 = 14$ and $a_{16} = 79$. Thus we would have to add d 13 times to get from 14 to 79. That is,

$$14 + 13d = 79. \qquad a_3 \text{ and } a_{16} \text{ are 13 terms apart; } 16 - 3 = 13$$

Solving $14 + 13d = 79$, we obtain

$$13d = 65 \qquad \text{Subtracting 14 from both sides}$$
$$d = 5. \qquad \text{Dividing both sides by 13}$$

We subtract d twice from a_3 to get to a_1. Thus,

$$a_1 = 14 - 2 \cdot 5 = 4. \qquad a_1 \text{ and } a_3 \text{ are 2 terms apart; } 3 - 1 = 2$$

The sequence is 4, 9, 14, 19, ... Note that we could have subtracted d 15 times from a_{16} in order to find a_1.

In general, d should be subtracted $(n - 1)$ times from a_n in order to find a_1.

Sum of the First n Terms of an Arithmetic Sequence

When the terms of an arithmetic sequence are added, an **arithmetic series** is formed. To develop a formula for computing S_n when the series is arithmetic, we list the first n terms of the sequence as follows:

This is the next-to-last term. If you add d to this term, the result is a_n.

$$a_1, (a_1 + d), (a_1 + 2d), \ldots, (a_n - 2d), \overbrace{(a_n - d)}, a_n$$

This term is two terms back from the end. If you add d to this term, you get the next-to-last term, $a_n - d$.

Thus, S_n is given by

$$S_n = a_1 + (a_1 + d) + (a_1 + 2d) + \cdots + (a_n - 2d) + (a_n - d) + a_n.$$

Using a commutative law, we have a second equation:

$$S_n = a_n + (a_n - d) + (a_n - 2d) + \cdots + (a_1 + 2d) + (a_1 + d) + a_1.$$

Adding corresponding terms on each side of the above equations, we get

$$2S_n = [a_1 + a_n] + [(a_1 + d) + (a_n - d)] + [(a_1 + 2d) + (a_n - 2d)]$$
$$+ \cdots + [(a_n - 2d) + (a_1 + 2d)] + [(a_n - d) + (a_1 + d)] + [a_n + a_1].$$

This simplifies to

$$2S_n = [a_1 + a_n] + [a_1 + a_n] + [a_1 + a_n]$$
$$+ \cdots + [a_n + a_1] + [a_n + a_1] + [a_n + a_1]. \qquad \text{There are } n \text{ bracketed sums.}$$

Since $[a_1 + a_n]$ is being added n times, it follows that

$$2S_n = n[a_1 + a_n].$$

Dividing both sides by 2 leads to the following formula.

Student Notes _____

The formula for the sum of an arithmetic sequence is very useful, but remember that it does not work for sequences that are not arithmetic.

> **To Find S_n for an Arithmetic Sequence**
>
> The sum of the first n terms of an arithmetic sequence is given by
>
> $$S_n = \frac{n}{2}(a_1 + a_n).$$

EXAMPLE 5 Find the sum of the first 100 positive even numbers.

Solution The sum is

$$2 + 4 + 6 + \cdots + 198 + 200.$$

This is the sum of the first 100 terms of the arithmetic sequence for which

$$a_1 = 2, \quad n = 100, \quad \text{and} \quad a_n = 200.$$

Substituting in the formula

$$S_n = \frac{n}{2}(a_1 + a_n),$$

we get

$$S_{100} = \frac{100}{2}(2 + 200)$$

$$= 50(202) = 10{,}100.$$

The above formula is useful when we know the first and last terms, a_1 and a_n. To find S_n when a_n is unknown, but a_1, n, and d are known, we can use the formula $a_n = a_1 + (n - 1)d$ to calculate a_n and then proceed as in Example 5.

EXAMPLE 6 Find the sum of the first 15 terms of the arithmetic sequence $4, 7, 10, 13, \ldots$.

Solution Note that

$$a_1 = 4, \quad n = 15, \quad \text{and} \quad d = 3.$$

Before using the formula for S_n, we find a_{15}:

$$a_{15} = 4 + (15 - 1)3 \qquad \text{Substituting into the formula for } a_n$$

$$= 4 + 14 \cdot 3 = 46.$$

Thus, knowing that $a_{15} = 46$, we have

$$S_{15} = \tfrac{15}{2}(4 + 46) \qquad \text{Using the formula for } S_n$$

$$= \tfrac{15}{2}(50) = 375.$$

Problem Solving

In problem-solving situations, translation may involve sequences or series. As always, there is often a variety of ways in which a problem can be solved. You should use the approach that is best or easiest for you. In this chapter, however, we will try to emphasize sequences and series and their related formulas.

EXAMPLE 7   Hourly wages.   Chris accepts a job managing a music store, starting with an hourly wage of $14.60, and is promised a raise of 25¢ per hour every 2 months for 5 years. After 5 years of work, what will be Chris's hourly wage?

Solution

1. **Familiarize.**   It helps to write down the hourly wage for several two-month time periods.

   Beginning: 14.60,
   After two months: 14.85,
   After four months: 15.10,

   and so on.

   What appears is a sequence of numbers: 14.60, 14.85, 15.10, ... Since the same amount is added each time, the sequence is arithmetic.

   We list what we know about arithmetic sequences. The pertinent formulas are

   $$a_n = a_1 + (n - 1)d$$

   and

   $$S_n = \frac{n}{2}(a_1 + a_n).$$

   In this case, we are not looking for a sum, so it is probably the first formula that will give us our answer. We want to determine the last term in a sequence. To do so, we need to know a_1, n, and d. From our list above, we see that

   $$a_1 = 14.60 \quad \text{and} \quad d = 0.25.$$

   What is n? That is, how many terms are in the sequence? After 1 year, there have been 6 raises, since Chris gets a raise every 2 months. There are 5 years, so the total number of raises will be $5 \cdot 6$, or 30. Altogether, there will be 31 terms: the original wage and 30 increased rates.

2. **Translate.**   We want to find a_n for the arithmetic sequence in which $a_1 = 14.60$, $n = 31$, and $d = 0.25$.

3. **Carry out.**   Substituting in the formula for a_n gives us

   $$a_{31} = 14.60 + (31 - 1) \cdot 0.25$$
   $$= 22.10.$$

4. **Check.**   We can check by redoing the calculations or we can calculate in a slightly different way for another check. For example, at the end of a year, there will be 6 raises, for a total raise of $1.50. At the end of 5 years, the total raise will be $5 \times \$1.50$, or $7.50. If we add that to the original wage of $14.60, we obtain $22.10. The answer checks.

5. **State.**   After 5 years, Chris's hourly wage will be $22.10.

EXAMPLE 8   Telephone pole storage.   A stack of telephone poles has 30 poles in the bottom row. There are 29 poles in the second row, 28 in the next row, and so on. How many poles are in the stack if there are 5 poles in the top row?

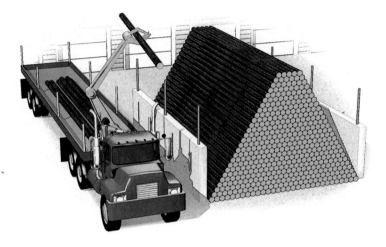

Solution

1. **Familiarize.**   The following figure shows the ends of the poles. There are 30 poles on the bottom and one fewer in each successive row. How many rows will there be?

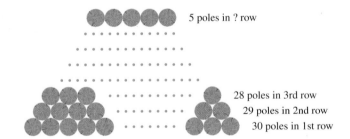

5 poles in ? row

28 poles in 3rd row
29 poles in 2nd row
30 poles in 1st row

Note that there are $30 - 1 = 29$ poles in the 2nd row, $30 - 2 = 28$ poles in the 3rd row, $30 - 3 = 27$ poles in the 4th row, and so on. The pattern leads to $30 - 25 = 5$ poles in the 26th row.

The situation is represented by the equation

$30 + 29 + 28 + \cdots + 5.$ There are 26 terms in this series.

Thus we have an arithmetic series. We recall the formula

$$S_n = \frac{n}{2}(a_1 + a_n).$$

2. **Translate.**   We want to find the sum of the first 26 terms of an arithmetic sequence in which $a_1 = 30$ and $a_{26} = 5$.

3. **Carry out.**   Substituting into the above formula gives us

$$S_{26} = \frac{26}{2}(30 + 5)$$
$$= 13 \cdot 35 = 455.$$

4. Check. In this case, we can check the calculations by doing them again. A longer, harder way would be to do the entire addition:

$$30 + 29 + 28 + \cdots + 5.$$

5. State. There are 455 poles in the stack.

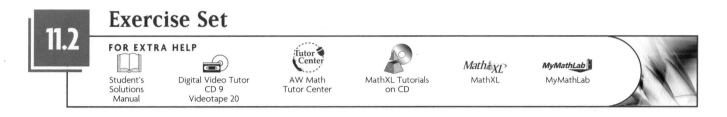

Exercise Set

11.2

FOR EXTRA HELP

Student's Solutions Manual   Digital Video Tutor CD 9 Videotape 20   AW Math Tutor Center   MathXL Tutorials on CD   MathXL   MyMathLab

↪ *Concept Reinforcement*   *Classify each statement as either true or false.*

1. In an arithmetic sequence, the difference between any two consecutive terms is always the same.

2. In an arithmetic sequence, if $a_9 - a_8 = 4$, then $a_{13} - a_{12} = 4$ as well.

3. In an arithmetic sequence containing 17 terms, the common difference is $a_{17} - a_1$.

4. In an arithmetic sequence, if $a_1 = 8$ and $a_3 = 12$, then a_6 must be 16.

5. The sum of the first 20 terms of an arithmetic sequence can be found by knowing just a_1 and a_{20}.

6. The sum of the first 30 terms of an arithmetic sequence can be found by knowing just a_1 and d, the common difference.

7. The notation S_5 means $a_1 + a_5$.

8. For any arithmetic sequence, $S_9 = S_8 + d$, where d is the common difference.

Find the first term and the common difference.

9. $2, 6, 10, 14, \ldots$

10. $1.06, 1.12, 1.18, 1.24, \ldots$

11. $7, 3, -1, -5, \ldots$

12. $-8, -5, -2, 1, \ldots$

13. $\frac{3}{2}, \frac{9}{4}, 3, \frac{15}{4}, \ldots$

14. $\frac{3}{5}, \frac{1}{10}, -\frac{2}{5}, \ldots$

15. $\$5.12, \$5.24, \$5.36, \$5.48, \ldots$

16. $\$214, \$211, \$208, \$205, \ldots$

17. Find the 15th term of the arithmetic sequence $7, 10, 13, \ldots$.

18. Find the 17th term of the arithmetic sequence $6, 10, 14, \ldots$.

19. Find the 18th term of the arithmetic sequence $8, 2, -4, \ldots$.

20. Find the 14th term of the arithmetic sequence $3, \frac{7}{3}, \frac{5}{3}, \ldots$.

21. Find the 13th term of the arithmetic sequence $\$1200, \$964.32, \$728.64, \ldots$.

22. Find the 10th term of the arithmetic sequence $\$2345.78, \$2967.54, \$3589.30, \ldots$.

23. In the sequence of Exercise 17, what term is 82?

24. In the sequence of Exercise 18, what term is 126?

25. In the sequence of Exercise 19, what term is -328?

26. In the sequence of Exercise 20, what term is -27?

27. Find a_{17} when $a_1 = 2$ and $d = 5$.

28. Find a_{20} when $a_1 = 14$ and $d = -3$.

29. Find a_1 when $d = 4$ and $a_8 = 33$.

30. Find a_1 when $d = 8$ and $a_{11} = 26$.

31. Find n when $a_1 = 5$, $d = -3$, and $a_n = -76$.

32. Find n when $a_1 = 25$, $d = -14$, and $a_n = -507$.

33. For an arithmetic sequence in which $a_{17} = -40$ and $a_{28} = -73$, find a_1 and d. Write the first five terms of the sequence.

34. In an arithmetic sequence, $a_{17} = \frac{25}{3}$ and $a_{32} = \frac{95}{6}$. Find a_1 and d. Write the first five terms of the sequence.

Aha! **35.** Find a_1 and d if $a_{13} = 13$ and $a_{54} = 54$.

36. Find a_1 and d if $a_{12} = 24$ and $a_{25} = 50$.

37. Find the sum of the first 20 terms of the arithmetic series $1 + 5 + 9 + 13 + \cdots$.

38. Find the sum of the first 14 terms of the arithmetic series $11 + 7 + 3 + \cdots$.

39. Find the sum of the first 250 natural numbers.

40. Find the sum of the first 400 natural numbers.

41. Find the sum of the even numbers from 2 to 100, inclusive.

42. Find the sum of the odd numbers from 1 to 99, inclusive.

43. Find the sum of all multiples of 6 from 6 to 102, inclusive.

44. Find the sum of all multiples of 4 that are between 15 and 521.

45. An arithmetic series has $a_1 = 4$ and $d = 5$. Find S_{20}.

46. An arithmetic series has $a_1 = 9$ and $d = -3$. Find S_{32}.

Solve.

47. *Band formations.* The South Brighton Drum and Bugle Corps has 7 marchers in the front row, 9 in the second row, 11 in the third row, and so on, for 15 rows. How many marchers are in the last row? How many marchers are there altogether?

48. *Gardening.* A gardener is planting bulbs near an entrance to a college. She has 39 plants in the front row, 35 in the second row, 31 in the third row, and so on. If the pattern is consistent, how many plants will be in the last row? How many plants will there be altogether?

49. *Archaeology.* Many ancient Mayan pyramids were constructed over a span of several generations. Each layer of the pyramid has a stone perimeter, enclosing a layer of dirt or debris on which a structure once stood. One drawing of such a pyramid indicates that the perimeter of the bottom layer contains 36 stones, the next level up contains 32 stones, and so on, up to the top row which contains 4 stones. How many stones are in the pyramid?

50. *Telephone pole piles.* How many poles will be in a pile of telephone poles if there are 50 in the first layer, 49 in the second, and so on, until there are 6 in the top layer?

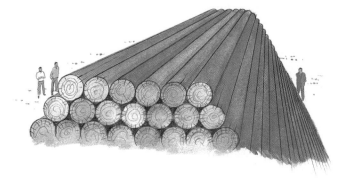

51. *Accumulated savings.* If 10¢ is saved on October 1, another 20¢ on October 2, another 30¢ on October 3, and so on, how much is saved during October? (October has 31 days.)

52. *Accumulated savings.* Renata saves money in an arithmetic sequence: $700 for the first year, another $850 the second, and so on, for 20 yr. How much does she save in all (disregarding interest)?

53. *Auditorium design.*   Theaters are often built with more seats per row as the rows move toward the back. The Sanders Amphitheater has 20 seats in the first row, 22 in the second, 24 in the third, and so on, for 19 rows. How many seats are in the amphitheater?

54. *Accumulated savings.*   Shirley sets up an investment such that it will return $5000 the first year, $6125 the second year, $7250 the third year, and so on, for 25 yr. How much in all is received from the investment?

55. It is said that as a young child, the mathematician Karl F. Gauss (1777–1855) was able to compute the sum $1 + 2 + 3 + \cdots + 100$ very quickly in his head. Explain how Gauss might have done this and present a formula for the sum of the first n natural numbers. (*Hint*: $1 + 99 = 100$.)

56. Is it true that if every number in a sequence is doubled and then added, the result is the same as if the numbers were first added and the sum then doubled? Why or why not?

SKILL MAINTENANCE

Simplify. [6.2]

57. $\dfrac{3}{10x} + \dfrac{2}{15x}$

58. $\dfrac{2}{9t} + \dfrac{5}{12t}$

Convert to an exponential equation. [9.3]

59. $\log_a P = k$

60. $\ln t = a$

Find an equation of the circle satisfying the given conditions. [10.1]

61. Center $(0, 0)$, radius 9

62. Center $(-2, 5)$, radius $3\sqrt{2}$

SYNTHESIS

63. Write a problem for a classmate to solve. Devise the problem so that its solution requires computing S_{17} for an arithmetic sequence.

64. The sum of the first n terms of an arithmetic sequence is also given by

$$S_n = \frac{n}{2}[2a_1 + (n - 1)d].$$

Use the earlier formulas for a_n and S_n to explain how this equation was developed.

65. A frog is at the bottom of a 100-ft well. With each jump, the frog climbs 4 ft, but then slips back 1 ft. How many jumps does it take for the frog to reach the top of the hole?

66. Find a formula for the sum of the first n consecutive odd numbers starting with 1:

$$1 + 3 + 5 + \cdots + (2n - 1).$$

67. In an arithmetic sequence, $a_1 = \$8760$ and $d = -\$798.23$. Find the first 10 terms of the sequence.

68. Find the sum of the first 10 terms of the sequence given in Exercise 67.

69. Prove that if p, m, and q are consecutive terms in an arithmetic sequence, then

$$m = \frac{p + q}{2}.$$

70. *Straight-line depreciation.*   A company buys a color copier for $5200 on January 1 of a given year. The machine is expected to last for 8 yr, at the end of which time its *trade-in*, or *salvage*, *value* will be $1100. If the company figures the decline in value to be the same each year, then the trade-in values, after t years, $0 \le t \le 8$, form an arithmetic sequence given by

$$a_t = C - t\left(\frac{C - S}{N}\right),$$

where C is the original cost of the item, N the years of expected life, and S the salvage value.

a) Find the formula for a_t for the straight-line depreciation of the copier.

b) Find the salvage value after 0 yr, 1 yr, 2 yr, 3 yr, 4 yr, 7 yr, and 8 yr.

c) Find a formula that expresses a_t recursively.

71. Use your answer to Exercise 39 to find the sum of all integers from 501 through 750.

11.3 Geometric Sequences and Series

Geometric Sequences • Sum of the First *n* Terms of a
Geometric Sequence • Infinite Geometric Series •
Problem Solving

In an arithmetic sequence, a certain number is added to each term to get the
next term. When each term in a sequence is *multiplied* by a certain fixed number to get the next term, the sequence is **geometric**. In this section, we examine geometric sequences (or progressions) and *geometric series*.

Geometric Sequences

Consider the sequence

$$2, 6, 18, 54, 162, \ldots .$$

If we multiply each term by 3, we obtain the next term. The multiplier is called
the *common ratio* because it is found by dividing any term by the preceding
term.

Geometric Sequence

A sequence is *geometric* if there exists a number r, called the
common ratio, for which

$$\frac{a_{n+1}}{a_n} = r, \quad \text{or} \quad a_{n+1} = a_n \cdot r \quad \text{for any integer } n \geq 1.$$

EXAMPLE 1   For each geometric sequence, find the common ratio.

a) $4, 20, 100, 500, 2500, \ldots$

b) $3, -6, 12, -24, 48, -96, \ldots$

c) $\$5200, \$3900, \$2925, \$2193.75, \ldots$

Solution

Sequence	*Common Ratio*	
a) $4, 20, 100, 500, 2500, \ldots$	5	$\frac{20}{4} = 5, \frac{100}{20} = 5$, and so on
b) $3, -6, 12, -24, 48, -96, \ldots$	-2	$\frac{-6}{3} = -2, \frac{12}{-6} = -2$, and so on
c) $\$5200, \$3900, \$2925, \$2193.75, \ldots$	0.75	$\frac{\$3900}{\$5200} = 0.75, \frac{\$2925}{\$3900} = 0.75$

To develop a formula for the general, or *n*th, term of a geometric sequence, let a_1 be the first term and let r be the common ratio. We write out the first few terms as follows:

a_1,

$a_2 = a_1 r$,

$a_3 = a_2 r = (a_1 r)r = a_1 r^2$,   Substituting $a_1 r$ for a_2

$a_4 = a_3 r = (a_1 r^2)r = a_1 r^3$.   Substituting $a_1 r^2$ for a_3

Note that the exponent is 1 less than the subscript.

Generalizing, we obtain the following.

> *To Find a_n for a Geometric Sequence*
>
> The *n*th term of a geometric sequence with common ratio r is given by
>
> $$a_n = a_1 r^{n-1}, \quad \text{for any integer } n \geq 1.$$

EXAMPLE 2   Find the 7th term of the geometric sequence 4, 20, 100, ...

Solution   First, we note that

$$a_1 = 4 \quad \text{and} \quad n = 7.$$

To find the common ratio, we can divide any term (other than the first) by the term preceding it. Since the second term is 20 and the first is 4,

$$r = \frac{20}{4}, \quad \text{or 5.}$$

The formula

$$a_n = a_1 r^{n-1}$$

gives us

$$a_7 = 4 \cdot 5^{7-1} = 4 \cdot 5^6 = 4 \cdot 15{,}625 = 62{,}500.$$

EXAMPLE 3   Find the 10th term of the geometric sequence

$$64, -32, 16, -8, \ldots.$$

Solution   First, we note that

$$a_1 = 64, \qquad n = 10, \quad \text{and} \quad r = \frac{-32}{64} = -\frac{1}{2}.$$

Then, using the formula for the *n*th term of a geometric sequence, we have

$$a_{10} = 64 \cdot \left(-\frac{1}{2}\right)^{10-1} = 64 \cdot \left(-\frac{1}{2}\right)^9 = 2^6 \cdot \left(-\frac{1}{2^9}\right) = -\frac{1}{2^3} = -\frac{1}{8}.$$

The 10th term is $-\frac{1}{8}$.

Sum of the First n Terms of a Geometric Sequence

We next develop a formula for S_n when a sequence is geometric:

$$a_1, \; a_1r, \; a_1r^2, \; a_1r^3, \ldots, a_1r^{n-1}, \ldots.$$

The **geometric series** S_n is given by

$$S_n = a_1 + a_1r + a_1r^2 + \cdots + a_1r^{n-2} + a_1r^{n-1}. \qquad (1)$$

Multiplying both sides by r gives us

$$rS_n = a_1r + a_1r^2 + a_1r^3 + \cdots + a_1r^{n-1} + a_1r^{n}. \qquad (2)$$

When we subtract corresponding sides of equation (2) from equation (1), the color terms drop out, leaving

$$S_n - rS_n = a_1 - a_1r^n$$
$$S_n(1 - r) = a_1(1 - r^n), \qquad \text{Factoring}$$

or

$$S_n = \frac{a_1(1 - r^n)}{1 - r}. \qquad \text{Dividing both sides by } 1 - r$$

> **To Find S_n for a Geometric Sequence**
>
> The sum of the first n terms of a geometric sequence with common ratio r is given by
>
> $$S_n = \frac{a_1(1 - r^n)}{1 - r}, \text{ for any } r \neq 1.$$

EXAMPLE 4   Find the sum of the first 7 terms of the geometric sequence $3, 15, 75, 375, \ldots$.

Solution   First, we note that

$$a_1 = 3, \quad n = 7, \quad \text{and} \quad r = \frac{15}{3} = 5.$$

Then, substituting in the formula $S_n = \dfrac{a_1(1 - r^n)}{1 - r}$, we have

$$S_7 = \frac{3(1 - 5^7)}{1 - 5} = \frac{3(1 - 78{,}125)}{-4}$$
$$= \frac{3(-78{,}124)}{-4}$$
$$= 58{,}593.$$

Infinite Geometric Series

Suppose we consider the sum of the terms of an infinite geometric sequence, such as $3, 6, 12, 24, 48, \ldots$. We get what is called an **infinite geometric series**:

$$3 + 6 + 12 + 24 + 48 + \cdots.$$

Here, as n increases, the sum of the first n terms, S_n, increases without bound. There are also infinite series that get closer and closer to some specific number. Here is an example:

$$\frac{1}{2} + \frac{1}{4} + \frac{1}{8} + \frac{1}{16} + \cdots + \frac{1}{2^n} + \cdots.$$

Let's consider S_n for the first four values of n:

$$S_1 = \tfrac{1}{2} \qquad\qquad\qquad = \tfrac{1}{2} = 0.5,$$
$$S_2 = \tfrac{1}{2} + \tfrac{1}{4} \qquad\qquad = \tfrac{3}{4} = 0.75,$$
$$S_3 = \tfrac{1}{2} + \tfrac{1}{4} + \tfrac{1}{8} \qquad = \tfrac{7}{8} = 0.875,$$
$$S_4 = \tfrac{1}{2} + \tfrac{1}{4} + \tfrac{1}{8} + \tfrac{1}{16} = \tfrac{15}{16} = 0.9375.$$

> The denominator of each sum is 2^n, where n is the subscript of S. The numerator is $2^n - 1$.

Thus, for this particular series, we have

$$S_n = \frac{2^n - 1}{2^n} = \frac{2^n}{2^n} - \frac{1}{2^n} = 1 - \frac{1}{2^n}.$$

Note that as n gets larger and larger, the value of $1/2^n$ gets closer to 0 and the value of S_n gets closer to 1. We say that 1 is the **limit** of S_n and that 1 is the sum of this infinite geometric series. An infinite geometric series is denoted S_∞. It can be shown (but we will not do it here) that the sum of the terms of an infinite geometric sequence exists if and only if $|r| < 1$ (that is, the common ratio's absolute value is less than 1).

To find a formula for the sum of an infinite geometric series, we first consider the sum of the first n terms:

$$S_n = \frac{a_1(1 - r^n)}{1 - r} = \frac{a_1 - a_1 r^n}{1 - r}. \qquad \text{Using the distributive law}$$

For $|r| < 1$, it follows that the value of r^n gets closer to 0 as n gets larger. (Check this by selecting a number between -1 and 1 and finding larger and larger powers on a calculator.) As r^n gets closer to 0, so too does $a_1 r^n$. Thus, S_n gets closer to $a_1/(1 - r)$.

> **The Limit of an Infinite Geometric Series**
>
> When $|r| < 1$, the limit of an infinite geometric series is given by
>
> $$S_\infty = \frac{a_1}{1 - r}. \qquad \text{(For } |r| \geq 1, \text{ no limit exists.)}$$

EXAMPLE 5 Determine whether each series has a limit. If a limit exists, find it.

a) $1 + 3 + 9 + 27 + \cdots$ \qquad\qquad **b)** $-35 + 7 - \tfrac{7}{5} + \tfrac{7}{25} + \cdots$

Solution

a) Here $r = 3$, so $|r| = |3| = 3$. Since $|r| \not< 1$, the series does *not* have a limit.

b) Here $r = -\frac{1}{5}$, so $|r| = |-\frac{1}{5}| = \frac{1}{5}$. Since $|r| < 1$, the series *does* have a limit. We find the limit by substituting into the formula for S_∞:

$$S_\infty = \frac{-35}{1 - \left(-\frac{1}{5}\right)} = \frac{-35}{\frac{6}{5}} = -35 \cdot \frac{5}{6} = \frac{-175}{6} = -29\frac{1}{6}.$$

EXAMPLE 6   Find fraction notation for 0.63636363....

Solution   We can express this as

$$0.63 + 0.0063 + 0.000063 + \cdots.$$

This is an infinite geometric series, where $a_1 = 0.63$ and $r = 0.01$. Since $|r| < 1$, this series has a limit:

$$S_\infty = \frac{a_1}{1 - r} = \frac{0.63}{1 - 0.01} = \frac{0.63}{0.99} = \frac{63}{99}.$$

Thus fraction notation for 0.63636363... is $\frac{63}{99}$, or $\frac{7}{11}$.

Problem Solving

For some problem-solving situations, the translation may involve geometric sequences or series.

EXAMPLE 7   *Daily wages.*   Suppose someone offered you a job for the month of September (30 days) under the following conditions. You will be paid $0.01 for the first day, $0.02 for the second, $0.04 for the third, and so on, doubling your previous day's salary each day. How much would you earn? (Would you take the job? Make a guess before reading further.)

Solution

1. **Familiarize.**   You earn $0.01 the first day, $0.01(2)$ the second day, $0.01(2)(2)$ the third day, and so on. Since each day's wages are a constant multiple of the previous day's wages, a geometric sequence is formed.

2. **Translate.**   The amount earned is the geometric series

$$\$0.01 + \$0.01(2) + \$0.01(2^2) + \$0.01(2^3) + \cdots + \$0.01(2^{29}),$$

where $a_1 = \$0.01$, $n = 30$, and $r = 2$.

3. **Carry out.**   Using the formula

$$S_n = \frac{a_1(1 - r^n)}{1 - r},$$

we have

$$S_{30} = \frac{\$0.01(1 - 2^{30})}{1 - 2}$$

$$= \frac{\$0.01(-1,073,741,823)}{-1} \qquad \text{Using a calculator}$$

$$= \$10,737,418.23.$$

4. **Check.**   The calculations can be repeated as a check.

5. **State.**   The pay exceeds $10.7 million for the month. Most people would probably take the job!

EXAMPLE 8

Loan repayment.   Francine's student loan is in the amount of $6000. Interest is to be 9% compounded annually, and the entire amount is to be paid after 10 yr. How much is to be paid back?

Solution

1. **Familiarize.**   Suppose we let P represent any principal amount. At the end of one year, the amount owed will be $P + 0.09P$, or $1.09P$. That amount will be the principal for the second year. The amount owed at the end of the second year will be $1.09 \times$ New principal $= 1.09(1.09P)$, or 1.09^2P. Thus the amount owed at the beginning of successive years is as follows:

$$\overset{①}{\downarrow} \qquad \overset{②}{\downarrow} \qquad \overset{③}{\downarrow} \qquad \overset{④}{\downarrow}$$
$$P, \qquad 1.09P, \qquad 1.09^2P, \qquad 1.09^3P, \quad \text{and so on.}$$

We have a geometric sequence. The amount owed at the beginning of the 11th year will be the amount owed at the end of the 10th year.

2. **Translate.**   We have a geometric sequence with $a_1 = 6000$, $r = 1.09$, and $n = 11$. The appropriate formula is

$$a_n = a_1 r^{n-1}.$$

3. **Carry out.**   We substitute and calculate:

$$a_{11} = \$6000(1.09)^{11-1} = \$6000(1.09)^{10}$$
$$\approx \$14{,}204.18. \qquad \begin{array}{l}\text{Using a calculator and rounding} \\ \text{to the nearest hundredth}\end{array}$$

4. **Check.**   A check, by repeating the calculations, is left to the student.

5. **State.**   Francine will owe $14,204.18 at the end of 10 yr.

EXAMPLE 9

Bungee jumping.   A bungee jumper rebounds 60% of the height jumped. Clyde's bungee jump is made using a cord that stretches to 200 ft.

a) After jumping and then rebounding 9 times, how far has Clyde traveled upward (the total rebound distance)?

b) Approximately how far will Clyde have traveled upward (bounced) before coming to rest?

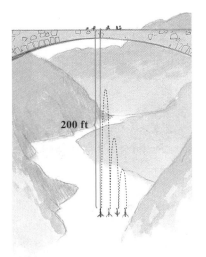

200 ft

Solution

1. Familiarize.   Let's do some calculations and look for a pattern.

First fall:	200 ft
First rebound:	0.6×200, or 120 ft
Second fall:	120 ft, or 0.6×200
Second rebound:	0.6×120, or $0.6(0.6 \times 200)$, which is 72 ft
Third fall:	72 ft, or $0.6(0.6 \times 200)$
Third rebound:	0.6×72, or $0.6(0.6(0.6 \times 200))$, which is 43.2 ft

The rebound distances form a geometric sequence:

① ② ③ ④

$120,\quad 0.6 \times 120,\quad 0.6^2 \times 120,\quad 0.6^3 \times 120, \ldots .$

2. Translate.
  a) The total rebound distance after 9 bounces is the sum of a geometric sequence. The first term is 120 and the common ratio is 0.6. There will be 9 terms, so we can use the formula

$$S_n = \frac{a_1(1 - r^n)}{1 - r}.$$

  b) Theoretically, Clyde will never stop bouncing. Realistically, the bouncing will eventually stop. To approximate the actual distance bounced, we consider an infinite number of bounces and use the formula

$$S_\infty = \frac{a_1}{1 - r}.$$ Since $r = 0.6$ and $|0.6| < 1$, we know that S_∞ exists.

3. Carry out.
  a) We substitute into the formula and calculate:

$$S_9 = \frac{120[1 - (0.6)^9]}{1 - 0.6} \approx 297.$$ Using a calculator

  b) We substitute and calculate:

$$S_\infty = \frac{120}{1 - 0.6} = 300.$$

4. Check.   We can do the calculations again.

5. State.
  a) In 9 bounces, Clyde will have traveled upward a total distance of about 297 ft.
  b) Clyde will have traveled upward a total of about 300 ft before coming to rest.

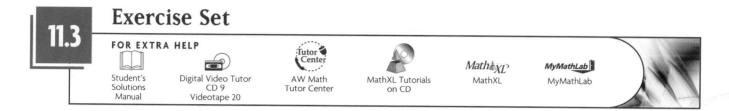

Exercise Set

11.3

FOR EXTRA HELP

Student's Solutions Manual | Digital Video Tutor CD 9 Videotape 20 | AW Math Tutor Center | MathXL Tutorials on CD | MathXL | MyMathLab

⌒ *Concept Reinforcement   Classify each of the following as an arithmetic sequence, a geometric sequence, an arithmetic series, a geometric series, or none of these.*

1. $2, 6, 18, 54, \ldots$

2. $3, 5, 7, 9, \ldots$

3. $1, 6, 11, 16, 21, \ldots$

4. $5, 15, 45, 135, 405, \ldots$

5. $4 + 20 + 100 + 500 + 2500 + 12{,}500$

6. $10 + 12 + 14 + 16 + 18 + 20$

7. $3 - \frac{3}{2} + \frac{3}{4} - \frac{3}{8} + \frac{3}{16} - \cdots$

8. $1 + \frac{1}{2} + \frac{1}{3} + \frac{1}{4} + \frac{1}{5} + \frac{1}{6} + \cdots$

Find the common ratio for each geometric sequence.

9. $7, 14, 28, 56, \ldots$

10. $2, 6, 18, 54, \ldots$

11. $6, -0.6, 0.06, -0.006, \ldots$

12. $-5, -0.5, -0.05, -0.005, \ldots$

13. $\frac{1}{2}, -\frac{1}{4}, \frac{1}{8}, -\frac{1}{16}, \ldots$

14. $\frac{2}{3}, -\frac{4}{3}, \frac{8}{3}, -\frac{16}{3}, \ldots$

15. $75, 15, 3, \frac{3}{5}, \ldots$

16. $12, -4, \frac{4}{3}, -\frac{4}{9}, \ldots$

17. $\frac{1}{m}, \frac{6}{m^2}, \frac{36}{m^3}, \frac{216}{m^4}, \ldots$

18. $4, \frac{4m}{5}, \frac{4m^2}{25}, \frac{4m^3}{125}, \ldots$

Find the indicated term for each geometric sequence.

19. $3, 6, 12, \ldots$; the 7th term

20. $2, 8, 32, \ldots$; the 9th term

21. $7, 7\sqrt{2}, 14, \ldots$; the 10th term

22. $4, 4\sqrt{3}, 12, \ldots$; the 8th term

23. $-\frac{8}{243}, \frac{8}{81}, -\frac{8}{27}, \ldots$; the 14th term

24. $\frac{7}{625}, \frac{-7}{125}, \frac{7}{25}, \ldots$; the 13th term

25. $\$1000, \$1080, \$1166.40, \ldots$; the 12th term

26. $\$1000, \$1070, \$1144.90, \ldots$; the 11th term

Find the nth, or general, term for each geometric sequence.

27. $1, 5, 25, 125, \ldots$

28. $2, 4, 8, \ldots$

29. $1, -1, 1, -1, \ldots$

30. $\frac{1}{4}, \frac{1}{16}, \frac{1}{64}, \ldots$

31. $\frac{1}{x}, \frac{1}{x^2}, \frac{1}{x^3}, \ldots$

32. $5, \frac{5m}{2}, \frac{5m^2}{4}, \ldots$

For Exercises 33–40, use the formula for S_n to find the indicated sum for each geometric series.

33. S_9 for $6 + 12 + 24 + \cdots$

34. S_6 for $16 - 8 + 4 - \cdots$

35. S_7 for $\frac{1}{18} - \frac{1}{6} + \frac{1}{2} - \cdots$

Aha! **36.** S_5 for $7 + 0.7 + 0.07 + \cdots$

37. S_8 for $1 + x + x^2 + x^3 + \cdots$

38. S_{10} for $1 + x^2 + x^4 + x^6 + \cdots$

39. S_{16} for $\$200, \$200(1.06), \$200(1.06)^2, \ldots$

40. S_{23} for $\$1000, \$1000(1.08), \$1000(1.08)^2, \ldots$

Determine whether each infinite geometric series has a limit. If a limit exists, find it.

41. $16 + 4 + 1 + \cdots$

42. $8 + 4 + 2 + \cdots$

43. $7 + 3 + \frac{9}{7} + \cdots$

44. $12 + 9 + \frac{27}{4} + \cdots$

45. $3 + 15 + 75 + \cdots$

46. $2 + 3 + \frac{9}{2} + \cdots$

47. $4 - 6 + 9 - \frac{27}{2} + \cdots$

48. $-6 + 3 - \frac{3}{2} + \frac{3}{4} - \cdots$

49. $0.43 + 0.0043 + 0.000043 + \cdots$

50. $0.37 + 0.0037 + 0.000037 + \cdots$

51. $\$500(1.02)^{-1} + \$500(1.02)^{-2} + \$500(1.02)^{-3} + \cdots$

52. $\$1000(1.08)^{-1} + \$1000(1.08)^{-2} + \$1000(1.08)^{-3} + \cdots$

Find fraction notation for each infinite sum. (Each can be regarded as an infinite geometric series.)

53. $0.7777\ldots$

54. $0.2222\ldots$

55. $8.3838\ldots$

56. $7.4747\ldots$

57. $0.15151515\ldots$

58. $0.12121212\ldots$

Solve. Use a calculator as needed for evaluating formulas.

59. *Rebound distance.* A ping-pong ball is dropped from a height of 20 ft and always rebounds one-fourth of the distance fallen. How high does it rebound the 6th time?

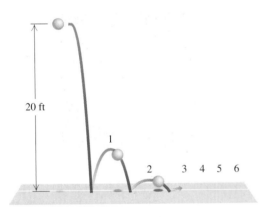

60. *Rebound distance.* Approximate the total of the rebound heights of the ball in Exercise 59.

61. *Population growth.* Yorktown has a current population of 100,000, and the population is increasing by 3% each year. What will the population be in 15 yr?

62. *Amount owed.* Gilberto borrows $15,000. The loan is to be repaid in 13 yr at 8.5% interest, compounded annually. How much will be repaid at the end of 13 yr?

63. *Shrinking population.* A population of 5000 fruit flies is dying off at a rate of 4% per minute. How many flies will be alive after 15 min?

64. *Shrinking population.* For the population of fruit flies in Exercise 63, how long will it take for only 1800 fruit flies to remain alive? (*Hint*: Use logarithms.) Round to the nearest minute.

65. *Housing units.* Approximately 534,000 new apartments and houses were built in the United States in 1991. Since then, the number has grown by about 5.35% per year. How many new apartments and houses were built in the United States from 1991 through 2004?
Sources: Based on data from the U.S. Bureau of the Census and the U.S. Department of Housing and Urban Development

66. *Housing units.* Approximately 144,000 new apartments and houses were built in the western United States in 1991. Since then, the number has grown by about 5.0% per year. How many new houses and apartments were built in the western United States from 1991 through 2004?
Sources: Based on data from the U.S. Bureau of the Census and the U.S. Department of Housing and Urban Development

67. *Rebound distance.* A superball dropped from the top of the Washington Monument (556 ft high) rebounds three-fourths of the distance fallen. How far (up and down) will the ball have traveled when it hits the ground for the 6th time?

68. *Rebound distance.* Approximate the total distance that the ball of Exercise 67 will have traveled when it comes to rest.

69. *Stacking paper.* Construction paper is about 0.02 in. thick. Beginning with just one piece, a stack is doubled again and again 10 times. Find the height of the final stack.

70. *Monthly earnings.*   Suppose you accepted a job for the month of February (28 days) under the following conditions. You will be paid $0.01 the first day, $0.02 the second, $0.04 the third, and so on, doubling your previous day's salary each day. How much would you earn?

Aha! **71.** Under what circumstances is it possible for the 5th term of a geometric sequence to be greater than the 4th term but less than the 7th term?

72. When r is negative, a series is said to be *alternating*. Why do you suppose this terminology is used?

SKILL MAINTENANCE

Multiply. [5.2]

73. $(x + y)(x^2 + 2xy + y^2)$

74. $(a - b)(a^2 - 2ab + b^2)$

Solve the system.

75. $5x - 2y = -3,$
$2x + 5y = -24$ [3.2]

76. $x - 2y + 3z = 4,$
$2x - y + z = -1,$
$4x + y + z = 1$ [3.4]

SYNTHESIS

77. Write a problem for a classmate to solve. Devise the problem so that a geometric series is involved and the solution is "The total amount in the bank is $900(1.08)^{40}$, or about \$19,550."

78. The infinite series
$$S_\infty = 2 + \frac{1}{2} + \frac{1}{2 \cdot 3} + \frac{1}{2 \cdot 3 \cdot 4} + \frac{1}{2 \cdot 3 \cdot 4 \cdot 5}$$
$$+ \frac{1}{2 \cdot 3 \cdot 4 \cdot 5 \cdot 6} + \cdots$$
is not geometric, but it does have a sum. Using S_1, S_2, S_3, S_4, S_5, and S_6, make a conjecture about the value of S_∞ and explain your reasoning.

Calculate each of the following sums.

79. $\sum_{k=1}^{\infty} 6(0.9)^k$

80. $\sum_{k=1}^{\infty} 5(-0.7)^k$

81. Find the sum of the first n terms of
$$x^2 - x^3 + x^4 - x^5 + \cdots.$$

82. Find the sum of the first n terms of
$$1 + x + x^2 + x^3 + \cdots.$$

83. The sides of a square are each 16 cm long. A second square is inscribed by joining the midpoints of the sides, successively. In the second square we repeat the process, inscribing a third square. If this process is continued indefinitely, what is the sum of all of the areas of all the squares? (*Hint*: Use an infinite geometric series.)

84. Show that $0.999\ldots$ is 1.

85. Using Example 5 and Exercises 41–52, explain how the graph of a geometric sequence can be used to determine whether a geometric series has a limit.

86. To compare the *graphs* of an arithmetic and a geometric sequence, we plot n on the horizontal axis and a_n on the vertical axis. Graph Example 1(a) of Section 11.2 and Example 1(a) of Section 11.3 on the same set of axes. How do the graphs of geometric sequences differ from the graphs of arithmetic sequences?

COLLABORATIVE

CORNER

Bargaining for a Used Car

Focus: Geometric series

Time: 30 minutes

Group size: 2

Materials: Graphing calculators are optional.

ACTIVITY*

1. One group member ("the seller") has a car for sale and is asking $3500. The second ("the buyer") offers $1500. The seller splits the difference ($2000 ÷ 2 = $1000) and lowers the price to $2500. The buyer then splits the difference again ($1000 ÷ 2 = $500) and counters with $2000. Continue in this manner and stop when you are able to agree on the car's selling price to the nearest penny.

2. What should the buyer's initial offer be in order to achieve a purchase price of $2000? (Check several guesses to find the appropriate initial offer.)

---

*This activity is based on the article, "Bargaining Theory, or Zeno's Used Cars," by James C. Kirby, *The College Mathematics Journal*, **27**(4), September 1996.

3. The seller's price in the bargaining above can be modeled recursively (see Exercises 81, 82, and 90 in Section 11.1) by the sequence

$$a_1 = 3500, \qquad a_n = a_{n-1} - \frac{d}{2^{2n-3}},$$

where d is the difference between the initial price and the first offer. Use this recursively defined sequence to solve parts (1) and (2) above either manually or by using the SEQ MODE and the TABLE feature of a graphing calculator.

4. The first four terms in the sequence in part (3) can be written as

$$a_1, \quad a_1 - \frac{d}{2}, \quad a_1 - \frac{d}{2} - \frac{d}{8},$$

$$a_1 - \frac{d}{2} - \frac{d}{8} - \frac{d}{32}.$$

Use the formula for the limit of an infinite geometric series to find a simple algebraic formula for the eventual sale price, P, when the bargaining process from above is followed. Verify the formula by using it to solve parts (1) and (2) above.

11.4 The Binomial Theorem

Binomial Expansion Using Pascal's Triangle •
Binomial Expansion Using Factorial Notation

CONNECTING THE CONCEPTS

Sequences and series occur in many settings, some of which were mentioned in Sections 11.1–11.3. Although you may not have viewed it this way before, the expression $(x + y)^2$ can be regarded as a series: $x^2 + 2xy + y^2$.

In Chapter 5, we found that the expansion of $(x + y)^n$, for powers greater than 2, can be quite time-consuming. The reason for this extends all the way back to Chapter 1 and the rules for the order of operations and the properties of exponents: $(x + y)^n \neq x^n + y^n$. Since the terms in the expansion of $(x + y)^n$ have many uses, we devote this section to two methods that streamline the expansion of this important algebraic expression.

Binomial Expansion Using Pascal's Triangle

Consider the following expanded powers of $(a + b)^n$:

$$(a + b)^0 = 1$$
$$(a + b)^1 = a + b$$
$$(a + b)^2 = a^2 + 2a^1b^1 + b^2$$
$$(a + b)^3 = a^3 + 3a^2b^1 + 3a^1b^2 + b^3$$
$$(a + b)^4 = a^4 + 4a^3b^1 + 6a^2b^2 + 4a^1b^3 + b^4$$
$$(a + b)^5 = a^5 + 5a^4b^1 + 10a^3b^2 + 10a^2b^3 + 5a^1b^4 + b^5.$$

Each expansion is a polynomial. There are some patterns to be noted:

1. There is one more term than the power of the binomial, n. That is, there are $n + 1$ terms in the expansion of $(a + b)^n$.

2. In each term, the sum of the exponents is the power to which the binomial is raised.

3. The exponents of a start with n, the power of the binomial, and decrease to 0 (since $a^0 = 1$, the last term has no factor of a). The first term has no factor of b, so powers of b start with 0 and increase to n.

4. The coefficients start at 1, increase through certain values, and then decrease through these same values back to 1.

Let's study the coefficients further. Suppose we wish to expand $(a + b)^8$. The patterns we noticed above indicate 9 terms in the expansion:

$$a^8 + c_1a^7b + c_2a^6b^2 + c_3a^5b^3 + c_4a^4b^4 + c_5a^3b^5 + c_6a^2b^6 + c_7ab^7 + b^8.$$

Study Skills _____

Make the Most of Each Minute

As your final exam nears, it is essential to make wise use of your time. If it is possible to reschedule haircuts, dentist appointments, and the like until after your exam(s), consider doing so. If you can, consider writing out your hourly schedule of daily activities leading up to the exam(s).

How can we determine the values for the *c*'s? One method seems very simple, but it has some drawbacks. It involves writing down the coefficients in a triangular array as follows. We form what is known as **Pascal's triangle**:

$$
\begin{array}{llccccccc}
(a + b)^0: & & & & & 1 & & & \\
(a + b)^1: & & & & 1 & & 1 & & \\
(a + b)^2: & & & 1 & & 2 & & 1 & \\
(a + b)^3: & & 1 & & 3 & & 3 & & 1 \\
(a + b)^4: & 1 & & 4 & & 6 & & 4 & & 1 \\
(a + b)^5: & 1 & 5 & & 10 & & 10 & & 5 & & 1 \\
\end{array}
$$

There are many patterns in the triangle. Find as many as you can.

Perhaps you discovered a way to write the next row of numbers, given the numbers in the row above it. There are always 1's on the outside. Each remaining number is the sum of the two numbers above:

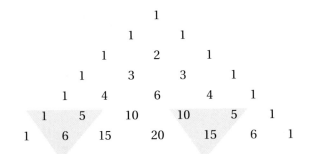

We see that in the bottom (seventh) row

the 1st and last numbers are 1;

the 2nd number is 1 + 5, or 6;

the 3rd number is 5 + 10, or 15;

the 4th number is 10 + 10, or 20;

the 5th number is 10 + 5, or 15; and

the 6th number is 5 + 1, or 6.

Thus the expansion of $(a + b)^6$ is

$$(a + b)^6 = 1a^6 + 6a^5b + 15a^4b^2 + 20a^3b^3 + 15a^2b^4 + 6ab^5 + 1b^6.$$

To expand $(a + b)^8$, we complete two more rows of Pascal's triangle:

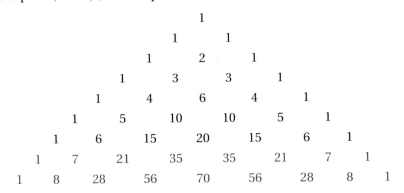

Thus the expansion of $(a + b)^8$ has coefficients found in the 9th row above:

$$(a + b)^8 = 1a^8 + 8a^7b + 28a^6b^2 + 56a^5b^3 + 70a^4b^4 + 56a^3b^5 + 28a^2b^6 + 8ab^7 + 1b^8.$$

We can generalize our results as follows:

The Binomial Theorem (Form 1)

For any binomial $a + b$ and any natural number n,

$$(a + b)^n = c_0a^nb^0 + c_1a^{n-1}b^1 + c_2a^{n-2}b^2 + \cdots + c_{n-1}a^1b^{n-1} + c_na^0b^n,$$

where the numbers $c_0, c_1, c_2, \ldots, c_n$ are from the $(n + 1)$st row of Pascal's triangle.

A proof of the binomial theorem is beyond the scope of this text.

EXAMPLE 1   Expand: $(u - v)^5$.

Solution   Using the binomial theorem, we have $a = u$, $b = -v$, and $n = 5$. We use the 6th row of Pascal's triangle: 1   5   10   10   5   1. Thus,

$$\begin{aligned}
(u - v)^5 &= [u + (-v)]^5 \qquad \text{Rewriting } u - v \text{ as a sum} \\
&= 1(u)^5 + 5(u)^4(-v)^1 + 10(u)^3(-v)^2 + 10(u)^2(-v)^3 \\
&\quad + 5(u)^1(-v)^4 + 1(-v)^5 \\
&= u^5 - 5u^4v + 10u^3v^2 - 10u^2v^3 + 5uv^4 - v^5.
\end{aligned}$$

Note that the signs of the terms alternate between $+$ and $-$. When $-v$ is raised to an odd power, the sign is $-$; when the power is even, the sign is $+$.

EXAMPLE 2   Expand: $\left(2t + \dfrac{3}{t}\right)^6$.

Solution   Note that $a = 2t$, $b = 3/t$, and $n = 6$. We use the 7th row of Pascal's triangle: 1   6   15   20   15   6   1. Thus,

$$\begin{aligned}
\left(2t + \frac{3}{t}\right)^6 &= 1(2t)^6 + 6(2t)^5\left(\frac{3}{t}\right)^1 + 15(2t)^4\left(\frac{3}{t}\right)^2 + 20(2t)^3\left(\frac{3}{t}\right)^3 \\
&\quad + 15(2t)^2\left(\frac{3}{t}\right)^4 + 6(2t)^1\left(\frac{3}{t}\right)^5 + 1\left(\frac{3}{t}\right)^6 \\
&= 64t^6 + 6(32t^5)\left(\frac{3}{t}\right) + 15(16t^4)\left(\frac{9}{t^2}\right) + 20(8t^3)\left(\frac{27}{t^3}\right) \\
&\quad + 15(4t^2)\left(\frac{81}{t^4}\right) + 6(2t)\left(\frac{243}{t^5}\right) + \frac{729}{t^6} \\
&= 64t^6 + 576t^4 + 2160t^2 + 4320 + 4860t^{-2} + 2916t^{-4} \\
&\quad + 729t^{-6}.
\end{aligned}$$

Binomial Expansion Using Factorial Notation

The drawback to using Pascal's triangle is that we must compute all the preceding rows in the table to obtain the row needed for the expansion in which we are interested. The following method avoids this difficulty. It will also enable us to find a specific term—say, the 8th term—without computing all the other terms in the expansion. This method is useful in such courses as finite mathematics, calculus, and statistics.

To develop the method, we need some new notation. Products of successive natural numbers, such as $6 \cdot 5 \cdot 4 \cdot 3 \cdot 2 \cdot 1$ and $8 \cdot 7 \cdot 6 \cdot 5 \cdot 4 \cdot 3 \cdot 2 \cdot 1$, have a special notation. For the product $6 \cdot 5 \cdot 4 \cdot 3 \cdot 2 \cdot 1$, we write $6!$, read "6 factorial."

Factorial Notation

For any natural number n,

$$n! = n(n-1)(n-2) \cdots (3)(2)(1).$$

Here are some examples:

$$6! = 6 \cdot 5 \cdot 4 \cdot 3 \cdot 2 \cdot 1 = 720,$$
$$5! = 5 \cdot 4 \cdot 3 \cdot 2 \cdot 1 = 120,$$
$$4! = 4 \cdot 3 \cdot 2 \cdot 1 = 24,$$
$$3! = 3 \cdot 2 \cdot 1 = 6,$$
$$2! = 2 \cdot 1 = 2,$$
$$1! = 1 = 1.$$

We also define $0!$ to be 1 for reasons explained shortly.

To simplify expressions like

$$\frac{8!}{5! \, 3!},$$

note that

$$8! = 8 \cdot 7 \cdot 6 \cdot 5 \cdot 4 \cdot 3 \cdot 2 \cdot 1 = 8 \cdot 7! = 8 \cdot 7 \cdot 6! = 8 \cdot 7 \cdot 6 \cdot 5!,$$

and so on.

Caution! $\dfrac{6!}{3!} \neq 2!$ To see this, note that

$$\frac{6!}{3!} = \frac{6 \cdot 5 \cdot 4 \cdot \cancel{3} \cdot \cancel{2} \cdot \cancel{1}}{\cancel{3} \cdot \cancel{2} \cdot \cancel{1}} = 6 \cdot 5 \cdot 4.$$

EXAMPLE 3   Simplify: $\dfrac{8!}{5!\,3!}$.

Student Notes _____

It is important to recognize factorial notation as representing a product with descending factors. Thus, $7!$, $7 \cdot 6!$, and $7 \cdot 6 \cdot 5!$ all represent the same product.

Solution

$$\frac{8!}{5!\,3!} = \frac{8 \cdot 7 \cdot 6 \cdot 5!}{5! \cdot 3 \cdot 2 \cdot 1} = 8 \cdot 7 \qquad \text{Removing a factor equal to 1:}$$
$$\frac{6 \cdot 5!}{5! \cdot 3 \cdot 2} = 1$$
$$= 56$$

The following notation is used in our second formulation of the binomial theorem.

$\dbinom{n}{r}$ *Notation*

For n, r nonnegative integers with $n \geq r$,

$$\binom{n}{r}, \quad \text{read ``n choose r,''} \quad \text{means} \quad \frac{n!}{(n-r)!\,r!}.^{*}$$

EXAMPLE 4   Simplify: **(a)** $\dbinom{7}{2}$; **(b)** $\dbinom{9}{6}$; **(c)** $\dbinom{6}{6}$.

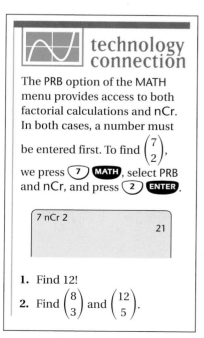

technology connection

The PRB option of the MATH menu provides access to both factorial calculations and nCr. In both cases, a number must be entered first. To find $\dbinom{7}{2}$, we press ⑦ **MATH**, select PRB and nCr, and press ② **ENTER**.

> 7 nCr 2
>   21

1. Find $12!$

2. Find $\dbinom{8}{3}$ and $\dbinom{12}{5}$.

Solution

a) $\dbinom{7}{2} = \dfrac{7!}{(7-2)!\,2!}$

$\qquad = \dfrac{7!}{5!\,2!} = \dfrac{7 \cdot 6 \cdot 5!}{5! \cdot 2 \cdot 1} = \dfrac{7 \cdot 6}{2}$ We can write $7!$ as $7 \cdot 6 \cdot 5!$ to aid our simplification.

$\qquad = 7 \cdot 3$

$\qquad = 21$

b) $\dbinom{9}{6} = \dfrac{9!}{3!\,6!}$

$\qquad = \dfrac{9 \cdot 8 \cdot 7 \cdot 6!}{3 \cdot 2 \cdot 1 \cdot 6!} = \dfrac{9 \cdot 8 \cdot 7}{3 \cdot 2}$ Writing $9!$ as $9 \cdot 8 \cdot 7 \cdot 6!$ to help with simplification

$\qquad = 3 \cdot 4 \cdot 7$

$\qquad = 84$

---

*In many books and for many calculators, the notation $_{n}C_{r}$ is used instead of $\dbinom{n}{r}$.

c) $\binom{6}{6} = \dfrac{6!}{0!\,6!} = \dfrac{6!}{1 \cdot 6!}$ Since $0! = 1$

$\quad\quad = \dfrac{6!}{6!}$

$\quad\quad = 1$

Now we can restate the binomial theorem using our new notation.

The Binomial Theorem (Form 2)

For any binomial $a + b$ and any natural number n,

$$(a + b)^n = \binom{n}{0}a^n + \binom{n}{1}a^{n-1}b + \binom{n}{2}a^{n-2}b^2 + \cdots + \binom{n}{n}b^n.$$

EXAMPLE 5   Expand: $(3x + y)^4$.

Solution   We use the binomial theorem (Form 2) with $a = 3x$, $b = y$, and $n = 4$:

$$(3x + y)^4 = \binom{4}{0}(3x)^4 + \binom{4}{1}(3x)^3y + \binom{4}{2}(3x)^2y^2 + \binom{4}{3}(3x)y^3 + \binom{4}{4}y^4$$

$$= \frac{4!}{4!\,0!}3^4x^4 + \frac{4!}{3!\,1!}3^3x^3y + \frac{4!}{2!\,2!}3^2x^2y^2 + \frac{4!}{1!\,3!}3xy^3 + \frac{4!}{0!\,4!}y^4$$

$$= 1 \cdot 81x^4 + 4 \cdot 27x^3y + 6 \cdot 9x^2y^2 + 4 \cdot 3xy^3 + y^4 \left.\vphantom{\begin{array}{c}1\\1\\1\end{array}}\right\}$$ Simplifying

$$= 81x^4 + 108x^3y + 54x^2y^2 + 12xy^3 + y^4.$$

EXAMPLE 6   Expand: $(x^2 - 2y)^5$.

Solution   In this case, $a = x^2$, $b = -2y$, and $n = 5$:

$$(x^2 - 2y)^5 = \binom{5}{0}(x^2)^5 + \binom{5}{1}(x^2)^4(-2y) + \binom{5}{2}(x^2)^3(-2y)^2$$

$$+ \binom{5}{3}(x^2)^2(-2y)^3 + \binom{5}{4}(x^2)(-2y)^4 + \binom{5}{5}(-2y)^5$$

$$= \frac{5!}{5!\,0!}x^{10} + \frac{5!}{4!\,1!}x^8(-2y) + \frac{5!}{3!\,2!}x^6(-2y)^2$$

$$+ \frac{5!}{2!\,3!}x^4(-2y)^3 + \frac{5!}{1!\,4!}x^2(-2y)^4 + \frac{5!}{0!\,5!}(-2y)^5$$

$$= x^{10} - 10x^8y + 40x^6y^2 - 80x^4y^3 + 80x^2y^4 - 32y^5.$$

Note that in the binomial theorem (Form 2), $\binom{n}{0}a^n b^0$ gives us the first term, $\binom{n}{1}a^{n-1}b^1$ gives us the second term, $\binom{n}{2}a^{n-2}b^2$ gives us the third term, and so on. This can be generalized to give a method for finding a specific term without writing the entire expansion.

> ### Finding a Specific Term
>
> When $(a + b)^n$ is expanded and written in descending powers of a, the $(r + 1)$st term is
>
> $$\binom{n}{r}a^{n-r}b^r.$$

EXAMPLE 7   Find the 5th term in the expansion of $(2x - 3y)^7$.

Solution   First, we note that $5 = 4 + 1$. Thus, $r = 4$, $a = 2x$, $b = -3y$, and $n = 7$. Then the 5th term of the expansion is

$$\binom{7}{4}(2x)^{7-4}(-3y)^4, \text{ or } \frac{7!}{3!\,4!}(2x)^3(-3y)^4, \text{ or } 22{,}680x^3y^4.$$

It is because of the binomial theorem that $\binom{n}{r}$ is called a *binomial coefficient*. We can now explain why 0! is defined to be 1. In the binomial theorem,

$$\binom{n}{0} \text{ must equal 1 when using the definition } \binom{n}{r} = \frac{n!}{(n - r)!\,r!}.$$

Thus we must have

$$\binom{n}{0} = \frac{n!}{(n - 0)!\,0!} = \frac{n!}{n!\,0!} = 1.$$

This is satisfied only if 0! is defined to be 1.

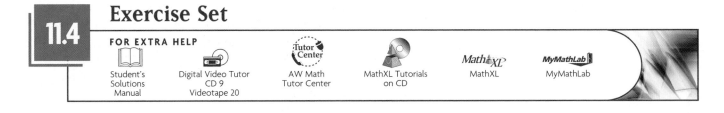

Exercise Set

11.4

FOR EXTRA HELP

Student's Solutions Manual | Digital Video Tutor CD 9 Videotape 20 | Tutor Center AW Math Tutor Center | MathXL Tutorials on CD | Math XL MathXL | MyMathLab MyMathLab

✎ *Concept Reinforcement   Complete each of the following.*

1. The last term in the expansion of $(x + 2)^5$ is _____.

2. The expansion of $(x + y)^7$, when simplified, contains a total of _____ terms.

3. In the expansion of $(a + b)^9$, the exponents in each term add to _____.

4. The expression _____ represents $4 \cdot 3 \cdot 2 \cdot 1$.

5. The expression _____ represents $\dfrac{8!}{3! \, 5!}$.

6. In the expansion of $(a + b)^{10}$, the coefficient of $a^8 b^2$ is _____.

7. In the expansion of $(x + y)^9$, the coefficient of $x^2 y^7$ is the same as the coefficient of _____.

8. The notation $\binom{10}{4}$ is read _____.

Simplify.

9. $9!$

10. $8!$

11. $11!$

12. $10!$

13. $\dfrac{8!}{6!}$

14. $\dfrac{7!}{4!}$

15. $\dfrac{9!}{5!}$

16. $\dfrac{10!}{7!}$

17. $\binom{7}{4}$

18. $\binom{8}{2}$

19. $\binom{9}{5}$

20. $\binom{10}{6}$

21. $\binom{30}{3}$

22. $\binom{20}{18}$

23. $\binom{40}{38}$

24. $\binom{35}{2}$

Expand. Use both of the methods shown in this section.

25. $(a - b)^4$

26. $(m + n)^5$

27. $(p + q)^7$

28. $(x - y)^6$

29. $(3c - d)^7$

30. $(x^2 - 3y)^5$

31. $(t^{-2} + 2)^6$

32. $(3c - d)^6$

33. $(x - y)^5$

34. $(x - y)^3$

35. $\left(3s + \dfrac{1}{t}\right)^9$

36. $\left(x + \dfrac{2}{y}\right)^9$

37. $(x^3 - 2y)^5$

38. $(a^2 - b^3)^5$

39. $\left(\sqrt{5} + t\right)^6$

40. $\left(\sqrt{3} - t\right)^4$

41. $\left(\dfrac{1}{\sqrt{x}} - \sqrt{x}\right)^6$

42. $(x^{-2} + x^2)^4$

Find the indicated term for each binomial expression.

43. 3rd, $(a + b)^6$

44. 6th, $(x + y)^7$

45. 12th, $(a - 3)^{14}$

46. 11th, $(x - 2)^{12}$

47. 5th, $\left(2x^3 + \sqrt{y}\right)^8$

48. 4th, $\left(\dfrac{1}{b^2} + c\right)^7$

49. Middle, $(2u + 3v^2)^{10}$

50. Middle two, $\left(\sqrt{x} + \sqrt{3}\right)^5$

Aha! **51.** 9th, $(x - y)^8$

52. 13th, $\left(a - \sqrt{b}\right)^{12}$

📖 **53.** Maya claims that she can calculate mentally the first two and the last two terms of the expansion of $(a + b)^n$ for any whole number n. How do you think she does this?

📖 **54.** Without performing any calculations, explain why the expansions of $(x - y)^8$ and $(y - x)^8$ must be equal.

SKILL MAINTENANCE

Solve. [9.6]

55. $\log_2 x + \log_2(x - 2) = 3$

56. $\log_3(x + 2) - \log_3(x - 2) = 2$

57. $e^t = 280$

58. $\log_5 x^2 = 2$

SYNTHESIS

59. Explain how someone can determine the x^2-term of the expansion of $\left(x - \dfrac{3}{x}\right)^{10}$ without calculating any other terms.

60. Devise two problems requiring the use of the binomial theorem. Design the problems so that one is solved more easily using Form 1 and the other is solved more easily using Form 2. Then explain what makes one form easier to use than the other in each case.

61. Show that there are exactly $\dbinom{5}{3}$ ways of choosing a subset of size 3 from $\{a, b, c, d, e\}$.

62. *Baseball.*   At one point in July 2004, Barry Bonds of the San Francisco Giants had a batting average of .360. At that time, if someone were to randomly select 5 of his "at-bats," the probability of Bonds getting exactly 3 hits would be the 3rd term of the binomial expansion of $(0.360 + 0.640)^5$. Find that term and use a calculator to estimate the probability.
Source: www.mlb.com

63. *Widows or divorcees.*   The probability that a woman will be either widowed or divorced is 85%. If 8 women are randomly selected, the probability that exactly 5 of them will be either widowed or divorced is the 6th term of the binomial expansion of $(0.15 + 0.85)^8$. Use a calculator to estimate that probability.

64. *Baseball.*   In reference to Exercise 62, the probability that Bonds will get *at most* 3 hits is found by adding the last 4 terms of the binomial expansion of $(0.360 + 0.640)^5$. Find these terms and use a calculator to estimate the probability.

65. *Widows or divorcees.*   In reference to Exercise 63, the probability that *at least* 6 of the women will be widowed or divorced is found by adding the last three terms of the binomial expansion of $(0.15 + 0.85)^8$. Find these terms and use a calculator to estimate the probability.

66. Find the term of
$$\left(\frac{3x^2}{2} - \frac{1}{3x}\right)^{12}$$
that does not contain x.

67. Prove that
$$\binom{n}{r} = \binom{n}{n - r}$$
for any whole numbers n and r. Assume $r \le n$.

68. Find the middle term of $(x^2 - 6y^{3/2})^6$.

69. Find the ratio of the 4th term of
$$\left(p^2 - \frac{1}{2}p\sqrt[3]{q}\right)^5$$
to the 3rd term.

70. Find the term containing $\dfrac{1}{x^{1/6}}$ of
$$\left(\sqrt[3]{x} - \frac{1}{\sqrt{x}}\right)^7.$$

Aha! **71.** Multiply: $(x^2 + 2xy + y^2)(x^2 + 2xy + y^2)^2(x + y)$.

72. What is the degree of $(x^3 + 2)^4$?

11 Study Summary

An ordered list of numbers, such as

$$5, 7, 8, 11, 17 \quad \text{or} \quad 6, 9, 12, 15, 18, 21, \ldots$$

is called a **sequence**, or **progression** (p. 704). The first sequence above ends so it is considered **finite**, whereas the second sequence is **infinite** (p. 704). When the difference between consecutive terms is constant—as in the second sequence above — the sequence is **arithmetic** (p. 711). If d represents the **common difference**, we have

$$a_n = a_1 + (n-1)d, \qquad \text{To find the } n\text{th term}$$

$$a_{n+1} = a_n + d, \qquad \text{To find the next term when the } n\text{th term is known}$$

$$S_n = \frac{n}{2}(a_1 + a_n). \qquad \text{The sum of the first } n \text{ terms. This is an \textbf{arithmetic series}.}$$

When each term in a sequence is multiplied by a certain fixed number to get the next term, the sequence is **geometric** (p. 720). If r represents this fixed number, or **common ratio**, we have

$$a_n = a_1 r^{n-1}, \qquad \text{To find the } n\text{th term}$$

$$a_{n+1} = a_n \cdot r, \qquad \text{To find the next term when the } n\text{th term is known}$$

$$S_n = \frac{a_1(1 - r^n)}{1 - r}, \qquad \text{The sum of the first } n \text{ terms. This is a \textbf{geometric series}.}$$

$$S_\infty = \frac{a_1}{1 - r}, \ |r| < 1. \qquad \text{The limit of an infinite geometric series with } |r| < 1$$

The expression $\sum\limits_{k=1}^{n} a_k$ indicates a **partial sum**, or **finite series**, using **sigma** or **summation notation** (p. 707). The letter k serves as the **index of summation**.

It is possible to regard expressions of the form $(a + b)^n$ as series for which **Pascal's triangle** can be used to determine the **binomial coefficients** (pp. 732, 737). These coefficients can also be found using the **binomial theorem** (pp. 733, 736). Use of the binomial theorem generally requires understanding **factorial notation** (p. 734).

$$n! = n(n-1)(n-2) \cdots 3 \cdot 2 \cdot 1 \qquad \text{Factorial notation}$$

$$\binom{n}{r} = \frac{n!}{(n-r)!\, r!} \qquad \text{Binomial coefficient}$$

$$(a + b)^n = \binom{n}{0}a^n + \binom{n}{1}a^{n-1}b + \binom{n}{2}a^{n-2}b^2 + \cdots + \binom{n}{n}b^n \quad \text{Binomial theorem}$$

$$\binom{n}{r}a^{n-r}b^r \qquad \qquad (r+1)\text{st term of } (a + b)^n$$

🐌 *Concept Reinforcement*   *Classify each of the following as either true or false.*

1. The next term in the arithmetic sequence 10, 15, 20, . . . is 35. [11.2]

2. The next term in the geometric sequence 2, 6, 18, 54, . . . is 162. [11.3]

3. $\sum\limits_{k=1}^{3} k^2$ means $1^2 + 2^2 + 3^2$. [11.1]

4. If $a_n = 3n - 1$, then $a_{17} = 19$. [11.1]

5. The nth term of an arithmetic sequence with common difference d is $a_n = a_1 + (n - 1)d$, for any integer $n \geq 1$. [11.2]

6. The nth term of a geometric sequence with common ratio r is given by $a_n = a_1 r^{n-1}$, for any integer $n \geq 1$. [11.3]

7. For any natural number n, $n! = n(n - 1)$. [11.4]

8. When simplified, the expansion of $(x + y)^{17}$ has 19 terms. [11.4]

Find the first four terms; the 8th term, a_8; and the 12th term, a_{12}. [11.1]

9. $a_n = 4n - 3$

10. $a_n = \dfrac{n - 1}{n^2 + 1}$

Predict the general term. Answers may vary. [11.1]

11. 7, 14, 21, 28, 35, . . .

12. $-1, 3, -5, 7, -9, . . .$

Write out and evaluate each sum. [11.1]

13. $\sum\limits_{k=1}^{5} (-2)^k$

14. $\sum\limits_{k=2}^{7} (1 - 2k)$

Rewrite using sigma notation. [11.1]

15. $4 + 8 + 12 + 16 + 20$

16. $\dfrac{-1}{2} + \dfrac{1}{4} + \dfrac{-1}{8} + \dfrac{1}{16} + \dfrac{-1}{32}$

17. Find the 14th term of the arithmetic sequence $-6, 1, 8, ...$ [11.2]

18. An arithmetic sequence has $a_1 = 11$ and $a_{13} = 43$. Find the common difference, d. [11.2]

19. An arithmetic sequence has $a_8 = 20$ and $a_{24} = 40$. Find the first term, a_1, and the common difference, d. [11.2]

20. Find the sum of the first 17 terms of the arithmetic series $-8 + (-11) + (-14) + \cdots.$ [11.2]

21. Find the sum of all the multiples of 6 from 12 to 318, inclusive. [11.2]

22. Find the 20th term of the geometric sequence $2, 2\sqrt{2}, 4, ...$ [11.3]

23. Find the common ratio of the geometric sequence $2, \frac{4}{3}, \frac{8}{9}, ...$ [11.3]

24. Find the nth term of the geometric sequence $-2, 2, -2, ...$ [11.3]

25. Find the nth term of the geometric sequence $3, \frac{3}{4}x, \frac{3}{16}x^2, ...$ [11.3]

26. Find S_6 for the geometric series
$$3 + 12 + 48 + \cdots. \text{ [11.3]}$$

27. Find S_{12} for the geometric series
$$3x - 6x + 12x - \cdots. \text{ [11.3]}$$

Determine whether each infinite geometric series has a limit. If a limit exists, find it. [11.3]

28. $6 + 3 + 1.5 + 0.75 + \cdots$

29. $7 - 4 + \frac{16}{7} - \cdots$

30. $2 + (-2) + 2 + (-2) + \cdots$

31. $0.04 + 0.08 + 0.16 + 0.32 + \cdots$

32. $\$2000 + \$1900 + \$1805 + \$1714.75 + \cdots$

33. Find fraction notation for $0.555555 . . . .$ [11.3]

34. Find fraction notation for $1.39393939 . . . .$ [11.3]

35. Adam took a job working in a convenience store starting with an hourly wage of $11.50. He was promised a raise of 40¢ per hour every 3 mos for 8 yr. At the end of 8 yr, what will be his hourly wage? [11.2]

36. A stack of poles has 42 poles in the bottom row. There are 41 poles in the second row, 40 poles in the third row, and so on, ending with 1 pole in the top row. How many poles are in the stack? [11.2]

37. Stacey's student loan is in the amount of $12,000. Interest is 4%, compounded annually, and the amount is to be paid off in 7 yr. How much is to be paid back? [11.3]

38. Find the total rebound distance of a ball, given that it is dropped from a height of 12 m and each rebound is one-third of the preceding one. [11.3]

Simplify. [11.4]

39. $7!$

40. $\binom{8}{3}$

41. Find the 3rd term of $(a + b)^{20}$. [11.4]

42. Expand: $(x - 2y)^4$. [11.4]

SYNTHESIS

43. What happens to a_n in a geometric sequence with $|r| < 1$, as n gets larger? Why? [11.3]

44. Compare the two forms of the binomial theorem given in the text. Under what circumstances would one be more useful than the other? [11.4]

45. Find the sum of the first n terms of the geometric series $1 - x + x^2 - x^3 + \cdots$. [11.3]

46. Expand: $(x^{-3} + x^3)^5$. [11.4]

11 Chapter Test

1. Find the first five terms and the 12th term of a sequence with general term $a_n = 6n - 5$.

2. Predict the general term of the sequence
$$\frac{4}{3}, \frac{4}{9}, \frac{4}{27}, \ldots.$$

3. Write out and evaluate:
$$\sum_{k=2}^{6} (3 - 2^k).$$

4. Rewrite using sigma notation:
$$1 + (-8) + 27 + (-64) + 125.$$

5. Find the 13th term, a_{13}, of the arithmetic sequence $9, 4, -1, \ldots.$

Assume arithmetic sequences for Questions 6 and 7.

6. Find the common difference d when $a_1 = 7$ and $a_7 = 9\frac{1}{4}$.

7. Find a_1 and d when $a_5 = 16$ and $a_{10} = -3$.

8. Find the sum of all the multiples of 12 from 24 to 240, inclusive.

9. Find the 6th term of the geometric sequence $72, 18, 4\frac{1}{2}, \ldots.$

10. Find the common ratio of the geometric sequence $22\frac{1}{2}, 15, 10, \ldots.$

11. Find the nth term of the geometric sequence $3, 9, 27, \ldots.$

12. Find the sum of the first nine terms of the geometric series
$$(1 + x) + (2 + 2x) + (4 + 4x) + \cdots.$$

Determine whether each infinite geometric series has a limit. If a limit exists, find it.

13. $0.5 + 0.25 + 0.125 + \cdots$

14. $0.5 + 1 + 2 + 4 + \cdots$

15. $\$1000 + \$80 + \$6.40 + \cdots$

16. Find fraction notation for $0.85858585\ldots$.

17. An auditorium has 31 seats in the first row, 33 seats in the second row, 35 seats in the third row, and so on, for 18 rows. How many seats are in the 17th row?

18. Lindsay's Uncle Ken gave her $100 for her first birthday, $200 for her second birthday, $300 for her third birthday, and so on, until her eighteenth birthday. How much did he give her in all?

19. Each week the price of a $15,000 boat will be reduced 5% of the previous week's price. If we assume that it is not sold, what will be the price after 10 weeks?

20. Find the total rebound distance of a ball that is dropped from a height of 18 m, with each rebound two thirds of the preceding one.

21. Simplify: $\dbinom{12}{9}$.

22. Expand: $(x^2 - 3y)^5$.

23. Find the 4th term in the expansion of $(a + x)^{12}$.

SYNTHESIS

24. Find a formula for the sum of the first n even natural numbers:
$$2 + 4 + 6 + \cdots + 2n.$$

25. Find the sum of the first n terms of
$$1 + \frac{1}{x} + \frac{1}{x^2} + \frac{1}{x^3} + \cdots.$$

1-11 Cumulative Review

Simplify.

1. $(-7x^2y^3)(5x^4y^{-7})$ [1.6]

2. $|-3.5 + 9.8|$ [1.2]

3. $y - [3 - 4(5 - 2y) - 3y]$ [1.3]

4. $(10 \cdot 8 - 9 \cdot 7)^2 - 54 \div 9 - 3$ [1.2]

5. Evaluate
$$\frac{ab - ac}{bc}$$
for $a = -2$, $b = 3$, and $c = -4$. [1.1], [1.2]

Perform the indicated operations to create an equivalent expression. Be sure to simplify your result if possible.

6. $(5a^2 - 3ab - 7b^2) - (2a^2 + 5ab + 8b^2)$ [5.1]

7. $(-3x^2 + 4x^3 - 5x - 1) + (9x^3 - 4x^2 + 7 - x)$ [5.1]

8. $(2a - 1)(3a + 5)$ [5.2]

9. $(3a^2 - 5y)^2$ [5.2]

10. $\dfrac{1}{x - 2} - \dfrac{4}{x^2 - 4} + \dfrac{3}{x + 2}$ [6.2]

11. $\dfrac{x^2 - 6x + 8}{4x + 12} \cdot \dfrac{x + 3}{x^2 - 4}$ [6.1]

12. $\dfrac{3x + 3y}{5x - 5y} \div \dfrac{3x^2 + 3y^2}{5x^3 - 5y^3}$ [6.1]

13. $\dfrac{x - \dfrac{a^2}{x}}{1 + \dfrac{a}{x}}$ [6.3]

Factor, if possible, to form an equivalent expression.

14. $4x^2 - 12x + 9$ [5.5]

15. $27a^3 - 8$ [5.6]

16. $a^3 + 3a^2 - ab - 3b$ [5.3]

17. $15y^4 + 33y^2 - 36$ [5.7]

18. For the function described by
$$f(x) = 3x^2 - 4x,$$
find $f(-2)$. [2.2]

19. Divide:

$$(7x^4 - 5x^3 + x^2 - 4) \div (x - 2). \; [6.6]$$

Solve.

20. $8(x - 1) - 3(x - 2) = 1$ [1.3]

21. $\dfrac{6}{x} + \dfrac{6}{x + 2} = \dfrac{5}{2}$ [6.4]

22. $2x + 1 > 5 \; or \; x - 7 \le 3$ [4.2]

23. $5x + 3y = -2,$
$3x + 5y = 2$ [3.2]

24. $x + y - z = 0,$
$3x + y + z = 6,$
$x - y + 2z = 5$ [3.4]

25. $3\sqrt{x - 1} = 5 - x$ [7.6]

26. $x^4 - 29x^2 + 100 = 0$ [8.5]

27. $x^2 + y^2 = 8,$
$x^2 - y^2 = 2$ [10.4]

28. $4^x = 7$ [9.6]

29. $\log (x^2 - 25) - \log (x + 5) = 3$ [9.6]

30. $\log_5 x = -2$ [9.6]

31. $7^{2x+3} = 49$ [9.6]

32. $|2x - 1| \le 5$ [4.3]

33. $7x^2 + 14 = 0$ [8.1]

34. $x^2 + 4x = 3$ [8.2]

35. $y^2 + 3y > 10$ [8.9]

36. Let $f(x) = x^2 - 2x$. Find a such that $f(a) = 48$. [5.8]

37. If $f(x) = \sqrt{-x + 4} + 3$ and $g(x) = \sqrt{x - 2} + 3$, find a such that $f(a) = g(a)$. [7.6]

Solve.

38. The perimeter of a rectangular sign is 34 ft. The length of a diagonal is 13 ft. Find the dimensions of the sign. [10.4]

39. A music club offers two types of membership. Limited members pay a fee of $10 a year and can buy CDs for $10 each. Preferred members pay $20 a year and can buy CDs for $7.50 each. For what numbers of annual CD purchases would it be less expensive to be a preferred member? [4.1]

40. Find three consecutive integers whose sum is 198. [1.4]

41. A pentagon with all five sides the same size has a perimeter equal to that of an octagon in which all eight sides are the same size. One side of the pentagon is 2 less than three times one side of the octagon. What is the perimeter of each figure? [3.3]

42. Roger's Organics mixes herbs that cost $2.68 an ounce with herbs that cost $4.60 an ounce to create a seasoning that costs $3.80 an ounce. How many ounces of each herb should be mixed together to make 24 oz of the seasoning? [3.3]

43. An airplane can fly 190 mi with the wind in the same time it takes to fly 160 mi against the wind. The speed of the wind is 30 mph. How fast can the plane fly in still air? [6.5]

44. Bianca can tap the sugar maple trees in Southway Park in 21 hr. Delia can tap the trees in 14 hr. How long would it take them, working together, to tap the trees? [6.5]

45. The centripetal force F of an object moving in a circle varies directly as the square of the velocity v and inversely as the radius r of the circle. If $F = 8$ when $v = 1$ and $r = 10$, what is F when $v = 2$ and $r = 16$? [8.6]

46. The Brighton recreation department plans to fence in a rectangular park next to a river. (Note that no fence will be needed along the river.) What is the area of the largest region that can be fenced in with 200 ft of fencing? [8.8]

Graph.

47. $3x - y = 7$ [2.4]

48. $y - 4 = -\frac{2}{3}(x - 1)$ [2.5]

49. $x^2 + y^2 = 100$ [10.1]

50. $\dfrac{x^2}{36} - \dfrac{y^2}{9} = 1$ [10.3]

51. $y = \log_2 x$ [9.3]

52. $f(x) = 2^x - 3$ [9.2]

53. $2x - 3y < -6$ [4.4]

54. Graph: $f(x) = -2(x - 3)^2 + 1$. [8.7]
   a) Label the vertex.
   b) Draw the axis of symmetry.
   c) Find the maximum or minimum value.

55. Solve $V = P - Prt$ for r. [1.5]

56. Solve $I = \dfrac{R}{R + r}$ for R. [6.8]

57. Find a linear equation whose graph has a y-intercept of $(0, -8)$ and is parallel to the line whose equation is $3x - y = 6$. [2.5]

Find the domain of each function.

58. $f(x) = \sqrt{6 - 8x}$ [4.2]

59. $g(x) = \dfrac{x - 4}{x^2 - 2x + 1}$ [5.8]

60. Multiply $(8.9 \times 10^{-17})(7.6 \times 10^4)$. Write scientific notation for the answer. [1.7]

61. Multiply and simplify: $\sqrt{8x}\,\sqrt{8x^3y}$. [7.3]

62. Simplify: $(25x^{4/3}y^{1/2})^{3/2}$. [7.2]

63. Divide and simplify:
$$\dfrac{\sqrt[3]{25x}}{\sqrt[3]{5y^2}}. \text{ [7.5]}$$

64. Write an equivalent expression by rationalizing the denominator:
$$\dfrac{1 - \sqrt{x}}{1 + \sqrt{x}}. \text{ [7.5]}$$

65. Multiply these complex numbers:
$$(3 + 2i)(4 - 7i). \text{ [7.8]}$$

66. Write a quadratic equation whose solutions are $5\sqrt{2}$ and $-5\sqrt{2}$. [8.4]

67. Find the center and the radius of the circle given by
$$x^2 + y^2 - 4x + 6y - 23 = 0. \text{ [10.1]}$$

68. Write an equivalent expression that is a single logarithm:
$$\tfrac{2}{3} \log_a x - \tfrac{1}{2} \log_a y + 5 \log_a z. \text{ [9.4]}$$

69. Write an equivalent exponential equation:
$\log_a c = 5$. [9.3]

Use a calculator to find each of the following. [9.5]

70. $\log 5677.2$

71. $10^{-3.587}$

72. $\ln 5677.2$

73. $e^{-3.587}$

74. The number of personal computers in Mexico has grown exponentially from 0.12 million in 1985 to 5.0 million in 2002. [9.7]
   Source: Center for International Development
   a) Find the exponential growth rate, k, to three decimal places and write an exponential function describing the number of personal computers in Mexico t years after 1985.
   b) Predict the number of personal computers in Mexico in 2010.

75. Find the distance between the points $(-1, -5)$ and $(2, -1)$. [10.1]

76. Find the 21st term of the arithmetic sequence $19, 12, 5, \ldots$ [11.2]

77. Find the sum of the first 25 terms of the arithmetic series $-1 + 2 + 5 + \cdots$. [11.2]

78. Find the general term of the geometric sequence $16, 4, 1, \ldots$ [11.3]

79. Find the 7th term of $(a - 2b)^{10}$. [11.4]

80. Find the sum of the first nine terms of the geometric series $x + 1.5x + 2.25x + \cdots$. [11.3]

81. On Elyse's 9th birthday, her grandmother opened a savings account for her with $500. The account pays 3% interest, compounded annually. If Elyse neither adds to nor withdraws any money from the bank, how much will be in the account on her 18th birthday? [11.3]

SYNTHESIS

Solve.

82. $\dfrac{9}{x} - \dfrac{9}{x + 12} = \dfrac{108}{x^2 + 12x}$ [6.4]

83. $\log_2 (\log_3 x) = 2$ [9.6]

84. y varies directly as the cube of x and x is multiplied by 0.5. What is the effect on y? [8.6]

85. Divide these complex numbers:
$$\dfrac{2\sqrt{6} + 4\sqrt{5}i}{2\sqrt{6} - 4\sqrt{5}i}. \text{ [7.8]}$$

86. Diaphantos, a famous mathematician, spent $\tfrac{1}{6}$ of his life as a child, $\tfrac{1}{12}$ as an adolescent, and $\tfrac{1}{7}$ as a bachelor. Five years after he was married, he had a son who died 4 years before his father at half his father's final age. How long did Diaphantos live? [3.5]

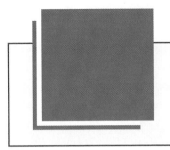

The Graphing Calculator

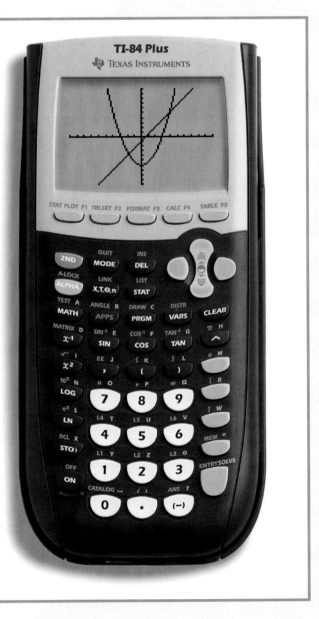

G raphing calculators and computer software make possible the quick graphing of equations. This appendix discusses the graphing calculator. Sections I.1 and I.2 of this appendix introduce the calculator, while the other sections are each linked to a corresponding section in the text.

Different calculators may require different keystrokes to use the same feature. In this appendix, exact keystrokes referenced and screens shown apply to the Texas Instruments TI-84® Plus calculator. If your calculator does not have a key by the name shown, consult the user's manual that came with your calculator to learn how to perform the procedure. Often it is helpful to simply try different combinations of keystrokes.

I.1   Introduction to the Graphing Calculator

A typical screen on a graphing calculator holds up to 8 lines of text and is 16 columns wide. Screens are composed of *pixels*, or dots. After the calculator has been turned on, a blinking rectangle or line appears on the screen. This rectangle or line is called the *cursor*. It indicates your current position on the screen. In the graphing mode, the cursor may appear as a cross. The four arrow keys are used to move the cursor.

The *contrast* controls how dark the characters appear on the screen. Once the graphing calculator is on, you can adjust the contrast by pressing **2ND** and either the up or down arrow key.

Entering and Editing Expressions

The screen allows you to view an expression in its entirety before it is calculated. Pressing the operation keys ($+$, $-$, $\times$, $\div$) does not automatically perform the operation; you must also press **ENTER**.

EXAMPLE 1   Calculate: $4 + 3(7 - 10)$.

Solution   We first enter the expression as it is written and check our typing. If it is correct, pressing **ENTER** will perform all the operations in the correct order. The answer will appear on the right side of the screen, as shown below.

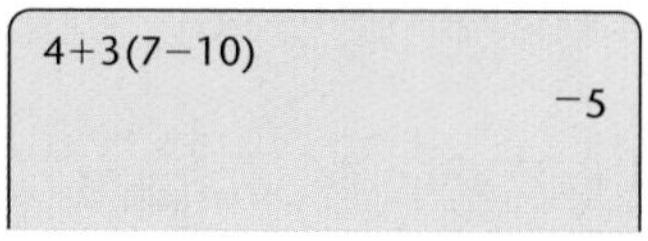

```
4+3(7−10)
                          −5
```

Thus, $4 + 3(7 - 10) = -5$.

If, before pressing **ENTER**, you see that your keystrokes are not correct, use the arrow keys to move through the expression and correct any errors. Pressing **DEL** will delete the character under the cursor. To replace the character under the cursor with another character, simply press the desired key. If you left out a character, press **2ND** INS. Characters can then be inserted to the left of the cursor.

After pressing **ENTER**, you can still edit, or change, the previous expression. To do so, press **2ND** ENTRY to copy the previous expression on the screen. You can then change it as described above.

A graphing calculator has many features not written on the keys. Written above each key are color-coded secondary features, which are performed by first pressing **2ND** and then the key below the desired feature. Above many keys, there is also an *alpha character*. These are accessed by first pressing **ALPHA** and then the key below the character. **ALPHA** can be locked on by pressing **2ND** A-LOCK. Press **ALPHA** again to unlock the key.

Many features are available from *menus*. When certain keys are pressed, a menu, or list of options, appears on the screen. You can select an item from the menu by pressing the number of the item or by highlighting the item and then pressing **ENTER**. Sometimes a menu has a *secondary menu* from which to choose. For example, pressing **MATH** yields the menu shown below. The secondary menu is listed across the top of the screen.

```
MATH NUM CPX PRB
1:▶ Frac
2:▶ Dec
3: 3
4: ³√ (
5: ˣ√
6: fMin(
7↓fMax(
```

The screen in which you perform calculations is called the *home screen*. To return to the home screen, simply press **2ND** QUIT.

Functions and Grouping Symbols

On a graphing calculator, most operations can be entered in the order in which you would write them on paper, except that you may need to add parentheses. In particular, a fraction bar is entered as a division symbol, $/$, and does not serve as a grouping symbol, so numerators and denominators may need to be enclosed in parentheses. It is important to remember that the subtraction symbol and the negative symbol are two different keys. To enter a negative number, we press (-), and to subtract, we press −.

EXAMPLE 2 Calculate $\sqrt{3 \cdot 5 - 3(6 - 4)}$ and $\dfrac{4 + 1.5}{3.2 - 1}$.

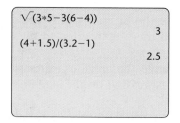

$\sqrt{(3*5-3(6-4))}$
$\qquad\qquad 3$
$(4+1.5)/(3.2-1)$
$\qquad\qquad 2.5$

Solution The keystrokes and the results of the calculations can be seen in the screen at left. Note that the entire radicand, as well as the numerator and the denominator of the rational expression, require parentheses.

Other functions, like absolute value, abs, work in a similar way. On some graphing calculators, the left parenthesis is supplied automatically.

Any number can be squared by typing the number, pressing **x²**, and then pressing **ENTER**. For powers other than 2, press **^** followed by the exponent. If the exponent is itself an expression, enclose it within parentheses. An exponent that is an integer or decimal does not need to be enclosed in parentheses. But an exponent written as a fraction must be in parentheses. (For more on rational exponents, see Chapter 7.)

EXAMPLE 3 Calculate: $(3 + 5)^{-2}$, 7^{8-1}, and $8^{2/3}$.

Solution The results of the calculations are shown in the screen below.

$(3+5)^{\wedge}-2$
$\qquad\qquad .015625$
$7^{\wedge}(8-1)$
$\qquad\qquad 823543$
$8^{\wedge}(2/3)$
$\qquad\qquad 4$

Exercise Set

I.1

Short exercise sets appear throughout this appendix, providing practice with the skills just discussed. Answers can be found at the back of the book.

Use a graphing calculator to evaluate each expression.

1. $\sqrt{-7 + 2(10 - 2)}$

2. $|-9|$

3. $|9 - (5 + (-3(6.4 - 19)))|$

4. $\dfrac{-4 - 3|13 - 5.3| + \sqrt{2.25}}{5.6 \div 7 - 55 \div 10}$

5. $(3.4 - 5.6(7.3 - 8.79))^2$

6. $2.6^{2(3-1)}$

7. $\left(\dfrac{3}{4}\right)^{-2}$

8. $16^{3/4}$

I.2   Variables and Functions

Graphing calculators have memory available in which values of variables, equations of functions, and programs can be stored. To store a numerical value, type the number, press **STO›**, and then type the name of the variable using an alpha character. To recall that value, press **2ND** [RCL] and the variable name or press **ALPHA** followed by the variable name. This can be done in the middle of a calculation.

To store an expression for a function, press (Y=). We enter the expression the same way we would write it on paper. To enter the variable x, we use the **X,T,Θ,n** key.

EXAMPLE 1   Enter $y_1 = x^2 - x - 6$, $y_2 = 2x - 6$, and $y_3 = 2x^4 - 5x^3 + x^2 - 10$.

Solution   Press (Y=) **X,T,Θ,n** **x²** (−) **X,T,Θ,n** (−) (6) **ENTER** to enter the first equation. You will now be on the line for Y2 and can press (2) **X,T,Θ,n** (−) (6) **ENTER**. Enter y_3 in a similar way. Your screen should look like the one below.

```
Plot1   Plot2   Plot3
\Y1■X²−X−6
\Y2■2X−6
\Y3■2X^4−5X^3+X²
−10
\Y4=
\Y5=
\Y6=
```

Remember: After you have worked on a screen other than the home screen, press **2ND** (QUIT) to return to the home screen.

Once entered, a function can be graphed or used in equations. Graphing functions and solving equations are discussed throughout the text and in the remainder of this appendix.

To evaluate a function for a particular value of x, you can use function notation or store the value to the variable X. Shown below are the keystrokes for each method of finding $f(2)$, where $f(x) = y_1 = x^2 - x - 6$.

Function Notation

VARS ( ) (1) (1) (( ) (2) ( )) **ENTER**

Storing the Value to X

(2) **STO›** **X,T,Θ,n** **ENTER**
VARS ( ) (1) (1) **ENTER**

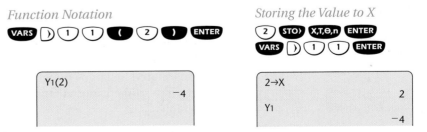

```
Y1(2)
              −4
```

```
2→X
               2
Y1
              −4
```

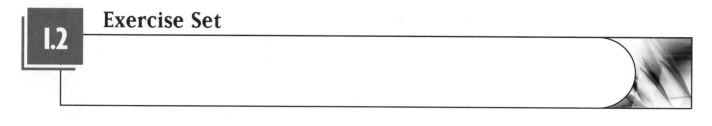

Exercise Set
1.2

1. Find the circumference and the area of a circle with radius 10.3452 cm using the following steps.
   a) Store the value 10.3452 to the variable R.
   b) Calculate the circumference using the formula $C = 2\pi R$. Use the π key and the stored value for R. (You can type simply $2\pi R$ and press **ENTER**.) Round to four decimal places.
   c) Calculate the area using the formula $A = \pi R^2$. Round to four decimal places.

2. The formula $I = PRT$ gives the amount of simple interest earned on a principal P at an interest rate R (in decimal form) for a length of time T (in years).

You have $1452.39 to invest for 7 months. Store the values for P and T and use them to calculate the interest earned when the interest rate is each of the following.
   a) 10%   b) 6%
   c) 4.375%   d) 8.65%

Evaluate each polynomial for the given values.

3. $y = 3x^4 - 2x^3 + 9x - 17$, for $x = 4, -1, 36,$ and 98.

4. $y = 1.5x^6 - 2.8x^4 + 0.1x^2$, for $x = 1.23, -1.23, 107,$ and -107.

Graphs
2.1

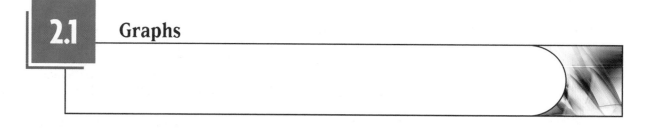

When we draw a graph with a graphing calculator, it appears in a *window*. The window is described by four numbers: the minimum x-value shown (Xmin), the maximum x-value (Xmax), the minimum y-value (Ymin), and the maximum y-value (Ymax). These dimensions are often written in the form [L, R, B, T], or Left and Right endpoints of the x-axis and Bottom and Top endpoints of the y-axis. The number of tick marks on each axis is determined by setting the x-scale (Xscl) and the y-scale (Yscl). The *scale* is the distance between the tick marks. The graphs in this appendix are labeled to indicate the viewing window used. If Xscl or Yscl is not 1, we indicate that as well.

Choosing an appropriate viewing window is important. The window dimensions should be large enough to include important features of the graph. However, if the dimensions are too large, it may be difficult to see the curvature of the graph.

EXAMPLE 1   Graph $y = x^3 + 3x^2 - x + 1$ using the windows

$$[0, 10, 0, 10], \quad [-100, 100, -100, 100], \quad \text{and} \quad [-10, 10, -10, 10].$$

Solution   We first enter the equation using ⎛Y=⎞ and then set the window dimensions using ⎛WINDOW⎞. A good choice of scale might be 1 for the first and last graphs and 10 for the second graph. (Scales need not be the same for both axes.) We then graph the equation by pressing ⎛GRAPH⎞ after setting each window.

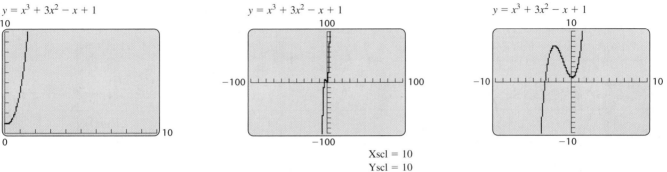

Which viewing window is the best? The answer depends on what we need to see from the graph. In Example 1, the third window would probably be the best choice since it shows where the x-axis is crossed, as well as the overall curvature.

Some window choices can be set quickly using the ZOOM option. The **standard viewing window** $[-10, 10, -10, 10]$ can be set up directly by pressing ⎛ZOOM⎞ ⎛6⎞. The ZOOM menu can also be used to magnify a portion of the graph (Zoom In) and to set up an appropriate window for a set of data (ZoomStat).

The VALUE option in the ⎛CALC⎞ menu gives a function value for a given value of x. Alternatively, pressing ⎛TRACE⎞ displays the coordinates of various points on the graph. The points traced by the cursor depend on the window. If we are interested in evaluating a function for a particular x-value, most graphing calculators allow us to simply enter that x-value.

EXAMPLE 2   Graph $y = x^2 - 4x + 2$ using the standard viewing window. Then use TRACE to find the value of y when x is 1.1.

Solution   We enter the equation and then press ⎛ZOOM⎞ ⎛6⎞ to set the window and graph the equation. (See the graph on the left below.) Next, we press ⎛ZOOM⎞ ⎛4⎞. The equation is graphed again in a smaller window. Pressing ⎛TRACE⎞ and entering 1.1, we find that y is -1.19 when x is 1.1.

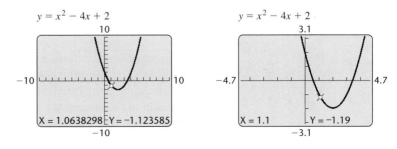

On most graphing calculators, two or more equations can be graphed one after the other (*sequentially*) or at the same time (*simultaneously*). If equations are graphed simultaneously, all *y*-values are calculated and plotted for one *x*-value, then all *y*-values are calculated for the next *x*-value, and so on. Often a graphing calculator can be set to SEQUENTIAL or SIMULTANEOUS after pressing **MODE**. To *select* the equation(s) that is graphed, locate the cursor on the equals sign of the desired equation and press **ENTER**. When the equals sign is highlighted, the equation is selected and can be graphed. Pressing **ENTER** again deselects the equation.

2.1 Exercise Set

Graph each equation or pair of equations using the given viewing windows. Use an Xscl and a Yscl of 1 unless otherwise indicated.

1. $y = 2x - 5$
   a) $[-10, 10, -10, 10]$
   b) $[0, 10, 0, 10]$

2. $y = 2/x$
   a) $[-100, 100, -100, 100]$
       $Xscl = 10$ $Yscl = 10$
   b) $[-10, 10, -10, 10]$

3. $y_1 = 3x - 2, \; y_2 = x^2 + x + 1$
   $[-10, 10, -10, 10]$
   a) Graph sequentially.
   b) Graph simultaneously.

4. Use the graph of $y = 3x^2 - x - 1$ to find the value of y when $x = -1.2$.

2.5 Other Equations of Lines

The graph of an equation can be used to visually check what type of equation it represents. If the graph is not a straight line, the equation is not linear. If the graph appears to be a line, the equation is *probably* linear. However, a graph appearing to be linear may have a different shape when viewed in a different window. Algebraic checks for linearity are more accurate than graphing.

EXAMPLE 1   Determine whether $y = 3x^2 + 5x - 1$ is linear.

Solution   For the sake of illustration, the graph of the equation is shown using two different viewing windows. Note that in the figure on the left, the graph appears to be a line.

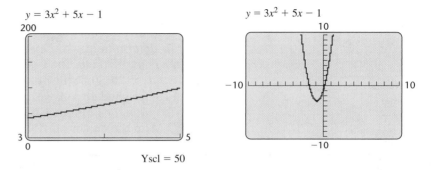

From the figure on the right, we see that the equation is not linear. Algebraically, it is not linear because it cannot be written in the standard form of a linear equation.

If a set of data is graphed and seems to lie on a straight line, we say that the data appear to be linear. We can find the equation of a line through two points. If more than two data points are known, the line that *best* describes the data may not actually go through any of the given points. The most commonly used method for fitting an equation to data is called *regression*. Most graphing calculators offer regression as a way of fitting a line or curve to a set of data. If the equation is a line, the method is called *linear regression*.

EXAMPLE 2   The number of shopping centers in the United States has grown in recent years, as shown in the following table (*Source*: International Council of Shopping Centers).

Years After 1990	Number of Centers (in thousands)
0	36.515
2	38.966
5	41.235
8	43.661
10	45.016
12	46.336

a) Plot the data and determine whether they appear to be linear.

b) If the data appear to be linear, fit a linear equation to the data and graph the line.

c) Use the equation found in part (b) to predict the number of shopping centers in the United States in 2008.

Solution

L1	L2	L3
0	36.515	--------
2	38.966	
5	41.235	
8	43.661	
10	45.016	
12	46.336	
--------	--------	

$L1(1) = 0$

a) We enter the data as lists of numbers using the EDIT menu. To do so, we press **STAT** and choose the EDIT option. Then we clear any data already in the lists by moving the cursor to the list name (L_1, L_2) and pressing **CLEAR** **ENTER**.

Next, we enter the years (the first column in the table at left) as L1 and the number of shopping centers as L2, and type the data in the correct row and column and press **ENTER**.

To determine the dimensions of the viewing window, we look at the numbers in the table. The years are between 0 and 12, so we set Xmin = 0 and Xmax = 20. Since the number of shopping centers, in thousands, is between 36.515 and 46.336, we can set Ymin = 0, Ymax = 75, and Yscl = 5.

To graph the points, we turn the STAT PLOT on by pressing **2ND** (STAT PLOT) **ENTER**. With the cursor on ON, we press **ENTER** again. Using the arrow keys and **ENTER**, we choose the first graph Type. If necessary, we set Xlist to L_1 and Ylist to L_2 using the functions above the ①and ② keys on the keypad.

Finally, we clear or deselect any equations stored in the graphing calculator and plot the points. The data appear to be linear—that is, they appear to lie in a straight line.

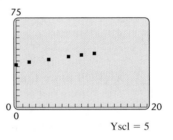

Yscl = 5

b) To calculate the linear regression line, we press **STAT**. Next, we choose the CALC submenu from the top of the screen and choose option **4: LinReg (ax + b)**. In order to be able to graph the equation, we copy the regression equation as Y1 by pressing **VARS** ▷①① and then **ENTER**.

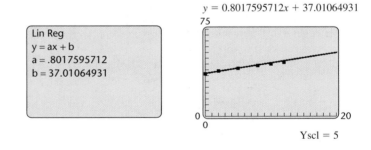

$y = 0.8017595712x + 37.01064931$

Lin Reg
$y = ax + b$
$a = .8017595712$
$b = 37.01064931$

Yscl = 5

The equation is given in slope–intercept form $y = ax + b$. (The calculator uses a instead of m.) Substituting the given values of a and b, we have the equation $y = 0.8017595712x + 37.01064931$. We press (GRAPH) to see the graph of the equation.

c) To predict the number of shopping centers in the United States in 2008, we evaluate the linear equation for $x = 18$. When $x = 18$, $y = 51.442322$. The number of shopping centers in 2008 will be about 51,442.

Exercise Set

2.5

Determine whether each equation is linear.

1. $y = 2.98x + 1.307$

2. $y = |3x - 6.5|$

3. $y = 0.002x^3$

4. The following table shows the life expectancy of males born in the United States in selected years.
Source: *Statistical Abstract of the United States*, 2003

Years Since 1970	Life Expectancy, y (in years)
0	67.1
10	70.0
20	71.8
30	74.3

a) Use linear regression to find a linear function that can be used to predict a man's life expectancy as a function of the year in which he was born, where x represents the number of years since 1970.
b) Predict the life expectancy in 2010.

5. The following table shows the estimated annual expenditures on a child by a family with an annual income between $39,100 and $65,800.
Source: *Statistical Abstract of the United States*, 2003

Age of Child	Annual Expenditures
1	$ 9,030
4	9,260
7	9,260
13	9,940
16	10,140

a) Use linear regression to find a linear function that can be used to predict the annual expenditure on a child as a function of the age of the child.
b) Estimate the annual expenditure on a 10-year-old child.

6. The following table gives the total waste generated in the United States in millions of tons per year.
Source: Characterization of MSW in the US, US EPA, Washington, DC

Years Since 1960	Total Waste Generated (in millions of tons)
0	88.1
10	121.1
20	151.6
30	196.9
40	221.7

a) Use linear regression to find a linear function that can be used to predict the total waste generated as a function of the number of years since 1960.
b) Estimate the total waste generated in the United States in 2010.

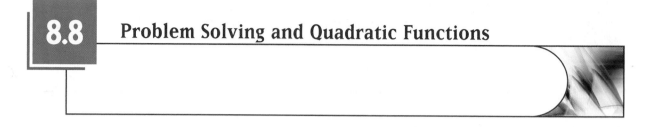

8.8   Problem Solving and Quadratic Functions

The point at which a maximum or minimum value of a function occurs can often be seen directly from its graph. To help identify this value, many graphing calculators can calculate the maximum or minimum value of a function over a specified interval.

EXAMPLE 1   Estimate the minimum value of the function $g(x) = 1.65x^2 - 3.7x - 2.95$.

Solution   We can see from the graph that $g(x)$ has a minimum value somewhere near $x = 1$.

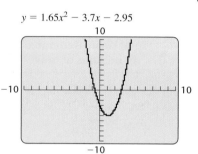

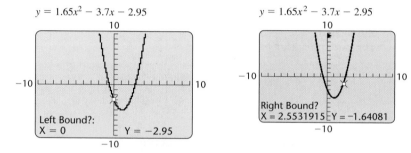

The $\boxed{\text{CALC}}$ menu, which is the 2nd function of the $\boxed{\text{TRACE}}$ key, lists some information that can be calculated from the graph. When we select option **3: minimum**, the graph is displayed, along with the question "Left Bound?" as shown in the left figure above. We position the cursor on the graph somewhere to the left of the minimum and press **ENTER**. We can also simply enter a value to the left of the minimum.

Next, we are asked for a right bound, as shown on the right above. We position the cursor to the *right* of the minimum and press **ENTER** or simply enter a value that is to the right of the minimum.

Finally, we are asked for a guess, as shown on the left below. We move the cursor closer to the minimum and press **ENTER**.

At the bottom of the graph on the right below, we can now read the coordinates of the minimum point on the graph. To the nearest hundredth, the minimum value of the function is -5.02.

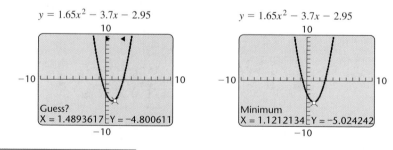

Exercise Set

8.8

Find the maximum or minimum value of each function and the value of x at which it occurs. Round to the nearest hundredth.

1. $f(x) = -2.15x^2 + 3.8x - 6.9$

2. $f(x) = 8.3x^2 + 6.7x + 3.9$

3. $g(x) = 0.016 + 2.3x + 0.003x^2$

Photo Credits

Answers

CHAPTER 1

Exercise Set 1.1, pp. 11–13

1. Constant 3. Factors 5. Left 7. Rational
9. Terminating 11. Let n represent the number; $n - 6$
13. Let t represent the number; $12t$ 15. Let x represent the number; $0.65x$, or $\frac{65}{100}x$ 17. Let y represent the number; $2y + 9$ 19. Let s represent the number; $0.1s + 8$, or $\frac{10}{100}s + 8$ 21. Let m and n represent the numbers; $m - n - 1$ 23. $90 \div 4$, or $\frac{90}{4}$ 25. 25
27. 8 29. 27 31. 0 33. 5 35. 25 37. 225
39. 33 41. 7 43. 17.5 sq ft 45. 11.2 sq m
47. {a, e, i, o, u}, or {a, e, i, o, u, y}
49. {1, 3, 5, 7, ...} 51. {5, 10, 15, 20, ...}
53. {$x \mid x$ is an odd number between 10 and 20}
55. {$x \mid x$ is a whole number less than 5}
57. {$n \mid n$ is a multiple of 5 between 7 and 79}
59. (a) 0, 6; (b) -3, 0, 6; (c) -8.7, -3, 0, $\frac{2}{3}$, 6; (d) $\sqrt{7}$;
(e) -8.7, -3, 0, $\frac{2}{3}$, $\sqrt{7}$, 6 61. (a) 0, 3; (b) -17, 0, 3;
(c) -17, -4.13, 0, $\frac{5}{4}$, 3; (d) $\sqrt{77}$; (e) -17, -4.13, 0, $\frac{5}{4}$, 3, $\sqrt{77}$
63. False 65. True 67. False 69. True 71. True
73. False 75. 📓 77. 📓 79. Let a and b represent the
numbers; $\dfrac{a + b}{a - b}$ 81. Let r and s represent the numbers;
$\dfrac{1}{2}(r^2 - s^2)$, or $\dfrac{r^2 - s^2}{2}$ 83. {0} 85. {5, 10, 15, 20, ...}
87. {1, 3, 5, 7, ...} 89.

Technology Connection, p. 16

1. 📊 2. 0.97

Exercise Set 1.2, pp. 22–24

1. True 3. True 5. False 7. False 9. True 11. 9
13. 6 15. 6.2 17. 0 19. $1\frac{7}{8}$ 21. 4.21 23. -6 is less than or equal to -2; true 25. -9 is greater than 1; false
27. 3 is greater than or equal to -5; true 29. -8 is less than -3; true 31. -4 is greater than or equal to -4; true
33. -5 is less than -5; false 35. 12 37. -11 39. -1.2
41. $-\frac{11}{35}$ 43. -9.06 45. $\frac{5}{9}$ 47. -4.5 49. 0 51. -6.4
53. -3.14 55. $4\frac{1}{3}$ 57. 0 59. -9 61. 2.7 63. -1.79
65. 0 67. 3 69. -3 71. 7 73. -19 75. -3.1
77. $-\frac{11}{10}$ 79. 2.9 81. 7.9 83. -30 85. 36 87. -21
89. $-\frac{3}{7}$ 91. 0 93. 5.44 95. 5 97. -5 99. -73
101. 0 103. $\frac{1}{4}$ 105. $-\frac{1}{9}$ 107. $\frac{3}{2}$ 109. $-\frac{11}{3}$ 111. $\frac{5}{6}$
113. $-\frac{6}{5}$ 115. $\frac{1}{36}$ 117. 1 119. 25 121. $-\frac{6}{11}$
123. Undefined 125. $\frac{11}{43}$ 127. 31 129. -3
131. 15 133. $7b + 4a; a \cdot 4 + b \cdot 7; 4a + b \cdot 7$
135. $y(7x); (x \cdot 7)y$ 137. $3(xy)$ 139. $(x + 2y) + 5$
141. $2a + 10$ 143. $4x - 4y$ 145. $-10a - 15b$
147. $9ab - 9ac + 9ad$ 149. $5(x + 3)$ 151. $3(p - 3)$
153. $7(x - 3y + 2z)$ 155. $17(15 - 2b)$ 157. 📓
159. 16; 16 160. 11; 11 161. 📓
163. $(8 - 5)^3 + 9 = 36$ 165. $5 \cdot 2^3 \div (3 - 4)^4 = 40$
167. -6.2 169. 📓

Exercise Set 1.3, pp. 30–31

1. Equivalent expressions 3. Equivalent equations
5. Equivalent equations 7. Equivalent expressions
9. Equivalent equations 11. Equivalent
13. Not equivalent 15. Not equivalent 17. 16.3 19. 9
21. 18 23. 25 25. 24 27. $10x$ 29. $-2rt$ 31. $10t^2$
33. $11a$ 35. $-7n$ 37. $10x$ 39. $7x - 2x^2$
41. $22x + 18$ 43. $-5t^2 + 2t + 4t^3$ 45. $5a - 5$
47. $-5m + 2$ 49. $5d - 12$ 51. $-7x + 14$

53. $44a - 22$ **55.** $-100a - 90$ **57.** $-12y - 145$ **59.** 6
61. 12   **63.** 4   **65.** 3   **67.** 3   **69.** 7   **71.** 5   **73.** 2
75. 2   **77.** $\frac{49}{9}$ **79.** $\frac{4}{5}$ **81.** $\frac{19}{5}$ **83.** $-\frac{1}{2}$
85. $\varnothing$; contradiction   **87.** $\{0\}$; conditional
89. $\varnothing$; contradiction   **91.** $\mathbb{R}$; identity   **93.**   **95.** Let n represent the number; $2n + 9$, or $9 + 2n$

96. Let n represent the number; $0.42\left(\dfrac{n}{2}\right)$ **97.**

99. -4.176190476 **101.** 8   **103.** $\frac{224}{29}$ **105.**

Exercise Set 1.4, pp. 39–41

1. Let x and $x + 7$ represent the numbers; $x + (x + 7) = 65$
3. Let t represent the time, in hours, it will take the swimmer to swim 1.8 km upstream; $2.7t = 1.8$
5. Let t represent the time, in seconds, it takes Alida to walk the length of the sidewalk; $9t = 300$
7. Let x, $x + 1$, and $x + 2$ represent the angle measures; $x + (x + 1) + (x + 2) = 180$ **9.** Let w represent the wholesale price; $w + 0.5w + 1.50 = 22.50$
11. Let t represent the number of minutes spent climbing; $8000 + 3500t = 29{,}000$ **13.** Let x represent the measure of the second angle, in degrees; $3x + x + (2x - 12) = 180$
15. Let x represent the first even number; $2x + 3(x + 2) = 76$ **17.** Let s represent the length, in centimeters, of a side of the smaller triangle; $3s + 3 \cdot 2s = 90$ **19.** Let c represent Brian's calls on his next shift; $\dfrac{5 + 2 + 1 + 3 + c}{5} = 3$ **21.** $97
23. Approximately 11.15 million cases   **25.** 22 tickets
27. Length: 45 cm; width: 15 cm   **29.** Length: 52 m; width: 13 m   **31.** $\frac{2}{3}$ hr   **33.** 100°, 25°, 55°   **35.** $150
37. $14.00   **39.**   **41.** $\frac{9}{2}$ **42.** 36   **43.** 19   **44.** $\frac{4}{3}$
45.   **47.** 10 points   **49.** $110,000

Exercise Set 1.5, pp. 47–50

1. Equation   **3.** Circumference   **5.** $A = bh$
7. Subscripts   **9.** $r = \dfrac{d}{t}$ **11.** $a = \dfrac{F}{m}$ **13.** $I = \dfrac{W}{E}$
15. $h = \dfrac{V}{lw}$ **17.** $k = Ld^2$ **19.** $n = \dfrac{G - w}{150}$
21. $l = p - 2w - 2h$ **23.** $n = \dfrac{k}{l + m}$ **25.** $x = \dfrac{w}{y + z}$
27. $y = \dfrac{C - Ax}{B}$ **29.** $F = \dfrac{9}{5}C + 32$ **31.** $r^3 = \dfrac{3V}{4\pi}$
33. $a = \dfrac{d}{b + c}$ **35.** $y = \dfrac{w}{x - z}$ **37.** $n = \dfrac{q_1 + q_2 + q_3}{A}$
39. $t = \dfrac{d_2 - d_1}{v}$ **41.** $d_1 = d_2 - vt$ **43.** $m = \dfrac{r}{1 + np}$
45. $a = \dfrac{y}{c^2 + b}$ **47.** 12%   **49.** 6 cm   **51.** About 235 lb

53. About 1504.6 g   **55.** 9 ft   **57.** 1 yr   **59.** 1205
61. 5 ft 7 in.   **63.** 61   **65.** About 8.5 cm
67. 34 appointments   **69.**
71. $(7 \cdot 3)(a \cdot a)$; $a \cdot 3 \cdot a \cdot 7$ **72.** $4 \cdot x \cdot y \cdot y$; $(y \cdot 4)(x \cdot y)$

73.   **75.** About 10.9 g   **77.**   **79.** $l = \dfrac{A - w^2}{4w}$

81. $T_1 = \dfrac{P_1 V_1 T_2}{P_2 V_2}$ **83.** $d = \dfrac{me^2}{f}$ **85.** $t = \dfrac{1}{s}$

Technology Connection, p. 56

1. Answers may vary; (2) (x^y) (5) (−) (=), (2) (∧) (−) (5) ENTER, (2) (x^y) (−x) (5) ENTER   **2.** Compute $1 \div (2 \times 2 \times 2 \times 2 \times 2)$.

Exercise Set 1.6, pp. 58–60

1. The power rule   **3.** Raising a product to a power
5. The product rule   **7.** Raising a quotient to a power
9. The quotient rule   **11.** 2^{12} **13.** 5^{10} **15.** m^9
17. $18x^7$ **19.** $16m^{13}$ **21.** $x^{10}y^{10}$ **23.** a^6 **25.** $3t^5$
27. m^5n^4 **29.** $4x^6y^4$ **31.** $-4x^8y^6z^6$ **33.** -1 **35.** 1
37. 16   **39.** -16 **41.** $\frac{1}{16}$ **43.** $-\frac{1}{16}$ **45.** $-\frac{1}{16}$ **47.** $-\frac{1}{64}$
49. $\dfrac{1}{a^3}$ **51.** 125   **53.** $\dfrac{8}{x^3}$ **55.** $\dfrac{3a^8}{b^6}$ **57.** $\dfrac{1}{3x^5z^4}$ **59.** $\dfrac{y^7z^4}{x^2}$
61. 8^{-4} **63.** $(-12)^{-2}$ **65.** $\dfrac{1}{x^{-5}}$ **67.** $\dfrac{4}{x^{-2}}$ **69.** $(5y)^{-3}$
71. $\dfrac{y^{-4}}{3}$ **73.** 8^{-6}, or $\dfrac{1}{8^6}$ **75.** b^{-3}, or $\dfrac{1}{b^3}$ **77.** a^3
79. $-8m^4n^5$ **81.** $-14x^{-11}$, or $-\dfrac{14}{x^{11}}$ **83.** $10a^{-6}b^{-2}$, or
$\dfrac{10}{a^6b^2}$ **85.** 10^{-9}, or $\dfrac{1}{10^9}$ **87.** 2^{-2}, or $\dfrac{1}{2^2}$, or $\dfrac{1}{4}$ **89.** y^9
91. $-3ab^2$ **93.** $-\dfrac{7}{4}a^{-4}b^2$, or $-\dfrac{7b^2}{4a^4}$ **95.** $-\dfrac{1}{4}x^3y^{-2}z^{11}$,
or $-\dfrac{x^3z^{11}}{4y^2}$ **97.** x^{12} **99.** 9^{-12}, or $\dfrac{1}{9^{12}}$ **101.** t^{40}
103. $25x^2y^2$ **105.** $a^{12}b^4$ **107.** x^{-8}, or $\dfrac{1}{x^8}$ **109.** $32a^{-4}$,
or $\dfrac{32}{a^4}$ **111.** 1   **113.** $\dfrac{9x^8y^9}{2}$ **115.** $\dfrac{625}{256}x^{-20}y^{24}$, or $\dfrac{625y^{24}}{256x^{20}}$
117. $\dfrac{16}{9}a^{-4}b^{-12}$, or $\dfrac{16}{9a^4b^{12}}$ **119.** 1   **121.**   **123.** 35.1
124. 44   **125.**   **127.** $4a^{-x-4}$ **129.** $2^{-2a-2b+ab}$
131. 3^{a^2+2a} **133.** 7^{6b-2ab} **135.** $\frac{2}{27}$ **137.** $\dfrac{a^{-14ac}}{b^{27ac}}$

Exercise Set 1.7, pp. 66–68

1. Positive power of 10   **3.** Negative power of 10
5. Positive power of 10   **7.** 0.0005   **9.** 973,000,000

11. 0.0000000004923   **13.** 90,300,000,000
15. 0.00000004037   **17.** 7,010,000,000,000
19. 8.3×10^{10} **21.** 8.63×10^{17} **23.** 1.6×10^{-8}
25. 7×10^{-11} **27.** 8.03×10^{11} **29.** 9.04×10^{-7}
31. 4.317×10^{11} **33.** 9.7×10^{-5} **35.** 1.3×10^{-11}
37. 1.4×10^{11} **39.** 4.6×10^{3} **41.** 6.0   **43.** 1.5×10^{3}
45. 3.0×10^{-5} **47.** 4.0×10^{-16} **49.** 3.00×10^{-22}
51. 2.00×10^{26} **53.** 7.8×10^{-9} **55.** 1.2×10^{24}
57. 2.0×10^{9} neutrinos   **59.** Approximately
4.5×10^{-16} in³   **61.** Approximately 5×10^{2} in³, or
3×10^{-1} ft³   **63.** 8.00 light years   **65.** 3.08×10^{26} Å
67. 1×10^{22} cu Å, or 1×10^{-8} m³   **69.** 7.9×10^{7} bacteria
71. 4.49×10^{4} km/h   **73.** 🖩   **75.** 8   **76.** 32   **77.** 🖩
79. 2.5×10^{-10} oz to 9.5×10^{-10} oz   **81.** $8 \cdot 10^{-90}$ is
larger by 7.1×10^{-90} **83.** 8   **85.** 8×10^{18} grains

Review Exercises: Chapter 1, pp. 71–72

1. (e)   **2.** (g)   **3.** (j)   **4.** (a)   **5.** (i)   **6.** (b)   **7.** (f)
8. (c)   **9.** (d)   **10.** (h)   **11.** Let x and y represent the
numbers; $\dfrac{x}{y} - 5$ **12.** 22   **13.** $\{1, 3, 5, 7, 9\}$;
$\{x \mid x \text{ is an odd natural number between 0 and 10}\}$
14. 1750 sq cm   **15.** 9.3   **16.** 4.09   **17.** 0   **18.** -10.2
19. $-\frac{23}{35}$ **20.** $\frac{7}{15}$ **21.** -11.5 **22.** $-\frac{1}{6}$ **23.** -5.4
24. 12.6   **25.** $-\frac{5}{12}$ **26.** -4.8 **27.** 6   **28.** -9.1 **29.** $-\frac{21}{4}$
30. 4.01   **31.** $a + 9$ **32.** $y \cdot 7$ **33.** $x \cdot 5 + y$, or $y + 5x$
34. $4 + (a + b)$ **35.** $x(y \cdot 3)$ **36.** $7m(n + 2)$
37. $4x^3 - 6x^2 + 5$ **38.** $47x - 60$ **39.** 11.6   **40.** $\frac{27}{2}$
41. $-\frac{4}{11}$ **42.** $\mathbb{R}$; identity   **43.** $\varnothing$; contradiction
44. Let x represent the number; $2x + 15 = 21$ **45.** 48
46. 90°, 30°, 60°   **47.** $m = PS$ **48.** $x = \dfrac{c}{m - r}$ **49.** 4 cm
50. $-12a^6b^8$ **51.** $4xy^6$ **52.** 1, 28.09, -28.09
53. 3^5, or 243   **54.** 5^3a^6, or $125a^6$ **55.** $-\dfrac{a^9}{8b^6}$ **56.** $\dfrac{z^8}{x^4y^6}$
57. $\dfrac{b^{16}}{16a^{20}}$ **58.** $\frac{3}{7}$ **59.** 0   **60.** 1.03×10^{-7}
61. 3.086×10^{13} **62.** 3.7×10^{7} **63.** 2.0×10^{-6}
64. 1.4×10^{4} mm³, or 1.4×10^{-5} m³   **65.** 🖩 To write an
equation that has no solution, begin with a simple
equation that is false for any value of x, such as $x = x + 1$.
Then add or multiply by the same quantities on both sides
of the equation to construct a more complicated equation
with no solution.   **66.** 🖩 Use the distributive law to
rewrite a sum of like terms as a single term by first writing
the sum as a product. For example, $2a + 5a = (2 + 5)a = 7a$.   **67.** 0.0000003%   **68.** $\frac{25}{24}$ **69.** The 17-in. pizza is a
better deal. It costs about 5¢ per square inch; the 13-in.
pizza costs about 6¢ per square inch.   **70.** 729 cm³
71. $z = y - \dfrac{x}{m}$ **72.** $3^{-2a+2b-8ab}$ **73.** $88.\overline{3}$ **74.** -39
75. $-40x$ **76.** $a \cdot 2 + cb + cd + ad = ad + a \cdot 2 + cb + cd = a(d + 2) + c(b + d)$ **77.** $\sqrt{5}/4$; answers may vary

Test: Chapter 1, pp. 73–74

1. [1.1] Let m and n represent the numbers; $mn + 4$
2. [1.1], [1.2] -47 **3.** [1.1] 3.75 sq cm   **4.** [1.2] -31
5. [1.2] -3.7 **6.** [1.2] -5.11 **7.** [1.2] -14.2
8. [1.2] -43.2 **9.** [1.2] -33.92 **10.** [1.2] $-\frac{19}{12}$ **11.** [1.2] $\frac{5}{49}$
12. [1.2] 6   **13.** [1.2] $-\frac{4}{3}$ **14.** [1.2] $-\frac{5}{2}$
15. [1.2] $y + 7x$; $x \cdot 7 + y$; answers may vary   **16.** [1.3] $-y$
17. [1.3] $a^2b - 4ab^2 + 2$ **18.** [1.3] $-4x + 8$ **19.** [1.3] -2
20. [1.3] $\mathbb{R}$; identity   **21.** [1.5] $P_2 = \dfrac{P_1 V_1 T_2}{T_1 V_2}$ **22.** [1.4] 94
23. [1.4] 17, 19, 21   **24.** [1.3] $-8x - 1$ **25.** [1.3] $24b - 9$
26. [1.6] $-\dfrac{42}{x^{10}y^6}$ **27.** [1.6] $-\dfrac{1}{3^2}$, or $-\dfrac{1}{9}$ **28.** [1.6] $\dfrac{y^8}{36x^4}$
29. [1.6] $\dfrac{x^6}{4y^8}$ **30.** [1.6] 1   **31.** [1.7] 2.01×10^{-7}
32. [1.7] 2.0×10^{10} **33.** [1.7] 3.8×10^{2}
34. [1.7] 4.7×10^{8} km   **35.** [1.6] $8^c x^{9ac} y^{3bc+3c}$
36. [1.6] $-9a^3$ **37.** [1.6] $\dfrac{4}{7y^2}$

CHAPTER 2

Technology Connection, p. 79

1. $y = -4x + 3$ $(-1.5, 9), (1, -1)$
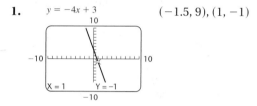

Technology Connection, p. 81

1. $y = 5x - 3$

2. $y = x^2 - 4x + 3$

3. $y = (x + 4)^2$
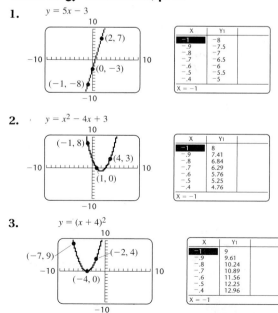

4. $y = \sqrt{x} + 2$

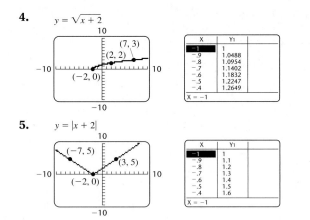

5. $y = |x + 2|$

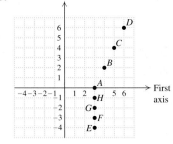

Exercise Set 2.1, pp. 82–84

1. Negative   **3.** Axes   **5.** Solutions   **7.** $(5, 3)$, $(-4, 3)$, $(0, 2)$, $(-2, -3)$, $(4, -2)$, and $(-5, 0)$

9.

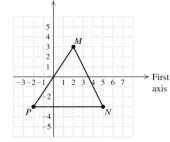

11.

Triangle, 21 units^2

13. II   **15.** III   **17.** I   **19.** IV   **21.** Yes   **23.** No   **25.** Yes
27. Yes   **29.** Yes   **31.** Yes   **33.** No   **35.** No

37.   **39.**

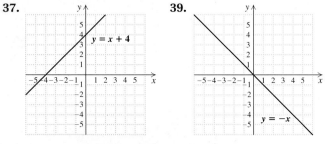

41.   **43.**

45.   **47.**

49.   **51.**

53.   **55.**

57.   **59.** ▨   **61.** -2

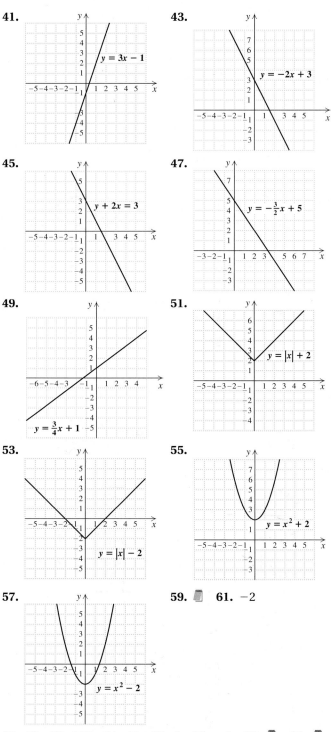

62. 13   **63.** 100   **64.** 49   **65.** 0   **66.** -3 **67.** ▨   **69.** ▨
71. **(a)** IV; **(b)** III; **(c)** I; **(d)** II   **73.** (a), (d)
75. $(2, 4)$, $(-5, -3)$; 49 sq units

77.

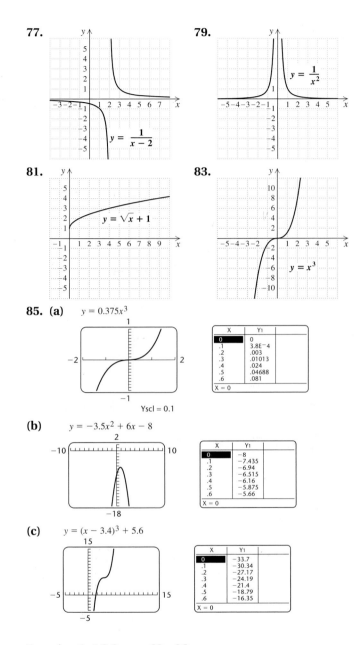

79.

$y = \dfrac{1}{x-2}$

$y = \dfrac{1}{x^2}$

81.

$y = \sqrt{x} + 1$

83.

$y = x^3$

85. (a) $y = 0.375x^3$

Yscl = 0.1

(b) $y = -3.5x^2 + 6x - 8$

(c) $y = (x - 3.4)^3 + 5.6$

(d) $\{1, 2, 3, 4\}$ **35.** Yes   **37.** Yes   **39.** No   **41.** No
43. (a) 5; **(b)** 1; **(c)** -2; **(d)** 13; **(e)** $a + 7$; **(f)** $a + 7$
45. (a) 0; **(b)** 1; **(c)** 57; **(d)** $5t^2 + 4t$; **(e)** $20a^2 + 8a$; **(f)** 48
47. (a) $\frac{3}{5}$; **(b)** $\frac{1}{3}$; **(c)** $\frac{4}{7}$; **(d)** 0; **(e)** $\dfrac{x - 1}{2x - 1}$
49. (a) $\{x \mid x \text{ is a real number } and \ x \neq 3\}$;
(b) $\{x \mid x \text{ is a real number } and \ x \neq 6\}$; **(c)** $\mathbb{R}$; **(d)** $\mathbb{R}$;
(e) $\left\{x \mid x \text{ is a real number } and \ x \neq \frac{5}{2}\right\}$; **(f)** $\mathbb{R}$
51. $4\sqrt{3} \text{ cm}^2 \approx 6.93 \text{ cm}^2$ **53.** $36\pi \text{ in}^2 \approx 113.10 \text{ in}^2$
55. $1\frac{20}{33}$ atm; $1\frac{10}{11}$ atm; $4\frac{1}{33}$ atm   **57.** 159.48 cm   **59.** 14°F
61. 75 heart attacks per 10,000 men   **63.** 56%
65. 60 watts; 140 watts

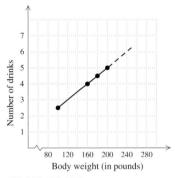

67. 3.5 drinks; 6 drinks

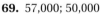

69. 57,000; 50,000

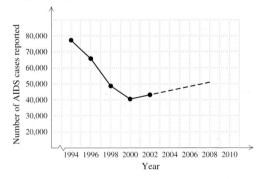

Exercise Set 2.2, pp. 93–98

1. Domain   **3.** Exactly   **5.** Horizontal   **7.** "f of 3," "f at 3," or "the value of f at 3"   **9.** Yes   **11.** Yes   **13.** Yes
15. No   **17.** Function   **19.** Function
21. (a) -1; **(b)** $\{x \mid -4 \leq x \leq 3\}$; **(c)** -3; **(d)** $\{y \mid -2 \leq y \leq 5\}$
23. (a) 3; **(b)** $\{x \mid -1 \leq x \leq 4\}$; **(c)** 3; **(d)** $\{y \mid 1 \leq y \leq 4\}$
25. (a) 3; **(b)** $\{x \mid -4 \leq x \leq 3\}$; **(c)** 0; **(d)** $\{y \mid -5 \leq y \leq 4\}$
27. (a) 3; **(b)** $\{x \mid -4 \leq x \leq 3\}$; **(c)** -3; **(d)** $\{y \mid -2 \leq y \leq 5\}$
29. (a) 1; **(b)** $\{-3, -1, 1, 3, 5\}$; **(c)** 3; **(d)** $\{-1, 0, 1, 2, 3\}$
31. (a) 4; **(b)** $\{x \mid -3 \leq x \leq 4\}$; **(c)** $-1, 3$; **(d)** $\{y \mid -4 \leq y \leq 5\}$
33. (a) 2; **(b)** $\{x \mid -4 \leq x \leq 4\}$; **(c)** $\{x \mid 0 < x \leq 2\}$;

71. $257,000; $306,000

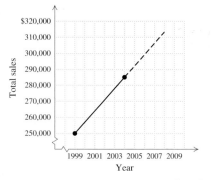

73. 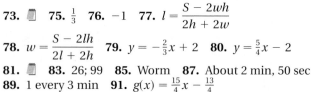   **75.** $\frac{1}{3}$ **76.** -1 **77.** $l = \dfrac{S - 2wh}{2h + 2w}$

78. $w = \dfrac{S - 2lh}{2l + 2h}$ **79.** $y = -\frac{2}{3}x + 2$ **80.** $y = \frac{5}{4}x - 2$

81. 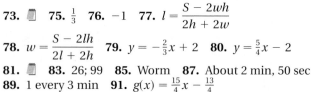   **83.** 26; 99   **85.** Worm   **87.** About 2 min, 50 sec
89. 1 every 3 min   **91.** $g(x) = \frac{15}{4}x - \frac{13}{4}$

Exercise Set 2.3, pp. 107–112

1. (e)   **3.** (c)   **5.** (a)
7.

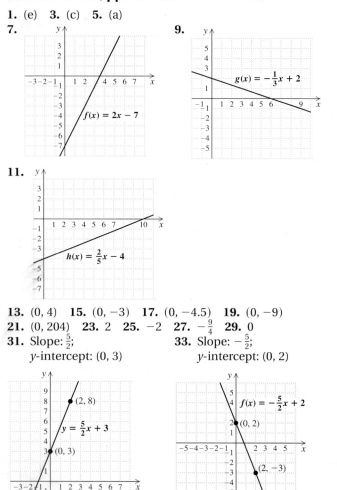

9.

11.

13. $(0, 4)$ **15.** $(0, -3)$ **17.** $(0, -4.5)$ **19.** $(0, -9)$
21. $(0, 204)$ **23.** 2   **25.** -2 **27.** $-\frac{9}{4}$ **29.** 0
31. Slope: $\frac{5}{2}$;
   y-intercept: $(0, 3)$
33. Slope: $-\frac{5}{2}$;
   y-intercept: $(0, 2)$

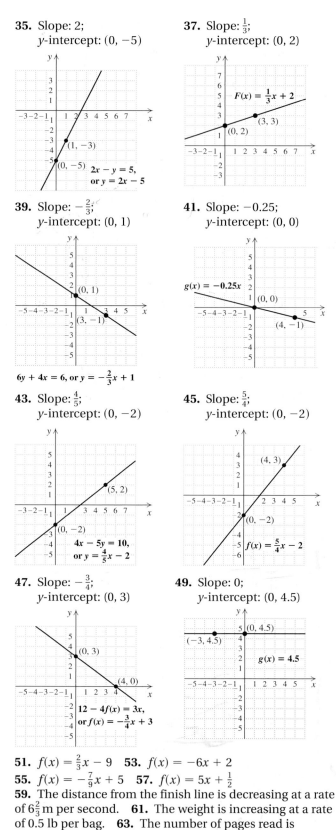

35. Slope: 2;
   y-intercept: $(0, -5)$
37. Slope: $\frac{1}{3}$;
   y-intercept: $(0, 2)$

39. Slope: $-\frac{2}{3}$;
   y-intercept: $(0, 1)$
41. Slope: -0.25;
   y-intercept: $(0, 0)$

43. Slope: $\frac{4}{5}$;
   y-intercept: $(0, -2)$
45. Slope: $\frac{5}{4}$;
   y-intercept: $(0, -2)$

47. Slope: $-\frac{3}{4}$;
   y-intercept: $(0, 3)$
49. Slope: 0;
   y-intercept: $(0, 4.5)$

51. $f(x) = \frac{2}{3}x - 9$ **53.** $f(x) = -6x + 2$
55. $f(x) = -\frac{7}{9}x + 5$ **57.** $f(x) = 5x + \frac{1}{2}$
59. The distance from the finish line is decreasing at a rate of $6\frac{2}{3}$ m per second.   **61.** The weight is increasing at a rate of 0.5 lb per bag.   **63.** The number of pages read is

increasing at a rate of 75 pages per day.   **65.** The average SAT math score is increasing at a rate of 1 point per thousand dollars of family income.   **67.** 7.5 mph   **69.** $\frac{5}{96}$ of the house per hour   **71.** 300 ft/min   **73.** (a) II; (b) IV; (c) I; (d) III   **75.** 25 signifies that the cost per person is $25; 75 signifies that the setup cost for the party is $75. **77.** $\frac{1}{2}$ signifies that Oscar's hair grows $\frac{1}{2}$ in. per month; 1 signifies that his hair is 1 in. long when cut. **79.** $\frac{1}{7}$ signifies that the life expectancy of American women increases $\frac{1}{7}$ yr per year, for years after 1970; 75.5 signifies that the life expectancy in 1970 was 75.5 yr.   **81.** 0.227 signifies that the price increases $0.227 per year, for years since 1995; 4.29 signifies that the average cost of a movie ticket in 1995 was $4.29.   **83.** 2 signifies that the cost per mile of a taxi ride is $2; 2.5 signifies that the minimum cost of a taxi ride is $2.50.   **85.** (a) −5000 signifies that the depreciation is $5000 per year; 90,000 signifies that the original value of the truck was $90,000; (b) 18 yr; (c) $\{t \mid 0 \le t \le 18\}$ **87.** (a) −150 signifies that the depreciation is $150 per winter of use; 900 signifies that the original value of the snowblower was $900; (b) after 4 winters of use; (c) $\{n \mid 0 \le n \le 6\}$ **89.** ▨   **91.** $-\frac{8}{5}$ **92.** −5   **93.** −25   **94.** 3   **95.** $-\frac{9}{2}$ **96.** $\frac{3}{4}$

97. ▨   **99.** Slope: $-\dfrac{r}{r+p}$; y-intercept: $\left(0, \dfrac{s}{r+p}\right)$

101. Since (x_1, y_1) and (x_2, y_2) are two points on the graph of $y = mx + b$, then $y_1 = mx_1 + b$ and $y_2 = mx_2 + b$. Using the definition of slope, we have

Slope $= \dfrac{y_2 - y_1}{x_2 - x_1}$

$= \dfrac{(mx_2 + b) - (mx_1 + b)}{x_2 - x_1}$

$= \dfrac{m(x_2 - x_1)}{x_2 - x_1}$

$= m.$

103. False   **105.** False   **107.** (a) III; (b) IV; (c) I; (d) II

109. (a) $-\dfrac{5c}{4b}$; (b) undefined; (c) $\dfrac{a+d}{f}$ **111.** ▨

Exercise Set 2.4, pp. 121–124

1. Horizontal   **3.** Vertical   **5.** 0; x   **7.** Intersection **9.** Linear   **11.** 0   **13.** Undefined   **15.** 0   **17.** Undefined **19.** Undefined   **21.** 0   **23.** Undefined   **25.** Undefined **27.** $-\frac{2}{3}$

29.   **31.**

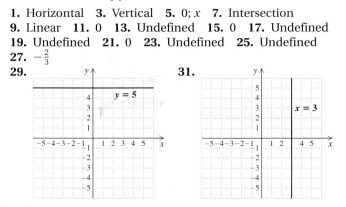

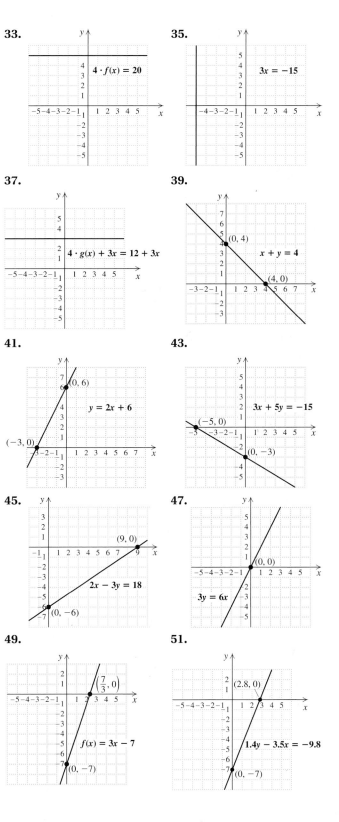

53.

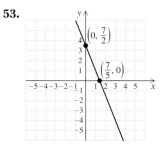

$$5x + 2g(x) = 7$$

55. 8   **57.** 1   **59.** 4   **61.** $-\frac{3}{2}$ **63.** 4   **65.** 2   **67.** $10,500 over $100   **69.** 5 months   **71.** 2 hr 15 min   **73.** 180 lb   **75.** Linear; $\frac{5}{3}$ **77.** Linear; 0   **79.** Not linear   **81.** Linear; $\frac{14}{3}$ **83.** Not linear   **85.** Linear; 5   **87.** 📝   **89.** -1 **90.** -1 **91.** $-5x - 15$ **92.** $-2x - 8$ **93.** $\frac{2}{3}x - \frac{2}{3}$ **94.** $-\frac{3}{2}x - \frac{12}{5}$ **95.** 📝   **97.** $4x - 5y = 20$ **99.** Linear   **101.** Linear   **103.** The slope of equation B is $\frac{1}{2}$ the slope of equation A.   **105.** $a = 7, b = -3$ **107.**

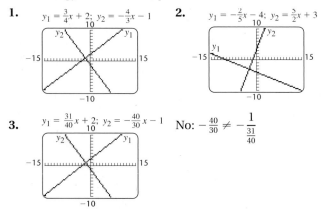

109. $0.\overline{6}$, or $\frac{2}{3}$ **111.** 2.6   **113.** 249 shirts

Technology Connection, p. 128

1. $y_1 = \frac{3}{4}x + 2; \ y_2 = -\frac{4}{3}x - 1$

2. $y_1 = -\frac{2}{5}x - 4; \ y_2 = \frac{5}{2}x + 3$

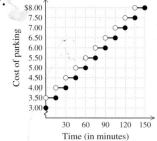

3. $y_1 = \frac{31}{40}x + 2; \ y_2 = -\frac{40}{30}x - 1$ No: $-\frac{40}{30} \neq -\frac{1}{\frac{31}{40}}$

Although the lines appear to be perpendicular, they are not, because the product of their slopes is not -1:

$$\frac{31}{40}\left(-\frac{40}{30}\right) = -\frac{1240}{1200} \neq -1.$$

Exercise Set 2.5, pp. 129–133

1. True   **3.** False   **5.** False   **7.** True   **9.** True
11.   **13.**

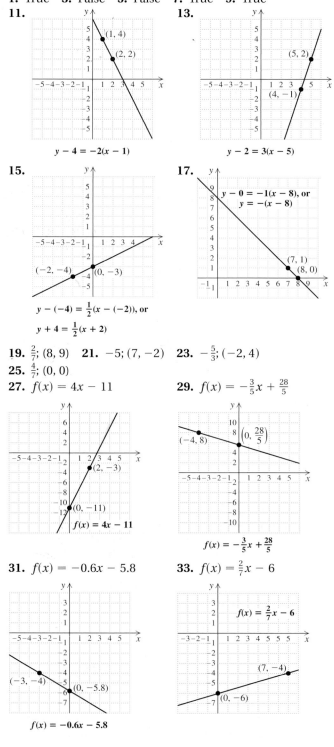

$$y - 4 = -2(x - 1)$$ $$y - 2 = 3(x - 5)$$

15.   **17.**

$$y - (-4) = \tfrac{1}{2}(x - (-2)), \text{ or}$$
$$y + 4 = \tfrac{1}{2}(x + 2)$$

19. $\frac{2}{7}$; $(8, 9)$ **21.** -5; $(7, -2)$ **23.** $-\frac{5}{3}$; $(-2, 4)$
25. $\frac{4}{7}$; $(0, 0)$
27. $f(x) = 4x - 11$ **29.** $f(x) = -\frac{3}{5}x + \frac{28}{5}$

$$f(x) = 4x - 11$$ $$f(x) = -\tfrac{3}{5}x + \tfrac{28}{5}$$

31. $f(x) = -0.6x - 5.8$ **33.** $f(x) = \frac{2}{7}x - 6$

$$f(x) = -0.6x - 5.8$$ $$f(x) = \tfrac{2}{7}x - 6$$

35. $f(x) = \frac{3}{5}x + \frac{42}{5}$

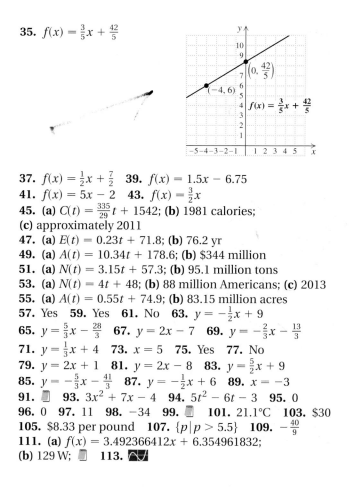

37. $f(x) = \frac{1}{2}x + \frac{7}{2}$ **39.** $f(x) = 1.5x - 6.75$
41. $f(x) = 5x - 2$ **43.** $f(x) = \frac{3}{2}x$
45. (a) $C(t) = \frac{335}{29}t + 1542$; (b) 1981 calories;
(c) approximately 2011
47. (a) $E(t) = 0.23t + 71.8$; (b) 76.2 yr
49. (a) $A(t) = 10.34t + 178.6$; (b) $344 million
51. (a) $N(t) = 3.15t + 57.3$; (b) 95.1 million tons
53. (a) $N(t) = 4t + 48$; (b) 88 million Americans; (c) 2013
55. (a) $A(t) = 0.55t + 74.9$; (b) 83.15 million acres
57. Yes   **59.** Yes   **61.** No   **63.** $y = -\frac{1}{2}x + 9$
65. $y = \frac{5}{3}x - \frac{28}{3}$ **67.** $y = 2x - 7$ **69.** $y = -\frac{2}{3}x - \frac{13}{3}$
71. $y = \frac{1}{3}x + 4$ **73.** $x = 5$ **75.** Yes   **77.** No
79. $y = 2x + 1$ **81.** $y = 2x - 8$ **83.** $y = \frac{5}{2}x + 9$
85. $y = -\frac{5}{3}x - \frac{41}{3}$ **87.** $y = -\frac{1}{2}x + 6$ **89.** $x = -3$
91. 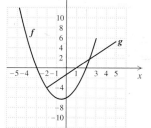   **93.** $3x^2 + 7x - 4$ **94.** $5t^2 - 6t - 3$ **95.** 0
96. 0   **97.** 11   **98.** −34   **99.**   **101.** 21.1°C   **103.** $30
105. $8.33 per pound   **107.** $\{p\,|\,p > 5.5\}$ **109.** $-\frac{40}{9}$
111. (a) $f(x) = 3.492366412x + 6.354961832$;
(b) 129 W;   **113.**

Exercise Set 2.6, pp. 139–142

1. Domain   **3.** Evaluate   **5.** Excluding   **7.** 1   **9.** −41
11. 12   **13.** $\frac{13}{18}$ **15.** 5   **17.** $x^2 - 3x + 3$ **19.** $x^2 - x + 3$
21. 23   **23.** 5   **25.** 56   **27.** $\frac{x^2 - 2}{5 - x}, x \neq 5$ **29.** $\frac{2}{7}$
31. 4%   **33.** $1.3 + 2.2 = 3.5$ million   **35.** About
50 million; the number of passengers using Newark
Liberty and LaGuardia in 1998   **37.** About 81 million; the
number of passengers using the three airports in 2002
39. About 51 million; the number of passengers using
LaGuardia and Newark Liberty in 2002   **41.** $\mathbb{R}$
43. $\{x\,|\,x$ is a real number and $x \neq 3\}$
45. $\{x\,|\,x$ is a real number and $x \neq 0\}$
47. $\{x\,|\,x$ is a real number and $x \neq 1\}$
49. $\{x\,|\,x$ is a real number and $x \neq 2$ and $x \neq 4\}$
51. $\{x\,|\,x$ is a real number and $x \neq 3\}$
53. $\{x\,|\,x$ is a real number and $x \neq 4\}$
55. $\{x\,|\,x$ is a real number and $x \neq 4$ and $x \neq 5\}$
57. $\{x\,|\,x$ is a real number and $x \neq -1$ and $x \neq -\frac{5}{2}\}$
59. 4; 3   **61.** 5; −1   **63.** $\{x\,|\,0 \leq x \leq 9\}$; $\{x\,|\,3 \leq x \leq 10\}$;
$\{x\,|\,3 \leq x \leq 9\}$; $\{x\,|\,3 \leq x \leq 9\}$

65.

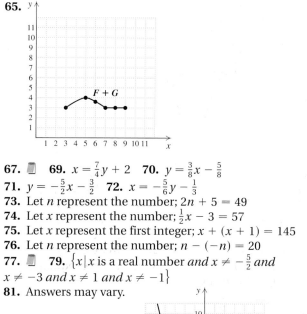

67.   **69.** $x = \frac{7}{4}y + 2$ **70.** $y = \frac{3}{8}x - \frac{5}{8}$
71. $y = -\frac{5}{2}x - \frac{3}{2}$ **72.** $x = -\frac{5}{6}y - \frac{1}{3}$
73. Let n represent the number; $2n + 5 = 49$
74. Let x represent the number; $\frac{1}{2}x - 3 = 57$
75. Let x represent the first integer; $x + (x + 1) = 145$
76. Let n represent the number; $n - (-n) = 20$
77.   **79.** $\{x\,|\,x$ is a real number and $x \neq -\frac{5}{2}$ and
$x \neq -3$ and $x \neq 1$ and $x \neq -1\}$
81. Answers may vary.

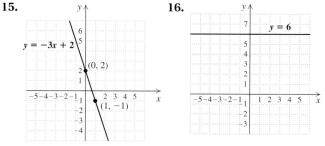

83. $\{x\,|\,x$ is a real number and $-1 < x < 5$ and $x \neq \frac{3}{2}\}$
85. Answers may vary. $f(x) = \dfrac{1}{x + 2}, g(x) = \dfrac{1}{x - 5}$ **87.**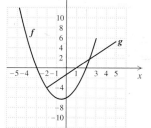

Review Exercises: Chapter 2, pp. 144–147

1. True   **2.** False   **3.** False   **4.** False   **5.** False   **6.** True
7. True   **8.** True   **9.** True   **10.** True   **11.** Yes   **12.** No
13. Yes   **14.** No
15.

16.

17.

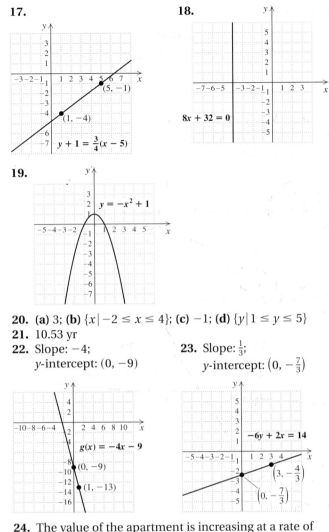

18.

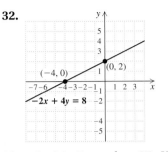

32.

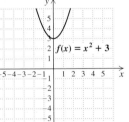

19.

33. -1 **34.** 7 months **35.** Yes **36.** Yes **37.** No
38. No **39.** $y - 4 = -2(x - (-3))$, or $y - 4 = -2(x + 3)$
40. $f(x) = \frac{4}{3}x + \frac{7}{3}$ **41.** Perpendicular **42.** Parallel
43. (a) $R(t) = -0.0215t + 19.75$; (b) 19.21 sec; 19.11 sec
44. $y = \frac{3}{5}x - \frac{31}{5}$ **45.** $y = -\frac{5}{3}x - \frac{5}{3}$ **46.** -6 **47.** 26
48. 102 **49.** -17 **50.** $-\frac{9}{2}$ **51.** $3a + 3b - 6$ **52.** $\mathbb{R}$
53. $\{x \mid x$ is a real number $and\ x \neq 2\}$ **54.** For a
function, every member of the domain corresponds to
exactly one member of the range. Thus, for any function,
each member of the domain corresponds to *at least one*
member of the range. Therefore, a function is a relation. In
a relation, every member of the domain corresponds to *at
least one*, but not necessarily *exactly one*, member of the
range. Therefore, a relation may or may not be a function.
55. The slope of a line is the rise between two points
on the line divided by the run between those points. For a
vertical line, there is no run between any two points, and
division by 0 is undefined; therefore, the slope is
undefined. For a horizontal line, there is no rise between
any two points, so the slope is 0/run, or 0. **56.** -9
57. $-\frac{9}{2}$ **58.** $f(x) = 3.09x + 3.75$
59. (a) III; (b) IV; (c) I; (d) II

Test: Chapter 2, pp. 147–148

1. [2.1] No **2.** [2.1] Yes
3. [2.1], [2.3]   **4.** [2.1], [2.2]

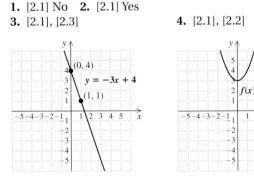

20. (a) 3; (b) $\{x \mid -2 \leq x \leq 4\}$; (c) -1; (d) $\{y \mid 1 \leq y \leq 5\}$
21. 10.53 yr
22. Slope: -4;   **23.** Slope: $\frac{1}{3}$;
  y-intercept: $(0, -9)$ y-intercept: $\left(0, -\frac{7}{3}\right)$

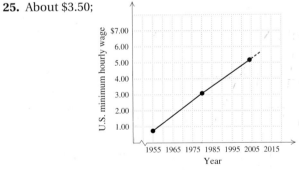
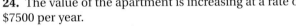

24. The value of the apartment is increasing at a rate of
$7500 per year.
25. About $3.50;

26. About $5.60 **27.** $\frac{4}{7}$ **28.** Undefined
29. 159,475 homes per month **30.** 645 signifies that
tuition is increasing at a rate of $645 a year;
9800 represents tuition costs in 1997.
31. $f(x) = \frac{2}{7}x - 6$

5. [2.5]   **6.** [2.4]

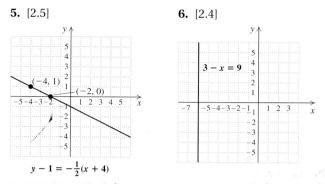

$$y - 1 = -\tfrac{1}{2}(x + 4)$$

7. [2.2] **(a)** 1; **(b)** $\{x \mid -3 \le x \le 4\}$; **(c)** 3; **(d)** $\{y \mid -1 \le y \le 2\}$
8. **(a)** [2.2] $50 billion; **(b)** [2.3] 1.7 signifies that the rate of increase in U.S. book sales was $1.7 billion per year; 26.2 signifies that U.S. book sales were $26.2 billion in 1995.
9. [2.2] About 44 million international visitors

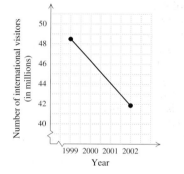

10. [2.3] Slope: $-\tfrac{3}{5}$; y-intercept: $(0, 12)$
11. [2.3] Slope: $-\tfrac{2}{5}$; y-intercept: $\left(0, -\tfrac{7}{5}\right)$ **12.** [2.3] $\tfrac{5}{8}$
13. [2.3] 0   **14.** [2.3] World population is increasing at a rate of 70,000,000 people per year.
15. [2.3] $f(x) = -5x - 1$
16. [2.4]

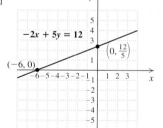

17. [2.4] 3   **18.** [2.4] (a), (c)
19. [2.5] $y - (-4) = 4(x - (-2))$, or $y + 4 = 4(x + 2)$
20. [2.5] $f(x) = -x + 2$ **21.** [2.5] Parallel
22. [2.5] Perpendicular   **23.** [2.5] $y = \tfrac{2}{5}x + \tfrac{16}{5}$
24. [2.5] $y = -\tfrac{5}{2}x - \tfrac{11}{2}$ **25.** **(a)** [2.2] 5; **(b)** [2.6] -130;
(c) [2.6] $\{x \mid x$ is a real number *and* $x \ne -\tfrac{4}{3}\}$
26. [2.5] **(a)** $C(m) = 0.3m + 25$; **(b)** $175
27. [2.2], [2.3] **(a)** 30 mi; **(b)** 15 mph
28. [2.5] $s = -\dfrac{3}{2}r + \dfrac{27}{2}$, or $\dfrac{27 - 3r}{2}$
29. [2.6] $h(x) = 7x - 2$

CHAPTER 3

Technology Connection, p. 155
1. $(1.53, 2.58)$ **2.** $(-0.26, 57.06)$ **3.** $(2.23, 1.14)$
4. $(0.87, -0.32)$

Exercise Set 3.1, pp. 156–158
1. True   **3.** True   **5.** True   **7.** False   **9.** Yes   **11.** No
13. Yes   **15.** Yes   **17.** $(4, 1)$ **19.** $(2, -1)$ **21.** $(4, 3)$
23. $(-3, -2)$ **25.** $(-3, 2)$ **27.** $(3, -7)$ **29.** $(7, 2)$
31. $(4, 0)$ **33.** No solution   **35.** $\{(x, y) \mid y = 3 - x\}$
37. All except 33   **39.** 35   **41.** Let x represent the first number and y the second number; $x + y = 50, x = 0.25y$
43. Let m represent the number of ounces of mineral oil and v the number of ounces of vinegar;
$m + v = 16, m = 2v + 4$ **45.** Let x and y represent the angles; $x + y = 180, x = 2y - 3$ **47.** Let x represent the number of two-point shots and y the number of foul shots;
$x + y = 64, 2x + y = 100$ **49.** Let x represent the number of $8.50 brushes sold and y the number of $9.75 brushes sold; $x + y = 45, 8.50x + 9.75y = 398.75$
51. Let h represent the number of vials of Humulin sold and n the number of vials of Novolin Velosulin;
$h + n = 50, 27.06h + 34.39n = 1565.57$ **53.** Let l represent the length, in feet, and w the width, in feet; $2l + 2w = 288, l = w + 44$ **55.** 📓   **57.** 15   **58.** $\tfrac{19}{12}$
59. $\tfrac{9}{20}$ **60.** $\tfrac{13}{3}$ **61.** $y = -\tfrac{3}{4}x + \tfrac{7}{4}$ **62.** $y = \tfrac{2}{5}x - \tfrac{9}{5}$ **63.** 📓
65. Answers may vary. **(a)** $x + y = 6, x - y = 4$;
(b) $x + y = 1, 2x + 2y = 3$; **(c)** $x + y = 1, 2x + 2y = 2$
67. $A = -\tfrac{17}{4}, B = -\tfrac{12}{5}$ **69.** Let x and y represent the number of years that Lou and Juanita have taught at the university, respectively; $x + y = 46, x - 2 = 2.5(y - 2)$
71. Let s and v represent the number of ounces of baking soda and vinegar needed, respectively; $s = 4v, s + v = 16$
73. $(0, 0), (1, 1)$ **75.** $(0.07, -7.95)$ **77.** $(0.00, 1.25)$

Exercise Set 3.2, pp. 165–166
1. (d)   **3.** (a)   **5.** (c)   **7.** $(2, -3)$ **9.** $(-4, 3)$ **11.**
$(2, -2)$ **13.** $\{(x, y) \mid 2x - 3 = y\}$ **15.** $(-2, 1)$ **17.** $\left(\tfrac{1}{2}, \tfrac{1}{2}\right)$
19. $\left(\tfrac{25}{23}, -\tfrac{11}{23}\right)$ **21.** No solution   **23.** $(1, 2)$ **25.** $(2, 7)$
27. $(-1, 2)$ **29.** $\left(\tfrac{128}{31}, -\tfrac{17}{31}\right)$ **31.** $(6, 2)$ **33.** No solution
35. $\left(\tfrac{110}{19}, -\tfrac{12}{19}\right)$ **37.** $(3, -1)$ **39.** $\{(x, y) \mid -4x + 2y = 5\}$
41. $\left(\tfrac{140}{13}, -\tfrac{50}{13}\right)$ **43.** $(-2, -9)$ **45.** $(30, 6)$
47. $\{(x, y) \mid x = 2 + 3y\}$ **49.** No solution   **51.** $(140, 60)$
53. $\left(\tfrac{1}{3}, -\tfrac{2}{3}\right)$ **55.** 📓   **57.** 4 mi   **58.** 86
59. $11\tfrac{2}{3}$ billion on bathrooms; $23\tfrac{1}{3}$ billion on kitchens
60. 30 m, 90 m, 360 m   **61.** 450.5 mi   **62.** 460.5 mi
63. 📓   **65.** $m = -\tfrac{1}{2}, b = \tfrac{5}{2}$ **67.** $a = 5, b = 2$
69. $\left(-\tfrac{32}{17}, \tfrac{38}{17}\right)$ **71.** $\left(-\tfrac{1}{5}, \tfrac{1}{10}\right)$ **73.** 📓

Exercise Set 3.3, pp. 177–181

1. 10, 40 **3.** Mineral oil: 12 oz; vinegar: 4 oz **5.** 119°, 61°
7. Two-point shots: 36; foul shots: 28 **9.** $8.50-brushes:
32; $9.75-brushes: 13 **11.** Humulin vials: 21; Novolin
Velosulin vials: 29 **13.** Width: 50 ft; length: 94 ft
15. Nonrecycled sheets: 38; recycled sheets: 112
17. General Electric bulbs: 60; SLi bulbs: 140 **19.** HP
cartridges: 15; Apple cartridges: 35 **21.** Kenyan: 8 lb;
Sumatran: 12 lb **23.** 10 lb of each **25.** Deep Thought:
12 lb; Oat Dream: 8 lb **27.** $7500 at 6%; $4500 at 9%
29. Arctic Antifreeze: 12.5 L; Frost No-More: 7.5 L
31. 87-octane: 4 gal; 93-octane: 8 gal **33.** Whole milk:
$169\frac{3}{13}$ lb; cream: $30\frac{10}{13}$ lb **35.** 375 km **37.** 24 mph
39. About 1489 mi **41.** Length: 265 ft; width: 165 ft
43. Simon: 122 properties; DeBartolo: 61 properties
45. 30-sec commercials: 4; 60-sec commercials: 8
47. Quarters: 17; fifty-cent pieces: 13 **49.** ▨ **51.** 16
52. 11 **53.** -28 **54.** -10 **55.** $\frac{49}{12}$ **56.** $\frac{13}{10}$ **57.** ▨
59. 0%: 20 reams; 30%: 40 reams **61.** 1.8 L
63. 180 members **65.** Brown: 0.8 gal; neutral: 0.2 gal
67. City: 261 mi; highway: 204 mi **69.** $P(x) = \dfrac{0.1 + x}{1.5}$
(This expresses the percent as a decimal quantity.)

Exercise Set 3.4, pp. 187–189

1. True **3.** False **5.** True **7.** No **9.** $(4, 0, 2)$
11. $(2, -2, 2)$ **13.** $(3, -2, 1)$ **15.** No solution
17. $(2, 1, 3)$ **19.** $(2, -5, 6)$ **21.** The equations are
dependent. **23.** $\left(\frac{1}{2}, 4, -6\right)$ **25.** $\left(\frac{1}{2}, \frac{1}{3}, \frac{1}{6}\right)$ **27.** $\left(\frac{1}{2}, \frac{2}{3}, -\frac{5}{6}\right)$
29. $(15, 33, 9)$ **31.** $(3, 4, -1)$ **33.** $(10, 23, 50)$ **35.** No
solution **37.** The equations are dependent. **39.** ▨
41. Let x and y represent the numbers; $x = 2y$ **42.** Let x
and y represent the numbers; $x + y = 3x$ **43.** Let x
represent the first number; $x + (x + 1) + (x + 2) = 45$
44. Let x and y represent the numbers; $x + 2y = 17$
45. Let x, y, and z represent the numbers; $x + y = 5z$
46. Let x and y represent the numbers; $xy = 2(x + y)$
47. ▨ **49.** $(1, -1, 2)$ **51.** $(-3, -1, 0, 4)$
53. $\left(-\frac{1}{2}, -1, -\frac{1}{3}\right)$ **55.** 14 **57.** $z = 8 - 2x - 4y$

Exercise Set 3.5, pp. 193–196

1. 16, 19, 22 **3.** 8, 21, -3 **5.** 32°, 96°, 52° **7.** Individual
adult: $64; spouse: $57; child: $43 **9.** Bran muffin: 1.5 g;
banana: 3 g; 1 cup Wheaties: 3 g **11.** Basic price: $24,695;
4WD: $1970; sunroof: $800 **13.** Elrod: 20 ft/hr; Dot:
24 ft/hr; Wendy: 30 ft/hr **15.** 12-oz cups: 17; 16-oz cups:
25; 20-oz cups: 13 **17.** Small: 10; medium: 25; large: 5
19. Roast beef: 2 servings; baked potato: 1 serving;
broccoli: 2 servings **21.** Asia: 4.8 billion; Africa: 1.8
billion; rest of world: 2.5 billion **23.** Two-point field
goals: 32; three-point field goals: 5; foul shots: 13
25. ▨ **27.** -8 **28.** 33 **29.** -55 **30.** -71
31. $-14x + 21y - 35z$ **32.** $-24a - 42b + 54c$ **33.** $-5a$

34. $11x$ **35.** ▨ **37.** Applicant: $102; spouse: $58; first
child: $43; second child: $40 **39.** 20 yr **41.** 35 tickets

Exercise Set 3.6, pp. 200–201

1. Horizontal; columns **3.** Entry **5.** Multiple
7. $\left(-\frac{1}{3}, -4\right)$ **9.** $(-4, 3)$ **11.** $\left(\frac{3}{2}, \frac{5}{2}\right)$ **13.** $\left(2, \frac{1}{2}, -2\right)$
15. $(2, -2, 1)$ **17.** $\left(4, \frac{1}{2}, -\frac{1}{2}\right)$ **19.** $(1, -3, -2, -1)$
21. Dimes: 18; nickels: 24 **23.** $4.05-granola: 5 lb;
$2.70-granola: 10 lb **25.** $400 at 7%; $500 at 8%;
$1600 at 9% **27.** ▨ **29.** 13 **30.** -22 **31.** 37 **32.** 422
33. ▨ **35.** 1324

Exercise Set 3.7, pp. 205–206

1. True **3.** False **5.** False **7.** 18 **9.** 36 **11.** 27
13. -3 **15.** -5 **17.** $(-3, 2)$ **19.** $\left(\frac{9}{19}, \frac{51}{38}\right)$
21. $\left(-1, -\frac{6}{7}, \frac{11}{7}\right)$ **23.** $(2, -1, 4)$ **25.** $(1, 2, 3)$ **27.** ▨
29. $\frac{333}{245}$ **30.** -12 **31.** One piece: 20.8 ft; other piece: 12 ft
32. Scientific calculators: 18; graphing calculators: 27
33. Mazzas: 28 rolls; Kranepools: 8 rolls **34.** Buckets:
17; dinners: 11 **35.** ▨ **37.** 12 **39.** 10

Exercise Set 3.8, pp. 211–213

1. (b) **3.** (e) **5.** (h) **7.** (g)
9. (a) $P(x) = 20x - 300,000$; (b) (15,000 units, $975,000)
11. (a) $P(x) = 50x - 120,000$; (b) (2400 units, $144,000)
13. (a) $P(x) = 45x - 22,500$; (b) (500 units, $42,500)
15. (a) $P(x) = 18x - 16,000$; (b) (889 units, $35,560)
17. (a) $P(x) = 50x - 100,000$; (b) (2000 units, $250,000)
19. ($70, 300) **21.** ($22, 474) **23.** ($50, 6250)
25. ($10, 1070) **27.** (a) $C(x) = 125,300 + 450x$;
(b) $R(x) = 800x$; (c) $P(x) = 350x - 125,300$; (d) $90,300
loss, $14,700 profit; (e) (358 computers, $286,400)
29. (a) $C(x) = 16,404 + 6x$; (b) $R(x) = 18x$;
(c) $P(x) = 12x - 16,404$; (d) $19,596 profit, $4404 loss;
(e) (1367 dozen caps, $24,606) **31.** ▨ **33.** 12 **34.** 15
35. $\frac{8}{3}$ **36.** 4 **37.** $\frac{9}{2}$ **38.** $\frac{1}{3}$ **39.** ▨ **41.** ($5, 300 yo-yo's)
43. (a) $8.74; (b) 24,509 units

Review Exercises: Chapter 3, pp. 216–217

1. Substitution **2.** Elimination **3.** Approximate
4. Dependent **5.** Inconsistent **6.** Infinite **7.** Parallel
8. Square **9.** Determinant **10.** Zero **11.** $(-2, 1)$
12. $(3, 2)$ **13.** $\left(-\frac{11}{15}, -\frac{43}{30}\right)$ **14.** No solution **15.** $\left(-\frac{4}{5}, \frac{2}{5}\right)$
16. $\left(\frac{37}{17}, \frac{53}{17}\right)$ **17.** $\left(\frac{76}{17}, -\frac{2}{119}\right)$ **18.** $(2, 2)$
19. $\{(x, y) \mid 3x + 4y = 6\}$ **20.** DVD: $17; videocassette: $14
21. 4 hr **22.** 8% juice: 10 L; 15% juice: 4 L
23. $(4, -8, 10)$ **24.** The equations are dependent.
25. $(2, 0, 4)$ **26.** No solution **27.** $\left(\frac{8}{9}, -\frac{2}{3}, \frac{10}{9}\right)$
28. A: 90°; B: 67.5°; C: 22.5°
29. Oil: $21\frac{1}{3}$ oz; lemon juice: $10\frac{2}{3}$ oz

30. Lumber: 29 pallets; plywood: 13 pallets   **31.** $\left(55, -\frac{89}{2}\right)$
32. $(-1, 1, 3)$ **33.** 2   **34.** 9   **35.** $(6, -2)$ **36.** $(-3, 0, 4)$
37. ($3, 81)   **38. (a)** $C(x) = 0.75x + 9000$;
(b) $R(x) = 5.25x$; **(c)** $P(x) = 4.5x - 9000$; **(d)** $2250 loss;
$13,500 profit; **(e)** (2000 pints of honey, $10,500)
39. ▨ To solve a problem involving four variables, go
through the *Familiarize* and *Translate* steps as usual. The
resulting system of equations can be solved using the
elimination method just as for three variables but likely
with more steps.   **40.** ▨ A system of equations can be
both dependent and inconsistent if it is equivalent to a
system with fewer equations that has no solution. An
example is a system of three equations in three unknowns
in which two of the equations represent the same plane,
and the third represents a parallel plane.   **41.** 8000 pints
42. $(0, 2), (1, 3)$
43. $a = -\frac{2}{3}, b = -\frac{4}{3}, c = 3; f(x) = -\frac{2}{3}x^2 - \frac{4}{3}x + 3$

Test: Chapter 3, p. 218

1. [3.1] $(2, 4)$ **2.** [3.2] $\left(3, -\frac{11}{3}\right)$ **3.** [3.2] $\left(\frac{15}{7}, -\frac{18}{7}\right)$
4. [3.2] $\left(-\frac{3}{2}, -\frac{3}{2}\right)$ **5.** [3.2] No solution   **6.** [3.3] Length:
30 units; width: 18 units   **7.** [3.3] Pepperidge Farm
Goldfish: 120 g; Rold Gold Pretzels: 500 g   **8.** [3.4] The
equations are dependent.   **9.** [3.4] $\left(2, -\frac{1}{2}, -1\right)$
10. [3.4] No solution   **11.** [3.4] $(0, 1, 0)$
12. [3.6] $\left(\frac{34}{107}, -\frac{104}{107}\right)$ **13.** [3.6] $(3, 1, -2)$ **14.** [3.7] 34
15. [3.7] 133   **16.** [3.7] $\left(\frac{13}{18}, \frac{7}{27}\right)$ **17.** [3.5] Electrician:
3.5 hr; carpenter: 8 hr; plumber: 10 hr   **18.** [3.8] ($3, 55)
19. [3.8] **(a)** $C(x) = 25x + 40,000$; **(b)** $R(x) = 70x$;
(c) $P(x) = 45x - 40,000$; **(d)** $26,500 loss, $500 profit;
(e) (889 hammocks, $62,230)
20. [2.3], [3.3] $m = 7, b = 10$ **21.** [3.5] Adults' tickets:
1346; senior citizens' tickets: 335; children's tickets: 1651

Cumulative Review: Chapters 1–3, pp. 219–220

1. 14.87   **2.** -22 **3.** -42.9 **4.** 20   **5.** -56 **6.** 6
7. -5 **8.** $\frac{10}{9}$ **9.** $-\frac{32}{5}$ **10.** $\frac{18}{17}$ **11.** x^{11} **12.** $-\frac{40x}{y^5}$
13. $-288x^4y^{18}$ **14.** y^{10} **15.** $-\frac{2a^{11}}{5b^{33}}$ **16.** $\frac{81x^{36}}{256y^8}$
17. 1.12×10^6 **18.** 4.00×10^6 **19.** $b = \frac{2A}{h} - t$, or
$\frac{2A - ht}{h}$ **20.** Yes

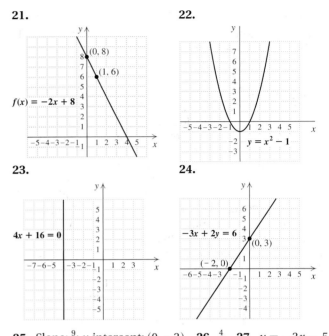

21. **22.** **23.** **24.**

25. Slope: $\frac{9}{4}$; y-intercept: $(0, -3)$ **26.** $\frac{4}{3}$ **27.** $y = -3x - 5$
28. $y = -\frac{1}{10}x + \frac{12}{5}$ **29.** Parallel   **30.** $y = -2x + 5$
31. $\{-5, -3, -1, 1, 3\}; \{-3, -2, 1, 4, 5\}; -2; 3$
32. $\{x \mid x$ is a real number *and* $x \neq \frac{1}{2}\}$ **33.** -31 **34.** 3
35. 7   **36.** $8a^2 + 4a - 4$ **37.** $(1, 1)$ **38.** $(-2, 3)$
39. $\left(-3, \frac{2}{5}\right)$ **40.** $(-3, 2, -4)$ **41.** $(0, -1, 2)$ **42.** 14
43. 0   **44.** 8, 18   **45.** 33 recovery plans per year
46. 8.8 signifies that the number of U.S. domestic trips is
increasing by 8.8 million each year; 954 signifies that there
were 954 million U.S. domestic trips in 1994.
47. **(a)** $A(t) = 0.2125t + 7.5$; **(b)** 10.9 million departures
48. Soakem: $48\frac{8}{9}$ oz; Rinsem: $71\frac{1}{9}$ oz   **49.** 3, 5, 7   **50.** 90
51. Length: 10 cm; width: 6 cm   **52.** Nickels: 19; dimes: 15
53. $120   **54.** Wins: 23; losses: 33; ties: 8   **55.** Cookie: 90;
banana: 80; yogurt: 165   **56.** $-12x^{2a}y^{b+y+3}$ **57.**
$151,000
58. $m = -\frac{5}{9}, b = -\frac{2}{9}$

CHAPTER 4

Exercise Set 4.1, pp. 230–233

1. Equivalent inequalities   **3.** Equivalent equations
5. Not equivalent   **7.** Equivalent expressions   **9.** Not
equivalent   **11.** No, no, yes, yes   **13.** No, yes, yes, no
15. $(-\infty, 6), \{y \mid y < 6\}$
17. $[-4, \infty), \{x \mid x \geq -4\}$
19. $(-3, \infty), \{t \mid t > -3\}$
21. $(-\infty, -7], \{x \mid x \leq -7\}$
23. $\{x \mid x > -6\}$, or $(-6, \infty)$

25. $\{a \mid a \le -20\}$, or $(-\infty, -20]$

27. $\{x \mid x \le 17\}$, or $(-\infty, 17]$

29. $\{y \mid y > -9\}$, or $(-9, \infty)$

31. $\{y \mid y \le 14\}$, or $(-\infty, 14]$

33. $\{t \mid t < -9\}$, or $(-\infty, -9)$

35. $\{x \mid x < -60\}$, or $(-\infty, -60)$

37. $\{x \mid x \le 0.9\}$, or $(-\infty, 0.9]$

39. $\left\{x \mid x \le \frac{5}{6}\right\}$, or $\left(-\infty, \frac{5}{6}\right]$

41. $\{x \mid x < -26\}$, or $(-\infty, -26)$

43. $\left\{t \mid t \ge -\frac{13}{3}\right\}$, or $\left[-\frac{13}{3}, \infty\right)$

45. $\{x \mid x \ge 6\}$, or $[6, \infty)$

47. $\{x \mid x \ge 2\}$, or $[2, \infty)$

49. $\left\{x \mid x > \frac{2}{3}\right\}$, or $\left(\frac{2}{3}, \infty\right)$

51. $\left\{x \mid x \ge \frac{1}{2}\right\}$, or $\left[\frac{1}{2}, \infty\right)$

53. $\left\{y \mid y \le -\frac{53}{6}\right\}$, or $\left(-\infty, -\frac{53}{6}\right]$ **55.** $\left\{t \mid t < \frac{29}{5}\right\}$, or $\left(-\infty, \frac{29}{5}\right)$
57. $\left\{m \mid m > \frac{7}{3}\right\}$, or $\left(\frac{7}{3}, \infty\right)$ **59.** $\{x \mid x \ge 2\}$, or $[2, \infty)$
61. $\{y \mid y < 5\}$, or $(-\infty, 5)$ **63.** $\left\{x \mid x \le \frac{4}{7}\right\}$, or $\left(-\infty, \frac{4}{7}\right]$
65. Mileages less than or equal to 150 mi   **67.** $11,500 or
more   **69.** For 1175 min or more   **71.** More than
25 checks   **73.** Gross sales greater than $7000
75. Parties of more than 80   **77.** At least 625 people
79. **(a)** $\left\{x \mid x < 8181\frac{9}{11}\right\}$, or $\{x \mid x \le 8181\}$;
(b) $\left\{x \mid x > 8181\frac{9}{11}\right\}$, or $\{x \mid x \ge 8182\}$ **81.**
83. $\{x \mid x$ is a real number $and \ x \ne 2\}$
84. $\{x \mid x$ is a real number $and \ x \ne -3\}$
85. $\left\{x \mid x$ is a real number $and \ x \ne \frac{7}{2}\right\}$
86. $\left\{x \mid x$ is a real number $and \ x \ne \frac{9}{4}\right\}$ **87.** $7x + 10$
88. $22x - 7$ **89.**   **91.** $\left\{x \mid x \le \dfrac{2}{a-1}\right\}$
93. $\left\{y \mid y \ge \dfrac{2a+5b}{b(a-2)}\right\}$ **95.** $\left\{x \mid x > \dfrac{4m-2c}{d-(5c+2m)}\right\}$
97. False; $2 < 3$ and $4 < 5$, but $2 - 4 = 3 - 5$.   **99.**
101. $\mathbb{R}$
103. $\{x \mid x$ is a real number $and \ x \ne 0\}$

105.

Exercise Set 4.2, pp. 241–244

1. (h)   **3.** (f)   **5.** (e)   **7.** (b)   **9.** (c)   **11.** $\{9, 11\}$
13. $\{0, 5, 10, 15, 20\}$ **15.** $\{b, d, f\}$ **17.** $\{r, s, t, u, v\}$
19. $\varnothing$ **21.** $\{3, 5, 7\}$
23. $(3, 7)$

25. $[-6, -2]$

27. $(-\infty, -1) \cup (4, \infty)$

29. $(-\infty, -2] \cup (1, \infty)$

31. $(-2, 4]$

33. $(-2, 4)$

35. $(-\infty, 5) \cup (7, \infty)$

37. $(-\infty, -4] \cup [5, \infty)$

39. $[-3, 7)$

41. $[3, 7)$

43. $(-\infty, 5)$

45. $\{t \mid -3 < t < 7\}$, or $(-3, 7)$

47. $\{x \mid -1 < x \le 4\}$, or $(-1, 4]$

49. $\{a \mid -2 \le a < 2\}$, or $[-2, 2)$

51. $\mathbb{R}$, or $(-\infty, \infty)$

53. $\{x \mid 7 < x < 23\}$, or $(7, 23)$

55. $\{x \mid -3 \le x \le 2\}$, or $[-3, 2]$

57. $\{x \mid 1 \le x \le 3\}$, or $[1, 3]$

59. $\left\{x \mid -\frac{7}{2} < x \le 7\right\}$, or $\left(-\frac{7}{2}, 7\right]$

61. $\{x \mid x \le 1 \ or \ x \ge 3\}$, or $(-\infty, 1] \cup [3, \infty)$

63. $\{x \mid x < 2 \ or \ x > 6\}$, or $(-\infty, 2) \cup (6, \infty)$

65. $\left\{a \mid a < \frac{7}{2}\right\}$, or $\left(-\infty, \frac{7}{2}\right)$

67. $\{a \mid a < -5\}$, or $(-\infty, -5)$

69. $\mathbb{R}$, or $(-\infty, \infty)$

71. $\{t \mid t \le 6\}$, or $(-\infty, 6]$

73. $(-\infty, -8) \cup (-8, \infty)$ **75.** $[6, \infty)$
77. $(-\infty, 4) \cup (4, \infty)$ **79.** $\left[-\frac{7}{2}, \infty\right)$ **81.** $(-\infty, 4]$
83.

85.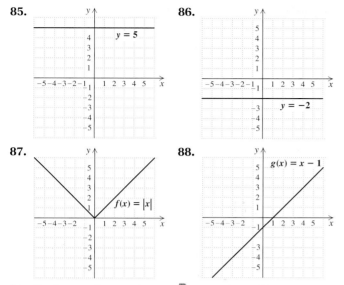
86.
87.
88.

89. $(8, 5)$ **90.** $(-5, -3)$ **91.** 📝 **93.** $(-1, 6)$
95. From 2000 to 2025 **97.** Sizes between 6 and 13
99. $1965 \le y \le 1981$ **101.** Between 12 and 240 trips
103. $\left\{m \mid m < \frac{6}{5}\right\}$, or $\left(-\infty, \frac{6}{5}\right)$

105. $\left\{x \mid -\frac{1}{8} < x < \frac{1}{2}\right\}$, or $\left(-\frac{1}{8}, \frac{1}{2}\right)$

107. False **109.** True **111.** $\left(-\infty, -7\right) \cup \left(-7, \frac{3}{4}\right]$
113. 📈 **115.** 📈

Technology Connection, p. 246

1. The graphs of $y_1 = \text{abs}(4 - 7x)$ and $y_2 = -8$ do not intersect.

Technology Connection, p. 250

1. The x-values on the graph of $y_1 = |4x + 2|$ that are *below* the line $y = 6$ solve the inequality $|4x + 2| < 6$. The x-values on the graph of $y_1 = |4x + 2|$ that are *on* the line $y = 6$ solve the equation $|4x + 2| = 6$. **2.** The x-values on the graph of $y_1 = |3x - 2|$ that are below the line $y = 4$ are in the interval $\left(-\frac{2}{3}, 2\right)$.

Exercise Set 4.3, pp. 251–253

1. True **3.** False **5.** True **7.** False **9.** $\{-7, 7\}$ **11.** $\varnothing$
13. $\{0\}$ **15.** $\{-5.5, 5.5\}$ **17.** $\left\{-\frac{1}{2}, \frac{7}{2}\right\}$ **19.** $\varnothing$ **21.** $\{-4, 8\}$
23. $\{2, 8\}$ **25.** $\{-2, 16\}$ **27.** $\{-8, 8\}$ **29.** $\left\{-\frac{11}{7}, \frac{11}{7}\right\}$
31. $\{-7, 8\}$ **33.** $\{-12, 2\}$ **35.** $\left\{-\frac{1}{3}, 3\right\}$ **37.** $\{-7, 1\}$
39. $\{-8.7, 8.7\}$ **41.** $\left\{-\frac{8}{3}, 4\right\}$ **43.** $\{1, 11\}$ **45.** $\left\{-\frac{1}{2}\right\}$
47. $\left\{-\frac{3}{5}, 5\right\}$ **49.** $\mathbb{R}$ **51.** $\{1\}$ **53.** $\left\{32, \frac{8}{3}\right\}$

55. $\{a \mid -9 \le a \le 9\}$, or $[-9, 9]$
57. $\{x \mid x < -8 \text{ or } x > 8\}$, or $(-\infty, -8) \cup (8, \infty)$

59. $\{t \mid t < 0 \text{ or } t > 0\}$, or $(-\infty, 0) \cup (0, \infty)$

61. $\{x \mid -3 < x < 5\}$, or $(-3, 5)$
63. $\{x \mid -8 \le x \le 4\}$, or $[-8, 4]$
65. $\{x \mid x < -2 \text{ or } x > 8\}$, or $(-\infty, -2) \cup (8, \infty)$

67. $\mathbb{R}$, or $(-\infty, \infty)$
69. $\left\{a \mid a \le -\frac{2}{3} \text{ or } a \ge \frac{10}{3}\right\}$, or $\left(-\infty, -\frac{2}{3}\right] \cup \left[\frac{10}{3}, \infty\right)$

71. $\{y \mid -9 < y < 15\}$, or $(-9, 15)$

73. $\{x \mid x \le -8 \text{ or } x \ge 0\}$, or $(-\infty, -8] \cup [0, \infty)$

75. $\left\{y \mid y < -\frac{4}{3} \text{ or } y > 4\right\}$, or $\left(-\infty, -\frac{4}{3}\right) \cup (4, \infty)$

77. $\varnothing$
79. $\left\{x \mid x \le -\frac{2}{15} \text{ or } x \ge \frac{14}{15}\right\}$, or $\left(-\infty, -\frac{2}{15}\right] \cup \left[\frac{14}{15}, \infty\right)$

81. $\{m \mid -9 \le m \le 3\}$, or $[-9, 3]$

83. $\{a \mid -6 < a < 0\}$, or $(-6, 0)$
85. $\left\{x \mid -\frac{1}{2} \le x \le \frac{7}{2}\right\}$, or $\left[-\frac{1}{2}, \frac{7}{2}\right]$

87. $\left\{x \mid x \le -\frac{7}{3} \text{ or } x \ge 5\right\}$, or $\left(-\infty, -\frac{7}{3}\right] \cup [5, \infty)$

89. $\{x \mid -4 < x < 5\}$, or $(-4, 5)$
91. 📝 **93.** $\left(-\frac{16}{13}, -\frac{41}{13}\right)$ **94.** $(-2, -3)$ **95.** $(10, 4)$
96. $(-1, 7)$ **97.** $(1, 4)$ **98.** $(24, -41)$ **99.** 📝
101. $\left\{t \mid t \ge \frac{5}{3}\right\}$, or $\left[\frac{5}{3}, \infty\right)$ **103.** $\mathbb{R}$, or $(-\infty, \infty)$ **105.** $\left\{-\frac{1}{7}, \frac{7}{3}\right\}$
107. $|x| < 3$ **109.** $|x| \ge 6$ **111.** $|x + 3| > 5$
113. $|x - 7| < 2$, or $|7 - x| < 2$ **115.** $|x - 3| \le 4$
117. $|x + 4| < 3$ **119.** Between 80 ft and 100 ft
121. $\{x \mid 1 \le x \le 5\}$, or $[1, 5]$ **123.** 📈 **125.** 📝, 📈

Technology Connection, p. 257

1. $y > x + 3.5$

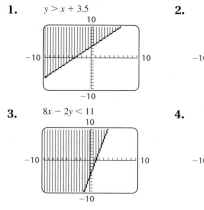

2. $7y \leq 2x + 5$

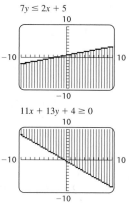

3. $8x - 2y < 11$

4. $11x + 13y + 4 \geq 0$

Technology Connection, p. 260

1. $y_1 \leq 4 - x, \ y_2 > x - 4$

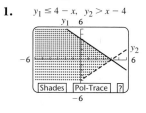

Exercise Set 4.4, pp. 262–264

1. (e)   **3.** (d)   **5.** (b)   **7.** Yes   **9.** No

11. $y > \frac{1}{2}x$

13. $y \geq x - 3$

15. $y \leq x + 5$

17. $x - y \leq 4$

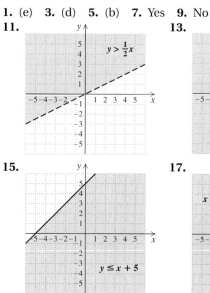
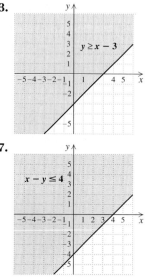

19. $2x + 3y < 6$

21. $2y - x \leq 4$

23. $2x - 2y \geq 8 + 2y$

25. $y \geq 3$

27. $x \leq 6$

29. $-2 < y < 7$

31. $-4 \leq x \leq 2$

33. $0 \leq y \leq 3$

35.

37.

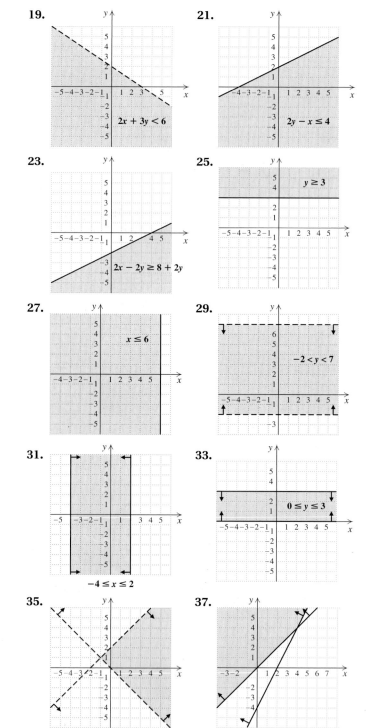

39.
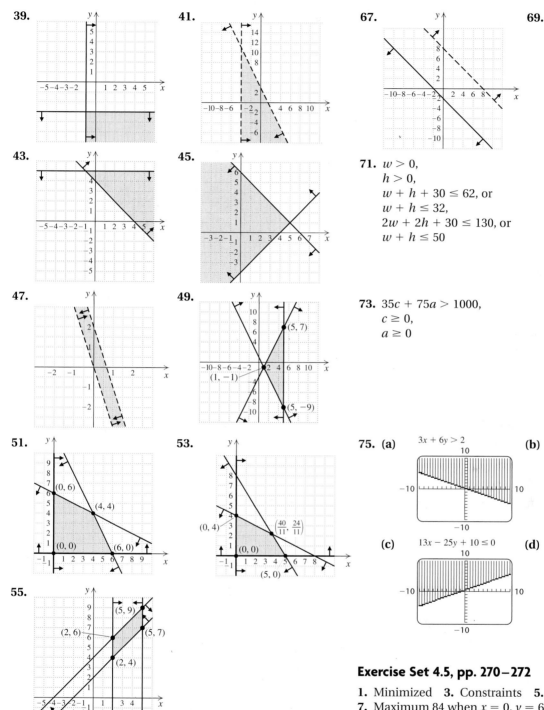

41.

43.

45.

47.

49.

51.

53.

55.

57. ▨   **59.** Peanuts: $6\frac{2}{3}$ lb; fancy nuts: $3\frac{1}{3}$ lb
60. Hendersons: 10 bags; Savickis: 4 bags   **61.** Activity-card holders: 128; noncard holders: 75   **62.** Students: 70; adults: 130   **63.** 25 ft   **64.** 2.75%   **65.** ▨

67.

69.

71. $w > 0,$
$h > 0,$
$w + h + 30 \le 62,$ or
$w + h \le 32,$
$2w + 2h + 30 \le 130,$ or
$w + h \le 50$

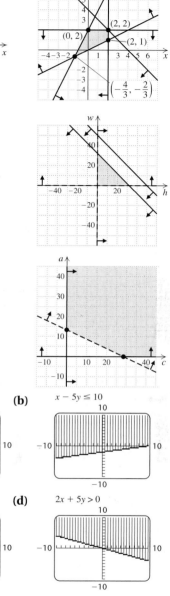

73. $35c + 75a > 1000,$
$c \ge 0,$
$a \ge 0$

75. (a) $3x + 6y > 2$ **(b)** $x - 5y \le 10$

(c) $13x - 25y + 10 \le 0$ **(d)** $2x + 5y > 0$

Exercise Set 4.5, pp. 270–272

1. Minimized   **3.** Constraints   **5.** Feasible
7. Maximum 84 when $x = 0, y = 6$; minimum 0 when $x = 0, y = 0$ **9.** Maximum 76 when $x = 7, y = 0$; minimum 16 when $x = 0, y = 4$ **11.** Maximum 5 when $x = 3, y = 7$; minimum -15 when $x = 3, y = -3$
13. Car: 9 gal; moped: 3 gal; maximum: 480 mi
15. Lumber: 100 units; plywood: 300 units
17. Corporate bonds: $22,000; municipal bonds: $18,000; maximum: $3110

19. Matching: 5; essay: 15; maximum: 425   **21.** Merlot: 80 acres; Cabernet: 160 acres   **23.** Knit suits: 2; worsted suits: 4   **25.** 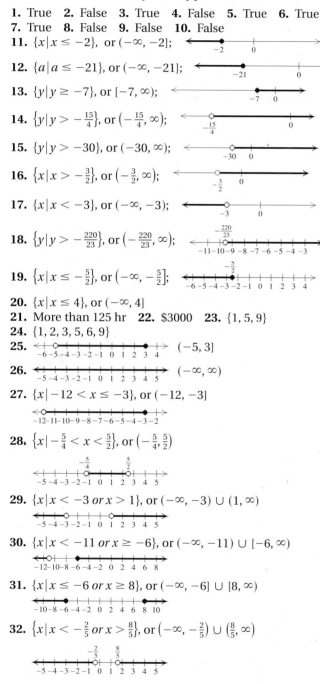   **27.** −40   **28.** 46   **29.** 10x + 5   **30.** 26t + 20   **31.** 3x − 6   **32.** 2t + 2   **33.** 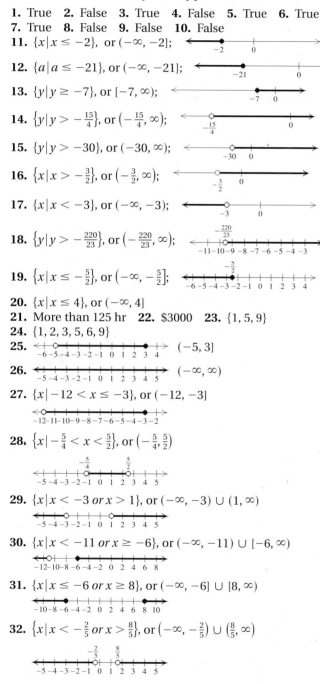   **35.** T3's: 30; S5's: 10   **37.** Chairs: 25; sofas: 9

Review Exercises: Chapter 4, pp. 274–275

1. True   **2.** False   **3.** True   **4.** False   **5.** True   **6.** True
7. True   **8.** False   **9.** False   **10.** False
11. $\{x \mid x \le -2\}$, or $(-\infty, -2]$;
12. $\{a \mid a \le -21\}$, or $(-\infty, -21]$;
13. $\{y \mid y \ge -7\}$, or $[-7, \infty)$;
14. $\{y \mid y > -\frac{15}{4}\}$, or $\left(-\frac{15}{4}, \infty\right)$;
15. $\{y \mid y > -30\}$, or $(-30, \infty)$;
16. $\{x \mid x > -\frac{3}{2}\}$, or $\left(-\frac{3}{2}, \infty\right)$;
17. $\{x \mid x < -3\}$, or $(-\infty, -3)$;
18. $\{y \mid y > -\frac{220}{23}\}$, or $\left(-\frac{220}{23}, \infty\right)$;
19. $\{x \mid x \le -\frac{5}{2}\}$, or $\left(-\infty, -\frac{5}{2}\right]$;
20. $\{x \mid x \le 4\}$, or $(-\infty, 4]$
21. More than 125 hr   **22.** $3000   **23.** $\{1, 5, 9\}$
24. $\{1, 2, 3, 5, 6, 9\}$
25. $(-5, 3]$
26. $(-\infty, \infty)$
27. $\{x \mid -12 < x \le -3\}$, or $(-12, -3]$
28. $\{x \mid -\frac{5}{4} < x < \frac{5}{2}\}$, or $\left(-\frac{5}{4}, \frac{5}{2}\right)$
29. $\{x \mid x < -3 \text{ or } x > 1\}$, or $(-\infty, -3) \cup (1, \infty)$
30. $\{x \mid x < -11 \text{ or } x \ge -6\}$, or $(-\infty, -11) \cup [-6, \infty)$
31. $\{x \mid x \le -6 \text{ or } x \ge 8\}$, or $(-\infty, -6] \cup [8, \infty)$
32. $\{x \mid x < -\frac{2}{5} \text{ or } x > \frac{8}{5}\}$, or $\left(-\infty, -\frac{2}{5}\right) \cup \left(\frac{8}{5}, \infty\right)$

33. $(-\infty, 8) \cup (8, \infty)$ **34.** $[-5, \infty)$ **35.** $\left(-\infty, \frac{8}{3}\right]$
36. $\{-5, 5\}$ **37.** $\{t \mid t \le -3.5 \text{ or } t \ge 3.5\}$, or $(-\infty, -3.5] \cup [3.5, \infty)$ **38.** $\{-4, 10\}$
39. $\{x \mid -\frac{17}{2} < x < \frac{7}{2}\}$, or $\left(-\frac{17}{2}, \frac{7}{2}\right)$
40. $\{x \mid x \le -\frac{11}{3} \text{ or } x \ge \frac{19}{3}\}$, or $\left(-\infty, -\frac{11}{3}\right] \cup \left[\frac{19}{3}, \infty\right)$
41. $\{-14, \frac{4}{3}\}$ **42.** $\varnothing$ **43.** $\{x \mid -16 \le x \le 8\}$, or $[-16, 8]$
44. $\{x \mid x < 0 \text{ or } x > 10\}$, or $(-\infty, 0) \cup (10, \infty)$ **45.** $\varnothing$
46.   **47.**

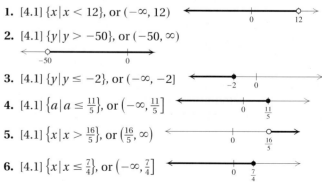

48.

49. Maximum 40 when $x = 7$, $y = 15$; minimum 10 when $x = 1$, $y = 3$ **50.** Ohio plant: 120; Oregon plant: 40
51. 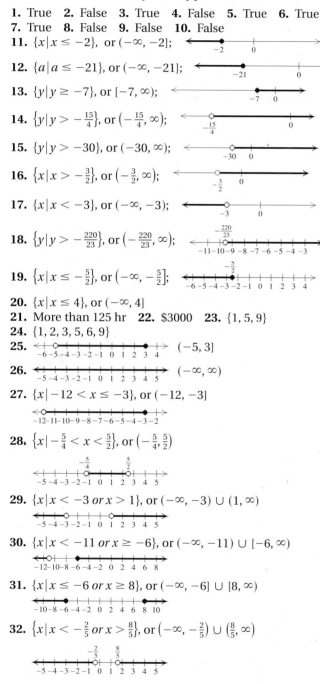 The equation $|X| = p$ has two solutions when p is positive because X can be either p or $-p$. The same equation has no solution when p is negative because no number has a negative absolute value.   **52.** 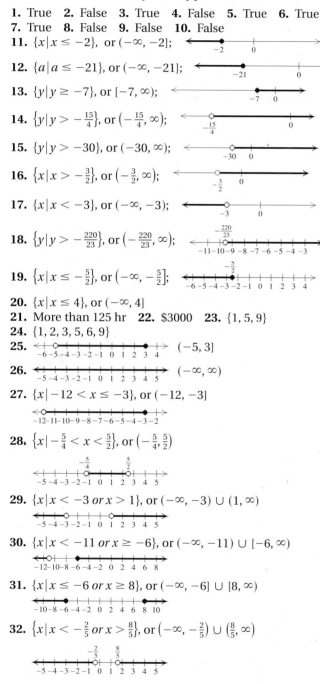 The solution set of a system of inequalities is all ordered pairs that make *all* the individual inequalities true. This consists of ordered pairs that are common to all the individual solution sets, or the intersection of the graphs.
53. $\{x \mid -\frac{8}{3} \le x \le -2\}$, or $\left[-\frac{8}{3}, -2\right]$ **54.** False; $-4 < 3$ is true, but $(-4)^2 < 9$ is false.   **55.** $|d - 1.1| \le 0.03$

Test: Chapter 4, p. 276

1. [4.1] $\{x \mid x < 12\}$, or $(-\infty, 12)$
2. [4.1] $\{y \mid y > -50\}$, or $(-50, \infty)$
3. [4.1] $\{y \mid y \le -2\}$, or $(-\infty, -2]$
4. [4.1] $\{a \mid a \le \frac{11}{5}\}$, or $\left(-\infty, \frac{11}{5}\right]$
5. [4.1] $\{x \mid x > \frac{16}{5}\}$, or $\left(\frac{16}{5}, \infty\right)$
6. [4.1] $\{x \mid x \le \frac{7}{4}\}$, or $\left(-\infty, \frac{7}{4}\right]$

7. [4.1] $\{x \mid x > 1\}$, or $(1, \infty)$ **8.** [4.1] More than $166\frac{2}{3}$ mi
9. [4.1] Less than or equal to 2.5 hr   **10.** [4.2] $\{3, 5\}$
11. [4.2] $\{1, 3, 5, 7, 9, 11, 13\}$ **12.** [4.2] $(-\infty, 4]$
13. [4.2] $\{x \mid 1 < x < 8\}$, or $(1, 8)$

14. [4.2] $\left\{t \mid -\frac{2}{5} < t \leq \frac{9}{5}\right\}$, or $\left(-\frac{2}{5}, \frac{9}{5}\right]$

15. [4.2] $\{x \mid x < 3 \text{ or } x > 6\}$, or $(-\infty, 3) \cup (6, \infty)$

16. [4.2] $\left\{x \mid x < -4 \text{ or } x > -\frac{5}{2}\right\}$, or $(-\infty, -4) \cup \left(-\frac{5}{2}, \infty\right)$

17. [4.2] $\left\{x \mid 4 \leq x < \frac{15}{2}\right\}$, or $\left[4, \frac{15}{2}\right)$

18. [4.3] $\{-13, 13\}$

19. [4.3] $\{a \mid a < -7 \text{ or } a > 7\}$, or $(-\infty, -7) \cup (7, \infty)$

20. [4.3] $\left\{x \mid -2 < x < \frac{8}{3}\right\}$, or $\left(-2, \frac{8}{3}\right)$

21. [4.3] $\left\{t \mid t \leq -\frac{13}{5} \text{ or } t \geq \frac{7}{5}\right\}$, or $\left(-\infty, -\frac{13}{5}\right] \cup \left[\frac{7}{5}, \infty\right)$

22. [4.3] $\varnothing$
23. [4.2] $\left\{x \mid x < \frac{1}{2} \text{ or } x > \frac{7}{2}\right\}$, or $\left(-\infty, \frac{1}{2}\right) \cup \left(\frac{7}{2}, \infty\right)$

24. [4.3] $\{1\}$
25. [4.4]   **26.** [4.4]

27. [4.5] Maximum 57 when $x = 6$, $y = 9$; minimum 5 when $x = 1$, $y = 0$ **28.** [4.5] Manicures: 35; haircuts: 15; maximum: $690   **29.** [4.3] $[-1, 0] \cup [4, 6]$ **30.** [4.2] $\left(\frac{1}{5}, \frac{4}{5}\right)$
31. [4.3] $|x + 3| \leq 5$

CHAPTER 5

Technology Connection, p. 283

1. Correct   **2.** Incorrect   **3.** Correct   **4.** Incorrect

Exercise Set 5.1, pp. 285–289

1. (g)   **3.** (a)   **5.** (b)   **7.** (j)   **9.** (f)   **11.** 5, 3, 2, 1, 0; 5
13. 3, 7, 6, 0; 7   **15.** 5, 6, 2, 1, 0; 6
17. $-18y^4 + 11y^3 + 6y^2 - 5y + 3$; $-18y^4$; -18
19. $-a^7 + 8a^5 + 5a^3 - 19a^2 + a$; $-a^7$; -1
21. $-9 + 6x - 5x^2 + 3x^4$
23. $-9xy + 5x^2y^2 + 8x^3y^2 - 5x^4$ **25.** -23 **27.** -12
29. 9   **31.** -13; 11   **33.** 282; -9 **35.** About 25.5 mph
37. About 29.6 ft   **39.** 6840   **41.** 26.4 million
43. 6.4 million   **45.** 14; 55 oranges
47. About 2.3 mcg/mL   **49.** 2.3   **51.** 56.5 in^2
53. $18,750   **55.** $8375   **57.** $2a^3 - a + 4$
59. $-6a^2b - 3b^2$ **61.** $10x^2 + 2xy + 15y^2$
63. $16x + 7y - 5z$ **65.** $-2x^2 + x - xy - 1$
67. $6x^2y - 4xy^2 + 5xy$ **69.** $9r^2 + 9r - 9$
71. $-\frac{5}{24}xy - \frac{27}{20}x^3y^2 + 1.4y^3$
73. $-(5x^3 - 7x^2 + 3x - 9)$, $-5x^3 + 7x^2 - 3x + 9$
75. $-(-12y^5 + 4ay^4 - 7by^2)$, $12y^5 - 4ay^4 + 7by^2$
77. $10x - 9$ **79.** $-4x^2 - 3x + 13$ **81.** $6a - 6b + 5c$
83. $-2a^2 + 12ab - 7b^2$ **85.** $8a^2b + 16ab + 3ab^2$
87. $x^4 - x^2 - 1$ **89.** $4x^2 + 2x + 6$ **91.** $13r^2 - 8r - 1$
93. $3x^2 - 9$ **95.** $9700   **97.** 🖩   **99.** x^{10} **100.** t^{12}
101. $25x^6$ **102.** $49t^8$ **103.** x^9 **104.** a^8 **105.** 🖩
107. $68x^5 - 81x^4 - 22x^3 + 52x^2 + 2x + 250$
109. $45x^5 - 8x^4 + 208x^3 - 176x^2 + 116x - 25$
111. 494.55 cm^3 **113.** $5x^2 - 8x$ **115.** $8x^{2a} + 7x^a + 7$
117. $x^{5b} + 4x^{4b} + x^{3b} - 6x^{2b} - 9x^b$ **119.** 〰️

Technology Connection, p. 296

1. ↘ $y_1 = x^2 - 9 - (x - 3)(x + 3)$

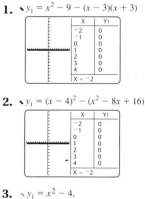

2. ↘ $y_1 = (x - 4)^2 - (x^2 - 8x + 16)$

3. ↘ $y_1 = x^2 - 4$,
↘ $y_2 = (x + 2)(x - 2)$

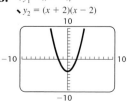

If $y_3 = y_2 - y_1$, the graph of y_3 should be the x-axis.

Exercise Set 5.2, pp. 298–300

1. True **3.** True **5.** False **7.** True **9.** $40a^3$
11. $-20x^3y$ **13.** $-20x^5y^6$ **15.** $21x - 7x^2$
17. $20c^3d^2 - 25c^2d^3$ **19.** $x^2 + 8x + 15$
21. $t^2 + 5t - 14$ **23.** $8a^2 + 14a + 3$
25. $10x^2 - 3xy - 4y^2$ **27.** $x^3 - x^2 - 5x + 2$
29. $t^3 - 3t^2 - 13t + 15$ **31.** $a^4 + 5a^3 - 2a^2 - 9a + 5$
33. $x^3 + 27$ **35.** $a^3 - b^3$ **37.** $t^2 - \frac{7}{12}t + \frac{1}{12}$
39. $r^3 + 4r^2 + r - 6$ **41.** $x^2 + 10x + 25$
43. $3t^2 + 1.5st - 15s^2$ **45.** $4y^2 - 28y + 49$
47. $25a^2 - 30ab + 9b^2$ **49.** $4a^6 - 12a^3b^2 + 9b^4$
51. $x^6y^8 + 10x^3y^4 + 25$
53. $12x^4 - 21x^3 - 17x^2 + 35x - 5$ **55.** $25x^2 - 20x + 4$
57. $4x^2 - \frac{4}{3}x + \frac{1}{9}$ **59.** $c^2 - 49$ **61.** $16x^2 - 1$
63. $9m^2 - 4n^2$ **65.** $x^6 - y^2z^2$ **67.** $-m^2n^2 + m^4$, or
$m^4 - m^2n^2$ **69.** $14x + 58$ **71.** $3m^2 + 4mn - 5n^2$
73. $a^2 + 2ab + b^2 - 1$ **75.** $4x^2 + 12xy + 9y^2 - 16$
77. $A = P + 2Pi + Pi^2$ **79. (a)** $t^2 - 2t + 6$; **(b)** $2ah + h^2$;

(c) $2ah - h^2$ **81.** 🖩 **83.** $a = \dfrac{d}{b+c}$ **84.** $y = \dfrac{w}{x+z}$

85. $m = \dfrac{p}{n+1}$ **86.** $s = \dfrac{t}{r+1}$ **87.** Dimes: 5; nickels: 2;

quarters: 6 **88.** 8 weekdays **89.** 🖩 **91.** $x^4 - y^{2n}$
93. $5x^{n+2}y^3 + 4x^2y^{n+3}$ **95.** $t^{3n} + t^{2n} - 5t^n + 3$
97. $a^2 + 2ac + c^2 - b^2 - 2bd - d^2$
99. $\frac{4}{9}x^2 - \frac{1}{9}y^2 - \frac{2}{3}y - 1$ **101.** $x^{4a} - y^{4b}$ **103.** $x^{a^2-b^2}$
105. 0 **107.** $2a + h$ **109.** (b) and (c) are identities.

Technology Connection, p. 304

1. A table should show that $y_3 = 0$ for any value of x.

Exercise Set 5.3, pp. 306–308

1. False **3.** True **5.** True **7.** True **9.** $3y(y + 2)$
11. $x(x + 9)$ **13.** $x^2(x + 8)$ **15.** $4y^2(2 + y^2)$
17. $5x^2y^2(y + 3x)$ **19.** $5(x^2 - x + 3)$
21. $2x(4y + 5z - 7w)$ **23.** $3x^2y^4z^2(3xy^2 - 4x^2z^2 + 5yz)$
25. $-5(x + 8)$ **27.** $-8(t - 9)$ **29.** $-2(x^2 - 6x - 20)$
31. $-7(-x + 8y)$, or $-7(8y - x)$ **33.** $-5(-r + 2s)$, or
$-5(2s - r)$ **35.** $-(p^3 + 4p^2 - 11)$
37. $-(m^3 + m^2 - m + 2)$ **39.** $(b - 5)(a + c)$
41. $(x + 7)(2x - 3)$ **43.** $(x - y)(a^2 - 5)$
45. $(y + z)(x + w)$ **47.** $(y - 1)(y^2 + 3)$
49. $(t + 6)(t^2 - 2)$ **51.** $3a^2(4a^2 - 7a - 3)$
53. $y(y - 1)(y^2 + 1)$ **55.** $(2 - x)(xy - 3)$
57. (a) $h(t) = -8t(2t - 9)$; **(b)** $h(1) = 56$ ft
59. $R(n) = n(n - 1)$ **61.** $P(t) = t(t - 5)$
63. $R(x) = 0.4x(700 - x)$ **65.** $P(n) = \frac{1}{2}(n^2 - 3n)$
67. $H(n) = \frac{1}{2}n(n - 1)$ **69.** 🖩 **71.** -26 **72.** -1
73. -30 **74.** -19 **75.** 56, 58, 60 **76.** A: 75 suitcases;
B: 84 suitcases; C: 63 suitcases **77.** 🖩
79. $x^5y^4 + x^4y^6 = x^3y(x^2y^3 + xy^5)$

81. $(x^2 - x + 5)(r + s)$ **83.** $(x^4 + x^2 + 5)(a^4 + a^2 + 5)$
85. $x^{-9}(x^3 + 1 + x^6)$ **87.** $x^{1/3}(1 - 5x^{1/6} + 3x^{5/12})$
89. $x^{-5/2}(1 + x)$ **91.** $x^{-7/5}(x^{3/5} - 1 + x^{16/15})$
93. $3a^n(a + 2 - 5a^2)$ **95.** $y^{a+b}(7y^a - 5 + 3y^b)$ **97.** 📈

Technology Connection, p. 311

1. They should coincide. **2.** The x-axis
3. Let $y_1 = x^3 - x^2 - 30x$, $y_2 = x(x + 5)(x - 6)$, and
$y_3 = y_2 - y_1$. The graphs of y_1 and y_2 should coincide; the
graph of y_3 should be the x-axis.
4. Let $y_1 = 2x^2 + x - 15$, $y_2 = (2x + 5)(x - 3)$, and
$y_3 = y_2 - y_1$. The graphs of y_1 and y_2 do not coincide; the
graph of y_3 is not the x-axis.

Exercise Set 5.4, pp. 317–318

1. True **3.** False **5.** True **7.** False **9.** $(x + 1)(x + 5)$
11. $(y + 3)(y + 9)$ **13.** $(t - 5)(t + 3)$
15. $2(a - 4)(a - 4)$, or $2(a - 4)^2$ **17.** $x(x + 9)(x - 6)$
19. $(y + 8)(y + 4)$ **21.** $(p - 8)(p + 5)$
23. $(a - 4)(a - 7)$ **25.** $(x + 3)(x - 2)$
27. $5(y + 1)(y + 7)$ **29.** $(8 - y)(4 + y)$
31. $x(8 - x)(7 + x)$ **33.** $y^2(y + 12)(y - 7)$ **35.** Prime
37. $(x + 3y)(x + 9y)$ **39.** $(x - 7y)(x - 7y)$, or $(x - 7y)^2$
41. $x^2(x - 1)(x - 49)$ **43.** $x^4(x - 7)(x + 9)$
45. $(3x + 2)(x - 6)$ **47.** $x(3x - 5)(2x + 3)$
49. $(3a - 4)(a - 2)$ **51.** $(3a + 2)(3a + 4)$
53. $2(5x + 3)(3x - 1)$ **55.** $6(3x - 4)(x + 1)$
57. $t^6(t + 7)(t - 2)$ **59.** $2x^2(5x - 2)(7x - 4)$
61. $2(6a + 5)(a - 2)$ **63.** $(3x + 1)(3x + 4)$
65. $(x + 3)(4x + 3)$ **67.** $-2(2t - 3)(2t + 5)$
69. $xy(6y + 5)(3y - 2)$ **71.** $(24x + 1)(x - 2)$
73. $3x(7x + 3)(3x + 4)$ **75.** $2x^2(6x + 5)(4x - 3)$
77. $(4a - 3b)(3a - 2b)$ **79.** $(2x - 3y)(x + 2y)$
81. $(2x - 7y)(3x - 4y)$ **83.** $(3x - 5y)(3x - 5y)$, or
$(3x - 5y)^2$ **85.** $(9xy - 4)(xy + 1)$ **87.** 🖩 **89.** $27t^3$
90. $125a^{12}$ **91.** $64s^9$ **92.** $-8x^6$ **93.** -24 **94.** 880 ft;
960 ft; 1024 ft; 624 ft; 240 ft **95.** 🖩
97. $5(4x^4y^3 + 1)(3x^4y^3 + 1)$ **99.** $\left(y + \frac{4}{7}\right)\left(y - \frac{2}{7}\right)$
101. $(2x^a + 1)(2x^a - 3)$ **103.** $(x^a + 8)(x^a - 3)$
105. $a(2r + s)(r + s)$ **107.** $(x - 4)(x + 8)$
109. 76, -76, 28, -28, 20, -20 **111.** Since
$ax^2 + bx + c = (mx + r)(nx + s)$, from FOIL we know
that $a = mn$, $c = rs$, and $b = ms + rn$. If $P = ms$ and
$Q = rn$, then $b = P + Q$. Since $ac = mnrs = msrn$, we
have $ac = PQ$. **113.** 📈 **115.** 🖩

Exercise Set 5.5, pp. 323–325

1. Difference of two squares **3.** Perfect-square trinomial
5. None of these **7.** Polynomial having a common factor
9. Perfect-square trinomial **11.** $(t + 3)^2$ **13.** $(a - 7)^2$
15. $4(a - 2)^2$ **17.** $(y + 6)^2$ **19.** $y(y - 9)^2$

21. $2(x - 10)^2$ **23.** $(1 - 4d)^2$ **25.** $y(y + 4)^2$
27. $(0.5x + 0.3)^2$ **29.** $(p - q)^2$ **31.** $(5a + 3b)^2$
33. $5(a - b)^2$ **35.** $(y + 10)(y - 10)$ **37.**
$(m + 8)(m - 8)$ **39.** $(pq + 5)(pq - 5)$ **41.**
$8(x + y)(x - y)$
43. $7x(y^2 + z^2)(y + z)(y - z)$ **45.** $a(2a + 7)(2a - 7)$
47. $3(x^4 + y^4)(x^2 + y^2)(x + y)(x - y)$
49. $a^2(3a + 5b^2)(3a - 5b^2)$ **51.** $\left(\frac{1}{7} + x\right)\left(\frac{1}{7} - x\right)$
53. $(a + b + 3)(a + b - 3)$ **55.** $(x - 3 + y)(x - 3 - y)$
57. $(t + 8)(t + 1)(t - 1)$ **59.** $(r - 3)^2(r + 3)$
61. $(m - n + 5)(m - n - 5)$ **63.** $(6 + x + y)(6 - x - y)$
65. $(r - 1 + 2s)(r - 1 - 2s)$ **67.** $(4 + a + b)(4 - a - b)$
69. $($ $+ 5)(x + 2)(x - 2)$ **71.** $(a - 2)(a + b)(a - b)$
73.   **75.** $8a^{12}b^{15}$ **76.** $125x^6y^{12}$
77. $x^3 + 3x^2y + 3xy^2 + y^3$ **78.** $a^3 + 3a^2 + 3a + 1$
79. $(2, -1, 3)$ **80.** $\left\{x \mid x \leq -\frac{4}{7} \text{ or } x \geq 2\right\}$, or
$\left(-\infty, -\frac{4}{7}\right] \cup [2, \infty)$ **81.** $\left\{x \mid -\frac{4}{7} \leq x \leq 2\right\}$, or $\left[-\frac{4}{7}, 2\right]$
82. $\left\{x \mid x < \frac{14}{19}\right\}$, or $\left(-\infty, \frac{14}{19}\right)$ **83.**   **85.** $-\frac{1}{54}(4r + 3s)^2$
87. $(0.3x^4 + 0.8)^2$, or $\frac{1}{100}(3x^4 + 8)^2$
89. $(r + s + 1)(r - s - 9)$ **91.** $(x^{2a} + y^b)(x^{2a} - y^b)$
93. $(5y^a + x^b - 1)(5y^a - x^b + 1)$
95. $(a - 3 - 4)^2$, or $(a - 7)^2$ **97.** $(m + 2n)(m + 2n + 5)$
99. $5(c^{50} + 4d^{50})(c^{25} + 2d^{25})(c^{25} - 2d^{25})$
101. $($ $^y + 1)^2$ **103.** $h(2a + h)(2a^2 + 2ah + h^2)$
105.

Exercise Set 5.6, pp. 328–329

1. Difference of two cubes   **3.** Difference of two squares
5. Sum of two cubes   **7.** None of these   **9.** Both a
difference of two cubes *and* a difference of two squares
11. $(x + 4)(x^2 - 4x + 16)$ **13.** $(z - 1)(z^2 + z + 1)$
15. $(x - 3)(x^2 + 3x + 9)$ **17.** $(3x + 1)(9x^2 - 3x + 1)$
19. $(4 - 5x)(16 + 20x + 25x^2)$
21. $(3y + 4)(9y^2 - 12y + 16)$ **23.** $(x - y)(x^2 + xy + y^2)$
25. $\left(a + \frac{1}{2}\right)\left(a^2 - \frac{1}{2}a + \frac{1}{4}\right)$ **27.** $8(t - 1)(t^2 + t + 1)$
29. $2(3x + 1)(9x^2 - 3x + 1)$ **31.** $a(b + 5)(b^2 - 5b + 25)$
33. $5(x - 2z)(x^2 + 2xz + 4z^2)$
35. $(x + 0.1)(x^2 - 0.1x + 0.01)$
37. $8(2x^2 - t^2)(4x^4 + 2x^2t^2 + t^4)$
39. $2y(y - 4)(y^2 + 4y + 16)$
41. $(z + 1)(z^2 - z + 1)(z - 1)(z^2 + z + 1)$
43. $(t^2 + 4y^2)(t^4 - 4t^2y^2 + 16y^4)$
45. $(x^4 - yz^4)(x^8 + x^4yz^4 + y^2z^8)$ **47.**   **49.** 228 ft^2
50. -2 **51.** Slope: $\frac{4}{3}$; y-intercept: $\left(0, -\frac{8}{3}\right)$
52. 24 ft; 288 ft; 320 ft; 288 ft; 128 ft   **53.** $\frac{5}{3}$ **54.** $-\frac{7}{2}$
55.   **57.** $(x^{2a} - y^b)(x^{4a} + x^{2a}y^b + y^{2b})$
59. $2x(x^2 + 75)$ **61.** $5\left(xy^2 - \frac{1}{2}\right)\left(x^2y^4 + \frac{1}{2}xy^2 + \frac{1}{4}\right)$
63. $-(3x^{4a} + 3x^{2a} + 1)$ **65.** $(t - 8)(t - 1)(t^2 + t +$
67. $h(2a + h)(a^2 + ah + h^2)(3a^2 + 3ah + h^2)$ **69.**

Exercise Set 5.7, pp. 333–334

1. Greatest common factor   **3.** Sum; cubes
5. FOIL; grouping   **7.** $4(m^2 + 5)(m^2 - 5)$
9. $(a + 9)(a - 9)$ **11.** $(4x + 1)(2x - 5)$ **13.** $(a + 5)^2$
15. $3(x + 12)(x - 7)$ **17.** $(5x + 3y)(5x - 3y)$
19. $(t^2 + 1)(t^4 - t^2 + 1)$ **21.** $(x + y + 3)(x - y + 3)$
23. $(7x + 3y)(49x^2 - 21xy + 9y^2)$ **25.** $2t(t + 12)(t - 2)$
27. $6x^4(1 + 2x)(1 - 2x)$ **29.** $(m^3 + 10)(m^3 - 2)$
31. $(a + d)(c - b)$ **33.** $(2c - d)^2$ **35.** $(8 + 3t^2)(3 + t)$
37. $2(x + 3)(x + 2)(x - 2)$
39. $2(3a - 2b)(9a^2 + 6ab + 4b^2)$ **41.** $(6y - 5)(6y + 7)$
43. $4(m^2 + 4n^2)(m + 2n)(m - 2n)$
45. $ab(a + 4b)(a - 4b)$ **47.** $2t(17t^2 - 3)$
49. $2(a - 3)(a + 3)$ **51.** $7a(a^3 - 2a^2 + 3a - 1)$
53. $(9ab + 2)(3ab + 4)$ **55.** $-5t(2t^2 - 3)$
57. $-2x(3x^3 - 4x^2 + 6)$ **59.** $p(1 - 4p)(1 + 4p + 16p^2)$
61. $(a - b - 3)(a + b + 3)$ **63.**   **65.** $\frac{9}{5}$ **66.** $-\frac{13}{7}$
67.   **68.**

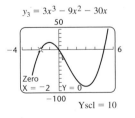

69. Correct: 55; incorrect: 20   **70.** $\frac{80}{7}$ units   **71.**
73. $a(7a - bc)(4a - 3bc)$ **75.** $x(x - 2p)$
77. $y(y - 1)^2(y - 2)$ **79.** $(x - 1)^3(x^2 + 1)(x + 1)$
81. $-x(x - 1)^3(x^2 - 3x + 3)$, or $x(1 - x)^3(x^2 - 3x + 3)$
83. $a(a^w + 1)^2$
85. $(b + a)(1 + y^4)(1 + y^2)(1 + y)(1 - y)$
87. $3(a + b + c + d)(a + b - c - d)$ **89.** $-2(3m^2 + 1)$

Technology Connection, p. 339

1. The graphs intersect at $(-2, 36)$, $(0, 0)$, and $(5, 225)$, so
the solutions are -2, 0, and 5.
2. The zeros are -2, 0, and 5.

$y_3 = 3x^3 - 9x^2 - 30x$

Zero
X = -2   Y = 0
Yscl = 10

Exercise Set 5.8, pp. 342–346

1. True   **3.** False   **5.** True   **7.** $\{-4, 3\}$ **9.** $\{-3, 2\}$
11. $\left\{0, \frac{3}{2}\right\}$ **13.** $\left\{0, \frac{7}{5}\right\}$ **15.** $\left\{-\frac{5}{2}, 7\right\}$ **17.** $\{2, 5\}$ **19.** $\{0, 10\}$
21. $\left\{\frac{1}{3}, \frac{5}{4}\right\}$ **23.** $\left\{0, \frac{5}{2}\right\}$ **25.** $\{-2, 8\}$ **27.** $\{-4, 7\}$ **29.** $\{4\}$
31. $\{-8\}$ **33.** $\{-5, -3\}$ **35.** $\{-3, 3\}$ **37.** $\{-7, 0, 9\}$
39. $\{-7, 7\}$ **41.** $\{-6, 6\}$ **43.** $\left\{\frac{1}{3}, \frac{4}{3}\right\}$ **45.** $\left\{-\frac{3}{4}, -\frac{1}{2}, 0\right\}$
47. $\{-3, 1\}$ **49.** $\left\{-\frac{7}{4}, \frac{4}{3}\right\}$ **51.** $\left\{-\frac{1}{8}, \frac{1}{8}\right\}$ **53.** $\{-5, -1, 1, 5\}$
55. $\{-8, -4\}$ **57.** $\left\{-4, \frac{3}{2}\right\}$ **59.** $\{-9, -3\}$ **61.** $\left\{\frac{1}{4}, \frac{5}{3}\right\}$
63. $\left\{-5, 0, \frac{3}{2}\right\}$
65. $\left\{x \mid x \text{ is a real number } and \ x \neq -1 \text{ and } x \neq 5\right\}$
67. $\left\{x \mid x \text{ is a real number } and \ x \neq -3 \text{ and } x \neq 3\right\}$
69. $\left\{x \mid x \text{ is a real number } and \ x \neq 0 \text{ and } x \neq \frac{1}{2}\right\}$
71. $\left\{x \mid x \text{ is a real number } and \ x \neq 0 \text{ and } x \neq 2 \text{ and } x \neq 5\right\}$
73. Length: 12 cm; width: 7 cm   **75.** 3 m   **77.** 3 cm
79. 10 ft   **81.** 16, 18, 20   **83.** Height: 6 ft; base: 4 ft
85. Height: 16 m; base: 7 m   **87.** 24 ft   **89.** Length: 12 m;
width: 9 m   **91.** 6 video cameras   **93.** 5 sec
95. $1\frac{1}{2}$ yr, 6 yr   **97.** 📓   **99.** $-\frac{5}{8}$ **100.** 2   **101.** Faster
car: 54 mph; slower car: 39 mph   **102.** Conventional: 36;
surround-sound: 42   **103.** $\left\{x \mid x > \frac{10}{3}\right\}$, or $\left(\frac{10}{3}, \infty\right)$
104. $(2, -2, -4)$ **105.** 📓   **107.** $\left\{-\frac{11}{8}, -\frac{1}{4}, \frac{2}{3}\right\}$
109. $\{-3, 1\}; \{x \mid -4 \leq x \leq 2\}$, or $[-4, 2]$
111. Answers may vary. $g(x) = 3x^3 - 9x^2 - 39x + 45$
113. Length: 28 cm; width: 14 cm   **115.** About 5.7 sec
117. 〰️   **119.** $\{6.90\}$ **121.** $\{3.48\}$

Review Exercises: Chapter 5, pp. 348–349

1. (g)   **2.** (e)   **3.** (j)   **4.** (h)   **5.** (c)   **6.** (i)   **7.** (b)
8. (f)   **9.** (a)   **10.** (d)   **11.** 7, 11, 3, 0; 11
12. $-5x^3 + 2x^2 + 3x + 9$; $-5x^3$; -5
13. $-3x^2 + 2x^3 + 8x^6y - 7x^8y^3$ **14.** 0; -6 **15.** 4
16. $-2a^3 + a^2 - 3a - 4$ **17.** $-x^2y - 2xy^2$
18. $-2x^3 + 2x^2 + 5x + 3$ **19.** $-3x^4 + 3x^3 - x + 16$
20. $-5xy^2 - 2xy + x^2y$ **21.** $14x - 7$
22. $-2a + 6b + 7c$ **23.** $6x^2 - 7xy + 3y^2$ **24.** $-18x^3y^4$
25. $x^8 - x^6 + 5x^2 - 3$ **26.** $8a^2b^2 + 2abc - 3c^2$
27. $4x^2 - 25y^2$ **28.** $9x^2 - 24xy + 16y^2$
29. $2x^2 + 5x - 3$ **30.** $x^4 + 8x^2y^3 + 16y^6$
31. $5t^2 - 42t + 16$ **32.** $x^2 - \frac{1}{2}x + \frac{1}{18}$ **33.** $x(7x + 6)$
34. $3y^2(3y^2 - 1)$ **35.** $3x(5x^3 - 6x^2 + 7x - 3)$
36. $(a - 9)(a - 3)$ **37.** $(3m + 2)(m + 4)$ **38.** $(5x + 2)^2$
39. $4(y + 2)(y - 2)$ **40.** $x(x - 2)(x + 7)$
41. $(a + 2b)(x - y)$ **42.** $(y + 2)(3y^2 - 5)$
43. $(a^2 + 9)(a + 3)(a - 3)$ **44.** $4(x^4 + x^2 + 5)$
45. $(3x - 2)(9x^2 + 6x + 4)$
46. $(0.4b - 0.5c)(0.16b^2 + 0.2bc + 0.25c^2)$
47. $y(y^4 + 1)$ **48.** $2z^6(z^2 - 8)$
49. $2y(3x^2 - 1)(9x^4 + 3x^2 + 1)$ **50.** $4(3x - 5)^2$
51. $(3t + p)(2t + 5p)$ **52.** $(x + 3)(x - 3)(x + 2)$
53. $(a - b + 2t)(a - b - 2t)$ **54.** $\{8\}$ **55.** $\left\{\frac{2}{3}, \frac{3}{2}\right\}$

56. $\left\{0, \frac{7}{4}\right\}$ **57.** $\{-4, 4\}$ **58.** $\{-3, 0, 7\}$ **59.** $\{-1, 6\}$
60. $\{-4, 11\}$
61. $\left\{x \mid x \text{ is a real number } and \ x \neq -7 \text{ and } x \neq \frac{2}{3}\right\}$
62. 5 units   **63.** 3, 5, 7; $-7, -5, -3$
64. Length: 8 in.; width: 5 in.   **65.** 17 ft
66. 📓 The roots of a polynomial function are the
x-coordinates of the points at which the graph of the
function crosses the x-axis.   **67.** 📓 The principle of zero
products states that if a product is equal to 0, at least one
of the factors must be 0. If a product is nonzero, we cannot
conclude that any one of the factors is a particular value.
68. $2(2x - y)(4x^2 + 2xy + y^2)(2x + y)(4x^2 - 2xy + y^2)$
69. $-2(3x^2 + 1)$ **70.** $a^3 - b^3 + 3b^2 - 3b + 1$
71. z^{5n^5} **72.** $\left\{-1, -\frac{1}{2}\right\}$

Test: Chapter 5, p. 350

1. [5.1] 9   **2.** [5.1] $5x^5y^4 - 9x^4y - 14x^2y + 8xy^3$
3. [5.1] $-5a^3$ **4.** [5.1] 4; 2   **5.** [5.2] $2ah + h^2 - 3h$
6. [5.1] $4xy + 3xy^2$ **7.** [5.1] $-3x^3 + 3x^2 - 6y - 7y^2$
8. [5.1] $7m^3 + 2m^2n + 3mn^2 - 7n^3$ **9.** [5.1] $5a - 8b$
10. [5.1] $5y^2 + 5y + y^3$ **11.** [5.2] $64x^3y^3$
12. [5.2] $12a^2 - 4ab - 5b^2$ **13.** [5.2] $x^3 - 2x^2y + y^3$
14. [5.2] $4x^6 - 16x^3 + 16$ **15.** [5.2] $25t^2 - 90t + 81$
16. [5.2] $x^2 - 4y^2$ **17.** [5.3] $5x^2(3 - x^2)$
18. [5.5] $(y + 5)(y + 2)(y - 2)$ **19.** [5.4] $(p - 14)(p + 2)$
20. [5.4] $(6m + 1)(2m + 3)$ **21.** [5.5] $(3y + 5)(3y - 5)$
22. [5.6] $3(r - 1)(r^2 + r + 1)$ **23.** [5.5] $(3x - 5)^2$
24. [5.5] $(x^4 + y^4)(x^2 + y^2)(x + y)(x - y)$
25. [5.5] $(y + 4 + 10t)(y + 4 - 10t)$
26. [5.5] $5(2a - b)(2a + b)$ **27.** [5.4] $2(4x - 1)(3x - 5)$
28. [5.6] $2ab(2a^2 + 3b^2)(4a^4 - 6a^2b^2 + 9b^4)$
29. [5.3] $4xy(y^3 + 9x + 2x^2y - 4)$ **30.** [5.8] $\{-3, 10\}$
31. [5.8] $\{-5, 5\}$ **32.** [5.8] $\left\{-7, -\frac{3}{2}, 0\right\}$ **33.** [5.8] $\left\{-\frac{1}{3}, 0\right\}$
34. [5.8] $\{0, 5\}$
35. [5.8] $\{x \mid x \text{ is a real number } and \ x \neq -1\}$
36. [5.8] Length: 8 cm; width: 5 cm   **37.** [5.8] $4\frac{1}{2}$ sec
38. (a) [5.2] $x^5 + x + 1$;
(b) [5.2], [5.7] $(x^2 + x + 1)(x^3 - x^2 + 1)$
39. [5.4] $(3x^n + 4)(2x^n - 5)$

CHAPTER 6

Technology Connection, p. 357

1. Let $y_1 = (7x^2 + 21x)/(14x)$, $y_2 = (x + 3)/2$, and
$y_3 = y_1 - y_2$ (or $y_2 - y_1$). A table or the TRACE feature can
be used to show that, except when $x = 0$, y_3 is always 0. As
an alternative, let $y_1 = (7x^2 + 21x)/(14x) - (x + 3)/2$ and
show that, except when $x = 0$, y_1 is always 0.
2. Let $y_1 = (x + 3)/x$, $y_2 = 3$, and $y_3 = y_1 - y_2$ (or $y_2 - y_1$).
Use a table or the TRACE feature to show that y_3 is not
always 0. As an alternative, let $y_1 = (x + 3)/x - 3$ and
show that y_1 is not always 0.

Exercise Set 6.1, pp. 360–363

1. (e)   **3.** (g)   **5.** (i)   **7.** (a)   **9.** (d)   **11.** $\frac{40}{13}$ hr, or $3\frac{1}{13}$ hr

13. $\frac{2}{3}$; 28; $\frac{163}{10}$ **15.** $-\frac{9}{4}$; does not exist; $-\frac{11}{9}$ **17.** $\dfrac{4x(x-3)}{4x(x+2)}$

19. $\dfrac{(t-2)(-1)}{(t+3)(-1)}$ **21.** $\dfrac{3}{x}$ **23.** $\dfrac{2}{3t^4 s^7}$ **25.** $a-5$

27. $\dfrac{3}{5a-6}$ **29.** $\dfrac{x-4}{x+5}$ **31.** $f(x)=\dfrac{3}{x}, x\neq -7, 0$

33. $g(x)=\dfrac{x-3}{5}, t\neq -3$ **35.** $h(x)=-\dfrac{1}{5}, x\neq 4$

37. $f(t)=\dfrac{t+4}{t-4}, t\neq 4$ **39.** $g(t)=-\dfrac{7}{3}, t\neq 3$

41. $h(t)=\dfrac{t+4}{t-9}, t\neq -1, 9$ **43.** $f(x)=3x+2, x\neq \frac{2}{3}$

45. $g(t)=\dfrac{4+t}{4-t}, t\neq 4$ **47.** $\dfrac{7b^2}{6a^4}$ **49.** $\dfrac{8x^2}{25}$

51. $\dfrac{(x+4)(x-4)}{x(x+3)}$ **53.** $-\dfrac{a+1}{2+a}$ **55.** 1   **57.** $c(c-2)$

59. $\dfrac{a^2+ab+b^2}{3(a+2b)}$ **61.** $6x^4 y^7$ **63.** $\dfrac{5}{x^5}$ **65.** $-\dfrac{5x+2}{x-3}$

67. $-\dfrac{1}{y^3}$ **69.** $\dfrac{(x+4)(x+2)}{3(x-5)}$ **71.** $\dfrac{x^2+4x+16}{(x+4)^2}$

73. $f(t)=\dfrac{t+4}{4}, t\neq -3, 4$

75. $g(x)=\dfrac{(x+5)(2x+3)}{7x}, x\neq 0, \frac{3}{2}, 7$

77. $f(x)=\dfrac{(x+2)(x+4)}{x^7}, x\neq -4, 0, 2$

79. $h(n)=\dfrac{n(n^2+3)}{(n+3)(n-2)}, n\neq -7, -3, 2, 3$

81. $\dfrac{3(x-3y)}{2(2x-y)(2x-3y)}$ **83.** $\dfrac{(2a-b)(a-1)}{(a-b)(a+1)}$ **85.**

87. $-\frac{7}{30}$ **88.** $-\frac{13}{40}$ **89.** $\frac{5}{14}$ **90.** $\frac{1}{35}$

91. $4x^3-7x^2+9x-5$ **92.** $-2t^4+11t^3-t^2+10t-3$

93.   **95.** $2a+h$ **97.** (a) $\dfrac{2x+2h+3}{4x+4h-1}$; (b) $\dfrac{2x+3}{8x-9}$;

(c) $\dfrac{x+5}{4x-1}$ **99.** $\dfrac{4s^2}{(r+2s)^2(r-2s)}$ **101.** $\dfrac{6t^2-26t+30}{8t^2-15t-21}$

103. $\dfrac{a^2+2}{a^2-3}$ **105.** $\dfrac{(u^2-uv+v^2)^2}{u-v}$

107. (a) $\dfrac{16(x+1)}{(x-1)^2(x^2+x+1)}$; (b) $\dfrac{x^2+x+1}{(x+1)^3}$;

(c) $\dfrac{(x+1)^3}{x^2+x+1}$ **109.**   **111.**

Exercise Set 6.2, pp. 370–374

1. True   **3.** False   **5.** True   **7.** True   **9.** $\dfrac{4}{y}$ **11.** $\dfrac{1}{3m^2 n^2}$

13. 2   **15.** $\dfrac{2t+4}{t-4}$ **17.** $f(x)=\dfrac{45x^2+4x+12}{5x(x+3)}, x\neq -3, 0$

19. $f(x)=\dfrac{3}{x^2-9}, x\neq \pm 3$

21. $f(x)=\dfrac{13x^2-7}{3x^2(x-1)(x+1)}, x\neq 0, \pm 1$

23. $f(x)=\dfrac{1}{1-x}, x\neq 1$ **25.** $\dfrac{1}{x-7}$ **27.** $\dfrac{-1}{a+3}$

29. $-(s+r)$ **31.** $\dfrac{7}{a}$ **33.** $-\dfrac{1}{y+5}$ **35.** $\dfrac{1}{y^2+9}$

37. $\dfrac{1}{r^2+rs+s^2}$ **39.** $\dfrac{2a^2-a+14}{(a-4)(a+3)}$ **41.** $\dfrac{5x+1}{x+1}$

43. $\dfrac{x+y}{x-y}$ **45.** $\dfrac{3x^2+7x+14}{(2x-5)(x-1)(x+2)}$

47. $\dfrac{8x+1}{(x+1)(x-1)}$ **49.** $\dfrac{-x+34}{20(x+2)}$ **51.** $\dfrac{-a^2+7ab-b^2}{(a-b)(a+b)}$

53. $\dfrac{x-5}{(x+5)(x+3)}$ **55.** $\dfrac{y}{(y-2)(y-3)}$ **57.** $\dfrac{7x+1}{x-y}$

59. $\dfrac{3y^2-3y-29}{(y-3)(y+8)(y-4)}$ **61.** $\dfrac{-y}{(y+3)(y-1)}$

63. $-\dfrac{2y}{2y+1}$ **65.** $f(x)=\dfrac{3(x+4)}{x+3}, x\neq \pm 3$

67. $f(x)=\dfrac{(x-7)(2x-1)}{(x-4)(x-1)(x+3)}, x\neq -4, -3, 1, 4$

69. $f(x)=\dfrac{-2}{(x+1)(x+2)}, x\neq -3, -2, -1$ **71.**

73. $\dfrac{3y^6 z^7}{7x^5}$ **74.** $\dfrac{7b^{11}c^7}{9a^2}$ **75.** $\dfrac{17s^{12}r^{40}}{5t^{60}}$ **76.** $y=\frac{5}{4}x+\frac{11}{2}$

77. Dimes: 2 rolls; nickels: 5 rolls; quarters: 5 rolls

78. 30-min tapes: 4; 60-min tapes: 8   **79.**

81. 420 days   **83.** 12 parts

85. $x^4(x^2+1)(x+1)(x-1)(x^2+x+1)(x^2-x+1)$

87. $8a^4, 8a^4 b, 8a^4 b^2, 8a^4 b^3, 8a^4 b^4, 8a^4 b^5, 8a^4 b^6, 8a^4 b^7$

89. $\dfrac{x^4+6x^3+2x^2}{(x+2)(x-2)(x+5)}$ **91.** $\dfrac{x^5}{(x^2-4)(x^2+3x-10)}$

93. $\dfrac{9x^2+28x+15}{(x-3)(x+3)^2}$ **95.** $\dfrac{1}{2x(x-5)}$ **97.** $-4t^4$ **99.**

Technology Connection, p. 378

1. $-1, 0, \frac{2}{3}, 2$

Exercise Set 6.3, pp. 380–383

1. (e)   **3.** (f)   **5.** (b)   **7.** $\dfrac{(x+2)(x-3)}{(x-1)(x+4)}$ **9.** $\dfrac{5b-4a}{2b+3a}$

11. $\dfrac{2z+3y}{4y-z}$ **13.** $\dfrac{a+b}{a}$ **15.** $\dfrac{3}{3x+2}$ **17.** $\dfrac{1}{x-y}$

19. $\dfrac{1}{a(a-h)}$ **21.** $\dfrac{(a-2)(a-7)}{(a+1)(a-6)}$ **23.** $\dfrac{x+2}{x+3}$ **25.** $\dfrac{1+2y}{1-3y}$

27. $\dfrac{y^2 + 1}{y^2 - 1}$ **29.** $\dfrac{4x - 7}{7x - 9}$ **31.** $\dfrac{a^2 - 3a - 6}{a^2 - 2a - 3}$ **33.** $\dfrac{a + 1}{2a + 5}$

35. $\dfrac{-1 - 3x}{8 - 2x}$, or $\dfrac{3x + 1}{2x - 8}$ **37.** $\dfrac{1}{y + 5}$ **39.** $-y$

41. $\dfrac{6a^2 + 30a + 60}{3a^2 + 2a + 4}$ **43.** $\dfrac{(2x + 1)(x + 2)}{2x(x - 1)}$

45. $\dfrac{(2a - 3)(a + 5)}{2(a - 3)(a + 2)}$ **47.** -1 **49.** $\dfrac{2x^2 - 11x - 27}{2x^2 + 21x + 13}$

51. 🖩   **53.** 1   **54.** $\frac{11}{4}$ **55.** $y = \dfrac{t - rs}{r}$ **56.** $\{-2, 5\}$

57. First frame: 15 cm; second frame: 29 cm   **58.** $34

59. 🖩   **61.** $\dfrac{5(y + x)}{3(y - x)}$ **63.** $\dfrac{8c}{17}$ **65.** $\dfrac{-3}{x(x + h)}$

67. $\dfrac{2}{(1 + x + h)(1 + x)}$

69. $\{x \,|\, x$ is a real number $and\ x \neq \pm 1$ $and\ x \neq \pm 4$

$and\ x \neq \pm 5\}$ **71.** $\dfrac{2 + a}{3 + a}, a \neq -2, -3$

73. $\dfrac{x^4}{81}$; $\{x \,|\, x$ is a real number $and\ x \neq 3\}$ **75.** 📈

77. $168.61

Exercise Set 6.4, pp. 388–390

1. Equation   **3.** Expression   **5.** Equation   **7.** Equation
9. Expression   **11.** $\frac{42}{5}$ **13.** -8 **15.** $-\frac{11}{9}$ **17.** 5   **19.** -3
21. No solution   **23.** $-4, -1$ **25.** $-2, 6$
27. No solution   **29.** -5 **31.** $-\frac{10}{3}$ **33.** -1 **35.** 2, 3
37. -145 **39.** No solution   **41.** -1 **43.** 4   **45.** $-6, 5$
47. $-\frac{3}{2}, 5$ **49.** 14   **51.** $\frac{3}{4}$ **53.** $\frac{5}{14}$ **55.** $\frac{3}{5}$ **57.** 🖩
59. Multiple-choice: 25; true–false: 30; fill-in: 15
60. Child's: 118; adult's: 132   **61.** 24,640 m²
62. 16 and 18   **63.** (a) Inconsistent; (b) consistent
64. $\{x \,|\, x < -1\ or\ x > 5\}$, or $(-\infty, -1) \cup (5, \infty)$ **65.** 🖩
67. $\frac{1}{5}$ **69.** $-\frac{7}{2}$ **71.** 0.0854697   **73.** Yes   **75.** 📈   **77.** 📈

Exercise Set 6.5, pp. 397–400

1. 2   **3.** $-3, -2$ **5.** 6 and 7, -7 and -6 **7.** $3\frac{3}{14}$ hr
9. $8\frac{4}{7}$ hr   **11.** $19\frac{4}{5}$ min   **13.** Canon: 36 min; HP: 72 min
15. Blueair 402: 30 min; Panasonic F-P20HU1: 40 min
17. Erickson Air-Crane: 10 hr; S-58T: 40 hr
19. Zsuzanna: $\frac{4}{3}$ hr; Stan: 4 hr   **21.** 8 hr   **23.** 7 mph
25. 4.3 ft/sec   **27.** Freight: 66 mph; passenger: 80 mph
29. Express: 45 mph; local: 38 mph   **31.** 9 km/h
33. 2 km/h   **35.** 25 mph   **37.** 20 mph   **39.** 🖩
41. $5a^4b^6$ **42.** $5x^6y^4$ **43.** $4s^{10}t^8$
44. $-2x^4 - 7x^2 + 11x$ **45.** $-8x^3 + 28x^2 - 25x + 19$
46. $11x^4 + 7x^3 - 2x^2 - 4x - 10$ **47.** 🖩   **49.** $49\frac{1}{2}$ hr
51. 11.25 min   **53.** 2250 people per hour   **55.** $14\frac{7}{8}$ mi
57. Page 278   **59.** $8\frac{2}{11}$ min after 10:30   **61.** $51\frac{3}{7}$ mph

Exercise Set 6.6, pp. 405–407

1. True   **3.** True   **5.** False   **7.** $\frac{16}{3}x^4 + 3x^3 - \frac{9}{2}$
9. $3a^2 + a - \dfrac{3}{7} - \dfrac{2}{a}$ **11.** $-6t^3 + 5 - \dfrac{7}{t}$

13. $-4z + 2y^2z^3 - 3y^4z^2$ **15.** $8y - \dfrac{9}{2} - \dfrac{4}{y}$

17. $-5x^5 + 7x^2 + 1$ **19.** $1 - ab^2 - a^3b^4$ **21.** $x + 3$

23. $a - 12 + \dfrac{32}{a + 4}$ **25.** $x - 5 + \dfrac{1}{x - 4}$ **27.** $y - 5$

29. $y^2 - 2y - 1 + \dfrac{-8}{y - 2}$ **31.** $2x^2 - x + 1 + \dfrac{-5}{x + 2}$

33. $a^2 + 4a + 15 + \dfrac{70}{a - 4}$ **35.** $2y^2 + 2y - 1 + \dfrac{8}{5y - 2}$

37. $2x^2 - x - 9 + \dfrac{3x + 12}{x^2 + 2}$ **39.** $4x^2 + 6x + 9, x \neq \frac{3}{2}$
41. $2x - 5, x \neq -\frac{2}{3}$ **43.** $x^2 + 1, x \neq -5, x \neq 5$
45. $2x^3 - 3x^2 + 5, x \neq -1, x \neq 1$ **47.** 🖩
49. $c = \dfrac{ab - k}{d}$ **50.** $z = \dfrac{xy - t}{w}$ **51.** 12, 13, 14
52. $-54a^3$ **53.** $\{x \,|\, x < -2\ or\ x > 5\}$, or $(-\infty, -2) \cup (5, \infty)$
54. $\{x \,|\, -\frac{7}{3} < x < 3\}$, or $(-\frac{7}{3}, 3)$ **55.** 🖩   **57.** $a^2 + ab$
59. $a^6 - a^5b + a^4b^2 - a^3b^3 + a^2b^4 - ab^5 + b^6$ **61.** $-\frac{3}{2}$
63. 🖩   **65.** 📈

Exercise Set 6.7, pp. 411–412

1. True   **3.** True   **5.** True   **7.** True

9. $x^2 - x + 1 + \dfrac{-6}{x - 1}$ **11.** $a + 5 + \dfrac{-4}{a + 3}$

13. $x^2 - 9x + 5 + \dfrac{-7}{x + 2}$ **15.** $3x^2 - 2x + 2 + \dfrac{-3}{x + 3}$

17. $y^2 + 2y + 1 + \dfrac{12}{y - 2}$ **19.** $x^4 + 2x^3 + 4x^2 + 8x + 16$

21. $3x^2 + 6x - 3 + \dfrac{2}{x + \frac{1}{3}}$ **23.** 6   **25.** 1   **27.** 54

29. 🖩   **31.** $b = \dfrac{a - 9}{c + 1}$ **32.** $a = \dfrac{bd - 8}{c - b}$

33. $\{x \,|\, x$ is a real number $and\ x \neq -5\ and\ x \neq 5\}$
34. $\{x \,|\, x$ is a real number $and\ x \neq -\frac{9}{2}\ and\ x \neq 1\}$
35.   **36.**

37. 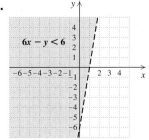   **39. (a)** The degree of R must be less than 1, the degree of $x - r$; **(b)** Let $x = r$. Then
$$P(r) = (r - r) \cdot Q(r) + R$$
$$= 0 \cdot Q(r) + R$$
$$= R.$$
41. $0; -3, -\frac{5}{2}, \frac{3}{2}$ **43.**   **45.** 0

Exercise Set 6.8, pp. 420–425

1. LCD   **3.** Factor   **5.** Inverse   **7.** Direct   **9.** Inverse
11. $d = \dfrac{L}{f}$ **13.** $v_1 = \dfrac{2s}{t} - v_2$, or $\dfrac{2s - tv_2}{t}$ **15.** $b = \dfrac{at}{a - t}$
17. $R = \dfrac{2V}{I} - 2r$, or $\dfrac{2V - 2Ir}{I}$ **19.** $g = \dfrac{Rs}{s - R}$
21. $n = \dfrac{IR}{E - Ir}$ **23.** $q = \dfrac{pf}{p - f}$ **25.** $t_1 = \dfrac{H}{Sm} + t_2$, or
$\dfrac{H + Smt_2}{Sm}$ **27.** $r = \dfrac{Re}{E - e}$ **29.** $r = 1 - \dfrac{a}{S}$, or $\dfrac{S - a}{S}$
31. $a + b = \dfrac{f}{c^2}$ **33.** $r = \dfrac{A}{P} - 1$, or $\dfrac{A - P}{P}$
35. $t_2 = \dfrac{d_2 - d_1}{v} + t_1$, or $\dfrac{d_2 - d_1 + t_1 v}{v}$ **37.** $b^2 = \dfrac{a^2 y^2}{a^2 - x^2}$
39. $Q = \dfrac{2Tt - 2AT}{A - q}$ **41.** $k = 7; y = 7x$
43. $k = 1.7; y = 1.7x$ **45.** $k = 6; y = 6x$
47. $k = 60; y = \dfrac{60}{x}$ **49.** $k = 112; y = \dfrac{112}{x}$
51. $k = 9; y = \dfrac{9}{x}$ **53.** 241,920,000 cans   **55.** 6 amperes
57. 3.5 hr   **59.** 32 kg   **61.** 50 min   **63.** About 21 min
65. 122,269,230 tons   **67.** $y = \frac{2}{3}x^2$ **69.** $y = \dfrac{54}{x^2}$
71. $y = 0.3xz^2$ **73.** $y = \dfrac{4wx^2}{z}$ **75.** About 2.9 sec
77. 308 cm^3 **79.** About 57.42 mph   **81.**   **83.** $\mathbb{R}$
84. $\mathbb{R}$
85.

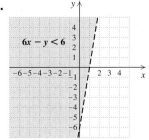

86. $8a^3 - 2a$ **87.** $(t + 2b)(t^2 - 2bt + 4b^2)$
88. $-\frac{5}{3}, \frac{7}{2}$ **89.**   **91.** 567 mi   **93.** Ratio is $\dfrac{a + 12}{a + 6}$;
percent increase is $\dfrac{6}{a + 6} \cdot 100\%$, or $\dfrac{600}{a + 6}\%$

95. $t_1 = t_2 + \dfrac{(d_2 - d_1)(t_4 - t_3)}{a(t_4 - t_2)(t_4 - t_3) + d_3 - d_4}$
97. The intensity is halved.   **99.** About 1.697 m
101. $d(s) = \dfrac{28}{s}$; 70 yd

Review Exercises: Chapter 6, pp. 429–431

1. True   **2.** False   **3.** False   **4.** True   **5.** False   **6.** False
7. True   **8.** False   **9.** True   **10.** True   **11. (a)** $-\frac{2}{9}$;
(b) $-\frac{3}{4}$; **(c)** 0   **12.** $48x^3$ **13.** $(x + 5)(x - 2)(x - 4)$
14. $x + 3$ **15.** $\dfrac{b^2 c^6 d^2}{a^5}$ **16.** $\dfrac{15np + 14m}{18m^2 n^4 p^2}$
17. $\dfrac{(x - 2)(x + 5)}{x - 5}$ **18.** $\dfrac{(x^2 + 4x + 16)(x - 6)}{(x + 4)(x + 2)}$
19. $\dfrac{x - 3}{(x + 1)(x + 3)}$ **20.** $\dfrac{x - y}{x + y}$ **21.** $2(x + y)$
22. $\dfrac{-y}{(y + 4)(y - 1)}$ **23.** $f(x) = \dfrac{1}{x - 1}, x \neq 1, 4$
24. $f(x) = \dfrac{2}{x - 8}, x \neq -8, -5, 8$
25. $f(x) = \dfrac{3x - 1}{x - 3}, x \neq -3, -\frac{1}{3}, 3$ **26.** $\frac{4}{9}$
27. $\dfrac{a^2 b^2}{2(b^2 - ba + a^2)}$ **28.** $\dfrac{(y + 11)(y + 5)}{(y - 5)(y + 2)}$
29. $\dfrac{(14 - 3x)(x + 3)}{2x^2 + 16x + 6}$ **30.** 2   **31.** 6   **32.** No solution
33. 0   **34.** $-1, 4$ **35.** $5\frac{1}{7}$ hr   **36.** Celeron: 45 sec;
Pentium 4: 30 sec   **37.** 24 mph   **38.** Motorcycle: 62 mph;
car: 70 mph   **39.** $6s^2 + 5s - 4rs^2$ **40.** $y^2 - 5y + 25$
41. $4x + 3 + \dfrac{-9x - 5}{x^2 + 1}$ **42.** $x^2 + 6x + 20 + \dfrac{54}{x - 3}$
43. 341   **44.** $s = \dfrac{Rg}{g - R}$ **45.** $m = \dfrac{H}{S(t_1 - t_2)}$
46. $c = \dfrac{b + 3a}{2}$ **47.** $t_1 = \dfrac{-A}{vT} + t_2$, or $\dfrac{-A + vTt_2}{vT}$
48. About 22 lb   **49.** 64 L   **50.** $y = \dfrac{\frac{3}{4}}{x}$
51. The least common denominator was used to add and subtract rational expressions, to simplify complex rational expressions, and to solve rational equations.
52. A rational *expression* is a quotient of two polynomials. Expressions can be simplified, multiplied, or added, but they cannot be solved for a variable. A rational *equation* is an equation containing rational expressions. In a rational equation, we often can solve for a variable.
53. All real numbers except 0 and 13   **54.** 45   **55.** $9\frac{9}{19}$ sec

Test: Chapter 6, pp. 431–432

1. [6.1] $\dfrac{3}{4(t-1)}$ **2.** [6.1] $\dfrac{x^2-3x+9}{x+4}$

3. [6.2] $(x-3)(x+8)(x-9)$ **4.** [6.2] $\dfrac{25x+x^3}{x+5}$

5. [6.2] $3(a-b)$ **6.** [6.2] $\dfrac{a^3-a^2b+4ab+ab^2-b^3}{(a-b)(a+b)}$

7. [6.2] $\dfrac{-2(2x^2+5x+20)}{(x-4)(x+4)(x^2+4x+16)}$

8. [6.2] $f(x)=\dfrac{x-4}{(x+3)(x-2)}, x\neq -3, 2$

9. [6.1] $f(x)=\dfrac{(x-1)^2(x+1)}{x(x-2)}, x\neq -2, 0, 1, 2$

10. [6.3] $\dfrac{a(2b+3a)}{5a+b}$ **11.** [6.3] $\dfrac{(x-9)(x-6)}{(x+6)(x-3)}$

12. [6.3] $\dfrac{2x^2-7x+1}{3x^2+10x-17}$ **13.** [6.4] $-\frac{21}{4}$ **14.** [6.4] 15

15. [6.1] 5; 0   **16.** [6.4] $\frac{5}{3}$ **17.** [6.5] $1\frac{31}{32}$ hr

18. [6.6] $\dfrac{4b^2c}{a}-\dfrac{5bc^2}{2a}+3bc$ **19.** [6.6] $y-14+\dfrac{-20}{y-6}$

20. [6.6] $6x^2-9+\dfrac{5x+22}{x^2+2}$

21. [6.7] $x^2+9x+40+\dfrac{153}{x-4}$ **22.** [6.7] 449

23. [6.8] $b_1=\dfrac{2A}{h}-b_2$, or $\dfrac{2A-b_2h}{h}$ **24.** [6.5] 5 and 6;

−6 and −5   **25.** [6.5] $3\frac{3}{11}$ mph   **26.** [6.8] 30 workers

27. [6.8] 637 in^2 **28.** [6.4] $-\frac{19}{3}$

29. [6.4] $\{x\,|\,x$ is a real number $and\ x\neq 0\ and\ x\neq 15\}$

30. [6.3], [6.4] x-intercept: $(11, 0)$; y-intercept: $\left(0, -\frac{33}{5}\right)$

31. [6.5] Hans: 56 lawns; Franz: 42 lawns

Cumulative Review: Chapters 1–6, pp. 433–434

1. 10   **2.** 3.91×10^8 **3.** Slope: $\frac{7}{4}$; y-intercept: $(0, -3)$

4. $y=-\frac{10}{3}x+\frac{11}{3}$ **5.** $(-3, 4)$ **6.** $(-2, -3, 1)$

7. Small: 16; large: 29   **8.** 12, $\frac{1}{2}$, $7\frac{1}{2}$

9. \$28.36 million per year   **10.** (a) $V(t)=\frac{5}{18}t+15.4$;

(b) about 19.8 min   **11.** (a) $-\frac{1}{2}$;

(b) $\{x\,|\,x$ is a real number $and\ x\neq 5\}$ **12.** $\frac{1}{4}$

13. $-\frac{12}{7}, \frac{12}{7}$ **14.** $\{x\,|\,x>-3\}$, or $(-3, \infty)$ **15.** $\{x\,|\,x\geq -1\}$,

or $[-1, \infty)$ **16.** $\{x\,|\,-10<x<13\}$, or $(-10, 13)$

17. $\left\{x\,\middle|\,x<-\frac{4}{3}\ or\ x>6\right\}$, or $\left(-\infty, -\frac{4}{3}\right)\cup(6, \infty)$

18. $\{x\,|\,x<-6.4\ or\ x>6.4\}$, or $(-\infty, -6.4)\cup(6.4, \infty)$

19. $\left\{x\,\middle|\,-4\leq x\leq \frac{16}{3}\right\}$, or $\left[-4, \frac{16}{3}\right]$ **20.** 2   **21.** −1

22. No solution   **23.** $\frac{1}{3}$ **24.** 1, $\frac{7}{3}$ **25.** $[9, \infty)$

26. $n=\dfrac{m-12}{3}$ **27.** $a=\dfrac{Pb}{4-P}$

28. 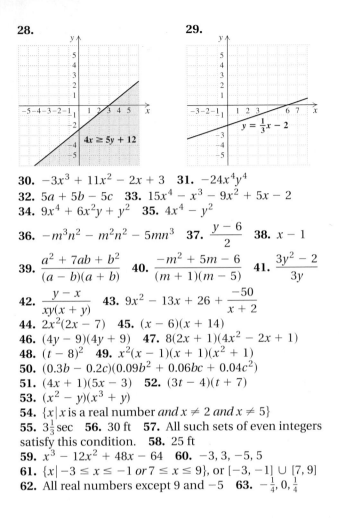   **29.**

30. $-3x^3+11x^2-2x+3$ **31.** $-24x^4y^4$

32. $5a+5b-5c$ **33.** $15x^4-x^3-9x^2+5x-2$

34. $9x^4+6x^2y+y^2$ **35.** $4x^4-y^2$

36. $-m^3n^2-m^2n^2-5mn^3$ **37.** $\dfrac{y-6}{2}$ **38.** $x-1$

39. $\dfrac{a^2+7ab+b^2}{(a-b)(a+b)}$ **40.** $\dfrac{-m^2+5m-6}{(m+1)(m-5)}$ **41.** $\dfrac{3y^2-2}{3y}$

42. $\dfrac{y-x}{xy(x+y)}$ **43.** $9x^2-13x+26+\dfrac{-50}{x+2}$

44. $2x^2(2x-7)$ **45.** $(x-6)(x+14)$

46. $(4y-9)(4y+9)$ **47.** $8(2x+1)(4x^2-2x+1)$

48. $(t-8)^2$ **49.** $x^2(x-1)(x+1)(x^2+1)$

50. $(0.3b-0.2c)(0.09b^2+0.06bc+0.04c^2)$

51. $(4x+1)(5x-3)$ **52.** $(3t-4)(t+7)$

53. $(x^2-y)(x^3+y)$

54. $\{x\,|\,x$ is a real number $and\ x\neq 2\ and\ x\neq 5\}$

55. $3\frac{1}{3}$ sec   **56.** 30 ft   **57.** All such sets of even integers

satisfy this condition.   **58.** 25 ft

59. $x^3-12x^2+48x-64$ **60.** $-3, 3, -5, 5$

61. $\{x\,|\,-3\leq x\leq -1\ or\ 7\leq x\leq 9\}$, or $[-3, -1]\cup[7, 9]$

62. All real numbers except 9 and −5   **63.** $-\frac{1}{4}, 0, \frac{1}{4}$

CHAPTER 7

Technology Connection, p. 439

1. False   **2.** True   **3.** False

Exercise Set 7.1, pp. 443–444

1. Two   **3.** Positive   **5.** Irrational   **7.** Nonnegative

9. 7, −7   **11.** 12, −12   **13.** 20, −20   **15.** 30, −30

17. $-\frac{6}{7}$ **19.** 21   **21.** $-\frac{4}{9}$ **23.** 0.2   **25.** −0.05

27. $p^2+4; 2$ **29.** $\dfrac{x}{y+4}; 3$ **31.** $\sqrt{20}$; 0; does not exist;

does not exist   **33.** −3; −1; does not exist; 0

35. 1; $\sqrt{2}$; $\sqrt{101}$ **37.** 1; does not exist; 6   **39.** $6|x|$

41. $6|b|$ **43.** $|8-t|$ **45.** $|y+8|$ **47.** $|2x+7|$

49. 4   **51.** −1   **53.** $-\frac{2}{3}$ **55.** $|x|$ **57.** $6|a|$ **59.** 6

61. $|a+b|$ **63.** $|a^{11}|$ **65.** Cannot be simplified

67. $4x$ **69.** $3t$ **71.** $a+1$ **73.** $2(x+1)$, or $2x+2$

75. $3t - 2$ **77.** 3 **79.** $2x$ **81.** -6 **83.** $5y$ **85.** t^9
87. $(x - 2)^4$ **89.** 2; 3; -2; -4 **91.** 2; does not exist; does not exist; 3 **93.** $\{x \mid x \ge 6\}$, or $[6, \infty)$ **95.** $\{t \mid t \ge -8\}$, or $[-8, \infty)$ **97.** $\{x \mid x \ge 5\}$, or $[5, \infty)$ **99.** $\mathbb{R}$ **101.** $\{z \mid z \ge -\frac{2}{5}\}$, or $\left[-\frac{2}{5}, \infty\right)$ **103.** $\mathbb{R}$ **105.** 📓 **107.** $a^9 b^6 c^{15}$ **108.** $10a^{10}b^9$
109. $\dfrac{a^6 c^{12}}{8b^9}$ **110.** $\dfrac{x^6 y^2}{25z^4}$ **111.** $2x^4 y^5 z^2$ **112.** $\dfrac{5c^3}{a^4 b^7}$
113. 📓 **115.** (a) 34.6 lb; (b) 10.0 lb; (c) 24.9 lb; (d) 42.3 lb
117. $\{x \mid x \ge -5\}$, or $[-5, \infty)$ **119.** $\{x \mid x \ge 0\}$, or $[0, \infty)$

121. $\{x \mid -3 \le x < 2\}$, or $[-3, 2)$
123. $\{x \mid x < -1 \text{ or } x > 6\}$, or $(-\infty, -1) \cup (6, \infty)$ **125.** 📈

Technology Connection, p. 447

1. Without parentheses, the expression entered would be $\dfrac{7^2}{3}$. **2.** For $x = 0$ or $x = 1$, $y_1 = y_2 = y_3$; on $(0, 1)$, $y_1 > y_2 > y_3$; on $(1, \infty)$, $y_1 < y_2 < y_3$.

Technology Connection, p. 449

1. Most graphing calculators do not have keys for radicals of index 3 or higher. On those graphing calculators that offer $\sqrt[x]{\ }$ in a MATH menu, rational exponents still require fewer keystrokes.

Exercise Set 7.2, pp. 449–452

1. (g) **3.** (e) **5.** (a) **7.** (b) **9.** $\sqrt[6]{x}$ **11.** 4 **13.** 3
15. 3 **17.** $\sqrt[3]{xyz}$ **19.** $\sqrt[5]{a^2 b^2}$ **21.** $\sqrt[5]{t^2}$ **23.** 8 **25.** 81
27. $27\sqrt[4]{x^3}$ **29.** $125x^6$ **31.** $20^{1/3}$ **33.** $17^{1/2}$ **35.** $x^{3/2}$
37. $m^{2/5}$ **39.** $(cd)^{1/4}$ **41.** $(xy^2 z)^{1/5}$ **43.** $(3mn)^{3/2}$
45. $(8x^2 y)^{5/7}$ **47.** $\dfrac{2x}{z^{2/3}}$ **49.** $\dfrac{1}{x^{1/3}}$ **51.** $\dfrac{1}{(2rs)^{3/4}}$ **53.** 8
55. $2a^{3/5}c$ **57.** $\dfrac{5y^{4/5}z}{x^{2/3}}$ **59.** $\dfrac{a^3}{3^{5/2} b^{7/3}}$ **61.** $\left(\dfrac{3c}{2ab}\right)^{5/6}$
63. $\dfrac{6a}{b^{1/4}}$ **65.** $7^{7/8}$ **67.** $3^{3/4}$ **69.** $5.2^{1/2}$ **71.** $10^{6/25}$
73. $a^{23/12}$ **75.** 64 **77.** $\dfrac{m^{1/3}}{n^{1/8}}$ **79.** $\sqrt[3]{x^2}$ **81.** a^3
83. a^2 **85.** $x^2 y^2$ **87.** $\sqrt{7a}$ **89.** $\sqrt[4]{8x^3}$ **91.** $\sqrt[18]{a}$
93. $x^3 y^3$ **95.** $a^6 b^{12}$ **97.** $\sqrt[12]{xy}$ **99.** 📓

101. $11x^4 + 14x^3$ **102.** $-3t^6 + 28t^5 - 20t^4$
103. $15a^2 - 11ab - 12b^2$ **104.** $49x^2 - 14xy + y^2$
105. \$93,500 **106.** 0, 1 **107.** 📓 **109.** $\sqrt[6]{x^5}$
111. $\sqrt[6]{p + q}$ **113.** $2^{7/12} \approx 1.498 \approx 1.5$ **115.** 53.0%
117. About 7.937×10^{-13} to 1 **119.** 📈

Technology Connection, p. 454

1. The graphs differ in appearance because the domain of y_1 is the intersection of $[-3, \infty)$ and $[3, \infty)$, or $[3, \infty)$. The domain of y_2 is $(-\infty, -3] \cup [3, \infty)$.

Exercise Set 7.3, pp. 457–459

1. True **3.** False **5.** True **7.** $\sqrt{35}$ **9.** $\sqrt[3]{14}$ **11.** $\sqrt[4]{18}$
13. $\sqrt{26xy}$ **15.** $\sqrt[5]{80y^4}$ **17.** $\sqrt{y^2 - b^2}$ **19.** $\sqrt[3]{0.21y^2}$
21. $\sqrt[5]{(x - 2)^3}$ **23.** $\sqrt{\dfrac{7s}{11t}}$ **25.** $\sqrt[7]{\dfrac{5x - 15}{4x + 8}}$ **27.** $3\sqrt{2}$
29. $3\sqrt{3}$ **31.** $2\sqrt{2}$ **33.** $3\sqrt{22}$ **35.** $6a^2\sqrt{b}$ **37.** $2x\sqrt[3]{y^2}$
39. $-2x^2\sqrt[3]{2}$ **41.** $f(x) = 5x\sqrt[3]{x^2}$ **43.** $f(x) = |7(x - 3)|$, or $7|x - 3|$ **45.** $f(x) = |x - 1|\sqrt{5}$ **47.** $a^3 b^3\sqrt{b}$
49. $xy^2 z^3\sqrt[3]{x^2 z}$ **51.** $-2ab^2\sqrt[5]{a^2 b}$ **53.** $x^2 y z^3\sqrt[5]{x^3 y^3 z^2}$
55. $-2a^4\sqrt[3]{10a^2}$ **57.** $3\sqrt{2}$ **59.** $3\sqrt{35}$ **61.** 3 **63.** $18a^3$
65. $a^3\sqrt{10}$ **67.** $2x^3\sqrt{5x}$ **69.** $s^2 t^3\sqrt[3]{t}$ **71.** $(x + 5)^2$
73. $2ab^3\sqrt[4]{5a}$ **75.** $x(y + z)^2\sqrt[5]{x}$ **77.** 📓
79. $\dfrac{12x^2 + 5y^2}{64xy}$ **80.** $\dfrac{2a + 6b^3}{a^4 b^4}$ **81.** $\dfrac{-7x - 13}{2(x - 3)(x + 3)}$
82. $\dfrac{-3x + 1}{2(x - 5)(x + 5)}$ **83.** $3a^2 b^2$ **84.** $3ab^5$ **85.** 📓
87. 175.6 mi **89.** (a) $-3.3°C$; (b) $-16.6°C$; (c) $-25.5°C$; (d) $-54.0°C$ **91.** $25x^5\sqrt[3]{25x}$ **93.** $a^{10}b^{17}\sqrt{ab}$
95.

$f(x) = h(x); f(x) \ne g(x)$

97. $\{x \mid x \le 2 \text{ or } x \ge 4\}$, or $(-\infty, 2] \cup [4, \infty)$ **99.** 6
101. 📈 , 📓

Exercise Set 7.4, pp. 465–467

1. (e) **3.** (f) **5.** (h) **7.** (a) **9.** $\dfrac{6}{5}$ **11.** $\dfrac{4}{3}$ **13.** $\dfrac{7}{y}$
15. $\dfrac{6y\sqrt{y}}{x^2}$ **17.** $\dfrac{3a\sqrt[3]{a}}{2b}$ **19.** $\dfrac{2a}{bc^2}$ **21.** $\dfrac{ab^2}{c^2}\sqrt[4]{\dfrac{a}{c^2}}$
23. $\dfrac{2x}{y^2}\sqrt[5]{\dfrac{x}{y}}$ **25.** $\dfrac{xy}{z^2}\sqrt[6]{\dfrac{y^2}{z^3}}$ **27.** $\sqrt{5}$ **29.** 3 **31.** $y\sqrt{5y}$
33. $2\sqrt[3]{a^2 b}$ **35.** $\sqrt{2ab}$ **37.** $2x^2 y^3\sqrt[4]{y^3}$

39. $\sqrt[3]{x^2 + xy + y^2}$ **41.** $\dfrac{\sqrt{6}}{2}$ **43.** $\dfrac{2\sqrt{15}}{21}$ **45.** $\dfrac{2\sqrt[3]{6}}{3}$

47. $\dfrac{\sqrt[3]{75ac^2}}{5c}$ **49.** $\dfrac{y\sqrt[3]{180x^2y}}{6x^2}$ **51.** $\dfrac{\sqrt[3]{2xy^2}}{xy}$ **53.** $\dfrac{\sqrt{14a}}{6}$

55. $\dfrac{3\sqrt{5y}}{10xy}$ **57.** $\dfrac{\sqrt{5b}}{6a}$ **59.** $\dfrac{5}{\sqrt{35x}}$ **61.** $\dfrac{2}{\sqrt{6}}$ **63.** $\dfrac{52}{3\sqrt{91}}$

65. $\dfrac{7}{\sqrt[3]{98}}$ **67.** $\dfrac{7x}{\sqrt{21xy}}$ **69.** $\dfrac{2a^2}{\sqrt[3]{20ab}}$ **71.** $\dfrac{x^2y}{\sqrt{2xy}}$ **73.**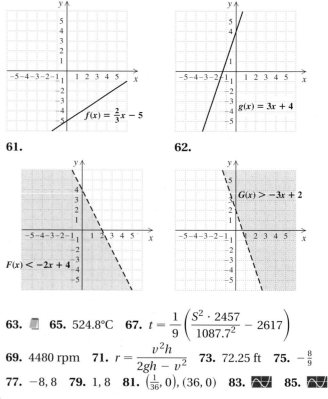

75. $\dfrac{3(x-1)}{(x-5)(x+5)}$ **76.** $\dfrac{7(x-2)}{(x+4)(x-4)}$ **77.** $\dfrac{a-1}{a+7}$

78. $\dfrac{t+11}{t+2}$ **79.** $125a^9b^{12}$ **80.** $225x^{10}y^6$ **81.**

83. (a) 1.62 sec; (b) 1.99 sec; (c) 2.20 sec **85.** $9\sqrt[3]{9n^2}$

87. $\dfrac{-3\sqrt{a^2-3}}{a^2-3}$, or $\dfrac{-3}{\sqrt{a^2-3}}$ **89.** Step 1: $\sqrt[n]{a}=a^{1/n}$, by

definition; Step 2: $\left(\dfrac{a}{b}\right)^n = \dfrac{a^n}{b^n}$, raising a quotient to a

power; Step 3: $a^{1/n}=\sqrt[n]{a}$, by definition
91. $(f/g)(x) = 3x$, where x is a real number and $x > 0$
93. $(f/g)(x) = \sqrt{x+3}$, where x is a real number and $x > 3$

Exercise Set 7.5, pp. 472–474

1. Radicands, indices **3.** Bases **5.** Numerator, conjugate
7. $9\sqrt{5}$ **9.** $2\sqrt[3]{4}$ **11.** $10\sqrt[3]{y}$ **13.** $7\sqrt{2}$
15. $13\sqrt[3]{7} + \sqrt{3}$ **17.** $9\sqrt{3}$ **19.** $23\sqrt{5}$ **21.** $9\sqrt[3]{2}$
23. $(1+6a)\sqrt{5a}$ **25.** $(x+2)\sqrt[3]{6x}$ **27.** $3\sqrt{a-1}$
29. $(x+3)\sqrt{x-1}$ **31.** $4\sqrt{3}+3$ **33.** $15-3\sqrt{10}$
35. $6\sqrt{5}-4$ **37.** $3-4\sqrt[3]{63}$ **39.** $a+2a\sqrt[3]{3}$
41. $4+3\sqrt{6}$ **43.** $\sqrt{6}-\sqrt{14}+\sqrt{21}-7$ **45.** 4
47. -2 **49.** $2-8\sqrt{35}$ **51.** $7+4\sqrt{3}$ **53.** $5-2\sqrt{6}$
55. $2t+5+2\sqrt{10t}$ **57.** $14+x-6\sqrt{x+5}$
59. $6\sqrt[4]{63}+4\sqrt[4]{35}-3\sqrt[4]{54}-2\sqrt[4]{30}$ **61.** $\dfrac{20+5\sqrt{3}}{13}$
63. $\dfrac{12-2\sqrt{3}+6\sqrt{5}-\sqrt{15}}{33}$ **65.** $\dfrac{a-\sqrt{ab}}{a-b}$ **67.** -1
69. $\dfrac{12-3\sqrt{10}-2\sqrt{14}+\sqrt{35}}{6}$ **71.** $\dfrac{3}{5\sqrt{7}-10}$
73. $\dfrac{2}{14+2\sqrt{3}+3\sqrt{2}+7\sqrt{6}}$ **75.** $\dfrac{x-y}{x+2\sqrt{xy}+y}$
77. $\dfrac{1}{\sqrt{a+h}+\sqrt{a}}$ **79.** $a\sqrt[4]{a}$ **81.** $b\sqrt[10]{b^9}$
83. $xy\sqrt[6]{xy^5}$ **85.** $3a^2b\sqrt[4]{ab}$ **87.** $a^2b^2c^2\sqrt[6]{a^2bc^2}$
89. $\sqrt[12]{a^5}$ **91.** $\sqrt[12]{x^2y^5}$ **93.** $\sqrt[10]{ab^9}$ **95.** $\sqrt[20]{(3x-1)^3}$
97. $\sqrt[15]{(2x+1)^4}$ **99.** $x\sqrt[6]{xy^5}-\sqrt[15]{x^{13}y^{14}}$
101. $2m^2 + m\sqrt[4]{n} + 2m\sqrt[3]{n^2} + \sqrt[12]{n^{11}}$ **103.** $\sqrt[4]{2x^2}-x^3$
105. x^2-7 **107.** $27+10\sqrt{2}$ **109.** $8-2\sqrt{15}$ **111.**
113. 8 **114.** $\frac{15}{2}$ **115.** $\frac{1}{5}, 1$ **116.** $\frac{1}{7}, 1$ **117.** $-5, 4$
118. Length: 20 units; width: 5 units **119.**

121. $f(x) = -6x\sqrt{5+x}$ **123.** $f(x) = (x+3x^2)\sqrt[4]{x-1}$
125. $ac^2\left[(3a+2c)\sqrt{ab} - 2\sqrt[3]{ab}\right]$
127. $9a^2(b+1)\sqrt[6]{243a^5(b+1)^5}$ **129.** $1-\sqrt{w}$
131. $\left(\sqrt{x}+\sqrt{5}\right)\left(\sqrt{x}-\sqrt{5}\right)$
133. $\left(\sqrt{x}+\sqrt{a}\right)\left(\sqrt{x}-\sqrt{a}\right)$ **135.** $2x-2\sqrt{x^2-4}$

Technology Connection, p. 477

1. The x-coordinates of the points of intersection should approximate the solutions of the examples.

Exercise Set 7.6, pp. 479–481

1. False **3.** True **5.** True **7.** $\frac{51}{5}$ **9.** $\frac{25}{3}$ **11.** 168
13. 56 **15.** 3 **17.** 82 **19.** 0, 9 **21.** 64 **23.** -27
25. 125 **27.** No solution **29.** $\frac{80}{3}$ **31.** 57 **33.** $-\frac{5}{3}$
35. 1 **37.** $\frac{106}{27}$ **39.** 4 **41.** 3, 7 **43.** $\frac{80}{9}$ **45.** -1
47. No solution **49.** 2, 6 **51.** 2 **53.** 4 **55.**
57. Height: 7 in.; base: 9 in. **58.** 8 60-sec commercials
59.

$f(x) = \frac{2}{3}x - 5$

60.

$g(x) = 3x + 4$

61.

$F(x) < -2x + 4$

62.

$G(x) > -3x + 2$

63. **65.** 524.8°C **67.** $t = \dfrac{1}{9}\left(\dfrac{S^2 \cdot 2457}{1087.7^2} - 2617\right)$

69. 4480 rpm **71.** $r = \dfrac{v^2h}{2gh - v^2}$ **73.** 72.25 ft **75.** $-\frac{8}{9}$

77. $-8, 8$ **79.** 1, 8 **81.** $\left(\frac{1}{36}, 0\right), (36, 0)$ **83.** **85.**

Exercise Set 7.7, pp. 487–490

1. Right, hypotenuse **3.** Square roots
5. 30°, 60°, 90°, leg **7.** $\sqrt{34}$; 5.831 **9.** $9\sqrt{2}$; 12.728
11. 5 **13.** 4 m **15.** $\sqrt{19}$ in.; 4.359 in. **17.** 1 m

19. 250 ft   **21.** $\sqrt{8450}$, or $65\sqrt{2}$ ft; 91.924 ft   **23.** 12 in.
25. $\left(\sqrt{340} + 8\right)$ ft; 26.439 ft   **27.** $\left(110 - \sqrt{6500}\right)$ paces;
29.377 paces   **29.** Leg = 5; hypotenuse = $5\sqrt{2} \approx 7.071$
31. Shorter leg = 7; longer leg = $7\sqrt{3} \approx 12.124$
33. Leg = $5\sqrt{3} \approx 8.660$; hypotenuse = $10\sqrt{3} \approx 17.321$
35. Both legs = $\dfrac{13\sqrt{2}}{2} \approx 9.192$
37. Leg = $14\sqrt{3} \approx 24.249$; hypotenuse = 28
39. $3\sqrt{3} \approx 5.196$ **41.** $13\sqrt{2} \approx 18.385$
43. $\dfrac{19\sqrt{2}}{2} \approx 13.435$ **45.** $\sqrt{10561}$ ft ≈ 102.767 ft
47. $h = 2\sqrt{3}$ ft ≈ 3.464 ft   **49.** $(0, -4), (0, 4)$ **51.** 🗒
53. -47 **54.** 5   **55.** $x(x - 3)(x + 3)$
56. $7a(a - 2)(a + 2)$ **57.** $\left\{-\frac{2}{3}, 4\right\}$ **58.** $\left\{-\frac{4}{3}, 10\right\}$
59. 🗒   **61.** $36\sqrt{3}$ cm^2; 62.354 cm^2 **63.** $d = s + s\sqrt{2}$
65. 5 gal. The total area of the doors and windows is
134 ft^2 or more.   **67.** 60.28 ft by 60.28 ft

Exercise Set 7.8, pp. 497–498

1. False   **3.** True   **5.** True   **7.** False   **9.** $6i$ **11.** $i\sqrt{13}$,
or $\sqrt{13}i$ **13.** $3i\sqrt{2}$, or $3\sqrt{2}i$ **15.** $i\sqrt{3}$, or $\sqrt{3}i$ **17.** $9i$
19. $-10i\sqrt{3}$, or $-10\sqrt{3}i$ **21.** $6 - 2i\sqrt{21}$, or $6 - 2\sqrt{21}i$
23. $\left(-2\sqrt{19} + 5\sqrt{5}\right)i$ **25.** $\left(3\sqrt{2} - 10\right)i$ **27.** $11 + 10i$
29. $4 + 5i$ **31.** $2 - i$ **33.** $-12 - 5i$ **35.** -42 **37.** -24
39. -18 **41.** $-\sqrt{10}$ **43.** $-3\sqrt{14}$ **45.** $-30 + 10i$
47. $-28 - 21i$ **49.** $1 + 5i$ **51.** $38 + 9i$ **53.** $2 - 46i$
55. $-11 - 16i$ **57.** $13 - 47i$ **59.** $12 - 16i$
61. $-5 + 12i$ **63.** $-5 - 12i$ **65.** $\frac{28}{17} - \frac{7}{17}i$ **67.** $\frac{6}{13} + \frac{4}{13}i$
69. $\frac{3}{17} + \frac{5}{17}i$ **71.** $-\frac{5}{6}i$ **73.** $-\frac{3}{4} - \frac{5}{4}i$ **75.** $1 - 2i$
77. $-\frac{23}{58} + \frac{43}{58}i$ **79.** $\frac{19}{29} - \frac{4}{29}i$ **81.** $\frac{6}{25} - \frac{17}{25}i$ **83.** $-i$ **85.** 1
87. -1 **89.** i **91.** -1 **93.** $-125i$ **95.** 0   **97.** 🗒
99.   **100.**

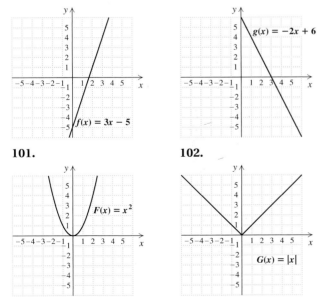

101.   **102.**

103. $-\frac{4}{3}, 7$ **104.** $\left\{x \mid -\frac{29}{3} < x < 5\right\}$, or $\left(-\frac{29}{3}, 5\right)$ **105.** 🗒
107. $-9 - 27i$ **109.** $50 - 120i$ **111.** $\frac{250}{41} + \frac{200}{41}i$ **113.** 8
115. $\frac{3}{5} + \frac{9}{5}i$ **117.** 1

Review Exercises: Chapter 7, pp. 502–503

1. True   **2.** False   **3.** True   **4.** True   **5.** True   **6.** True
7. True   **8.** False   **9.** $\frac{7}{3}$ **10.** -0.5 **11.** 5
12. $\left\{x \mid x \geq \frac{7}{2}\right\}$, or $\left[\frac{7}{2}, \infty\right)$ **13.** $5|t|$ **14.** $|c + 8|$
15. $|x - 3|$ **16.** $|2x + 1|$ **17.** -2 **18.** $-\dfrac{4x^2}{3}$
19. $|x^3 y^2|$, or $|x^3| y^2$ **20.** $2x^2$ **21.** $(5ab)^{4/3}$
22. $8a^4\sqrt{a}$ **23.** $x^3 y^5$ **24.** $\sqrt[3]{x^2 y}$ **25.** $\dfrac{1}{x^{2/5}}$ **26.** $7^{1/6}$
27. $f(x) = 5|x - 6|$ **28.** $\sqrt{6xy}$ **29.** $3a\sqrt[3]{a^2 b^2}$
30. $-6x^5 y^4\sqrt[3]{2x^2}$ **31.** $y\sqrt[3]{6}$ **32.** $\dfrac{5\sqrt{x}}{2}$ **33.** $\dfrac{2a^2\sqrt[4]{3a^3}}{c^2}$
34. $7\sqrt[3]{x}$ **35.** $\sqrt{3}$ **36.** $(2x + y^2)\sqrt[3]{x}$ **37.** $15\sqrt{2}$
38. $\sqrt{15} + 4\sqrt{6} - 6\sqrt{10} - 48$ **39.** $\sqrt[4]{x^3}$ **40.** $\sqrt[12]{x^5}$
41. $a^2 - 2a\sqrt{2} + 2$ **42.** $-4\sqrt{10} + 4\sqrt{15}$
43. $\dfrac{20}{\sqrt{10} + \sqrt{15}}$ **44.** 19   **45.** -126 **46.** 4   **47.** 14
48. $5\sqrt{2}$ cm; 7.071 cm   **49.** $\sqrt{32}$ ft; 5.657 ft
50. Short leg = 10; long leg = $10\sqrt{3} \approx 17.321$
51. $-2i\sqrt{2}$, or $-2\sqrt{2}i$ **52.** $-2 - 9i$ **53.** $6 + i$ **54.** 29
55. -1 **56.** 3   **57.** $-\frac{2}{5} + \frac{9}{10}i$ **58.** $9 - 12i$ **59.** $\frac{13}{25} - \frac{34}{25}i$
60. 🗒 A complex number $a + bi$ is real when $b = 0$. It is
imaginary when $b \neq 0$.
61. 🗒 An absolute-value sign must be used to simplify
$\sqrt[n]{x^n}$ when n is even, since x may be negative. If x is
negative while n is even, the radical expression cannot be
simplified to x, since $\sqrt[n]{x^n}$ represents the principal, or
positive, root. When n is odd, there is only one root, and it
will be positive or negative depending on the sign of x.
Thus there is no absolute-value sign when n is odd.
62. $\dfrac{2i}{3i}$; answers may vary

Test: Chapter 7, p. 504

1. [7.3] $5\sqrt{2}$ **2.** [7.4] $-\dfrac{2}{x^2}$ **3.** [7.1] $9|a|$
4. [7.1] $|x - 4|$ **5.** [7.3] $x^2 y\sqrt[5]{x^2 y^3}$ **6.** [7.4] $\left|\dfrac{5x}{6y^2}\right|$, or $\dfrac{5|x|}{6y^2}$
7. [7.3] $\sqrt[3]{15y^2 z}$ **8.** [7.4] $\sqrt[5]{x^2 y^2}$ **9.** [7.5] $xy\sqrt[4]{x}$
10. [7.5] $\sqrt[20]{a^3}$ **11.** [7.5] $6\sqrt{2}$ **12.** [7.5] $(x^2 + 3y)\sqrt{y}$
13. [7.5] $14 - 19\sqrt{x} - 3x$ **14.** [7.2] $(7xy)^{1/2}$
15. [7.2] $\sqrt[6]{(4a^3 b)^5}$ **16.** [7.1] $\{x \mid x \geq 5\}$, or $[5, \infty)$
17. [7.5] $27 + 10\sqrt{2}$ **18.** [7.5] $\dfrac{5\sqrt{3} - \sqrt{6}}{23}$ **19.** [7.6] 7

20. [7.6] No solution **21.** [7.7] $7\sqrt{3}$ cm or 12.124 cm (if the shorter leg is 7 cm); $\dfrac{7}{\sqrt{3}}$ cm or 4.041 cm (if the longer leg is 7 cm) **22.** [7.7] $\sqrt{10{,}600}$ ft; 102.956 ft
23. [7.8] $5i\sqrt{2}$, or $5\sqrt{2}i$ **24.** [7.8] $12 + 2i$ **25.** [7.8] -24
26. [7.8] $15 - 8i$ **27.** [7.8] $-\frac{11}{34} - \frac{7}{34}i$ **28.** [7.8] i
29. [7.6] 3 **30.** [7.8] $-\frac{17}{4}i$ **31.** [7.7] The isosceles right triangle is larger by 1.206 ft^2.

CHAPTER 8

Technology Connection, p. 515

1. The right-hand x-intercept should be an approximation of $4 + \sqrt{23}$. **2.** x-intercepts should be approximations of $-3 + \sqrt{5}$ and $-3 - \sqrt{5}$ for Example 7; approximations of $\left(-5 + \sqrt{37}\right)/2$ and $\left(-5 - \sqrt{37}\right)/2$ for Example 9(b)
3. Most graphing calculators can give only rational-number approximations of the two irrational solutions. An *exact* solution cannot be found with a graphing calculator.
4. The graph of $y = x^2 - 6x + 11$ has no x-intercepts.

Exercise Set 8.1, pp. 515–517

1. $\sqrt{k}$; $-\sqrt{k}$ **3.** $t + 3$; $t + 3$ **5.** 25; 5 **7.** $\pm\sqrt{5}$ **9.** $\pm\frac{4}{3}i$
11. $\pm\sqrt{\dfrac{7}{5}}$, or $\pm\dfrac{\sqrt{35}}{5}$ **13.** $-6, 8$ **15.** $13 \pm 3\sqrt{2}$
17. $-1 \pm 3i$ **19.** $-\dfrac{3}{4} \pm \dfrac{\sqrt{17}}{4}$, or $\dfrac{-3 \pm \sqrt{17}}{4}$
21. $-3, 13$ **23.** $1, 9$ **25.** $-4 \pm \sqrt{13}$ **27.** $-14, 0$
29. $x^2 + 16x + 64 = (x + 8)^2$
31. $t^2 - 10t + 25 = (t - 5)^2$ **33.** $x^2 + 3x + \frac{9}{4} = \left(x + \frac{3}{2}\right)^2$
35. $t^2 - 9t + \frac{81}{4} = \left(t - \frac{9}{2}\right)^2$ **37.** $x^2 + \frac{2}{5}x + \frac{1}{25} = \left(x + \frac{1}{5}\right)^2$
39. $t^2 - \frac{5}{6}t + \frac{25}{144} = \left(t - \frac{5}{12}\right)^2$ **41.** $-7, 1$ **43.** $4, 6$
45. $-9, -1$ **47.** $-4 \pm \sqrt{19}$
49. $\left(-3 - \sqrt{2}, 0\right), \left(-3 + \sqrt{2}, 0\right)$
51. $\left(-6 - \sqrt{11}, 0\right), \left(-6 + \sqrt{11}, 0\right)$
53. $\left(5 - \sqrt{47}, 0\right), \left(5 + \sqrt{47}, 0\right)$ **55.** $-\frac{4}{3}, -\frac{2}{3}$ **57.** $-\frac{1}{3}, 2$
59. $-\dfrac{2}{5} \pm \dfrac{\sqrt{19}}{5}$, or $\dfrac{-2 \pm \sqrt{19}}{5}$
61. $\left(-\dfrac{1}{4} - \dfrac{\sqrt{13}}{4}, 0\right), \left(-\dfrac{1}{4} + \dfrac{\sqrt{13}}{4}, 0\right)$, or $\left(\dfrac{-1 - \sqrt{13}}{4}, 0\right), \left(\dfrac{-1 + \sqrt{13}}{4}, 0\right)$
63. $\left(\dfrac{3}{4} - \dfrac{\sqrt{17}}{4}, 0\right), \left(\dfrac{3}{4} + \dfrac{\sqrt{17}}{4}, 0\right)$, or $\left(\dfrac{3 - \sqrt{17}}{4}, 0\right), \left(\dfrac{3 + \sqrt{17}}{4}, 0\right)$ **65.** 10% **67.** 18.75%
69. 4% **71.** About 8.1 sec **73.** About 9.5 sec **75.** 📝
77. 28 **78.** -92 **79.** $3\sqrt[3]{10}$ **80.** $4\sqrt{5}$ **81.** 5 **82.** 7

83. 📝 **85.** ±18 **87.** $-\frac{7}{2}, -\sqrt{5}, 0, \sqrt{5}, 8$
89. Barge: 8 km/h; fishing boat: 15 km/h
91. 📈 **93.** 📝, 📈

Exercise Set 8.2, pp. 522–524

1. True **3.** False **5.** False **7.** $-\dfrac{7}{2} \pm \dfrac{\sqrt{61}}{2}$ **9.** $3 \pm \sqrt{7}$
11. $-\dfrac{1}{2} \pm \dfrac{\sqrt{3}}{2}i$ **13.** $2 \pm 3i$ **15.** $3 \pm \sqrt{5}$
17. $-\dfrac{4}{3} \pm \dfrac{\sqrt{19}}{3}$ **19.** $-\dfrac{1}{2} \pm \dfrac{\sqrt{17}}{2}$ **21.** $-\dfrac{3}{8} \pm \dfrac{\sqrt{129}}{24}$
23. $\frac{2}{5}$ **25.** $-\dfrac{11}{8} \pm \dfrac{\sqrt{41}}{8}$ **27.** 5, 10 **29.** $\dfrac{13}{10} \pm \dfrac{\sqrt{509}}{10}$
31. $2 \pm \sqrt{5}i$ **33.** $2, -1 \pm \sqrt{3}i$ **35.** $\frac{2}{3}, 1$ **37.** $5 \pm \sqrt{53}$
39. $\dfrac{7}{2} \pm \dfrac{\sqrt{85}}{2}$ **41.** $\frac{3}{2}, 6$ **43.** $-5.31662479, 1.31662479$
45. $0.7639320225, 5.236067978$ **47.** -1.265564437,
2.765564437 **49.** 📝 **51.** Kenyan: 30 lb; Kona: 20 lb
52. Cream-filled: 46; glazed: 44 **53.** $9a^2b^3\sqrt{2a}$
54. $4a^2b^3\sqrt{6}$ **55.** $\dfrac{3(x + 1)}{3x + 1}$ **56.** $\dfrac{4b}{3ab^2 - 4a^2}$ **57.** 📝
59. $(-2, 0), (1, 0)$ **61.** $4 - 2\sqrt{2}, 4 + 2\sqrt{2}$
63. $-1.1792101, 0.3392101$ **65.** $\dfrac{-5\sqrt{2} \pm \sqrt{34}}{4}$ **67.** $\frac{1}{2}$
69. 📈

Exercise Set 8.3, pp. 528–531

1. First part: 60 mph; second part: 50 mph **3.** 40 mph
5. Cessna: 150 mph, Beechcraft: 200 mph; or
Cessna: 200 mph, Beechcraft: 250 mph
7. To Hillsboro: 10 mph; return trip: 4 mph
9. About 14 mph **11.** 12 hr **13.** About 3.24 mph
15. $r = \dfrac{1}{2}\sqrt{\dfrac{A}{\pi}}$ **17.** $r = \dfrac{-\pi h + \sqrt{\pi^2 h^2 + 2\pi A}}{2\pi}$
19. $s = \sqrt{\dfrac{kQ_1Q_2}{N}}$ **21.** $g = \dfrac{4\pi^2 l}{T^2}$
23. $c = \sqrt{d^2 - a^2 - b^2}$ **25.** $t = \dfrac{-v_0 + \sqrt{v_0^2 + 2gs}}{g}$
27. $n = \dfrac{1 + \sqrt{1 + 8N}}{2}$ **29.** $h = \dfrac{V^2}{12.25}$
31. $t = \dfrac{-b \pm \sqrt{b^2 - 4ac}}{2a}$ **33. (a)** 10.1 sec; **(b)** 7.49 sec;
(c) 272.5 m **35.** 2.9 sec **37.** 0.968 sec **39.** 2.5 m/sec
41. 7% **43.** 📝 **45.** -104 **46.** $2i\sqrt{11}$ **47.** $\dfrac{x + y}{2}$
48. $\dfrac{a^2 - b^2}{b}$ **49.** $\dfrac{1 + \sqrt{5}}{2}$ **50.** $\dfrac{1 - \sqrt{7}}{5}$ **51.** 📝
53. $t = \dfrac{-10.2 + 6\sqrt{-A^2 + 13A - 39.36}}{A - 6.5}$

55. $\pm\sqrt{2}$ **57.** $l = \dfrac{w + w\sqrt{5}}{2}$

59. $n = \pm\sqrt{\dfrac{r^2 \pm \sqrt{r^4 + 4m^4r^2p - 4mp}}{2m}}$

61. $A(S) = \dfrac{\pi S}{6}$

Technology Connection, p. 534

1. $(-0.4, 0)$ is the other x-intercept of $y = 5x^2 - 13x - 6$.
2. The x-intercepts of $y = x^2 - 175$ are $(-13.22875656, 0)$ and $(13.22875656, 0)$, or $(-5\sqrt{7}, 0)$ and $(5\sqrt{7}, 0)$.
3. The x-intercepts of $y = x^3 + 3x^2 - 4x$ are $(-4, 0)$, $(0, 0)$, and $(1, 0)$.

Exercise Set 8.4, pp. 535–537

1. Discriminant **3.** Two **5.** Rational **7.** Two irrational
9. Two imaginary **11.** Two irrational **13.** Two rational
15. Two imaginary **17.** One rational **19.** Two rational
21. Two rational **23.** Two irrational **25.** Two imaginary
27. Two irrational **29.** $x^2 + 4x - 21 = 0$
31. $x^2 - 6x + 9 = 0$ **33.** $x^2 + 4x + 3 = 0$
35. $4x^2 - 23x + 15 = 0$ **37.** $8x^2 + 6x + 1 = 0$
39. $x^2 - 2x - 0.96 = 0$ **41.** $x^2 - 3 = 0$
43. $x^2 - 20 = 0$ **45.** $x^2 + 16 = 0$
47. $x^2 - 4x + 53 = 0$ **49.** $x^2 - 6x - 5 = 0$
51. $3x^2 - 6x - 4 = 0$ **53.** $x^3 - 4x^2 - 7x + 10 = 0$
55. $x^3 - 2x^2 - 3x = 0$ **57.** 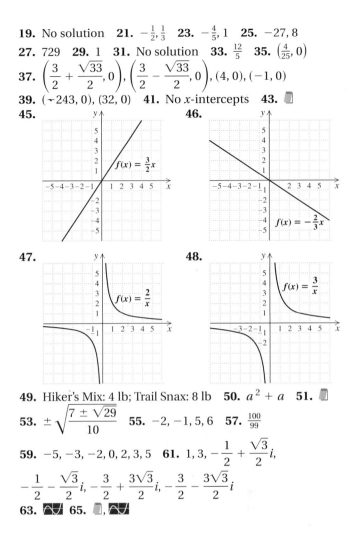 **59.** $81a^8$ **60.** $16x^6$
61. $(-1, 0), (8, 0)$ **62.** $(2, 0), (4, 0)$ **63.** 6 commercials
64.

$y = -\dfrac{3}{7}x + 4$

65. 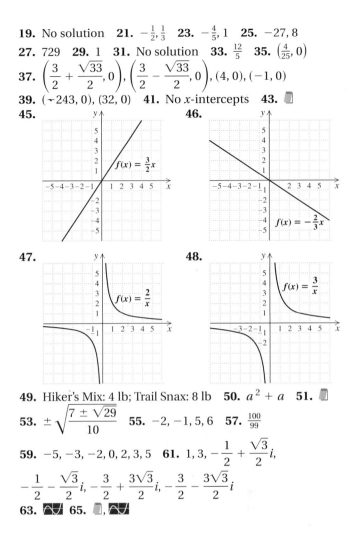 **67.** $a = 1, b = 2, c = -3$ **69. (a)** $-\dfrac{3}{5}$; **(b)** $-\dfrac{1}{3}$
71. (a) $9 + 9i$; **(b)** $3 + 3i$
73. The solutions of $ax^2 + bx + c = 0$ are
$x = \dfrac{-b \pm \sqrt{b^2 - 4ac}}{2a}$. When there is just one solution,
$b^2 - 4ac$ must be 0, so $x = \dfrac{-b \pm 0}{2a} = \dfrac{-b}{2a}$.

75. $a = 8, b = 20, c = -12$
77. $x^4 - 8x^3 + 21x^2 - 2x - 52 = 0$ **79.**

Exercise Set 8.5, pp. 541–543

1. (f) **3. (h)** **5. (g)** **7. (e)** **9.** $\pm 1, \pm 2$ **11.** $\pm\sqrt{5}, \pm 2$
13. $\pm\dfrac{\sqrt{3}}{2}, \pm 2$ **15.** $8 + 2\sqrt{7}$ **17.** $\pm 2\sqrt{2}, \pm 3$

19. No solution **21.** $-\dfrac{1}{2}, \dfrac{1}{3}$ **23.** $-\dfrac{4}{5}, 1$ **25.** $-27, 8$
27. 729 **29.** 1 **31.** No solution **33.** $\dfrac{12}{5}$ **35.** $\left(\dfrac{4}{25}, 0\right)$
37. $\left(\dfrac{3}{2} + \dfrac{\sqrt{33}}{2}, 0\right), \left(\dfrac{3}{2} - \dfrac{\sqrt{33}}{2}, 0\right), (4, 0), (-1, 0)$
39. $(-243, 0), (32, 0)$ **41.** No x-intercepts **43.**
45. **46.**

47. **48.**

49. Hiker's Mix: 4 lb; Trail Snax: 8 lb **50.** $a^2 + a$ **51.**
53. $\pm\sqrt{\dfrac{7 \pm \sqrt{29}}{10}}$ **55.** $-2, -1, 5, 6$ **57.** $\dfrac{100}{99}$
59. $-5, -3, -2, 0, 2, 3, 5$ **61.** $1, 3, -\dfrac{1}{2} + \dfrac{\sqrt{3}}{2}i,$
$-\dfrac{1}{2} - \dfrac{\sqrt{3}}{2}i, -\dfrac{3}{2} + \dfrac{3\sqrt{3}}{2}i, -\dfrac{3}{2} - \dfrac{3\sqrt{3}}{2}i$
63.   **65.** ,

Technology Connection, p. 544

1. The graphs of $y_1, y_2,$ and y_3 open upward. The graphs of $y_4, y_5,$ and y_6 open downward. The graph of y_1 is wider than the graph of y_2. The graph of y_3 is narrower than the graph of y_2. Similarly, the graph of y_4 is wider than the graph of y_5, and the graph of y_6 is narrower than the graph of y_5. **2.** If A is positive, the graph opens upward. If A is negative, the graph opens downward. Compared with the graph of $y = x^2$, the graph of $y = Ax^2$ is wider if $|A| < 1$ and narrower if $|A| > 1$.

Technology Connection, p. 546

1. Compared with the graph of $y = ax^2$, the graph of $y = a(x - h)^2$ is shifted left or right. It is shifted left if h is negative and right if h is positive. **2.** The value of A makes the graph wider or narrower, and makes the graph open downward if A is negative. The value of B shifts the graph left or right.

Technology Connection, p. 547

1. The graph of y_2 looks like the graph of y_1 shifted up 2 units, and the graph of y_3 looks like the graph of y_1 shifted down 4 units.   **2.** Compared with the graph of $y = a(x - h)^2$, the graph of $y = a(x - h)^2 + k$ is shifted up or down. It is shifted down if k is negative and up if k is positive.   **3.** The value of A makes the graph wider or narrower, and makes the graph open downward if A is negative. The value of B shifts the graph left or right. The value of C shifts the graph up or down.

Exercise Set 8.6, pp. 549–551

1. (h)   **3.** (f)   **5.** (b)   **7.** (e)

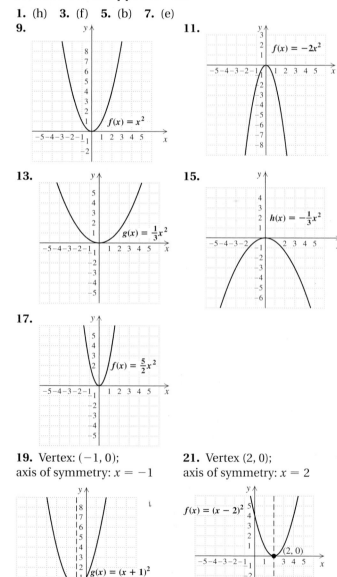

9.

11.

13.

15.

17.

19. Vertex: $(-1, 0)$; axis of symmetry: $x = -1$

21. Vertex $(2, 0)$; axis of symmetry: $x = 2$

23. Vertex: $(3, 0)$; axis of symmetry: $x = 3$

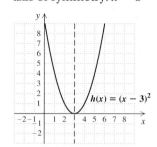

25. Vertex: $(-1, 0)$; axis of symmetry: $x = -1$

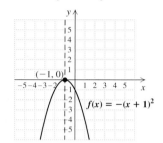

27. Vertex: $(2, 0)$; axis of symmetry: $x = 2$

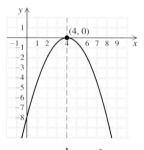

29. Vertex: $(-1, 0)$; axis of symmetry: $x = -1$

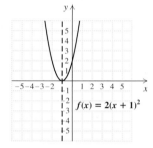

31. Vertex: $(4, 0)$; axis of symmetry: $x = 4$

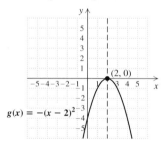

$$h(x) = -\tfrac{1}{2}(x - 4)^2$$

33. Vertex: $(1, 0)$; axis of symmetry: $x = 1$

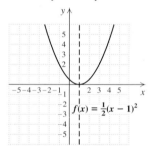

35. Vertex: $(-5, 0)$; axis of symmetry: $x = -5$

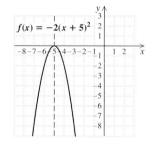

37. Vertex: $\left(\tfrac{1}{2}, 0\right)$; axis of symmetry: $x = \tfrac{1}{2}$

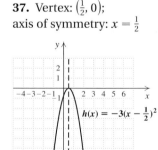

39. Vertex: (5, 2);
axis of symmetry: $x = 5$;
minimum: 2

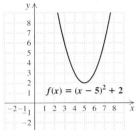

$f(x) = (x - 5)^2 + 2$

41. Vertex: $(-1, -3)$;
axis of symmetry: $x = -1$;
minimum: -3

$f(x) = (x + 1)^2 - 3$

43. Vertex: $(-4, 1)$;
axis of symmetry: $x = -4$;
minimum: 1

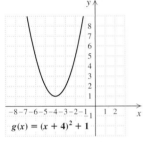

$g(x) = (x + 4)^2 + 1$

45. Vertex: $(1, -3)$;
axis of symmetry: $x = 1$;
maximum: -3

$h(x) = -2(x - 1)^2 - 3$

47. Vertex: $(-4, 1)$; axis of
symmetry: $x = -4$;
minimum: 1

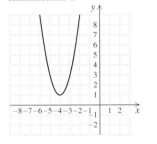

$f(x) = 2(x + 4)^2 + 1$

49. Vertex: (1, 4); axis of
symmetry: $x = 1$;
maximum: 4

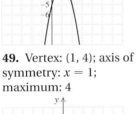

$g(x) = -\frac{3}{2}(x - 1)^2 + 4$

51. Vertex: (8, 7); axis of symmetry: $x = 8$; minimum: 7
53. Vertex: $(-6, 11)$; axis of symmetry: $x = -6$;
maximum: 11   **55.** Vertex: $\left(-\frac{1}{4}, -13\right)$; axis of
symmetry: $x = -\frac{1}{4}$; minimum: -13
57. Vertex: $(-4.58, 65\pi)$; axis of
symmetry: $x = -4.58$; minimum: 65π **59.**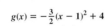
61.

62.

63. $(-5, -1)$ **64.** $(-1, 2)$
65. $x^2 + 5x + \frac{25}{4} = \left(x + \frac{5}{2}\right)^2$
66. $x^2 - 9x + \frac{81}{4} = \left(x - \frac{9}{2}\right)^2$ **67.**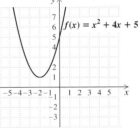
69. $f(x) = \frac{3}{5}(x - 4)^2 + 1$ **71.** $f(x) = \frac{3}{5}(x - 3)^2 - 1$
73. $f(x) = \frac{3}{5}(x + 2)^2 - 5$ **75.** $f(x) = 2(x - 2)^2$
77. $g(x) = -2x^2 + 3$ **79.** $F(x) = 3(x - 5)^2 + 1$
81.

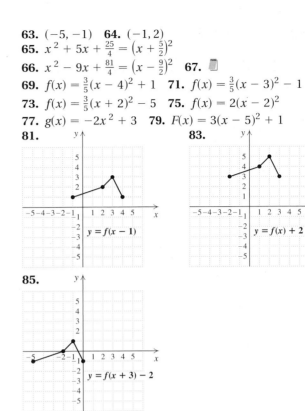

$y = f(x - 1)$

83.

$y = f(x) + 2$

85.

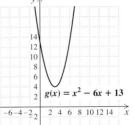

$y = f(x + 3) - 2$

87. 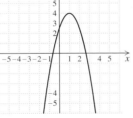   **89.**

Exercise Set 8.7, pp. 556–557

1. 9   **3.** 9   **5.** 3   **7.** $\frac{5}{2}, (-4)$
9. (a) Vertex: $(-2, 1)$; axis of symmetry: $x = -2$;
(b)

$f(x) = x^2 + 4x + 5$

11. (a) Vertex: (3, 4); axis of symmetry: $x = 3$;
(b)

$g(x) = x^2 - 6x + 13$

13. (a) Vertex: $(-4, 4)$; axis of symmetry: $x = -4$;
(b)

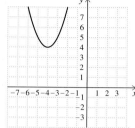

$$f(x) = x^2 + 8x + 20$$

15. (a) Vertex: $(4, -7)$; axis of symmetry: $x = 4$;
(b)

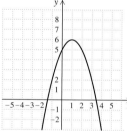

$$h(x) = 2x^2 - 16x + 25$$

17. (a) Vertex: $(1, 6)$; axis of symmetry: $x = 1$;
(b)

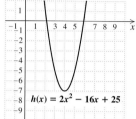

$$f(x) = -x^2 + 2x + 5$$

19. (a) Vertex: $\left(-\frac{3}{2}, -\frac{49}{4}\right)$; axis of symmetry: $x = -\frac{3}{2}$;
(b)

$$g(x) = x^2 + 3x - 10$$

21. (a) Vertex: $(4, 2)$; axis of symmetry: $x = 4$;
(b)

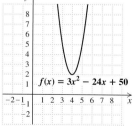

$$f(x) = 3x^2 - 24x + 50$$

23. (a) Vertex: $\left(-\frac{7}{2}, -\frac{49}{4}\right)$; axis of symmetry: $x = -\frac{7}{2}$;
(b)

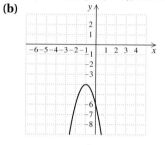

$$h(x) = x^2 + 7x$$

25. (a) Vertex: $(-1, -4)$; axis of symmetry: $x = -1$;
(b)

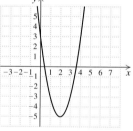

$$f(x) = -2x^2 - 4x - 6$$

27. (a) Vertex: $(2, -5)$; axis of symmetry: $x = 2$;
(b)

$$g(x) = 2x^2 - 8x + 3$$

29. (a) Vertex: $\left(\frac{5}{6}, \frac{1}{12}\right)$; axis of symmetry: $x = \frac{5}{6}$;
(b)

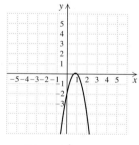

$$f(x) = -3x^2 + 5x - 2$$

31. (a) Vertex: $\left(-4, -\frac{5}{3}\right)$; axis of symmetry: $x = -4$;
(b)

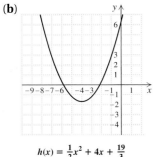

$$h(x) = \tfrac{1}{2}x^2 + 4x + \tfrac{19}{3}$$

33. $(3 - \sqrt{6}, 0), (3 + \sqrt{6}, 0); (0, 3)$ **35.** $(-1, 0), (3, 0);$
$(0, 3)$ **37.** $(0, 0), (9, 0); (0, 0)$ **39.** $(2, 0); (0, -4)$ **41.** No
x-intercept; $(0, 6)$ **43.** **45.** $(2, -2)$ **46.** $(7, 1)$
47. $(3, 2, 1)$ **48.** $(1, -3, 2)$ **49.** 5 **50.** 4 **51.** 🗒
53. (a) Minimum: -6.953660714; (b) $(-1.056433682, 0),$
$(2.413576539, 0); (0, -5.89)$ **55.** (a) $-2.4, 3.4;$
(b) $-1.3, 2.3$ **57.** $f(x) = m\left(x - \dfrac{n}{2m}\right)^2 + \dfrac{4mp - n^2}{4m}$
59. $f(x) = \frac{5}{16}x^2 - \frac{15}{8}x - \frac{35}{16}$, or $f(x) = \frac{5}{16}(x - 3)^2 - 5$
61.   **63.**

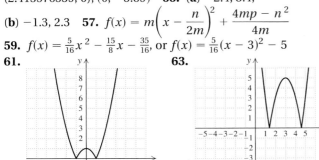

$f(x) = |x^2 - 1|$ $f(x) = |2(x - 3)^2 - 5|$

Technology Connection, p. 562

1. 32.7%

Exercise Set 8.8, pp. 563–568

1. (e) **3.** (c) **5.** (d) **7.** 11 days after the concert was
announced; about 62 tickets **9.** \$120/Dobro; 350 Dobros
11. 32 in. by 32 in. **13.** 450 ft^2; 15 ft by 30 ft (The house
serves as a 30-ft side.) **15.** 3.5 in. **17.** 81; 9 and 9
19. -16; 4 and -4 **21.** 25; -5 and -5
23. $f(x) = mx + b$ **25.** $f(x) = ax^2 + bx + c, a < 0$
27. $f(x) = ax^2 + bx + c, a > 0$
29. $f(x) = ax^2 + bx + c, a < 0$
31. $f(x) = ax^2 + bx + c, a > 0$ **33.** $f(x) = mx + b$
35. $f(x) = 2x^2 + 3x - 1$ **37.** $f(x) = -\frac{1}{4}x^2 + 3x - 5$
39. (a) $A(s) = \frac{3}{16}s^2 - \frac{135}{4}s + 1750$; (b) about 531 accidents
41. $h(d) = -0.0068d^2 + 0.8571d$ **43.** 🗒
45. $\dfrac{x - 9}{(x + 9)(x + 7)}$ **46.** $\dfrac{(x - 3)(x + 1)}{(x - 7)(x + 3)}$

47. $\dfrac{(t + 2)(t + 8)}{(t - 3)(t + 1)}$ **48.** $\dfrac{2t(t + 2)}{(t - 7)(t - 3)(t + 7)}$
49. $\{x \mid x < 8\}$, or $(-\infty, 8)$ **50.** $\{x \mid x \geq 10\}$, or $[10, \infty)$
51. 🗒 **53.** 158 ft **55.** \$15 **57.** The radius of the
circular portion of the window and the height of the
rectangular portion should each be $\dfrac{24}{\pi + 4}$ ft.
59. (a) $c(x) = 261.875x^2 - 882.5642857x + 2134.571429;$
(b) 55,053 cars

Technology Connection, p. 573

1. $\{x \mid -0.78 \leq x \leq 1.59\}$, or $[-0.78, 1.59]$
2. $\{x \mid x \leq -0.21 \text{ or } x \geq 2.47\}$, or $(-\infty, -0.21] \cup [2.47, \infty)$
3. $\{x \mid x < -1.26 \text{ or } x > 2.33\}$, or $(-\infty, -1.26) \cup (2.33, \infty)$
4. $\{x \mid x > -1.37\}$, or $(-1.37, \infty)$

Exercise Set 8.9, pp. 576–577

1. True **3.** True **5.** True **7.** False
9. $(-4, 3)$, or $\{x \mid -4 < x < 3\}$
11. $(-\infty, -7] \cup [2, \infty)$, or $\{x \mid x \leq -7 \text{ or } x \geq 2\}$
13. $(-\infty, -1) \cup (2, \infty)$, or $\{x \mid x < -1 \text{ or } x > 2\}$ **15.** $\varnothing$
17. $(-2, 6)$, or $\{x \mid -2 < x < 6\}$
19. $(-\infty, -2) \cup (0, 2)$, or $\{x \mid x < -2 \text{ or } 0 < x < 2\}$
21. $[-2, 1] \cup [4, \infty)$, or $\{x \mid -2 \leq x \leq 1 \text{ or } x \geq 4\}$
23. $[-2, 2]$, or $\{x \mid -2 \leq x \leq 2\}$ **25.** $(-1, 2) \cup (3, \infty)$, or
$\{x \mid -1 < x < 2 \text{ or } x > 3\}$ **27.** $(-\infty, 0] \cup [2, 5]$, or
$\{x \mid x \leq 0 \text{ or } 2 \leq x \leq 5\}$ **29.** $(-\infty, -5)$, or $\{x \mid x < -5\}$
31. $(-\infty, -1] \cup (3, \infty)$, or $\{x \mid x \leq -1 \text{ or } x > 3\}$
33. $(-\infty, -6)$, or $\{x \mid x < -6\}$ **35.** $(-\infty, -1] \cup [2, 5)$, or
$\{x \mid x \leq -1 \text{ or } 2 \leq x < 5\}$ **37.** $(-\infty, -3) \cup [0, \infty)$, or
$\{x \mid x < -3 \text{ or } x \geq 0\}$ **39.** $(0, \infty)$, or $\{x \mid x > 0\}$
41. $(-\infty, -4) \cup [1, 3)$, or $\{x \mid x < -4 \text{ or } 1 \leq x < 3\}$
43. $\left(-\frac{3}{4}, \frac{5}{2}\right]$, or $\left\{x \mid -\frac{3}{4} < x \leq \frac{5}{2}\right\}$ **45.** $(-\infty, 2) \cup [3, \infty)$, or
$\{x \mid x < 2 \text{ or } x \geq 3\}$ **47.** 🗒 **49.** $8a^9b^6c^{12}$ **50.** $25a^8b^{14}$
51. $\frac{1}{32}$ **52.** $\frac{1}{81}$ **53.** $3a^2 + 6a + 3$ **54.** $5a + 7$ **55.** 🗒
57. $\left(-1 - \sqrt{6}, -1 + \sqrt{6}\right)$, or
$\left\{x \mid -1 - \sqrt{6} < x < -1 + \sqrt{6}\right\}$
59. $\{0\}$ **61.** (a) $(10, 200)$, or $\{x \mid 10 < x < 200\}$;
(b) $[0, 10) \cup (200, \infty)$, or $\{x \mid 0 \leq x < 10 \text{ or } x > 200\}$
63. $\{n \mid n \text{ is an integer } and \ 12 \leq n \leq 25\}$ **65.** $f(x) = 0$ for
$x = -2, 1, 3; f(x) < 0$ for $(-\infty, -2) \cup (1, 3)$, or
$\{x \mid x < -2 \text{ or } 1 < x < 3\}; f(x) > 0$ for $(-2, 1) \cup (3, \infty)$, or
$\{x \mid -2 < x < 1 \text{ or } x > 3\}$ **67.** $f(x)$ has no zeros; $f(x) < 0$
for $(-\infty, 0)$, or $\{x \mid x < 0\}; f(x) > 0$ for $(0, \infty)$, or $\{x \mid x > 0\}$
69. $f(x) = 0$ for $x = -1, 0; f(x) < 0$ for
$(-\infty, -3) \cup (-1, 0)$,
or $\{x \mid x < -3 \text{ or } -1 < x < 0\}; f(x) > 0$ for
$(-3, -1) \cup (0, 2) \cup (2, \infty)$, or
$\{x \mid -3 < x < -1 \text{ or } 0 < x < 2 \text{ or } x > 2\}$ **71.** 📈

Review Exercises: Chapter 8, pp. 580–581

1. False   **2.** True   **3.** True   **4.** True   **5.** True   **6.** False
7. True   **8.** True   **9.** True   **10.** True   **11.** $\pm\frac{3}{2}$ **12.** $0, -\frac{3}{4}$

13. $3, 9$ **14.** $2 \pm 2i$ **15.** $3, 5$ **16.** $-\frac{9}{2} \pm \frac{\sqrt{85}}{2}$

17. $-0.3722813233, 5.3722813233$ **18.** $-\frac{1}{4}, 1$
19. $x^2 - 12x + 36 = (x - 6)^2$
20. $x^2 + \frac{3}{5}x + \frac{9}{100} = \left(x + \frac{3}{10}\right)^2$ **21.** $3 \pm 2\sqrt{2}$
22. 10%   **23.** 6.7 sec   **24.** About 153 mph   **25.** 6 hr
26. Two irrational   **27.** Two imaginary   **28.** $x^2 - 5 = 0$
29. $x^2 + 8x + 16 = 0$ **30.** $(-3, 0), (-2, 0), (2, 0), (3, 0)$
31. $-5, 3$ **32.** $\pm\sqrt{2}, \pm\sqrt{7}$
33.

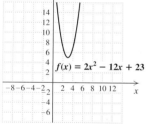

$f(x) = -3(x + 2)^2 + 4$
Maximum: 4

34. (a) Vertex: $(3, 5)$; axis of symmetry: $x = 3$;
(b)

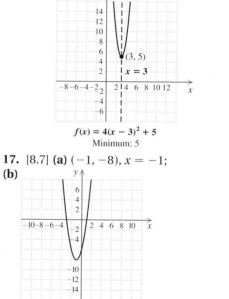

$f(x) = 2x^2 - 12x + 23$

35. $(2, 0), (7, 0); (0, 14)$ **36.** $p = \dfrac{9\pi^2}{N^2}$

37. $T = \dfrac{1 \pm \sqrt{1 + 24A}}{6}$ **38.** Quadratic   **39.** Linear
40. 225 ft²; 15 ft by 15 ft
41. (a) $f(x) = -42x^2 + 167x + 281$;
(b) 277 million books
42. $(-1, 0) \cup (3, \infty)$, or $\{x \mid -1 < x < 0 \ or \ x > 3\}$
43. $(-3, 5]$, or $\{x \mid -3 < x \le 5\}$
44. ▨ The x-coordinate of the maximum or minimum
point lies halfway between the x-coordinates of the
x-intercepts.
45. ▨ Yes; if the discriminant is a perfect square, then the
solutions are rational numbers, p/q and r/s. (Note that if
the discriminant is 0, then $p/q = r/s$.) Then the equation
can be written in factored form, $(qx - p)(sx - r) = 0$.
46. ▨ Four; let $u = x^2$. Then $au^2 + bu + c = 0$ has at
most two solutions, $u = m$ and $u = n$. Now substitute x^2

for u and obtain $x^2 = m$ or $x^2 = n$. These equations yield
the solutions $x = \pm\sqrt{m}$ and $x = \pm\sqrt{n}$. When $m \ne n$, the
maximum number of solutions, four, occurs.
47. ▨ Completing the square was used to solve quadratic
equations and to graph quadratic functions by rewriting
the function in the form $f(x) = a(x - h)^2 + k$.
48. $f(x) = \frac{7}{15}x^2 - \frac{14}{15}x - 7$ **49.** $h = 60, k = 60$
50. 18, 324

Test: Chapter 8, p. 582

1. [8.1] $\pm\dfrac{\sqrt{11}}{2}$ **2.** [8.2] 2, 9   **3.** [8.2] $-1 \pm \sqrt{2}i$

4. [8.2] $1 \pm \sqrt{6}$ **5.** [8.5] $-2, \frac{2}{3}$
6. [8.2] $-4.192582404, 1.192582404$ **7.** [8.2] $-\frac{3}{4}, \frac{7}{3}$
8. [8.1] $x^2 - 16x + 64 = (x - 8)^2$
9. [8.1] $x^2 + \frac{2}{7}x + \frac{1}{49} = \left(x + \frac{1}{7}\right)^2$ **10.** [8.1] $-5 \pm \sqrt{10}$
11. [8.3] 16 km/h   **12.** [8.3] 2 hr   **13.** [8.4] Two imaginary
14. [8.4] $3x^2 + 5x - 2 = 0$ **15.** [8.5] $(-4, 0), (4, 0)$
16. [8.6]

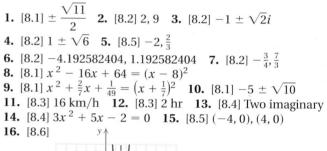

$f(x) = 4(x - 3)^2 + 5$
Minimum: 5

17. [8.7] **(a)** $(-1, -8), x = -1$;
(b)

$f(x) = 2x^2 + 4x - 6$

18. [8.7] $(-2, 0), (3, 0); (0, -6)$ **19.** [8.3] $r = \sqrt{\dfrac{3V}{\pi} - R^2}$

20. [8.8] Neither   **21.** [8.8] Minimum: \$129/cap when
325 caps are built   **22.** [8.8] $f(x) = \frac{1}{5}x^2 - \frac{3}{5}x$
23. [8.9] $(-6, 1)$, or $\{x \mid -6 < x < 1\}$
24. [8.9] $[-1, 0) \cup [1, \infty)$, or $\{x \mid -1 \le x < 0 \ or \ x \ge 1\}$
25. [8.4] $\frac{1}{2}$ **26.** [8.4] $x^4 - 14x^3 + 67x^2 - 114x + 26 = 0$;
answers may vary
27. [8.4] $x^6 - 10x^5 + 20x^4 + 50x^3 - 119x^2 - 60x + 150 = 0$; answers may vary

CHAPTER 9

Technology Connection, p. 586

1. To check $(f \circ g)(x)$, we use a table to show that $y_2 = y_3$.

$y_1 = x - 1$; $y_2 = \sqrt{y_1}$;
$y_3 = \sqrt{x - 1}$

X	Y2	Y3
1	0	0
2	1	1
3	1.4142	1.4142
4	1.7321	1.7321
5	2	2
6	2.2361	2.2361
7	2.4495	2.4495
X =		

A similar table shows that for $y_2 = \sqrt{x}$ and $y_4 = y_2(y_1)$, we have $y_3 = y_4$. The check for $(g \circ f)(x)$ is similar. A graph can also be used.

Technology Connection, p. 592

1. Graph each pair of functions in a square window along with the line $y = x$ and determine whether the first two functions are reflections of each other across $y = x$. For further verification, examine a table of values for each pair of functions.   **2.** Yes; most graphing calculators do not require that the inverse relation be a function.

Exercise Set 9.1, p. 593–595

1. True   **3.** False   **5.** False   **7.** True
9. $(f \circ g)(1) = 2$; $(g \circ f)(1) = 1$;
$(f \circ g)(x) = 4x^2 - 12x + 10$;
$(g \circ f)(x) = 2x^2 - 1$ **11.** $(f \circ g)(1) = -8$; $(g \circ f)(1) = 1$;
$(f \circ g)(x) = 2x^2 - 10$; $(g \circ f)(x) = 2x^2 - 12x + 11$
13. $(f \circ g)(1) = 8$; $(g \circ f)(1) = \frac{1}{64}$;
$(f \circ g)(x) = \frac{1}{x^2} + 7$; $(g \circ f)(x) = \frac{1}{(x + 7)^2}$
15. $(f \circ g)(1) = 2$; $(g \circ f)(1) = 4$;
$(f \circ g)(x) = \sqrt{x + 3}$; $(g \circ f)(x) = \sqrt{x} + 3$
17. $(f \circ g)(1) = 2$; $(g \circ f)(1) = \frac{1}{2}$; $(f \circ g)(x) = \sqrt{\frac{4}{x}}$;
$(g \circ f)(x) = \frac{1}{\sqrt{4x}}$ **19.** $(f \circ g)(1) = 4$; $(g \circ f)(1) = 2$;
$(f \circ g)(x) = x + 3$; $(g \circ f)(x) = \sqrt{x^2 + 3}$
21. $f(x) = x^2$; $g(x) = 7 + 5x$
23. $f(x) = \sqrt{x}$; $g(x) = 2x + 7$ **25.** $f(x) = \frac{2}{x}$; $g(x) = x - 3$
27. Yes   **29.** No   **31.** Yes   **33.** No   **35.** (a) Yes;
(b) $f^{-1}(x) = x - 4$ **37.** (a) Yes; (b) $f^{-1}(x) = \frac{x}{2}$
39. (a) Yes; (b) $g^{-1}(x) = \frac{x + 1}{3}$ **41.** (a) Yes;
(b) $f^{-1}(x) = 2x - 2$ **43.** (a) No   **45.** (a) Yes;
(b) $h^{-1}(x) = \frac{x - 4}{-2}$ **47.** (a) Yes; (b) $f^{-1}(x) = \frac{1}{x}$

49. (a) No   **51.** (a) Yes; (b) $f^{-1}(x) = \frac{3x - 1}{2}$ **53.** (a) Yes;
(b) $f^{-1}(x) = \sqrt[3]{x + 5}$ **55.** (a) Yes; (b) $g^{-1}(x) = \sqrt[3]{x} + 2$
57. (a) Yes; (b) $f^{-1}(x) = x^2, x \geq 0$

59.   **61.**

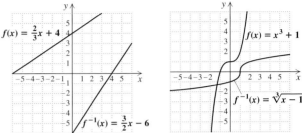

63.   **65.**

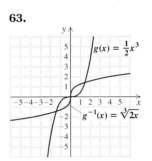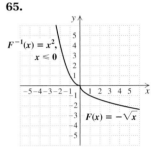

67.

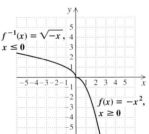

69. **(1)** $(f^{-1} \circ f)(x) = f^{-1}(f(x))$
$= f^{-1}(\sqrt[3]{x - 4}) = (\sqrt[3]{x - 4})^3 + 4$
$= x - 4 + 4 = x$;
(2) $(f \circ f^{-1})(x) = f(f^{-1}(x))$
$= f(x^3 + 4) = \sqrt[3]{x^3 + 4 - 4}$
$= \sqrt[3]{x^3} = x$

71. **(1)** $(f^{-1} \circ f)(x) = f^{-1}(f(x)) = f^{-1}\left(\frac{1 - x}{x}\right)$
$= \cfrac{1}{\left(\cfrac{1 - x}{x}\right) + 1}$
$= \cfrac{1}{\cfrac{1 - x + x}{x}}$
$= x$;

(2) $(f \circ f^{-1})(x) = f(f^{-1}(x)) = f\left(\dfrac{1}{x+1}\right)$

$$= \dfrac{1 - \left(\dfrac{1}{x+1}\right)}{\left(\dfrac{1}{x+1}\right)}$$

$$= \dfrac{\dfrac{x+1-1}{x+1}}{\dfrac{1}{x+1}} = x$$

73. **(a)** 40, 44, 52, 60; **(b)** $f^{-1}(x) = (x - 24)/2$;
(c) 8, 10, 14, 18 **75.** 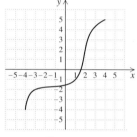 **77.** $a^{13}b^{13}$ **78.** $x^{10}y^{12}$

79. 81 **80.** 125 **81.** $y = \frac{3}{2}(x + 7)$ **82.** $y = \dfrac{10 - x}{3}$
83.

85.

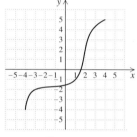

87. $g(x) = \dfrac{x}{2} + 20$ **89.**

91. Suppose that $h(x) = (f \circ g)(x)$. First, note that for
$I(x) = x$, $(f \circ I)(x) = f(I(x)) = f(x)$ for any function f.
(i) $((g^{-1} \circ f^{-1}) \circ h)(x) = ((g^{-1} \circ f^{-1}) \circ (f \circ g))(x)$
$\qquad\qquad = ((g^{-1} \circ (f^{-1} \circ f)) \circ g)(x)$
$\qquad\qquad = ((g^{-1} \circ I) \circ g)(x)$
$\qquad\qquad = (g^{-1} \circ g)(x) = x$
(ii) $(h \circ (g^{-1} \circ f^{-1}))(x) = ((f \circ g) \circ (g^{-1} \circ f^{-1}))(x)$
$\qquad\qquad = ((f \circ (g \circ g^{-1})) \circ f^{-1})(x)$
$\qquad\qquad = ((f \circ I) \circ f^{-1})(x)$
$\qquad\qquad = (f \circ f^{-1})(x) = x.$
Therefore, $(g^{-1} \circ f^{-1})(x) = h^{-1}(x).$ **93.** Yes **95.** No
97. **(1)** C; **(2)** A; **(3)** B; **(4)** D **99.**

Technology Connection, p. 598

1. $y_1 = \left(\frac{5}{2}\right)^x$; $y_2 = \left(\frac{2}{5}\right)^x$ **2.** $y_1 = 3.2^x$; $y_2 = 3.2^{-x}$

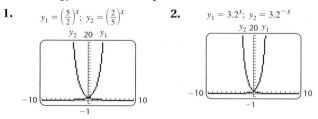

3. $y_1 = \left(\frac{3}{7}\right)^x$; $y_2 = \left(\frac{7}{3}\right)^x$ **4.** $y_1 = 5000(1.08)^x$; $y_2 = 5000(1.08)^{x-3}$

$\text{Xscl} = 5, \text{Yscl} = 1000$

Exercise Set 9.2, pp. 602–604

1. True **3.** True **5.** False
7. $y = f(x) = 3^x$ **9.** $y = 6^x$

11. $y = 2^x + 1$ **13.** $y = 3^x - 2$

15. $y = 2^x$ $y = 2^x - 5$ **17.** $y = 2^{x-2}$

19. $y = 2^{x+1}$ **21.** $y = \left(\frac{1}{4}\right)^x$

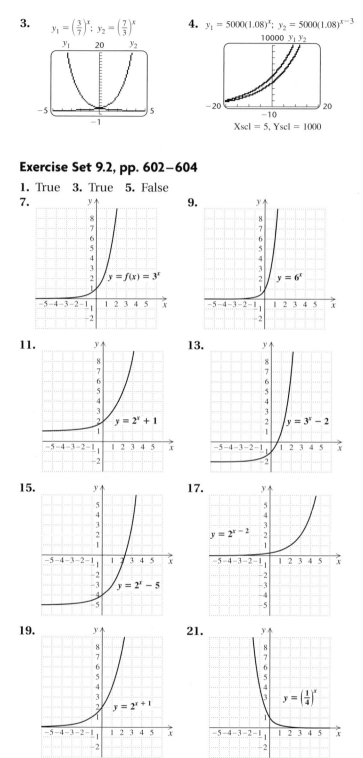

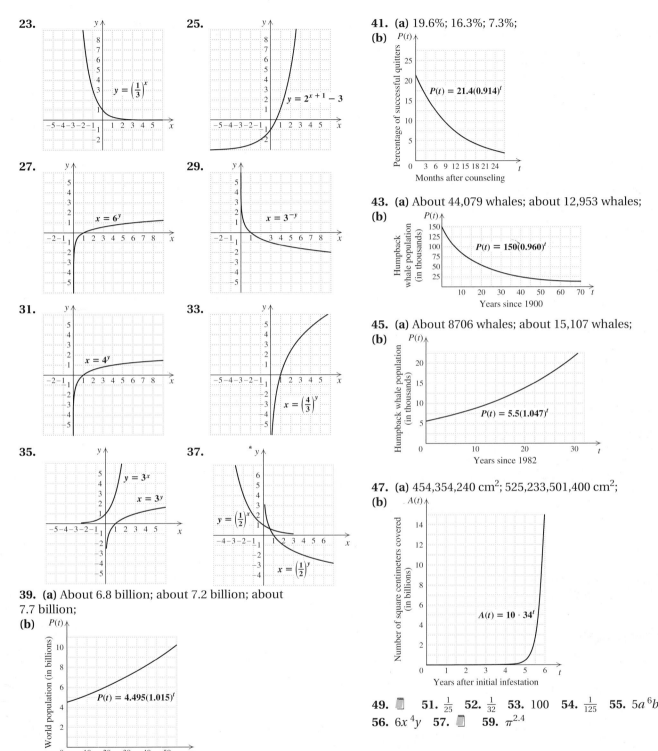

23. $y = \left(\frac{1}{3}\right)^x$

25. $y = 2^{x+1} - 3$

27. $x = 6^y$

29. $x = 3^{-y}$

31. $x = 4^y$

33. $x = \left(\frac{4}{3}\right)^y$

35. $y = 3^x$; $x = 3^y$

37. $y = \left(\frac{1}{2}\right)^x$; $x = \left(\frac{1}{2}\right)^y$

39. (a) About 6.8 billion; about 7.2 billion; about 7.7 billion;

(b) $P(t) = 4.495(1.015)^t$

41. (a) 19.6%; 16.3%; 7.3%;

(b) $P(t) = 21.4(0.914)^t$

43. (a) About 44,079 whales; about 12,953 whales;

(b) $P(t) = 150(0.960)^t$

45. (a) About 8706 whales; about 15,107 whales;

(b) $P(t) = 5.5(1.047)^t$

47. (a) 454,354,240 cm²; 525,233,501,400 cm²;

(b) $A(t) = 10 \cdot 34^t$

49. 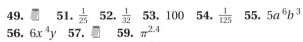   **51.** $\frac{1}{25}$ **52.** $\frac{1}{32}$ **53.** 100   **54.** $\frac{1}{125}$ **55.** $5a^6b^3$

56. $6x^4y$ **57.** 🖩   **59.** $\pi^{2.4}$

61. **63.**

65. **67.**

69.

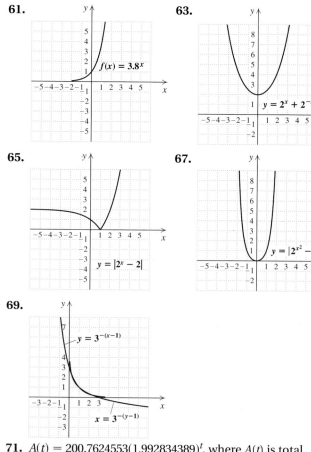

71. $A(t) = 200.7624553(1.992834389)^t$, where $A(t)$ is total sales, in millions of dollars, t years after 1997; $395,244,465,700

73. (a)

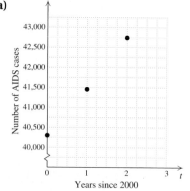

(b) linear; the points appear to lie on a straight line.
75.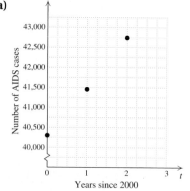

Exercise Set 9.3, pp. 610–611

1. (g) **3.** (a) **5.** (b) **7.** (e) **9.** 3 **11.** 4 **13.** 4
15. -2 **17.** -1 **19.** 4 **21.** 1 **23.** 0 **25.** 5 **27.** -2
29. $\frac{1}{2}$ **31.** $\frac{3}{2}$ **33.** $\frac{2}{3}$ **35.** 7

37. **39.**

41. **43.**

45.

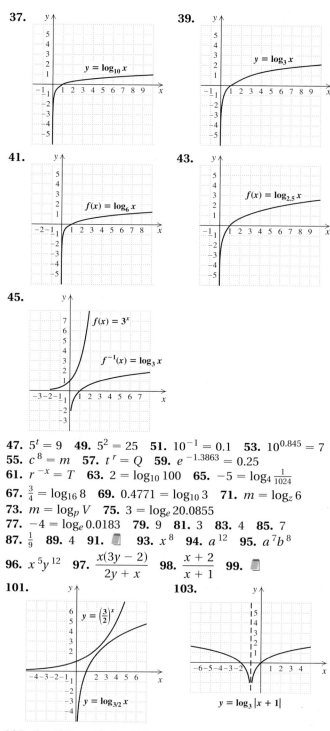

47. $5^t = 9$ **49.** $5^2 = 25$ **51.** $10^{-1} = 0.1$ **53.** $10^{0.845} = 7$
55. $c^8 = m$ **57.** $t^r = Q$ **59.** $e^{-1.3863} = 0.25$
61. $r^{-x} = T$ **63.** $2 = \log_{10} 100$ **65.** $-5 = \log_4 \frac{1}{1024}$
67. $\frac{3}{4} = \log_{16} 8$ **69.** $0.4771 = \log_{10} 3$ **71.** $m = \log_z 6$
73. $m = \log_p V$ **75.** $3 = \log_e 20.0855$
77. $-4 = \log_e 0.0183$ **79.** 9 **81.** 3 **83.** 4 **85.** 7
87. $\frac{1}{9}$ **89.** 4 **91.** ▨ **93.** x^8 **94.** a^{12} **95.** $a^7 b^8$
96. $x^5 y^{12}$ **97.** $\dfrac{x(3y-2)}{2y+x}$ **98.** $\dfrac{x+2}{x+1}$ **99.** ▨

101. **103.**

105. 6 **107.** $-25, 4$ **109.** -2 **111.** 0 **113.** Let $b = 0$, and suppose that $x_1 = 1$ and $x_2 = 2$. Then $0^1 = 0^2$, but $1 \neq 2$. Then let $b = 1$, and suppose that $x_1 = 1$ and $x_2 = 2$. Then $1^1 = 1^2$, but $1 \neq 2$.

Exercise Set 9.4, pp. 617–619

1. (e)   **3.** (a)   **5.** (c)   **7.** $\log_3 81 + \log_3 27$
9. $\log_4 64 + \log_4 16$ **11.** $\log_c r + \log_c s + \log_c t$
13. $\log_a (5 \cdot 14)$, or $\log_a 70$ **15.** $\log_c (t \cdot y)$ **17.** $8 \log_a r$
19. $6 \log_c y$ **21.** $-3 \log_b C$ **23.** $\log_2 25 - \log_2 13$
25. $\log_b m - \log_b n$ **27.** $\log_a \frac{17}{6}$ **29.** $\log_b \frac{36}{4}$, or $\log_b 9$
31. $\log_a \frac{7}{18}$ **33.** $\log_a x + \log_a y + \log_a z$
35. $3 \log_a x + 4 \log_a z$ **37.** $2 \log_a x - 2 \log_a y + \log_a z$
39. $4 \log_a x - 3 \log_a y - \log_a z$
41. $\log_b x + 2 \log_b y - \log_b w - 3 \log_b z$
43. $\frac{1}{2}(7 \log_a x - 5 \log_a y - 8 \log_a z)$
45. $\frac{1}{3}(6 \log_a x + 3 \log_a y - 2 - 7 \log_a z)$ **47.** $\log_a (x^8 z^3)$
49. $\log_a x$ **51.** $\log_a \dfrac{y^5}{x^{3/2}}$ **53.** $\log_a (x - 2)$ **55.** 1.953
57. -0.369 **59.** -1.161 **61.** $\frac{3}{2}$ **63.** Cannot be found
65. 7   **67.** m **69.**
71.   **72.**

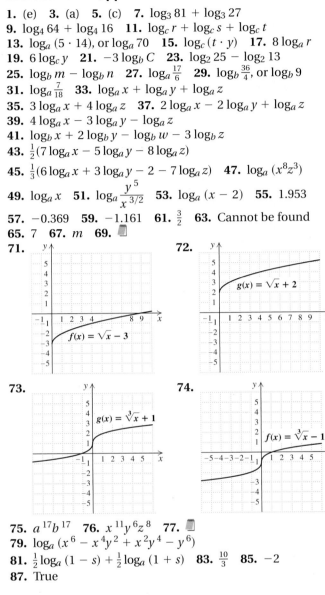

75. $a^{17}b^{17}$ **76.** $x^{11}y^6 z^8$ **77.**
79. $\log_a (x^6 - x^4 y^2 + x^2 y^4 - y^6)$
81. $\frac{1}{2} \log_a (1 - s) + \frac{1}{2} \log_a (1 + s)$ **83.** $\frac{10}{3}$ **85.** -2
87. True

Technology Connection, p. 620

1. ⬛LOG ⬤7 ⬤) ⬤÷ ⬛LOG ⬤3 ⬤) ⬛ENTER

Technology Connection, p. 621

1. As x gets larger, the value of y_1 approaches
2.7182818284.... **2.** For large values of x, the graphs of y_1
and y_2 will be very close or appear to be the same line,
depending on the window chosen. **3.** Using ⬤TRACE, no
y-value is given for $x = 0$. Using a table, an error message
appears for y_1 when $x = 0$. The domain does not include 0
because division by 0 is undefined.

Technology Connection, p. 624

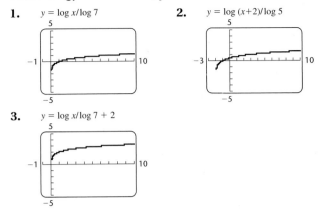

1. $y = \log x / \log 7$
2. $y = \log (x+2)/\log 5$
3. $y = \log x / \log 7 + 2$

Exercise Set 9.5, pp. 624–626

1. True   **3.** True   **5.** True   **7.** True   **9.** 0.7782
11. 1.8621   **13.** 3   **15.** -0.2782 **17.** 1.7986
19. 199.5262   **21.** 1.4894   **23.** 0.0011   **25.** 1.6094
27. 4.0431   **29.** -5.0832 **31.** 96.7583   **33.** 15.0293
35. 0.0305   **37.** 109.9472   **39.** 2.5237   **41.** 6.6439
43. 2.1452   **45.** -2.3219 **47.** -2.3219 **49.** 3.5471
51.   Domain: $\mathbb{R}$; range: $(0, \infty)$

$f(x) = e^x$

53.   Domain: $\mathbb{R}$; range: $(3, \infty)$

$f(x) = e^x + 3$

55.   Domain: $\mathbb{R}$; range: $(-2, \infty)$

$f(x) = e^x - 2$

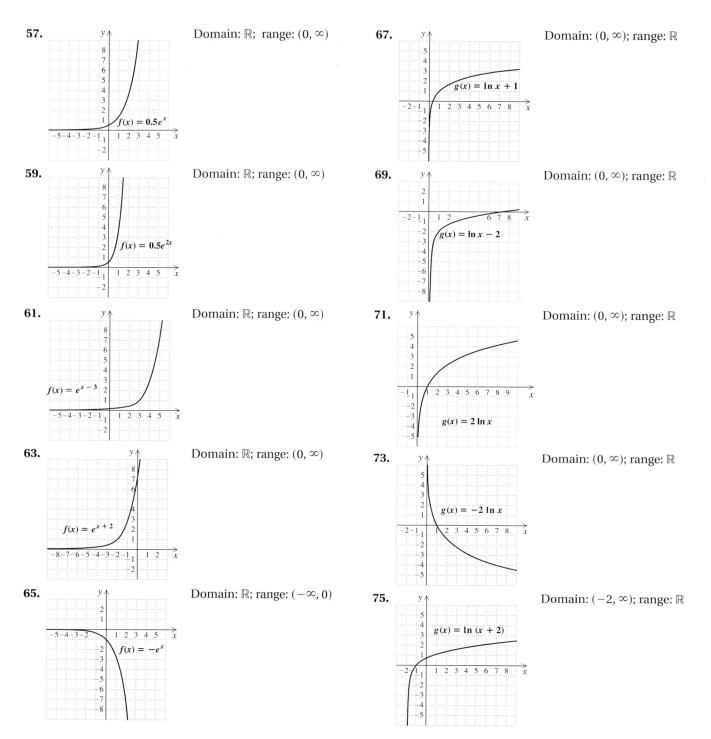

57. Domain: $\mathbb{R}$; range: $(0, \infty)$
$f(x) = 0.5e^x$

59. Domain: $\mathbb{R}$; range: $(0, \infty)$
$f(x) = 0.5e^{2x}$

61. Domain: $\mathbb{R}$; range: $(0, \infty)$
$f(x) = e^{x-3}$

63. Domain: $\mathbb{R}$; range: $(0, \infty)$
$f(x) = e^{x+2}$

65. Domain: $\mathbb{R}$; range: $(-\infty, 0)$
$f(x) = -e^x$

67. Domain: $(0, \infty)$; range: $\mathbb{R}$
$g(x) = \ln x + 1$

69. Domain: $(0, \infty)$; range: $\mathbb{R}$
$g(x) = \ln x - 2$

71. Domain: $(0, \infty)$; range: $\mathbb{R}$
$g(x) = 2 \ln x$

73. Domain: $(0, \infty)$; range: $\mathbb{R}$
$g(x) = -2 \ln x$

75. Domain: $(-2, \infty)$; range: $\mathbb{R}$
$g(x) = \ln (x + 2)$

77.

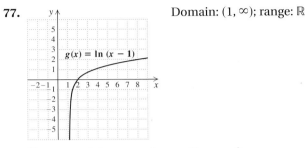

Domain: $(1, \infty)$; range: $\mathbb{R}$

79.   **81.** $-\frac{5}{2}, \frac{5}{2}$ **82.** $0, \frac{7}{5}$ **83.** $\frac{15}{17}$ **84.** $\frac{9}{13}$ **85.** 16, 256
86. $\frac{1}{4}, 9$ **87.**   **89.** 2.452 **91.** 1.442

93. $\log M = \dfrac{\ln M}{\ln 10}$ **95.** 1086.5129 **97.** 4.9855

99. **(a)** Domain: $\{x \mid x > 0\}$, or $(0, \infty)$;
range: $\{y \mid y < 0.5135\}$, or $(-\infty, 0.5135)$; **(b)** $[-1, 5, -10, 5]$;
(c)

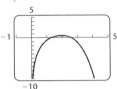

$y = 3.4 \ln x - 0.25e^x$

101. **(a)** Domain: $\{x \mid x > 0\}$, or $(0, \infty)$;
range: $\{y \mid y > -0.2453\}$, or $(-0.2453, \infty)$;
(b) $[-1, 5, -1, 10]$;
(c)

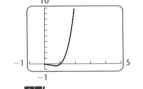

$y = 2x^3 \ln x$

103.

Technology Connection, p. 631

1. 0.38 **2.** -1.96 **3.** 0.90 **4.** -1.53 **5.** 0.13, 8.47
6. $-0.75, 0.75$

Exercise Set 9.6, pp. 632–633

1. (e) **3.** (f) **5.** (b) **7.** (g) **9.** $\dfrac{\log 19}{\log 2} \approx 4.248$

11. $\dfrac{\log 17}{\log 8} + 1 \approx 2.362$ **13.** $\ln 1000 \approx 6.908$

15. $\dfrac{\ln 5}{0.03} \approx 53.648$ **17.** $\dfrac{\log 5}{\log 3} - 1 \approx 0.465$ **19.** 1

21. $\dfrac{\log 87}{\log 4.9} \approx 2.810$ **23.** $\dfrac{\ln\left(\frac{19}{2}\right)}{4} \approx 0.563$ **25.** $\dfrac{\ln 2}{5} \approx 0.139$

27. 81 **29.** $\frac{1}{8}$ **31.** $e^5 \approx 148.413$ **33.** 2 **35.** $\dfrac{e^3}{4} \approx 5.021$

37. $10^{2.5} \approx 316.228$ **39.** $\dfrac{e^4 - 1}{2} \approx 26.799$ **41.** $e \approx 2.718$

43. $e^{-3} \approx 0.050$ **45.** -4 **47.** 10 **49.** No solution
51. $\frac{83}{15}$ **53.** 1 **55.** 6 **57.** 1 **59.** 5 **61.** $\frac{17}{2}$ **63.** 4

65.   **67.** $y = 9x$ **68.** $y = \dfrac{21.35}{x}$ **69.** $L = \dfrac{8T^2}{\pi^2}$

70. $c = \sqrt{\dfrac{E}{m}}$ **71.** $1\frac{1}{5}$ hr **72.** $9\frac{3}{8}$ min **73.**   **75.** $\frac{12}{5}$

77. $\sqrt[3]{3}$ **79.** -1 **81.** $-3, -1$ **83.** $-625, 625$
85. $\frac{1}{2}, 5000$ **87.** $-3, -1$ **89.** $\frac{1}{100,000}, 100,000$ **91.** $-\frac{1}{3}$
93. 38 **95.** 1

Exercise Set 9.7, pp. 641–646

1. **(a)** About 2001; **(b)** 1 yr **3.** **(a)** 146,293; **(b)** 51
5. **(a)** 6.4 yr; **(b)** 23.4 yr **7.** **(a)** 2005; **(b)** 2018
9. **(a)** 2019; **(b)** 15.1 yr **11.** 4.9 **13.** 10^{-7} moles per liter
15. 65 dB **17.** $10^{-1.5}$ W/m^2 **19.** **(a)** $P(t) = P_0 e^{0.025t}$;
(b) \$5126.58; \$5256.36; **(c)** 27.7 yr
21. **(a)** $P(t) = 292.80 e^{0.009t}$, where t is the number of years
after 2004 and $P(t)$ is in millions; **(b)** 295.45 million;
(c) 2016 **23.** 6.7 months
25. **(a)** About 2010; **(b)** about 2019;
(c)

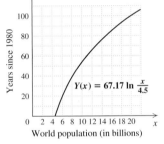

27. **(a)** 68%; **(b)** 54%, 40% **(d)** 6.9 months
(c)

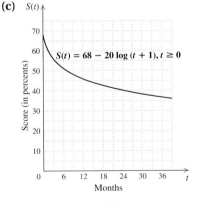

29. **(a)** $N(t) = 17 e^{0.534t}$, where t is the number of years
since 2000; **(b)** about 419
31. **(a)** $k \approx 0.004$; $P(t) = 987 e^{-0.004t}$, where t is the
number of years after 1990 and $P(t)$ is in millions;
(b) 918 million acres; **(c)** about 2043 **33.** About 2103 yr

35. About 7.2 days   **37.** 69.3% per year
39. **(a)** $k \approx 0.099$; $V(t) = 451{,}000e^{0.099t}$, where t is the number of years after 1991; **(b)** about \$1.99 million; **(c)** 7.0 yr; **(d)** 2010   **41.**
43.

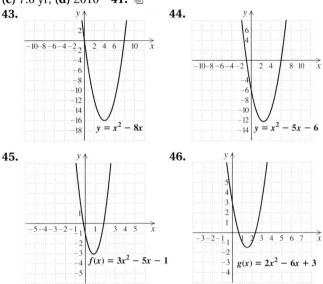

44.

45.

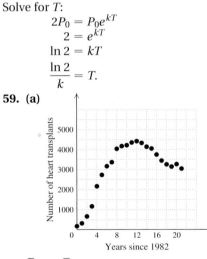

46.

47. $4 \pm \sqrt{23}$ **48.** $-5 \pm \sqrt{31}$ **49.** 🗒   **51.** \$18.9 million
53. $P(t) = 100 - 63.03(0.95)^t$ **55.** About 80,922 yr, or with rounding of k, about 80,792 yr   **57.** Consider an exponential growth function $P(t) = P_0e^{kt}$. At time T,
$P(T) = 2P_0$.
Solve for T:
$$2P_0 = P_0e^{kT}$$
$$2 = e^{kT}$$
$$\ln 2 = kT$$
$$\frac{\ln 2}{k} = T.$$
59. (a)

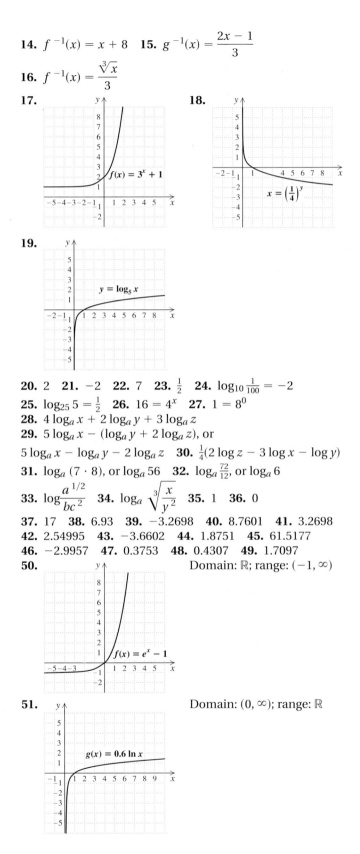

(b) 🗒; **(c)** 🗒

Review Exercises: Chapter 9, pp. 651–652

1. True   **2.** True   **3.** True   **4.** False   **5.** False   **6.** True
7. False   **8.** False   **9.** True   **10.** False
11. $(f \circ g)(x) = 4x^2 - 12x + 10$; $(g \circ f)(x) = 2x^2 - 1$
12. $f(x) = \sqrt{x}$; $g(x) = 3 - x$ **13.** No
14. $f^{-1}(x) = x + 8$ **15.** $g^{-1}(x) = \dfrac{2x - 1}{3}$
16. $f^{-1}(x) = \dfrac{\sqrt[3]{x}}{3}$
17.

18.

19.

20. 2   **21.** -2 **22.** 7   **23.** $\frac{1}{2}$ **24.** $\log_{10}\frac{1}{100} = -2$
25. $\log_{25} 5 = \frac{1}{2}$ **26.** $16 = 4^x$ **27.** $1 = 8^0$
28. $4 \log_a x + 2 \log_a y + 3 \log_a z$
29. $5 \log_a x - (\log_a y + 2 \log_a z)$, or
$5 \log_a x - \log_a y - 2 \log_a z$ **30.** $\frac{1}{4}(2 \log z - 3 \log x - \log y)$
31. $\log_a (7 \cdot 8)$, or $\log_a 56$ **32.** $\log_a \frac{72}{12}$, or $\log_a 6$
33. $\log \dfrac{a^{1/2}}{bc^2}$ **34.** $\log_a \sqrt[3]{\dfrac{x}{y^2}}$ **35.** 1   **36.** 0
37. 17   **38.** 6.93   **39.** -3.2698 **40.** 8.7601   **41.** 3.2698
42. 2.54995   **43.** -3.6602 **44.** 1.8751   **45.** 61.5177
46. -2.9957 **47.** 0.3753   **48.** 0.4307   **49.** 1.7097
50.   Domain: $\mathbb{R}$; range: $(-1, \infty)$

51.   Domain: $(0, \infty)$; range: $\mathbb{R}$

52. 5 **53.** −2 **54.** $\frac{1}{81}$ **55.** 2 **56.** $\frac{1}{1000}$

57. $e^{-2} \approx 0.1353$ **58.** $\frac{1}{2}\left(\dfrac{\log 19}{\log 4} + 5\right) \approx 3.5620$

59. −5, 1 **60.** $\dfrac{\log 8.3}{\log 4} \approx 1.5266$ **61.** $\dfrac{\ln 0.03}{-0.1} \approx 35.0656$

62. $e^{-3} \approx 0.0498$ **63.** 4 **64.** 8 **65.** 20 **66.** $\sqrt{43}$
67. (a) 82; **(b)** 66.8; **(c)** 35 months **68. (a)** 6.6 yr; **(b)** 3.1 yr
69. (a) $k \approx 0.128$; $C(t) = 10e^{0.128t}$, where t is the number
of years after 1999 and $C(t)$ is in thousands; **(b)** $35,966;
(c) 2009 **70.** 23.105% per year **71.** 16.5 yr **72.** 3463 yr
73. 6.6 **74.** 90 dB
75. 🗒 Negative numbers do not have logarithms because
logarithm bases are positive, and there is no exponent to
which a positive number can be raised to yield a negative
number. **76.** 🗒 Taking the logarithm on each side of an
equation produces an equivalent equation because the
logarithm function is one-to-one. If two quantities are
equal, their logarithms must be equal, and if the
logarithms of two quantities are equal, the quantities must
be the same. **77.** e^{e^3} **78.** −3, −1 **79.** $\left(\frac{8}{3}, -\frac{2}{3}\right)$

Test: Chapter 9, pp. 653–654

1. [9.1] $(f \circ g)(x) = 2 + 6x + 4x^2$;

$(g \circ f)(x) = 2x^2 + 2x + 1$ **2.** [9.1] $f(x) = \dfrac{1}{x}$;

$g(x) = 2x^2 + 1$ **3.** [9.1] No **4.** [9.1] $f^{-1}(x) = \dfrac{x - 4}{3}$

5. [9.1] $g^{-1}(x) = \sqrt[3]{x} - 1$
6. [9.2]   **7.** [9.3]

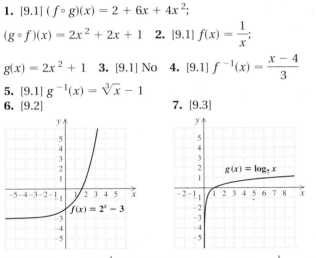

8. [9.3] 3 **9.** [9.3] $\frac{1}{2}$ **10.** [9.3] 18 **11.** [9.3] $\log_4 \frac{1}{64} = -3$
12. [9.3] $\log_{256} 16 = \frac{1}{2}$ **13.** [9.3] $49 = 7^m$
14. [9.3] $81 = 3^4$ **15.** [9.4] $3 \log a + \frac{1}{2}\log b - 2 \log c$
16. [9.4] $\log_a (z^2\sqrt[3]{x})$ **17.** [9.4] 1 **18.** [9.4] 23
19. [9.4] 0 **20.** [9.4] 1.146 **21.** [9.4] 0.477
22. [9.4] 1.204 **23.** [9.5] 1.0899 **24.** [9.5] 0.1585
25. [9.5] −3.3524 **26.** [9.5] 121.5104 **27.** [9.5] 2.4022

28. [9.5]

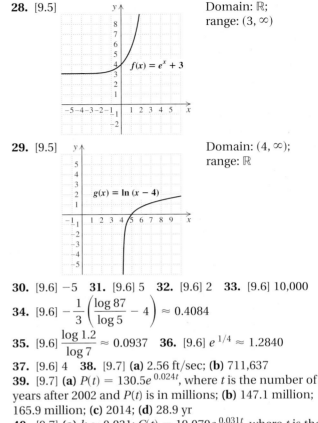

Domain: $\mathbb{R}$;
range: $(3, \infty)$

29. [9.5]   Domain: $(4, \infty)$;
range: $\mathbb{R}$

30. [9.6] −5 **31.** [9.6] 5 **32.** [9.6] 2 **33.** [9.6] 10,000
34. [9.6] $-\dfrac{1}{3}\left(\dfrac{\log 87}{\log 5} - 4\right) \approx 0.4084$

35. [9.6] $\dfrac{\log 1.2}{\log 7} \approx 0.0937$ **36.** [9.6] $e^{1/4} \approx 1.2840$

37. [9.6] 4 **38.** [9.7] **(a)** 2.56 ft/sec; **(b)** 711,637
39. [9.7] **(a)** $P(t) = 130.5e^{0.024t}$, where t is the number of
years after 2002 and $P(t)$ is in millions; **(b)** 147.1 million;
165.9 million; **(c)** 2014; **(d)** 28.9 yr
40. [9.7] **(a)** $k \approx 0.031$; $C(t) = 19,070e^{0.031t}$, where t is the
number of years after 1997; **(b)** $28,535; **(c)** 2028
41. [9.7] 4.6% **42.** [9.7] 4684 yr **43.** [9.7] $10^{-4.5}\ \text{W/m}^2$
44. [9.7] 7.0 **45.** [9.6] −309, 316 **46.** [9.4] 2

Cumulative Review: Chapters 1–9, pp. 654–656

1. 2 **2.** 6 **3.** $\dfrac{y^{12}}{16x^8}$ **4.** $\dfrac{20x^6z^2}{y}$ **5.** $\dfrac{-y^4}{3z^5}$ **6.** $-2x - 1$

7. 25 **8.** $\frac{14}{5}$ **9.** $(3, -1)$ **10.** $(1, -2, 0)$ **11.** $-2, 5$
12. $\frac{9}{2}$ **13.** $\frac{5}{8}$ **14.** $\frac{3}{4}$ **15.** $\frac{1}{2}$ **16.** $\pm 5i$ **17.** 9, 25
18. $\pm 2, \pm 3$ **19.** 7 **20.** 6 **21.** $\frac{3}{2}$ **22.** $\dfrac{\log 7}{5 \log 3} \approx 0.3542$

23. $\dfrac{8e}{e - 1} \approx 12.6558$ **24.** $(-\infty, -5) \cup (1, \infty)$, or
$\{x \mid x < -5\ or\ x > 1\}$ **25.** $-3 \pm 2\sqrt{5}$
26. $\{x \mid x \le -2\ or\ x \ge 5\}$, or $(-\infty, -2] \cup [5, \infty)$

27. $a = \dfrac{Db}{b - D}$ **28.** $q = \dfrac{pf}{p - f}$ **29.** $B = \dfrac{3M - 2A}{2}$, or

$B = \frac{3}{2}M - A$ **30.** 2 **31.** 3
32. $\{x \mid x$ is a real number $and\ x \ne -\frac{1}{3}\ and\ x \ne 2\}$

33. (a) $\dfrac{1.7 \text{ million barrels}}{13 \text{ yr}}$, or approximately

130,769 barrels/yr; **(b)** $g(t) = \dfrac{1.7}{13}t + 7.2$, where $g(t)$ is in

millions of barrels; **(c)** $G(t) = 7.2e^{0.016t}$, where t is the
number of years after 1990 and $G(t)$ is in millions of barrels
34. Length: 36 m; width: 20 m   **35.** A: 15°; B: 45°; C: 120°
36. $5\frac{5}{11}$ min   **37.** Thick and Tasty: 6 oz;
Light and Lean: 9 oz   **38.** $2\frac{7}{9}$ km/h   **39.** -49; -7 and 7
40. 78   **41.** 67.5   **42.** $P(t) = 33.7e^{0.01t}$
43. 35.8 million; 38.0 million   **44.** 69.3 yr   **45.** 18
46. $7p^2q^3 + pq + p - 9$ **47.** $8x^2 - 11x - 1$
48. $9x^4 - 12x^2y + 4y^2$ **49.** $10a^2 - 9ab - 9b^2$

50. $\dfrac{(x + 4)(x - 3)}{2(x - 1)}$ **51.** $\dfrac{1}{x - 4}$ **52.** $\dfrac{a + 2}{6}$

53. $\dfrac{7x + 4}{(x + 6)(x - 6)}$ **54.** $x(y + 2z - w)$

55. $(2 - 5x)(4 + 10x + 25x^2)$ **56.** $2(3x - 2y)(x + 2y)$
57. $(x^3 + 7)(x - 4)$ **58.** $2(m + 3n)^2$
59. $(x - 2y)(x + 2y)(x^2 + 4y^2)$ **60.** -12

61. $x^3 - 2x^2 - 4x - 12 + \dfrac{-42}{x - 3}$ **62.** 1.8×10^{-1}

63. $2y^2\sqrt[3]{y}$ **64.** $14xy^2\sqrt{x}$ **65.** $81a^8b\sqrt[3]{b}$

66. $\dfrac{6 + \sqrt{y} - y}{4 - y}$ **67.** $\sqrt[10]{(x + 5)^3}$ **68.** $18 - 2\sqrt{3}i$

69. $13 - i$ **70.** $f^{-1}(x) = \dfrac{x - 9}{-2}$, or $f^{-1}(x) = \dfrac{9 - x}{2}$

71. $f(x) = -10x + 8$ **72.** $y = \frac{1}{2}x + 7$
73.   **74.**

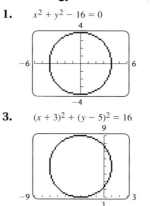

$5x = 15 + 3y$

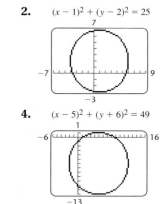

$y = 2x^2 - 4x - 1$

75.   **76.**

$y = \log_3 x$

$y = 3^x$

77.

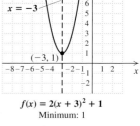

$-2x - 3y \le 12$

78.

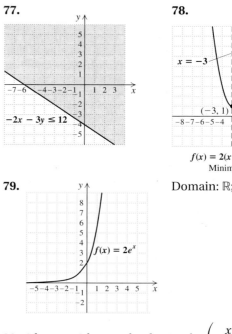

$x = -3$

$(-3, 1)$

$f(x) = 2(x + 3)^2 + 1$
Minimum: 1

Domain: $\mathbb{R}$; range: $(0, \infty)$

79.

$f(x) = 2e^x$

80. $2 \log a + 3 \log c - \log b$ **81.** $\log\left(\dfrac{x^3}{y^{1/2}z^2}\right)$

82. $a^x = 5$ **83.** $\log_x t = 3$ **84.** -1.2545 **85.** 776.2471
86. 2.5479   **87.** 0.2466   **88.** All real numbers except
1 and -2 **89.** $\frac{1}{3}, \frac{10{,}000}{3}$ **90.** 35 mph

CHAPTER 10

Technology Connection, p. 665

1. $x^2 + y^2 - 16 = 0$ **2.** $(x - 1)^2 + (y - 2)^2 = 25$

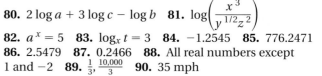

3. $(x + 3)^2 + (y - 5)^2 = 16$ **4.** $(x - 5)^2 + (y + 6)^2 = 49$

Exercise Set 10.1, pp. 666–670

1. (f) **3.** (g) **5.** (c) **7.** (d)

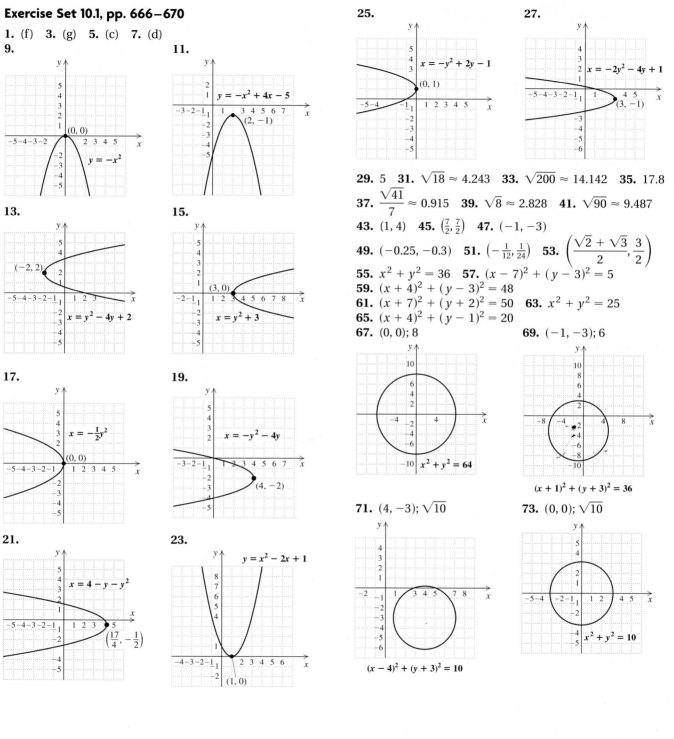

9.

11.
$y = -x^2 + 4x - 5$
$(2, -1)$

13.
$(-2, 2)$
$x = y^2 - 4y + 2$

15.
$(3, 0)$
$x = y^2 + 3$

17.
$x = -\frac{1}{2}y^2$
$(0, 0)$

19.
$x = -y^2 - 4y$
$(4, -2)$

21.
$x = 4 - y - y^2$
$\left(\frac{17}{4}, -\frac{1}{2}\right)$

23.
$y = x^2 - 2x + 1$
$(1, 0)$

25.
$x = -y^2 + 2y - 1$
$(0, 1)$

27.
$x = -2y^2 - 4y + 1$
$(3, -1)$

29. 5 **31.** $\sqrt{18} \approx 4.243$ **33.** $\sqrt{200} \approx 14.142$ **35.** 17.8

37. $\dfrac{\sqrt{41}}{7} \approx 0.915$ **39.** $\sqrt{8} \approx 2.828$ **41.** $\sqrt{90} \approx 9.487$

43. $(1, 4)$ **45.** $\left(\frac{7}{2}, \frac{7}{2}\right)$ **47.** $(-1, -3)$

49. $(-0.25, -0.3)$ **51.** $\left(-\frac{1}{12}, \frac{1}{24}\right)$ **53.** $\left(\dfrac{\sqrt{2} + \sqrt{3}}{2}, \dfrac{3}{2}\right)$

55. $x^2 + y^2 = 36$ **57.** $(x - 7)^2 + (y - 3)^2 = 5$

59. $(x + 4)^2 + (y - 3)^2 = 48$

61. $(x + 7)^2 + (y + 2)^2 = 50$ **63.** $x^2 + y^2 = 25$

65. $(x + 4)^2 + (y - 1)^2 = 20$

67. $(0, 0)$; 8

69. $(-1, -3)$; 6

$x^2 + y^2 = 64$

$(x + 1)^2 + (y + 3)^2 = 36$

71. $(4, -3)$; $\sqrt{10}$

73. $(0, 0)$; $\sqrt{10}$

$(x - 4)^2 + (y + 3)^2 = 10$

$x^2 + y^2 = 10$

75. $(5, 0); \frac{1}{2}$

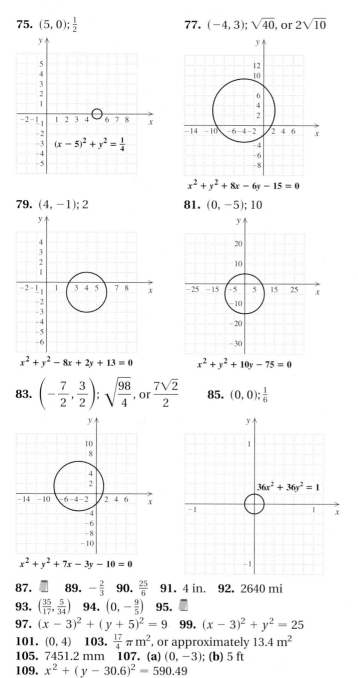

$(x - 5)^2 + y^2 = \frac{1}{4}$

77. $(-4, 3); \sqrt{40}$, or $2\sqrt{10}$

$x^2 + y^2 + 8x - 6y - 15 = 0$

79. $(4, -1); 2$

$x^2 + y^2 - 8x + 2y + 13 = 0$

81. $(0, -5); 10$

$x^2 + y^2 + 10y - 75 = 0$

83. $\left(-\frac{7}{2}, \frac{3}{2}\right); \sqrt{\frac{98}{4}}$, or $\frac{7\sqrt{2}}{2}$

$x^2 + y^2 + 7x - 3y - 10 = 0$

85. $(0, 0); \frac{1}{6}$

$36x^2 + 36y^2 = 1$

87.   **89.** $-\frac{2}{3}$ **90.** $\frac{25}{6}$ **91.** 4 in.   **92.** 2640 mi
93. $\left(\frac{35}{17}, \frac{5}{34}\right)$ **94.** $\left(0, -\frac{9}{5}\right)$ **95.**
97. $(x - 3)^2 + (y + 5)^2 = 9$ **99.** $(x - 3)^2 + y^2 = 25$
101. $(0, 4)$ **103.** $\frac{17}{4}\pi \, \text{m}^2$, or approximately $13.4 \, \text{m}^2$
105. 7451.2 mm   **107. (a)** $(0, -3)$; **(b)** 5 ft
109. $x^2 + (y - 30.6)^2 = 590.49$

111.

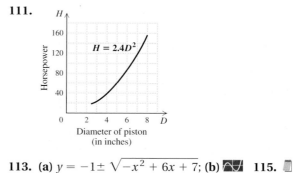

$H = 2.4D^2$

Horsepower

Diameter of piston
(in inches)

113. (a) $y = -1 \pm \sqrt{-x^2 + 6x + 7}$; **(b)**   **115.**

Exercise Set 10.2, pp. 674–676

1. True   **3.** True   **5.** False   **7.** True
9.

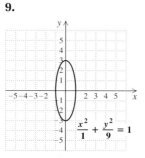

$\frac{x^2}{1} + \frac{y^2}{9} = 1$

11.

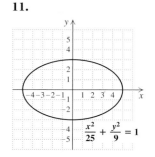

$\frac{x^2}{25} + \frac{y^2}{9} = 1$

13.

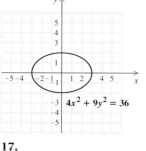

$4x^2 + 9y^2 = 36$

15.

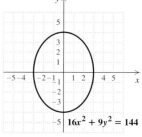

$16x^2 + 9y^2 = 144$

17.

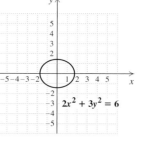

$2x^2 + 3y^2 = 6$

19.

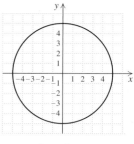

$5x^2 + 5y^2 = 125$

21.

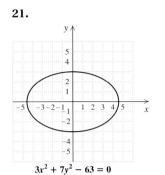

$$3x^2 + 7y^2 - 63 = 0$$

23.

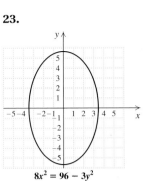

$$8x^2 = 96 - 3y^2$$

25.

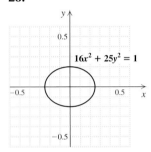

$$16x^2 + 25y^2 = 1$$

27.

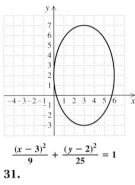

$$\frac{(x-3)^2}{9} + \frac{(y-2)^2}{25} = 1$$

29.

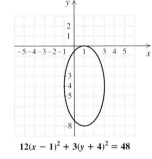

$$\frac{(x+4)^2}{16} + \frac{(y-3)^2}{49} = 1$$

31.

$$12(x-1)^2 + 3(y+4)^2 = 48$$

33.

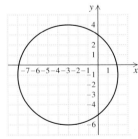

$$4(x+3)^2 + 4(y+1)^2 - 10 = 90$$

35.   **37.** $\frac{16}{9}$ **38.** $-\frac{19}{8}$ **39.** $5 \pm \sqrt{3}$ **40.** $3 \pm \sqrt{7}$

41. $\frac{3}{2}$ **42.** No solution **43.**   **45.** $\dfrac{x^2}{81} + \dfrac{y^2}{121} = 1$

47. $\dfrac{(x-2)^2}{16} + \dfrac{(y+1)^2}{9} = 1$ **49.** $\dfrac{x^2}{9} + \dfrac{y^2}{25} = 1$

51. (a) Let $F_1 = (-c, 0)$ and $F_2 = (c, 0)$. Then the sum of the distances from the foci to P is $2a$. By the distance formula,

$$\sqrt{(x+c)^2 + y^2} + \sqrt{(x-c)^2 + y^2} = 2a, \text{ or}$$
$$\sqrt{(x+c)^2 + y^2} = 2a - \sqrt{(x-c)^2 + y^2}.$$

Squaring, we get

$$(x+c)^2 + y^2 = 4a^2 - 4a\sqrt{(x-c)^2 + y^2} + (x-c)^2 + y^2,$$

or

$$x^2 + 2cx + c^2 + y^2 = 4a^2 - 4a\sqrt{(x-c)^2 + y^2}$$
$$+ x^2 - 2cx + c^2 + y^2.$$

Thus

$$-4a^2 + 4cx = -4a\sqrt{(x-c)^2 + y^2}$$
$$a^2 - cx = a\sqrt{(x-c)^2 + y^2}.$$

Squaring again, we get

$$a^4 - 2a^2cx + c^2x^2 = a^2(x^2 - 2cx + c^2 + y^2)$$
$$a^4 - 2a^2cx + c^2x^2 = a^2x^2 - 2a^2cx + a^2c^2 + a^2y^2,$$

or

$$x^2(a^2 - c^2) + a^2y^2 = a^2(a^2 - c^2)$$
$$\frac{x^2}{a^2} + \frac{y^2}{a^2 - c^2} = 1.$$

(b) When P is at $(0, b)$, it follows that $b^2 = a^2 - c^2$. Substituting, we have

$$\frac{x^2}{a^2} + \frac{y^2}{b^2} = 1.$$

53. 5.66 ft

55.

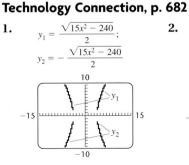

$$\frac{(x-2)^2}{16} + \frac{(y+1)^2}{4} = 1$$

57.

Technology Connection, p. 682

1.
$$y_1 = \frac{\sqrt{15x^2 - 240}}{2};$$
$$y_2 = -\frac{\sqrt{15x^2 - 240}}{2}$$

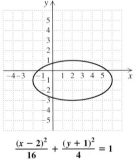

2.
$$y_1 = \sqrt{\frac{16x^2 - 64}{3}};$$
$$y_2 = -\sqrt{\frac{16x^2 - 64}{3}}$$

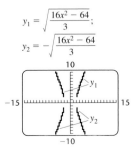

3.
$$y_1 = \frac{\sqrt{5x^2 + 320}}{4};$$
$$y_2 = -\frac{\sqrt{5x^2 + 320}}{4}$$

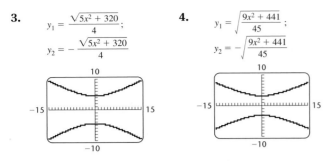

4.
$$y_1 = \sqrt{\frac{9x^2 + 441}{45}};$$
$$y_2 = -\sqrt{\frac{9x^2 + 441}{45}}$$

21.

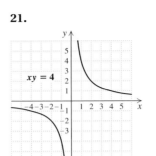

$xy = 4$

23.

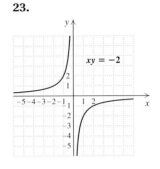

$xy = -2$

25.

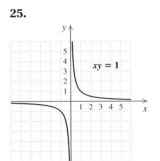

$xy = 1$

Exercise Set 10.3, pp. 687–688

1. (d)   **3.** (h)   **5.** (g)   **7.** (c)

9.

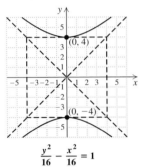

$$\frac{y^2}{16} - \frac{x^2}{16} = 1$$

11.

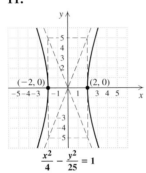

$$\frac{x^2}{4} - \frac{y^2}{25} = 1$$

27. Circle   **29.** Ellipse   **31.** Hyperbola   **33.** Circle
35. Ellipse   **37.** Hyperbola   **39.** Parabola
41. Hyperbola   **43.** Circle   **45.** Ellipse   **47.** ▨
49. $\left(-\frac{22}{3}, \frac{37}{9}\right)$ **50.** $(9, -4)$ **51.** $-3, 3$ **52.** $-1, 1$
53. \$35   **54.** 69   **55.** ▨   **57.** $\dfrac{y^2}{36} - \dfrac{x^2}{4} = 1$

59. C: $(5, 2)$; V: $(-1, 2)$, $(11, 2)$;
asymptotes: $y - 2 = \frac{5}{6}(x - 5)$, $y - 2 = -\frac{5}{6}(x - 5)$

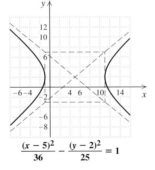

$$\frac{(x - 5)^2}{36} - \frac{(y - 2)^2}{25} = 1$$

13.

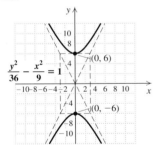

$$\frac{y^2}{36} - \frac{x^2}{9} = 1$$

15.

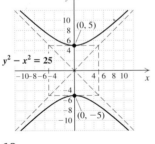

$y^2 - x^2 = 25$

17.

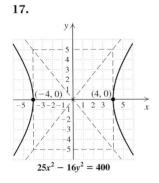

$25x^2 - 16y^2 = 400$

19.

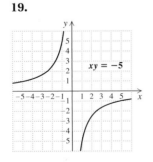

$xy = -5$

61. $\dfrac{(y + 3)^2}{4} - \dfrac{(x - 4)^2}{16} = 1$; C: $(4, -3)$; V: $(4, -5)$, $(4, -1)$;
asymptotes: $y + 3 = \frac{1}{2}(x - 4)$, $y + 3 = -\frac{1}{2}(x - 4)$

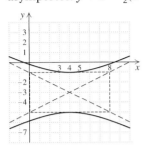

$8(y + 3)^2 - 2(x - 4)^2 = 32$

63. $\dfrac{(x + 3)^2}{1} - \dfrac{(y - 2)^2}{4} = 1$; C: $(-3, 2)$; V: $(-4, 2)$, $(-2, 2)$;
asymptotes: $y - 2 = 2(x + 3)$, $y - 2 = -2(x + 3)$

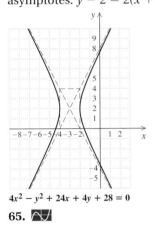

$4x^2 - y^2 + 24x + 4y + 28 = 0$

65.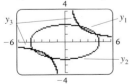

Technology Connection, p. 691

1. $(-1.50, -1.17)$; $(3.50, 0.50)$
2. $(-2.77, 2.52)$; $(-2.77, -2.52)$

Technology Connection, p. 693

1. $y_1 = \sqrt{(20 - x^2)/4}$; $y_2 = -\sqrt{(20 - x^2)/4}$; $y_3 = 4/x$

Exercise Set 10.4, pp. 696–698

1. True   **3.** False   **5.** True   **7.** $(-4, -3)$, $(3, 4)$
9. $(0, 2)$, $(3, 0)$ **11.** $(-2, 1)$

13. $\left(\dfrac{5 + \sqrt{70}}{3}, \dfrac{-1 + \sqrt{70}}{3}\right)$, $\left(\dfrac{5 - \sqrt{70}}{3}, \dfrac{-1 - \sqrt{70}}{3}\right)$
15. $\left(4, \frac{3}{2}\right)$, $(3, 2)$ **17.** $\left(\frac{7}{3}, \frac{1}{3}\right)$, $(1, -1)$ **19.** $\left(\frac{11}{4}, -\frac{5}{4}\right)$, $(1, 4)$
21. $\left(\dfrac{7 - \sqrt{33}}{2}, \dfrac{7 + \sqrt{33}}{2}\right)$, $\left(\dfrac{7 + \sqrt{33}}{2}, \dfrac{7 - \sqrt{33}}{2}\right)$
23. $(3, -5)$, $(-1, 3)$ **25.** $(-5, -8)$, $(8, 5)$ **27.** $(0, 0)$, $(1, 1)$,
$\left(-\dfrac{1}{2} + \dfrac{\sqrt{3}}{2}i, -\dfrac{1}{2} - \dfrac{\sqrt{3}}{2}i\right)$, $\left(-\dfrac{1}{2} - \dfrac{\sqrt{3}}{2}i, -\dfrac{1}{2} + \dfrac{\sqrt{3}}{2}i\right)$
29. $(-3, 0)$, $(3, 0)$ **31.** $(-4, -3)$, $(-3, -4)$, $(3, 4)$, $(4, 3)$
33. $\left(\dfrac{16}{3}, \dfrac{5\sqrt{7}}{3}i\right)$, $\left(\dfrac{16}{3}, -\dfrac{5\sqrt{7}}{3}i\right)$, $\left(-\dfrac{16}{3}, \dfrac{5\sqrt{7}}{3}i\right)$,
$\left(-\dfrac{16}{3}, -\dfrac{5\sqrt{7}}{3}i\right)$ **35.** $(-3, -\sqrt{5})$, $(-3, \sqrt{5})$, $(3, -\sqrt{5})$,
$(3, \sqrt{5})$ **37.** $(4, 2)$, $(-4, -2)$, $(2, 4)$, $(-2, -4)$
39. $(4, 1)$, $(-4, -1)$, $(2, 2)$, $(-2, -2)$ **41.** $(2, 1)$, $(-2, -1)$
43. $\left(2, -\frac{4}{5}\right)$, $\left(-2, -\frac{4}{5}\right)$, $(5, 2)$, $(-5, 2)$
45. $\left(-\sqrt{2}, \sqrt{2}\right)$, $\left(\sqrt{2}, -\sqrt{2}\right)$
47. Length: 8 cm; width: 6 cm
49. Length: 5 in.; width: 4 in.
51. Length: 12 ft; width: 5 ft   **53.** 6 and 10; -6 and -10
55. 24 ft, 16 ft   **57.** 13 and 12   **59.** 🖩   **61.** -16
62. -32 **63.** 1   **64.** $-\frac{1}{4}$ **65.** 44   **66.** 28   **67.** 🖩
69. $(x + 2)^2 + (y - 1)^2 = 4$
71. $(-2, 3)$, $(2, -3)$, $(-3, 2)$, $(3, -2)$ **73.** Length: 55 ft;
width: 45 ft   **75.** 10 in. by 7 in. by 5 in.
77. Length: 61.02 in.; height: 34.32 in.   **79.** 📈

Review Exercises: Chapter 10, pp. 700–701

1. True   **2.** False   **3.** False   **4.** False   **5.** True   **6.** True
7. False   **8.** True   **9.** 4   **10.** 5   **11.** $\sqrt{90.1} \approx 9.492$
12. $\sqrt{9 + 4a^2}$ **13.** $\left(\frac{9}{2}, -1\right)$ **14.** $(-3, 7)$
15. $\left(\dfrac{3}{4}, \dfrac{\sqrt{3} - \sqrt{2}}{2}\right)$ **16.** $\left(\frac{1}{2}, 2a\right)$ **17.** $(-3, 2)$, $\sqrt{7}$
18. $(5, 0)$, 7   **19.** $(3, 1)$, 3   **20.** $(-4, 3)$, $\sqrt{35}$
21. $(x + 4)^2 + (y - 3)^2 = 48$
22. $(x - 7)^2 + (y + 2)^2 = 20$
23. Circle   **24.** Ellipse

$5x^2 + 5y^2 = 80$ $9x^2 + 2y^2 = 18$

25. Parabola

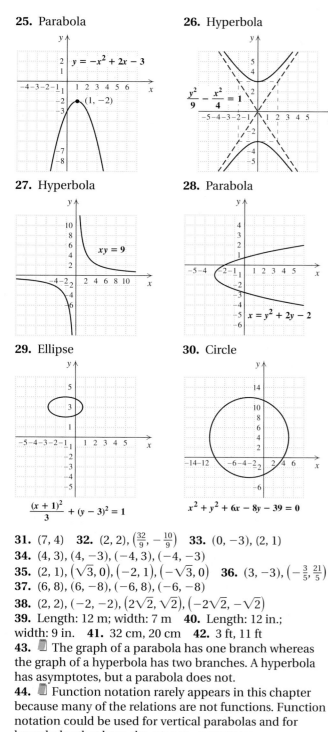

$y = -x^2 + 2x - 3$

$(1, -2)$

26. Hyperbola

$$\dfrac{y^2}{9} - \dfrac{x^2}{4} = 1$$

27. Hyperbola

$xy = 9$

28. Parabola

$x = y^2 + 2y - 2$

29. Ellipse

$$\dfrac{(x + 1)^2}{3} + (y - 3)^2 = 1$$

30. Circle

$x^2 + y^2 + 6x - 8y - 39 = 0$

31. $(7, 4)$ **32.** $(2, 2), \left(\dfrac{32}{9}, -\dfrac{10}{9}\right)$ **33.** $(0, -3), (2, 1)$
34. $(4, 3), (4, -3), (-4, 3), (-4, -3)$
35. $(2, 1), \left(\sqrt{3}, 0\right), (-2, 1), \left(-\sqrt{3}, 0\right)$ **36.** $(3, -3), \left(-\dfrac{3}{5}, \dfrac{21}{5}\right)$
37. $(6, 8), (6, -8), (-6, 8), (-6, -8)$
38. $(2, 2), (-2, -2), \left(2\sqrt{2}, \sqrt{2}\right), \left(-2\sqrt{2}, -\sqrt{2}\right)$
39. Length: 12 m; width: 7 m   **40.** Length: 12 in.;
width: 9 in.   **41.** 32 cm, 20 cm   **42.** 3 ft, 11 ft
43. The graph of a parabola has one branch whereas
the graph of a hyperbola has two branches. A hyperbola
has asymptotes, but a parabola does not.
44. Function notation rarely appears in this chapter
because many of the relations are not functions. Function
notation could be used for vertical parabolas and for
hyperbolas that have the axes as asymptotes.
45. $\left(-5, -4\sqrt{2}\right), \left(-5, 4\sqrt{2}\right), \left(3, -2\sqrt{2}\right), \left(3, 2\sqrt{2}\right)$
46. $(0, 6), (0, -6)$ **47.** $(x - 2)^2 + (y + 1)^2 = 25$
48. $\dfrac{x^2}{81} + \dfrac{y^2}{25} = 1$ **49.** $\left(\dfrac{9}{4}, 0\right)$

Test: Chapter 10, p. 702

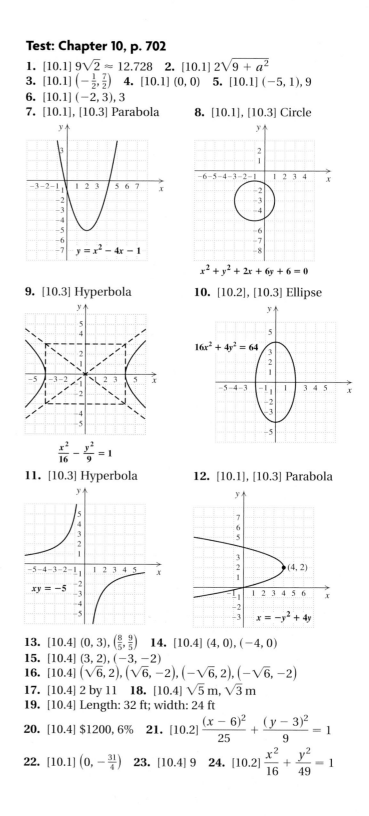

1. [10.1] $9\sqrt{2} \approx 12.728$ **2.** [10.1] $2\sqrt{9 + a^2}$
3. [10.1] $\left(-\dfrac{1}{2}, \dfrac{7}{2}\right)$ **4.** [10.1] $(0, 0)$ **5.** [10.1] $(-5, 1), 9$
6. [10.1] $(-2, 3), 3$
7. [10.1], [10.3] Parabola   **8.** [10.1], [10.3] Circle

$y = x^2 - 4x - 1$

$x^2 + y^2 + 2x + 6y + 6 = 0$

9. [10.3] Hyperbola   **10.** [10.2], [10.3] Ellipse

$$\dfrac{x^2}{16} - \dfrac{y^2}{9} = 1$$

$16x^2 + 4y^2 = 64$

11. [10.3] Hyperbola   **12.** [10.1], [10.3] Parabola

$xy = -5$

$(4, 2)$

$x = -y^2 + 4y$

13. [10.4] $(0, 3), \left(\dfrac{8}{5}, \dfrac{9}{5}\right)$ **14.** [10.4] $(4, 0), (-4, 0)$
15. [10.4] $(3, 2), (-3, -2)$
16. [10.4] $\left(\sqrt{6}, 2\right), \left(\sqrt{6}, -2\right), \left(-\sqrt{6}, 2\right), \left(-\sqrt{6}, -2\right)$
17. [10.4] 2 by 11   **18.** [10.4] $\sqrt{5}$ m, $\sqrt{3}$ m
19. [10.4] Length: 32 ft; width: 24 ft
20. [10.4] $1200, 6%   **21.** [10.2] $\dfrac{(x - 6)^2}{25} + \dfrac{(y - 3)^2}{9} = 1$
22. [10.1] $\left(0, -\dfrac{31}{4}\right)$ **23.** [10.4] 9   **24.** [10.2] $\dfrac{x^2}{16} + \dfrac{y^2}{49} = 1$

CHAPTER 11

Exercise Set 11.1, pp. 708–710

1. (f) **3.** (d) **5.** (c) **7.** 5, 7, 9, 11; 23; 33
9. 3, 6, 11, 18; 102; 227 **11.** $0, \frac{3}{5}, \frac{4}{5}, \frac{15}{17}, \frac{99}{101}, \frac{112}{113}$
13. $1, -\frac{1}{2}, \frac{1}{4}, -\frac{1}{8}; -\frac{1}{512}; \frac{1}{16{,}384}$ **15.** $-4, 5, -6, 7; 13; -18$
17. $0, 7, -26, 63; 999; -3374$ **19.** 13 **21.** 364
23. -23.5 **25.** -363 **27.** $\frac{441}{400}$ **29.** $2n$ **31.** $(-1)^{n+1}$
33. $(-1)^n \cdot n$ **35.** $2n + 1$ **37.** $(-1)^n \cdot 2 \cdot (3)^{n-1}$
39. $\frac{n}{n + 1}$ **41.** 5^n **43.** $(-1)^n \cdot n^2$ **45.** 4 **47.** 30
49. $\frac{1}{2} + \frac{1}{4} + \frac{1}{6} + \frac{1}{8} + \frac{1}{10} = \frac{137}{120}$
51. $3^0 + 3^1 + 3^2 + 3^3 + 3^4 = 121$
53. $2 + \frac{3}{2} + \frac{4}{3} + \frac{5}{4} + \frac{6}{5} + \frac{7}{6} + \frac{8}{7} = \frac{1343}{140}$
55. $(-1)^2 2^1 + (-1)^3 2^2 + (-1)^4 2^3 + (-1)^5 2^4 + (-1)^6 2^5 + (-1)^7 2^6 + (-1)^8 2^7 + (-1)^9 2^8 = -170$
57. $(0^2 - 2 \cdot 0 + 3) + (1^2 - 2 \cdot 1 + 3) + (2^2 - 2 \cdot 2 + 3) + (3^2 - 2 \cdot 3 + 3) + (4^2 - 2 \cdot 4 + 3) + (5^2 - 2 \cdot 5 + 3) = 43$
59. $\frac{(-1)^3}{3 \cdot 4} + \frac{(-1)^4}{4 \cdot 5} + \frac{(-1)^5}{5 \cdot 6} = -\frac{1}{15}$ **61.** $\sum_{k=1}^{5} \frac{k+1}{k+2}$
63. $\sum_{k=1}^{6} k^2$ **65.** $\sum_{k=2}^{n} (-1)^k k^2$ **67.** $\sum_{k=1}^{\infty} 5k$
69. $\sum_{k=1}^{\infty} \frac{1}{k(k+1)}$ **71.** 🖩 **73.** 77 **74.** 23
75. $x^3 + 3x^2y + 3xy^2 + y^3$ **76.** $a^3 - 3a^2b + 3ab^2 - b^3$
77. $8a^3 - 12a^2b + 6ab^2 - b^3$
78. $8x^3 + 12x^2y + 6xy^2 + y^3$ **79.** 🖩
81. 1, 3, 13, 63, 313, 1563 **83.** $5200, $3900, $2925, $2193.75, $1645.31, $1233.98, $925.49, $694.12, $520.59, $390.44 **85.** $S_{100} = 0; S_{101} = -1$ **87.** $i, -1, -i, 1, i; i$
89. 11th term

Exercise Set 11.2, pp. 717–719

1. True **3.** False **5.** True **7.** False **9.** $a_1 = 2, d = 4$
11. $a_1 = 7, d = -4$ **13.** $a_1 = \frac{3}{2}, d = \frac{3}{4}$
15. $a_1 = $5.12, d = 0.12 **17.** 49 **19.** -94
21. $-$1628.16 **23.** 26th **25.** 57th **27.** 82 **29.** 5
31. 28 **33.** $a_1 = 8; d = -3; 8, 5, 2, -1, -4$
35. $a_1 = 1; d = 1$ **37.** 780 **39.** 31,375 **41.** 2550
43. 918 **45.** 1030 **47.** 35 marchers; 315 marchers
49. 180 stones **51.** $49.60 **53.** 722 seats **55.** 🖩
57. $\frac{13}{30x}$ **58.** $\frac{23}{36t}$ **59.** $a^k = P$ **60.** $e^a = t$
61. $x^2 + y^2 = 81$ **62.** $(x + 2)^2 + (y - 5)^2 = 18$
63. 🖩 **65.** 33 jumps **67.** $8760, $7961.77, $7163.54, $6365.31, $5567.08; $4768.85, $3970.62, $3172.39, $2374.16,

$1575.93 **69.** Let $d = $ the common difference. Since p, m, and q form an arithmetic sequence, $m = p + d$ and $q = p + 2d$. Then $\frac{p + q}{2} = \frac{p + (p + 2d)}{2} = p + d = m$.
71. 156,375

Exercise Set 11.3, pp. 727–729

1. Geometric sequence **3.** Arithmetic sequence
5. Geometric series **7.** Geometric series **9.** 2
11. -0.1 **13.** $-\frac{1}{2}$ **15.** $\frac{1}{5}$ **17.** $\frac{6}{m}$ **19.** 192
21. $112\sqrt{2}$ **23.** 52,488 **25.** $2331.64 **27.** $a_n = 5^{n-1}$
29. $a_n = (-1)^{n-1}$, or $a_n = (-1)^{n+1}$
31. $a_n = \frac{1}{x^n}$, or $a_n = x^{-n}$ **33.** 3066 **35.** $\frac{547}{18}$ **37.** $\frac{1 - x^8}{1 - x}$, or $(1 + x)(1 + x^2)(1 + x^4)$ **39.** $5134.51 **41.** $\frac{64}{3}$ **43.** $\frac{49}{4}$
45. No **47.** No **49.** $\frac{43}{99}$ **51.** $25,000 **53.** $\frac{7}{9}$ **55.** $\frac{830}{99}$
57. $\frac{5}{33}$ **59.** $\frac{5}{1024}$ ft **61.** 155,797 **63.** 2710 flies
65. 10,723,491 apartments and houses **67.** 3100.35 ft
69. 20.48 in. **71.** 🖩 **73.** $x^3 + 3x^2y + 3xy^2 + y^3$
74. $a^3 - 3a^2b + 3ab^2 - b^3$ **75.** $\left(-\frac{63}{29}, -\frac{114}{29}\right)$
76. $(-1, 2, 3)$ **77.** 🖩 **79.** 54 **81.** $\frac{x^2[1 - (-x)^n]}{1 + x}$
83. 512 cm² **85.** 🖩, 📈

Technology Connection, p. 735

1. 479,001,600 **2.** 56; 792

Exercise Set 11.4, pp. 738–739

1. 2^5, or 32 **3.** 9 **5.** $\binom{8}{5}$ **7.** x^7y^2 **9.** 362,880
11. 39,916,800 **13.** 56 **15.** 3024 **17.** 35 **19.** 126
21. 4060 **23.** 780 **25.** $a^4 - 4a^3b + 6a^2b^2 - 4ab^3 + b^4$
27. $p^7 + 7p^6q + 21p^5q^2 + 35p^4q^3 + 35p^3q^4 + 21p^2q^5 + 7pq^6 + q^7$
29. $2187c^7 - 5103c^6d + 5103c^5d^2 - 2835c^4d^3 + 945c^3d^4 - 189c^2d^5 + 21cd^6 - d^7$
31. $t^{-12} + 12t^{-10} + 60t^{-8} + 160t^{-6} + 240t^{-4} + 192t^{-2} + 64$
33. $x^5 - 5x^4y + 10x^3y^2 - 10x^2y^3 + 5xy^4 - y^5$
35. $19{,}683s^9 + \frac{59{,}049s^8}{t} + \frac{78{,}732s^7}{t^2} + \frac{61{,}236s^6}{t^3} + \frac{30{,}618s^5}{t^4} + \frac{10{,}206s^4}{t^5} + \frac{2268s^3}{t^6} + \frac{324s^2}{t^7} + \frac{27s}{t^8} + \frac{1}{t^9}$
37. $x^{15} - 10x^{12}y + 40x^9y^2 - 80x^6y^3 + 80x^3y^4 - 32y^5$
39. $125 + 150\sqrt{5}t + 375t^2 + 100\sqrt{5}t^3 + 75t^4 + 6\sqrt{5}t^5 + t^6$
41. $x^{-3} - 6x^{-2} + 15x^{-1} - 20 + 15x - 6x^2 + x^3$

43. $15a^4b^2$ **45.** $-64{,}481{,}508a^3$ **47.** $1120x^{12}y^2$
49. $1{,}959{,}552u^5v^{10}$ **51.** y^8 **53.** 🔲 **55.** 4 **56.** $\frac{5}{2}$
57. 5.6348 **58.** ± 5 **59.** 🔲
61. List all the subsets of size 3: $\{a, b, c\}$, $\{a, b, d\}$, $\{a, b, e\}$, $\{a, c, d\}$, $\{a, c, e\}$, $\{a, d, e\}$, $\{b, c, d\}$, $\{b, c, e\}$, $\{b, d, e\}$, $\{c, d, e\}$.

There are exactly 10 subsets of size 3 and $\binom{5}{3} = 10$, so

there are exactly $\binom{5}{3}$ ways of forming a subset of size 3

from $\{a, b, c, d, e\}$.

63. $\binom{8}{5}(0.15)^3(0.85)^5 \approx 0.084$

65. $\binom{8}{6}(0.15)^2(0.85)^6 + \binom{8}{7}(0.15)(0.85)^7 +$

$\binom{8}{8}(0.85)^8 \approx 0.89$

67. $\binom{n}{n-r} = \dfrac{n!}{[n-(n-r)]!\,(n-r)!}$

$= \dfrac{n!}{r!\,(n-r)!} = \binom{n}{r}$

69. $\dfrac{-\sqrt[3]{q}}{2p}$ **71.** $x^7 + 7x^6y + 21x^5y^2 + 35x^4y^3 + 35x^3y^4 + 21x^2y^5 + 7xy^6 + y^7$

Review Exercises: Chapter 11, pp. 741–742

1. False **2.** True **3.** True **4.** False **5.** True **6.** True
7. False **8.** False **9.** 1, 5, 9, 13; 29; 45
10. $0, \frac{1}{5}, \frac{1}{5}, \frac{3}{17}; \frac{7}{65}; \frac{11}{145}$ **11.** $a_n = 7n$
12. $a_n = (-1)^n(2n-1)$
13. $-2 + 4 + (-8) + 16 + (-32) = -22$
14. $-3 + (-5) + (-7) + (-9) + (-11) + (-13) = -48$
15. $\displaystyle\sum_{k=1}^{5} 4k$ **16.** $\displaystyle\sum_{k=1}^{5}\frac{1}{(-2)^k}$ **17.** 85 **18.** $\frac{8}{3}$
19. $a_1 = \frac{45}{4}, d = \frac{5}{4}$ **20.** -544 **21.** 8580 **22.** $1024\sqrt{2}$
23. $\frac{2}{3}$ **24.** $a_n = 2(-1)^n$ **25.** $a_n = 3\left(\frac{x}{4}\right)^{n-1}$ **26.** 4095
27. $-4095x$ **28.** 12 **29.** $\frac{49}{11}$ **30.** No **31.** No
32. \$40,000 **33.** $\frac{5}{9}$ **34.** $\frac{46}{33}$ **35.** \$24.30 **36.** 903 poles
37. \$15,791.18 **38.** 6 m **39.** 5040 **40.** 56
41. $190a^{18}b^2$ **42.** $x^4 - 8x^3y + 24x^2y^2 - 32xy^3 + 16y^4$
43. 🔲 For a geometric sequence with $|r| < 1$, as n gets larger, the absolute value of the terms gets smaller, since $|r^n|$ gets smaller.
44. 🔲 The first form of the binomial theorem draws the coefficients from Pascal's triangle; the second form uses factorial notation. The second form avoids the need to compute all preceding rows of Pascal's triangle, and is generally easier to use when only one term of an

expression is needed. When several terms of an expansion are needed and n is not large (say, $n \le 8$), it is often easier to use Pascal's triangle.

45. $\dfrac{1 - (-x)^n}{x + 1}$

46. $x^{-15} + 5x^{-9} + 10x^{-3} + 10x^3 + 5x^9 + x^{15}$

Test: Chapter 11, pp. 742–743

1. [11.1] 1, 7, 13, 19, 25; 67 **2.** [11.1] $a_n = 4\left(\frac{1}{3}\right)^n$
3. [11.1] $-1 + (-5) + (-13) + (-29) + (-61) = -109$
4. [11.1] $\displaystyle\sum_{k=1}^{5}(-1)^{k+1}k^3$ **5.** [11.2] -51 **6.** [11.2] $\frac{3}{8}$
7. [11.2] $a_1 = 31.2; d = -3.8$ **8.** [11.2] 2508 **9.** [11.3] $\frac{9}{128}$
10. [11.3] $\frac{2}{3}$ **11.** [11.3] 3^n **12.** [11.3] $511 + 511x$
13. [11.3] 1 **14.** [11.3] No **15.** [11.3] $\frac{\$25{,}000}{23} \approx \$1086.96$
16. [11.3] $\frac{85}{99}$ **17.** [11.2] 63 seats **18.** [11.2] \$17,100
19. [11.3] \$8981.05 **20.** [11.3] 36 m **21.** [11.4] 220
22. [11.4] $x^{10} - 15x^8y + 90x^6y^2 - 270x^4y^3 + 405x^2y^4 - 243y^5$ **23.** [11.4] $220a^9x^3$ **24.** [11.2] $n(n+1)$

25. [11.3] $\dfrac{1 - \left(\dfrac{1}{x}\right)^n}{1 - \dfrac{1}{x}}$, or $\dfrac{x^n - 1}{x^{n-1}(x - 1)}$

Cumulative Review: Chapters 1–11, pp. 743–745

1. $-35x^6y^{-4}$, or $\dfrac{-35x^6}{y^4}$ **2.** 6.3 **3.** $-4y + 17$ **4.** 280
5. $\frac{7}{6}$ **6.** $3a^2 - 8ab - 15b^2$ **7.** $13x^3 - 7x^2 - 6x + 6$
8. $6a^2 + 7a - 5$ **9.** $9a^4 - 30a^2y + 25y^2$ **10.** $\dfrac{4}{x + 2}$
11. $\dfrac{x - 4}{4(x + 2)}$ **12.** $\dfrac{(x + y)(x^2 + xy + y^2)}{x^2 + y^2}$ **13.** $x - a$
14. $(2x - 3)^2$ **15.** $(3a - 2)(9a^2 + 6a + 4)$
16. $(a + 3)(a^2 - b)$ **17.** $3(y^2 + 3)(5y^2 - 4)$ **18.** 20
19. $7x^3 + 9x^2 + 19x + 38 + \dfrac{72}{x - 2}$ **20.** $\frac{3}{5}$ **21.** $-\frac{6}{5}, 4$
22. $\mathbb{R}$, or $(-\infty, \infty)$ **23.** $(-1, 1)$ **24.** $(2, -1, 1)$ **25.** 2
26. $\pm 2, \pm 5$ **27.** $\left(\sqrt{5}, \sqrt{3}\right), \left(\sqrt{5}, -\sqrt{3}\right), \left(-\sqrt{5}, \sqrt{3}\right),$
$\left(-\sqrt{5}, -\sqrt{3}\right)$ **28.** 1.4037 **29.** 1005 **30.** $\frac{1}{25}$ **31.** $-\frac{1}{2}$
32. $\{x \mid -2 \le x \le 3\}$, or $[-2, 3]$ **33.** $\pm i\sqrt{2}$
34. $-2 \pm \sqrt{7}$ **35.** $\{y \mid y < -5 \text{ or } y > 2\}$, or
$(-\infty, -5) \cup (2, \infty)$ **36.** $-6, 8$ **37.** 3 **38.** 5 ft by 12 ft
39. More than 4 purchases **40.** 65, 66, 67 **41.** $11\frac{3}{7}$
42. \$2.68 herb: 10 oz; \$4.60 herb: 14 oz **43.** 350 mph
44. $8\frac{2}{5}$ hr, or 8 hr 24 min **45.** 20 **46.** 5000 ft^2

47.

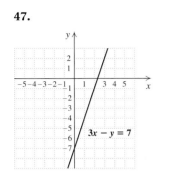

$3x - y = 7$

48.

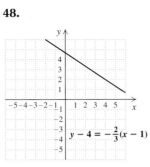

$y - 4 = -\frac{2}{3}(x - 1)$

49.

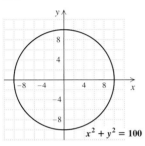

$x^2 + y^2 = 100$

50.

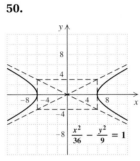

$\frac{x^2}{36} - \frac{y^2}{9} = 1$

51.

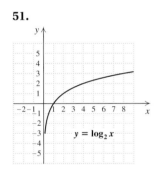

$y = \log_2 x$

52.

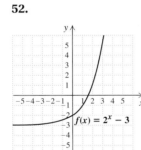

$f(x) = 2^x - 3$

53.

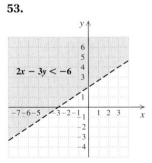

$2x - 3y < -6$

54.

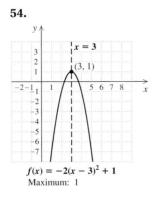

$f(x) = -2(x - 3)^2 + 1$
Maximum: 1

55. $r = \dfrac{V - P}{-Pt}$, or $\dfrac{P - V}{Pt}$ **56.** $R = \dfrac{Ir}{1 - I}$ **57.** $y = 3x - 8$

58. $\{x \mid x \le \frac{3}{4}\}$, or $\left(-\infty, \frac{3}{4}\right]$

59. $\{x \mid x$ is a real number and $x \ne 1\}$, or $(-\infty, 1) \cup (1, \infty)$

60. 6.8×10^{-12} **61.** $8x^2\sqrt{y}$ **62.** $125x^2y^{3/4}$ **63.** $\dfrac{\sqrt[3]{5xy}}{y}$

64. $\dfrac{1 - 2\sqrt{x} + x}{1 - x}$ **65.** $26 - 13i$ **66.** $x^2 - 50 = 0$

67. $(2, -3); 6$ **68.** $\log_a \dfrac{\sqrt[3]{x^2} \cdot z^5}{\sqrt{y}}$ **69.** $a^5 = c$

70. 3.7541   **71.** 0.0003   **72.** 8.6442   **73.** 0.0277

74. (a) $k \approx 0.219$; $C(t) = 0.12e^{0.219t}$; **(b)** about 29 million
computers   **75.** 5   **76.** −121   **77.** 875

78. $16\left(\frac{1}{4}\right)^{n-1}$ **79.** $13{,}440a^4b^6$ **80.** $74.88671875x$

81. \$652.39   **82.** All real numbers except 0 and −12

83. 81   **84.** y gets divided by 8   **85.** $-\dfrac{7}{13} + \dfrac{2\sqrt{30}}{13}i$

86. 84 yr

APPENDIX

Exercise Set I.1, p. 750

1. 3   **2.** 9   **3.** 33.8   **4.** 5.446808511   **5.** 137.921536
6. 45.6976   **7.** 1.777777778   **8.** 8

Exercise Set I.2, p. 752

1. $C = 65.0008$ cm; $A = 336.2232$ cm^2 **2. (a)** \$84.72;
(b) \$50.83; **(c)** \$37.07; **(d)** \$73.29

3. 659; −21; 4,945,843; 274,828,929

4. −1.06329696; −1.06329696; 2.250728506 × 10^{12};
2.250728506 × 10^{12}

Exercise Set 2.1, p. 754

1.

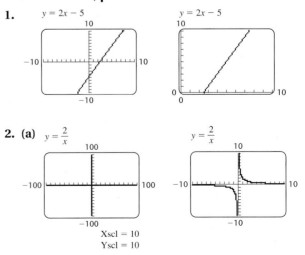

3. $y_1 = 3x - 2;\ y_2 = x^2 + x + 1$

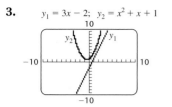

(a) The graphing calculator will graph the entire equation $y_1 = 3x - 2$ first, and then $y_2 = x^2 + x + 1$.   **(b)** The graphing calculator will appear to draw both graphs at the same time.   **4.** If we use ZDecimal and TRACE, y is 4.52 when $x = -1.2$.

Exercise Set 2.5, p. 757

1. Yes   **2.** No   **3.** No   **4. (a)** $f(x) = 0.234x + 67.29$;
(b) 76.65   **5. (a)** $f(x) = 76.12403101x + 8901.782946$;
(b) about \$9663.02   **6. (a)** $f(x) = 3.43x + 87.28$; **(b)** about
258.8 million tons

Exercise Set 8.8, p. 759

1. Maximum of -5.22 at $x = 0.88$ **2.** Minimum of 2.55 at
$x = -0.40$ **3.** Minimum of -440.82 at $x = -383.33$

Glossary

A

Absolute value [1.2]   The distance that a number is from 0 on the number line.

Additive inverse [1.2]   A number's opposite. Two numbers are additive inverses of each other if their sum is zero.

Algebraic expression [1.1]   A number or variable or a collection of numbers and variables on which the operations $+, -, \cdot, \div, (\ )^n$, or $\sqrt[n]{(\ )}$ are performed.

Arithmetic sequence [11.2]   A sequence in which the difference between any two successive terms is constant.

Arithmetic series [11.2]   A series for which the associated sequence is arithmetic.

Ascending order [5.1]   A polynomial in one variable written with the terms arranged according to degree, from least to greatest.

Associative law of addition [1.2]   The statement that when three numbers are added, regrouping the addends gives the same sum.

Associative law of multiplication [1.2]   The statement that when three numbers are multiplied, regrouping the factors gives the same product.

Asymptote [10.3]   A line that a graph approaches more and more closely as x increases or as x decreases.

Average [1.4]   Most commonly, the mean of a set of numbers.

Axes [2.1]   Two perpendicular number lines used to identify points in a plane.

Axis of symmetry [8.6]   A line that can be drawn through a graph such that the part of the graph on one side of the line is an exact reflection of the part on the opposite side.

B

Bar graph [1.1]   A graphic display of data using bars proportional in length to the numbers represented.

Base [1.1]   In exponential notation, the number being raised to a power.

Binomial [5.1]   A polynomial composed of two terms.

Branches [10.3]   The two curves that comprise a hyperbola.

Break-even point [3.8]   In business, the point of intersection of the revenue function and the cost function.

C

Circle [10.1]   A set of points in a plane that are a fixed distance r, called the radius, from a fixed point (h, k), called the center.

Circle graph   A graphic display of data using sectors of a circle to represent percents.

Circumference [1.5]   The distance around a circle.

Closed interval [a, b] [4.1]   The set of all numbers x for which $a \leq x \leq b$. Thus, $[a, b] = \{x \mid a \leq x \leq b\}$.

Coefficient [5.1]   The numerical multiplier of a variable.

Combined variation [6.8]   A mathematical relationship in which a variable varies directly and/or inversely, at the same time, with more than one other variable.

Common logarithm [9.5]   A logarithm with base 10.

Commutative law of addition [1.2]   The statement that when two numbers are added, changing the order in which the numbers are added does not affect the sum.

Commutative law of multiplication [1.2]   The statement that when two numbers are multiplied, changing the order in which the numbers are multiplied does not affect the product.

Completing the square [8.1]   Adding a particular constant to an expression so that the resulting sum is a perfect square.

Complex number [7.8]   Any number that can be written as $a + bi$, where a and b are real numbers.

Complex rational expression [6.3]   A rational expression that has one or more rational expressions within its numerator and/or denominator.

Complex-number system [7.8]   A number system that contains the real-number system and is designed so that negative numbers have square roots.

Composite function [9.1]   A function in which a quantity depends on a variable that, in turn, depends on another variable.

Composite number   A natural number, other than 1, that is not prime.

Compound inequality [4.2]   A statement in which two or more inequalities are combined using the word *and* or the word *or*.

Compound interest [8.1]   Interest computed on the sum of an original principal and the interest previously accrued by that principal.

Conditional equation [1.3]   An equation that is true for some replacements and false for others.

Conic section [10.1]   A curve formed by the intersection of a plane and a cone.

Conjugates [7.5]   Pairs of radical terms, like $\sqrt{a} + \sqrt{b}$ and $\sqrt{a} - \sqrt{b}$, for which the product does not have a radical term.

Conjunction [4.2]   A sentence in which two statements are joined by the word *and*.

Consecutive numbers [1.4]   Integers that are one unit apart.

Consistent system of equations [3.1]   A system of equations that has at least one solution.

Constant [1.1]   A known number.

Constant function [2.4]   A function given by an equation of the form $f(x) = b$, where b is a real number.

Constant of proportionality [6.8]   The constant in an equation of direct or inverse variation.

Constraint [4.5]   A requirement imposed on a problem.

Contradiction [1.3]   An equation that is never true.

Coordinates [2.1]   The numbers in an ordered pair.

Cube root [7.1]   The number c is called the cube root of a if $c^3 = a$.

D

Data point [8.8]   A given ordered pair of a function, usually found experimentally.

Degree of a polynomial [5.1]   The degree of the term of highest degree in a polynomial.

Degree of a term [5.1]   The number of variable factors in a term.

Demand function [3.8]   A function modeling the relationship between the price of a good and the quantity of that good demanded.

Denominator   The number below the fraction bar in a fraction.

Dependent equations [3.1]   The equations in a system are dependent if one equation can be removed without changing the solution set.

Descending order [5.1]   A polynomial in one variable written with the terms arranged according to degree, from greatest to least.

Determinant [3.7]   The determinant of a two-by-two matrix $\begin{bmatrix} a & c \\ b & d \end{bmatrix}$ is denoted $\begin{vmatrix} a & c \\ b & d \end{vmatrix}$ and represents $ad - bc$.

Difference of squares [5.5]   An expression that can be written in the form $a^2 - b^2$.

Direct variation [6.8]   A situation that translates to an equation of the form $y = kx$, with k a constant.

Discriminant [8.4]   The expression $b^2 - 4ac$ from the quadratic formula.

Disjunction [4.2]   A sentence in which two statements are joined by the word *or*.

Distributive law [1.2]   The statement that multiplying a factor by the sum of two numbers gives the same result as multiplying the factor by each of the two numbers and then adding.

Domain [2.2]   The set of all first coordinates of the ordered pairs in a function.

Doubling time [9.7]   The time necessary for a population to double in size.

E

Elimination method [3.2]   An algebraic method that uses the addition principle to solve a system of equations.

Ellipse [10.2]   The set of all points in a plane for which the sum of the distances from two fixed points F_1 and F_2 is constant.

Equation [1.1]   A number sentence with the verb $=$.

Equation of variation [6.8]   An equation used to represent direct, inverse, or combined variation.

Equilibrium point [3.8]   The point of intersection between the demand function and the supply function.

Equivalent equations [2.1]   Equations with the same solutions.

Equivalent expressions [1.2]   Expressions that have the same value for all allowable replacements.

Equivalent inequalities [4.1]   Inequalities that have the same solution set.

Evaluate [1.1]   To substitute a value for each occurrence of a variable in an expression.

Exponent [1.1]   In expressions of the form a^n, the number n is an exponent. For n a natural number, a^n represents n factors of a.

Exponential decay [9.7]   A decrease in quantity over time that can be modeled by an exponential equation of the form $P(t) = P_0 e^{-kt}$, $k > 0$.

Exponential equation [9.6]   An equation in which a variable appears as an exponent.

Exponential function [9.2]   A function that can be described by an exponential equation.

Exponential growth [9.7]   An increase in quantity over time that can be modeled by an exponential function of the form $P(t) = P_0 e^{kt}$, $k > 0$.

Exponential notation [1.1]   A representation of a number using a base raised to a power.

Extrapolation [2.2]   The process of predicting a future value on the basis of given data.

F

Factor [1.2]   *Verb*: to write an equivalent expression that is a product. *Noun*: a multiplier.

Finite sequence [11.1]   A function having for its domain a set of natural numbers: $\{1, 2, 3, 4, 5, \ldots, n\}$, for some natural number n.

Fixed costs [3.8]   In business, costs that are incurred whether or not a product is produced.

Focus [10.2]   One of two fixed points that determine the points of an ellipse.

FOIL [5.2]   To multiply two binomials by multiplying the First terms, the Outside terms, the Inside terms, and the Last terms, and then adding the results.

Formula [1.5]   An equation that uses numbers or letters to represent a relationship between two or more quantities.

Fraction notation   A number written using a numerator and a denominator.

Function [2.2]   A correspondence that assigns to each member of a set called the domain exactly one member of a set called the range.

G

General term of a sequence [11.1]   The nth term, denoted a_n.

Geometric sequence [11.3]   A sequence in which the ratio of every pair of successive terms is constant.

Geometric series [11.3]   A series for which the associated sequence is geometric.

Graph [2.1]   A picture or diagram of the data in a table. A line, curve, or collection of points that represents all the solutions of an equation.

Greatest common factor [5.3]   The common factor of a polynomial with the largest possible coefficient and the largest possible exponent(s).

H

Half-life [9.7]   The amount of time necessary for half of a quantity to decay.

Half-open interval [4.1]   An interval that includes exactly one of two endpoints.

Horizontal-line test [9.1]   If it is impossible to draw a horizontal line that intersects the graph of a function more than once, then that function is one-to-one.

Hyperbola [10.3]   The set of all points P in the plane such that the difference of the distance from P to two fixed points is constant.

Hypotenuse [5.8]   In a right triangle, the side opposite the right angle.

I

Identity [1.3]   An equation that is always true.

Identity property of 0   The statement that the sum of a number and 0 is always the original number.

Identity property of 1   The statement that the product of a number and 1 is always the original number.

Imaginary number [7.8]   A number that can be written in the form $a + bi$, where a and b are real numbers and $b \neq 0$.

Imaginary number i [7.8]   The square root of -1. That is, $i = \sqrt{-1}$ and $i^2 = -1$.

Inconsistent system of equations [3.1]   A system of equations for which there is no solution.

Independent equations [3.1]   Equations that are not dependent.

Index [7.1]   In the radical $\sqrt[n]{a}$, the number n is called the index.

Inequality [1.2]   A mathematical sentence using $<, >, \leq, \geq$, or $\neq$.

Infinite geometric series [11.3]   The sum of the terms of an infinite geometric sequence.

Infinite sequence [11.1]   A function having for its domain the set of natural numbers: $\{1, 2, 3, 4, 5, \ldots\}$.

Input [2.2]   A member of the domain of a function.

Integers [1.1]   The whole numbers and their opposites.

Interpolation [2.2]   The process of estimating a value between given values.

Intersection of two sets [4.2]   The set of all elements that are common to both sets.

Interval notation [4.1]   The use of a pair of numbers inside parentheses and brackets to represent the set of all numbers between those two numbers. *See also* Closed and Open intervals.

Inverse relation [9.1]   The relation formed by interchanging the members of the domain and the range of a relation.

Inverse variation [6.8]   A situation that translates to an equation of the form $y = k/x$, with k a constant.

Irrational number [1.1]   A real number that cannot be named as a ratio of two integers.

Isosceles right triangle [7.7]   A right triangle in which both legs have the same length.

J

Joint variation [6.8]   A situation that translates to an equation of the form $y = kxz$, with k a constant.

L

Leading coefficient [5.1]   The coefficient of the term of highest degree in a polynomial.

Leading term [5.1]   The term of highest degree in a polynomial.

Least common denominator [6.2]   The least common multiple of the denominators.

Legs [5.8]   In a right triangle, the two sides that form the right angle.

Like radicals [7.5]   Radical expressions that have a common radical factor.

Like terms [1.3]   Terms that have exactly the same variable factors.

Line graph   A graph in which quantities are represented as points connected by straight-line segments.

Linear equation [2.1]   Any equation that can be written in the form $y = mx + b$, or $Ax + By = C$, where x and y are variables.

Linear function [2.3]   A function that can be described by an equation of the form $f(x) = mx + b$, where m and b are constants.

Linear inequality [4.4]   An inequality whose related equation is a linear equation.

Linear programming [4.5]   A branch of mathematics involving graphs of inequalities and their constraints.

Logarithmic equation [9.6]   An equation containing a logarithmic expression.

Logarithmic function, base a [9.3]   The inverse of an exponential function with base a.

M

Matrix [3.6]   A rectangular array of numbers.

Maximum value [8.6]   The largest function value (output) achieved by a function.

Minimum value [8.6]   The smallest function value (output) achieved by a function.

Monomial [5.1]   A constant, a variable, or a product of a constant and one or more variables.

Motion problem [6.5]   A problem that deals with distance, speed, and time.

Multiplicative inverses [1.2]   Reciprocals; two numbers whose product is 1.

Multiplicative property of zero   The statement that the product of 0 and any real number is 0.

N

Natural logarithm [9.5]   A logarithm with base e.

Natural numbers [1.1]   The counting numbers: 1, 2, 3, 4, 5, ...

Nonlinear equation [2.1]   An equation whose graph is not a straight line.

Numerator   The number above the fraction bar in a fraction.

O

Objective function [4.5]   In linear programming, the function in which the expression being maximized or minimized appears.

One-to-one function [9.1]   A function for which different inputs have different outputs.

Open interval (a, b) [4.1]   The set of all numbers x for which $a < x < b$. Thus, $(a, b) = \{x \mid a < x < b\}$.

Opposite [1.2]   The opposite, or additive inverse, of a number a is written $-a$. Opposites are the same distance from 0 on the number line but on different sides of 0.

Ordered pair [2.1]   A pair of numbers of the form (h, k) for which the order in which the numbers are listed is important.

Origin [2.1]   The point on a graph where the two axes intersect.

Output [2.2]   A member of the range of a function.

P

Parabola [8.6]   A graph of a quadratic function.

Parallel lines [2.5]   Lines that extend indefinitely in the same plane without intersecting.

Pascal's triangle [11.4]   A triangular array of coefficients of the expansion $(a + b)^n$ for $n = 0, 1, 2, \ldots$.

Perfect square [7.1]   A rational number for which there exists a number a for which $a^2 = p$.

Perfect-square trinomial [5.5]   A trinomial that is the square of a binomial.

Perpendicular lines [2.5]   Lines that form a right angle.

Point–slope equation [2.5]   An equation of the type $y - y_1 = m(x - x_1)$, where x and y are variables.

Polynomial [5.1]   A monomial or a sum of monomials.

Polynomial equation [5.8]   An equation in which two polynomials are set equal to each other.

Polynomial inequality [8.9]   An inequality that is equivalent to an inequality with a polynomial as one side and 0 as the other.

Prime factorization [5.3]   The factorization of a whole number into a product of its prime factors.

Prime number   A natural number that has exactly two different factors: the number itself and 1.

Principal square root [7.1]   The nonnegative square root of a number.

Pure imaginary number [7.8]   A complex number of the form $a + bi$, with $a = 0$ and $b \neq 0$.

Pythagorean theorem [5.8]   In any right triangle, if a and b are the lengths of the legs and c is the length of the hypotenuse, then $a^2 + b^2 = c^2$.

Q

Quadrants [2.1]   The four regions into which the axes divide a plane.

Quadratic equation [5.8]   An equation equivalent to one of the form $ax^2 + bx + c = 0$, where $a \neq 0$.

Quadratic formula [8.2]   The solutions of $ax^2 + bx + c, a \neq 0$, are given by the equation
$$x = \frac{-b \pm \sqrt{b^2 - 4ac}}{2a}.$$

Quadratic function [8.1]   A second-degree polynomial function in one variable.

Quadratic inequality [8.9]   A second-degree polynomial inequality in one variable.

R

Radical equation [7.6]   An equation in which a variable appears in a radicand.

Radical expression [7.1]   An algebraic expression in which a radical sign appears.

Radical sign [7.1]   The symbol $\sqrt{}$.

Radical term [7.5]   A term in which a radical sign appears.

Radicand [7.1]   The expression under the radical sign.

Radius [10.1]   The distance from the center of a circle to a point on the circle. Also, a segment connecting the center to a point on the circle.

Range [2.2]   The set of all second coordinates of the ordered pairs in a function.

Ratio [2.3]   The ratio of a to b is a/b, also written $a:b$.

Rational equation [6.4]   An equation containing one or more rational expressions.

Rational expression [6.1]   A quotient of two polynomials.

Rational inequality [8.9]   An inequality containing a rational expression.

Rational number [1.1]   A number that can be written in the form $\frac{a}{b}$, where a and b are integers and $b \neq 0$.

Rationalizing the denominator [7.4]   A procedure for finding an equivalent expression without a radical in the denominator.

Rationalizing the numerator [7.4]   A procedure for finding an equivalent expression without a radical in the numerator.

Real number [1.1]   Any number that is either rational or irrational.

Reciprocal [1.2]   A multiplicative inverse. Two numbers are reciprocals if their product is 1.

Reflection [8.6]   The mirror image of a graph.

Relation [2.2]   A correspondence between the domain and the range such that each member of the domain corresponds to at least one member of the range.

Repeating decimal [1.1]   A decimal in which a number pattern repeats indefinitely.

Right triangle [5.8]   A triangle that includes a right angle.

Row-equivalent operations [3.6]   Operations used to produce equivalent systems of equations.

S

Scientific notation [1.7]   A number written in the form $N \times 10^m$, where m is an integer, $1 \leq N < 10$, and N is expressed in decimal notation.

Sequence [11.1]   A function for which the domain is a set of consecutive positive integers beginning with 1.

Series [11.1]   The sum of specified terms in a sequence.

Set [1.1]   A collection of objects.

Set-builder notation [1.1]   The naming of a set by describing basic characteristics of the elements in the set.

Sigma notation [11.1]   The naming of a sum using the Greek letter Σ (sigma) as part of an abbreviated form.

Similar triangles   Triangles in which corresponding sides are proportional.

Simplify [1.3]   To rewrite an expression in an equivalent, abbreviated, form.

Slope [2.3]   The ratio of the rise to the run for any two points on a line.

Slope–intercept equation [2.3]   An equation of the form $y = mx + b$, where x and y are variables.

Solution [1.1]   A replacement or substitution that makes an equation or inequality true.

Solution set [4.1]   The set of all solutions of an equation, an inequality, or a system of equations or inequalities.

Solve [1.1]   To find all solutions of an equation, an inequality, or a system of equations or inequalities; to find the solution(s) of a problem.

Speed [6.5]   The ratio of distance traveled to the time required to travel that distance.

Square matrix [3.7]   A matrix with the same number of rows and columns.

Square root [7.1]   The number c is a square root of a if $c^2 = a$.

Substitute [1.1]   To replace a variable with a number.

Substitution method [3.2]   An algebraic method for solving systems of equations.

Supply function [3.8]   A function modeling the relationship between the price of a good and the quantity of that good supplied.

System of equations [3.1]   A set of two or more equations that are to be solved simultaneously.

T

Term [1.3]   A number, a variable, or a product or a quotient of numbers and/or variables.

Terminating decimal [1.1]   A decimal that can be written using a finite number of decimal places.

Total cost [3.8]   The amount spent to produce a product.

Total profit [3.8]   The amount taken in less the amount spent, or total revenue minus total cost.

Total revenue [3.8]   The amount taken in from the sale of a product.

Trinomial [5.1]   A polynomial that is composed of three terms.

U

Union of *A* and *B* [4.2]   The set of all elements belonging to either *A* or *B*.

V

Value [1.1]   The numerical result after a number has been substituted into an expression.

Variable [1.1]   A letter that represents an unknown number.

Variable costs [3.8]   In business, costs that vary according to the amount of products produced.

Variable expression [1.1]   An expression containing a variable.

Vertex [8.6]   The point at which the graph of a quadratic equation crosses its axis of symmetry.

Vertical-line test [2.2]   The statement that a graph represents a function if it is impossible to draw a vertical line that intersects the graph more than once.

W

Whole numbers [1.1]   The natural numbers and 0: 0, 1, 2, 3, . . .

X

***x*-intercept** [2.4]   The point at which a graph crosses the *x*-axis.

Y

***y*-intercept** [2.4]   The point at which a graph crosses the *y*-axis.

Z

Zeros [8.9]   The x-values for which $f(x)$ is 0, for any function f.

Index

A

Absolute value, 13–14, 16
  equations with, 245–248, 273
    absolute-value principle for, 246–248, 273
    inequalities with, 248–250, 273
    principles for solving, 249
Absolute-value principle for equations, 246–248, 273
Absolute-value signs, 439
Addition, 3, 69. *See also* Sums
  of complex numbers, 492–493
  of functions, 134–135, 144
  of polynomials, 282–283
  of radical expressions, 467–468
  of rational expressions
    with different denominators, 365–370
    with like denominators, 354–365
  of real numbers, 14–15
  series and, 706, 740
Addition principle
  for equations, 25, 69
  for inequalities, 224–225, 227, 273
Additive inverses
  of equations, 27
  laws of, 15, 69
  of polynomials, 283–284
Algebraic expressions, 2–3, 69
  evaluating, 4–5
  polynomials and, 278–280
  translating to, 3–4
Approximating solutions and functions, 519

Area formulas
  for circle, 70
  for parallelogram, 70
  for rectangle, 43, 70
  for trapezoid, 44, 70
  for triangle, 4–5, 70
Arithmetic sequences, 711–714, 740
  sum of first n terms of, 713–714
Ascending order in polynomials, 280
Associative laws, 20–21
Asymptotes, 679
AUTO function, 283
Axes
  of graphs, 76, 143
  of hyperbolas, 678
Axis of symmetry, 544, 579

B

Base e, 620–621
Bases, logarithmic, changing, 622–623
Binomial expansion
  with factorial notation, 734–737, 740
  with Pascal's triangle, 731–733, 740
Binomials, 279
  multiplying monomials and, 291
  squaring, 294–295
Binomial theorem, 733, 736, 740
Boundary of graph of linear inequality, 254, 274
Branches of hyperbolas, 678
Break-even analysis, 207–209, 214
Break-even point, 209, 214

C

Calculators
  common logarithms on, 619–620
  graphing. *See* Graphing calculator
Canceling, 357
Cartesian coordinate system, 76, 143
Center
  of ellipses, 671
  of hyperbolas, 678
Change-of-base formula, 622–623
Checking selections, 35
Circles, 663–665, 683, 699
  area of, formula for, 70
  center of, 663
  circumference of, formula for, 70
  equation of, 664–665, 699
  radius of, 663
Circumference of a circle, formula for, 70
Closed interval, 223
Coefficients
  leading, 279
  of terms, 279
Collaborative Corner, 42, 68, 142, 167, 189, 234, 244, 264, 290, 301, 383, 400, 426, 452, 482, 569, 604, 647, 677, 730
Collecting like terms, 26–28, 69
Columns, of matrices, 196, 215
Combined variation, 420, 429
Combining like terms, 26–28, 69
Common factors, 302–304, 347
  greatest (largest), 302
Common logarithms, 619–620, 649
Commutative laws, 20, 69

INDEX OF APPLICATIONS

Selected Keys of the Scientific Calculator

This secondary function takes the square root of number displayed.

Squares number displayed.

Activates secondary functions printed above certain keys. Also denoted INV or 2nd.

Used when entering numbers in scientific notation. Also denoted EXP.

Finds reciprocal of number displayed.

Used to raise any base to a power. Also denoted y^x, a^x, or ^.

Stores number displayed in memory. Also denoted MIN or M.

Recalls number stored in memory. Also denoted MR.

This secondary function raises 10 to any power entered.

Clears all preceding numbers and operations. Also used to turn calculator on.

Used as an approximation for pi.

Used to perform indicated operation.

Used to control order in which certain operations are performed.

Clears last number displayed but not preceding operations.

Used when entering decimal notation.

Used to change sign of number displayed.

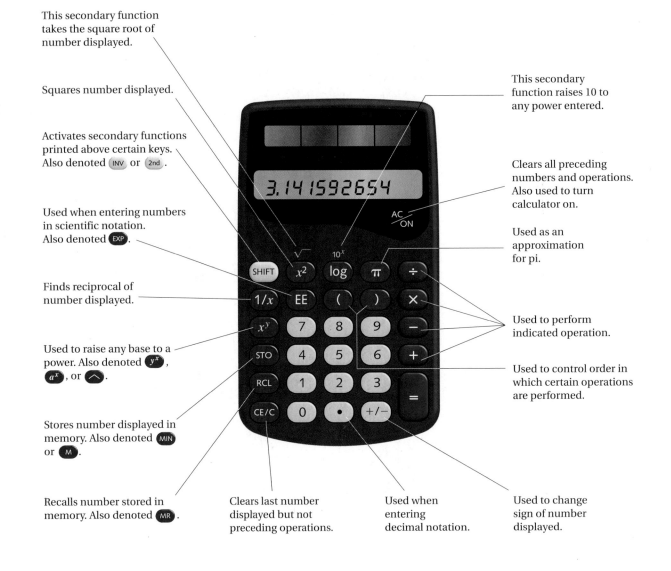

Selected Keys of the Graphing Calculator*

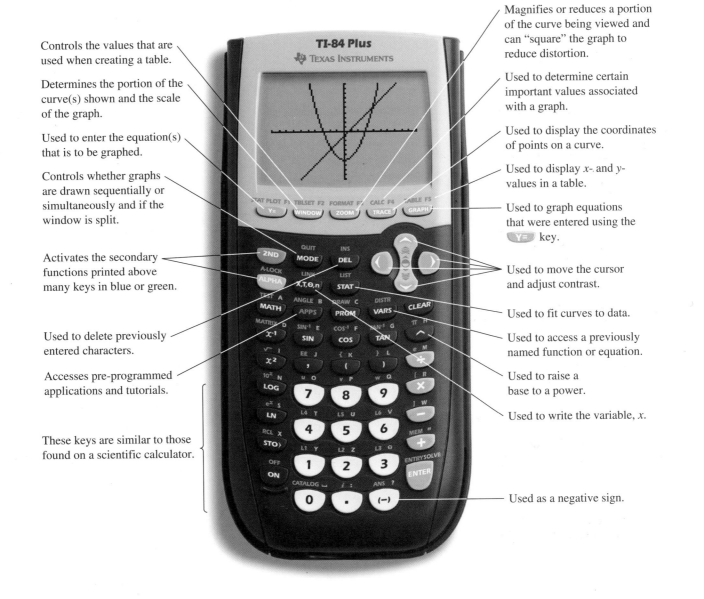

Magnifies or reduces a portion of the curve being viewed and can "square" the graph to reduce distortion.

Used to determine certain important values associated with a graph.

Used to display the coordinates of points on a curve.

Used to display x- and y- values in a table.

Used to graph equations that were entered using the Y= key.

Used to move the cursor and adjust contrast.

Used to fit curves to data.

Used to access a previously named function or equation.

Used to raise a base to a power.

Used to write the variable, x.

Used as a negative sign.

Controls the values that are used when creating a table.

Determines the portion of the curve(s) shown and the scale of the graph.

Used to enter the equation(s) that is to be graphed.

Controls whether graphs are drawn sequentially or simultaneously and if the window is split.

Activates the secondary functions printed above many keys in blue or green.

Used to delete previously entered characters.

Accesses pre-programmed applications and tutorials.

These keys are similar to those found on a scientific calculator.

*Key functions and locations are the same for the TI-83 Plus.

Frequently Used Symbols and Formulas

SYMBOLS

$=$	Is equal to
$\approx$	Is approximately equal to
$>$	Is greater than
$<$	Is less than
$\geq$	Is greater than or equal to
$\leq$	Is less than or equal to
$\in$	Is an element of
$\subseteq$	Is a subset of
$\|x\|$	The absolute value of x
$\{x \mid x...\}$	The set of all x such that $x...$
$-x$	The opposite of x
$\sqrt{x}$	The square root of x
$\sqrt[n]{x}$	The nth root of x
LCM	Least Common Multiple
LCD	Least Common Denominator
π	Pi
i	$\sqrt{-1}$
$f(x)$	f of x, or f at x
$f^{-1}(x)$	f inverse of x
$(f \circ g)(x)$	$f(g(x))$
e	Approximately 2.7
Σ	Summation
$n!$	Factorial notation

FORMULAS

$m = \dfrac{y_2 - y_1}{x_2 - x_1}$	Slope of a line
$y = mx + b$	Slope–intercept form of a linear equation
$y - y_1 = m(x - x_1)$	Point–slope form of a linear equation
$(A + B)(A - B) = A^2 - B^2$	Product of the sum and difference of the same two terms
$\left.\begin{array}{l}(A + B)^2 = A^2 + 2AB + B^2, \\ (A - B)^2 = A^2 - 2AB + B^2\end{array}\right\}$	Square of a binomial
$d = rt$	Formula for distance traveled
$\dfrac{1}{a} \cdot t + \dfrac{1}{b} \cdot t = 1$	Work principle
$s = 16t^2$	Free-fall distance
$y = kx$	Direct variation
$y = \dfrac{k}{x}$	Inverse variation
$x = \dfrac{-b \pm \sqrt{b^2 - 4ac}}{2a}$	Quadratic formula
$P(t) = P_0 e^{kt}, \; k > 0$	Exponential growth
$P(t) = P_0 e^{-kt}, \; k > 0$	Exponential decay
$d = \sqrt{(x_2 - x_1)^2 + (y_2 - y_1)^2}$	Distance formula
$\dbinom{n}{r} = \dfrac{n!}{(n - r)! \, r!}$	$\dbinom{n}{r}$ notation